Design of Feedback Control Systems

THE OXFORD SERIES IN ELECTRICAL AND COMPUTER ENGINEERING

ADEL S. SEDRA, Series Editor

Allen and Holberg, *CMOS Analog Circuit Design*
Bobrow, *Elementary Linear Circuit Analysis, 2nd Edition*
Bobrow, *Fundamentals of Electrical Engineering, 2nd Edition*
Burns and Roberts, *An Introduction to Mixed-Signal IC Test and Measurement*
Campbell, *The Science and Engineering of Microelectronic Fabrication, 2nd Edition*
Chen, *Analog & Digital Control System Design*
Chen, *Digital Signal Processing*
Chen, *Linear System Theory and Design, 3rd Edition*
Chen, *System and Signal Analysis, 2nd Edition*
Comer, *Digital Logic and State Machine Design, 3rd Edition*
Cooper and McGillem, *Probabilistic Methods of Signal and System Analysis, 3rd Edition*
DeCarlo and Lin, *Linear Circuit Analysis, 2nd Edition*
Dimitrijev, *Understanding Semiconductor Devices*
Fortney, *Principles of Electronics: Analog & Digital*
Franco, *Electric Circuits Fundamentals*
Granzow, *Digital Transmission Lines*
Guru and Hiziroğlu, *Electric Machinery and Transformers, 3rd Edition*
Hoole and Hoole, *A Modern Short Course in Engineering Electromagnetics*
Jones, *Introduction to Optical Fiber Communication Systems*
Krein, *Elements of Power Electronics*
Kuo, *Digital Control Systems, 3rd Edition*
Lathi, *Modern Digital and Analog Communications Systems, 3rd Edition*
Martin, *Digital Integrated Circuit Design*
McGillem and Cooper, *Continuous and Discrete Signal and System Analysis, 3rd Edition*
Miner, *Lines and Electromagnetic Fields for Engineers*
Parhami, *Computer Arithmetic*
Roberts and Sedra, *SPICE, 2nd Edition*
Roulston, *An Introduction to the Physics of Semiconductor Devices*
Sadiku, *Elements of Electromagnetics, 3rd Edition*
Santina, Stubberud and Hostetter, *Digital Control System Design, 2nd Edition*
Sarma, *Introduction to Electrical Engineering*
Schaumann and Van Valkenburg, *Design of Analog Filters*
Schwarz, *Electromagnetics for Engineers*
Schwarz and Oldham, *Electrical Engineering: An Introduction, 2nd Edition*
Sedra and Smith, *Microelectronic Circuits, 4th Edition*
Stefani, Savant, Shahian, and Hostetter, *Design of Feedback Control Systems, 4th Edition*
Van Valkenburg, *Analog Filter Design*
Warner and Grung, *MOSFET Theory and Design*
Warner and Grung, *Semiconductor Device Electronics*
Wolovich, *Automatic Control Systems*
Yariv, *Optical Electronics in Modern Communications, 5th Edition*

Design of Feedback Control Systems

FOURTH EDITION

Raymond T. Stefani
California State University, Long Beach

Bahram Shahian
California State University, Long Beach

Clement J. Savant, Jr.

Gene H. Hostetter

New York Oxford
OXFORD UNIVERSITY PRESS
2002

Oxford University Press

Oxford New York
Athens Auckland Bangkok Bogotá Buenos Aires Calcutta
Cape Town Chennai Dar es Salaam Delhi Florence Hong Kong Istanbul
Karachi Kuala Lumpur Madrid Melbourne Mexico City Mumbai
Nairobi Paris São Paulo Shanghai Singapore Taipei Tokyo Toronto Warsaw

and associated companies in
Berlin Ibadan

Published by Oxford University Press, Inc.
198 Madison Avenue, New York, New York, 10016
http://www.oup-usa.org

Library of Congress Cataloging-in-Publication Data

Design of feedback control systems / Raymond T. Stefani ... [et al.].-- 4th ed.
p. cm. -- (Oxford series in electrical and computer engineering)
Includes bibliographical references and index.
ISBN 0-19-514249-7
1. Feedback control systems. I. Stefani, Raymond T. II. Series.

TJ 216 .D417 2001
629.8'3--dc21 00-058913

Printing number: 9 8 7 6 5 4 3 2 1
Printed in the United States of America
on acid-free paper

TO

Ted, Rick, and my Inspiration

Saleh and Mahin; Farahnaz, Bita and Nima

Barbara and the Savant family in memory of Clement

Donna and the Hostetter family in memory of Gene

Contents

Preface xv

CHAPTER 1 Continuous-Time System Description 1

1.1 Preview 1
1.2 Basic Concepts 2
1.2.1 Control System Terminology 2
1.2.2 The Feedback Concept 4
1.3 Modeling 7
1.4 System Dynamics 9
1.5 Electrical Components 10
1.5.1 Mesh Analysis 11
1.5.2 State Variables 13
1.5.3 Node Analysis 15
1.5.4 Analyzing Operational Amplifier Circuits 18
1.5.5 Operational Amplifier Applications 21
1.6 Translational Mechanical Components 25
1.6.1 Free-Body Diagrams 25
1.6.2 State Variables 29
1.7 Rotational Mechanical Components 32
1.7.1 Free-Body Diagrams 32
1.7.2 Analogies 35
1.7.3 Gear Trains and Transformers 37

1.8 Electromechanical Components 40
1.9 Aerodynamics 45
1.9.1 Nomenclature 46
1.9.2 Dynamics 46
1.9.3 Lateral and Longitudinal Motion 50
1.10 Thermal Systems 52
1.11 Hydraulics 54
1.12 Transfer Function and Stability 55
1.12.1 Transfer Functions 55
1.12.2 Response Terms 57
1.12.3 Multiple Inputs and Outputs 67
1.12.4 Stability 69
1.13 Block Diagrams 73
1.13.1 Block Diagram Elements 73
1.13.2 Block Diagram Reduction 75
1.13.3 Multiple Inputs and Outputs 78
1.14 Signal Flow Graphs 79
1.14.1 Comparison with Block Diagrams 79
1.14.2 Mason's Rule 83
1.15 A Positioning Servo 91
1.16 Controller Model of the Thyroid Gland 94
1.17 Stick–Slip Response of an Oil Well Drill 96
1.18 Summary 101
References 103
Problems 105

CHAPTER 2 Continuous-Time System Response 119

2.1 Preview 119
2.2 Response of First-Order Systems 120
2.3 Response of Second-Order Systems 126
2.3.1 Time Response 126
2.3.2 Overdamped Response 127
2.3.3 Critically Damped Response 128
2.3.4 Underdamped Response 128
2.3.5 Undamped Natural Frequency and Damping Ratio 129
2.3.6 Rise Time, Overshoot, and Settling Time 136
2.4 Higher-Order System Response 141
2.5 Stability Testing 143
2.5.1 Coefficient Tests 143
2.5.2 Routh–Hurwitz Testing 145
2.5.3 Significance of the Array Coefficients 147

2.5.4 Left-Column Zeros 148
2.5.5 Row of Zeros 150
2.5.6 Eliminating a Possible Odd Divisor 154
2.5.7 Multiple Roots 155

2.6 Parameter Shifting 159
2.6.1 Adjustable Systems 159
2.6.2 Kharitonov's Theorem 163

2.7 An Insulin Delivery System 165

2.8 Analysis of an Aircraft Wing 168

2.9 Summary 171

References 173

Problems 174

CHAPTER 3 Performance Specifications 183

3.1 Preview 183

3.2 Analyzing Tracking Systems 184
3.2.1 Importance of Tracking Systems 184
3.2.2 Natural Response, Relative Stability, and Damping 187

3.3 Forced Response 189
3.3.1 Steady State Error 189
3.3.2 Initial and Final Values 190
3.3.3 Steady State Errors to Power-of-Time Inputs 192

3.4 Power-of-Time Error Performance 198
3.4.1 System Type Number 198
3.4.2 Achieving a Given Type Number 200
3.4.3 Unity Feedback Systems 201
3.4.4 Unity Feedback Error Coefficients 204

3.5 Performance Indices and Optimal Systems 208

3.6 System Sensitivity 215
3.6.1 Calculating the Effects of Changes in Parameters 215
3.6.2 Sensitivity Functions 216
3.6.3 Sensitivity to Disturbance Signals 220

3.7 Time Domain Design 223
3.7.1 Process Control 224
3.7.2 Ziegler–Nichols Compensation 224
3.7.3 Chien–Hrones–Reswick Compensation 225

3.8 An Electric Rail Transportation System 231

3.9 Phase-Locked Loop for a CB Receiver 234

3.10 Bionic Eye 237

3.11 Summary 240

References 242

Problems 244

CHAPTER 4 Root Locus Analysis 254

4.1 Preview 254
4.2 Pole–Zero Plots 255
4.2.1 Poles and Zeros 255
4.2.2 Graphical Evaluation 256
4.3 Root Locus for Feedback Systems 260
4.3.1 Angle Criterion 260
4.3.2 High and Low Gains 261
4.3.3 Root Locus Properties 262
4.4 Root Locus Construction 263
4.5 More About Root Locus 272
4.5.1 Root Locus Calibration 272
4.5.2 Computer-Aided Root Locus 284
4.6 Root Locus for Other Systems 286
4.6.1 Systems with Other Forms 286
4.6.2 Negative Parameter Ranges 288
4.6.3 Delay Effects 293
4.7 Design concepts (Adding Poles and Zeros) 295
4.8 A Light-Source Tracking System 300
4.9 An Artificial Limb 302
4.10 Control of a Flexible Spacecraft 308
4.11 Bionic Eye 310
4.12 Summary 313
References 314
Problems 314

CHAPTER 5 Root Locus Design 327

5.1 Preview 327
5.2 Shaping a Root Locus 328
5.3 Adding and Canceling Poles and Zeros 329
5.3.1 Adding a Pole or Zero 329
5.3.2 Canceling a Pole or Zero 330
5.4 Second-Order Plant Models 334
5.5 An Uncompensated Example System 338
5.6 Cascade Proportional Plus Integral (PI) 341
5.6.1 General Approach to Compensator Design 341
5.6.2 Cascade PI Compensation 343
5.7 Cascade Lag Compensation 347
5.8 Cascade Lead Compensation 351
5.9 Cascade Lag–Lead Compensation 355

5.10 Rate Feedback Compensation (PD) 357
5.11 Proportional-Integral-Derivative Compensation 361
5.12 Pole Placement 365
5.12.1 Algebraic Compensation 366
5.12.2 Selecting the Transfer Function 367
5.12.3 Incorrect Plant Transmittance 370
5.12.4 Robust Algebraic Compensation 373
5.12.5 Fixed-Structure Compensation 378
5.13 An Unstable High-Performance Aircraft 381
5.14 Control of a Flexible Space Station 385
5.15 Control of a Solar Furnace 388
5.16 Summary 393
References 394
Problems 395

CHAPTER 6 Frequency Response Analysis 405

6.1 Preview 405
6.2 Frequency Response 406
6.2.1 Forced Sinusoidal Response 406
6.2.2 Frequency Response Measurement 407
6.2.3 Response at Low and High Frequencies 410
6.2.4 Graphical Frequency Response Methods 412
6.3 Bode Plots 420
6.3.1 Amplitude Plots in Decibels 420
6.3.2 Real Axis Roots 424
6.3.3 Products of Transmittance Terms 428
6.3.4 Complex Roots 433
6.4 Using Experimental Data 446
6.4.1 Finding Models 446
6.4.2 Irrational Transmittances 447
6.5 Nyquist Methods 449
6.5.1 Generating the Nyquist (polar) Plot 450
6.5.2 Interpreting the Nyquist Plot 456
6.6 Gain Margin 464
6.7 Phase Margin 469
6.8 Relations Between Closed-Loop and Open-Loop Frequency Response 475
6.9 Frequency Response of a Flexible Spacecraft 480
6.10 Summary 485
References 488
Problems 488

CHAPTER 7 Frequency Response Design 501

7.1 Preview 501
7.2 Relation Between Root Locus, Time Domain, and Frequency Domain 501
7.3 Compensation Using Bode Plots 505
7.4 Uncompensated System 507
7.5 Cascade Proportional Plus Integral (PI) and Cascade Lag Compensations 509
7.6 Cascade Lead Compensation 514
7.7 Cascade Lag–Lead Compensation 517
7.8 Rate Feedback Compensation 520
7.9 Proportional-Integral-Derivative Compensation 523
7.10 An Automobile Driver as a Compensator 525
7.11 Summary 529
References 530
Problems 530

CHAPTER 8 State Space Analysis 535

8.1 Preview 535
8.2 State Space Representation 536
8.2.1 Phase-Variable Form 537
8.2.2 Dual Phase-Variable Form 540
8.2.3 Multiple Inputs and Outputs 542
8.2.4 Physical State Variables 547
8.2.5 Transfer Functions 551
8.3 State Transformations and Diagonalization 554
8.3.1 Diagonal Forms 558
8.3.2 Diagonalization Using Partial Fraction Expansion 562
8.3.3 Complex Conjugate Characteristic Roots 564
8.3.4 Repeated Characteristic Roots 567
8.4 Time Response from State Equations 575
8.4.1 Laplace Transform Solution 575
8.4.2 Time Domain Response of First-Order Systems 576
8.4.3 Time Domain Response of Higher-Order Systems 577
8.4.4 System Response Computation 579
8.5 Stability 584
8.5.1 Asymptotic Stability 584
8.5.2 BIBO Stability 585
8.5.3 Internal Stability 587
8.6 Controllability and Observability 589
8.6.1 The Controllability Matrix 592
8.6.2 The Observability Matrix 594
8.6.3 Controllability, Observability, and Pole–Zero Cancellation 595

8.6.4 Causes of Uncontrollability 596
8.7 Inverted Pendulum Problems 603
8.8 Summary 610
References 612
Problems 614

CHAPTER 9 State Space Design 626
9.1 Preview 626
9.2 State Feedback and Pole Placement 626
9.2.1 Stabilizability 630
9.2.2 Choosing Pole Locations 632
9.2.3 Limitations of State Feedback 635
9.3 Tracking Problems 637
9.3.1 Integral Control 638
9.4 Observer Design 640
9.4.1 Control Using Observers 644
9.4.2 Separation Property 646
9.4.3 Observer Transfer Function 647
9.5 Reduced-Order Observer Design 650
9.5.1 Separation Property 653
9.5.2 Reduced-Order Observer Transfer Function 654
9.6 A Magnetic Levitation System 657
9.7 Summary 667
References 668
Problems 669

CHAPTER 10 Advanced State Space Methods 675
10.1 Preview 675
10.2 The Linear Quadratic Regulator Problem 676
10.2.1 Properties of the LQR Design 680
10.2.2 Return Difference Inequality 680
10.2.3 Optimal Root Locus 682
10.3 Optimal Observers—the Kalman Filter 685
10.4 The Linear Quadratic Gaussian (LQG) Problem 687
10.4.1 Critique of LQG 690
10.5 Robustness 692
10.5.1 Feedback Properties 693
10.5.2 Uncertainty Modeling 695
10.5.3 Robust Stability 698
10.6 Loop Transfer Recovery (LTR) 705

10.7 H_∞ Control 709
10.7.1 A Brief History 709
10.7.2 Some Preliminaries 710
10.7.3 H_∞ Control: Solution 713
10.7.4 Weights in H_∞ Control Problems 715
10.8 Summary 722
References 723
Problems 724

CHAPTER 11 Digital Control 733

11.1 Preview 733
11.2 Computer Processing 734
11.2.1 Computer History and Trends 734
11.3 A/D and D/A Conversion 737
11.3.1 Analog-to-Digital Conversion 737
11.3.2 Sample and Hold 739
11.3.3 Digital-to-Analog Conversion 741
11.4 Discrete-Time Signals 741
11.4.1 Representing Sequences 741
11.4.2 z-Transformation and Properties 744
11.4.3 Inverse z Transform 749
11.5 Sampling 751
11.6 Reconstruction of Signals from Samples 753
11.6.1 Representing Sampled Signals with Impulses 753
11.6.2 Relation Between the z Transform and the Laplace Transform 756
11.6.3 The Sampling Theorem 757
11.7 Discrete-Time Systems 760
11.7.1 Difference Equations and Response 760
11.7.2 z-Transfer Functions 762
11.7.3 Block Diagrams and Signal Flow Graphs 763
11.7.4 Stability and the Bilinear Transformation 764
11.7.5 Computer Software 768
11.8 State-Variable Descriptions of Discrete-Time Systems 771
11.8.1 Simulation Diagrams and Equations 771
11.8.2 Response and Stability 774
11.8.3 Controllability and Observability 777
11.9 Digitizing Control Systems 779
11.9.1 Step-Invariant Approximation 779
11.9.2 z-Transfer Functions of Systems with Analog Measurements 782
11.9.3 A Design Example 785
11.10 Direct Digital Design 788
11.10.1 Steady State Response 788

11.10.2 Deadbeat Systems 789
11.10.3 A Design Example 790
11.11 Summary 798
References 800
Problems 802

APPENDIX A Matrix Algebra 812

A.1 Preview 812
A.2 Nomenclature 812
A.3 Addition and Subtraction 812
A.4 Transposition 813
A.5 Multiplication 813
A.6 Determinants and Cofactors 814
A.7 Inverse 816
A.8 Simultaneous Equations 817
A.9 Eigenvalues and Eigenvectors 819
A.10 Derivative of a Scalar with Respect to a Vector 821
A.11 Quadratic Forms and Symmetry 823
A.12 Definiteness 824
A.13 Rank 826
A.14 Partitioned Matrices 827
Problems 830

APPENDIX B Laplace Transform 834

B.1 Preview 834
B.2 Definition and Properties 834
B.3 Solving Differential Equations 835
B.4 Partial Fraction Expansion 837
B.5 Additional Properties of the Laplace Transform 841
B.5.1 Real Translation 842
B.5.2 Second Independent Variable 842
B.5.3 Final-Value and Initial-Value Theorems 843
B.5.4 Convolution Integral 844

Index 845

Preface

As the new millennium begins, we look back in gratitude to the many faculty and students who have used the three earlier editions of this textbook and made many helpful suggestions to the authors. In those earlier editions we introduced comprehensive design examples, drill problems, and wide margins with notes. Other texts followed our lead and emulated those items. What other texts cannot emulate, we believe, is the clear and understandable exposition we bring to the field of control system science. Throughout this book we try to make complicated methodology accessible to a spectrum of students with widely varying backgrounds. Detail is there for those who want to know "why." Summaries and marginal comments are there for those who simply want to know "how."

Revisions

The most obvious change in this edition is the comprehensive keying of this text to MATLAB. We created sections of "Computer-Aided Learning" by which each student can learn how the MATLAB platform can be used to verify all figures and tables included in the text. We selected a small group of MATLAB commands to efficiently focus the use of that computational package. In a basic course such as this, it is essential that every student use the computer as an aid to learning and not as the primary source of information. The student should learn all basics and should be able to sketch (albeit roughly) time response plots, root locus plots, and Bode/Nyquist plots manually. MATLAB (or any other computer tool) may then be used to fine-tune understanding and to obtain results of high accuracy. But, those results must be critically reviewed by a knowledgeable user; otherwise the computer becomes the master and the user becomes the slave.

Chapter 1 has been substantially revised. Linearization is introduced by which models may be generated. Operational amplifier applications are included for the various types of compensator designed later in the text. Substantive coverage is made of aerodynamics, thermal systems, and hydraulic systems. Drill problems cover those topics. Stability is covered in more detail. Signal flow graphs are better compared to block diagrams. Design examples are added for the human thyroid gland as a controller and for oil well drill dynamics.

For Chapter 2, we include the significance of Routh array coefficients and the stability implication of multiple roots occurring as even divisors. An example of Kharitonov's theorem is added.

Hurwitz determinants are now presented in Chapter 3. It is now shown how coefficients of the transfer function may be selected to force a given type number to occur. An interesting biomedical design example is added, that of a bionic eye for the blind. Time response examples are added to illustrate time domain design.

The main change to Chapter 4 is inclusion of computer-aided means for calculating breakaway points, entry points, departure angles, and approach angles. The MATLAB command `rltool` is introduced. Delay effects are evaluated as a function of $1/T$ where T is the delay in seconds. The bionic eye example is again used, this time to illustrate use of the root locus.

Chapter 5 is revised comprehensively. Root locus design methods are now more general and more flexible. The effect of adding or canceling poles or zeros is covered in detail. The MATLAB command `rltool` is suggested as a primary computer aid in that the effect of each root locus design point may be evaluated in terms of step response and the Bode plot. A new design example is introduced for a solar furnace.

Chapter 6 now begins with an introduction to all frequency response plots. It is argued that frequency response data are complex vectors, hence can be plotted in a variety of ways resulting in Bode, Nyquist, and Nichols plots. There is a new section that discusses the relation between open-loop and closed-loop frequency response plots. Closed-loop frequency response data such as bandwidth and peak resonance are introduced more formally. Nichols plots, Nichols charts, and constant loci M and N circles are also discussed. Chapter 7 on frequency domain design remains unchanged.

Chapter 8 now includes a design example of the classic inverted pendulum problem and several variations. This famous problem has become a benchmark for testing novel control design techniques and provides an excellent tool for introducing the important concepts of controllability, observability, pole–zero cancellation, and practical issues such as sensor placement. Appropriate MATLAB commands for state space modeling, transformation, analysis, and simulation are also discussed.

Chapters 9–11 have minor corrections along with the introduction of MATLAB commands for digital control.

Use of This Textbook

The text can be divided into six areas:

Classical analysis including modeling (Chapters 1–4, 6)
Classical design (Chapters 5 and 7)
State-variable analysis (Chapter 8)
State-variable design (Chapter 9)
Advanced topics (Chapter 10)
Digital control (Chapters 11)

These six areas represent building blocks to construct a course. We have purposely included more material than a three-semester unit course or a four-quarter unit course would normally cover. The extra material is intended to give the instructor flexibility in structuring a course to meet the needs of the program, the university, and the community served. We suggest that it is better to cover a smaller number of units well than to cover a larger number poorly.

For example, a two-course sequence could be created where the first course covers classical analysis (Chapters 1–4 and 6) followed by a second course including state variables, design, and advanced topics (Chapters 5 and 7–10). Chapter 11 is often used as reference material, introducing the student to digital control and providing a comparison with analog methods. The possibilities are endless.

Raymond T. Stefani
Bahram Shahian
Clement J. Savant Jr. (late)
Gene H. Hostetter (late)

CHAPTER

Continuous-Time System Description

1

1.1 Preview

The first conscious use of feedback control of a physical system by mankind lives in prehistory. Possibly it was a spillway in an irrigation network, where excess water was automatically drained. Development of a mathematical framework for the description, analysis, and design of control systems dates from the introduction of James Watt's flyball governor (1760), which was used to regulate the speed of steam engines, and the subsequent work by James Clerk Maxwell (ca. 1868) and others to improve the design and extend its applicability.

Since that era, the theory and practice of control system design advanced rapidly. Important new concepts and tools were developed in connection with telephone and radio communications in the 1920s and 1930s. Rather poorly performing electronic devices, including amplifiers and modulators, were dramatically improved by feedback. World War II further accelerated the development of classical control theory and practice. Heavy guns had to be rapidly and accurately positioned. Precise navigation and target tracking were increasingly important, and aircraft performance was improved greatly with the incorporation of complex control systems to aid the pilot. Latter, *automation* became a household word as industry began to depend more and more upon automatically controlled machinery.

Today, feedback control systems are pervasive in industry and in our everyday lives. They range from governmental regulation (such as that governing monetary policy) to automated and highly flexible manufacturing plants to sophisticated automobiles, household appliances, and entertainment systems. It is our purpose to learn to design feedback control systems for a wide variety of applications.

Control system designers find that block diagrams provide a particularly useful way to visualize the interconnections of system components, thus revealing the system structure. Successful design begins by creating a mathematical model of the system to be stabilized. Next, the contents of the blocks within a diagram must be identified. Finally, values must be selected for those parameters that are adjustable, and sometimes additional components must be added to provide acceptable performance.

This chapter begins by defining basic control system terminology. Since design requires a model of each system of interest, the behaviors of many typical electrical, mechanical, and electromechanical systems are described. The resulting differential equations must be rendered into a form useful to the controls engineer. The goal can be accomplished by Laplace-transforming each differential equation and then generating a relationship, the transmittance, between the input and output of each block of the control system block diagram. In Appendix B, a summary of the Laplace transform method is presented.

The block diagram can be reduced to just one input–output relationship, the system overall transfer function. By converting the block diagram into an equivalent form, the signal flow graph can be developed. Subsequent chapters will describe the design steps that follow once the block diagram has been defined and the transfer function has become available.

All the chapters of this text conclude with examples that are intended to reinforce the key points of the chapter in an interesting and informative manner. Chapter 1 concludes with discussion of a positioning servo, analysis of the thyroid gland, and design of an oil well drilling system.

While the material in the first chapter involves subjects already known to the reader from previous experience, the text provides a coherent review. The emphasis here is on using rather than proving results.

1.2 Basic Concepts

1.2.1 Control System Terminology

Control systems influence each facet of modern life. Automatic washers and dryers, microwave ovens, chemical processing plants, navigation and guidance systems, space satellites, pollution control, mass transit, and economic regulation are a few examples. In the broadest sense, a control system is any interconnection of components to provide a desired function.

The plant (process), inputs, and outputs are defined.

The portion of a system that is to be controlled is called the **plant** or the **process**. It is affected by applied signals, called **inputs**, and produces signals of particular interest, called **outputs**, as indicated in Figure 1.1(a). The plant is fixed insofar as the control system designer is concerned. Whether the plant is an automobile engine, an electrical generator, or a nuclear reactor, it is the designer's job to ensure that the plant operates as required. Other components must be specially created and connected as a means to an end.

Controller and open-loop control are defined.

A **controller** may be used to produce a desired behavior of the plant, as shown in Figure 1.1(b). The controller generates plant input signals designed to produce

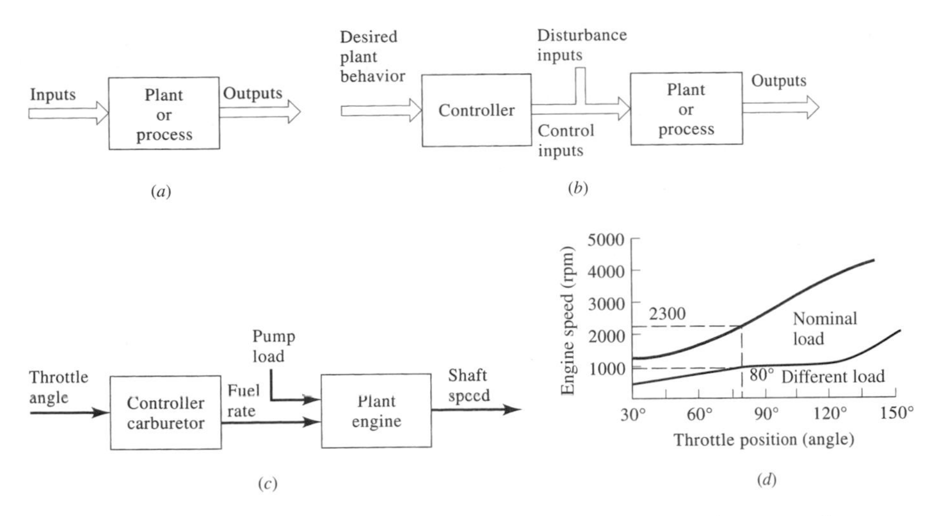

Figure 1.1 (a) A plant or process to be controlled. (b) An open-loop control system. (c) Example of an open-loop control system. (d) Engine speed versus throttle angle curves.

desired outputs. Some of the plant inputs are accessible to the designer and some are generally not available. The inaccessible input signals are often disturbances to the plant. The double lines in the figure indicate that several signals of each type may be involved. Arrows indicate direction of flow. This system is termed **open-loop** because the control inputs are not influenced by the plant outputs: that is, there is no feedback around the plant.

Such an open-loop control system has the advantage of simplicity, but its performance is highly dependent upon the properties of the plant, which may vary with time. The disturbances to the plant may also create an unwanted response, which it would be desirable to reduce.

Open-loop examples are presented.

As an example, suppose that a gasoline engine is used to drive a large pump, as depicted in Figure 1.1(c). The carburetor and the engine comprise a common type of control system wherein a large-power output is controlled with a small-power input. The carburetor is the controller in this case, and the engine is the plant. The desired plant output, a certain engine shaft speed, may be obtained by adjusting the throttle angle.

Two plots of engine speed versus throttle angle are shown in Figure 1.1(d). If the nominal curve is used, a throttle angle of 80° produces an engine speed of 2300 rpm. Suppose that a disturbances occurs, consisting of a change in engine load. For the new curve, a throttle angle of 80° produces an engine speed of only 1000 rpm. In some cases open-loop control may be acceptable. In other cases, it may not be acceptable to have system output change when other values change. In these more critical cases, the closed-loop procedure of the next section may be needed.

Table 1.1 shows five examples. The first two examples are for open-loop systems in that no measurements are taken to adjust controller influence on the plant. Each of the two controllers is specified when a manual setting is made of temperature and speed respectively. Hair dampness and the type of material being drilled are disturbances affecting desired performance. In these two cases, the user simply alters the total time until the job is done. In the case of the hair dryer, output air temperature remains constant while drying time for hair will vary according to wetness. In the case of the drill, output speed may vary while the drilling requirement remains constant. Figure 1.1(b) describes these systems.

1.2.2 The Feedback Concept

Closed-loop control is distinguished from open-loop control.

If the requirements of the system cannot be satisfied with an open-loop control system, a **closed-loop** or **feedback** system is desirable. A path (or loop) is provided from the output back to the controller. Some or all of the system outputs are measured and used by the controller, as indicated in Figure 1.2(a). The controller may then compare a desired plant output with the actual output and act to reduce the difference between the two.

Let us return to Table 1.1 and consider the third and fourth examples. Temperatures and speed are the system outputs, as was the situation for the first two examples, but now measurements are used to keep the outputs constant in the presence of disturbances. If outside temperatures drops, a thermostat determines that the room is becoming too cold. The thermostat causes furnace heat to increase which, in turn, causes the room temperature to increase to the predetermined value.

Changes in driving conditions represent disturbances affecting an automobile's speed. One possible feedback control configuration is shown in Figure 1.2(b). A tachometer produces a voltage proportional to the engine shaft speed. The input voltage, which is proportional to the desired speed, is set with a potentiometer. The tachometer voltage is subtracted from the input voltage, giving an error voltage that is proportional to the difference between the actual speed and the desired speed.

The error voltage is then amplified and used to position the throttle. The throttle actuator could be a reversible electric motor, geared to the throttle arm. When the engine shaft speed is equal to the desired speed (when the difference or *error* is zero), the throttle remains fixed. If a change in load or a change in the engine components

Table 1.1 *Examples of Open-Loop and Closed-Loop Systems*

Input	Controller	Plant	Disturbance	Output	Measurement
Heat setting	Dial	Hair dryer	Hair dampness	Hot air temperature	None
Speed setting	Dial	Drill	Type of material	Rotating drill bit speed	None
Desired temperature	Thermostat	Furnace	Outside temperature	Hot air temperature	Room temperature
Desired speed	Cruise control	Auto engine	Driving conditions	Car speed	Engine rpm
Desired performance	Electorate	President	Economy	Decisions	Evaluation

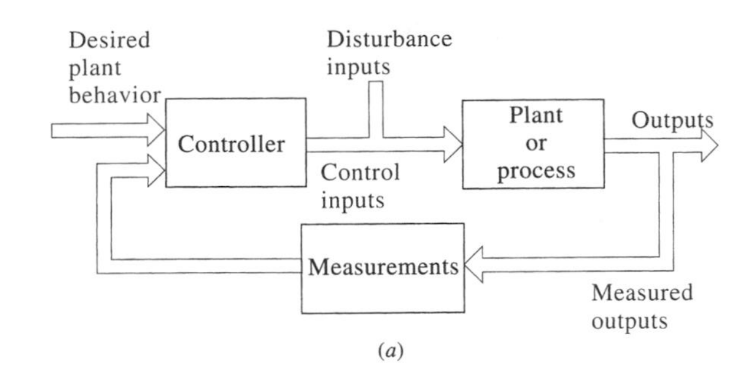

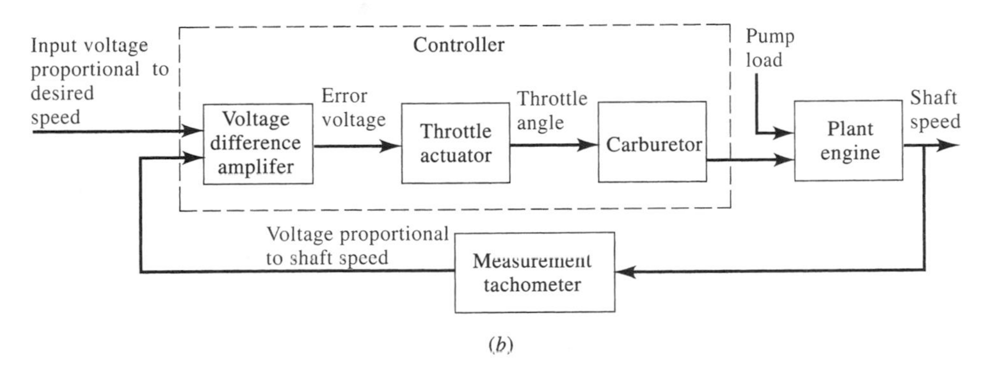

Figure 1.2 (a) Closed-loop or feedback control. (b) A closed-loop engine control system.

should occur in the system, and the actual speed is no longer equal to the desired speed, the error voltage becomes nonzero, causing the throttle setting to change so that the actual speed approaches the desired speed. The controller here consists of the voltage difference amplifier, the throttle actuator, and the carburetor.

The last example of Table 1.1 differs in a subtle yet very important way from the first four examples. Consider the task of affecting U.S. presidential decisions. To fit the control system model of Figure 1.2(a), those decisions would be the output. The electorate functions as a controller. As the economy changes, those changes become a disturbance to decisions tentatively made by the president. Collectively, the voters have some shared desired performance of government. These voters measure the president's decisions against desired performance and vote accordingly. It is interesting (and perhaps a bit surprising) that after 54 presidential elections from 1788 through 2000, the Democratic party and its predecessors have won 26 elections while the Republican party and its predecessors have won 28, a remarkably balanced control. The important difference from the first four examples is that here there is a time delay between a president's decisions and the next election. Action in the first four examples is continuous. In some control applications, the presence of delay can destabilize a system. Imagine driving a car with eyes being open for 30 s, then closed for 30 s, and so on. Activity by nearby traffic during the 30 s that the eyes are closed could easily require corrective action, which could not occur in time. Delay effects are considered later in this text.

Some of the advantages that feedback control offers to the designer are:

Advantages of feedback.

1. **Increased accuracy.** The closed-loop system may be designed to drive the error (difference between desired and measured response) to zero.
2. **Reduced sensitivity to changes in components.** As in the preceding examples, the system may be designed to seek zero error despite changes in the plant.
3. **Reduced effects of disturbances.** The effects of disturbances to the system may be greatly attenuated.
4. **Increased speed of response and bandwidth.** Feedback may be used to increase the range of frequencies over which a system will respond and to make it respond more desirably. A satellite booster rocket, for example, has aerodynamics resembling those of a giant broomstick. It may, with feedback, behave with beauty and grace.

❑ DRILL PROBLEMS

D1.1 Using the format of Table 1.1, list five additional examples of control systems that do not employ feedback.

D1.2 Using the format of Table 1.1, list five additional examples of control systems that employ feedback.

D1.3 Identify the input, controller, and output for each of the following control systems. Which are open loop and which are closed loop?

(a) A heater with thermostat

(b) A toaster

(c) A human being reaching to touch an object

(d) A human being piloting an aircraft

(e) A hydroelectric generator

Ans.

	Input	**Controller**	**Output**	**Open/Closed Loop**
a.	Temperature	Thermostat	Hot air temperature	Closed
b.	Darkness	Dial	Heat	Open
c.	Position	Brain	Position	Closed
d.	Destination	Human	Speed, heading	Closed
e.	Desired flow rate	Pipes/nozzles	Power	open

D1.4 What measurements and changes should be taken so the open-loop systems of Drill Problem D1.3 can become closed-loop systems?

Ans. (b) measure dryness of the toast and adjust the dial setting of heat;
(e) measure power demand and adjust the flow through the pipes and nozzles

D1.5 Draw diagrams similar to Figure 1.2 for the following systems:

(a) Control of human skin temperature by sweating

(b) Control of a nuclear reactor

(c) The learning process with feedback, assuming that available study time is a disturbance

1.3 Modeling

Start with a process.

Control engineers must be able to analyze and design systems of many kinds. For example, to design a speed control system for an automobile, it is necessary to understand how the vacuum pressure of an engine affects throttle setting (pneumatics), how temperature and pressure within a cylinder affect the power out as the gas–air mixture from the carburetor explodes (thermodynamics), how the car will respond to the power applied by the pistons in the cylinders (mechanics), and how electrical devices may be created to measure and store important variables like temperature and vacuum pressure (electrical circuits).

Create a model.

In each case it is necessary to create a mathematical model that behaves similarly to the actual system within some operating range. The result is the description of a plant for which a controller and measurement device may then be designed. For example, certain values of a spring-mass-damper may be able to simulate the motion of a car within some range of power applied while other values are needed for different powers applied.

The process of linearization may be used to construct a model that is valid for some range of operating conditions. For example, suppose a system output y (maybe speed) depends on some input x (perhaps power), as represented by Figure 1.3, in which

$$y = f(x) = x^2 \quad [1.1]$$

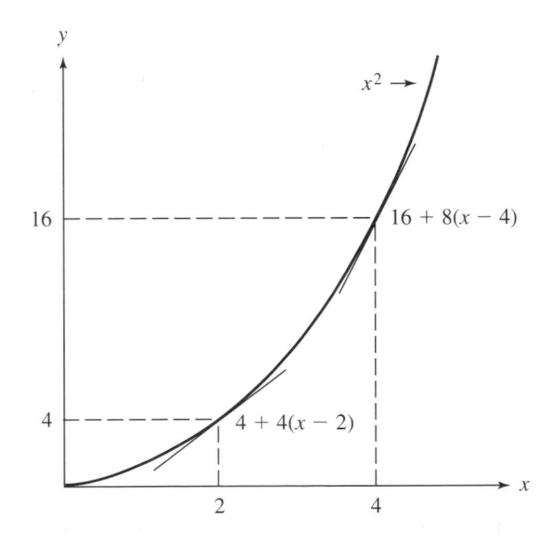

Figure 1.3 Two approximations for $y = x^2$.

Table 1.2 *Two Approximations to* $y = x^2$

x	x^2	$4 + 4(x - 2)$	$16 + 8(x - 4)$
2	4.00	4.00	0.00
2.1	4.41	4.40	0.80
2.2	4.84	4.80	1.60
3.0	9.00	8.00	8.00
4.1	16.81	12.40	16.80

Linearization.

Instead of this nonlinear equation, it may be more useful to create a linear model that operates near some value of x called x_0. A Taylor series approximation to $f(x)$ at the point x_0 is given by the following where $f^1(x_0)$ means that $f(x)$ is differentiated with respect to x and then evaluated when x equals x_0. The tilde symbol ($\sim$) implies an approximation

$$y \sim y_0 + f^1(x_0)(x - x_0) \qquad \textbf{[1.2]}$$

$$y \sim x_0^2 + 2x_0(x - x_0) \qquad \textbf{[1.3]}$$

If we choose x_0 to be 2, then the approximation of Equation (1.3) becomes

$$y \sim 4 + 4(x - 2) \qquad \textbf{[1.4]}$$

Table 1.2 and Figure 1.3 show values of Equation (1.4) near $x_0 = 2$ and also results farther away from $x_0 = 2$.

Notice that Equation (1.4) is good approximation to x^2 for values of x near 2 but that the approximation becomes worse for x values which are higher than 2. For example, if x moves to the vicinity of 4, then Equation (1.3) becomes $16 + 8(x - 4)$, which yields an approximate value of 16.80 at $x = 4.1$, very close to the true value of 16.81 and much better than the value 12.40 that we would get using the other approximation. Even Ohm's famous law that $v = iR$ is good only for some range of voltage versus current. In Figure 1.4 there is a linear region where the slope of v versus i is constant (and Ohm's law applies) and other regions where the slope is not constant (and Ohm's law does not apply).

❑ DRILL PROBLEMS

D1.6 Approximate $y = \sqrt{x}$ for values of $x = 2.2, 2.4, 2.6, 2.8$, and 3.0 by linearizing $\sqrt{x}$ about $x_0 = 2$, using Equation (1.2). Compare approximate values with true values.

Ans. $y \approx 1.414 + 0.354(x - 2)$

x	Approximate	True
2.2	1.485	1.483
2.4	1.556	1.549
2.6	1.626	1.612
2.8	1.697	1.673
3.0	1.768	1.732

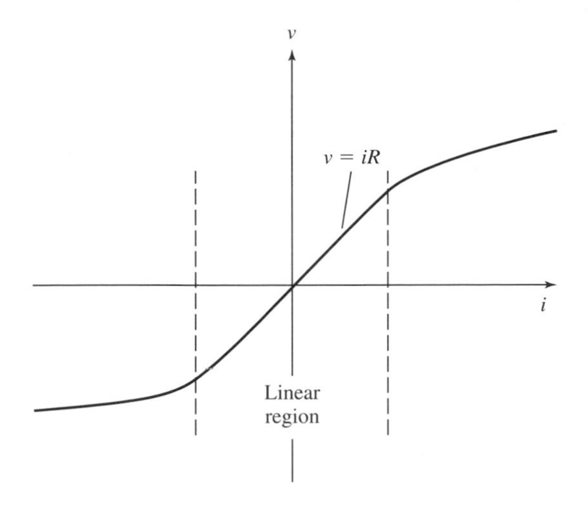

Figure 1.4 Linear region where $v = iR$.

D1.7 If x is in radians, use Equation (1.2) to approximate $y = \sin x$ for values of $x = 0.1, 0.2, 0.3$, and 0.4, by linearizing $\sin x$ about $x_0 = 0$. Compare approximate values with true values.

Ans. $y \approx x$

x (rad)	x (deg)	Approximate	True
0.1	5.7	0.100	0.100
0.2	11.5	0.200	0.199
0.3	17.2	0.300	0.296
0.4	22.9	0.400	0.389

1.4 System Dynamics

A controls engineer usually works from the Laplace-transformed description of a system. Each application has its own unique properties. Some systems are purely electrical while others may employ electrical, hydraulic, and mechanical subsystems, all tied together in a coordinated effort to maintain some desired performance. We shall examine methods for analyzing components of the following types:

Applications.

Electrical (mesh analysis, node analysis, state variables, operational amplifier applications)
Mechanical translational (free-body diagrams and state variables)
Mechanical rotational (free-body diagrams)
Electromechanical
Aerodynamic
Hydraulic
Thermodynamic

For each of these system types, the basic dynamic equations are shown and then converted to Laplace transforms.

1.5 Electrical Components

Electrical networks are governed by two Kirchhoff laws:

1. The algebraic sum of voltages around a closed loop equals zero.
2. The algebraic sum of currents flowing into a circuit node equals zero.

Basic laws.

Network element models include resistors, capacitors, inductors, voltage sources, and current sources. The voltage–current relations for these are summarized in Figure 1.5. The voltage–current relations of Figure 1.5 can be Laplace-transformed by using the techniques of Appendix B. With all initial conditions equal to zero, the relations are Laplace-transformed as follows:

$$V_R(s) = RI(s) \qquad V_L(s) = sLI(s) \qquad V_C(s) = (1/sC)I(s) \tag{1.5}$$

$$I(s) = (1/R)V_R(s) \qquad I(s) = 1/(sL)V_L(s) \qquad I(s) = sCV_C(s) \tag{1.6}$$

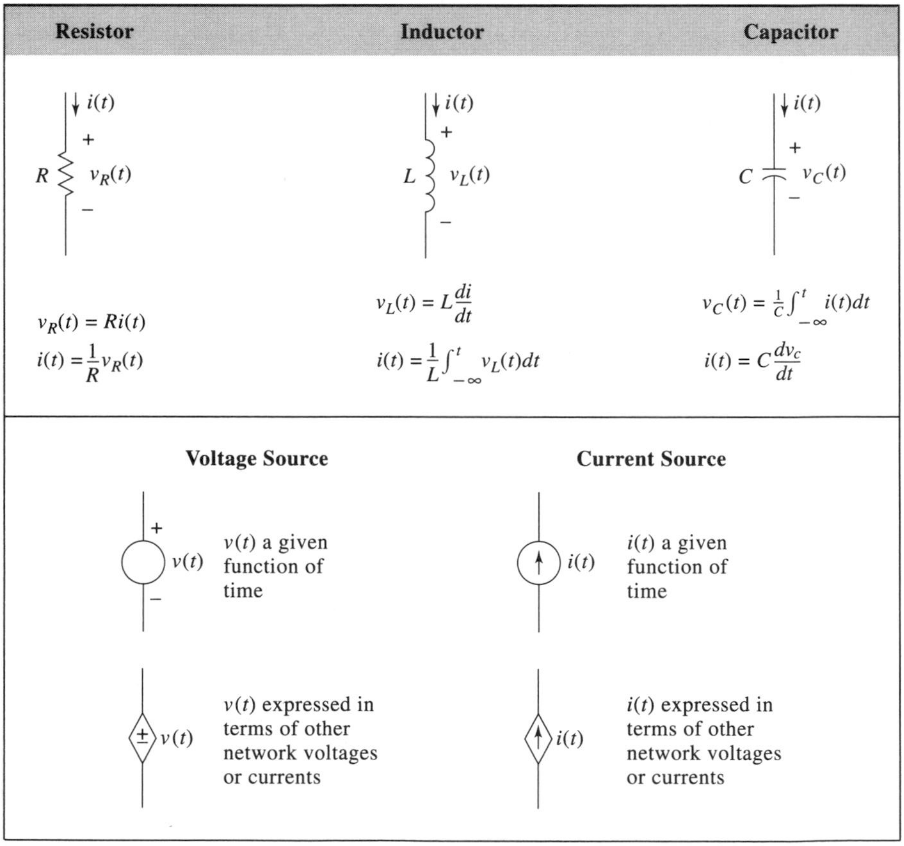

Resistor

$i(t)$, $+$, R, $v_R(t)$, $-$

$v_R(t) = Ri(t)$

$i(t) = \frac{1}{R}v_R(t)$

Inductor

$i(t)$, $+$, L, $v_L(t)$, $-$

$v_L(t) = L\frac{di}{dt}$

$i(t) = \frac{1}{L}\int_{-\infty}^{t} v_L(t)dt$

Capacitor

$i(t)$, $+$, C, $v_C(t)$, $-$

$v_C(t) = \frac{1}{C}\int_{-\infty}^{t} i(t)dt$

$i(t) = C\frac{dv_c}{dt}$

Voltage Source

$+$ $v(t)$ $-$: $v(t)$ a given function of time

$v(t)$: $v(t)$ expressed in terms of other network voltages or currents

Current Source

$i(t)$: $i(t)$ a given function of time

$i(t)$: $i(t)$ expressed in terms of other network voltages or currents

Figure 1.5 Electrical element voltage–current relations.

Note that uppercase letters are used for voltage and current to indicate that these variables are functions of s and not t. When these functions are Laplace-transformed, it is easy to consider an inductance as an impedance of value sL and a capacitor as an impedance of value $1/sC$.

Figure 1.5 illustrates voltage and current sources of two types—independent and dependent.

1.5.1 Mesh Analysis

One systematic method of network analysis consists of defining a loop current in each mesh of a network, equating the algebraic sums of the voltages around each mesh to zero. For example, in the network of Figure 1.6 the application of Kirchhoff's voltage law around the i_1 mesh gives

Example of network analysis.

$$2i_1 + 3\frac{di_1}{dt} + 4(i_1 - i_2) + \frac{1}{5}\int_{-\infty}^{t}(i_1 - i_2)\,dt = 8\cos 9t$$

Similarly, around the i_2 mesh, we have

$$6\frac{di_2}{dt} + 7i_2 - 4(i_1 - i_2) - \frac{1}{5}\int_{-\infty}^{t}(i_1 - i_2)dt = 0$$

Collecting terms, the following simultaneous integrodifferential equations in i_1 and i_2 result:

$$3\frac{di_1}{dt} + 6i_1 + \frac{1}{5}\int_{-\infty}^{t} i_1 dt - 4i_2 - \frac{1}{5}\int_{-\infty}^{t} i_2 dt = 8\cos 9t$$

$$-4i_1 - \frac{1}{5}\int_{-\infty}^{t} i_1 dt + 6\frac{di_2}{dt} + 11i_2 + \frac{1}{5}\int_{-\infty}^{t} i_2 dt = 0$$

Of course, each of these equations can be differentiated to eliminate integrals if a purely differential equation is desired.

These equations can be Laplace-transformed, assuming zero initial conditions, with the result:

$$\left(3s + 6 + \frac{1}{5s}\right) I_1(s) + \left(-4 - \frac{1}{5s}\right) I_2(s) = \frac{8s}{s^2 + 9^2} \qquad [1.7]$$

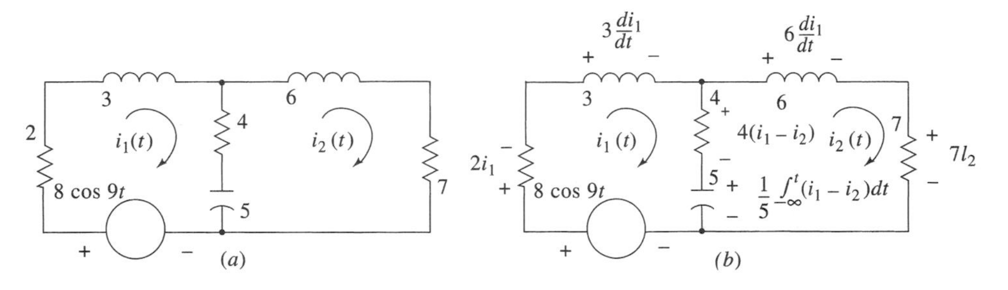

Figure 1.6 Writing simultaneous mesh equations for an electrical network. (a) Electrical network example for mesh analysis. (b) Network with element voltages expressed in terms of the mesh currents.

and for the second mesh,

$$\left(-4 - \frac{1}{5s}\right) I_1(s) + \left(6s + 11 + \frac{1}{5s}\right) I_2(s) = 0 \qquad [1.8]$$

Notice that since these are functions of s, uppercase variables are used. Equations (1.7) and (1.8) can be written in a standard form as follows:

Using a matrix to summarize mesh equations.

$$Z_{11}(s)I_1(s) + Z_{12}(s)I_2(s) = E_1(s) \qquad [1.9]$$

$$Z_{21}(s)I_1(s) + Z_{22}(s)I_2(s) = E_2(s) \qquad [1.10]$$

where

$Z_{11}(s) =$ sum of all impedances around the I_1 mesh, in this case equal to $[3s + 6 + (1/5s)]$

$Z_{12}(s) = Z_{21}(s) =$ sum of all impedances common to both mesh I_1 and mesh I_2, in this case equal to $[-4 - (1/5s)]$; these terms are negative when all mesh currents are taken in a clockwise direction

$Z_{22}(s) =$ sum of all impedances around the I_2 mesh, in this case equal to $[6s + 11 + (1/5s)]$

$E_1(s) =$ independent voltage source driving mesh I_1, in this case equal to $8s/(s^2 + 9^2)$

$E_2(s) =$ independent voltage source driving mesh I_2, in this case equal to 0

The solution of Equations (1.9) and (1.10) is best accomplished with Cramer's rule. In this example, $I_2(s)$ is to be found as follows:

Using Cramer's rule.

$$I_2(s) = \frac{\begin{vmatrix} Z_{11} & E_1(s) \\ Z_{21} & E_2(s) \end{vmatrix}}{\begin{vmatrix} Z_{11} & Z_{12} \\ Z_{21} & Z_{22} \end{vmatrix}} \qquad [1.11]$$

when the values of this impedance and the driving functions are given above. The denominator of Equation (1.11) is expanded as follows:

$$\Delta = \begin{vmatrix} Z_{11} & Z_{12} \\ Z_{21} & Z_{22} \end{vmatrix} = Z_{11}Z_{22} - Z_{12}Z_{21}$$

where the (s) symbol is understood for all Z_{ij}.

The symbol Δ is used to represent the determinant of the impedance functions. When these functions are substituted into Equation (1.11), the following expression results:

$$I_2(s) = \frac{[8s/(s^2 + 9^2)][4 + (1/5s)]}{[3s + 6 + (1/5s)][6s + 11 + (1/5s)] - [4 + (1/5s)]^2} \qquad [1.12]$$

1.5.2 State Variables

Another way to handle circuit problem is to write differential equations in terms of the "energy storage" variables such as inductor currents and capacitor voltages. The result is a set of first-order differential equations. For example, we could choose the inductor currents i_1 and i_2 and the capacitor voltage v_c

$$v_c = \frac{1}{5}\int_{-\infty}^{t}(i_1 - i_2)dt$$

We have

$$di_1/dt = -2i_1 + 4/3i_2 - 1/3v_c + 8/3\cos 9t$$

$$di_2/dt = 4/6i_1 - 11/6i_2 + 1/6v_c$$

$$dv_c/dt = 1/5i_1 - 1/5i_2$$

These differential equations can be written in matrix form

$$d/dt\begin{bmatrix} i_1 \\ i_2 \\ v_c \end{bmatrix} = \begin{bmatrix} -2 & 4/3 & -1/3 \\ 2/3 & -11/6 & 1/6 \\ 1/5 & -1/5 & 0 \end{bmatrix}\begin{bmatrix} i_1 \\ i_2 \\ v_c \end{bmatrix} + \begin{bmatrix} v_s(t)/3 \\ 0 \\ 0 \end{bmatrix} \quad [1.13]$$

Time domain.

where $v_s(t) = 8\cos 9t$.

In Laplace form, Equation (1.13) becomes

$$\begin{bmatrix} s+2 & -4/3 & 1/3 \\ -2/3 & s+11/6 & -1/6 \\ -1/5 & 1/5 & s \end{bmatrix}\begin{bmatrix} I_1(s) \\ I_2(s) \\ V_c(s) \end{bmatrix} = \begin{bmatrix} V_s(s)/3 \\ 0 \\ 0 \end{bmatrix} \quad [1.14]$$

Lapalace domain.

Next, Equation (1.14) can be solved for $I_2(s)$ using Cramer's rule:

$$I_2(s) = \frac{\begin{vmatrix} s+2 & 1/3 & 1/3 \\ -2/3 & 0 & -1/6 \\ 1/5 & 0 & s \end{vmatrix}}{\begin{vmatrix} s+2 & -4/3 & 1/3 \\ -2/3 & s+11/6 & -1/6 \\ -1/5 & 1/5 & s \end{vmatrix}} V_s(s) \quad [1.15]$$

Using Cramer's rule.

The result of solving Equation (1.15) is

$$I_2(s) = \frac{2/9(s+1/20)\left[8s/(s^2+9^2)\right]}{s^3 + 23/6s^2 + 259/90\,s + 1/10} \quad [1.16]$$

Equation (1.16) equals equation (1.12). Notice that the system is obviously of order 3 using equation (1.14), while the block diagram approach awaited the cancellation of some terms before the system order became known, which means that the system order is less obvious from the block diagram approach. A set of equations such as (1.13) and (1.14) are called "state variable" equations in that these variables represent the minimum set needed to evaluate the "state" of the system.

❑ Computer-Aided Learning

Throughout this textbook, we encourage the understanding of basic concepts. Once understood, we then encourage the use of computer packages to carry out difficult calculations. The pairing of basic concepts with computer-aided learning should be an equal partnership without overemphasis on the computer as the primary mode of understanding, but neither should there be theory without practice. This text is keyed to the use of the student version of MATLAB, including the Control Systems Toolbox. MATLAB has become the preeminent program of choice among control engineers. Of course, other computer packages may be applied to the computer-aided learning sections of this text.

In this section, the following MATLAB commands are introduced:

```
syms s
inv(matrix)
simplify
```

One strength of MATLAB is its ability to perform symbolic mathematical operations. Polynomials in "s" can be created and multiplied. The "syms" command causes any indicated variable (such as "s") to be used symbolically (not as a numerical value). Other useful commands are "simplify" which causes the computer to combine terms and cancel where needed and "inv(m)" which causes the inverse of the matrix m to be taken. Rows of m are separated by ";" and columns are separated by spaces or commas. The "inv" command makes it possible to solve a number of difficult problems just presented in the text.

C1.1 Write MATLAB commands to evaluate and simplify Equation (1.12), which gives I_2(s).

Ans.
```
Sym s
t1=8*s/(s^2+81)
t2=4+1/5/s
t3=3*s+6+1/5/s
t4=6*s+11+1/5/s
t5=t2
t112=t1*t2/(t3*t4-t5^2)
simplify(t112)
```

C1.2 Write MATLAB commands to solve Equations (1.7) and (1.8) to obtain the I_2(s) of Equation (1.16),

Ans.
```
syms s
m=[3*s+6+1/5/s -4-1/5/s; -4-1/5/s 6*s+11+1/5/s]
e=[8*s/(s^2+81); 0]
I=inv(m)*e
```

Note: $I_2(s)$ is $I(2, 1)$.

❑ DRILL PROBLEMS

D1.8 For the circuit of Figure D1.8, Write an integrodifferential mesh equations in terms of i_1 and i_2. Use Laplace transforms to obtain the form of Equations (1.9) and (1.10).

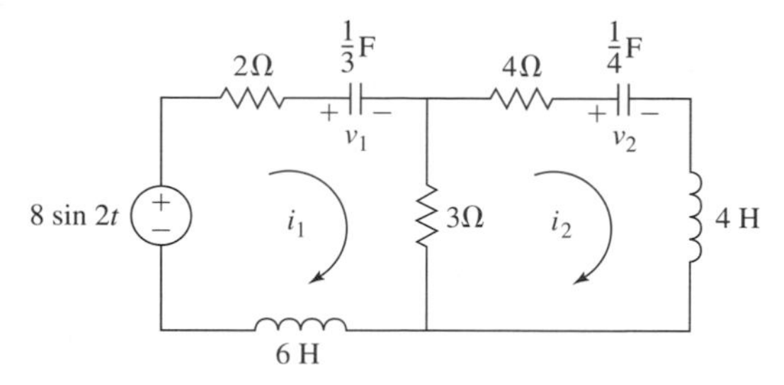

Figure D1.8

Ans.

$$\left(5 + 6s + \frac{3}{s}\right) I_1(s) - 3I_2(s) = \frac{16}{s^2 + 4}$$

$$-3I_1(s) + \left(7 + 4s + \frac{4}{s}\right) = 0$$

D1.9 For the circuit of Figure D.1.8, choose one state variable for each energy storage variable; that is, choose i_1, i_2, v_1 and v_2. Write four first-order differential equations and convert to Laplace form of Equation (1.14).

Ans.

$$\begin{bmatrix} \left(s + \frac{5}{6}\right) & -\frac{1}{2} & \frac{1}{6} & 0 \\ -\frac{3}{4} & s + \frac{7}{4} & 0 & \frac{1}{4} \\ -3 & 0 & s & 0 \\ 0 & -4 & 0 & s \end{bmatrix} \begin{bmatrix} I_1(s) \\ I_2(s) \\ v_1(s) \\ v_2(s) \end{bmatrix} = \begin{bmatrix} \frac{3}{4}\left(2/(s^2+4)\right) \\ 0 \\ 0 \\ 0 \end{bmatrix}$$

Characteristic polynomial $= 24s^2 + 62s^3 + 62s^2 + 41s + 12$

1.5.3 Node Analysis

In the nodal method of network analysis, one node in the network is chosen as the reference node and voltages between the reference node and each other node are defined. Expressing the element currents in terms of node voltages and applying Kirchhoff's current law at each node except the reference node gives the same number of independent simultaneous integrodifferential equations as there are node voltages. For the network of Figure 1.7 (a), the node voltages are labeled v_1(t) and $v_2(t)$. In Figure 1.7 (b) the branch currents are expressed in terms of these node voltages. Applying Kirchhoff's current law at node 1 gives:

Node analysis methodology.

$$\frac{1}{2}\int_{-\infty}^{t} v_1 dt + \frac{1}{3}v_1 + 4\frac{d}{dt}(v_1 - v_2) + \frac{1}{5}(v_1 - v_2) = 12$$

At node 2.

$$\frac{1}{6}v_2 - 4\frac{d}{dt}(v_1 - v_2) - \frac{1}{5}(v_1 - v_2) = -\sin t$$

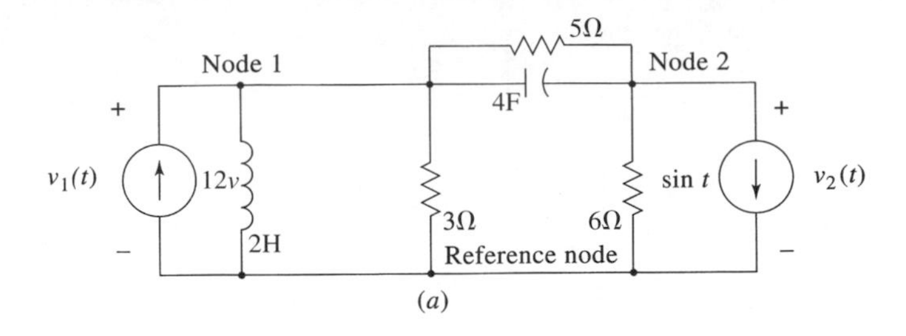

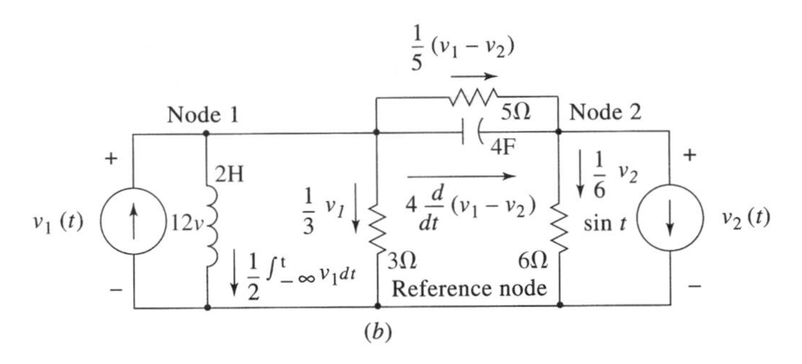

Figure 1.7 Writing simultaneous node equations for an electrical network. (a) Electrical network example for nodal analysis. (b) Network with element currents expressed in terms of the node voltages.

Collecting terms gives the following two simultaneous integrodifferential equations in $v_1(t)$ and $v_2(t)$:

$$4\frac{dv_1}{dt} + \frac{8}{15}v_1 + \frac{1}{2}\int_{-\infty}^{t} v_1 dt - 4\frac{dv_2}{dt} - \frac{1}{5}v_2 = 12$$

$$-4\frac{dv_1}{dt} - \frac{1}{5}v_1 + 4\frac{dv_2}{dt} + \frac{11}{30}v_2 = -\sin t$$

These equations are Laplace-transformed, again with zero initial conditions, with the result:

$$\left(4s + \frac{8}{15} + \frac{1}{2s}\right) V_1(s) + \left(-4s - \frac{1}{5}\right) V_2(s) = \frac{12}{s} \qquad [1.17]$$

$$\left(-4s - \frac{1}{5}\right) V_1(s) + \left(4s + \frac{11}{30}\right) V_2(s) = \frac{-1}{s^2+1} \qquad [1.18]$$

Again, since these are now functions of s rather than t, uppercase variables are used. Equations (1.17) and (1.18) are written in standard form as follows:

Using matrix to summarize node equations.

$$Y_{11}V_1(s) + Y_{12}V_2(s) = I_1(s) \qquad [1.19]$$

$$Y_{21}V_1(s) + Y_{22}V_2(s) = I_2(s) \qquad [1.20]$$

where again the (s) notation is understood for Y_{ij}, and

Y_{11} = sum of all admittances attached to node 1, in this case equal to $4s + 8/15 + (1/2s)$

$Y_{12} = Y_{21}$ = sum of all admittances connected between node 1 and node 2, in this case equal to $(-4s - 1/5)$

Y_{22} = sum of all admittances attached to node 2, in this case equal to $(4s + 11/30)$

$I_1(s)$ = independent current source driving node 1, in this case equal to $12/s$

$I_2(s)$ = independent current source driving node 2, in this case equal to $-1/(s^2+1)$

The solution for $V_2(s)$ of Equations (1.19) and (1.20) is found with Cramer's rule, as follows:

$$V_2(s) = \frac{\begin{bmatrix} Y_{11} & I_1(s) \\ Y_{21} & I_2(s) \end{bmatrix}}{\begin{bmatrix} Y_{11} & Y_{12} \\ Y_{21} & Y_{22} \end{bmatrix}} \qquad [\mathbf{1.21}]$$

Using Cramer's rule.

where the values of the admittances and independent driving currents are given above. Notice that this solution has the same form as that for the loop analysis shown in Equation (1.11).

Simple models for other common electrical and electronic devices are summarized in Figure 1.8. Of special importance is the operational amplifier shown in Figure 1.8(c).

❑ Computer-Aided Learning

C1.3 Write MATLAB commands to obtain $V_1(s)$ and $V_2(s)$ from Equations [1.17] and [1.18].

Ans.
```
syms s
m=[4*s+8/15+1/2/s -4*s-1/5;-4*s-1/5 4*s+11/30]
i=[12/s;-1/(s^2 +1)]
V=inv(m)*i
```

❑ DRILL PROBLEM

D1.10 Write simultaneous Laplace nodal equations for the electrical network shown in Figure D1.10 in terms of the indicated node voltages.

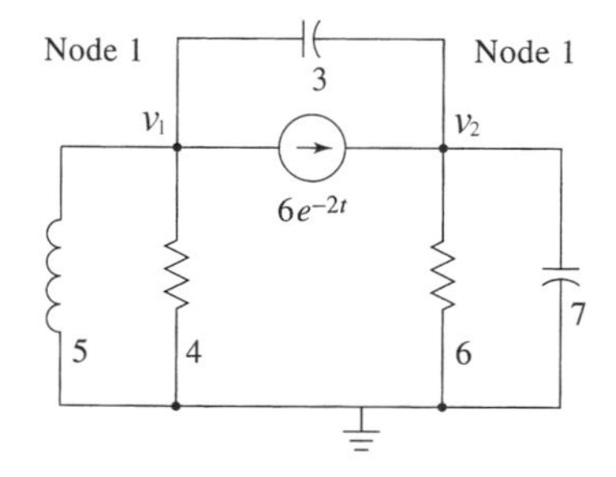

Figure D1.10

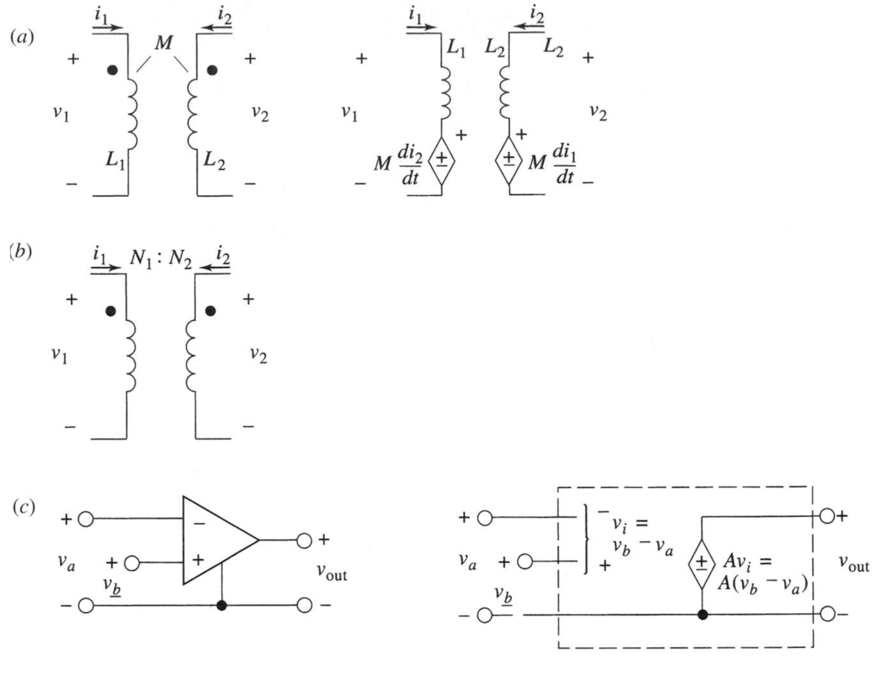

Figure 1.8 Simple models of some electrical and electronic devices. (a) *Transformer* with core in linear region. Resistance of the coil wires can be included as additional resistors at each port. (b) *Ideal transformer*. The ideal transformer models a transformer with perfect magnetic coupling, $M = \sqrt{L_1 L_2}$. For the ideal transformer,

$$v_2 = \frac{N_2}{N_1} v_1 \quad i_2 = -\frac{N_1}{N_2} i_1$$

where N_1 and N_2 are the number of turns of the L_1 and L_2 coils, respectively. (c) *Operational amplifier*. The idealized operational amplifier produces an output voltage that is proportional to the difference between two input voltages.

Ans.

$$V_1(s)\left(\frac{1}{4} + \frac{1}{5s} + 3s\right) - 3s\,V_2(s) = -\frac{6}{s+2}$$

$$-3s\,V_1(s) + V_2(s)\left(\frac{1}{6} + 10s\right) = \frac{6}{s+2}$$

1.5.4 Analyzing Operational Amplifier Circuits

We present here an analysis method that easily yields the mathematical model for "ideal" operational amplifier circuits. An ideal operational amplifier (op-amp) has the following characteristics:

1. Input resistance into the v_a and v_b terminals approaches infinity, so no current enters those terminals.

2. The output resistance is nearly zero, so amplifiers can be cascaded without "loading" effects.
3. The open-loop voltage gain A approaches infinity, so $v_a = v_b$.

Models of electrical and electronic devices are presented.

As a result of such large voltage gain A, the output voltage, from Figure 1.8 (c)

$$v_{\text{out}} = A(v_a - v_b)$$

so

$$v_a - v_b = \frac{v_{\text{out}}}{A}$$

As A is allowed to approach infinity. $v_a - v_b = 0$ and

$$v_a = v_b \qquad [1.22]$$

Several examples are presented to explain the method of ideal op-amp analysis. The analysis of a network containing an op-amp is accomplished by following this procedure:

1. Write a node equation for node v_a, assuming that no current enters the amplifier.
2. Write a node equation for node v_b, assuming that no current enters the amplifier.
3. Equate v_a to v_b and solve.

Example 1: Find the closed-loop gain for the noninverting amplifier of Figure 1.9. First, the node equation is written for node v_a,

Non-inverting op amp.

$$\frac{v_a - 0}{R_a} + \frac{v_a - v_{\text{out}}}{R_F} = 0$$

and

$$v_a = \frac{R_a}{R_a + R_F} v_{\text{out}}$$

Second, write the node equation for node v_b. Obviously, $v_b = v_{\text{in}}$.

Third, set $v_a = v_b$, since A approaches infinity and

$$v_a = \frac{R_a v_{\text{out}}}{R_a + R_F}$$

Hence the closed-loop gain is

$$\frac{v_{\text{out}}}{v_{\text{in}}} = 1 + \frac{R_F}{R_a} \qquad [1.23]$$

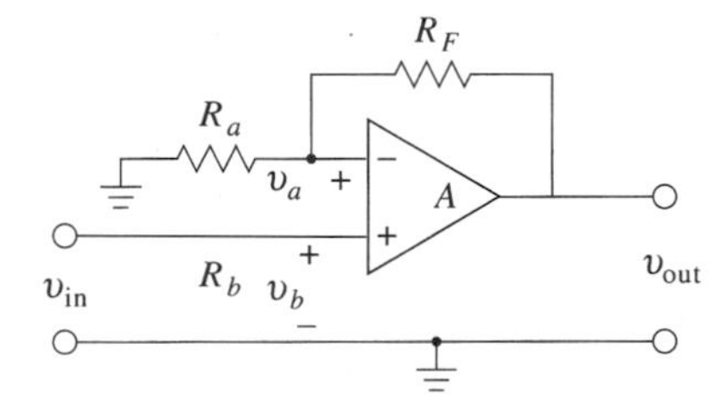

Figure 1.9 Noninverting amplifier.

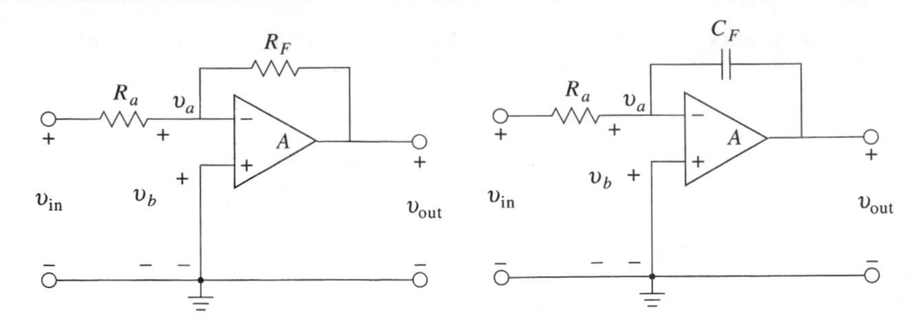

Figure 1.10 Inverting amplifier.

Figure 1.11 Integrating amplifier.

Example 2: The closed-loop gain for the inverting amplifier shown in Figure 1.10 is now determined using the procedure shown in Example 1. First write the node equations of v_a:

$$\frac{v_a - v_{\text{in}}}{R_a} + \frac{v_a - v_{\text{out}}}{R_F} = 0$$

Next write the node equation for the node at v_b. Here $v_b = 0$ and $v_a = v_b = 0$. Solving for the closed-loop gain.

$$\frac{v_{\text{out}}}{v_{\text{in}}} = \frac{-R_F}{R_a} \qquad [1.24]$$

Example 3: Consider the integrating amplifier of Figure 1.11. Replace R_F in Example 2 with the impedance of the capacitor $1/sC_F$. The node equation for v_a is

$$\frac{V_a - V_{\text{in}}}{R_a} + (V_a - V_{\text{out}})sC_F = 0$$

The node equation at v_b is $V_b = 0$, and

$$V_a = V_b \quad \text{so} \quad V_a = 0$$

Uppercase letters indicate that the variables are functions of s. Solving for the closed-loop gain.

$$\frac{V_{\text{out}}(s)}{V_{\text{in}}(s)} = \frac{-1}{sR_aC_F} = \left(-\frac{1}{s}\right)\frac{1}{R_aC_F} \qquad [1.25]$$

which is the Laplace transform for a negative integrator, due to the $-1/s$.

❑ DRILL PROBLEMS

D1.11 For the operational amplifier circuit of Figure D1.11, do the following.

(a) Find v_{out}.

(b) Find i.

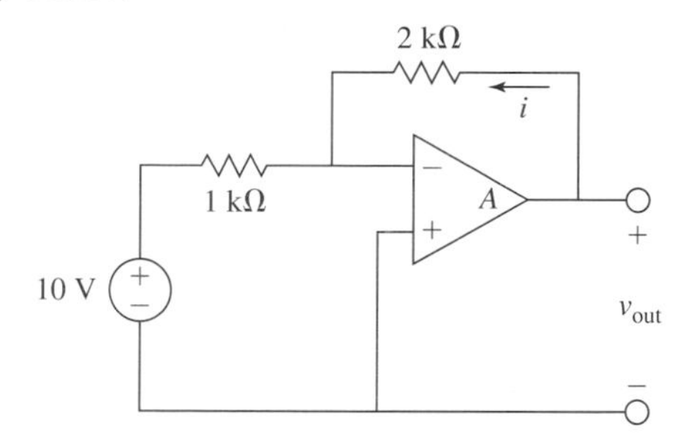

Figure D1.11

Ans. (a) $v_{\text{out}} = -20$ V; (b) $i = -10$ mA

D1.12 For the operational amplifier circuit of Figure D1.12, assuming $v_{\text{out}} = 0$ for $t = 0$:

(a) Find $v_{\text{out}}(t)$ if $v_{\text{in}}(t) = 10$ V.

(b) Find $v_{\text{out}}(t)$ if $v_{\text{in}}(t) = 20 \cos 100t$.

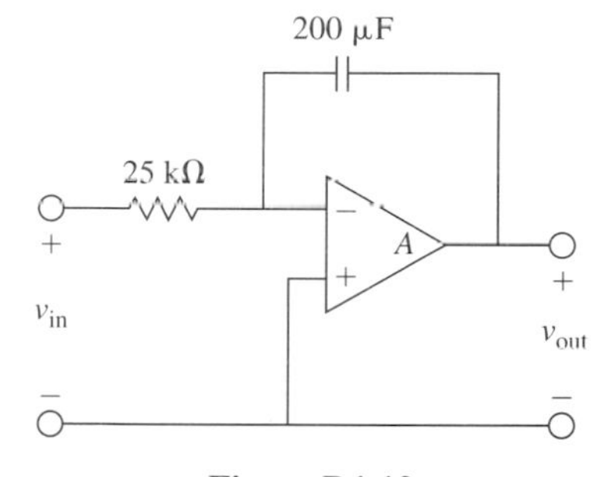

Figure D1.12

Ans. (a) $-2t$; (b) $-\frac{1}{25} \sin 100t$.

1.5.5 Operational Amplifier Applications

The result of Example 2 can be extended to a wide range of other applications by replacing R_a in Figure 1.10 with an impedance Z_a and by replacing R_f in Figure 1.10 with an impedance Z_f (which was $1/sC_F$ in Example 3). The result is Figure 1.12. Using Equation (1.24), it follows easily that

$$\frac{V_{\text{out}}(s)}{V_{\text{in}}(s)} = -\frac{Z_f(s)}{Z_a(s)} \qquad [1.26]$$

Suppose the circuit of Figure 1.13 is used in either the input or feedback branch of Figure 1.12. It is useful to obtain the impedance of the circuit of Figure 1.13. Since

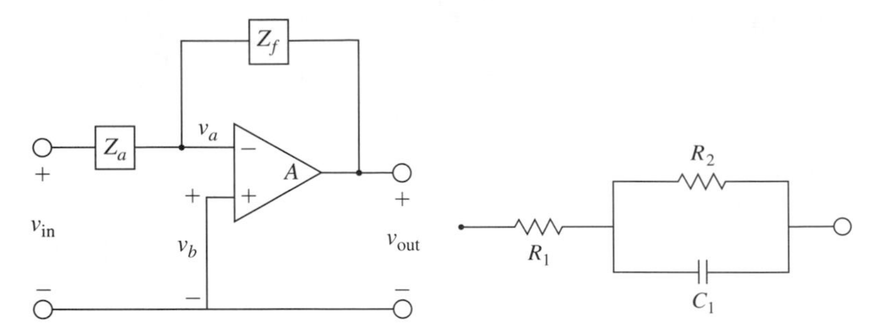

Figure 1.12 Inverting amplifier with arbitrary impedances.

Figure 1.13 Lag impedance circuit.

the impedance of the capacitor is $1/sC_1$, the circuit impedance is given by

$$Z(s) = R_1 + \frac{R_2/sC_1}{R_2 + 1/sC_1} = R_1 + \frac{R_2}{sR_2C_1 + 1}$$

$$= \frac{sR_1R_2C_1 + R_1 + R_2}{sR_2C_1 + 1} = \frac{R_1\left[s + \frac{1 + R_2/R_1}{R_2C_1}\right]}{s + 1/R_2C_1} \quad [\mathbf{1.27}]$$

The impedance of Equation (1.27) can be written compactly as

$$Z(s) = \frac{k(s+a)}{s+b} \quad [\mathbf{1.28}]$$

where $k = R_1$, $b = 1/R_2C_1$, and $a = (1 + R_2/R_1)b$. Thus a must be larger than b so the pole at $-b$ is located to the right of the zero at $-a$.

Figure 1.14 shows nine operational amplifier applications for various choices of Z_a and Z_F. For convenience, the negative of the transfer function is shown. We have already seen the inverter of Figure 1.14(a) in Figure 1.10 and the integrator of Figure 1.14(b) in Figure 1.11. To create a differentiator in Figure 1.14(c), a capacitor is placed in the input path while a resistor appears in the feedback path, a configuration exactly the opposite of that for the integrator. For the differentiator, the output voltage becomes the negative derivative of the input voltage. If the input voltage has noisy components, the differentiator output may have unwanted large high-frequency components. Often a filter is employed with the differentiator to eliminate the high-frequency spikes.

Differentiator and integrator.

In Figure 1.14(d), the circuit of Figure 1.13 is placed in the feedback path. The resulting circuit transfer function has a zero to the left of the pole, which results in negative phase at each frequency, referred to as phase lag; hence, the circuit of Figure 1.14(d) is called a lag circuit. The circuit of Figure 1.14(e) creates an inverse transfer function compared to the lag circuit; it is called a lead circuit owing to the positive phase angle of the transfer function where the zero is to the right of the pole.

Lag and lead circuits.

Suppose that two versions of the circuit of Figure 1.13 (having different components but the same form) are placed in the input path and in the feedback path.

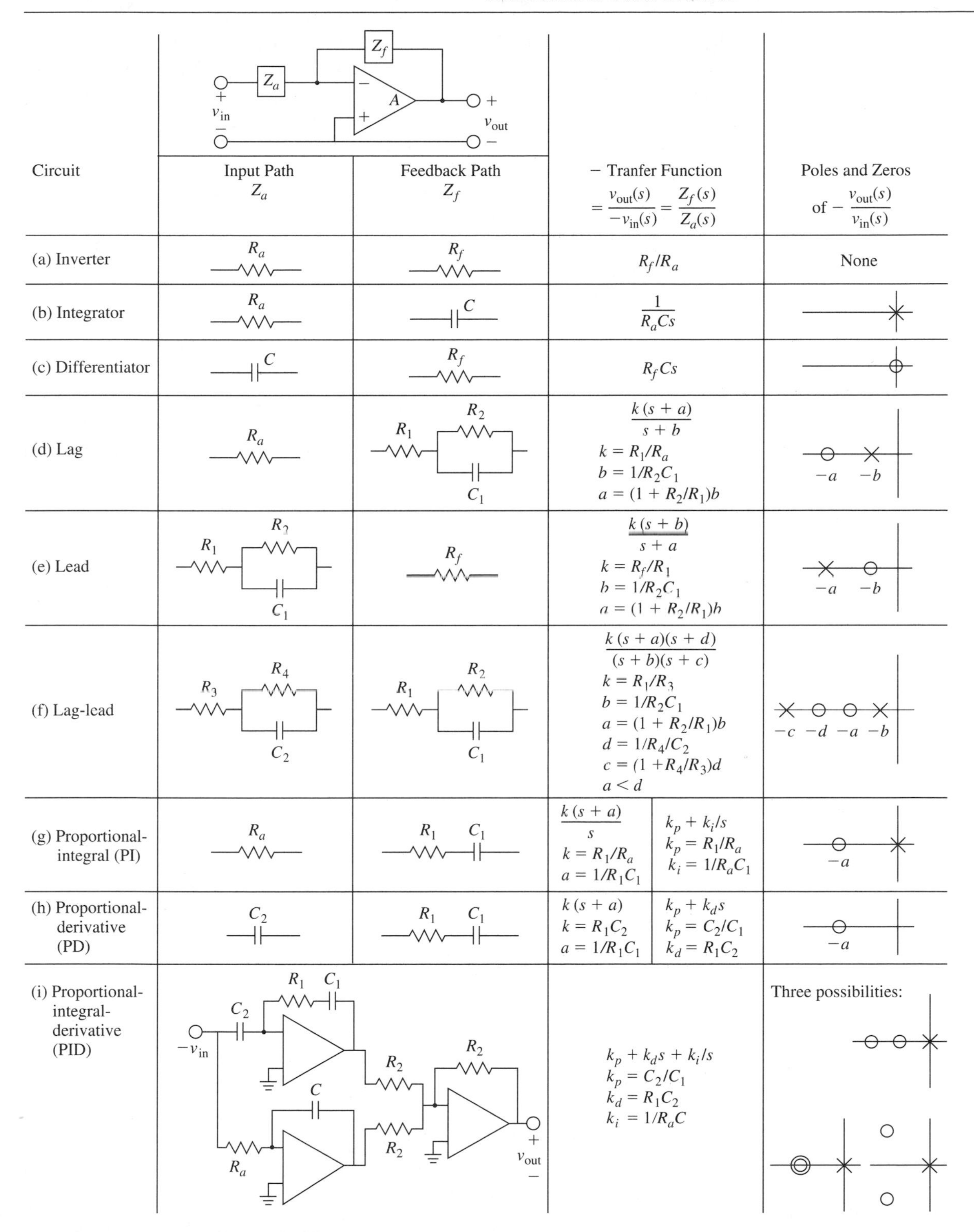

Figure 1.14 Operational amplifier applications.

The resulting circuit's transfer function of Figure 1.14(f) is the product of the transfer functions of Figure 1.14(d,e), causing lag (negative phase) and lead (positive phase) to both be present. If the lag pole and zero are placed closer to the origin than the lead pole and zero, the phase lag occurs at lower frequencies than does the phase lead. Appropriately, the circuit of Figure 1.14(f) is called a lag–lead circuit. The circuits of Figure 1.14(a–e) are special cases of the lag–lead circuit evolving by selecting appropriate values for the lag–lead components. For example, if $C_1 = \infty$, $R_2 = 0$ and $R_1 = R_f$, the lag–lead circuit becomes the lead circuit of Figure 1.14(e).

PI, PD, and PID circuits.

The circuits of Figures 1.14(g)–(i) are named because of a proportional effect (P, with a constant transmittance), derivative effect (D, with an s in the numerator of transmittance) and/or integral effect (I, with an s in denominator of the transmittance). These circuits provide PI, PD, and PID operations, respectively. Each circuit transmittance realizes a proportional gain k_p, an integral gain k_i and/or a derivative gain k_d. The PID circuit output is the sum of an integrator output and a PD output. For the PID circuit and PD circuit (as was the case for the differentiator), it may be desirable to include a low-pass filter to eliminate high-frequency spikes due to differentiating possibly noisy signals. The zeros of a PID circuit can be real and different, double real, or complex conjugate, as desired.

The circuits of Figure 1.14 will be very useful later in this text when compensation is explored, in which circuits are added to a system to improve system behavior, primarily PI, PD and PID, lag, lead, and lag–lead circuits.

As an example of circuit design, suppose we want to create a circuit with

$$-\frac{V_{\text{out}}(s)}{V_{\text{in}}(s)} = \frac{20(s+10)}{s+50}$$

Since the zero is to the right of the pole, choose the lead circuit of Figure 1.14(e). Then $k = 20$, $a = 50$, and $b = 10$. There are four dependent circuit elements, R_1, R_2, C_1 and R_f but only three independent constraints, k, a, and b. Thus, one circuit element may be selected arbitrarily. Suppose we want resistors in the kilohm range. Then if we select R_1 to be 1 kΩ,

$$k = R_f/R_1 = 20$$

$$R_f = kR_1 = 20\,\text{k}\Omega$$

$$a = 50 = (1 + R_2/R_1)b = (1 + R_2/R_1)10$$

$$R_2/R_1 = 5 - 1 = 4$$

$$R_2 = 4R_1 = 4\,\text{k}\Omega$$

$$b = 1/R_2C_1 = 10$$

$$C_1 = 1/10R_2 = 1/(40 \times 10^3) = 25\,\mu\text{F}$$

Figure 1.15 shows the final circuit.

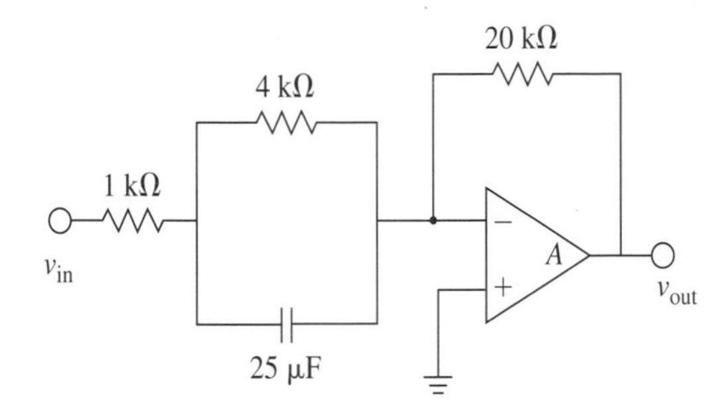

Figure 1.15 Operational amplifier circuit for $-V_{out}(s)/V_{in}(s) = [20(s + 10)/(s + 50)]$.

❑ **DRILL PROBLEMS**

D1.13 Design an operational amplifier circuit with $R_1 = 1\text{k}\,\Omega$ and

$$-\frac{V_{out}(s)}{V_{in}(s)} = \frac{20(s+50)}{s+10}$$

Ans. Lag circuit with $R_1 = 1\,\text{k}\Omega$, $R_a = 50\,\Omega$, $R_2 = 4\,\text{k}\Omega$, $C_1 = 25\,\mu\text{F}$.

D1.14 Design an operational amplifier circuit with $R_3 = 1\,\text{k}\Omega$ and

$$-\frac{V_{out}(s)}{V_{in}(s)} = \frac{10(s+10)(s+20)}{(s+2)(s+80)}$$

Ans. Lag–lead circuit with $R_1 = 10\,\text{k}\Omega$, $R_2 = 40\,\text{k}\Omega$, $C_1 = 12.5\,\mu\text{F}$, $R_3 = 1\,\text{k}\Omega$, $R_4 = 3\,\text{k}\Omega$, $C_2 = 16.7\,\mu\text{F}$.

D1.15 Design an operational amplifier circuit with $R_a = 1\,\text{k}\Omega$, $R_1 = 10\,\text{k}\Omega$ and

$$-\frac{V_{out}(s)}{V_{in}(s)} = 10 + 0.002s + \frac{300}{s}$$

Ans. PID circuit with $R_a = 1\,\text{k}\Omega$, $R_1 = 10\,\text{k}\Omega$, $C = 3.33\,\mu\text{F}$, $C_1 = .02\,\mu\text{F}$, $C_2 = .2\,\mu\text{F}$.

1.6 Translational Mechanical Components

1.6.1 Free-Body Diagrams

The force and position relations for the translational mechanical mass, spring, and damper elements are given in Figure 1.16. As in a free-body diagram, the forces shown are those applied to the element. The force produced by a spring is proportional to the translation of this spring. For example, in Figure 1.16, as the mass is moved a position displacement of amount x, the spring produces a force on the mass in the negative x direction in the amount of kx. Similarly, as the mass is moved in the positive dx/dt direction, the linear viscous damper produces a force on the mass in the negative dx/dt direction in the amount $B(dx/dt)$. A method of analysis for translational mechanical systems involving these elements is as follows:

1. Define positions with directional senses for each mass in the system.

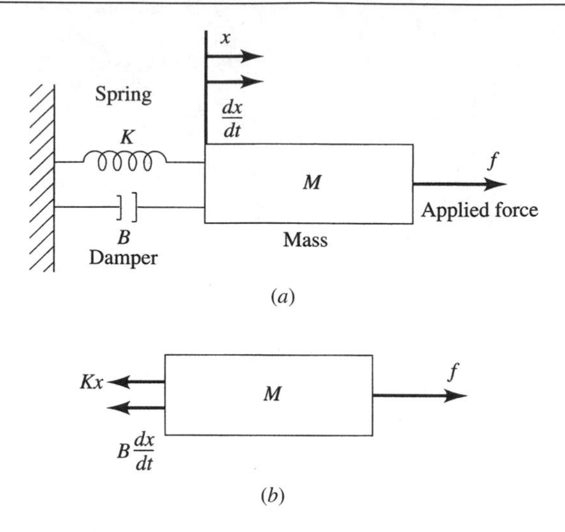

Figure 1.16 Translational mechanical network force—position relations. (a) Spring-mass-damper system. At equilibrium (with no applied force), x is zero. (b) Free-body diagram.

Procedure for translational systems.

2. Draw free-body diagrams for each of the masses, expressing the forces on them in terms of mass position and velocity.
3. Write an equation for each mass, equating the algebraic sum of forces acting in the same direction to $M(d^2x/dt^2)$.

This procedure is applied to the system of Figure 1.17(a), where mass positions have been defined. In Figure 1.17(b), free-body diagrams for the two masses are shown.

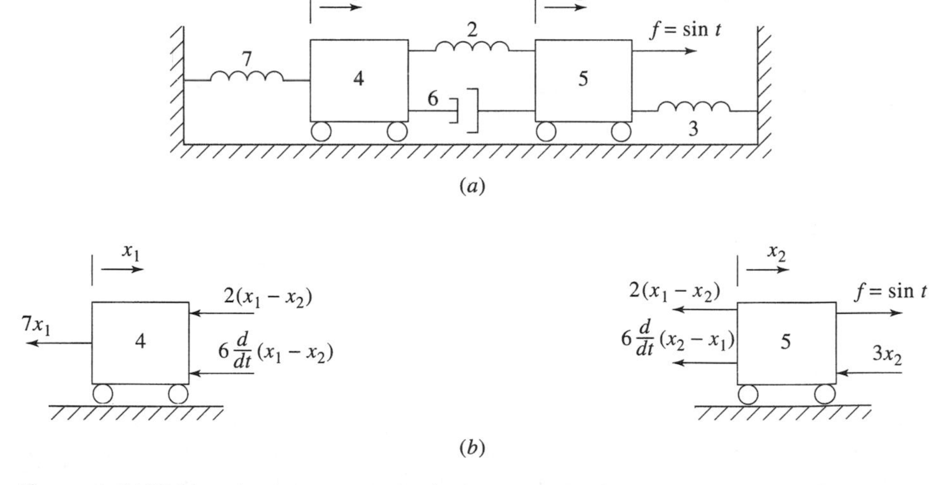

Figure 1.17 Writing simultaneous mechanical network equations. (a) A translational mechanical network. (b) Free-body diagrams for the masses.

Equating forces for the first mass gives:

$$4\frac{d^2x_1}{dt^2} = -7x_1 - 2(x_1 - x_2) - 6\frac{d}{dt}(x_1 - x_2)$$

The forces on the second mass, of the amount 5, are summed as follows:

$$5\frac{d^2x_2}{dt^2} = f - 3x_2 - 2(x_2 - x_1) - 6\frac{d}{dt}(x_2 - x_1)$$

where $f = \sin t$.

Collecting terms. We find that the simultaneous differential equations in x_1 and x_2 for the translational system are

$$4\frac{d^2x_1}{dt^2} + 6\frac{dx_1}{dt} + 9x_1 - 6\frac{dx_2}{dt} - 2x_2 = 0$$

$$-6\frac{dx_1}{dt} - 2x_1 + 5\frac{d^2x_2}{dt^2} + 6\frac{dx_2}{dt} + 5x_2 = \sin t$$

These terms are Laplace-transformed (see Appendix B) assuming zero conditions, with the result

$$(4s^2 + 6s + 9)X_1(s) + (-6s - 2)X_2(s) = 0 \qquad [1.29]$$

$$(-6s - 2)X_1(s) + (5s^2 + 6s + 5)X_2(s) = \frac{1}{(s^2 + 1)} \qquad [1.30]$$

The X variables are uppercase since they are functions of s rather then functions of t. Equations (1.29) and (1.30) can be written in standard form as follows:

Using a matrix to summarize translational equations.

$$W_{11}X_1(s) + W_{12}X_2(s) = F_1(s) \qquad [1.31]$$

$$W_{21}X_1(s) + W_{22}X_2(s) = F_2(s) \qquad [1.32]$$

where the (s) is omitted from W_{ij}, and

W_{11} = inertial force M_1s^2 plus all the other forces attached to M_1 (variable X_1); in this case $W_{11} = 4s^2 + 6s + 9$

$W_{12} = W_{21}$ = sum of all forces connected between $X_1(s)$ and $X_2(s)$; in this case $W_{12} = W_{21} = (-6s - 2)$

W_{22} = inertial force M_2s^2 plus all other forces attached to M_2 (variable X_2); in this case, $W_{22} = (5s^2 + 6s + 5)$

$F_1(s)$ = independent driving force on M_1; in this case, $F_1(s) = 0$

$F_s(s)$ = independent driving force on M_2; in this case, $F_2(s) = 1/(s^2 + 1)$

To solve for $X_1(s)$, use Cramer's rule as follows:

Using Cramer's rule.

$$X_1(s) = \frac{\begin{vmatrix} F_1(s) & W_{12} \\ F_2(s) & W_{22} \end{vmatrix}}{\begin{vmatrix} W_{11} & W_{12} \\ W_{21} & W_{22} \end{vmatrix}} \qquad [1.33]$$

Notice the similarity between Equations (1.33), (1.21), (1.11). All these equations represent linear differential equations with constant coefficients.

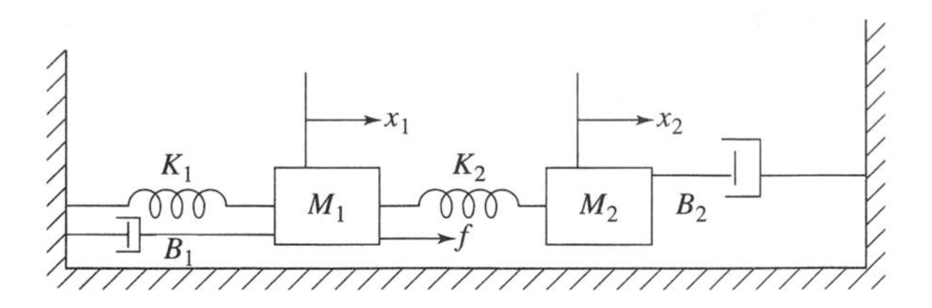

Figure 1.18 Vibration–damper example.

As an interesting example, consider the vibration damper shown in the schematic of Figure 1.18. The positive direction for both x_1 and x_2 is to the right. Newton's equations for this translation system are

$$M_1 \frac{d^2x_1}{dt^2} = f - K_1x_1 - K_2(x_1 - x_2) - B_1\frac{dx_1}{dt}$$

$$M_2 \frac{d^2x_2}{dt^2} = -K_2(x_2 - x_1) - B_2\frac{dx_2}{dt}$$

These equations are Laplace-transformed, with zero initial conditions, and then rearranged as follows:

$$(M_1s^2 + B_1s + K_1 + K_2)X_1(s) + (-K_2)X_2(s) = F(s) \qquad [1.34]$$

$$(-K_2)X_1(s) + (M_2s^2 + B_2s + K_2)X_2(s) = 0 \qquad [1.35]$$

Solving for $X_1(s)$.

The transforms of the variable and driving functions are capitalized to indicate that they are functions of s. Equations (1.34) and (1.35) are solved for $X_1(s)$ under the condition of low damping, so that $B_1 = 0 = B_2$. Solving with Cramer's rule. We have

$$X_1(s) = \frac{\begin{vmatrix} F(s) & (-K_2) \\ 0 & M_2s^2 + K_2 \end{vmatrix}}{\begin{vmatrix} M_1s^2 + K_1 + K_2 & (-K_2) \\ (-K_2) & M_2s^2 + K_2 \end{vmatrix}} \qquad [1.36]$$

which simplifies to:

$$\frac{X_1(s)}{F(s)} = \frac{M_2s^2 + K_2}{(M_1M_2)s^4 + [M_1K_2 + M_2(K_1 + K_2)]s^2 + K_1K_2} \qquad [1.37]$$

The system is a vibration damper, which can be "tuned" to substantially reduce a particular vibration frequency.

❑ DRILL PROBLEM

D1.16 Write simultaneous differential equations for the translational mechanical networks shown in Figure D1.16 in terms of the indicated mass positions. The mass positions are defined so that when all positions are zero, the spring forces are zero. Use Laplace to obtain the form of Equations (1.31) and (1.32). Find $X_1(s)$.

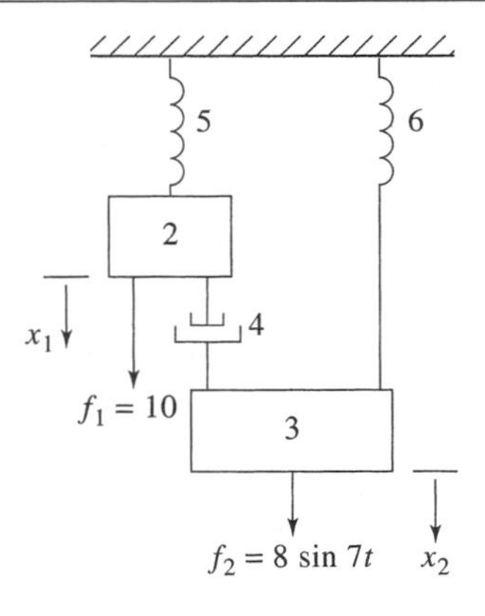

Figure D1.16

Ans.

$$\frac{2d^2x_1}{dt^2} = 10 - 5x_1 - 4\frac{d}{dt}[x_1 - x_2]$$

$$\frac{3d^2x_2}{dt^2} = 8\sin 7t - 6x_2 - 4\frac{d}{dt}[x_2 - x_1]$$

$$\begin{bmatrix} (2s^2 + 4s + 5) & -4s \\ -4s & (3s^2 + 4s + 6) \end{bmatrix} \begin{bmatrix} X_1(s) \\ X_2(s) \end{bmatrix} = \begin{bmatrix} \dfrac{10}{s} \\ \dfrac{56}{s^2 + 49} \end{bmatrix}$$

$$X_1(s) = \frac{30s^2 + 40s + 60}{SD(s)} + \frac{224s}{(s^2 + 49)D(s)}$$

$$D(s) = (2s^2 + 4s + 5)(3s^2 + 4s + 6) - 16s^2$$

1.6.2 State variables

As we did in section 1.5.2, we can select the system variables in such a way that a set of first-order differential equations may be used to describe the dynamic behavior. For the system of Figure 1.18, two second-order differential equations result, one in terms of the second derivative of x_1 and one in terms of the second derivative of x_2. Since the second derivative is the first derivative of the first derivative, state variables dx_1/dt and dx_2/dt should be used, in which case two first-order differential equations result:

$$\frac{d^2x_1}{dt^2} = \frac{d}{dt}\frac{dx_1}{dt}$$

$$\frac{d^2x_2}{dt^2} = \frac{d}{dt}\frac{dx_2}{dt}$$

Two other state variables could be x_1 and x_2, whose first derivatives are dx_1/dt and dx_2/dt. If we select state variables, dx_1/dt, x_1, dx_2/dt, and x_2, a set of four first-order differential equations results, two of which simply are identities.

Selecting state variables.

The result is

Time domain matrices.

$$\frac{d}{dt}\begin{bmatrix}\dfrac{dx_1}{dt}\\ x_1\\ \dfrac{dx_2}{dt}\\ x_2\end{bmatrix}=\begin{bmatrix}\dfrac{-B_1}{M_1} & -\left(\dfrac{K_1}{M_1}+\dfrac{K_2}{M_1}\right) & 0 & \dfrac{K_2}{M_1}\\ 1 & 0 & 0 & 0\\ 0 & \dfrac{K_2}{M_2} & \dfrac{-B_2}{M_2} & \dfrac{-K_2}{M_2}\\ 0 & 0 & 1 & 0\end{bmatrix}\begin{bmatrix}\dfrac{dx_1}{dt}\\ x_1\\ \dfrac{dx_2}{dt}\\ x_2\end{bmatrix}+\begin{bmatrix}\dfrac{f}{M_1}\\ 0\\ 0\\ 0\end{bmatrix} \quad [1.38]$$

As in Section 1.5.2, the Laplace transform is taken with the initial conditions equal to zero.

Laplace domain matrices.

$$\begin{bmatrix}s+\dfrac{B_1}{M_1} & \dfrac{(K_1+K_2)}{M_1} & 0 & \dfrac{-K_2}{M_1}\\ -1 & s & 0 & 0\\ 0 & \dfrac{-K_2}{M_2} & s+\dfrac{B_2}{M_2} & \dfrac{K_2}{M_2}\\ 0 & 0 & -1 & s\end{bmatrix}\begin{bmatrix}sX_1(s)\\ X_1\\ sX_2(s)\\ X_2(s)\end{bmatrix}=\begin{bmatrix}\dfrac{F(s)}{M_1}\\ 0\\ 0\\ 0\end{bmatrix} \quad [1.39]$$

If Cramer's rule is applied to Equation (1.39) to get $X_1(s)$ with $B_1 = B_2 = 0$, then

$$X_1(s)=\frac{\begin{bmatrix}s & F(s)/M_1 & 0 & -K_2/M_1\\ -1 & 0 & 0 & 0\\ 0 & 0 & s & K_2/M_2\\ 0 & 0 & -1 & s\end{bmatrix}}{\begin{bmatrix}s & (K_1+K_2)/M_1 & 0 & -K_2/M_1\\ -1 & s & 0 & 0\\ 0 & -K_2/M_2 & s & K_2/M_2\\ 0 & 0 & -1 & s\end{bmatrix}} \quad [1.40]$$

so that

$$\frac{X_1(s)}{F(s)}=\frac{s^2/M_1+K_2/M_1M_2}{s^4+s^2\left(K_2/M_2+(K_1+K_2)/M_1\right)+K_1K_2/M_1M_2} \quad [1.41]$$

Equation (1.41) is equivalent to Equation (1.37) if the numerator and denominator of Equation (1.37) are divided by M_1M_2.

The selection of state variables can proceed for any system. The result is a set of n, first-order differential equations, where n is the highest power of s appearing in the denominators of equations like (1.16) and (1.41). We will have much more to say about ratios such as $X_1(s)/F(s)$ and polynomials like the denominator of Equation (1.41).

❑ DRILL PROBLEM

D1.17 Write simultaneous differential equations for the translational mechanical networks shown in Figure D1.17 in terms of the indicated mass positions. The mass positions are defined so that when all positions are zero, the spring forces are zero. Convert your equations to the state variable form of Equation (1.38). Find $X_1(s)$.

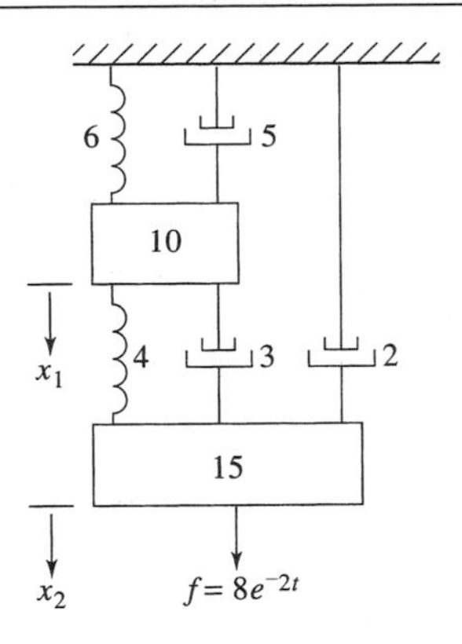

Figure D1.17

Ans.

$$10\frac{d^2x_1}{dt^2} = -6x_1 - 4(x_1 - x_2) - 5\frac{dx_1}{dt} - 3\frac{d}{dt}(x_1 - x_2)$$

$$15\frac{d^2x_2}{dt^2} = 8e^{-2t} - 4(x_2 - x_1) - 3\frac{d}{dt}(x_2 - x_1) - 2\frac{dx_2}{dt}$$

$$\frac{d}{dt}\begin{bmatrix} dx_1/dt \\ x_1 \\ dx_2/dt \\ x_2 \end{bmatrix} = \begin{bmatrix} -0.8 & -1 & 0.3 & 0.4 \\ 1 & 0 & 0 & 0 \\ 0.2 & \frac{4}{15} & -\frac{1}{3} & -\frac{4}{15} \\ 0 & 0 & 1 & 0 \end{bmatrix}\begin{bmatrix} dx_1/dt \\ x_1 \\ dx_2/dt \\ x_2 \end{bmatrix} + \begin{bmatrix} 0 \\ 0 \\ (\frac{8}{15})e^{-2t} \\ 0 \end{bmatrix}$$

$$x_1(s) = \frac{0.16s + 0.213}{(s+2)(s^4 + 1.133s^3 + 1.423s^2 + 0.387s + 0.160)}$$

❑ Computer-Aided Learning

C1.4 Write MATLAB commands to solve Equations (1.29) and (1.30) for $X_1(s)$ and $X_2(s)$.

Ans.
```
syms s
m=[(4*s^2+6*s+9)(-6*s-2);(-6*s-2)(5*s^2+6*s+5)]
f=[0; 1/(s^2+1)]
X=inv(m)*f
```

C1.5 Write MATLAB commands to obtain Equation (1.41) from Equation (1.39) when $B_1 = B_2 = 0$.

Ans.
```
syms s m1 m2 k1 k2
m=[s (k1+k2)/m1 0 -k2/m1;-1 s 0 0;
    0 -k2/m2 s k2/m2; 0 0 -1 s]
f=[ 1/m1; 0; 0; 0]
y=inv(m)*f
```

Note: $y(2, 1)$ is $X_1(s)/F(s)$.

C1.6 Write MATLAB commands to solve Drill Problem D1.17.

Ans.
```
Syms S
m=[(s+0.8) 1 -0.3 -0.4; -1 s 0 0;
    -0.2 -4/15 (s+1/3) 4/15; 0 0 -1 s]
f=[0; 0; (8/15)/(s+2); 0]
y=inv(m)*f
```

Note: $y(2, 1)$ is $X_1(s)$.

1.7 Rotational Mechanical Components

1.7.1 Free-Body Diagrams

Differential equations for angular motion are obtained in the same way as those for translational motion. Torque–position relations for rotational elements are summarized in Figure 1.19. The torques shown are those applied to the element. An analysis procedure is as follows:

Procedure for rotational systems.

1. Define angular positions with directional senses for each rotational mass.
2. Draw free-body diagrams for each of the rotational masses, expressing each torque in terms of the angular positions of the masses.
3. Write an equation for each rotational mass, equating the algebraic sum of torques on it to $J(d^2\theta/dt^2)$.

The procedure is applied to the rotational system of Figure 1.20(a). Free-body diagrams for this system are drawn in Figure 1.20(b). For the first rotational mass,

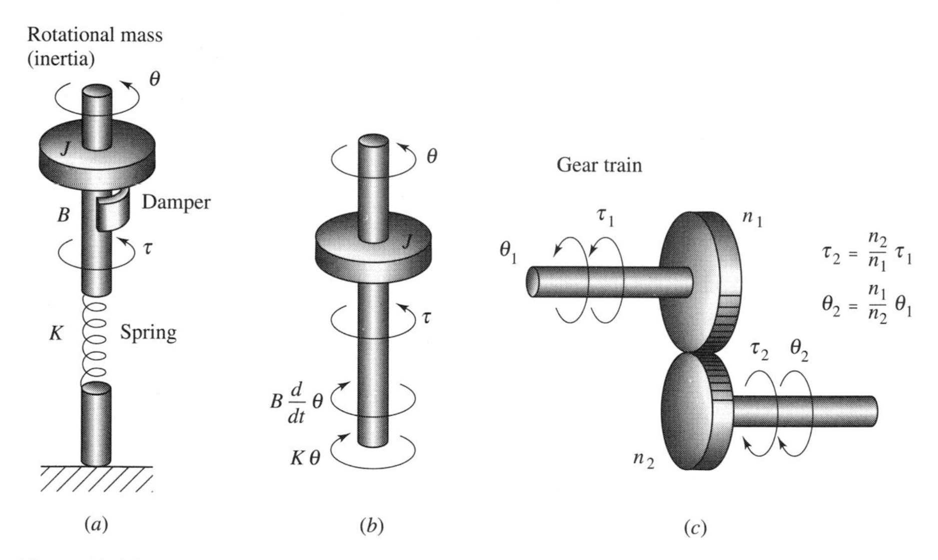

Figure 1.19 Rotational mechanical element torque–angle relations. (a) Rotational spring-mass-damper systems. At equilibrium (with no applied torque), the angle θ is zero. (b) Free-body diagram. (c) Gear train.

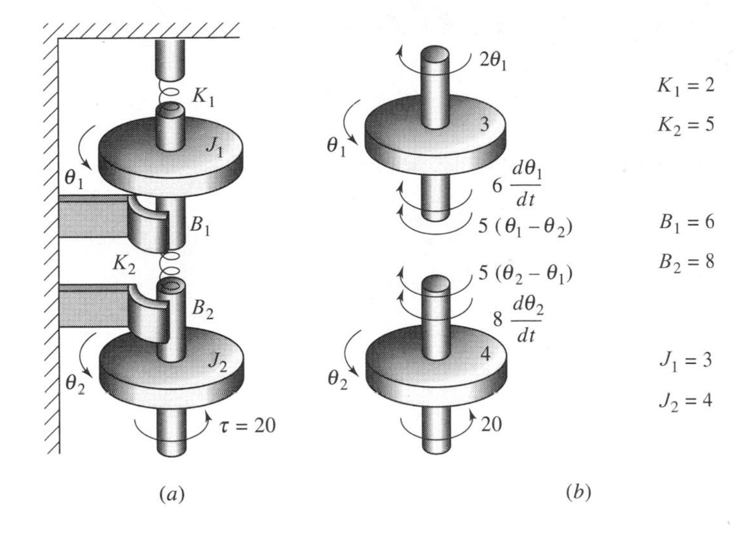

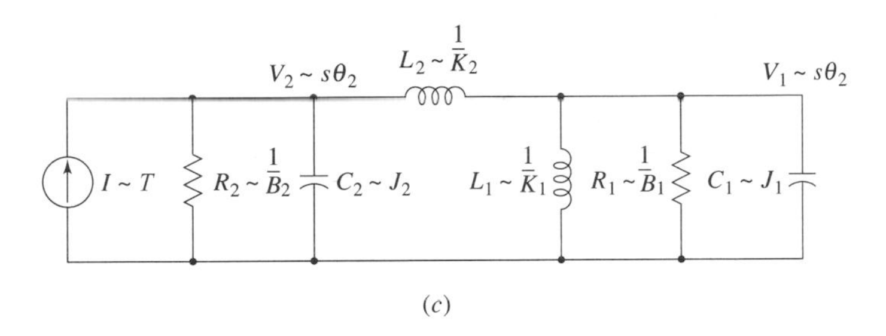

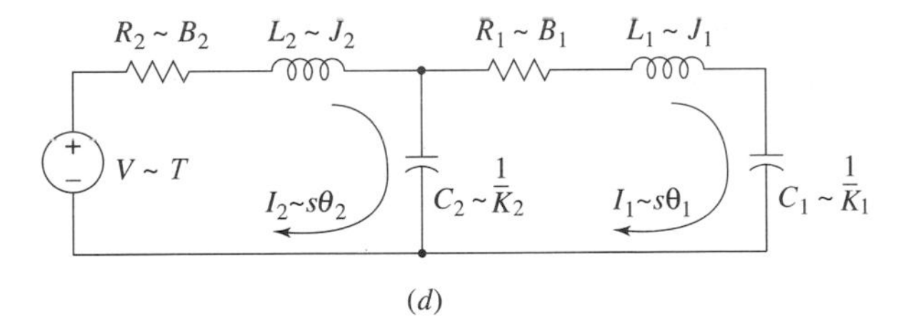

Figure 1.20 Writing simultaneous rotational mechanical network equations. (a) A rotational mechanical network. (b) Free-body diagram for the inertias with numerical values. (c) Nodal analog. (d) Mesh analog.

equating torques gives

$$3\frac{d^2\theta_1}{dt^2} = -2\theta_1 - 5(\theta_1 - \theta_2) - 6\frac{d\theta_1}{dt}$$

The applied torque, $\tau = 20$ is applied in the direction of $+\theta_2$, so

$$4\frac{d^2\theta_2}{dt^2} = -5(\theta_2 - \theta_1) - 8\frac{d\theta_2}{dt} + 20$$

With terms collected, the two simultaneous differential equations in θ_1 and θ_2 are

$$3\frac{d^2\theta_1}{dt^2} + 6\frac{d\theta_1}{dt} + 7\theta_1 - 5\theta_2 = 0$$

$$-5\theta_1 + 4\frac{d^2\theta_2}{dt^2} + 8\frac{d\theta_2}{dt} + 5\theta_2 = 20$$

$$(3s^2 + 6s + 7)\theta_1(s) + (-5)\theta_2(s) = 0 \quad [1.42]$$

$$(-5)\theta_1(s) + (4s^2 + 8s + 5)\theta_2(s) = \frac{20}{s} \quad [1.43]$$

Here the θ variables are uppercase since they are functions of s rather functions of time. Equations (1.42) and (1.43) can be written in standard form as follows:

Using a matrix to summarize rotational equation.

$$U_{11}\theta_1(s) + U_{12}\theta_2(s) = T_1(s) \quad [1.44]$$

$$U_{21}\theta_1(s) + U_{22}\theta_2(s) = T_2(s) \quad [1.45]$$

where the function (s) is understood for U_{ij}, and

U_{11} = initial torque $J_1 s^2$ plus all the other torques attached to J_1 (variable θ_1); in this case $U_{11} = 3s^2 + 6s + 7$

$U_{12} = U_{21}$ = sum of all torques connected between θ_1 and θ_2; in this case $U_{12} = U_{21} = -5$

U_{22} = initial torque $J_2 s^2$ plus all of the other torques attached to J_1 (variable θ_2); in this case $U_{22} = 4s^2 + 8s + 5$

$T_1(s)$ = independent driving torque on J_1; in this case $T_1(s) = 0$

$T_2(s)$ = independent driving torque on J_2; in this case $T_2(s) = 20/s$

Again, Equations (1.44) and (1.45) are solved with Crammer's rule.

Applying Cramer's rule.

$$\theta_2(s) = \frac{\begin{vmatrix} U_{11} & T_1(s) \\ U_{21} & T_2(s) \end{vmatrix}}{\begin{vmatrix} U_{11} & U_{12} \\ U_{21} & U_{22} \end{vmatrix}} \quad [1.46]$$

❑ DRILL PROBLEM

D1.18 Write simultaneous differential equations for the rotational mechanical networks shown in Figure D1.18 in terms of the indicated rotational mass angles. The angles are defined so that when all angles are zero, the spring torques are zero.

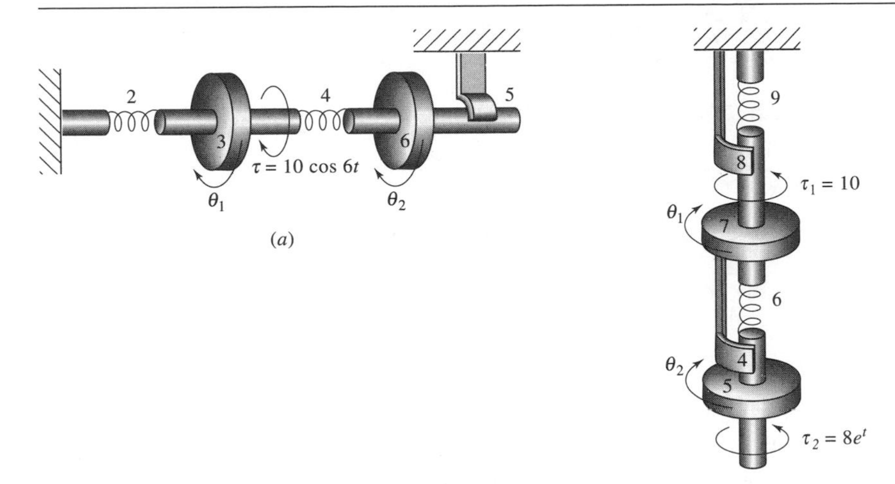

Figure D1.18

Ans. (a) $3\dfrac{d^2\theta_1}{dt^2} = 10\cos 6t - 2\theta_1 - 4(\theta_1 - \theta_2)$

$$6\frac{d^2\theta_2}{dt^2} = -5\frac{d\theta_2}{dt} - 4(\theta_2 - \theta_1);$$

(b) $7\dfrac{d^2\theta_1}{dt^2} = -10 - 9\theta_1 - 6(\theta_1 - \theta_2) - 8\dfrac{d\theta_1}{dt} - 4\dfrac{d}{dt}(\theta_1 - \theta_2)$

$$5\frac{d^2\theta_2}{dt^2} = -8e^t - 6(\theta_2 - \theta_1) - 4\frac{d}{dt}(\theta_2 - \theta_1)$$

1.7.2 Analogies

Equations of motion are derived similarly for all lumped-parameter linear systems, whether the systems are electrical, mechanical, or electromechanical. The procedures, the use of basic laws, and the solutions are much the same. We introduce the electrical analogies because frequently certain students can more readily understand the electrical analog for a mechanical circuit. It is possible to construct electrical analogies, both nodal and mesh, for mechanical circuits. The electrical analog is found by deriving an electrical network having a differential equation of the same form as that of the mechanical system. The following procedure is suggested:

1. Write the mechanical system equations.
2. Substitute electrical quantities using electrical network constants and variables.
3. Interpret these equations to yield the analog network.

Procedure for obtaining analogies.

Nodal Analog

To obtain a nodal analog for a mechanical equation, compare the equations of the mechanical circuit

$$\left(sM_{ii} + B_{ii} + \frac{K_{ii}}{s}\right) V_i \cdots = F_i$$

$$\left(sJ_{ii} + B_{ii} + \frac{K_{ii}}{s}\right)\Omega_i \cdots = T_i$$

with the electrical admittance of a node circuit,

$$\left(sC_{ii} + \frac{1}{R_{ii}} + \frac{1}{sL_{ii}}\right)V_i \cdots = I_i$$

Note that the variable is linear velocity V_i or angular velocity Ω_i rather than linear displacement X_i [Equation (1.36)] or angular displacement θ_i [Equation (1.46)]. Comparison of the coefficients of these equations indicates the following analogous quantities:

Mass M or moment of inertia J	Analogous to capacitance C
Damping constant B	Analogous to conductance $1/R$
Spring constant K	Analogous to inverse inductance $1/L$
Force or torque	Analogous to current
Linear velocity V or angular velocity Ω	Analogous to voltage

Summarizing,

$$\left.\begin{matrix} M \\ J \end{matrix}\right\} \sim C \qquad B \sim \frac{1}{R} \qquad K \sim \frac{1}{L} \qquad \left.\begin{matrix} F \\ T \end{matrix}\right\} \sim I \qquad \left.\begin{matrix} V \\ \Omega \end{matrix}\right\} \sim E$$

Mesh Analog

Compare the operational mechanical admittance with the electrical impedance.

$$\left(sL_{ii} + R_{ii} + \frac{1}{sC_{ii}}\right)I_i \cdots = E_i$$

which yields the following analogies listed:

Mass M or moment of inertia J	Analogous to inductance L
Damping constant B	Analogous to resistance R
Spring constant K	Analogous to inverse capacitance $1/C$
Forces or torques	Analogous to voltages
Linear velocity V or angular velocity Ω	Analogous to current

Summarizing, we write

$$\left.\begin{matrix} M \\ J \end{matrix}\right\} \sim L \qquad B \sim R \quad K \sim \frac{1}{C} \qquad \left.\begin{matrix} F \\ T \end{matrix}\right\} \sim E \qquad \left.\begin{matrix} V \\ \Omega \end{matrix}\right\} \sim I$$

There is a possibility of more than one electrical analog for a given mechanical system.

As an example, an analog for the mechanical system of Figure 1.20(a) is found. Figure 1.20(c) shows the nodal analog and Figure 1.20(d) is the mesh analog. The two electrical circuits of Figure 1.20(c) and (d) are termed "duals." The two networks thus derived are duals, since node pair voltages in Figure 1.20(c) are analogous to loop currents in Figure 1.20(d).

❑ DRILL PROBLEM

D1.19 Obtain the electrical nodal and mesh analogies to the mechanical system of Drill Problem D1.18(a).

Ans.

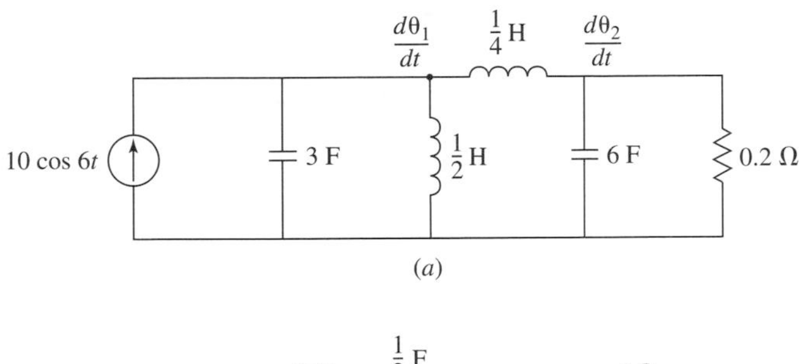

(a)

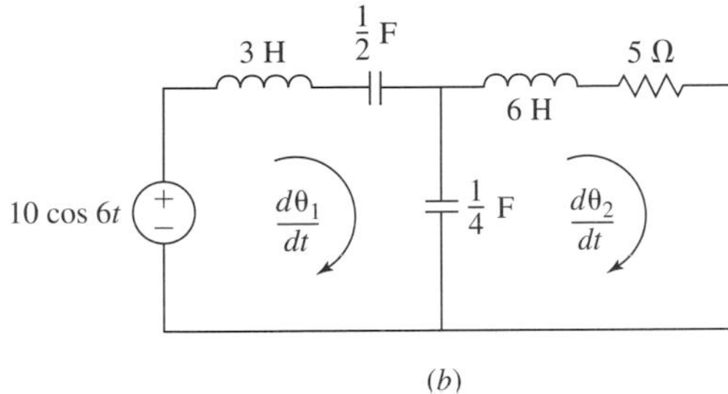

(b)

Figure D1.19

1.7.3 Gear Trains and Transformers

The mechanical network of Figure 1.21(a) involves a gear train. It has rotational equations from Figure 1.19. If we examine the torque acting to the left of the gear with n_1 teeth, the torque τ_1 equals the input torque τ minus the inertial and damper torques. Thus

$$\tau_1 = \tau - J_1 \frac{d^2\theta_1}{dt^2} - B_1 \frac{d\theta_1}{dt} \qquad [\mathbf{1.47}]$$

or

$$J_1 \frac{d^2\theta_1}{dt^2} + B_1 \frac{d\theta_1}{dt} + \tau_1 = \tau \qquad [\mathbf{1.48}]$$

The torque τ_2 applied to the load connected to the n_2 teeth gear equals the inertial and damper torque, so

$$\tau_2 = J_2 \frac{d^2\theta_2}{dt} + B_2 \frac{d\theta_2}{dt}$$

or

$$J_2 \frac{d^2\theta_2}{dt^2} + B_2 \frac{d\theta_2}{dt} - \tau_2 = 0 \qquad [\mathbf{1.49}]$$

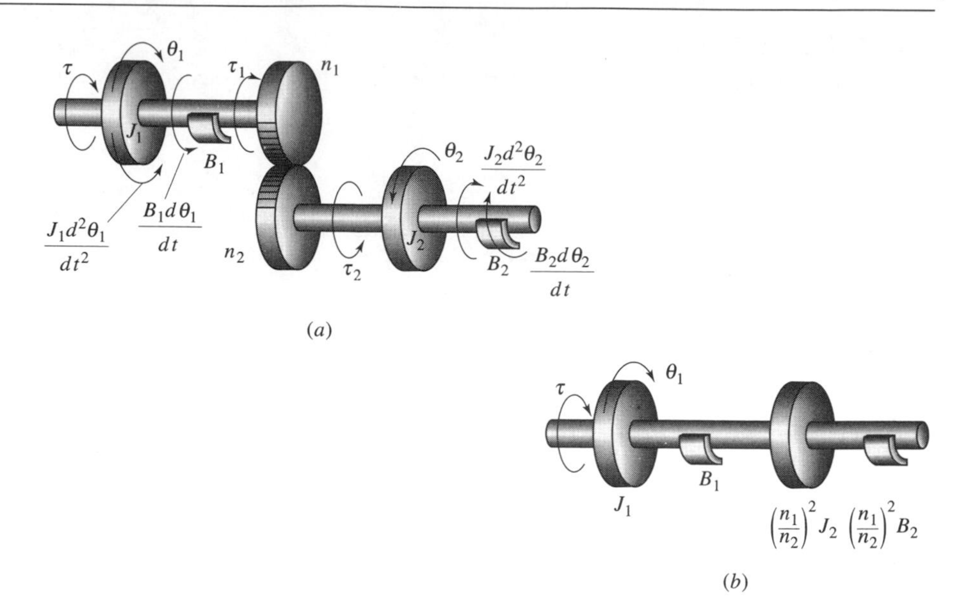

Figure 1.21 A mechanical network coupled by gears. (a) Gear system. (b) Equivalent system.

The gear relations from Figure 1.19 are

$$\tau_2 = \frac{n_2}{n_1}\tau_1 \quad [1.50]$$

$$\theta_2 = \frac{n_1}{n_2}\theta_1 \quad [1.51]$$

substituting the gear relations of Equations (1.50) and (1.51) into Equation (1.49) gives

$$\tau_1 = \left(\frac{n_1}{n_2}\right)^2 \left(J_2\frac{d^2\theta_1}{dt^2} + B_2\frac{d\theta_1}{dt}\right)$$

Eliminating τ_1 in Equation (1.48) gives

$$J_1\frac{d^2\theta_1}{dt^2} + B_1\frac{d\theta_1}{dt} + \left(\frac{n_1}{n_2}\right)^2 J_2\frac{d^2\theta_1}{dt^2} + \left(\frac{n_1}{n_2}\right)^2 B_2\frac{d\theta_1}{dt} = \tau$$

in which the gear load has been reflected to the left side of the gear train. An equivalent mechanical network is given in Figure 1.21(b).

The electrical analog of a gear train is a transformer. The electrical network of Figure 1.22(a) involves an ideal transformer. The transformer windings are labeled "primary" and "secondary," with the primary winding closer to the voltage source $v(t)$. Simultaneous mesh equations for the network are

Reflecting a load to the primary side.

$$L_1\frac{di_1}{dt} + R_1 i_1 + v_1 = v \quad [1.52]$$

$$L_2\frac{di_2}{dt} + R_2 i_2 - v_2 = 0 \quad [1.53]$$

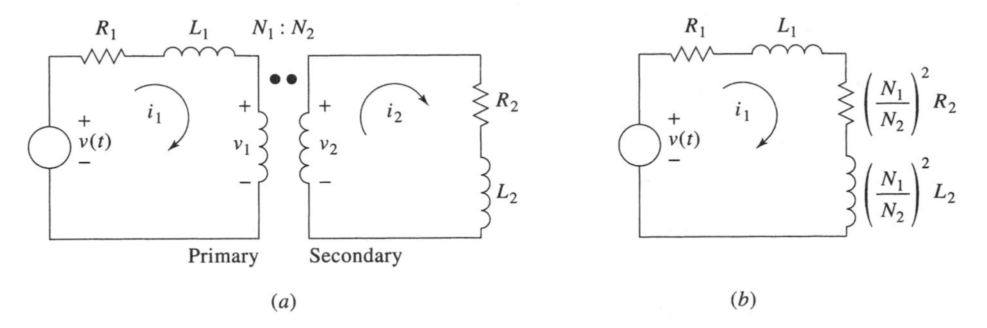

Figure 1.22 An electrical network involving a transformer.

where the ideal transformer voltages and currents are related by the turns ratio:

$$v_2 = \frac{N_2}{N_1} v_1 \quad [1.54]$$

$$i_2 = \frac{N_1}{N_2} i_1 \quad [1.55]$$

Substituting the transformer relations of Equations (1.54) and (1.55) into Equation (1.53) gives

$$v_1 = \left(\frac{N_1}{N_2}\right)^2 L_2 \frac{di_1}{dt} + \left(\frac{N_1}{N_2}\right)^2 R_2 i_1$$

Eliminating v_1 in Equation (1.52), we write:

$$L_1 \frac{di_1}{dt} + R_1 i_1 + \left(\frac{N_1}{N_2}\right)^2 L_2 \frac{di_1}{dt} + \left(\frac{N_1}{N_2}\right)^2 R_2 i_1 = v$$

The secondary load, consisting of L_2 and R_2, is said to be reflected to the primary side of the transformer, through the square of the transformer turns ratio, as in the equivalent circuit of Figure 1.22(b).

This electrical network and the mechanical network of Figure 1.21 are analogous to each other. They are described by equations of the same form, with the following equivalent quantities:

$$i_1 \sim \frac{d\theta_1}{dt} \qquad L_1 \sim J_1$$

$$R_1 \sim B_1 \qquad \frac{N_1}{N_2} \sim \frac{n_1}{n_2}$$

$$i_2 \sim \frac{d\theta_2}{dt} \qquad L_2 \sim J_2$$

$$v \sim \tau \qquad R_2 \sim B_2$$

Other analogies are possible if an alternate electrical circuit is used. For example, a circuit with a current source as a forcing function would provide an analogy between current and torque. All other analogies would change accordingly.

❑ **DRILL PROBLEM**

D1.20 Find $\theta(t)$. Mass and damping effects in this system are negligible. (See Figure D1.20).

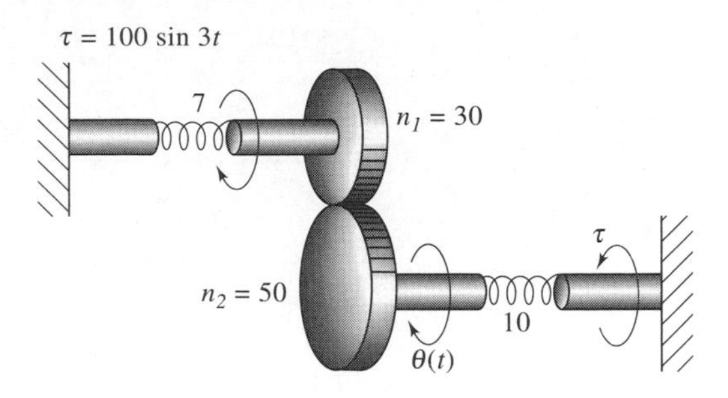

Figure D1.20

Ans. $$\left[7+\left(\frac{30}{50}\right)^2 10\right]\theta = -\left(\frac{30}{50}\right)^2 100 \sin 3t$$

$$\theta = -5.66 \sin 3t$$

1.8 Electromechanical Components

Many electromechanical devices are encountered in engineering and scientific applications. Solenoids, actuators, motors, generators, gyroscopes, accelerometers, and loudspeakers are just a few of these. For many control systems it is necessary to deal with equations for a combination of electrical and mechanical components. Figure 1.23 gives idealized equations for several common electromechanical devices. These devices operate as indicated over a suitable range of parameters and conditions.

Potentiometer.

A **potentiometer** contains a slider that moves along a resistance element as in Figure 1.23(a). The potentiometer has a voltage V applied across the entire resistance, while a fraction of that voltage appears across the output. That fraction depends on the ratio of the angle subtended by the slider compared to the maximum angle.

Dc motor.

Figure 1.23(b) contains a model for a **dc motor with a fixed field**. An input voltage causes a current in the armature of the dc motor. If that current creates a large enough field to interact with the fixed field, the armature begins to turn. The armature turns because a torque is applied to the inertia and friction of the motor and to the inertia and friction of the load connected to the motor. That torque is proportional to the armature current. Control motors (both ac and dc) are manufactured to obtain a linear relation between the torque and the armature current. Practically, this curve is not linear, however, a straight-line approximation is used to develop these equations. This is a good approximation, since control motors operate near the zero speed point where both the torque–speed curve and the torque–current curve are most linear.

As the armature spins, a reverse voltage is induced so as to oppose the input voltage to the armature circuit. To improve linear performance, a dc control motor is

(*a*) **Potentiometer**

θ_{max} is the maximum angular position θ. For $\theta = \theta_{max}$, $R_1 = R$.

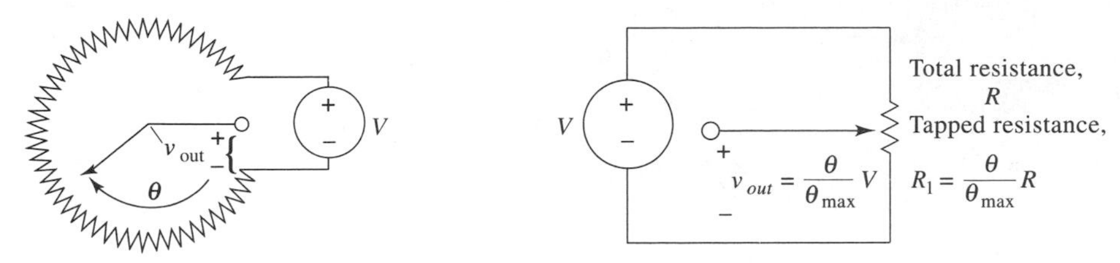

(*b*) **Dc Control Monitor Driving an Inertial Load**

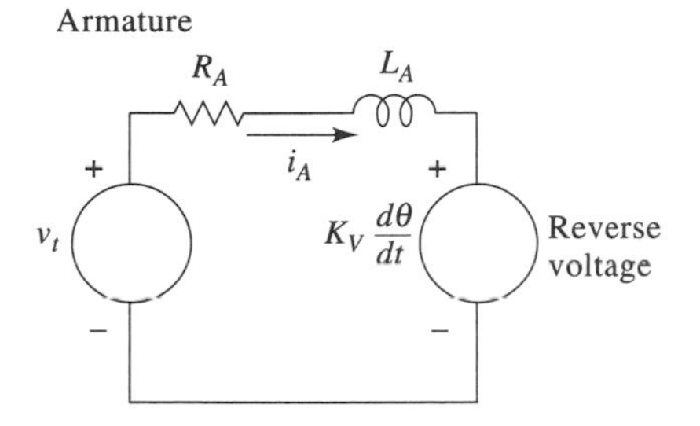

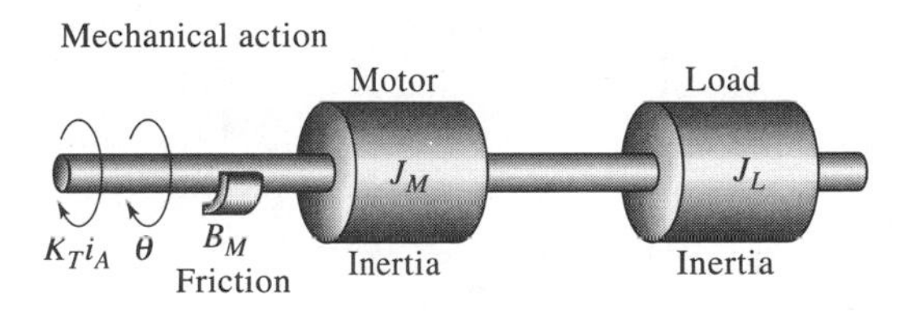

(*c*) **Dc Generator (Constant Field)**

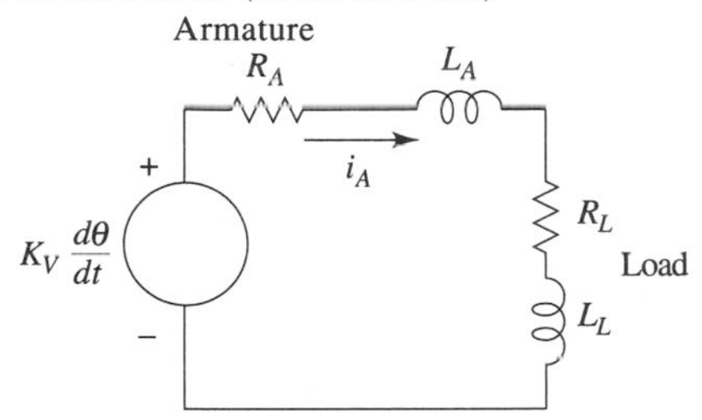

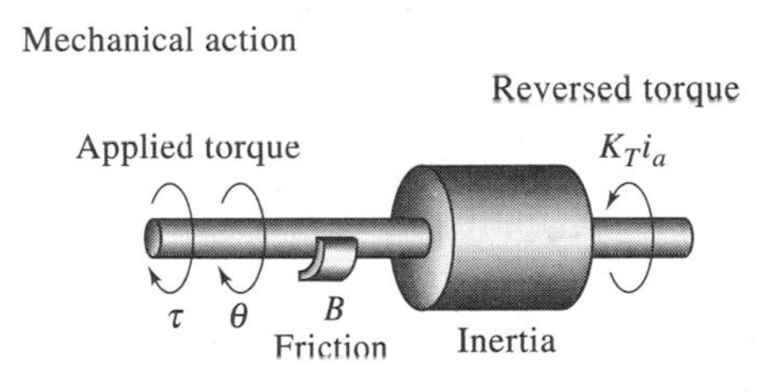

(*d*) **Tachometer**

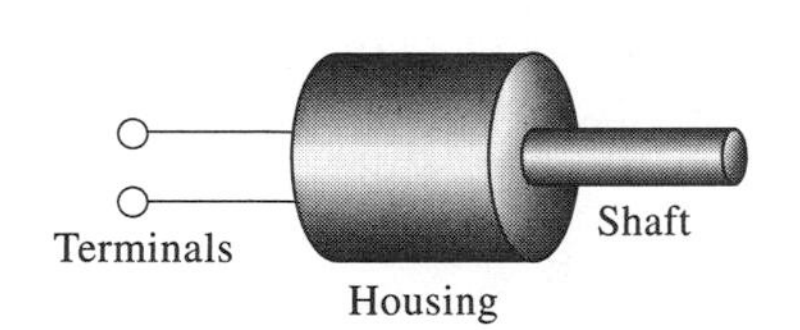

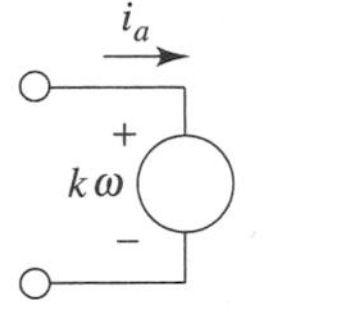

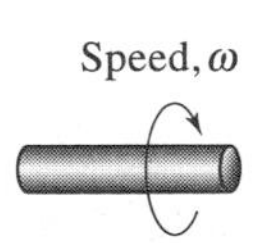

Figure 1.23 Simple models for common electromechanical elements.

(*e*) **linear Actuator Solenoids**

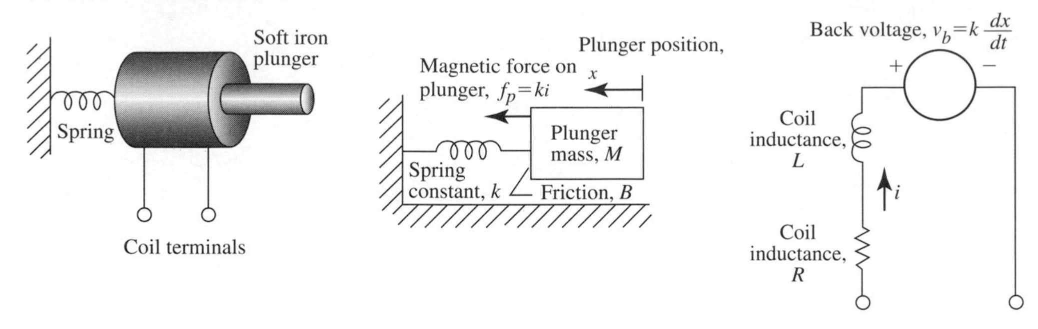

(*f*) **Gyroscope**

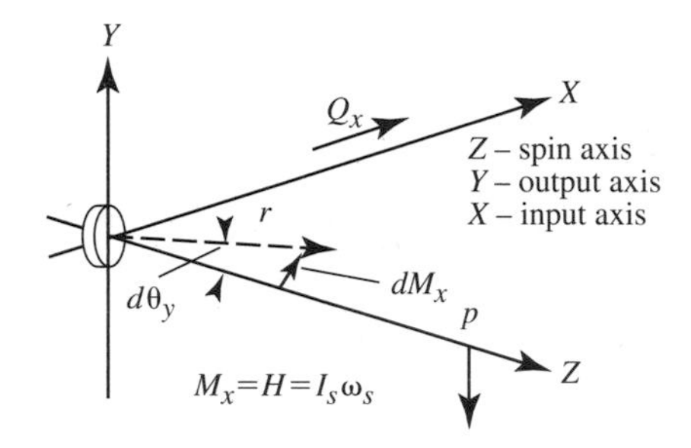

Figure 1.23 (*continued*) Simple models for common electromechanical elements.

designed with a large value for armature resistance R_A and a small value for armature inductance L_A. In addition, the internal damping B_m of the motor is designed to be quite small. Two equations are written for the dc control motor driving an inertia load. The electric circuit is

$$v_i = R_A i_A + L_A \frac{di}{dt} + k_v \frac{d\theta}{dt}$$

When L_A approaches zero, then

$$v_i = R_A i_A + k_v \frac{d\theta}{dt} \qquad [1.56]$$

Notice the reverse voltage $k_v(d\theta/dt)$ that is generated to oppose the armature current. The mechanical equation [when B_m is neglected and $J = (J_L + J_m)$] is

$$(J_L + J_m)\frac{d^2\theta}{dt^2} = k_T i_A \qquad [1.57]$$

Equations (1.56) and (1.57) are Laplace-transformed (see Appendix B) with zero initial conditions,

$$V_i(s) = R_A I_A(s) + k_v s\Theta(s)$$
$$0 = -k_T I_A(s) + Js^2\Theta(s)$$

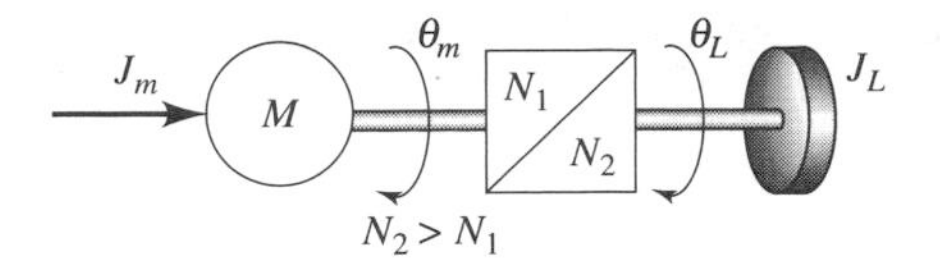

Figure 1.24 Motor driving an identical load through a gear train.

And solving for $\Theta(s)$,

Simplifying the dynamics.

$$\frac{\Theta(s)}{V_i(s)} = \frac{1/k_v}{s\,[1 + (R_A J/k_T k_v s]} = \frac{k_m}{s(\tau s + 1)} \qquad [1.58]$$

where the motor gain constant $k_m = 1/k_v$ and the motor time constant $\tau = R_A J/(k_T k_v)$. Notice that the mathematical model Equation (1.58) is also a good approximation for an ac control motor.

It is important to remember that Equation (1.58) applies only to a motor driving an inertia load. If another load is on the motor, the torque opposing $k_T i_A$ in Equation (1.57) must be for that load.

Gear effects.

As an example, consider a control motor driving an inertia load through a gear train, as shown in the schematic diagram of Figure 1.24. A gear train is required in most cases, since a control motor operates at high speed with low torque, whereas the load, such as an antenna, operates at low speed with high torque. The transfer function desired is between the input voltage V_i and the output angle θ_2. The load inertia is reflected across the gear train as $(N_1/N_2)^2 J_L$. This is added to the motor inertia J_m to form an equivalent inertia, $J_e = J_m + (N_1/N_2)^2 J_L$. This is inserted into Equation (1.57) as follows:

$$k_T i_A = J_e \frac{d^2\theta_m}{dt^2} \qquad [1.59]$$

and θ_m is related to θ_L as follows:

$$\theta_L = \theta_m \frac{N_1}{N_2} \qquad [1.60]$$

When Equations (1.59) and (1.60) are used. The transfer function becomes:

$$\frac{\Theta_L(s)}{V_i(s)} = \frac{k_m N_1/N_2}{s(\tau_L s + 1)} \qquad [1.61]$$

where $\tau_L = R_A/(k_T k_v)[J_m + (N_1 + N_2)^2 J_L]$. Various other types of control motor exist. However, they are usually designed to have a linear speed–torque characteristic. For this reason, the transfer function of Equation (1.58) is a good approximation for control motors of other types.

Dc generator.

Figure 1.23(c) contains a model for a **dc generator with a fixed field**. Some external means is employed to cause the generator to rotate (e.g., hydrological forces or steam pressure). As the armature rotates in the fixed field, a voltage is induced across the armature coils. That voltage can be applied to a resistive–inductive load. For example, the field created by the armature current will create a reverse torque, which opposes the applied torque. The dc generator model is governed by

$$(L_A + L_L)\frac{di_A}{dt} + (R_A + R_L)\,i_A - k_v\frac{d\theta}{dt} = 0$$

$$J\frac{d^2\theta}{dt^2} + B\frac{d\theta}{dt} + k_T i_A = \tau$$

Tachometer.

Figure 1.23(d) is a **tachometer**, which develops an output voltage proportional to shaft angular velocity. The tachometer output voltage is used for damping in position control systems and as an output comparison unit for a speed control system.

The tachometer is a special case of a dc generator in which the field is replaced by a permanent magnet, which is equivalent to having a constant field current. It is normally used with a small electrical load so that i_a is nearly zero. Friction and inertia are typically made as small as is practical. An equation that expresses the performance of a tachometer is

$$E_t(s) = k_t s\Theta(s)$$

where k_t is the voltage gradient in volts per radian per second and θ is the shaft angular position in radians. A tachometer is usually connected to a rotating device, to permit measurement of its rate of rotation without hindering the operation of the device being monitored.

Solenoid.

The solenoid in Figure 1.23(e) converts electrical energy into linear motion. The magnetic force of attraction created by current flowing in an RL circuit (e.g., in the ignition system of an automobile where the solenoid acts as a switch) causes the soft iron plunger to move. The constant k that appears in two places has the same numerical value if consistent units are used. The k associated with force could have units of newtons per ampere. In the electrical circuit, k has units of volts per meters per second. Voltage has units of joules per coulomb, the same as units of newton-meters/coulomb. Therefore,

$$\frac{\text{volts}}{\text{meters/second}} = \frac{\text{newton-meters}}{\text{coulomb-meters/second}}$$

is the same as units of newtons/(coulomb/second) or newtons/ampere. The two values of k are the same.

Gyroscope operation.

Figure 1.23(f) shows a simplified model for a **gyroscope**. A gyroscope consists of a wheel mounted on a shaft and arranged to be spun at high angular velocity. Frequently, the wheel is mounted in a system of gimbals that permits complete freedom of movement of all three axes.

The most useful characterstic of the gyroscope is its tendency to maintain its spin axis in a fixed direction in space. This phenomenon is best explained by a consideration of rotation dynamics. A simplified model is used to gain insight into the operation of the gyroscope. In this the model the effects of moments of inertia of the wheel and gimbal system about axes other than the spin axis are neglected. The resulting equation will fail to show certain characteristics. For many purposes, however, the results are adequate. For this derivation the following nomenclature is needed:

M_x, M_y, M_z = components of angular momentum about x, y, and z axes
I_s = moment of inertia of wheel about spin axis
ω_s = angular velocity of wheel
$H = I_s\omega_s$ = angular momentum of wheel

These quantities are identified in Figure 1.23(f).

At time $t = 0$, suppose the torque Q_x is applied about the x axis by pressing down on the gyro housing at point p. Initially, $M_z = I_s\omega_s = H$, $M_x = M_y = 0$, and the angular momentum of spin lies along OZ and has a magnitude H. Since the rate of change of angular momentum of a system is equal to the applied torque, the following expression can be written:

$$Q_x = \frac{dM_x}{dt}$$

This is expressed in different form:

$$dM_x = Q_x dt$$

If this term is added vectorically to the initial angular momentum, a new value is obtained. $M_z + dM_x$ is separated from the initial value by an angle $d\theta_y$ which is given by

$$d\theta_y = \frac{dM_x}{H} = \frac{Q_x dt}{H}$$

and is written as

$$\omega_y = \frac{d\theta_y}{dt} = \frac{Q_x}{H}$$

The gyro is thus rotating about the OY axis with a velocity ω_y. This is the fundamental gyroscopic law: A torque about any axis other than the spin axis produces a velocity about the axis that is orthogonal to the applied-torque axis. Because of this property, the gyro is an important instrument for measuring torques ($Q_x = H\omega_y$) by measuring this orthogonal or precession velocity.

Alternatively, a large gyroscope can be used to obtain a stabilizing countertorque. For applied torques about the x or y axis, the gyro supplies an equal countertorque that prevents motion in the direction of the applied torque as long as the gyro can precess. Once the precession angle θ has reached 90°, the gyroscope is in a state of "gimbal lock" and ceases to function as described. In the state of gimbal lock the OZ axis has precessed into the OX axis, about which the torque is being applied. With a torque applied about the OX, or spin, axis, the gyro ceases to produce a countertorque. Thus the gyro tends to rotate to align its spin axis in the direction of applied torque.

1.9 Aerodynamics

Control engineers work on a wide range of aerodynamic applications such as passenger aircraft, fighter aircraft, and missile systems. Submarines also have equations of motion very similar to those of aerodynamic vehicles. Aircraft and submarines simply move through fluids with different densities but the basic forces and torques behave similarly.

1.9.1 Nomenclature

Figure 1.25(a) shows typical aerodynamic nomenclature with an x, y, z coordinate system attached to the airplane. The x axis points forward along the vehicle centerline. The y-axis points along one wing, while the z axis points downward following a "right-hand" coordiante system. If you rotate the fingers of your right hand from the x axis to the y axis, your thumb points to the z axis. A similar rotation from y to z yields x and from z to x yields y as the thumb position.

Figure 1.25(b) shows a number of velocities, angles, and angular rates.

Definition of terms.

u = x-axis velocity
v = y-axis velocity
w = z-axis velocity

ϕ = roll angle about the x axis
θ = pitch angle about the y axis
ψ = yaw angle around the z axis

p = roll rate = $d\phi/dt$
q = pitch rate = $d\theta/dt$
r = yaw rate = $d\psi/dt$

The aircraft has a velocity vector V as shown in Figure 1.25(c) with components u, v, w. There are three angles between the velocity vector V and the x, y, z axes, which are, respectively:

α = pitch angle
β = side-slip angle
γ = bank angle

1.9.2 Dynamics

To develop the aircraft equations of motion, six equations (degrees of freedom) are usually derived for the three applied forces along each of the three axes and for the three applied torques around each of the three axes. In vector form we can write

Compact matrix description.

$$\text{Forces applied} = m\dot{V} + \Omega \times mV \qquad [\mathbf{1.62}]$$

$$\text{Torques applied} = I\dot{\Omega} + \Omega \times I\Omega \qquad [\mathbf{1.63}]$$

where

$$V = \begin{bmatrix} u \\ v \\ w \end{bmatrix} \quad \Omega = \begin{bmatrix} p \\ q \\ r \end{bmatrix} \quad \text{Forces} = \begin{bmatrix} F_x \\ F_y \\ F_z \end{bmatrix} \quad \text{Torques} = \begin{bmatrix} T_x \\ T_y \\ T_z \end{bmatrix}$$

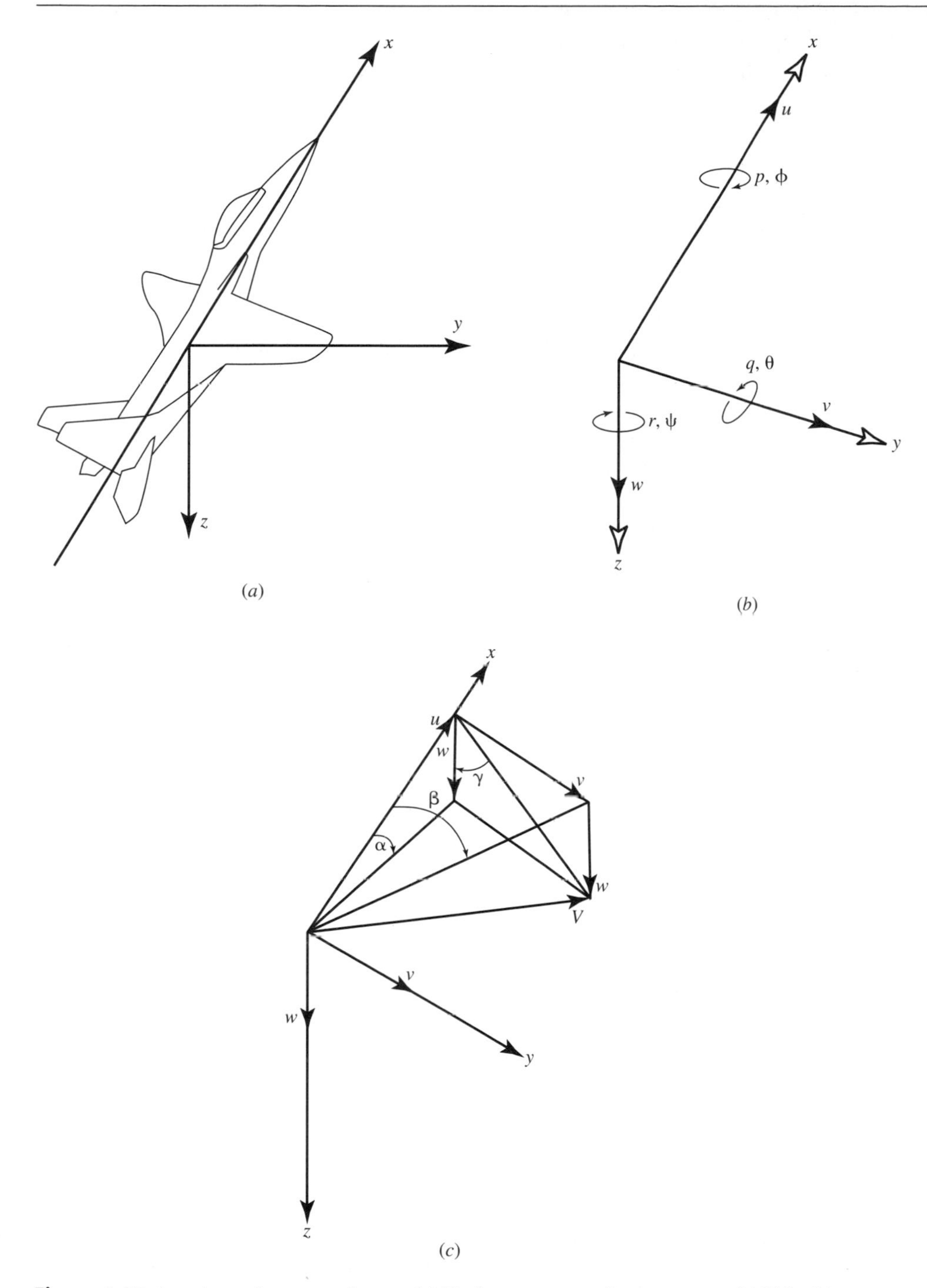

Figure 1.25 Aerodynamic nomenclature. (a) Basic x, y, z coordinate system. (b) Velocities (u, v, w), angles (ϕ, θ, ψ), angular rates (p, q, r). (c) Angle of attack (α), side-slip angle (β), bank angle (γ), and velocity vector (V).

The symbol "×" represents the cross product. The two cross products can be conveniently calculated using i, j, k to represent unit vectors along the x, y, z axes, respectively.

$$\omega \times mV = \begin{vmatrix} i & j & k \\ p & q & r \\ mu & mv & mw \end{vmatrix}$$
$$= im(qw - rv) + jm(ru - pw) + km(pv - qu) \quad [1.64]$$

Here m is the mass of the vehicle. The factors that multiply the unit vector i represent the x components of the cross product. Similarly the multipliers of j are the y-axis components and the multipliers of k are the z-axis components.

For convenience, the inertial matrix is simplified to include inertias taken about each axis. For some aircraft applications, a more complicated inertial matrix may be desired for more precise dynamical modeling.

$$\Omega \times I\omega = \begin{vmatrix} i & j & k \\ p & q & r \\ I_x p & I_y q & I_z r \end{vmatrix} \quad [1.65]$$

Each inertia is taken about the axis indicated by the subscript.

Equation (1.64) may be used to evaluate Equation (1.62). For convenience, the result is divided by m to yield translational accelerations along each axis.

Translational accelerations.

$$\begin{aligned} a_x &= \dot{u} + qw - rv \\ a_y &= \dot{v} + ru - pw \\ a_z &= \dot{w} + pv - qu \end{aligned} \quad [1.66]$$

Similarly, Equation (1.63) may be evaluated by using Equation (1.65) to yield rotational torque equations around each axis.

Rotational torques.

$$\begin{aligned} T_x &= I_x\dot{p} + qr(I_z - I_y) \\ T_y &= I_y\dot{q} + pr(I_x - I_z) \\ T_z &= I_z\dot{r} + pq(I_y - I_x) \end{aligned} \quad [1.67]$$

Simplifications.

For modeling, design, and simulation, acceleration in the x direction is usually considered small. Since the velocity u is much larger than v and w, a small-angle approximation may be used for α and β. Thus, α in radians is approximately equal to w/u and β in radians is approximately equal to v/u. Accelerations and torques are due to the shape and motion of the vehicle, so those effects must be accounted for. The symbols X, Y, and Z give the accelerations along the designated axes, including a subscript denoting the source of the acceleration. For example, Y_β represents the acceleration along the y axis due to the side-slip angle β. Torques about the x, y, z axes respectively, may be represented by the symbols L, M, and N with the appropriate

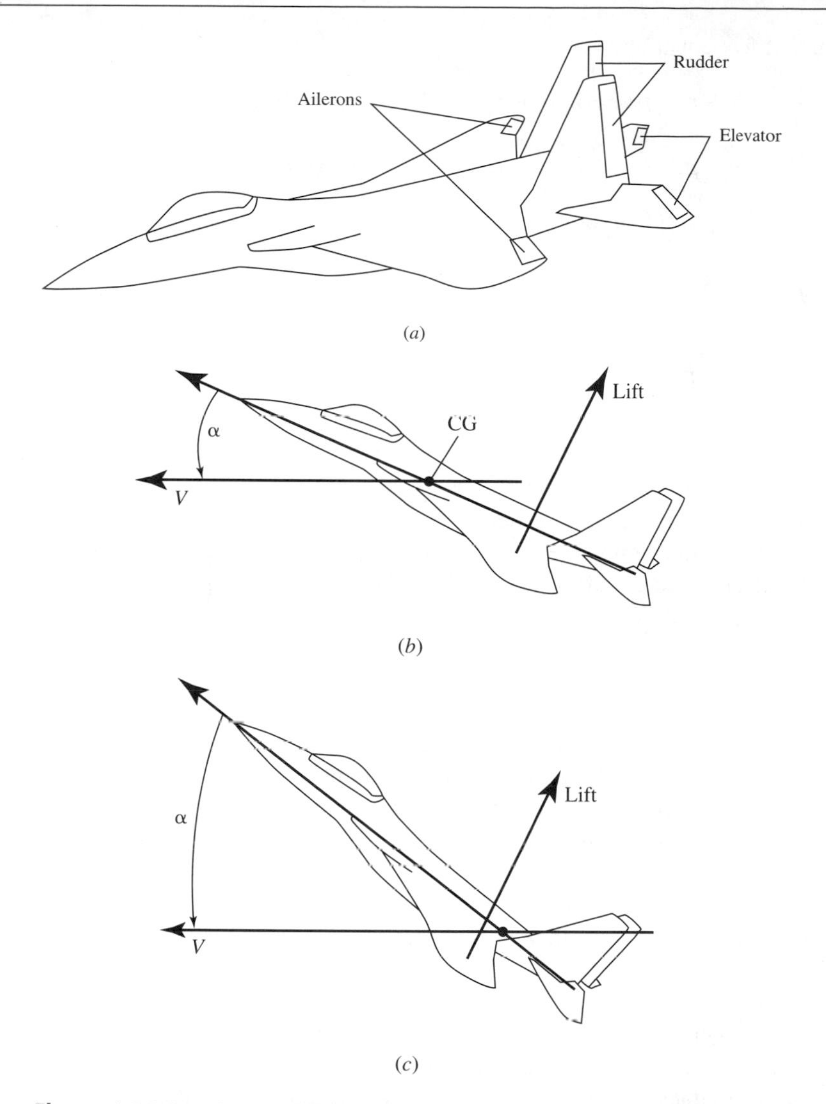

Figure 1.26 Rotational and lifting effects. (a) Control surfaces. (b) Center of lift is aft of the CG; α will decrease (V is the velocity vector). (c) Center of lift is forward of the CG; α will increase (V is the velocity vector).

subscript. For example the torque about the y axis due to angle of attack α is given by M_{α}.

Moving one or more of the three control surfaces shown in Figure 1.26(a) generates other forces and torques. By moving the aileron located on the main wing through an angle δ_a, rolling about the x axis is induced. By moving another control surface called an elevator on the rear wing through an angle δ_e, pitching along the y axis is induced. Similarly, moving the rudder on the vertical stabilizer through an angle δ_r induces yawing motion about the z axis. The downward force of gravity, g, is resolved into y and z components via angles ϕ and θ. Ignoring x axis acceleration results in five equations.

$$\frac{a_y}{u} = \dot{\beta} + r - p\alpha = Y_\beta \beta + Y_p p + Y_r r + Y_{d_r} \delta_r + (g/u) \sin\phi \cos\theta$$

$$\frac{a_z}{u} = \dot{\alpha} + p\beta - q = Z_\alpha \alpha + Z_q q + Z_{\delta_e} \delta_e + (g/u) \cos\phi \cos\theta$$

$$\frac{T_x}{I_x} = \dot{p} + \frac{qr(I_z - I_y)}{I_x} = L_\beta \beta + L_p p + L_r r + L_{\delta_a} \delta_a + L_{\delta_r} \delta_r \quad [1.68]$$

$$\frac{T_y}{I_y} = \dot{q} + \frac{pr(I_x - I_z)}{I_y} = M_\alpha \alpha + M_q q + M_{\delta_e} \delta_e$$

$$\frac{T_z}{I_z} = \dot{r} + \frac{pq(I_y - I_x)}{I_z} = N_p p + N_r r + N_\beta \beta + N_{\delta_a} \delta_a + N_{\delta_r} \delta_r$$

For convenience, u is absorbed into the Y and Z coefficients. Similarly, inertias are absorbed into the L, M and N coefficients. The Y, Z, L, M, and N coefficients are found by placing a scale model into a wind tunnel or by using a computer simulation, which is becoming the most common tool today. Typical flight conditions are simulated, and then a coefficient such as Z_a is broken down into components such as

$$Z_a = C_{Z_\alpha} S(0.5\rho |V|^2)$$

where the term $0.5\rho|V|^2$ is called the dynamic pressure, $|V|$ is the magnitude of the velocity, ρ is the density of the air being displaced, S is the cross-sectional area of the air being displaced, and $C_{z\alpha}$ is called an aerodynamic "derivative," which may be plotted and stored versus the flight conditions simulated.

Some coefficients may be negative. For example, a positive angle of attack a will cause the aircraft to move in the negative-z direction, that is, upward; hence Z_a is negative.

Two possible rotational moment effects due to M_a are shown in Figure 1.26(b,c). If the center of lift is behind the center of gravity (CG, around which the plane rotates), as in Figure 1.26(b), then a positive angle of attack will create a lift that tends to reduce that angle of attack and stabilize the aircraft. In that case M_a is considered to be negative. Quite a different situation is depicted in Figure 1.26(c). There, a positive angle of attack creates a lift that is forward on the CG, causing a further increase in that angle of attack. The system becomes unstable and usually that M_a is considered to be positive. In such a case, the pilot would need to be especially attentive or a computer would be employed. The pilot essentially makes a request to the computer that actually flies the plane to ensure safety.

1.9.3 Lateral and Longitudinal Motion

Often, motion is restricted to a few axes to simplify the task of design. Two such simplified motions are longitudinal motion in the x-z plane and lateral motion in the x-y plane.

Longitudinal dynamics.

To illustrate longitudinal motion, suppose we are interested only in motion in the x-z plane where the wings remain level and the vehicle does not roll. In that case, we can assume y axis velocity v, x axis rotational rate p and z axis rotational rate r

are zero. That is, there is neither sideways velocity nor rotation about the x or z axes, only rotation about the y axis. Assume only elevator deflection δ_e, so δ_r and δ_a are zero. Two equations remain where perturbations are taken with respect to gravity, eliminating gravity from the equations:

$$\begin{aligned} \dot{\alpha} - q &= Z_\alpha \alpha + Z_q q + Z_{\delta_e} \delta_e \\ \dot{q} &= M_\alpha \alpha + M_q q + M_{\delta_e} \delta_e \end{aligned} \tag{1.69}$$

The Laplace transform is

$$\begin{bmatrix} s - Z_\alpha & -(1 + Z_q) \\ -M_\alpha & s - M_q \end{bmatrix} \begin{bmatrix} \alpha(s) \\ Q(s) \end{bmatrix} = \begin{bmatrix} Z_{\delta_e} \delta_e(s) \\ M_{\delta_e} \delta_e(s) \end{bmatrix} \tag{1.70}$$

$\alpha(s)$ can be found by using Cramer's rule:

$$\begin{aligned} \alpha(s) &= \frac{\begin{vmatrix} Z_{\delta_e} \delta_e(s) & -(1 + Z_q) \\ M_{\delta_e} \delta_e(s) & s - M_q \end{vmatrix}}{\begin{vmatrix} s - Z_\alpha & -(1 + Z_q) \\ -M_\alpha & s - M_q \end{vmatrix}} \\ &= \frac{\left[Z_{\delta_e} s + M_{\delta_e}(1 + Z_q) - M_q Z_{\delta_e}\right] \delta_e(s)}{s^2 + s(-M_q - Z_\alpha) + M_q Z_\alpha - M_\alpha(1 + Z_q)} \end{aligned} \tag{1.71}$$

Generally, Z_α and M_q are negative. Notice that M_α must also be negative to cause all coefficients of the second-order characteristic polynomial [denominator of Equation (1.71)] to be positive, creating left half-plane roots, corresponding to the stabilized aircraft of Figure 1.26(b).

Lateral dynamics.

For lateral motion, the aircraft may yaw and roll but not pitch. The centerline of the aircraft remains level (not tipped up or down) so motion is restricted to the x-y plane. Only the rudder controls the aircraft. Here δ_e, δ_a, α, w, θ and q are all zero. A small angle approximation $\sin\phi = \phi$ may be used for the gravity term in Equation (1.68). Three of the five equations from Equation (1.68) remain. If we want a set of first-order differential equations in the variables β, p, r, and ϕ, then the differential equation $d\phi/dt = p$ must be added to those three.

$$\dot{\beta} - Y_\beta \beta - Y_p p + r(1 - Y_r) - \frac{\phi g}{u} = Y_{d_r} \delta_r$$

$$\dot{p} - L_\beta \beta - L_p p - L_r r = L_{\delta_r} \delta_r$$

$$\dot{r} - N_\beta \beta - N_p p - N_r r = N_{\delta_r} \delta r$$

$$\dot{\phi} - p = 0$$

Using Laplace transforms, we have

$$\begin{bmatrix} s - Y_\beta & -Y_p & (1 - Y_r) & -g/u \\ -L_\beta & s - L_p & -L_r & 0 \\ -N_\beta & -N_p & s - N_r & 0 \\ 0 & -1 & 0 & s \end{bmatrix} \begin{bmatrix} \beta(s) \\ P(s) \\ R(s) \\ \phi(s) \end{bmatrix} = \begin{bmatrix} Y_{d_r} \\ L_{\delta_r} \\ N_{\delta_r} \\ 0 \end{bmatrix} \delta_r \tag{1.72}$$

The fourth-order characteristic polynomial roots for Equation (1.72) include a pair of lightly damped complex conjugates. The ensuing step response of that complex conjugate pair shows coupled yaw–roll response in which the aircraft bobs and weaves in what is called "Dutch roll," supposedly because the bobbing and weaving resembles the upper body movement of a skater moving along a frozen canal in the Netherlands. Designing a controller to damp out the Dutch roll response is a challenging task.

Dutch roll.

1.10 Thermal Systems

It is often necessary to control temperature for complex thermodynamic processes. Under some simplifying assumptions, the flow of heat through various media may be modeled with simple electrical analogs. Table 1.3 shows analogs between heat flow and an electrical RC circuit.

Figure 1.27 shows examples of heat flowing through glass from a fluid at temperature θ_2 toward a fluid at temperature θ_1. This could represent the escape of heat from the air of a warm house through a window to cooler air outside. For Figure 1.27(a), with one pane of glass, using nomenclature from Table 1.2, we write

Heat flow.

$$q = \frac{\theta_2 - \theta_1}{R} = q_0 \qquad \textbf{[1.73a]}$$

This rate of heat flow is analogous to current flowing through an electrical circuit given by

$$i = \frac{v_2 - v_1}{R}$$

Suppose a room has two similar windows (each with thermal resistance R) as in Figure 1.27(b). These two windows provide for parallel heat flow; thus the equivalent thermal resistance is the combination of the two individual window resistances using the same rule used for two equal parallel electrical resistances.

Thermal resistance.

$$q = \frac{\theta_2 - \theta_1}{R/2} = 2q_0$$

Thus, the rate of heat flow doubles compared to Figure 1.27(a).

Instead, suppose one window with double-thick glass is present, as in Figure 1.27(c). The thermal resistance doubles, as does the combination of two

Table 1.3 *Thermal-Electrical Analogs*

Thermal Symbol	Thermal Quantity	Electrical Symbol	Electrical Quality
q	Rate of heat flow	i	Current
θ	Temperature	v	Voltage
R	Thermal resistance	R	Electrical resistance
C	Thermal capacitance	C	Electrical capacitance

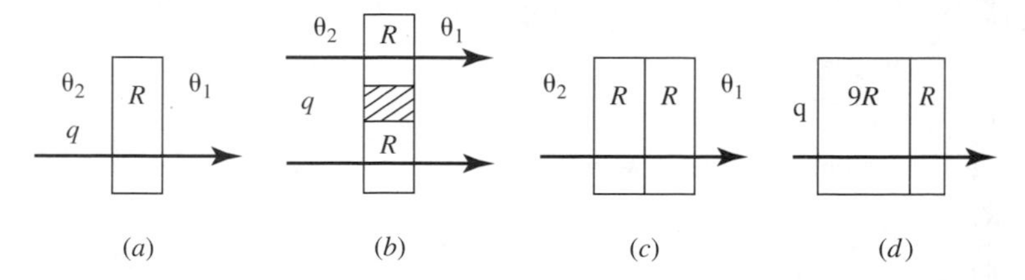

Figure 1.27 Examples of heat flow. (a) Flow through a window with one pane of glass. (b) Flow through two windows. (c) Flow through a double-thick window. (d) Flow through a thermally insulated window.

electrical resistors in series. The heat flow becomes

$$q = \frac{\theta_2 - \theta_1}{2R} = \frac{q_0}{2}$$

The heat flow is halved compared to Figure 1.27(a).

Special thermally insulting material having thermal resistance $9R$ could be coated onto the glass having thermal resistance R. The total thermal resistance acts like two series electrical resistors of $9R$ and R, respectively, for a total thermal resistance of $10R$. Then

$$q = \frac{\theta_2 - \theta_1}{10R} = \frac{q_0}{10}$$

for even less heat loss.

Temperature of a fluid changes in a manner analogous to the change in voltage across an electrical capacitor owing to the flow of heat. For the thermal system

$$q = C\, d\theta/dt \qquad \textbf{[1.73b]}$$

while the electrical analog is

$$i = C\, dv/dt$$

Here, C can represent either thermal or electrical capacitance.

Suppose an otherwise perfectly insulted room can lose (or gain) heat via some heat-exchanging material as in Figure 1.28 (a). The room temperature θ evolves per

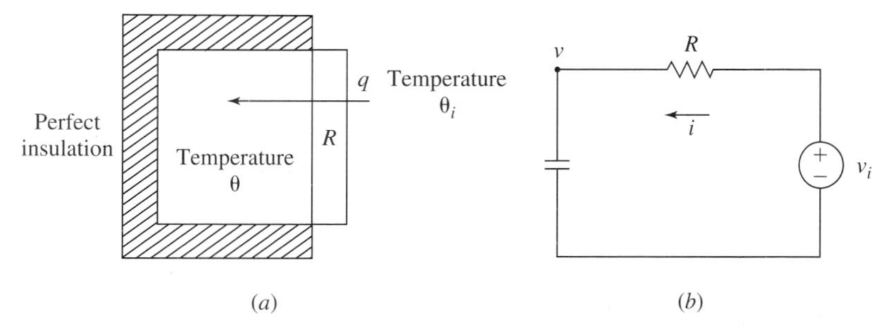

Figure 1.28 Electrical model of heat flow. (a) Heat flow into a room. (b) electrical analog: $v_s = \theta_s$, $v = \theta$, $i = q$ (rate of heat flow).

Equations 1.73(a,b)

$$C\frac{d\theta}{dt} + \frac{(\theta - \theta_i)}{R} = 0$$

analogous to an electrical circuit with

$$C\frac{dv}{dt} + \frac{(v - v_i)}{R} = 0$$

The Laplace transform for the thermal system is

$$\theta(s) = \frac{\theta(0^-)}{s + 1/RC} + \frac{(1/RC)\theta_i(s)}{s + 1/RC}$$

Suppose $\theta_i(s)$ represents a step of temperature θ_2 and thus we have $\theta_i(s) = \theta_2/s$. The step response of temperature versus time is analogous to the voltage of an RC circuit.

$$\theta(t) = \theta(0^-)e^{-t/RC} + \theta_2(1 - e^{-t/RC}) \qquad \textbf{[1.74]}$$

Thus, the temperature changes exponentially from the initial inside temperature $\theta(0^-)$ to the outside temperature θ_2. This model provides a simple way of analyzing many thermal systems.

1.11 Hydraulics

There are many systems that employ the movement of a fluid that is incompressible or nearly so. Figure 1.29, for example, shows a hydraulic amplifier. A pressure p_i is applied to a tube entering the sealed chamber. If the pilot valve moves through a distance x_i, fluid enters the pilot tube producing an overpressure on the left-hand portion at the output piston. That piston moves to the right, causing an equal displacement of fluid to exit through the right-side output tube at a pressure p_o.

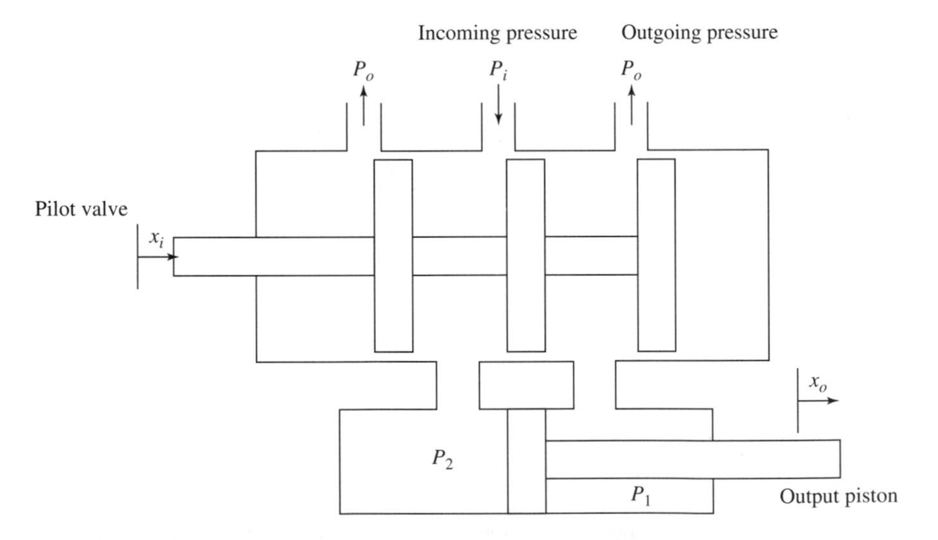

Figure 1.29 Hydraulic actuator.

The flow rate into the left output piston chamber equals the flow rate out. If the incoming flow has resistance R_i and the outgoing flow has resistance R_o, then

$$\rho A\dot{x}_0 = \frac{x_i}{R_i}(p_i - p_2)^{0.5} = \frac{x_i}{R_0}(p_1 - p_0)^{0.5} \quad [1.75]$$

Fluid flow.

for a small horizontal movement x_i of the pilot valve and a small movement x_o of the output piston. In Equation (1.75), ρ is the density of the fluid, A is the area of the piston, and p represents pressure.

The pressure difference $p_2 - p_1$ drives the output piston. If that piston has mass m and exerts a force F on some load, then

$$m\ddot{x}_0 = A(p_2 - p_1) - F \quad [1.76]$$

Equations (1.75) and (1.76) form a pair of nonlinear differential equations; however, under a few simplifying assumptions, these equations are easy to solve. Let's assume that the output piston pushes on one of the control surfaces of an aircraft as in Section 1.10. If the piston moves with constant velocity, then $\ddot{x}_0$ is zero and Equation (1.76) indicates that

$$p_2 - p_1 = F/A \quad [1.77]$$

If $R_i = R_o = R$ in Equation (1.75), then

$$p_1 - p_0 = p_i - p_2 \quad [1.78]$$

By solving Equations (1.77) and (1.78), we have

$$p_2 = \frac{p_i + p_0 + F/A}{2}$$

This result can be used to solve Equation 1.75.

$$\dot{x}_0 = \frac{x_i}{\rho AR\sqrt{2}}(p_i - p_o - F/A)^{0.5} \quad [1.79]$$

Thus, a small movement x_i of the pilot valve may produce a velocity $\dot{x}_o$ of the output piston moving what may be a very large load if sufficient pressure is applied.

1.12 Transfer Function and Stability

1.12.1 Transfer Functions

One of the most powerful tools of control system analysis and design is the transfer function representation, which is a generalization of the impedance concept in electrical and mechanical networks. For a single-input, single-output system with input $r(t)$ and output $y(t)$, the transfer function (also called the transmittance) relating the output to the input is defined as follows:

The transfer function is defined for zero-valued initial conditions.

$$T(s) = \left.\frac{Y(s)}{R(s)}\right|_{\text{when all initial conditions are zero}}$$

To find the transfer function, Laplace-transform the system equations, with zero initial conditions, and form the ratio of output transform to input transform.

Consider the system described by

$$\frac{d^2y}{dt^2} + 6\frac{dy}{dt} + 8y = -\frac{dr}{dt} + 5r \qquad [1.80]$$

Laplace-transforming and collecting terms gives

$$s^2Y(s) - sy(0^-) - y'(0^-) + 6[sY(s) - y(0^-)] + 8Y(s)$$
$$= -[sR(s) - r(0^-)] + 5R(s)$$
$$Y(s)[s^2 + 6s + 8] = sy(0^-) + y'(0^-) + 6y(0^-) + r(0^-) + R(s)[-s + 5] \qquad [1.81]$$

In Equation (1.81), the polynomial multiplying $Y(s)$ is called the characteristic polynomial. On the right side of Equation (1.81) are two distinct sets of values, one created by initial conditions and one created by the input $R(s)$. According to the definition of a transfer function, the initial conditions are set equal to zero, yielding

$$T(s) = \frac{Y(s)}{R(s)} = \frac{-s + 5}{s^2 + 6s + 8} \qquad [1.82]$$

In systems described by linear, constant-coefficient integrodifferential equations, every Laplace-transformed signal is related to every other such signal by a transfer function.

❑ Computer-Aided Learning

In this section, the following MATLAB commands are introduced:

```
den=[coefficients]
roots(den)
tf(num, den)
zpk([num roots], [den roots], gain k)
```

Suppose we want to use MATLAB to enter the following transfer function into the computer.

$$T(s) = \frac{5(s + 6)}{s^2 + 6s + 25}$$

We have learned how to input a function like $T(s)$ using "s" as a symbolic variable; but, a number of MATLAB functions require the use of either of two other mechanism for inputting a polynomial ratio in s. In one mechanism, we must create a vector having the coefficients of the numerator and denominator for descending powers of s. Since the numerator is $5s + 30$, the commands

```
num=[5  30]
den=[1  6  25]
```

code the coefficients of each polynomial. The transfer function may then be created by

```
t=tf(num, den)
```

The roots of a polynomial follow from

```
Roots(den)
```

Which are -3+j4 and -3-j4.

The other approach is to use the zero-pole-gain or zpk form. In that case, numerator and denominator roots are placed in respective vectors and an empty vector [ ] denotes lack of a root when a constant is present, usually for the numerator. For the system above, we can create $T(s)$ by inputting the zero, poles, and gain k.

```
t=zpk(-6,[-3+j*4 -3-j*4],5]
```

C1.7 Write MATLAB commands using the "tf" command to create

(a) $T_a(s) = \dfrac{8s}{s^2 + 9^2}$

(b) $T_b(s) = \dfrac{(2/9)(s + 1/20)}{s^3 + (23/6)s^2 + (259/90)s + 1/10}$

Ans. (a)
```
numa=[8 0]
dena=[1 0 81]
ta=tf(numa,dena)
```

(b)
```
numb=[2/9 2/180]
denb=[1 23/6 259/90 1/10]
tb=tf(numb,denb)
```

C1.8 Write MATLAB commands to obtain the roots of the denominator of the transfer function of C1.7(b)

Ans.
```
denb=[1 23/6 259/90 1/10]
r=roots(denb)
```

C1.9 Suppose the roots from C1.8 are contained in the vector r, write a MATLAB command to create the transfer function of C1.7(b) by using the "zpk" command.

Ans.
```
tc3=zpk(-1/20,r,2/9)
```

1.12.2 Response Terms

The type of time function corresponding to each partial fraction expansion term for a Laplace-transformed signal depends upon the term's root location in the complex plane and upon whether the root is repeated. Figure 1.30 shows representative time functions associated with various transform denominator root locations.

Response Components

The system output when the initial conditions are all zero is termed the *zero-state* response component. Its Laplace transform is given simply by the product of the

Zero-state response is defined.

transfer function and the input transform. If the system of Equation (1.80) has zero initial conditions and if the input is

$$r(t) = 7e^{-3t}$$

the system output is given by

$$Y_{\text{zero-state}}(s) = T(s)R(s) = \frac{7(-s+5)}{(s^2+6s+8)(s+3)} \qquad \textbf{[1.83]}$$

If the system initial conditions are not zero, there is an additional output component present. The *zero-input* part of the response. For this example, assume $y(0^-) = 0$, $y'(0^-) = 1$, and $r(0^-) = 7$. Then we have

$$Y_{\text{zero-input}}(s) = \frac{8}{s^2+6s+8}$$

Denominator Root Locations **Time Function**

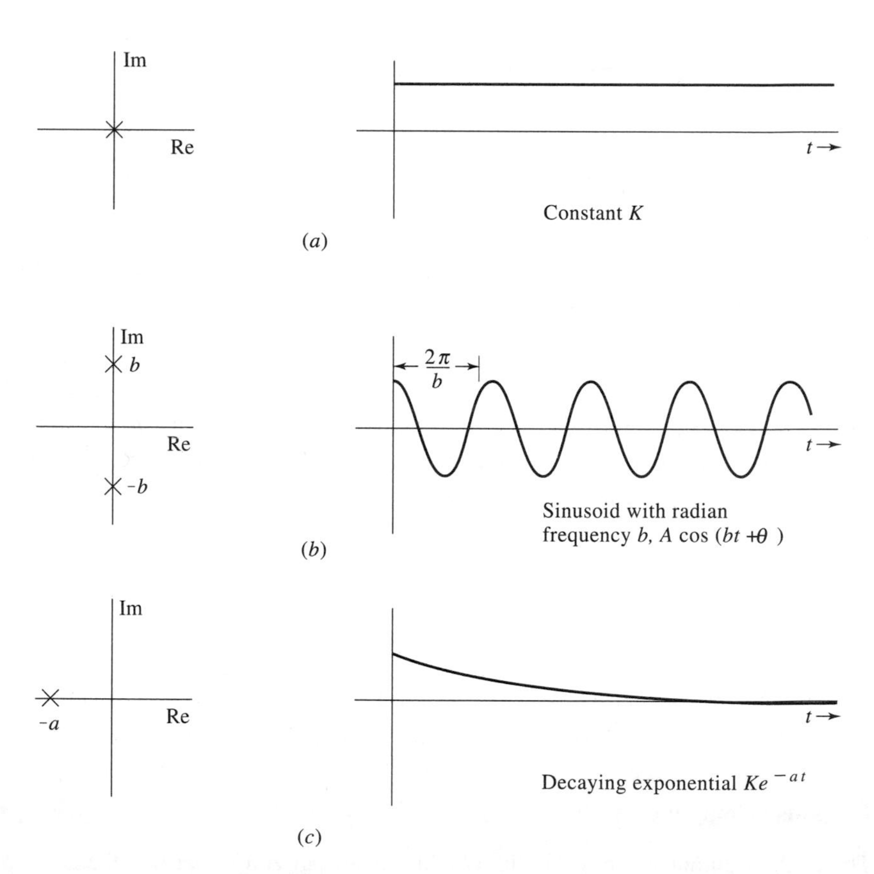

Figure 1.30 Laplace-transform denominator root locations and corresponding time functions.

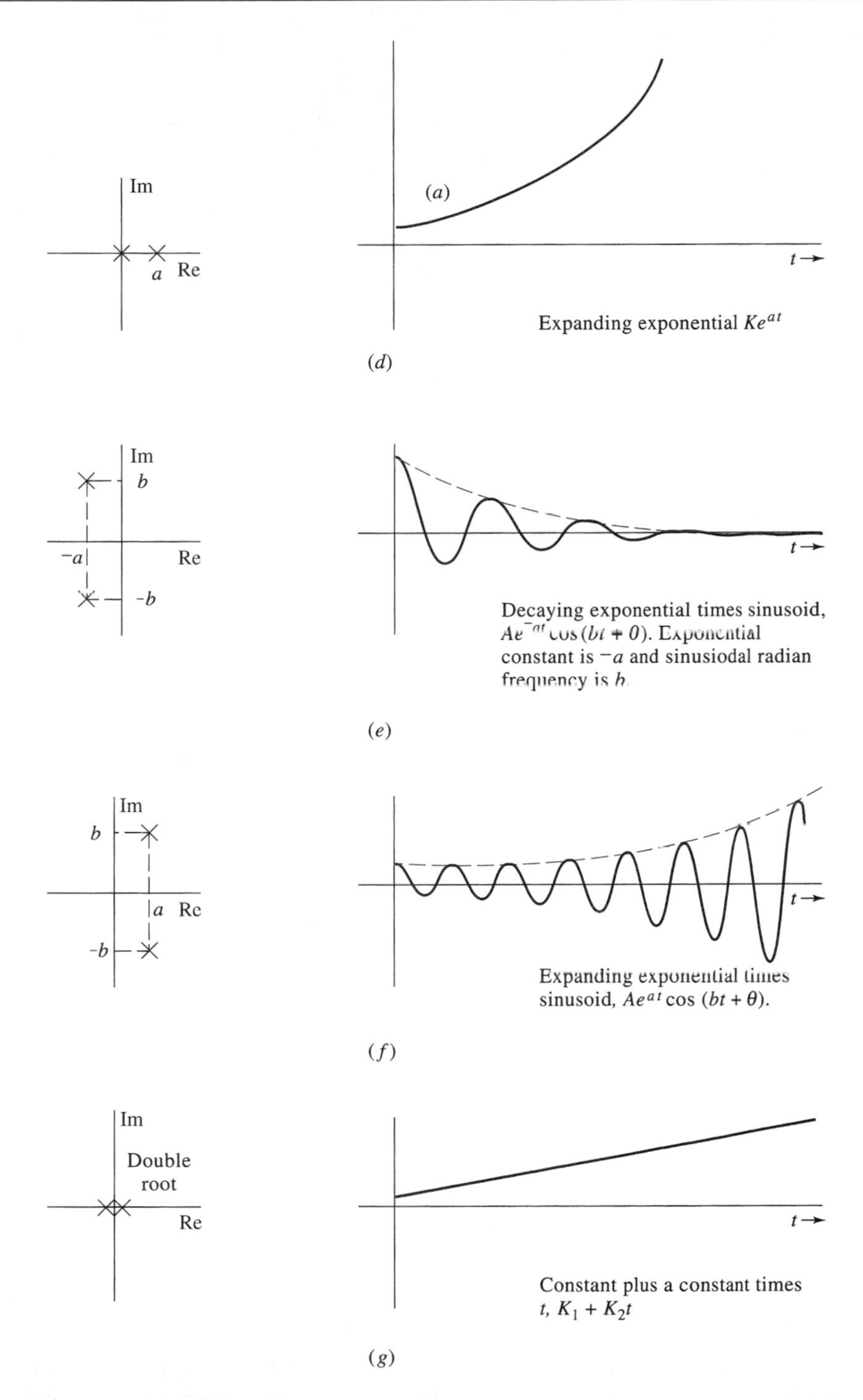

Figure 1.30 *(continued)*

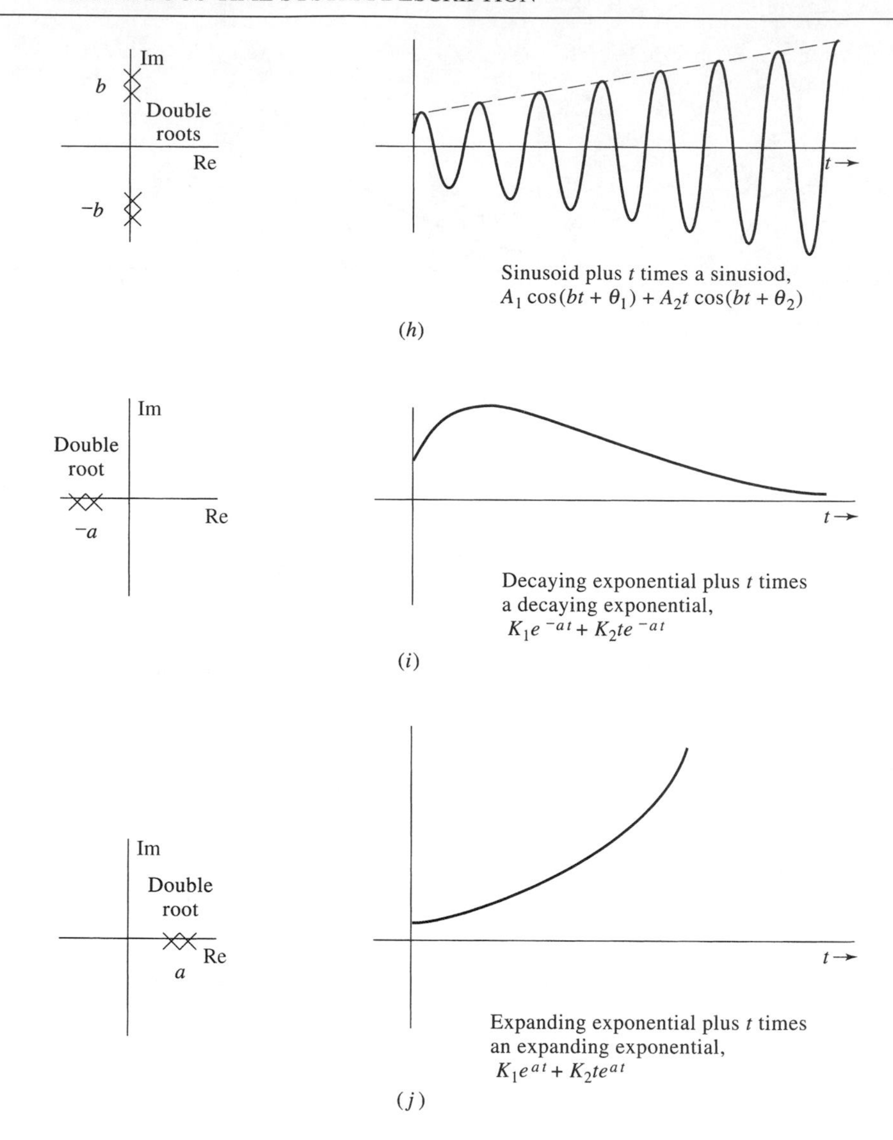

Figure 1.30 *(continued)*

The complete response

$$Y(s) = Y_{\text{zero-state}}(s) + Y_{\text{zero-input}}(s)$$
$$= \frac{7(-s+5)}{(s^2+6s+8)(s+3)} + \frac{8}{s^2+6s+8} \qquad [\mathbf{1.84}]$$

Zero-input response is defined.

Although $T(s)$ assumes zero initial conditions, the zero-input response can be found from the transfer function, provided there have been no unwarranted cancellations of terms made between numerator and denominator of the transfer function. It is possible, although perhaps unlikely, that system parameters are just the right numbers so that a factor in the transfer function numerator cancels a denominator factor, causing the corresponding term in the system's zero-input response to be overlooked.

For a transfer function

$$T(s) = \frac{b_m s^m + b_{m-1}s^{m-1} + \cdots + b_1 s + b_0}{a_n s^n + a_{n-1}s^{n-1} + \cdots + a_1 s + a_0} \qquad [1.85]$$

a differential equation relating the output $y(t)$ to the input $r(t)$ is

$$a_n \frac{d^n y}{dt^n} + a_{n-1}\frac{d^{n-1}y}{dt^{n-1}} + \cdots + a_1 \frac{dy}{dt} + a_0 y$$
$$= b_m \frac{d^m r}{dt^m} + b_{m-1}\frac{d^{m-1}r}{dt^{m-1}} + \cdots + b_1 \frac{dr}{dt} + b_0 r$$

For rational systems, $n > m$.

The differential equation above can be Laplace-transformed again so that initial condition terms are restored. Including the initial condition terms in the Laplace-transformed equations gives

$$a_n[s^n y(s) - s^{n-1}y(0^-) - s^{n-2}y'(0^-) - \cdots]$$
$$+ a_{n-1}[s^{n-1}Y(s) \quad s^{n-2}y(0^-) - s^{n-3}y'(0^-) - \cdots] + \cdots + a_0 Y(s)$$
$$= b_m[s^m R(s) - s^{m-1}r(0^-) - s^{m-2}r'(0^-) - \cdots] + \cdots + b_0 R(s)$$
$$(a_n s^n + a_{n-1}s^{n-1} + \cdots + a_1 s + a_0)Y(s) = (b_m s^m + \cdots + b_0)R(s)$$
$$+ \text{ polynomial in } s \text{ with coefficients dependent upon initial conditions}$$

$$Y(s) = \underbrace{T(s)R(s)}_{\text{zero-state component}} + \underbrace{\frac{\text{polynomial in } s \text{ with coefficients dependent upon initial conditions}}{a_n s^n + a_{n-1}s^{n-1} + \cdots + a_1 s + a_0}}_{\text{zero-input component}} \qquad [1.86]$$

Total response.

The zero-input component transform has the same denominator polynomial as $T(s)$.

The transfer function denominator polynomial is the system *characteristic polynomial*, and the roots of the characteristic polynomial, that is, the solutions of the characteristic equation

$$a_n s^n + a_{n-1}s^{n-1} + \cdots + a_1 s + a_0 = 0 \qquad [1.87]$$

are known as the system's *characteristic roots*.

Natural and forced Components are defined.

An alternative to the zero input/zero-state system decomposing of system response is to separate the response into *natural* and *forced* parts. The natural component consists of all the characteristic root terms in the partial fraction expansion for the response. The forced response component is the remainder of the response and is composed of the terms associated with the input transform. For example, for the system of Equation (1.80) the characteristic roots are $s = -2$ and $s = -4$. The response of Equation (1.84) expands in partial fractions as

The zero-state response contributes the forced response and part of the natural response. The zero input response contributes only to the natural response.

$$Y(s) = \underbrace{\frac{-7s + 35}{(s+2)(s+4)(s+3)}}_{\text{zero-state component}} + \underbrace{\frac{8}{(s+2)(s+4)}}_{\text{zero-input component}}$$

$$= \underbrace{\frac{-56}{s+3}}_{\text{forced component}} + \underbrace{\frac{28.5}{s+2} + \frac{27.5}{s+4}}_{\text{natural component}} \qquad [1.88]$$

Both the zero-input and zero-state response components generally contribute to the natural response component.

If desired, however, system initial conditions can be considered to be system inputs, and transfer functions can be found that relate the outputs to the initial condition inputs. For a system described by

$$\frac{d^2y}{dt^2} + 3\frac{dy}{dt} + 2y = 4r(t)$$

Laplace-transforming with the initial conditions included gives

$$s^2Y(s) - sy(0^-) - y'(0^-) + 3\left[sY(s) - y(0^-)\right] + 2Y(s) = 4R(s)$$

$$Y(s) = \left[\frac{4}{s^2 + 3s + 2}\right]R(s) + \left[\frac{s+3}{s^2 + 3s + 2}\right]y(0^-) + \left[\frac{1}{s^2 + 3s + 2}\right]y'(0^-) \qquad [1.89]$$

$$= T_1(s)R(s) + T_2(s)y\left(0^-\right) + T_3(s)y'(0^-)$$

❑ Computer-Aided Learning

You should thoroughly understand the basis of Laplace and inverse Laplace transforming. To enhance that understanding, and not to replace basic knowledge, MATLAB provides the ability to carry out tedious symbolic calculations. In this section, the following commands are covered:

```
syms
laplace(time function)
ilaplace(Laplace function)
```

If we want the inverse Laplace transform of

$$Y(s) = \frac{s+8}{s^2 + 6s + 36}$$

then we can get $y(t)$ by the commands

```
syms s t
Y=(s+8)/(s^2+6*s+36)
y=ilaplace(Y)
```

C1.10 Use MATLAB to get the inverse Laplace transforms of

(a) $Y(s) = \dfrac{8}{(s+2)(s+4)}$

(b) $Y(s) = \dfrac{s+13}{s(s^2+4s+13)}$

(c) $Y(s) = \dfrac{9}{s(s+3)^2}$

(d) $Y(s) = \dfrac{100\,(s+12)}{\left(s^2+9\right)\left(s^2+s+5\right)^2}$

Ans. (a)
```
syms s t
Ya=8/(s+2)/(s+4)
ya=ilaplace(Ya)
```

(b)
```
syms s t
Yb=(s+13)/s/s^2+4*s+13
yb=ilaplace(Yb)
```

(c)
```
syms s t
Yc=9/s/(s+3)^2
yc=ilaplace(Yc)
```

(d)
```
syms s t
Yd=100*(s+12)/(s^2+9)/(s^2+s+5)^2
yd=ilaplace(Yd)
```

C1.11 Write commands to get the following Laplace transforms.

(a) $y(t) = 1 + 4e^{-2t}\cos(3t)$

(b) $y(t) = 4 - 4e^{-6t}$

(c) $y(t) = 3t^3 + 4t^2 + 6e^{-4t}$

(d) $y(t) = 3t^2e^{-2t}\cos(4t)$

Ans. (a)
```
syms t s
ya=1+4*exp(-2*t)*cos(3*t)
Ya=laplace(ya)
```

(b)
```
syms t s
yb=4-4*exp(-6*t)
Yb=laplace(yb)
```

(c)
```
syms t s
yc=3*t^3+4*t^2+6*exp(-4*t)
Yc=laplace(yc)
```

(d)
```
syms t s
yd=3*t^2*exp(-2*t)*cos(4*t)
Yd=laplace(yd)
```

C1.12 Use MATLAB to find the inverse Laplace transform of $Y(s) = T(s)R(s)$ when $r(t)$ is $e^{-3t}u(t)$ and $T(s) = 4/(s+6)$.

Ans.
```
Syms t s
T=4/(s+6)
r=exp(-3*t)
R=laplace(r)
Y=T*R
y=ilaplace(Y)
```

❑ DRILL PROBLEMS

D1.21 For systems with input $r(t)$ and output $y(t)$ that are described by the following equations, find the system transfer functions:

(a)

$$\frac{d^2y}{dt^2} + 3\frac{dy}{dt} + 7y = 6r$$

Ans. $6/\left(s^2 + 3s + 7\right)$

(b)

$$\frac{d^3y}{dt^3} + 6\frac{d^2y}{dr^2} + 2\frac{dy}{dt} + 4y = -5\frac{d^2r}{dt^2} + 8\frac{dr}{dt}$$

Ans. $\left(-5s^2 + 8s\right) / \left(s^3 + 6s^2 + 2s + 4\right)$

(c)

$$y(t) = 8r(t-3)$$

Ans. $8e^{-3s}$

(d)

$$\begin{cases} \dfrac{dx_1}{dt} = -3x_1 + x_2 + 4r \\ \dfrac{dx_2}{dt} = -2x_1 - r \\ y = x_1 - 2x_2 \end{cases}$$

Ans. $(6s + 21)/(s^2 + 3s + 2)$

D1.22 For the following systems with outputs y and inputs r, find the transfer functions:

(a)

$$\frac{d^2y}{dt^2} + 3\frac{dy}{dt} + 7y = 6r_1 + 5r_2 - 4\frac{dr_2}{dt}$$

Ans. $T_{11}(s) = 6/(s^2 + 3s + 7)$; $T_{12}(s) = (-4s + 5)/(s^2 + 3s + 7)$

(b)

$$\begin{cases} \dfrac{d^2y_1}{dt^2} + 6\dfrac{dy_1}{dt} + 2y_1 = \dfrac{dr}{dt} - 3r \\ \dfrac{dy_2}{dt} + 6y_2 = 4\dfrac{dr}{dt} \end{cases}$$

Ans. $T_{11}(s) = (s - 3)/(s^2 + 6s + 2)$; $T_{21}(s) = 4s/(s + 6)$

(c)

$$\begin{cases} \dfrac{d^3y_1}{dt^3} + 7\dfrac{d^2y_1}{dt^2} + 6\dfrac{dy_1}{dt} + y_1 = \dfrac{d^2r_1}{dt^2} + 3r_1 + r_2 \\ \dfrac{d^3y_2}{dt^3} + 7\dfrac{d^2y_2}{dt^2} + 6\dfrac{dy_2}{dt} + y_2 = 4\dfrac{dr_2}{dt} \end{cases}$$

Ans. $T_{11}(s) = (s^3 + 3)/(s^3 + 7s^2 + 6s + 1)$;
$T_{12}(s) = 1/(s^3 + 7s^2 + 6s + 1)$;
$T_{21}(s) = 0$;
$T_{22}(s) = 4s/(s^3 + 7s^2 + 6s + 1)$

D1.23 Find the zero-state response for $t \geq 0$ of the systems with the following transfer functions and inputs:

(a)

$$T(s) = \frac{4}{s+3}$$

$$r(t) = u(t)$$

Ans. $\frac{4}{3} - \frac{4}{3}e^{-3t}$

(b)

$$T(s) = \frac{3s}{s+2}$$

$$r(t) = \delta(t)$$

Ans. $3\delta(t) - 6e^{-2t}$

(c)

$$T(s) = \frac{-5s}{s^2 + 4s + 3}$$

$$r(t) = 6u(t)e^{-2t}$$

Ans. $15e^{-t} - 60e^{-2t} + 45e^{-3t}$

(d)

$$T(s) = \frac{4}{s + 3}$$

$$r(t) = 3u(t)\cos 2t$$

Ans. $-\frac{36}{13}e^{-3t} + \frac{36}{13}\cos 2t + \frac{24}{13}\sin 2t = -\frac{36}{13}e^{-3t} + 3.32\cos(2t - 33.7°)$

D1.24 Find the complete response for $t \geq 0$ of each of the following systems with the indicated input and initial conditions. Identify the zero-state and zero-input response components and the forced and natural response components.

(a)

$$T(s) = \frac{4}{s + 3}$$

$$r(t) = u(t)$$

$$y(0^-) = -2$$

Ans. $\frac{4}{3} - \frac{10}{3}e^{-3t}$; $\frac{4}{3} - \frac{4}{3}e^{-3t}$; $-2e^{-3t}$; $\frac{4}{3}$; $-\frac{10}{3}e^{-3t}$

(b)

$$T(s) = \frac{10}{s + 4}$$

$$r(t) = \delta(t)$$

$$y(0^-) = 0$$

Ans. $10e^{-4t}$; $10e^{-4t}$; 0; 0; $10e^{-4t}$

(c)

$$T(s) = \frac{s - 5}{s^2 + 3s + 2}$$

$$r(t) = u(t)$$

$$y(0^-) = -3$$

$$y(0^-) = 4$$

Ans. $-\frac{5}{2} + 4e^{-t} - \frac{9}{2}e^{-2t}$; $-\frac{5}{2} - \frac{7}{2}e^{-2t} + 6e^{-t}$; $-2e^{-t} - e^{-2t}$; $-\frac{5}{2}$; $4e^{-t} - \frac{9}{2}e^{-2t}$

1.12.3 Multiple Inputs and Outputs

If a system has several inputs $r_1(t), r_2(t), \ldots$, and/or several outputs $y_1(t), y_2(t), \ldots$, there is a transfer function that relates each one of the outputs to each one of the inputs, when all other inputs are zero:

$$T_{ij}(s) = \left. \frac{Y_i(s)}{R_j(s)} \right|_{\substack{\text{When all initial conditions are zero} \\ \text{when all inputs except } R_j \text{ are zero}}}$$

(i — output number; j — input number)

In general, when the system initial conditions are zero, the outputs are given by

$$\begin{aligned} Y_1(s) &= T_{11}(s)R_1(s) + T_{12}(s)R_2(s) + T_{13}(s)R_3(s) + \cdots \\ Y_2(s) &= T_{21}(s)R_1(s) + T_{22}(s)R_2(s) + T_{23}(s)R_3(s) + \cdots \\ Y_3(s) &= T_{31}(s)R_1(s) + T_{32}(s)R_2(s) + T_{33}(s)R_3(s) + \cdots \end{aligned} \quad [1.90]$$

In vector form we have

$$Y(s) = T(s)R(s)$$

where $Y(s)$ and $R(s)$ are column vectors and $T(s)$ is a matrix of transfer functions.

Suppose that a two-input, two-output system is described by the following differential equations, where r_1 and r_2 are the inputs and y_1 and y_2 are the outputs:

$$\frac{dy_1}{dt} + 2y_1 = r_1 + 5r_2$$

$$y_1 + \frac{dy_2}{dt} + 3y_2 = 4r_2 + \frac{dr_2}{dt}$$

Laplace-transforming with zero initial conditions gives

$$\begin{aligned} (s+2)Y_1(s) &= R_1(s) + 5R_2(s) \\ Y_1(s) + (s+3)Y_2(s) &= (s+4)R_2(s) \end{aligned} \quad [1.91]$$

To find the transfer function that relates Y_1 to R_1, set R_2 to zero and solve for Y_1:

$$\begin{aligned} (s+2)Y_1(s) &= R_1(s) \\ Y_1(s) + (s+3)Y_2(s) &= 0 \end{aligned} \qquad Y_1(s) = \frac{1}{s+2}R_1(s)$$

Then we have

$$T_{11} = \frac{Y_1(s)}{R_1(s)} = \frac{1}{s+2} \quad [1.92]$$

Solving, instead, for Y_2 gives

$$Y_2(s) = \frac{-1}{(s+2)(s+3)}R_1(s)$$

$$T_{21}(s) = \frac{Y_2(s)}{R_1(s)} = \frac{-1}{(s+2)(s+3)} \quad [1.93]$$

Similarly, setting all inputs but R_2 to zero in Equation (1.91) gives

$$\begin{cases} (s+2)Y_1(s) = 5R_2(s) \\ Y_1(s) + (s+3)Y_2(s) = (s+4)R_2(s) \end{cases}$$

$$Y_1(s) = \frac{5}{s+2}R_2(s) \quad T_{12}(s) = \frac{Y_1(s)}{R_2(s)} = \frac{5}{s+2} \qquad [1.94]$$

and

$$Y_2(s) = \frac{s^2+6s+3}{(s+2)(s+3)}R_2(s)$$

$$T_{22}(s) = \frac{Y_2(s)}{R_2(s)} = \frac{s^2+6s+3}{(s+2)(s+3)} \qquad [1.95]$$

For zero initial conditions and inputs

$$r_1(t) = 6\sin 4t$$

$$r_2(t) = 10$$

the outputs are given by

$$Y_1(s) = T_{11}(s)R_1(s) + T_{12}(s)R_2(s) = \left(\frac{1}{s+2}\right)\left(\frac{24}{s^2+16}\right) + \left(\frac{5}{s+2}\right)\left(\frac{10}{s}\right) \qquad [1.96]$$

and

$$Y_2(s) = T_{21}(s)R_1(s) + T_{22}(s)R_2(s)$$

$$= \frac{-1}{(s+2)(s+3)}\left(\frac{24}{s^2+16}\right) + \frac{s^2+6s+3}{(s+2)(s+3)}\left(\frac{10}{s}\right) \qquad [1.97]$$

The natural part of the output y_1 will have an $\exp(-2t)$ contribution from T_{11} and another $\exp(-2t)$ contribution from T_{12}, and so is of the form

$$y_{1_{\text{natural}}} = K_1 e^{-2t}$$

The natural response in y_2 has $\exp(-2t)$ and $\exp(-3t)$ terms from each of T_{21} and T_{22} so it is of the form

$$y_{2_{\text{natural}}} = K_2 e^{-2t} + K_3 e^{-3t}$$

❑ DRILL PROBLEM

D1.25 A three-input, two-output system has the following transfer functions and inputs. Find the two outputs for $t \geq 0$ if all system initial conditions are zero:

$$T_{11}(s) = \frac{3}{s+2} \qquad T_{22}(s) = \frac{-6s+4}{s^2+5s+6}$$

$$T_{12}(s) = \frac{s}{s+3} \qquad T_{23}(s) = \frac{s+2}{s+3}$$

$$T_{13}(s) = \frac{s+7}{s^2+5s+6} \qquad r_1(t) = u(t)$$

$$r_2(t) = \delta(t)$$

$$T_{21}(s) = \frac{10}{(s+2)(s+3)} \qquad r_3(t) = 6u(t)\,e^{-2t}$$

Ans. $\delta(t) + \frac{3}{2} - \frac{51}{2}e^{-2t} + 21e^{-3t} + 30te^{-2t}$;

$\frac{10}{6} + 11e^{-2t} - \frac{38}{3}e^{-3t}$

1.12.4 Stability

The stability of a system is usually evaluated from two different viewpoints, which may give two different results. The viewpoints are internal and external behaviors. External behavior may be affected by cancellation of some transfer function poles by some transfer function zeros, while internal behavior includes the effect of all characteristic polynomial roots.

By definition, for a system to have ***internal (asymptotic)*** stability, the zero-input response decays to zero, as time approaches infinity, for all possible initial conditions. In this case, all of the characteristic polynomial roots influence the response. This type of stability is ensured if all the characteristic polynomial roots are located in the left half-plane (LHP).

By definition, for ***external (bounded-input, bounded-output, or BIBO)*** stability, the zero-state response is bounded, as time approaches infinity, for all bounded inputs. For a bounded input, we can at least expect the forced response(which has the same form as the input) to be bounded. For the system to be BIBO, the natural response of the output should also be bounded. If the output natural response decays to zero as time approaches infinity, for a bounded input, the system is clearly BIBO. A simple way to require the natural response of the output to decay to zero is to require the zero-state impulse response of the output (the inverse Laplace transform of the transfer function) to decay to zero.

If all characteristic polynomial roots are LHP, then the natural response of all variables (both internal and output) do decay to zero and the system is stable from asymptotic and BIBO viewpoints. Another possibility for BIBO stability exists, other than having all characteristic polynomial roots in the LHP. Suppose the characteristic polynomial [denominator of $T(s)$] contains some right half-plane roots that are therefore system poles; however, RHP zeros of $T(s)$ cancel all RHP poles from the transfer function and from the zero-state impulse response. In that case, at least one term in the zero-input response will go to infinity as time goes to infinity, while the zero-state

Table 1.4 *Types of Stability and Conditions for Each Type*

Test	To Be Stable	To Be Unstable	To Be Marginally Stable
Response as $t \to \infty$ $y_{zs,impulse}$ (BIBO) y_{zi} (asymptotic)	Goes to zero	Goes to infinity	Goes neither to zero nor to infinity
Roots of Denominator of $T(s)$ including cancellation (BIBO) Denominator of $T(S)$ excluding cancellation (asymptotic)	All LHP	At least one RHP and/or repeated imaginary axis	Simple imaginary axis

impulse response decays to zero. That system would be asymptotically unstable but BIBO stable. Thus, the two types of stability differ when all RHP poles of $T(s)$ are canceled by RHP zeros. In general, we will evaluate BIBO stability by examining the zero-state impulse response, which allows for pole-zero cancellation. For asymptotic stability we will examine the zero-input response. Table 1.4 summarizes the stability tests.

Consider the following three system transfer functions:

$$T_1(s) = \frac{s+1}{(s+2)(s+3)} = \frac{2}{s+3} - \frac{1}{s+2}$$

$$T_2(s) = \frac{s-7}{(s-2)(s+3)} = \frac{2}{s+3} - \frac{1}{s-2}$$

$$T_2(s) = \frac{2(s-2)}{(s-2)(s+3)} = \frac{2}{s+3} + \frac{0}{s-2}$$

The system of $T_1(s)$ will have terms in the zero-input response of the form

$$y_{zi1}(t) = K_1 e^{-3t} + K_2 e^{-2t}$$

K_1 and K_2 are determined by initial conditions. Consider the zero-state impulse response when the input $r(t) = \delta(t)$, a unit impulse function. Here $R(s)$ is simply one and $Y_{zs}(S)$ is $T(s)$, the transfer function, as in Figure 1.31. Then

$$y_{zs,\text{impulse}1} = 2e^{-3t} - e^{-2t}$$

which decays to zero so system 1 is both asymptotically stable and BIBO stable.

For the second system

$$y_{zi2}(s) = K_1 e^{-3t} + K_2 e^{2t}$$

$$y_{zs,\text{impulse}2}(t) = 2e^{-3t} - e^{2t}$$

System 2 is asymptotically unstable because e^{2t} will approach infinity and BIBO is unstable owing to the same term.

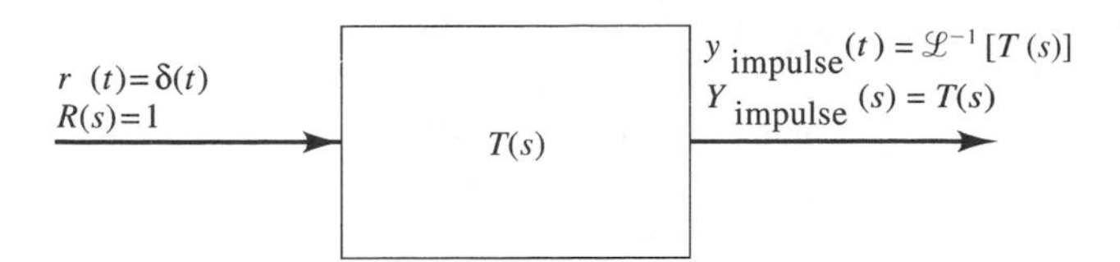

Figure 1.31 The transfer function of a single-output system is the Laplace transform of its unit impulse response.

For the third system

$$y_{zi3}(t) = K_1 e^{-3t} + K_2 e^{2t}$$

$$y_{zs,\text{impulse}3}(t) = 2e^{-3t}$$

System 3 is asymptotically unstable due to the e^{2t} term [caused by the RHP pole of $T(s)$ at $+2$] but BIBO stable due to cancellation of the RHP pole by the RHP zero also located at $+2$.

Importance of asymptotic stability.

The most conservative design approach is to demand asymptotic stability. Suppose the zero- state response of each of the three systems provides the external position of an aircraft while the zero-input response provides the response of some internal electrical device. Clearly, the first system has an acceptable response both for the aircraft and for the electrical device, while the second system has an unacceptable response for both . Although the zero-state response of the position of the third aircraft is stable, the electrical device is probably destroyed (only visible in the zero-input response), which is obviously unacceptable. Perhaps that electrical device plays an important role in a future activity of the aircraft and its destruction could imperil a later flight. It is best to accept only asymptotic (internal) stability, which ensures viability of all components, rather than to be satisfied with BIBO stability, which relates only to the output variable.

Let us examine the time functions of Figure 1.30 and classify those responses. If each response is a zero-state impulse response, then we are classifying BIBO stability. If each response is the zero-input response, we are classifying asymptotic stability. In Figure 1.30, the systems of (c), (e), and (i) are stable owing to LHP denominator roots of $T(s)$. The systems of (d), (f), (g), (h), and (j) are unstable because the responses rise toward infinity as time increases. Notice the RHP denominator $T(s)$ roots in (d), (f), and (j), a case we have already discussed as implying instability. An unstable case we have not yet considered occurs in (g) and (h) where there are repeated (multiple) imaginary axes (IA). See Table 1.4 for a summary of the stable and unstable cases we have considered.

There remains one situation we have not yet considered. In Figures 1.30(a,b) the responses are bounded but do not approach zero. The denominator roots of $T(s)$ are IA but are simple (not repeated). The systems of (a) and (b) are referred to as being marginally stable in that the zero-state response may display a bounded output for some bounded inputs and unbounded outputs for other bounded inputs. For example, the system $T(s) = 1/s$ is classified as being marginally stable in Table 1.4, owing to the pole at $s = 0$. If the input is a unit impulse function $r(t) = \delta(t)$, then $Y_{zs,\text{impulse}}(s) = T(s) = 1/s$ and $y_{zs,\text{impulse}}(t)$ is just $u(t)$, which is obviously a bounded output. Conversely if $r(t) = u(t)$, which is obviously a bounded input, $Y_{zs,\text{input}}(s) = 1/s^2$ and $y_{zs}(t) = tu(t)$, which is an unbounded output. Table 1.4 summarizes all three types of stability for both BIBO and asymptotic cases.

An important problem exists when $T(s)$ and $R(s)$ share lightly damped poles. The zero-state response for this situation, although stable because the poles are in the left half-plane, can cause damage because of the closeness of the poles to the j axis in the s plane. Examples of this condition are the 1939 problem with wind stress on the bridge over Tacoma Narrows and the more recent San Francisco earthquake that caused a freeway to collapse.

$$Y_{\text{zero-state}}(s) = T(s)R(s) = \frac{2}{(s+1)^2+4} \cdot \frac{s}{(s+1)^2+4}$$

Because the damping is so small, replace $(s+1)$ with s and approximate the zero-state response with

$$Y_{\text{zero-state}}(s) = \frac{2s}{(s^2+4)^2}$$

By using Table B.1 to inverse-Laplace-transform the expression, we obtain

$$y_{\text{zero-state}}(t) = 0.5t \sin 2t \qquad \textbf{[1.98]}$$

This $y_{\text{zero-state}}$ component is a linearly increasing sinusoid that is unstable. If the 0.1 damping term is included, the $y_{\text{zero-state}}$ takes the form

$$y_{\text{zero-state}}(t) = Ate^{-.1t} \sin(2t + \Psi_1) + be^{-.1t} \sin(2t + \Psi_2)$$

Since the damping term is so small, the time function approximates Equation (1.98) for small times, and much damage can be done to the structure before the $e^{-.1t}$ can reduce the increasing sinusoid ($t \sin 2t$).

❑ DRILL PROBLEM

D1.26 Classify the systems with the following transfer functions as being stable, marginally stable, or unstable, from asymptotic and BIBO viewpoint.

(a)

$$T(s) = \frac{10(s-3)}{(s+1)^2(s-3)}$$

Ans. unstable, stable

(b)

$$T(s) = \frac{s^2-4s}{(s^2+9)(s+10)}$$

Ans. marginally stable, marginally stable

(c)

$$T(s) = \frac{10}{s^2+2s+10}$$

Ans. stable, stable

(d)

$$T(s) = \frac{10s + 1}{s(s^2 - 2s + 10)}$$

Ans. unstable, unstable

(e)

$$T(s) = \frac{4(s + 2)s^2}{s^2(s + 6)}$$

Ans. unstable, stable

(f)

$$T(s) = \frac{3s}{(s^2 + 4)^2(s + 6)}$$

Ans. unstable, unstable

1.13 Block Diagrams

1.13.1 Block Diagram Elements

A block diagram contains blocks, summers and junctions pickoff points.

Block diagrams are used to describe the component parts of systems. They offer an alternative to dealing directly with equations. A *block* is used to indicate a proportional relationship between two Laplace-transformed signals. The proportionality functions called a transfer function or *transmittance*, relates incoming and outgoing signals and is indicated within the block. A *summer* is used to show additions and subtractions of signals. A summer can have any number of incoming signals, but only one outgoing signal. The algebraic signs to be used in the summation are indicated next to the arrowhead for each incoming signal. A *junction* (sometimes termed a "pickoff point") indicates that the same signal is to go several places. Examples of each of these elements are shown in Figure 1.32.

For example, a system that satisfies the second-order linear differential equation

$$\frac{d^2y}{dt^2} + 4\frac{dy}{dt} + 13y = 4r$$

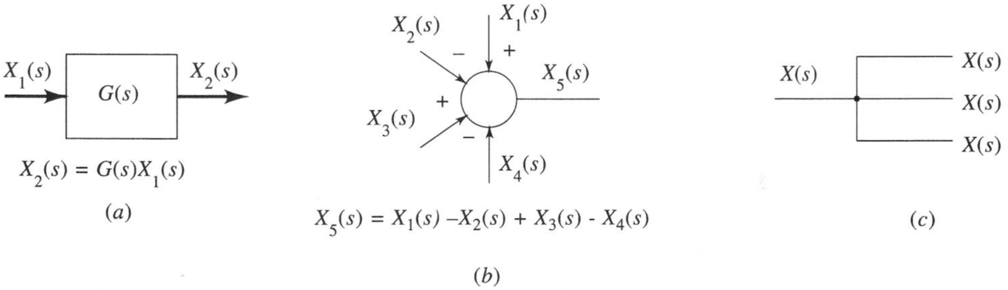

$X_1(s)$
$G(s)$
$X_2(s)$
$X_2(s) = G(s)X_1(s)$
(a)
$X_2(s)$
$X_1(s)$
−
+
$X_5(s)$
+
$X_3(s)$
−
$X_4(s)$
$X_5(s) = X_1(s) - X_2(s) + X_3(s) - X_4(s)$
(b)
$X(s)$
$X(s)$
$X(s)$
$X(s)$
(c)

Figure 1.32 Elements of block diagrams.

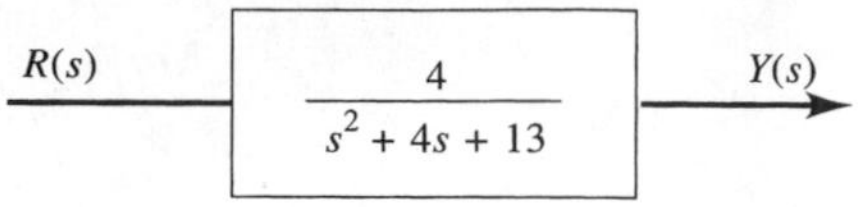

Figure 1.33 Transmittance of a second-order system.

has the transmittance

$$T(s) = \left.\frac{Y(s)}{R(s)}\right|_{\text{initial conditions}=0} = \frac{4}{s^2 + 4s + 13} \qquad [1.99]$$

so that Figure 1.33 contains the transmittance in a block.

A temperature control system provides a more complicated example. If x represents the heat applied to some object, suppose the object's temperature y satisfies

$$\frac{dy}{dt} + by = bx$$

creating the transmittance

$$T_1(s) = \left.\frac{Y(s)}{X(s)}\right|_{\text{initial conditions}=0} = \frac{b}{s + b} \qquad [1.100]$$

which becomes one of the blocks in Figure 1.34. If the system compares the desired temperature r to the actual temperature y, the error e results, where

$$e = r - y$$
$$E(s) = R(s) - Y(s)$$

The difference is created in Figure 1.34 by using a junction to produce the measurement of y for comparison with r and then using a summer with one negative sign to produce e.

Finally, suppose the oven operates upon e so as to modify the heat applied to the object, where

$$\frac{dx}{dt} + ax = ae$$

creating the transmittance

$$T_2(s) = \left.\frac{X(s)}{E(s)}\right|_{\text{initial conditions}=0} = \frac{a}{s + a} \qquad [1.101]$$

which completes Figure 1.34. Figure 1.34 contains two blocks, one summer, and one junction.

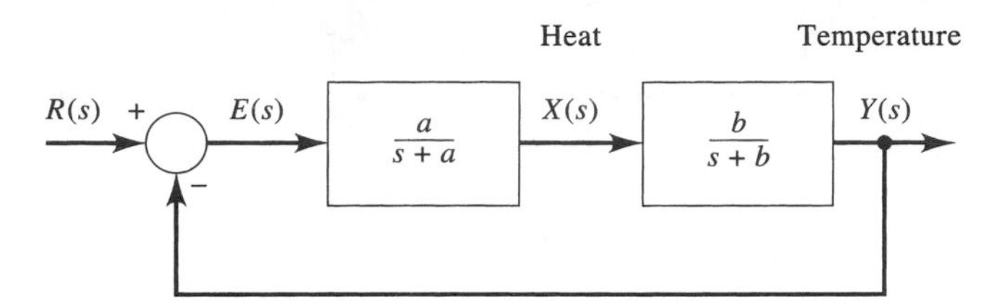

Figure 1.34 Block diagram of a temperature control system.

1.13.2 Block Diagram Reduction

Rearranging system block diagrams to effect simplification or special structures is termed **block diagram algebra**. Since the block diagrams represent Laplace-transformed system equations, manipulating a diagram is equivalent to algebraic manipulation of the original equations. Diagram manipulation provides better physical insight about the structure of the system. For a single-input, single-output block diagram, *reduction* means simplifying the composite diagram to the point where it is a single block, displaying the transfer function relating the output to the input. In reducing a block diagram, it is helpful to proceed step by step, always maintaining the same overall relationship between input and output.

Cascade and tandem-block can be combined.

Some useful simplifications are the following. Two blocks in *cascade* (or *series*), with no additional connections between them, are equivalent so far as the incoming and outgoing signals are concerned to a single "product of transmittances" block, as indicated in Figure 1.35(a). Two blocks in *tandem* (or *parallel*). Figure 1.35(b) are equivalent to a single "sum of transmittances" block. This result is modified if there are other signs besides pluses on the summer.

The feedback configuration is analyzed.

For two blocks in a *feedback* configuration. Figure 1.36, two possible algebraic signs on the summer are considered. $G(s)$ is termed the *forward transmittance* and

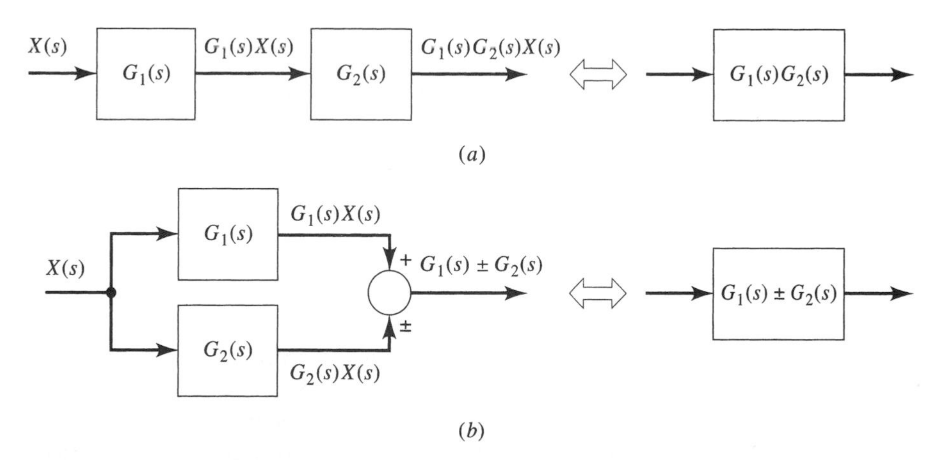

Figure 1.35 Equivalents of blocks. (a) In cascade. (b) In tandem with positive or negative summer signs.

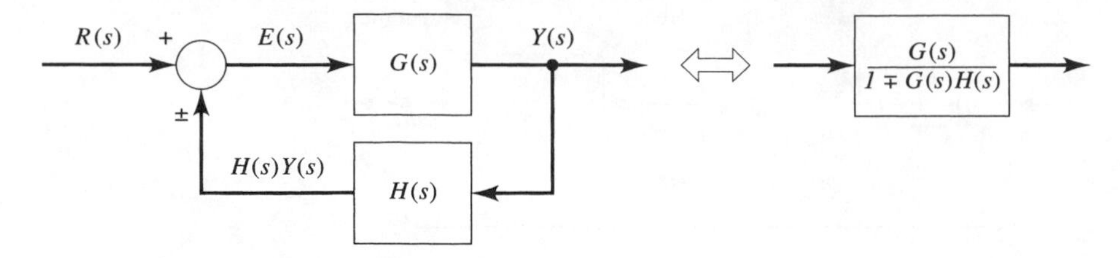

Figure 1.36 The feedback configuration.

$H(s)$ is the *feedback transmittance* in this arrangement. The relationship between the signals is

$$Y(S) = G(s)E(s)$$
$$E(s) = R(s) \pm H(s)Y(s)$$

Solving for $Y(s)$ in terms of $R(s)$ by eliminating $E(s)$, we write

$$Y(s) = G(s)\,[R(s) \pm H(s)Y(s)] = G(s)R(s) \pm G(s)H(s)Y(s)$$

there results

$$T(s) = \frac{Y(s)}{R(s)} = \frac{G(s)}{1 \mp G(s)H(s)} \qquad \textbf{[1.102]}$$

A negative sign on the feedback summation in Figure 1.36 results in a positive algebraic sign in the denominator of $T(s)$; a plus sign on the summer gives a minus sign.

Negative feedback is preferable to positive feedback.

Figure 1.36 is fundamental to control engineering because it reveals the effect of applying feedback to a system. Let's examine the practical meaning of the two possible summations in Figure 1.36: positive and negative. Imagine that $G(s)$ is the transmittance for the combination of a cruise controller and the dynamics of an automobile, while $H(s)$ is the transmittance of a velocity-measuring device. Here, $R(s)$ could represent the Laplace transform of the desired velocity while $Y(s)$ similarly represents the actual velocity. For example, the desired velocity could be 65 mi/h and the actual velocity could be 55 mi/h. Using the negative sign on the summer causes a correction or error equal to (65 − 55) mi/h or 10 mi/h, to be represented by $E(s)$, which would cause the car to speed up by 10 mi/h until the actual velocity reaches 65 mi/h, clearly a logical choice.

Conversely, a positive sign on the summer creates an error or correction equal to (65 + 55) mi/h, or 110 mi/h, obviously the wrong choice. At the same time, a positive sign on the summer induces negative coefficients in the denominator of $T(s)$. The negative coefficients cause some right half-plane roots to evolve for the characteristic polynomial. For example, a first-order polynomial with negative sign like $s - 2$ has a root of $s = +2$ and the system would be unstable due to the RHP root. Clearly

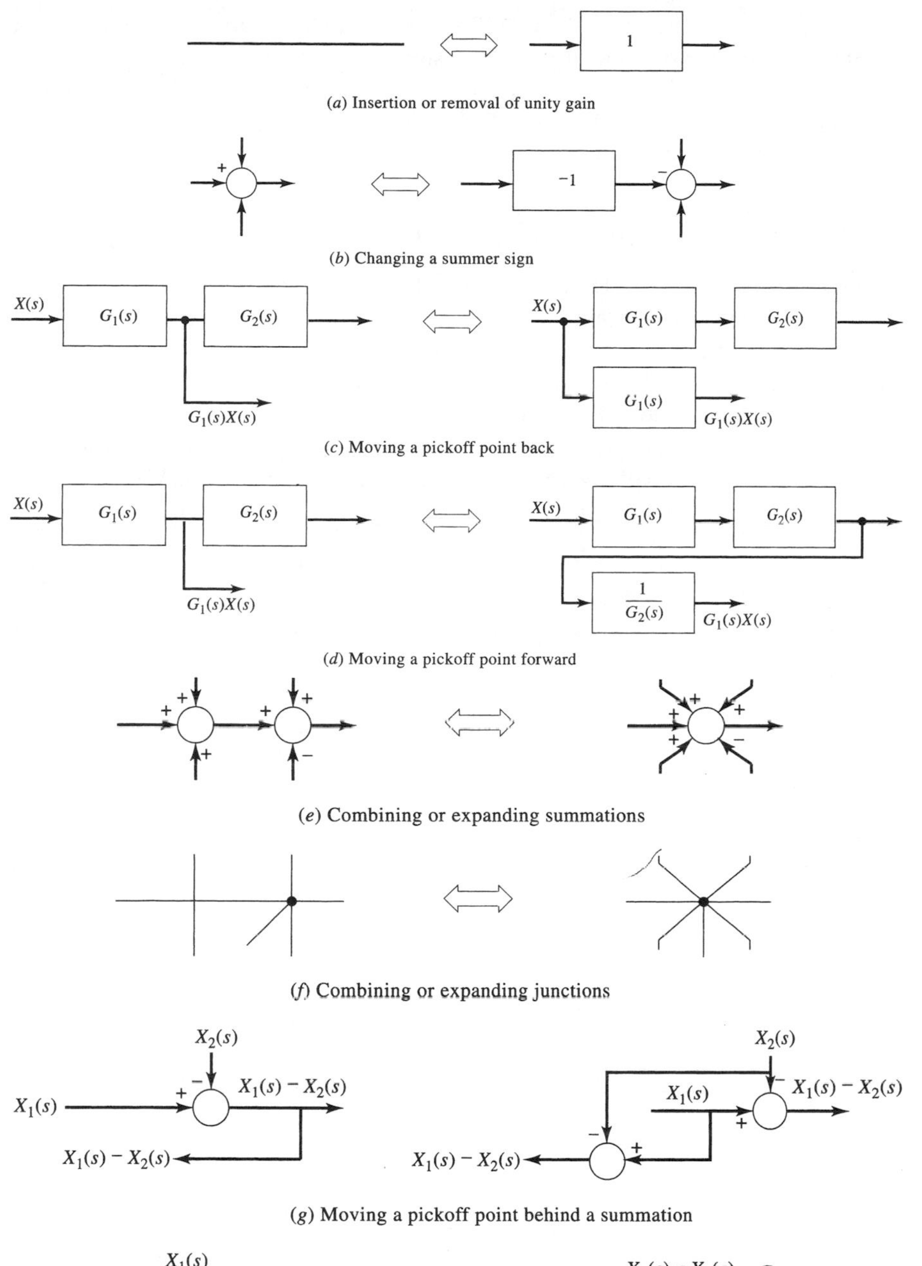

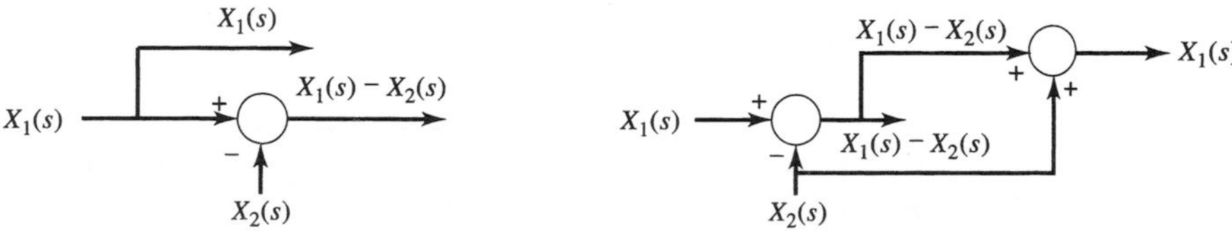

(h) Moving a pickoff point forward of a summation

Figure 1.37 Other useful block diagram equivalence.

the positive sign on the summer should be avoided. In summary, negative feedback should be employed, resulting in a positive sign for Equation (1.102).

Other useful equivalences are given in Figure 1.37. In Figure 1.38, several of the equivalences are used to reduce a block diagram of a single-input, single-output system, to find its transfer function.

1.13.3 Multiple Inputs and Outputs

For a multiple-input, multiple-output system, block diagram reduction involves finding each of the system transfer functions. This is done by considering only one output at a time and setting all but one input to zero to determine the transfer function relating that output to that input. For example, see the four transfer functions for the two-input, two-output system found in Figure 1.39.

The input and output signals in a two-input system are related as shown in the canonical block diagram of Figure 1.40. Every block diagram is reducible to a similar equivalent form, where the transfer functions are placed in the indicated blocks.

❑ DRILL PROBLEMS

D1.27 Reduce the block diagrams shown in Figure D1.27 to obtain the system transfer function.

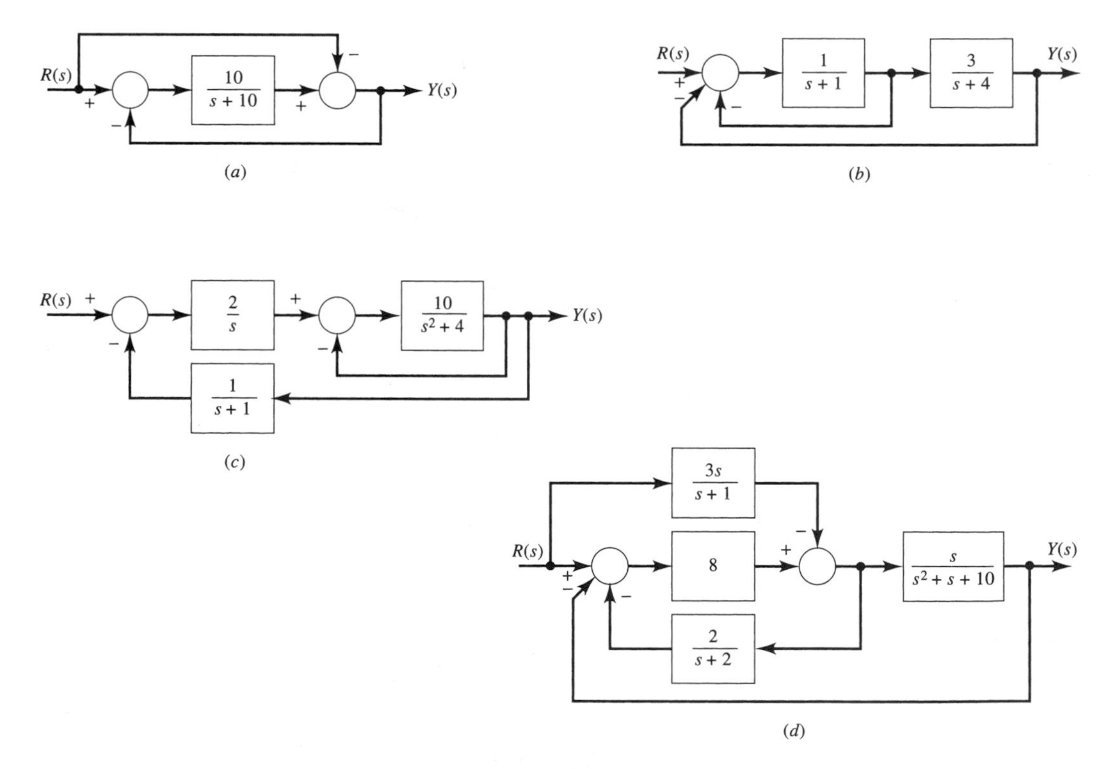

Figure D1.27

Ans. (a) $-s/(s+20)$; (b) $3/(s^2+6s+11)$;
(c) $20(s+1)/(s^4+s^3+14s^2+14s+20)$; (d) $[-3s^2(s+2)$
$+8s(s+2)(s+1)]/(s^4+28s^3+71s^2+224s+180)$

D1.28 Find the six system transfer functions of the system shown in Figure D1.28.

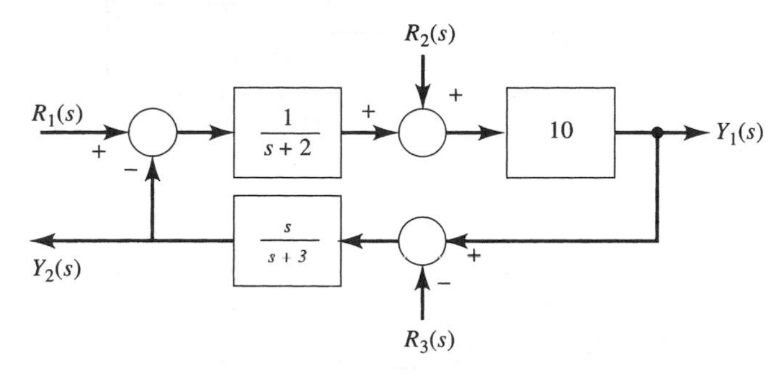

Figure D1.28

Ans. $10(s+3)/(s^2+15s+6)$; $10(s+2)(s+3)/(s^2+15s+6)$;
$10s/(s^2+15s+6)$; $10s/(s^2+15s+6)$;
$10s(s+2)/(s^2+15s+6)$; $-s(s+2)/(s^2+15s+6)$

D1.29 If all the system initial conditions are zero, find the Laplace transform of the output for the given system inputs shown in Figure D1.29.

$r_1(t) = 3e^{-t}$

$r_2(t) = 4u(t)$, where $u(t)$ is the unit step function

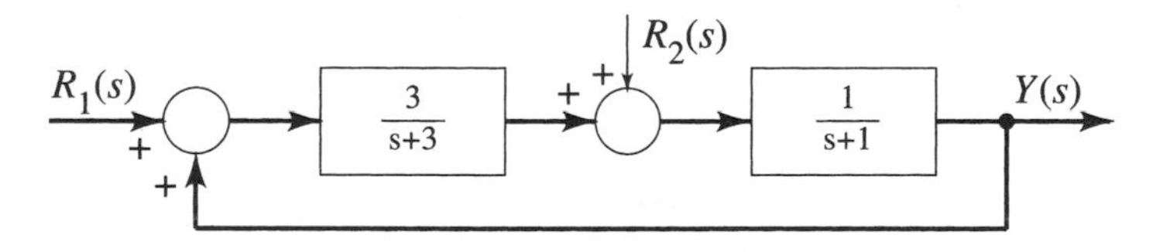

Figure D1.29

Ans. $(4s^2+25s+12)/[s^2(s+4)(s+1)]$

1.14 Signal Flow Graphs

1.14.1 Comparison with Block Diagrams

Comparison.

Another way of depicting relations between Laplace-transformed signals is to employ a signal flow graph. The advantage of using a signal flow graph for complicated systems is that a straightforward procedure is available for finding the transfer function in which it is not necessary to move pickoff points around or to redraw the system several times as with block diagram manipulations. Figure 1.41 shows the relationship

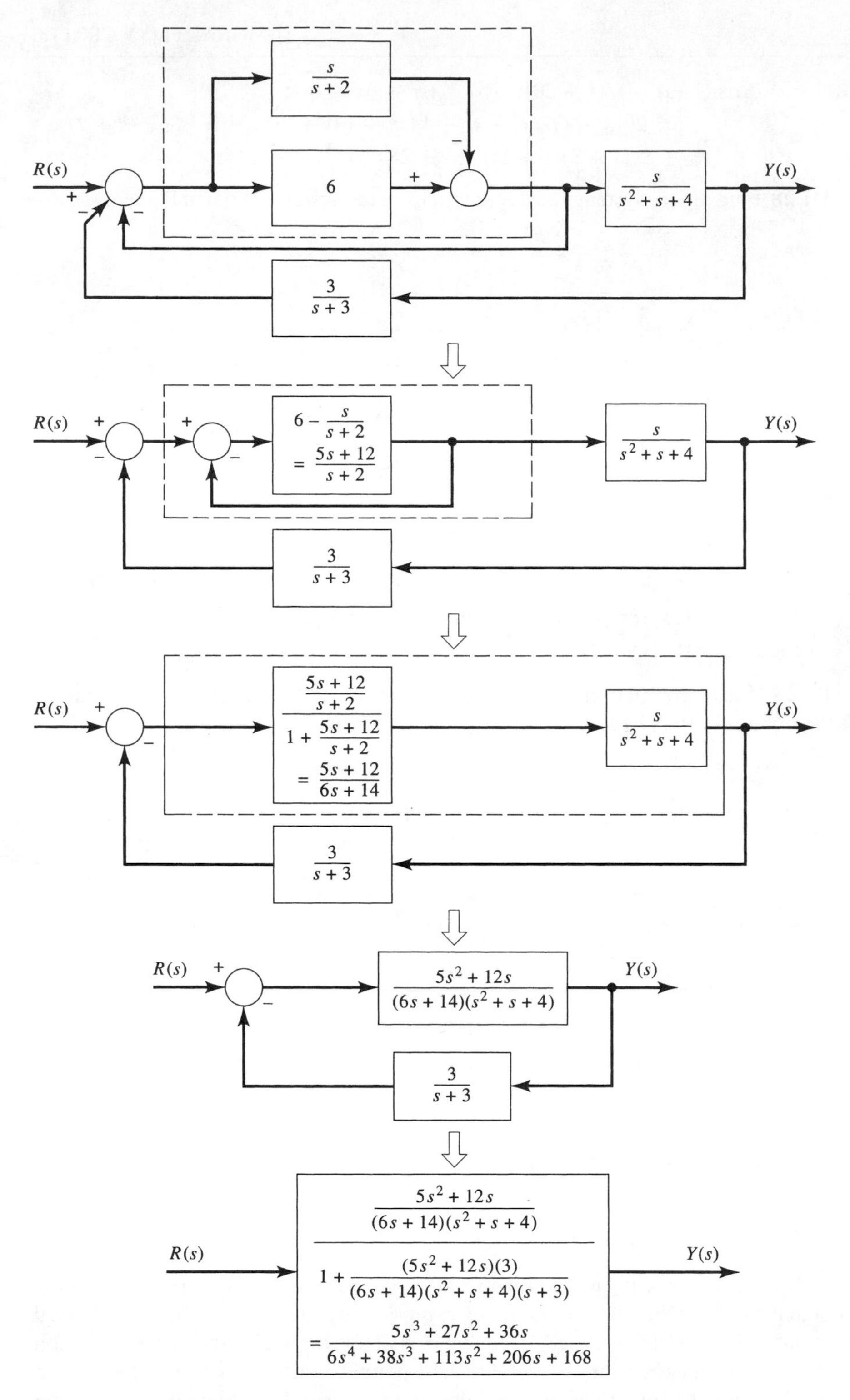

Figure 1.38 Example of block diagram reduction for a single-input, single-output system.

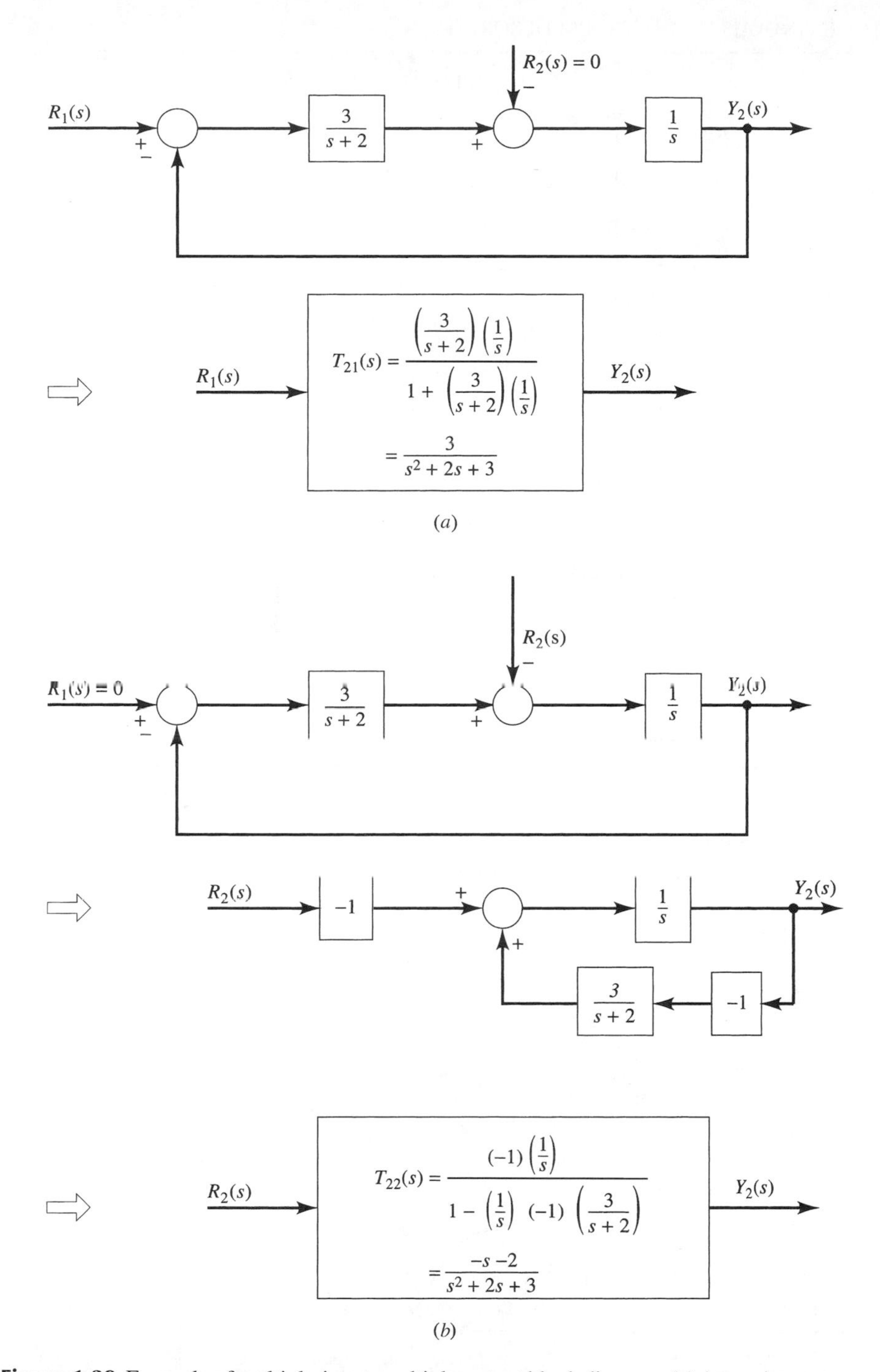

Figure 1.39 Example of multiple-input, multiple-output block diagram. (a) A two- input, two-output system. (b) Block diagram reduction to find $T_{11}(s)$. Input R_2 is set to zero and output Y_2 is ignored. (c) Reduction to find $T_{12}(s)$. R_1 is set to zero and Y_2 is ignored. (d) Reduction to find $T_{21}(s)$. R_2 is set to zero and Y_1 is ignored. (e) Reduction to find $T_{22}(s)$. R_1 is set zero and Y_1 is ignored.

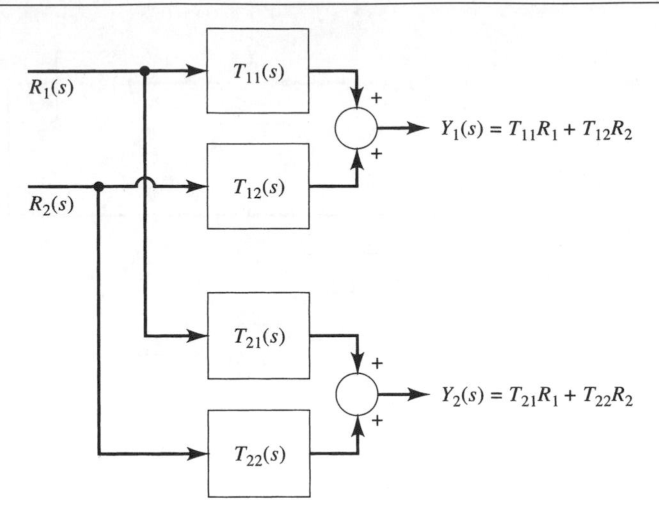

Figure 1.40 Canonical block diagram for a two-input, two-output system.

between block diagrams and signal flow graphs. Here are the steps for converting a block diagram to a signal flow graph.

1. Replace every block in a block diagram with a branch in a signal flow graph.
2. Replace each combination of a summer and pickoff point in a block diagram with a node in a signal flow graph. All sums are assumed to be positive for each node, so signs are not shown on a signal flow graph. For each negative sum, a negative sign must be included with the branch. Label the node with the variable assigned to the pickoff point.

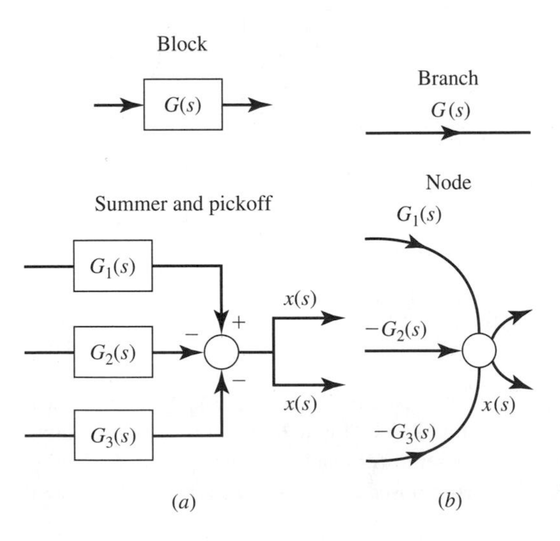

Figure 1.41 Comparison of block diagram and signal flow elements.

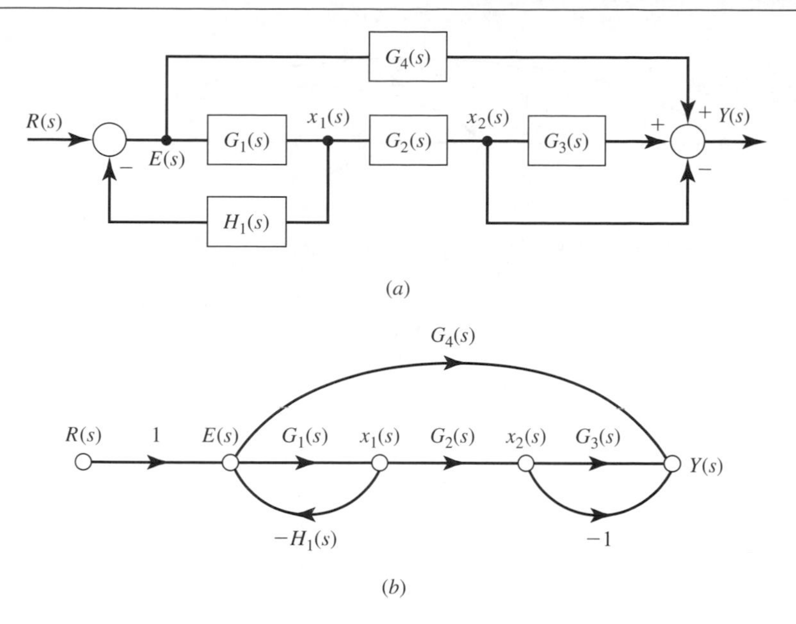

Figure 1.42 Converting a block diagram (a) to an equivalent signal flow graph (b).

3. Replace each solitary pickoff point (not connected to a summer) with a node labeled with the variable assigned to the pickoff point.
4. For each input, show a node labeled with the variable assigned to the input.
5. Add unity branches as needed for clarity or to make connections.

Notice that there will be a node in a signal flow graph for each block diagram summer–pickoff combination, each solitary pickoff point, and each input. Some unity branches may need to be added for clarity or to make some connections.

Nodes.

For example, the block diagram of Figure 1.42(a) contains five blocks, $G_1(s)$, $G_2(s)$, $G_3(s)$, $G_4(s)$, $H_1(s)$, and two unity transmittances conveying $X_2(s)$ and $R(s)$. The signal flow graph of Figure 1.42(b) shows seven branches, one for each transmittance. The block diagram has two summer–pickoff combinations, $E(s)$ and $Y(s)$, two solitary pickoffs not having a summer, $X_1(s)$ and $X_2(s)$, and the input $R(s)$. The signal flow graph accordingly has five nodes with appropriate labels. The two negative summations of Figure 1.42(a) cause negative signs to be assigned to the transmittance of one operating on $X_2(s)$ and to the transmittance of $H_1(s)$, operating on $X_1(s)$.

Example system.

1.14.2 Mason's Rule

the fact that a signal flow graph has fewer elements than a block diagrams (in particular, that a node replaces a summer and pickoff point) results in a straightforward method called Mason's rule for determining the transfer function of a system directly from the signal flow graph. To explain the application of Mason's rule, some terms will first be defined and illustrated with the example signal flow graph of Figure 1.43.

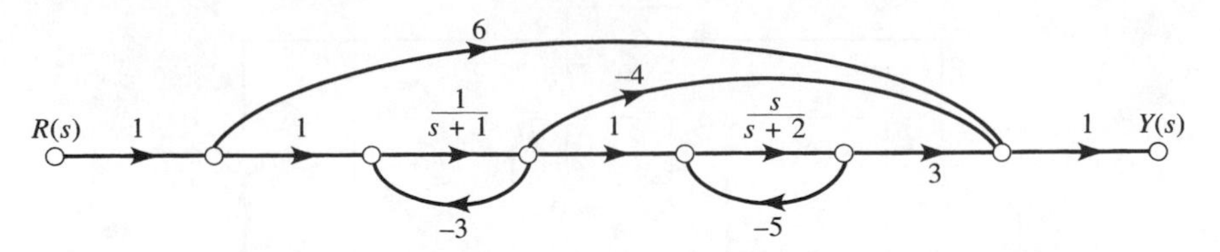

Figure 1.43 A single-input, single-output signal flow graph.

Path.

A *path* is any succession of branches, from input to output, in the direction of the arrows, that does not pass any node more than once, as indicated for the example in Figure 1.43. The *path gain* is the product of the transmittances of the branches comprising the path. For example,

An application of Mason's rule.

$$P_1 = 6$$

$$P_2 = \frac{-4}{s+1}$$

$$P_3 = \frac{3s}{(s+1)(s-2)}$$

Loop.

A **loop** is any closed succession of branches in the direction of the arrows that does not pass any node more than once, as in Figure 1.43. The **loop gain** is the product of the transmittances of the branches comprising the loop. For the example,

$$L_1 = \frac{-3}{s+1}$$

$$L_2 = \frac{-5s}{s+2}$$

Touching.

Two loops are said to be *touching* if they have any node in common. Otherwise, they are nontouching. Similarly, a loop and a path are touching if they have any node in common.

Determinant.

The **determinant** of a signal flow graph is

$$\Delta = 1-(\text{sum of all loop gains}) + (\text{sum of products of gains of all combinations if 2 nontouching loops}) - (\text{sum of products of gains of all combinations of 3 nontouching loops}) + \cdots$$

For the example,

$$\Delta = 1 - (L_1 + L_2) + L_1 L_2$$

Cofactor.

The **cofactor** of a path is the determinant of the signal flow graph formed by deleting all loops touching the path; in this example,

$$\Delta_1 = 1 - (L_1 + L_2) + L_1 L_2$$

$$\Delta_2 = 1 - L_2$$

$$\Delta_3 = 1$$

Mason's gain rule is as follows: The transfer function of a system with single-input, single-output signal flow graph is

$$T(s) = \frac{P_1\Delta_1 + P_2\Delta_2 + P_3\Delta_3 + \cdots}{\Delta}$$

Getting T(s).

For the example shown in Figure 1.43, we have

$$T(s) = \frac{P_1\Delta_1 + P_2\Delta_2 + P_3\Delta_3}{\Delta}$$

$$= \frac{6\left[1 + \frac{3}{s+1} + \frac{5s}{s+2} + \frac{15s}{(s+1)(s+2)}\right] + \left(\frac{-4}{s+1}\right)\left(1 + \frac{5s}{s+2}\right) + \left(\frac{3}{s+1}\right)\left(\frac{s}{s+2}\right)}{1 + \frac{3}{s+1} + \frac{5s}{s+2} + \frac{15s}{(s+1)(s+2)}}$$

$$= \frac{36s^2 + 135s + 40}{6s^2 + 26s + 8} \qquad [1.103]$$

Consider a second example, Figure 1.44 where the products of the transmittances for the paths are

A second example is presented.

$$P_1 = \left(\frac{1}{s+1}\right)\left(\frac{1}{s^2+s}\right)(10)\left(\frac{1}{s}\right)\left(\frac{1}{s}\right)$$

$$P_2 = \left(\frac{1}{s^2+4}\right)\left(\frac{8}{s+8}\right)\left(\frac{1}{s}\right)\left(\frac{1}{s}\right)$$

Path gains.

There are five loops for Figure 1.44:

$$L_1 = \frac{-4}{s^2+s}$$

$$L_2 = -s$$

$$L_3 = \frac{-56}{s+8}$$

$$L_4 = \frac{-6}{s}$$

$$L_5 = (10)\left(\frac{1}{s}\right)\left(\frac{1}{s}\right)\left(\frac{3}{s+3}\right)\left(\frac{s}{s+2}\right)$$

Loop gains.

The determinant for Figure 1.44 is

$$\Delta = 1 - (L_1 + L_2 + L_3 + L_4 + L_5) + (L_1L_2 + L_1L_3 + L_1L_4 + L_2L_3 + L_2L_4) - (L_1L_2L_3 + L_1L_2L_4)$$

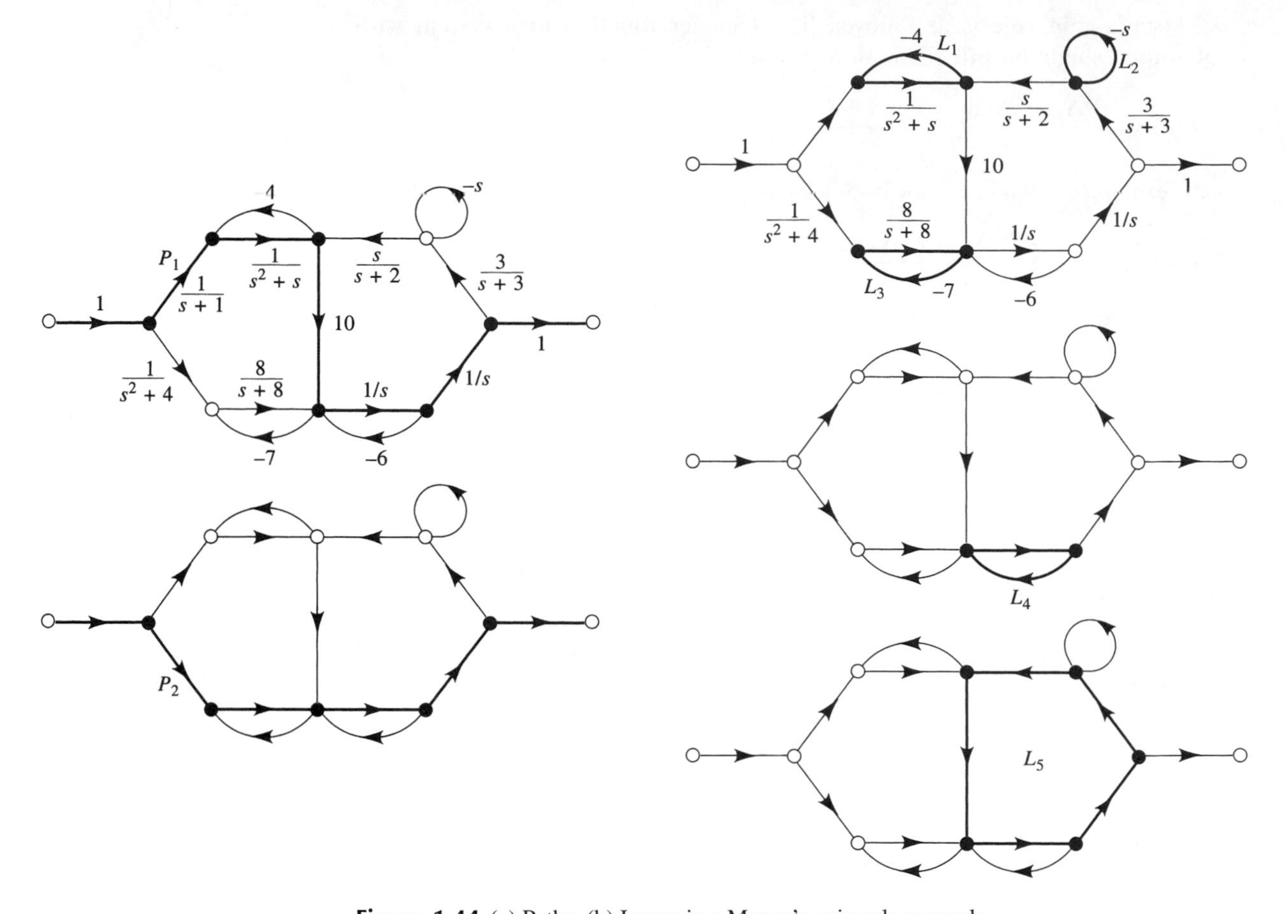

Figure 1.44 (a) Paths. (b) Loops in a Mason's gain rule example.

The cofactors are

$$\Delta_1 = 1 - L_2 = 1 + s$$

$$\Delta_2 = 1 - (L_1 + L_2) + (L_1 L_2) = 1 + \frac{4}{s^2+s} + s + \frac{4s}{s^2+s}$$

Since there are two paths, the transfer function is

$$T(s) = \frac{P_1\Delta_1 + P_2\Delta_2}{\Delta}$$

where P_1, P_2, Δ, Δ_1, and Δ_2 are given above.

The block diagram of Drill Problem D1.27(b) is used to compare the block diagram method with Mason's rule. This system is reduced with block diagram algebra as shown in Figure 1.45. A flow diagram is obtained from the block diagram by setting forth the nodes, x_1 and x_2 in Figure D1.27(b), and drawing the functions connecting

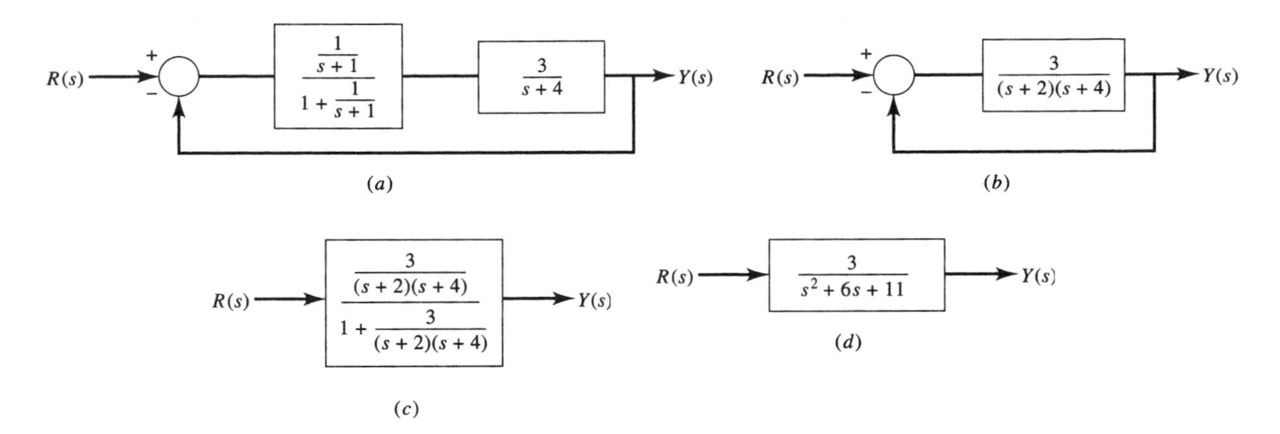

Figure 1.45 Example of block diagram reduction.

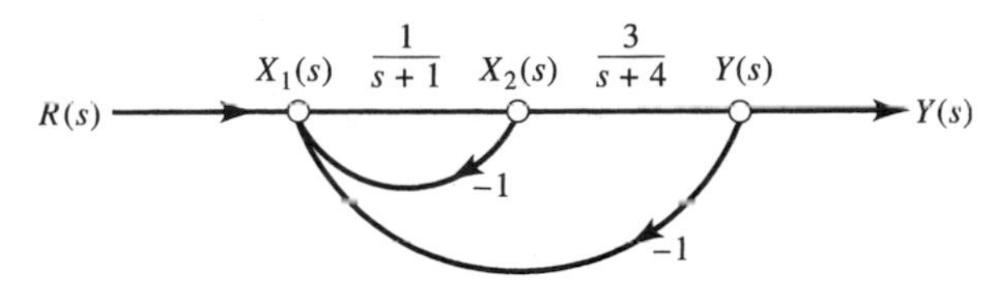

Figure 1.46 Flow diagram for Figure D1.27(b).

these nodes, as shown in Figure 1.46. Mason's rule is applied for one path

$$P_1 = \frac{1}{s+1} \cdot \frac{3}{s+4}$$

and two loops

$$L_1(s) = \frac{-1}{s+1}$$

$$L_2(s) = \frac{-3}{(s+1)(s+4)}$$

The determinant is

$$\Delta = 1 - [L_1(s) + L_2(s)] = 1 + \frac{1}{s+1} + \frac{3}{(s+1)(s+4)}$$

and the cofactor is

$$\Delta_1 = 1$$

Then from Table 1.5,

$$T(s) = \frac{P_1 \Delta_1}{\Delta} = \frac{\dfrac{3}{(s+1)(s+4)}}{1 + \dfrac{1}{s+1} + \dfrac{3}{(s+1)(s+4)}} = \frac{3}{s^2 + 6s + 11} \qquad \textbf{[1.104]}$$

which of course agrees with the final block diagram of Figure 1.45(d).

Table 1.5 *Signal Flow Graph Definitions for Mason's Rule*

Path	A succession of branches, from input to output, in the direction of the arrows, that does not pass any node more than once.
Path gain	Product of the transmittances of the branches of the path. For the ith path, the path gain is denoted by P_i.
Loop	A closed succession of branches, in the direction of the arrows, that does not pass any node more than once.
Loop gain	Product of the transmittances of the branches of the loop.
Touching	Loops with one or more nodes in common are termed *touching*. A loop and a path are touching if they have a common node.
Determinant	The determinant of a signal flow graph is $\Delta = 1 -$ (sum of all loop gains) + (sum of products of gains of all combinations of 2 nontouching loops) − (sum of products of gains of all combinations of 3 nontouching loops) $+ \cdots$
Cofactor	The cofactor of the ith path, denoted by Δ_i, is the determinant of the signal flow graph formed by deleting all loops touching path i.
Mason's gain rule	$T(s) = (P_1\Delta_1 + P_2\Delta_2 + \cdots)/\Delta$

T(s) can be found by simultaneous solutions, which are avoided using Mason's rule.

As was mentioned earlier, Mason's rule evolved from his intuitive look at using Cramer's rule to solve simultaneous node equations for a signal flow graph. It is instructive to show that Equation (1.104) can also be derived by using Cramer's rule. The nodes of Figure 1.46 are described by the following matrices:

$$\begin{bmatrix} 1 & 1 & 1 \\ -\left(\dfrac{1}{s+1}\right) & 1 & 0 \\ 0 & -\left(\dfrac{3}{s+4}\right) & 1 \end{bmatrix} \begin{bmatrix} X_1(s) \\ X_2(s) \\ Y(s) \end{bmatrix} = \begin{bmatrix} 1 \\ 0 \\ 0 \end{bmatrix} R(s)$$

Using Cramer's rule, we write

$$Y(s) = \frac{\begin{vmatrix} 1 & 1 & R(s) \\ -\left(\dfrac{1}{s+1}\right) & 1 & 0 \\ 0 & -\left(\dfrac{3}{s+4}\right) & 0 \end{vmatrix}}{\begin{vmatrix} 1 & 1 & 1 \\ -\left(\dfrac{1}{s+1}\right) & 1 & 0 \\ 0 & -\left(\dfrac{3}{s+4}\right) & 1 \end{vmatrix}}$$

$$= \frac{\left(\frac{1}{s+1}\right)\left(\frac{3}{s+4}\right)R(s)}{1+\left(\frac{1}{s+1}\right)+\left(\frac{1}{s+1}\right)\left(\frac{3}{s+4}\right)} = \frac{3}{s^2+6s+11}R(s) \qquad \textbf{[1.105]}$$

As expected, Equation (1.105) agrees with Equation (1.104). If you look carefully at the calculations required to get Equation (1.105), you can see how the loop gains affect the denominator and how the path and loop gains affect the numerator. The properties touching and nontouching affect the presence or lack of zeros in the determinants, and therefore affect which terms are zeroed out and which terms appear in the final answer.

For multiple-input, multiple-output systems, Mason's gain rule is simply applied repeatedly for each different combination of output and input. The signal flow graph of Figure 1.47(a) is of a system with two inputs and two outputs. The corresponding single-input, single-output signal flow graphs for calculating the four associated transfer functions are shown in Figure 1.47(b)–(e). Using Mason's gain rule, the system transfer functions are as follows:

Multiple transfer functions.

$$T_{11}(s) = \frac{Y_1(s)}{R_1(s)} = \frac{\left(\frac{s}{s+2}\right)(1)+\frac{10}{s(s+3)}(1)}{1-\left(\frac{-4s}{s+2}\right)-\left[\frac{-40}{s(s+3)}\right]} = \frac{s^3+3s^2+10s+20}{5s^3+17s^2+46s+80}$$

$$T_{21}(s) = \frac{Y_2(s)}{R_1(s)} = \frac{\left[\frac{1}{s(s+3)}\right](1)}{1-\left(\frac{-4s}{s+2}\right)-\left[\frac{-40}{s(s+3)}\right]} = \frac{s+2}{5s^3+17s^2+46s+80}$$

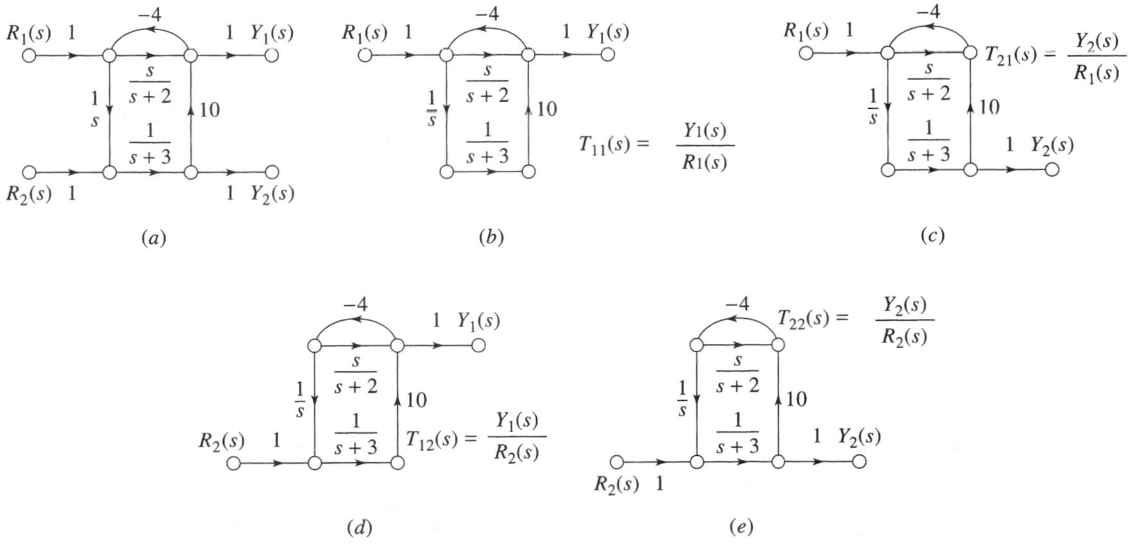

Figure 1.47 A multiple-input, multiple-output signal flow graph and its transfer functions.

$$T_{12}(s) = \frac{Y_1(s)}{R_2(s)} = \frac{\left(\dfrac{10}{s+3}\right)(1)}{1 - \left(\dfrac{-4s}{s+2}\right) - \left[\dfrac{-40}{s(s+3)}\right]} = \frac{10s^2 + 20s}{5s^3 + 17s^2 + 46s + 80}$$

$$T_{22}(s) = \frac{Y_2(s)}{R_2(s)} = \frac{\left(\dfrac{1}{s+3}\right)\left[1 - \left(\dfrac{-4s}{s+2}\right)\right]}{1 - \left(\dfrac{-4s}{s+2}\right) - \left[\dfrac{-40}{s(s+3)}\right]} = \frac{5s^2 + 2s}{5s^3 + 17s^2 + 46s + 80}$$

❑ DRILL PROBLEMS

D1.30 Write a set of simultaneous Laplace-transformed equations for the signal flow graph of Figure D1.30.

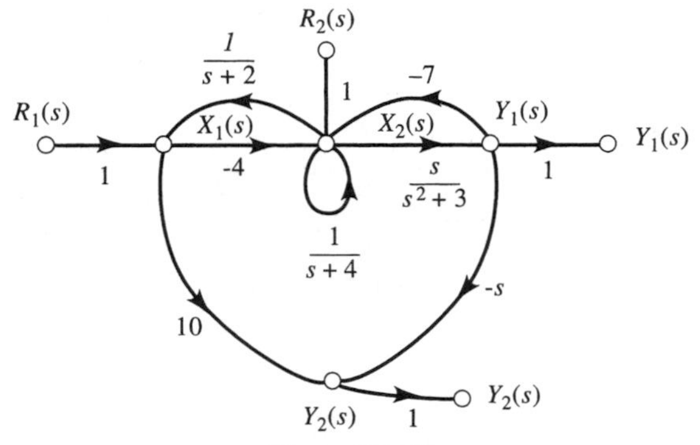

Figure D1.30

Ans. $X_1(s) = R_1(s) + (1/(s+2))X_2(s)$; $X_2(s) = -4X_1(s) + R_2(s) - 7Y_1(s) + (1/(s+4))X_2(s)$; $Y_1(s) = (s/(s^2+3))X_2(s)$; $Y_2(s) = 10X_1(s) - sY_1(s)$

D1.31 Use Mason's gain rule to find the transfer function of each system of Figure D1.31.

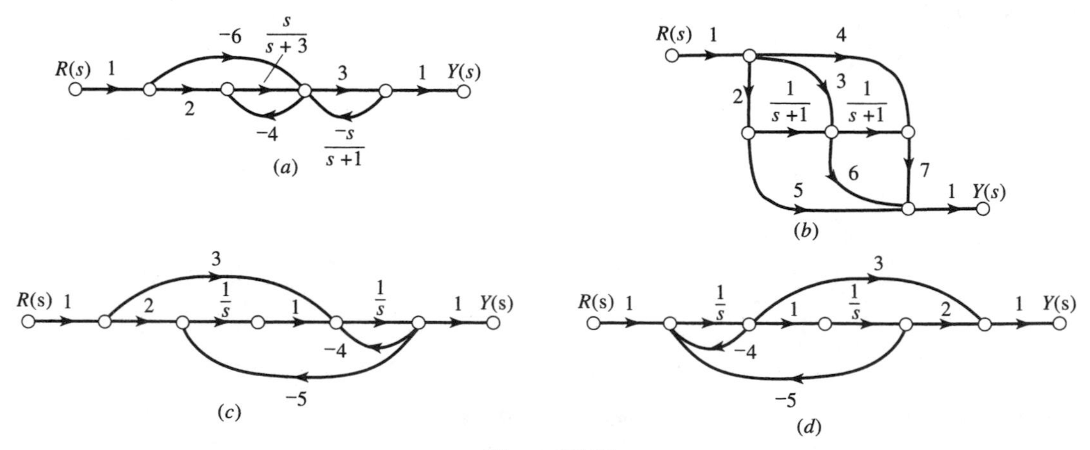

Figure D1.31

Ans. (a) $(-12s^2 - 66s - 54)/(8s^2 + 17s + 3)$;

(b) $(56s^2 + 145s + 103)/(s + 1)^2$; (c) $(3s + 2)/(s^2 + 4s + 5)$;

(d) $(3s + 2)/(s^2 + 4s + 5)$

D1.32 Use Mason's gain rule to find the six transfer functions of the system shown in Figure D1.32.

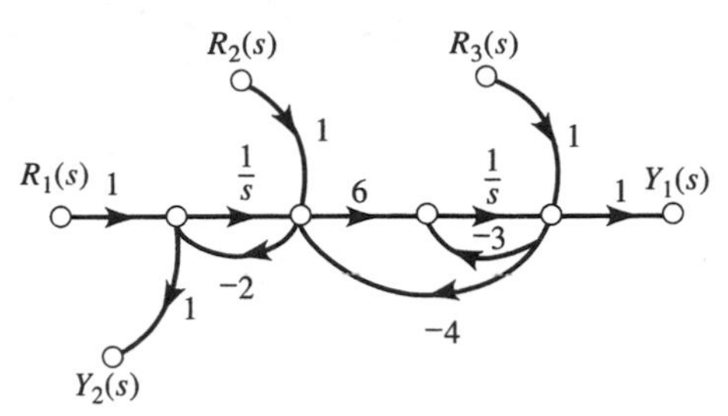

Figure D1.32

Ans. $6/(s^2 + 29s + 6)$; $6s/(s^2 + 29s + 6)$;
$s(s + 2)/(s^2 + 29s + 6)$; $s(s + 27)/(s + 29s + 6)$;
$-s(2s + 6)/(s^2 + 29s + 6)$; $8s^2/(s^2 + 29s + 6)$

D1.33 If all system initial conditions are zero, find the Laplace transforms of the outputs for the given system inputs for Figure D1.33.

$r_1(t) = 4 \sin t$

$r_2(t) = 3\delta(t)$, where $\delta(t)$ is the unit impulse

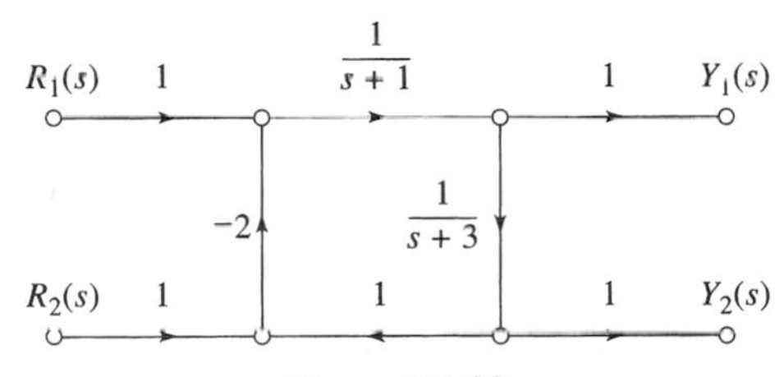

Figure D1.33

Ans. $Y_1(s) = (-6s^2 - 2)(s + 3)/(s^2 + 1)(s^2 + 4s + 5)$;
$Y_2(s) = (-6s^2 - 2)/(s^2 + 1)(s^2 + 4s + 5)$

1.15 A Positioning Servo

A simple but practical feedback control system is diagrammed in Figure 1.48. It is a positioning system or *position servo* for a large video satellite antenna modeled as a mass having a large moment of inertia, J. An output potentiometer measures the output shaft position, converting the position to a proportional voltage according to

$$v_0 = K_p\theta$$

where θ is the output shaft angle in radians and v_0 is the output potentiometer voltage; K_p, the constant of proportionality between shaft position and potentiometer voltage,

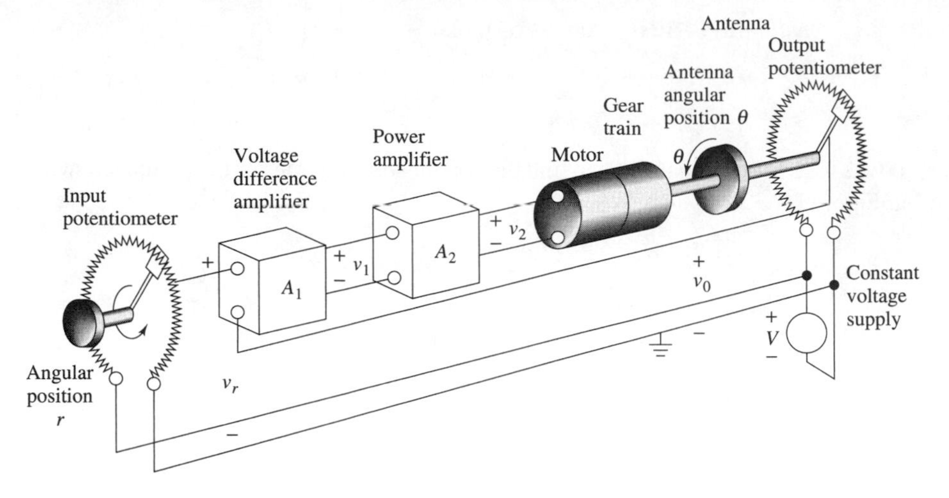

Figure 1.48 A position servo.

is the total voltage V divided by the maximum rotation of the potentiometer:

$$K_p = \frac{V}{\theta_{\max}} \text{ volts/radian}$$

The elements of the block diagram are derived from Figure 1.48.

The input potentiometer slider position r is converted to a voltage with a potentiometer identical to the output potentiometer:

$$v_r = K_p r$$

The difference between the two potentiometer signals is then amplified with gain A_1,

$$v_1 = A_1(v_r - v_0) = A_1 K_p(r - \theta)$$

where v_1 is the error voltage output of the difference amplifier. This voltage is then further amplified with gain A_2 and is applied to the motor terminals,

$$v_2 = A_2 v_1 = A_1 A_2 K_p(r - \theta) \qquad \textbf{[1.106]}$$

where v_2 is the motor voltage. The second amplifier is the power amplifier, which is capable of supplying the electrical power necessary to drive the motor. The motor is coupled to the antenna with a gear train, of ratio

$$\theta = \frac{N_1}{N_2}\theta_m$$

where θ_m is the motor shaft angle and θ is the angular position of the antenna with moment of inertia J. N_1 is much smaller than N_2 since the high-speed shaft of the motor must drive the antenna at low speed but with a much higher torque. The transmittance of the motor driving an inertia load is given by Equation (1.61) in transformed form. The control motor has negligible armature inductance and negligible internal damping. The equation is repeated here,

$$\frac{\Theta(s)}{V_2(s)} = \frac{K_m(N_1/N_2)}{s(\tau_L s + 1)} \qquad [1.107]$$

Control motor dynamics.

where $\tau_L = R_A/(K_T K_v)[J_m + (N_1/N_2)^2 J_L]$ and $K_m = 1/K_v$. Equation (1.106) is Laplace-transformed to

$$V_2(s) = A_1 A_2 K_p[R(s) - \Theta(s)]$$

and combined with Equation (1.107), to obtain

$$\Theta(s) = \frac{(N_1/N_2)K_m A_1 A_2 K_p[R(s) - \Theta(s)]}{s(\tau_L s + 1)}$$

Some of the equation coefficients, and thus some of the system properties, can be selected by the designer by appropriately choosing the control components. Other parameters, such as the moment of inertia of the load J cannot be changed.

The transfer function relating the input position $R(s)$ to the output position $\Theta(s)$ is given by

$$T(s) = \frac{(N_1/N_2)K_m A_1 A_2 K_p}{\tau_L s^2 + s + (N_1/N_2)K_m A_1 A_2 K_p} \qquad [1.108]$$

This transfer function can also be derived using a block diagram, shown in Figure 1.49(a). Block diagram reduction results in Equation (1.108)

The signal flow graph, shown in Figure 1.49(b) is reduced with the aid of Mason's rule as follows:

$$P_1(s) = K_p A_1 A_2 \frac{N_1}{N_2} \frac{1}{\tau_L s} \frac{1}{s} K_m$$

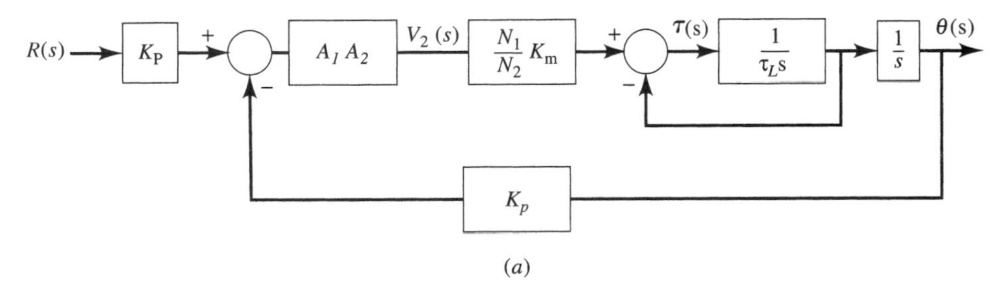

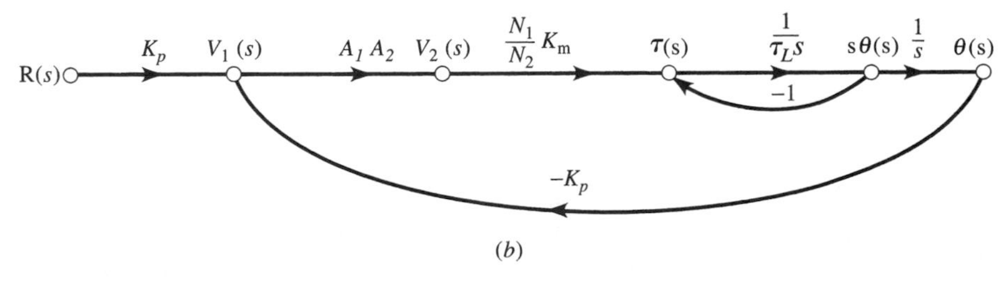

Figure 1.49 (a) Block diagram for the system of Figure 1.48. (b) Signal flow graph for the system of Figure 1.48.

$$\Delta_1 = 1$$

$$L_1(s) = A_1 A_2 \frac{N_1}{N_2} K_m \frac{1}{\tau_L s} \frac{1}{s}(-K_p)$$

$$= \frac{-A_1 A_2 (N_1/N_2) K_m K_p}{\tau_L s^2}$$

$$L_2(s) = \frac{-1}{\tau_L s}$$

and the transfer function is

$$T(s) = \frac{P_1 \Delta_1}{1 - (L_1 + L_2)}$$

$$= \frac{K_p A_1 A_2 (N_1/N_2) K_m}{\tau_L s^2 + s + (N_1/N_2) K_m A_1 A_2 K_p}$$

which of course agrees with Equation (1.108).

The signal flow graph makes the interrelations among variables easy to visualize; in addition, the transfer functions are easy to derive when the paths and loops are clearly identified. This simple example illustrates the importance of signal flow graphs and block diagrams in control system analysis. Design can then follow where adjustable values are selected to meet specifications.

1.16 Controller Model of the Thyroid Gland

The thyroid as a controller.

The human thyroid gland operates as part of a controller, regulating a specific thyroid hormone. The operation of the thyroid gland can be modeled using the methods of this chapter. In Figure 1.50, the hypothalamus (a part of the brain) and the pituitary gland (beneath the brain), collectively designated H/P, compare the desired level of a thyroid hormone with the actual level of the same hormone. If the actual level is lower than the desired level, a thyroid-stimulating hormone is produced which is intended to cause the functioning of the thyroid to speed up, resulting in secretion of more of its hormonal output.

Sometimes, the thyroid gland cannot secrete enough hormonal output. In that case, the H/P continues to produce thyroid-stimulating hormone, to compensate for the thyroid's continued underactivity. If the thyroid is permanently disabled, and therefore permanently underactive, its continued (and unsuccessful) efforts to speed up in response to the thyroid-stimulating hormone can eventually cause the thyroid to

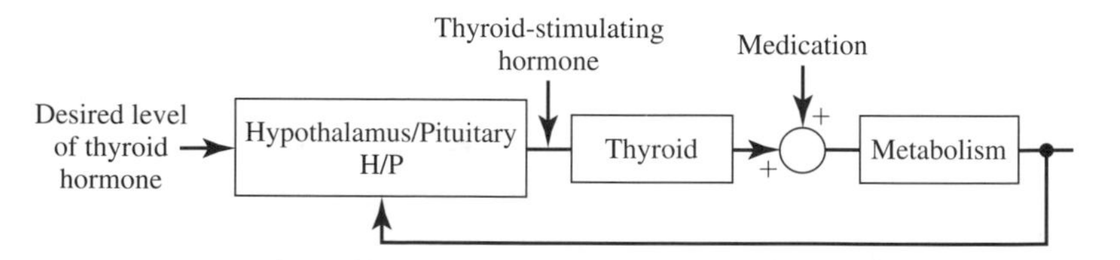

Figure 1.50 Thyroid hormone control system.

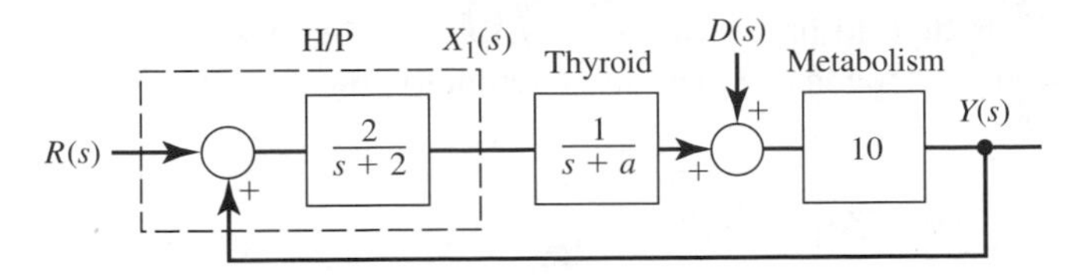

Figure 1.51 Block diagram values for Figure 1.50.

swell and possibly require surgical removal. Medication may be prescribed that can be metabolized to create additional thyroid hormone. Then, the H/P should recognize that the additional thyroid hormone is present in the bloodstream and greatly reduce the levels of thyroid-stimulating hormone. The thyroid thus believes its output is satisfactory and ceases attempts to speed up its functioning.

To simulate these activities, Figure 1.51 includes a summer and the transmittance $2/(s+2)$ to simulate the H/P, $X_1(s)$ to represent the thyroid-stimulating hormone, 10 to simulate metabolism, $1/(s+a)$ to simulate the thyroid, and $Y(s)$ to represent the thyroid hormone. Medication (oral or by injection) is represented by $D(s)$ while the desired level of thyroid hormone is represented by $R(s)$. The output $Y(s)$, the actual value of the thyroid hormone, is given by

Simulation.

$$Y(s) = R(s)\left[\frac{20}{s^2+(a+2)s+2a+20}\right] + D(s)\left[\frac{10(s+2)(s+a)}{s^2+(a+2)s+2a+20}\right] \quad [1.109]$$

while $X_1(s)$, the thyroid-stimulating hormone, is

$$X_1(s) = R(s)\left[\frac{2(s+a)}{s^2+(a+2)s+2a+20}\right] - D(s)\left[\frac{20(s+a)}{s^2+(a+2)s+2a+20}\right] \quad [1.110]$$

Suppose $R(s) = r/s$ and $D(s) = d/s$. If the final-value theorem is applied,

Steady state responses.

$$y(\infty) = r\left[\frac{20}{2a+20}\right] + d\left[\frac{20a}{2a+20}\right] \quad [1.111]$$

$$x_1(\infty) = r\left[\frac{2a}{2a+20}\right] - d\left[\frac{20a}{2a+20}\right] \quad [1.112]$$

When the thyroid operates properly, $a = 0$ and no medication is required ($d = 0$). In that case, $y(\infty) = r$, the desired level, while $x_1(\infty) = 0$, and thyroid receives no additional stimulation.

Defective thyroid.

To simulate an underactive (defective) thyroid where no medication is being taken, let $a = 5$, and $d = 0$. Notice that $y(\infty) = (2/3)r$, indicating that the thyroid produces two-thirds of what is needed. Meanwhile, $x_1(\infty)$ is $(1/3)r$. Thus the H/P will continue to attempt to stimulate the thyroid, even though the thyroid cannot produce more hormone. Eventually, the thyroid may be damaged.

Suppose a dosage $d = r/10$ is taken orally. With $a = 5$, we now have $y(\infty) = r$ where $(2/3)r$ is due to the thyroid and $(1/3)r$ is due to the medication. In total, no steady state thyroid-stimulating hormone $x_1(\infty)$ is secreted, since $x_1(\infty)$ is the sum of $r/3$ due to the underactivity of the thyroid and a compensating $-r/3$ due to the

Therapy.

medication. The thyroid produces only $(2/3)r$, but the lack of thyroid-stimulating hormone means that the thyroid will not attempt to produce any additional hormone.

1.17 Stick–Slip Response of an Oil Well Drill

The dynamic behavior of an oil well drill provides an interesting response to analyze and with it a difficult design problem to solve. The main problem is caused by nonlinear bit friction. Figure 1.52 shows a simplified version of an oil well derrick and drill. Torque is applied to the rotating table on the surface, causing the drill shaft to begin to rotate. At the bottom of the hole are drill collars securing the drill shaft to the drill bit. A nonlinear reverse torque is applied to the bit by the material being drilled.

Figure 1.53(a) contains a spring-mass-damper model for Figure 1.52. The lumped inertia J_2 includes the above-ground rotary table, attached DC motor, and gearbox. B_2 is due to the damping operating on J_2. The inertia of the below-ground drilling

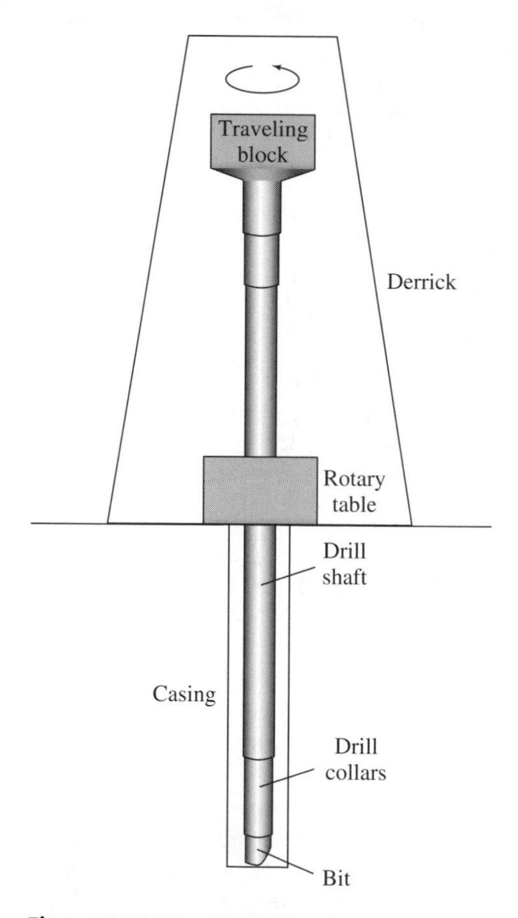

Figure 1.52 Simplified oil well drilling system.

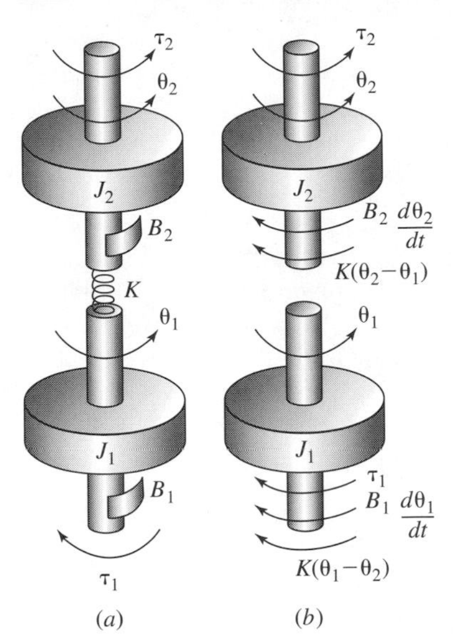

Figure 1.53 Oil well drill.
(a) Spring-mass-damper model. (b) Free-body diagram.

column, collars, and bit are lumped into J_1, while B_1 contains damping effects for below ground dynamics.

Torque applied to the rotary table is represented by T_2 while the nonlinear reverse torque of the drilled material acting on the bit is given by T_1. The free-body diagram of Figure 1.53(b) allows us to write the following two equations.

Dynamics.

$$\frac{d^2\theta_1}{dt^2} = -\frac{B_1}{J_1}\frac{d\theta_1}{dt} + \frac{K}{J_1}\theta_2 - \frac{K}{J_1}\theta_1 - \frac{T_1}{J_1} \qquad [\mathbf{1.113}\text{a}]$$

$$\frac{d^2\theta_2}{dt^2} = -\frac{B_2}{J_2}\frac{d\theta_2}{dt} - \frac{K}{J_2}\theta_2 + \frac{K}{J_2}\theta_1 + \frac{T_2}{J_2} \qquad [\mathbf{1.113}\text{b}]$$

These equations convert easily, using Laplace transforms with zero initial conditions, to the signal flow graph of Figure 1.54. Suppose we want four transfer functions relating each of two inputs $T_1(s)$ and $T_2(s)$ to each of two outputs $\theta_1(s)$ and $\theta_2(s)$, where

$$\begin{bmatrix} \theta_1(s) \\ \theta_2(s) \end{bmatrix} = \begin{bmatrix} T_{11}(s) & T_{12}(s) \\ T_{21}(s) & T_{22}(s) \end{bmatrix} \begin{bmatrix} T_1(s) \\ T_2(s) \end{bmatrix} \qquad [\mathbf{1.114}]$$

Mason's gain rule can be used for each $T_{ij}(s)$, $i = 1, 2$ and $j = 1, 2$. Paths terminate at $\theta_1(s)$ and $\theta_2(s)$. From Figure 1.54, the loop gains and path gains are

Mason's rule is applied.

L_1 (does not touch L_3, L_4) $= -(B_1/J_1)(1/s)$

L_2 (does not touch L_3, L_4) $= -(K/J_1)(1/s^2)$

L_3 (does not touch L_1, L_2) $= -(B_2/J_2)(1/s)$

L_4 (does not touch L_1, L_2) $= -(K/J_2)(1/s^2)$

L_5 (touches all) $= (K^2/J_1J_2)(1/s^4)$

P_1 (touches L_1, L_2, L_5) $= -(1/J_1)(1/s^2)$

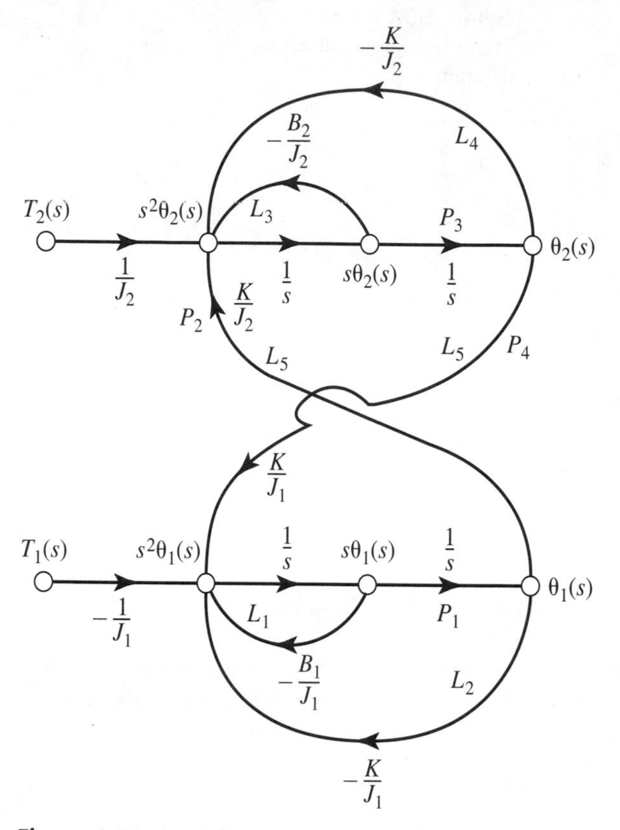

Figure 1.54 signal flow graph for oil well drill.

P_2 (touches all loops) $= -(1/J_1)(1/s^4)(K/J_2)$

P_3 (touches L_3, L_4, L_5) $= (1/J_2)(1/s^2)$

P_4 (touches all loops) $= (1/J_2)(1/s^4)(K/J_1)$

The determinant and cofactors follow easily

$$\begin{aligned}
\Delta &= 1-(L_1+L_2+L_3+L_4+L_5)+L_1L_3+L_1L_4+L_2L_3+L_2L_4 \\
&= \left\{s^4+s^3\left[\frac{B_1}{J_1}+\frac{B_2}{J_2}\right]+s^2\left[\frac{K}{J_1}+\frac{K}{J_2}+\frac{B_1}{J_1}\frac{B_2}{J_2}\right]+s\left[\frac{K(B_1+B_2)}{J_1J_2}\right]\right\}\frac{1}{s^4} \\
\Delta_1 &= 1-(L_3+L_4) = \frac{s^2+(B_2/J_2)s+K/J_2}{s^2} \\
\Delta_2 &= 1 \\
\Delta_3 &= 1-(L_1+L_2) = \frac{s^2+(B_1/J_1)s+K/J_1}{s^2} \\
\Delta_4 &= 1
\end{aligned} \qquad [1.115]$$

Finally, the four transfer functions are

$$T_{11}(s) = \left.\frac{\theta_1(s)}{T_1(s)}\right|_{T_2(s)=0} = \frac{P_1\Delta_1}{\Delta} = -\frac{(1/J_1)\left[s^2 + (B_2/J_2)s + (K/J_2)\right]}{s^4\Delta}$$

$$T_{12}(s) = \left.\frac{\theta_1(s)}{T_2(s)}\right|_{T_1(s)=0} = \frac{P_4\Delta_4}{\Delta} = \frac{(1/J_2)(K/J_1)}{s^4\Delta}$$

$$T_{21}(s) = \left.\frac{\theta_2(s)}{T_1(s)}\right|_{T_2(s)=0} = \frac{P_2\Delta_2}{\Delta} = -\frac{(1/J_1)(K/J_2)}{s^4\Delta}$$

$$T_{22}(s) = \left.\frac{\theta_2(s)}{T_2(s)}\right|_{T_1(s)=0} = \frac{P_3\Delta_3}{\Delta} = \frac{(1/J_2)\left[s^2 + (B_1/J_1)s + (K/J_1)\right]}{s^4\Delta}$$

An equivalent way to obtain the four transfer functions is to write Equations (1.113a) and (1.113b) in matrix form, using four "state" variables selected to allow Equations (1.113a) and (1.113b) to be written in terms of first-order differential equations.

$$\frac{d}{dt}\begin{bmatrix} \theta_1(t) \\ \dfrac{d\theta_1(t)}{dt} \\ \theta_2(t) \\ \dfrac{d\theta_2(t)}{dt} \end{bmatrix} = \begin{bmatrix} 0 & 1 & 0 & 0 \\ -\dfrac{K}{J_1} & -\dfrac{B_1}{J_1} & \dfrac{K}{J_1} & 0 \\ 0 & 0 & 0 & 1 \\ \dfrac{K}{J_2} & 0 & -\dfrac{K}{J_2} & -\dfrac{B_2}{J_2} \end{bmatrix} \begin{bmatrix} \theta_1(t) \\ \dfrac{d\theta_1(t)}{dt} \\ \theta_2(t) \\ \dfrac{d\theta_2(t)}{dt} \end{bmatrix} + \begin{bmatrix} 0 & 0 \\ -\dfrac{1}{J_1} & 0 \\ 0 & 0 \\ 0 & \dfrac{1}{J_2} \end{bmatrix} \begin{bmatrix} T_1(t) \\ T_2(t) \end{bmatrix} \quad [1.116]$$

Time domain matrices

Converting to Laplace, we write

$$\begin{bmatrix} s & -1 & 0 & 0 \\ K/J_1 & s + (B_1/J_1) & -K/J_1 & 0 \\ 0 & 0 & s & -1 \\ -K/J_2 & 0 & K/J_2 & s + (B_2/J_2) \end{bmatrix} \begin{bmatrix} \theta_1(s) \\ s\theta_1(s) \\ \theta_2(s) \\ s\theta_2(s) \end{bmatrix} = \begin{bmatrix} 0 & 0 \\ -1/J_1 & 0 \\ 0 & 0 \\ 0 & 1/J_2 \end{bmatrix} \begin{bmatrix} T_1(s) \\ T_2(s) \end{bmatrix} \quad [1.117]$$

Laplace domain matrices.

The four transfer functions can be found using Cramer's rule. The determinant of the matrix on the left side of Equation (1.117) is $s^4\Delta$ where Δ is given in Equation (1.115).

To get $T_{11}(s)$, for example, we write

Using Cramer's rule.

$$T_{11}(s) = \frac{\theta_1(s)}{T_1(s)} = \frac{\begin{vmatrix} 0 & -1 & 0 & 0 \\ -1/J_1 & s + B_1/J_1 & -K/J_1 & 0 \\ 0 & 0 & s & 1 \\ 0 & 0 & K/J_2 & s + B_2/J_2 \end{vmatrix}}{s^4 \Delta}$$

$$= \frac{\begin{vmatrix} -1/J_1 & -K/J_1 & 0 \\ 0 & s & -1 \\ 0 & K/J_2 & s + B_2/J_2 \end{vmatrix}}{s^4 \Delta}$$

which results in the value given earlier.

Figure 1.55 shows the time response of $\theta_1(t)$ and $\theta_2(t)$, where the response is assumed to be linear and $T_1(t)$ and $T_2(t)$ are both step functions. Parameter values are chosen from published values:

$$J_1 = 374\,\text{kg} \cdot \text{m}^2$$

$$J_2 = 2122\,\text{kg} \cdot \text{m}^2$$

$$B_1 = 2.5\,\text{N} \cdot \text{m/(rad/s)}$$

$$B_2 = 425\,\text{N} \cdot \text{m/(rad/s)}$$

$$K = 473\,\text{N} \cdot \text{m/rad}$$

$$T_1 = 1000\,\text{N} \cdot \text{m}$$

$$T_2 = 3000\,\text{N} \cdot \text{m}$$

In Figure 1.55, notice that $\theta_1(t)$ for the drill bit is much more oscillatory than $\theta_2(t)$ for the rotary table. In a real system, the reverse torque $T_1(t)$ is not linear and neither is the time response. Generally, the drill bit cannot go in reverse, so when the system attempts to drive $\theta_1(t)$ negative, the actual bit "sticks" in place.

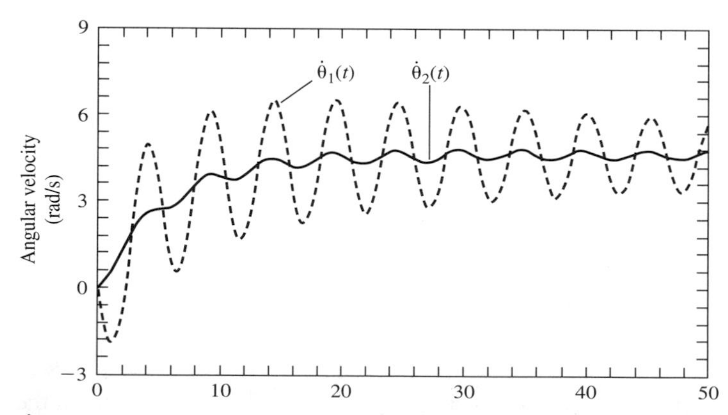

Figure 1.55 Linear motion of the drill bit, $\theta_1(t)$, and rotary table, $\theta_2(t)$.

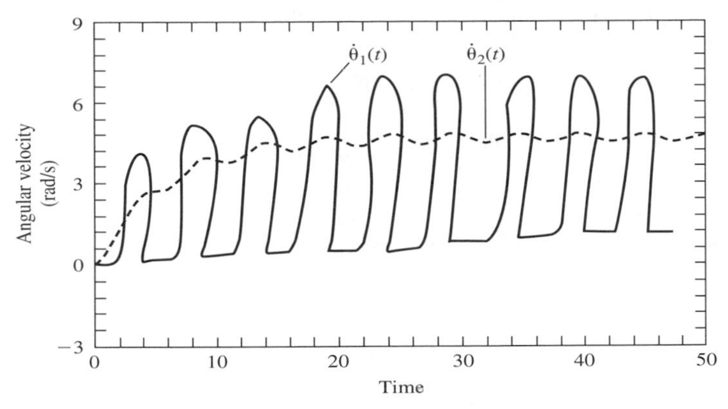

Figure 1.56 Nonlinear stick–slip Response.

When this happens, acceleration builds up, creating a large positive value for $\theta_1(t)$, which in turn causes the drill bit to "slip" in a wild positive rotation, followed by a reverse motion causing the bit to "stick" again. Figure 1.56 illustrates the alternating stick and slip modes. It is possible to use advanced methods (beyond the scope of this chapter) to select $T_2(t)$ to counteract the stick–slip problem.

1.18 SUMMARY

The goal of a control system is to cause the output of the system to behave in some desirable way, perhaps maintaining some desired level of performance or following some desired path. A control system comprises a controller (e.g., a thermostat), which causes the plant (e.g., a furnace) to change its output value. In an open-loop system, the setting of the controller does not change in spite of changes in the plant output. A closed-loop system includes measurement devices (e.g., temperature measurements) that evaluate the plant output, and the measurements, in turn, cause the controller to change setting (e.g., the thermostat calls for a higher output when the furnace temperature is too low). Feedback systems can result in improved accuracy, faster response, and less dependence of the output on disturbances. A disturbance is some outside influence (like outside temperature) that is not controllable by the system.

It is necessary to model controller, plant, and measurement. Often engineers approximate the response of a device by linearizing that response about some operating condition. The linearized response can be expressed in Laplace transform variables to obtain the output/input with zero initial conditions as a polynomial ratio called the transfer function $T(s)$ or in time domain state variable form (so chosen that a set of first-order linear differential equations results).

Among the types of subsystem commonly found as part of a control system are electrical networks (described by node or mesh methods) including operational amplifier circuits, which can be used to realize integrators, differentiators, and lag–lead and lead–lead transfer functions. Other subsystems include mechanical rotating systems, mechanical translating systems, electromechanical systems, aerodynamic systems (including longitudinal pitch and lateral yaw–roll simplifications), thermal systems, and hydraulic systems.

The response of a given system may be separated into components in two ways:

1. Zero-state and zero-input components
2. Forced and natural components

The zero-state response component is the system response when all initial conditions are zero

$$Y_{\text{zero-state}}(s) = T(s)R(s)$$

where $T(s)$ is the transfer function and $R(s)$ is the Laplace transform of the input. The zero-input response component is the system response when the input is zero. The natural response component contains terms due to the roots of the characteristic polynomial, while the forced response component contains those terms, due to the inputs, that do not correspond to the roots of the characteristic polynomial.

System stability is evaluated from either asymptotic or BIBO (bounded-input, bounded-output) viewpoints. Asymptotic stability is influenced by all roots of the characteristic polynomial [poles of $T(s)$] while BIBO stability is influenced by poles of $T(s)$ not canceled by zeros of $T(s)$. A system is stable when all appropriate response terms decay to zero as time approaches infinity. All appropriate roots must be in the LHP (left half-plane). A system is unstable when the appropriate response goes toward infinity, in which case some of the appropriate roots are in the RHP (right half-plane) and/or are repeated along the imaginary axis (IA). A system is marginally stable when the appropriate response goes neither to zero nor to infinity, in which case there are no RHP roots and some roots are along the IA and are not repeated.

A control system can be graphically displayed as the interconnection of blocks, summers, and junctions where each block contains the transfer function of that subsystem. To reduce the block diagram elements of a single-input, single-output system to just one block, three basic configurations may be simplified wherever present. Cascade configurations are replaced by the product of the blocks; tandem configurations are replaced by the sum or difference of the blocks, and feedback configurations are replaced by

$$T(s) = \frac{G(s)}{1 + G(s)H(s)}$$

where $T(s)$ refers to the closed-loop system, $G(s)$ refers to the forward path, $H(s)$ refers to the feedback path, and negative feedback is assumed. In some cases it may be necessary to move pickoff points and to split or combine summers and pickoff points. For more involved systems, the use of a signal flow graph and Mason's rule is suggested. A signal flow graph is the interconnection of branches (equivalent to blocks) and nodes (equivalent to summers and/or junctions). Mason's rule results in the calculation of the system transfer function directly from the system by

$$T(s) = \frac{\sum_i P_i \Delta_i}{\Delta}$$

where the P_i terms are the path gains, Δ_i is the cofactor of the ith path, Δ is the signal flow graph determinant. Mason's rule is applied to multi-input, multi-output systems by taking one input–output pair at a time. Once we have the system transfer function, the roots of the denominator indicate the stability of the overall system.

The chapter concluded with three example systems: a positioning servo, a controller model of the thyroid gland, and analysis of stick–slip in an oil well drilling system.

Tedious calculations, including symbolic mathematics, can be avoided, by using MATLAB, assuming that the user fully understands the underlying theory. Commands were presented by which transfer functions may be entered into a computer, roots of polynomials may be taken, and both Laplace and inverse Laplace transforms may be calculated, including the solution of simultaneous symbolic equations. These commands include the following:

```
den=[coeff vector]
ilaplace(Laplace function)
inv(matrix)
laplace(time function)
roots(den)
simplify
syms variables
zpk([num roots],[den roots], gain)
```

REFERENCES

The references given here and in the following chapters trace the history of topics presented in the text. While by no means comprehensive, they give a series of milestones in the development and understanding of these ideas. This, too, is our way of acknowledging those works from which we learned and to which we all owe a great deal.

Feedback

Black, H. S., "Inventing the Negative Feedback Amplifier." *IEEE Spectrum* (December 1977).

Blackman, R. B., "Effect of Feedback on Impedance." *Bell Syst. Tech. J.*, 22 (October 1943).

Bode, H. W., "Feedback—The History of an Idea," in *Selected Papers on Mathematical Trends in Control Theory.* New York: Dover, 1964.

Fuller, A. T., "The Early Development of Control Theory." *Trans. ASME, J. Dyn. Syst., Meas. Control*, 96G (June 1976).

———, "The Early Development of Control Theory II." *Trans. ASME J. Dyn. Syst. Meas. Control* 98G (September 1976).

Maxwell, J. C., "On Governors." *Proc. R. Soc.*, 16(1868).

Mayr, O., "The Origins of Feedback Control." *Sci. Am.* (October 1970).

Nyquist, H., "Regeneration Theory." *Bell Syst. Tech. J.*, 11 (January 1932).

Wolf, A. *A History of Science, Technology and Philosophy in the Eighteenth Century*. New York: McGraw-Hill, 1939.

System Equations

Cannon, R. H., Jr., *Dynamics of Physical Systems.* New York: McGraw-Hill, 1967.

Close, C. M., and Frederick, D. K., *Modeling and Analysis of Dynamic Systems.* Boston: Houghton Mifflin, 1978.

Luenberger, D. G., *Introduction to Dynamic Systems.* New York: Wiley, 1979.

Perkins, W. R., and Cruz, J. B., Jr., *Engineering of Dynamic Systems.* New York: Wiley, 1969.

Van Valkenburg, M. E., *Network Analysis*. Englewood Cliffs, NJ: Prentice-Hall, 1974.

Single Flow Graphs

Dertouzos, M. L ., Athans, M., Spann, R. N., and Mason, S. J., *Systems, Networks and Computation.* New York: McGraw-Hill, 1973.

Mason, S. J., "Feedback Theory: Some Properties of Signal Flow Graphs." *Proc. IRE*, 41 (September 1953).

———, "Feedback Theory: Further Properties of Signal Flow Graphs." *Proc. IRE*, 44 (July 1956).

Automobile Control Systems

Jurgen, R.K., "Drivers Get More Options in 1983." *IEEE Spectrum* (November 1982): 30–36.

———, "Detroit Unveils Sophisticated Electronics." *IEEE Spectrum* (October 1983): 33–39.

———, "More Electronics in Detroit's 1985 Models." *IEEE Spectrum* (October 1984): 54–60.

———, "Detroit's 1987 Models: New Electronic Inroads." *IEEE Spectrum* (October 1986): 68–72.

Generating and Controlling Power

Fischetti, M., and Zorpette, G., "Power and Energy." *IEEE Spectrum* (January 1986).

Gaushell, D. J., "Automating and Power Grid." *IEEE Spectrum* (October 1985): 39–45.

Zorpette, G., "HVDC: Wheeling Lots of Power." *IEEE Spectrum* (June 1985): 30–36.

Aerodynamics

Pachter, M, et al., "*Aerospace Centrols*," in *The Control Handbook*. Boca Raton, FL: CRC Press. 1996, pp. 1287–1295.

Thyroid Gland

Claymon, C. B, (ed.), The American Medical Association Encyclopedia of Medicine. New York: Random House, 1989.

Oil Well Slip–Stick

Serrarens, A. et al., "*H*-Infinity Control for Suppressing Stick-Slip in Oil Well Drillstrings." *IEEE Control Syst. Mag.* April 1998, pp. 19–31.

PROBLEMS

1. Linearize each function about the indicated x_0 and express your answer in the form $y \approx y_0 + m\,(x - x_0)$.

   $(a)\ y = x^3, \qquad x_0 = 1$

   $(b)\ y = 2x^{1/3}, \qquad x_0 = 1$

   $(c)\ y = 2\cos 2x, \qquad x_0 = \dfrac{\pi}{4}$, x is in radians

2. Write simultaneous Laplace-transformed loop equations for the electrical networks of Figure P1.2. All initial conditions are zero.

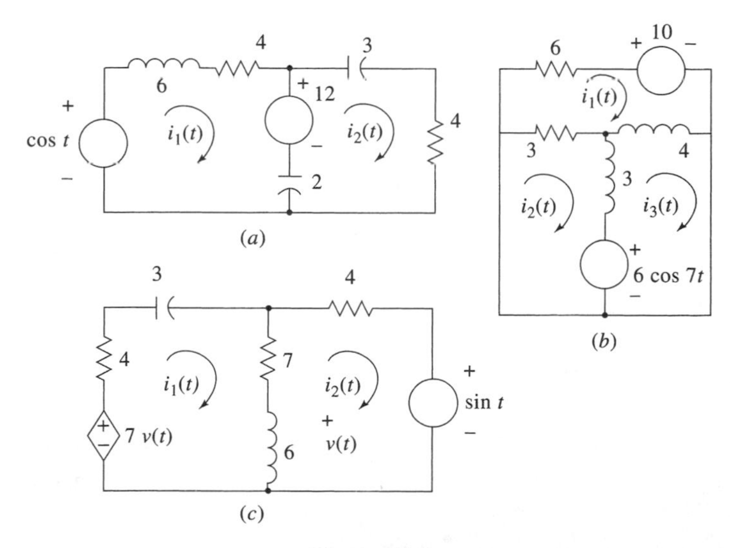

Figure P1.2

3. If all initial conditions of the networks of Figure P1.3 are zero, find $v(t)$, $t \geq 0$.

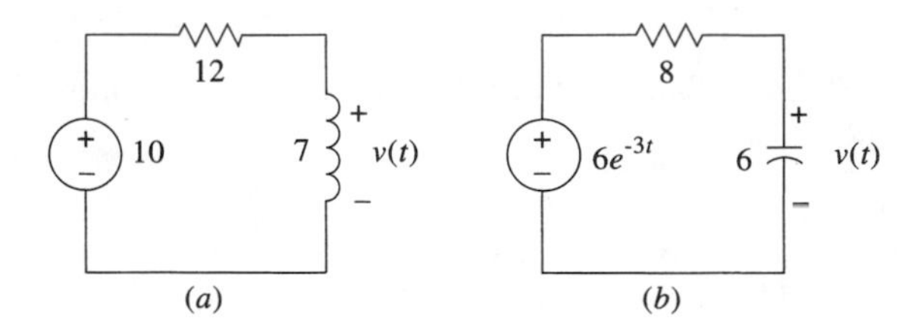

Figure P1.3

Ans. (a) $v(t) = 10e^{-(12/7)t}u(t)$;
(b) $v(t) = (0.041e^{-t/48} - 0.041e^{-3t})u(t)$

4. Write simultaneous Laplace-transformed nodal equations for the electrical networks of Figure P1.4. All initial conditions are zero.

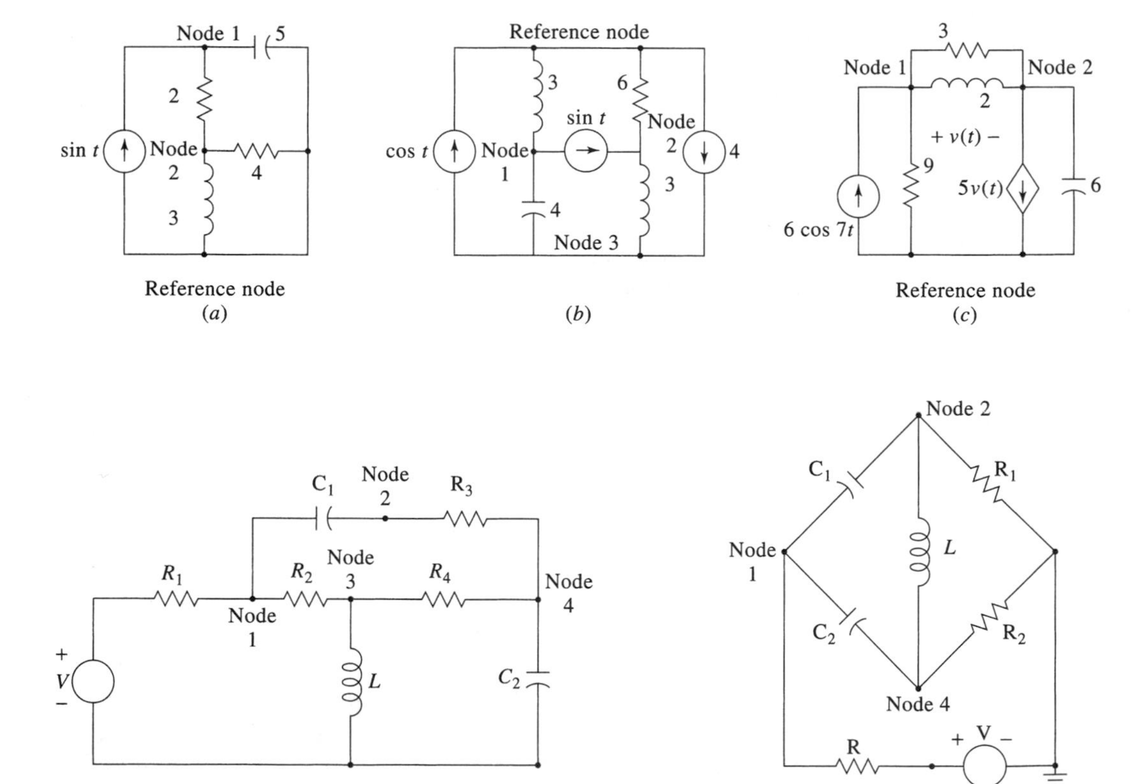

Figure P1.4

5. All initial conditions in the networks of Figure P1.5 are zero. Find $I(s)$.

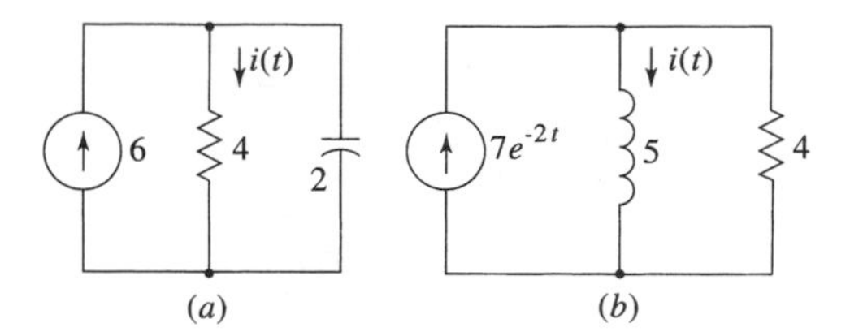

Figure P1.5

6. Write simultaneous integrodifferential loop equations for the electrical networks shown in Figure P1.6 in terms of the indicated mesh currents. Then use the derivative of each mesh current and the derivative of each capacitor voltage to convert your equations to state variable form.

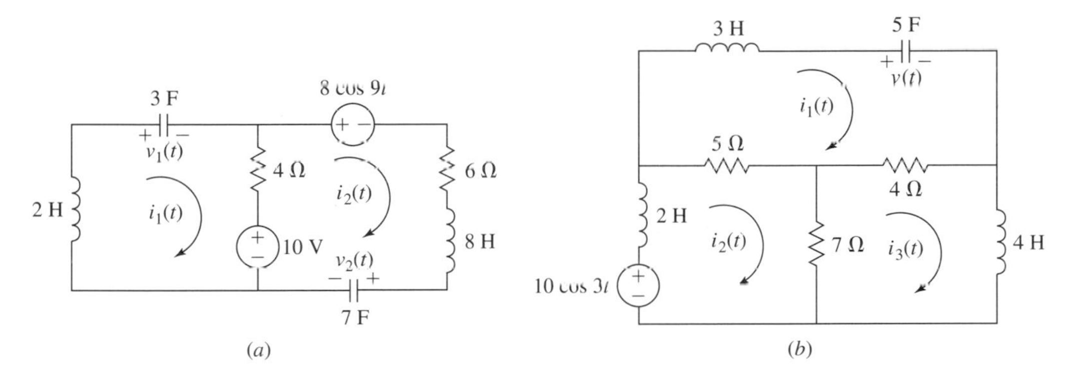

Figure P1.6

7. Find $-v_{\text{out}}(s)/v_{\text{in}}(s)$ for each of the operational amplifier Circuits in Figure P1.7.

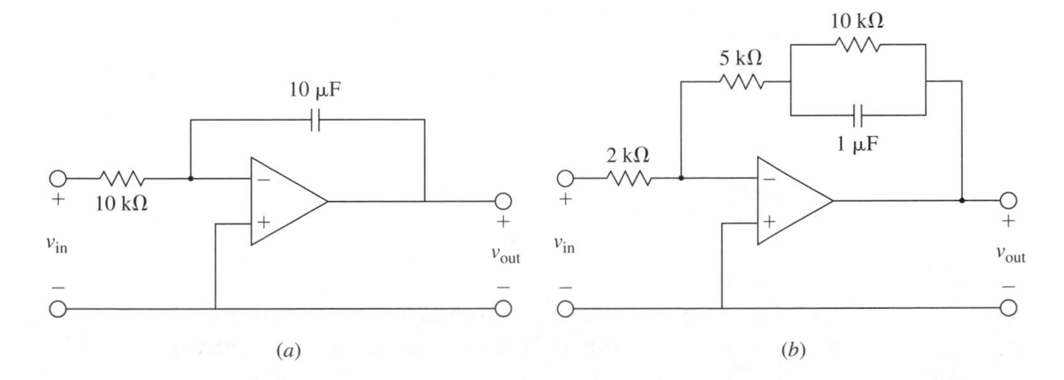

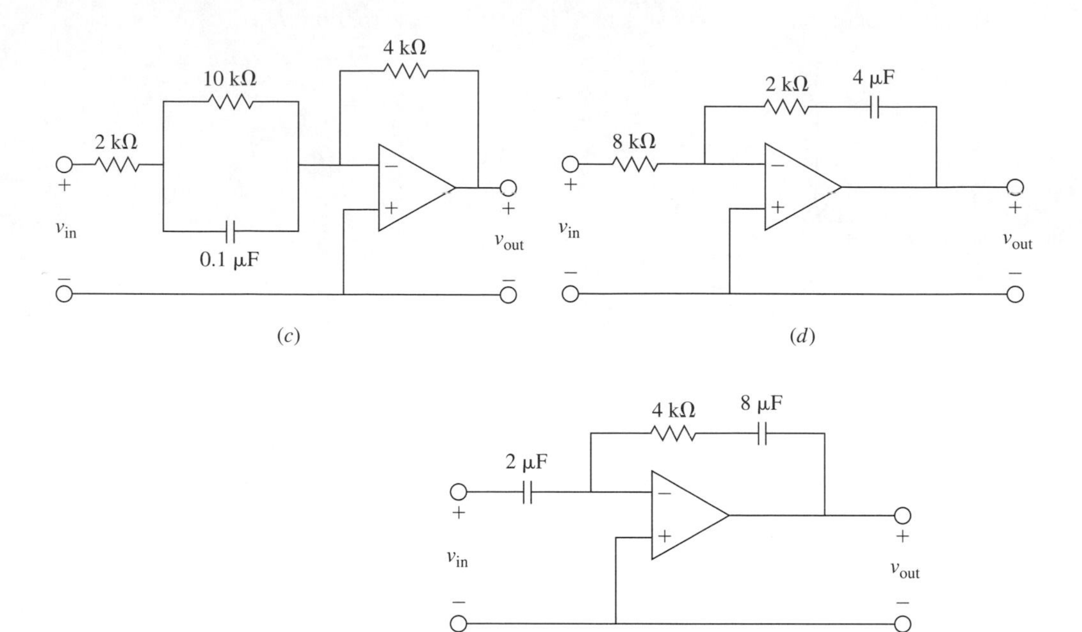

Figure P1.7

8. Select an operational amplifier circuit to realize each of the following values of $-v_{\text{out}}(s)/v_{\text{in}}(s)$. Some values are given. Refer to Figure 1.14.

(a) $0.001s,\ R_f = 2\,\text{k}\Omega$

(b) $\dfrac{5(s+6)}{s+60},\ R_1 = 1\,\text{k}\Omega$

(c) $\dfrac{4(s+21)}{s+3},\ R_1 = 10\,\text{k}\Omega$

(d) $10\left(\dfrac{s+40}{s+4}\right)\left(\dfrac{s+50}{s+200}\right),\ R_1 = 10\,\text{k}\Omega$

(e) $20\,(s+5)\,,\ R_1 = 1\,\text{k}\Omega$

(f) $\dfrac{10(s^2+10s+100)}{s},\quad C_2 = 1\,\mu\text{F},\ R_a = 1k\Omega$

9. Write simultaneous Laplace-transformed differential equations for the translational mechanical networks of Figure P1.9. All initial conditions are zero.

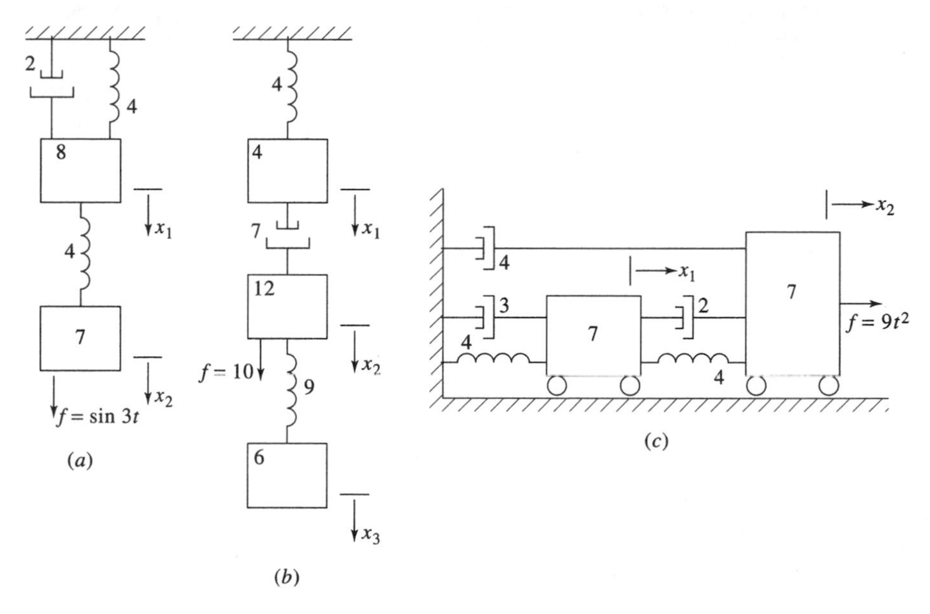

Figure P1.9

10. Convert the simultaneous differential equations for Figure P1.9 to the state variable forms of Equations (1.38) and (1.39) by using the following state variables.

    (a) $dx_1/dt, x_1, dx_2/dt, x_2$

    (b) $dx_1/dt, x_1, dx_2/dt, x_2, dx_3/dt, x_3$

    (c) $dx_1/dt, x_1, dx_2/dt, x_2$

11. Find $X(s)$ and $x(t)$, $t \geq 0$ for the networks of Figure P1.11, for which all initial conditions are zero.

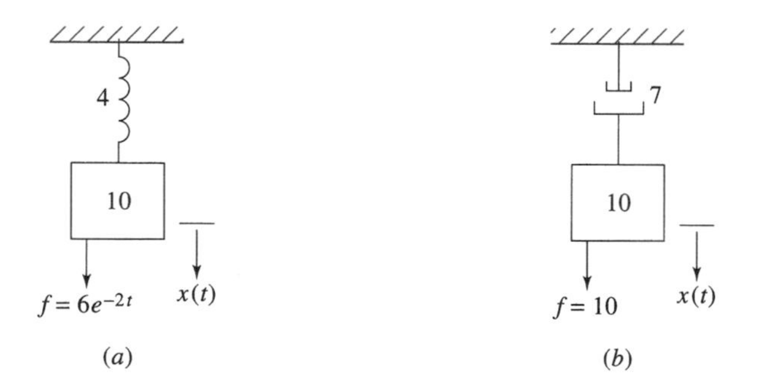

Figure P1.11

Ans. (a) $(0.44\cos(0.63t - 107°) - 0.13e^{-2t})u(t)$;

(b) $(1.43t + 2.04e^{-0.7t} - 2.04)u(t)$

12. In vibration studies, the human body is often modeled by springs, masses, and dampers. For the model of a seated body with applied force f (Figure P1.12), find the Laplace-transformed system equations.

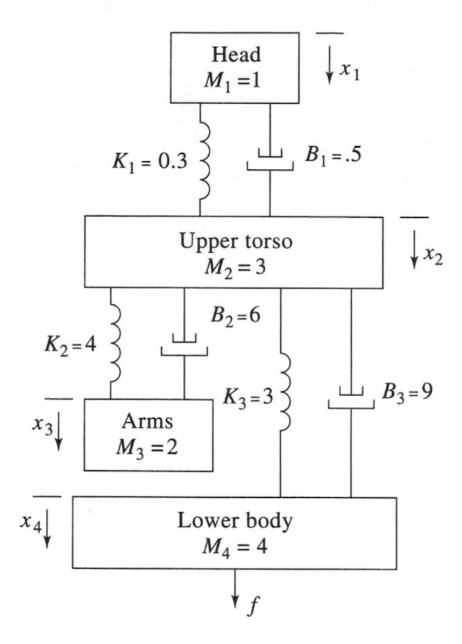

Figure P1.12

13. A shock absorber in an automobile or truck can alter the ride comfort by providing anything from a "soft" ride appropriate for the highway to a "stiff" ride appropriate for a vehicle that may traverse rough terrain. Figure P1.13 simulates the suspension of an automobile with M, B, and K values. Suppose $F = 250$ and $M = 10$. Select values of B and K for the indicated characteristic polynomial roots. Find and plot the step responses.

    (a) $-10, \; -2.5$ (stiff ride)

    (b) $-5, \; -5$ (less stiff ride)

    (c) $-3 + j4, \; -3 - j4$ (soft ride)

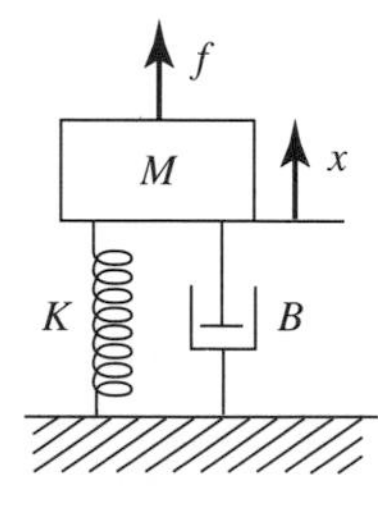

Figure P1.13

14. Find the transmittance $X(s)/F(s)$ for the system in Figure P1.14.

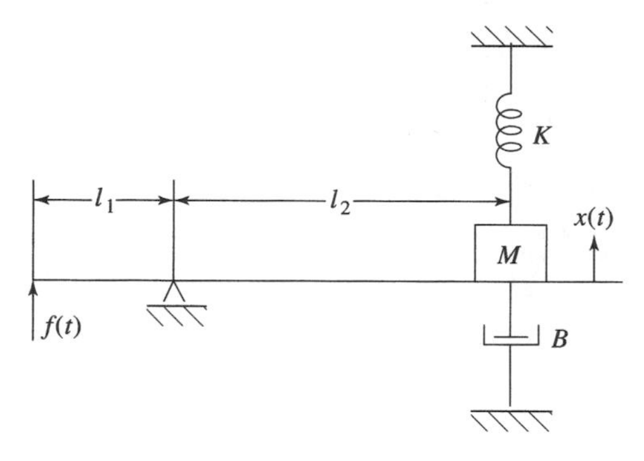

Figure P1.14

Ans. $\dfrac{X(s)}{F(s)} = \dfrac{-l_1/l_2}{Ms^2 + Bs + K}$

15. Write simultaneous Laplace-transformed differential equations for the rotational mechanical networks of Figure P1.15. All the initial conditions are zero.

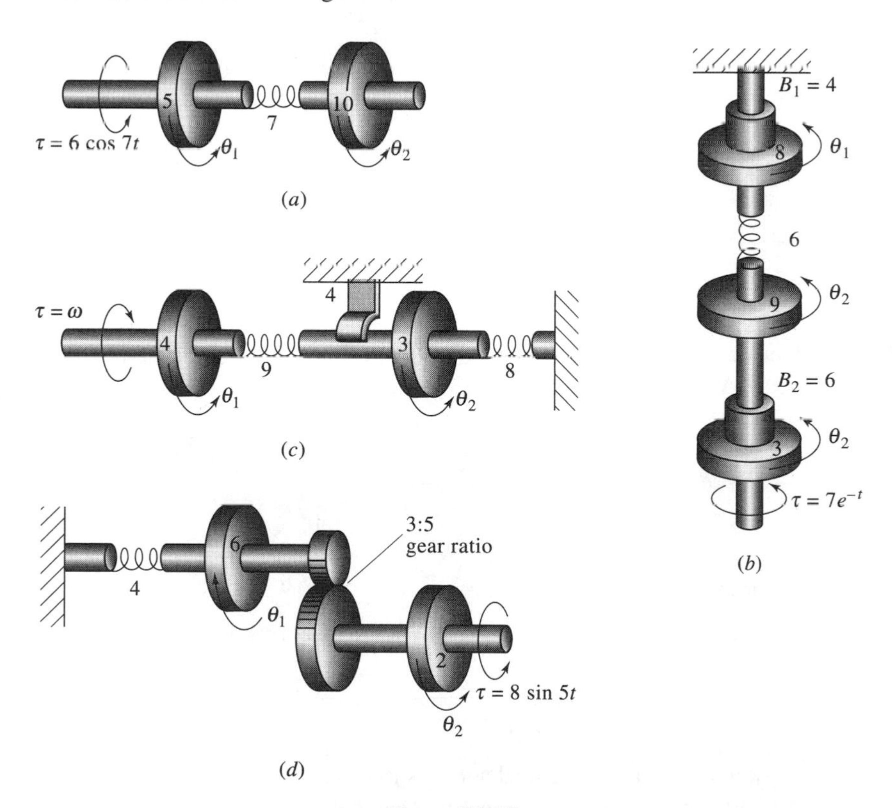

Figure P1.15

16. Find $\Theta(s)$ and $\theta(t)$, $t \geq 0$, for the networks of Figure P1.16, for which all initial conditions are zero.

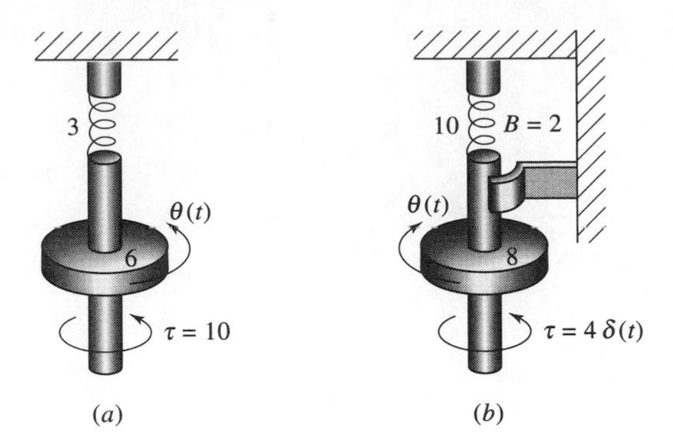

Figure P1.16

Ans. (a) $(3.33 - 3.34\cos 0.707t)u(t)$; (b) $(0.46e^{-1.25t}\sin 1.1t)u(t)$

17. Draw an electircal network that is analogous to the translational mechanical network of Figure P1.17.

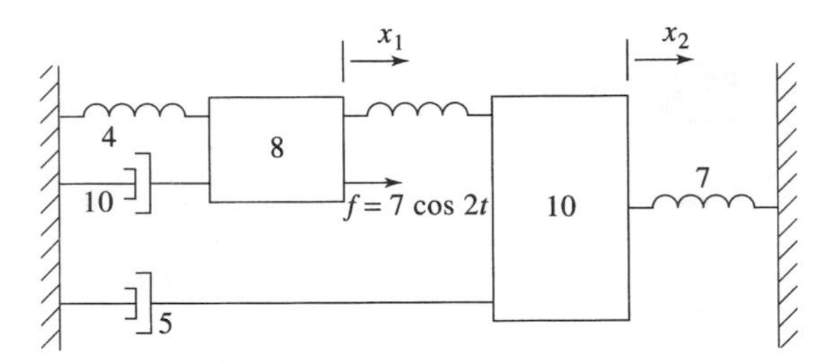

Figure P1.17

18. (a) Draw a translational mechanical network that is analogoue to the electrical network of Figure P1.18.

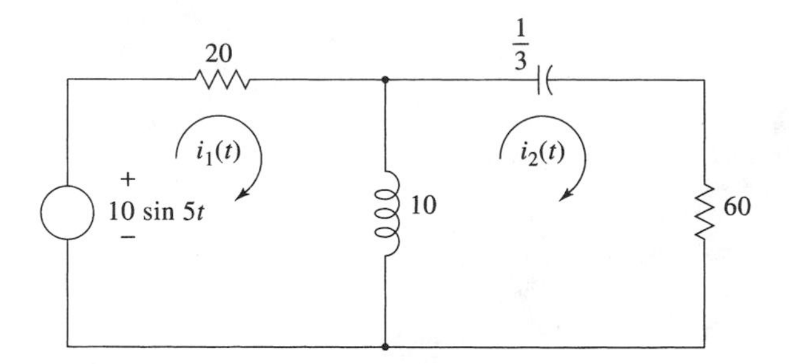

Figure P1.18

(b) Draw a rotational mechanical network that is analogous to the same electrical network (and to the translational mechanical network).

19. Equation (1.71) contains the transfer function for the longitudinal dynamics of an aircraft. Using that equation, determine whether the system is asymptotically

stable, marginally stable, or unstable if

(a) $M_q = -20;\ Z_\alpha = -30;\ Z_q = 5;\ M_\alpha = 100$

(b) $M_q = -30;\ Z_\alpha = -20;\ Z_q = 3;\ M_\alpha = -200$

(c) $M_q = -10;\ Z_\alpha = -40;\ Z_q = 2;\ M_\alpha = 200$

20. Figure D1.20 represents two rooms of equal size and thermal capacity C. Each room has two equal-sized panes of glass with thermal resistance R. In addition, room 2 has a coating on its window having an additional thermal resistance of $9R$. Suppose that each room is shuttered with a room temperature of 70°F. The shutters are opened when the outside temperature is 100°F. After $t = RC$ seconds, find the temperature in each room. Show the electrical equivalent circuit of each room.

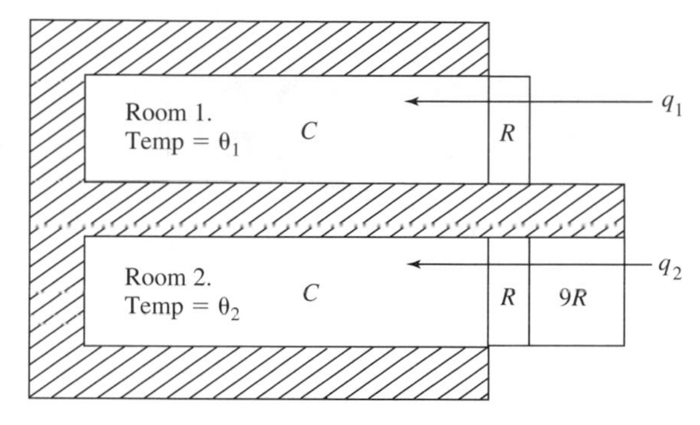

Figure P1.20

21. Find both the asymptotic stability and the BIBO stability of each system given the transfer function $T(s)$.

(a) $$T(s) = \frac{s+2}{s(s+2)(s+6)}$$

(b) $$T(s) = \frac{s^2}{s^2(s+1)^2(s+8)}$$

(c) $$T(s) = \frac{(s^2+4)(s+5)}{(s^2+4)^2(s+6)^2}$$

(d) $$T(s) = \frac{s(s+6)}{(s+2)^2(s-6)}$$

(e) $$T(s) = \frac{(s^2+6s+10)(s-6)}{(s^2+3s+20)(s-6)}$$

22. Indicate the type of asymptotic stability of each system given the zero-input responses of Figure P1.22.

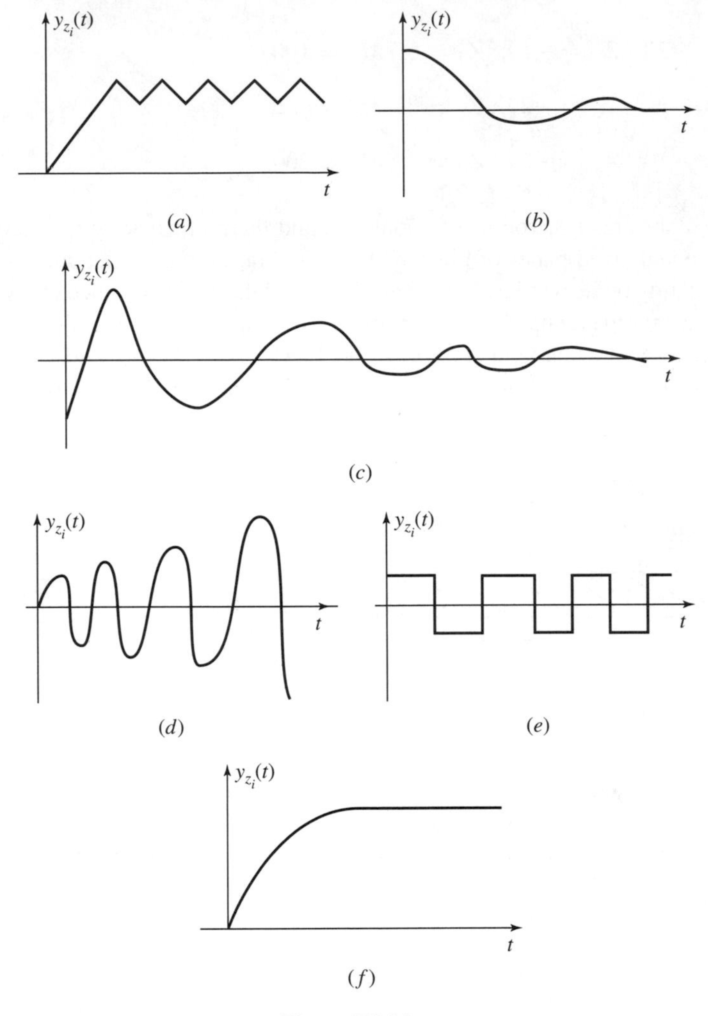

Figure P1.22

23. For the system of Figure P1.23, determine the asymptotic stability for

 (a) $K = 0.1$

 (b) $K = 2$

 (c) $K = 20$

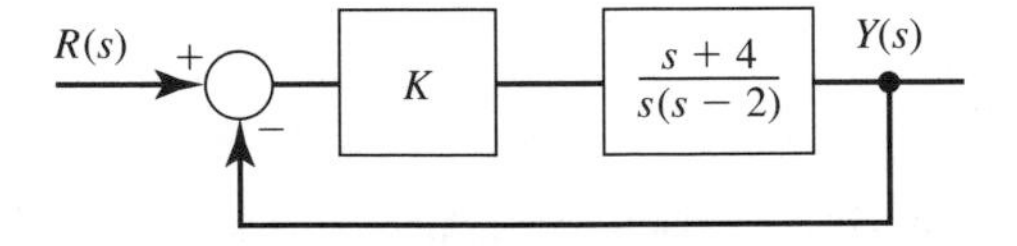

Figure P1.23

24. Draw block diagrams to represent the following Laplace-transformed equations. The R signals are inputs, and the Y signals are outputs. This is a synthesis problem and has, in each case, several possible solutions.

(a)
$$X_1(s) = \frac{3s}{s+1}R(s) + X_2(s)$$
$$X_2(s) = R(s) + 9X_1(s)$$
$$Y(s) = R(s) + \frac{3}{s^2+4}X_2(s)$$

(b)
$$X_1(s) = R(s) - KX_2(s)$$
$$X_2(s) = [3/(s+2)]X_1(s) - R(s) - (2/s)Y(s)$$
$$Y(s) = [4/(s^2+4)]X_2(s)$$

25. For the system equations of Problem 24, draw instead representative signal flow graphs.

26. Using equivalences, reduce the block diagrams in Figure P1.26 to single blocks or to a multiple-block canonical form, displaying the system transfer function (s).

Ans. (a) $T(s) = s(s+2)/(3s^2+3s+2)$;
(c) $T(s) = [2(s+2)]/(s^2+13s+24)$

27. Use Mason's gain rule to find the transfer function(s) relating output(s) to input(s) for each of the signal flow graphs of Figure P1.27.

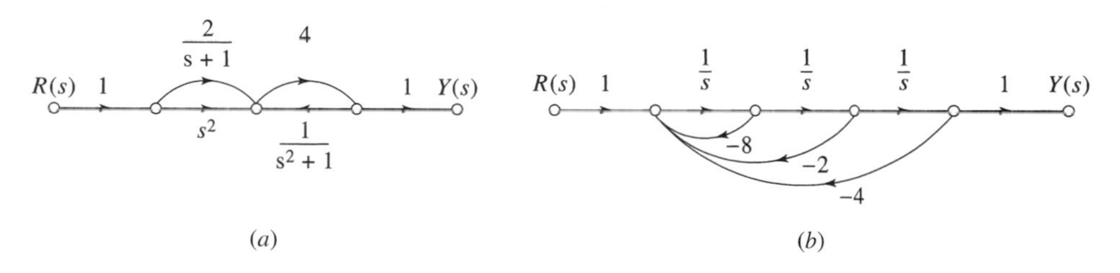

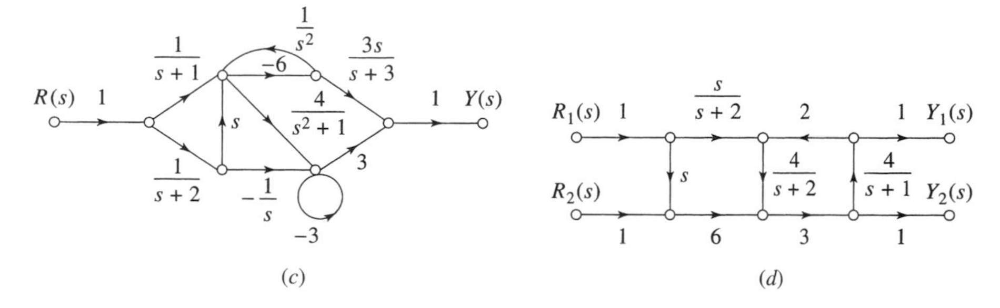

Figure P1.27

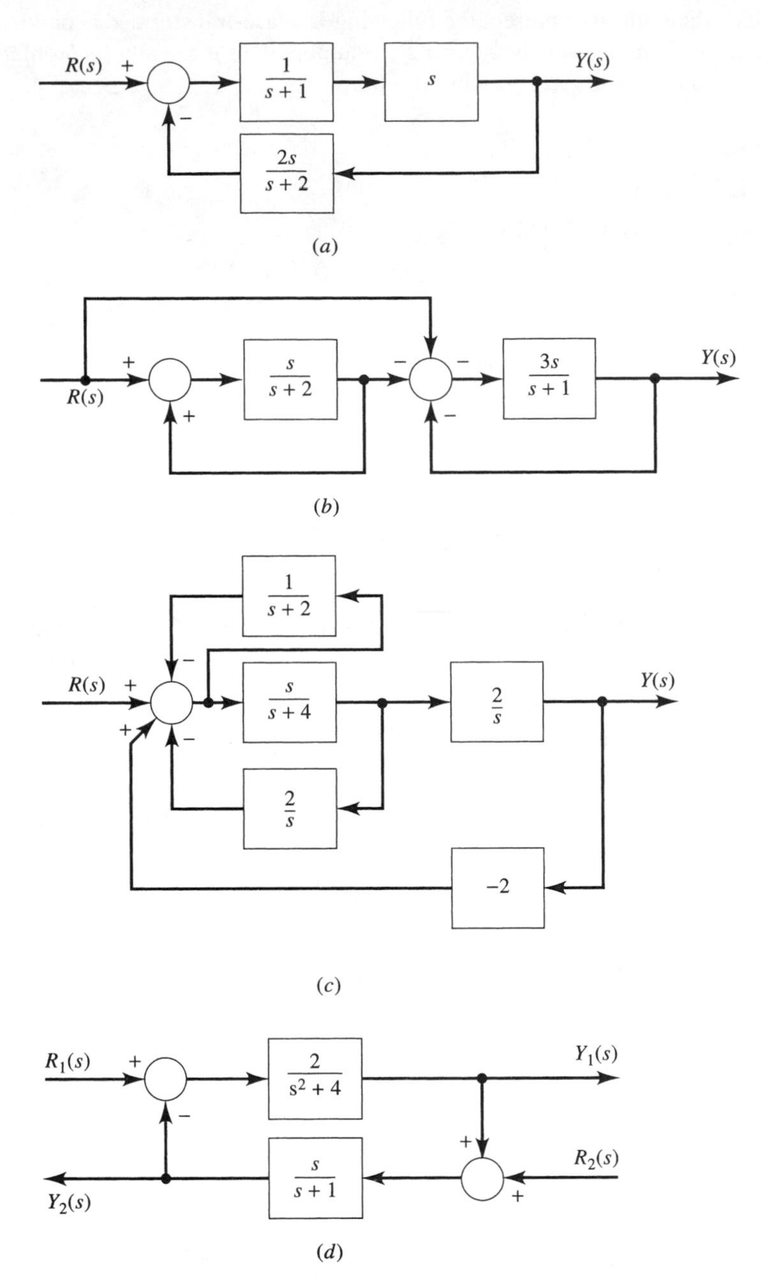
R(s)
+
−
1/(s + 1)
s
Y(s)
2s/(s + 2)
(a)
+
+
R(s)
s/(s + 2)
−
−
−
3s/(s + 1)
Y(s)
(b)
1/(s + 2)
−
R(s)
+
+
−
s/(s + 4)
2/s
Y(s)
2/s
−2
(c)
R1(s)
+
−
2/(s^2 + 4)
Y1(s)
+
R2(s)
+
s/(s + 1)
Y2(s)
(d)

Figure P1.26

28. The position control system for a spacecraft platform is governed by the following approximate equations:

$$\frac{d^2p}{dt^2} + \frac{dp}{dt} + 4p = \theta$$

$$v_1 = r - p$$

$$\frac{d\theta}{dt} = 0.4v_2$$

$$v_2 = 7v_1$$

where the variables are

$r(t)$ = desired platform position (input)

$p(t)$ = actual platform position (output)

$v_1(t)$ = amplifier input voltage

$v_2(t)$ = amplifier output voltage

$\theta(t)$ = motor shaft position

Draw a block diagram of the system, identifying the component parts and their transmittances. Then determine the system transfer function.

29. A simplified block diagram of an aircraft roll control is given in Figure P1.29. Find its transfer functions.

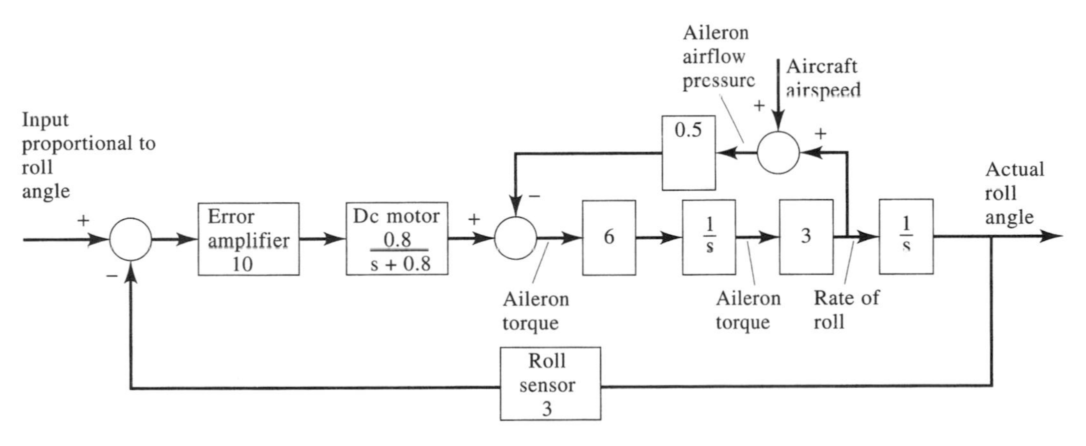

Figure P1.29

30. A system without an input that can be used to generate a sinusoidal output signal is shown in Figure P1.30. When the differential equation describing $y(t)$ has a characteristic equation with roots on the imaginary axis of the complex plane, the system's zero-input response is sinusoidal. Find the value of the adjustable constant K, in terms of a, for which the system has a sinusoidal output. Also find the frequency in hertz, of the oscillations in terms of the constant a.

Achieving characteristic roots precisely on the imaginary axis is impossible for inexact K. If the characteristic roots are slightly to the right of the imaginary axis, the oscillation amplitude will increase exponentially. If they are to the left of the imaginary axis, the oscillations will decay exponentially in time. In practice, another control system can be used to slowly adjust K to maintain nearly constant sinusoidal amplitude.

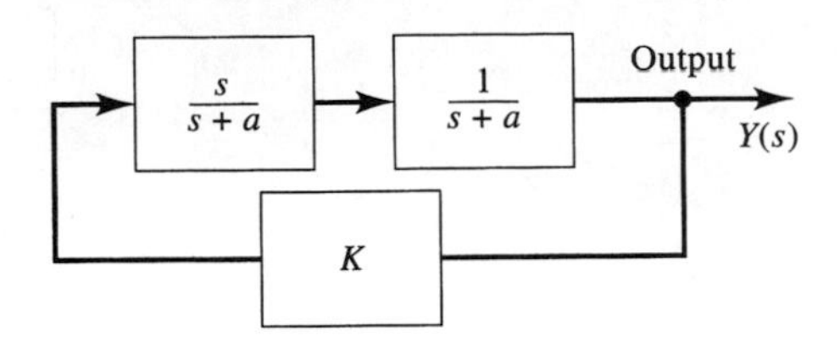

Figure P1.30

Ans. $K = 2a$; frequency $= a/2\pi$ Hz

CHAPTER 2

Continuous-Time System Response

2.1 Preview

Designing a control system for a practical application usually entails a series of steps. For example, we can use the methods of Chapter 1 to examine a variety of electrical and mechanical systems and to write a set of differential equations to describe their operation. In doing so, techniques developed over several hundred years are used. In the late 1660s, Newton and Leibniz developed methods for writing and solving differential equations. These and other related topics are now called *calculus*. Over 100 years later, Laplace developed a transform to aid in solving those equations.

In the 1890s, about 100 years after Laplace, steam engines represented a challenge to controls designers in that steam pressure and various rotational velocities had to be controlled. The efforts of a mathematician (Hurwitz) and an engineer (Routh) resulted in a test of performance using the coefficients of the characteristic polynomial, a polynomial resulting from the same methods presented in Chapter 1. Thus the legacy of the past is used to understand the behavior of today's technology using the methods developed about 100 years ago by Routh and Hurwitz (some 300 years after Newton and Leibniz).

This chapter begins by recognizing that a characteristic polynomial, as defined in Chapter 1, can be factored into first- and second-order terms with real-valued coefficients. If the behavior of first- and second-order systems is well understood, the behavior of higher-order systems follows as a combination of the first- and second-order building blocks. Next, it is useful to understand whether an automobile or aircraft described by similar equations could yield a comfortable and safe ride. Certain definitions are presented that clarify the quality of performance in terms of a system's stability.

Even though computing capability has grown enormously since the time of Routh and Hurwitz, their method remains a valuable tool for determining a range of values for an unknown parameter so that the stability of the resulting closed-loop system is ensured.

Chapter 2 concludes with examples of two systems that illustrate the power of the analytical methods. A delivery system is selected so that a desired flow of insulin can be induced to the bloodstream of a diabetic person. An application of a very different sort shows that the periodic vibration of turbine engines can cause excessive wing deflection in an aircraft.

2.2 Response of First-Order Systems

In a first-order system, the output $y(t)$ and input $r(t)$ are related by a differential equation of the form

$$\frac{dy}{dt} + a_0 y = b_0 r \qquad [2.1]$$

where the input terms, possibly involving derivatives of $r(t)$, form the driving function of the equation, and a_0 and b_0 are constants. The corresponding transfer function is

$$T(s) = \frac{b_0}{s + a_0}$$

Stability of a first-order system.

A system is considered to be stable if the natural response decays to zero. The denominator of $T(s)$ is the characteristic polynomial whose roots must all be in the left half-plane to force all the natural response terms to decay to zero and therefore to force the system to be stable. In the first-order case above, the system is stable if and only if $a_0 > 0$.

Laplace-transforming the input–output Equation (2.1), we have

$$sY(s) - y(0^-) + a_0 Y(s) = b_0 R(s)$$

$$Y(s) = \underbrace{\frac{b_0}{s + a_0} R(s)}_{\text{zero-state component}} + \underbrace{\frac{y(0^-)}{s + a_0}}_{\text{zero-input component}}$$

The **zero-state component** is the result of a driving function $R(s)$ with zero initial conditions. The **zero-input component** is the result of a zero driving function and only nonzero initial conditions driving the system.

For a step input signal

$$r(t) = Au(t) \quad R(s) = \frac{A}{s}$$

and zero initial conditions, $Y(s)$ contains only the zero-state component

Zero-state response.

$$Y(s) = T(s)R(s) = \frac{b_0 A}{s(s + a_0)} = \frac{b_0 A/a_0}{s} + \frac{-b_0 A/a_0}{s + a_0}$$

the time solution is found with partial fractions as reviewed in Appendix B:

$$y(t) = \left(\frac{b_0 A}{a_0} - \frac{b_0 A}{a_0} e^{-a_0 t} \right) u(t)$$

which is sketched in Figure 2.1. If the characteristic root $s = -a_0$ is negative (a_0 positive), the system is stable and the exponential natural (or "transient") response term decays with time, leaving the constant forced (or "steady state") term. For a positive characteristic root (negative a_0), the exponential term expands with time, and the system is unstable.

Total response.

If, instead, the initial conditions are not zero,

$$Y(s) = T(s)R(s) + \frac{y(0^-)}{s + a_0} = \frac{b_0 A}{s(s + a_0)} + \frac{y(0^-)}{s + a_0}$$
$$= \frac{(b_0 A / a_0)}{s} + \frac{y(0^-) - (b_0 A / a_0)}{s + a_0}$$

giving

$$y(t) = \left\{ \frac{b_0 A}{a_0} + \left[y(0^-) - \frac{b_0 A}{a_0} \right] e^{-a_0 t} \right\} u(t)$$

The amplitude of the exponential term is changed, as illustrated in Figure 2.2.

The response to other inputs is calculated similarly, using the Laplace transform of the input signal. The natural exponential term in the response consists of contributions from both the zero-state and the zero-input parts of the response. In general, the natural and zero-input responses die out in time when the system is stable.

Time constant is defined.

The *time constant* of a stable first-order system is

$$\tau = \frac{1}{a_0}$$

It is the time interval over which the exponential $K \exp(-a_0 t)$ decays by a factor of $e^{-1} = 0.37$. First-order system transfer functions (and similar terms in higher-order

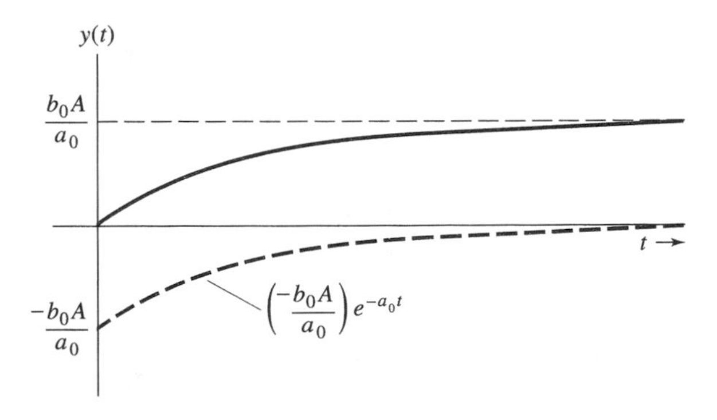

Figure 2.1 First-order system step response with zero initial conditions.

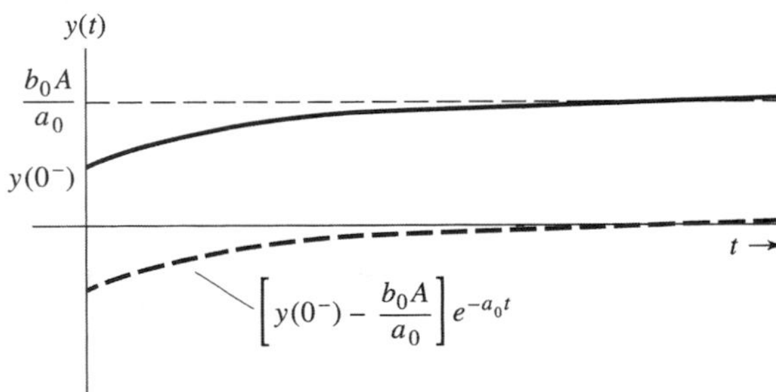

Figure 2.2 First-order system step response with nonzero initial conditions.

transfer functions) are sometimes placed in the form

$$T(s) = \frac{\text{numerator polynomial}}{s + a_0} = \frac{(\text{numerator polynomial})/a_0}{(1/a_0)s + 1}$$
$$= \frac{(\text{numerator polynomial})/a_0}{\tau s + 1}$$

to show the time constant explicitly. For the first-order system

$$\frac{dy}{dt} + a_0 y = b_1 \frac{dr}{dt} + b_0 r$$

we write

$$Y(s) = \underbrace{\frac{b_1 s + b_0}{s + a_0} R(s)}_{\substack{\text{zero-state} \\ \text{component}}} + \underbrace{\frac{y(0^-) - b_1 r(0^-)}{s + a_0}}_{\substack{\text{zero-input} \\ \text{component}}}$$

The transfer function relating $Y(s)$ to $R(s)$ is

$$T(s) = \frac{b_1 s + b_0}{s + a_0}$$

which can be expressed as follows:

$$T(s) = b_1 + \frac{(b_0 - a_0 b_1)}{s + a_0}$$

Such a system can be considered to be composed of a gain b_1 in tandem with a first-order system with constant transfer function numerator, as indicated in Figure 2.3. This system output thus contains a component proportional to the input, in addition to a component of the preceding type.

Typical first-order systems.

Numerous systems in nature are governed by first-order systems, with widely varying time constants. Electrical *RL* and *RC* circuits may result in time constants of a few milliseconds. The room heating system with an electromechanical thermostat described in Section 1.2 can result in a first-order system with a time constant of tens of minutes. A biological growth process may have a time constant of several weeks, and a radioactive decay can have a time constant of several years.

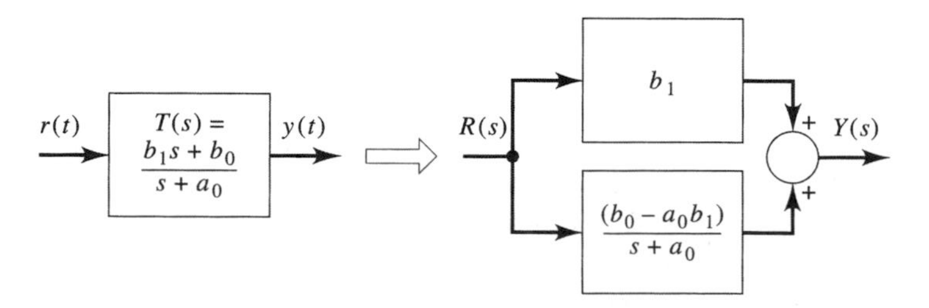

Figure 2.3 Tandem representation of a first-order system.

❑ DRILL PROBLEMS

D2.1 Find and also sketch the response of the systems with the following transfer functions, inputs, and initial conditions:

(a)

$$T(s) = \frac{3}{s+3}$$
$$r(t) = 6u(t)$$
$$y(0^-) = 10$$

Ans. $6 + 4e^{-3t}$

(b)

$$T(s) = \frac{1}{s+10}$$
$$r(t) = 3u(t)\cos 10t$$
$$y(0^-) = 0$$

Ans. $0.212\cos(10t - 45°) - 0.15e^{-10t}$

(c)

$$T(s) = \frac{s}{s+1000}$$
$$r(t) - 7u(t)$$
$$y(0^-) = 4$$

Ans. $11e^{-1000t}$

(d)

$$T(s) = \frac{20s}{s+300}$$
$$r(t) = 8u(t)\sin 100t$$
$$y(0^-) = -10$$

Ans. $50.6\cos(100t - 18.4°) - 58e^{-300t}$

(e)

$$T(s) = \frac{-4s+20}{s+300}$$
$$r(t) = 10u(t)$$
$$y(0^-) = 0$$

Ans. $0.67 - 40.67e^{-300t}$

D2.2 Find the time constants of the following systems:

(a)

$$T(s) = \frac{4s-1}{3s+2}$$

Ans. 3/2 s

(b)

$$\frac{dy}{dt} + 4y = -3\frac{dr}{dt}$$

Ans. 1/4 s

(c)

The system with a block diagram shown in Figure D2.2.

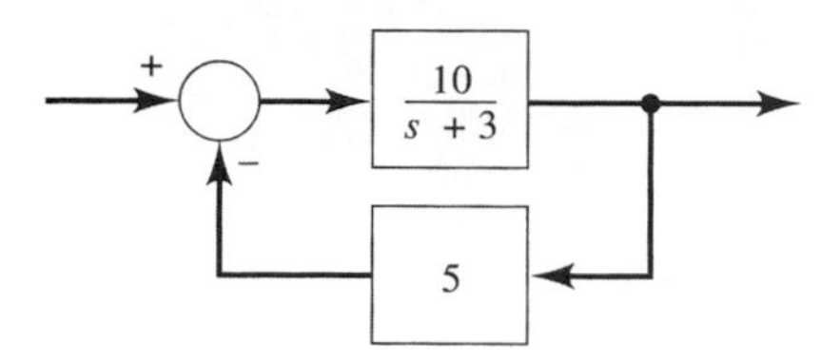

Figure D2.2

Ans. 1/53 s

❑ Computer-Aided Learning

Two methods for plotting the time response are presented here. The MATLAB commands are

```
step(Laplace transfer function)
ezplot(symbolic time function, [min time max time])
```

We will also use the command

```
conv([coeff1], [coeff2])
```

For the step command the transfer function coefficients are input in either vector form, followed by the "tf" command, or the "zpk" form may be used to define the transfer function. For example, suppose we want the step response for a system with transfer function

$$T(s) = \frac{10(s+1)}{s^2 + 5s + 6} = \frac{10(s+1)}{(s+2)(s+3)}$$

The step response using the vector form may be created by

```
num=[10 10]
den=[1 5 6]
t=tf(num, den)
step(t)
```

Another approach is to use the "zpk" command

```
t=zpk(-1, [-2 -3], 10)
step(t)
```

In either case, the computer, in effect, multiplies $T(s)$ by $1/s$ to get the step response. That response is then calculated and plotted.

Responses to other than a step response may be taken by employing a simple trick with the step command. If you want the time response of some $Y(s) = T(s)R(s)$ when $R(s)$ is not $1/s$, obtain the step response of $sY(s)$. The computer can be thought of as multiplying by $1/s$, then canceling the s, resulting in the time response of $Y(s)$.

Suppose we want the time response for the $T(s)$ when

$$r(t) = 6\cos 4t$$
$$R(s) = \frac{6s}{s^2 + 16}$$

Then

$$sY(s) = sT(s)R(s) = s\frac{10(s+1)}{(s+2)(s+3)}\frac{6s}{(s^2+16)}$$

To employ the step command in vector form, the "conv" (meaning convolve) command may be used to multiply polynomial vectors

```
snum=60*conv([1 0 0], [1 1])
den1=conv([1 2], [1 3])
den=conv(den1, [1 0 16])
sy=tf(snum, den)
step(sy)
```

Using the "zpk" form may sometimes be easier:

```
sy=zpk([0 0 -1], [-2 -3 j*4 -j*4], 60)
step(sy)
```

C2.1 Use the step command to get the step response of the following:

(a) $T(s) = \dfrac{5}{(s+4)^2}$

(b) $T(s) = \dfrac{5(s+6)}{(s+1+j4)(s+1-j4)}$

Ans. (a)
```
ta=zpk([], [-4 -4], 5)
step(ta)
```

(b)
```
num=[5 30]
den=[1 2 17]
tb=tf(num, den)
step(tb)
```

C2.2 Use the step command to get the time response of the following:

(a) $Y(s) = \dfrac{32}{(s+2)^2(s^2+16)}$

(b) $Y(s) = T(s)R(s)$, where

$$T(s) = \frac{48}{(s+3)(s^2+s+16)}$$
$$r(t) = 3e^{-3t}$$

Ans. (a)
```
sya=zpk([0], [-2 -2 -j*4 j*4],32)
step(sya)
```
(b)
```
snum=[48 0]*3
den1=conv([1 3], [1 1 16])
den=conv(den1, [1 3])
stb=tf(snum, den)
step(stb)
```

In Chapter 1, we used symbolic techniques to get Laplace and inverse Laplace transforms. These symbolic capabilities of MATLAB provide another way of using the "ezplot" command to get the time response. If we want to solve

$$Y(s) = T(s)R(s)$$

$$Y(s) = \frac{10(s+1)}{s^2+5s+6}$$

$$r(t) = 6\cos 4t$$

we can use symbolic commands

```
syms s t
T=10*(s+1)/(s^2+5*s+6)
R=laplace(6*cos(4*t))
Y=R*T
y=ilaplace (Y)
ezplot (y, [0 10])
```

We need to specify the time range (here from 0 to 10 s). That range may require some trial-and-error work.

C2.3 Find the time response of C2.2(a) and C2.2(b) using symbolic methods and the "ezplot" command.

Ans. (a)
```
syms s t
Ya=32/(s+2)^2/((s^2+16)
ya=ilaplace(Ya)
ezplot(ya, [0 15])
```
(b)
```
syms s t
T=48/(s+3)/(s^2+s+16)
r=3*exp(-3*t)
R=laplace (r)
Yb=T*R
yb=ilaplace (Yb)
ezplot (yb, [0 2])
```

2.3 Response of Second-Order Systems

2.3.1 Time Response

In a second-order system, the output $y(t)$ and the input $r(t)$ are related by a second-order linear differential equation in $y(t)$; $y(t)$ is the dependent variable and $r(t)$ is the independent variable, as follows:

$$\frac{d^2y}{dt^2} + a_1\frac{dy}{dt} + a_0 y = b_1\frac{dr}{dt} + b_0 r$$

As can be seen from Figure 1.30 in Chapter 1, stability implies that the natural response decays to zero. The denominator of $T(s)$ is the characteristic polynomial, whose roots must all be in the left half-plane to force all the natural response terms to decay to zero, and therefore to force the system to be stable. In the second-order case, both a_0 and a_1 must be greater than zero for stability

Stability of a second-order system.

$$T(s) = \frac{b_1 s + b_0}{s^2 + a_1 s + a_0}$$

and the system response is of the form

$$Y(s) = \underbrace{\frac{b_1 s + b_0}{s^2 + a_1 s + a_0} R(s)}_{\text{zero-state component}} + \underbrace{\frac{\begin{pmatrix}\text{first-degree numerator}\\ \text{polynomial dependent on}\\ \text{initial conditions}\end{pmatrix}}{s^2 + a_1 s + a_0}}_{\text{zero-input component}}$$

The characteristic polynomial of a second-order system is

$$s^2 + a_1 s + a_0 = (s - s_1)(s - s_2)$$

with roots s_1 and s_2 given by the quadratic formula

$$s_1, s_2 = \frac{-a_1 \pm \sqrt{a_1^2 - 4a_0}}{2}$$

2.3.2 Overdamped Response

If the characteristic roots s_1 and s_2 are real and distinct, the zero-state response to a unit impulse of the system

$$\frac{d^2y}{dt^2} + a_1\frac{dy}{dt} + a_0 y = b_1\frac{dr}{dt} + b_0 r \qquad [2.2]$$

is

$$Y_{\text{zero-state}}(s) = \frac{K_1}{s - s_1} + \frac{K_2}{s - s_2}$$

$$y_{\text{zero-state}}(t) = K_1 e^{s_1 t} + K_2 e^{s_2 t} \quad t \geqslant 0$$

which is the sum of two real exponential terms. Such a system is termed *overdamped* when s_1 and s_2 are both negative. For example, the overdamped second-order system with transfer function

$$T(s) = \frac{1}{s^2 + 3s + 2}$$

and unit step intput

$$R(s) = \frac{1}{s}$$

and zero initial conditions has response given by

$$Y(s) = \frac{1}{s(s^2+3s+2)} = \frac{1/2}{s} + \frac{-1}{s+1} + \frac{1/2}{s+2}$$

$$y(t) = \left[\tfrac{1}{2} - e^{-t} + \tfrac{1}{2}e^{-2t}\right]u(t)$$

This response is sketched in Figure 2.4. Nonzero initial conditions result in different amplitudes for the two exponential natural terms.

2.3.3 Critically Damped Response

If the two characteristic roots are equal, the second-order system of Equation (2.2) has

$$Y_{\text{zero-state}}(s) = \frac{\text{numerator polynomial}}{s^2+a_1s+a_0} = \frac{\text{numerator polynomial}}{(s-s_1)^2}$$

$$= \frac{K_1}{s-s_2} + \frac{K_2}{(s-s_1)^2}$$

for which the corresponding time function after $t = 0$ is

$$y_{\text{zero-state}}(t) = K_1e^{s_1t} + K_2te^{s_1t} \qquad t \geqslant 0$$

Such a second-order system is said to be *critically damped* where s_1 is negative. The critically damped system with transfer function

$$T(s) = \frac{10s+8}{s^2+4s+4}$$

for example, has unit step response, with zero initial conditions, given by

$$Y(s) = \frac{10s+8}{s(s^2+4s+4)} = \frac{2}{s} + \frac{-2}{s+2} + \frac{6}{(s+2)^2}$$

for which

$$y(t) = [2 - 2e^{-2t} + 6te^{-2t}]u(t)$$

This response is sketched in Figure 2.5. Other initial conditions result in different amplitudes for the $\exp(-2t)$ and $t\exp(-2t)$ natural terms, but the same character of response.

2.3.4 Underdamped Response

If the roots of the characteristic polynomial are complex numbers, they are complex conjugates of one another

$$s_1, s_2 = -a \pm j\omega$$

and the zero-state response to a unit impulse component is of the form

$$Y_{\text{zero-state}}(s) = \frac{\text{numerator polynomial}}{s^2+a_1s+a_0} = \frac{\text{numerator polynomial}}{(s+a-j\omega)(s+a+j\omega)}$$

$$= \frac{\text{numerator polynomial}}{(s+a)^2+\omega^2}$$

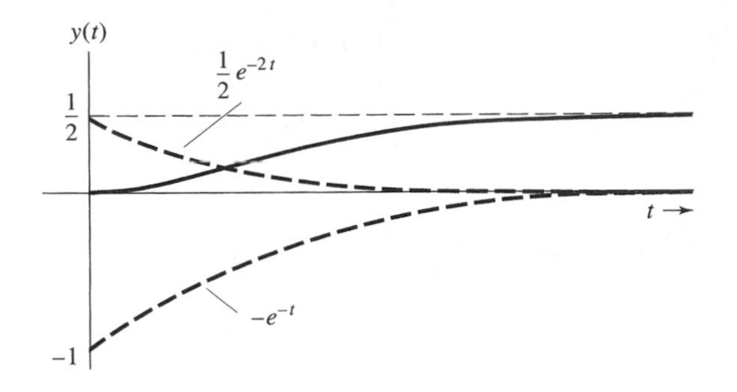

Figure 2.4 Step response of an overdamped second-order system.

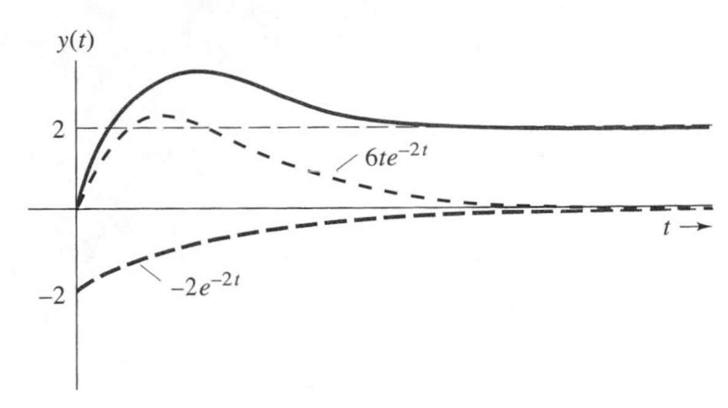

Figure 2.5 Step response of a critically damped second-order system.

corresponding to the time behavior

$$y_{\text{zero-state}}(t) = [Ae^{-at}\cos(\omega t + \theta)]u(t)$$

This type of second-order system is termed *underdamped* where a is positive. For example, the underdamped system with transfer function.

$$T(s) = \frac{-3s + 17}{s^2 + 2s + 17}$$

has unit step response

$$\begin{aligned} Y(s) &= \frac{-3s + 17}{s(s^2 + 2s + 17)} = \frac{1}{s} + \frac{-(s+5)}{(s+1)^2 + (4)^2} \\ &= \frac{1}{s} + \frac{Me^{j\theta}}{s + 1 - j4} + \frac{Me^{-j\theta}}{s + 1 + j4} \end{aligned}$$

By using the evaluation method to find M and θ, we find

$$Me^{j\theta} = \left.\frac{-s - 5}{s + 1 + j4}\right|_{s=-1+j4} = \frac{1 - j4 - 5}{j8} = -\frac{1}{2} + j\frac{1}{2} = \frac{\sqrt{2}}{2}e^{j135^\circ}$$

From Table B.1 in Appendix B, we have

$$\begin{aligned} y(t) &= [1 + \sqrt{2}e^{-t}\cos(4t + 135^\circ)]u(t) \\ &= [1 - \sqrt{2}e^{-t}\cos(4t - 45^\circ)]u(t) \end{aligned}$$

This response is sketched in Figure 2.6. Other initial conditions give different constants A and θ.

2.3.5 Undamped Natural Frequency and Damping Ratio

Underdamped second-order systems have a natural response that is described by a radian frequency of oscillation ω and an exponential constant σ, which are found

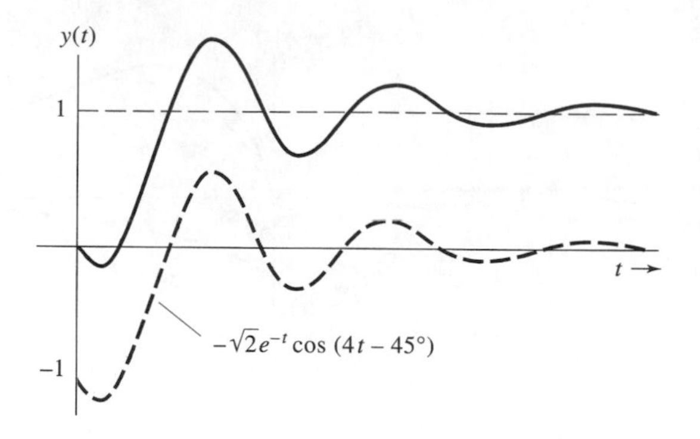

Figure 2.6 Step response of an underdamped second-order system.

from the system's characteristic polynomial:

$$s^2 + a_1 s + a_0 = (s + \sigma)^2 + \omega^2$$

With underdamped second-order system response, a more useful form is described by the **undamped natural frequency** ω_n and the **damping ratio** ζ:

$$s^2 + a_1 s + a_0 = s^2 + 2\zeta\omega_n s + \omega_n^2$$

The two sets of quantities are related by

$$\sigma = \zeta\omega_n$$
$$\omega = \omega_n\sqrt{1 - \zeta^2}$$

For ζ between 0 and 1, the characteristic roots lie on a circle of radius ω_n about the origin in the left half of the complex plane, as shown in Figure 2.7. For $\zeta = 0$, the roots are on the imaginary axis, and for $\zeta = 1$, both roots are on the negative real axis, repeated. The undamped natural frequency ω_n is the radian frequency at which the oscillations would occur if the damping ratio ζ were zero. If ζ were zero, the system zero-state response to a unit impulse would have the form

$$Y_{\text{zero-state}}(s) = \frac{\text{numerator polynomial}}{s^2 + \omega_n^2}$$

$$y_{\text{zero-state}}(t) = [A\cos(\omega_n t + \theta)]u(t)$$

Damping ratio is related to the damping angle.

The damping ratio is related to the *damping angle* ϕ in Figure 2.7 by

$$\zeta = \cos\phi$$

Consider the second-order system with transfer function

$$T(s) = \frac{100}{s^2 + 3s + 13}$$

The undamped natural frequency of the system is

$$\omega_n = \sqrt{13} = 3.6$$

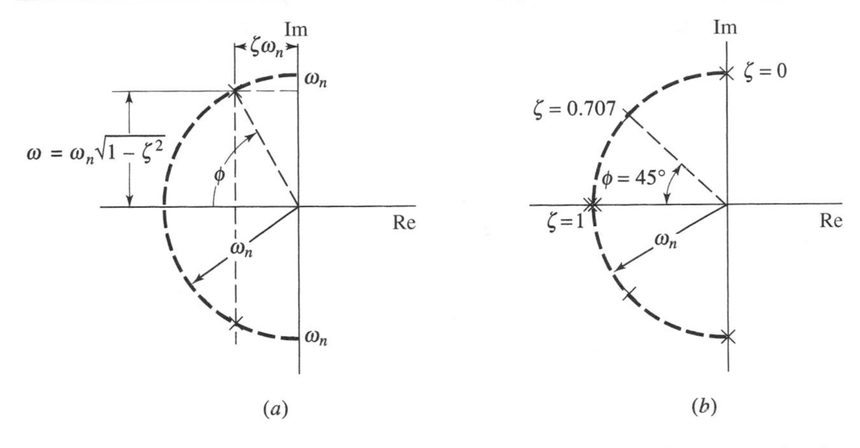

Figure 2.7 Underdamped natural frequency and damping ratio for a second-order response term. (a) Relation to characteristic root locations. (b) Damping ratios corresponding to various root locations.

and the damping ratio is given by

$$2\zeta\omega_n = 3 \quad \zeta = \frac{3}{2\omega_n} = \frac{3}{7.2} = 0.42$$

The true oscillation frequency is

$$\omega = \omega_n\sqrt{1-\zeta^2} = 3.6(0.9) = 3.27$$

Figure 2.8 shows a set of normalized step response curves for underdamped second-order systems of the form

Standard second-order T(s).

$$T(s) = \frac{\omega_n^2}{s^2 + 2\zeta\omega_n s + \omega_n^2}$$

For $\zeta = 0$, the oscillations continue forever. Larger values of ζ give more rapid decay of the oscillations but a slower rise of the response. For $\zeta = 1$, the system is critically damped.

At this point, the student should stop and study Figure 2.8 for the purpose of gaining an understanding of how system response depends on the values of ζ and ω.

❑ Computer-Aided Learning

C2.4 Write MATLAB commands to obtain the step response of

(a) Figure 2.4

(b) Figure 2.5

(c) Figure 2.6

Ans. (a)

```
numa=1
dena=[1 3 2]
ta=tf(numa, dena)
step(ta)
```

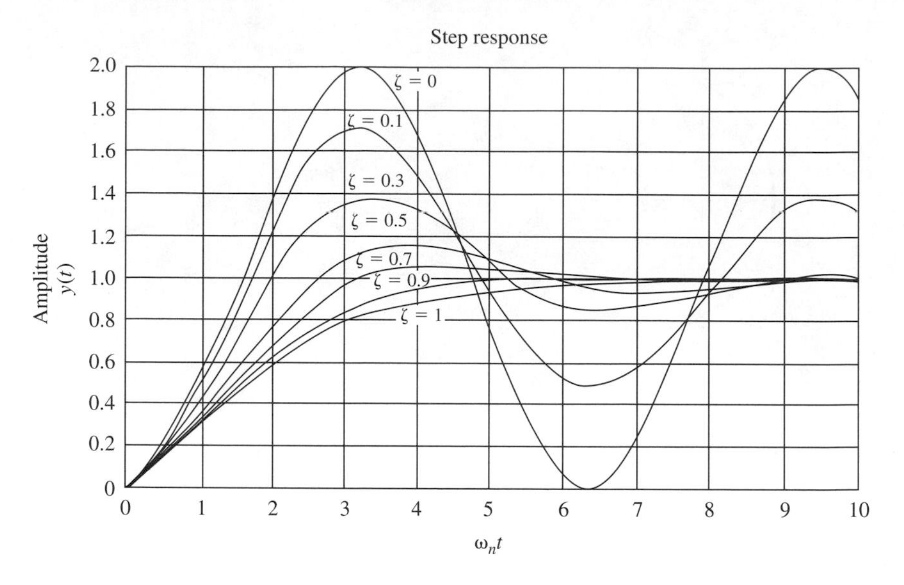

Figure 2.8 Normalized step response of a second-order system with constant transfer function numerator.

(b)
```
numb=[10 8]
denb=[1 4 4]
tb=tf(numb,denb)
step(tb)
```

(c)
```
numc=[-3 17]
denc=[1 2 17]
tc=tf(numc, denc)
step(tc)
```

C2.5 Use MATLAB Commands to obtain Figure 2.8. For convenience, let $\omega_n = 1$, so the time axis becomes $t = \omega_n t$.

Ans.
```
num=1
den0=[1 0 1]
den1=[1 .2 1]
den3=[1 .6 1]
den5=[1 1 1]
den7=[1 1.4 1]
den9=[1 1.8 1]
den10=[1 2 1]
t0=tf(num, den0)
t1=tf(num, den1)
t3=tf(num, den3)
t5=tf(num, den5)
```

```
t7=tf(num, den7)
t9=tf(num, den9)
t10=tf(num, den10)
step(t0, t1, t3, t5, t7, t9, t10)
```

❑ DRILL PROBLEMS

D2.3 Find and also sketch the response of the system with the following transfer functions, input, and initial conditions:

(a)

$$T(s) = \frac{s}{s^2 + 7s + 8}$$

$$r(t) = u(t)$$

$$y(0^-) = 7$$

$$y'(0^-) = -4$$

Ans. $8.63e^{-1.4t} - 1.62e^{-5.6t}$

(b)

$$T(s) = \frac{4}{s^2 + 4s + 4}$$

$$r(t) = 0$$

$$y(0^-) = -3$$

$$y'(0^-) = 2$$

Ans. $-3e^{-2t} - 4te^{-2t}$

(c)

$$T(s) = \frac{3}{s^2 + 0.5s + 4}$$

$$r(t) = 2u(t)$$

$$y(0^-) = 0$$

$$y'(0^-) = 0$$

Ans. $1.50 + 1.51e^{-0.25t}\cos(1.98t + 173°)$

(d)

$$T(s) = \frac{3}{s^2 + s + 8}$$

$$r(t) = 4u(t)$$

$$y(0^-) = 0$$

$$y'(0^-) = 0$$

Ans. $1.5 + 1.52e^{-0.5t}\cos(2.78t + 169.8°)$

(e)

$$T(s) = \frac{4s - 20}{s^2 + 4s + 29}$$
$$r(t) = 10\delta(t)$$
$$y(0^-) = 0$$
$$y'(0^-) = 6$$

Ans. $67.85e^{-2t}\cos(5t + 54°)$

(f)

$$T(s) = \frac{s^2}{s^2 + 2s + 17}$$
$$r(t) = 0$$
$$y(0^-) = 10$$
$$y'(0^-) = 0$$

Ans. $10.3e^{-t}\cos(4t - 14°)$

D2.4 Determine which of the following second-order systems are underdamped, which are critically damped, and which are overdamped.

(a) $T(s) = \dfrac{9s^2 + 3s + 10}{s^2 + 5s + 2}$

(b) $T(s) = \dfrac{s^2 - 2s}{s^2 + 6s + 9}$

(c) $T(s) = \dfrac{64}{3s^2 + 4s + 5}$

(d) $T(s) = \dfrac{19s - 20}{s^2 + s + 100}$

(e) $T(s) = \dfrac{s^2 + 2s + 100}{s^2 + 7s + 49}$

Ans. (a) overdamped; (b) critically damped; (c) underdamped; (d) underdamped; (e) underdamped

D2.5 For second-order systems with the following transfer functions, determine the undamped natural frequency, the damping ratio, and the oscillation frequency.

(a)

$$T(s) = \frac{100}{s^2 + s + 100}$$

Ans. 10, 0.05, 9.99

(b)

$$T(s) = \frac{3s - 49}{s^2 + 3s + 49}$$

Ans. 7, 0.214, 6.84

(c)

$$T(s) = \frac{s^2 + 9s}{s^2 + 4s + 10}$$

Ans. 3.16, 0.632, 2.45

(d)

$$T(s) = \frac{s^2 + 20}{s^2 + 2s + 20}$$

Ans. 4.47, 0.224, 4.36

(e)

$$T(s) = \frac{-3s + 0.7}{s^2 + 0.3s + 4}$$

Ans. 2, 0.075, 1.99

D2.6 Find the constant k for which the system with transfer function $T(s)$ has the given second-order response property.

(a)

$$T(s) = \frac{10}{s^2 + 40s + k}$$

$$\zeta = 0.7$$

Ans. 816

(b)

$$T(s) = \frac{ks + 6}{s^2 + ks + 49}$$

$$\omega = 4$$

Ans. 11.49

(c)

$$T(s) = \frac{20s}{3s^2 + 2s + k + 4}$$

$$\zeta = 0.1$$

Ans. 29.33

(d)

$$T(s) = \frac{s^2 - 6}{ks^2 + s + 6}$$

$$\omega_n = 2$$

Ans. 1.5

2.3.6 Rise Time, Overshoot, and Settling Time

The quality of the performance of a stable system is commonly characterized by the **rise time**, **peak time**, **overshoot**, and **settling time** of its response to a step input. As indicated in Figure 2.9, rise time is the interval of time required for the step response of a system to go from 10% to 90% of its final value. Peak time, T_p, is the time required for the step response of a system to reach the first peak. Overshoot is the percentage difference between the maximum and the steady

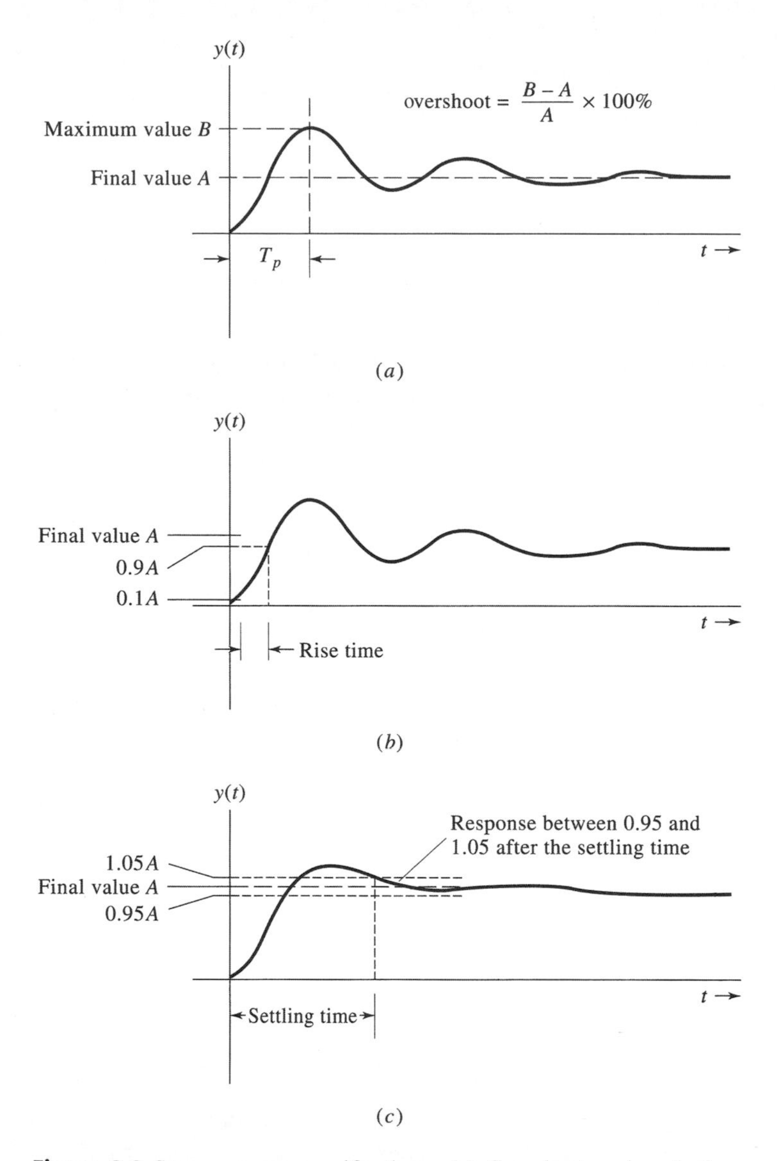

Figure 2.9 Step response specifications. (a) Overshoot and peak time. (b) Rise time. (c) Settling time.

state values of the response. Settling time is the minimum time required before the system response remains within ±5% of the final value. Other definitions of rise time (e.g., 5% to 95%) and settling time (±10%) are possible but are not used in this text.

Dominant roots are considered.

The real roots of most higher-order systems are often widely separated, so most terms in the response decay much more rapidly than the others. When all but two roots of a transfer function exhibit rapid decay, the response appears to be second-order, although the system may actually be third- or higher-order. This system is said to be *dominated by two least-damped roots.*

Rise time, peak time, overshoot, and settling time depend on the roots of the characteristic polynomial, the initial conditions, and the command. Considerable insight into these performance measures can be gained by considering only second-order underdamped systems with zero initial conditions and a unit step command. Other step commands would change the response values accordingly.

For an underdamped second-order system with transfer function of the form

$$T(s) = \frac{\omega_n^2}{s^2 + 2\zeta\omega_n s + \omega_n^2} \qquad [2.3]$$

the step response has Laplace transform

Zero-state response.

$$Y(s) = \frac{A}{s}T(s) = \frac{A\omega_n^2}{s(s^2 + 2\zeta\omega_n s + \omega_n^2)}$$

For a unit step command A is 1, and so

$$Y(s) = \frac{k_1}{s} + \frac{k_2}{s+\sigma-j\omega} + \frac{k_3}{s+\sigma+j\omega}$$
$$\sigma = \zeta\omega_n$$
$$\omega = \omega_n\sqrt{1-\zeta^2}$$
$$k_1 = 1$$
$$k_2 = \frac{1}{2z}\Big/ 90^\circ + \tan^{-1}\frac{\omega}{\sigma}$$
$$z = \sqrt{1-\zeta^2}$$

The time response is

$$y = 1 + \frac{1}{z}e^{-\zeta\omega_n t}\cos\left(z\omega_n t + \tan^{-1}\frac{z}{\zeta} + 90^\circ\right) \qquad [2.4]$$

The time response depends on the product of ω_n and t. It is common practice to normalize curves of this type by using $\omega_n t$ as the time axis. The value of time can always be computed later when the undamped natural frequency is known. For example, plots such as those in Figure 2.8 are created for each value of the damping ratio (with $\omega_n t$ being the horizontal axis).

Percent overshoot vs. damping ratio. Rise time vs. damping ratio.

The horizontal axis for Figures 2.10(a–c) is the damping ratio. From Figure 2.8, it is clear that as the damping ratio diminishes, the time response moves upward more quickly, so assuming that the undamped natural frequency remains constant, rise time [Figure 2.10(a)] diminishes as the damping ratio diminishes. However, as the damping ratio diminishes, the percentage overshoot [Figure 2.10(b)] increases. Rise time and percent overshoot move in opposite directions as a function of damping ratio. To find the peak time, dy/dt is set to zero resulting in

$$T_p = \pi/\omega$$

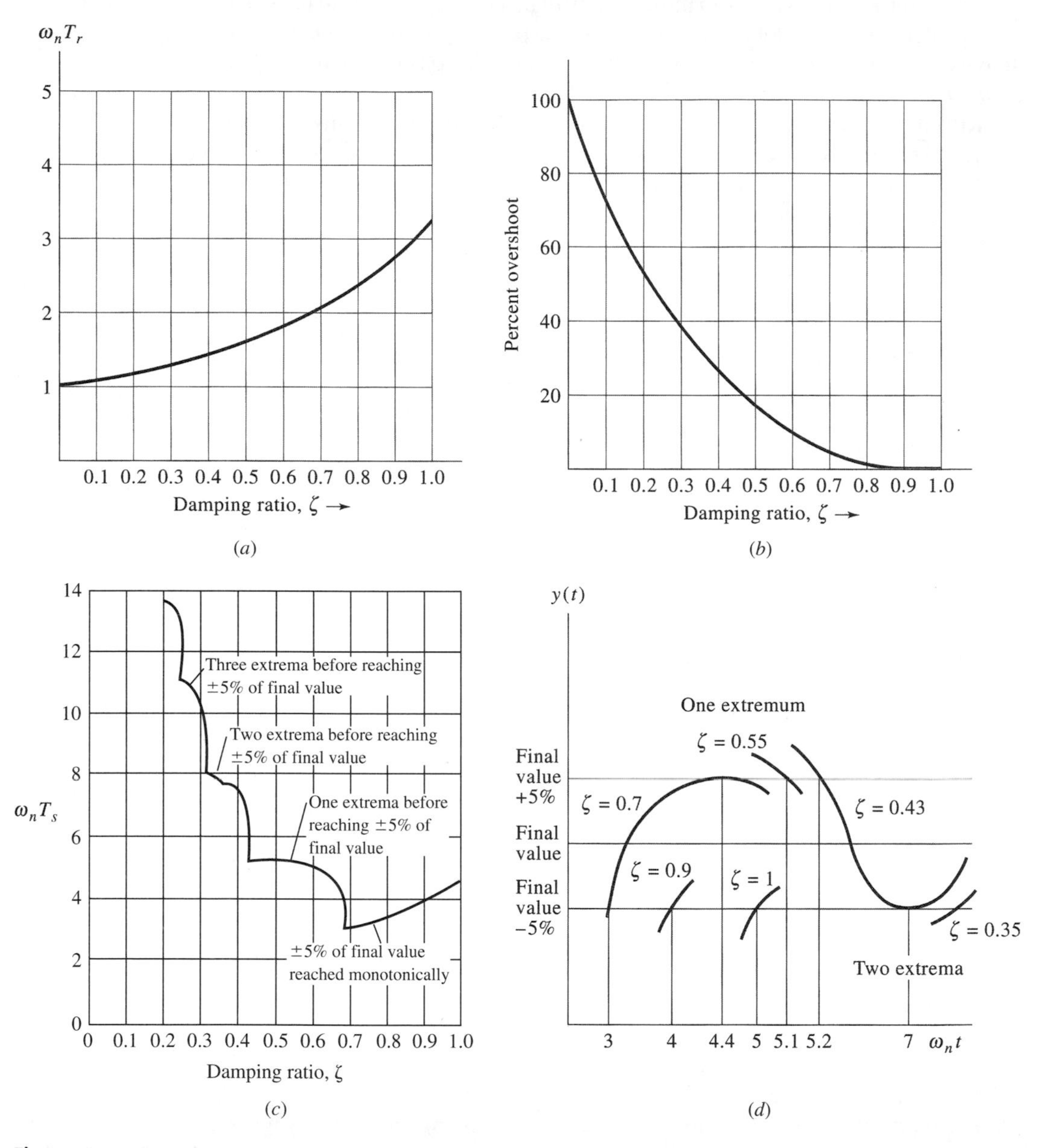

Figure 2.10 Step response performance of second-order systems with transfer functions having a constant numerator. (a) Rise time. (b) Overshoot. (c) Settling time. (d) Time response of $Y(t)$ for various damping ratios (not to scale).

Settling time provides a compromise in designing a second-order system, in that the speed of response measured by rise time is tempered with the undesirable feature of overshooting (or undershooting) the final value and then having to move back in the opposite direction.

In the following discussion, the product of undamped natural frequency and settling time will be called the *normalized settling time*. The product of undamped natural frequency and time is called *normalized time*. The normalized settling time versus damping ratio curve of Figure 2.10(c) can be understood by examining Figure 2.10(d), which has an exaggerated scale to facilitate this discussion. If the damping ratio is 1.0, the lower limit is crossed when the normalized time is 5, but the upper limit is never crossed. As the damping ratio is reduced to $\zeta = 0.9$, $\omega_n T_s$ reduces to 4. A damping ratio of 0.7 has the smallest $\omega_n T_s$, since the response curve crosses the -5% line at $\omega_n t = 3$, as seen in Figure 2.10(d). Therefore, normalized settling time decreases as damping ratio decreases from 1 down to 0.7. When the damping ratio is about 0.7, the time response overshoots about 5%, reaching the upper limit after about 4.4 units of normalized time. Normalized settling time therefore jumps from 3 to 4.4. As the damping ratio drops from 0.7 down to about 0.43, the upper limit is entered after the derivative has gone to zero once (one extremum), but the lower limit is not subsequently exceeded. The normalized settling time increases from 4.4 when the damping ratio is 0.7 to $\omega_n T_s = 5.1$ when $\zeta = 0.55$. At a damping ratio of $\zeta = 0.43$, the upper $+5\%$ value is crossed at 5.2. When the damping ratio is about 0.43, the lower limit is exceeded, so the normalized settling time jumps to about 7. For damping ratios from 0.43 down to about 0.3, it takes longer for the lower limit to be entered, so normalized settling time increases while $y(t)$ transverses two time points where the derivative is zero (two extrema).

Settling time vs. damping ratio is explained.

In general, the normalized settling time curve comprises an infinite number of segments, where, for each range of decreasing damping ratios, a $\pm 5\%$, limit is eventually exceeded, calling for one additional extremum (and another segment). Because the threshold is chosen to be $\pm 5\%$, settling time is a minimum for a damping ratio of about 0.7. That value is often used in control system design. Chapters 5 and 6 explore further uses of rise time and settling time in designing control systems.

❑ Computer-Aided Learning

C2.6 According to Figure 2.10, if a second-order system has a damping ratio of 0.5, there should be about 16% overshoot. If, in addition to $\zeta = 0.5$, $\omega_n = 1$, then the rise time T_r should be about 1.7 s and the settling time T_s should be about 5.4 s. That data refers to the step response of

$$T(s) = \frac{1}{s^2 + s + 1}$$

(a) Add a pole at $-10\ s_0$

$$T(s) = \left(\frac{10}{s + 10}\right)\left(\frac{1}{s^2 + s + 1}\right)$$

and use MATLAB to get the step response. Find the percent overshoot, T_r and T_s, where the third pole is not dominant (i.e., well to the left).

(b) Repeat (a), but assume that the third pole has the same real part as the second-order complex conjugate pair, so all three poles are now dominant. That is, the step response for

$$T(s) = \frac{0.5}{s + 0.5}\frac{1}{s^2 + s + 1}$$

Ans. (a)
```
numa=10
dena=conv([1 10], [1 1 1])
ta=tf(numa, dena)
step(ta)
```
(% overshoot = 16%, $T_r = 1.7$ s, $T_s = 5.4$ s)

(b)
```
numb=0.5
denb=conv([1 0.5], [1 1 1])
tb=tf(numb, denb)
step(tb)
```
(% overshoot = 0%, $T_r = 3.4$ s, $T_s = 5.5$ s)

❑ DRILL PROBLEMS

D2.7 Find the form of the natural response of systems with the following transfer functions:

(a)

$$T(s) = \frac{100}{\left(s^2 + 4s + 4\right)\left(s^2 + 4s + 5\right)}$$

Ans. $K_1 e^{-2t} + K_2 t e^{-2t} + K_3 e^{-2t}\cos(t + \theta)$

(b)

$$T(s) = \frac{3s - 12}{s^3 + 4s^2 + 13s}$$

Ans. $K_1 + K_2 e^{-2t}\cos(3t + \theta)$

(c)

$$T(s) = \frac{s^2}{3(s + 3)^2(s^2 + 2s + 10)}$$

Ans. $K_1 e^{-3t} + K_2 t e^{-3t} + K_3 e^{-t}\cos(3t + \theta)$

(d)

$$T(s) = \frac{5(s^2 + 2s + 1)(s^2 + 2s + 2)}{(s + 1)^2(s^2 + 4)(s + 8)}$$

Ans. $K_1 e^{-t} + K_2 t e^{-t} + K_3\cos(2t + \theta) + K_4 e^{-8t}$

(e)

$$T(s) = \frac{6s^3 - 4s^2 + 2s + 400}{(s^2 + s + 10)(s^2 + s + 20)}$$

Ans. $K_1 e^{-0.5t}\cos(3.12t + \theta_1) + K_2 e^{-0.5t}\cos(4.44t + \theta_2)$

D2.8 Using the curves given in the text, find approximately the percent overshoot, peak time, rise time, and settling time of the following systems when driven by a step input.

(a)

$$T(s) = \frac{100}{s^2 + 4s + 100}$$

Ans. 54%, 0.32, 0.12, 1.1

(b)

$$T(s) = \frac{49}{s^2 + 4s + 49}$$

Ans. 40%, 0.47, 0.18, 1.5

(c)

$$T(s) = \frac{60}{2s^2 + 8s + 30}$$

Ans. 18%, 0.95, 0.44, 1.29

(d)

$$T(s) = \frac{75}{s^2 + 3s + 20}$$

Ans. 32%, 0.75, 0.34, 1.7

It is important to realize the amount of information that can be obtained from the $T(s)$ function. In Drill Problem D2.8, the percent overshoot, peak time, rise time, and settling time are found from the Laplace-transformed form of $T(s)$. In most control system designs, it is not necessary to convert the Laplace transformed $T(s)$ function into the time solution. The system can be analyzed and designed by staying in the s plane. The location of the roots of the characteristic equation tells the designer the form of the time solution. Figure 1.30 shows the form of the time solution for various root locations. Figure 2.8 shows how the time response of a second-order system depends upon the damping ratio ζ and the undamped natural frequency ω_n.

A second observation must be made. The characteristic equation for the system is the denominator of $T(s)$, which influences the zero-state response with a unit impulse input. If initial conditions exist to find the zero input function, the same characteristic equation is used and no new stability information is obtained. As a result, most control system analysis and design sets all initial conditions to zero and the characteristic equation is found from the Laplace-transformed zero-state response with an impulse input. That response is $T(s)$, since the zero-state response is $T(s)R(s)$ with $R(s) = 1$.

2.4 Higher-Order System Response

The zero-state response to a unit impulse input of third- and higher-order systems consists of a sum of terms, one term for each characteristic root. For each distinct real characteristic root, there is a real exponential term in the system zero-state impulse response. For each pair of complex conjugate roots there is a pair of complex exponential terms, which are better expressed as an exponential times a sinusoid. Repeated roots give additional terms involving powers of time multiplied by the exponent. For example, a system with transfer function

$$T(s) = \frac{-8s^2+5}{s^4+9s^3+37s^2+81s+52} = \frac{-8s^2+5}{(s+1)(s+4)(s^2+4s+13)}$$

has a zero-state impulse response of the form

$$Y_{\text{zero-state}}(s) = \frac{K_1}{s+1} + \frac{K_2}{s+4} + \frac{K_3 s + K_4}{(s+2)^2+3^2}$$

$$y_{\text{zero-state}}(t) = K_1 e^{-t} + K_2 e^{-4t} + Ae^{-2t}\cos(3t+\theta) \qquad t \geqslant 0$$

A system with transfer function

$$T(s) = \frac{-7s^3+6s^2+9s+23}{(s+9)(s+12)(s^2+12s+40)(s^2+2s+37)} \qquad [2.5]$$

has a zero-state response to a unit impulse input of the form

$$\begin{aligned} Y_{\text{zero-state}}(s) &= \frac{K_1}{s+9} + \frac{K_2}{s+12} + \frac{K_3}{s+6+j2} \\ &\quad + \frac{K_3^*}{s+6-j2} + \frac{K_4}{s+1+j6} + \frac{K_4^*}{s+1-j6} \\ &= \frac{K_1}{s+9} + \frac{K_2}{s+12} + \frac{\text{first-order numerator}}{s^2+12s+40} \\ &\quad + \frac{\text{first-order numerator}}{s^2+2s+37} \\ y_{\text{zero-state}}(t) &= K_1 e^{-9t} + K_2 e^{-12t} + A_1 e^{-6t}\cos(2t+\theta_1) \\ &\quad + A_2 e^{-t}\cos(6t+\theta_2) \end{aligned} \qquad [2.6]$$

The time solution shows two exponential terms, one with a 1/9 s time constant and the other with a 1/12 s time constant. Two damped sinusoids are also included in Equation (2.6) as follows:

1. $\omega_{n1} = \sqrt{40} = 6.32$, $\zeta_1 = 12/[2(6.32)] = 0.95$
2. $\omega_{n2} = \sqrt{37} = 6.08$, $\zeta_2 = 2/[2(6.08)] = 0.16$

Notice that it is not necessary to solve for the time solution of Equation (2.6) to obtain stability information. Instead, the same information can be obtained from the Laplace-transformed $T(s)$ function of Equation (2.5).

The system of Equation (2.5) demonstrates the concept of "dominant roots." Notice that each of the first three terms in Equation (2.6) has a small time constant: 1/9 s, 1/12 s, and 1/6 s, whereas the fourth term is damped with a one-second time constant. This is significiant, since the first three terms of Equation (2.6) will decay much faster than the fourth term. In the s plane, the roots for the first three terms in

Dominant and nondominant roots.

Equation (2.5) are farther to the left of the j axis. The first three terms are termed *nondominant*, whereas the fourth term

$$A_2 e^{-t} \cos(6t + \theta_2) \qquad [2.7]$$

is termed *dominant*. This fourth pair of dominant roots determines the behavior of the system more than the nondominant roots. This is because the first three terms in Equation (2.6) will decay to zero in such a short period of time that the time response will be determined largely by the fourth term, shown in Equation (2.7). The system behavior can often be approximated by the dominant roots, or the "least-damped roots" ($\zeta_2 = 0.16$). This concept is important in higher-order systems, which are usually dominated by the pair of "least-damped roots" located closest to the j axis of the s plane. For roots to be dominant, the real parts of the complex conjugate roots should be approximately 20% of the real part of any other roots of the system. The relative magnitude of the residues (constant multiplier) of each root must also be considered. These residues, of course, depend upon the location of the zeros of the system function. The concept of dominant roots is useful to estimate the system response, but it should be used with care.

❑ **Computer-Aided Learning**

C2.7 Use MATLAB "ezplot" command to get the zero-state impulse response for the two transfer functions given in Section 2.4.

Ans. (a)

```
syms t s
Ya=(-8*s^2+5)/(s+1)/(s+4)/(s^2+4*5+13)
ya=ilaplace(Ya)
ezplot(ya,[0,4])
```

(b)

```
syms  t  s
Yb=(-7*s^3+6*s^2+9*s+23)/(s+9)/(s+12)
/(s^2+12*s+40)/(s^2+2*s+37)
yb=ilaplace(Yb)
ezplot(yb, [0,5])
```

2.5 Stability Testing

2.5.1 Coefficient Tests

For first- and second-order systems, stability (see Section 1.12) is determined by inspection of the characteristic polynomial. A first- or second-order polynomial has all roots in the left half of the complex plane if and only if all polynomial coefficients have the same algebraic sign. For example,

$$3s^2 + s + 10$$

is the characteristic polynomial of a stable system, while

$$3s^2 + s - 10$$

represents an unstable system.

First- and Second-order stable systems.

For higher-order polynomials, representing higher-order systems, the algebraic signs of the polynomial coefficients may or may not yield information as to stability. A polynomial with all roots in the left half-plane (LHP) has factors of the form

$$(s+a) \qquad a>0 \qquad \text{(real axis root in the LHP)}$$

and

$$(s^2+bs+c) \qquad b>0 \quad \text{and} \quad c>0 \qquad \text{(two LHP roots, perhaps complex conjugate)}$$

When multiplied out, such a polynomial must have all coefficients of the same algebraic sign, all positive or all negative. No coefficient can be zero ("missing") in a system with LHP roots because there are no minus signs involved and thus no way for a coefficient to be canceled.

If imaginary axis roots exist in the polynomial, factors of the following forms can be present, in addition to the others:

$$(s) \qquad \text{(root at the origin)}$$

and

$$(s^2+a) \qquad a>0 \qquad \text{(complex conjugate roots on the imaginary axis)}$$

With such roots present, all polynomial coefficients must be of the same algebraic sign, but some coefficients can be zero.

Right half-plane (RHP) roots involve factors of the form

Negative signs indicate RHP roots (unstable system).

$$(s-a) \qquad a>0 \qquad \text{(real axis root in the RHP)}$$

and

$$(s^2-as+b) \qquad a>0 \quad \text{and} \quad b>0 \qquad \text{(two roots in the RHP, perhaps complex conjugate)}$$

The presence of such factors may or may not cause differing algebraic signs of the coefficients and (by cancellation) zero coefficients.

Table 2.1 summarizes the information conveyed by these coefficient tests. For example, the polynomial

$$7s^6+5s^4-3s^3-2s^2+s+10$$

Table 2.1 *Polynomial Coefficient Tests*

Properties of the Polynomial Coefficients	Conclusion About Roots from the Coefficient Test
Differing algebraic signs	At least one RHP root
Zero-valued coefficients	Imaginary axis or RHP roots or both
All of the same algebraic sign, none zero	No direct information

has one or more RHP roots, indicated by the differing algebraic signs of the coefficients. Examination of the coefficient signs yields no information about root locations for the following polynomial:

$$8s^5 + 6s^4 + 3s^3 + 2s^2 + 7s + 10$$

The polynomial

$$s^6 + 3s^5 + 2s^4 + 8s^2 + 3s + 17$$

has imaginary axis roots or RHP roots or both, indicated by the missing s^3 term. Imaginary axis roots in the polynomial, if they are present, are complex conjugate, since if there were an imaginary axis root at $s = 0$, s would be a factor of the polynomial.

Implication of missing terms.

2.5.2 Routh–Hurwitz Testing

The Routh–Hurwitz test is a numerical procedure for determining the numbers of right half-plane (RHP) and imaginary axis (IA) roots of polynomial. The characteristic polynomial, which is the denominator of $T(s)$ is expressed as follows:

$$p(s) = a_n s^n + a_{n-1} s^{n-1} + \cdots + a_1 s + a_0 \qquad [2.8]$$

An array of numbers a_i is established as follows:

$$\begin{array}{c|cccc} s^n & a_n & a_{n-2} & a_{n-4} & \\ s^{n-1} & a_{n-1} & a_{n-3} & a_{n-5} & \\ s^{n-2} & b_1 & b_2 & b_3 & \\ s^{n-3} & c_1 & c_2 & c_3 & \cdots \\ s^{n-4} & d_1 & d_2 & d_3 & \\ \vdots & & & & \\ s^0 & & & & \end{array} \qquad [2.9]$$

Routh array format.

The first row starts with the highest-order constant a_n, and then every other constant is set forth. The second row starts with the second-highest-order constant a_{n-1} and then every other constant. The first two rows are formed from the coefficients of $p(s)$. The subsequent rows are formed as follows:

$$b_1 = \frac{-\begin{vmatrix} a_n & a_{n-2} \\ a_{n-1} & a_{n-3} \end{vmatrix}}{a_{n-1}} = \frac{a_{n-1}a_{n-2} - a_n a_{n-3}}{a_{n-1}} \qquad [2.10]$$

$$b_2 = \frac{-\begin{vmatrix} a_n & a_{n-4} \\ a_{n-1} & a_{n-5} \end{vmatrix}}{a_{n-1}} = \frac{a_{n-1}a_{n-4} - a_n a_{n-5}}{a_{n-1}} \qquad [2.11]$$

And the fourth row is formed as follows:

$$c_1 = \frac{-\begin{vmatrix} a_{n-1} & a_{n-3} \\ b_1 & b_2 \end{vmatrix}}{b_1} = \frac{b_1 a_{n-3} - b_2 a_{n-1}}{b_1} \quad [2.12]$$

$$c_2 = \frac{-\begin{vmatrix} a_{n-1} & a_{n-5} \\ b_1 & b_3 \end{vmatrix}}{b_1} = \frac{b_1 a_{n-5} - b_3 a_{n-1}}{b_1} \quad [2.13]$$

The other rows follow in the same form. Before we advance to the use of this procedure, we shall establish an array for the following polynomial:

$$p(s) = 2s^4 + 3s^3 + 5s^2 + 2s + 6 \quad [2.14]$$

First, the initial part of the array is formed. The powers of s are written to the left, and the polynomial coefficients are alternated between the first and second rows, as shown. It is helpful to imagine the rows to continue to the right with entries of zeros.

s^4	2	5	6
s^3	3	2	
s^2			
s^1			
s^0			

The array is completed by proceeding, row by row. Each element calcuated is derived from four elements in the above two rows, two of them at the left column and two in the column to the right of the element being calculated.

For the example of Equation (2.14), the first element of the s^2 row is

$$b_1 = \frac{3(5) - 2(2)}{3} = 11/3$$

And the second term of the s^2 row is

$$b_2 = \frac{3(6) - 2(0)}{3} = 6$$

The first element of the s^1 row is

Repeating pattern of calculations.

$$c_1 = \frac{(11/3)\,2 - 3(6)}{11/3} = -32/11$$

and the second element of the s^1 row is

$$c_2 = 0$$

The first element of the s^0 row is

$$d_1 = \frac{-\,(32/11)\,(6) - 0}{-32/11} = 6$$

This is shown in the shaded areas as follows:

$$\begin{array}{c|ccc} s^4 & 2 & 5 & 6 \\ s^3 & 3 & 2 & \\ s^2 & \frac{11}{3} & & \\ s^1 & & & \\ s^0 & & & \end{array} \qquad \begin{array}{c|ccc} s^4 & 2 & 5 & 6 \\ s^3 & 3 & 2 & 0 \\ s^2 & \frac{11}{3} & 6 & \\ s^1 & & & \\ s^0 & & & \end{array} \qquad \begin{array}{c|ccc} s^4 & 2 & 5 & 6 \\ s^3 & 3 & 2 & \\ s^2 & \frac{11}{3} & 6 & \\ s^1 & -\frac{32}{11} & & \\ s^0 & & & \end{array}$$

The completed Routh array is shown below. The number of RHP (right half-plane) roots of $p(s)$ is the number of algebraic sign changes in the elements of the left column of the array, proceeding from top to bottom. For this example, there are two sign changes in the left column, as indicated with the arrows; therefore $p(s)$ has two RHP roots.

$$\begin{array}{c|ccc} s^4 & 2 & 5 & 6 \\ s^3 & 3 & 2 & \\ s^2 & \frac{11}{3} & 6 & \\ s^1 & -\frac{32}{11} & & \\ s^0 & 6 & & \end{array}$$

If $p(s)$ is the denominator polynomial of a system's transfer function, that system is unstable.

2.5.3 Significance of the Array Coefficients

The Routh array contains coefficients for polynomials resulting from a sequence of polynomial divisions. The first row is an abbreviated form of the even powers of s present in the $p(s)$ of Equation (2.8):

$$E(s) = 2s^4 + 5s^2 + 6$$

The second row of the array, in effect, contains coefficients of the odd powers of s from $p(s)$:

$$D(s) = 3s^3 + 2s$$

Suppose we divide the lower-order polynomial $D(s)$ into the higher-order polynomial $E(s)$:

$$\begin{array}{r} \frac{2}{3}s \\ 3s^3 + 2s \overline{\smash{)}\, 2s^4 + 5s^2 + 6} \\ \underline{2s^4 + \frac{4}{3}s^2 } \\ \frac{11}{3}s^2 + 6 \end{array}$$

Notice that the remainder

$$R_1(s) = \tfrac{11}{3}s^2 + 6$$

defines the s^2 row of the Routh array. Now, if we divide $R_1(s)$ into $D(s)$

$$\begin{array}{r@{}l} & \frac{9}{11}s \\ \frac{11}{3}s^2 + 6s \big) & \overline{3s^3 + 2s} \\ & 3s^3 + \frac{54}{11}s \\ \hline & \quad -\frac{32}{11}s \end{array}$$

The remainder

$$R_2(s) = -\tfrac{32}{11}s$$

defines the s^1 row of the Routh array. Finally, the s^0 row of the Routh array is the remainder when $R_2(s)$ is divided into $R_1(s)$.

$$\begin{array}{r@{}l} & \frac{-121}{96}s \\ -\frac{32}{11}s \big) & \overline{\frac{11}{3}s^2 + 6} \\ & \frac{11}{3}s^2 \\ \hline & \qquad 6 \end{array}$$

Notice that if a remainder is zero, an entire row of the Routh array is zero. If, for example, $R_2(s)$ had been zero, then $R_1(s)$ would have been a factor of $p(s)$. That fact will be exploited when we consider a procedure for handling a row of zeros.

As another example, a system with transfer function

$$T(s) = \frac{5s^2 - 7s + 2}{s^4 + 2s^3 + 3s^2 + 4s + 1}$$

has characteristic polynomial

$$s^4 + 2s^3 + 3s^2 + 4s + 1$$

which has the Routh array of Equation (2.15). There are no algebraic sign changes in the left column, so the polynomial has no RHP roots.

$$\begin{array}{c|ccc} s^4 & 1 & 3 & \textcircled{1} \\ s^3 & 2 & 4 & \\ s^2 & 1 & \textcircled{1} & \\ s^1 & 2 & & \\ s^0 & \textcircled{1} & & \end{array} \qquad [2.15]$$

Each array has properties that serve as a partial check on its correct completion. The number of nonzero row entries is normally reduced by one every two rows, with just one nonzero element in the s^1 row and in the s^0 row. And the last coefficient of the polynomial appears periodically as the last nonzero entry in every other row, as shown in the array of Equation (2.15).

2.5.4 Left-Column Zeros

Sometimes the polynomial coefficients are such that a zero occurs in the left column of the array, so that the array cannot be completed. The situation of having a zero at the left of a row, although the entire row does not consist of zeros, is termed a *left-column zero*. For example, the polynomial

$$p(s) = 3s^4 + 6s^3 + 2s^2 + 4s + 5 \qquad [2.16]$$

has an array that begins as follows:

$$\begin{array}{c|ccc} s^4 & 3 & 2 & 5 \\ s^3 & 6 & 4 & \\ s^2 & 0 & 5 & \\ s^1 & & & \\ s^0 & & & \end{array}$$

The array cannot be completed in the usual way, because of the necessity to divide by zero. This zero in the first column can be eliminated by extending the array. The row where the first column 0 occurs is renamed the A row, and a new row, termed the B row, is formed from the A row. The B row is formed by sliding the A row to the left until the zeros disappear. The sign of the row is changed by $(-1)^n$ where n is the number of times this row is shifted to the left to eliminate the zeros. In the example, the A and B rows are as follows because $n = 1$,

Row manipulations solve the left-column zero problem.

$$\begin{array}{c|cc} A & 0 & 5 \\ B & -5 & 0 \end{array} \qquad [2.17]$$

The new s^2 row is formed by adding the A and B rows together, with the result

$$\begin{array}{c|cc} s^2 & -5 & 5 \end{array} \qquad [2.18]$$

After the remaining terms are calculated as before, the entire array becomes

$$\begin{array}{c|ccc} s^4 & 3 & 2 & 5 \\ s^3 & 6 & 4 & \\ s^2 & -5 & 5 & \\ s^1 & +10 & & \\ s^0 & 5 & & \end{array} \qquad [2.19]$$

From the array of Equation (2.19), there are two sign changes— +6 to −5 and −5 to +10—and hence there are two roots in the RHP.

Some years ago, another procedure was used when a left column zero was encountered, the so-called epsilon method. The new "row-shifting" method was found to work when the epsilon method would not work, so the epsilon method is not presented here. The article by Benidir and Picinbond cited at the end of this chapter describes both the epsilon method and the row-shifting method.

The epsilon method also solves the left-column zero problem.

Another example is presented, with the following polynomial:

$$p(s) = s^5 + s^4 + 2s^3 + 3s^2 + s + 4 \qquad [2.20]$$

The array is formed in the normal fashion as follows:

$$\begin{array}{c|ccc} s^5 & 1 & 2 & 1 \\ s^4 & 1 & 3 & 4 \\ s^3 & -1 & -3 & \\ s^2 & 0 & 4 & \end{array} \qquad [2.21]$$

The s^2 row now becomes the A row and the B row is formed by sliding the A row to the left one column and multiplying by –1, with the result

$$\begin{array}{c|cc} A & 0 & 4 \\ B & -4 & 0 \end{array} \qquad [2.22]$$

These are added to form the s^2 row and the array is completed in the usual fashion as follows:

$$\begin{array}{c|ccc} s^5 & +1 & 2 & 1 \\ s^4 & +1 & 3 & 4 \\ s^3 & -1 & -3 & \\ s^2 & -4 & 4 & \\ s^1 & -4 & 0 & \\ s^0 & +4 & & \end{array} \qquad [2.23]$$

With two sign changes, from $+1$ to -1 and from -4 to $+4$, the polynomial of Equation (2.20) has two RHP roots.

2.5.5 Row of Zeros

An entire row of zeros occurs in the Routh array, when $p(s)$ contains an even polynomial as a factor, so, no remainder polynomial occurs in an odd-ordered row. For example, the test of the polynomial

$$p(s) = s^5 + 2s^4 + 8s^3 + 11s^2 + 16s + 12 \qquad [2.24]$$

has a row of zeros at s^1 row:

$$\begin{array}{c|ccc} s^5 & 1 & 8 & 16 \\ s^4 & 2 & 11 & 12 \\ s^3 & \frac{5}{2} & 10 & \\ s^2 & 3 & 12 & \\ s^1 & 0 & 0 & \\ s^0 & & & \end{array} \qquad [2.25]$$

Significance of a row of zeros.

Since the row of zeros indicates an even polynomial divisor of the original polynomial, the coefficients of that polynomial are those given in the row above the row of zeros:

$$p_{\text{divisor}}(s) = 3s^2 + 12 = 3(s^2 + 4) = 3(s + 2j)(s - 2j)$$

The tested polynomial thus has two imaginary axis roots at $s = \pm j2$ and can be divided as follows:

$$\begin{array}{r|l}
 & s^3 + 2s^2 + 4s + 3 \\
\hline
s^2 + 4 & s^5 + 2s^4 + 8s^3 + 11s^2 + 16s + 12 \\
 & s^5 \qquad\quad + 4s^3 \\
\hline
 & \qquad 2s^4 + 4s^3 + 11s^2 + 16s + 12 \\
 & \qquad 2s^4 \qquad + 8s^2 \\
\hline
 & \qquad\qquad 4s^3 + 3s^2 + 16s + 12 \\
 & \qquad\qquad 4s^3 \qquad + 16s \\
\hline
 & \qquad\qquad\qquad 3s^2 \qquad + 12 \\
 & \qquad\qquad\qquad 3s^2 \qquad + 12 \\
\hline
 & \qquad\qquad\qquad\qquad 0
\end{array}$$

To complete the array, the row of zeros is replaced by the coefficients of the derivative of the divisor polynomial:

$$\frac{dp_{\text{divisor}}(s)}{ds} = 6s$$

$$\begin{array}{c|ccc}
s^5 & 1 & 8 & 16 \\
s^4 & 2 & 11 & 12 \\
s^3 & \frac{5}{2} & 10 & \\
\hdashline
s^2 & 3 & 12 & \\
s^1 & 6 & & \\
s^0 & 12 & &
\end{array}$$

We test the $(n = 2)$-order divisor by looking for any left-column sign changes below the dashed line. There are none, and there are two complex conjugate roots located at $s = \pm j2$ and no RHP or LHP roots. The remaining three roots are tested by using the three left-column signs, starting with the coefficient "1" and ending with the coefficient "3". Notice that the coefficient "3" is used twice: once to initiate a test of $E(s)$ and again to complete the test for the remaining roots. There are no sign changes, so the remaining roots must be in the LHP.

Test below the dashed line separately from the top of the array.

Even and odd polynomials have root locations that are symmetric about the imaginary axis. An odd polynomial always has s as a factor. Whether or not s is a factor is evident from a glance at the polynomial, and the remainder is an even polynomial. For any even polynomial, replacing s with $(-s)$ leaves the polynomial unchanged. Thus, besides occurring in conjugate pairs, the roots of an even polynomial are also symmetrical about the imaginary axis. There are three basic types of factor possible in an even polynomial. One type is

$$(s + ja)(s - ja) = (s^2 + a^2)$$

which consists of complex conjugate roots on the imaginary axis, as indicated in Figure 2.11(a). Another is

$$(s + a)(s - a) = (s^2 - a^2)$$

which are symmetrical roots on the real axis, one in the LHP and one in the RHP, as in Figure 2.11(b). the third type of factor, Figure 2.11(c), involves complex roots in quadrature, one pair in the LHP and one pair in the RHP:

$$(s + a + jb)\,(s + a - jb)\,(s - a + jb)\,(s - a - jb)$$
$$= s^4 + 2(b^2 - a^2)s^2 + (a^2 + b^2)^2$$

The additional symmetry of even polynomial roots about the imaginary axis allows determination of the number of imaginary axis roots. Each RHP root of an even polynomial must be matched by just one corresponding LHP root. Thus if an even polynomial is of sixth order and is known to have just just one RHP root, it has just one LHP root, and the remaining four roots must be on the imaginary axis. If an eighth-order polynomial has three RHP roots, it must have three LHP roots and two imaginary axis roots. For example, for the polynomial

$$s^6 + s^5 + 5s^4 + s^3 + 2s^2 - 2s - 8 \qquad \textbf{[2.26]}$$

the coefficient tests indicate the presence of at least one RHP root. The Routh–Hurwitz test begins as follows:

The row of zeros means there is an even divisor.

s^6	1	5	2	−8
s^5	1	1	−2	
s^4	4	4	−8	
s^3	0	0		

So

$$4s^4 + 4s^2 - 8$$

is a divisor of the original polynomial. Replacing the row of zeros by the coefficients of the derivative of the divisor polynomial, we write

$$\frac{d}{ds}\left(4s^4 + 4s^2 - 8\right) = 16s^3 + 8s$$

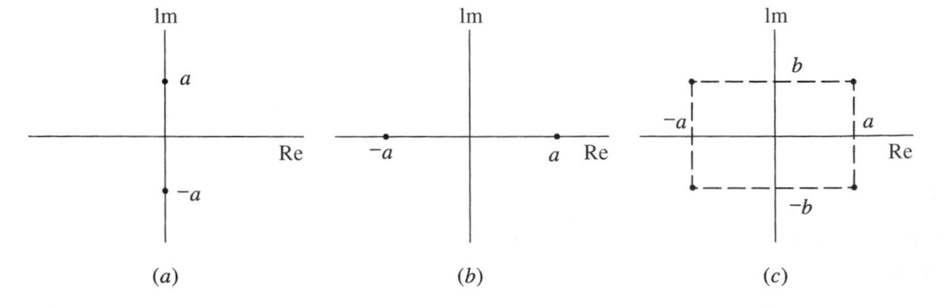

Figure 2.11 Even polynomial root locations.

The completed array is as follows:

s^6	1	5	2	−8
s^5	1	1	−2	
s^4	4	4	−8	
s^3	16	8		
s^2	2	−8		
s^1	72			
s^0	−8			

[2.27]

The test of the even polynomial divisor is that portion of the array below the dashed line.

Test below the dashed line separated from the top of the array.

Below the dashed line for the ($n = 4$)-order even polynomial, there is one root in the RHP and one root in the LHP. There are two roots on the imaginary axis. The remaining two roots are tested by using the top two sign changes determined by the top three rows (a maximum of two possible sign changes). Since there are no such sign changes, the remaining two roots are in the LHP, and we have the following numbers of the roots of various types:

RHP = 1

LHP = 3

IA = 2

Another array with an all-zero row is the following:

s^4	1	9	20
s^3	6	24	
s^2	5	20	
s^1	0	0	
s^0			

The s^1 is an all-zero row, so the system has a divisor found from the s^2 row as follows:

$$5s^2 + 20 = 0 = 5(s^2 + 4) \quad \textbf{[2.28]}$$

From Equation (2.28), the roots on the j axis are at $\pm j2$. Replacing the row of zeros by the coefficients of the derivative of the divisor polynomial, we obtain $d/ds(5s^2 + 20) = 10s$. The array is continued with the result

s^4	1	9	20
s^3	6	24	
s^2	5	20	
s^1	10		
s^0	20		

For the ($n = 2$) order even divisor, there are no sign changes below the dashed line, so no RHP roots. Two roots are on the imaginary axis located at $\pm j2$, hence there are no LHP roots. Since there are no sign changes above the dashed line, there must be two roots in the LHP. The roots are as follows:

$$\text{RHP} = 0$$
$$\text{LHP} = 2$$
$$\text{IA} = 2$$

The polynomial

$$s^5 + s^4 + 6s^3 + 6s^2 + 25s + 25$$

has an all-zero row at the s^3 row. The system is continued by differentiating the s^4 row with the result:

s^5	1	6	25
s^4	1	6	25
s^3	4	12	
s^2	3	25	
s^1	$-\frac{64}{3}$		
s^0	25		

the fourth-order even polynomial divisor has two RHP roots, and so its remaining two roots must be in the LHP. There can thus be no imaginary axis roots. The entire polynomial has two RHP roots, and the remaining three roots must be in the LHP.

2.5.6 Eliminating a Possible Odd Divisor

Could an odd divisor appear in a Routh array? That is, could a row of zeros occur in an even-ordered row, causing an odd-ordered row above it to contain a divisor of $p(s)$? If there is an odd divisor of $p(s)$, then s must be divisible into $p(s)$ so that $p(s)$ would lack an a_0 coefficient.

When the a_0 coefficient is zero, we need only factor out an appropriate number of s values and test what remains.

Suppose we want to test

$$p(s) = s^5 + 3s^4 + 2s^3 + 4s$$
$$= sp_1(s) = s[s^4 + 3s^3 + 2s^2 + 4]$$

Instead, test

$$p_1(s) = [s^4 + 3s^3 + 2s^2 + 4]$$

and remember that one additional root of $p(s)$ exists at $s = 0$. Once it has been agreed to remove any s factors, a row of zeros can happen only in an odd-ordered row of a Routh array, revealing an even divisor of $p(s)$ above that row of zeros.

2.5.7 Multiple Roots

Assuming that we have acted to exclude odd divisors, if multiple roots are present, the even divisor must, itself, have an even divisor, meaning that the Routh array must reveal a row of zeros twice. When there is one row of zeros, at best the system is marginally stable if some roots are in the IA and the rest are in the LHP. When multiple roots are present, either RHP roots are multiple or IA roots are multiple. In any case, if a row of zeros occurs more than once, the system sill be unstable.

Consider the polynomial

$$p(s) = s^6 + 5s^5 + 14s^4 + 40s^3 + 64s^2 + 80s + 96$$

We begin by forming the Routh array and proceed with row evaluations. Notice that all polynomial coefficients are positive; but no stability implications should be made.

s^6	1	14	64	96
s^5	5	40	80	
s^4	6	48	96	
s^3	0	0		

Since the s^3 row contains all zeros for the remainder polynomial, the system has an even divisor.

$$E_1(s) = 6s^4 + 48s^2 + 96 = 6[s^4 + 8s + 16]$$

To complete part of the rest of the array, replace the zeros with the coefficients of

$$\frac{dE_1(s)}{ds} = 24s^3 + 96$$

s^6	1	14	64	96
s^5	5	40	80	
s^4	6	48	96	
s^3	24	96		
s^2	24	96		
s^1	0			

Now, $E_1(s)$ must have a divisor given by

$$E_2(s) = 24s^2 + 96 = 24[s^2 + 4]$$

Since

$$\frac{dE_2(s)}{ds} = 48s$$

the final array is

s^6	1	14	64	96
s^5	5	40	80	
s^4	6	48	96	
s^3	24	96		
s^2	24	96		
s^1	48			
s^0	96			

To properly interpret the array, start with the test of the second-order divisor $E_2(S)$: that is, by testing the bottom three rows. Since there are no left-column sign changes down the last three rows, there are no LHP roots and no RHP roots. The two roots must thus be on the IA.

Next, test the fourth-order $E_1(s)$. For the s^4 row downward, there are no left-column sign changes, so there are no LHP roots, no RHP roots, and four IA roots (which comprise pairs of double conjugate IA roots). The remaining two roots are determined by examining the two potential sign changes going down the left column of the first three rows. Since there are no RHP roots among the last two, they must be in the LHP.

We conclude that there are two LHP roots and a pair of double conjugate IA roots. In fact, the roots are $-2, -3, j2, j2, -j2, -j2$. The system is unstable because of the repeated IA roots.

❑ DRILL PROBLEMS

D2.9 What can be determined about the roots of the following polynomials from the coefficient tests?

(a)

$$-3s^4 + 2s^3 + s + 10$$

Ans. at least one RHP root

(b)

$$4s^4 + 3s^3 + 10s^2 + 8s + 1$$

Ans. nothing

(c)

$$s^5 + 4s^3 + 8$$

Ans. imaginary axis (IA) or RHP roots or both

(d)

$$s^6 + 6s^4 + 3s^2 + 10$$

Ans. IA or RHP roots or both

D2.10 How many roots of each of the following polynomials are in the right half of the complex plane?

(a)

$$s^3 + 2s^2 + 3s + 4$$

Ans. 0

(b)

$$s^4 - 6s^3 + 7s^2 + 2s + 4$$

Ans. 2

(c)

$$0.3s^4 + 1.1s^3 + 0.7s^2 + s + 2.1$$

Ans. 2

(d)

$$s^5 + s^4 + 2s^3 + 3s^2 + \tfrac{1}{2}$$

Ans. 4

(e)

$$2s^5 + s^4 + 2s^3 + 4s^2 + s + 6$$

Ans. 2

D2.11 The Routh–Hurwitz tests for the following polynomials might involve left-column zeros. For each polynomial, use the array to find the number of roots in the right half of the complex plane.

(a)

$$s^3 + 2s + 3$$

Ans. 2

(b)

$$3s^4 + 6s^3 + 2s^2 + 4s + 5$$

Ans. 2

(c)

$$2s^4 + 2s^3 + s^2 + s - 3$$

Ans. 1

(d)

$$s^5 + s^4 + 3s^3 + 2s^2 + 4s + 2$$

Ans. 2

D2.12 The Routh–Hurwitz tests for the following polynomials might involve an all-zero row in the arrays. For each polynomial, complete the array and determine the number of roots in the right half of the complex plane.

(a)

$$s^4 + 8s^2 - 7$$

Ans. 1

(b)

$$s^4 + 2s^3 + 9s^2 + 4s + 14$$

Ans. 0

(c)

$$s^5 + s^3 + 2s$$

Ans. 2

(d)

$$s^5 + 3s^4 + 4s^3 + 7s^2 + 4s + 2$$

Ans. 0

D2.13 For each of the following polynomials, how many roots are in the LHP, how many are in the RHP, and how many are on the imaginary axis?

(a)

$$s^4 + 3s^2 + 4$$

Ans. 2 RHP, 2 LHP

(b)

$$s^4 + 2s^3 + 5s^2 - 4s - 14$$

Ans. 1 RHP, 3 LHP

(c)

$$s^5 + 2s^4 + 3s^3 + 6s^2 + 2s + 4$$

Ans. 1 LHP, 4 IA

(d)

$$3s^5 + 2s^3 + s$$

Ans. 2 RHP, 2 LHP, 1 IA

(e)

$$2s^5 + 4s^4 + s^3 + 2s^2 + 3s + 6$$

Ans. 3 LHP, 2 RHP

D2.14 Are the systems of Figure D2.14 stable?

Ans. (a) no; (b) yes; (c) yes

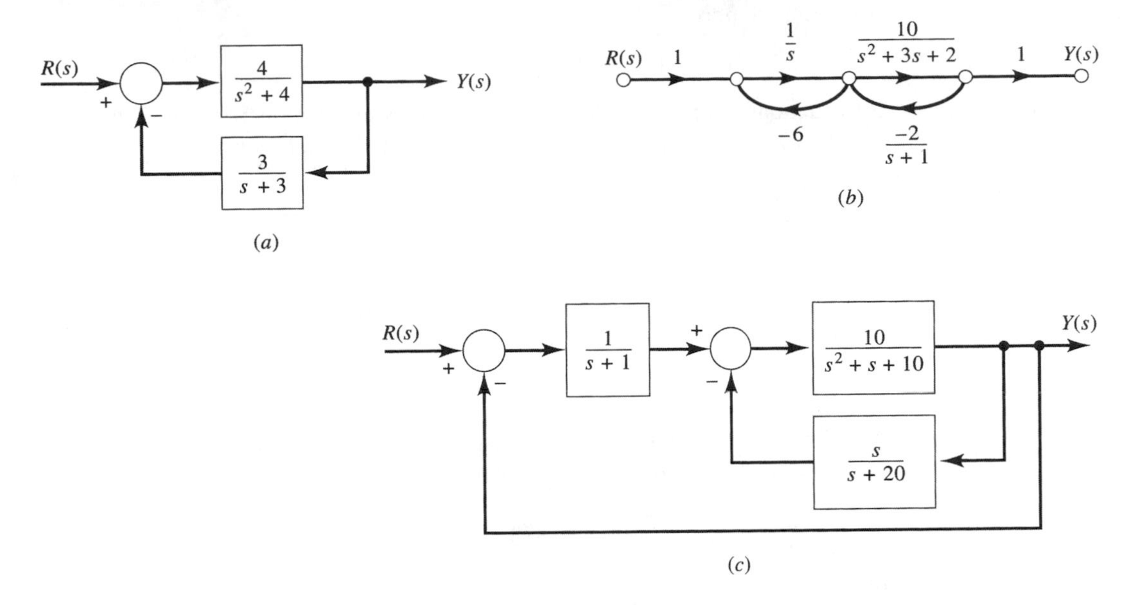

Figure D2.14

2.6 Parameter Shifting

2.6.1 Adjustable Systems

It is often desired to know for what range or ranges of an adjustable parameter K a system is stable. For example, suppose the transfer function of a system, in terms of K, is

$$T(s) = \frac{4}{s^2 + 2s + K}$$

This system is stable for all positive values of K because the roots of a quadratic are in the LHP if and only if all coefficients have the same algebraic sign. For $K = 0$,

$$T(s) = \frac{4}{s(s+2)}$$

and the system is marginally stable because of the imaginary axis characteristic root at $s = 0$. For negative values of K, the transfer function has a characteristic root in the RHP and the system is unstable. The system with transfer function

$$T(s) = \frac{2s + K}{s^2 + (2 + K)s + 4}$$

is, similarly, stable for $K > -2$.

For systems with characteristic polynomials of higher order, the Routh–Hurwitz test is a useful tool for determining stability in terms of a constant but adjustable parameter. Suppose

$$s^4 + 2s^3 + 4s^2 + 2s + K$$

Select K for no left-column sign changes.

is the denominator polynomial of a system transfer function, in terms of an adjustable constant K. The Routh–Hurwitz test in terms of K is as follows:

$$\begin{array}{c|ccc} s^4 & 1 & 4 & K \\ s^3 & 2 & 2 & \\ s^2 & 3 & K & \\ s^1 & \dfrac{6-2K}{3} & & \\ s^0 & K & & \end{array}$$

All of the left-column array entries must be of the same algebraic sign if the polynomial is to have no RHP roots. Thus, for system stability,

$$\frac{6-2K}{3} > 0 \quad \text{and} \quad K > 0$$

Range of values of K for stability.

or

$$0 < K < 3$$

For $K = 3$, the array contains a row of zeros, $3s^2 + 3$ is a factor of the polynomial, and the system is marginally stable. For $K = 0$, s is a factor of the polynomial and, again, the system is marginally stable.

For the adjustable polynomial

$$s^4 + 2s^3 + 4s^2 + Ks + 6$$

the Routh–Hurwitz array is the following:

$$\begin{array}{c|ccc} s^4 & 1 & 4 & 6 \\ s^3 & 2 & K & \\ s^2 & 4 - \dfrac{K}{2} & 6 & \\ s^1 & K - \dfrac{12}{4 - K/2} & & \\ s^0 & 6 & & \end{array}$$

If the polynomial is to have all LHP roots,

$$4 - \frac{K}{2} > 0, \quad K < 8, \quad \text{and} \quad K - \frac{12}{4 - K/2} > 0$$

Making use of the requirement that $4 - K/2$ be positive, from the first inequality, the second inequality can be multiplied by the positive number $4 - K/2$. If the inequality were multiplied by a negative number, its sense would be reversed. Then

$$\left(4 - \frac{K}{2}\right)\left(K - \frac{12}{4 - K/2}\right) = -\frac{K^2}{2} + 4K - 12 > 0$$

The quadratic function

$$-\frac{K^2}{2} + 4K - 12$$

is negative for large negative K and is negative for large positive K. To determine whether there are intermediate values of K for which the function is positive, we use the quadratic formula to find the roots of K:

$$K = \frac{-4 \pm \sqrt{16 - 24}}{2\left(-\frac{1}{2}\right)}$$

Since the function's roots are complex, it is concluded that the inequalities cannot be satisfied for any (real) K. The original polynomial thus has RHP roots for all K.

There is no range of K for which the system is stable.

As an interesting example of this technique, consider again the positioning system for a video satellite antenna shown in Figure 1.48. The system is changed slightly, in that a time constant is added to the voltage amplifier. The transmittance of the amplifier is now

$$\frac{A_1}{\tau_a s + 1}$$

The signal flow diagram is shown in Figure 2.12. Mason's gain rule is used to find the transfer function.

$$T = \frac{\theta}{R} = \frac{K}{\tau_a \tau_L s^3 + (\tau_a + \tau_L)\, s^2 + s + K} \qquad \textbf{[2.29]}$$

where

$K = (K_p N_1 K_m A_1 A_2)/N_2$

$K_p =$ potentiometer transmittance

$N_1/N_2 =$ step-down motor gear ratio

$K_m =$ motor constant $= 1/K_v$

$A_1 A_2 =$ product of amplifier gains

$\tau_a =$ voltage amplifier time constant

$\tau_L =$ motor time constant including gear train

$\quad = R_A K_m / K_T [J_m + (N_1/N_2)^2 J_L]$

When the amplifier time constant τ_a is zero, the transfer function of Equation (2.29) is governed by a second-order characteristic equation with positive constant

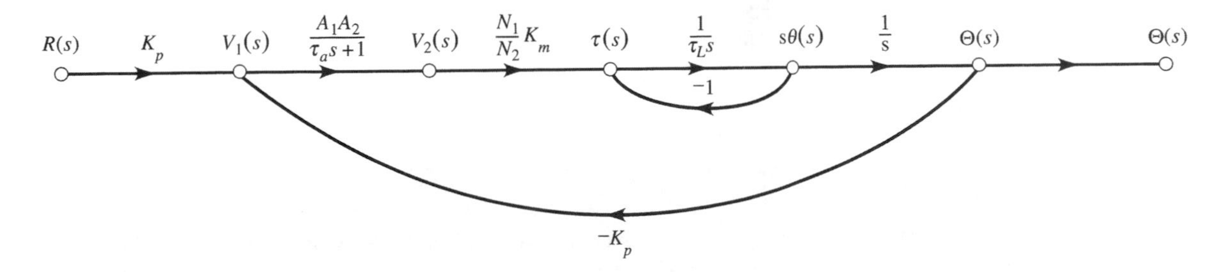

Figure 2.12 Signal flow graph for satellite antenna positioning system.

coefficients. However, with the addition of a variable τ_a the Routh–Hurwitz method is used to determine the effect on stability as τ_a varies. The array is established for the denominator of Equation (2.29):

$$\begin{array}{c|ll} s^3 & \tau_a \tau_L & 1 \\ s^2 & \tau_a + \tau_L & K \\ s^1 & (\tau_a + \tau_L - \tau_a \tau_L K)/(\tau_a + \tau_L) & 0 \\ s^0 & K & \end{array}$$

For marginal stability, the s^1 row must be all zeros, or

$$\frac{\tau_a + \tau_L - \tau_a \tau_L K}{\tau_a + \tau_L} = 0$$

which is rearranged in the form

$$K = \frac{1}{\tau_a} + \frac{1}{\tau_L} \qquad [2.30]$$

Equation (2.30) shows the effect of τ_a upon the system stability: the larger τ_a, the smaller K for a given degree of stability. As τ_a approaches zero, K approaches infinity, which verifies the second-order results with no τ_a.

❑ DRILL PROBLEMS

D2.15 For what range(s), if any, for the adjustable constant K are all roots of the following polynomials in the left half of the complex plane?

(a)

$$s^3 + (2 + K)s^2 + (8 + K)s + 6$$

Ans. $-1.12 < K$

(b)

$$2s^3 + (6 - 2K)s^2 + (4 + 3K)s + 10$$

Ans. $-\frac{1}{3} < K < 2$

(c)

$$s^4 + (10 + k)s^3 + 9s + 11$$

Ans. No value of K

(d)

$$s^4 + s^3 + 3s^2 + 2s + 4 + K$$

Ans. $-4 < K < -2$

D2.16 Find ranges of positive, constant K, if any, for which the systems shown in Figure D2.16 are stable.

Ans. (a) $0 < K < 32/9$; (b) $4/3 < K$

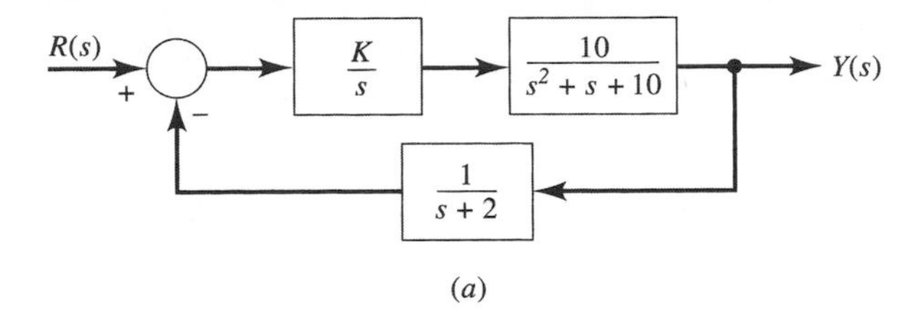

(a)

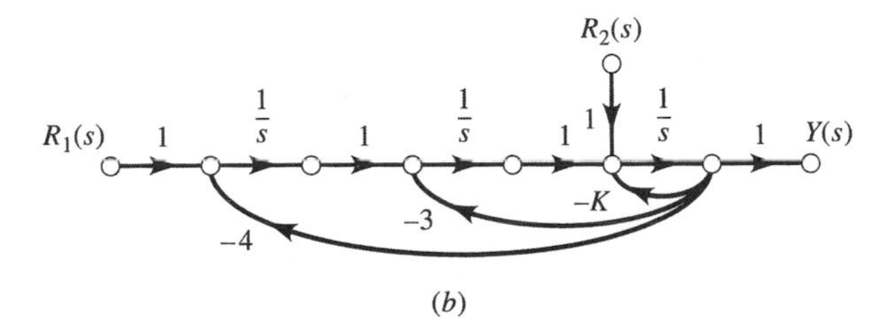

(b)

Figure D2.16

2.6.2 Kharitonov's Theorem

In the preceding sections, the Routh–Hurwitz method was used to determine whether any roots of the characteristic equation lie in the RHP. In recent years, however, much attention has been given to "robustness" analysis. Robustness analysis arises because the system coefficients of the characteristic equation are often uncertain, or are changing as a function of time. The characteristic polynomial is written as follows:

$$f(s,q) = q_0 + q_1 s + q_2 s^2 + \cdots + q_n s^n \qquad \textbf{[2.31]}$$

when q sets forth the uncertainty or the time variation of the coefficients of the polynomial of Equation (2.31). The uncertain polynomial of Equation (2.31) must still be tested to be certain that all the roots of $f(s,q)$ lie in the left half-plane for any variation of q.

A variation of the Routh–Hurwitz method, termed the *Kharitonov theorem*, is used for $f(s,q)$. References at the end of this chapter more fully describe this theorem, which enables the engineer to determine the stability of an uncertain system by defining four bounding polynomials and applying the Routh–Hurwitz test to each of them.

To obtain those four bounding polynomials, designate the lowest value of a coefficient with a superscript "−" and the highest value of a coefficient with a superscript "+." The four polynomials possess patterns of "−" and "+" that repeat every four coefficients. Suppose we have a third-order system.

$$p(s) = q_0 + q_1 s + q_2 s^2 + q_3 s^3$$

Suppose the ranges are

$$q_0 \varepsilon\, [0.5,\ 1.5] = \left[q_0^-,\, q_0^+\right]$$

$$q_1 \varepsilon\, [1.5,\ 2.5] = \left[q_1^-,\, q_1^+\right]$$

$$q_2 \varepsilon\, [2.5,\ 3.5] = [q_2^-, q_2^+]$$
$$q_3 \varepsilon\, [3.5,\ 4.5] = [q_3^-,\ q_3^+]$$

According to Kharitonov,

$$p_1(s) = q_0^+ + q_1^+ s + q_2^- s^2 + q_3^- s^3 + \cdots$$
$$p_2(s) = q_0^- + q_1^- s + q_2^+ s^2 + q_3^+ s^3 + \cdots$$
$$p_3(s) = q_0^+ + q_1^- s + q_2^- s^2 + q_3^+ s^3 + \cdots$$
$$p_4(s) = q_0^- + q_1^+ s + q_2^+ s^2 + q_3^- s^3 + \cdots$$

Using the ranges, we find

$$p_1(s) = 1.5 + 2.5s + 2.5s^2 + 3.5s^3$$
$$p_2(s) = 0.5 + 1.5s + 3.5s^2 + 4.5s^3$$
$$p_3(s) = 1.5 + 1.5s + 2.5s^2 + 4.5s^3$$
$$p_4(s) = 0.5 + 2.5s + 3.5s^2 + 3.5s^3$$

It is left as an exercise to test these four polynomials by using the Routh array. You will find that $p_1(s)$, $p_2(s)$, and $p_4(s)$ have all three roots in the LHP but that $p_3(s)$ has two RHP roots. This indicates that the system does not remain stable, given ranges for the coefficients. Notice how only four polynomials have to be tested even though an infinite number of polynomials exist for the ranges given.

❑ DRILL PROBLEMS

D2.17 A system has the third-order characteristic polynomial

$$p(s) = q_0 + q_1 s + q_2 s^2 + q_3 s^3$$

The coefficients are known to lie in the ranges

$$q_0 \varepsilon [2,\ 3] \qquad q_1 \varepsilon [3,\ 4]$$
$$q_2 \varepsilon [4,\ 5] \qquad q_5 \varepsilon [5,\ 6]$$

D2.18 Use Kharitonov's theorem to find the four polynomials, and determine the number of RHP roots of each. Is the system stable, given the range of values for the coefficients?

Ans. $p_1(s) = 3 + 4s + 4s^2 + 5s^3$ (0 RHP roots)
$p_2(s) = 2 + 3s + 5s^2 + 6s^3$ (0 RHP roots)
$p_3(s) = 3 + 3s + 4s^2 + 6s^3$ (2 RHP roots)
$p_4(s) = 2 + 4s + 5s^2 + 5s^3$ (0 RHP roots)

System is *not* stable, given the ranges.

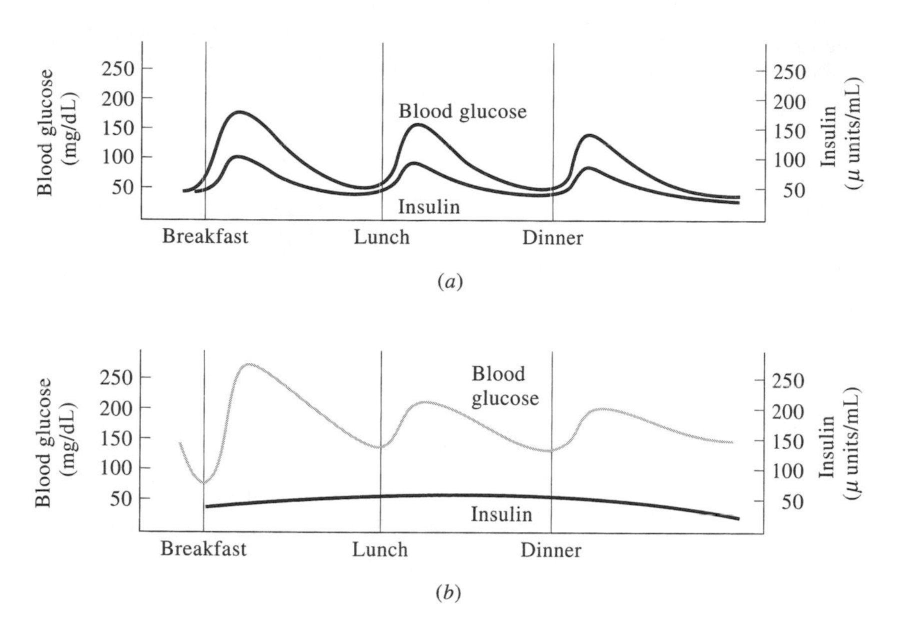

Figure 2.13 Typical blood sugar and insulin concentrations. (a) Normal person. (b) Diabetic with one daily insulin injection.

2.7 An Insulin Delivery System

Control system methods have been applied to the biomedical field to create an implantable insulin delivery system for diabetics. When food is eaten and digested, sugars, mainly glucose, are absorbed into the bloodstream. Normally, the pancreas secretes insulin into the bloodstream to metabolize the sugar. The pancreas of a diabetic person secretes insufficient insulin to metabolize blood sugar; blood sugar levels can thus become high enough to threaten damage to the body's organs.

One solution to this problem is for the diabetic to take one injection of insulin each day. In Figure 2.13(a), typical blood sugar and insulin concentration histories for a day are shown for a normal person. In Figure 2.13(b), the blood sugar and insulin concentration histories for a day are shown for a diabetic who takes one insulin injection in the morning. Notice that blood sugar is often higher than normal, but the sugar concentration is far less than would be the case if no insulin had been injected. A higher dose of insulin could be taken in the morning to counteract the low insulin residual after dinner, but blood sugar concentration might be driven unacceptably low in the morning (hypoglycemia), causing weakness, trembling, and possibly fainting. Three injections a day, one before each meal, is generally not a feasible regimen because of the damage to the veins and skin tissue.

One automatic control system of interest consists of a tiny insulin reservoir, control motor, and a pump that is implanted in the body below the diaphragm. This electronic pancreas delivers insulin into the peritoneum, using preprogrammed

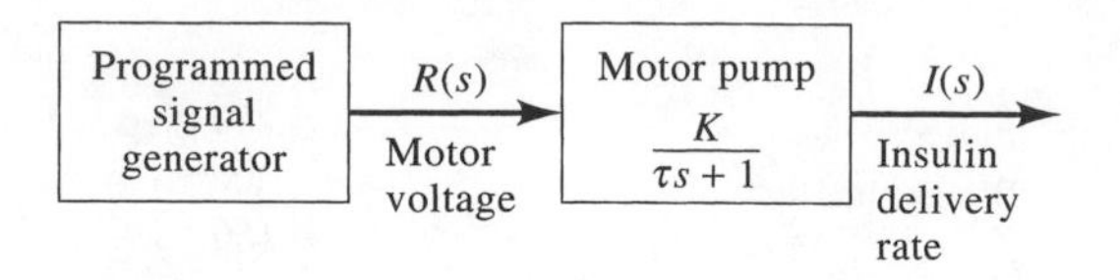

Figure 2.14 The implanted open-loop insulin delivery system.

commands intended to establish insulin levels close to the levels of a normal individual. The pump runs at higher rates after meals than otherwise. The patient must time meals to complement the behavior of the implanted system, but injections are required only every few weeks, to refill the insulin reservoir. Systems of this type have operated for many years without malfunction.

The methods of this chapter can be used to design such an insulin delivery system, which is the open-loop one shown in Figure 2.14. A signal generator is programmed to drive the motor pump in such a way that the insulin delivery rate $i(t)$ approximates a desired delivery rate $i_D(t)$.

Figure 2.15(a) shows an approximate desired insulin rate $I_D(t)$ in cm³/sec for one-third day, beginning with a meal. Figure 2.15(b) shows a similar function,

$$i(t) = Ate^{-at}u(t)$$

which has the particularly simple Laplace transform

$$I(s) = \mathcal{L}[i(t)] = \frac{A}{(s+a)^2}$$

A good approximation of $i_D(t)$ by $i(t)$ occurs when the constants A and a are selected so that $i(t)$ is maximum at $t = 3600$, as is $i_D(t)$, and so that the areas under the two curves are equal, with value 0.17 cm³:

$$\frac{di}{dt} = -aAte^{-at} + Ae^{-at} = A(1-at)e^{-at}$$

For the maximum of $i(t)$ to occur at $t = 3600$,

$$\left.\frac{di}{dt}\right|_{t=3600} = A(1-3600a)e^{-3600a} = 0 \qquad a = \frac{1}{3600} = 2.78 \times 10^{-4}$$

The area under the $i(t)$ curve after $t = 0$ is

$$\int_0^{\infty} Ate^{-at}\,dt = A\left[-\frac{1}{a}te^{-at} - \frac{1}{a^2}e^{-at}\right]_0^{\infty} = \frac{A}{a^2}$$

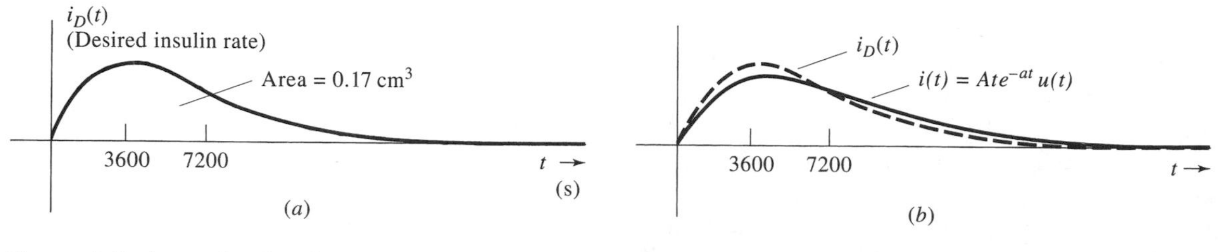

Figure 2.15 Approximating the required insulin rate with a time-weighted exponential. (a) Insulin rate required for a diabetic. (b) The function $i(t) = Ate^{-at}u(t)$.

Equating to the desired area of $0.17\ \text{cm}^3$ gives

$$\frac{A}{a^2} = (3600)^2 A = 0.17 \qquad A = 1.31 \times 10^{-8}$$

If $i(t)$ is to be produced by the system input $r(t)$, then

$$I(s) = \frac{A}{(s+a)^2} = \frac{K}{\tau s + 1} R(s)$$

giving the required programmed signal for each mealtime:

$$R(s) = \frac{A(\tau s + 1)/K}{(s+a)^2}$$

For a motor pump with

$$\tau = 5\ \text{s} \qquad K = 2.3 \times 10^{-6}\ \text{cm}^3/\text{volt-s}$$

and with the delivery rate for which

$$a = 2.78 \times 10^{-4}\ \text{s}^{-1} \qquad A = 1.31 \times 10^{-8}\ \text{cm}^3/\,\text{s}^2$$

then

$$R(s) = \frac{(1.31 \times 10^{-8})(5s+1)/2.3 \times 10^{-6}}{(s + 2.78 \times 10^{-4})^2}$$
$$= \frac{K_1}{(s + 2.78 \times 10^{-4})} + \frac{K_2}{(s + 2.78 \times 10^{-4})^2}$$

giving

$$K_1 = 28.5 \times 10^{-3} \qquad K_2 = 5.7 \times 10^{-3} - 2.78 \times 10^{-4} K_1$$
$$= 5.69 \times 10^{-3}$$

The programmed motor drive signal is thus to be

$$R(s) = \frac{28.5 \times 10^{-3}}{s + 2.78 \times 10^{-4}} + \frac{5.69 \times 10^{-3}}{(s + 2.78 \times 10^{-4})^2}$$

$$r(t) = [28.5 e^{-(2.78 \times 10^{-4})t} + 5.69 t e^{-(2.78 \times 10^{-4})t}] 10^{-3} u(t)\ \text{V}$$

which is sketched in Figure 2.16(a). Repetition, three times a day, of the motor drive signal will provide insulin delivery for periodic meals, as shown in Figure 2.16(b).

The insulin delivery system described in this section entered active use in 1981. A second-generation titanium insulin pump was in active service by the mid-1980's.

Programmed rate of infusion, amount of insulin in the reservoir, and battery charge can be read automatically and transmitted over the telephone to the patient's doctor. The doctor can reprogram the pump over the same telephone line. As before, the reservoir is refilled by injection through the skin and into the reservoir.

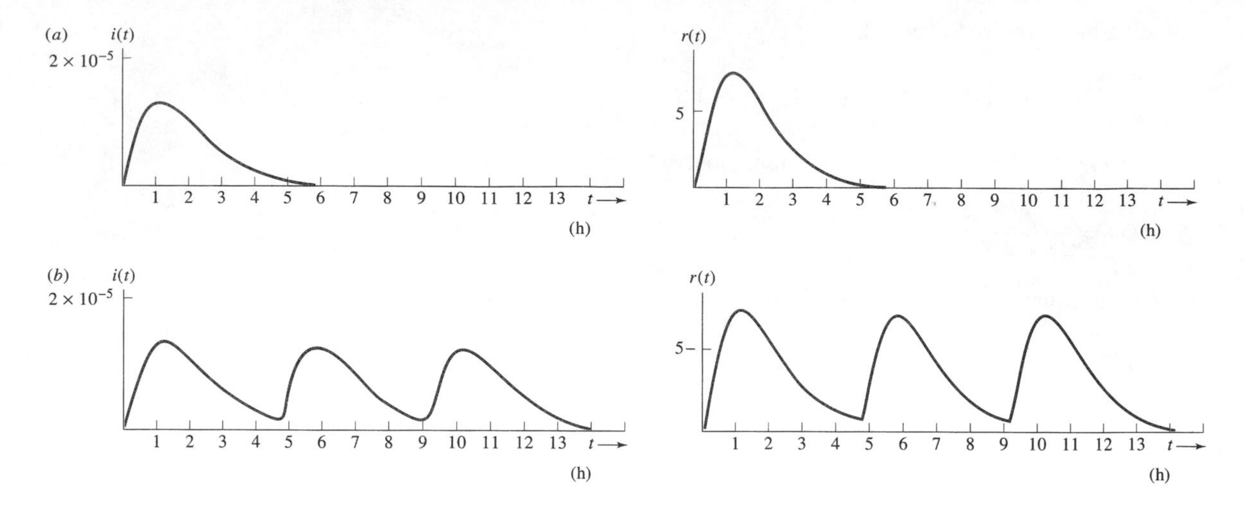

Figure 2.16 Response of an insulin delivery system. (a) Motor voltage and resulting insulin rate for one mealtime. (b) Repetitive motor voltage and insulin rate for three meals a day.

A third-generation insulin pump entered development after 1985. The insulin pump operates in a closed-loop mode so that direct measurement of the blood glucose level causes the pump to track the patient's insulin requirements. This third- generation system frees the patient from some aspects of dietary regimen.

2.8 Analysis of an Aircraft Wing

Physics students commonly perform an experiment in which a tuning fork is struck above an empty tube, resulting in a resonating tone. One can obtain the same effect by blowing into a bottle at just the right angle. This occurs because the blowing or the tuning fork is at a natural frequency of the hollow glass chamber, causing a reinforcement or resonating effect. This effect can be catastrophic in certain mechanical systems. For example, in 1939 a bridge over Puget Sound at the Tacoma Narrows began to sway and twist in a wind that whistled down the Narrows at just the right speed. In a mechanical sense the bridge was excited at a resonance, resulting in deflections that exceeded its structural limit and led to collapse. A similar effect must be considered in aircraft wing design. Structural failure of wings on certain turbine-driven jet aircraft was traced to a mechanical resonance excited by the jet turbine engines. The following example illustrates the problem of mechanical resonance for an aircraft wing.

Figure 2.17(a) depicts an aircraft wing with an applied force $f(t)$ and resulting deflection $x(t)$ at the wingtip. Aerodynamic and accelerative forces are ignored and we consider only a sinusoidal excitation of the wing caused by the turbine:

$$f(t) = D \cos \omega t$$

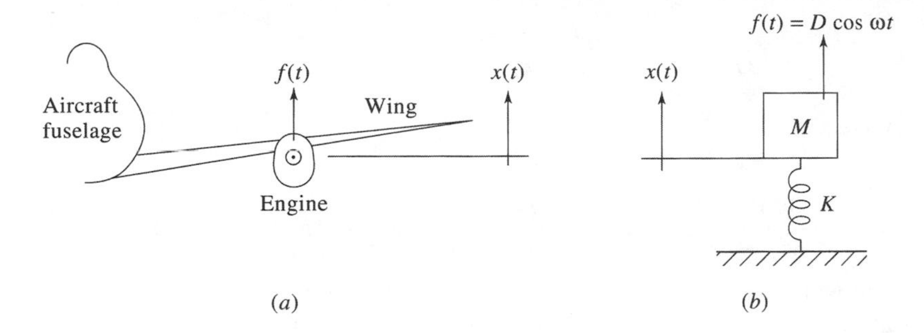

Figure 2.17 Deflection of an aircraft wing due to jet turbine vibrational force. (a) The aircraft wing. (b) Model of aircraft wing.

Figure 2.17(b) shows a simple mechanical model of the aircraft wing, a spring–mass system excited by the force $f(t)$. The differential equation that $f(t)$ to $x(t)$ is

$$M\frac{d^2x}{dt^2} + Kx = f(t) = D\cos\omega t$$

Laplace-transforming, we write

$$Ms^2X(s) - Msx(0^-) - Mx'(0^-) + KX(s) = \frac{Ds}{s^2+\omega^2}$$

$$X(s) = \underbrace{\frac{Ds}{(s^2+\omega^2)(Ms^2+K)}}_{\substack{\text{zero-state response}\\ \text{component}}} + \underbrace{\frac{Msx(0^-)+Mx'(0^-)}{Ms^2+K}}_{\substack{\text{zero-input response}\\ \text{component}}}$$

For zero initial conditions,

$$X(s) = \frac{(D/M)s}{(s^2+\omega^2)(s^2+K/M)} = \frac{K_1s+K_2}{s^2+\omega^2} + \frac{K_3s+K_4}{s^2+K/M}$$

$$= \frac{(K_1+K_3)s^3 + (K_2+K_4)s^2 + (K_1K/M + K_3\omega^2)s + (K_2K/M + K_4\omega^2)}{(s^2+\omega^2)(s^2+K/M)}$$

giving

$$\begin{aligned}
K_1 \quad + \quad K_3 \qquad\qquad &= 0\\
K_2 \qquad + \quad K_4 &= 0\\
\frac{K}{M}K_1 \quad + \omega^2 K_3 \qquad\qquad &= \frac{D}{M}\\
\frac{K}{M}K_2 \qquad + \omega^2 K_4 &= 0
\end{aligned}$$

These have solution

$$K_2 = K_4 = 0 \qquad K_1 = -K_3 = \frac{D/M}{(K/M)-\omega^2}$$

giving

$$X(s) = \frac{\dfrac{D/M}{(K/M) - \omega^2}s}{s^2 + \omega^2} + \frac{\dfrac{-D/M}{(K/M) - \omega^2}s}{s^2 + (K/M)}$$

$$x(t) = \frac{D/M}{(K/M) - \omega^2}\cos\omega t - \frac{D/M}{(K/M) - \omega^2}\cos\sqrt{K/M}\ t \qquad t \geqslant 0$$

The following data are for an aircraft wing; the frequency ω of the turbine and the natural frequency $\sqrt{K/M}$ are close to one another.

$M = 1000$ lb of mass

$D = 3000$ lb of force

$K = 400{,}800.4$ lb of force/ft of deflection

$\omega = 20$ rad/s

Deflections exceeding 5 ft are considered to be beyond the wing's structural limit. With these numbers,

$$x(t) = 3.748(\cos 20t - \cos 20.02t) \qquad t \geqslant 0$$

Maximum deflection will obviously occur at any time, when

$$\cos 20t = 1 \quad \text{and} \quad \cos 20.02t = -1$$

These are times for which

$$20t = n2\pi \qquad n \text{ an integer}$$

and

$$20.02t = \pi + m2\pi \qquad m \text{ an integer}$$

one such solution being $n = m = 500$, $t = 50\pi$. The amount of maximum deflection will be

$$x_{\text{max}} = 3.748(2) = 7.496 \text{ ft}$$

To reduce the deflection amplitude, the wing can be stiffened to increase K to obtain a natural frequency $\sqrt{K/M}$ much larger than ω, giving a much smaller value for the amplitudes

$$K_1 = -K_3 = \frac{D/M}{(K/M) - \omega^2}$$

Stiffening the wing slightly to give

$$K = 441{,}000 \text{ lb of force/ft}$$

of deflection results instead in

$$K_1 = -K_3 = \frac{D/M}{(K/M) - \omega^2} = 0.073$$

$$x(t) = 0.073(\cos\ 20t - \cos\ 21t)$$

for which

$$x_{max} = 0.073(2) = 0.146 \text{ ft}$$

This is the solution that was used to prevent wing fatigue.

2.9 SUMMARY

A first-order system has transfer function

$$T(s) = \frac{b_m s^m + \cdots + b_1 s + b_0}{s + a_0}$$

It is stable for any $a_0 > 0$. First-order system response to an input $r(t)$ is of the form

$$Y(s) = T(s)R(s) + \frac{\begin{pmatrix}\text{constant numerator dependent}\\ \text{on initial conditions}\end{pmatrix}}{s + a_0}$$

The natural component of the response is of the form

$$y_{natural}(t) = Ke^{-a_0 t}$$

and has time constant

$$\tau = \frac{1}{a_0}$$

A second-order system has transfer function

$$T(s) = \frac{b_m s^m + \cdots + b_1 s + b_0}{s^2 + a_1 s + a_0}$$

It is stable if and only if both a_1 and a_0 are positive. Second-order system response is of the form

$$Y(s) = T(s)R(s) + \frac{\begin{pmatrix}\text{first-degree numerator polynomial}\\ \text{dependent on initial conditions}\end{pmatrix}}{s^2 + a_1 s + a_0}$$

If the characteristic polynomial

$$s^2 + a_1 s + a_0 = (s - s_1)(s - s_2)$$

has roots s_1 and s_2 that are real and distinct, the natural response component is *overdamped*:

$$y_{natural}(t) = K_1 e^{s_1 t} + K_2 e^{s_2 t}$$

If the characteristic roots are equal, $s_1 = s_2$, the natural response is *critically damped*:

$$y_{natural}(t) = K_1 e^{s_1 t} + K_2 t e^{s_1 t}$$

If the characteristic roots are complex numbers,

$$s_1, s_2 = \sigma \pm j\omega$$

the natural response component is *underdamped*:

$$y_{\text{natural}}(t) = K_1 e^{s_1 t} + K_2 e^{s_2 t} = A e^{-\sigma t} \cos(\omega t + \theta)$$

The underdamped second-order system natural response is also described by the undamped natural frequency ω_n and damping ratio ζ. In terms of these, the characteristic polynomial is

$$s^2 + 2\zeta\omega_n s + \omega_n^2$$

and

$$\sigma = \zeta\omega_n$$
$$\omega = \omega_n\sqrt{1-\zeta^2}$$

For ζ between 0 and 1, the characteristic roots lie on a circle of radius ω_n about the origin in the left half of the complex plane. The *damping angle* ϕ is related to the damping ratio by

$$\zeta = \cos\phi$$

Higher-order systems have a natural response that consists of a sum of terms, one term for each characteristic root. For each real, distinct characteristic root, there is a real exponential natural response term. Repeated characteristic roots give response terms involving powers of time times an exponential. For each pair of complex conjugate roots, there are a pair of complex exponential terms which combine into the exponential-times-sinusoid form.

One set of measures of a stable system's performance is its *rise time*, *overshoot*, and *settling time* to a step input. These were plotted for specific second-order systems with

$$T(s) = \frac{\omega_n^2}{s^2 + 2\zeta\omega_n s + \omega_n^2}$$

in Figure 2.10.

A system is stable if and only if all its characteristic roots are in the left half of the complex plane (LHP). First- and second-order polynomials have all roots in the LHP if and only if all polynomial coefficients have the same algebraic sign. Occasionally, the presence of RHP and imaginary roots of a high-order polynomial can be detected by an examination of the algebraic signs of the polynomial coefficients, as summarized in Table 2.1.

The Routh–Hurwitz test provides a convenient, definitive method of ascertaining system stability in general. The number of algebraic sign changes in the left column of the Routh array equals the number of RHP roots of the polynomial tested. A left-column zero situation occurs when there is a zero entry at the left of a row but at least one nonzero entry in the rest of the row. To complete an array with a left-column zero, row shifting may be used.

An all-zero row of a Routh array is any row, through the s^0 row, that consists solely of zeros. An all-zero row indicates that the tested polynomial has an even polynomial divisor. The coefficients of the divisor polynomial are given by the entries in the row above the row of zeros. To complete the array, replace the row of zeros with the coefficients of the derivative with respect to s of the divisor polynomial and complete the array as usual. The completed array includes the test of the divisor polynomial which must have equal numbers of RHP and LHP roots. Any remaining roots of an even or odd divisor polynomial must be in the imaginary axis.

An adjustable polynomial has coefficients dependent upon a parameter K. The Routh–Hurwitz test can be performed in terms of K to determine the range(s) of K for which all polynomial roots are in the LHP. A defnition of "robustness" and Kharitonov's theorem is presented.

MATLAB may be used to obtain the time response. The "step" command allows the step response to be taken. Any other response may be obtained by multiplying the Laplace transform of the command by s. The "conv" command is used to multiply two polynomials. The "ezplot" command operates upon a symbolic inverse Laplace transform to obtain any desired time response.

Two design examples were presented: design of an insulin delivery system and design of an aircraft wing.

REFERENCES

Routh–Hurwitz Testing

Bendir, M., and Picinbond, B., "Extended Table for Eliminating the Singularities in Routh's Array." *IEEE Trans. Auto. Control* (February 1990).

Clark, R. "The Routh–Hurwitz Stability Criterion Revisited." *IEEE Control Syst. Mag.* (July 1992): 119–120.

Robustness

Kharitonov, C. L., "Asymptotic Stability of an Equilibrium Position of a Family of Linear Differential Equations." *Differensialnye Uravnewiya,* 14:1086–1088, 1978.

Tempo, R. and Blanchini, F. "Interval Polynomials: Kharitonov's Theorem and Value Set Geometry." *The Control Handbook.* CRC Press, Boca Raton, FL, 1996. pp 501, 502.

Insulin Delivery and Biomedical Applications

Albisser, A. M., "Review of Artificial Pancreas Research." *Arch. Intern. Med.*, 137 (May 1977):639–649.

Blackshear, P.J., "Implantable Drug-Delivery Systems." *Sci. Am.*, (December 1979): 66–73

Horgan, J., "Medical Electronics." *IEEE Spectrum* (January 1985); 89–94.

Spencer, W. J., "For Diabetics: An Electronic Pancreas." *IEEE Spectrum* (June 1978): 38–42.

Aircraft Dynamics

Etkin, B.B., *Dynamics of Atmospheric Flight*. New York: Wiley, 1972.

Kolk, R. W., *Modern Flight Dynamics*. Englewood Cliffs, NJ: Prentice-Hall, 1961.

PROBLEMS

1. Find the response $y(t), t \geqslant 0$, of the first-order systems with following transfer functions, inputs $r(t)$, and initial conditions. Indicate the natural part of the response and find its time constant.

(a)
$$T(s) = \frac{6}{s+3}$$
$$r(t) = \delta(t), \text{ an impulse}$$
$$y(0^-) = -2$$

(b)
$$T(s) = \frac{-4s}{3s+2}$$
$$r(t) = u(t), \text{ a step}$$
$$y(0^-) = 3$$

(c)
$$T(s) = \frac{s+1}{2s+3}$$
$$r(t) = u(t), \text{ a step}$$
$$y(0^-) = 4$$

(d)
$$T(s) = \frac{3}{s+7}$$
$$r(t) = 2\cos 4t$$
$$y(0^-) = 2$$

(e)
$$T(t) = \frac{2}{2s+1}$$
$$r(t) = 6e^{-(1/2)t}$$
$$y(0^-) = 3$$

2. When the heater for an industrial controlled-temperature chamber was turned off, the measured temperature decayed as shown in the table. What is the chamber's time constant? What is the temperature outside the oven?

Oven Temperature Decay Data

Time	Temperature (°C)
14:23:10	120
14:31:00	108
14:39:30	92
14:48:35	80.5
15:05:00	63
15:34:00	40.5

Ans. $\tau = 35.8$ s; temperature outside $= 40°C$

3. Find the response $y(t)$, $t \geq 0$, of the second-order systems with the following transfer functions, inputs $r(t)$, and zero initial conditions. Indicate the natural part of the response.

(a) $T(s) = \dfrac{2}{s^2 + 4s + 16}$

$r(t) = \delta(t)$, an impulse

(b) $T(s) = \dfrac{s - 2}{s^2 + 3s + 12}$

$r(t) = u(t)$, a step

(c) $T(s) = \dfrac{s + 1}{s^2 + 4s + 10}$

$r(t) = u(t)$, a step

(d) $T(s) = \dfrac{1}{s^2 + s + 6}$

$r(t) = 2t$

(e) $T(s) = \dfrac{s}{s^2 + 4s + 4}$

$r(t) = 4 \sin 2t$

4. For the underdamped systems with the following transfer functions, find the undamped natural frequency ω_n, the damping ratio ζ, the exponential constant $a = \zeta\omega_n$, and the oscillation frequency ω:

(a) $T(s) = \dfrac{-s^2}{s^2 + 4s + 25}$

(b) $T(s) = \dfrac{s^2 + 6s + 10}{4s^2 + 2s + 50}$

(c) $T(s) = \dfrac{20}{3s^2 + s + 10}$

(d) $T(s) = \dfrac{s-2}{s^2+16}$

(e) $T(s) = \dfrac{1}{2s^2+6s+20}$

Ans. (a) 5, 0.4, 2, 4.58; (c) 1.82, 0.09, 0.16, 1.8

5. Find a controller transmittance $G(s)$ such that the overall system of Figure P2.5 is second-order and critically damped. A solution to this problem is not unique.

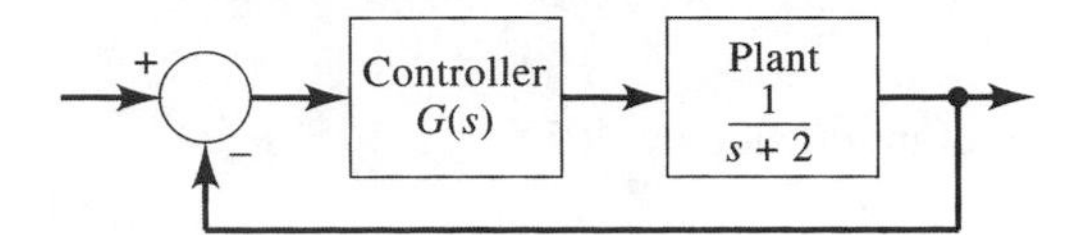

Figure P2.5

6. For the system of Figure P2.6 find the constant value of gain, K, for which the damping ratio of the overall system is 0.7. For this value of K, what is the system's undamped natural frequency?

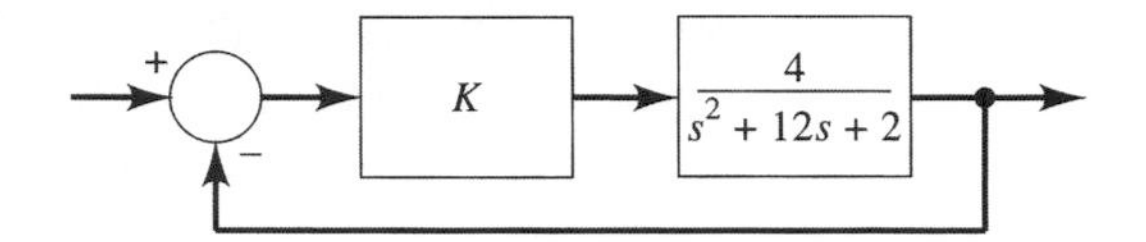

Figure P2.6

Ans. 17.87, 8.57

7. For the system of Figure P2.7 find a gain K for which the natural component of the output decays at least as rapidly as $\exp(-10t)$. A solution to this problem is not unique.

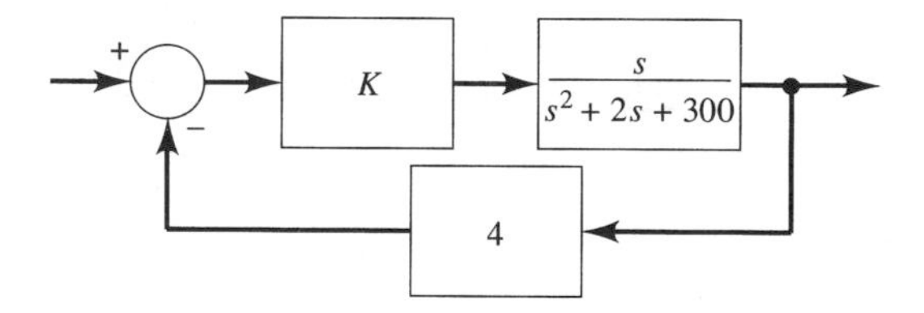

Figure P2.7

8. Draw "block" diagrams to represent the transfer function

$$T(s) = \frac{10(s-1)(s+4)}{(s+2)(s+3)(s+6)}$$

as follows:

(a) In terms of cascaded (end-to-end connection) first-order blocks.

(b) In terms of tandem (same input, outputs summed) first-order blocks.

(c) As a nontrival combination of cascaded and tandem first-order blocks.

9. Identify the type of natural response (overdamped, critically damped, or underdamped) associated with each of the following characteristic polynomials:

   (a) $s^2 + 8s + 8$

   (b) $s^2 + s + 7$

   (c) $2s^2 + 9s + 3$

   (d) $s^2 + s + 6$

   (e) $s^2 + 6s + 9$

   Ans. (a) overdamped; (c) overdamped; (e) critically damped

10. Use the Routh–Hurwitz test to show that all roots of a cubic,

$$s^3 + a_2 s^2 + a_1 s + a_0$$

are in the LHP if and only if a_2, a_1, and a_0 are positive and $a_1 a_2 > a_0$

11. Determine a range of values for K, if a range exists, for which each of the systems shown in Figure P2.11 is stable.

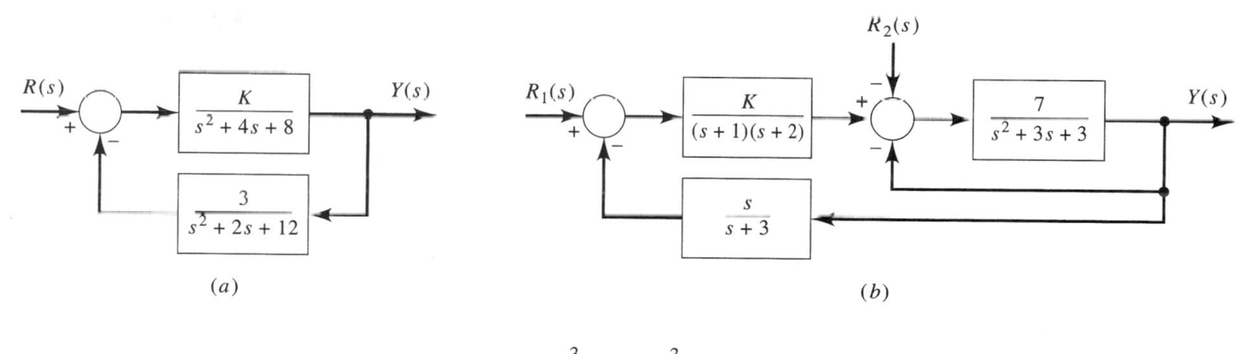

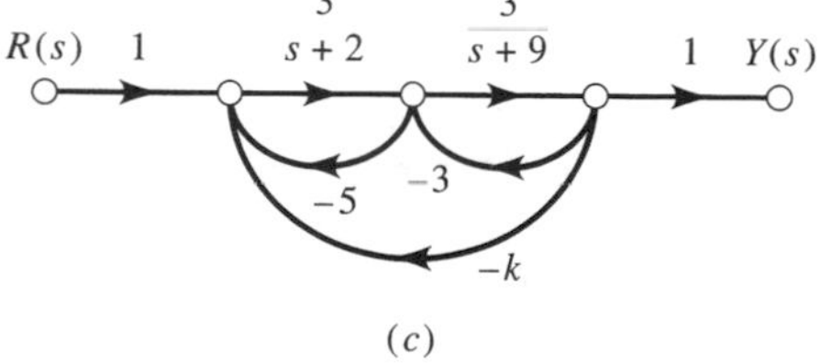

Figure P2.11

Ans. (a) $-32 < K < 29.6$; (c) $K > -19$

12. A plot such as the example sketch of Figure P2.12 shows the range of values of two parameters K_1 and K_2 for which a system is stable. It is called a *stability boundary diagram.* Draw such a diagram for a system with characteristic equation

$$s^2 + (6 + 0.5K_1)s + 3(K_1 + K_2) = 0$$

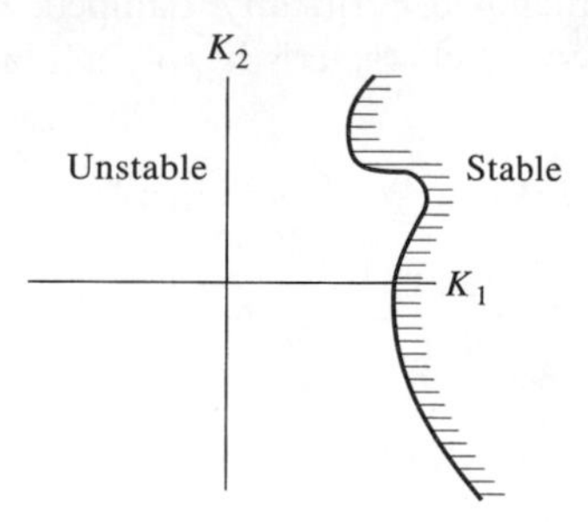

Figure P2.12

13. For what range(s) of the adjustable parameter K do the following polynomials have all roots in the LHP?

    (a) $Ks^3 + s^2 + 6s + 12$

    (b) $s^3 + Ks^2 + 3s + 12$

    (c) $s^3 + 7s^2 + Ks + K$

14. Find the range(s) of the adjustable parameter $K > 0$ for which the systems of Figure P2.14 are stable.

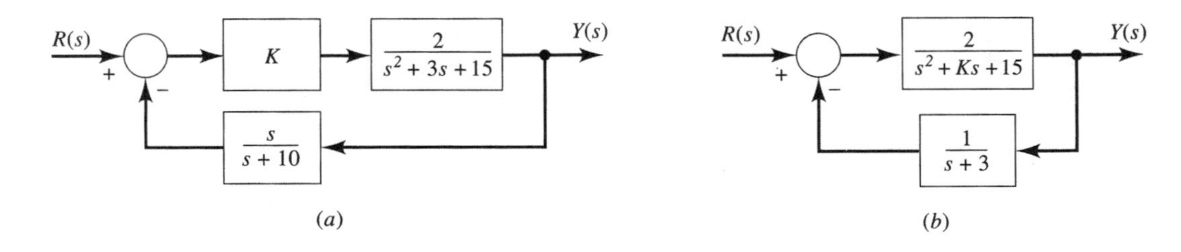

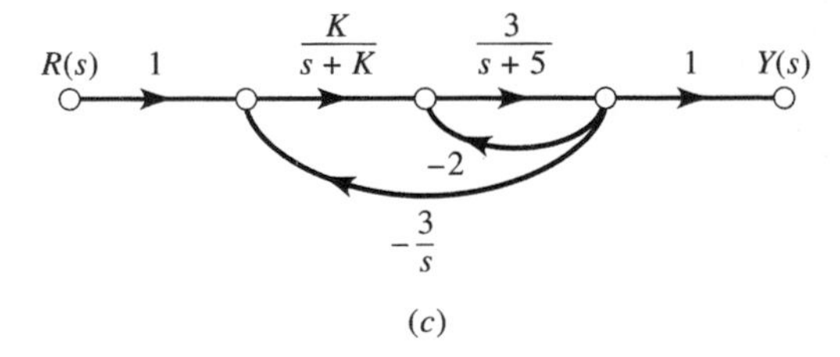

Figure P2.14

15. The required insulin delivery rate varies considerably from person to person because of differences in body chemistry. For the insulin delivery system (Section 2.7), suppose that a peak delivery of insulin must occur at 2 h (7200 s) instead of at 1 h. Let the total insulin volume to be delivered for each meal be 0.35 cm^3/s instead of 0.17 cm^3/s. Find a single cycle of the required motor drive signal.

16. For the aircraft wing problem (Section 2.8), instead of stiffening the wing with mass remaining constant, suppose the stiffness remains constant but the mass M of the wing varies. K_1 represents half the maximum wing deflection. Plot K_1 versus M for M between 990 and 1000 lb of mass.

17. For the original M, D, and K of the aircraft wing problem (Section 2.8), plot K_1 versus ω for values of ω from 19.8 to 20 rad/s.

18. A savings account with an initial deposit of \$1000 and 0.7% interest per month has a balance during the nth month of

$$b(n) = \$1000(1.007)^n \qquad n = 0, 1, 2, \ldots$$

This discrete function of the integer values n can be modeled by a continuos exponential function

$$f(t) = Ae^{at}$$

which has the same values as $b(n)$ for integer values of t, as indicated in Figure P2.18. For convenience, t can be measured in months. For this model, find the constants A and a.

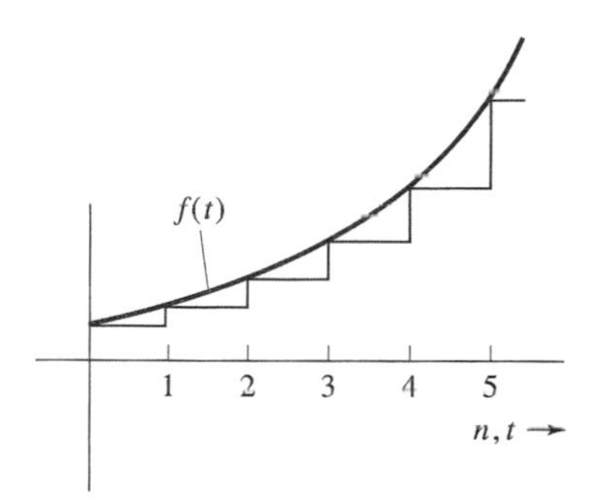

Figure P2.18

19. Power steering for an automobile is a feedback system that can be modeled as in Figure P2.19. For a unit step input $A(s)$, find the values of K_1 and K_2, if possible, for which the response $w(t)$ is critically damped and has a forced response of 0.4 unit. Repeat for a damping ratio of 0.707 and a forced response of 0.23 unit.

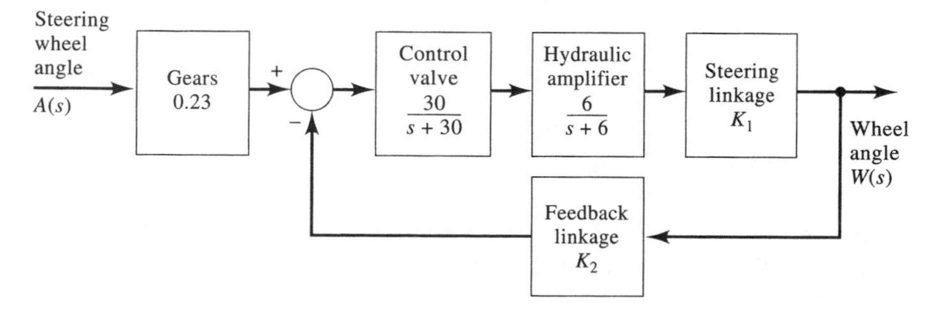

Figure P2.19

20. A feedback control system which is designed to maintain constant torque on a rotary shaft in Figure P2.20. The torque sensor monitors strain on a section of the shaft, which is nearly proportional to the applied shaft torque.

For a step input of desired torque, choose the constants k_1 and k_2 in the controller, if possible, so that the system response is oscillatory with a damping ratio of 0.7 and an undamped natural frequency of 6 rad/s.

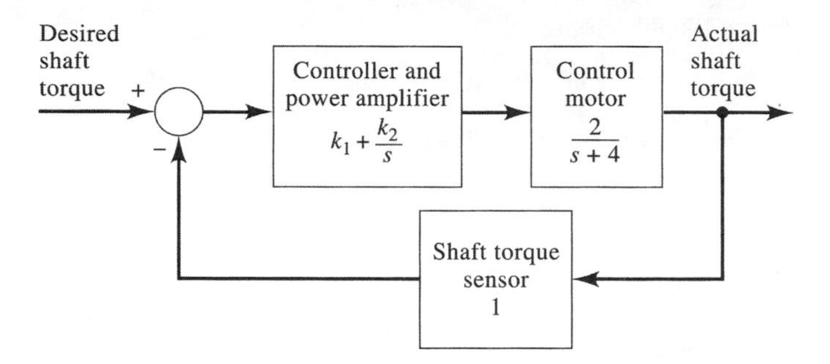

Figure P2.20

Ans. $k_1 = 2.2, k_2 = 18$

21. A simple model for the roll stabilizer on a large ship is given in Figure P2.21. Find the two system transfer functions in terms of the fin actuator gain K, and determine the range of $K > 0$ for which the system is stable.

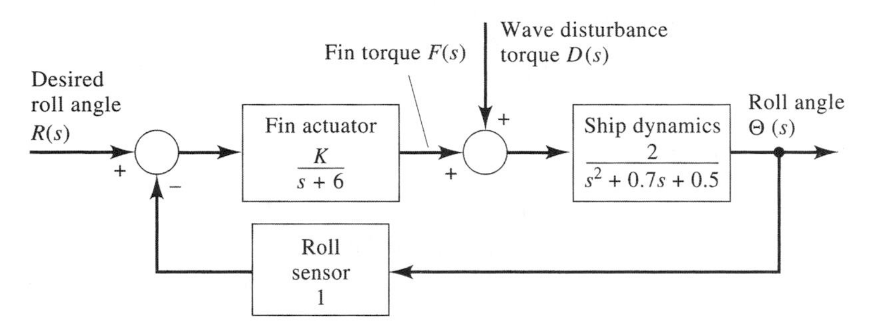

Figure P2.21

22. When paper is rolled in a paper mill, it is important to maintain a specific tension as the paper is wound. A model for the control of this process is given in Figure P2.22 in terms of various constants. Choose nonzero values for K_1, K_2, a_1, a_2, a_3, and a_4, and determine if the system model is stable.

23. A motor shaft velocity control system model is given in Figure P2.23. For what range of $K > 0$ is this system stable?

Ans. Any $K > 0$ will work.

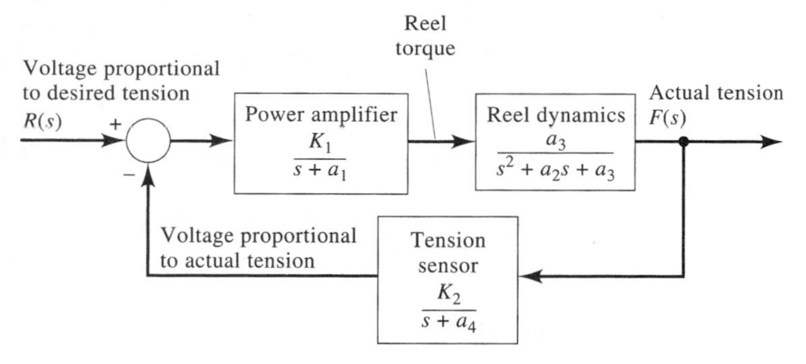

Figure P2.22

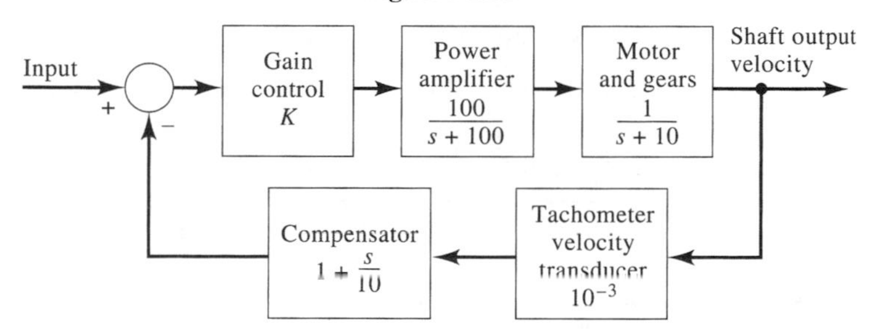

Figure P2.23

24. A motor shaft position control system is modeled in Figure P2.24. Find the relative stability of the system as a function of the positive gain K.

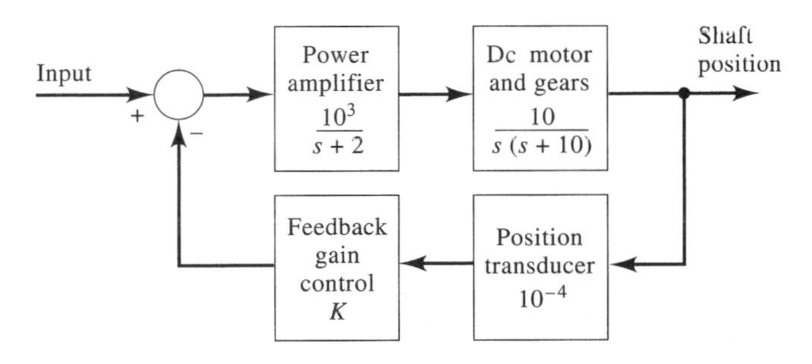

Figure P2.24

25. The following is the forward path transmittance for a unit feedback system. For what K is the closed-loop system marginally stable? At that K, where is the imaginary axis root?

$$G(s) = \frac{K(s + 8)}{s(s + 1)(s^2 + 2s + 1)}$$

26. The block diagram for a radar antenna tracking system is shown in Figure P2.26. K_p is 3.18. Set A so that the damping ratio of the complex conjugate closed-loop roots is 0.1.

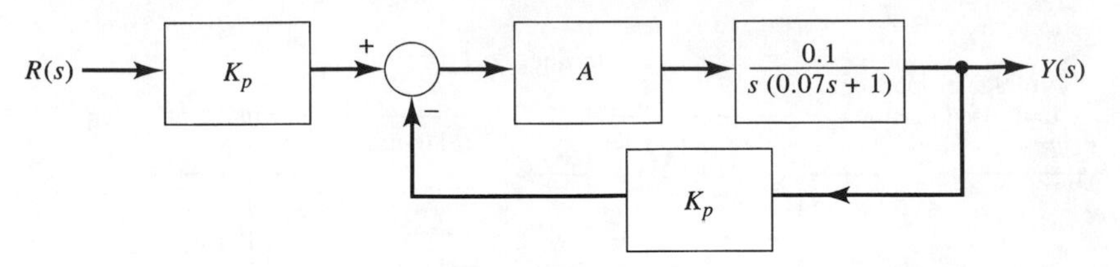

Figure P2.26

Ans. $A = 1124$

27. For the closed-loop system shown in Figure P2.27,

(a) For what values of K is the system stable?

(b) For what value of K is the system marginally stable?

(c) For the value of K in (b), what are the two imaginary roots?

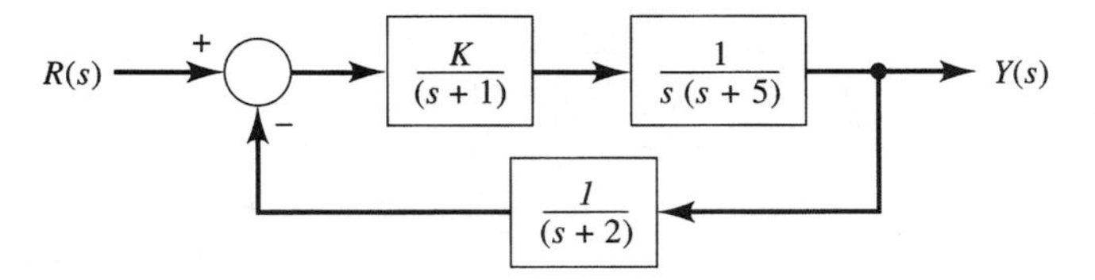

Figure P2.27

CHAPTER 3

Performance Specifications

3.1 Preview

Two aspects of performance are often considered when a control system is designed: the transient performance and the steady state performance. For example, consider an automobile that is being driven along a level, smooth highway. The driver pushes down on the accelerator pedal and holds it stationary. The automobile then slowly gathers speed. The time history of speed would be the transient performance. A high-performance car would be designed to be far more responsive than a car driven for transportation to work. When the automobile speedometer reads a constant velocity, steady state has been achieved. The final (steady state) velocity depends on how aerodynamic the car is, on road and wind conditions, and on the efficiency of the engine, to name just a few factors.

As another example, consider a radar surveillance operator trying to detect enemy aircraft. The radar dish should quickly and smoothly move to sweep a suspicious portion of the sky (the transient response) and eventually achieve a desired sweep rate (the steady state result). Sometimes the steady state result is a constant (the automobile velocity), and sometimes a signal that increases linearly with time (the angular position of the radar dish).

In this chapter we analyze the tendency of a closed-loop system to follow a desired command. Chapters 1 and 2 were dedicated to defining the differential equations, the transfer function, and the stability of a system. Stability is defined in terms of the natural response, so stability is a property of the transient aspect of performance.

In Chapter 3, the emphasis shifts to steady state performance, in particular to the tendency of the system to follow a desired command. Special attention is focused on commands that are powers of time, since most commands can be approximated by the sum of a step, ramp, and so forth.

Integral squared error is calculated and position-integral-derivative (PID) compensation is presented for the time domain. As system parameters change, the system transfer function may also change. The study of that cause-effect relationship, sensitivity, is also covered.

Time domain compensation is considered using the Zieger–Nichols compensation and the Chien–Hrones–Reswick (CHR) compensation. The chapter ends with two design examples: the design of accurate velocity and positioning control for an electric rail transportation system, and the acquisition characteristics of a phase-locked loop for a citizens band radio receiver.

3.2 Analyzing Tracking Systems

3.2.1 Importance of Tracking Systems

A control system is often part of a larger device containing a guidance system that creates the commands that the control system acts upon, as in Figure 3.1(a). The driver of an automobile is actually a sort of guidance system. The driver monitors environmental conditions and the destination. The driver issues commands to the automobile by depressing the pedals and by using the steering wheel. The driver expects prompt, smooth activation of the ever-changing commands as conditions change. A missile has an electronic guidance system to monitor the target motion and to compute corrective commands so that the missile will intercept the target. A guidance system causes the spacecraft to automatically fly near enough to the final destination for the crew to take over the terminal phases of the descent.

A guidance system provides the command to a control system.

From the viewpoint of the guidance system, the control system should quickly respond to the commands. From the viewpoint of what should be a very responsive control system, the guidance commands should appear to be fairly slowly varying functions of time. In fact, a command like that in Figure 3.1(b) can be approximated (as far as the control system is concerned) by the sum of steps, ramps, and so on.

Importance of steps and ramps.

In general, the input $r(t)$ can be written as a power series in terms of powers of t. For example

$$\begin{aligned} r(t) = {} & r(a) + \left.\frac{dr}{dt}\right|_{t=a}(t-a) + \frac{(d^2r/dt^2)\big|_{t=a}}{2!}(t-a)^2 \\ & + \frac{(d^3r/dt^3)\big|_{t=a}}{3!}(t-a)^3 + \cdots \\ = {} & A_0 + A_1 t + A_2 t^2 + A_3 t^3 + \cdots \end{aligned} \qquad \textbf{[3.1]}$$

The viewpoint of this chapter is to examine how a control system responds to commands—in particular those that are powers of t, because more complicated commands are expressible in terms of powers of t.

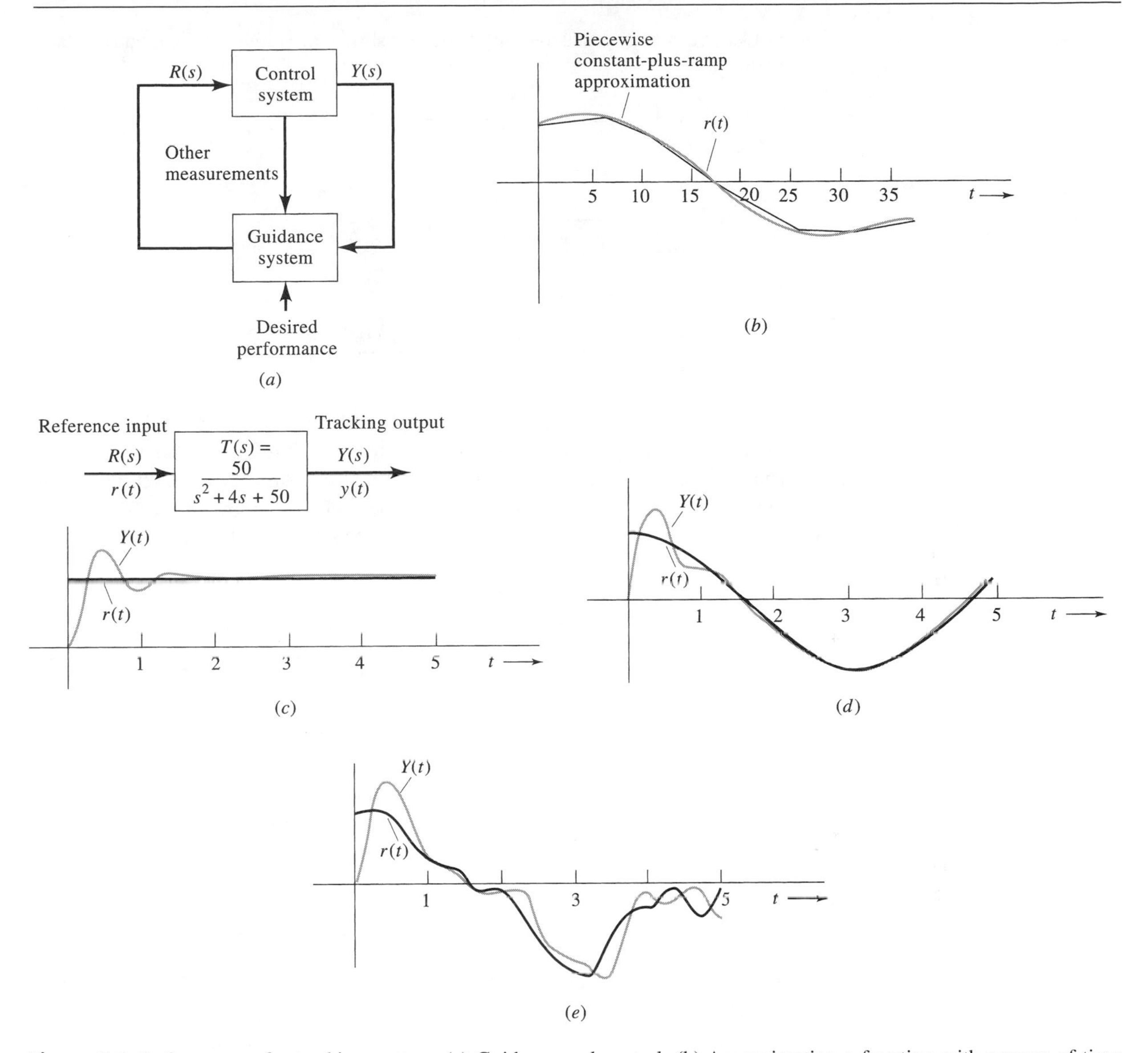

Figure 3.1 Performance of a tracking system. (a) Guidance and control. (b) Approximating a function with powers of time. (c) Unit step reference input. (d) Arbitrary slowly varying reference input. (e) Arbitrary more quickly varying input.

A tracking system is defined.

A tracking system is a control system that creates an output that tracks (follows) the input to some level of tolerance. Figure 3.1(c)–(e) shows how a typical tracking system responds to several reference inputs. The system shown is the elevation control system for a shipboard satellite dish antenna. The overall transfer function model is:

$$T(s) = \frac{50}{s^2 + 4s + 50}$$

The unit step response of this tracking system is shown in Figure 3.1(c). It has Laplace transform

$$Y(s) = T(s)\left(\frac{1}{s}\right) = \frac{50}{s(s^2+4s+50)}$$

$$= \underbrace{\frac{1}{s}}_{\substack{\text{forced}\\ \text{component}}} + \underbrace{\frac{-s-4}{s^2+4s+50}}_{\substack{\text{natural}\\ \text{component}}}$$

and is, as a function of time after $t = 0$,

$$y(t) = \underbrace{1}_{\substack{\text{forced}\\ \text{component}}} + \underbrace{1.04e^{-2t}\cos(6.78t + 163.6°)}_{\substack{\text{natural}\\ \text{component}}}$$

The natural component of the response dies out, with a time constant of 0.5 s leaving a forced response component that equals the constant unit reference input.

The tracking of a relatively slowly varying but otherwise arbitrary reference input is shown in Figure 3.1(d). The output does track the reference input pretty well. When the reference input for this system varies more quickly, Figure 3.1(e), the performance is poorer.

Qualities of a good tracking system.

A good tracking system has a natural response that decays rapidly and without excessive fluctuations, leaving the forced component of the response. The natural response component depends on the system initial conditions but otherwise is not affected by the specific reference input. The forced response component, which is the tracking system response after the natural component has decayed, should adequately track reference inputs of the class to be encountered. For example, in a steel rolling mill control system where step changes in the desired steel thickness are to be made, fast, smooth decay of the natural response and accuracy of the forced response for constant inputs are of primary importance. In a high-performance terrain-following aircraft altitude control system, not only must the natural response component decay quickly, the forced response should accurately track any reference input within the plane's safe physical limits.

The analysis and design of tracking systems can be separated into two parts:

1. Locating the characteristic roots (poles) of the transfer function. These determine the character of the system's natural response component. Normally, it is desired that the natural response decay rapidly and that any oscillatory terms be well damped.
2. Tracking of the reference input by the forced response of the system for those kinds of inputs to be encountered in practice.

In the remainder of this section we consider the natural response component and in subsequent sections, we analyze tracking system forced response. Additionally, the designer is concerned with

3. Performance when the plant model is inaccurate.
4. Tracking system response due to unwanted, inaccessible disturbance inputs.

A variation of the tracking system is the "regulator" system, which is intended to maintain the output of the system at some desired constant level. The speed control system of Chapter 1 (Figure 1.2) is designed to maintain the output shaft velocity at some prescribed level, and as such is a regulator.

3.2.2 Natural Response, Relative Stability, and Damping

The relative stability of a system is the distance into the left half of the complex plane from the imaginary axis to the nearest characteristic root or roots. For example, the system with the characteristic roots (or poles) shown in Figure 3.2(a) has a relative stability of 2 units. The most slowly decaying term in this system's natural response component decays as $\exp(-2t)$. For the natural response to decay at least as quickly as $\exp(-\sigma t)$, a system must have a relative stability of at least σ units. That is, its characteristic roots must be on or to the left of the line $\text{Re}(s) = -\sigma$ as indicated in Figure 3.2(b).

A pair of complex conjugate characteristic roots

$$s_1, s_2 = -a \pm j\omega$$

gives rise to a damped oscillatory term in the natural response component of the form

$$y_i(t) = Ae^{-at}\cos(\omega t + \theta)$$

where the constants A and θ depend on the initial conditions. The damping ratio of such a term is (Section 2.3)

$$\zeta = \cos\phi$$

where ϕ is the damping angle, as shown in Figure 3.3(a). A low damping ratio, which occurs for ϕ near 90°, is undesirable for most tracking systems because it means the output will likely exhibit large fluctuations as its natural response decays. It is

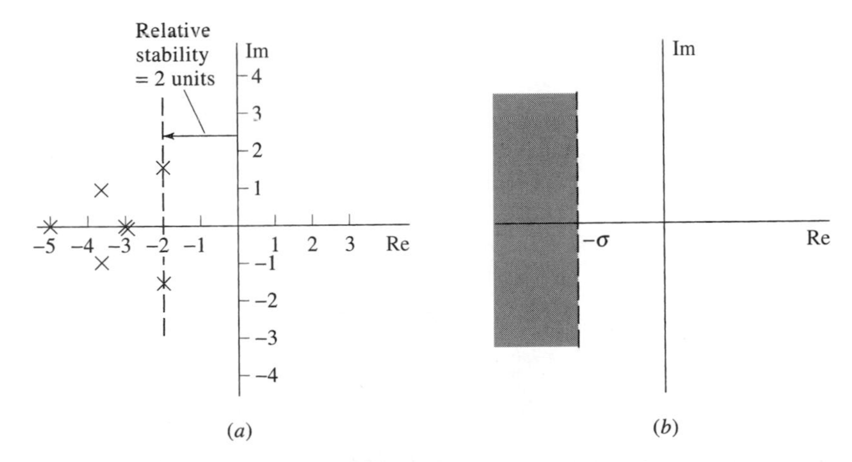

Figure 3.2 Relative stability. (a) An example, (b) Region of greater relative stability than σ.

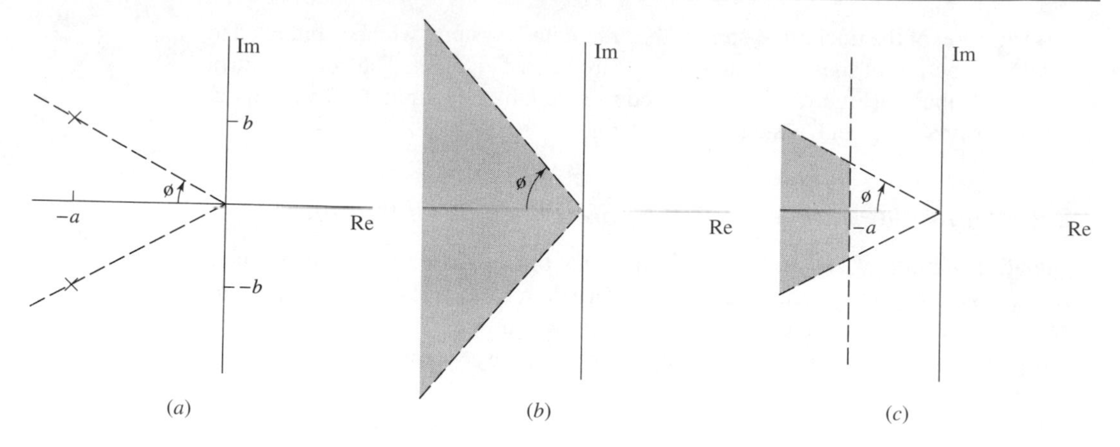

Figure 3.3 Damping ratio for response terms corresponding to complex conjugate transfer function characteristic root (pole) pairs. (a) Damping angle. (b) Region of more than a certain amount of damping. (c) Region of greater relative stability than a and more than a certain amount of damping.

Best damping ratio.

therefore usually important that the characteristic roots (poles) of a tracking system be within a region of more than a certain amount of damping, as shown in Figure 3.3(b). Often, the choice of maximum damping angle $\phi = 45°$ is made, corresponding to a minimum damping ratio of 0.707.

When the two requirements of relative stability and damping ratio are combined, the result is Figure 3.3(c). The characteristic roots of a tracking system should be to the left of the line $\text{Re}(s) = -\sigma$, where σ is the minimum desired relative stability. The natural component of the system's response then decays at least as quickly as $\exp(-\sigma t)$. The characteristic roots should be within a maximum damping angle ϕ so that the oscillatory terms in the natural component of the response are damped quickly.

❑ Computer-Aided Learning

The "roots" command may be used to obtain the relative stability of a system. The relative stability is the smallest real part of the LHP roots for a stable system.

C3.1 Use MATLAB to obtain the relative stability of the following:

(a) $T(s) = \dfrac{4}{(s+2)(s^2+5s+100)}$

(b) $T(s) = \dfrac{80}{(s^2+6s+20)(s^2+8s+90)(s+5)}$

(c) $T(s) = \dfrac{80(s+3)}{s^5+10.6s^4+95.8s^3+413.8s^2+1325s+1680}$

Ans. (a)
```
dena=conv([1 2], [1 5 100])
roots(dena)
```

(b)
```
denb=conv([1 5], [ 1 8 90])
denb=conv(denb, [1 6 20])
roots(denb)
```

(c)
```
denc=(1 10.6 95.8 413.8 1325 1680]
roots(denc)
```

❑ DRILL PROBLEM

D3.1 Determine the relative stability of systems with the following transfer functions. Also determine the damping ratios of any complex conjugate pairs of characteristic roots.

(a)

$$T(s) = \frac{5s^2 - 1}{(s+3)^2(s+4)(s+6)}$$

Ans. 3

(b)

$$T(s) = \frac{4s^2 - 3s + 1}{(s+2)(s^2+3s+10)^2}$$

Ans. 1.5, 0.47

(c)

$$T(s) = \frac{24}{(s^3+s+4)(s^2+2s+6)}$$

Ans. 0.5, 0.25, 0.41

(d)

$$T(s) = \frac{(s+4)^2(s-2)}{(s+2)(s^2+5s+20)(s+3)}$$

Ans. 2, 0.56

3.3 Forced Response

3.3.1 Steady State Error

The error between the output and input of a system is given by

$$E(s) = R(s) - Y(s) = R(s) - T(s)R(s) = [1 - T(s)]R(s) = T_E(s)R(s)$$

The error between the input and the output of the system has the transmittance

$$T_E(s) = 1 - T(s) = \left.\frac{E(s)}{R(s)}\right|_{\text{zero initial conditions}} \qquad [3.2]$$

$T_E(s)$ is defined.

which relates the error and input transforms. The poles of $T_E(s)$ are the same as the poles of $T(s)$. Like all system signals, the error signal is generally composed of natural and forced parts. The natural part is composed of terms, one for each pole of $T_E(s)$, or $T(s)$, with amplitudes dependent on initial conditions. For a system with relative stability σ, this natural response decays to zero at least as fast as $\exp(-\sigma t)$.

The forced part of the error signal is

$$e_{\text{forced}}(t) = r(t) - y_{\text{forced}}(t)$$

If the system output tracks a reference input $r(t)$ well, then $e_{\text{forced}}(t)$ will be small. For perfect tracking,

$$y_{\text{forced}}(t) = r(t) \qquad \text{and} \qquad e_{\text{forced}}(t) = 0$$

3.3.2 Initial and Final Values

The *initial value* of a function of time $y(t)$ is related to the function's Laplace transform by

$$y(0) = \lim_{s\to\infty} [sY(s)] \qquad \textbf{[3.3]}$$

For example, the initial value of the function with Laplace transform

$$Y(s) = \frac{-4s^4 + 3s^3 + s^2 - s + 1}{3s^5 - 2s^4 + s^3 - s + 10}$$

is

$$y(0) = \lim_{s\to\infty} [sY(s)] = -\tfrac{4}{3}$$

In general, the values $y(0^-)$, $y(0)$, and $y(0^+)$ for a Laplace-transformable function can differ. If there is an impulse $y(t)$ at $t = 0$, then $y(0)$ will be infinite.

For functions with rational transforms, the initial-value theorem is especially easy to visualize. The partial fraction expansion of a rational function can have the representative types of terms in Table 3.1. For each, and for other terms as well, multiplying by s and taking the limit as s goes to infinity gives the correct contribution to $y(0)$.

The *final value* (or *steady state value*) of a function $y(t)$ is

$$\lim_{t\to\infty} y(t)$$

If the limit exists and is finite, the final value is related to the function's Laplace transform by what is called the *final-value theorem*:

$$\lim_{t\to\infty} y(t) = \lim_{s\to 0}[sY(s)] \qquad \textbf{[3.4]}$$

Applicability of the final-value theorem.

For a final value of $y(t)$ to exist, all denominator roots of $Y(s)$ must be in the LHP except possibly for one root at $s = 0$. It is the root at $s = 0$, corresponding to a constant term in $y(t)$ after $t = 0$, that then contributes a nonzero final value.

Table 3.1 *Application of the Initial-Value and Final-Value Theorems to Representative Laplace Transform Terms*

Transform $Y(s)$	**Time Function $y(t), t \geq 0$**	$\lim_{s\to\infty} sY(s)$	$\lim_{s\to\infty} sY(s)$
A	$A\delta(t)$	∞	0
$\frac{A}{s}$	A	A	A
$\frac{A}{s^2}$	At	0	∞
$\frac{A}{s+a}$	Ae^{-at}	A	0
$\frac{Ab}{s^2+b^2}$	$A \sin bt$	0	0
$\frac{Ab}{(s+a)^2+b^2}$	$Ae^{-at} \sin bt$	0	0
$\frac{As}{s^2+b^2}$	$A \cos bt$	A	0
$\frac{A(s+a)}{(s+a)^2+b^2}$	$Ae^{-at} \cos bt$	A	0

Application of this result is simply a calculation of the residue of a *K*/*s* term in the partial fraction expansion of $Y(s)$. The function

$$Y(s) = \frac{-4s^3 - s^2 + 7s + 3}{s^3 + 9s^2 + 2s}$$

for example, has all denominator roots in the LHP except for the single $s = 0$ root. The final value of $y(t)$ is

$$\lim_{t\to\infty} y(t) = \lim_{s\to 0}[sY(s)] = \tfrac{3}{2}$$

If $sY(s)$ has multiple poles at the origin ($s = 0$), the final value is not mathematically applicable; however, the theorem does correctly predict the final value. For example, if

$$Y(s) = \frac{-4s^3 - s^2 + 7s + 3}{s(s^3 + 9s^2 + 2s)}$$

apply the final-value theorem of Equation (3.4) as follows:

$$\lim_{t\to\infty} y(t) = \lim_{s\to 0}[sY(s)] = \infty$$

This result is valid because $y(t)$ has a term in the time response of the form bt, which approaches infinity as t approaches infinity. Unfortunately, the final-value theorem gives answers even when a final value does not exist, as is demonstrated by the entries in Table 3.1. The transform

$$Y(s) = \frac{s-6}{s(s-1)[(s+1)^2+4]} = \frac{K_1}{s} + \frac{K_2}{s-1} + \frac{K_3 s + K_4}{(s+1)^2+2^2}$$

is of a time function of the form

$$y(t) = \left[K_1 + K_2 e^t + Ae^{-t}\cos(2t+\theta)\right]u(t)$$

Invalid use of the Final-Value theorem.

This does not have a finite final value because of the e^t term, which comes from the root at $+1$. However, application of the final-value theorem gives

$$\lim_{s\to 0}[sY(s)] = \tfrac{6}{5}$$

which is incorrect. The Routh–Hurwitz method can be used to see whether any roots are in the right half-plane (RHP).

For the Laplace transform

$$Y(s) = \frac{6s+7}{s(s^3+s^2+2s+3)}$$

the limit is

$$\lim_{s\to 0} sY(s) = \tfrac{7}{3}$$

However, the final value is infinite because owing to the third-order polynomial factor, $Y(s)$ has denominator roots in the RHP. This is determined with the Routh–Hurwitz test:

s^3	1	2
s^2	1	3
s^1	−1	
s^0	3	

Two sign changes indicate two RHP roots. It is therefore important to determine whether any roots lie in the RHP, which can be accomplished with the Routh–Hurwitz method before the final-value theorem is applied.

3.3.3 Steady State Errors to Power-of-Time Inputs

The standard ith degree power-of-time inputs have Laplace transforms

$$R_i(s) = \frac{1}{s^{i+1}}$$

The corresponding time functions are the unit step,

$$R_0(s) = \frac{1}{s} \qquad r(t) = u(t)$$

the unit ramp,

Power of time inputs of power $i = 0,1,2$.

$$R_1(s) = \frac{1}{s^2} \qquad r(t) = tu(t)$$

one-half the unit parabola,

$$R_2(s) = \frac{1}{s^3} \qquad r(t) = \frac{t^2 u(t)}{2}$$

and so on. If the input to a tracking system is a power-of-time input, the error signal is given by

$$E_i(s) = T_E(s)R_i(s) = \frac{1}{s^{i+1}}T_E(s)$$

As a function of time, the error will consist of natural response terms, one for each pole of $T_E(s)$, and a forced response consisting of power-of-time terms, through the ith degree term. For example, a system with transfer function

$$T(s) = \frac{6}{s^2 + 3s + 2}$$

has error transmittance

$$T_E(s) = 1 - T(s) = \frac{s^2 + 3s - 4}{s^2 + 3s + 2}$$

Its error to the standard ramp input

$$R(s) = \frac{1}{s^2} \qquad r(t) = tu(t)$$

is

$$E(s) = T_E(s)R(s) = \frac{s^2 + 3s - 4}{s^2(s^2 + 3s + 2)} = \underbrace{\frac{\frac{9}{2}}{s} + \frac{-2}{s^2}}_{\text{forced response component}} + \underbrace{\frac{-6}{s+1} + \frac{\frac{3}{2}}{s+2}}_{\text{natural response component}}$$

or

$$e(t) = \underbrace{\tfrac{9}{2} - 2t}_{\text{forced response component}} + \underbrace{-6e^{-t} + \tfrac{3}{2}e^{-2t}}_{\text{natural response component}}$$

after $t = 0$.

For a stable system driven by a power-of-time input, the forced component of the error can do only one of three things:

Three possible steady state behaviors.

1. The forced error can be zero, meaning that after the natural response has died out, the error is zero and the tracking system output equals the power-of-time reference input.
2. The forced error can be a constant, so that after the natural response has decayed to zero, the error is constant and the tracking system output and the reference input differ by a constant.

3. The forced error can involve a nonzero term proportional to t or a higher power of t, in which case the error grows without bound.

These three situations are easily distinguished, without calculating $e(t)$, by applying the final-value theorem to $E(s)$. For a stable system, if the final value of $e(t)$ is zero, then the situation must be that of situation 1, ideal tracking of the input:

$$y_{\text{forced}}(t) = r(t)$$

The final-value theorem does not indicate how quickly the final value is achieved, but it does indicate what the final value will be. If the final value of $e(t)$ is a finite nonzero constant, we have situation 2, where $y_{\text{forced}}(t)$ and $r(t)$ differ by that constant. If $E(s)$ has more than a single pole at $s = 0$ (the final-value theorem does not apply then, but the final value is correctly predicted), the final value of $e(t)$ approaches infinity and situation 3 exists.

For a system with a stable transfer function

$$T(s) = \frac{10}{s^2 + 3s + 10}$$

The system has zero steady state step error.

the error to a unit step input is given by

$$E(s) = [1 - T(s)]R(s) = \frac{s^2 + 3s}{s(s^2 + 3s + 10)}$$

Its final value is

$$\lim_{t\to\infty} e(t) = \lim_{s\to 0} sE(s) = \lim_{s\to 0} \frac{s^2 + 3s}{s^2 + 3s + 10} = 0$$

so that after this system's natural response has decayed to zero, it tracks any constant input with zero error. The error of his system to a unit ramp input is given by

$$E(s) = [1 - T(s)]R(s) = \frac{s^2 + 3s}{s^2(s^2 + 3s + 10)}$$

There is a steady state ramp error.

and has final value

$$\lim_{t\to\infty} e(t) = \lim_{s\to 0} sE(s) = \lim_{s\to 0} \frac{s + 3}{s^2 + 3s + 10} = \frac{3}{10}$$

After the system's natural response has died out, the output tracks any ramp input with an error of $\frac{3}{10}$. For a standard parabolic input

$$R(s) = \frac{1}{s^3} \qquad r(t) = \frac{t^2 u(t)}{2}$$

the error is given by

$$E(s) = [1 - T(s)]R(s) = \frac{s^2 + 3s}{s^3(s^2 + 3s + 10)}$$

and

$$\lim_{t\to\infty} e(t) \to \infty$$

❑ Computer-Aided Learning

We learned in Chapter 2 that the response to other than a step input can be obtained by using the MATLAB step command by multiplying the $Y(s)$ by s. In effect, the step response causes $sY(s)$ to be multiplied by $1/s$, resulting in the inverse Laplace transform of $Y(s)$. That technique can be employed to plot ramp response and ramp error response.

For example, suppose we want the ramp response for $R(s) = 1/s^2$. Then $Y(s) = T(s)R(s) = T(s)/s^2$. The inverse Laplace transform of $Y(s)$ may be found by using the step command applied to $sY(s)$, which is $T(s)/s$ here. The polynomial $R_1(s) = 1/s$ can be obtained by

```
r1=zpk([], 0, 1)
```

where [] is used when there is no zero.

Suppose we want the ramp response and the ramp error response for

$$T(s) = \frac{10}{s^2 + 3s + 10}$$

Let $Y(s)$ represent the Laplace transform of the ramp response while $E(s)$ represents the Laplace transform of the ramp error response obtained by using the step command

```
r1=zpk([],0,1)
numt=10
dent=[1 3 10]
t=tf(numt,dent)
sy=t*r1
step(sy)
te=1-t
se=te*r1
step(se)
```

Here step(sy) provides the ramp response of $Y(s)$ while step(se) provides the ramp error response of $E(s)$.

C3.2 Write MATLAB Commands to get the ramp response and ramp error response for

(a) $T(s) = \dfrac{-3s^2 + 5}{s^3 + 3s^2 + 2s + 5}$

(b) $T(s) = \dfrac{5s + 1}{s^3 + 4s^2 + 5s + 1}$

Ans. (a)

```
r1=zpk ([], 0, 1)
numta=[-3 0 5]
denta=[1 3 2 5]
ta=tf(numta, denta)
sya=ta*r1
```

```
step(sya)
tea=1-ta
sea=tea*r1
step(sea)
```

(b)
```
r1=zpk([], 0, 1])
numtb=[5 1]
dentb[1 4 5 1]
tb=tf(numtb, dentb)
syb=tb*r1
step(syb)
teb=1-tb
seb=teb*r1
step(seb)
```

C3.3 Use MATLAB to get the step and ramp responses of

(a) $T(s) = \dfrac{20}{s^2 + 5s + 25}$

(b) $T(s)\dfrac{9s + 20}{s^3 + 6s^2 + 9s + 20}$

Ans. (a)
```
numta=20
denta=[1 5 25]
ta=tf(numta, denta)
step(ta)
r1=zpk([], 0, 1)
sya=ta*r1
step(sya)
```

(b)
```
numtb=[9 20]
dentb=[1 6 9 20]
tb=tf(numtb, dentb)
step(tb)
r1=zpk([], 0,1)
syb=tb*r1
step(syb)
```

❑ DRILL PROBLEMS

D3.2 For each of the following Laplace-transformed signals, find the initial value $y(0)$.

(a)

$$Y(s) = \frac{4s - 1}{s^2 + 3s}$$

Ans. 4

(b)

$$Y(s) = \frac{5}{s^3 + 2s^2 + 11}$$

Ans. 0

(c)

$$Y(s) = \frac{6e^{-2s}}{s+4}$$

Ans. 0

(d)

$$Y(s) = \frac{30s}{s^3 + 2s^2 + 11s + 3}$$

Ans. 0

D3.3 For each of the following Laplace-transformed signals, find the final value if it exists. If a final value of the signal does not exist, so state.

(a)

$$Y(s) = \frac{6s^3 - 3s^2 + 2s - 4}{s^4 + 3s^3 + 2s^2 + s + 10}$$

Ans. does not exist

(b)

$$Y(s) = \frac{s^2 - 2s + 10}{(s^2 + 2s + 10)(s^2 + 3s)}$$

Ans. $\frac{1}{3}$

(c)

$$Y(s) = 7e^{-3s}$$

Ans. 0

(d)

$$Y(s) = \frac{6s^2 + 5}{s^3 + 2s^2 + 11s}$$

Ans. $\frac{5}{11}$

(e)

$$Y(s) = \frac{30}{s^3 + 2s^2 + 11s + 3}$$

Ans. 0

D3.4 For each of the systems in Figure D3.4 the input $R(s)$ is a unit step. Find the steady state value of the output signal $Y(s)$ if it exists.

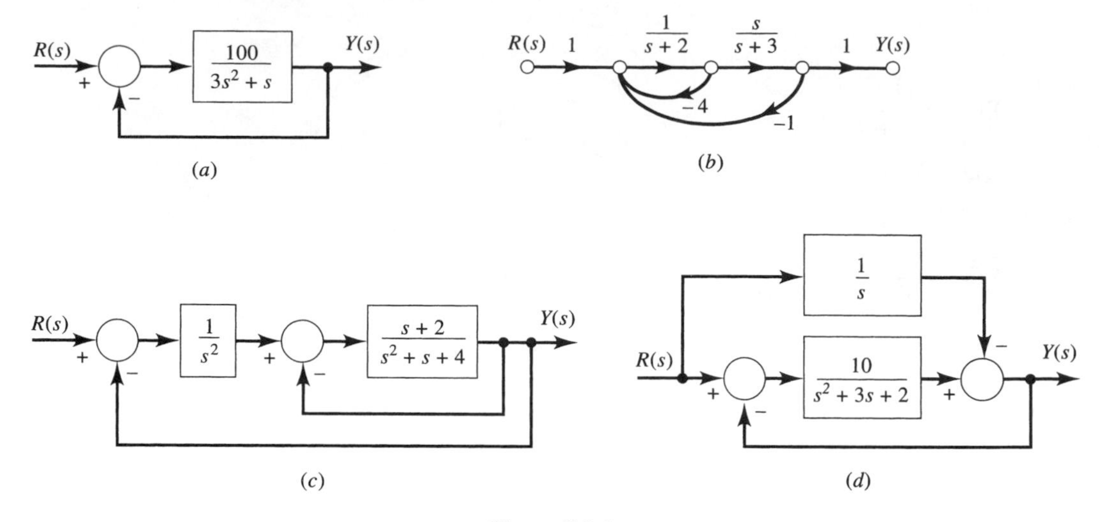

Figure D3.4

Ans. (a) 1; (b) 0; (c) 1; (d) does not exist

D3.5 For systems with the following transfer functions, find steady state errors, if they exist, between output and input for unit step, ramp, and parabolic inputs:

(a)

$$T(s) = \frac{-3s^2 + 5}{s^3 + 3s^2 + 2s + 5}$$

Ans. $0, \frac{2}{5}, \infty$

3.4 Power-of-Time Error Performance

3.4.1 System Type Number

System type number using $T_E(s)$.

Steady state error to power-of-time inputs is intimately related to the number of factors of s in the numerator of the error transmittance $T_E(s)$, the *type number* of the system. If $T_E(s)$ has no factor of s in its numerator, the type number is 0 and the steady state error to a step input

$$R(s) = \frac{A}{s}$$

is

$$\lim_{t\to\infty} e_{\text{step}}(t) = \lim_{s\to 0} sT_E(s)R(s) = \lim_{s\to 0} sT_E(s)\frac{A}{s} = AT_E(0)$$

which is finite. Figure 3.4 illustrates how a system can respond to give a constant steady state error. For no factors of s in the numerator of $T_E(s)$, higher power-of-t

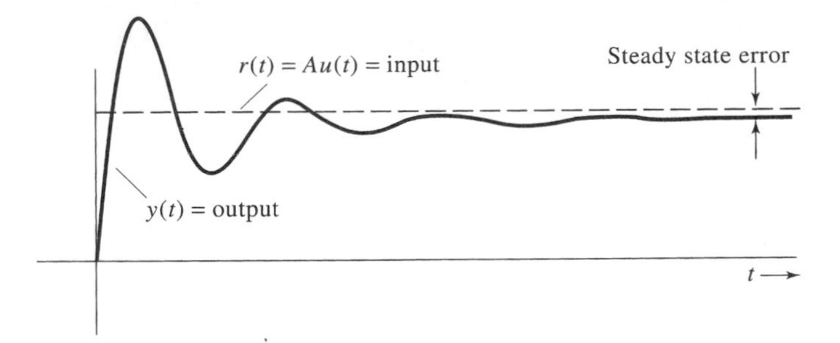

Figure 3.4 Type 0 system step response.

inputs give infinite steady state error. For a ramp input,

$$R(s) = \frac{A}{s^2}$$

the error

$$E(s) = T_E(s)\frac{A}{s^2}$$

has a repeated denominator root at $s = 0$, indicative of a ramp term in $e(t)$. The final-value theorem does not apply in this case because $e(t)$ does not approach a final value. Taking the limit

$$\lim_{s \to 0} sT_E(s)\frac{A}{s^2} = \infty$$

does give the correct answer, though.

If the system is stable and $T_E(s)$ has one factor of s in the numerator, the steady state error to a step is zero:

$$\lim_{t \to \infty} e_{\text{step}}(t) = \lim_{s \to 0} sT_E(s)\frac{A}{s} = AT_E(0) = 0$$

To a ramp, the error is constant:

$$\lim_{t \to \infty} e_{\text{ramp}}(t) = \lim_{s \to 0} sT_E(s)\frac{A}{s^2} = A \lim_{s \to 0} \frac{T_E(s)}{s}$$

The constant steady state error results by cancellation with the s factor in the numerator of $T_E(s)$. Figure 3.5 illustrates how a system can respond to a ramp input to give a finite steady state error. For higher power-of-t inputs, the error of such a system is infinite, since

$$E(s) = T_E(s)\frac{A}{s^n}$$

has a repeated $s = 0$ denominator root.

A factor of s^2 in the numerator of $T_E(s)$ means a type 2 system. For a type 2 system, there is zero steady state error to a step, zero steady state error to a ramp, finite error to a parabola (Figure 3.6), and ever-increasing error for higher power-of-t inputs. The behavior of various system types for power-of-time inputs is summarized in Table 3.2.

The steady state error is finite and nonzero when the system type number equals the power of t.

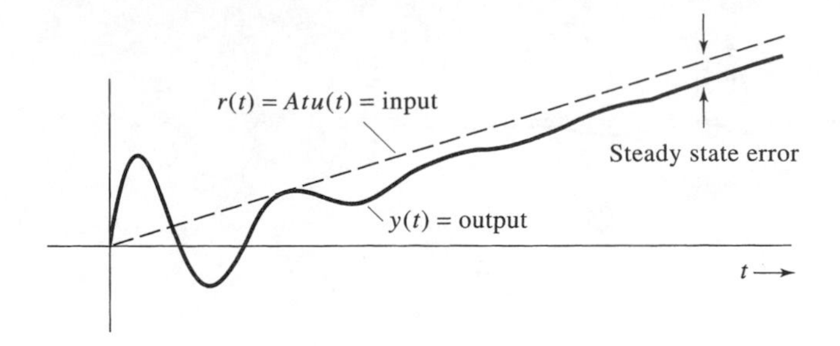

Figure 3.5 Type 1 system ramp response.

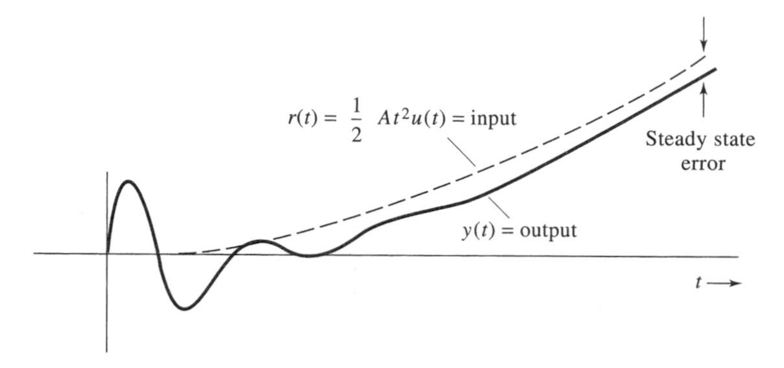

Figure 3.6 Type 2 system parabolic response.

Table 3.2 *Steady State Errors*

System Type (Number of $s = 0$ numerator roots of the error transmittance T_E)	Steady State Error to Step Input $r(t) = Au(t)$ $R(t) = A/s$	Steady State Error to Ramp Input $r(t) = Atu(t)$ $R(s) = A/s^2$	Steady State Error to Parabolic Input $r(t) = \frac{1}{2}At^2u(t)$ $R(s) = A/s^3$	Steady State Error to Input $r(t) = \frac{1}{6}At^3u(t)$ $R(s) = A/s^4$
0	$AT_E(0)$	∞	∞	∞
1	0	$A \lim_{s\to 0}\left[\frac{T_E(s)}{s}\right]$	∞	∞
2	0	0	$A \lim_{s\to 0}\left[\frac{T_E(s)}{s^2}\right]$	∞
3	0	0	0	$A \lim_{s\to 0}\left[\frac{T_E(s)}{s^2}\right]$
⋮				

3.4.2 Achieving a Given Type Number

To achieve a given type number, certain numerator and denominator coefficients of $T(s)$ must match exactly. For example, if we have a third-order system, the transfer function may be represented by

$$T(s) = \frac{b_2 s^2 + b_1 s + b_0}{s^3 + a_2 s^2 + a_1 s + a_0}$$

then $T_E(s)$ is

$$T_E(s) = 1 - T(s) = \frac{s^3 + s^2(a_2 - b_2) + s(a_1 - b_1) + a_0 - b_0}{s^3 + a_2 s^2 + a_1 s + a_0}$$

To have a type 1 system; one power of s must be factorable from the numerator of $T(s)$. Obviously, b_0 must equal a_0. Similarly, for type 2 behavior, we must have $b_1 = a_1$ and $b_0 = a_0$. For type 3 behavior, $b_2 = a_2$, $b_1 = a_1$ and $b_0 = a_0$. Thus, steady state accuracy requires matching as many numerator and denominator coefficients as possible.

The pattern of infinities, constants, and zeros in Table 3.2 may be predicted by thinking of the steady state error as resulting from a contest between the system (represented by its type number) and the input (represented by its power of time). When the power of time exceeds the type number, the input wins, in that the steady state error is infinite. When the type number exceeds the power of time, the system wins, and the steady state error is zero. When the system type number equals the power of time, there is a tie of sorts between the system and the input, resulting in a constant steady state error. For example, a ramp input has a power of time equal to one. When the system number is 0, the input wins 1-0, resulting in an infinite steady state error. When the system type number rises to 1 for a ramp input, the score is a 1-1 tie, and there is a constant steady state error. If the system type number rises to 2 for a ramp input, the system wins 2-1 and the ramp error is zero. The totality of Table 3.2 is predictable from the analogy of a contest between the input and the system.

3.4.3 Unity Feedback Systems

When a control system has unity feedback, the input r and output y are compared directly, as in Figure 3.7(a). The error signal drives the forward transmittance $G(s)$. System type is determined by the number of $s = 0$ numerator roots of

$$T_E(s) = 1 - T(s) = 1 - \frac{G(s)}{1 + G(s)} = \frac{1}{1 + G(s)} \qquad [3.5]$$

For a unity feedback system, the system type can be determined by the number of $s = 0$ denominator roots of the transmittance $G(s)$. If $G(s)$ consists of the ratio of polynomials

Using $G_E(s)$ to find System type number.

$$G(s) = \frac{N(s)}{D(s)}$$

then

$$T_E(s) = \frac{1}{1 + G(s)} = \frac{D(s)}{D(s) + N(s)}$$

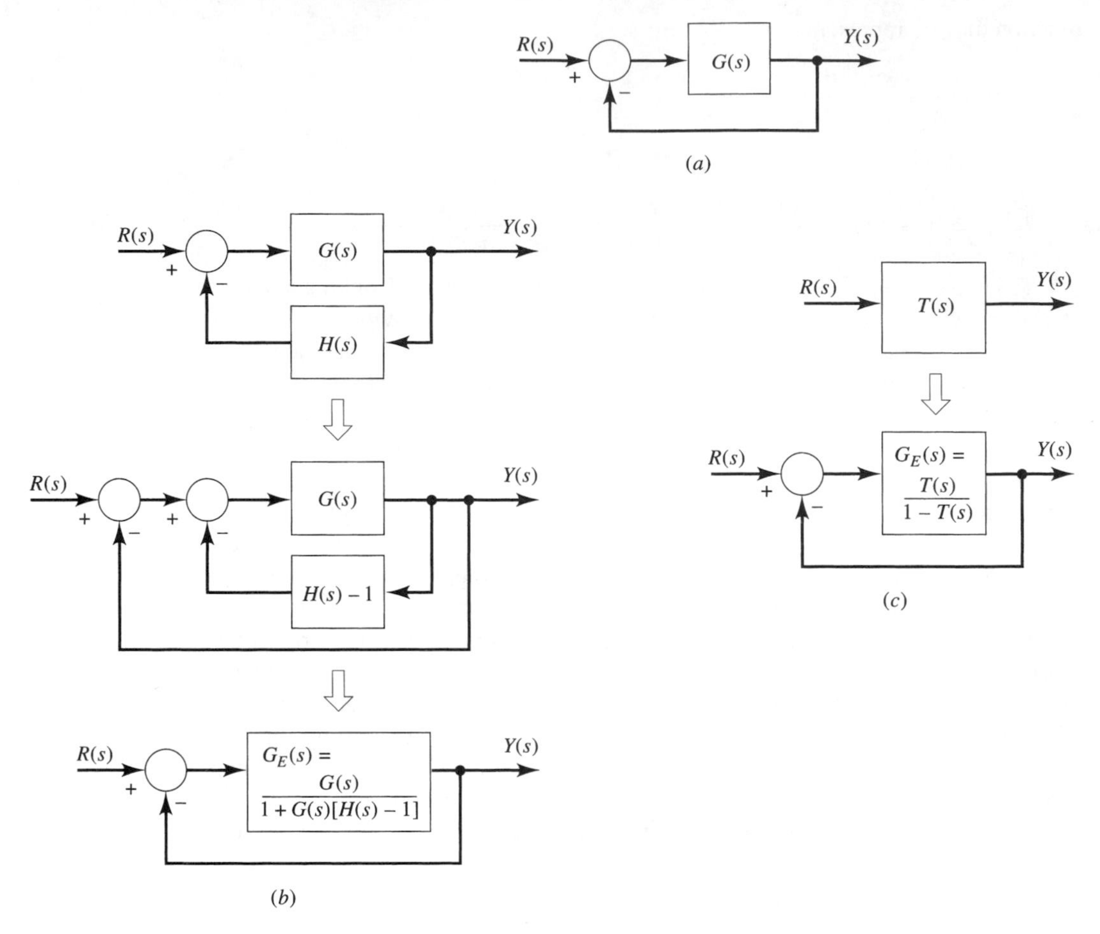

Figure 3.7 (a) A unity feedback system. (b) Converting a feedback system to an equivalent unity feedback system. (c) General conversion of a system to an equivalent unity feedback system.

Hence it is seen that adding factors of s to $D(s)$ raises the system type number, as long as the resulting system is stable (and the system is unity feedback).

A system that does not have unity feedback but does have a transfer function $T(s)$ possesses an equivalent unity feedback transmittance $G_E(s)$ as in Figure 3.7(c). For example, any system of the form of Figure 3.7(b) has the indicated $G_E(s)$. In general, a unity feedback system [for which $G_E(s)$ is just $G(s)$ and $H(s) = 1$] should be treated by the methods that follow. It will usually be easier to treat a nonunity feedback system using the $T_E(s)$ approach.

Consider the unity feedback system Figure 3.8. The difference between the input and the output is formed, amplified, and applied to the plant input in such a way as to reduce this difference. The system transfer function is

$$T(s) = \frac{Y(s)}{R(s)} = \frac{K/(s+2)}{1 + K/(s+2)} = \frac{K}{s+2+K}$$

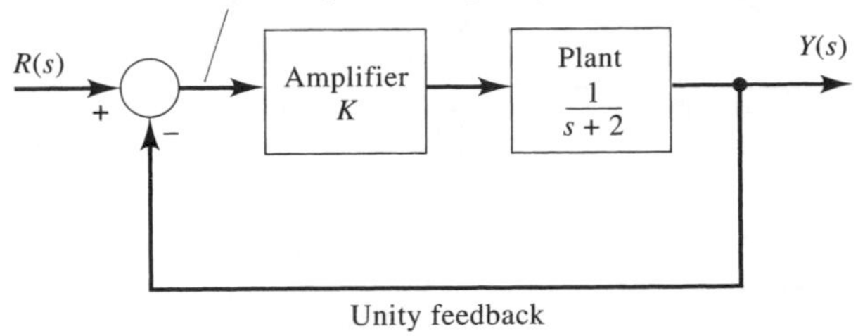

Figure 3.8 A unity feedback system.

For a step input, we write

$$R(s) = \frac{A}{s} \qquad Y(s) = T(s)R(s) = T(s)\frac{A}{s}$$

and the error between input and output is

$$E(s) = R(s) - Y(s) = \frac{A}{s}[1 - T(s)] = \frac{A}{s}\left(\frac{s+2}{s+2+K}\right)$$

The error reaches a steady state for any positive K (in fact, for any K larger than -2) and is

$$\lim_{t\to\infty} e(t) = \lim_{s\to\infty} sE(s) = \lim_{s\to\infty} A\left(\frac{s+2}{s+2+K}\right) = \frac{2A}{2+K}$$

which can be made arbitrarily small in this case by choosing a sufficiently large amplifier gain K.

For a ramp, parabolic, or higher power-of-t input, the error for the example system becomes infinite. For a ramp input

$$R(s) = \frac{A}{s^2}$$

we have

$$E(s) = \left(\frac{A}{s^2}\right)\left(\frac{s+2}{s+2+K}\right) = \frac{K_1}{s} + \frac{K_2}{s^2} + \frac{K_3}{s+2+K}$$

The term K_2/s^2 in the partial fraction expansion of $E(s)$ corresponds to the time function K_2t, which grows without bound.

A simple method of obtaining zero steady state error for a step input is to drive the plant with a signal proportional to the integral of the error. This places an $s = 0$ pole in the system's forward transmittance, raising the type number. The addition of an integrator to the preceding example system is shown in Figure 3.9. For this system,

Including an integrator raises type number.

$$T(s) = \frac{K/s(s+2)}{1 + K/s(s+2)} = \frac{K}{s^2 + 2s + K}$$

For a step input

$$R(s) = \frac{A}{s}$$

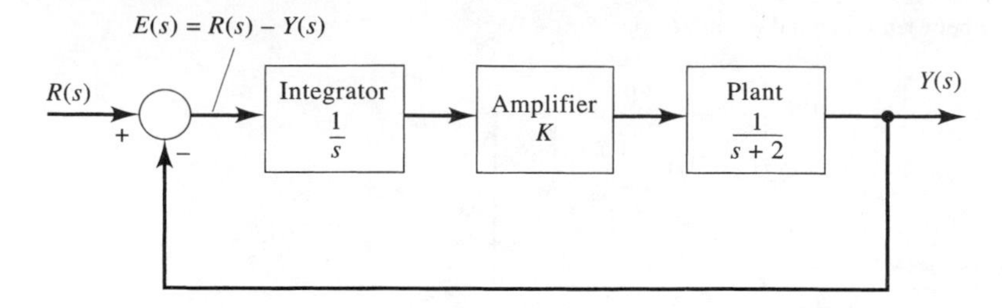

Figure 3.9 A tracking system with error integration.

the new error signal is given by

$$E(s) = R(s) - Y(s) = \frac{A}{s}[1 - T(s)] = \frac{A}{s}\left(\frac{s^2 + 2s}{s^2 + 2s + K}\right)$$

The steady state error for any positive K is

$$\lim_{t\to\infty} e(t) = \lim_{s\to 0} sE(s) = 0$$

This result occurs because of the factor of s in the numerator of $E(s)$. Without the integrator, the numerator of $E(s)$ contains no such factor and a finite steady state error results for a step input.

For a ramp input

$$R(s) = \frac{A}{s^2}$$

the steady state error is

$$\lim_{t\to\infty} e(t) = \lim_{s\to 0} \frac{A\,(s+2)}{s^2 + 2s + K} = \frac{2A}{K}$$

3.4.4 Unity Feedback Error Coefficients

The *steady state error coefficients* of a unity feedback system are

$$\kappa_i = \lim_{s\to 0} s^i G(s)$$

Since the error transmittance is

$$T_E(s) = 1 - T(s) = 1 - \frac{G(s)}{1 + G(s)} = \frac{1}{1 + G(s)}$$

the error to a power-of-t input with power $i - 1$

$$R(s) = \frac{A}{s^i}$$

is

$$E(s) = T_E(s)R(s) = \frac{A}{s^i[1 + G(s)]}$$

When the limits exist and are finite,

$$\begin{pmatrix}\text{Steady state error}\\ \text{to input } A/s^i\end{pmatrix} = \lim_{s\to 0} sE(s) = \lim_{s\to 0} \frac{A}{s^{i-1}[1+G(s)]}$$

The steady state error coefficient is finite and nonzero when the power of time equals the system type number.

For a step input, $i = 1$,

$$\begin{pmatrix}\text{Steady state error}\\ \text{to input } A/s\end{pmatrix} = \lim_{s\to 0} \frac{A}{1+G(s)} = \frac{A}{1+\kappa_0}$$

For a step input, $i = 2$,

$$\begin{pmatrix}\text{Steady state error}\\ \text{to input } A/s^2\end{pmatrix} = \lim_{s\to 0} \frac{A}{s[1+G(s)]} = \lim_{s\to 0} \frac{A}{sG(s)} = \frac{A}{\kappa_1}$$

For higher power-of-t inputs,

$$\begin{pmatrix}\text{Steady state error}\\ \text{to input } A/s^i\end{pmatrix} = \lim_{s\to 0} \frac{A}{s^{i-1}[1+G(s)]} = \lim_{s\to 0} \frac{A}{s^{i-1}G(s)}$$

$$= \frac{A}{\kappa_{i-1}} i = 2, 3, 4, \ldots$$

These relations are summarized in Table 3.3. For the unity feedback type 0 system of Figure 3.8,

$$G(s) = \frac{K}{s+2}$$

$$\kappa_0 = \lim_{s\to 0} G(s) = \frac{K}{2}$$

giving

$$\begin{pmatrix}\text{Steady state error}\\ \text{to input } A/s\end{pmatrix} = \frac{A}{1+\kappa_0} = \frac{2A}{2+K}$$

Table 3.3 *Steady State Error of Unity Feedback Systems in Terms of Error Coefficients*

System Type	Steady State Error to Step Input $R(s) = A/s$	Steady State Error to Ramp Input $R(s) = A/s^2$	Steady State Error to Parabolic Input $R(s) = A/s^3$	Steady State Error to Input $R(s) = A/s^4$
0	$\frac{A}{1+\kappa_0}$	∞	∞	∞
1	0	$\frac{A}{\kappa_1}$	∞	∞
2	0	0	$\frac{A}{\kappa_2}$	∞
3	0	0	0	$\frac{A}{\kappa_3}$
⋮				

as was found earlier. When an integrator is added to this system, as in Figure 3.9, to make it type 1,

$$\kappa_0 = \lim_{s\to 0} G(s) = \lim_{s\to 0} \frac{K}{s(s+2)} = \infty$$

and the step error is

$$\begin{pmatrix}\text{Steady state error} \\ \text{to input } A/s\end{pmatrix} = \frac{A}{1+\kappa_0} = 0$$

For a ramp input

$$\kappa_1 = \lim_{s\to 0} \frac{K}{s+2} = \frac{K}{2}$$

and

$$\begin{pmatrix}\text{Steady state error} \\ \text{to input } A/s^2\end{pmatrix} = \frac{A}{\kappa_1} = \frac{2A}{K}$$

It is clear from this example that having more poles located at the origin of $G(s)$ tend to reduce the steady state errors. This concept is useful in designing closed-loop systems. The number of poles at the origin of $G(s)$ can be purposely chosen to provide the desired type and number, hence the desired steady state error performance. If, for example, it is necessary to have zero error to a ramp input to the tracking system, then two poles must be purposely placed at the origin of the $G(s)$ function.

❑ Computer-Aided Learning

C3.4 Find the closed-loop step and ramp responses for the following open-loop transfer functions, assuming unity feedback

(a) $G(s) = \dfrac{20}{s(s+1)(s+2)}$

(b) $G(s) = \dfrac{6}{s^2+2s+6}$

(c) $G(s) = \dfrac{(s+2)}{s^2(s+4)}$

Ans. (a)
```
ga=zpk([], [0,-1,-2],20)
ta=ga/(1+ga)
step(ta)
r1=zpk([],0,1)
sya=ta*r1
step(sya)
```

(b)
```
numgb=6
dengb=[1 2 6]
gb=tf(numgb, dengb)
```

```
tb=gb/(1+gb)
step (tb)
r1=zpk([],0,1)
syb=tb*r1
step(syb)
```

(c)
```
gc=zpk(-2,[0,0,-4],1)
tc=gc/(1+gc)
step(tc)
r1=zpk([],0,1)
syc=tc*r1
step(syc)
```

C3.5 Use MATLAB to get the step error and ramp error responses for the system of C3.4(b), Assume that r1 and tb have been created.

Ans.
```
teb=1-tb
step(teb)
seb=teb*r1
step(seb)
```

❑ DRILL PROBLEMS

D3.6 Find output–input error transmittance of each of the systems in Figure D3.6. Then determine the type number of each system and, if the response reaches steady state, steady state errors to unit step and to unit ramp inputs.

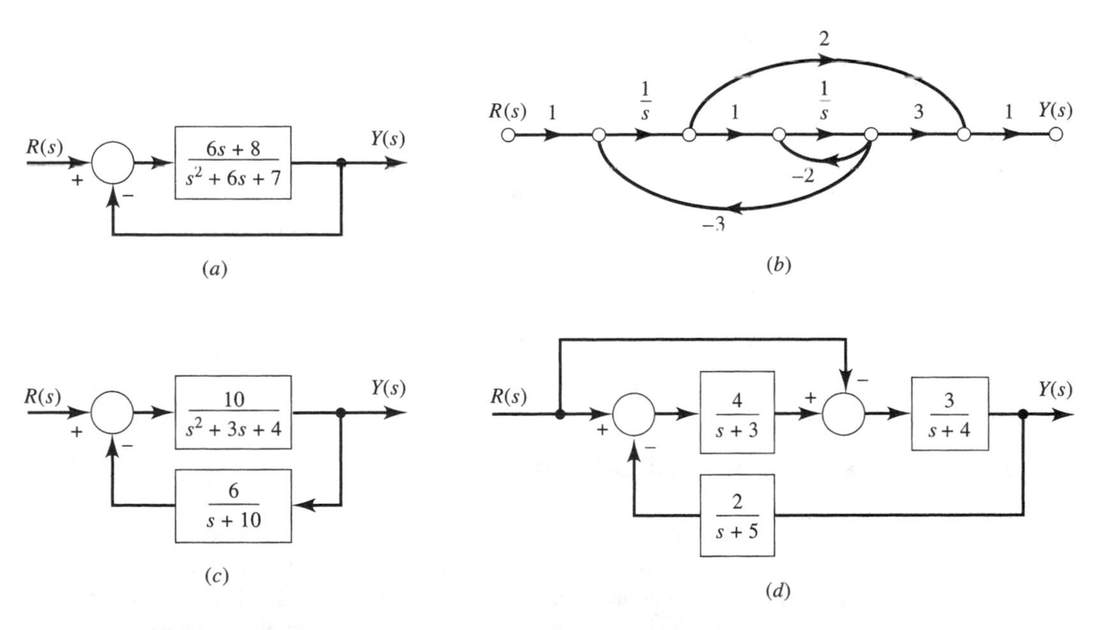

Figure D3.6

Ans. $0, \frac{7}{15}, \infty;\quad 0, -\ \frac{4}{3}, \infty;\quad 1, 0, \frac{6}{25}; 0, \frac{23}{28}, \infty$

D3.7 For unity feedback systems with the following forward transmittances, determine each system type number and, if the response reaches steady state, find steady state output–input errors to unit step and to unit ramp inputs.

(a)

$$G(s) = \frac{10}{s^3 + 8s^2 + 2s}$$

Ans. $1, 0, \frac{1}{5}$

(b)

$$G(s) = \frac{1}{s^3 + 2s^2 + s + 3}$$

Ans. 0, unstable

(c)

$$G(s) = \frac{4(s+1)}{(s+2)(s+3)(s^2+s+10)}$$

Ans. $0, \frac{15}{16}, \infty$

(d)

$$G(s) = \frac{2s+1}{2s^4 + 4s^3 + 4s}$$

Ans. 1, unstable

D3.8 Find the error coefficients for each of the systems in Figure D3.8. Then, if the responses reach steady state, find the steady state output–input errors to unit step and to unit ramp inputs.

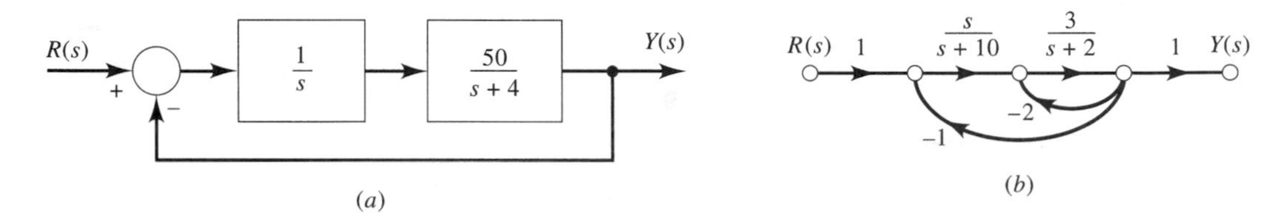

Figure D3.8

Ans. $\kappa_0 = \infty, \kappa_1 = 12.5, \kappa_2 = 0, 0, 0.08;\quad \kappa_0 = 0, \kappa_1 = 0, 1, \infty$

3.5 Performance Indices and Optimal Systems

A system design problem generally reaches the point at which one or more parameters are to be selected to give the best performance. If a measure or index of performance can be expressed mathematically, the problem can be solved for the best choice of the adjustable parameters. The resulting system is termed *optimal* with respect to the selection criteria. Optimization generally proceeds as in Figure 3.10. A given system provides a value for some given performance index. The adjustable parameters

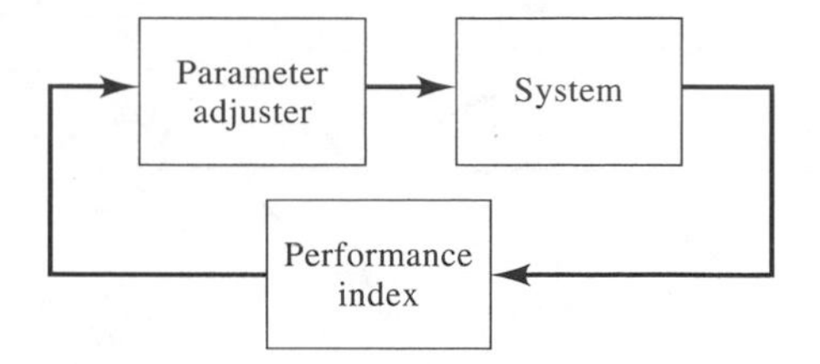

Figure 3.10 Optimal parameter selection.

are modified to maximize (or minimize, depending on the application) the performance index. The term *optimization* pertains to both maximization and minimization. Optimization can occur before the system begins operation (off-line optimization), or the parameters may vary as the system operates (on-line optimization).

Selection of an appropriate performance index is as much a part of the design process as fabricating the final system. An optimal value for an inappropriate performance measure may result in poor performance, by other standards. Imagine that a system minimizes rather than maximizes profit, for example.

A commonly used performance index J is the integral of the square of the error to a step input

$$J = I_{SE} = \int_0^{\infty} e^2_{\text{step}}(t)dt$$

Adjusting Parameters.

If the step error is expressed as a function of the adjustable parameters, the index can be minimized with respect to the parameters, yielding the optimal parameter values.

For this I_{SE} to be finite, the steady state value of $e_{\text{step}}(t)$ must be zero; otherwise the I_{SE} would be infinite. For finite I_{SE} the system must be type 1, causing $e_{\text{step}}(\infty)$ to be zero. In the following, $e(t)$ and $E(s)$ are assumed to be for a step input, each system is type 1, and J refers to I_{SE}.

Consider the adjustable system of Figure 3.11(a), for which it is desired to choose the parameter k to give minimum integral square error to a step input. The system transfer function, found by using Mason's gain rule on the system signal flow graph in Figure 3.11 (b), is

$$T(s) = \frac{2/s(s+3) - k/(s+3)}{1 + 2/s(s+3)} = \frac{2 - ks}{s^2 + 3s + 2} \qquad [\mathbf{3.6}]$$

The error transmittance is

$$T_E(s) = 1 - T(s) = \frac{s^2 + (3+k)s}{s^2 + 3s + 2}$$

and the error to a step input is given by

$$E(s) = \frac{A}{s}T_E(s) = A\frac{s+3+k}{s^2+3s+2} = A\left(\frac{k+2}{s+1} + \frac{-k-1}{s+2}\right) \qquad [\mathbf{3.7}]$$

As a function of time, the error signal is

$$e(t) = \mathcal{L}^{-1}[E(s)] = A[(k+2)e^{-t} - (k+1)e^{-2t}]u(t)$$

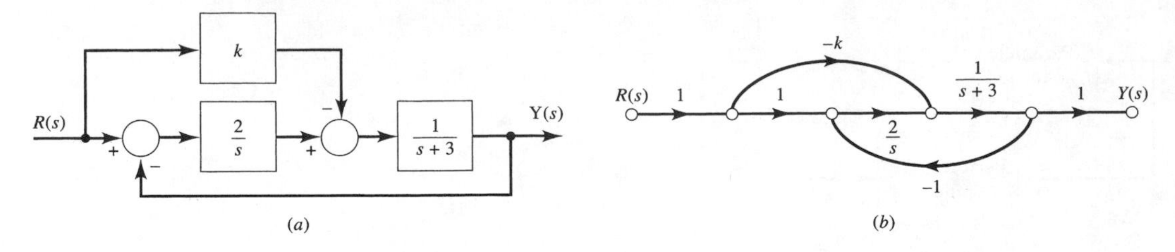

Figure 3.11 An adjustable feedback system. (a) System block diagram. (b) System signal flow graph.

The square of the error is

$$e^2(t) = A^2[(k+2)^2e^{-2t} - 2(k+1)(k+2)e^{-3t} + (k+1)^2e^{-4t}]u(t)$$

and the integral square error, in terms of k, is

$$\begin{aligned} J(k) &= \int_0^\infty e^2(t)dt = A^2\left[(k^2+4k+4)\left(\frac{e^{-2t}}{-2}\right)\right. \\ &\qquad \left.-2(k^2+3k+2)\left(\frac{e^{-3t}}{-3}\right)\right] + (k^2+2k+1)\left(\frac{e^{-4t}}{-4}\right)\Bigg]_0^\infty \\ &= A^2\left[(k^2+4k+4)\left(\tfrac{1}{2}\right) - 2(k^2+3k+2)\left(\tfrac{1}{3}\right)\right. \\ &\qquad \left.+(k^2+2k+1)\left(\tfrac{1}{4}\right)\right] \\ &= \frac{A^2}{12}[k^2+6k+11] \end{aligned}$$

As a function of k, $I_S(k)$ is a parabola, with minimum given by

$$\frac{dJ}{dk} = \frac{A^2}{12}[2k+6] = 0 \qquad k = -3$$

Thus the optimal system, in the sense of minimum integral square error to a step with all parameters but k fixed, is the one with this value of k.

An easier way.

Instead of the time domain calculation of this J, the complex frequency domain can be used. According to a theorem in mathematics by Parseval

$$J = I_{SE} = \int_0^\infty e^2(t)dt = \frac{1}{2\pi j}\int_{-j\infty}^{j\infty} E(s)E(-s)ds \qquad \textbf{[3.8]}$$

Where $E(s)$ can be expressed as follows:

$$E(S) = \frac{N_{n-1}s^{n-1} + \cdots + N_1 s + N_o}{D_n s^n + D_{n-1}s^{n-1} + \cdots + D_1 s + D_0} \qquad \textbf{[3.9]}$$

assuming type 1 behavior.

J follows from complex variable theory. To clarify the effect of system order, the subscript for J will be the system order. For an nth-order system,

$$J_n = (-1)^{n-1}\frac{B_n}{2D_nH_n} \qquad \textbf{[3.10]}$$

where H_n and B_n are determinants. H_n is the determinant of the $n \times n$ Hurwitz matrix. The first two rows of the Hurwitz matrix are formed from the coefficients of $D(s)$, while the remaining rows consist of right-shifted versions of the first two rows until the $n \times n$ matrix is formed. Thus we write

$$H_n = \begin{vmatrix} D_{n-1} & D_{n-3} \cdots\cdots & \\ D_n & D_{n-2} \cdots\cdots & \\ \emptyset & D_{n-1} & D_{n-3} \cdots \\ \emptyset & D_n & D_{n-2} \cdots \\ \cdots & \cdots & \cdots \end{vmatrix} \qquad [3.11]$$

The determinant B_n is found by first calculating

$$N(s)N(-s) = b_{2n-2}s^{2n-2} + \cdots + b_2 s^2 + b_0 \qquad [3.12]$$

Then the first row of the Hurwitz matrix is replaced by the coefficients of $N(s)N(-s)$ while the remaining rows are unchanged.

$$B_n = \begin{vmatrix} b_{2n-2} & \cdots\cdots & b_2 \cdots b_0 \\ D_n & D_{n-2} \cdots\cdots & \\ \emptyset & D_{n-1} & D_{n-3} \cdots \\ \emptyset & D_n & D_{n-2} \cdots \\ \cdots & \cdots & \cdots \end{vmatrix} \qquad [3.13]$$

For example, suppose the error has a second-order transform

$$E(s) = \frac{N_1 s + N_0}{D_2 s^2 + D_1 s + D_0}$$

so that

$$N(s)N(-s) = b_2 s^2 + b_0$$
$$= -N_1^2 s^2 + N_0^2$$

$$H_2 = \begin{vmatrix} D_1 & \emptyset \\ D_2 & D_0 \end{vmatrix} = D_1 D_0$$

$$B_2 = \begin{vmatrix} -N_1^2 & N_0^2 \\ D_2 & D_0 \end{vmatrix} = -N_1^2 D_0 - N_0^2 D_2$$

$$J_2 = (-1)\frac{B_2}{2D_2 H_2}$$
$$= \frac{N_1^2 D_0 + N_0^2 D_2}{2D_2 D_1 D_0}$$

Proceeding in this way, one can find any J_n. Table 3.4 contains J_1, J_2, and J_3. Remember that the index for J is the order of the denominator of $E(s)$. The product of $E(s)$ $E(-s)$ is of twice that order. We can rework the problem of Figue 3.11, where

$$E(s) = \frac{As + A(3+k)}{s^2 + 3s + 2} = \frac{N_1 s + N_0}{D_2 s^2 + D_1 s + D_0}$$

Table 3.4 *Values for the J integral*

n	J_n
1	$\dfrac{N_0^2}{2D_0D_1}$
2	$\dfrac{N_1^2D_0 + N_0^2D_2}{2D_0D_1D_2}$
3	$\dfrac{N_2^2D_0D_1 + (N_1^2 - 2N_0N_2)D_0D_3 + N_0^2D_2D_3}{2D_0D_3(-D_0D_3 + D_1D_2)}$

$$N_1 = A \qquad N_0 = A(3+k)$$
$$D_2 = 1 \qquad D_1 = 3 \qquad D_0 = 2$$

Then the value of J_2 follows from Equation (3.10) and Table 3.4.

$$J_2 = \frac{A^2(2) + A^2 + (3+k)^2}{2(2)(3)}$$
$$= \frac{A^2}{12}(k^2 + 6k + 11)$$

which verifies the earlier result, but the time domain is avoided; and, better yet, we have a general formula for J_2 that works for any stable system.

Determination of the "best" damping ratio for a second-order system offers another example of the use of the performance indices. Consider a stable second-order system with complex conjugate characteristic roots, with transfer function of the form

$$T(s) = \frac{\omega_n^2}{s^2 + 2\zeta\omega_n s + \omega_n^2}$$

A damping ratio of 0.5 minimizes the integral of the squared error for a step input.

For a unit step input, the error between output and input is given by

$$E(s) = \frac{1}{s}[1 - T(s)] = \frac{1}{s}\left[\frac{s^2 + 2\zeta\omega_n s}{s^2 + 2\zeta\omega_n s + \omega_n^2}\right] \tag{3.14}$$
$$= \frac{s + 2\zeta\omega_n}{s^2 + 2\zeta\omega_n s + \omega_n^2}$$

Then for Equation (3.14) we have

$$N_0 = 2\zeta\omega_n \qquad N_1 = 1$$
$$D_0 = \omega_n^2 \qquad D_1 = 2\zeta\omega_n \qquad D_2 = 1$$

$$J_2(\zeta) = \frac{N_1^2 D_0 + N_0^2 D_2}{2D_0 D_1 D_2} = \frac{\omega_n^2 + 4\zeta^2\omega_n^2}{4\zeta\omega_n^3}$$

$$= \frac{1 + 4\zeta^2}{4\zeta\omega_n} \qquad [3.15]$$

The best damping ratio ζ is the value that causes

$$\frac{dJ_2(\zeta)}{d\zeta} = 0 = \frac{4\zeta\omega_n\,(d/d\zeta)\,(1 + 4\zeta^2) - \left[(1 + 4\zeta^2)\,(d/d\zeta)\,(4\zeta\omega_n)\right]}{4\zeta\omega_n^2}$$

$$4\omega_n\left[8\zeta^2 - 4\zeta^2 - 1\right] = 0$$

$$4\zeta^2 = 1$$

$$\zeta^2 = \tfrac{1}{4}$$

$$\zeta = \tfrac{1}{2}$$

Thus, the lowest integral squared error occurs with a damping ratio of 0.5. Figure 3.12 shows $\omega_n I_{SE}$ versus damping ratio, where the minimum value is 1 for $\zeta = 0.5$.

Other useful performance indices can involve errors for other test signals, such as a ramp input, and include the integral of the magnitude of the error,

$$I_M = \int_0^\infty |e(t)|\,dt$$

and other performance measures

$$I_{TS} = \int_0^\infty te^2(t)\,dt$$

$$I_{TM} = \int_0^\infty t\,|e(t)|\,dt$$

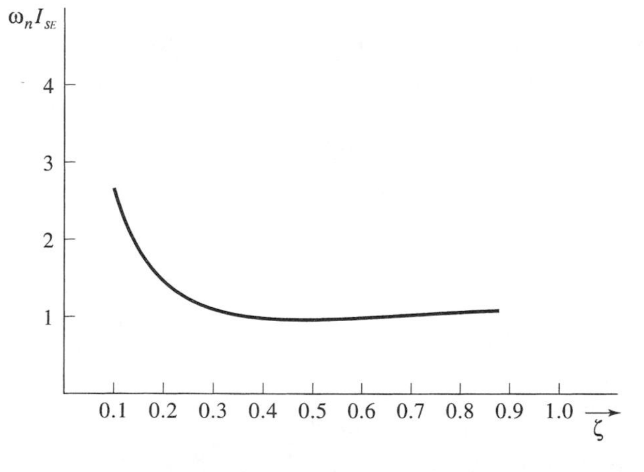

Figure 3.12 Integral square error performance measure for a certain second-order system with adjustable damping ratio.

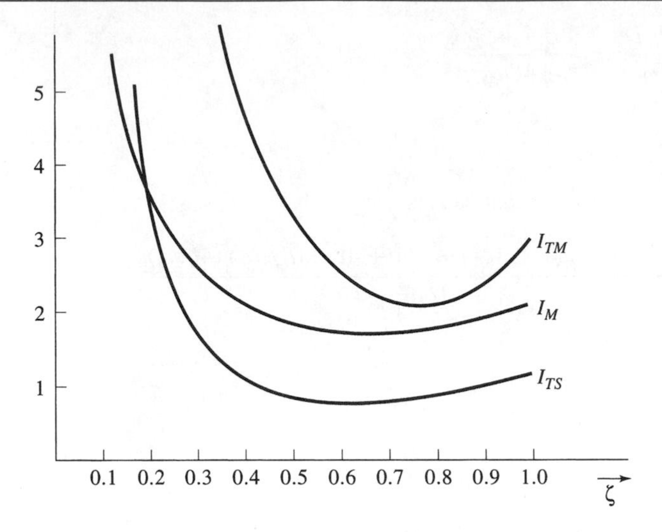

Figure 3.13 Other performance measures for a certain second-order system with adjustable damping ratio.

Figure 3.13 shows the performance I_M, I_{TS}, and I_{TM} for the second-order system of Equation (3.8). Clearly, the optimum value of ζ depends upon the definition of "goodness," the performance measure used. Minimum integral magnitude error I_M for the example system occurs for $\zeta = 0.67$. Minimum I_{TS} occurs for $\zeta = 0.6$, and minimum I_{TM} for $\zeta = 0.7$.

In some situations, such as in the control of a spacecraft for minimal fuel use, the performance index is clearly given by the design objectives. In others, an index, if used, must be chosen somewhat arbitrarily. In the latter case, choosing desired response characteristics such as rise time, overshoot and steady state error directly may be much more sensible than choosing a performance index.

❑ DRILL PROBLEMS

D3.9 Use Table 3.4 to determine the optimum choice of the parameter k for the given transfer function to obtain minimum I_{SE}.

(a)

$$T(s) = \frac{100}{s^2 + ks + 100}$$

Ans. 10

(b)

$$T(s) = \frac{k}{2s^2 + s + k}$$

Ans. 0.5

(c)

$$T(s) = \frac{k}{s^2 + ks + k}$$

Ans. 1.0

(d)

$$T(s) = \frac{10k}{s^2 + 8s + 10k}$$

Ans. 6.4

D3.10 Find the optimum choice of the parameter k, with a minimum square error step response performance measure, for a system with output–input error transmittance

$$T_E(s) = \frac{ks^2 + (1 - k)s}{s^2 + 3s + 2}$$

Ans. $\frac{1}{3}$

3.6 System Sensitivity

3.6.1 Calculating the Effects of Changes in Parameters

One of the major advantages of feedback is that it can be used to make the response of a system relatively independent of certain types of changes or inaccuracies in the plant model. For example, in the system of Figure 3.14(a), the nominal system transfer function is

$$T(s) = \frac{400/(s + 2)}{1 + 400/(s + 2)} = \frac{400}{s + 402}$$

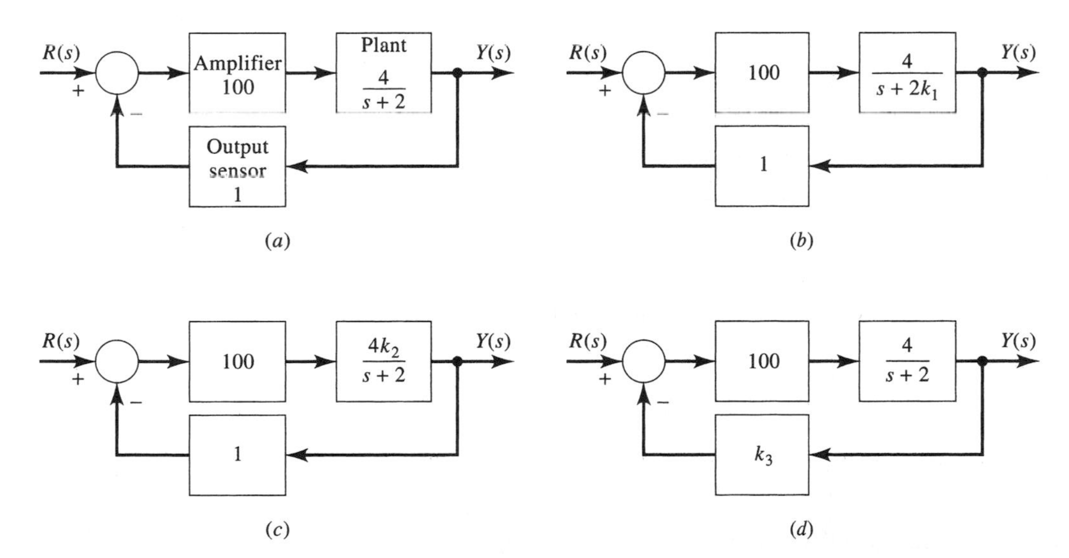

Figure 3.14 Determining the effects of changes in the parameters of a feedback system. (a) The nominal system. (b) Perturbation of plant pole. (c) Perturbation of plant gain. (d) Perturbation of the sensor gain.

Feedback reduces sensitivity to parameter changes.

Now suppose one of the plant parameters changes or is wrongly modeled, as in 3.14(b). For $k_1 = 1$, the plant is the nominal one; other values of k_1 correspond to perturbations from the nominal plant. In terms of k_1,

$$T(s) = \frac{400/(s+2k_1)}{1+400/(s+2k_1)} = \frac{400}{s+400+2k_1}$$

and it is seen that even 50% changes in the parameter, ranging from $k_1 = \frac{1}{2}$ to $k_1 = \frac{3}{2}$, result in a relatively minor change in $T(s)$. Even negative values of k_1 (say $k_1 = -1$), for which the plant is unstable, give much the same, stable, overall transfer function $T(s)$.

The system's steady state error to a unit step input is, in terms of k_1,

$$\lim_{s\to 0} s\left(\frac{1}{s}\right)[1 - T(s)] = \lim_{s\to 0} \frac{s+2k_1}{s+400+2k_1} = \frac{2k_1}{400+2k_1}$$

which is dominated by the factor of 400 and is nearly proportional to k_1.

For the parameter perturbed by k_2 in Figure 3.14(c),

$$T(s) = \frac{400k_2/(s+2)}{1+400k_2/(s+2)} = \frac{400k_2}{s+400k_2+2}$$

For this parameter, the system's steady state error (between output and input) to a unit step input is

$$\lim_{s\to 0} s\left(\frac{1}{s}\right)[1 - T(s)] = \lim_{s\to 0} \frac{s+2}{s+400k_2+2} = \frac{2}{400k_2+2}$$

This steady state error is dominated by the factor of $400k_2$ for moderate changes in k_2 from the nominal $k_2 = 1$ and is nearly inversely proportional to k_2. Changes from the nominal amplifier gain of 400 will produce the same effects on $T(s)$ and its step response.

If the sensor gain is perturbed, as in Figure. 3.14(d), we have

$$T(s) = \frac{400/(s+2)}{1+400k_3/(s+2)} = \frac{400}{s+400k_3+2}$$

The steady state error to a unit step input is

$$\lim_{s\to o} s\left(\frac{1}{s}\right)[1 - T(s)] = \frac{s+400(k_3-1)+2}{s+400k_3+2} = \frac{400(k_3-1)+2}{400k_3+2}$$

which can become quite large in comparison to the earlier expressions for comparable percentage parameter changes. In this case, the result is expected, since an error by the sensor in the perceived plant output is indistinguishable by the rest of the system from an actual output error.

3.6.2 Sensitivity Functions

Sensitivity is defined.

In general, the *sensitivity* of a single-output system transfer function to changes in a specific parameter a is defined as

$$S_a = \lim_{\Delta a\to 0} \frac{\Delta T/T}{\Delta a/a} = \lim_{\Delta a\to 0} \frac{a}{T}\frac{\Delta T}{\Delta a} = \frac{a}{T}\frac{\partial T}{\partial a}$$

It is the limiting ratio of the fractional change in the transfer function to the fractional change in the parameter. Generally, S_a may be reduced by using feedback.

For example, the feedback system with constant "block" transmittances in Figure 3.15 might represent a feedback amplifier over a range of frequencies. The transfer function of this system is the constant

$$T = \frac{G}{1+GH} = \frac{10}{1+\frac{10}{3}} = \frac{30}{13}$$

The sensitivity of T to changes in G is

$$S_G = \frac{G}{T}\frac{\partial T}{\partial G} = \frac{1}{1+GH}$$

which for $G = 10$, $H = \frac{1}{3}$ is

$$S_G = \frac{1}{1+\frac{10}{3}} = \frac{3}{13}$$

This is to say that, with the feedback, the transfer function changes only $\frac{3}{13}$ as much with small changes in G as it would without feedback. As long as H is greater than zero, S_G is less than 1 (the value without feedback).

The sensitivity of T to changes in H is

$$S_H = \frac{H}{T}\frac{\partial T}{\partial H} = \frac{-GH}{1+GH}$$

which for $G = 10$, $H = \frac{1}{3}$ is

$$S_H = \frac{-\frac{10}{3}}{1+\frac{10}{3}} = -\frac{10}{13}$$

The transfer function is affected by changes in H much more than by changes in G. The minus sign in S_H indicates that T decreases with an increase in H.

Feedback reduces the sensitivity.

For the feedback system of Figure 3.16, suppose that the plant parameter a is nominally $a = 2$ but is subject to small changes about the nominal value. The transfer function of the system is

$$T(s) = \frac{1/(s+a)}{1+K/(s+a)} = \frac{1}{s+a+K}$$

The sensitivity of $T(s)$ to changes in a is given by

$$S_a = \frac{a}{T}\frac{\partial T}{\partial a} = a(s+a+K)\frac{-1}{(s+a+K)^2} = \frac{-a}{s+a+K}$$

Sensitivities are generally functions of the complex variable s. For $a = 2$, the sensitivity is

$$S_a = \frac{-2}{s+2+K}$$

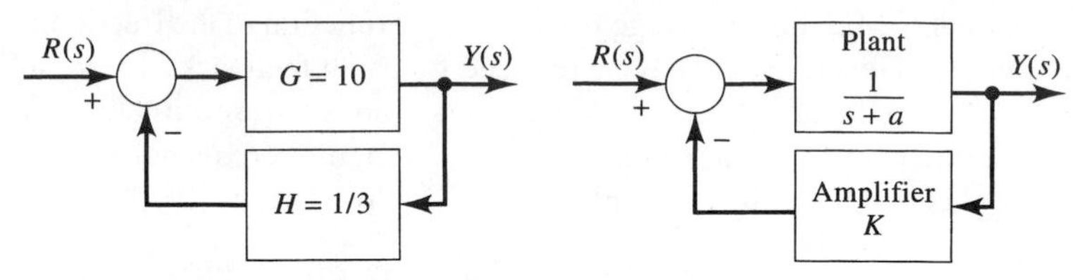

Figure 3.15 Finding the sensitivity of a feedback system with constant transmittances.

Figure 3.16 Finding the sensitivity of a feedback system to changes in a plant parameter.

Increasing K reduces sensitivity.

which can, for any, s, be reduced by making K sufficiently large. Without control, $K = 0$ and S_a becomes fixed as a function of s. Without control S_a has a value that becomes infinite for $s = -2$. In general, control provides a mechanism for reducing sensitivity to parameter variations in much the same way that control can reduce the influence of disturbances.

As a more involved example, consider the precision positioning system of Figure 3.17, for which the transfer function, as a function of the parameter a, is

$$T\,(s, a) = \frac{\left(\frac{s+3}{s}\right)\left(\frac{10}{s^2 + as + 10}\right)}{1 + \left(\frac{s+3}{s}\right)\left(\frac{10}{s^2 + as + 10}\right)} = \frac{10s + 30}{s^3 + as^2 + 20s + 30}$$

$$\frac{\partial T}{\partial a} = \frac{-s^2\,(10s + 30)}{\left(s^3 + as^2 + 20s + 30\right)^2}$$

and

$$S_a = \frac{a}{T}\frac{\partial T}{\partial a} = \left[\frac{a}{\left(\frac{10s + 30}{s^3 + as^2 + 20s + 30}\right)}\right]\left[\frac{-s^2\,(10s + 30)}{\left(s^3 + as^2 + 20s + 30\right)^2}\right]$$

$$= \frac{-as^2}{s^3 + as^2 + 20s + 30}$$

For the nominal value of $a = 2$

$$S_a = \frac{-2s^2}{s^3 + 2s^2 + 20s + 30}$$

For small changes in a about the nominal value of $a = 2$,

$$S_a \cong \frac{a}{T}\frac{\Delta T}{\Delta a}$$

$$\frac{\Delta T\,(s)}{T\,(s)} \cong \frac{\Delta a}{a} Sa = \frac{\Delta a}{a}\frac{-2s^2}{s^3 + 2s^2 + 20s + 30} \qquad [3.16]$$

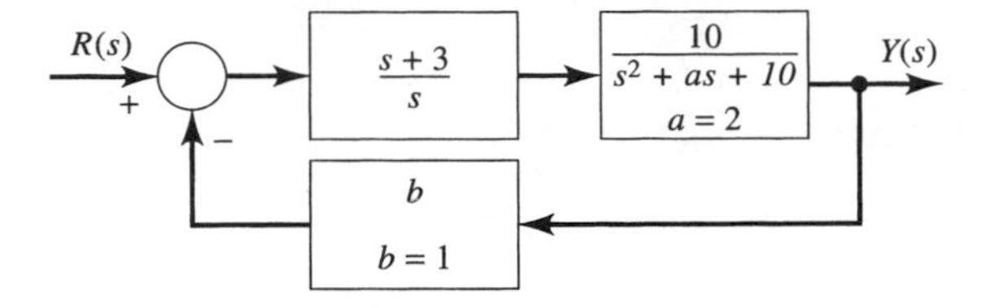

Figure 3.17 Calculating sensitivities to parameter changes for precision positioning system.

For a specific value of s, Equation (3.16) relates fractional changes in the transfer function to fractional small changes in the parameter. In calculating the response to a step input to the system, for example, the transfer function is evaluated at $s = 0$ to obtain the residue corresponding to the forced response. Equation (3.2) with $s = 0$ gives

$$\frac{\Delta T(0)}{T(0)} \cong 0 \qquad \Delta T(0)\, 0$$

for small changes Δa.

For the parameter b in the feedback path of the system of Figure 3.17,

$$T(s, b) = \frac{\left(\dfrac{s+3}{s}\right)\left(\dfrac{10}{s^2+2s+10}\right)}{1 + b\left(\dfrac{s+3}{s}\right)\left(\dfrac{10}{s^2+2s+10}\right)}$$

$$= \frac{10s + 30}{s^3 + 2s^2 + (10 + 10b)s + 30b}$$

so that

$$\frac{\partial T}{\partial b} = \frac{-(10s+30)(10s+30)}{\left(s^3 + 2s^2 + 10s + 10bs + 30b\right)^2}$$

For the nominal $b = 1$

$$\frac{\partial T}{\partial b} = \frac{-(10s+30)(10s+30)}{\left(s^3 + 2s^2 + 20s + 30\right)^2}$$

and

$$S_b = \frac{b}{T}\frac{\partial T}{\partial b} = \left[\frac{1}{\left(\dfrac{10s+30}{s^3+2s^2+20s+30}\right)}\right]\left[\frac{-(10s+30)(10s+30)}{(s^3+2s^2+20s+30)^2}\right]$$

$$= \frac{-10s - 30}{s^3 + 2s^2 + 20s + 30}$$

For small changes Δb,

$$\frac{\Delta T(s)}{T(s)} \cong \frac{\Delta b}{b} S_b = \left(\frac{\Delta b}{b}\right) \frac{-10s - 30}{s^3 + 2s^2 + 20s + 30}$$

At $s = 0$, for calculating the forced response to a step input,

$$\frac{\Delta T\ (0)}{T\ (0)} \cong \left(\frac{\Delta b}{b}\right) (-1)$$

Changes in T are proportional (in the opposite sense) to changes in b.

3.6.3 Sensitivity to Disturbance Signals

Another major advantage of feedback is that it can be used to reduce the effects of disturbance inputs upon system response. For example, for the thermal control system of Figure 3.18(a), a disturbance signal $D(s)$ affects the plant but is not accessible to the designer. The transfer function relating $Y(s)$ to $D(s)$ is

$$T_D(s) = \frac{1}{s + 2}$$

For a unit step disturbance input, the final value of the output due to the disturbance is given by

$$Y(s) = \frac{1}{s} \frac{1}{s + 2}$$

$$\lim_{t \to \infty} y\ (t) = \lim_{s \to 0} sY(s) = \tfrac{1}{2}$$

If the plant is driven in the feedback arrangement of Figure 3.18(b), the transfer function relating $Y(s)$ to $D(s)$ becomes, instead,

$$T_D(s) = \frac{1/\ (s + 2)}{1 - [-K/\ (s + 2)]} = \frac{1}{s + 2 + K}$$

For a unit step disturbance input to the feedback system, the resulting steady state output is given by

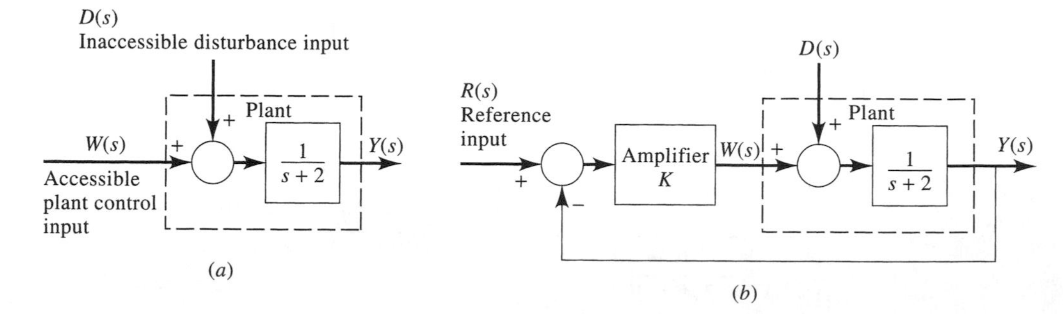

Figure 3.18 Disturbance input to a system with and without feedback. (a) Plant with control and disturbance inputs. (b) Plant with feedback.

$$Y(s) = \frac{1}{s}\left(\frac{1}{s+2+K}\right)$$

$$\lim_{t\to\infty} y(t) = \lim_{s\to 0} sY(s) = \frac{1}{2+K}$$

which can be made arbitrarily small by making K sufficiently large.

Another example system is shown in Figure 3.19(a). The disturbance signal $D(s)$ represents an unwanted, largely unknown effect upon the plant. The two-system transfer functions are

$$T_R(s) = \frac{Y(s)}{R(s)} = \frac{10(s+3)/s(s^2+2s+10)}{1+10(s+3)/s(s^2+2s+10)} = \frac{10(s+3)}{s^3+2s^2+20s+30}$$

$$T_D(s) = \frac{Y(s)}{D(s)} = \frac{10/(s^2+2s+10)}{1+10(s+3)/s(s^2+2s+10)} = \frac{10s}{s^3+2s^2+20s+30}$$

The system is stable, as is easily determined from a Routh–Hurwitz test:

s^3	1	20
s^2	2	30
s^1	5	
s^0	30	

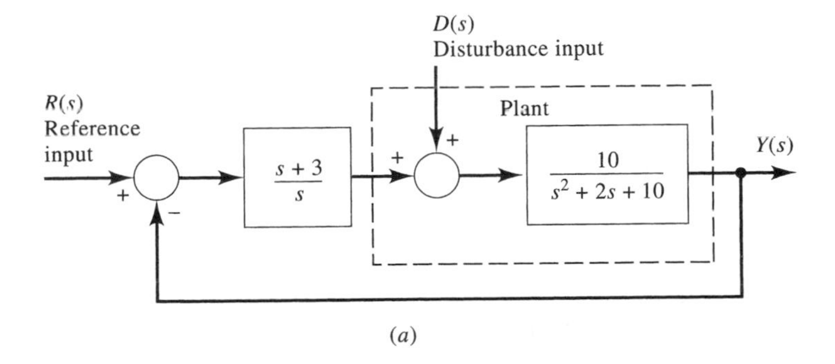

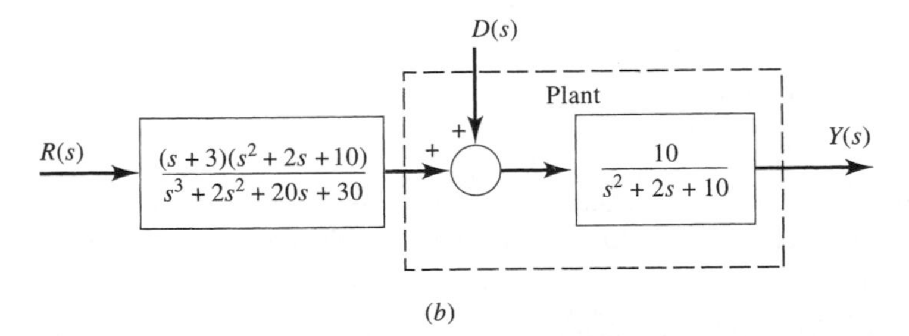

Figure 3.19 Reducing the effects of disturbance signals with feedback. (a) A feedback control system. (b) An open-loop system with the same relation between $Y(s)$ and $R(s)$.

A unit step disturbance to this system will produce zero contribution to the steady state output since

$$\lim_{s\to 0} s\left(\frac{1}{s}\right) T_D(s) = \lim_{s\to 0} \frac{10s}{s^3 + 2s^2 + 20s + 30} = 0$$

Now consider the open-loop (nonfeedback) system of Figure 3.19(b). It has the same transfer function relating $Y(s)$ and $R(s)$ as does the feedback system. For this system, however, the relationship between the output and the disturbance is not modified by feedback:

$$T_D(s) = \frac{Y(s)}{D(s)} = \frac{10}{s^2 + 2s + 10}$$

A unit step disturbance of the open-loop system will produce a unit contribution to the steady state output:

$$\lim_{s\to 0} s\left(\frac{1}{s}\right) T_D(s) = \lim_{s\to 0} \frac{10}{s^2 + 2s + 10} = 1$$

Feedback can improve disturbance rejection.

It is difficult to generalize about methods for improving disturbance rejection with feedback because most practical situations involve many specific structural constraints. For example, it may or may not be acceptable to sense (provide as an output) a certain plant signal or to supply a control signal to a certain part of the plant.

❑ DRILL PROBLEMS

D3.11 For the feedback system with constant "block" transmittances, find the sensitivity of each of the four transfer functions shown in Figure D3.11 to small changes in G_1 and to small changes in G_2.

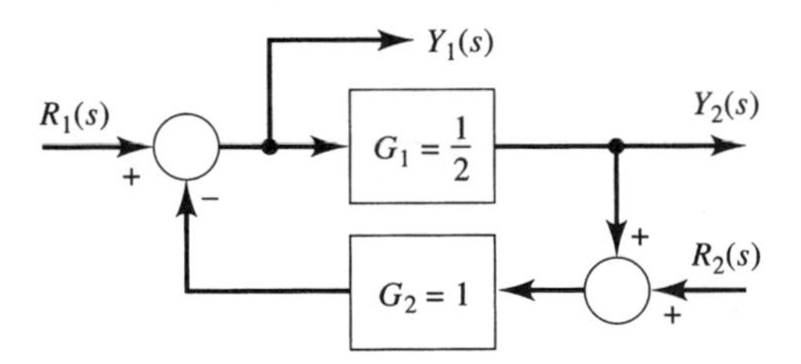

Figure D3.11

Ans. Sensitivity of T_{11} to G_1 is $-\frac{1}{3}$

D3.12 Find the sensitivities of the transfer functions of the systems shown in Figure D3.12 to small changes in k_1, k_2, and k_3 about the given nominal values.

Ans. (a) $(s+2)/(s+5)$, $-2/(s+5)$, $-3/(s+5)$;

(b) $(s^3 + 9s^2 + 26s + 24)/(s^3 + 9s^2 + 40s + 72)$,

$-3(s^2 + 6s + 16)/(s^3 + 9s^2 + 40s + 72)$,

$(3s^3 + 27s^2 + 128s + 240)/(s+4)(s^3 + 9s^2 + 40s + 72)$

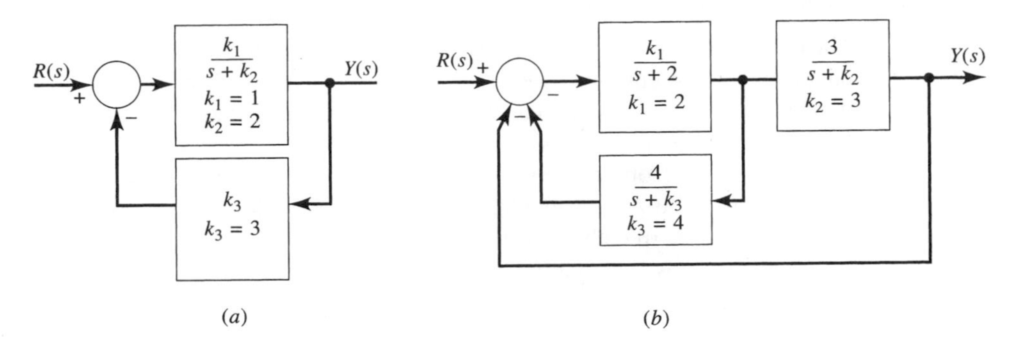

Figure D3.12

D3.13 For each of the systems shown in Figure D3.13, find the steady state error to a unit step input $R(s)$ and the steady state error to a unit step disturbance input $D(s)$.

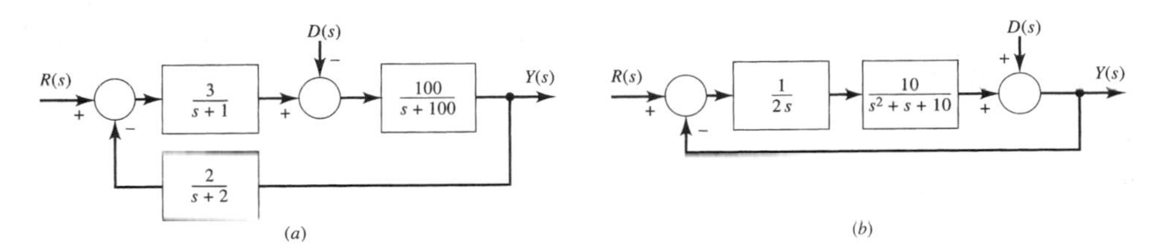

Figure D3.13

Ans. $\frac{1}{4}, \frac{5}{4};$ 0, 1

3.7 Time Domain Design

Most control system design uses frequency domain techniques, which are covered in subsequent chapters, as follows:

Nyquist methods, developed in 1932
Bode techniques, developed in 1945
Root locus methods, developed by Evans in 1948

Another significant body of knowledge, known as *time domain design,* exists, and its techniques are especially valuable for process control. Three such methods are presented in this chapter:

1. Integral squared error, introduced in Section 3.5
2. Ziegler–Nichols compensation
3. Chien–Hrones–Reswick (CHR) compensation

The latter two are introduced here.

3.7.1 Process Control

Process control is defined.

The term **process control** applies to a wide variety of industrial processes—including chemical and refining processes, steel and paper making—even automated milking machines. The output of an industrial process—for example, paper making—is continuously monitored for thickness, opacity, strength, and so on. As these monitored variables deviate from the preset values, the elements of the process, such as drying temperature, wood pulp, and alum are changed to yield the correct output. Process control plants are frequently overdamped, having real poles and zeros. Since these are type 0 systems, integral compensation is needed to eliminate steady state error for a step input.

3.7.2 Ziegler–Nichols compensation

As mentioned earlier, the typical process control plant has real poles and zeros and is type 0. Figure 3.20 shows a typical unity feedback system with plant $G_p(s)$ and a compensator $G_c(s)$. The compensator may be either proportional (P), proportional-integral (PI), or proportional-integral-derivative (PID). The form of the PID compensator is

$$G_c(s) = K_p + \frac{K_i}{s} + K_d s = \frac{K_d s^2 + K_p s + K_i}{s}$$

where K_p is the proportional gain, K_i is the integral gain, and K_d is the derivative gain. P or PI compensators have only the obvious components.

The Ziegler–Nichols method for tuning a compensator for process control was introduced in 1942. There are two steps, as follows:

The Ziegler–Nichols procedure.

1. A proportional compensator is applied so that

$$G_c(s) = K_p$$

The gain is adjusted until the system becomes marginally stable. This value of gain is designated K_{po}, and the period of oscillation is designated T_o.

2. The compensator is defined by

$$G_c(s) = K_p \left(1 + \frac{1}{T_i s} + T_d s\right)$$

Using the foregoing definition of a compensator, it follows that once values of T_i and T_d have been computed, then

$$K_i = \frac{K_p}{T_i}$$

$$K_d = K_p T_d$$

The design equations for Ziegler–Nichols compensators are shown in Table 3.5. There are P, PI, and PID versions.

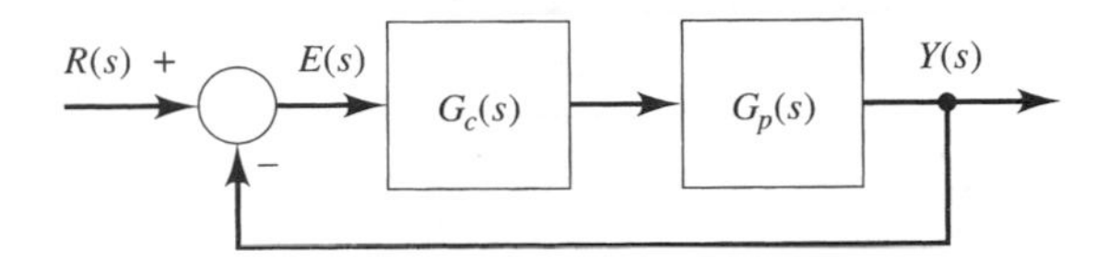

Figure 3.20 Process control system including compensator.

As an example of this method, consider the process control plant

$$G_p(s) = \frac{64}{s^3 + 14s^2 + 56s + 64} \quad [3.17]$$

The first step is to let $G_c(s) = K_p$ and find the value of $K_p = K_{po}$ such that $1 + G_c(s)G_p(s)$ is marginally stable, thus

$$T(s) = \frac{G_c(s)G_p(s)}{1 + G_c(s)G_p(s)} = \frac{64}{s^3 + 14s^2 + 56s + 64\left(1 + K_p\right)} \quad [3.18]$$

The Routh array is formed,

s^3	1	56
s^2	14	$64\left(1 + K_p\right)$
s^1	$(1/14)\left[784 - 64\left(1 + K_p\right)\right]$	0
s^0	$64\left(1 + K_p\right)$	

The value of K_{po} is found by setting the first term in the s^1 row to zero, with the result that $K_{po} = 11.25$. At this value of K_{po}, the complex conjugate roots are obtained from the s^2 row giving $14s^2 + 784$ or $14(s^2 + 56)$ with roots $\pm j7.483$. The characteristic polynomial of Equation (3.18) is divided by $(s^2 + 56)$ to find the remaining real root at -14. Hence, $T_0 = 2\pi/\omega_0 = 2\pi/7.483 = 0.8397$. The values for the Ziegler–Nichols compensators are easily found from Table 3.5 as follows:

Normal Ziegler–Nichols

P compensator	$K_p = 5.63$
PI compensator	$K_p = 5.06$
	$T_i = 0.697$
PID compensator	$K_p = 6.75$
	$T_i = 0.420$
	$T_d = 0.105$

3.7.3 Chien–Hrones–Reswick Compensation

Many engineers believe that the method developed in 1952 by Chien, Hrones, and Reswick (CHR) provides a better way to select a compensator for process control

Table 3.5 *Ziegler–Nichols Compensation*

Compensator	Values
P	$K_p = 0.5\,K_{po}$
PI	$K_p = 0.45\,K_{po}$ $T_i = 0.83\,T_o$
PID	$K_p = 0.6\,K_{po}$ $T_i = 0.5\,T_o$ $T_d = 0.125\,T_o$

applications. The method operates upon the shape of the step response of the open-loop plant. Since the plant is often type 0, the steady state value of the output of the open-loop plant driven by a unit step is given by the dc gain $G_p(0)$. That result follows from the final-value theorem. The procedure can be described in two steps.

1. As in Figure 3.21, a line is drawn through the linear portion of the open- loop plant unit srtep response just after $t = 0$ resulting in values for T_g and T_u.
2. The values for the compensator chosen depend on the ratio R,

$$R = T_g / T_u$$

The CHR method.

as shown in Table 3.6(a).

There are two types of CHR compensator, one that should provide overdamped closed-loop behavior and one that should provide 20% overshoot. These goals are not necessarily met exactly, but the response is usually close to the desired values. In Table 3.6(b), $K_g = G_p(0)$ is the dc gain of the open-loop plant.

As an example of this method, consider the process control plant of Equation (3.17), which was considered earlier. The first step in this process is to find the

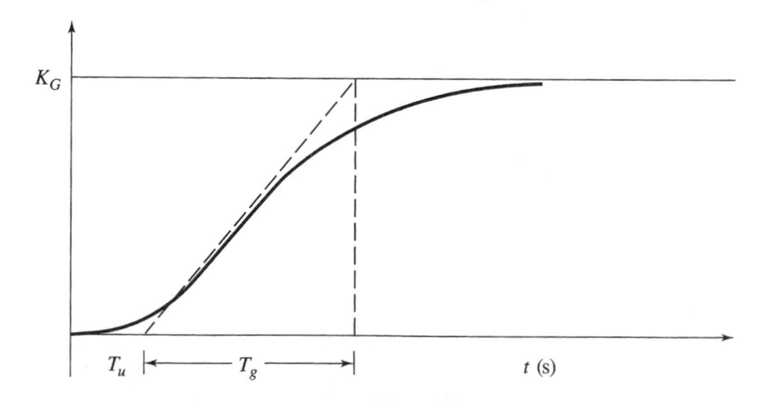

Figure 3.21 Unit step response of an open-loop plant.

Table 3.6 *Chien–Hrones–Reswick Compensator*

(a) Values for $R = T_g/T_u$

Compensator	R
P	$R > 10$
PI	$7.5 < R < 10$
PID	$3 < R < 7.5$
Higher order	$R < 3$

(b) CHR Compensation

Compensator	Overdamped	20% Overshoot
P	$K_p = 0.3R/K_g$	$K_p = 0.7R/K_g$
PI	$K_p = 0.35R/K_g$ $T_i = 1.2T_g$	$K_p = 0.6R/K_g$ $T_i = T_g$
PID	$K_p = 0.6R/K_g$ $T_i = T_g$ $T_d = 0.5T_u$	$K_p = 0.95R/K_g$ $T_i = 1.35T_g$ $T_d = 0.47T_u$

unit step response for the open-loop plant. A final time of 3.0 s is used. Figure 3.22 shows this time response. The line drawn through the linear portion of Figure 3.22 yield the following values:

$$T_u = 0.20 \text{ s}$$

$$T_g = 1.3 - 0.2 = 1.1 \text{ s}$$

$$R = \frac{1.1}{0.2} = 5.5$$

A PID compensator is selected from Table 3.6(a), and from Table 3.6(b), the values for K_p, T_i, and T_d are found for either the overdamped or the 20% overshoot case. Note that K_g is the dc gain of the open-loop plant, which for Equation (3.17) is unity. These values are easily calculated for the PID compensator.

PID compensator	
Over damped	$K_p = 3.31$ $T_i = 1.1$ $T_d = 0.10$
20% Overshoot	$K_p = 5.22$ $T_i = 1.49$ $T_d = 0.094$

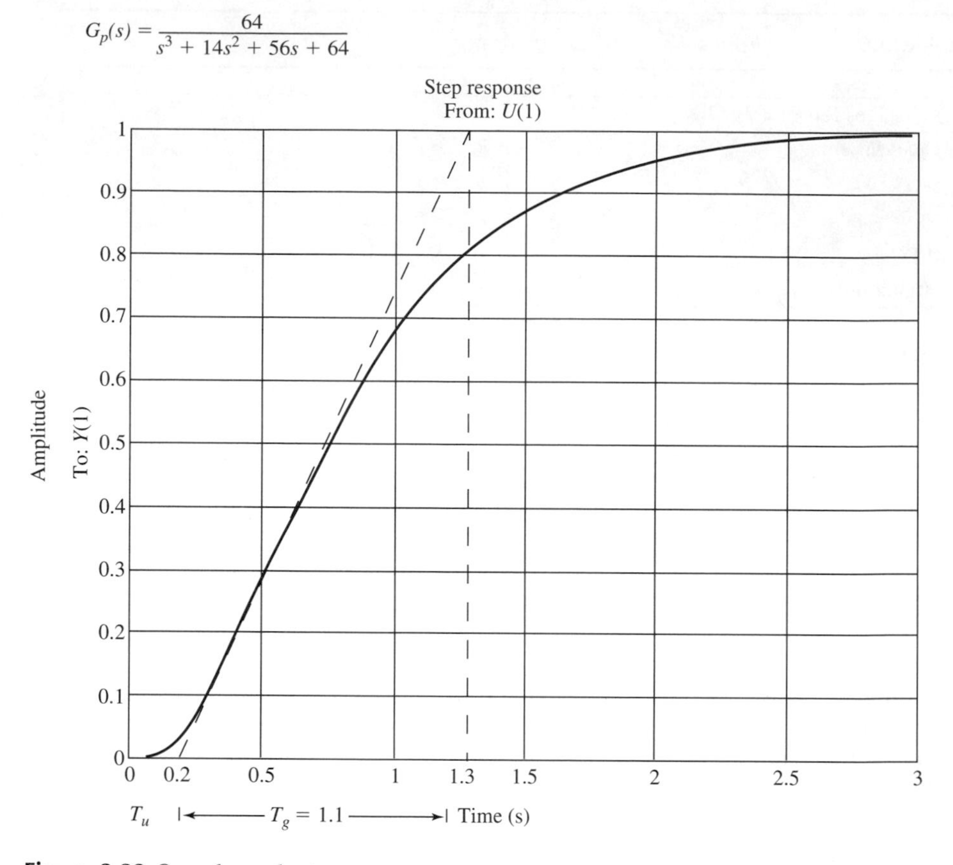

Figure 3.22 Open-loop plant response.

Figure 3.23 shows the unit step response for the three PID compensators we have designed. Notice that the greatest overshoot is caused by the Ziegler–Nichols design. The CHR 20% overshoot design does result in nearly 20% overshoot. The CHR overdamped result is somewhat underdamped but clearly has the least overshoot of the three resulting unit step responses.

The Ziegler–Nichols method and CHR method are intended for type 0 plants with real poles and zeros. Thus these methods do not work for type 1 plants or for those with complex conjugate poles or zeros.

❑ Computer-Aided Learning

Three features of MATLAB can be used to carry out the time domain design just discussed:

```
Programname.M
Pole (transfer function)
% non-executable comments
```

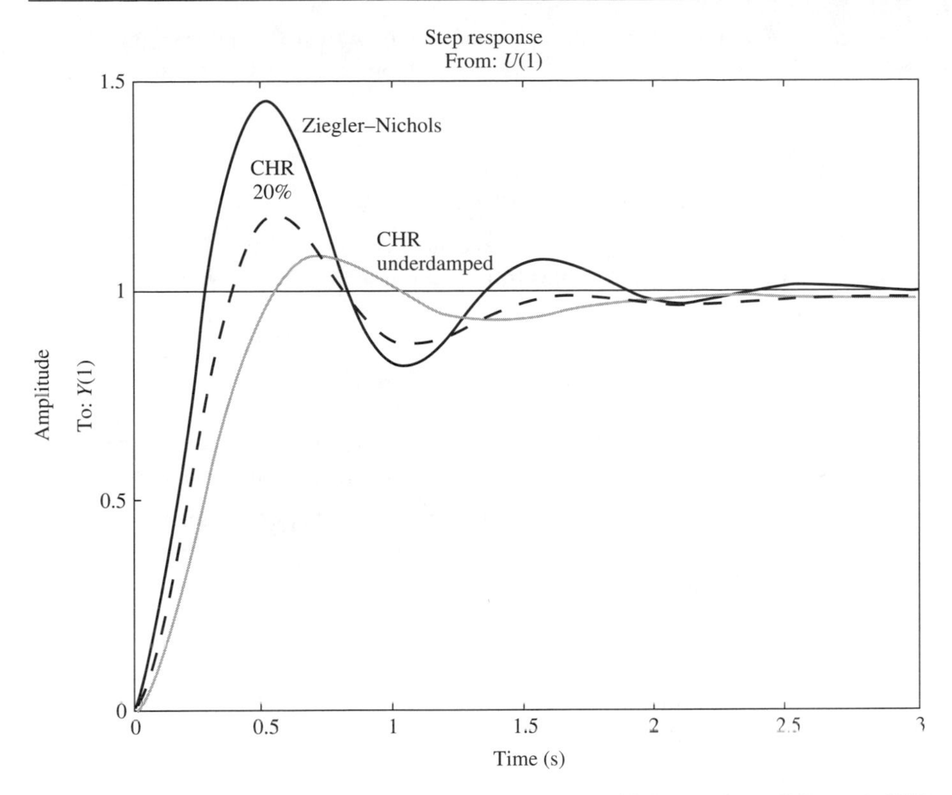

Figure 3.23 Response using Ziegler–Nichols PID, CHR 20% overshoot PID, and CHR underdamped PID.

Suppose we want to get the closed-loop poles for a system with an open- loop $G(s)$ equal to

$$G(s) = \frac{1}{s(s+1)(s+2)}$$

for gain values from 5.5 to 6.5 in steps of 0.1 (5.5, 5.6, 5.7, etc.) Begin by clicking on file, new, M files. In the window that opens, the sequence of commands

```
% get poles. M closed-loop pole calculations
gp=zpk ([], [0 -1 -2], 1)
k=kstart-diff
for i=1:11
   k=k+diff
   disp(pole(k*gp/(1+k*gp)))
end
```

beginning with a non-printable comment line creates a program that displays the gain and closed-loop poles. The "pole" command creates the poles of the indicated transfer function.

Here, these closed-loop poles are calculated for 11 gain values. The gain values would be from kstart to kstart+10 diff. After creating the program click on file; save as and type getpoles in the small window to store the program which will be stored as getpoles. If you want closed-loop poles for $k = 5.5$ to $k = 6.5$, then type

```
kstart=5.5
diff=0.1
getpoles
```

This will cause the program to be executed from $k = 5.5$ to $k = 6.5$. If you wanted 11 sets of closed-loop poles from $k = 6$ to $k = 6.1$, you would type

```
kstart=6.0
diff=0.01
getpoles
```

C3.6 Use MATLAB to find the gain causing the closed-loop system to be marginally stable for the plant of Equation (3.17). Use the trial-and-error approach.

Ans.

```
% get kpo. M find k for marginal stability
numgp=64
dengp=[1 14 56 64]
gp=tf (numgp, dengp)
k=kstart-diff
for i=1:11
     k=k+diff
     disp(pole(k*gp/(1+k*gp)))
end
(Save the program)
kstart=
diff=
getkpo
```

(repeat the last 3 commands by trial and error until you have k=kpo for marginal stability)

C3.7 Obtain the step response of the PID compensator for Ziegler–Nichols design.

Ans.

```
kp=6.75
Ti=0.420
Td=0.105
numgc=kp*[Td*Ti Ti 1]
dengc=[Ti 0]
gc=tf(numgc, dengc)
gf=gc*gp
t=gf/(1+gf)
step(t)
```

C3.8 Repeat C3.6 through C3.7 for

$$G_p(s) = \frac{64}{(s+4)^3}$$

Ans. $K_{po} = 8$, $K_p = 4.8$, $To = 0.907$, $Ti = 0.454$, $Td = 0.113$

C3.9 Obtain the step response for the open-loop plant of Equation (3.17) and Figure 3.22. Verify values of R, T_u, and T_g.

Ans.
```
numgp=64
dengp=[1 14 56 64]
gp=tf(numgp, dengp)
step(gp)
```

C3.10 Obtain the step response for the CHR designs applied to the plant of Equation (3.17)

(a) Underdamped CHR PID compensated system

(b) 20% overshoot CHR PID compensated system

Ans. (a)
```
Kp=3.31
Ti=1.1
Td=0.1
numgcchr=Kp*[Ti*Td Ti 1]
dengcchr=[Ti 0]
gcchr=tf(numgcchr, dengcchr)
gfchr=gp*gcchr
tchr=gfchr/(1+gfchr)
step(tchr)
```

(b)
```
Kp=5.22
Ti=1.49
Td=0.094
```
[repeat other instructions from (a)]

C3.11 Repeat C3.9 and C3.10 for the plant of C3.8. That is, carry out the CHR design of the two PID compensators for the plant of C3.8 and obtain the step responses.

Ans. $T_u = 0.2,\ T_g = 0.92,\ R = 4.6;$
Overdamped PID: $K_p = 2.76,\ T_i = 0.92,\ T_d = 0.1;$
20% overshoot PID: $K_p = 4.37,\ T_i = 1.242,\ T_d = 0.094$

3.8 An Electric Rail Transportation System

The control of transportation systems is an interesting design, area. Many European trains are electric, so control of electric reaction motors is common in that area of the world. Desired speed inputs to the system are made by the operator, with an override occurring in case of emergency. A similar system is used in Japan for the 100 mi/h Kyoto-to-Tokyo train. In San Francisco, the BART (Bay Area Rapid Transit) system is designed to automatically vary the speed as conditions warrant, without human intervention.

Manned aircraft and space flight are further examples of transportation systems where control inputs to the vehicle are generated by an autopilot. The same methodology can be applied to bus, car, and passenger train operation to improve performance. In this example we consider a velocity and position control system for a rail vehicle; the system is similar to those employed for passenger trains in Germany and Switzerland.

Figure 3.24 shows the relation between motor drive and velocity for an electric rail car. A unit step input $D(s)$ will produce a steady state car velocity given by

$$\lim_{s\to 0} s\left(\frac{1}{s}\right)\left(\frac{15}{s+0.1}\right) = 150\ \text{ft/s}$$

with a time constant of 10 s. In perhaps more familiar terms, in 10 s, the car will accelerate to

$$(150)(0.63) = 94.5\ \text{ft/s}$$

$$(94.5)(3600/5280) = 64\ \text{mi/h}$$

$$(94.5)(3600/3281) = 103.7\ \text{km/h}$$

Automobile racing enthusiasts would characterize this vehicle by saying that it can accelerate from 0 to 60 mi/h in 8.8 s or from 0 to 100 km/h in 9.3 s.

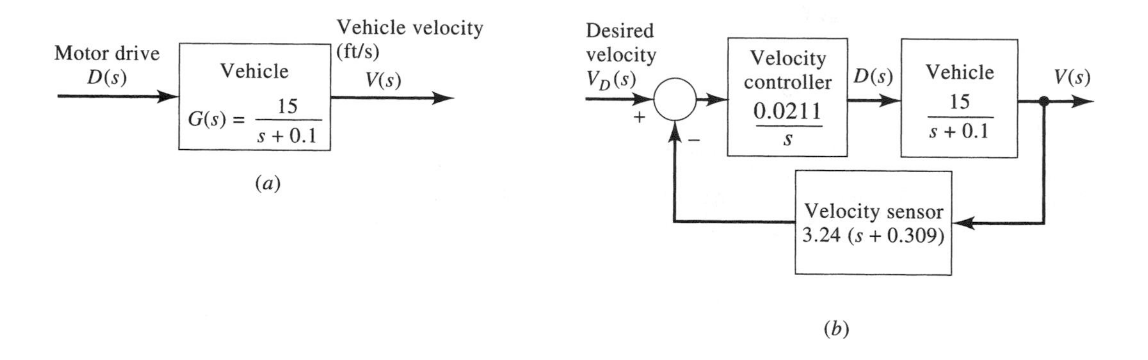

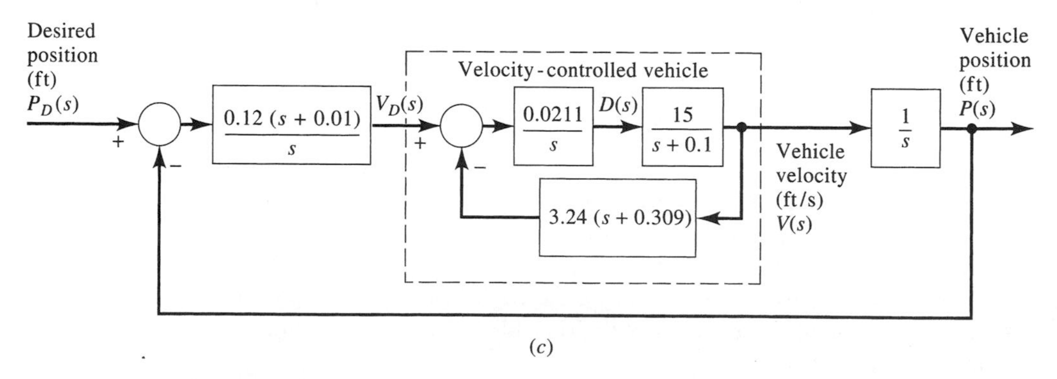

Figure 3.24 Controlling a transportation vehicle. (a) Vehicle model. (b) Vehicle with velocity control feed back loop. (c) Vehicle with velocity and position control.

In Figure 3.24(b) the vehicle is shown as part of a feedback system for velocity control, where V_D is the desired velocity input. This system has transfer function

$$T_V(s) = \frac{(0.0211/s)/(15/(s+0.1))}{1+(0.0211/s)(15/(s+0.1))(3.24)(s+0.309)} = \frac{0.317}{s^2+1.125s+0.317}$$

The error between desired and actual velocity is

$$E_V(s) = V_D(s) - V(s) = V_D(s) - T_V(s)V_D(s) = [1 - T_V(s)]V_D(s)$$

The error signal in this case is not the summer signal because this system does not have unity feedback. The error transmittance is

$$T_{VE}(s) = 1 - T_V(s) = \frac{s(s+1.125)}{s^2+1.125s+0.317}$$

Since the system is type 1, its steady state error to a step input will be zero:

$$\lim_{s\to 0} s\left(\frac{1}{s}\right)\left[\frac{s(s+1.125)}{s^2+1.125s+0.317}\right] = 0$$

Its normalized steady state error to a ramp input is

$$\lim_{s\to 0} s\left(\frac{1}{s^2}\right)\left[\frac{s(s+1.125)}{s^2+1.125s+0.317}\right] = \frac{1.125}{0.317} = 3.55 \text{ ft/s}$$

The velocity control feedback system has the characteristic equation

$$s^2 + 1.125s + 0.317 = 0$$

and repeated characteristics roots $s_1,\ s_2 = -0.56$. Its natural response dies out as fast as $\exp(-0.56t)$ compared to the natural response, $\exp(-0.1t)$, of the vehicle alone, Figure 3.24(a). To achieve control of the vehicle position, a second feedback loop has been added to the system in Figure 3.24(c). The transfer function of the complete system is

$$T(s) = \frac{\left[\dfrac{0.12(s+0.01)}{s}\right]\left(\dfrac{0.317}{s^2+1.125s+0.317}\right)\left(\dfrac{1}{s}\right)}{1+\left[\dfrac{0.12(s+0.01)}{s}\right]\left(\dfrac{0.317}{s^2+1.125s+0.317}\right)\left(\dfrac{1}{s}\right)} = \frac{0.038s+0.00038}{s^4+1.125s^3+0.317s^2+0.038s+0.00038}$$

The system is stable, as a Routh–Hurwitz test easily shows:

s^4	1	0.317	0.00038
s^3	1.125	0.038	
s^2	0.283	0.0038	
s^1	0.036		
s^0	0.00038		

The error between desired and actual position is

$$E(s) = P_D(s) - P(s) = [1 - T(s)]P_D(s)$$

giving the following error transmittance:

$$T_E(s) = 1 - T(s) = \frac{s^2(s^2 + 1.125s + 0.317)}{s^4 + 1.125s^3 + 0.317s^2 + 0.38s + 0.00038}$$

The system is type 2; it exhibits zero steady state error to a step and to a ramp input $P_D(s)$. A ramp desired position constitutes a step in desired velocity. In the steady state, this system will thus approach zero error in both position and velocity.

3.9 Phase-Locked Loop for a CB Receiver

The phase-locked loop (PLL) is an important component of many telecommunications systems. It is used to demodulate the stereo channel in FM broadcast receivers, to detect and maintain the color subcarrier in color television receivers, to generate precise frequencies in citizens band (CB) receivers, and in many other applications.

Figure 3.25(a) shows a diagram of the operation of a superheterodyne receiver for the citizens band. The signal from the receiver's antenna and a sinusoidal mixer signal generated in the receiver are mixed in a way that produces sums and differences of the antenna signal frequencies with the sinusoidal mixing frequency signal. Those difference frequencies which are centered at 10.7 MHz are passed by the band-pass filter and detected. By changing the mixing signal frequency, different incoming channels are translated to the 10.7 MHz pass band to be detected.

From a highly stable crystal-controlled oscillator, logic circuits produce a digital waveform at the proper mixing frequency of the desired channel to be detected. The mixer requires a smooth, sinusoidal mixing signal, and it is the purpose of the phase-locked loop to "lock" the frequency of a voltage-controlled oscillator to the digital rate. In some receivers, a PLL is also used to aid in the frequency synthesis, but that situation is not considered here.

A phase-locked loop model is given in Figure 3.25(b). The phase of a sinusoidal signal is proportional to the integral of its frequency. The difference in phase between the input and output signals is sensed and is used to control the frequency of an oscillator in such a way that difference in phase (and thus is frequency) is reduced. Commercial phase-locked loops may differ from this arrangement in that the phase difference detector is nonlinear and capable of sensing phase differences only as large as π or 2π rad. Nonetheless, this simple model adequately predicts PLL performance for most applications. The system block diagram of Figure 3.25(b) is unusual, compared with previous examples, because it relates frequencies and phases of signals, not the signals themselves.

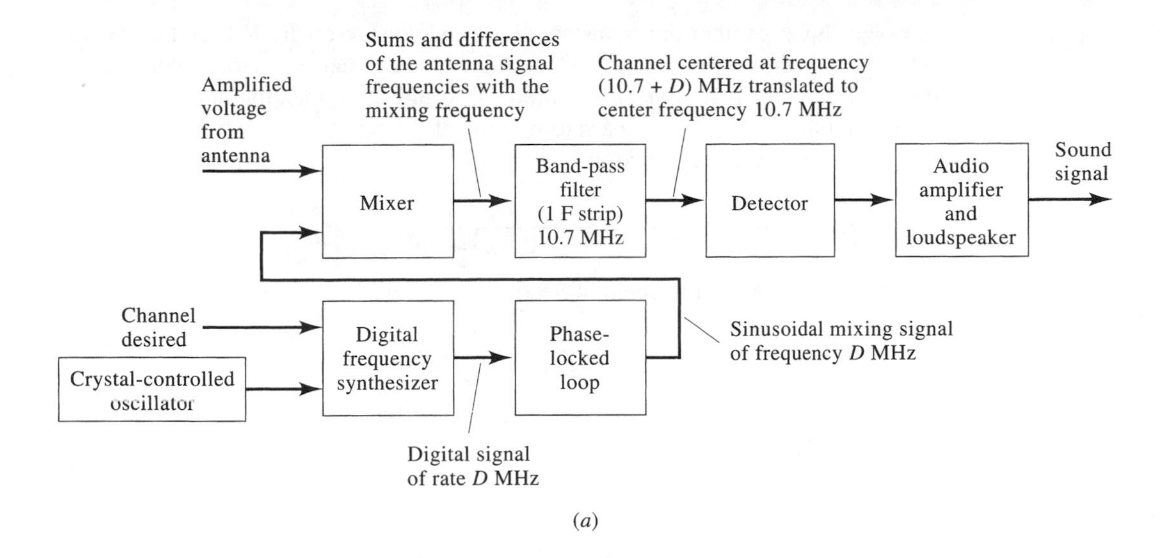

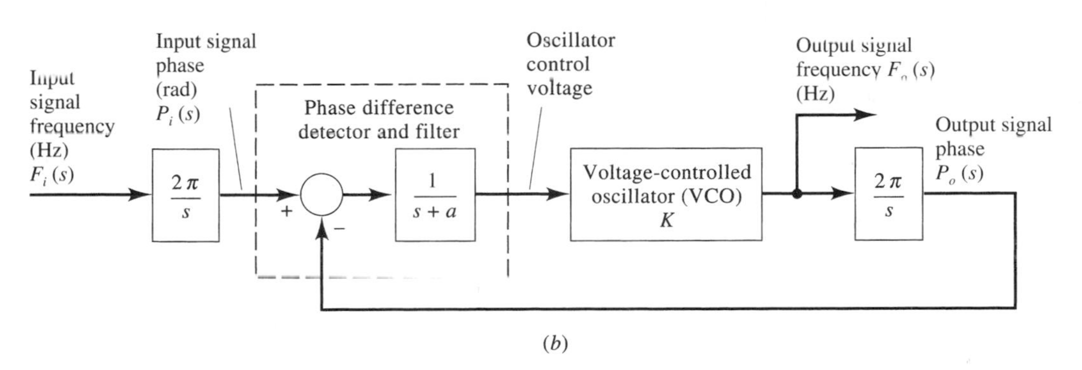

Figure 3.25 Phase-locked loop for a citizen band receiver. (a) Description of a CB receiver. (b) Phase-locked loop model.

The transfer function of the system that relates output frequency to input frequency is

$$T(s) = \frac{F_o(s)}{F_i(s)} = \frac{(2\pi/s)[1/(s+a)](K)}{1+[1/(s+a)](K)(2\pi/s)} = \frac{2\pi K}{s^2+as+2\pi K}$$

The error between input and output frequency is given by

$$E(s) = F_i(s) - F_o(s) = [1 - T(s)]F_i(s)$$

and has transmittance

$$T_E(s) = 1 - T(s) = \frac{s^2+as}{s^2+as+2\pi K}$$

The system is type 1 and has zero steady state error to a step input frequency change.

The choice of filter time constant $(1/a)$ and voltage-controlled oscillator gain K is based upon the desired characteristics of the system when responding to a change in desired frequency. A reasonable requirement would be for the minimum mean square error damping ratio of 0.5 and a settling time of 0.1 s.

The system transfer function

$$T(s) = \frac{2\pi K}{s^2 + as + 2\pi K} = \frac{\omega_n^2}{s^2 + 2\zeta\omega_n s + \omega_n^2}$$

is of the form analyzed in detail in Section 2.3, with

$$\omega_n = \sqrt{2\pi K} \qquad \text{and} \qquad \zeta = \frac{a}{2\omega_n} = \frac{a}{2\sqrt{2\pi k}} \tag{3.19}$$

Using Figure 2.10(c), which shows normalized settling time as a function of damping ratio, for a 0.5 damping ration, we can write

$$\omega_n T_s = 5.3 \tag{3.20}$$

For a settling time $T_s = 0.1$, Equation (3.20) gives

$$\omega_n = \frac{5.3}{0.1} = 53$$

From Equation (3.19) we find

$$K = \frac{\omega_n^2}{2\pi} = \frac{(53)^2}{6.28} = 447$$

and with $\zeta = 0.5$, we have

$$a = 2\omega_n\zeta = 2(53)(0.5) = 53$$

This preliminary design is now examined for "worst-case" behavior in changing from channel to channel. The lowest CB carrier frequency is 26.97 MHz, and the highest is 27.26 MHz, corresponding respectively to PLL input frequencies of

$$26.97 - 10.7 = 16.27 \text{ MHz} \qquad \text{and} \qquad 27.26 - 10.7 = 16.56 \text{ MHz}$$

The largest change in PLL input frequency between stations will be from 16.27 MHz to 16.56 MHz, a step change 290,000 Hz. By design, the frequency settles to within 5%,

$$(0.05)(290{,}000) = 14{,}500 \text{ Hz}$$

in 0.1 s, but it will take about three times that long for the frequency to settle to below about 50 Hz, which is necessary for intelligible reception. A 0.3 s maximum time to change stations is probably quite acceptable.

When the device is first turned on, however, it may have to respond to a step change of up to 16.56 MHz. Although the behavior of a PLL for such a change in input frequency will likely be nonlinear at first, the linear model will be used to predict an approximate initial length of time until the PLL output frequency has settled to within about 50 Hz. In the first 0.1 s, the response will settle to

$$(0.05)(16{,}560{,}000) = 828{,}000 \text{ Hz}$$

In as much as the envelope of the second-order oscillatory natural behavior is exponential, the response will settle to within 5% of this value in the next 0.1 s:

$$(0.05)\ (828{,}000) = 41{,}400\,\text{Hz}$$

Continuing, we see that

$$(0.05)\ (41400) = 2070\,\text{Hz}$$

$$(0.05)\ (2070) = 103.5\,\text{Hz}$$

so an acceptable "worst-case" initial tuning is predicted to be within about 0.5 s.

This system is type 1 insofar as a step change in frequency is concerned, so that zero steady state frequency error results. The transfer function relating $P_i(s)$ and $P_o(s)$ is also type 1. However, a constant frequency implies that phase (the integral of frequency) is a ramp, and thus some steady state phase error occurs. The error can be controlled by raising K or by increasing system type to 2.

3.10 Bionic Eye

Control system science and biomedical instrumentation form a partnership that has the promise of restoring sight to those without central vision. Figure 3.26(a) shows a block diagram of normal human vision. When a person wants to view an object, the brain causes the muscles to rotate the eye's centerline toward the object. The image from that object passes through the eye's lens to the retina, sending impulses back to the brain, closing the loop. In Figure 3.26(b), the bionic artificial eye system, the brain attempts to view the object by pointing the poorly functioning eye along some angle. An artificial eye orients a camera along the same angle as the eye. The image from the object enters the camera and is then routed to some components that elicit response in the brain, closing the loop.

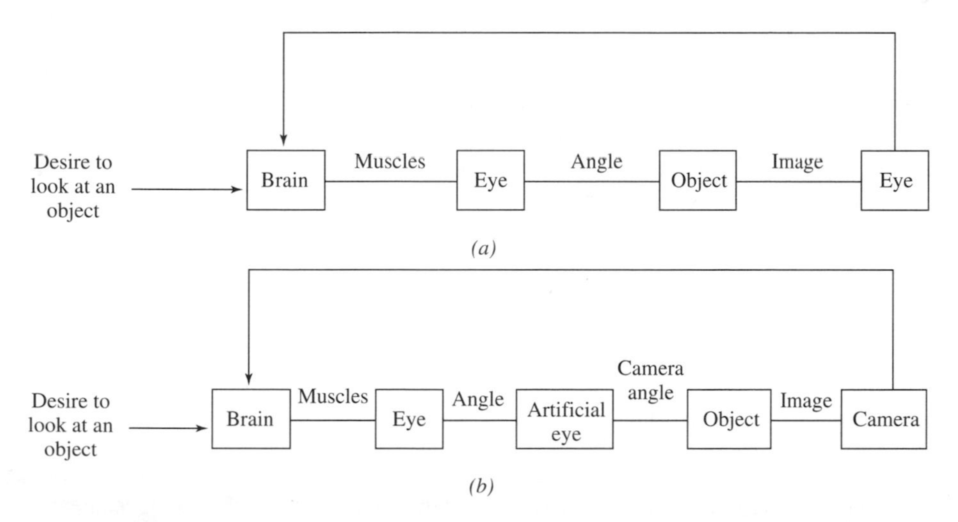

Figure 3.26 (a) Normal vision and (b) Bionic vision (with an artificial eye).

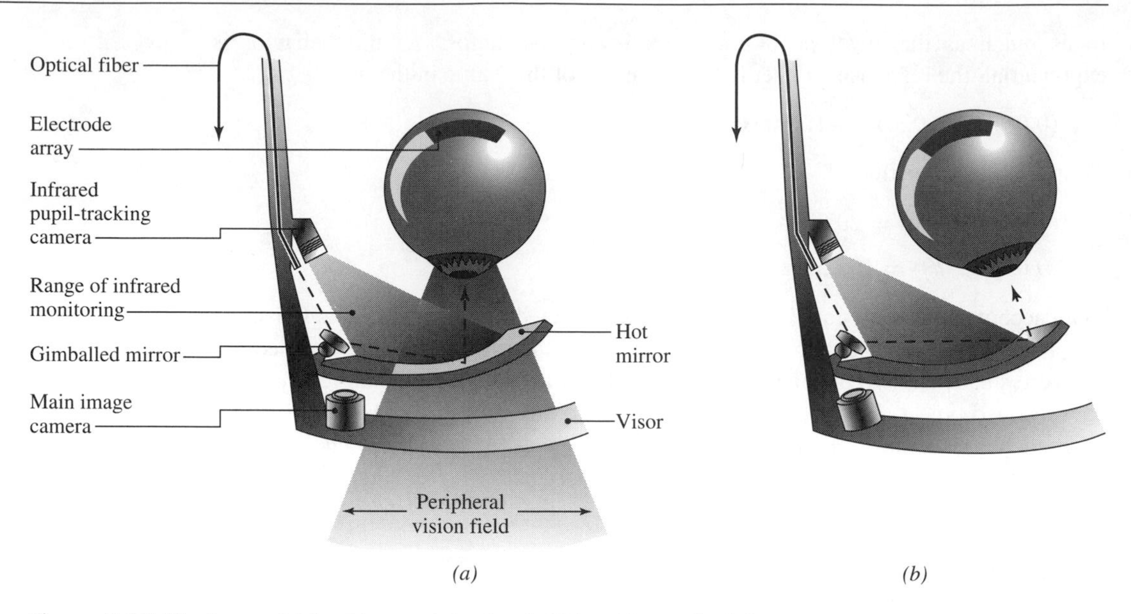

Figure 3.27 Bionic eye. (a) Looking straight ahead. (b) Looking to the left.

Figure 3.27 shows part of a special pair of glasses used to create the artificial eye portion of Figure 3.26(b), that is, the portion of the system having the eye's pointing angle as an input while the camera angle is the output. As the brain causes the eye to rotate, a mirror rotates via gimbals, deflecting an infra red (IR) beam toward the eye's lens with resulting impact on the electronic retinal detector that is implanted there. As the mirror gimbals in the correct direction, the detector accepts more light until an IR level equaling that of the reference IR beam is reached. The mirror is linked to the camera, so that the camera follows the eye's angle, creating artificial vision. Figure 3.28(a) is a block diagram of the control loop of the artificial eye. The compensator of Figure 3.28(b) simulates the lens/detector portion of Figure 3.28(a).

To design the compensator $G_c(s)$ and gimballed mirror $G_p(s)$, we need to think about the properties of normal vision this electronic eye should reproduce.

For example, the eye should be able to change focus from one object to another object without any steady state angular positioning error. Thus, there should be zero steady state step error. Similarly, the eye should be able to scan a scene or track a moving object with smooth response. That property will follow if the artificial eye provides zero steady state ramp error. Therefore, the control loop should be type 2. If $G_c(s)$ and $G_p(s)$ each have a pole at $s = 0$ as in Figure 3.28(c), then the steady state properties are assured in that the control loop is clearly type 2.

Next, we must select values for K, a, and b. The closed loop transfer function is

$$T(s) = \frac{G_c(s)G_p(s)}{1 + G_c(s)G_p(s)} = \frac{Kb(s+a)}{s^3 + bs^2 + Kbs + Kba}$$

Suppose we require the eye to refocus with a step response that achieves nearly zero step error in less than one second. If we choose the dominant closed-loop pole

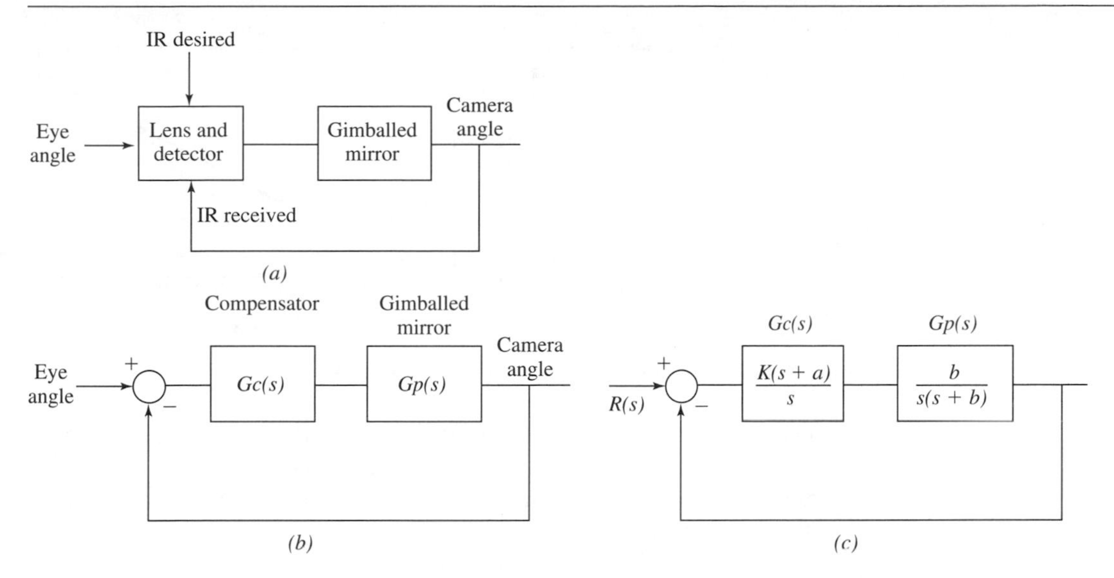

Figure 3.28 Artificial eye control loop dynamics. (a) The control loop. (b) Simulation of (a). (c) Selected components.

so that three time constants equal 0.6 s, then one time constant is 0.2 s, requiring the pole to be $1/0.2 = -5$. There are two other closed-loop poles. One should nearly cancel the closed-loop zero at $-a$. Choose $-1.01a$ for the closed-loop pole, while the other should be well to the left at, perhaps, -50.

To find K, a, and b, we equate

$$
\begin{aligned}
&(s + 1.01a)(s + 5)(s + 50) \\
&\quad = s^3 + (1.01a + 55)s^2 + (55.55a + 250)s + 252.5a \\
&\quad = s^3 + bs^2 + Kbs + Kba
\end{aligned}
$$

There are thus three equations:

$$Kba = 252.5a$$

$$Kb = 55.55a + 250$$

$$b = 1.01a + 55$$

Solving, we write

$$
\begin{aligned}
Kb &= 252.5 = 55.55a + 250 \\
a &= 0.045 \\
b &= 55.04545 \\
K &= 4.587
\end{aligned}
$$

$$T(s) = \frac{252.5(s + 0.045)}{s^3 + 55.04545s^2 + 252.55 + 11.3625}$$

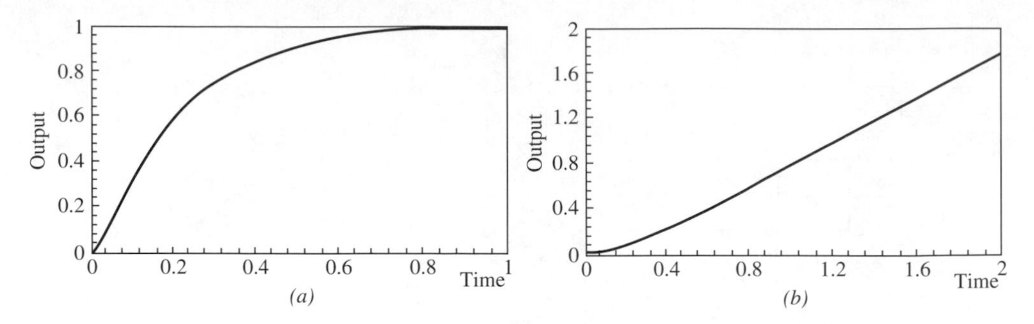

Figure 3.29 Response of the bionic eye. (a) Step response. (b) Ramp response.

The step response of Figure 3.29(a) is nearly first-order with a time constant of 0.2 s, while the ramp response of Figure 3.29(b) is smooth and linear after about three time constants or 0.6 s. The design meets the indicated goals. It is estimated that a system of this sort may be available by 2010.

3.11 SUMMARY

The concern of this chapter is the analysis of tracking system performance. In subsequent chapters, these analysis tools will be brought to bear on tracking system design. The analysis (and design) of tracking systems can be separated into two parts:

1. Location of the system characteristic roots to give natural responses that decay sufficiently rapidly and have acceptable damping ratios for oscillatory terms. It is usually desirable for the pole locations to be within a region on the complex plane of the shape shown in Figure 3.3(c). The amplitudes of the natural response terms depend only on the system's initial conditions, so the designer can normally choose the manner in which the natural response decays to zero.
2. A forced response component that tracks the reference input well for the class of inputs to be encountered. This can be characterized by the system's steady state error to power-of-time inputs.

For any function $y(t)$, the initial value is related to its Laplace transform by

$$y(0) = \lim_{s\to\infty} sY(s)$$

If $y(t)$ has a final value, it is given by

$$\lim_{t\to\infty} y(t) = \lim_{s\to 0} sY(s)$$

If the function $y(t)$ does not have a final value, the limit can give a misleading result.

Tracking systems are designed so that the forced component of the system output, as nearly as possible, equals the input. An important measure of performance in this regard is the steady state error between input and output for step, ramp, parabolic, and other power-of-time input signals. When the error has a final value, the steady state error to an input

$$r_i(t) = \frac{1}{i!}t^i u(t) \qquad R_i(s) = \frac{1}{s^{i+1}}$$

is given by the final-value theorem as follows:

$$\lim_{s \to 0} \frac{1}{s^i}[1 - T(s)]$$

System type number is the number of $s = 0$ numerator zeros in the error transmittance

$$T_E(s) = 1 - T(s)$$

Type number determines the steady state power-of-time error properties of a system, as summarized in Table 3.2. A system that has finite steady state error for the ith power-of-t input will have zero steady error for lower powers of t and infinite steady state error for higher powers of t.

Unity feedback systems have error transmittance

$$T_E(s) = 1 - T(s) = \frac{1}{1 + G(s)}$$

Their system type number, the number of $s = 0$ numerator roots in $T_E(s)$, is also the number of is $s = 0$ denominator roots of $G(s)$. Unity feedback system error coefficients are defined as follows:

$$\kappa_i = \lim_{s \to 0} s^i G_E(s)$$

and these are related to steady state errors in Table 3.3.

If the quality of performance of a system can be expressed with a performance measure I in terms of the adjustable system parameters, they can be selected by the mathematical process of finding the set of parameters that maximize or minimize I. A common performance index for a tracking system is the integral square error to a step input:

$$I_s = \int_0^\infty e_{\text{step}}^2(t)dt$$

The mathematics of finding extrema of functions I of many variables is formidable in most cases, and results can be quite dependent upon the specific performance measure used.

Feedback can be used to make system response relatively independent of inaccuracies in some of the system's parameters. A feedback system can thus be designed to perform well even when the controlled plant parameters are not known accurately or when they drift with time. In general, the sensitivity of a transfer function T to a change in a parameter a is

$$S_a = \lim_{\Delta a \to 0} \frac{\Delta T/T}{\Delta a/a} = \frac{a}{T}\frac{\partial T}{\partial a}$$

It is the fractional rate of change of T with fractional change in a. Sensitivities generally depend upon the complex variable s.

Feedback is also used to reduce the effects of disturbances signals upon a system's response. A tracking feedback arrangement that compares the output with a desired reference signal and works to drive the error between the two to zero tends to lessen disturbance effects.

Time domain compensation is presented with particular emphasis on process control systems. Two methods are presented:

1. Ziegler–Nichols compensation
2. Chien–Hrones–Reswick (CHR) compensation, which helps in the selection of an appropriate compensator

The velocity and position control system for a rail transportation system illustrates how one can use multiple feedback loops in tracking system design. In this example, an inner loop effected velocity control of the electric train. Then, an outer feedback loop involving error integration was need to achieve position control. Finally, step response was seen to be a highly useful and easily visualized test signal to use in analyzing the performance of a phase-locked loop for a CB receiver.

It may become possible to restore sight to someone who would otherwise lack central vision by using a clever combination of a mirror and a camera to follow a sightless eye's motion, stimulating the brain with electronic impulses received by the camera.

A MATLAB program may be created to facilitate trial and error aspects of time domain design. The "pole" command causes the close-loop poles to be displayed.

REFERENCES

Pole Placement and Steady State Error

Chestnut, H., and Mayer, R. W., *Servomechanisms and Regulating System Design*, vol. 1. New York: Willey, 1959.

DiStefano, J. J. III, Stubberud, A. R., and Williams I. J., *Feedback and Control Systems* (*Schaum's Outline*). New York: McGraw-Hill, 1967.

James, H. M., Nichols, N. B., and Phillips, R. S., *Theory of Servomechanisms* (MIT Radiation Laboratory Series, vol. 25). New York: McGraw-Hill, 1947.

Savant, C. J. Jr., *Basic Feedback Control System Design*. New York: McGraw-Hill, 1958.

Truxal, J. C., *Control System Synthesis*. New York: McGraw-Hill, 1955.

Sensitivity

Cruz, J. B. Jr., *Feedback Systems*. New York: McGraw-Hill, 1972.

———, ed. *System Sensitivity Analysis*. Stroudsburg, PA: Dowden, 1973.

Horowitz, I. M., *Synthesis of Feedback Systems.* New York: Academic Press, 1963.

Kreindler, E., "On the Definition and Application of the Sensitivity Function," *J. Franklin Inst.* 285 (January 1968).

Tomovic, R., *Sensitivity Analysis of Dynamic Systems.* New York: McGraw-Hill, 1963

Disturbance Rejection

Friedland, B., *Control System Design: An Introduction to State Space Methods.* New York: McGraw-Hill, 1986.

Hostetter, G. H., *Digital Control System Design.* New York: Holt, Rinehart & Winston, 1987.

Optimal Control

Athans, M., "The Status of Optimal Control Theory and Applications for Deterministic Systems," *IEEE Trans. Autom. Control* (July 1966).

McCausland, I., *Introduction to Optimal Control.* New York: Wiley, 1969.

Sage, A. P., and White, C. C. III., *Optimum Systems Control.* Englewood Cliffs, NJ: Prentice-Hall, 1977.

Schultz, D. G., and Melsa, J. L., *State Functions and Linear Control Systems.* New York: McGraw-Hill, 1967.

Time Domain Compensation

Cerr, M. *Instrumentation Industrelle*, vol. 2, Paris.

Chein, K. L., Hrones, J. A., and Reswick, J. B., "On the Automatic Control of Generalized Passsive Systems." *Trans ASME*, 74(1952). 175–185.

Herget, D., "A Closed Loop PID Control Program for Process Control." *Int. J. Appl. Eng. Educ.*, 5 no. 1 (1989): 83–88.

Ohta, T., et al., A New Optimization Method of PID Control Parameters for Automatic Tuning by Process Computer, in Computer-Aided Design of Control Systems, Proceedings of the IFAC Symposium, Zurich, 1979, pp. 133–138.

Ziegler, J. G., and Nichols, N. B., "Optimum Settings for Automatic Controllers." *Trans. ASME*, 64 (1942): 759–768

Transportation Systems

Friedlander, G. D., "Electronics and Swiss Railways," *IEEE Spectrum* (September 1974): 68–75.

Kaplan, G., "Microprocessors Monitor Rail Car Systems; Software Optimizes Headway in Miami's People Mover." *IEEE Spectrum* (January 1987): 59–61.

———, "Rail Transportation," *IEEE Spectrum* (January 1984): 82–85.

Stefani, R. T., "Design and Stimulation of an Automobile Guidance and Control System." *Comput. Educ. (COED)Trans. Am. Soc. Eng. Educ.*, (January 1978).

Artificial Eye

Dagnelic, G, and Massot R.W., "Toward an Artificial Eye." *IEEE Spectrum* (May 1996): 20–29.

PROBLEMS

1. Find the relative stability of the feedback systems of Figure P3.1. If there are complex conjugate pairs of characteristics roots, also find the damping ratios.

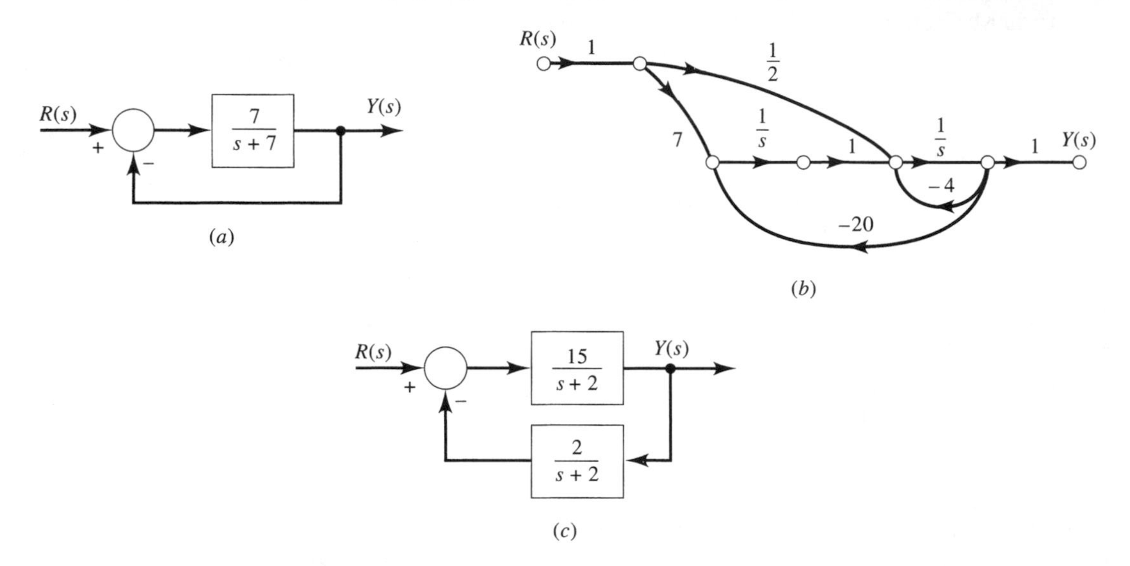

Figure P3.1

2. If possible, find a value of the adjustable constant k so that the system of Figure P3.2 has a relative stability of at least 2 units and a minimum damping ratio of 0.5.

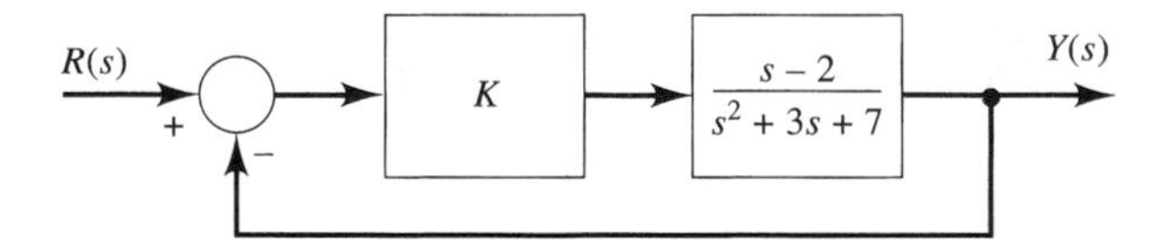

Figure P3.2

3. Find the initial values $y(0)$ of the signals with the following Laplace transforms.

(a) $Y(s) = \dfrac{-5s^3 + 3s^2 + 2}{s^3 + 4s^2 + 8s + 20}$

(b) $Y(s) = \dfrac{2s + 6se^{-4s}}{s^2 + 8s + 9}$

(c) $Y(s) = \dfrac{4s^2 - 3s + 9}{11s^4 + 4s^3 + 2s^2 + 3s}$

(d) $Y(s) = \dfrac{4s^2 + 3}{s^5 + 9s^4 - 4s^3}$

Ans. (a) ∞; (c) 0

4. Determine whether signals with the following Laplace transforms have a final value. If a finite value exists, use the final-value theorem to find it.

(a) $Y(s) = \dfrac{-3s^2 + 2s + 2}{s^3 + 7s^2 + 8s + 2}$

(b) $Y(s) = \dfrac{2s^2 + 8s + 16}{s^4 + 2s^3 + 4s^2 + 9s + 18}$

(c) $Y(s) = \dfrac{4e^{-2s}}{4s^4 + 6s^3 + 2s^3 + 2s + 9}$

(d) $Y(s) = \dfrac{10}{s^5 + 3s^4 + 7s^3 + 4s^2 + 4s}$

Ans. (a) exists, 0; (d) exists, $\frac{10}{4}$

5. For the systems of Figure P3.5 find the steady state error for a unit step input. The error signal is $r(t) - y(t)$.

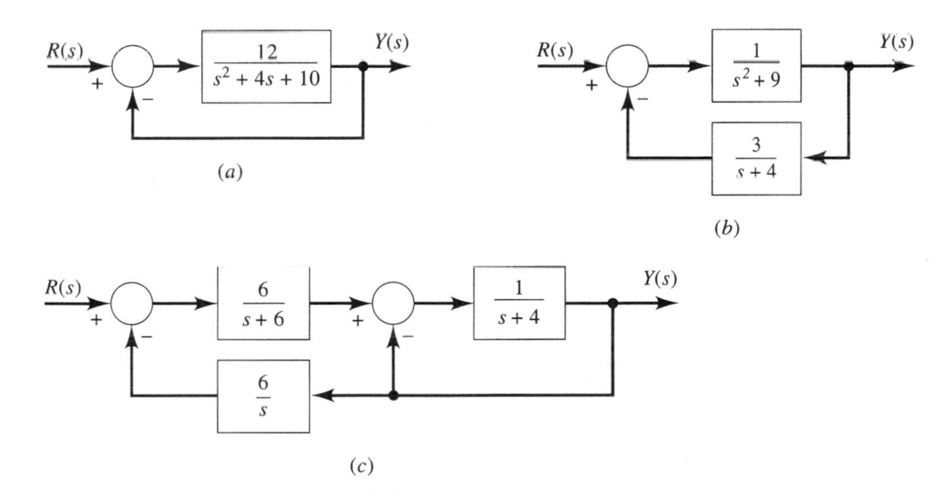

Figure P3.5

6. Find the type number and steady state errors with unit step, ramp, and parabolic inputs for the tracking systems with following transfer functions:

(a) $T(s) = \dfrac{7}{s^2 + 4s + 7}$

(b) $T(s) = \dfrac{4s + 7}{s^2 + 4s + 7}$

(c) $T(s) = \dfrac{5}{s^2 + 4s + 12}$

(d) $T(s) = \dfrac{s+4}{s^3 + 4s^2 + s + 4}$

(e) $T(s) = \dfrac{2s^2 + s + 4}{s^4 + 3s^3 + 10s^2 + s + 4}$

Ans. (a) $1, 0, \frac{4}{7}, \infty$; (c) $0, \frac{7}{12}, \infty, \infty$; (e) 2, unstable

7. The forward transmittances of unity feedback tracking systems are given below. For each, find the type number of the system, the steady state error coefficients κ_0, κ_1, and κ_2, and the steady state errors to unit step, ramp, and parabolic inputs.

(a) $G_E(s) = \dfrac{-3s + 7}{s^4 + 3s^3 + 2s^2 + 7s}$

(b) $G_E(s) = \dfrac{2}{s^3 + 2s^2 + 7s}$

(c) $G_E(S) = \dfrac{6s^2 - 3s + 6}{s^4}$

(d) $G_E(s) = \dfrac{3s + 9}{s^4 + 3s^3}$

(e) $G_E(s) = \dfrac{4s^2 + 9s + 11}{s^4 + 8s^3}$

Ans. (b) $1, \infty, \frac{2}{7}, 0, 0, \frac{7}{2}, \infty$; (d) 3, unstable

8. Choose a transmittance $G(s)$ for the block so that the overall system of Figure P3.8 is type 2. Many different choices for $G(s)$ are possible.

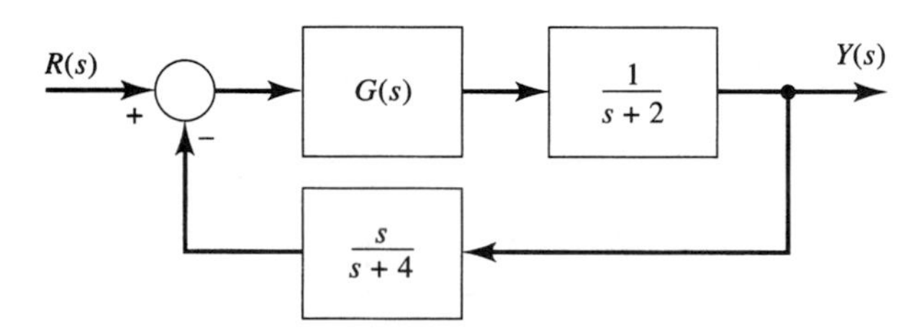

Figure P3.8

9. Find the sensitivities of the following transfer functions to small changes in the parameter a about the given nominal values:

(a) $T(s) = \dfrac{7a}{s + a}$ $\quad a = 3$

(b) $T(s) = \dfrac{6}{s^2 + as + 2}$ $\quad a = 2$

(c) $T(s) = \dfrac{as + 10}{s^2 + as + 10}$ $\quad a = 4$

Ans. (a) $s/(s+3)$; (b) $-2s/(s^2 + 2s + 2)$

10. It is sometimes possible to eliminate the effects of an inaccessible disturbance upon the output of a feedback system entirely. When this is done, the disturbance is said to be *decoupled* from the output. For the system of Figure P3.10 find a "block" transmittance $G(s)$, if possible, for which $D(s)$ is decoupled from $Y(s)$.

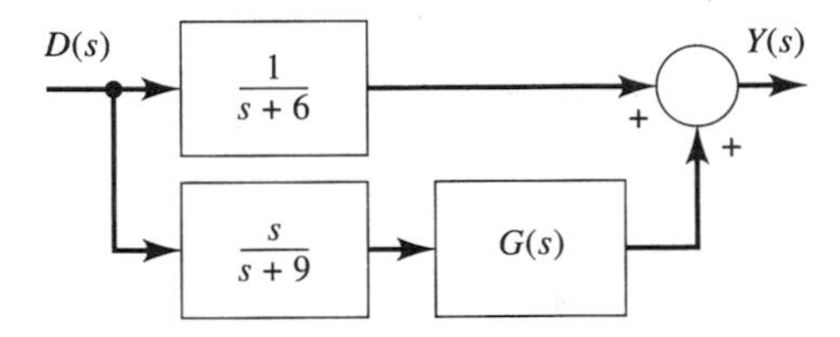

Figure P3.10

11. For the system of Figure P3.11, find the value of k, if it exists, such that for a step input $r(t)$, the integral square error between $y(t)$ and (t) is minimum.

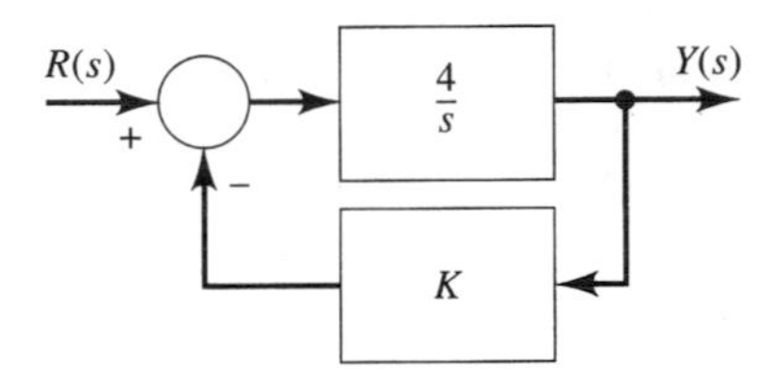

Figure P3.11

12. For signals with the following Laplace transforms, find values of the adjustable constants k, if possible, for which

$$I_S = \int_0^\infty e^2(t)dt$$

is smallest:

(a) $E(s) = \dfrac{ks + 2}{s^2 + 7s + 10}$

(b) $E(s) = \dfrac{-6s + k}{s^2 + 5s + 6}$

Ans. (a) $k = 0$.

13. Tracking systems such as the one in Figure P3.13, in which the reference input is zero (since it is missing), are called *regulators*. It is desired to keep the output as near to zero as is possible in the system, even when there are disturbances present. For what range of the adjustable constant K, if any, is the steady state value of $y(t)$ less than 0.1 when $d(t)$ is a unit step signal?

14. For the electric rail system of Section 3.8, suppose the vehicle transmittance is instead

$$G(s) = \frac{10}{s + 0.4}$$

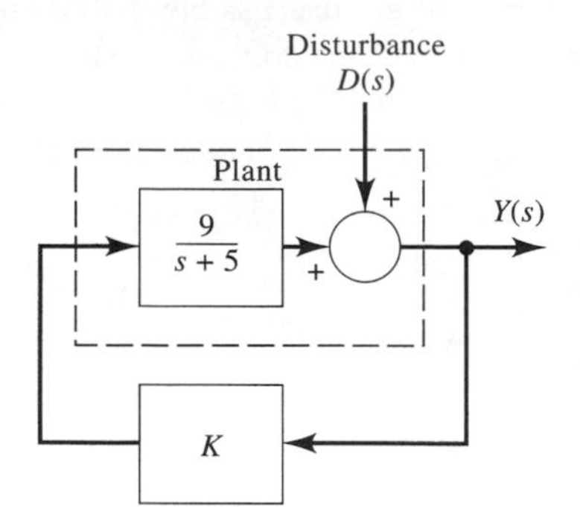

Figure P3.13

Find the steady state position error for a 10,000-ft step change in desired position.

Ans. 0

15. For the electric rail transportation system of Section 3.8 with the parameter values given in the text, suppose that an electrical malfunction causes a unit step to be added to $V_D(s)$ before the signal is applied to the velocity-controlled vehicle subsystem. Find the resulting steady state error in vehicle position.

16. For the phase-locked loop of Section 3.9, find a and K such that the system's natural response is critically damped and has suitable speed of response for the CB receiver application.

    Ans. $K = 168$, $a = 65$, rise time $= 0.1$ s

17. For the citizens band receiver of Section 3.9 with the values of K and a chosen in the text, suppose a listener is initially tuned to the station at 26.97 MHz. The listener suddenly tunes to the station at 27.26 MHz and remains tuned for 1 s. the listener is curious about the end of a message he had been tuned to, and the suddenly returns the receiver to the station at 26.97 MHz. Sketch $f_0(t)$.

18. A power plant frequency control system has the block diagram given in Figure P3.18. Find, in terms of K, relative stability, steady state errors due to power-of-t reference inputs, and steady state error due to a unit step change in load torque. Choose a value of K on the basis of your feeling of how a large power generator should be controlled. For this value of K, find the form of the system's natural response.

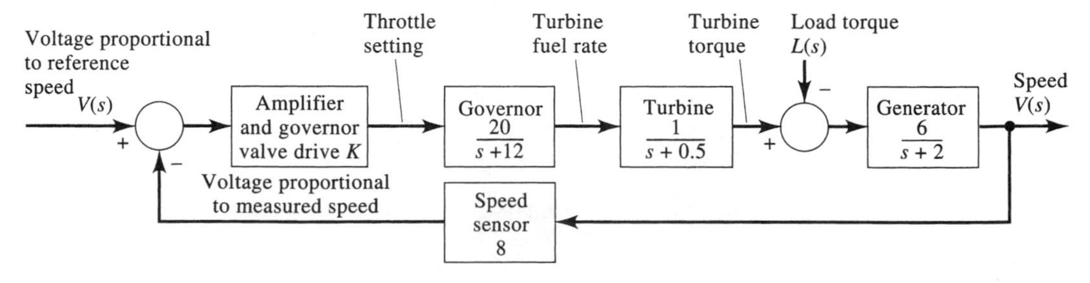

Figure P3.18

19. A simplified block diagram for a chemical process control system is shown in Figure P3.19. The controller is called a PID (*or three-term*) type, because it develops a signal that is a linear combination of terms that are proportional to, the derivative of, and the integral of the incoming signal $f(t)$. If possible, choose values for k_1, k_2, and k_3 such that the resulting system has a relative stability of at least 0.5 unit.

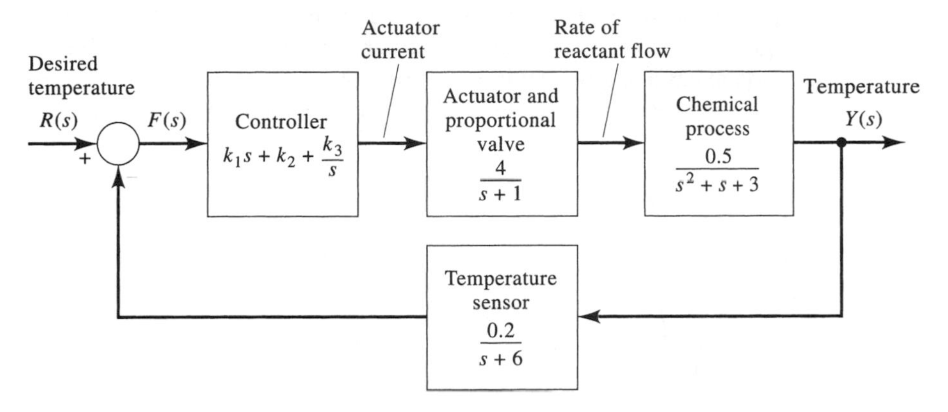

Figure P3.19

20. Figure P3.20 shows a simplified model of a depth control system for a submerged submarine. The ship-settling dynamics transmittance is

$$G(s) = \frac{10^4}{s^2 + 3 \times 10^3 s + 14 \times 10^6}$$

(a) Carefully explain the meaning of the "block" with transmittance $(1000/s)$. is this a component of the system in the same sense as an amplifier or a motor might be?

(b) Determine whether the system is stable.

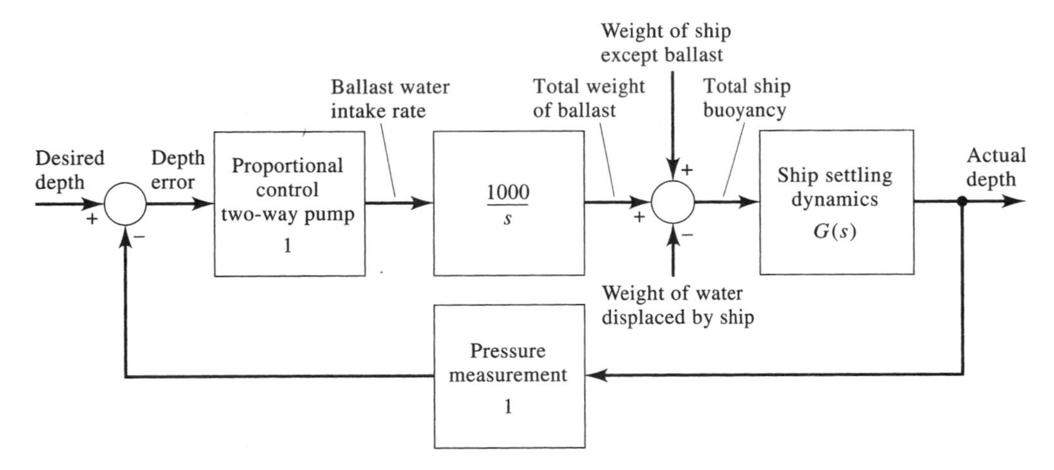

Figure P3.20

(c) Find the steady state change in actual depth due to a unit step change in desired depth.

(d) A sonar beacon is released by the ship, causing a unit step change in the weight of the vessel. Find the steady state change in actual depth.

Ans. (b) Stable; (c) 1; (d) 0

21. A tape loop positioning system for a digital computer tape drive is diagrammed in Figure P3.21. Find values of the constants k_1, k_2, and k_3, if possible, so that this system has a relative stability in excess of 3 units and a steady state error no greater than 50%. The motor time constant k_3 cannot be less than 0.2.

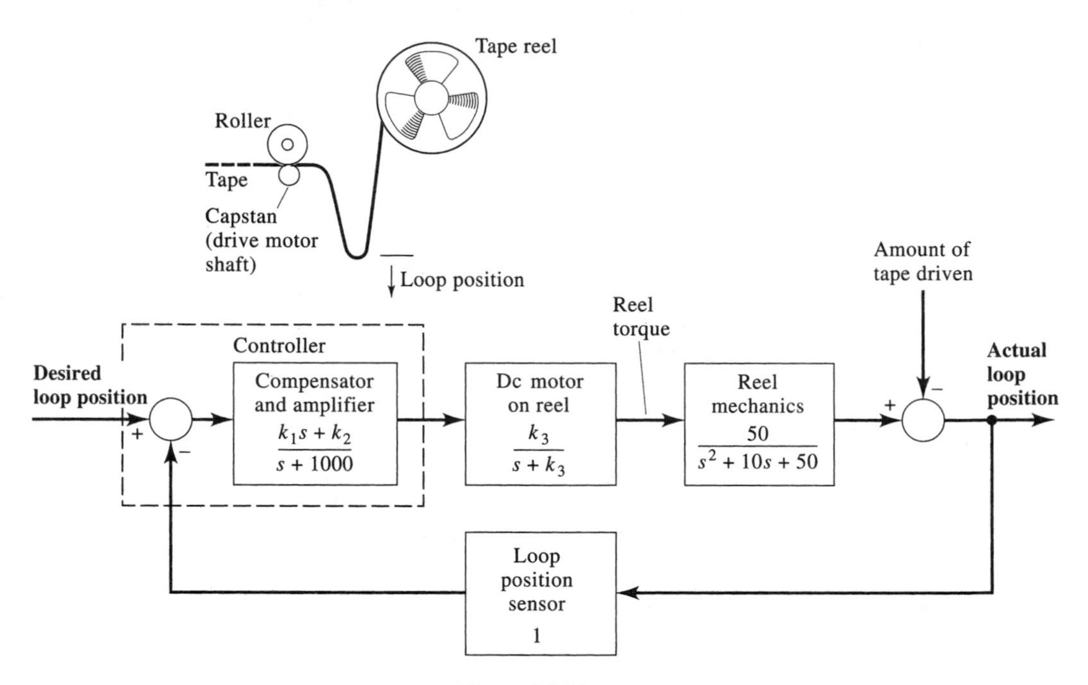

Figure P3.21

22. Table P3.22 gives approximate measurements of the output $y(t)$ of a system at various times t when the system input is a step,

$$r(t) = 6.3u(t)$$

(a) Using the data, find an approximate first-order, linear, time-invariant system model, specifying its transfer function.

(b) Repeat, but find an approximate second-order, linear, time-invariant system model. Specify the approximate system transfer function.

23. The forward transmittances of unity feedback tracking systems are given. For each, find the type number of the system, the steady state error coefficients κ_0, κ_1, and κ_2, and the steady state errors to unit step, ramp, and parabolic inputs.

Table P3.22 *An Input-Output Record for a System*

Time $t(s)$	Output Value $y(t)$
0.0	0.0
0.15	1.1
0.3	2.2
0.45	2.8
0.6	3.3
0.75	3.5
0.9	3.5
1.05	3.4
1.2	3.3
1.35	3.3

(a) $G_E(s) = \dfrac{K}{(s+6)(s+80)}$

(b) $G_F(s) = \dfrac{K}{s(s+6)(s+80)}$

Ans. $1, \infty K/480, 0, 0, 480/K, \infty$

(c) $G_E(s) = \dfrac{Ks}{(s+1)(s+6)(s+80)}$

Ans. $1, 0, 0, 0, 1, \infty, \infty$

(d) $G_E(s) = \dfrac{K}{s^2(s^2 + 2\zeta\omega_n s + \omega_n^2)}$

(e) $G_E(s) = \dfrac{100}{(s^2 + 2\zeta\omega_n s + \omega_n^2)}$

(f) $G_E(s) = \dfrac{70(s+10)(s+15)}{s^3(s^2+2s+10)}$

Ans. 3 (system is unstable)

24. Compute the step, ramp, and parabolic error coefficients for the non–unity feedback system of Figure P3.24.

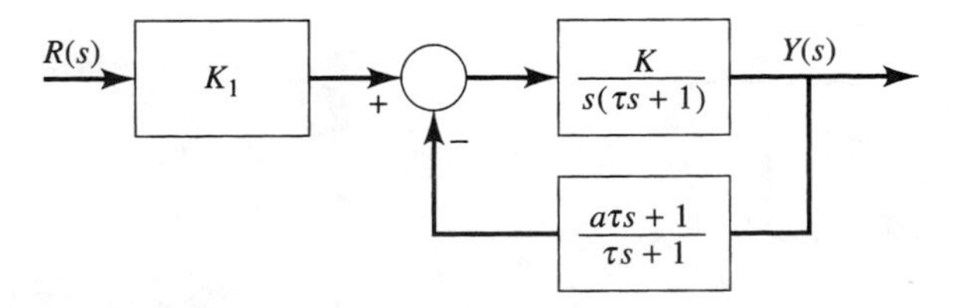

Figure P3.24

25. The following questions pertain to the system of Figure P3.25. A control motor with a linear torque–speed curve rotates a missile launcher while wind gusts provide a disturbance torque thwarting accurate targeting.

    (a) When the disturbance torque is zero, compute the output if the input $r(t)$ is a step and if the input $r(t)$ is a ramp.

    (b) If the disturbance torque $\tau_D(t)$ is a constant T_0, compute the steady state output due to that disturbance.

    Ans. $T_0/A_1A_2k_1$

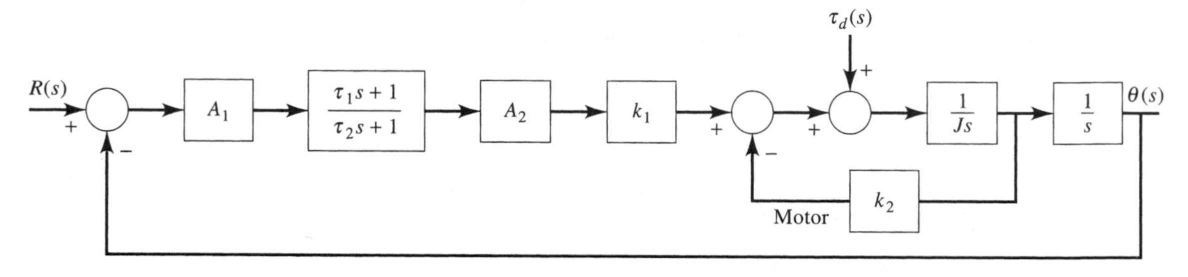

Figure P3.25

26. Repeat Problem 25(b) if the disturbance torque is bt.

27. Repeat Problem 26 where A_2 becomes K/s.

    Ans. b/A_1k_1K

28. Repeat Problem 26 where A_2 becomes Ks.

    Ans. infinity

29. The open-loop plant for a unity feedback system has the following $G(s)$ function

$$G(s)\frac{K(s+s_1)}{s^2(s+s_2)(s+s_4)}$$

    (a) State the type of system.

    (b) Find κ_0, κ_1, and κ_2

    Ans. (a) type 2, (b) $\kappa_0 = \infty = \kappa_1$, $\kappa_2 = Ks_1/s_2s_4$

30. In a position servo, addition of "rate feedback" increases the damping. The block diagram for the system is shown in Figure P3.30.

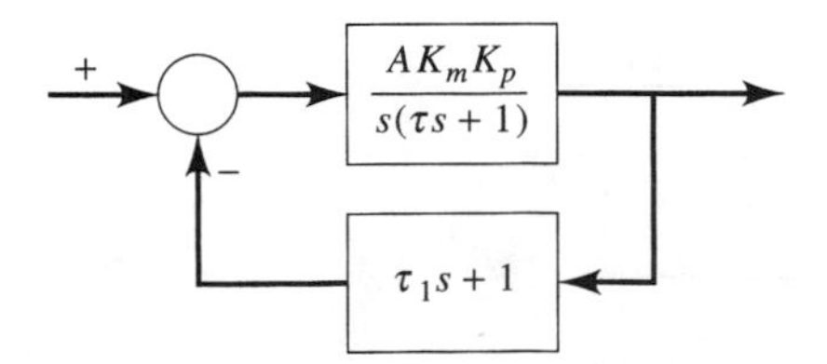

Figure P3.30

(a) Derive the damping ratio and the resonant frequency and show how they depend on τ_1.

(b) Calculate the steady state error in response to a unit ramp function input.

31. For a unity feedback process control system with

$$G_p(s) = \frac{4000}{(s+40)(s+10)^2}$$

design a Ziegler–Nichols PID compensator and plot the step response of the compensated system.

32. Repeat Problem 31 for CHR design of both underdamped and 20% overshoot compensator.

33. For the artificial eye control system of Figure 3.28(a), select K, a, and b so the closed-loop poles are at -8, $-10 + j10$, $-10 - j10$.

34. For the artificial eye control system of Figure 3.28(c), suppose $a = 0$, $K = 10$, $b = 20$. Find the steady state error if the input is $tu(t)$.

CHAPTER

4

Root Locus Analysis

4.1 Preview

In Chapter 1, we learned that differential equations could be written for electromechanical devices often associated with control system applications. The equations can be solved and the response can be divided into forced and natural components. Chapter 2 established definitions related to the natural response, especially definitions of types of stability. Stability is generally considered to be a property of the natural response, since the natural response is due to the system and not outside influences by which the control input evolves. The other response component was considered in Chapter 3: the steady state response—that is, the tendency of a device to follow (track) a command.

In Chapter 4, we return to stability, a property of the natural response, but we provide a much broader and more useful measure of that stability (as compared with Chapter 3). The Routh–Hurwitz approach developed in the 1890s, provides a generally yes or no answer to the stability question, as would be expected of a method predating modern digital computation. Although a range of values emerge for a variable gain (so that the closed-loop system is stable), there is no real advice about which of these gain values are preferable.

A logical approach to determining stability is to extract the roots of the characteristic polynomial as the adjustable gain varies (and indeed computational packages are readily available to do just that). Faced with poor root-solving capabilities in his era, Walter Evans in the mid-1940s developed a set of rules by which the path traced by the closed-loop characteristic equation roots can be sketched to reasonable accuracy as the gain varies. This plot is referred to as a *root locus.*

At this point in our discussion, it might seem logical to let today's computers do all the work, and to skip the sketching rules altogether. Suppose a beginning trigonometry student reaches a comparable conclusion, since trig functions are readily available on handheld calculators. That student might attempt to key in the cosine of 60° and write down the computer display of 1.732. A nearby student who had studied trig would conclude that the tangent of 60° was obtained instead. The student who owns the calculator would not otherwise recognize that an error had been made.

By all means, computational packages should be used to remove the drudgery of repetitive computation and plotting, but when using the sketching rules to follow, the user must be able to distinguish obviously erroneous results from those generally expected. If the user cannot critically evaluate the computer's numerical output, in a very real sense the student and the computer change roles in terms of which is the master and which is the slave, blindly following orders.

Root locus methods are so widely used in industry that two chapters are devoted to the topic. Chapter 4 is devoted to a basic understanding of root locus principles, while Chapter 5 deals with root locus compensation, a design application of the method.

The versatility of the root locus method is demonstrated by the three examples that terminate Chapter 4. A system is designed to track a light source. A control system is developed in which minute electrical muscle impulses are amplified to drive an artificial limb. A root locus analysis of a flexible spacecraft demonstrates that omitting flexible dynamics could cause the designer to overlook a potential cause of instability.

4.2 Pole–Zero Plots

4.2.1 Poles and Zeros

The *zeros* of a function are the values of the variable for which the function is zero. The *poles* of a function are the values of the variable for which the function is infinite, for which its inverse is zero. For a rational function, the zeros are the roots of the numerator polynomial and the poles are the roots of the denominator polynomial. The function

Poles, zeros, and multiplying constants are defined.

$$F(s) = \frac{-3s^3 + 6s^2 - 3s + 6}{2s^4 + 12s^3 + 36s^2 + 80s} = \frac{-3(s-2)(s+j)(s-j)}{2s(s+4)(s+1+3j)(s+1-3j)}$$

has zeros at $s = 2$ and $s = \pm j$. Its poles are at $s = 0$, $s = -4$, and $s = -1 \pm 3j$. In general, a rational function can be placed in the factored form

$$\begin{aligned} F(s) &= \frac{b_m s^m + b_{m-1}s^{m-1} + \cdots + b_1 s + b_0}{a_n s^n + a_{n-1}s^{n-1} + \cdots + a_1 s + a_0} \\ &= \frac{k(s-z_1)(s-z_2)\cdots(s-z_m)}{(s-p_1)(s-p_2)\cdots(s-p_n)} \end{aligned}$$

where its zeros, $z_1, z_2, \ldots, z_m$ and its poles, $p_1, p_2, \ldots, p_n$ are in evidence. The constant

$$k = \frac{b_m}{a_n}$$

is the *multiplying constant* of the function. The multiplying constant is chosen so that the coefficients for the highest powers of s in both the numerator and denominator both have a magnitude of one. If we write $F(s)$ with a multiplying constant as defined above, then there is a unique way to write each $F(s)$.

When the poles and zeros of a function are plotted on the complex plane, the result is a *pole–zero plot*, from which important properties of the function can be visualized. The zero locations are indicated by ○ on the plot and pole locations are indicated by × Figure 4.1 shows pole–zero plots for the following functions:

$$F_1(s) = \frac{4s+5}{s^3+4s^2+13s} = \frac{4(s+\frac{5}{4})}{s(s+2+3j)(s+2-3j)}$$

$$F_2(s) = \frac{-s^2-4}{2s^2+14s+24} = \frac{-\frac{1}{2}(s+2j)(s-2j)}{(s+3)(s+4)}$$

The multiplying constant is placed on the real axis.

We use the notation whereby the multiplying constant of a rational function is placed in a box at the right of the pole–zero plot. The rational function is then entirely given by the plot.

4.2.2 Graphical Evaluation

Of course a computer program can be used to find the poles and zeros of any polynomial ratio. We want to be able to predict some features of the closed-loop poles of a control system without actually using a program. In that way, the information that we will eventually obtain from a program can be checked against what we expect about the stability of the system. It is important to be able to double-check computer results because we may have put incorrect information into the computer, or perhaps our understanding of the system was not correct. For the analysis of a control system to be reliable, computer results and intuitive results must agree when the analysis is complete.

Stability can be checked using graphical procedures.

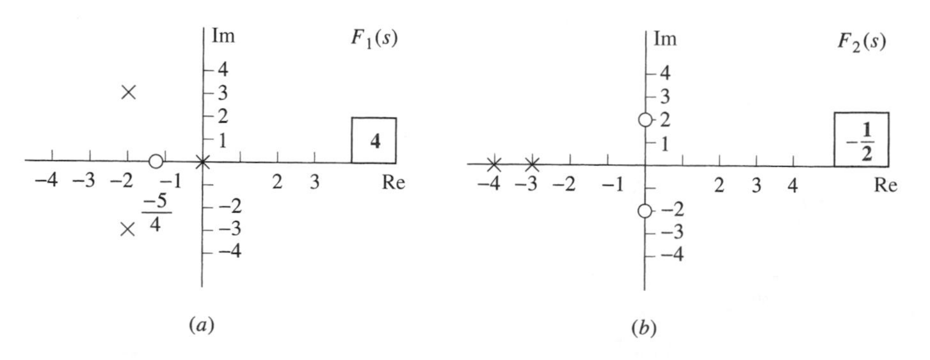

Figure 4.1 Pole–zero plots.

Shortly, we will learn that each value of s that is a closed-loop pole (root of the characteristic polynomial) must satisfy an angle criterion and a magnitude criterion, so we can identify closed-loop poles by their ability to satisfy those criteria. It is therefore useful to be able to find the magnitude and angle of a function $F(s)$ for a given value of s because that skill will be useful in predicting which values of s are actually closed-loop system poles.

A rational function

$$F(s) = \frac{k(s - z_1)(s - z_2)\cdots(s - z_m)}{(s - p_1)(s - p_2)\cdots(s - p_n)}$$

when evaluated at a specific value of the variable, $s = s_0$, is

$$F(s_0) = \frac{k(s_0 - z_1)(s_0 - z_2)\cdots(s_0 - z_m)}{(s_0 - p_1)(s_0 - p_2)\cdots(s_0 - p_n)}$$

On a pole–zero plot, suppose a directed line segment is drawn from the position of a pole, say p_1, to the value s_0 at which the function is to be evaluated. The segment has length $|s_0 - p_1|$ and makes the angle $\angle(s_0 - p_1)$ with the real axis, as indicated in Figure 4.2. Thus we have

$$F(s_0) = \frac{k(|s_0 - z_1|\, e^{j\angle(s_0 - z_1)})(|s_0 - z_2|\, e^{j\angle(s_0 - z_2)})\cdots}{(|s_0 - p_1|\, e^{j\angle(s_0 - p_1)})(|s_0 - p_2|\, e^{j\angle(s_0 - p_2)})\cdots}$$

$$|F(s_0)| = \frac{|k|\begin{pmatrix}\text{product of the lengths of the directed}\\ \text{line segments from the zeros to } s_0\end{pmatrix}}{\begin{matrix}\text{Product of the lengths of the directed}\\ \text{line segments from the poles to } s_0\end{matrix}}$$

$$\angle F(s_0) = \begin{matrix}\text{(sum of the angles of the directed line segments}\\ \text{from the zero to } s_0) - \text{(sum of the pole angles)}\\ +\ 180^\circ \text{ if } k \text{ is negative}\end{matrix}$$

If k is positive, the 180° is not added to the angle, and $|k| = k$.

For example, for $F(s)$ with the pole–zero plot of Figure 4.3, a graphical evaluation at $s_0 = -1 + j3$ gives

$$|F(s = -1 + j3)| = \frac{6(5)}{(3)(3)(5.4)(2.2)} = 0.28$$

$$\angle F(s = -1 + j3) = 143^\circ - 90^\circ - 90^\circ - 68^\circ - 27^\circ + 180^\circ = 48^\circ$$

❑ DRILL PROBLEMS

D4.1 Draw pole–zero plots for the following functions. Include the multiplying constant with the plot.

(a)

$$F(s) = \frac{3s - 1}{s^2 + 2s}$$

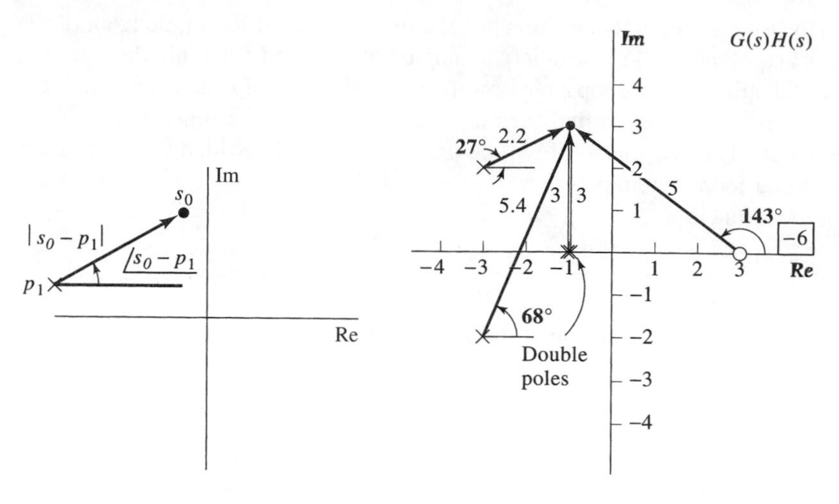

Figure 4.2 Evaluating a rational function at a point $s = s_0$.

Figure 4.3 Graphical evaluation of a rational function.

Ans. zero at $s = \frac{1}{3}$; poles at $s = 0$ and $s = -2$; multiplying constant 3

(b)

$$F(s) = \frac{9s^2 + 1}{(s^2 + 8s + 17)^2}$$

Ans. zero at $s = \pm j\frac{1}{3}$; repeated poles at $s = -4 \pm j$; multiplying constant 9

(b)

$$F(s) = \frac{-2s^2 + 6s + 3}{(s^2 + 3s + 8)(s^2 + 6s + 15)}$$

Ans. zeros at $s = 3.44$ and $s = -0.44$; poles at $-1.5 \pm j2.4$ and $-3 \pm j2.45$; multiplying constant -2

(c)

$$F(s) = \frac{(3s + 1)(2s + 1)}{(4s + 1)(7s + 1)^2}$$

Ans. zeros at $s = -\frac{1}{3}$ and $s = -\frac{1}{2}$; pole at $s = -\frac{1}{4}$ and repeated pole at $s = -\frac{1}{7}$; multiplying constant 3/98

D4.2 Find the rational functions represented by the pole–zero plots in Figure D4.2.

Ans. (a) $6(s + 3)(s^2 + 2s + 10)/s$; (b) $-3(s - 2000)/(s + 2000)(s^2 + 2000s + 5 \times 10^6)^2$; (c) $\frac{2}{3}(s^2 + 225)/(s + 30)^2 (s^2 + 20s + 500)$

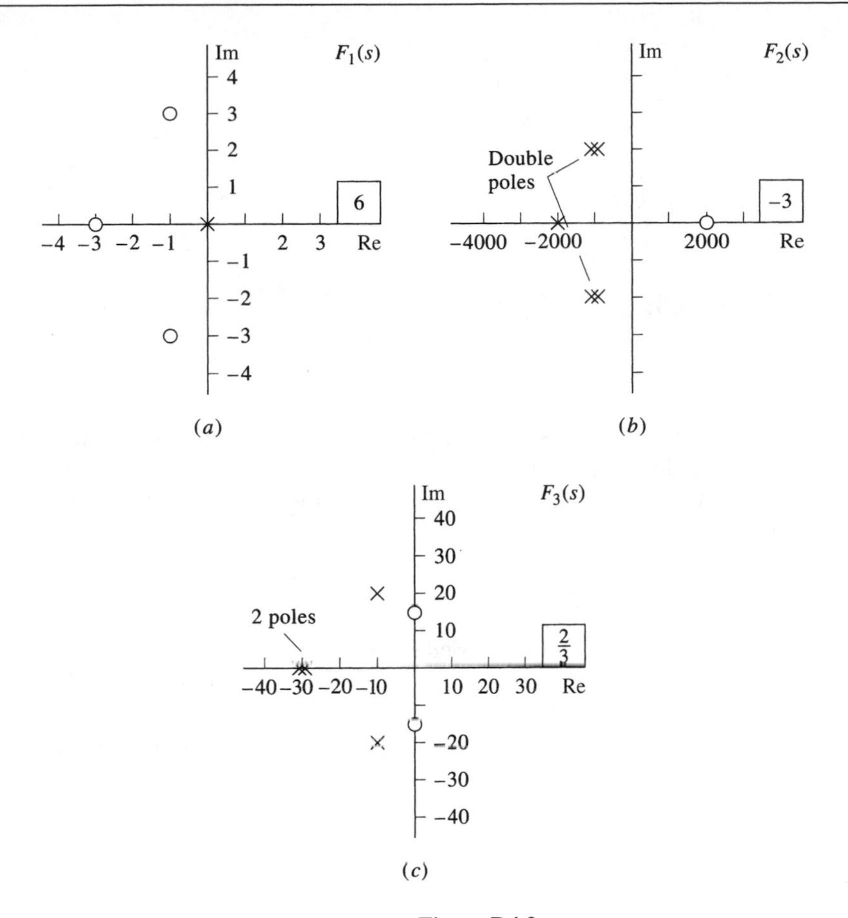

Figure D4.2

D4.3 Carefully sketch pole–zero plots for the following functions, then graphically evaluate the functions at the indicated value of the variable s using trigonometry.

(a)

$$F(s) = \frac{10(s-2)}{(s+1)(s+2)}$$

$$s = j3$$

Ans. $3.2e^{-j4^\circ}$

(b)

$$F(s) = \frac{4s^2+32}{(s^2+8s+20)(s+2)}$$

$$s = 2 + j$$

Ans. $0.3e^{-j11^\circ}$

(c)

$$F(s) = \frac{4(s^2 - 4s + 5)^2}{(s + 3)^2(s^2 + 6s + 10)}$$

$$s = j3$$

Ans. $2.0e^{-j34^\circ}$

4.3 Root Locus for Feedback Systems

4.3.1 Angle Criterion

The root locus for a feedback system is the path traced by the roots of the characteristic polynomial (the poles of the transfer function) as some system parameter is varied. Most control systems can be expressed in the form of Figure 4.4(a), where the transfer function $T(s)$ is

$$T(s) = \frac{KG(s)}{1 + KG(s)H(s)}$$

Closed- and open-loop transfer functions are defined.

The system parameter that is to be varied is the forward path gain K. It is normal practice to identify $T(s)$ as the *closed-loop transfer function* to distinguish it from another transfer function, which will now be introduced. In Figure 4.4(b), the feedback loop is broken at the variable called $Y_1(s)$. We can define a second transfer function, *the open- loop transfer function,* relating $Y_1(s)$ to $R(s)$. That second transfer function is

$$T_1(s) = KG(s)H(s) = \frac{Y_1(s)}{R(s)}$$

There are open-loop poles and zeros and closed-loop poles and zeros.

As always, each transfer function is defined with initial conditions equal to zero. Notice that the closed-loop transfer function depends heavily on the open-loop transfer function $KG(s)H(s)$. It is standard practice to define the zeros and poles of $G(s)H(s)$ as being *open-loop zeros* and *open-loop poles*, while the zeros and poles of $T(s)$ are called the *closed-loop zeros* and *closed-loop poles*. For convenience, the open-loop zeros and open-loop poles may be simply called *GH zeros* and *GH poles*. The term *roots* generally applies to the poles of $T(s)$ which are the roots of the characteristic polynomial.

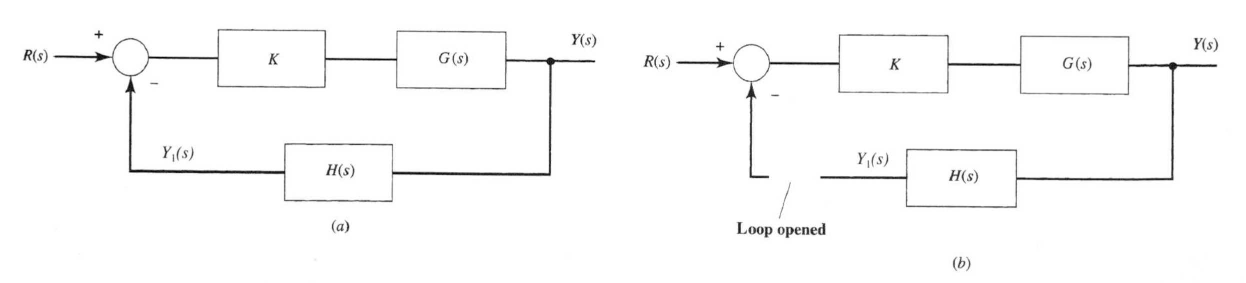

Figure 4.4 Feedback system with a variable gain K. (a) Closed-loop system. (b) Open-loop system with loop broken at $Y_1(s)$.

Since the closed-loop poles are the roots of the characteristic polynomial, then those closed-loop poles satisfy

$$1 + KG(s)H(s) = 0 \tag{4.1}$$

which is the same as requiring

$$KG(s)H(s) = -1 \tag{4.2}$$

In other words, any value of s that is a closed-loop pole must satisfy Equation (4.2), which is equivalent to satisfying both a magnitude criterion, which we get from the magnitude of Equation (4.2), and also an angle criterion, which we get from the angle of same equation. The magnitude criterion is

$$|K| = 1/\,|G(s)H(s)| \tag{4.3}$$

and the angle criterion is that

$$\angle KG(s)H(s) = \text{odd multiple of } 180^\circ \tag{4.4}$$

Importance of the angle criterion.

The angle criterion is the more important criterion because not all values of s that satisfy Equation (4.3) also satisfy Equation (4.4), while every s that satisfies Equation (4.4) can be used to find a value of K. A set of rules may be found to identify values of s satisfying Equation (4.4).

4.3.2 High and Low Gains

To understand how the roots depend on K, suppose we write the transmittances $G(s)$ and $H(s)$ with multiplying constants as we did in the last section.

$$G(s) = \frac{K_G N_G(s)}{D_G(s)}$$

$$H(s) = \frac{K_H N_H(s)}{D_H(s)}$$

where $N_G(s)$ is the numerator polynomial for $G(s)$ and $D_G(s)$ is the denominator polynomial for $G(s)$. Similar definitions apply to $N_H(s)$ and $D_H(s)$ for $H(s)$. The multiplying constants K_G and K_H are fixed by the process described by $G(s)$ and $H(s)$, so they cannot be changed; however, K can be varied.

The open-loop transfer function is therefore

$$KG(s)H(s) = \frac{KK_GK_HN_G(s)N_H(s)}{D_G(s)D_H(s)} = \frac{KN(s)}{D(s)} \tag{4.5}$$

where $D(s)$ is $D_G(s)D_H(s)$ and $N(s)$ equals $K_GK_HN_G(s)N_H(s)$. The closed-loop transfer function is

$$T(s) = \frac{KK_GN_G(s)/D_G(s)}{1 + \dfrac{KK_GK_HN_G(s)N_H(s)}{D_G(s)D_H(s)}} = \frac{KK_GN_G(s)D_H(s)}{D(s) + KN(s)} \tag{4.6}$$

The closed-loop zeros are those values of s for which the numerator of Equation (4.6) is zero, so the closed-loop zeros are the zeros of $G(s)$ and the poles of $H(s)$. The closed-loop poles are the roots of the denominator of Equation (4.6), and thus they depend on K. If K is small, then Equation (4.6) becomes

Effect of small K.

$$T(s) = \frac{KK_G N_G(s) D_H(s)}{D(s)}$$

so the roots for the root locus plot are located at the GH poles. By a similar reasoning, if K is large, then Equation (4.6) becomes

Effect of large K.

$$T(s) = \frac{KK_G N_G(s) D_H(s)}{KN(s)}$$

and the roots for the root locus plot are located at the GH zeros.

4.3.3 Root Locus Properties

Figure 4.5 shows an example root locus plot for a feedback system with open-loop transfer function

$$KG(s)H(s) = \frac{K(s+2)(s^2+2s+17)}{(s+4)(s^2+9)}$$

Some general comments about the root locus as K varies.

It consists of a pole–zero plot for $G(s)H(s)$, which is generally easy to construct because $G(s)$ and $H(s)$, being components of the system, are usually known in factored or partially factored from. Superimposed upon the pole–zero plot for $G(s)H(s)$ are the curves that are loci of the poles of $T(s)$ as K varies from zero to infinity. The locus segments are symmetrical about the real axis, and the sense of increasing K is usually indicated on each segment. As K approaches zero, the roots are located at the GH poles, and as K approaches $+\infty$ the roots are located at the GH zeros. For $0 \leqslant K < \infty$ the roots are located somewhere between the GH poles and the GH zeros.

To determine whether a given point s_0 is a point on the root locus for some value of K between zero and $+\infty$, it is only necessary to determine whether the angle of

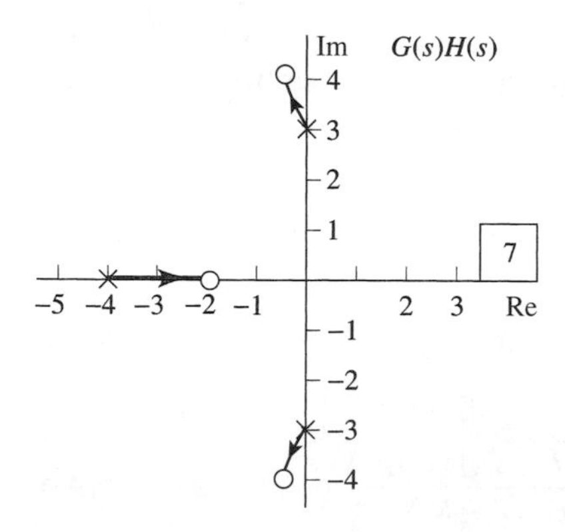

Figure 4.5 A root locus plot.

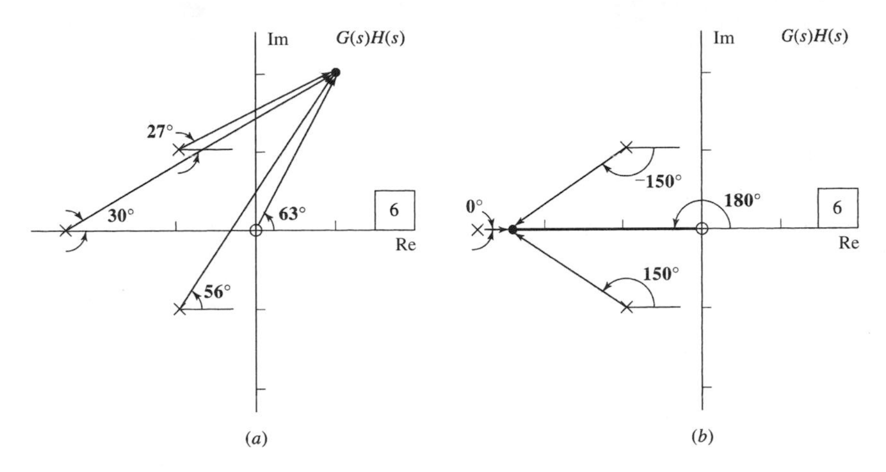

Figure 4.6 Testing an angle of $G(s)H(s)$ to determine whether a point is on the root locus.

$G(s)H(s)$ is 180°. This determination is easily made graphically, using directed line segments:

$$\begin{aligned}\angle G(s_0)H(s_0) &= \text{sum of zero angles to } s_0 - \text{ sum of pole angles to } s_0 \\ &\quad + 180^\circ \text{ if the multiplying constant is negative}\end{aligned}$$

For the GH product with pole–zero plot given in Figure 4.6(a), the angle of $G(s)H(s)$ for the indicated value of s is (approximately)

$$63^\circ - (56^\circ + 30^\circ + 27^\circ) = -50^\circ$$

Thus the indicated point is not on the root locus. For the point tested in Figure 4.6(b).

$$180^\circ - (0^\circ + 150^\circ - 150^\circ) = 180^\circ$$

Hence that point is on the root locus. There is a pole of $T(s)$ there for some positive value of K.

❑ DRILL PROBLEM

D4.4 Graphically find the angle of $G(s)H(s)$ at each of the points indicated In Figure D4.4.

Ans. (a) 180°; (b) 256°; (c) −117°

4.4 Root Locus Construction

At first, we consider only the most common case: the parameter K of interest is nonnegative and the multiplying constant of the $G(s)H(s)$ product is positive. There are several simple rules that allow approximate root locus sketches to be made easily and rapidly. Table 4.1 shows some examples of root locus plots. Remember that a computer could easily plot the roots for us, but since we want

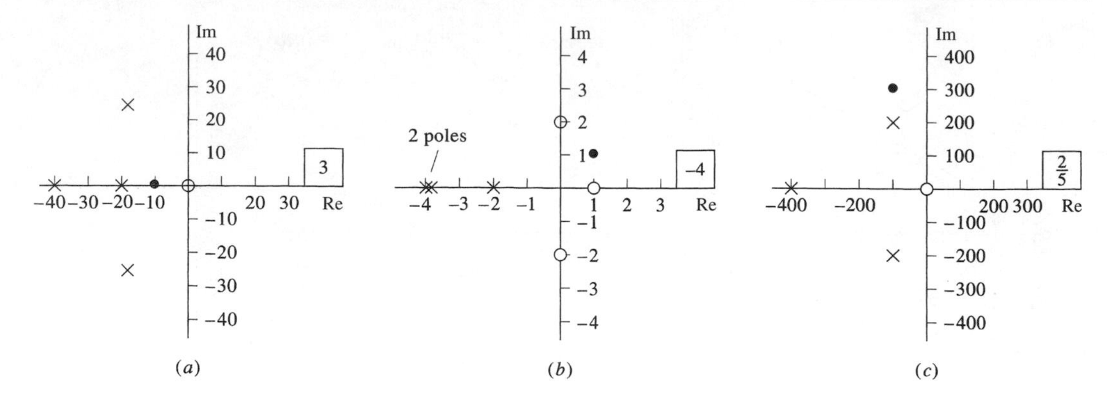

Figure D4.4

to be able to double-check the computer results and to analyze systems with some independence from software, we shall predict the rough shape of the root locus.

Because most transmittances encountered in practice have more poles than zeros, and because 180° and −180° are the same angle, the negative of the angle of the *GH* product is usually calculated, and the angle criterion becomes

$$\text{Sum of pole angles to } s_0 - \text{sum of zero angles to } s_0 = -\angle G(s_0)H(s_0) = 180^\circ$$

We will use this "reverse angle" evaluation in the development to follow.

Rule 1. Loci Branches

The branches of the locus are continuous curves that start at each of the *n* poles of GH, for $K = 0$. As K approaches $+\infty$, the locus branches approach the *m* zeros of GH. Locus branches for excess poles extend infinitely far from the origin; for excess zeros, locus segments extend from infinity.

Using Rule 1.

If $G(s)H(s)$ has more poles than zeros, some of the segments of the root locus, which start on poles (for $K = 0$), do not have a zero to end upon (for $K \to \infty$). These segments of the locus extend from the poles infinitely far from the origin of the complex plane. It is said that these loci extend "to infinity." If $G(s)H(s)$ has more poles n than zeros m, m segments of the locus extend from a pole to a zero, and $n - m$ excess segments each start at a pole and extend infinitely far from the origin. Segments never extend from a pole to infinity and then back from infinity to a zero. If $G(s)H(s)$ has more zeros than poles, the situation is similar, with n segments extending from a pole to a zero and $m - n$ excess segments coming from infinity to a zero. Some examples of roots locus plots are given in Table 4.2.

Rule 2. Real Axis Segments

The locus includes all points along the real axis to the left of an odd number of poles plus zeros of GH.

Table 4.1 *Basic Root Locus Principles*

1. The branches of the locus are continuous curves that start at each of the n poles of GH, for $K = 0$. As K approaches $+\infty$, the locus branches approach the m zeros of GH. Locus branches for excess poles extend infinitely far from the origin; for excess zeros, locus segments extend from infinity.
2. The locus includes all points along the real axis to the left of an odd number of poles plus zeros of GH.
3. As K approaches $+\infty$, the branches of the locus become asymptotic to straight lines with angles

$$\theta = \frac{180° + i360°}{n - m}$$

for $i = 0, \pm 1, \pm 2, \ldots$, until all $n - m$ or $m - n$ angles are obtained, where n is the number of poles and m is the number of zeros of GH.

4. The starting point of the asymptotes, the centroid of the pole–zero plot, is on the real axis at

$$\sigma = \frac{\sum \text{pole values of } GH - \sum \text{zero values of } GH}{n - m}$$

5. Loci leave the real axis at a gain K that is the maximum K in that region of the real axis. Loci enter the real axis at the minimum value of K in that region of the real axis. These points are termed *breakway points* and *entry points*, respectively. A pair of locus segments leave or enter the real axis at angles of $\pm 90°$.
6. The angle of departure ϕ of a locus branch from a complex pole is given by

$$\phi = -\sum \text{other } GH \text{ pole angles } + \sum GH \text{zero angles} + 180°$$

The angle of approach ϕ' of a locus branch to a complex zero is given by

$$\phi' = \sum GH \text{ pole angles} - \sum \text{other } GH \text{ zero angles} - 180°$$

where each GH pole angle and GH zero angle is calculated to the complex pole for ϕ and to the complex zero for ϕ'.

If the complex pole or zero is of order m, the m angles of departure and approach are given by

$$\phi = \frac{-\sum \text{other} GH \text{pole angles} + \sum GH \text{ zero angles} + (1 + 2i)180°}{m}$$

$$\phi' = \frac{\sum GH \text{ pole angles} - \sum \text{other } GH \text{ zero angles} - (1 + 2i)180°}{m}$$

for $i = 0, 1, 2, \ldots, (m - 1)$.

Table 4.2 *Some Root Locus Plots*

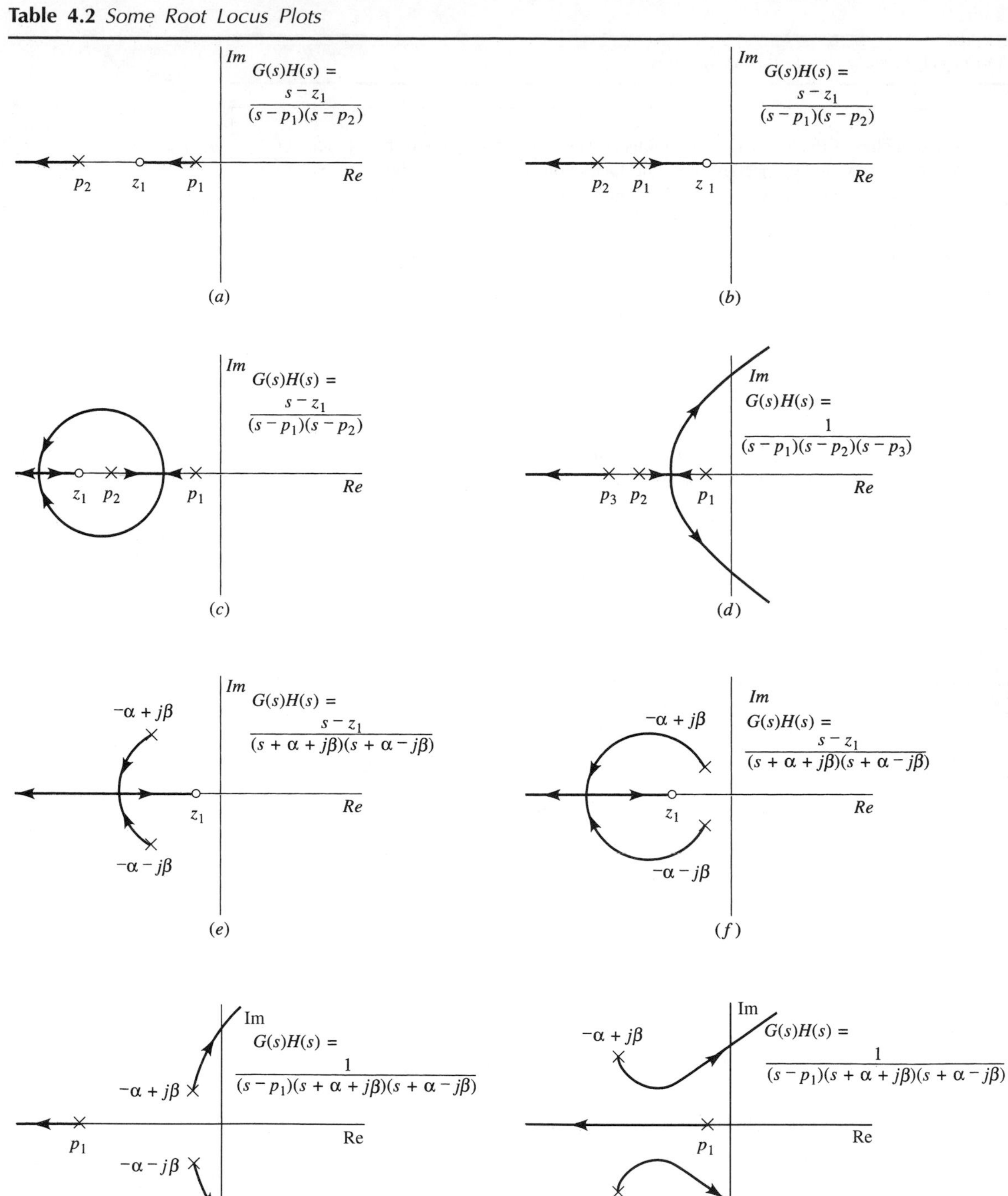

Table 4.2 *Some Root Locus Plots (Continued)*

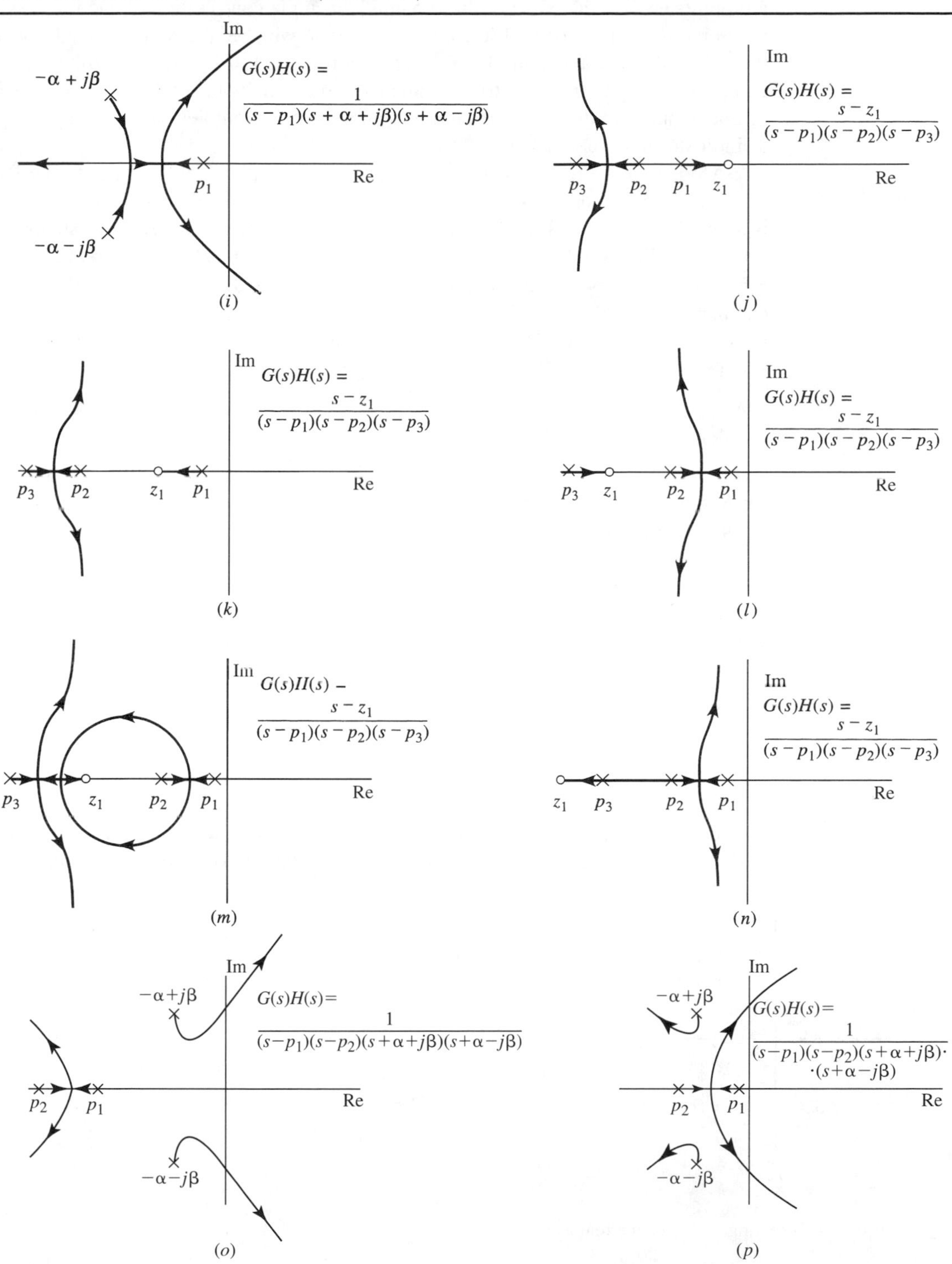

The easiest points on the complex plane to test to see if they are on the root locus are points on the real axis. For these points, the angle contribution of each real axis pole or zero is either 0° or 180°, depending upon whether the root is to right or to the left of the real axis point being tested, as indicated in Figure 4.7(a,b). A set of complex conjugate roots contributes angles to real axis points, which are negatives of one another, so the net contribution to an angle of a complex set of roots is zero, as indicated in Figure 4.7(c).

A point on the real axis is thus on the root locus if and only if it is to the left of an odd number of roots (poles and zeros), so that the angle of *GH* at that point is an odd multiple of 180°. The real axis root locus segments of several systems are sketched in Figure 4.8. In Figure 4.8(a), the loci are entirely along the real axis. In Figure 4.8(b), one segment extends from the double pole to the zero at $s = 0$ and one extends from the double pole to infinity. Two other root locus segments will extend from the complex poles; the sketching of these segments will be discussed later. In Figure 4.8(c), there are no real axis locus segments.

Rule 3. Asymptotic Angles

As K approaches $+\infty$, the branches of the locus become asymptotic to straight lines with angles

$$\theta = \frac{180^\circ + i360^\circ}{n - m}$$

for $i = 0, \pm 1, \pm 2, \ldots$, until all $(n - m)$ or $(m - n)$ angles are obtained; n is the number of poles and m is the number of zeros of GH.

At points on the complex plane very far from all the poles and zeros of $G(s)H(s)$, the reverse angle of *GH* is virtually

(Number of poles – number of zeros)θ

Finding asymptote angles.

where θ is the angle of the point itself, as indicated in Figure 4.9. At large distances from the cluster of poles and zeros, root loci extending to or from infinity approach

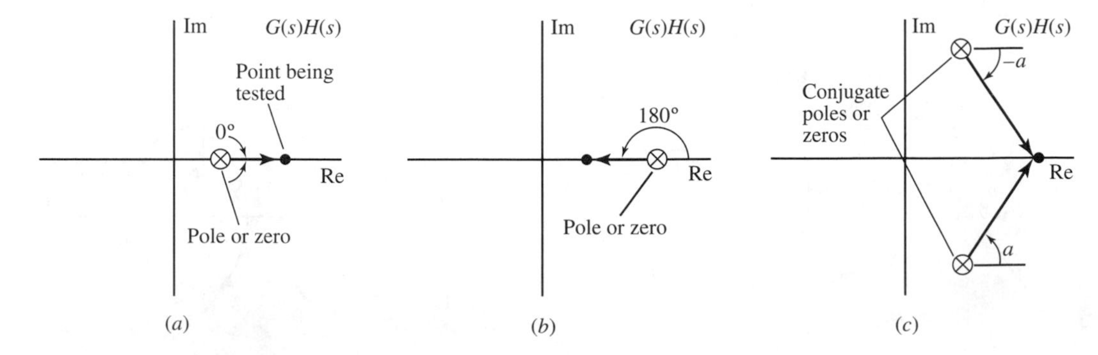

Figure 4.7 Testing points on the real axis.

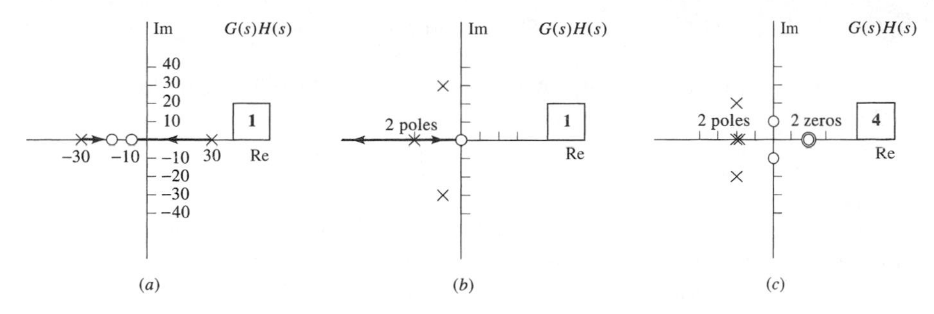

Figure 4.8 Example of a real axis root locus segments.

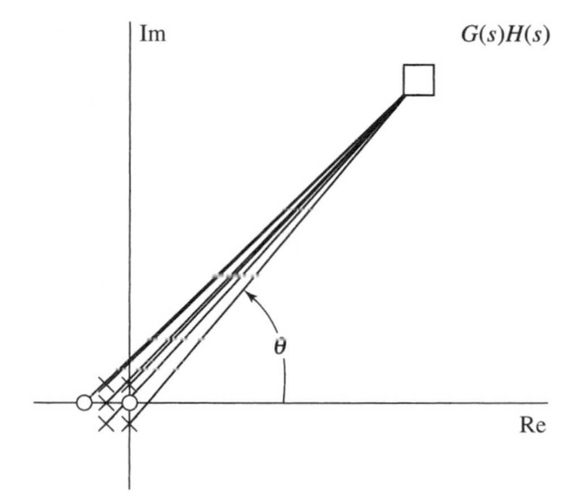

Figure 4.9 Angle contributions far from the poles of zeros of GH.

straight-line asymptotes at angles given by

$$(\text{Number of poles} - \text{number of zeros})\theta = 180^\circ \pm i360^\circ$$

It is important to include the multiples of 360° in this formulation because it is from this term that multiple solutions, where there is more than one asymptotic angle, arise.

If, for example, the GH product has two zeros and six poles, the asymptotic angles are given by

$$(6-2)\theta = 180^\circ \pm i360^\circ \quad \theta = \frac{180^\circ \pm i360^\circ}{4}$$

Substituting various integer values of i, say 0, 1, 2, 3, ..., we obtain the four different angles:

$$\begin{aligned}\theta &= +45^\circ, +135^\circ, 225^\circ, 315^\circ \\ &= \pm 45^\circ, \pm 135^\circ\end{aligned}$$

Substitutions of additional integers simply give repetitions of the same angles. There are $n - m = 4$ different asymptotic angles in this example.

Rule 4. Centroid of the Asymptotes

The starting point of the asymptotes, the centroid of the pole–zero plot, is on the real axis at

$$\sigma = \frac{\sum \text{poles values of } GH - \sum \text{zero values of } GH}{n - m}$$

The centroid is a sort of reverse vanishing point in that, as the closed-loop poles move to some distance away from the real axis, an observer looking backward would see the centroid as an apparent point of departure from the real axis. When $n-m$ is either zero or 1, the centroid is not used. For those values of $n - m$, closed-loop poles do not move off the real axis on the way toward infinity. When $n - m$ is zero, all the open-loop poles have an open- loop zero to approach as the gain rises toward infinity. When $n - m$ is 1, one closed-loop pole moves toward infinity along the negative real axis (moving at an asymptote angle of 180°).

Sometimes the centroid has no meaning ($n - m = 0$ or 1).

For large gain values, those s values on the root locus are approximately given by

$$s = \sigma + m(K)$$

where $m(K)$ is a large magnitude that grows with K. Since

$$s - \sigma = m(K)$$

each s in the equation

$$1 + KG(s)H(s) = 1 + \frac{KN(s)}{D(s)}$$

causes that equation to be

$$1 + KG(s)H(s) = 1 + \frac{K}{(s - \sigma)^{n-m}}$$

Because s is large, the polynomials $D(s)$ and $N(s)$ can be approximated by the first two terms as follows:

$$D(s) \approx s^n + a_{n-1}s^{n-1}$$

$$N(s) \approx s^m + b_{m-1}s^{m-1}$$

Using long division, we write

$$\frac{D(s)}{N(s)} \approx s^{n-m} + (a_{n-1} - b_{m-1})s^{n-m-1}$$

Using the binomial expansion, we write

$$(s - \sigma)^{n-m} \approx s^{n-m} + (n - m)\sigma s^{n-m-1}$$

The two expressions can be equated:

$$\sigma = \frac{a_{n-1} - b_{m-1}}{n - m}$$

The coefficient for the second highest power of s is always the sum of the roots of that polynomial; therefore, the numerator of the expression for σ is the sum of the

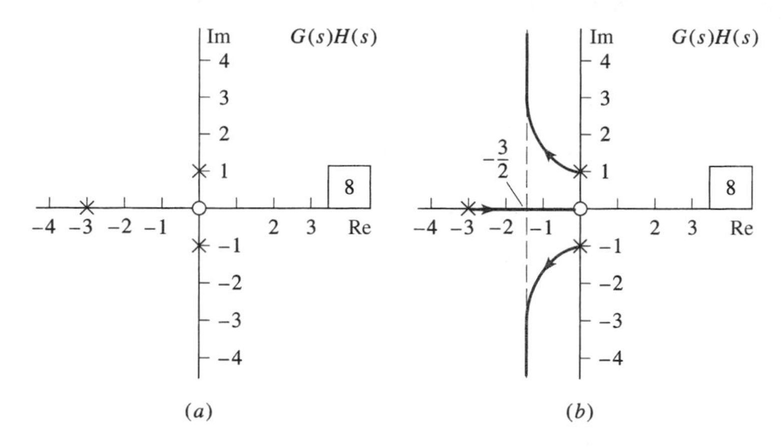

Figure 4.10 Root locus construction with asymptotes and centroid.

pole values of GH [the sum of the roots of $D(s)$] minus the sum of the zero values of GH [the sum of the roots of $N(s)$].

Consider the system with pole–zero plot given in Figure 4.10(a). A real axis segment of the locus extends from the real axis pole to the zero at $s = 0$. The other two locus segments extend from the imaginary axis poles to infinity. Their asymptotic angles are given by

Centroid example.

$$(3 - 1)\theta = 180^\circ \pm i360^\circ \qquad \theta = 90^\circ, -90^\circ$$

The centroid of asymptotes is

$$\sigma = \frac{(0 + j + 0 - j - 3) - 0}{3 - 1} = -\frac{3}{2}$$

The imaginary part contributions of conjugate sets of roots always cancel one another, so only the real parts of the root locations need to be included in the centroid calculation. The complete root locus is shown in Figure 4.10(b).

❑ DRILL PROBLEM

D4.5 Sketch root locus plots, for an adjustable constant K between 0 and $+\infty$, for systems with the following GH products. Find the asymptotic angles and centroid if applicable.

(a)

$$G(s)H(s) = \frac{3s}{(s + 2)(s^2 + 6s + 18)}$$

Ans. $\pm 90^\circ$; -4

(b)

$$G(s)H(s) = \frac{10}{(s + 6)(s^2 + 8s + 41)^2}$$

Ans. $\pm 36^\circ, \pm 108^\circ, -180^\circ$; -4.4

(c)

$$G(s)H(s) = \frac{1}{(s^2+8s+41)(s^2+2s+5)}$$

Ans. $\pm 45^\circ, \pm 135^\circ; -2.5$

(d)

$$G(s)H(s) = \frac{7}{(s+1)(s^2+10s+26)}$$

Ans. $\pm 60^\circ, -180^\circ; -\frac{11}{3}$

(e)

$$G(s)H(s) = \frac{2(s^2+4s+3)}{(s+2)^2(s+4)}$$

Ans. -180°; centroid not applicable

4.5 More About Root Locus

4.5.1 Root Locus Calibration

The values of the adjustable constant K corresponding to various points on a root locus can be found by applying the relation

$$|K| = \frac{1}{|G(s)H(s)|}$$

for points on the locus. The magnitude $|G(s)H(s)|$ can be found graphically for a point on the locus by using

$$|G(s)H(s)| = \frac{\begin{pmatrix}\text{magnitude of the} \\ \text{multiplying constant}\end{pmatrix}\begin{pmatrix}\text{product of} \\ \text{zero distances}\end{pmatrix}}{\text{product of pole distances}}$$

For the system of Figure 4.11, for example, the value of K corresponding to the imaginary axis points on the locus, where $T(s)$ is marginally stable, is given approximately by

Finding the largest K for stability.

$$|K| = \frac{1}{|G(s)H(s)|} = \frac{(3.3)(4.1)(5.0)}{10} \qquad K = 6.8$$

Thus for K greater than about this value, the overall system will be unstable. If the root locus is only approximate, this solution for K is approximate, too. More accuracy can easily be obtained, if needed, by testing the angles of some near the approximate locus and refining the solution for K.

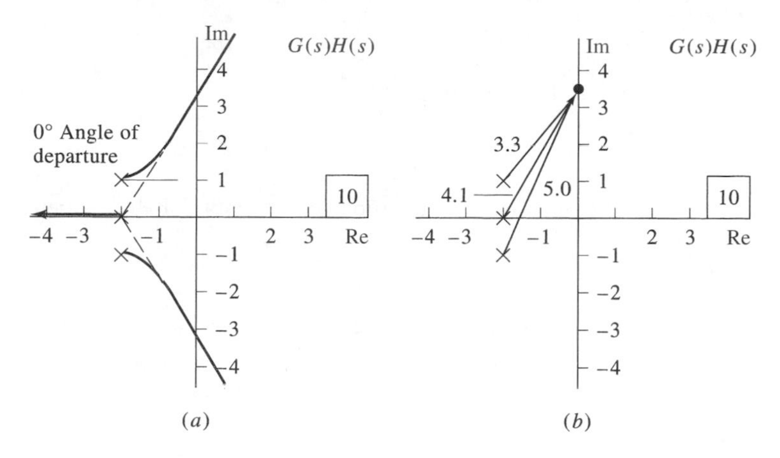

Figure 4.11 Calibrating a point on the root locus.

Rule 5. Breakaway and Entry Points

Loci leave the real axis at a gain K that is the maximum K in that region of the real axis. Loci enter the real axis at the minimum value of K in that region of the real axis. These points are termed breakaway points and entry points, respectively. Members of a pair of locus segments leave or enter the real axis at angles of ±90°.

From Rule 1 we know that the locus branches begin at the GH poles for $K = 0$. It follows that increasing K causes the locus branches to move away from the GH poles. In the case covered by Rule 5, there are pairs of GH poles from which these branches move until the branches eventually split and move away from the real axis. Therefore, the splitting occurs at some largest K for that part of the real axis. Any smaller K results in real axis roots that have not come together while any larger K results in roots that are not on the real axis.

Significance of the breakaway point.

By similar reasoning, locus branches approach GH zeros as K becomes large. Thus if there is a pair of real axis GH zeros, some roots must enter the real axis somewhere between the GH zeros at a K that is smallest for that part of the real axis. If K is smaller than that value, the roots will not be on the real axis; if K is larger, the roots will not be together.

Significance of the entry point.

There are two ways of finding each breakaway or entry point:

1. Differentiation of $K(s)$ with respect to s
2. Using $K(s)$ by trial and terror

The first procedure acknowledges that the value of s that maximizes K for breakaway or minimizes K for entry also causes the derivative of $K(s)$ taken with respect to s to become zero. In general, we know that any s on the root locus causes

$$1 + KG(s)H(s) = 1 + K\frac{N(s)}{D(s)} = 0 \qquad [4.7]$$

$$K = -\frac{D(s)}{N(s)} = K(s) \qquad [4.8]$$

If we differentiate Equation (4.8) with respect to s we have

$$\frac{dK}{ds} = -\frac{NdD/ds - DdN/ds}{N^2} = 0$$

Assuming that N is not zero, then we need the roots of a polynomial $P(s)$, where

$$P = N\frac{dD}{ds} - D\frac{dN}{ds} = 0 \qquad [4.9]$$

For the system with $G(s)H(s)$ as in Figure 4.12, there is a real axis root locus segment, but it is not complete because it does not extend from an open-loop pole to an open-loop zero. For this system, there are asymptotes with angles

$$\theta = \frac{180° + i360°}{2} = 90°, -90°$$

and centroid

$$\sigma = \frac{-3-1}{2} = -2$$

Since K varies fom zero to infinity, the poles of the overall transfer function begin as the poles of GH. For larger K, the overall transfer function $T(s)$ has two real axis poles, which become closer and closer together with increasing K. At some gain K there are two repeated roots which follow by solving Equation (4.9).
Here

$$N = 1$$

$$D = s^2 + 4s + 3$$

$$P = 1\frac{d\left(s^2+4s+3\right)}{ds} - \left(s^2+4s+3\right)\frac{dN}{dS}$$

$$= 2s + 4 = 0$$

Obviously s is -2 (the breakaway point). For that breakaway point, we write

$$K = -[s^2 + 4s + 3]_{s=-2} = -[4 - 8 + 3] = 1$$

For still larger values of K the roots are complex conjugates with larger and larger imaginary parts. In this simple case, the locus breaks away from the real axis at the centroid, $s = -2$.

In more involved systems, a breakaway of two root locus segments from the real axis is also at 90° angles, but the breakaway point is not necessarily midway between the real axis GH roots. For a segment extending from poles, as in the example of Figure 4.13(a), the real axis point at which the loci break away will correspond to the largest value of K for a root between -5 and 0. Here

$$6(s)H(s) = \frac{10(s+10)}{s(s+5)}$$

$$N(s) = 10(s+10)$$

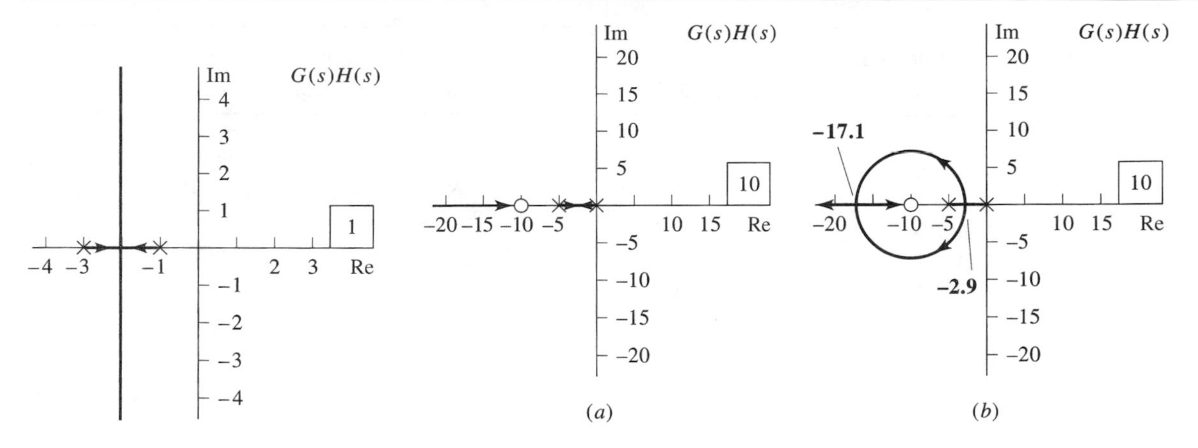

Figure 4.12 Breakaway of loci from the real axis.

Figure 4.13 Calculating breakaway and entry points.

$$D(s) = s(s+5) = s^2 + 5s$$

$$\begin{aligned} P &= (10s+100)\frac{d\left(s^2+5s\right)}{ds} - \left(s^2+5s\right)\frac{d\,(10s+100)}{ds} \\ &= (10s+100)(2s+5) - \left(s^2+5s\right)(10) \\ &= 10s^2 + 200s + 500 \\ &= 10\left(s^2+20s+50\right) \end{aligned}$$

The two roots of P are -2.9 (the breakaway point between 0 and -5) and -17.1 (the entry point to the left of -10). It is possible to have roots of P that correspond to entry and breakaway points of negative values of K. As long as we select only the roots along the parts of the real axis suggested by Rule 2, we easily select the entry and breakaway points for positive values of K. Of course we can also plug each s value into

$$K = -\frac{1}{G(s)H(s)} = -\frac{D(s)}{N(s)}$$

to determine the sign on K.

The second procedure mentioned earlier is to vary s by trial and error until we find the largest (or smallest) K value, using Equation (4.8):

$$K = -\frac{s(s+5)}{10(s+10)}$$

In Table 4.3 a search is performed by trial and error to find the breakaway point to two significant digits of s, with the corresponding K being rounded off to three significant digits, by seeking the largest K. The breakaway point is thus at $s = -2.9$ for $K = 0.0858$.

By proceeding in a similar manner, the entry point is sought in Table 4.4 by varying s to three significant digits to find the smallest K for roots to the left of -10. Thus the entry points is at -17.1 for K rounded off to five significant digits at $K = 2.9142$.

Entry example.

Table 4.3 *Breakaway Point Calculations for the Example of Figure 4.13*

	Varying
s	K
−2	0.075
−2.5	0.083
−2.8	0.0855
−2.9	0.0858
−3.0	0.0857
−3.5	0.081
−4.0	0.067

Table 4.4 *Entry Point Calculations for the Example of Figure 4.13*

	Varying
s	K
−15.0	3.000
−16.0	2.933
−16.5	2.919
−17.0	2.9143
−17.1	2.9142
−17.2	2.9144
−18.0	2.925

❑ Computer-Aided Learning

We shall soon discuss how to use MATLAB to obtain an entire root locus. For now, we consider just finding entry or breakaway points by employing Equation (4.9). To facilitate obtaining the roots of P, a macro can be written, assuming that n, d, and the derivatives of n and d are available. For convenience, we can define

$$G(s)H(s) = \frac{N(s)}{D(s)}$$

$$n_1 = \frac{dN(s)}{ds} \qquad n = N(s)$$

$$d_1 = \frac{dD(s)}{ds} \qquad d = D(s)$$

The script M file is stored as breakentry

```
% breakentry
p=conv(n,d1)-conv(d,n1)
roots(p)
```

For the example of Figure 4.13 use the commands

```
n=[10 100]
n1=10
d=[1 5 0]
d1=[2 5]
breakentry
```

The M file returns p and the roots of p, which are

```
p=[ 10 20 500]
roots -17.1, -2.9
```

There is one caution to be used with the breakentry program: the dimensions of `n*d1` and `d*n1` must be equal, which will not occur when $N(s)$ is a constant (since dimensions of n and n1 are both zero). In that case [when $N(s)$ is a constant], multiply both $N(s)$ and $D(s)$ by s *before* taking the derivative.

C4.1 Use MATLAB to find the entry point of

$$G(s)H(s) = \frac{s}{(s + j2)\,(s - j2)}$$

The entry point must be to the left of $s = 0$.

Ans.
```
n=[1 0]
n1=[1]
d=[1 0 4]
d1=[2 0]
breakentry
```
(answer is -2)

C4.2 Use MATLAB to find the entry point of

$$G(s)H(s) = \frac{s + 1}{(s + 2 + j6)\,(s + 2 - j6)}$$

The entry point must be to the left of $s = -1$.

Ans.
```
n=[1 1]
n1=[1]
d=[1 4 40]
d1=[2 4]
breakentry
```
(answer is -7.08)

C4.3 Use MATLAB to find the breakaway point and entry point of

$$G(s)H(s) = \frac{s + 10}{(s + 1)\,(s + 5)}$$

The breakaway point is between -1 and -5. The entry point is to the left of $s = -10$.

Ans.
```
n=[1 10]
n1=[1]
d=[1 6 5]
d1=[2 6]
breakentry
```

(The answers are -3.29 and -16.7.)

C4.4 Find the two breaking points of

$$G(s)H(s) = \frac{10}{(s + 4)(s + 6)(s + 10)(s + 40)}$$

One breakaway point is between -4 and -6 while the other is between -10 and -40.

Ans. First, use multiple convolution to get d. It is necessary to multiply numerator and denominator by s because the numerator is a constant.

```
n=[10 0]
n1=[10]
d4=conv([1 4], [1 6])
d3=conv([1 10], [1 40])
d2=conv(d4,d3)
d=conv(d2,[1 0])
  = [1 60 924 5200 9600 0]
d1=[5 240 2772 10400 9600]
breakentry
```

(The answers are −4.88, −31.7.)

Rule 6. Angles of Departure and Approach

The angle of departure Φ of a locus branch from a complex pole is given by

$$\phi = -\sum \text{other } GH \text{ pole angles} + \sum GH \text{ zero angles} + 180^\circ$$

The angle of approach ϕ' of a locus branch to a complex pole is given by

$$\phi' = \sum GH \text{ pole angles} - \sum \text{other } GH \text{ zero angles} - 180^\circ$$

where each *GH* pole angle and *GH* zero angle is calculated to the complex pole for ϕ and to the complex zero for ϕ'.

If the complex pole or zero is of order m, the m angles of departure and approach are given by

$$\phi = \frac{-\sum \text{other } GH \text{ pole angles } + \sum GH \text{ zero angles } + (1+2i)180^\circ}{m}$$

$$\phi' = \frac{\sum GH \text{ pole angles } - \sum \text{ other } GH \text{ zero angles } - (1+2i)180^\circ}{m}$$

for $i = 0, 1, 2, \ldots, (m-1)$.

For a set of complex conjugate poles of $G(s)H(s)$, the angle at which the root locus leaves one of the poles is found by considering a point on the locus branch very close to the pole, as in the example of Figure 4.14. Since the point is very close to the pole under consideration, the angles to the point are virtually the angles to the pole itself. Solving

Finding the angle of departure.

$$\text{(Sum of other pole angles to the pole under consideration)}$$
$$+\,\phi - \text{(Sum of zero angles to the pole)} = 180^\circ$$

will give the angle of departure, ϕ. Of course, the angle of departure from the lower pole of the conjugate set is the negative of the upper pole's angle of departure. Angles of approach to complex zeros are found similarly:

Finding the angle of approach.

$$\text{(Sum of pole angles to the zero under consideration)}$$
$$-\text{(sum of other zero angles to the zero)} - \phi' = 180^\circ$$

Figure 4.14 Finding an angle of departure.

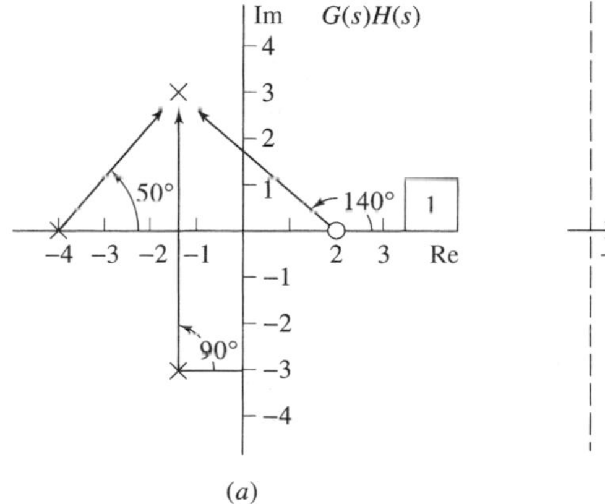

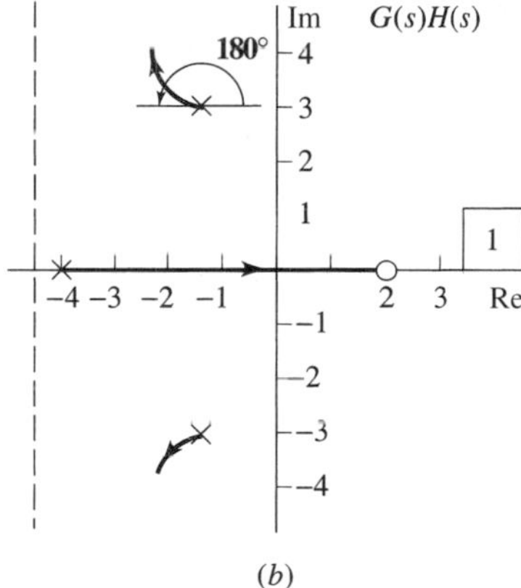

Figure 4.15 Using root locus construction to find angles of departure.

Consider the system with GH product given in Figure 4.15(a). For a point on the root locus near the top complex pole, approximately,

Angle of departure example for Figure 4.15.

$$\phi = -\left[90^\circ + 50^\circ\right] + 140^\circ + 180^\circ \qquad \phi = 180^\circ$$

where ϕ is the angle of departure of the locus from the pole. There is a complete real axis segment of the locus between the real axis pole and zero. The other two locus segments extend from the complex poles to infinity, with asymptotic angles given by

$$(3 - 1)\theta = 180^\circ \pm i360^\circ \qquad \theta = \pm 90^\circ$$

and centroid

$$\sigma = \frac{\left(-4 - \frac{3}{2} - \frac{3}{2}\right) - 2}{3 - 1} = -\frac{9}{2}$$

A completed root locus sketch is given in Figure 4.15(b).

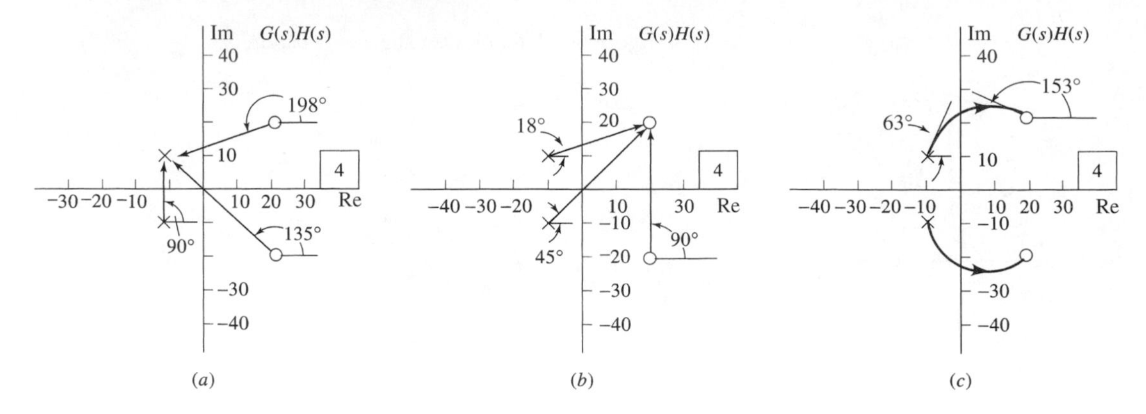

Figure 4.16 Root locus construction involving angles of departure and angles of approach.

Example using Figure 4.16.

Consider the system of Figure 4.16(a). For a point on the root locus near the top complex pole, approximately

$$\phi = -90^\circ + [198^\circ + 135^\circ] + 180^\circ \qquad \phi = 423^\circ = 63^\circ$$

where ϕ is the angle of departure of the locus from the top pole. For a point near the top complex zero, Figure 4.16(b), the angle ϕ' of arrival of the locus is given approximately by

$$\phi' = +\left[18^\circ + 45^\circ\right] - 90^\circ - 180^\circ \qquad \phi' = -207^\circ = 153^\circ$$

A complete root locus sketch is shown in Figure 4.16(c).

At repeated complex roots of GH, more than one locus segment begins or ends at the root location, so more than one angle of departure or approach will be found, a different angle for each locus segment. For the system of Figure 4.17(a), the angle contribution to a point near the top double pole is given approximately by

$$\phi = \left[-(90^\circ + 90^\circ + 108^\circ) + 124^\circ + (1 + 2i)180^\circ\right]/2$$
$$= -82^\circ + (1 + 2i)90^\circ \qquad \phi = 8^\circ (i = 0),\ 188^\circ (i = 1)$$

It is easy to see that multiple departure (or arrival) angles will be evenly spaced around the multiple roots. To complete the root locus sketch as in Figure 4.17(b), the real axis locus segment is drawn and the asymptotic angles and centroid are found:

$$(5 - 1)\theta = 180^\circ \pm i360^\circ \qquad \theta = \pm 45^\circ,\ \pm 135^\circ$$

$$\sigma = \frac{-20 - 20 - 20 - 20 - 10}{5 - 1} = -22.5$$

Another Example

As an example of application of the six root locus construction rules, consider the open-loop transfer function

$$KG(s)H(s) = \frac{K}{s(s + 3)(s^2 + 6s + 64)}$$

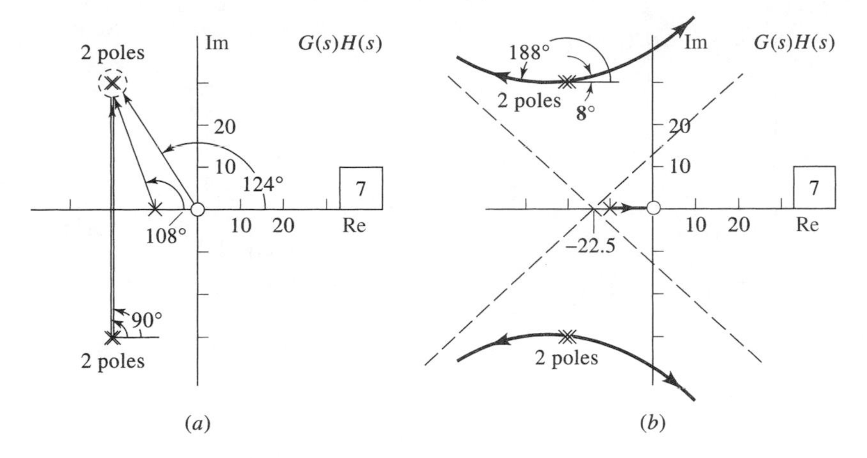

Figure 4.17 Angles of departure from multiple poles.

A step-by-step procedure is as follows.

Locate the open-loop poles and zeros and plot them (Rule 1). There are no zeros of GH. The poles of GH are located at 0, -3, $-3 + j7.4$ and $-3 - j7.4$.

Locate the real axis portions of the locus (Rule 2). The real axis segment between $s = 0$ and $s = -3$ is on the root locus. The root locus diagram so far is given in Figure 4.18(a).

Determine the angles of the asymptotes (Rule 3). The asymptotic angles are given by

$$\theta = \frac{180° + i360°}{4 - 0} = 45°,\ 135°,\ -135°,\ -45°$$

Determine the centroid of the asymptotes (Rule 4).

$$\sigma = \frac{0 - 3 - 3 + j7.4 - 3 - j7.4}{4 - 0} = \frac{-9}{4} = -2.25$$

Find the real axis breakway point (Rule 5). The values of K corresponding to various real axis points on the locus between $s = 0$ and $s = -3$ are found by using

$$K = \left| \frac{1}{G(s)H(s)} \right| = |s(s + 3)(s^2 + 6s + 64)|$$

With the aid of a calculator by trial and error, we find the following values:

s	K
−1.40	128.93
−1.43	129.01
−1.44	129.02
−1.45	129.01
−1.50	128.81

So the breakway point is at −1.44.

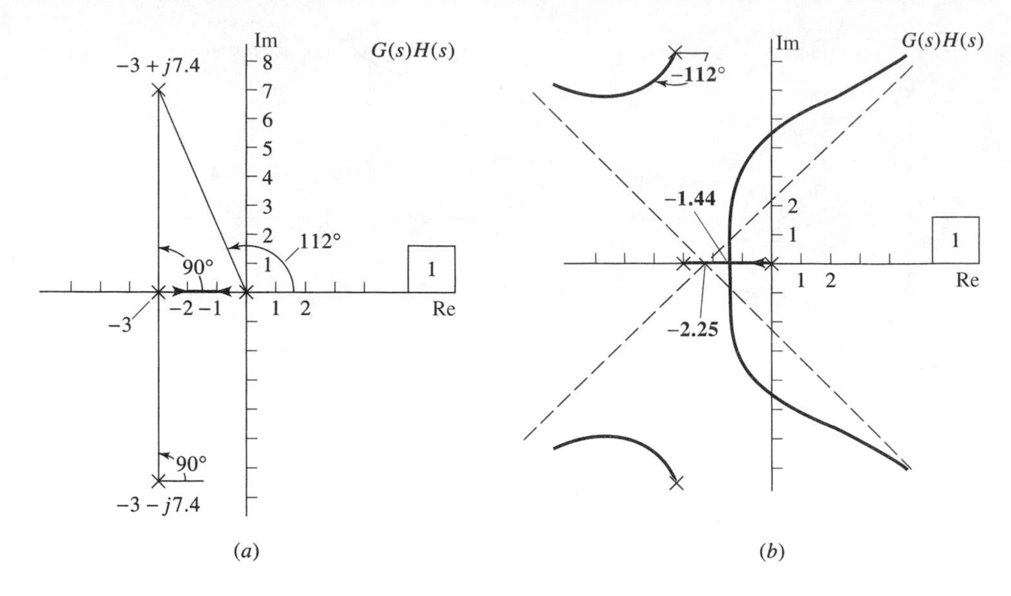

Figure 4.18 Applying the root locus construction rules.

Determine the angle of departure from the top pole (Rule 6). Using the construction of Figure 4.18(a),

$$\phi = -(112^\circ + 90^\circ + 90^\circ) + 180^\circ = -112^\circ$$

The completed root locus diagram is given in Figure 4.18(b). Of course, in many situations, only certain of the construction rules apply.

❑ Computer-Aided Learning

Angles of departure and approach may be calculated by using the MATLAB commands

```
ATAN2(y,x)
```

which returns the angle in radians for a vector terminating at the point (x, y).

To find the departure angle for an open-loop pole p_1 first eliminate that pole from $G(s)H(s)$ by taking

$$G_1(s) = (s - p_1)\, G(s)H(s)$$

If the point s is equal to $p_1 = a + jb$, then the angle of departure may be calculated, assuming $G_1(s)$ is given:

```
s=a+j*b
g1=
departure=180+180/pi*ATAN2(imag(g1), real(g1))
```

The factor 180/pi converts from radians to degrees. The y terminating point of the vector is given by the imaginary part of g1 while the x terminating point is given by the real part of g1.

For the example of Figure 4.16, the departure angle at the pole $-10+j10$ may be found by using the commands

```
s=-10+j*10
g1=4*(s-20+j*20)*(s-20-j*20)/(s+10+j*10)
departure=180+180/pi*ATAN2(imag(g1), real(g1))
```

To find an approach angle to an open-loop zero z_1, first eliminate that zero from $G(s)H(s)$:

$$G_2(s) = G(s)H(s)/(s - z_1)$$

If the point s is equal to the zero $z_1 = a + jb$, then the approach angle may be calculated, where we define $G_3(s) = 1/G_2(s)$:

```
s=a+j*b
g3=
approach=-180+180/pi*ATAN2(imag(g3), real(g3))
```

For the example of Figure 4.16, the approach angle at the zero $20 + j20$ may be found by

```
s=20+j*20
g3=(s+10+j*10)*(s+10-j*10)/4/(s-20+j*20)
approach=-180+180/pi*ATAN2(imag(g3), real(g3))
```

C4.5 Find the departure angle of

$$G(s)H(s) = \frac{s+1}{(s+2+j6)(s+2-j6)}$$

from the open-loop pole at $s = -2 + j6$.

Ans.
```
s=-2+j*6
g1=(s+1)/(s+2+j*6)
departure=180+180/pi*ATAN2(imag(g1), real(g1))
```
(189.5°)

C4.6 Find the approach angle of

$$G(s)H(s) = \frac{(s+2+j2)(s+2-j2)}{s(s+5)(s+10)}$$

to the open-loop zero at $s = -2 + j2$.

Ans.
```
s=-2+j*2
g3=s*(s+5)*(s+10)/(s+2+j*2)
approach=-180+180/pi*ATAN2(imag(g3), real(g3))
```
(-87.3°)

C4.7 For the system

$$G(s)H(s) = \frac{(s+2+j2)(s+2-j2)}{s(s+4+j6)(s+4-j6)}$$

Find the departure angle from $-4 + j6$ and the approach angle to $-2 + j2$.

Ans.
```
s=-4+j*6
g1=(s+2+j*2)*(s+2-j*2)/s/(s+4+j6)
departure=180+180/pi*ATAN2(imag(g1), real(g1))
s=-2+j*2
g3=s*(s+4+j*6)*(s+4-j*6)/(s+2+j*2)
approach=-180+180/pi*ATAN2((imag(g3), real(g3))
```
(angles are 186.9°, −122.5°)

C4.8 (a) Use MATLAB to find the breakaway point for the example of Figure 4.18.
(b) Use MATLAB to find the departure angle from the open-loop pole at $-3 + j7.4$.

Ans. (a) Multiply numerator and denominator of $G(s)H(s)$ by s:
```
n=[1 0]
d=[1 9 82 192 0 0]
n1=[1]
d1=[5 36 246 384 0]
breakentry
```
(b)
```
s=-3+j*7.4
g1=1/s/(s+3)/(s+3+j*7.4)
dep=180+180/pi*ATAN2(imag(g1), real(g1))
```

4.5.2 Computer-Aided Root Locus

The six rules permit the sketching of the general shape of a root locus manually. That endeavor displays a great deal of information about the system performance, as influenced by changing system gain. It is imperative that each engineer be able to sketch the general shape of a root locus by hand. Once that general shape has been obtained, a computer may be used to determine a more exact plot and to obtain accurate values of entry points, breakaway points, departure angles, and approach angles.

Of what value are entry and breakaway points, for example? In many cases the entry or breakaway point provides root locations with the greatest relative stability, that is, roots with the greatest real part. Other roots are closer to the vertical axis. As such, entry or breakaway points may provide a design point for system performance.

A number of computer programs are available for calculating and plotting a root locus. Although the commands differ, there are three basic pieces of information that a computer requires to generate a root locus. These are the numerator of *GH*, the denominator of *GH*, and the range of gains to be used. For example, suppose *GH* is

$$G(s)H(s) = \frac{s}{(s+2+j3)^2(s+2-j3)^2(s+6)}$$

The user can input the coefficients of the numerator of *GH* or the roots of the numerator of *GH*, followed by similar data for the *GH* denominator. At this point there are usually two options. If the user wants to investigate roots in some part of the complex plane under prompting from the computer, an interactive root locus may be sought. Another option is to select only one set of gains for which the root locus is to be found. That option is noninteractive in that the process terminates with the plotting of the root locus.

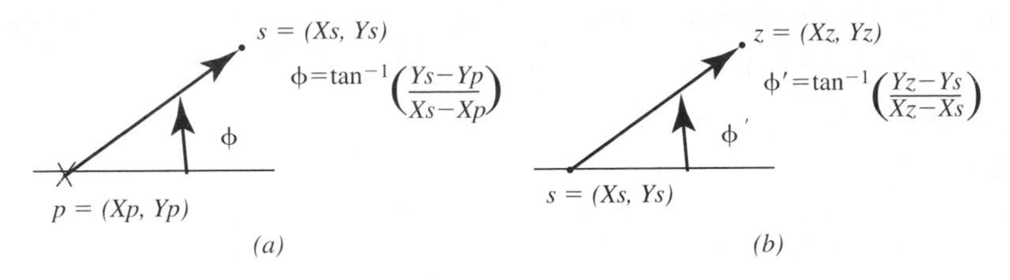

Figure 4.19 Determining (a) departure and (b) approach angles approximating, given a root locus.

For example, if an interactive root locus is available, each departure angle and each approach angle can be found by getting the x and y coordinates (real and imaginary values) for a point on the root locus very close to each GH pole [Figure 4.19(a)] and GH zero [Figure 4.19(b)]. Trigonometry may then be used to approximate each departure angle and each approach angle, where

$$\phi = \tan^{-1}\left(\frac{y_s - y_p}{x_s - x_p}\right) \quad [4.10]$$

$$\phi' = \tan^{-1}\left(\frac{y_z - y_s}{x_z - x_s}\right) \quad [4.11]$$

and the quadrant is given by the sign of each difference.

In this text, MATLAB is emphasized because of its widespread acceptance in the control community.

❑ Computer-Aided Learning

MATLAB provides the tools for root locus analysis:

```
rlocus(open loop transfer function)
rlocfind(open loop transfer function)
rltool(open loop transfer function)
```

The command rlocus, which displays the root locus only, is a noninteractive tool. The command rlocfind displays the root locus and a crosshair target, which can be located anywhere on the root locus by mcans of mouse. By clicking on a point, the user can display the gain and root values.

By far the most versatile and interactive tool is the “rltool” command. It is possible to move points anywhere on the root locus. The gain value is displayed, and the tools button provides the option of having closed-loop poles displayed. It is possible to obtain the step response for the chosen root locus gain. That capability allows a designer to see time domain implications of each gain choice. Further, poles and zeros can be added and then moved around to display the effect of each choice upon the root locus and the time response.

For example, to obtain the root locus for

$$G(s)H(s) = \frac{s}{(s + 2 + j3)^2(s + 2 - j3)^2(s + 6)}$$

The interactive root locus may be obtained by

```
gh=zpk([0],[-2+j*3 -2+j*3 -2-j*3 -2-j*3 -6],1)
rltool(gh)
```

C4.9 Use the "rltool" command to obtain each root locus and to verify values obtained previously. Use Equation (4.10) to verify departure angles and using Equation (4.11) to verify approach angles.

(a) Obtain the root locus and verify the entry point for Problem C4.1.
(b) Similarly solve C4.2.
(c) Obtain the root locus and verify the breakaway point and entry point for C4.3.
(d) Obtain the root locus and verify breakaway points for C4.4.
(e) Obtain the root locus and verify the departure angle for C4.5.
(f) Obtain the root locus and verify the approach angle for C4.6.
(g) Obtain the root locus and verify the departure and approach angle for C4.7.
(h) Obtain the root locus and verify values for C4.8.

❑ DRILL PROBLEMS

D4.6 Sketch root locus plots for the systems in Figure D4.6. Find asymptotic angles, centroid, approximate breakaway points, angles of departure, and angles of approach where applicable.

Ans. (a) $\pm 60°$, $-180°$; $-\frac{5}{3}$; none; $\mp 18°$; none; (b) $\pm 90°$; -20; -15.3; none; none; (c) $\pm 90°$; -6; -6.87; none, none; (d) $\pm 60°$, $-180°$; $-\frac{7}{3}$; none; ∓ 33.7; none.

D4.7 For systems with the root locus plots shown in Figure D4.7, find the value of the adjustable constant K for which the overall transfer function has a pole at the location indicated by the dot.

Ans. (a) $K = 1.4$; (b) $K = \frac{26}{7}$; (c) $K = 1000$

4.6 Root Locus for Other Systems

4.6.1 Systems with Other Forms

Problems other than basic one which has been considered exclusively here to this point, can be cast in the root locus form,

Standard root locus form.

$$1 + K\frac{N(s)}{D(s)} = 0 \qquad G(s)H(s) = \frac{N(s)}{D(s)}$$

where $N(s)$ and $D(s)$ are known polynomials. For example, the adjustable system of Figure 4.20(a), where K is adjustable, has an overall transfer function

Obtaining the standard form for adjustable K.

$$T(s) = \frac{s/(s+2)}{1 + (s/(s+2))(K/(1+K))} = \frac{s(s+K)}{s^2 + 2s + K(2s+2)}$$

$$= \frac{s(s+K)/(s^2+2s)}{1 + K[(2s+2)/(s^2+2s)]}$$

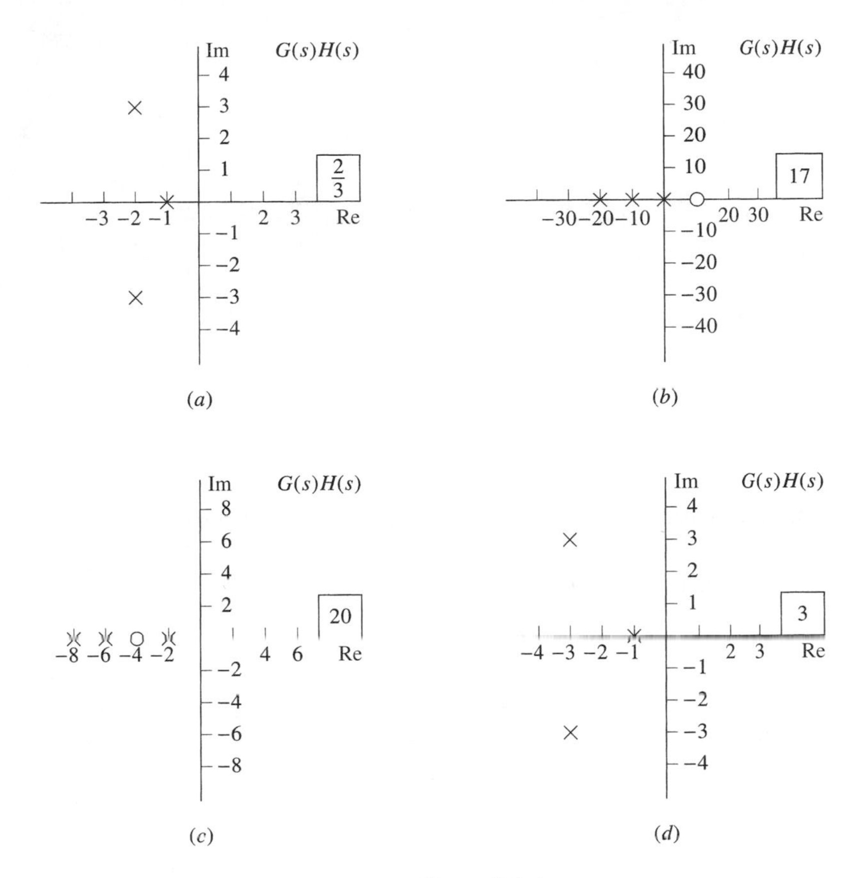

Figure D4.6

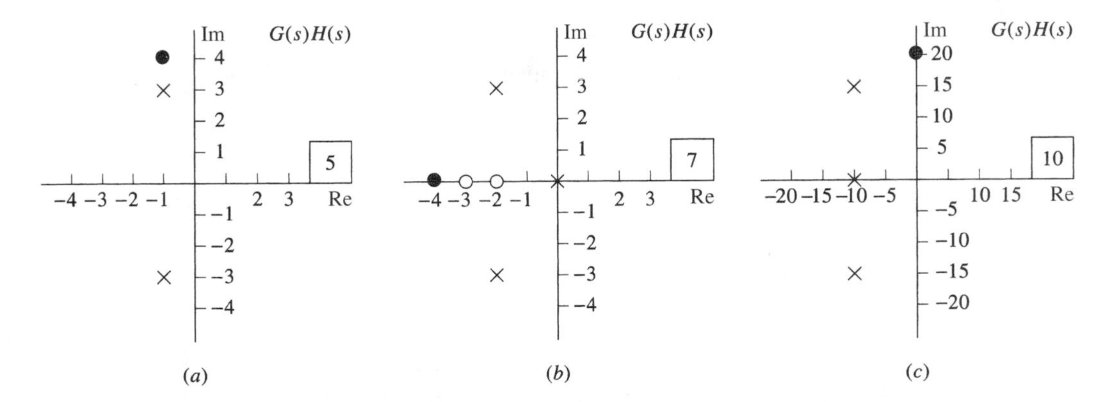

Figure D4.7

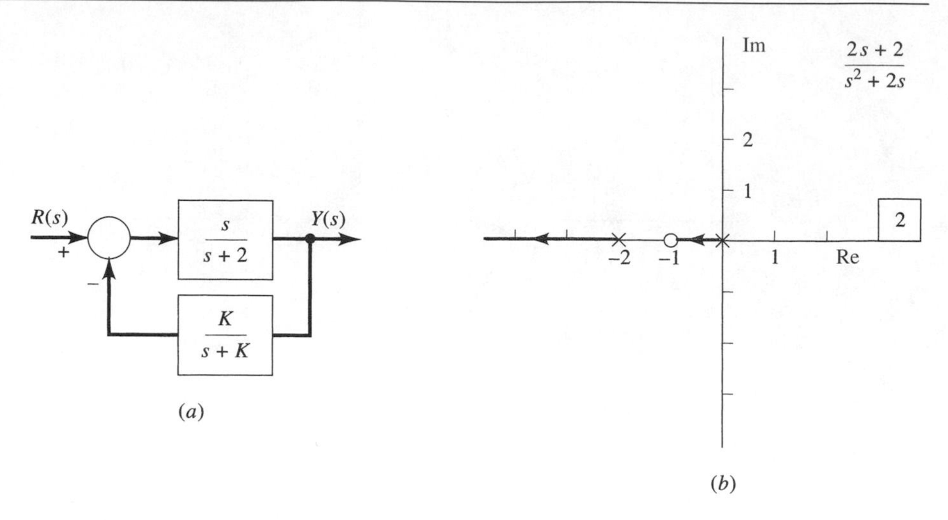

Figure 4.20 Root locus for a system with an adjustable time constraint.

Here the equivalent GH can be taken to be

$$G(s)H(s) = \frac{2(s+1)}{s^2+2s} = \frac{N(s)}{D(s)}$$

The root locus for K ranging from zero to infinity is shown in Figure 4.20(b). For the unity feedback system with adjustable damping in Figure 4.21(a), we have

Obtaining the standard form for adjustable ζ.

$$T(s) = \frac{(2s-4)/(s^2+6\zeta s+9)}{1+(2s-4)/(s^2+6\zeta s+9)} = \frac{2s-4}{s^2+2s+5+6\zeta s}$$

$$= \frac{(2s-4)/(s^2+2s+5)}{1+\zeta[6s/(s^2+2s+5)]}$$

The equivalent GH product can be taken to be

$$G(s)H(s) = \frac{6s}{s^2+2s+5}$$

The root locus for ζ in the range from zero to infinity is shown in Figure 4.21(b).

4.6.2 Negative Parameter Ranges

When it is desired to determine the root locus for a parameter K that ranges from zero to minus infinity, the applicable relations are

$$\begin{cases} \angle G(s)H(s) = 0^\circ \pm i360^\circ \\ |K| = 1/|G(s)H(s)| \end{cases}$$

The angle condition is changed from the case where $K > 0$ in that 360° replaces 180° in any rule related to an angle. The Rule 2 (real axis roots), Rule 3 (asymptote angles), and Rule 6 (angles of departure and arrival) change. There is no change to Rule 1 (number of branches and low and high gain effects) and Rule 4 (centroid). The

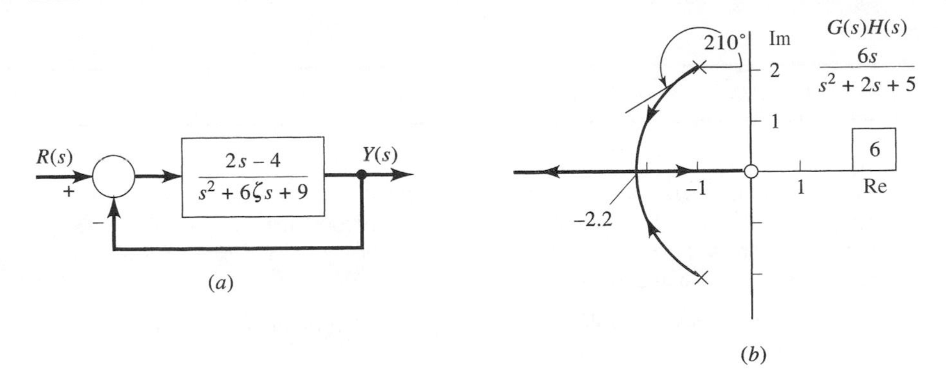

Figure 4.21 Root locus of a system with adjustable damping.

wording of Rule 5 (entry and breakaway points) does not change, but the part of the real axis to be searched does change to agree with Rule 2.

The rules change when $K < 0$, (versus $K > 0$ as earlier).

Table 4.5 summarizes the basic root locus principles for negative parameter ranges.

For example, the system with GH product given in Figure 4.22(a) has the loci shown for negative K. The asymptotic angles are given by

$$(4-1)\theta = 0^\circ \pm i360^\circ$$

$$\theta = 0^\circ, 120^\circ, -120^\circ$$

The centroid is

$$\sigma = \frac{-1+j2-1-j2+2-(-3)}{4-1} = 1$$

The angle of departure ϕ from the top pole is given by

$$\phi = -[90^\circ + 146.3^\circ + 116.6^\circ] + 45^\circ + 0^\circ$$

$$\phi = -307.9^\circ = 52.1^\circ$$

Real axis roots are to the left of an even number of poles plus zeros of GH.

The completed root locus plot is shown in Figure 4.22(b). The same 0° locus situation occurs, too, for positive K if the multiplying constant of $G(s)H(s)$ is negative. Then, despite the requirement that

$$\angle G(s)H(s) = 180^\circ$$

the angle contributions from the poles and zeros of GH must total 0°, since 180° is contributed to the angle by the negative multiplying constant.

❑ Computer-Aided Learning

Since the MATLAB software assumes that the gain K is positive, to obtain a negative-gain root locus, a negative sign must be introduced into the open-loop transfer function. For example, a negative-gain root locus for the system of Figure 4.22 may be obtained by defining

```
gh=zpk([-3],[-1+j*2 -1-j2 3 0],-3)
```

Table 4.5 *Basic Root Locus Principles for Negative Parameters*

1. The branches of the locus are continuous curves that start at each of the n poles of GH, for $K = 0$. As K approaches $-\infty$, the locus branches approach the m zeros of GH. Locus branches for excess poles extend infinitely far from the origin; for excess zeros, locus segments extend from infinity.
2. The locus includes all points along the real axis to the left of an even number of poles plus zeros of GH.
3. As K approaches $-\infty$, the branches of the locus become asymptotic to straight lines with angles

$$\theta = \frac{i360^\circ}{n - m}$$

for $i = 0, \ \pm 1, \ \pm 2, \ldots,$ until all $n - m$ or $m - n$ angles are obtained, where n is the number of poles and m is the number of zeros of GH.

4. The starting point of the asymptotes, the centroid of the pole–zero plot, is on the real axis at

$$\sigma = \frac{\sum \text{pole values of } GH - \sum \text{zero values of } GH}{n - m}$$

5. Loci leave the real axis at a gain K that is the maximum K in that region of the real axis. Loci enter the real axis at the minimum value of K in the region of the real axis. These points are termed *breakaway points* and *entry points*, respectively. A pair of locus segments leave or enter the real axis at angles of $\pm 90^\circ$.
6. The angel of departure ϕ of a locus branch from a complex zero is given by

$$\phi = -\sum \text{other } GH \text{ pole angles } + \sum GH \text{ zero angles} + 0^\circ$$

The angle of approach ϕ' of a locus branch to a complex pole is given by

$$\phi' = \sum GH \text{ pole angles } - \sum \text{ other } GH \text{ zero angles} - 0^\circ$$

where each GH pole angle and GH zero angle is calculated to the complex pole for ϕ and to the complex zero for ϕ'.

If the complex pole or zero is of order m, the m angles of departure and approach are given by

$$\phi = -\frac{\sum \text{other } GH \text{ pole angles } + \sum GM \text{ zero angle} + i360^\circ}{m}$$

$$\phi' = \frac{\sum GH \text{ pole angles } - \sum \text{ other } GH \text{ zero angle} - i360^\circ}{m}$$

for $i = 0, 1, 2, \ldots, (m - 1)$.

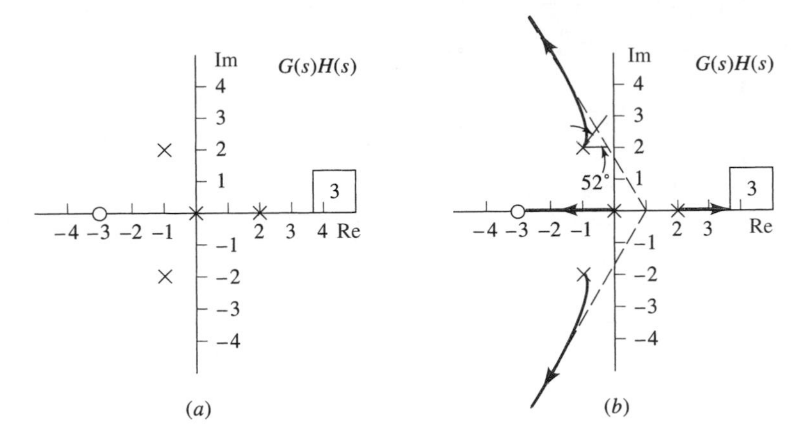

Figure 4.22 Root locus construction for negative K.

or by defining

```
n=[-3 -9]
d=[1 -1 -1 -15 0]
gh=tf(n,d)
```

and then using

```
rltool(gh)
```

C4.10 Use MATLAB to get the negative-gain root locus for the system of Figure 4.21.

Ans.
```
n=[-6 0]
d=[1 2 5]
gh=tf[n,d]
rltool(gh)
```

❑ DRILL PROBLEMS

D4.8 Develop root locus plots for the system in Figure D4.8 for K ranging from 0 to $+\infty$. Find asymptotic angles, centroid, approximate breakaway points, angles of departure, and angles of approach where applicable.

Ans. (a) $G(s)H(s) = (s + 2 + j2)(s + 2 - j2)/s(s + 2 + j1)$ $(s + 2 - j1)$; 180°; none; none; $\mp 63.4°$; $\pm 45°$

(b) $G(s)H(s) = 1/s(s^2 + s + 4)$; $\pm 60°$; 180°; $-\frac{2}{3}$; none; $\mp 30°$; none

(c) $G(s)H(s) = 2/s(s + 4)(s^2 + s + 4)$; $\pm 45°$; $\pm 135°$; -1.25; -2.8; $\mp 43.4°$; none

(d) $G(s)H(s) = (s + 4)/s(s^2 + 5s + 6)$; $\pm 90°$; $-\frac{1}{2}$; -0.9; none; none

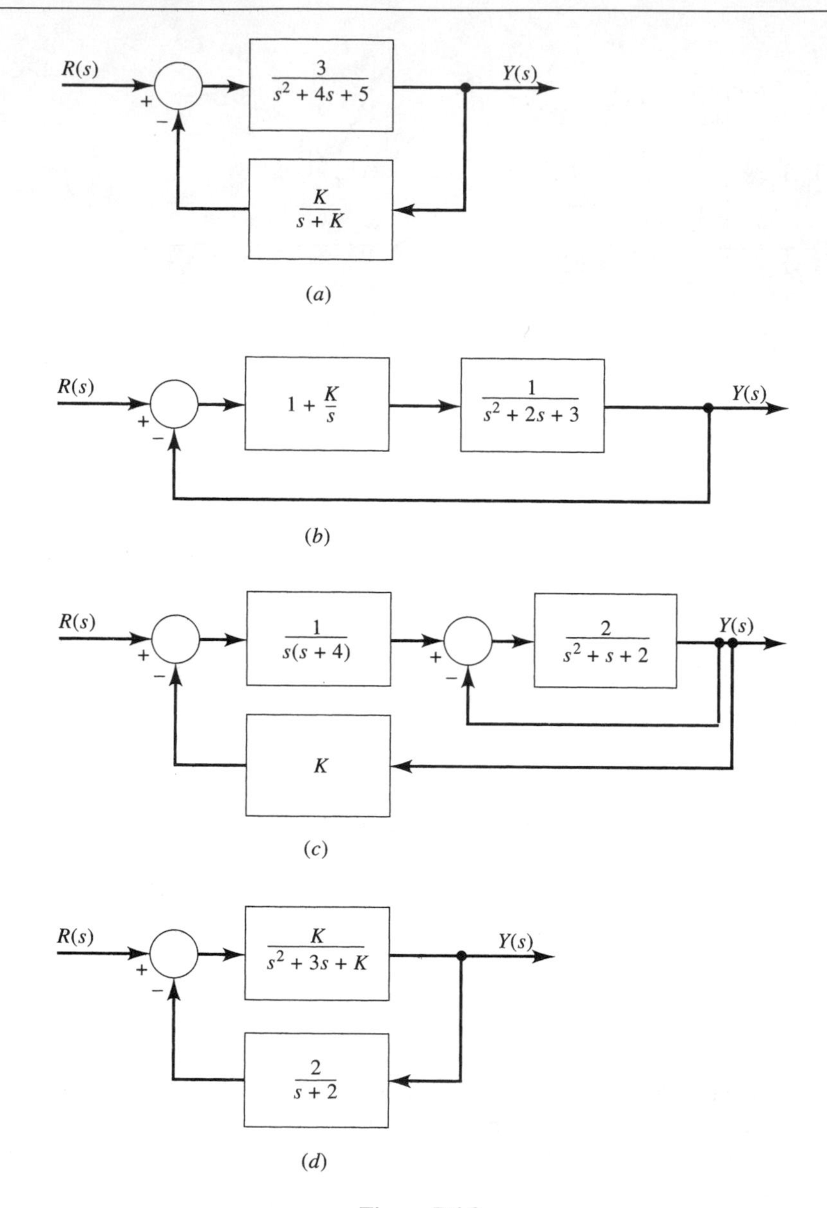

Figure D4.8

D4.9 Develop root locus plots for systems with the following overall transfer functions, for the indicated range of the adjustable parameter K.

(a)

$$T(s) = \frac{6Ks + 7}{s^3 + Ks^2 + (2K + 9)s + K}$$

$$0 < K < \infty$$

Ans. $G(s)H(s) = (s+1)^2/s(s^2+9)$

(b)

$$T(s) = \frac{10}{(1-K)s^2+3s+2}$$

$0 < K < \infty$

Ans. $G(s)H(s) = -[s^2/(s+1)(s+2)]$

(c)

$$T(s) = \frac{K[s/(s+4)]}{1+K[s(s-3)/(s+2)^2(s+4)]}$$

$-\infty < K < 0$

Ans. $G(s)H(s) = s(s-3)/(s+2)^2(s+4)$

(d)

$$T(s) = \frac{6s^2+Ks+2}{s^3+2s^2+(5+2K)s+2K}$$

$-\infty < K < \infty$

Ans. $G(s)H(s) = 2(s+1)/s(s^2+2s+5)$

4.6.3 Delay Effects

Examples of delay.

Many systems include delays throughout the control loop. For example, a chemical process may require that measurements be taken as part of some quality control scheme. Since the measurements take a finite amount of time, there is a delay in implementing the results. Certainly, any computer control system has a delay equal to one or more computational cycles, so the computer output pertains to data taken earlier. A delay can make an otherwise stable system less stable.

Figure 4.23 shows a control system with a delay of $G_d(s)$ in the forward path and a plant $G_p(s)$. The Laplace transform for a true delay of T seconds is e^{-Ts}. We cannot apply root locus rules to analyze a true delay because the root locus rules require rational transmittances (polynomial ratios) and the true delay e^{-Ts} is irrational. It is possible to use what is called a Padé approximation to approximate the delay with a polynomial ratio.

Padé approximation.

$$G_d(s) = \frac{e^{-Ts/2}}{e^{+Ts/2}} \approx \frac{1-Ts/2}{1+Ts/2} = \frac{-(s-2/T)}{s+2/T}$$

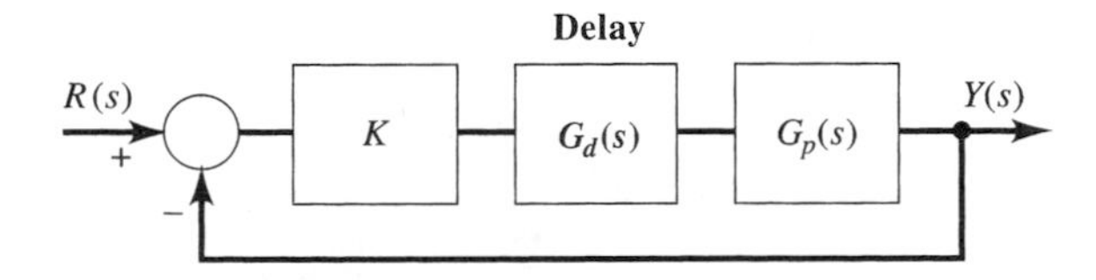

Figure 4.23 Control system including a delay.

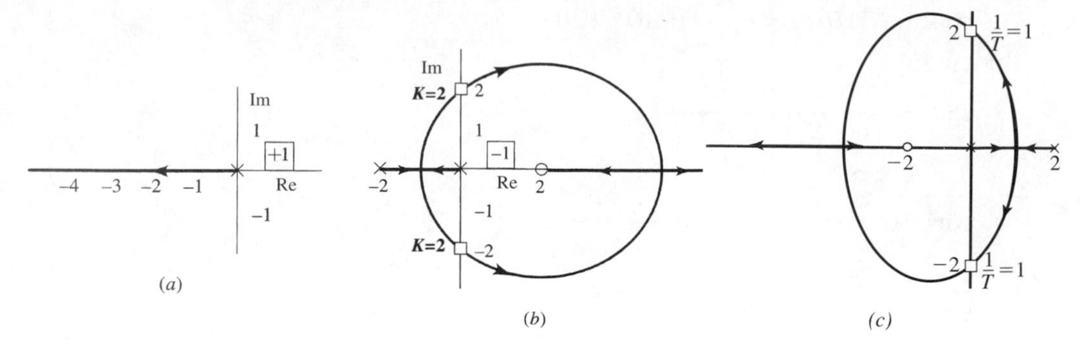

Figure 4.24 Delay example. (a) Root locus without delay ($T = 0,\ G_p(s) = 1/s$). (b) Root locus with delay ($T = 1,\ K$ varies).. (c) Roots locus with delay ($K = 2,\ 1/T$ varies).

Thus $e^{-Ts/2}$ and $e^{+Ts/2}$ are each approximated by a two-term Taylor series.

As an example, suppose the plant is $G_p(s) = 1/s$ and a delay is present as in Figure 4.23. Figure 4.24(a) shows the root locus for no delay, $T = 0$, so that $KG_d(s)G_p(s) = K/s$. The system is stable for all $K > 0$.

If there is a nonzero delay, the root locus can be found for

$$KG_d(s)G_p(s) = \frac{-K(s - 2/T)}{s(s + 2/T)}$$

which requires a negative gain root locus for $K > 0$. Figure 4.24(b) shows the root locus for a delay $T = 1$ and $K > 0$. With the delay included, the transfer function becomes

$$\begin{aligned} T(s) &= \frac{KG_d(s)G_p(s)}{1 + KG_d(s)G_p(s)} \\ &= \frac{-K[s - 2/T]}{s^2 + s[2/T - K] + 2K/T} \end{aligned}$$

This second-order system is stable for $(2/T - K) > 0$ and $K < 2/T\cdot$ Thus T and K are related for stability.

If the delay is small, then K can be large. Conversely, a large delay requires a small values of K. Since $T = 1$ in Figure 4.24(b), the system is marginally stable with $K = 2$ and stable for $K < 2$.

With K fixed, a root locus can be obtained for variable T. Usually, it is convenient to obtain a root locus in terms of $1/T$. We can write $T(s)$ as follows

$$\begin{aligned} T(s) &= \frac{-K(s - 2/T)}{s^2 - Ks + 2/T(s + K)} \\ &= \frac{-K(s - 2/T)/s(s + K)}{1 + 2/T[(s + K)/s(s - K)]} \end{aligned}$$

The effective $G(s)H(s)$ is

$$\frac{2(s + K)}{s(s - K)}$$

with $K = 2$ the root locus with $1/T$ varying is shown in Figure 4.24(c). The system is marginally stable when $1/T = 1$ $(T = 1)$ and stable for $1/T > 1$ $(T < 1)$.

In general, as a system is driven at larger gains (faster response), the value of delay must be smaller for stability. An analogy is driving in the rain with windshield wipers running. As it rains more heavily, you can see clearly only as the wiper moves past your line of sight (the delay). You then drive more slowly (effectively reducing K) until the activity not seen clearly is a less serious threat. When the rain diminishes (less delay in seeing clearly), you speed up appropriately (increasing K).

❑ DRILL PROBLEMS

D4.10 For each of the following $G_p(s)$ transmittances in Fig. 4.23, use the delay approximation

$$G_d(s) = \frac{-(s - 2/T)}{s + (2/T)}$$

and a root locus or root-solving computer program to find the maximum K for stability for no delay ($T = 0$, $G_d(s) = 1$) and for a one-second delay ($T = 1$).

(a) $G_p(s) = \dfrac{1}{(s+2)(s+3)}$

(b) $G_p(s) = \dfrac{s+3}{s(s+4)}$

(c) $G_p(s) = \dfrac{1}{s(s+1)(s+3)}$

Ans. (a) ∞, 11.11; (b) ∞, 2.78; (c) 12, 2.80

4.7 Design concepts (Adding Poles and Zeros)

The root locus technique is useful for both analysis and design. In analysis, we want to evaluate the performance of a system with known component values. If root locus is used to analyze a system, the GH poles and zeros are known as well as the gain K. Then the closed-loop poles are just one set of points along the root locus branches.

Analysis is distinguished from design.

The root locus technique can also be used for design, the purposeful selection of parameter values to meet objectives. If the GH poles and zeros are fixed, but K is adjustable as a design parameter, the designer can select the value of K for which the closed-loop poles meet design objectives. Sometimes more components are added to provide additional parameters for use in meeting design objectives.

In general, the addition of a pole makes a system less stable, while the addition of a zero makes a system more stable. These poles and zeros are often added to the system in the form of a compensator. Although a pole makes a system less stable, steady state error can be reduced.

For example, in Figure 4.25(a), a *compensator* $G_c(s)$ is added to the forward path in *cascade* with the plant $G_p(s)$. Suppose the plant has a transmittance

$$G_p(s) = 1/(s+1)^2$$

If the compensator is

$$G_s(s) = K$$

then we have what is called a proportional compensator. Since

$$G_c(s)G_p(s) = \frac{K}{(s+1)^2}$$

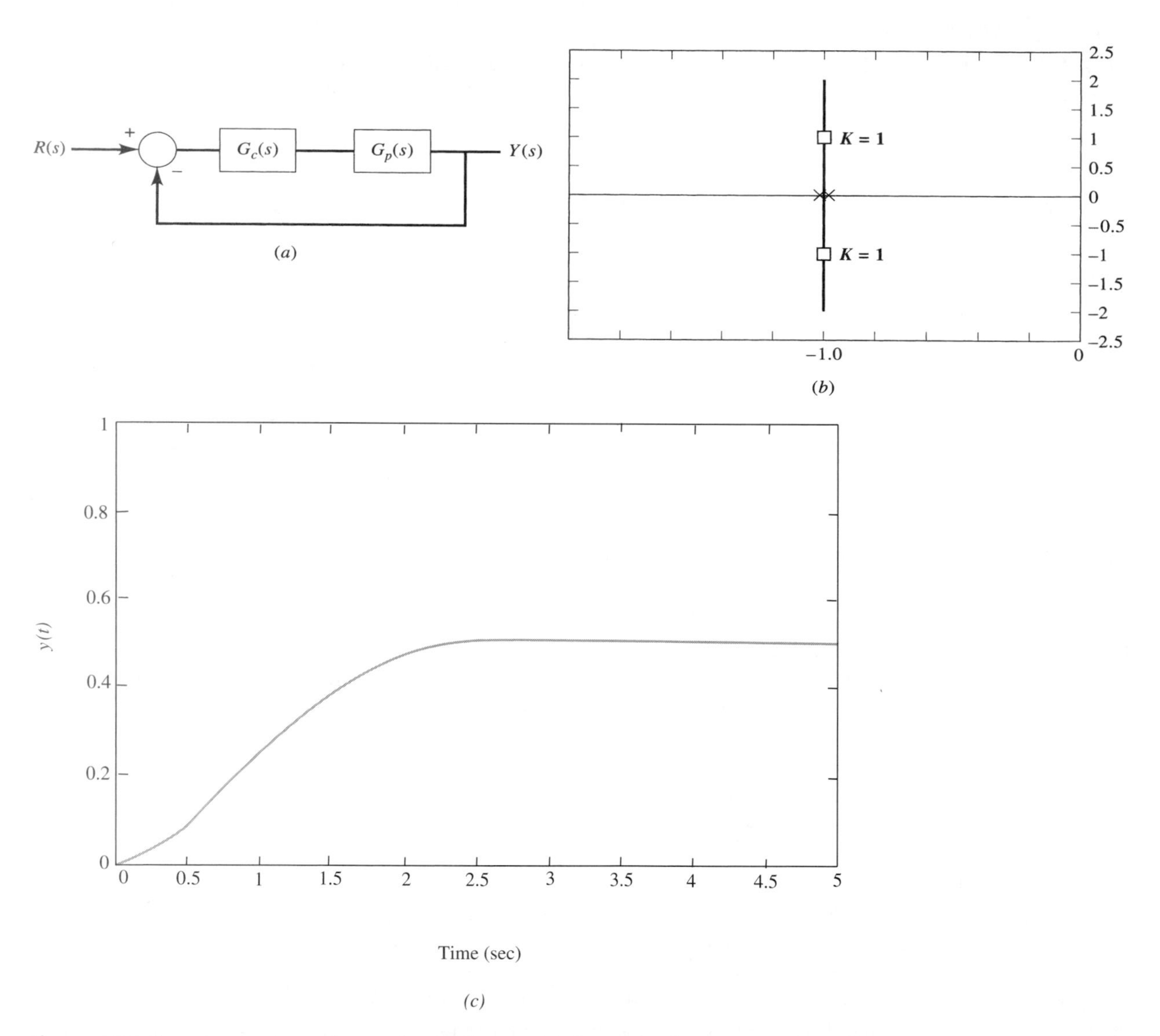

Figure 4.25 Cascade-compensated example. (a) System. (b) Proportional-compensated root locus. (c) Step response for (b) with $K = 1$.

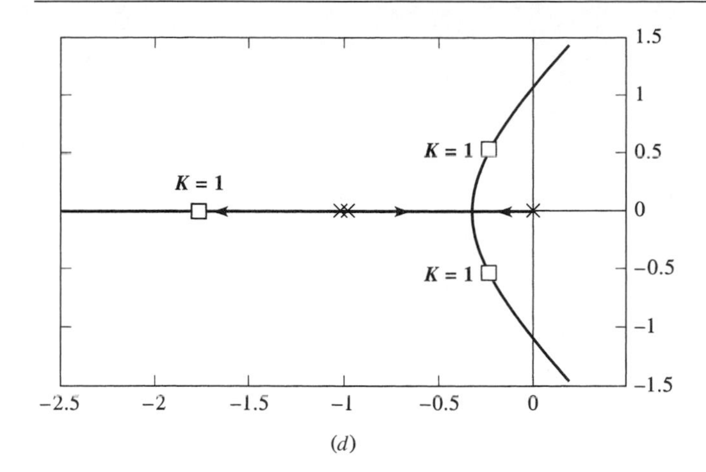

(d)

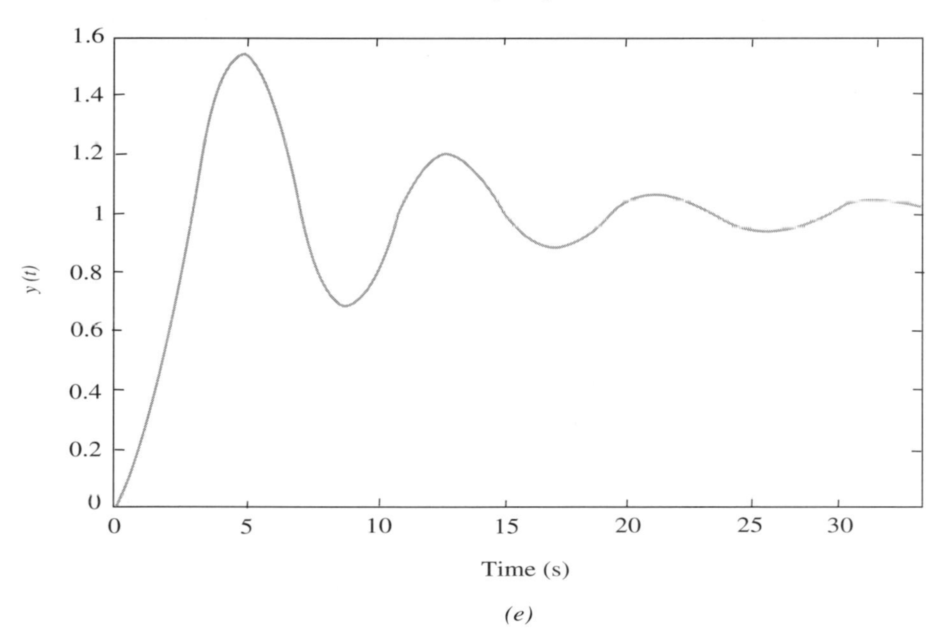

(e)

Figure 4.25 (*Continued*). (d) Integral-compensated root locus. (e) Step response for (d) with $K = 1$.

we have a type 0 plant. The step error coefficient is

$$K_0 = G_0(0)G_p(0) = K$$

For a unit step input, the steady state error is

$$e_{\text{step}}(\infty) = \frac{1}{1 + K_0} = \frac{1}{1 + K}$$

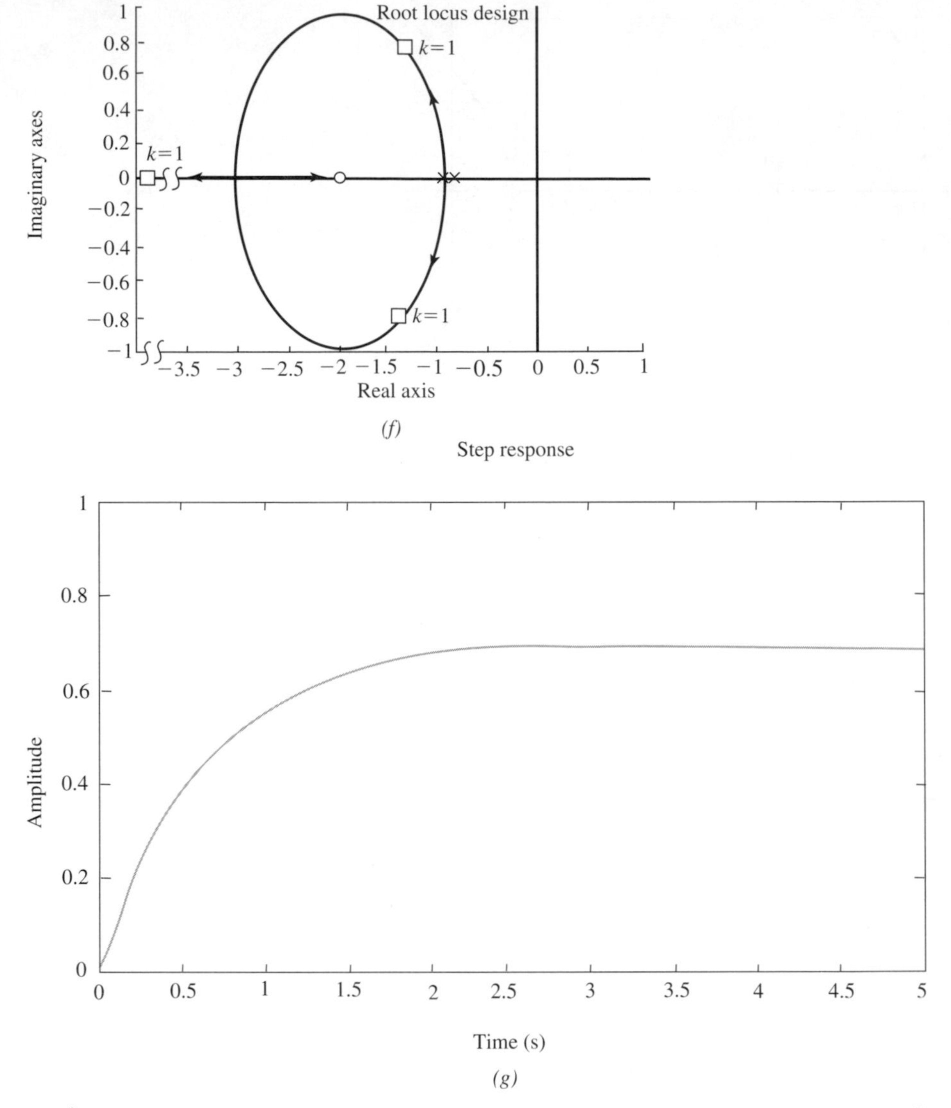

Figure 4.25 (*Continued*). (f) Proportional-plus-derivative-compensated root locus. (g) Step response for (f) with $K = 1$.

The closed-loop transfer function is

$$T(s)\frac{G_c(s)G_p(s)}{1+G_c(s)G_p(s)} = \frac{N(s)/D(s)}{1+N(s)/D(s)} = \frac{N(s)}{D(s)+N(s)}$$

where $N(s) = K$ and $D(s) = (s+1)^2$. Thus we have

$$T(s) = \frac{K}{s^2+2s+K+1} = \frac{K}{s^2+2\zeta\omega s+\omega_n^2}$$

It is possible to choose K for a 0.707 value of the damping ratio, causing minimum settling time. Then

$$\omega_n = \sqrt{K+1}$$

$$2\zeta\omega_n = 2 = 2(0.707)\sqrt{K+1}$$

$$K + 1 = 2$$

$$K = 1$$

In that case, we have

$$e_{\text{step}}(\infty) = \frac{1}{1+K} = 0.5$$

The resulting step response is shown in Figure 4.25(c). The steady state value of y is 0.5 (an error of 0.5), while the overshoot is about 5%.

Suppose we choose

$$G_c(s) = K/s$$

This is called an integral compensator because the $1/s$ Laplace function implies that the compensator creates integration in the time domain. Now we have

$$G_c(s)G_p(s) = \frac{K}{s(s+1)^2}$$

The system becomes type 1 where $K_0 = \infty$ and $e_{\text{step}}(\infty) = 0$; that is, the step error is removed. However, the root locus of Figure 4.25(d) moves to the right (less stable response) in relation to the root locus of Figure 4.25(b). The resulting step response of Figure 4.25 (with $K = 1$) is clearly less stable than that of Figure 4.25(c), but the final value becomes 1, eliminating step error. Here.

$$N(s) = K$$

$$D(s) = s(s+1)^2$$

$$T(s) = \frac{K}{s^3 + 2s^2 + s + K}$$

with $K = 1$, the closed-loop poles are $-0.122 \pm j0.745$ and -1.75.

Suppose we employ a cascade compensator with a zero instead of a pole. That is,

$$G_c(s) = K(s+2)$$

Usually, this is called a proportional-plus-derivative compensator because an s in the numerator implies differentiation. The root locus of Figure 4.25(f) moves to the left of that in Figure 4.25(b), making the system more stable. Here

$$N(s) = K(s+2)$$

$$D(s) = (s+1)^2$$

$$T(s) = \frac{K(s+2)}{s^2 + (2+K)s + 2K}$$

If we choose $K = 1$ to facilitate a meaningful comparison, the closed-loop poles are at $-1.5 \pm j0.87$. Also $K_0 = G_c(0)G_p(0) = 2K$

$$e_{\text{step}}(\infty) = \frac{1}{1+2K} = \frac{1}{3}$$

The step response of Figure 4.25(g) has less steady state error and is faster and less oscillatory (more stable) than, the one in Figure 4.25(c).

In general a designer selects the compensator poles, zeros, and gain K to meet objectives. The next chapter is devoted entirely to design.

❑ Computer-Aided Learning

C4.11 Obtain the step responses and root loci of section 4.7. Use

```
gp=zpk([],[-1 -1],1)
```

for that plant. Start with

```
rltool(gp)
```

You can obtain the step response for $K = 1$. Then, edit the compensator by adding a pole at $s = 0$ or a zero at $s = -2$. The step response can be obtained for any K, especially $K = 1$.

The root loci and step responses should also be obtained for the design examples that follow.

4.8 A Light-Source Tracking System

This example system is designed to follow, in one dimension, a moving light source. As pictured in Figure 4.26(a), when equal light intensities are detected by the two photodiodes, the electrical bridge is balanced, and zero voltage is applied to the drive motor. When one photodiode receives more light than the other, the bridge is unbalanced, and a nonzero voltage is amplified and applied to the drive motor, which then moves the photodiodes toward the equal-light-intensity position. Similar systems are used for precision machine tool alignment, where the light is reflected from a calibrated scale or transmitted through a tiny hole in the tool or the work. Variations of this system are used to track the sun or another star in navigation systems, to follow aircraft in collision avoidance systems, and to track the recording path on optical videodisks.

For small signals, a block diagram of the system is shown in Figure 4.26(b). The system transfer function is, in terms of the gain constant K,

$$T(s) = \frac{0.1K/s(s+2)}{1+0.1K/s(s+2)} = \frac{0.1K}{s^2+2s+0.1K}$$

which is stable for all $K > 0$. A relative stability of two units for a system means that the natural component of system response decays with time as $\exp(-2t)$, that is, with a 0.5 s time constant. This degree of stability cannot be achieved with the system. This is evident from the root locus plot for Figure 4.27(a), where it is seen that the system's relative stability (the distance from the imaginary axis to the nearest pole) is always equal to one unit. System responses to a unit step change in

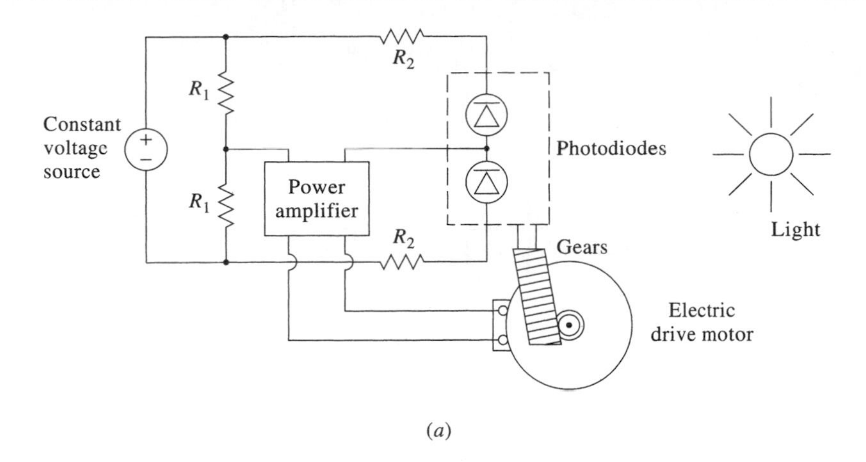

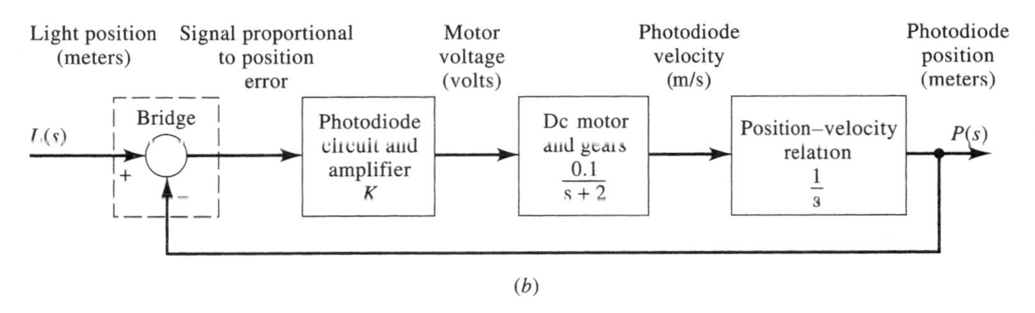

Figure 4.26 Light-source tracking system. (a) Physical arrangement. (b) Block diagram model.

light position, for various representative values of K, are shown in Figure 4.27(b). For each, the settling time is relatively long, in consequence of the small degree of relative stability.

Adding position and velocity feedback.

The performance of this system can be improved substantially by the addition of velocity feedback as well as the position feedback. A tachometer coupled to the drive motor shaft will produce a voltage nearly proportional to the motor speed, which in turn is proportional to the photodiode velocity. Adding a fraction of this voltage to the bridge voltage (which is amplified to drive the motor) results in the block diagram of Figure 4.28(a). Using Mason's gain rule on the system's signal flow graph in Figure 4.28(b), we write

$$T'(s) = \frac{(0.1K/(s+2))\,1/s}{1-(-4K')0.1K/(s+2)-(-1)\,(0.1K/(s+2))\,(1/s)}$$

$$= \frac{0.1K}{s^2+(2+0.4KK')s+0.1K}$$

Here it is seen that both coefficients of the characteristic polynomial can be chosen at will by the designer by selecting appropriate values of K and K'.

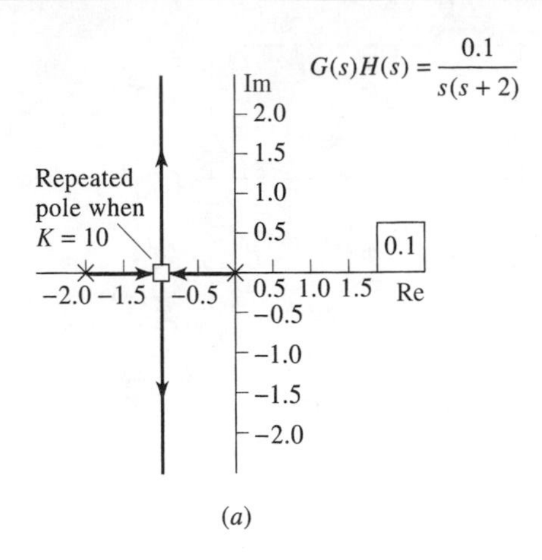

(*a*)

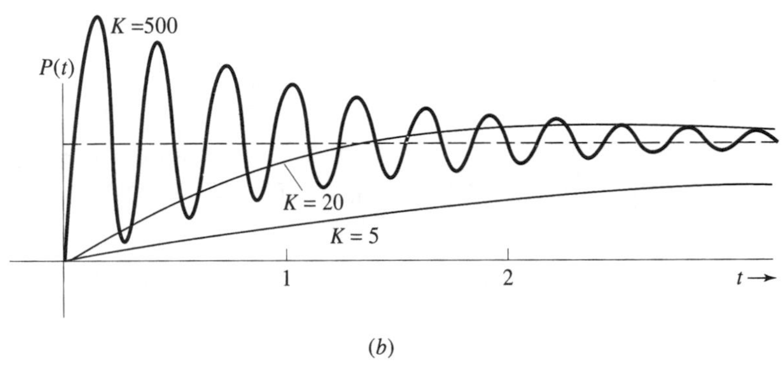

(*b*)

Figure 4.27 Root locus and typical responses for the original light source tracking system.

For example, a characteristic polynomial

$$(s+5)^2 = s^2 + 10s + 25 = s^2 + (2 + 0.4KK')s + 0.1K$$

is achieved with

$$K = 250 \qquad K' = 0.08$$

With these values of K and K', the system's step response is critically damped and has a relative stability of 5 units. This step response is shown in Figure 4.28(c).

4.9 An Artificial Limb

Advances in artificial limb design.

Development of prosthetics has paralleled wars and natural calamities. Until the 1980s all but a very few artificial limbs were nonactive devices connected to what remained of the limb. Lower-arm devices required the user to have an elbow joint, and lower-leg devices required a usable knee joint. The "Boston arm" was an early active artificial limb that used a torque motor and velocity feedback. The microprocessor revolution of the 1980s spawned more advanced artificial limbs capable of greater dexterity

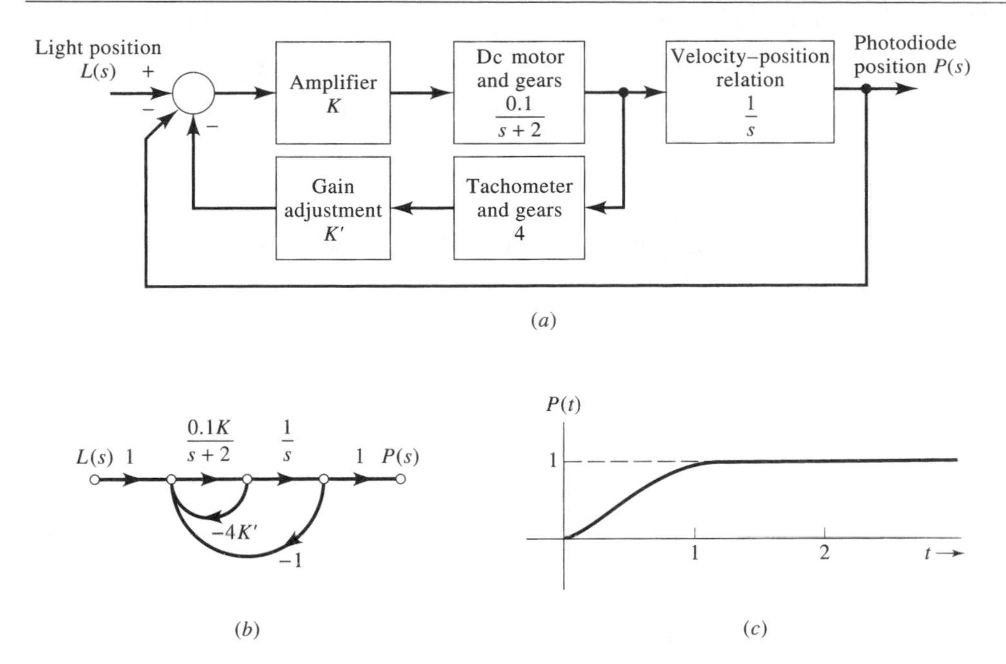

Figure 4.28 Light-source tracking system with both position and rate feedback. (a) Block diagram. (b) Signal flow graph. (c) Unit step response with $K - 250$, $K' = 0.08$.

and more precise operation. Advanced research included creation of a system that bypasses a damaged spinal cord so that muscle-electric (myoelectric) signals from the upper body cause a balance control system to facilitate movement of the legs and locomotion for an otherwise paralyzed patient. The example that follows exhibits typical design techniques for an artificial limb, although no actual working system is intended.

The human body has a time delay between decisions of the brain and reception of signals by the muscles, so rational limits must be placed upon the capabilities of an automatic artificial limb. If the device is too strong and sensitive, it could operate too rapidly for the brain and body control, making the human–machine team a bionic pretzel. On the other hand, too little sensitivity could result in a bionic statue. Between the two extremes lie useful designs.

Figure 4.29(a) shows a block diagram for a bionic arm in a closed-loop system with the body. For simplicity, motion is considered in one dimension only. The brain monitors the desired position and the sensed position, generating an error signal to the nervous system. Special sensors pick up the myoelectric impulses, and an amplifier produces a voltage that drives a dc control motor. The motor circuit involves tachometer feedback, as shown. The output of the motor circuit is the velocity of the limb in one dimension, which, when integrated, is the limb position.

When a control system is to be designed for a complicated system such as this, some course of action must be planned. In this case, a simplified block diagram will be created and the control will evolve by stabilizing the loops, starting with the inner loops, then progressing outward until all parameters have been selected.

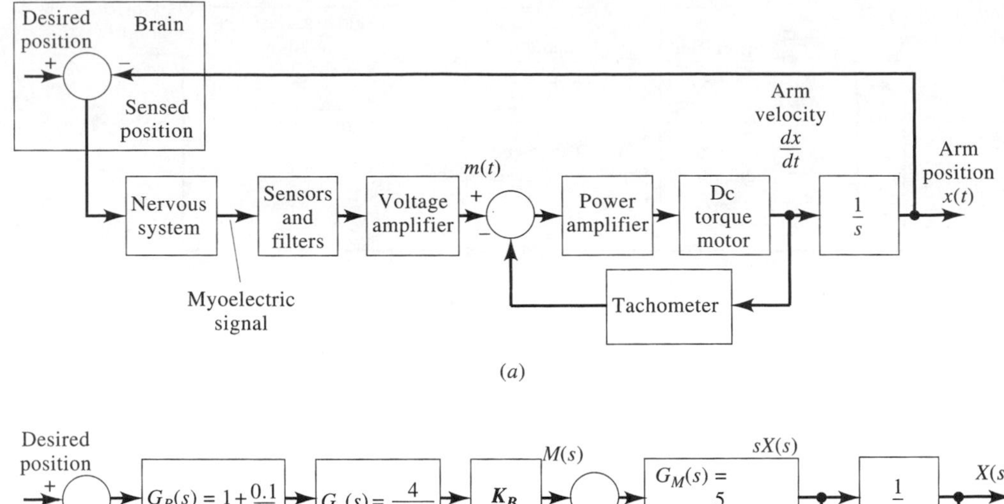

Desired position $G_B(s) = 1 + \frac{0.1}{s}$ $G_n(s) = \frac{4}{s+4}$ K_B $M(s)$ $G_M(s) = \frac{5}{s^2 + 11s + 10}$ $sX(s)$ $\frac{1}{s}$ $X(s)$ K_T

(b)

Figure 4.29 Prosthetic control system model. (a) Prosthetic and human control system model. (b) Simplified model for the system.

A simplified model for the system is given in Figure 4.29(b). The action of the brain is approximated by the transmittance

$$G_B(s) = 1 + \frac{0.1}{s}$$

which involves consideration of both the position error and its integral. The nervous system is modeled by the first-order system with transmittance

$$G_N(s) = \frac{1/T}{s + 1/T} = \frac{4}{s+4}$$

where $T = \frac{1}{4}$, the time constant, is approximately the neuromuscular delay time. The myoelectric signal is sensed and amplified with a gain K_B to form the amplified voltage $m(t)$.

Power amplifier, control motor, and mechanical load have second-order transmittance, relating motor control voltage $m(t)$ to arm velocity, modeled as follows:

$$G_M(s) = \frac{5}{s^2 + 11s + 10} = \frac{5}{(s+1)(s+10)}$$

This block exhibits time constants of 1 s, associated with mechanical inertia, and 1/10 s, due primarily to the motor itself. The step response of $G_M(s)$ is sketched in Figure 4.30(a). The tachometer, with constant transmittance K_T, provides feedback for the motor and arm, giving the following overall transmittance of those components:

$$G_T(s) = \frac{G_M(s)}{1 + K_T G_M(s)} = \frac{5/(s^2 + 11s + 10)}{1 + 5K_T/(s^2 + 11s + 10)}$$
$$= \frac{5}{s^2 + 11s + (10 + 5K_T)}$$

Selecting K_T for critical damping.

A root locus plot for this motor–arm subsystem, $G_T(s)$, in terms of the adjustable gain K_T is shown in Figure 4.31. The transmittance $G_T(s)$ has critical damping when its denominator polynomial is

$$s^2 + 11s + (10 + 5K_T) = (s + 5.5)^2 = s^2 + 11s + 30.25$$

which is obtained for

$$10 + 5K_T = 30.25 \qquad K_T = 4.05$$

For this value of K_T, the subsystem has a maximum relative stability of 5.5 units; larger values of K_T reduce the damping but do not increase relative stability. This local feedback makes the subsystem act as if it were a better motor, with faster response. It is best not to overdo a good thing, however. Huge improvements in response speed, when they are possible, generally require that the device being controlled be driven with very large input signals, as is the case if one wants a motor to quickly speed up. Very large inputs can cause damage and usually will drive the device into a nonlinear region of operation.

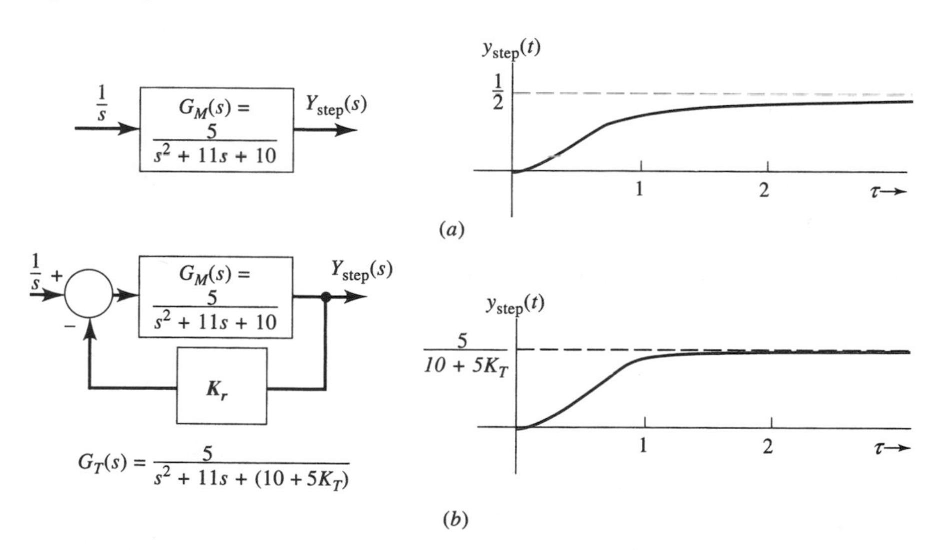

Figure 4.30 Improving the motor–arm performance with tachometer feedback and critical damping. (a) Step response of motor and arm without tachometer feedback. (b) Step response with the tachometer feedback.

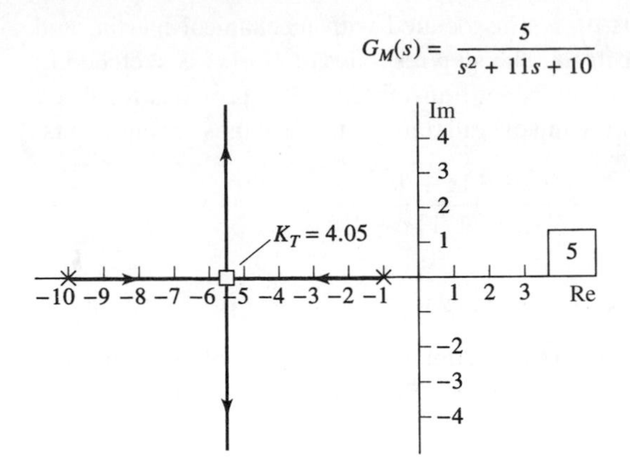

Figure 4.31 Root locus of the motor–arm subsystem.

The entire system is redrawn in Figure 4.32(a) for the tachometer gain $K_T = 4.05$, which gives critical damping of the motor–arm block. The transfer function of the entire system is

$$T(s) = \frac{G_B(s)G_N(s)K_B G_T(s)(1/s)}{1 + G_B(s)G_N(s)K_B G_T(s)(1/s)}$$

$$= \frac{\dfrac{20(s+0.1)K_B}{s^2(s+4)(s^2+11s+30.25)}}{1 + \dfrac{20(s+0.1)K_B}{s^2(s+4)(s^2+11s+30.25)}}$$

$$= \frac{20(s+0.1)K_B}{s^5 + 15s^4 + 74.25s^3 + 121s^2 + 20K_B s + 2K_B}$$

Stability range on K_B with K_T fixed at 4.05.

A root locus plot in terms of K_B is shown in Figure 4.32(b). These are two poles at the origin and a zero at $s = -0.1$ in the open-loop transmittance $G_B(s)G_N(s)G_T(s)(1/s)$. However, except very near to these roots, the loci are virtually the same as if a single pole were in that region.

A Routh–Hurwitz test for the stability of $T(s)$ is follows:

s^5	1	74	$20K_B$
s^4	15	121	$2K_B$
s^3	65.9	$19.86K_B$	
s^2	$121 - 4.52K_B$	$2K_B$	
s^1	$\dfrac{2271K_B - 89.76K_B^2}{121 - 4.54K_B}$		
s^0	$2K_B$		

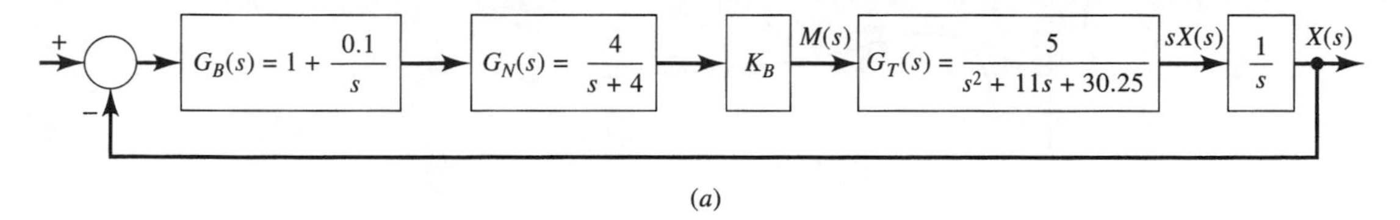

(a)

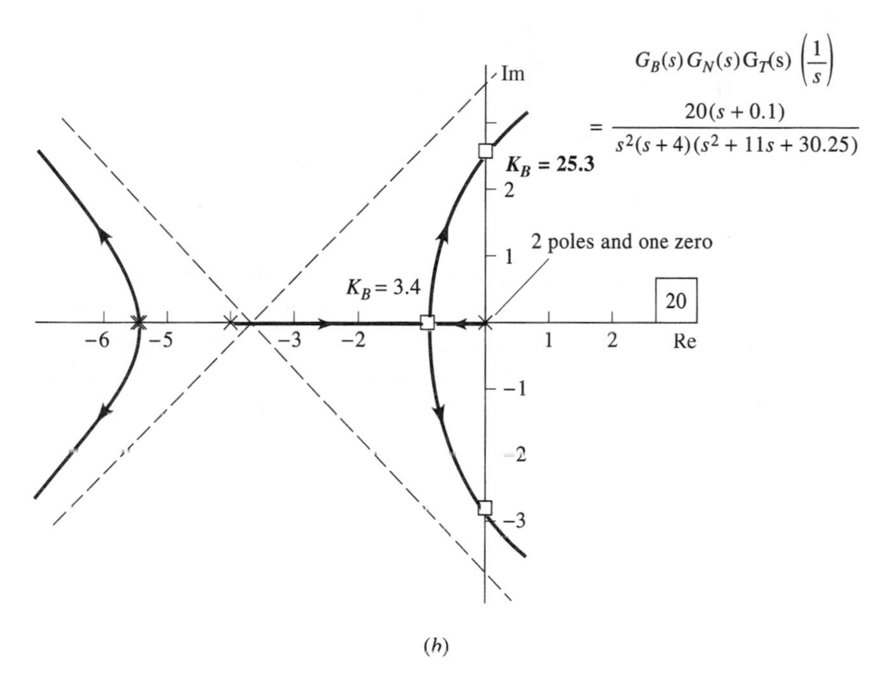

(b)

Figure 4.32 Root locus for the complete system. (a) Block diagram. (b) The root locus.

For stability, we write

$$\begin{cases} 121 - 4.52K_B > 0 \\ K_B(2271 - 89.76K_B) > 0 \\ 2K_B > 0 \end{cases}$$

or

$$\begin{cases} K_B < 26.8 \\ K_B < 25.3 \quad \text{for positive } K_B \\ K_B > 0 \end{cases}$$

Thus the system is stable for

$$0 < K_B < 25.3$$

and the root locus must cross the imaginary axis at $K_B = 25.3$.

The maximum relative stability of the closed-loop system occurs when $K_B = 3.4$ so this choice will be made. The overall system transfer function is then

$$T(s) = \frac{3.4(s+0.1)}{s^5 + 15s^4 + 74.25s^3 + 121s^2 + 68s + 6.8}$$

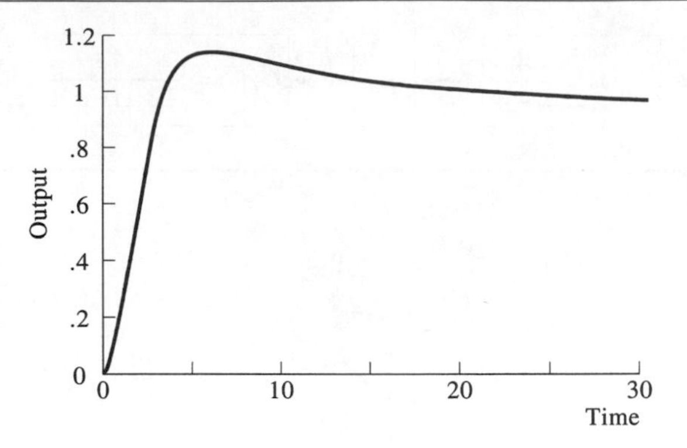

Figure 4.33 Unit step response for $K_B = 3.4$.

Figure 4.33 shows the unit step response with $K_B = 3.4$. The response is smooth. The output is one at steady state, so there is no steady state error. If other values of K_B are chosen, the response is degraded. In Figure 4.34(a) K_B is increased to 10, which causes more overshoot in the response, an undesirable attribute for a human response. In Figure 4.34(b) K_B is reduced to 1, in which case the response is obviously slower. The response for $K_B = 3.4$ seems to be a good choice; however, the design procedure can be repeated for other values of K_T and K_B until the final response is considered to be acceptable.

4.10 Control of a Flexible Spacecraft

A person observing an aircraft sees an apparently smooth motion through the sky. If the same person observes a satellite through a telescope, the satellite seems to follow a smooth trajectory. Actually, each spacecraft has a much more complicated motion, which can influence the design of an automatic control system.

Rigid and flexible motions are defined.

Very large satellites often extend solar panels to generate and store electrical energy. As a result, the satellites are very flexible. When a thruster fires to cause such a satellite to change position, the entire vehicle moves (rigid motion), but parts will bend and oscillate (flexible motion) in the same way that a taut guitar string oscillates when plucked.

Figure 4.35 shows the block diagram of a controlled flexible spacecraft. Often the controller output causes the rigid and flexible behaviors to move in opposite directions. Ignoring the flexible behavior can be a tragic mistake, especially for a manned spacecraft or airplane.

As a simple example of designing a control system that considers only rigid behavior, Figure 4.36 causes

$$KG(s)H(s) = \frac{K2(s+4)}{s^2}$$

Considering only rigid motion.

This system uses rate and position feedback to stabilize the system. The root locus of Figure 4.37 seems to indicate that the system is stable for all positive values of K.

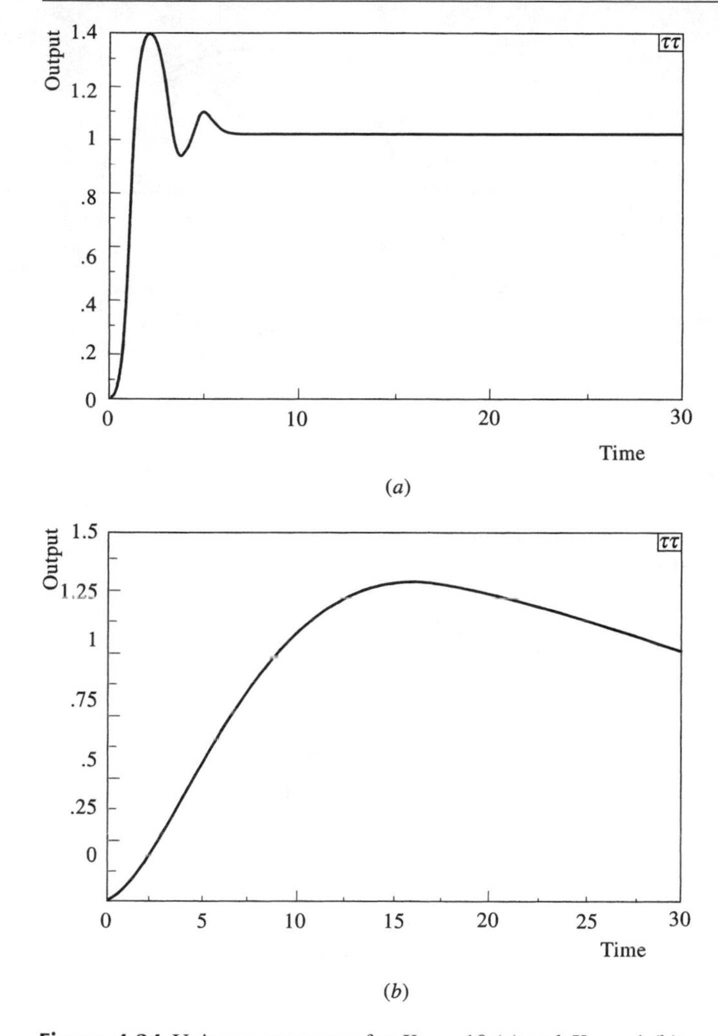

Figure 4.34 Unit step responses for $K_B = 10$ (a) and $K_B = 1$ (b).

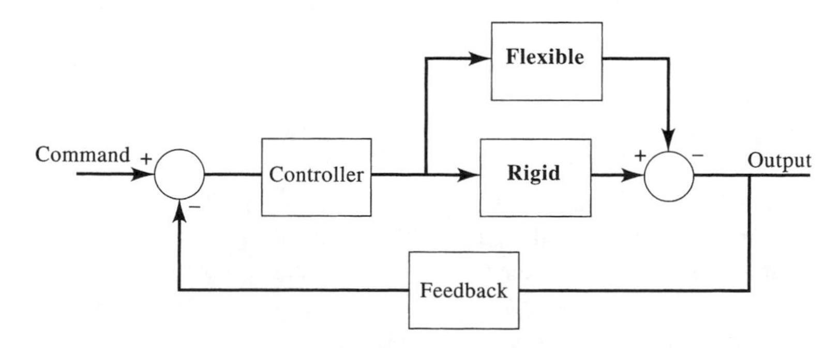

Figure 4.35 Block diagram of a flexible spacecraft.

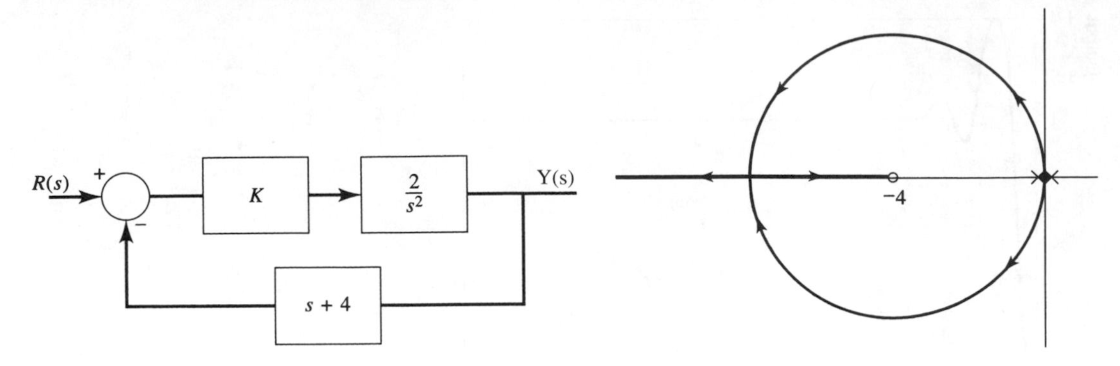

Figure 4.36 Rigid behavior example.

Figure 4.37 Root locus for Figure 4.36.

Suppose the spacecraft is actually quite flexible (including at least one mode of vibration) and Figure 4.38 results. Actually, there may be many higher-order vibration modes (resonances as in a musical instrument). From Figure 4.38, the two forward-path transmittances must be combined.

$$G(s) = \frac{2}{s^2} - \frac{1}{s^2 + s + 1} = \frac{s^2 + 2s + 2}{s^2(s^2 + s + 1)}$$

$$KG(s)H(s) = \frac{K(s + 1 + j1)(s + 1 - j1)(s + 4)}{s^2(s + 0.5 + j0.866)(s + 0.5 - j0.866)}$$

Considering both rigid and flexible motions.

Figure 4.39 shows the root locus where the one flexible mode is included. It now appears that the system can become unstable for some positive values of K, a conclusion that would not result if the flexible mode were ignored.

A designer can select values of K in Figure 4.39 to ensure stability. It is also possible to redesign the spacecraft to make it more rigid. For example, the poles and zeros due to the flexible mode (poles and zeros will always appear in pairs) can be moved away from the vertical axis to render the system stable over a larger gain range (perhaps for all positive K).

4.11 Bionic Eye

In accordance with the principle of artificial vision introduced in section 3.10, a gimballed mirror rotates to follow the motion of an otherwise blind eye. The mirror, in turn, causes a camera to detect light from the direction toward which the eye is pointed. That light proceeds to an artificial retina, to stimulate brain activity. Figure 4.40 reproduces Figure 3.28(c), a simulation of the artificial vision type 2 system.

Three values must be selected: the gain K, the parameter a, which places an open-loop zero at $-a$, and the parameter b, which places an open-loop pole at $-b$. Then we write

$$G(s)H(s) = \frac{b(s + a)}{s^2(s + b)}$$

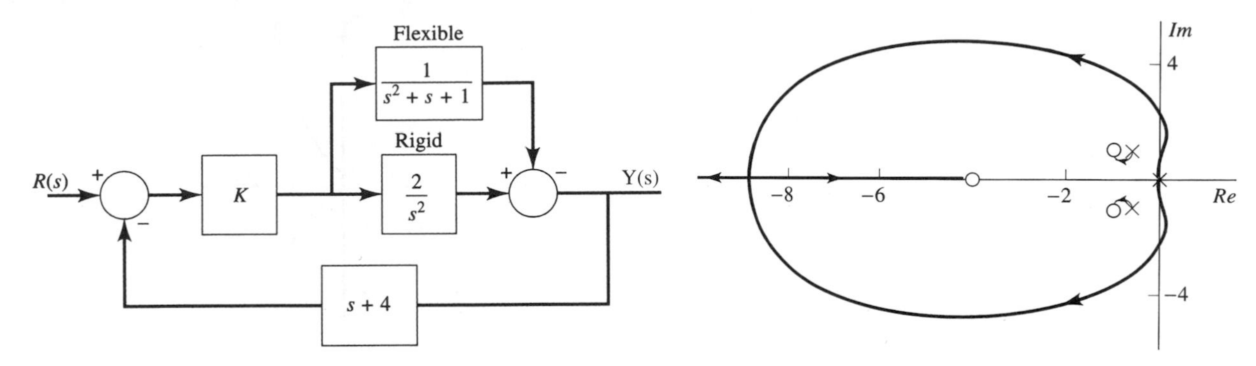

Figure 4.38 Flexible and rigid behavior example.

Figure 4.39 Root locus for Figure 4.38.

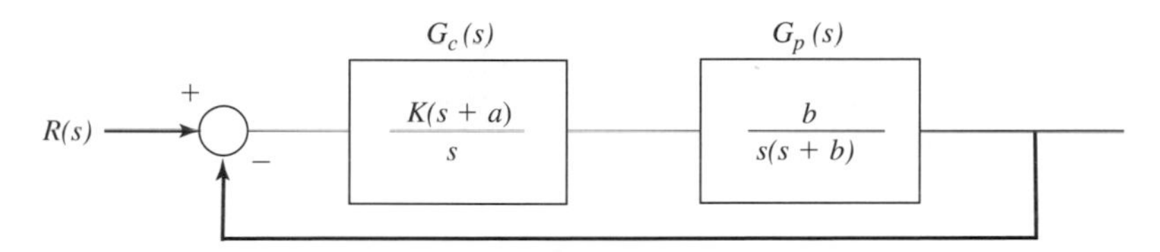

Figure 4.40 Artificial vision system.

As discussed in section 3.10 the bionic eye should track the desired visual target with a first-order step response having zero steady state error and a time constant of 0.2 s. The ramp response should be smooth and linear, with zero steady state error. The zero errors are ensured by the type 2 $G(s)H(s)$. The desired first-order behavior may be ensured if the closed-loop system has one closed-loop pole at -5, a closed-loop pole that nearly cancels, the closed-loop zero at $-a$ and one other closed-loop pole well to the left of -5 (near -50).

The open-loop pole at $-b$ and the open-loop zero at $-a$ should "draw" the root locus toward the desired closed-loop poles. If we choose $a > b$ as in Figure 4.41(a) then the centroid

$$\sigma = \frac{-b - [-a]}{3 - 1} = \frac{a - b}{2}$$

is positive. The root locus is unstable, so obviously we should try $a < b$, not $a > b$. In Figure 4.41(b), we have $a < b$ but a is relatively close to b and the dominant roots are complex conjugates. Since we want a dominant real root at -5, the situation of Figure 4.41(b) is not acceptable.

If we make a considerably less than b as in Figure 4.41(c), it is possible to select a gain K creating three real-valued, closed-loop poles. If one closed-loop pole is to be somewhere around -50, then $-b$ must be to the left of -50, perhaps at -60. If the closed-loop zero is to be canceled by a closed-loop pole to the right of -5, we can try selecting the closed-loop zero at -0.1. Next, we adjust K so one closed-loop

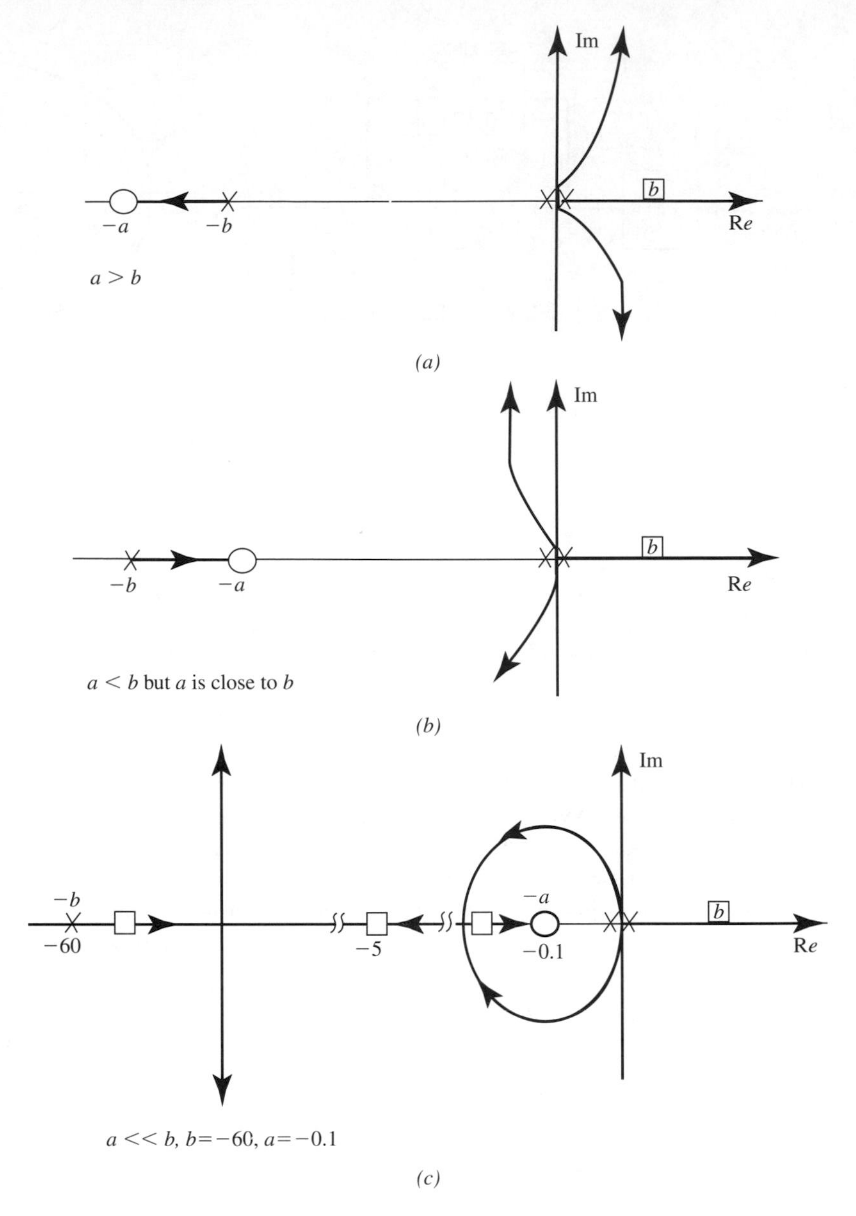

Figure 4.41 Root locus design of the artificial eye. (a) $a > b$. (b) $a < b$ but a is close to b. (c) $a \ll b$, $b = -60$, $a = -0.1$.

pole is at −5. If one closed-loop pole is near −0.1 and the other is near −50, the step and ramp responses should be dominated by the closed-loop pole at −5.

By choosing $K = 4.68$, we can ensure that the closed-loop poles are located at −0.102 (effectively canceling the closed-loop zero at −0.1), −5 (as desired), and −54.9 (well to the left of −5).

The resulting step and ramp responses are not shown because they are indistinguishable from the responses of section 3.10, Figure 3.29.

4.12 Summary

A pole–zero plot of a rational function consists of indications on the complex plane of the locations of the function's poles and zeros. The addition of the multiplying constant of the function to the diagram completely specifies the function.

To graphically evaluate a rational function $F(s)$ at a particular value $s = s_0$, directed line segments are drawn from each zero and each pole of $F(s)$ to the point s_0. The magnitude of the evaluated function is the product of the magnitude of the function's multiplying constant and the lengths of the directed line segments from the zeros, divided by the product of the lengths of the directed line segments from the poles. The angle of the evaluated function is the sum of the angles of the directed line segments from the zeros minus the sum of the angles of the directed line segments from the poles. An additional 180° must be added to or subtracted from the angle of the function if the multiplying constant is negative.

A root locus plot is a pole–zero plot of the rational open-loop system transmittance $G(s)H(s)$, upon which is superimposed the locus of the roots

$$1 + KG(s)H(s) = 0$$

as K is varied from zero to $+\infty$. The root locus is symmetrical about the real axis. A value of s for which the angle of $G(s)H(s)$ is 180° (plus or minus any multiple of 360°) is a point on the root locus, corresponding to some value of K. The six rules for root locus plot construction are summarized in Table 4.1, and Table 4.2 gives many root locus examples. The value of the constant K corresponding to a specific point on the locus is found using

$$|G(s)H(s)| = \frac{1}{|K|} = \frac{\begin{pmatrix}\text{magnitude of the} \\ \text{multiplying constant}\end{pmatrix}\begin{pmatrix}\text{product of} \\ \text{zero distances}\end{pmatrix}}{\text{product of pole distances}}$$

Root locus methods can be applied to other systems besides those of the standard form, and it is straightforward to extend the technique to systems with negative adjustable parameters, as shown in the chapter. A delay is present in many applications. A pure delay can be approximated by a first-order transmittance, and it can be shown that many systems that are stable when no delay is present become much less stable.

As poles are added to a system, the root locus plot shows that the system is less stable, while adding zeros tends to make a system more stable. These effects are used in the design procedures that appear in Chapter 5.

A light-source tracking system example illustrated some of the power of root locus for design. A control system for an artificial limb involved the design of subsystems as part of an overall system design. A root locus analysis of a flexible spacecraft demonstrated that omitting flexible dynamics can cause a designer to overlook a potential cause of instability. The chapter ends with a root-locus-based design of an artificial eye that tracks a target by means of the motion of an otherwise blind eye.

A camera directs an image to an implanted electronic retina, providing simulated vision for a blind person.

REFERENCES

Root Locus

Evans, W. R., "Graphical Analysis of Control Systems." *Trans. AIEE*, 67 (1948): 547–551.

———, "Control System Synthesis by the Root Locus Method." *Trans. AIEE*, 69 (1950): 67–69.

———, *Control System Dynamics.* New York: McGraw-Hill, 1954.

Truxal, J. G. *Control System Synthesis.* New York: McGraw-Hill, 1955.

Prosthetics

Allan, R., "Electronics Aids the Disabled." *IEEE Spectrum,* (November 1976): 36–40.

"Brain Controls Use of Artificial Arm." *Prod. Eng.*, October 7, 1968, p. 23.

"Novel Artificial Arm Gives Wearer six Degrees of Freedom." *Prod. Eng.,* October 20, 1969, p. 15.

Horgan, J.,"Medical Electronics." *IEEE Spectrum* (January 1984): 90–93.

Raibert, M. A., and Sutherland, I. E., "Machines That Walk." *Sci. Am.* (January 1983): 44–53.

Rauch, H., ed., "Dextrous Robotic Hand." *IEEE Control Systems. Mag.* (December 1986).

Control of Flexible Spacecraft

Larson, V., and Likins, P.W., "Optimal Estimation and Control of Elastic Spacecraft," in *Advances in Control and Dynamic Systems*. New York: Academic Press, 1977.

Martin, G. D., and Bryson, A. E., "Attitude Control of a Flexible Spacecraft," *J Guidance Control* (January–February 1980): 37–41.

Artificial Vision

Dagnelie, G., and Massof, R. W. "Toward an Artificial Eye." *IEEE Spectrum* (May 1996): 20–29.

PROBLEMS

1. Draw pole–zero plots for the following rational functions:

(a) $$F(s) = \frac{-3s^2 + 8}{(8s+1)(s+7)^2}$$

(b) $$F(s) = \frac{8\left(s^2 - 4s + 10\right)(1 + 2s)}{s^3\left(s^2 + 4s + 10\right)}$$

(c) $F(s) = \dfrac{100s\left(s^2 - 4s + 10\right)^2}{(s+3)^2\left(s^2 + 4s + 10\right)^2}$

(d) $F(s) = \dfrac{3s^3 + 9s}{\left(s^2 + 2s + 10\right)\left(s^2 + 2s + 8\right)}$

2. For each of the following GH products, construct root locus sketches. Find the range of the positive adjustable gain K in

$$T(s) = \frac{KG(s)}{1 + KG(s)H(s)}$$

for which each overall system is stable.

(a) $G(s)H(s) = \dfrac{1}{s(4s+1)}$

(b) $G(s)H(s) = \dfrac{3s+1}{s(4s+1)}$

(c) $G(s)H(s) = \dfrac{s+1}{s(4s+1)}$

(d) $G(s)H(s) = \dfrac{1}{s(s+1)(4s+1)}$

(e) $G(s)H(s) = \dfrac{1}{s^2 + 10s + 40}$

(f) $G(s)H(s) = \dfrac{s+4}{s^2 + s + 6}$

(g) $G(s)H(s) = \dfrac{s}{(s+2)(s+10)}$

(h) $G(s)H(s) = \dfrac{1}{s(s+2)\left(s^2 + 2s + 50\right)}$

Ans.

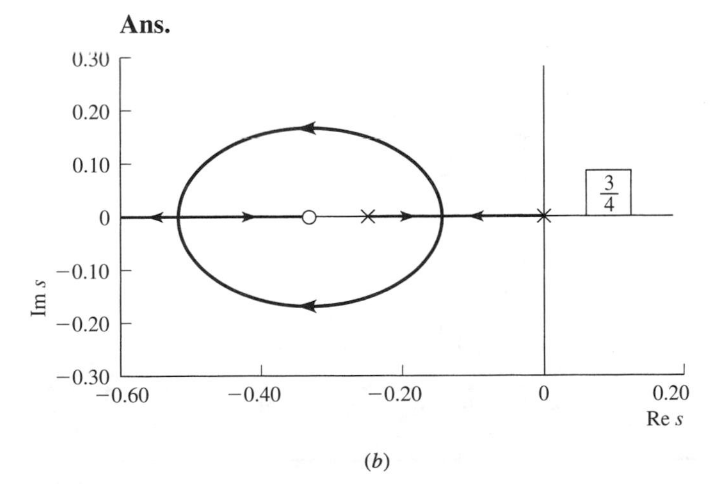

(*b*)

Figure P4.2

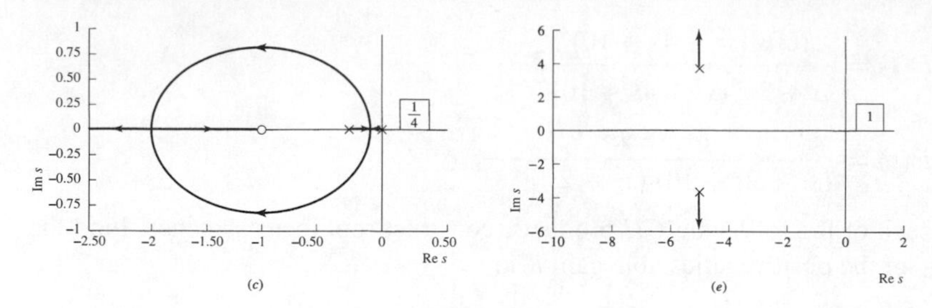

Figure P4.2 (Continued)

3. Sketch root locus plots for the systems of Figure P4.3 for K between zero and $+\infty$. Find asymptotes, centroid, angles of departure, angles of approach, and approximate breakaway points where applicable.

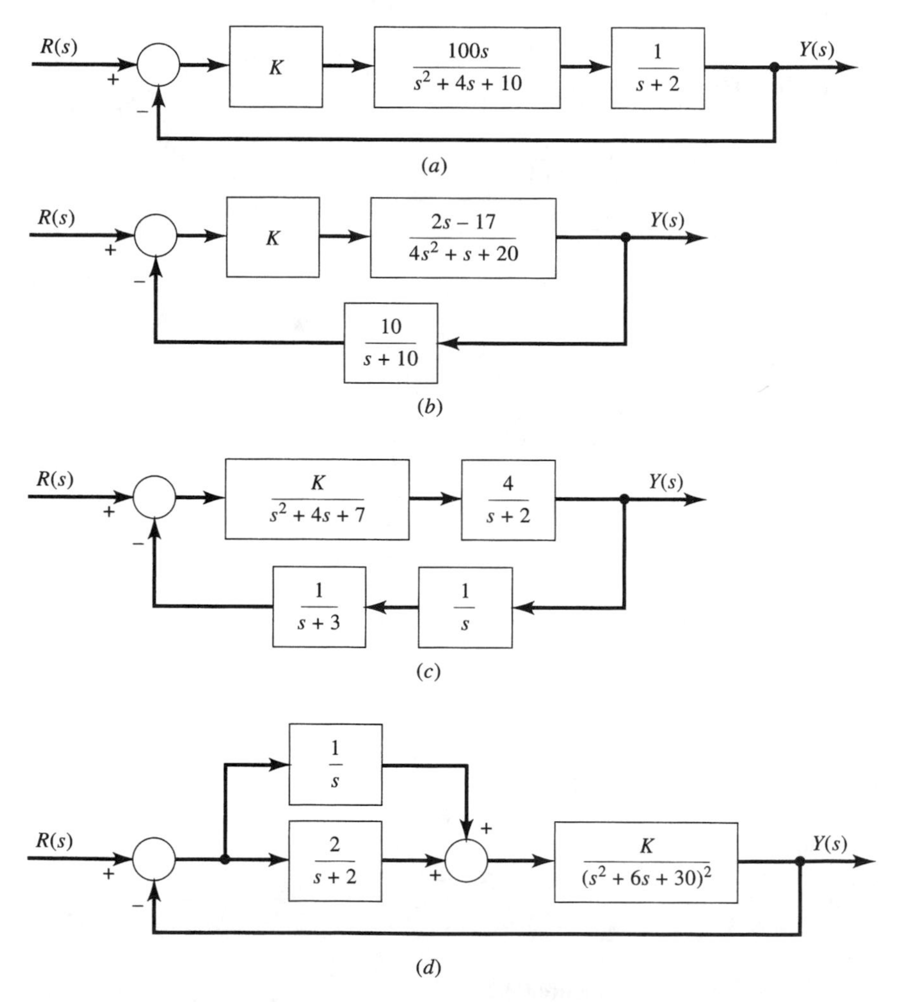

Figure P4.3

4. Sketch root locus plots for the systems with pole–zero plots given in Figure P4.4. Find asymptotes, centroid, angles of departure, angles of approach, and approximate breakaway points where applicable.

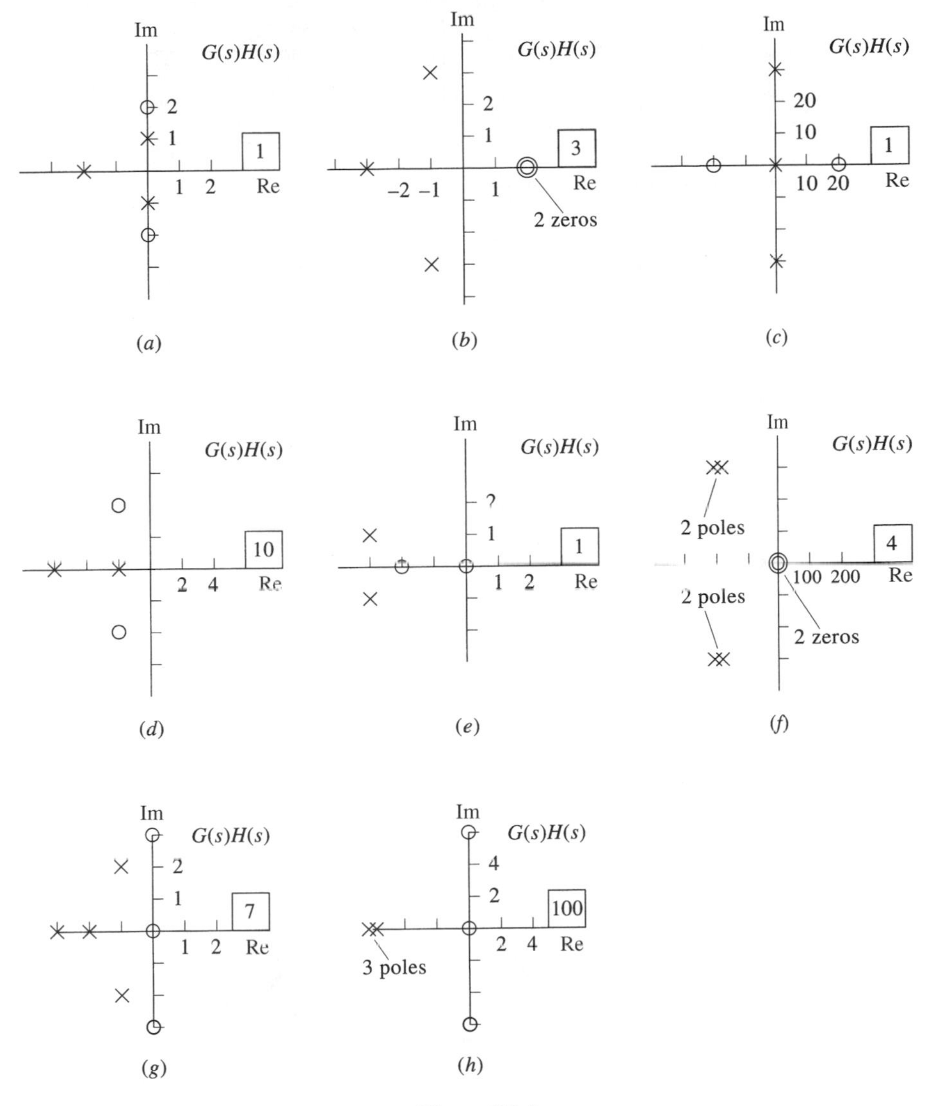

Figure P4.4

5. For the systems with the following GH products, use graphical root locus methods to find an approximate value of adjustable gain K for which (if possible) there is a complex conjugate set of roots of the overall system with damping ratio $\zeta = 0.2$.

(a) $\dfrac{1}{s\,(s+1)\,(0.2s+1)}$

Ans.

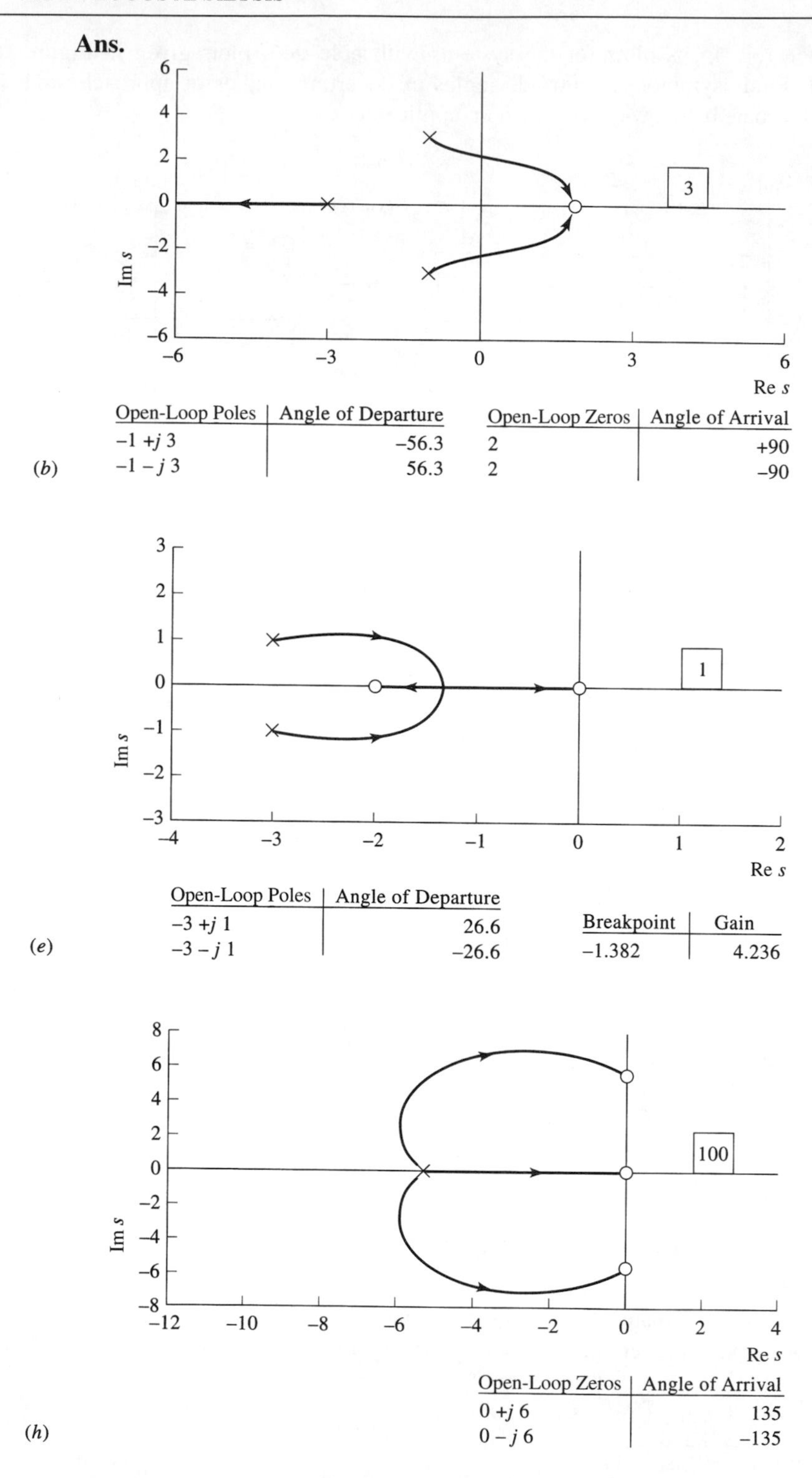

(b)

Open-Loop Poles	Angle of Departure
−1 +j 3	−56.3
−1 − j 3	56.3

Open-Loop Zeros	Angle of Arrival
2	+90
2	−90

(e)

Open-Loop Poles	Angle of Departure
−3 +j 1	26.6
−3 − j 1	−26.6

Breakpoint	Gain
−1.382	4.236

(h)

Open-Loop Zeros	Angle of Arrival
0 +j 6	135
0 − j 6	−135

Figure P4.4 (Continued)

(b) $\dfrac{100\,(s+0.2)}{s\,(s+1)\,(0.2s+1)\,(s+20)}$

(c) $\dfrac{100\,(s+2)}{s\,(s+1)\,(0.2s+1)\,(s+200)}$

(d) $\dfrac{100(s+5)}{s\,(s+1)\,(0.2s+1)\,(s+500)}$

Ans.

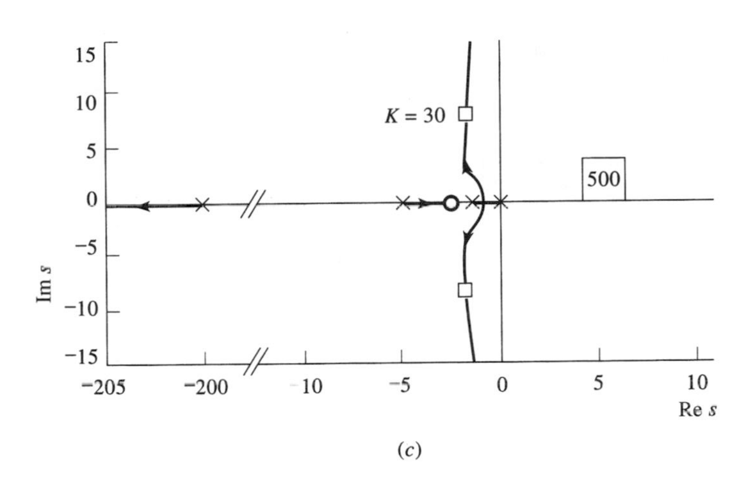

(c)

Figure P4.5

6. On the complex plane, sketch the locus of the roots of the following functions, as the parameter K is varied from zero to $+\infty$:

(a) $T(s) = \dfrac{6}{s^2+K}$

(b) $T(s) = \dfrac{s^2+7s-4}{Ks^2+7s+4}$

(c) $T(s) = \dfrac{100}{s^2+Ks+10}$

(d) $T(s) = \dfrac{-10}{s^2+(K-2)\,s+8+K}$

7. Sketch the root locus, for K from zero to $+\infty$, for a system with

$$G(s)H(s) = -\frac{s+4}{(s+1)^2\,(s^2+4s+9)}$$

Note the negative algebraic sign.

8. Sketch the root locus for K in the range from zero to *minus* infinity for system with

$$G(s)H(s) = \frac{2(s-4)^2}{(s+5)\left(s^2+4s+10\right)}$$

Using this root locus, approximately locate the poles of

$$T(s) = \frac{G(s)}{1+KG(s)H(s)}$$

for $K = -2$.

Ans.

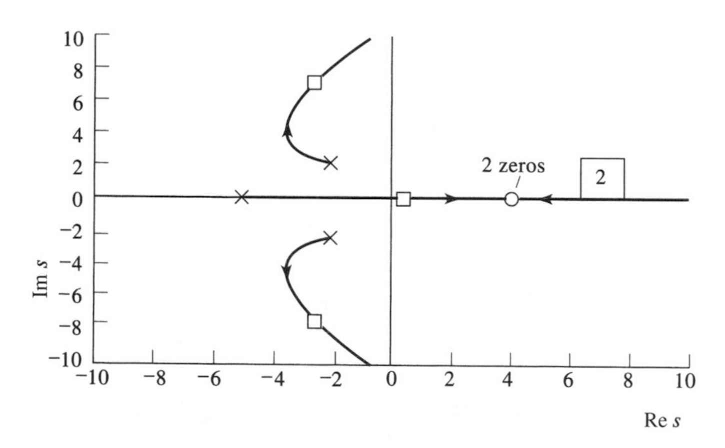

Figure P4.8

9. On the complex plane, sketch the locus of the poles of the function

$$T(s) = \frac{-10s+2}{4s^2+2s+Ks-10K}$$

as the parameter K is varied through the range $-\infty$ to $+\infty$.

10. For the light-source tracking system (Section 4.8), suppose that an electrical compensator network is added instead of the tachometer, as in Figure P4.10. Find gain constants K_1 and K_2, if possible, such that the system has a relative stability of at least 5 units.

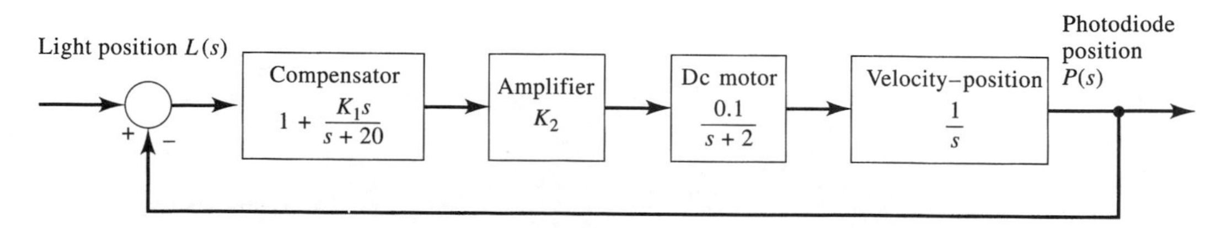

Figure P4.10

11. Sketch the root locus for

$$G(s)H(s) = \frac{1}{s(s+1)(s+2)(s+10)}$$

For $K = 200$, are there any characteristic equation roots in the right half-plane?

Ans. Yes

12. For the artificial limb system (Section 4.9) with $T = 0.25$ s, suppose that more tachometer feedback is used so that $K_T = 10$. Use root locus methods to find an acceptable value of K_B. Find the maximum value of K_B for which the overall system is stable.

13. A simple block diagram of a control rod positioning system for a nuclear power plant is given in Figure P4.13. Sketch a root locus plot for the system. Graphically test points along the imaginary axis to accurately determine where the loci cross into the RHP. Then determine graphically the value of K for which the system is marginally stable.

For a system of relatively low order such as this one, the same results are perhaps more easily obtained by Routh–Hurwitz testing in terms of K. With greater system complexity, a Routh–Hurwitz test becomes hopelessly involved, but root locus methods remain very suitable.

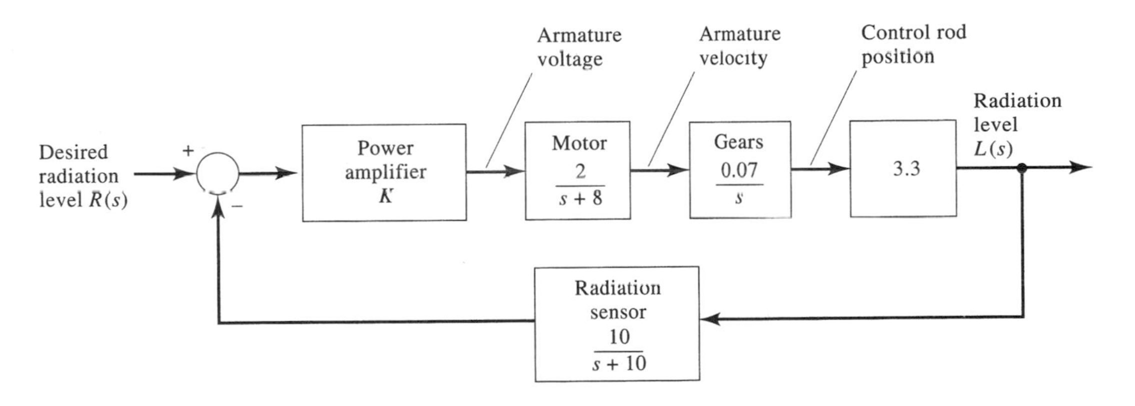

Figure P4.13

14. A linearized inventory control system model is given in Figure P4.14. Based upon the difference between desired and actual inventory levels, management sets the production quota, which in turn determines the production rate. Using graphical root locus methods, determine the production time constant

$$\tau = \frac{1}{K}$$

in the range

$$1 < \tau < 10$$

which gives greatest relative stability of the system.

Ans.

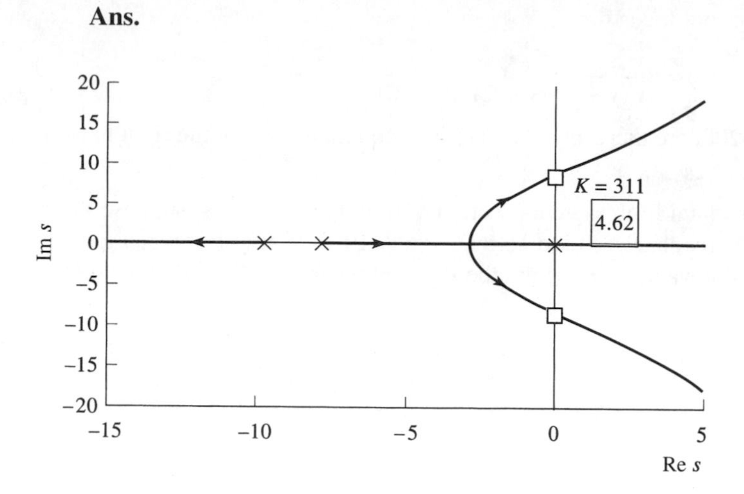

Figure P4.13

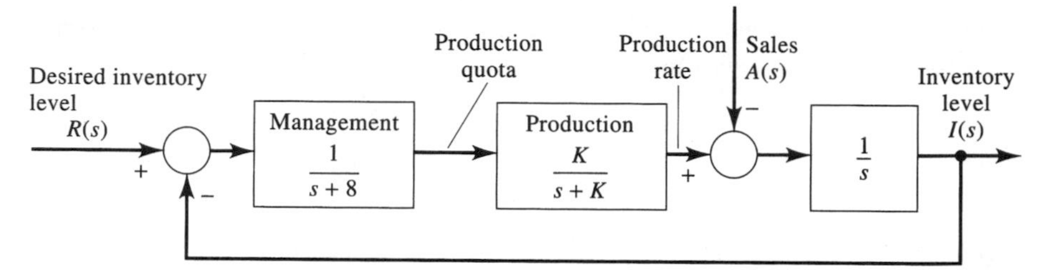

Figure P4.14

15. A simplified model of an automobile pollution control system is shown in Figure P4.15. Sketch the system root locus for $K > 0$, paying particular attention to the manner in which the loci leave the repeated poles. Use the root locus to determine, approximately, the value of K, if any, for which the system's natural response will decay at least as rapidly as $\exp(-0.4t)$.

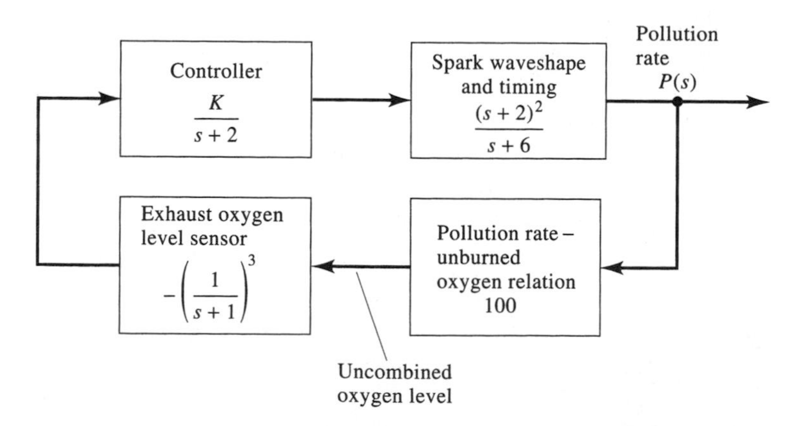

Figure P4.15

Ans.

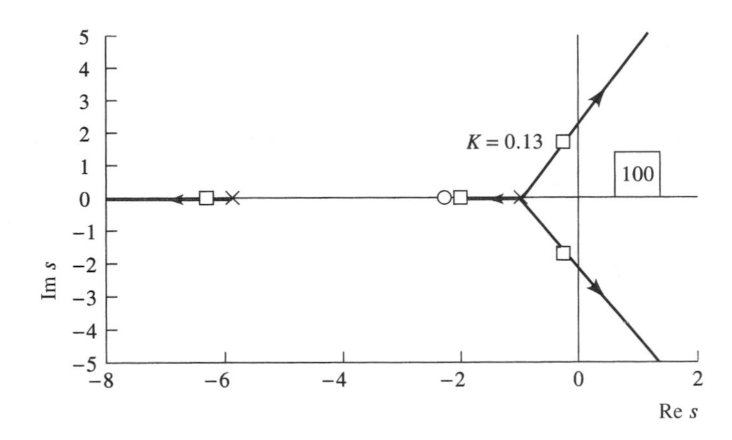

Figure P4.15

16. A linearized model of a frequency-locked loop is given in Figure P4.16. The frequency difference between an incoming sinusoidal signal and an oscillator is sensed and integrated to produce a voltage that drives the voltage-controlled oscillator. The oscillator's frequency is proportional to the control voltage. When there is zero frequency difference, the integrator input is zero, its output is constant, and the oscillator frequency equals the incoming frequency. For incoming signals that are distorted and corrupted by noise, the frequency-locked loop produces a nearly pure sinusoid, locked in frequency to the incoming signal. The effect is as if the incoming signal had been processed by a very sophisticated filter that had removed virtually all its noise and distortion. Use root locus procedures to select a value of positive K such that the damping ratio of the dominant roots is 0.2.

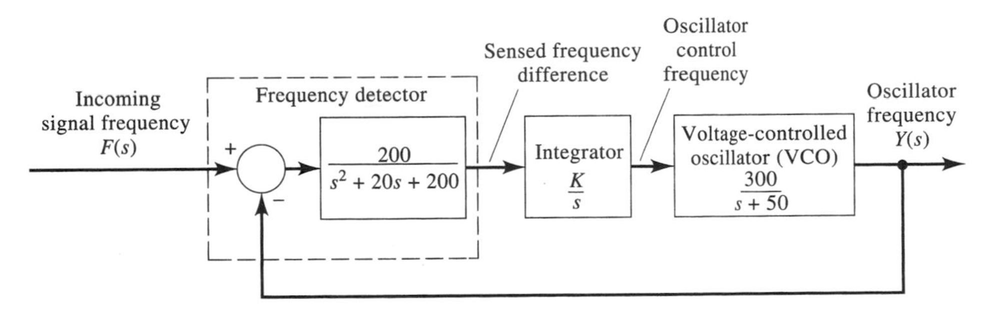

Figure P4.16

17. A position servosystem model is given in Figure P4.17. The compensator is an imperfect attempt to integrate the sensed velocity to obtain position feedback. Use root locus procedures to find positive values of the adjustable parameter K for which the system which relates shaft velocity to the input is stable.

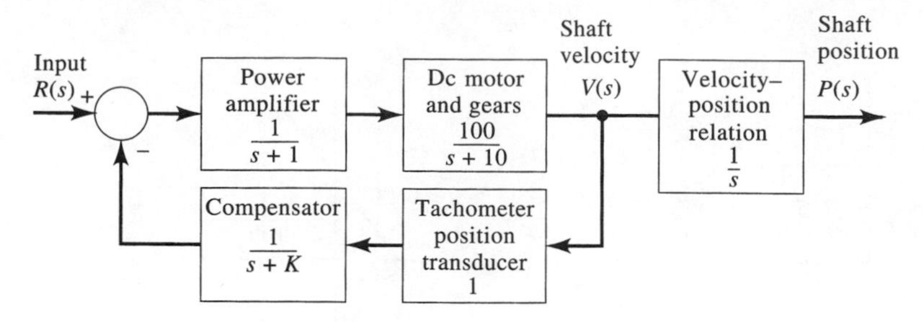

Figure P4.17

18. Use a computer to find root locus diagrams for

(a) $$G(s)H(s) = \frac{2(s+10)}{(s+2)^2\left(s^2+10s+125\right)}$$

(b) $$G(s)H(s) = \frac{2\,(s+10.4)}{(s+2)^2\left(s^2+10s+125\right)}$$

19. Sketch the root locus for

$$G(s)H(s) = \frac{1}{\left(s^2+1\right)\left(s^2-4\right)}$$

20. Determine the initial angles of departure from the complex poles and sketch the root locus diagrams for

$$G(s)H(s) = \frac{10s}{s^2+s+40}$$

21. A simple position servo is expressed in block diagram form as shown in Figure P4.21.

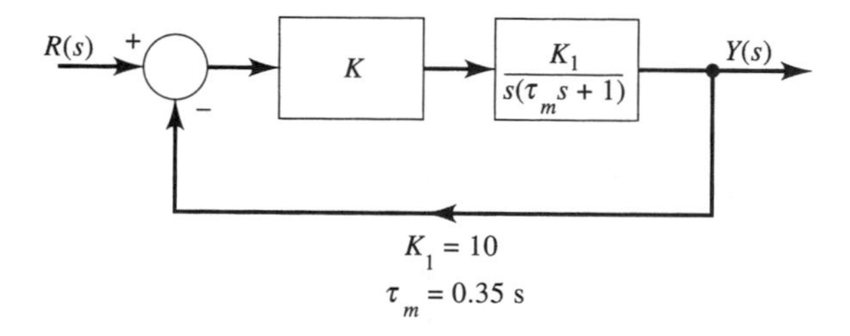

Figure P4.21

(a) Is the system stable for all values of gain?

(b) Upon construction of the servo of Figure P4. 21, the device is found to have zero damping for a gain $K = 30$. The instability is due to parasitic time lags in the synchros, amplifier, and motor. Approximate lumping these effects together by using an additional time lag in the forward transfer function—that is, by a factor $1/(1+\tau_a s)$. What is the value of τ_a?

Ans. $\tau_a = 0.00336$

22. Sketch root locus diagrams for the following systems:

(a) $\mathrm{GH} = \dfrac{1}{s^4 + 16}$

(b) $\mathrm{GH} = \dfrac{1}{s^4 - 16}$

(c) $\mathrm{GH} = \dfrac{1}{(s^2 + 4)(s^2 - 1)}$

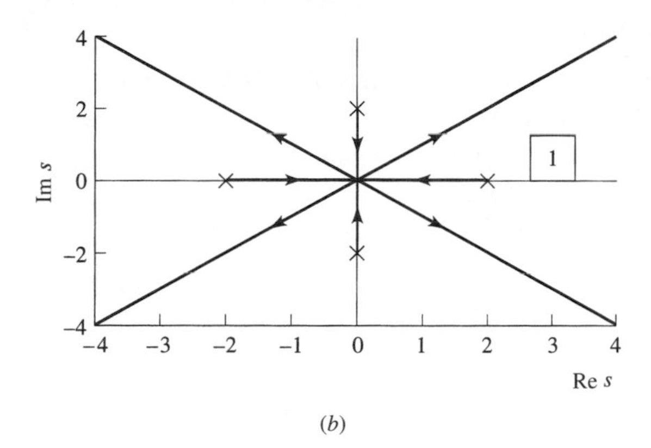

(b)

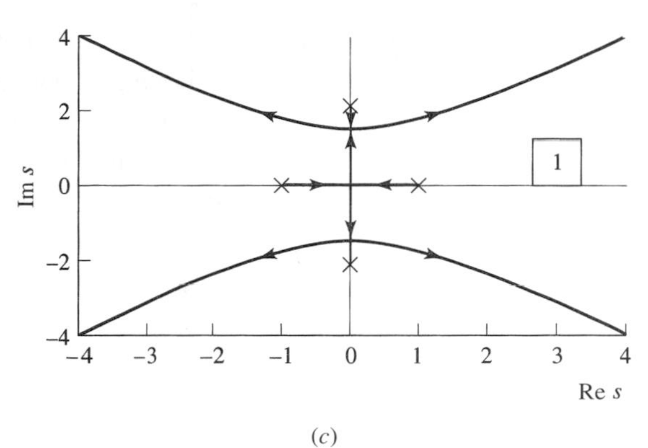

(c)

Figure P4.22

23. Sketch the root locus for the system with the following GH product. Determine all angles of departure and arrival, asymptote angles, and the centroid.

$$G(s)H(s) = \frac{\left[(s+2)^2 + 4\right]^2}{s^3(s+5)^2}$$

24. For Problem 23, find the value of the gain K so the dominant closed-loop roots have a damping ratio of 0.1.

25. For the system of Figure P4.25, determine the gain K so that the dominant closed-loop roots have a damping ratio of 0.2.

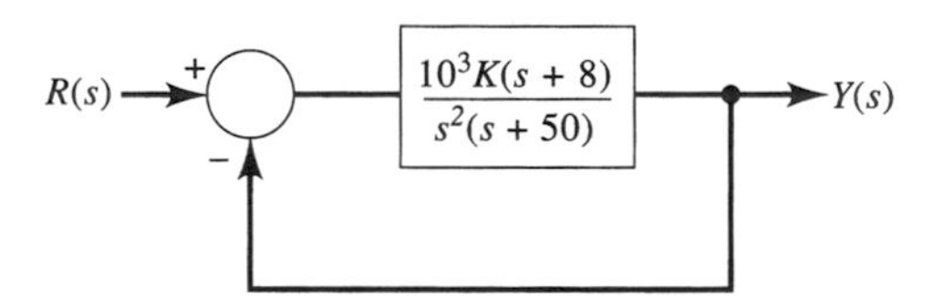

Figure P4.25

26. Find the gain K so that the system with the following GH product has dominant closed-loop roots with a damping ratio of 0.2.

$$G(s)H(s) = \frac{K(s+6)}{s(s^2 + 4s + 13)}$$

Ans. $K = 6$

27. For the system of Figure P4.27:

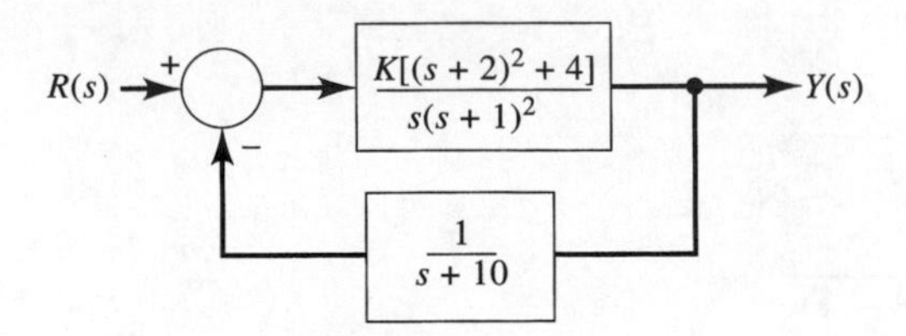

Figure P4.27

(a) Sketch the root locus and compute all angles of arrival, angles of departure, break-in points, entry points, and the centroid.

(b) Determine the gain K so the system has closed-loop dominant roots with a damping ratio of 0.4.

CHAPTER

Root Locus Design

5

5.1 Preview

Engineering practice in general and control system techniques in particular may be divided into two related areas: analysis and design. Analysis answers the question: "How does it work?" Design responds to the demand: "Make it work!" Most of the material in the preceding chapters provides the control system engineer with analytical tools. For example, given a differential equation, the Laplace transform may be used to generate the time response, from which the time constant of a first-order system or the damping ratio and undamped natural frequency of an underdamped second-order system may be obtained. The root locus demonstrates the effect that a variation in some gain would have upon the closed-loop characteristic equation roots. These are analytical tools, but each can be used to evaluate a design.

Design begins by using a simple unity feedback system (the uncompensated system) in which the gain is adjusted in an effort to provide good steady state performance (where does the system go?) and good stability (How does it get there?).

If the uncompensated system is unable to meet the design goals, compensators must be added. We considered forward path and feedback path compensators, which alter the root locus by adding poles and zeros or by canceling poles and zeros. System gain may be readjusted to create the desired performance (measured in terms of steady state response and stability).

To facilitate a comparison of the various compensators, and to aid in design, each is applied to a second-order model that possesses the dominant closed-loop poles. We consider lag, lead, lag–lead, PI, PD, and PID compensators. We also consider the exact placement of closed-loop poles. Many design examples are presented.

In the root loci that follow, the symbol × denotes an open-loop pole, the symbol ○ denotes an open-loop zero, and a box denotes a closed-loop transfer function pole.

5.2 Shaping a Root Locus

As K changes, the closed-loop poles move.

We have learned how to sketch the root locus of the system shown in Figure 5.1. As an analysis tool, the closed-loop poles and the time response are easily available for a given plant $G_p(s)$ and a given gain K. Root locus methods are even more valuable as a design tool when we are free to choose K. The root locus reveals the effect that K will have on the location of the closed-loop poles (called roots for simplicity) and thus on the time response.

The gain K is adjusted to provide acceptable performance, if possible. Typically, the designer is interested in the closed-loop system's stability and its steady state error performance. The forward path gain K has an output that is proportional to the amplifier input; hence, that sort of device is often referred to as being *proportional.*

Compensators are introduced.

Perhaps the gain K cannot be chosen so that the controlled closed-loop system has adequate performance (in terms of stability and steady state performance). In that case it is necessary to alter the root locus by including additional dynamic elements called compensators (see Figure 5.2 for various ways to implement a compensator).

In Figure 5.2(a), the compensator $G_c(s)$ is inserted into the system's forward path. In Figure 5.2(b), the compensator $H_c(s)$ is placed in the feedback path around the plant, forming a *feedback* compensator. A system involving both feedback and cascade compensation is given in Figure 5.2(c).

Proportional compensation is referred to as the uncompensated system.

Each of these devices will generally also have a gain K that has to be selected. To avoid confusing the proportional gain that will have to be selected for each more complicated compensator, the system of Figure 5.1 will be referred to as being *uncompensated* and the resulting gain will be called K_u. The gain associated with more complicated compensators will be referred to as just K. Of course, it may be desirable to set the gain of a compensator to the uncompensated value, in which case $K = K_u$.

Root locus designer's tools.

There are two basic tools in the root locus designer's toolbox: (1) a compensator may be chosen to add poles and zeros and (2) a compensator may be chosen to cancel poles and zeros. In either case, the root locus is reshaped purposefully. In some cases, closed-loop poles are placed approximately to meet requirements. In other cases, closed-loop poles are placed exactly.

Six compensators will be considered, all of which may be realized using Figure 1.14. Three compensators have real, non-zero poles and zeros: lag, lead and

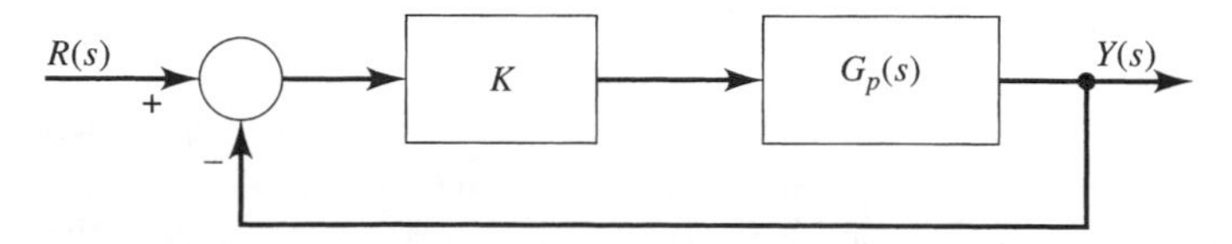

Figure 5.1 Unity feedback system with proportional amplifier (uncompensated system).

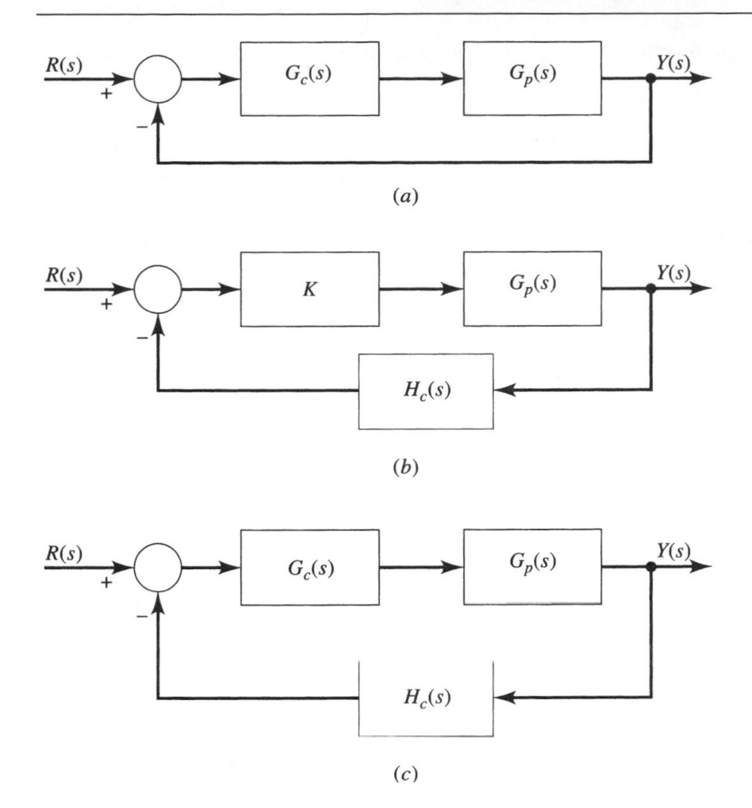

Figure 5.2 Compensator configurations. (a) Cascade-compensated system. (b) Feedback-compensated system. (c) System with feedback and cascade compensation.

lag–lead compensators. Three compensators are formed from components having a pole or zero at $s = 0$: PI, PD and PID compensators.

It is essential for the designer to understand the implication of adding poles and zeros and canceling poles and zeros; hence, we begin by evaluating those implications.

5.3 Adding and Canceling Poles and Zeros

5.3.1 Adding a Pole or Zero

In Section 4.7, we learned what happens when a cascade compensator [as in Figure 5.2(a)] contains a pole or a zero. When

$$G_c(s) = \frac{K}{s}$$

the added pole tends to move the root locus to the right, rendering the system less stable. System type number rises, which improves steady state performance.

Adding a pole makes the system less stable.

When

$$G_c(s) = K(s + a)$$

Adding a zero makes the system more stable.

a zero is added to the root locus, causing the root locus to move to the left, rendering the system more stable.

Remember that the final value theorem requires us to set $s = 0$. We can easily see the effect that any $G_c(s)$ will have for a unity feed back system by looking at $G_c(0)$. For the uncompensated case, $G_c(0) = K_u$. Suppose we agree to use K_u for each compensator. When we add a pole at $s = 0$, $G_c(0)$ is infinity, increasing some error coefficient K_i and reducing the error to zero for a power-of-time input

$$t^i u(t)$$

Effect on steady state accuracy.

When we add a zero, $G_c(0)$ is $K_u a$. For a given K_u, the error coefficient K_i increases for $a > 1$ and decreases for $a < 1$ (compared to the uncompensated case having K_u only), clearly the addition of a zero has nowhere near the steady state effect of adding a pole. Thus, we conclude that adding a pole primarily is a tool for improving steady state error, while adding a zero is primarily a tool for improving stability.

5.3.2 Canceling a Pole or Zero

There may be cases in which canceling a pole or zero offers a chance of improving performance in some way. For example, consider an uncompensated system in which only K is included in $G_c(s)$.

$$G_c(s) = K; \qquad G_p(s) = \frac{1}{s(s+2)}$$

The response may be speeded up by open-loop pole cancellation.

The root locus is shown in Figure 5.3(a). If $K = 1$, the system is critically damped with two closed-loop poles at -1. The step response for that critically damped case appears in Figure 5.3 (b). If that response is considered to be too slow, and overshoot is not desired, the response can be speeded up by canceling the open-loop pole at -2 by an open-loop zero at -2. Then, a new open-loop pole could be introduced at -10. Let us choose

$$G_c(s) = \frac{K(s+2)}{s+10}$$

$$G_p(s) = \frac{1}{s(s+2)}$$

$$G_c(s)G_p(s) = \frac{K}{s(s+10)}$$

The root locus of Figure 5.3(c) exhibits critical damping for $K = 25$. The resulting step response at Figure 5.3 (d) is exactly five times faster than that of Figure 5.3(b) because the closed-loop poles are now at -5. Canceling the open-loop pole at -2 quickened response considerably. In material to follow, we will call this compensator a lead compensator.

Approximate cancellation of a closed-loop zero.

The cancellation of the open-loop pole at -2 was exact. In other cases, we can approximately cancel a closed-loop zero by increasing K. Sometimes, a closed-loop zero can degrade performance, and it may be desirable to approximately cancel that closed-loop zero with a nearby closed-loop pole. As an example, suppose that we have

$$G_c(s) = K$$

$$G_p(s) = \frac{s + 0.5}{s^2}$$

According to the root locus of Figure 5.3(e), if $K = 2$ the system is critically damped with two closed-loop poles at $s = -1$, the same location as for Figure 5.3(a) with $K = 1$. Now, however, the step response of Figure 5.3(f) shows an overshoot because the transfer function is

A closed-loop zero can cause overshoot.

$$T(s) = \frac{2(s + 0.5)}{(s + 1)^2}$$

whereas for Figure 5.3(b) it is

$$T(s) = \frac{1}{(s + 1)^2}$$

The closed-loop zero at $s = -0.5$ is responsible for the overshoot. If we increase the gain K as in Figure 5.3(g), one closed-loop pole moves very close to the closed-loop zero located at -0.5. Specifically, with

$$K = 10$$

$$T(s) = \frac{10(s + 0.5)}{(s + 0.53)(s + 9.47)}$$

The closed-loop poles are -0.53 and -9.47. The closed-loop pole at -0.53 nearly cancels the closed-loop zero at -0.5 so we can approximate the transfer function by

Pole–zero cancellation.

$$T(s) \approx \frac{9.47}{s + 9.47}$$

The actual step response of Figure 5.3(h), does in fact, show very little overshoot. In keeping with a first-order approximation having a time constant

$$\tau(s) \approx \frac{1}{9.47} = 0.11$$

The step response reaches nearly 0.632 at $t = 0.11$, justifying the approximate $T(s)$.

❑ Computer-Aided Learning

The "rltool" command of MATLAB should be used to reproduce the results given so far. For example, if we use the commands

```
gp=zpk([], [0 -2],1)
rltool (gp)
```

We can obtain the root locus of Figure 5.3(a) and the step response of Figure 5.3(b). Then, any of the compensators just described can be implemented by editing and we can obtain closed-loop poles and step responses for various compensated systems and various K values.

The drill problems and design examples to follow should be implemented by using rltool. Perhaps you can improve on the example designs as you acquire knowledge of the design methods.

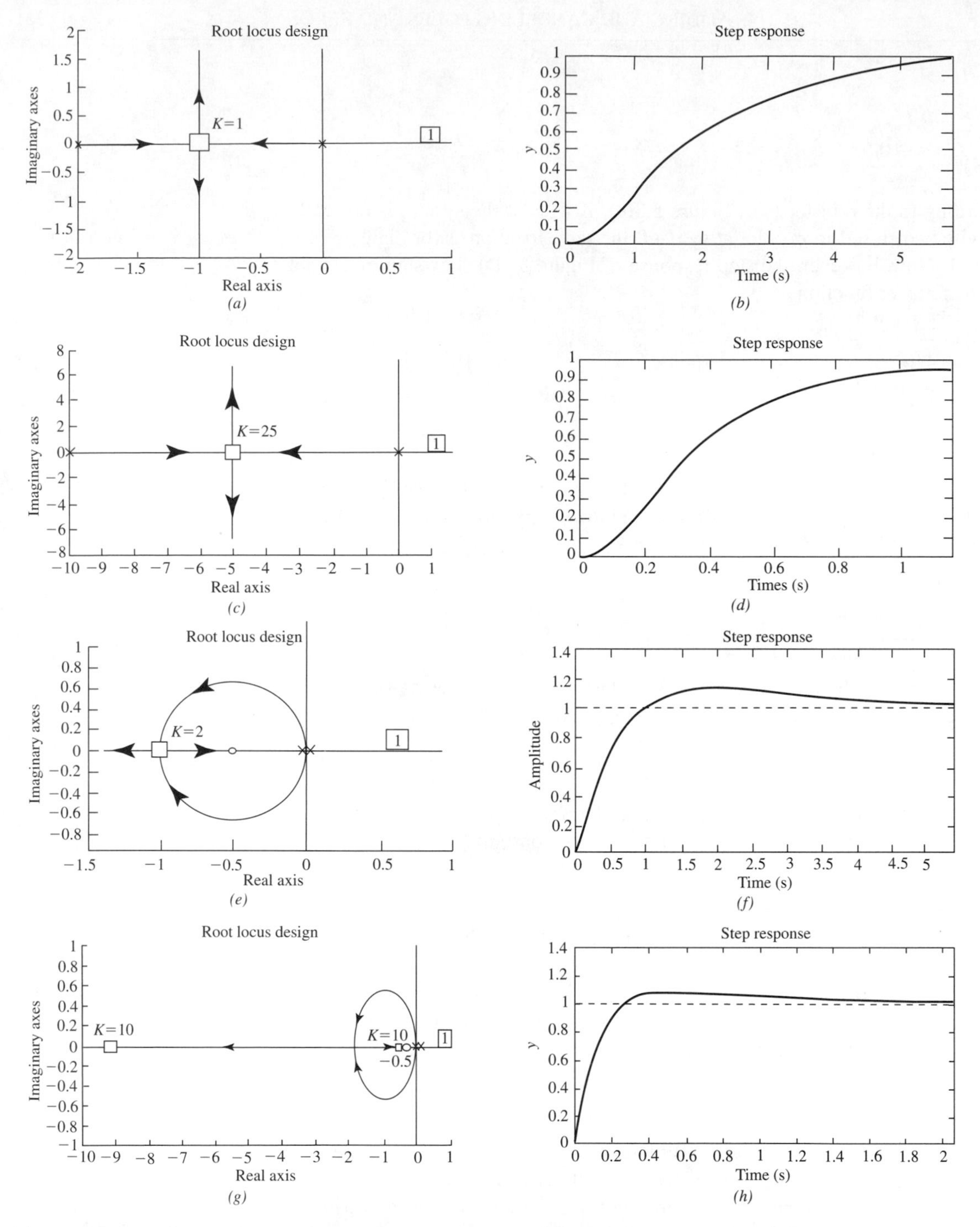

Figure 5.3 Canceling a pole or a zero. (a) Root locus for $G_c(s) = K$, $G_p(s) = 1/s(s+2)$. (b) Step response for (a) with $K = 1$. (c) Root locus for $G_c(s) = K(s+2)/(s+10)$, $G_p(s) = 1/s(s+2)$. (d) Step response for (c) with $K = 25$. (e) Root locus for $G_c(s) = K$, $G_p(s) = (s+0.5)/s^2$. (f) Step response for (e) with $K = 2$. (g) System of (e) with $K = 10$. (h) Step response of (g) with $K = 10$.

❑ DRILL PROBLEMS

D5.1 Use the root locus to select K so that the closed-loop poles have 0.707 damping for

$$G_c(s) = K$$

$$G_p(s) = \frac{1}{s(s+4)}$$

Find ω_n for the closed-loop poles. Use your value of K to obtain the step response. Find the percent overshoot and rise time.

Ans. 8, 2.83 s; 5%, 0.740 s

D5.2 For the system of Drill Problem D5.1, add an open-loop pole at –8, so now

$$G_c(s) = \frac{K}{s+8}$$

$$G_p(s) = \frac{1}{s(s+4)}$$

Repeat Drill problem D5.1 to find K, ω_n, percent overshoot, and rise time.

Ans. 41.7, 2.16 s; 5%, 1.02 s

D5.3 For the system of Drill Problem 5.1, add an open-loop zero at −8 so now

$$G_c(s) = K(s+8)$$

$$G_p(s) = \frac{1}{s(s+4)}$$

Repeat Drill Problem D5.1 to find K, ω_n, percent overshoot, and rise time,

Ans. 4.27, 5.84 s; 5%, 0.27 s

D5.4 For the system of Drill Problem D5.1, cancel the open-loop pole at −4 and replace with an open-loop pole at −8. Repeat Drill Problem D5.1 to find K, ω_n, percent overshoot, and rise time.

$$G_c(s) = K\frac{(s+4)}{(s+8)}$$

$$G_p(s) = \frac{1}{s(s+4)}$$

Ans. 32, 5.66 s; 5 %, 0.37 s

D5.5 In this drill problem, as K increases, one closed-loop pole approaches a closed-loop zero, effectively canceling that zero and causing the closed-loop step response to resemble that of a second-order system. Suppose

$$G_p(s) = \frac{s+0.5}{s(s+4)^2}$$

$$G_c(s) = K$$

For each K, find the damping ratio of the closed-loop complex conjugate poles, the percent overshoot from Figure 2.10 (b) that would occur for a purely second-order step response, and the percent overshoot of the actual step response.

(a) $K = 20$

(b) $K = 60$

(c) $K = 100$

(d) $K = 400$

Ans. (a) 0.66, 8%, 0%; (b) 0.44, 22%, 1.6%; (c) 0.36, 30%, 17%; (d) 0.18, 51%, 51%.

5.4 Second-Order Plant Models

We have explored some general implications of adding poles and zeros and some general implications of canceling poles and zeros. Next, we will cover the use of six specific types of compensators. There are an infinite number of plants that we could examine and for each unique plant, we might try to find some general design rules. Clearly, that would be a formidable task, completely beyond the scope of this textbook.

Generalized compensator design using a model.

To generalize compensator design, we will investigate only how each compensator affects the dominant roots. This calls for creation of a model that has the same set of second-order roots for the same uncompensated gain K_u as for the actual system. The manner in which each compensator affects the two roots of the model should represent an approximation of how the actual system roots are affected.

Remember that an uncompensated system is what we call a system like that of Figure 5.1, having a proportional gain K as the only compensating device. The uncompensated gain K_u is the value of K for Figure 5.1.

For example, suppose the uncompensated system is type 1. The closed-loop transfer function containing the dominant closed-loop poles (with damping ratio ζ and undamped natural frequency ω_n) is

$$T(s) = \frac{\omega_n^2}{s^2 + 2\zeta\omega_n s + \omega_n^2} \qquad [5.1]$$

An appropriate plant model would be

Type 1 plant model.

$$G_p(s) = \frac{\omega_n^2/K_u}{s(s + 2\zeta\omega_n)}$$

If the type 1 plant model is used in Figure 5.1 with $K = K_u$, then $T(s)$ is given by Equation (5.1) and the system of Figure 5.1 has the same roots as those that are dominant for the actual system.

Suppose the plant is type 0. The transfer function containing the dominant roots, but having a step error coefficient of K_0 would be

$$T(s) = \frac{K_0\omega_n^2/(K_0 + 1)}{s^2 + 2\zeta\omega_n s + \omega_n^2} \qquad [5.2]$$

The steady state error for a unit step input would be

$$\text{error} = \lim_{s \to 0} 1 - T(s) = \frac{1}{K_0 + 1}$$

We need a plant for Figure 5.1 so that when $K = K_u$ the transfer function is given by Equation (5.2). The following plant will work:

$$G_p(s) = \frac{K_0\omega_n^2/(K_0+1)K_u}{s^2 + 2\zeta\omega_n s + \omega_n^2/(K_0+1)}$$

Type 0 plant model.

Each compensator can now be applied to the type 1 or type 0 model to approximate how each compensator affects the dominant roots of the actual system.

To design a control system using root locus methods, one should understand the relationship between the root locus and the resulting time response. Suppose we have a type 1 $G_p(s)$ resulting in the closed-loop $T(s)$ of Equation (5.1).

If the damping ratio is between zero and one, complex conjugate poles of $T(s)$ result.

$$\begin{aligned} s^2 + 2\zeta\omega_n s + \omega_n^2 &= (s+\sigma+j\omega)(s+\sigma-j\omega) \\ &= s^2 + 2\sigma s + \sigma^2 + \omega^2 \end{aligned}$$

Figure 5.4 shows the complex conjugate closed-loop poles $T(s)$ denoted by two boxes. From the figure and the characteristic polynomial, the damping ratio and the angle are related.

$$\zeta = \frac{\sigma}{\omega_n} = \cos\phi$$

Notice that ω_n is the length of the vector from the origin to the complex conjugate poles. Figure 5.5 shows loci of constant ζ with increasing ω_n (straight lines projecting outward) and loci of constant ω_n with increasing ζ (semicircular arcs).

Relationships between closed-loop poles, rise time, and settling time.

Ultimately, a system performs in the time domain; hence, it is logical that root locus design should be considered a means to provide a good time response. When the closed-loop poles are moved as in Figure 5.5, the rise time and settling time of

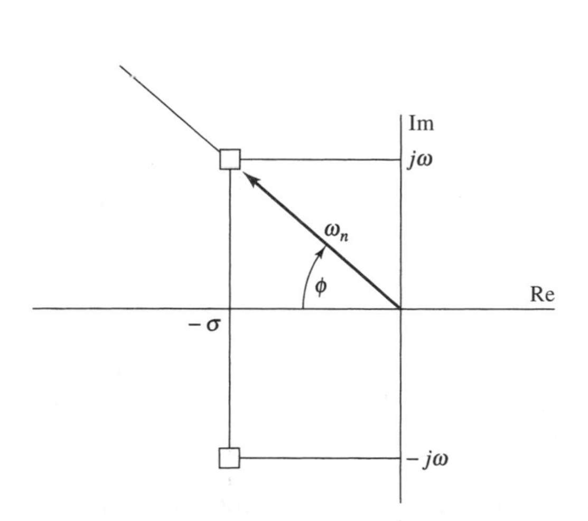

Figure 5.4 Complex conjugate poles.

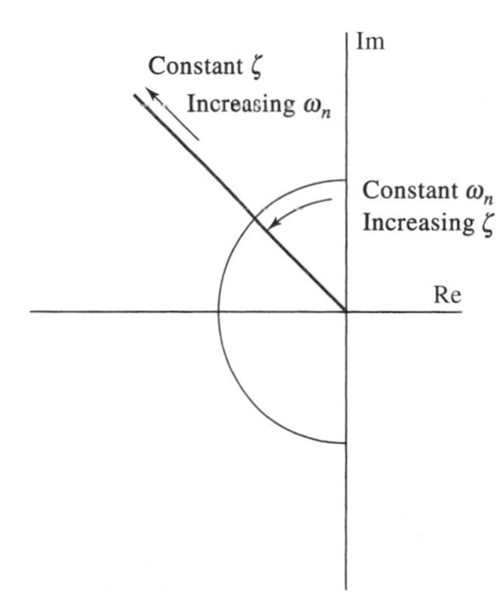

Figure 5.5 Changes in damping ratio and undamped natural frequency.

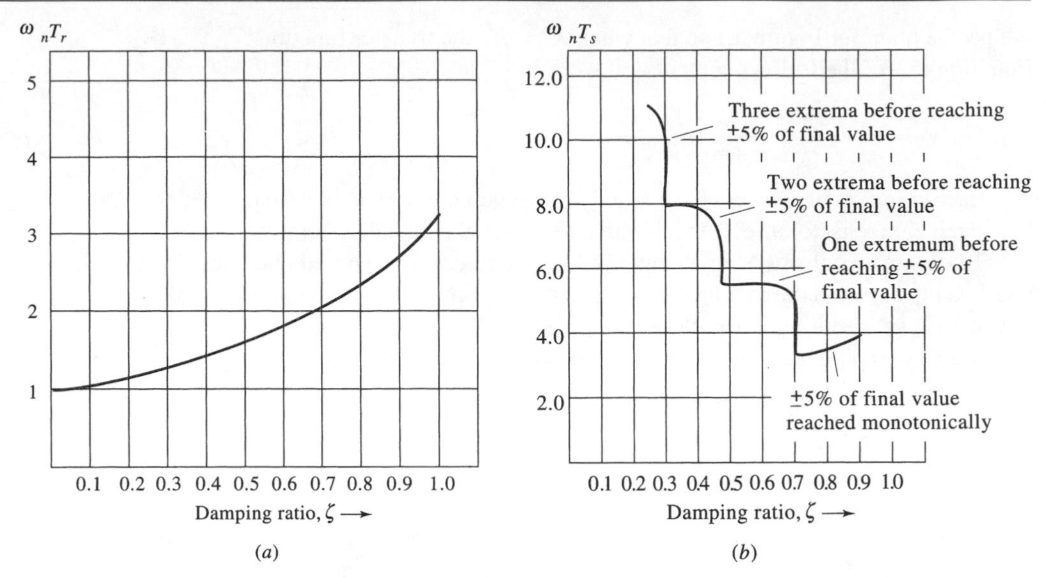

Figure 5.6 Rise time and settling time. (a) Rise time versus damping ratio. (b) Settling time versus damping ratio.

the dominant roots change as in Figure 5.6 (a) and (b), respectively. These figures predict that for a constant damping ratio, the rise time and settling time are inversely proportional to ω_n, since the vertical axes depend on ω_n For example, Figure 5.7(a) shows the unit step- response of a second-order underdamped system with $\zeta = 0.5$ and $\omega_n = 2.0$. From Figure 5.6(a) we can correctly predict that $T_r = 0.8\ s$, and from Figure 5.6(b) we can correctly predict that $T_s = 2.6\ s$. If the damping ratio remains at 0.5 while ω_n doubles, then rise time and settling time are halved as in Figure 5.7(b). Therefore, if the damping ratio is fixed and we want to reduce rise time and settling time, the root locus design must cause ω_n, and therefore the relative stability (distance from the vertical axis to the dominant closed-loop poles), to be as large as possible.

Selecting damping ratio for minimum settling time.

Suppose that ω_n is constant, but the damping ratio varies as in Figure 5.5. Figure 5.6(b) suggests that settling time can be minimized by selecting the damping ratio to be 0.707. Figure 5.7(c) has the same undamped natural frequency, as Figure 5.7(a), but the damping ratio rises from 0.5 to 0.707 so that the settling time drops from 2.6 s to 1.5 s, the minimum possible. If the damping ratio rises to 0.9, then the settling time rises to 2.0 s as in Figure 5.7(d).

In summary, the design strategy to be followed in this chapter regarding stability is to maximize the relative stability of the dominant poles when that is possible and to make the damping ratio 0.707 for the dominant poles when that choice is possible. Steady state performance will be shaped by modifying the error coefficients.

Other design strategies are possible, since design must be specific to the needs of each product and since economic and other practical considerations may result in other design requirements. For example, it may be necessary to design for some specific damping ratio to ensure some percentage overshoot and some

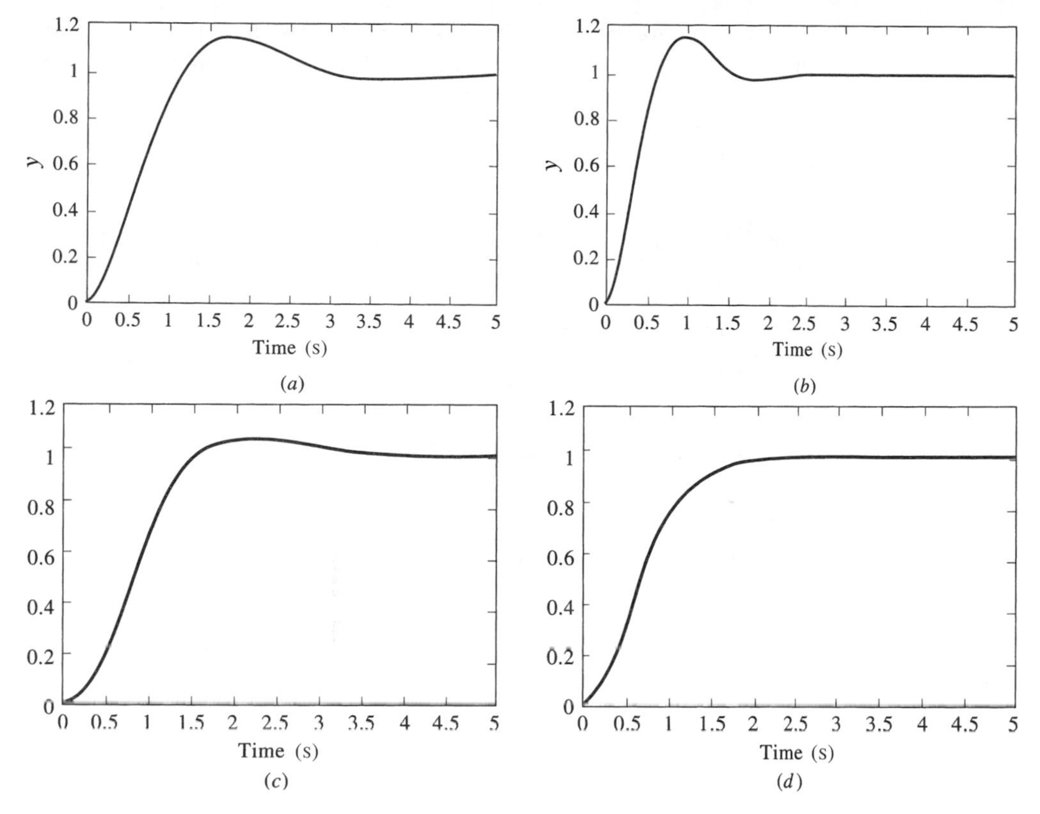

Figure 5.7 Unit step responses. (a) $\zeta = 0.5$, $\omega_n = 2$, $T_r = 0.8$ s, $T_s = 2.6$ s; (b) $\zeta = 0.5$, $\omega_n = 4$, $T_r = 0.4$ s, $T_s = 1.3$ s; (c) $\zeta = 0.707$, $\omega_n = 2$, $T_s = 1.5$ s, (d) $\zeta = 0.9$, $\omega_n = 2$, $T_s = 2.0$ s.

rise-time-to-settling-time relationship. We cannot possibly cover all design options; so we will focus on the objectives identified in the preceding paragraph.

❑ DRILL PROBLEMS

D5.6 (a) For the following characteristic polynomial, indicate the damping ratio and undamped natural frequency. Using Figures 5.6 and 5.7, estimate the rise time and settling time.

$$s^2 + 2s + 4$$

Ans. 0.5, 2, 0.85, 2.6

(b) Repeat for $s^2 + 3.2s + 4$

Ans. 0.8, 2, 1.2, 1.6

(c) Repeat for $s^2 + 4s + 16$

Ans. 0.5, 4, 0.425, 1.3

D5.7 (a) For a second-order system it is found that the rise time is 0.5 s and the settling time is 2.6 s. Use Figures 5.6 and 5.7 to estimate the damping ratio and undamped natural frequency.

Ans. 0.4, 3

(b) Repeat for a rise time of 0.3 s and a settling time of 1.56 s.

Ans. 0.4, 5

D5.8 (a) For the closed-loop system in Figure D5.8, what are the rise time and settling time for $K = 4$?

Ans. 0.85, 2.5

(b) Repeat Drill Problem D5.6 for $K = 16$.

Ans. 0.3, 2.8

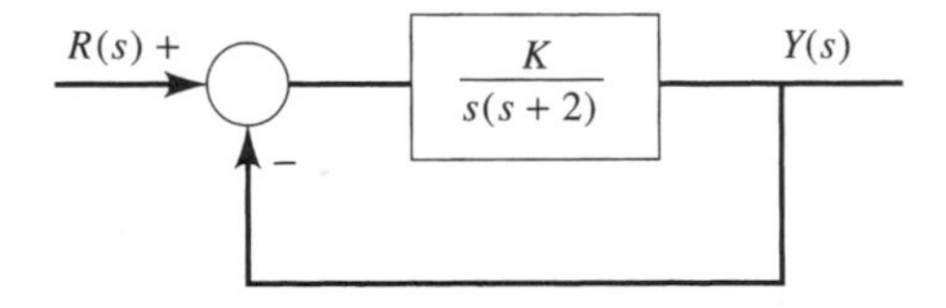

Figure D5.8

D5.9 (a) What range of values for K causes the rise time to be less than 0.8 s for the system of Drill Problem D5.8?

Ans. $K > 4$

(b) What value for K causes the minimum settling time for the system of Drill Problem D5.8?

Ans. $K = 2$

5.5 An Uncompensated Example System

In comparing the benefits of the different compensators, we shall use one example system throughout this chapter. The plant transmittance is

$$G_p(s) = \frac{1}{s(s+2+j)(s+2-j)} = \frac{1}{s(s^2+4s+5)}$$

A variety of computer programs such as MATLAB exist with which the root locus can be found as in Figure 5.8(a).

Motion of the dominant roots as K increases.

The dominant roots of Figure 5.8 move generally upward and then toward the right half-plane as K increases. For increased K the damping ratio of the dominant roots decreases from 1 to 0. For each K the ramp error coefficient is found by

$$K_1 = \lim_{s \to 0} sKG_p(s) = K/5$$

As K increases, the ramp error coefficient increases, the ramp error decreases, and the dominant roots become more lightly damped. Section 5.3 suggests that it is good design to select the damping ratio to be 0.707 (minimizing settling time) for a given ω_n when that is appropriate—which is the case here, since the damping ratio changes while ω_n is reasonably constant. To select K for 0.707 damping, a computer can be

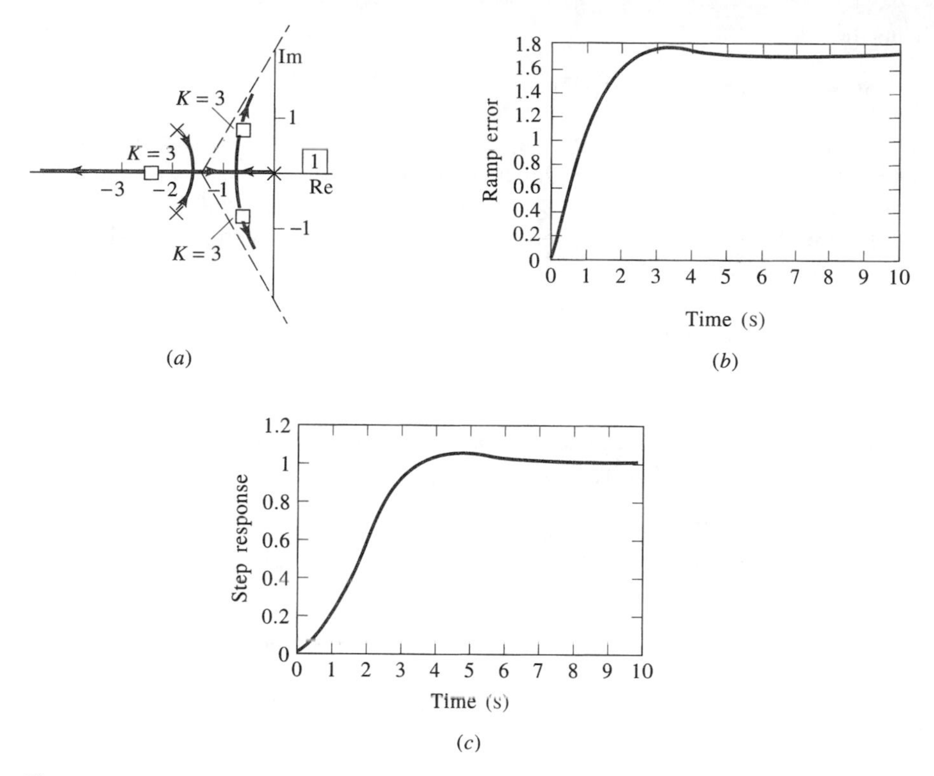

Figure 5.8 Uncompensated system performance. (a) Root locus. (b) Ramp error. (c) Step response.

used to solve the roots of the characteristic polynomial. The transfer function is

$$T(s) = \frac{KG_p(s)}{1 + KG_p(s)} = \frac{K}{s^3 + 4s^2 + 5s + K}$$

so the characteristic polynomial is $s^3 + 4s^2 + 5s + K$. For various K, the dominant roots and damping ratio can be found as follows:

K	Roots	ζ	ω_n
2	$-2, -1, -1$	1.00	1.00
2.5	$-2.29, -0.85 \pm j0.60$	0.82	1.04
2.8	$-2.41, -0.80 \pm j0.73$	0.74	1.08
2.9	$-2.44, -0.78 \pm j0.76$	0.72	1.09
3.0	$-2.47, -0.77 \pm j0.79$	0.70	1.10

The value of $K = 3$ provides a damping ratio closest to 0.707, hence the design of the uncompensated system ends with $K = 3$, which we shall designate as K_u in later sections of this chapter, and the resulting ramp error coefficient K_1 is $K/5 = 3/5 = 0.6$.

Select $K = 3$.

Relative stability = 0.707.

Thus the

Ramp error is 1.67.

$$\text{Steady state error to a ramp input} = \frac{1}{K_1} = 1.67$$

Ramp error.

A larger value of K would result in a larger error coefficient at the penalty of a smaller damping ratio and thus a larger settling time. It is not possible to increase K_1 and also reduce settling time without using a more complicated compensator.

Figure 5.8(b) shows the error versus time for a ramp input with $K = 3$ where the error $e = r - y$ so that

$$T_E(s) = 1 - T(s) = \frac{s(s^2 + 4s + 5)}{s^3 + 4s^2 + 5s + K}$$

$$E(s) = T_E(s)/s^2$$

If a computational package does not allow for a ramp response, it is possible to use the step response of $T'_E(s)$.

$$T'_E(s) = \frac{(s^2 + 4s + 5)}{s^3 + 4s^2 + 5s + K} = \frac{T_E(s)}{s}$$

$$E(s) = T'_E(s)/s$$

Step error.

Figure 5.8(c) shows the response of the uncompensated system when the input $r(t)$ is a unit step. There is no steady state error for a step input, since the uncompensated system is type 1. The damping ratio of 0.707 provides minimum settling time.

The following sections cover the design of more complex compensators. Rather than limit the discussion of compensation to this one example $G_p(s)$, we shall now examine how each compensator influences the dominant second-order closed-loop poles emerging from any uncompensated system design. The design rules are thus generalized and applicable to any system having such dominant roots. If a plant has open-loop zeros, it is assumed that those zeros are canceled by appropriately added open-loop poles.

❑ DRILL PROBLEM

D5.10 Each of the following uncompensated systems are of the form of Figure 5.1. Use a computer programs such as MATLAB to select K so that the dominant closed-loop poles are 0.707 damped. Identify K, the closed-loop poles, K_0 (the step error coefficient), K_1 (the ramp error coefficient), and the settling time for the step response.

(a)

$$G_p(s) = \frac{1}{s(s+2)}$$

Ans. $2; -1 \pm j1; \infty; 1; 3.2$

(b)

$$G_p(s) = \frac{1}{s(s+2)(s+4)}$$

Ans. 5.2; $-0.76 \pm j0.76$, -4.5; ∞; 0.65; 4.4

(c)

$$G_p(s) = \frac{s+8}{(s+1)(s+2)(s+6)}$$

Ans. 1.64; $-1.43 \pm j1.43$, -6.1; 1.09; 0; 2.3.

5.6 Cascade Proportional Plus Integral (PI)

5.6.1 General Approach to Compensator Design

Tables 5.1 and 5.2 contain the basic features of a number of compensators. These compensators will be examined one by one, as applied to a second-order model. The design procedures we learn are then applied to the third-order plant introduced in Section 5.5. The designs of the various compensators are summarized in Table 5.3.

How shall we proceed to design the various compensators? That is, can we find a systematic approach to choose parameter values? The following approach is used in this chapter.

1. Select $K = K_u$ for the uncompensated system. K_u usually is chosen to trade off steady state and stability characteristics.

Table 5.1 *Common Types of Compensator*

Compensator	Transmittance	Typical Effect on Steady State Errors	Typical Effect on Relative Stability
Cascade PI	$G_c(s) = \frac{K(s+a)}{s}$	Greatly improved	Reduced
Cascade lag	$G_c(s) = \frac{K(s+a)}{s+b}$ $b < a$	Improved	Reduced
Cascade lead	$G_c(s) = \frac{K(s+a)}{(s+b)}$ $a < b$	Somewhat improved or somewhat worse	Increased
Cascade lag–lead	$G_c(s) = K\left(\frac{s+a}{s+b}\right)_{\text{lag}} \times \left(\frac{s+a}{s+b}\right)_{\text{lead}}$	Improved	Increased
Rate feedback (PD)	$H_c(s) = 1 + As$	Somewhat improved or somewhat worse	Increased
Proportional-integral-derivative (PID)	$G_c(s) = \frac{K(s+a)}{s} + KAs$	Greatly improved	Increased

Table 5.2 *Common Types of Compensator*

Compensator	Change in $n - m$	Change in Centroid if $n - m$ is Constant	Change in Type Number	$\frac{K_i\,(\text{comp})}{K_i\,(\text{uncomp})}$
Cascade PI	0	$\frac{a}{n-m}$	+1	∞
Cascade lag ($a > b$)	0	$\frac{a-b}{n-m}$	0	$\frac{K}{K_u}\left(\frac{a}{b}\right)$
Cascade lead ($b > a$)	0	$\frac{-(b-a)}{n-m}$	0	$\frac{K}{K_u}\left(\frac{a}{b}\right)$
Cascade lag–lead	0	$\frac{(a-b)_{\text{lag}} - (b-a)_{\text{lead}}}{n-m}$	0	$\frac{K}{K_u}\left(\frac{a}{b}\right)_{\text{lag}}\left(\frac{a}{b}\right)_{\text{lead}}$
Rate feedback (PD)	−1		0	$\frac{1}{K_u/K + AK_1\,(\text{uncomp})}$
Proportional-integral-derivative	−1		+1	∞

Table 5.3 *Summary of Designs*

Compensator	Transmittance	Ramp Error Coefficient	Closed-Loop Zeros	Closed-Loop Poles	Settling Time	% Over-shoot
Uncompensated (P)	$K = 3$	0.6	None	$-0.767 \pm j0.793$, -2.47	3.0	4
Cascade PI	$G_c(s) = \frac{4.9(s+0.236)}{s}$	∞	-0.236	-0.313, $-0.452 \pm j1.06$, -2.78	5.3	4.8
Cascade lag	$G_c(s) = \frac{s(s+0.251)}{s+0.0251}$	10	-0.251	-0.325, $-0.450 \pm j1.8$ -2.80	5.1	47
Cascade lead	$G_c(s) = \frac{40(s+3.47)}{s+34.7}$	0.8	-3.47	-1.90, $-1.03 \pm j1.02$ -34.7	2.5	3
Cascade lag–lead	$G_c(s) = \frac{60(s+3.47)}{s+34.7} \times \frac{s+0.251}{s+0.0251}$	12	-0.251 -3.47	$-0.315, -2.14$ $-0.759 \pm j1.29$ -34.8	4.8	31
Rate feedback (PD)	$H_c(s) = 0.51(s+1.96)$ $K = 6$	0.72	None	-1.33 $-1.33 \pm j1.65$	2.5	0
PID	$G_c(s) = \frac{12(s+0.274)(s+1.69)}{s}$	∞	-0.274, -1.69	-0.322 -1.282 $-1.20 \pm j2.33$	4.6	24

2. Choose a compensator that can improve some undesirable trait of the uncompensated system.
3. With appropriate K obtain the root locus as a parameter of the compensator varies. Choose that parameter to trade off steady state performance and stability characteristics.
4. With the parameter chosen, obtain the root locus as K varies. Obtain the time response for the design value of K. Vary K to see if time response can be improved.
5. Vary the parameter(s) and K as needed to meet design goals, repeating steps 3 and 4.

Five-step design process.

5.6.2 Cascade PI compensation

We begin with design of the first compensator listed in Table 5.1 and 5.2

The cascade proportional plus integral (PI) compensator has a transmittance of the form

$$G_c(s) = A_p + \frac{A_i}{s} = \frac{K(s+a)}{s}$$

The A_p term creates an output from the compensator that is *proportional* to the compensator input, while the A_i/s term creates an output from the compensator that is the *integral* of the compensator input, hence the name *PI compensator.*

The PI compensator is defined.

The second form for the PI compensator uses a gain K and the polynomial factor a. The PI compensator thus adds a pole at $s = 0$ and a zero at $s = -a$ to the open-loop transmittance $G_c(s)G_p(s)$. For purposes of design we will select K and a.

System type increases by 1 because of the added open-loop pole at the origin. Hence, for a stable design, steady state error performance is improved. This compensator, having both a pole and a zero, does not change the difference between the number of open-loop poles and open-loop zeros. Thus the compensated system has the same root locus asymptotic angles as does the uncompensated one.

Effect on steady state error.

The centroid of the asymptotes does change, however. For the compensated system with open loop transmittance $G_c(s)G_p(s)$,

$$\sigma = \frac{\Sigma \text{ poles of } G_cG_p - \Sigma \text{ zeros of } G_cG_p}{\text{number of poles of } G_cG_p - \text{number of zeros of } G_cG_p}$$

As the compensator contributes a pole at $s = 0$ and a zero at $s = -a$,

$$\sigma = \frac{\Sigma \text{ poles of } G_p - \Sigma \text{ zeros of } G_p + a}{\text{number of poles of } G_p - \text{ number of zeros of } G_p}$$

$$= \sigma_u + \frac{a}{\text{number of poles of } G_p(s) - \text{ number of zeros of } G_p(s)}$$

Effect of PI compensator on the centroid.

where σ_u is the centroid for the uncompensated system. For a positive a, σ is moved to the right, tending to reduce relative stability from that of the uncompensated feedback system by an amount proportional to a.

For the cascade PI compensator, there are two adjustable parameters, K and a, and the designer may alternate between adjusting one parameter then the other until satisfactory design results.

Selecting K and a for type 1 model.

To select K and a using some generally applicable rules, suppose we apply the PI compensator to the type 1 model from Section 5.4 as in Figure 5.9(a). It is assumed that a value of K_u has been selected for the uncompensated system which creates closed-loop dominant poles with a damping ratio ζ and undamped natural frequency ω_n. If $K = K_u$ and $a = 0$, then the transfer function becomes that of the uncompensated system dominant poles given by Equation (5.1). For fixed K, the variable a is adjusted for stability. Here the compensator pole at the origin raises the system type number, thus eliminating step error for a type 0 plant, ramp error for a type 1 plant, and so on. For step 3, we can leave K at the value K_u and focus on selecting a for stability.

If we let $K = K_u$ while a varies, the transfer function becomes

$$T(s) = \frac{(s+a)\omega_n^2}{s^3 + 2\zeta\omega_n s^2 + \omega_n^2 s + \omega_n^2 a}$$

$$= \frac{\text{numerator}}{1 + a[\omega_n^2/s(s^2 + 2\zeta\omega_n s + \omega_n^2)]}$$

Root locus for variable a, type 1 model.

The equivalent $G(s)H(s)$ required to obtain the root locus for variable a is

$$aG(s)H(s) = \frac{a\omega_n^2}{s(s^2 + 2\zeta\omega_n s + \omega_n^2)} \quad \textbf{[5.3]}$$

The root locus for variable a is shown in Figure 5.9(b). For $a = 0$ there are closed-loop poles at

$$-\zeta\omega_n \pm j\omega_n\sqrt{1-\zeta^2}, 0$$

and there is a closed-loop zero at the compensator zero (which is zero for $a = 0$). After the closed-loop pole and zero at the origin have been canceled, the dominant closed-loop poles remain at the uncompensated values. As a increases, one closed-loop pole moves to the left while the complex conjugate closed-loop poles move to the right.

Choose a for maximum relative stability for a type 1 system.

A good design choice for a would be that value of a for which the closed-loop poles have maximum relative stability. For the uncompensated system, the relative stability is $\zeta\omega_n$. The maximum relative stability for the compensated case would be a value that would cause all three closed-loop poles to have the same real part $(-c)$. If a is increased beyond the value for a relative stability of c, the relative stability of the complex conjugate roots would be less than c. Also, if a is decreased, the relative stability of the real root would be less than c.

To determine the best value of a with $K = K_u$, equate the characteristic polynomial

$$s^3 + 2\zeta\omega_n s^2 + \omega_n^2 s + \omega_n^2 a$$

to the value of the characteristic polynomial if the roots are $-c$ and $-c \pm jd$

$$(s+c)(s+c+jd)(s+c-jd) = s^3 + 3cs^2 + (3c^2+d^2)s + c(c^2+d^2)$$

By equating the s^2 coefficients, we obtain

$$c = \frac{2}{3}\zeta\omega_n$$

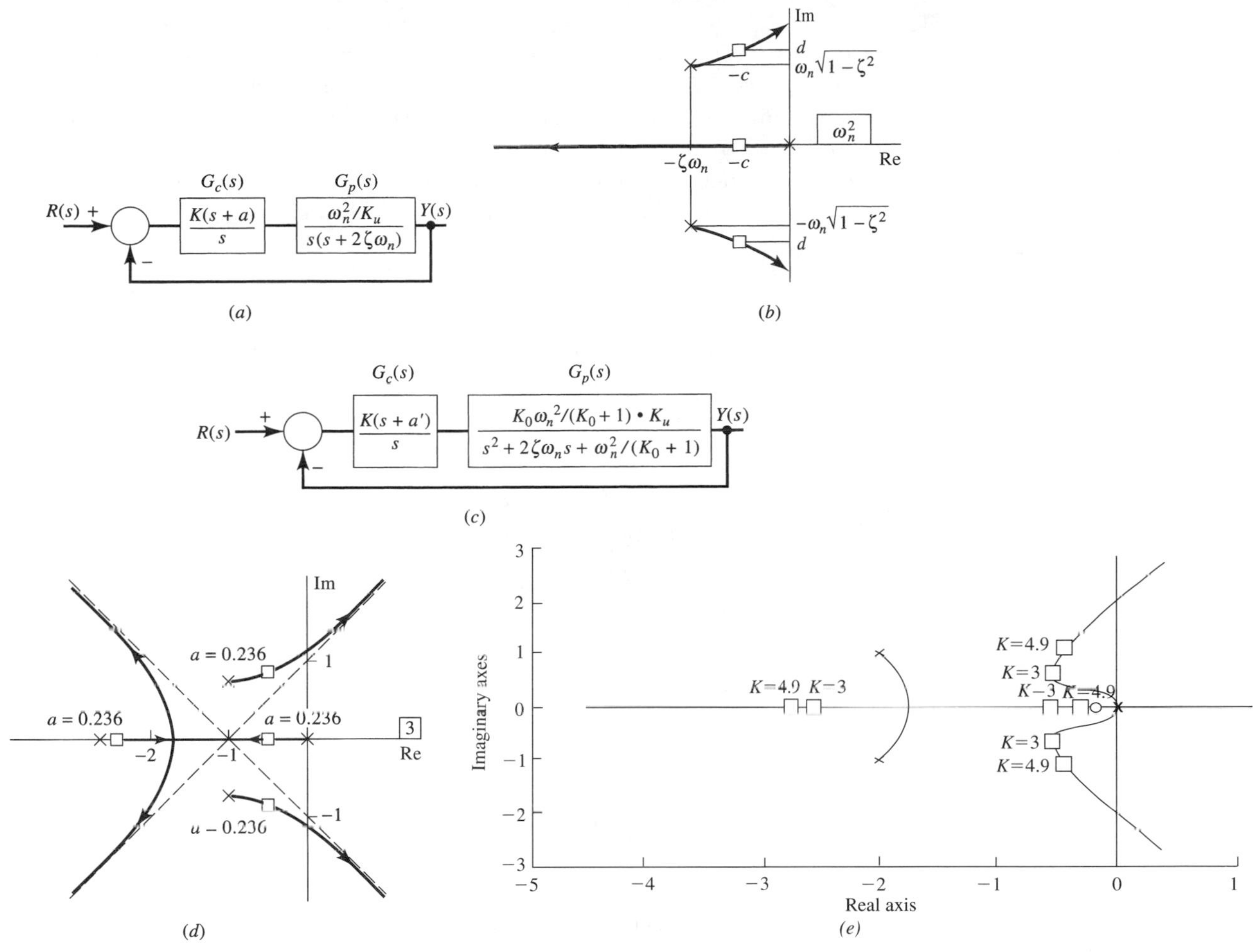

Figure 5.9 Proportional plus integral (PI) compensation. (a) PI compensation of the dominant roots (type 1 system). (b) Root locus of the model system of (a) for $K = K_u$, variable a. (c) PI compensation of the dominant roots (type 0 system). (d) Root locus of the example system for $K = 3$, variable a. (e) Root locus of the example system for $a = 0.236$, variable K.

This means that the relative stability becomes $\left(\frac{2}{3}\right)\zeta\omega_n$ a reduction by $\frac{1}{3}$ from the uncompensated value (however, the ramp error coefficient is now ∞ because the system is type 2).

Relative stability is $\frac{1}{3}$ less than for uncompensated systems.

Figure 5.9(c) applies a PI compensator to a type 0 model. If $K = K_u$, a root locus for variable a' can be constructed using the effective $G(s)H(s)$

$$a'G(s)H(s) = \frac{a'\omega_n^2 K_0/(K_0+1)}{s(s^2+2\zeta\omega_n s+\omega_n^2)} \qquad [5.4]$$

If Equations (5.3) and (5.4) are compared, then

$$a' = \frac{a(K_0+1)}{K_0} \qquad [5.5]$$

Best a'.

for the same closed-loop poles. As a result, a root locus for variable a, resulting from a type 1 system is exactly the same as the root locus for a type 0 system except that a' (for type 0) differs from a (for type 1) by a factor depending on K_0. As K_0 approaches infinity (approaching type 1 operation) then $(K_0 + 1)/K_0$ approaches one and a' approaches a, the value for type 1 operation.

PI compensator is applied to example system.

We can now apply a PI compensator to the example system from Section 5.5. If the gain $K = K_u = 3$, then the compensator for variable a is

$$G_c(s) = \frac{3(s+a)}{s}$$

giving an overall system transfer function

$$\begin{aligned} T(s) &= \frac{G_c(s)G_p(s)}{1 + G_c(s)G_p(s)} \\ &= \frac{3(s+a)/s[1/s(s^2+4s+5)]}{1 + 3(s+a)/s[1/s(s^2+4s+5)]} \\ &= \frac{3(s+a)}{s^2(s^2+4s+5) + 3s + 3a} \\ &= \frac{\text{numerator}}{1 + [3a/(s^2(s^2+4s+5)+3s)]} \\ &= \frac{\text{numerator}}{1 + [3a/s(s^3+4s^2+5s+3)]} \\ &= \frac{\text{numerator}}{1 + [3a/s(s+0.767+j0.793)(s+0.767-j0.793)(s+2.47)]} \end{aligned} \quad \text{[5.6]}$$

With K equal to the uncompensated value of 3 and a adjustable, two of the open-loop poles are the dominant closed-loop poles of the uncompensated system (the poles at $-0.767 \pm j0.793$). As a increases in the root locus of Figure 5.9(d) those two dominant roots migrate into the right half-plane.

Choose $a = 0.236$. Relative stability is 0.543 when $K = 3$.

The value of a for maximum relative stability is $a = 0.236$. The closed-loop poles are at -0.543, $-0.543 \pm j0.506$, and -2.37. A closed-loop zero is at -0.236 due to the forward-path cascade compensator zero. The relative stability is 0.54 (30% less than for the uncompensated case).

Vary K to improve performance.

Choose $K = 4.9$. Settling time is 5.3 s. Ramp error is zero.

Step 4 of the design process from Section 5.6.1 requires the root locus as K varies, given the chosen value of a. See Figure 5.9(e). If we take the step response with $K = 3$, the settling time is 9.0, quite a bit longer than the settling time of 3.0 s for the uncompensated system. If K is increased to 4, the settling time drops to 6.3 s. When $K = 4.9$, the settling time drops to 5.3 s. Above $K = 4.9$, the settling time increases. We could try an additional iteration upon a and K but the process will be ended here (with $K = 4.9$), so we can explore other compensators. The ramp error response for $K = 4.9$ is shown in Figure 5.10(a), The step response is shown in Figure 5.10(b). The closed-loop poles are the design values, summarized in Table 5.3.

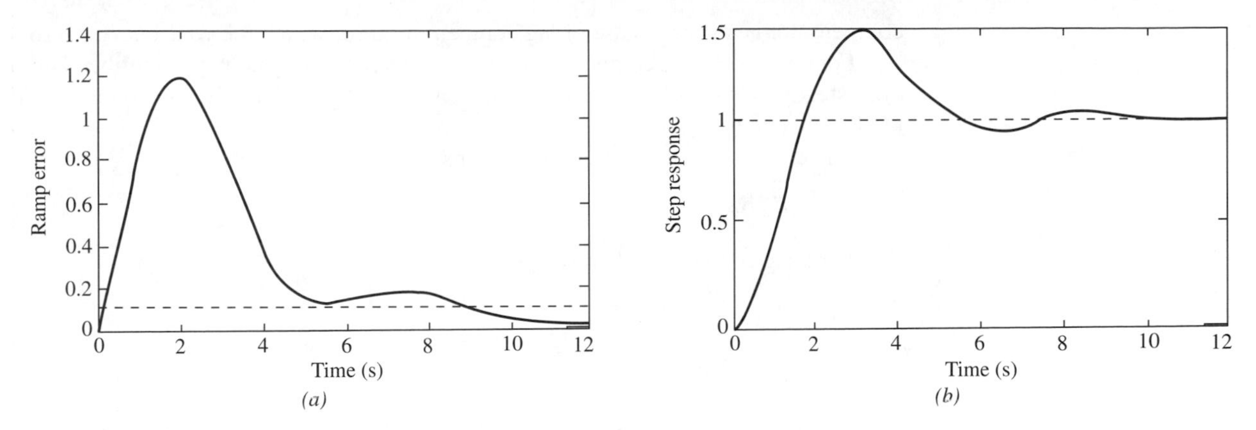

Figure 5.10 Example system with PI compensation. (a) Ramp error. (b) Step response.

❑ **DRILL PROBLEM**

D5.11 For each of the plants from Drill Problem D5.10, follow the design procedure for a cascade PI compensator and use a computer program to select a for maximum relative stability. Identify a, the closed-loop poles, K_0 (the step error coefficient), K_1 (the ramp error coefficient), and the settling time for the step response.

(a)

$$G_p(s) = \frac{1}{s(s+2)} \qquad K = K_u = 2$$

Ans. 0.37; $0.67 \pm j0.82$, -0.67; ∞; ∞; 5.8

(b)

$$G_p(s) = \frac{1}{s(s+2)(s+4)} \qquad K = K_u = 5.2$$

Ans. 0.251; $-0.52 \pm j0.55$, -0.52, -4.45; ∞; ∞; 8.5

(c)

$$G_p(s) = \frac{s+8}{(s+1)(s+2)(s+6)} \qquad K = K_u = 1.64$$

Ans. 0.98; $-0.96 \pm j1.12$, -0.96, -6.12; ∞; 1.07; 3.5

5.7 Cascade Lag Compensation

A cascade lag compensator has transmittance of the form

$$G_c(s) = \frac{K(s+a)}{s+b}$$

where K, a, and b are positive constants and $a > b$ so that the compensator zero is to the left of the compensator pole on the complex plane. The proportional-plus-integral

The cascade lag compensator is defined.

compensator is a special case of lag compensation for which the constant b is zero.

For nonzero b, this compensator does not increase the system type number. However, steady state error performance can be improved over that of the uncompensated feedback system.

The error coefficient for an uncompensated system is

$$K_i = K_u \lim_{s\to 0} s^i G_p(s)$$

For a lag compensated system, we write

$$K_i = \lim_{s\to 0} s^i G_c(s) G_p(s)$$

$$= K \left(\frac{a}{b}\right) \lim_{s\to 0} s^i G_p(s)$$

Effect of lag compensator on steady state accuracy.

The ratio of the error coefficient of the uncompensated system to the error coefficient for the cascade lag compensated system is

$$\frac{K_i\,(\text{compensated})}{K_i\ (\text{uncompensated})} = \frac{K}{K_u}\left(\frac{a}{b}\right)$$

The centroid for a cascade lag compensated system moves to the right by

$$\frac{a-b}{n-m}$$

One design approach is to select the ratio a/b to equal the factor by which the error coefficient is to be increased. If that factor is d, then

$$b = \frac{a}{d} \qquad d > 1$$

$$K = K_u$$

The error coefficient increases by the factor d.

$$K_i(\text{comp}) = K_i(\text{uncomp})\, d$$

If d approaches infinity, the cascade lag compensator becomes cascade PI; however, a large value of d creates a problem in fabricating a reliable RC circuit with a zero and a pole widely separated. In practice, the value of d is chosen around 10.

To realize a lag compensator, d should be limited to 10.

In chapter 1, Figure 1.14(d) shows an operational amplifier realization of a lag compensator employing one capacitor and three resistors. The ratio d is equal to $1 + R_2/R_1$. If we choose $d = 10$, then R_2/R_1 is 9. If we choose $d = 100$, then $R_2/R_1 = 99$. The larger is d, the more orders of magnitude exist among resistors in the lag circuit. Since it is desirable (from the standpoint of reliability) to keep similar components within one order of magnitude, d is limited to 10.

Selecting K, a, and d for a second-order model.

As for the PI compensator, the lag compensator can be applied to a second-order model to create general design rules for selecting K, a, and b. Since d is usually large while b is small, the design rules for a PI compensator (with $b = 0$ and $K = K_u$) are quite close to what we would get if the procedures in Section 5.6 were repeated with $K = K_u$ and d large. That is, we select a for maximum relative stability if d is fixed (perhaps at 10) and K is fixed (perhaps at K_u). Then, we examine the root locus and step response with K being the variable while d and a are fixed.

Selecting K, a, and d for the example system.

We can now consider the type 1 example system. With $K = K_u = 3$ for the lag compensated system, the effective $G(s)H(s)$ for variable a (with $b = a/d$) is found from $T(s)$:

$$T(s) = \frac{G_c(s)G_p(s)}{1 + G_c(s)G_p(s)}$$

$$aG(s)H(s) = \frac{a[(1/d)s^3 + (4/d)s^2 + (5/d)s + 3]}{s(s^3 + 4s^2 + 5s + 3)} \qquad \textbf{[5.7]}$$

Suppose we want to increase the error coefficient by a factor of 10 over the uncompensated case. Then we should choose $K = K_u = 3$ and $d = 10$.

Thus

$$aG(s)H(s) = \frac{a(0.1)(s^3 + 4s^2 + 5s + 30)}{s(s^3 + 4s^2 + 5s + 3)}$$

$$= \frac{a(0.1)(s + 0.20 + j2.6)(s + 0.20 - j2.6)(s + 4.4)}{s(s + 0.767 + j0.793)(s + 0.767 - j0.793)(s + 2.47)}$$

Results for the example.

The root locus for variable a is shown in Figure 5.11(a). The value of a for maximum relative stability is $a = 0.251$. The closed-loop poles are at 0.550 ± 0.523, -0.550, and -2.38. The root locus for $a = 0.251$ with variable K is shown in Figure 5.11(b).

Choose $a = 0.251$.

With $K = 5$, settling time is 5.1 s.

Following step 4 of the design procedure of section 5.6.1, we can vary K and evaluate the step response with $K = K_u = 3$, the settling time is 8.7 s. Upon increasing K to 5, the settling time drops to 5.1 s. Above $K = 5$, T_s increases, so we choose $K = 5$ as the design value. See Table 5.3 for the closed-loop poles. The ramp error coefficient is

$$K_1 = \frac{Kd}{5} = 10$$

Ramp error is 0.1.

The ramp error (steady state error = 0.1) is shown in Figure 5.11(c), while the step response is shown in Figure 5.11(d). Compared to the uncompensated system, the ramp error has been reduced by a factor of 16.6, while the step response exhibits less stability (higher rise time and percent overshoot).

The five steps of Section 5.6.1 may be summarized for this lag compensator design

1. Choose $K = K_u$ for 0.7 damping for the uncompensated system.
2. Select a lag compensator to improve steady state accuracy by a factor of about d.
3. Set $d = 10$. Find a for maximum relative stability.
4. Vary K (given d and a) to see if step response can be improved.
5. Repeat steps 3 and 4 as needed.

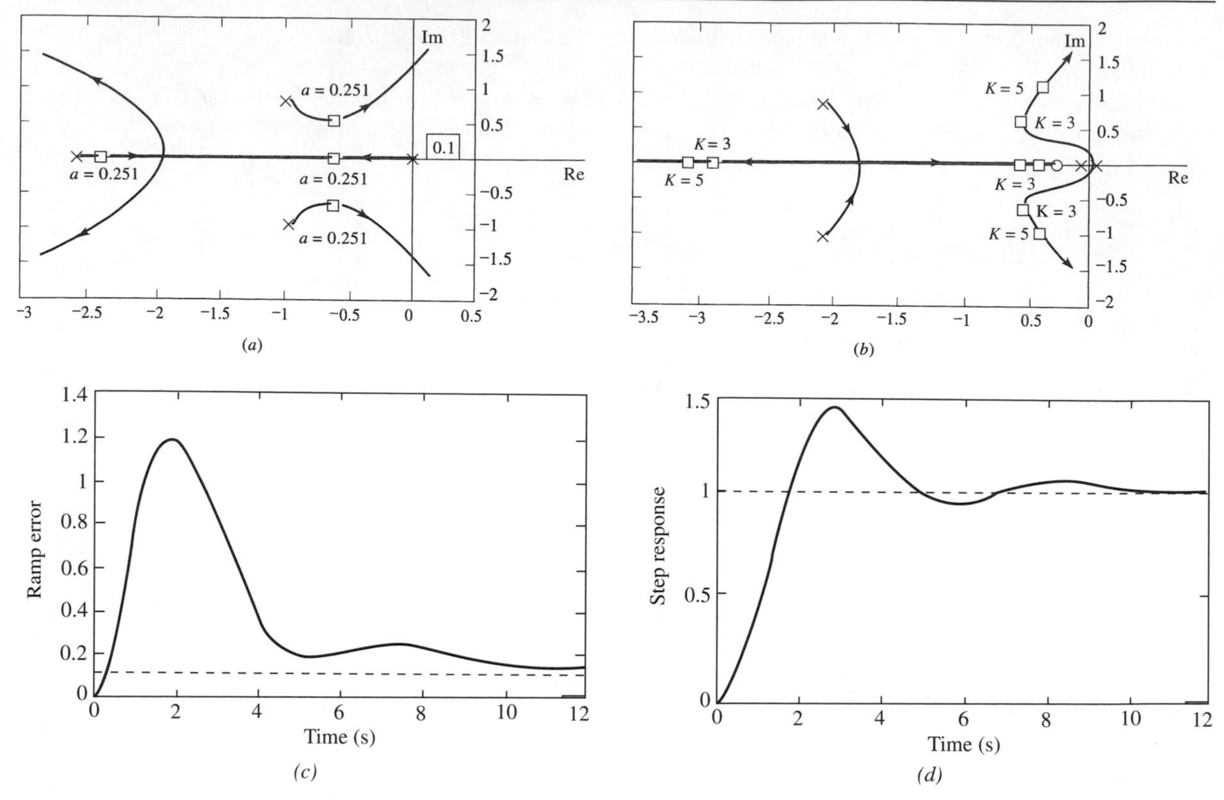

Figure 5.11 Example system with lag compensation. (a) Root locus for variable a, $K = K_u = 3$. (b) Root locus for variable K, $a = 0.251$. (c) Ramp error. (d) Step response.

❑ DRILL PROBLEM

D5.12 For each of the plants from Drill Problem D5.10, follow the design procedure for a cascade lag compensator to increase the appropriate error coefficient by a factor of 10. Use a computer program to select a for maximum relative stability. Identify a, the closed-loop poles, K_0 (the step error coefficient), K_1 (the ramp error coefficient), and the settling time for the step response.

(a)

$$G_p(s) = \frac{1}{s(s+2)} \qquad K = K_u = 2$$

Ans. $0.392; -0.68 \pm j0.83, -0.68; \infty; 10; 5.6$

(b)

$$G_p(s) = \frac{1}{s(s+2)(s+4)} \qquad K = K_u = 5.2$$

Ans. $0.267; -0.525 \pm j0.564, -0.525, -4.45; \infty; 6.5; 8.3$

(c)

$$G_p(s) = \frac{s+8}{(s+1)(s+2)(s+6)} \qquad K = K_u = 1.64$$

Ans. $0.995; -0.99 \pm j1.17, -0.99, -6.12; 10.9; 0; 3.4$

5.8 Cascade Lead Compensation

The cascade lead compensator is defined.

A cascade lead compensator has the same form as a cascade lag compensator

$$G_c(s) = \frac{K(s+a)}{s+b}$$

except that $b > a$. The compensator pole is to the left of the zero. The centroid moves to the left by

$$\frac{-(b-a)}{n-m}$$

Effect on asymptotic angles.

Since $n - m$ remains the same, the asymptote angles are unchanged. The root locus moves to the left. The lead compensator primarily influences system stability. The performance of the cascade lead compensator is more easily evaluated by replacing b with ad, where $d > 1$. The ith error coefficient of the compensated system is

$$K_i = \lim_{s\to 0} s^i G_c(s)G_p(s)$$
$$= \frac{K}{d}\lim_{s\to 0} s^i G_p(s)$$

The ratio of the error coefficient for the uncompensated system to the error coefficient for the cascade lead compensated system is

$$\frac{K_i(\text{compensated})}{K_i(\text{uncompensated})} = \frac{K/d}{K_u}$$

The value of K for the compensated system must be at least d times the value of K_u for the uncompensated system is order for the error coefficient to remain the same.

Three values must be selected: K, a, and d. Following Section 5.6.1, designer could select a value for K and d and then obtain a root locus for variable a. With a selected, other values of K and d could be tried until an acceptable design had been found.

Using the type 1 model to select K, a, and d.

We can get some help in understanding how to select K, a, and d by applying a lead compensator to the second-order model of a type 1 system. The lead compensated system is shown in Figure 5.12(a). The system reverts to the uncompensated form if $K = K_u d$ (the error coefficient is unchanged), and $a = \infty$, in which case $T(s)$ is

$$\frac{Y(s)}{R(s)} = \frac{\omega_n^2}{s^2 + 2\zeta\omega_n s + \omega_n^2}$$

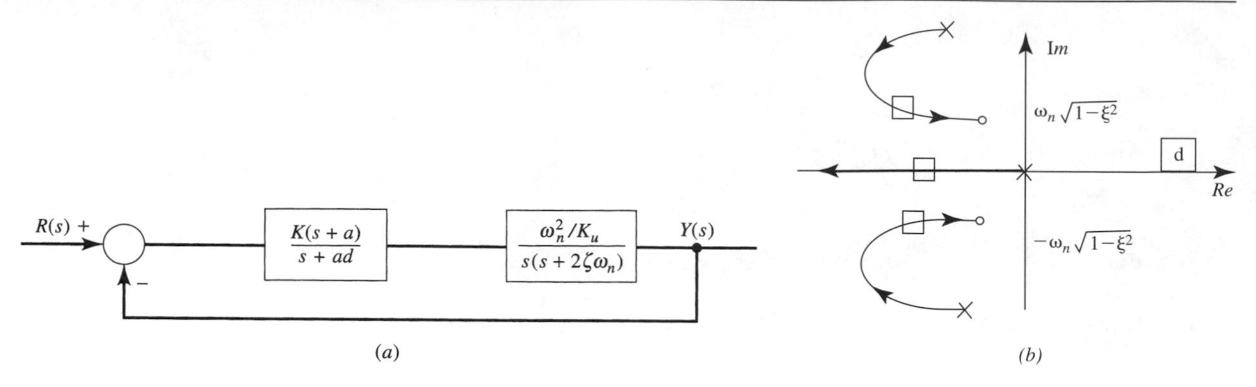

Figure 5.12 (a) Lead compensation of the uncompensated dominant roots. (b) Root locus for variable *a*.

In general, the transfer function can be written in terms of K, a, and d. To simplify the transfer function, it is convenient to define

$$K' = \frac{K}{K_u}$$

so that the transfer function becomes

$$\frac{Y(s)}{R(s)} = \frac{K'(s+a)\omega_n^2}{s^3 + s^2(2\zeta\omega_n + ad) + s(ad2\zeta\omega_n + K'\omega_n^2) + K'a\omega_n^2}$$

If K' and d are fixed, the root locus for adjustable a follows by factoring.

$$\frac{Y(s)}{R(s)} = \frac{\text{numerator}}{1 + \dfrac{ad[s^2 + 2\zeta\omega_n s + (K'/d)\omega_n^2]}{s(s^2 + 2\zeta\omega_n s + K'\omega_n^2)}}$$

The purpose of lead compensation is to improve the stability of a system—that is, to move the closed-loop roots to the left. Since the design of a lag compensator is supposed to improve steady state accuracy while minimizing the loss of relative stability, it is logical that the lead compensator should be designed to maintain the same steady state accuracy while improving stability. That is, a lag compensator is applied to a system with acceptable stability but poor steady state accuracy, while a lead compensator is applied to a system with acceptable steady state accuracy and poor stability. In this case steady state accuracy remains the same if we select $K' = d$. The root locus for adjustable a requires plotting

Effect on Steady state accuracy.

$$aG(s)H(s) = \frac{ad(s^2 + 2\zeta\omega_n s + \omega_n^2)}{s(s^2 + 2\zeta\omega_n s + d\omega_n^2)} \qquad [5.8]$$

The best choice for a should cause the dominant roots to be as far as possible to the left of the closed-loop uncompensated poles, which are the same as the open-loop zeros when $K' = d$ for variable a. In Figure 5.12(b) we can move the dominant roots to the left of the open-loop zeros and achieve maximum relative stability by placing the closed-loop poles as shown.

The design method we shall follow for lead compensation has five steps:

Design method.

1. Select $K = K_u$ (as before) for 0.7 damping for the uncompensated system.
2. Select a lead compensator to improve stability.
3. Select $K' = K/K_u = d$. This maintains steady state accuracy compared to the uncompensated case. A large value of d requires resistors differing by orders of magnitude as required by Figure 1.14(e). The pole/zero ratio is $1 + R_2/R_1$. Normally, d is in the range of 5 to 20. Choose $d = 10$ and find a for maximum relative stability.
4. Vary K (given a and d) to see whether step response can be improved.
5. Repeat steps 3 and 4 as needed.

Selecting K, a and d for the example system.

Lead compensation can now be applied to the example system. The closed-loop transfer function is

$$T(s) = \frac{G_c(s)G_p(s)}{1 + G_c(s)G_p(s)}$$

$$= \frac{K(s+a)}{s^4 + (4 + ad)s^3 + (5 + 4ad)s^2 + (K + 5ad)s + Ka}$$

With a as the variable $T(s)$ can be written

$$T(s) = \frac{\text{numerator}}{1 + \dfrac{ad(s^3 + 4s^2 + 5s + K/d)}{s(s^3 + 4s^2 + 5s + K)}}$$

For variable a the root locus follows by plotting

$$aG(s)H(s) = \frac{ad(s^3 + 4s^2 + 5s + K/d)}{s(s^3 + 4s^2 + 5s + K)} \qquad [5.9]$$

Choose d = 10.

We can now follow the design procedure just outlined. K_u has been selected to be 3. If we choose $d = 10$, as we did for the lag compensator, then to maintain the same ramp error coefficient as for the uncompensated case ($K_1 = 0.6$), $K' = K/K_u = d = 10$, so $K = 10K_u = 30$. The root locus for variable a follows from

$$aG(s)H(s) = \frac{ad(s^3 + 4s^2 + 5s + 3)}{s(s^3 + 4s^2 + 5s + 30)}$$

Vary K to improve performance.

Choose a = 3.47.

With K = 40, settling time is 2.5 s.

Ramp error is 1.25.

The root locus is shown in Figure 5.13(a). The value of a for maximum relative stability is 3.47. The closed-loop poles are at -1.33, $-1.33 \pm j0.712$, and -34.7. The relative stability of 1.33 is 73% higher than for the uncompensated case, in which the relative stability is 0.77. The root locus for variable K with $a = 3.47$ is shown in Figure 5.13(b), with $K = 30$; the settling time of the step response is 3.6 s with 0% overshoot. If we increase K to 40, settling time drops to 2.5 s, while the amount of overshoot is 3%. For higher K, the settling time rises (as does percent overshoot), so we can terminate design with the values shown in Table 5.3. With $K = 40$, we have

$$K_1 = \frac{K}{5d} = \frac{40}{50} = 0.8$$

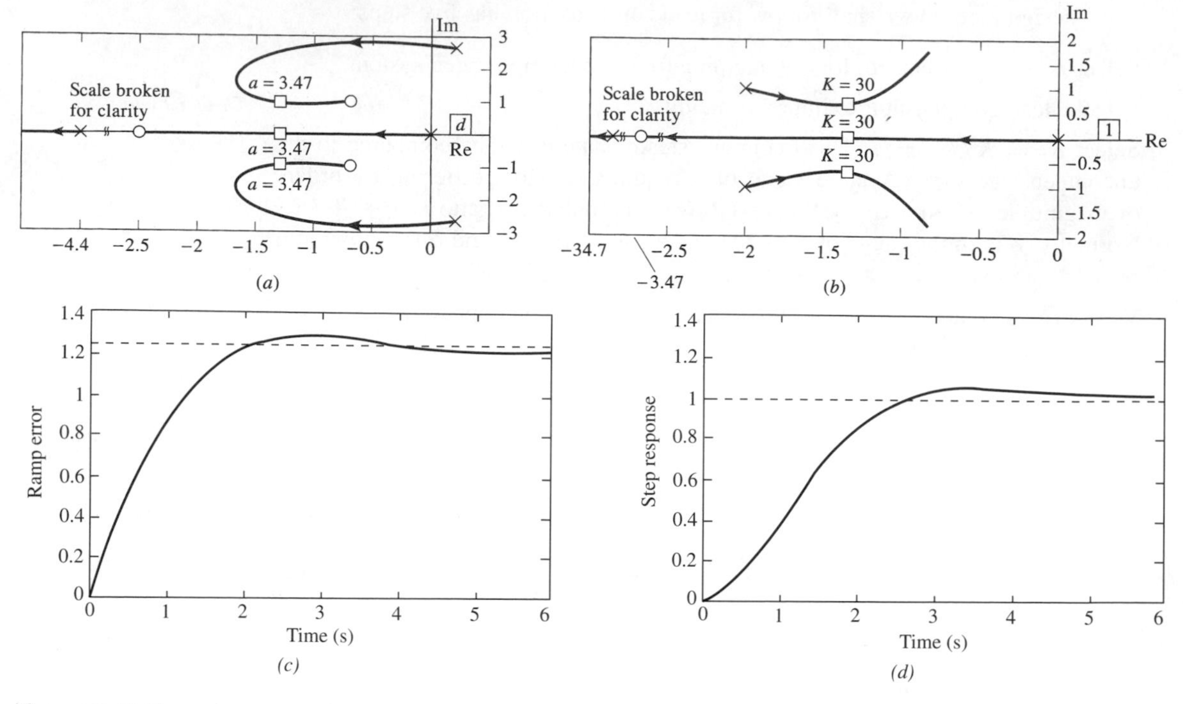

Figure 5.13 Example system with lead compensation (a) Root locus for variable a, $K = K_u d = 30$, $d = 10$. (b) Root locus for variable K, $a = 3.47$, $d = 10$. (c) Ramp error ($K = 40$). (d) Step response ($K = 40$).

The ramp error coefficient is 33% higher than for the uncompensated case, while settling time is less. The second-order closed-loop poles have a relative stability of 1.03, while that figure was 0.767 for the uncompensated system. The goal of improving stability is met. Figure 5.13(c) shows the ramp error for $K = 40$, while Figure 5.13(d) shows the step response for $K = 40$.

❑ DRILL PROBLEM

D5.13 For each of the plants from Problem D5.10, follow the design procedure for a cascade lead compensator with $d = 10$ and $K = K_u d$. Use a computer program to select a for maximum relative stability. Identify a, the closed-loop poles, K_0 (the step error coefficient), K_1 (the ramp error coefficient), and the settling time for the step response.

(a)

$$G_p(s) = \frac{1}{s(s+2)} \qquad K = K_u d = 20$$

Ans. 2.32; −1.44, −1.44, −22.3; ∞; 1; 2.8

(b)

$$G_p(s) = \frac{1}{s(s+2)(s+4)} \qquad K = K_u d = 52$$

Ans. 2.36; −1.23, −1.23, −3.44, −23.7; ∞; 0.65; 3.8

(c)

$$G_p(s) = \frac{s+8}{(s+1)(s+2)(s+6)} \qquad K = K_u d = 16.4$$

Ans. 2.03; −2.21, −2.21, −5.33, −19.5; 1.09; 0; 1.4

5.9 Cascade Lag–Lead Compensation

The cascade lag–lead compensator is defined.

A cascade lag compensator improves steady state accuracy, whereas a cascade lead compensator improves relative stability. The best attributes of both compensators can be combined, where

$$G_c(s) = K \left[\frac{s+a}{s+b}\right]_{\text{lag}} \left[\frac{s+a}{s+b}\right]_{\text{lead}}$$

The design is simplified if the pole-zero ratios are set by

$$\left[\frac{b}{a}\right]_{\text{lead}} = \left[\frac{a}{b}\right]_{\text{lag}} = d$$

The ratio of error coefficients becomes

$$\frac{K_i(\text{compensated})}{K_i(\text{uncompensated})} = \frac{K}{K_u}$$

Thus K must be greater than K_u to increase the error coefficient. When a lead compensator is cascaded with a lag compensator, stability can be improved over that of a lag compensated system while maintaining (or perhaps improving) the steady state accuracy available when a lag compensator is used alone.

If the pole zero ratios are as above, the compensator becomes (for $d > 1$)

$$G_c(s) = \frac{K(s + a_{\text{lag}})(s + a_{\text{lead}})}{[s + (a_{lag}/d)]\,(s + a_{\text{lead}}d)} \qquad [5.10]$$

Selecting a_{lag}, a_{lead}, and d.

To design the lag–lead compensator, a selection must be made for K, a_{lag}, a_{lead}, and d. We can simplify the design process by combining the previous lag and lead designs, followed by a final adjustment step to select K.

Lag and lead designs.

For the lag compensator, we chose $a_{\text{lag}} = 0.251$, $d = 10$, $K_{\text{lag}} = 5$, $K_{1\ \text{lag}} = 10$. For the lead compensator we chose $a_{\text{lead}} = 3.47$, $d = 10$, $K_{\text{lead}} = 40$, $K_{1\ \text{lead}} = 0.8$.

At this point in the design process, we have

$$G_c(s) = \frac{K(s+0.251)(s+3.47)}{(s+0.0251)(s+34.7)}$$

Varying K to improve performance.

The ramp error coefficient improved by a factor of $10/0.6 = 16.7$ for the lag compensator and by a factor of $0.8/0.6 = 1.33$ for the lead compensator. We can cascade those improvements and achieve an improvement of $16.7 \times 1.3 = 22$ by

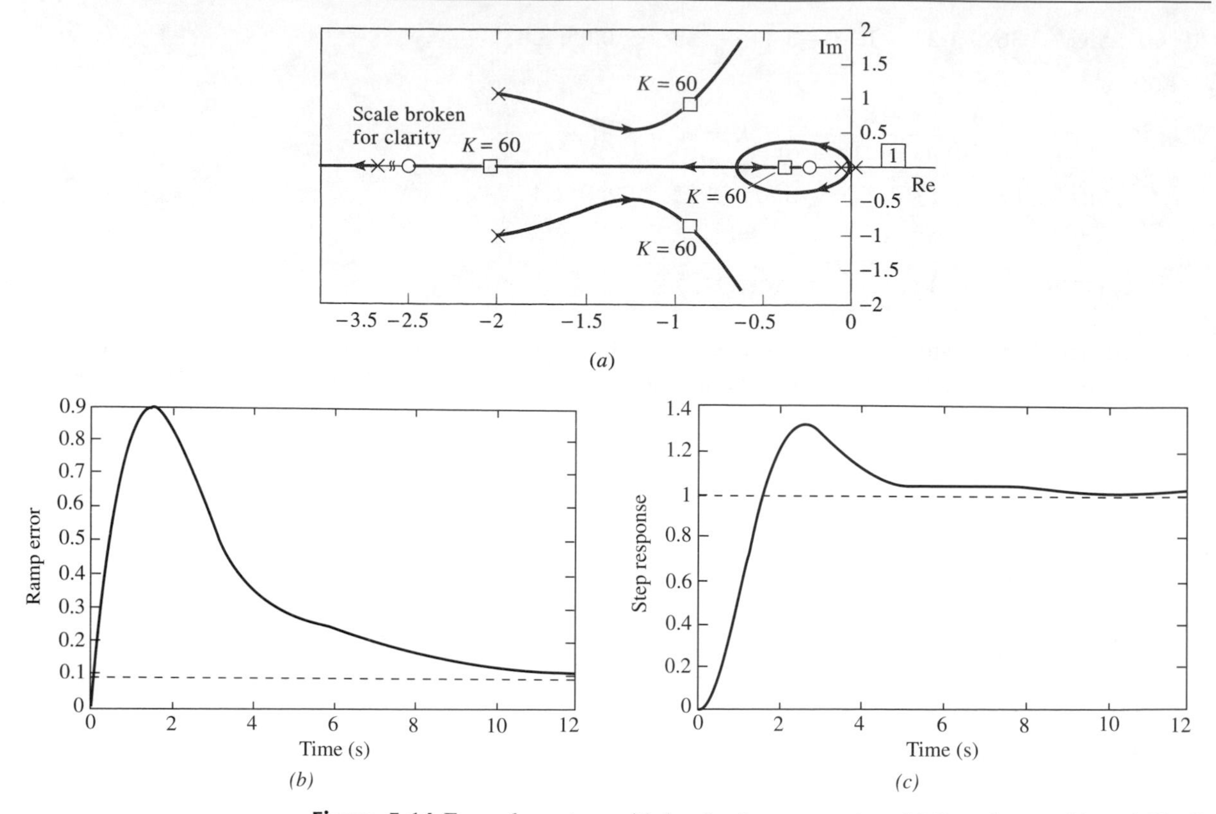

Figure 5.14 Example system with lag–lead compensation. (a) Root locus with variable K, $a_{\text{lag}} = 0.251$, $a_{\text{lead}} = 3.47$. (b) Ramp error with $K = 60$. (c) Step response with $K = 60$.

choosing

$$K = K_u \left[\frac{K_{\text{lag}}}{K_u}\right]\left[\frac{K_{\text{lead}}}{K_u}\right] = K_{\text{lag}}\left[\frac{K_{\text{lead}}}{K_u}\right] \quad [5.11]$$

With $K = 60$, settling time is 4.8 s.

Ramp error is 0.08.

Design goals are met.

or choose K to be $5(40/3) = 66.7$. Let's round off K to 60 and explore the result. Then $K_1 = K/5 = 12$ and thus K_1 improves by a factor of 20 over the uncompensated case. The root locus for variable K (given d, d_{lag}, and d_{lead}) is shown in Figure 5.14(a). With $K = 60$, the closed-loop poles (see Table 5.3) are at -0.315, -2.14, -34.8, and $-0.759 \pm j1.29$. The ramp error of Figure 5.14 (b) has less steady state error than for the lag-compensated system (and much less than for either the lead-compensated system or for the uncompensated system). The step response of Figure 5.14 (c) has a settling time of 4.8 s, which is less than for the lag-compensated system. Percent overshoot drops from 47 % for the lag-compensated system to 31 % for lag–lead compensation. The design goals are thus met by this very useful compensator. We can realize (create) such a compensator as in Figure 1.14(f).

❑ **DRILL PROBLEMS**

D5.14 For each of the plants from Problem D5.10, follow the design procedure for a cascade lag–lead compensator that must increase the error coefficient by a factor of $d = 10$. For design step 6, let $K = K_u d$ and show the compensator, the closed-loop poles, K_0 (the step error coefficient), K_1 (the ramp error coefficient), and the settling time for the step response. Then for $K = 2K_u d$ recalculate the closed-loop poles, K_0 (the step error coefficient), K_1 (the ramp error coefficient), and the settling time for the step response.

(a)

$$G_p(s) = \frac{1}{s(s+2)}$$

Ans. $G_c(s) = 20(s+0.392)(s+2.32)/(s+0.0392)(s+23.2)$;
$-0.569 \pm j0.370,\ -1.77,\ -22.3;\ \infty;\ 10;\ 6.6$

$G_c(s) = 40(s+0.392)(s+2.32)/(s+0.0392)(s+23.2)$;
$-0.50,\ -1.67 \pm j0.78, -21.4;\ \infty;\ 20;\ 4.2$

(b)

$$G_p(s) = \frac{1}{s(s+2)(s+4)}$$

Ans. $G_c(s) = 52(s+0.267)(s+2.36)/(s+0.0267)(s+23.6)$;
$-0.417 \pm j0.271, -1.61, -3.48, -23.7; \infty; 6.5;\ 9.4$
$G_c(s) = 104(s+0.267)(s+2.36)/(s+0.0267)(s+23.6)$;
$-0.338, -1.23 \pm j1.09, -3.0, -23.8; \infty; 13; 5.8$

(c)

$$G_p(s) = \frac{(s+8)}{(s+1)(s+2)(s+6)}$$

Ans. $G_c(s) = 16.4\,(s+0.995)(s+2.03)/(s+0.0995)(s+20.3)$;
$-0.982, -1.43, -1.92, -5.49, -19.6;\ 10.9;\ 0;\ 2.4$
$G_c(s) = 32.8(s+0.995)(s+2.03)/(s+0.0995)(s+20.3)$;
$-0.992, -2.11, -3.74 \pm j0.31, -18.8;\ 21.8;\ 0;\ 1.2$

5.10 Rate Feedback Compensation (PD)

A feedback-compensated system such as Figure 5.2(b), with feedback rate compensation has

The rate feedback compensator is defined.

$$H_c(s) = 1 + As$$

This compensator has an output that is proportional to its input and an output that is the derivative of its input. This type of device is referred to as a *proportional derivative (PD) compensator.*

Choosing K and A.

The designer must select K and A. The zero added by the rate feedback device is at $-1/A$, since

$$H_c(s) = A\left(s + \frac{1}{A}\right)$$

The value of $n - m$ decreases by 1 because one open-loop zero is added. Asymptote angles increase, generally improving stability.

An equivalent forward path transmittance for unity feedback would be

$$G_E(s) = \frac{KG_p(s)}{1 + AKsG_p(s)}$$

The ith error coefficient is

$$K_i(\text{comp}) = \lim_{s\to 0} s^i G_E(s)$$

$$= \lim_{s\to 0} \frac{s^i KG_p(s)}{1 + AKsG_p(s)} \quad [5.12]$$

Effect on steady state error.

To derive the ratio of K_i (comp) to K_i (uncomp), it is convenient to multiply the numerator and denominator of Equation (5.12) by the ratio K_u/K. Thus we have

$$K_i(\text{comp}) = \lim_{s\to 0} \frac{s^i K_u G_p(s)}{K_u/K + A\left[K_u s G_p(s)\right]}$$

$$= \frac{K_i(\text{uncomp})}{K_u/K + AK_1(\text{uncomp})}$$

$$\frac{K_i(\text{comp})}{K_i(\text{uncomp})} = \frac{1}{K_u/K + AK_1(uncomp)} \quad [5.13]$$

Thus, the error coefficient of the compensated system may be larger or smaller than for the uncompensated system, depending on the denominator of Equation (5.13). Certainly K should be larger than K_u if there is to be any chance of increasing the error coefficient with the compensator included.

A rate feedback compensator is primarily intended to improve the relative stability of a system—that is, to move the closed-loop poles to the left of their locations for the uncompensated case. To get some help in understanding how to select K and A, a rate feedback compensator can be applied to the type 1 second-order model of Figure 5.15(a). The closed-loop transfer function is

Selecting K and A for the type 1 model.

$$T(s) = \frac{\left(K/K_u\right)\omega_n^2}{s^2 + \left[2\zeta\omega_n + \left(K/K_u\right)A\omega_n^2\right]s + \left(K/K_u\right)\omega_n^2}$$

To get the root locus for variable A we must plot

$$AG(s)H(s) = \frac{sA\left(K/K_u\right)\omega_n^2}{s^2 + 2\zeta\omega_n s + \left(K/K_u\right)\omega_n^2} \quad [5.14]$$

That root locus is plotted in Figure 5.15(b), where it is assumed that $K/K_u > 1$ so that steady state accuracy will not be degraded by the rate feedback compensator.

From the root locus it appears that maximum relative stability occurs if the closed-loop poles are critically damped: that is, the damping ratio is one and there are double

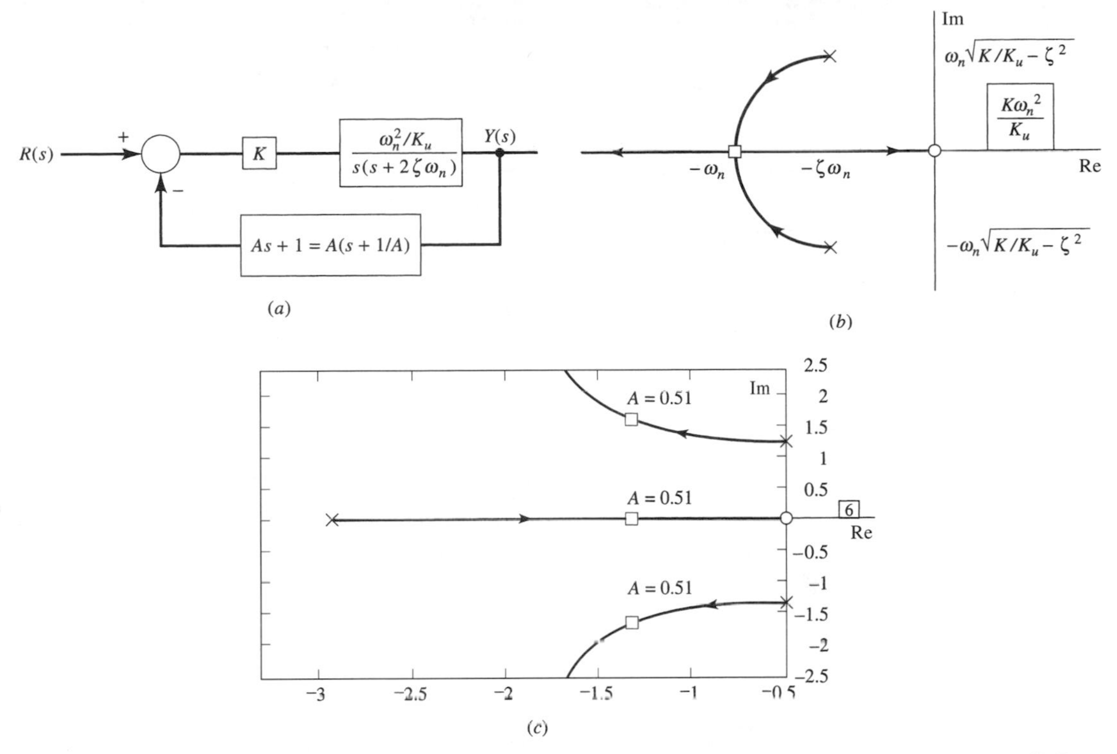

Figure 5.15 Rate feedback. (a) Rate feedback compensation of the dominant roots for a type 1 system. (b) Root locus for variable A for the system of (a). (c) Root locus for the example system for variable A with $K = 6$.

real closed-loop poles. For a type 0 model a similar value of A can be derived for maximum relative stability. In general, we can select A and K for a rate feedback compensator by following the procedure of Section 5.6.1.

Design procedure.

1. Select K_u for the uncompensated system to get 0.7 damping from the dominant closed-loop poles.
2. Select a rate feedback compensator to improve stability.
3. Select appropriate $K/K_u > 1$ to maintain an error coefficient compared to the uncompensated system. For that K, obtain the root locus as A varies. Select A, possibly requiring maximum relative stability.
4. For the given A, get the root locus as K varies. If necessary, vary K to improve step response.
5. Repeat steps 3 and 4 as required to meet design requirements.

The rate feedback design procedure can now be applied to the example system. The form of the system is that of Figure 5.2(b). Then

Selecting K and A for the example system.

$$T(s) = \frac{KG_p(s)}{1 + KH_c(s)G_p(s)}$$

$$= \frac{K}{s^3 + 4s^2 + (5 + KA)s + K}$$

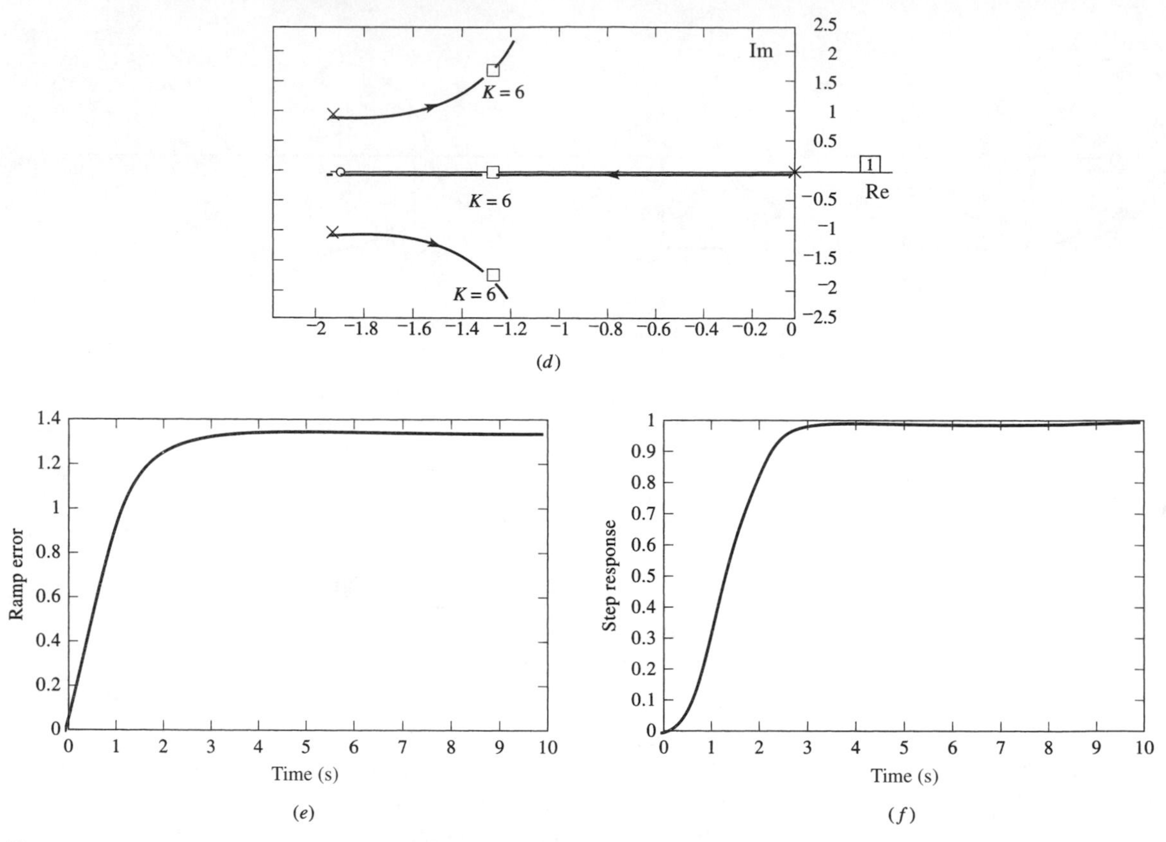

Figure 5.15 (*Continued*) (d) Root locus for the example system for variable K with $A = 0.51$. (e) Ramp error for the rate feedback–compensated system. (f) Step response for the rate feedback–compensated system.

The root locus for variable A requires using an effective $G(s)H(s)$, where

$$AG(s)H(s) = \frac{AKs}{s^3 + 4s^2 + 5s + K}$$

The ramp error coefficient can be found by using $G_E(s)$

$$G_E(s) = \frac{KG_p(s)}{1 + AKsG_p(s)}$$

$$= \frac{K}{s(s^2 + 4s + 5) + KAs}$$

$$K_1 = \frac{K}{5 + KA}$$

Example system results.

After step 1, K_u is 3. For step 3, suppose we choose $K/K_u = 2$ so that $K = 6$. Then the root locus for variable A requires plotting:

$$AG(s)H(s) = \frac{A6s}{s^3 + 4s^2 + 5s + 6}$$

Figure 5.15(c) shows the root locus for variable A with $K = 6$. By using the closed-loop pole data from a computer program, maximum relative stability occurs for $A = 0.51$. The closed-loop poles are then -1.33 and $-1.33 \pm j1.65$. The step error coefficient is

With $K = 6$, choose $A = 0.5$.

$$K_1 = \frac{6}{5 + (6 \times 0.51)} = 0.74.$$

The root locus for variable K when $A = 0.51$ is shown in Figure 5.15(d). In that case

Settling time is 2.5 s.

$$KH_c(s)G_p(s) = \frac{K(0.51)(s + 1.96)}{s(s^2 + 4s + 5)}$$

The ramp error is in Figure 5.15(e), while the step response is in Figure 5.15(f). The settling time for the step response is 2.5 s with no overshoot, an improvement over the uncompensated system. The steady state error and settling time are close to those of the cascade lead-compensated system. The rate feedback design has indeed improved stability and so the design ends with

Ramp error is 1.39.

$$K = 6 \qquad H_c(s) = 0.51(s + 1.96)$$

❑ **DRILL PROBLEM**

D5.15 For each of the plants from Drill Problem D5.10, follow the design procedure for a rate feedback compensator with the values of K given below. Use computer programs to select A for maximum relative stability. Identify A, the closed-loop poles, K_0 (the step error coefficient), K_1 (the ramp error coefficient), and the settling time for the step response.

(a)

$K = 2K_u = 4$

Ans. 0.50; $-2, -2$; ∞; 1; 2.4

(b)

$K = 2K_u = 10.4$

Ans. 0.50; $-2, -2 \pm j1.10$; ∞; 0.79 ; 2.6

(c)

$K = 2K_u = 3.28$

Ans. 0.453; $-2.79, -2.79, -4.9$; 2.19; 0; 1.9

5.11 Proportional-Integral-Derivative Compensation

A cascade proportional-integral-derivative (PID) compensator has a transmittance of the form

$$G_c(s) = A_p + \frac{A_i}{s} + A_d s$$

Controllers of this type are widely used in process control applications in industry. The A_p term provides proportional (P) action, where the compensator output is

The PID compensator is defined.

proportional to the compensator input. In a similar manner, the A_i/s provides integrator (I) action, and the $A_d s$ term provides derivative (D) action; hence, the compensator is referred to as PID.

Instead of designing values for the three coefficients A_p, A_i, and A simultaneously, it is convenient to rewrite the compensator as follows:

The variables become K, a, and A.

$$G_c(s) = \frac{K(s+a)}{s} + KAs \qquad \mathbf{[5.15]}$$

The variables are K, a, and A. Equation (5.15) shows clearly that the compensator combines PI action (where $A = 0$) with PD action (where $a = 0$). Here the derivative action is in the forward path, while with a rate feedback compensator, the derivative action is in the feedback path. Thus the PID compensator is seen to combine the steady state accuracy improvement of the PI compensator (with some loss of stability) with the improvement in stability of a PD compensator. We can therefore combine the design procedure of the PI compensator with that of the PD compensator in much the same way as we designed the cascade lag–lead compensator by combining the lag design, which improves steady state accuracy, with the lead design, which improves stability.

The PID compensator is designed as the combination of PI and PD designs.

The PID compensator design progresses by choosing a for the PI compensator (given K with $A = 0$) and by choosing A for the PD compensator (given K with $a = 0$). The value of K can then be changed (given a and A) to improve response. The values of a, A, and K can be iterated as needed. For the design example we chose $a = -0.236$ for the maximum relative stability using the PI compensator for $K = 3$ with $A = 0$. We chose $A = 0.51$ for maximum relative stability using the PD compensator for $K = 6$ with $a = 0$. Let's use $K = 6$ to evaluate the design.

Chose $A = 0.51$ and $a = -0.236$.

At this point in the design process,

$$G_c(s) = \frac{KA\left[s^2 + (1/A)s + a/A\right]}{s}$$

$$= \frac{K(0.51)(s^2 + 1.961s + 0.463)}{s}$$

$$= \frac{K(0.51)(s + 0.274)(s + 1.69)}{s}$$

With $K = 6$, the closed-loop poles are at -0.489, -0.647, and $-1.43 \pm j1.56$. Figure 5.16(a) shows the ramp error for $K = 6$. The steady state error is zero, since the system is now type 2. Figure 5.16(b) shows the step response with $K = 6$. The settling time is 7.5 s with 17 % overshoot. If we increase K to 12, the closed-loop poles become -0.322 (close to the closed-loop zero at -0.274), -1.282, and $-1.20 \pm j2.33$. Figure 5.16(c) shows the root locus for variable K. Figure 5.16(d) shows the ramp error for $K = 12$. The steady state error is again zero. Figure 5.16(e) shows the step response with $K = 12$. The settling time is 4.6 s, with 24 % overshoot. From Table 5.3, notice that for the PID compensated system, both the step response rise time and the step response percentage overshoot values are between those of the PI and PD compensated systems. Thus the stability is better than for PI compensation with the same steady state error, so the design goals of PID compensation are met

Results for $K = 6$.

Results for $K = 12$.

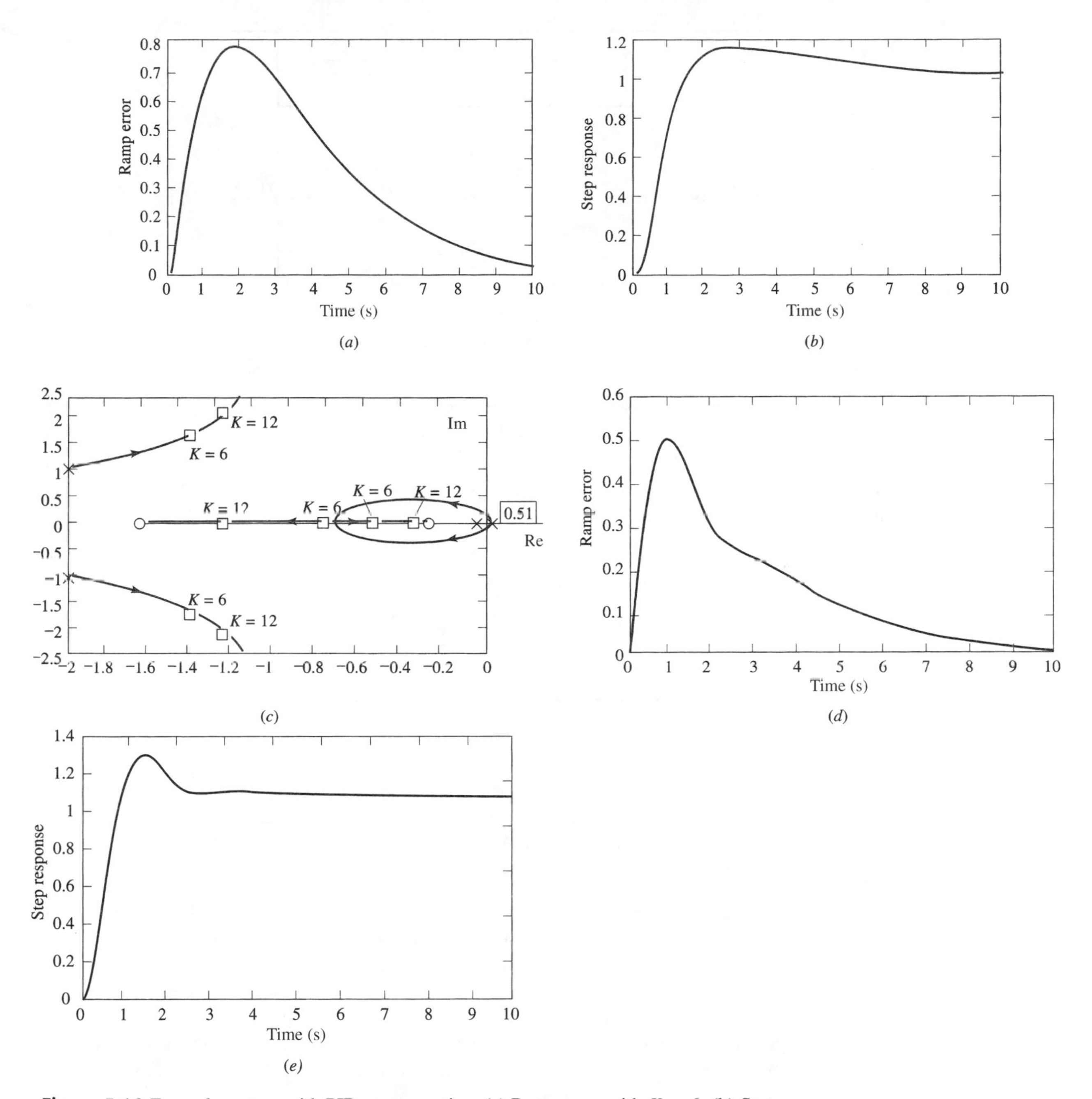

Figure 5.16 Example system with PID compensation. (a) Ramp error with $K = 6$. (b) Step response with $K = 6$. (c) Root locus with variable $K, a = 0.236,\ A = 0.51$. (d) Ramp error with $K = 12$. (e) Step response with $K = 12$.

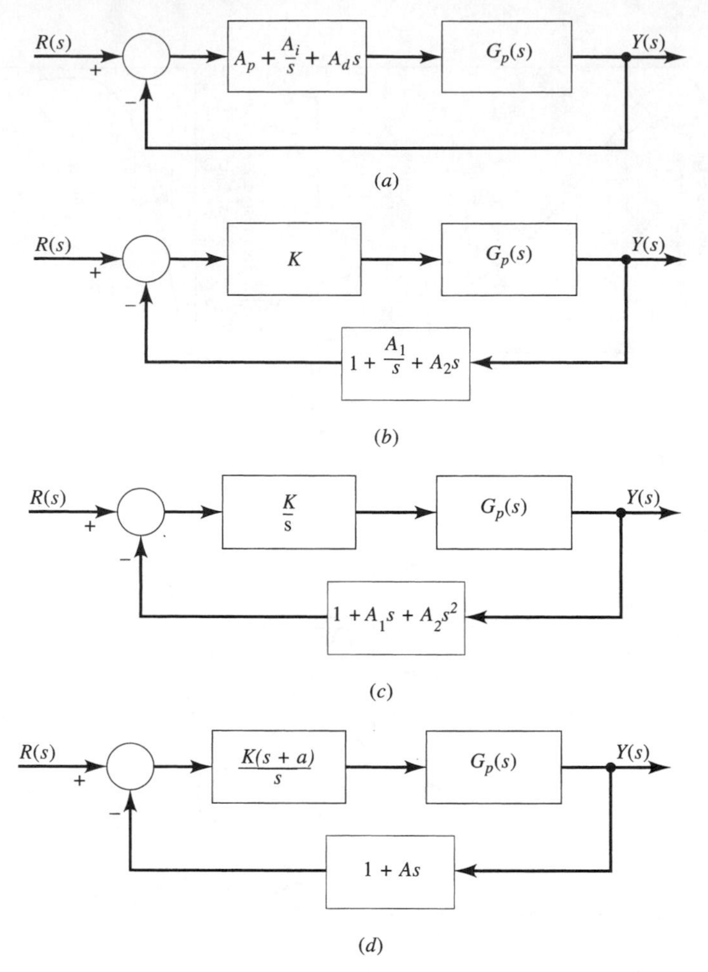

Figure 5.17 Some PID compensation arrangements. (a) Cascade compensation. (b) Feedback compensation. (c) A combination of cascade and feedback compensation. (d) Another combination of cascade and feedback compensation.

and no further changes to the design are necessary. Notice that the PID and lag–lead compensators provide similar settling times and percent overshoots.

PiD configurations.

Figure 5.17 shows a number of different feedback system configurations for which the open-loop system transmittance contains PID compensator terms. All these systems share the same root locus plot, but their steady state error performances and other characteristics differ somewhat. A designer selects a system configuration based upon the ease of developing and generating needed signals and the required performance characteristics.

❑ DRILL PROBLEMS

D5.16 For each of the plants for Drill Problem D5.10, follow the design procedure for cascade PID compensation. Use a from Drill Problem D5.11 and use $K = 2K_u$

and A from Drill Problem D5.15. For design step 6, let $K = 2K_u d$ and show the closed-loop poles, K_0 (the step error coefficient), K_1 (the ramp error coefficient), and the settling time for the step response. Then for $K = 4K_u d$ recalculate the closed-loop poles, K_0 (the step error coefficient), K_1 (the ramp error coefficient), and the settling time for the step response.

(a)

$$G_p(s) = \frac{1}{s(s+2)} \qquad a = 0.37 \qquad A = 0.50$$

$$G_c(s) = \frac{K(0.50)(s^2 + 2s + 0.74)}{s}$$

Ans. $K = 4;\ -0.632 \pm j0.376,\ -2.736;\ \infty;\ \infty;\ 5.2$
$K = 8;\ -0.664,\ -1.04,\ -4.30;\ \infty;\ \infty;\ 3.2$

(b)

$$G_p(s) = \frac{1}{s(s+2)(s+4)} \qquad a = 0.251 \quad A = 0.50$$

$$G_c(s) = \frac{K(0.50)(s^2 + 2s + 0.502)}{s}$$

Ans. $K = 10.4;\ -0.521,\ -0.727,\ -2.38 \pm j1.12;\ \infty;\ \infty;\ 7.0$
$K = 20.8;\ -0.345,\ -1.47,\ -2.09 \pm j2.43;\ \infty;\ \infty;\ 4.1$

(c)

$$G_p(s) = \frac{s+8}{(s+1)(s+2)(s+6)} \qquad a = 1.64 \quad A = 0.453$$

$$G_c(s) = \frac{K(0.452)(s^2 + 2.208s + 3.62)}{s}$$

Ans. $K = 3.28;\ -0.572 \pm j1.26,\ -4.67 \pm j0.87;\ \infty;\ 3.59;\ 3.3$
$K = 6.56;\ -0.731 \pm j1.46,\ -5.25 \pm j2.16;\ \infty;\ 7.17;\ 2.4$

D5.17 For each of the PID compensators of Drill Problem D5.16, use the same a and A with $K = 2K_u$. Suppose each compensator is realized with K and a in the forward path but with the feedback compensator being $1 + As$ as in Figure 5.17(d). Find K_0 (the step error coefficient) and K_1 (the ramp error coefficient).

Ans. (a) 0; 2; (b) 0; 2; (c) 0; 1.37

5.12 Pole Placement

The root locus design methods considered so far provide a modest ability to place closed-loop poles at desirable locations in the complex plane. When only the gain K is varied, closed-loop poles may reside only along the path of the root locus. When the various compensators we have considered so far are included, the new poles and zeros provide more help in repositioning the closed-loop poles as compared to using only the adjustable gain.

Instead of moving closed-loop poles into regions of the complex plane, it is possible to place one or more closed-loop poles at exactly the places desired. There are two basic types of compensation scheme for placing closed-loop poles: *algebraic* compensation and *fixed-structure* compensation. An *algebraic* compensator is used to place all the closed-loop poles and zeros as desired. The number of poles and zeros of each compensator will vary from problem to problem to fit the dimensional requirements of each problem. For example, a fifth-order compensator might be needed for a fifth-order system, while fourth-order compensator might work for a fourth-order system. A *fixed-structure* compensator has a fixed number of poles and zeros, and that same structure must be used for all problems. For example, a PID compensator provides three variables, the multiplying factors for proportional, integral, and derivative terms. Those same three variables would be all that are allowed when applied to a fourth-order or fifth-order system. A fixed-structure compensator cannot fix as many poles as an algebraic compensator, although the fixed-structure compensator would be easier to build.

Algebraic and fixed-structure compensators are defined and compared.

These pole placement compensators require knowledge of the plant transmittance. If the plant transmittance as used in calculating the compensator is incorrect, the resulting closed-loop transfer function will also be incorrect. A compensator that can operate fairly well when there is uncertainity regarding the plant is called a *robust* compensator. We will consider a robust algebraic compensation scheme.

Robust compensation is defined.

5.12.1 Algebraic Compensation

Figure 5.18 shows a typical cascade-compensated system. The compensator is $G_c(s)$. That compensator can be calculated using simple algebra to create a desired closed-loop $T(s)$.

$$T(s) = \frac{G_c(s)G_p(s)}{1 + G_c(s)G_p(s)} \qquad \textbf{[5.16]}$$

If $T(s)$ is given, the required compensator can be found.

$$T(s)\left[1 + G_c(s)G_p(s)\right] = G_c(s)G_p(s)$$

$$G_c(s)G_p(s)\left[1 - T(s)\right] = T(s)$$

$$G_c(s) = \frac{T(s)}{G_p(s)\left[1 - T(s)\right]} \qquad \textbf{[5.17]}$$

The polynomials that comprise $G_c(s)$ can be evaluated from

$$T(s) = \frac{N_t(s)}{D_t(s)} \qquad G_p(s) = \frac{N_p(s)}{D_p(s)}$$

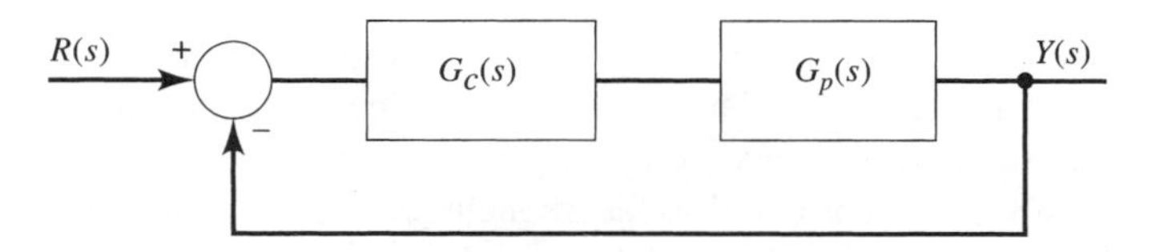

Figure 5.18 Cascade compensated system.

So that

$$G_c(s) = \frac{N_t(s)D_p(s)}{N_p(s)\,[D_t(s) - N_t(s)]} \qquad \textbf{[5.18]}$$

Algebraic compensator.

Assume that the plant $G_p(s)$ is known correctly and that a root locus is sought for the algebraic compensated system. A gain K can be included with $G_c(s)$, where $K = 1$ for the correct pole placement. The open-loop transfer function would be

$$G_c(s)G_p(s) = \frac{KT(s)}{1 - T(s)} \qquad \textbf{[5.19]}$$

Standard root locus form.

The root locus depends only on $T(s)$ and not on the plant, which cancels from the root locus.

5.12.2 Selecting the Transfer Function

Selecting desired closed-loop poles.

The closed-loop poles that generate the denominator of $T(s)$ should be selected for desired transient properties such as time constant (for a first-order term) and settling time (for a second-order system). Steady state (tracking) properties may be selected by specifying the numerator of $T(s)$.

Suppose $T(s)$ is given by

$$T(s) = \frac{b_2 s^2 + b_1 s + b_0}{s^3 + a_2 s^2 + a_1 s + a_0}$$

Then, for steady state operation we usually examine

$$\begin{aligned} T_E(s) &= 1 - T(s) \\ &= \frac{s^3 + (a_2 - b_2)\,s^2 + (a_1 - b_1)s + (a_0 - b_0)}{s^3 + a_2 s^2 + a_1 s + a_0} \end{aligned}$$

Selecting numerator coefficients for steady state accuracy.

For the system to become type 1, the numerator of $T_E(s)$ must have one factor of s. It follows that b_0 must equal a_0. If the system is to be type 2, then $b_0 = a_0$ and $b_1 = a_1$. In general, certain numerator coefficients must equal the corresponding denominator coefficient for type 1, type 2, and so on, behavior to be forced.

Suppose that the closed-loop poles are to be at -2 and -5 and that type 1 behavior is required. Then

$$T(s) = \frac{10}{(s+2)(s+5)} = \frac{10}{s^2 + 7s + 10}$$

Example System 1 (algebraic design)

Suppose the second-order system

$$G_p(s) = \frac{1}{(s+1)(s+2)} = \frac{N_p(s)}{D_p(s)}$$

Desired closed-loop poles.

is to be compensated so that the closed-loop poles become -3 and -4 and the system becomes type 1. The $T(s)$ would be

$$T(s) = \frac{12}{(s+3)(s+4)}$$
$$= \frac{12}{s^2+7s+12} = \frac{N_t(s)}{D_t(s)}$$

The system becomes type 1 because the s^0 terms in the numerator and denominator are the same. From Equation (5.18) the compensator must be

Algebraic compensator.

$$G_c(s) = \frac{12(s+1)(s+2)}{s^2+7s+12-12}$$
$$= \frac{12(s+1)(s+2)}{s(s+7)}$$

A root locus can be created as in Equation (5.19), where $K = 1$ generates the desired closed-loop pole locations.

Standard root locus form.

$$G_c(s)G_p(s) = \frac{12K}{s(s+7)}$$

The root locus is shown in Figure 5.19. Notice that the open-loop poles of $G_c(s)G_p(s)$ are placed so that at $K = 1$ the two closed-loop poles are at -3 and -4.

Example System 2 (algebraic design)

Consider the system used throughout the previous compensator design examples, that is,

$$G_p(s) = \frac{1}{s(s^2+4s+5)} = \frac{N_p(s)}{D_p(s)}$$

Desired closed-loop poles.

Suppose that the closed-loop poles are to become $-1 \pm j1$ and -1. To obtain type 1 operation, the numerator and denominator s^0 coefficients must be equal, so

$$T(s) = \frac{2}{(s+1)(s+1+j1)(s+1-j1)}$$
$$= \frac{2}{s^3+3s^2+4s+2} = \frac{N_t(s)}{D_t(s)}$$

From Equation (5.18), the compensator would be

Algebraic compensator.

$$G_c(s) = \frac{2s(s^2+4s+5)}{s^3+3s^2+4s+2-2}$$
$$= \frac{2(s^2+4s+5)}{s^2+3s+4}$$

Standard root locus form.

The root locus follows by including a gain K, which is 1 at the desired closed-loop pole locations

$$G_c(s)G_p(s) = \frac{2K}{s(s^2+3s+4)}$$

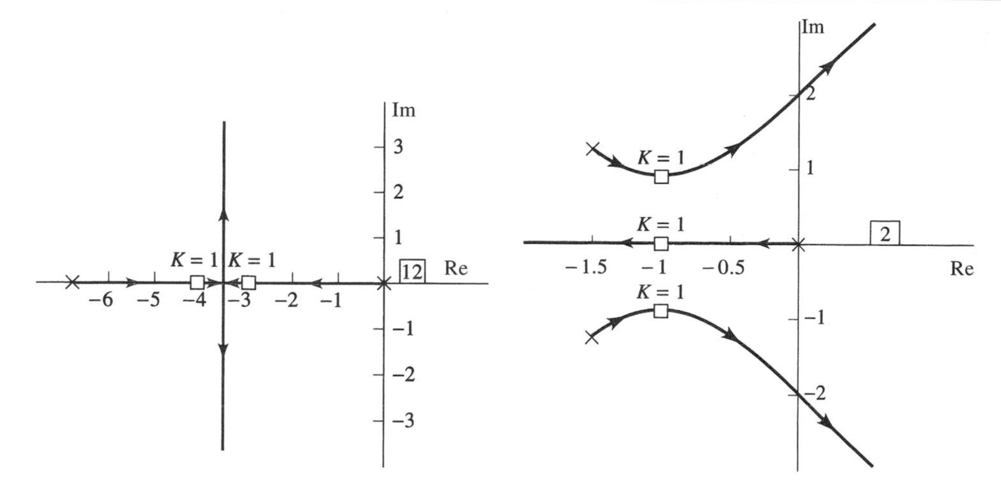

Figure 5.19 Root locus for example system 1. **Figure 5.20** Root locus for example system 2.

Figure 5.20 contains the root locus. The three open-loop poles at 0 and $-1.5 \pm j1.32$ cause the three closed-loop poles to be at $-1 \pm j1$ and -1 when $K = 1$.

❑ DRILL PROBLEM

D5.18 For each of the following plants, select the transfer function to meet specifications, design an algebraic compensator, and sketch the root locus for variable K, locating the desired closed-loop poles for $K = 1$.

(a) The closed-loop system should be second-order with two closed-loop poles having a damping ratio of 0.5, an undamped natural frequency of 2 rad/s, and zero steady state error for a unit step input.

$$G_p(s) = \frac{1}{s(s+3)}$$

(b) The closed-loop system should be second-order with closed-loop poles at -40 and -80, a closed-loop zero at -2, and zero steady state error for a unit step input

$$G_p(s) = \frac{6(s+2)}{(s+4)(s+8)}$$

(c) The closed-loop system should be third-order with one closed-loop pole at -50. The other two closed-loop poles should be 0.7 damped with an undamped natural frequency of 3 rad/s. There should be a closed-loop zero at -10, and there should be zero steady state error for a unit step input.

$$G_p(s) = \frac{(s+10)}{s^2(s+4)}$$

Ans. (a) $T(s) = \dfrac{4}{s^2 + 2s + 4}$

$$G_c(s) = \frac{4(s+3)}{(s+2)}$$

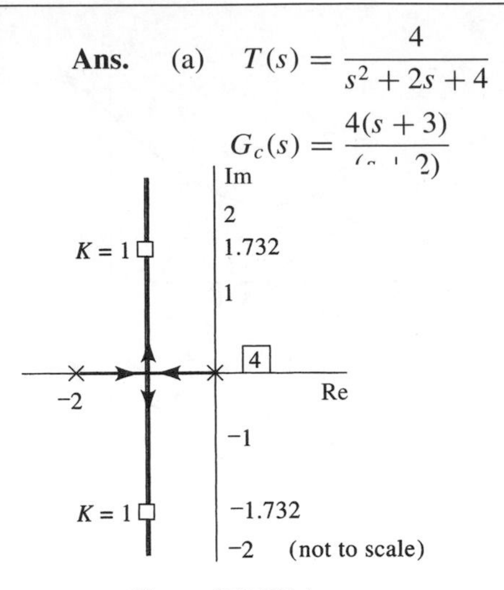

Figure D5.18(a)

(b) $T(s) = \dfrac{1600(s+2)}{(s+40)(s+80)}$

$$G_c(s) = \frac{267(s+4)(s+8)}{s(s-1480)}$$

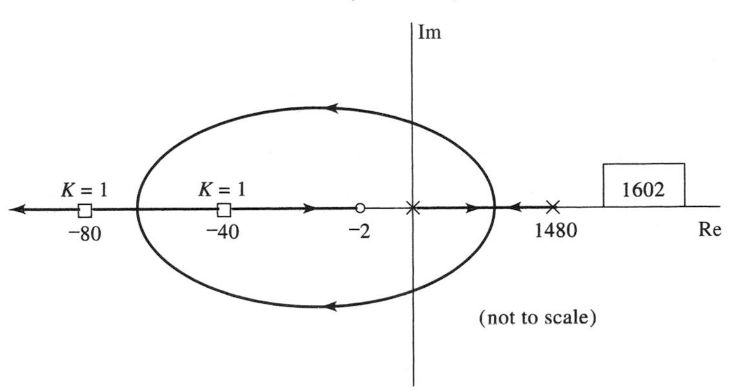

Figure D5.18(b)

(c) $T(s) = \dfrac{45(s+10)}{(s+50)(s^2+4.2s+9)}$

$$G_c(s) = \frac{45s(s+4)}{(s^2+54.2s+179)}$$

5.12.3 Incorrect Plant Transmittance

Sources of error may cause $G_p(s)$ to be incorrectly known.

The value of $G_p(s)$ results from the modeling of processes that can be subject to inaccuracies. For example, the actual system may be of very high order, whereas the model is only a low-order approximation. The measurement of aircraft parameters is

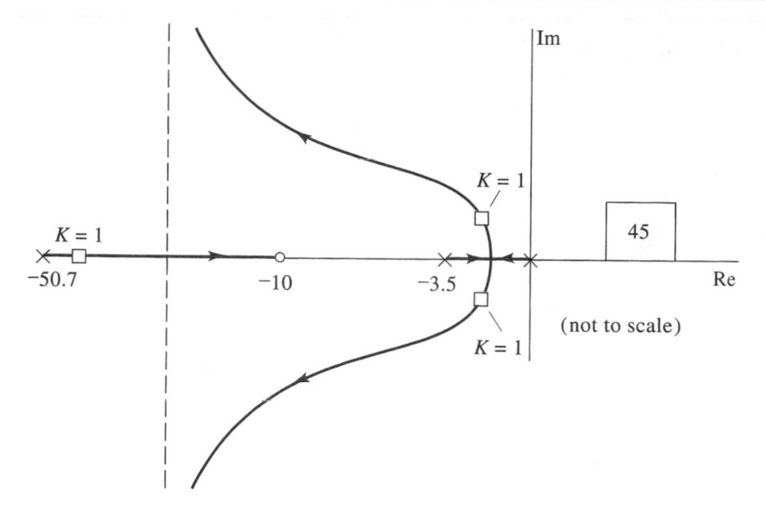

Figure D5.18(c)

subject to random errors (e.g., wind disturbances). The presence of errors in $G_p(s)$ means that the compensator is not correct, because it relies on $G_p(s)$ and the resulting $T(s)$ would not have the correct closed-loop poles and steady state accuracy.

Suppose the plant used to calculate the compensator is erroneous. Let $G'_p(s)$ represent the actual plant, which differs from the value$G_p(s)$ used to calculate the compensator. Let $T'(s)$ represent the actual transfer function using the actual plant and the erroneous compensator. Let $T(s)$ represent the desired transfer function used to calculate the compensator. From Equation (5.16), replacing $T(s)$ with $T'(s)$ and $G_p(s)$ and $G'_p(s)$ and using Equation (5.17) as written,

$$T'(s) = \frac{T(s)G'_p(s)}{G_p(s) + T(s)\left[G'_p(s) - G_p(s)\right]} \qquad \textbf{[5.20]}$$

Resulting incorrect T(s).

Notice that if $G_p(s) = G'_p(s)$, then $T'(s) = T(s)$, otherwise the actual closed-loop transfer function does not agree with the desired $T(s)$ and the actual $T'(s)$ could be unstable.

Suppose that example system 1 has a plant that is actually

$$G'_p(s) = \frac{1}{s(s+2)}$$

Actual plant is not $G_p(s)$.

but the earlier value of $G_p(s)$ is used to calculate the compensator and the same desired closed-loop transfer function is used. From Equation (5.20) the actual closed-loop transfer function is

$$T'(s) = \frac{\dfrac{12}{(s^2+7s+12)}\,\dfrac{1}{s(s+2)}}{\dfrac{1}{(s+1)(s+2)} + \dfrac{12}{s^2+7s+12}\left[\dfrac{1}{s(s+2)} - \dfrac{1}{(s+1)(s+2)}\right]}$$

$$= \frac{12(s+1)}{s^3+7s^2+12s+12}$$

Table 5.4 *Pole Placement Compensation (closed-loop roots)*

	Algebraic Using $G_p(s)$		Algebraic Using $G'_p(s)$		Algebraic Robust Using $G'_p(s)$		PID	
	Zeros	**Poles**	**Zeros**	**Poles**	**Zeros**	**Poles**	**Zeros**	**Poles**
Example 1	None	$-3, -4$	-1	$-.9 \pm j1.2, -5.1$	-1	$-2, -2, -3$	-3.12 $\pm j0.48$	$-3, -4,$ -20
Example 2	None	$-1 \pm j1,$ -1	None	$-1 \pm j1.4, -2$	None	$-1 \pm j1,$ -2	$-2.62,$ -0.38	$-1, -1,$ $-1 \pm j1$

Resulting incorrect T(s).

$$= \frac{12(s+1)}{(s+5.1)(s+0.9+j1.2)(s+0.9-j1.2)}$$

The closed-loop poles are no longer at the desired values of -3 and -4. There are now three closed-loop poles, at $-0.9 \pm j1.2$, and -5.1. In addition there is now a closed-loop zero at -1 (see Table 5.4).

Suppose that the plant of example system 2 is actually

Actual plant is not $G_p(s)$.

$$G'_p(s) = \frac{1}{(s+1)(s+4s+5)}$$

but the same desired $T(s)$ and the same $G_p(s)$ is used to calculate the compensator as before. Then, using Equation (5.20), we write

$$T'(s) = \frac{2}{s^3+4s^2+7s+6}$$

Resulting incorrect T(s).

$$T'(s) = \frac{2}{(s+1.0+j1.4)(s+1.0-j1.4)(s+2)}$$

The closed-loop poles have moved from the desired values of $-1 \pm j1$ and -1.

❑ DRILL PROBLEM

D5.19 For each of the plants of Problem D5.18, suppose that the actual plants are as follows but that the compensator remains at the value calculated in the earlier problem. Find the actual transfer function $T'(s)$ and compare the ideal with the actual closed-loop roots.

(a) $G'_p(s) = \dfrac{1}{(s+1)(s+3)}$

(b) $G'_p(s) = \dfrac{6(s+2)}{(s+1)(s+2)}$

(c) $G'_p(s) = \dfrac{(s+10)}{s\,(s+1)\,(s+4))}$

Ans. (a) $T'(s) = 4/(s^2+3s+6)$, CL poles are $-1.5 \pm j1.93$. They should be $-1 \pm j1.732$.

(b) $T'(s) = [1600(s+2)(s+4)]/(s^3 + 121s^2 + 8120s + 12{,}800)$, CL zeros are -2 and -4. There should be one zero at -2. CL poles are -1.6 and $-59.6 \pm j66.1$. They should be -40, -80.

(c) $T'(s) = [45(s+10)]/(s^3 + 55.2s^2 + 273.2s + 615)$, CL zero is correctly at -10. CL poles are -50 and $-2.6 \pm j2.34$. They should be -50 and $-2.1 \pm j2.14$.

5.12.4 Robust Algebraic Compensation

A number of schemes exist to render the system less sensitive to errors in the plant transmittance. Systems with reduced sensitivity are referred to as *robust*. The following procedure was suggested by Mafezzoni et al. in 1990. The idea is to separate those parts of the plant that are uncertain and also to select certain parts of $T(s)$ that should not be changed when the plant changes from the value used to calculate the compensator. A robust scheme requires a higher-order compensation procedure. Figure 5.21 shows a robust system with three compensators. The plant is broken into two parts so that

Achieving robust design.

$$G_p(s) = G_{p1}(s)G_{p2}(s)$$

The plant is separated into two parts, one of which ($G_{p1}(s)$) may be wrong.

where $G_{p1}(s)$ contains the part of the plant that may be incorrect. While $G_{p2}(s)$ is most likely to be correct. The desired closed-loop transfer function is divided into two parts:

$$T(s) = T_1(s)T_2(s)$$

where $T_1(s)$ contains roots that may move under uncertainty while $T_2(s)$ contains roots that should not change. The compensator $G_{c1}(s)$ is chosen to force the inner most loop to have the closed-loop transfer function $T_1(s)$. Using Equation (5.17), we write

$$G_{c1}(s) = \frac{T_1(s)}{G_{p1}(s)[1 - T_1(s)]} \qquad [5.21]$$

Robust compensator so $T_2(s)$ is correct.

The other two compensators are chosen to be

Other two compensators.

$$H_1(s) = T_2(s) \qquad [5.22]$$

$$G_{c2}(s) = \frac{T_2(s)}{G_{p2}(s)} \qquad [5.23]$$

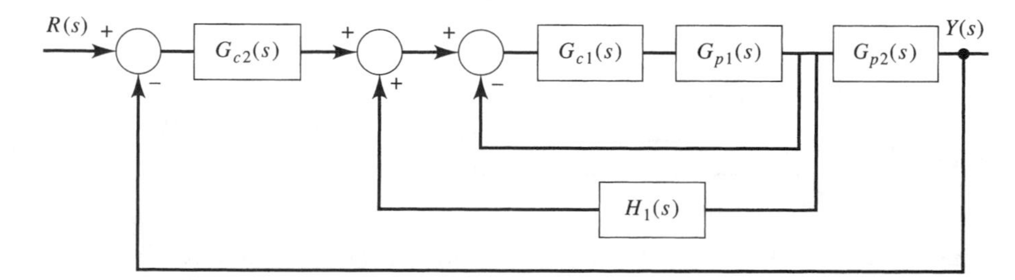

Figure 5.21 Robust algebraic compensator.

The block diagram manipulations in Figure 5.22 prove that

$$T(s) = T_1(s)T_2(s)$$

A root locus could be plotted from Figure 5.22(c) by including a K that is 1 at the desired $T(s)$. The root locus would use

$$\frac{KT_1(s)T_2(s)}{1 - T_1(s)T_2(s)} = \frac{KT(s)}{1 - T(s)}$$

The root locus is the same as for the previous algebraic compensator $G_c(s)$. However, when the plant is different from the value used to create the compensation scheme, the value of $T'(s)$ is different from Equation (5.20).

Suppose the actual plant is

$G_{p1}(s)$ is wrong.

$$G_p(s) = G'_{p1}(s)G_{p2}(s)$$

The innermost loop has a closed-loop transfer function similar to Equation (5.20).

Only $T_1(s)$ changes.

$$T'_1(s) = \frac{T_1(s)G'_{p1}(s)}{G_{p1}(s) + T_1(s)\left[G'_{p1}(s) - G_{p1}(s)\right]} \qquad [5.24]$$

The block diagram manipulations in Figure 5.23 prove that

$$T'(s) = T'_1(s)T_2(s)$$

so that all changes to $T(s)$ are confined to $T'_1(s)$.

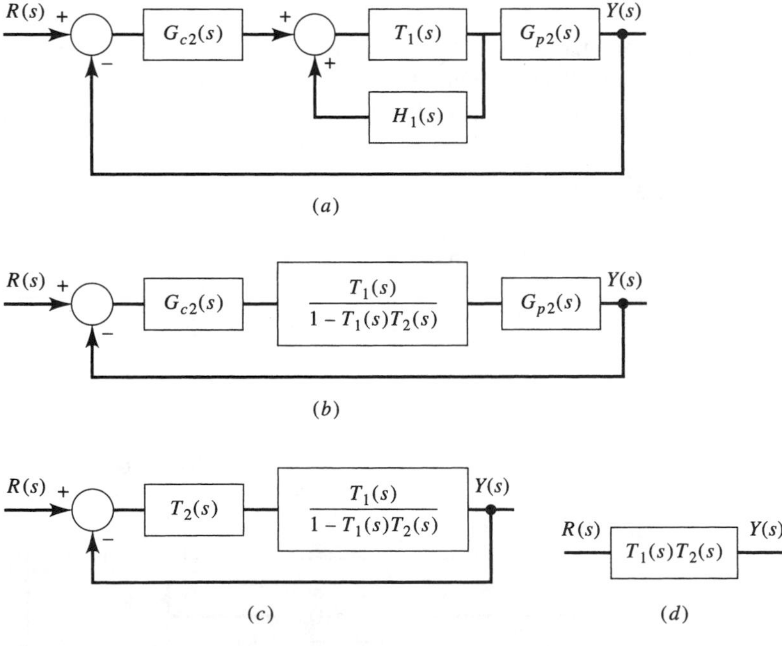

Figure 5.22 Block diagram manipulations of Figure 5.21. (a) Inner most loop closed. (b) $H_1(s)$ included in inner loop. (c) $G_{c2}(s)$ combined with $G_{p2}(s)$. (d) Closed-loop transfer function.

Example System 1 (robust design)

The system of example 1 is

$$G_p(s) = \frac{1}{(s+1)(s+2)}$$

and the desired closed-loop transfer function is

$$T(s) = \frac{12}{(s+3)(s+4)}$$

Suppose a designer is confident that one open-loop plant pole is at -2, but less confident that the other pole is actually at -1. Further, it is important to ensure that one closed-loop pole is at -3 while the other desired closed-loop pole should be at -4, but that pole could vary if the plant pole at -1 is incorrect. Then

$$G_{p1}(s) = \frac{1}{s+1} \qquad G_{p2}(s) = \frac{1}{s+2}$$

$$T_1(s) = \frac{4}{s+4} \qquad T_2(s) = \frac{3}{s+3}$$

From Equations (5.21) to (5.23), the compensators should be

Robust compensators for example system 1.

$$G_{c1}(s) = \frac{4(s+1)}{s} \qquad G_{c2}(s) = \frac{3(s+2)}{s+3} \qquad H_1(s) = \frac{3}{s+3}$$

The root locus is the same as in Figure 5.19 when the plant is actually $G_p(s)$. If $G_{p2}(s)$ is correct but

$G_{p1}(s)$ is wrong.

$$G'_{p1}(s) = \frac{1}{s}$$

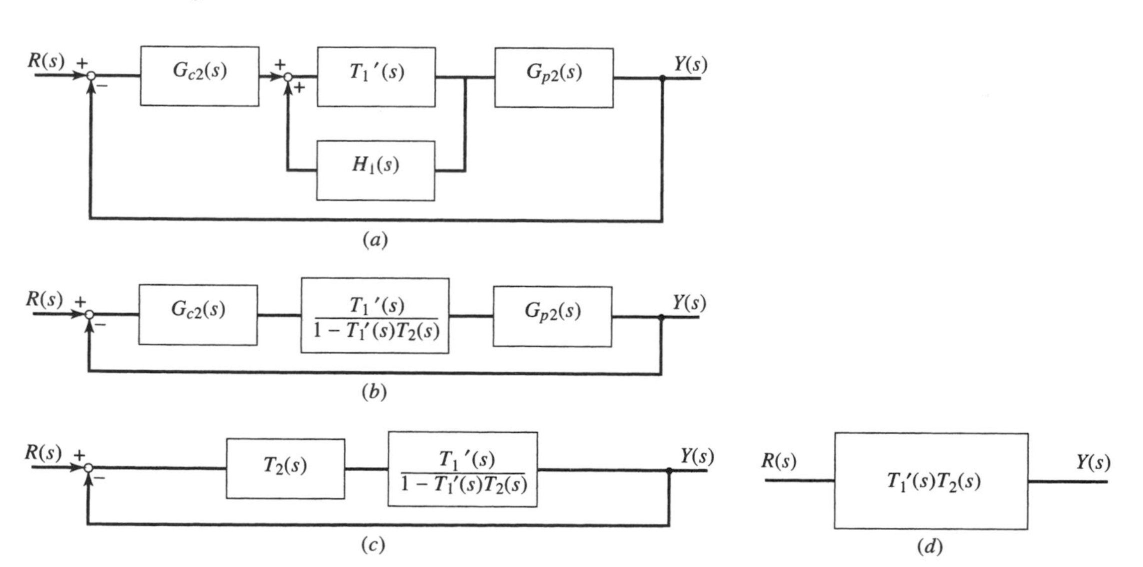

Figure 5.23 Robust system with nonideal plant. (a) Innermost loop closed. (b) $H_1(s)$ included in inner loop. (c) $G_{c2}(s)$ combined with $G_{p2}(s)$. (d) Closed-loop transfer function.

Only $T_1(s)$ changes.

then from Equation (5.24) we have

$$T_1'(s) = \frac{4(s+1)}{(s+2)^2}$$

$T_2(s)$ is not changed.

so that

$$T'(s) = \frac{12(s+1)}{(s+3)(s+2)^2}$$

and the desired closed-loop pole at -3 is unchanged, while the desired closed-loop pole at -4 is replaced by two closed-loop poles at -2 and one closed-loop zero at -1 (see Table 5.4).

Example system 2 (robust design)

The system of Example 2 is

$$G_p(s) = \frac{1}{s(s^2+4s+5)}$$

$$= \frac{1}{s(s+2+j1)(s+2-j1)}$$

and the desired closed-loop transfer function is

$$T(s) = \frac{2}{(s+1)(s+1+j1)(s+1-j1)}$$

Suppose a designer is confident in the location of the complex conjugate open-loop plant poles, but less confident that the other pole is actually at 0. Further, it is important to ensure that the complex conjugate closed-loop poles are at $-1 \pm j1$, while the other desired closed-loop pole should be at -1, but that pole could vary if the plant pole at 0 is incorrect. Then

$$G_{p1}(s) = \frac{1}{s} \qquad G_{p2}(s) = \frac{1}{s^2+4s+5}$$

$$T_1(s) = \frac{1}{s+1} \qquad T_2(s) = \frac{2}{s^2+2s+2}$$

From Equations (5.21) to (5.23), the compensators should be

Robust compensators for Example System 2.

$$G_{c1}(s) = 1 \quad G_{c2}(s) = \frac{2(s^2+4s+5)}{s^2+2s+2}$$

$$H_1(s) = \frac{2}{s^2+2s+2}$$

The root locus is the same as in Figure 5.20 when the plant is actually $G_p(s)$. If $G_{p2}(s)$ is correct but

$G_{p1}(s)$ is wrong.

$$G'_{p1}(s) = \frac{1}{s+1}$$

then from Equation (5.24) we write

Only $T_1(s)$ changes.

$$T_1'(s) = \frac{1}{(s+2)}$$

so that

$$T'(s) = \frac{2}{(s+2)(s^2+2s+2)}$$

$$= \frac{2}{(s+2)(s+1+j1)(s+1-j1)}$$

$T_2(s)$ is not changed.

and the desired complex closed-loop poles are unchanged while the desired closed-loop at -1 is replaced by a closed-loop pole at -2 (see Table 5.4).

❑ DRILL PROBLEM

D5.20 For each of the plants of Drill Problem 5.18, design a robust variable-structure compensator, where $G_{p1}(s)$ is the part of the plant that is uncertain and $T_1(s)$ is the part of the desired transfer function that will vary if $G_{p1}(s)$ is incorrect. For each of the following, determine $G_{c1}(s)$, $G_{c2}(s)$, and $H_1(s)$

(a) $$G_{p1}(s) = \frac{1}{s} \qquad G_{p2}(s) = \frac{1}{s+3}$$

$$T_1(s) = 4 \qquad T_2(s) = \frac{1}{s^2+2s+4}$$

(b)

$$G_{p1}(s) = \frac{1}{s+1} \qquad G_{p2}(s) = \frac{6(s+2)}{s+8}$$

$$T_1(s) = \frac{40}{s+40} \qquad T_2(s) = \frac{40(s+2)}{s+80}$$

(c)

$$G_{p1}(s) = \frac{1}{s} \qquad G_{p2}(s) = \frac{s+10}{s(s+4)}$$

$$T_1(s) = \frac{50}{s+50} \qquad T_2(s) = \frac{0.9(s+10)}{s^2+4.2s+9}$$

Ans. (a) $$G_{c1}(s) = -\frac{4}{3}s \qquad G_{c2}(s) = \frac{s+3}{s^2+2s+4}$$

$$H_1(s) = \frac{1}{s^2+2s+4}$$

(b) $$G_{c1}(s) = \frac{40(s+1)}{s} \qquad G_{c2}(s) = \frac{(20/3)(s+8)}{s+80}$$

$$H_1(s) = \frac{40(s+2)}{s+80}$$

(c) $$G_{c1}(s) = 50 \qquad G_{c2}(s) = \frac{0.9(s+4)}{s^2+4.2s+90}$$

$$H_1(s) = \frac{0.9(s+10)}{s^2+4.2s+9}$$

5.12.5 Fixed-Structure Compensation

The fixed-structure compensator is defined.

A *fixed-structure compensator* has a fixed dimension and form regardless of the application. As compared to an algebraic compensator, a fixed-structure compensator can place a smaller number of closed-loop poles (CLPs) but is generally easier to fabricate. If the numerator and denominator of the plant and cascade compensator in Figure 5.18 are

$$G_c(s) = \frac{N_c(s)}{D_c(s)} \quad G_p(s) = \frac{N_p(s)}{D_p(s)}$$

then the closed-loop transfer function is

$$T(s) = \frac{N_c(s)N_p(s)}{D_c(s)D_p(s) + N_c(s)N_p(s)}$$

and the characteristic polynomial $P(s)$ is

$$P(s) = D_c(s)D_p(s) + N_c(s)N_p(s) \qquad [5.25]$$

The number of CLPs that can be fixed is n.

The order of the closed-loop system is the order of the plant plus the order of the compensator. The number of the CLPs that can be placed is equal to the number of multiplying factors in the compensator. If the closed-loop system is fifth-order and the compensator has three multiplying factors, then three CLPs can be placed anywhere desired. The remaining two CLPs and the three multiplying factors of the compensator become dependent variables that must be computed from equations for coefficients for each of five powers of s in the characteristic polynomial—except for the highest power of s, which always has a unity coefficient.

In general, if the closed-loop system is of order n and the compensator has m multiplying factors, then the designer can place m CLPs as desired. The remaining $n-m$ CLPs and the m compensator multiplying factors are calculated from n equations arising from the characteristic polynomial. These equations are always linear and may possibly be simultaneous.

In the following designs, a PID (proportional-integral-derivative) compensator is used so that

A PID compensation is used as a fixed structure example.

$$G_c(s) = K_d s + K_p + K_i/s$$
$$= \frac{K_d s^2 + K_p s + K_i}{s} = \frac{N_c(s)}{D_c(s)}$$

where K_d causes the compensator output to depend on the derivative of the compensator input, K_p creates proportional behavior and K_i generates integral behavior.

Example System 1 (PID pole placement)

The system of Example 1 is

$$G_p(s) = \frac{1}{(s+1)(s+2)}$$

The plant is second-order, so the closed-loop system is third-order ($n = 3$), since $D_c(s)$ is s, thus raising the order by one and raising the type to 1. A PID compensator has three multiplying factors, so here $m = 3$ and all three CLPs can be placed. Suppose the preceding two CLPs are selected at -3 and -4, while the remaining pole is placed at -20, hence is nondominant. Then from Equation (5.25) we find

Three CLPs can be fixed.

$$
\begin{aligned}
P(s) &= s(s+1)(s+2) + K_d s^2 + K_p s + K_i \\
&= (s+3)(s+4)(s+20) \\
s^3 + 27s^2 + 152s + 240 &= s^3 + (3 + K_d)s^2 + (2 + K_p)s + K_i
\end{aligned}
$$

The three multiplying factors of the PID compensator can be calculated by equating the three power-of-s coefficients

$$
\begin{aligned}
3 + K_d &= 27 \quad K_d = 24 \\
2 + K_p &= 152 \quad K_p = 150 \\
K_i &= 240
\end{aligned}
$$

The PID compensator is

Resulting PID compensator.

$$
\begin{aligned}
G_c(s) &= \frac{24s^2 + 150s + 240}{s} \\
&= \frac{24(s^2 + 6.25s + 10)}{s} \\
&= \frac{24(s + 3.125 + j0.48)(s + 3.125 - j0.48)}{s}
\end{aligned}
$$

The root locus for the compensated system can be found for

$$
G_c(s)G_p(s) = \frac{24K(s + 3.125 + j0.48)(s + 3.125 - j0.48)}{s(s+1)(s+2)}
$$

where the desired closed-loop poles occur for $K = 1$. See Table 5.4.

Example System 2 (PID pole placement)

The plant is

$$
G_p(s) = \frac{1}{s(s^2 + 4s + 5)}
$$

Including the PID compensator, the closed-loop system becomes fourth-order ($n = 4$). Since there are only $m = 3$ multiplying constants in the PID compensator, only three of the four CLPS can be placed. The remaining ($n - m = 1$) CLP and the three PID multiplying factors are the four unknowns to be determined from four coefficients for four powers of s. Suppose the same three closed-loop poles are to be placed as before. It is convenient to call the other CLP $-a$ so that

Three CPLs can be fixed. The fourth pole and the PID multiplying factors are unknown.

$$
\begin{aligned}
P(s) &= s^2(s^2 + 4s + 5) + K_d s^2 + K_p s + K_i \\
&= (s + a)(s + 1 + j1)(s + 1 - j1)(s + 1) \\
s^4 + (3 + a)s^3 + (3a + 4)s^2 + (4a + 2)s + 2a \\
&= s^4 + 4s^3 + (5 + K_d)s^2 + K_p s + K_i
\end{aligned}
$$

Equating power-of-s coefficients, we have

$$3 + a = 4 \qquad a = 1$$
$$3a + 4 = 5 + K_d = 7 \quad K_d = 2$$
$$4a + 2 = K_p = 6$$
$$2a = k_i = 2$$

The fourth CLP is at -1, and the PID compensator is

Resulting PID compensator and fourth CLP.

$$G_c(s) = \frac{2s^2 + 6s + 2}{s} = \frac{2(s^2 + 3s + 1)}{s} = \frac{2(s + 2.62)(s + 0.38)}{s}$$

The root locus for the compensated system can be found for

$$G_c(s)G_p(s) = \frac{2K(s + 2.62)(s + 0.38)}{s^2(s^2 + 4s + 5)}$$

where the desired closed-loop poles occur for $K = 1$. See Table 5.4.

❑ DRILL PROBLEM

D5.21 For each of the plants of Problem 5.18, design a PID compensator as indicated and compute a closed-loop pole as indicated. Sketch the root locus for variable K and show the desired closed-loop poles for $K = 1$.

(a) Find K_d, K_p, and K_i, where

$$P(s) = (s + 10)(s^2 + 2s + 4)$$
$$G_p(s) = \frac{1}{s(s + 3)}$$
$$G_c(s) = \frac{K_d s^2 + K_p s + K_i}{s}$$

(b) Find K_p, K_i and a where

$$P(s) = (s + a)(s + 40)(s + 80)$$
$$G_p(s) = \frac{6(s + 2)}{(s + 4)(s + 8)}$$
$$G_c(s) = \frac{K_p s + K_i}{s}$$

(c) Find K_d, K_p, K_i, and a where

$$P(s) = (s + a)(s + 50)(s^2 + 4.2s + 9)$$
$$G_p(s) = \frac{(s + 10)}{s^2(s + 4)}$$

$$G_c(s) = \frac{K_d s^2 + K_p s + K_i}{s}$$

Ans. (a) $K_d = 9,\ K_p = 24,\ K_i = 40$

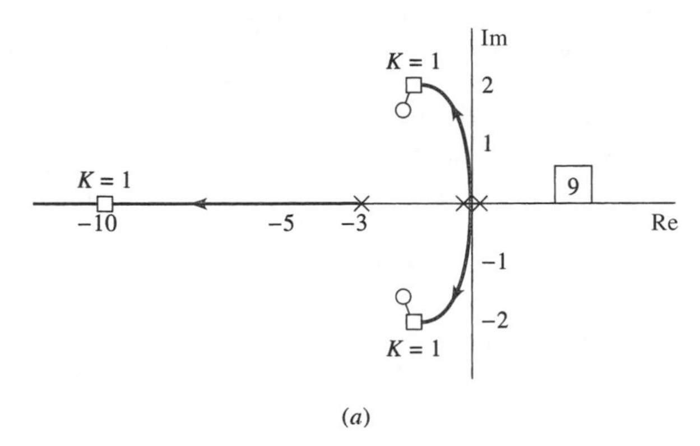

Figure D5.21(a)

Ans. (b) $a = 0.93,\ K_p = 108.93,\ K_i = 2964$

(c) $a = 12.2,\ K_d = 62.4,\quad K_p = 257.3,\quad K_i = 549$

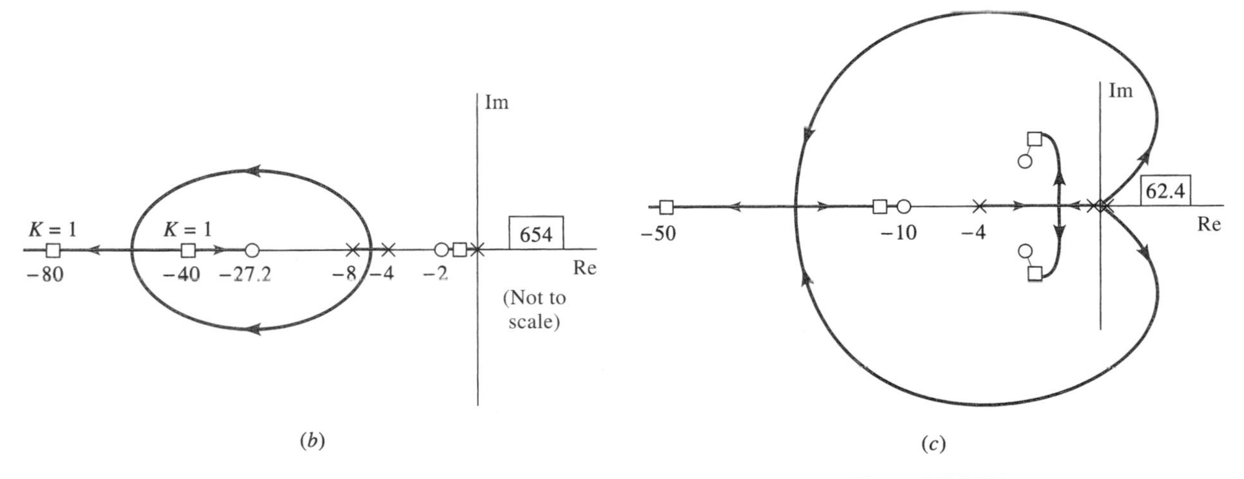

Figure D5.21(b) **Figure D5.21(c)**

5.13 An Unstable High-Performance Aircraft

Control system engineers may find it interesting that the design of high-performance aircraft has come full circle since the Wright brothers. Far from being experimental tinkerers, the Wright brothers performed careful analysis and wind tunnel testing prior to their first successful powered flight with a human at the controls.

The Wright brothers' aircraft was purposely open-loop unstable.

The German Otto Lilienthal carried out extensive experimentation using kites and human-controlled gliders from 1891 until his death in a stall-induced crash in 1899 (see Figures 5.24 and 5.25). The Wright brothers found errors in Lilienthal's aerodynamic data. Determined to design a very responsive aircraft, one in which a potential stall could be quickly eradicated, they purposely designed the famous Flyer of 1903 to be open-loop unstable. Some historians suggest that their experience building bicycles [which must be stabilized by constant (often unrealized) weight shifting by the rider] led naturally to creation of an aircraft that required weight shifting by the pilot to provide stability.

The F-6 is also open-loop unstable.

Subsequent designers learned enough about control surfaces to design aircraft that were responsive enough to avoid stalls but were stable (forgiving) enough to be more easily flown. From 1903 through the mid-1980s, open-loop stable operation was commonplace for passenger aircraft, fighter aircraft, and satellites. To provide greater responsiveness in combat, the F-16 was designed (in the mid-1970s) to be open-loop unstable, thus returning to the 80-year-old Wright brothers' scheme. Unlike the Wright Flyer, the pilot cannot actually control the F-16, and thus an auto-pilot must be used constantly. In the remainder of this section, design of piloted aircraft will be considered from the viewpoint of the root locus principles in this chapter.

Figure 5.26 shows a typical aircraft control system. The actuator consists of those electrical, mechanical, and hydraulic devices that move the flaps, elevators, fuel flow controllers, and other devices that cause the aircraft to vary its flight. Sensors provide information on velocity, heading, rate of rotation, and other flight data. This information is combined with the desired flight characteristics (commands) on displays visible to the pilot or electronically available to the autopilot. The autopilot should be able to fly the aircraft on a heading and under conditions set by the pilot. The command often consists of a predetermined heading or precoded warning strategies that are displayed as mentioned above. In combat the command might represent a radar-generated image of an enemy aircraft or of a ground-based target.

Figure 5.27 shows a simple version of Figure 5.26. Flight would be in one coordinate plane, where a coordinate plane refers to that part of our three-dimensional

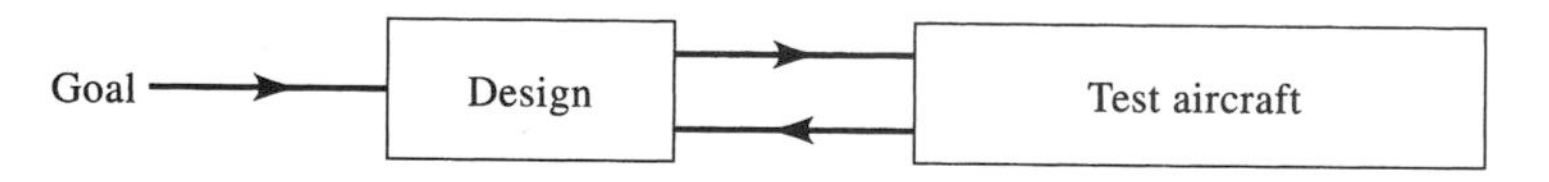

Figure 5.24 Aircraft design before Lilienthal.

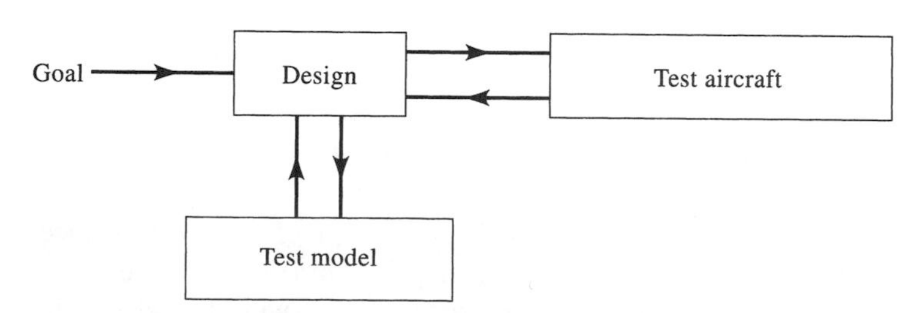

Figure 5.25 Aircraft design after Lilienthal.

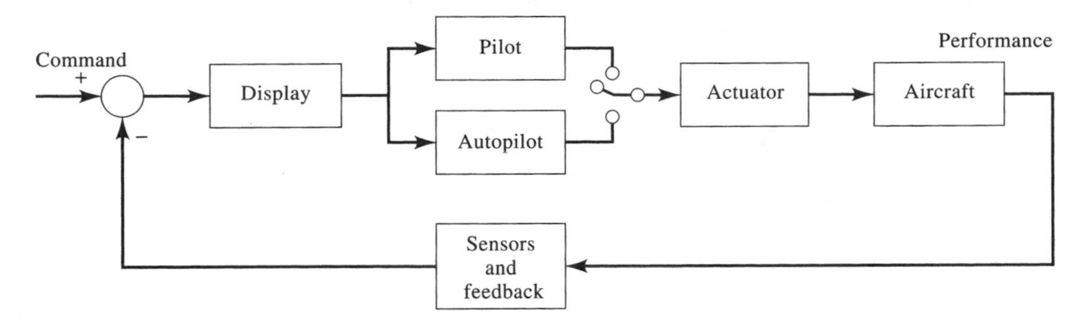

Figure 5.26 An aircraft control system.

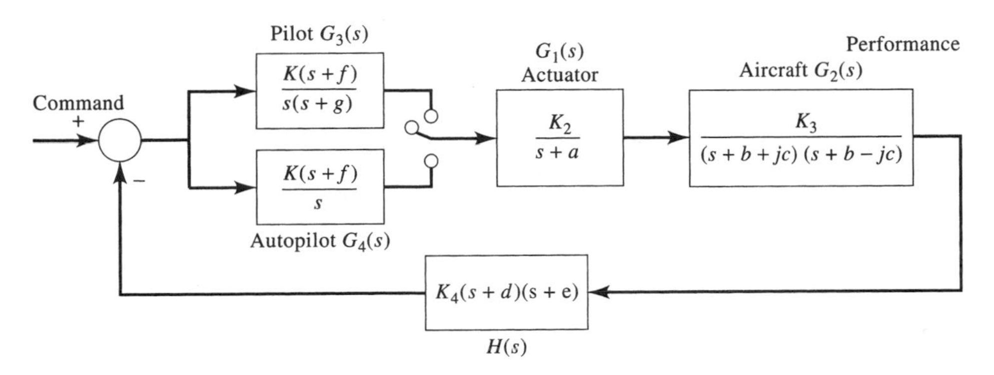

Figure 5.27 Aircraft control system dynamics.

world that is modeled. Often design focuses on a forward-moving aircraft that moves somewhat up or down without moving right or left and without rolling (rotating the wingtips). Such a study is called *pitch-plane design.* Figure 5.27 is typical of pitch-plane dynamics.

Simple pitch-plane model of the F-16.

The aircraft is modeled with a second-order transfer function having underdamped poles. The actuator may be approximated by a first-order $G_1(s)$. Where position, rate, and acceleration feedback occurs, $H(s)$ would contain two zeros. The pilot is assumed to provide integral plus proportional compensation. The pole at $-g$ represents time delay in the pilot's audiovisual–brain–neuromuscular system. The display is assumed to be the difference between the command and the output of $H(s)$.

It is assumed that K is adjustable by the pilot or by the autopilot but that K_2, K_3, and K_4 are fixed. For a piloted aircraft the root locus (for adjustable K) results from

$$KG(s)H(s) = K\left[\frac{s+f}{s(s+g)}G_1(s)G_2(s)\right]H(s)$$

For an autopiloted aircraft, we have

Root locus forms for piloted and autopiloted aircraft.

$$KG(s)H(s) = K\left[\frac{s+f}{s}G_1(s)G_2(s)\right]H(s)$$

The two systems differ because the autopiloted system does not have the delay represented by the term with a pole at $-g$.

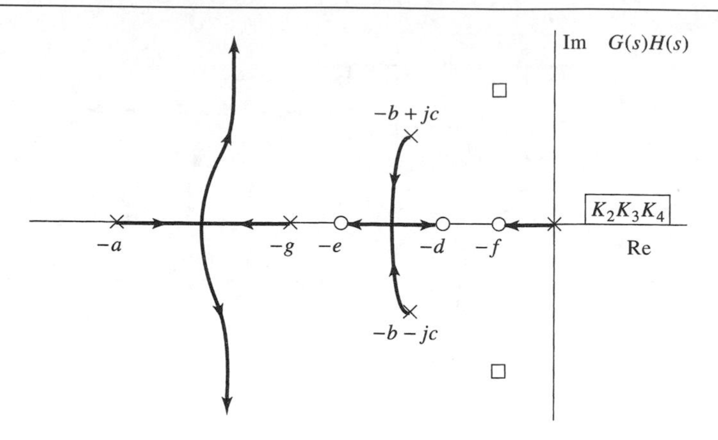

Figure 5.28 Pilot-controlled stable aircraft.

An open-loop stable aircraft is discussed.

Figure 5.28 contains the root locus for a pilot-controlled stable aircraft. The system has a closed-loop zero located at $-f$ since $-f$ is a forward path zero. The remaining coefficients show typical relative locationships. The closed- loop pole near $-f$ is nearly canceled by the closed-loop zero; hence the dominant poles (those most influencing the transient response) are the next two closed-loop poles further from the vertical axis (those leaving $-b \pm jc$). The pilot closes the loop by reacting to various cues and by moving the control stick accordingly. An inactive pilot opens the loop, creating about the same action as setting K to zero. The aircraft then operates due to its natural response, that is, due to values of $a,\ b,\ c,\ g,\ K_2,\ K_3$, and K_4. Inattentiveness by the pilot does not result in instability. First flown in late 1935, the DC-3 was particularly stable (forgiving). Many of these craft are still in service.

If the autopilot controls the system as in Figure 5.29, any selected value for K determines a stable response. The pole at $-g$ is eliminated, hence large values for K do not create a lightly damped pole pair to the left of $-g$ as in Figure 5.28.

Suppose the enemy has a plane with less rise time (more responsive performance) than the aircraft just discussed. To counteract this advantage, the closed-loop dominant poles should be moved to a point identified by the box in Figures 5.28 and 5.29. Clearly no value of K can create this more rapid response.

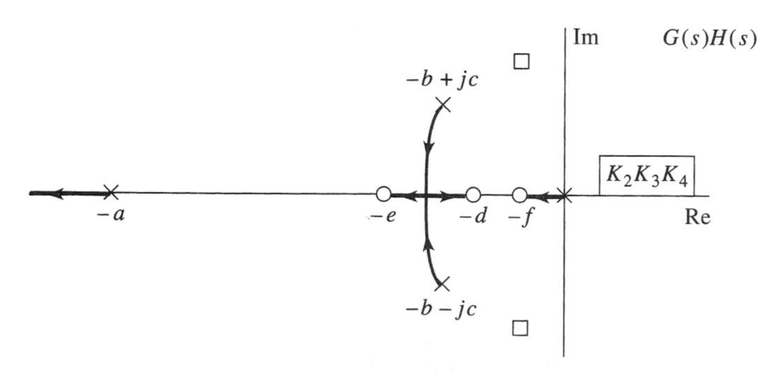

Figure 5.29 Autopilot-controlled stable aircraft.

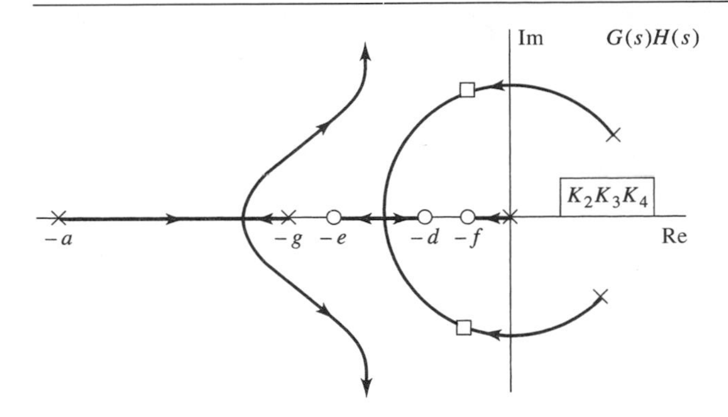

Figure 5.30 Pilot-controlled unstable aircraft.

Recall from Chapter 2 that a second-order characteristic polynomial can be written as follows:

$$s^2 + 2\zeta\omega_n s + \omega_n^2 = (s + \sigma + j\omega)(s + \sigma - j\omega)$$

The damping ratio is

$$\zeta = \frac{\sigma}{\omega_n^2} = \frac{\sigma}{\sqrt{\sigma^2 + \omega^2}}$$

For dominant complex conjugate roots at $-\sigma \pm j\omega$, if ω_n is held constant, a reduced damping ratio implies a lower value for ζ and a resulting reduction in rise time. In Figures 5.28 and 5.29, the desired dominant pole locations imply less rise time and (by that definition of performance) better response.

In Figure 5.30 the aircraft has been redesigned and is open-loop unstable [the poles of $G_2(s)$ are in the right half-plane]. With proper attention, the pilot can generate a value of K that provides the desired performance. The pilot must always maintain active control (as a rider does on a bicycle); otherwise any disturbance causes the system to go out of control. Pilot error or fatigue renders the design of Figure 5.30 unreliable. Only the autopilot-controlled system of Figure 5.31 provides a safe and reliable performance.

The pilot of an aircraft like the F-16 uses a standard-appearing stick, rudder, and other controls. However, the pilot is providing information to the autopilot (Figure 5.26), which actually moves all control surfaces and stabilizes the craft against disturbances that would otherwise grow so large as to compromise the survivability of the aircraft.

The partnership of autopilot, pilot, and open-loop unstable aircraft creates a closed-loop system with the desired dominant, combat-competitive behavior.

5.14 Control of a Flexible Space Station

When a large space station is placed into orbit around the earth, it is built from many parts placed into orbit during a sequence of launches. These components are then be assembled into living quarters, laboratories, and solar arrays for generating the

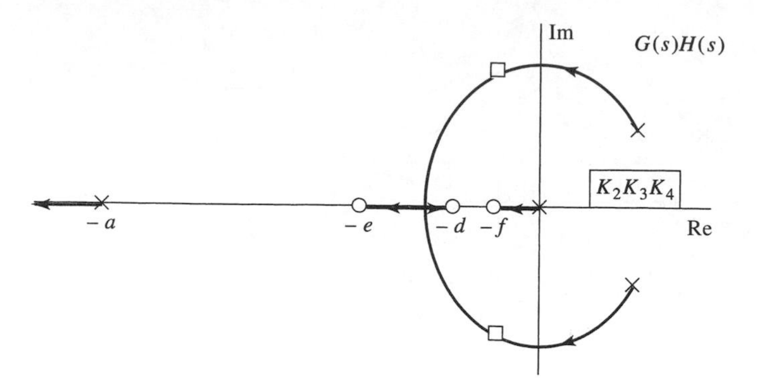

Figure 5.31 Autopilot-controlled unstable aircraft.

electrical power needed for the space station. Many of the components are relatively long and slender. The relative lack of an atmosphere in high earth orbit means that there is negligible aerodynamic drag to damp out oscillations when a command is sent to the space station to change attitude.

The space station is flexible.

There are often several resonances (frequencies at which a space station vibrates with negligible damping.) Figure 5.32 shows a block diagram of a space station with several of these resonances (modes). Suppose there are two modes, as in Figure 5.33. The plant is then

$$G_p(s) = \frac{1}{s^2+4} + \frac{1}{s^2+16}$$
$$= \frac{2(s^2+10)}{(s^2+4)(s^2+16)}$$

The uncompensated system oscillates.

The output is the attitude angle of the space station. Figure 5.34 shows the step response for the plant. The attitude would oscillate and the space station would be marginally stable (and very uncomfortable to live in).

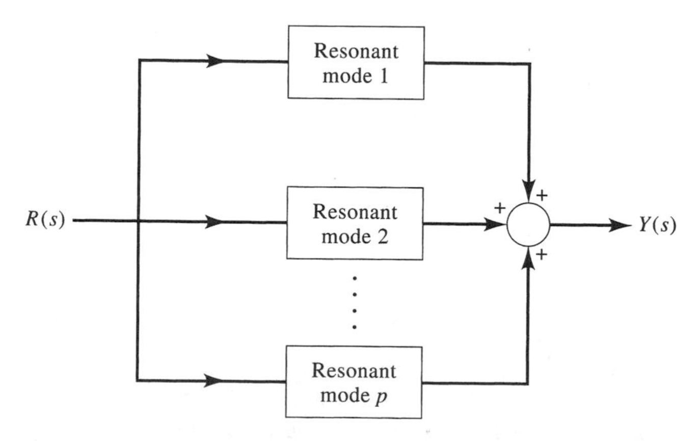

Figure 5.32 Space station with lightly damped resonant modes.

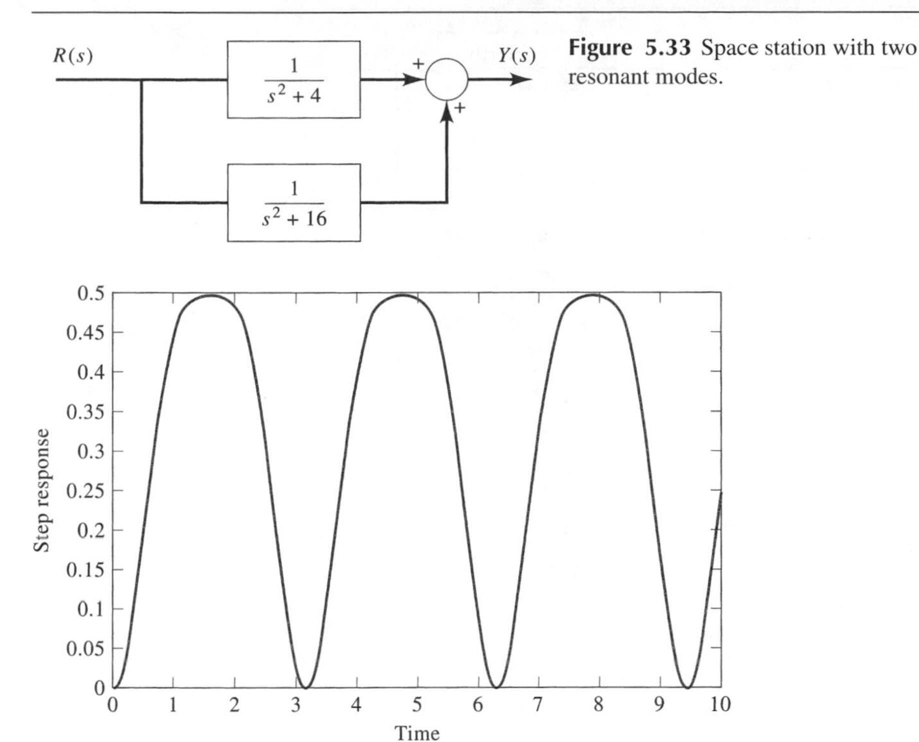

Figure 5.33 Space station with two resonant modes.

Figure 5.34 Step response for the plant of Figure 5.33.

PD compensation is used.

Rate and position feedback (PD compensation) can make stability much better. In this case proportional compensation means feeding back a signal proportional to the output position. In Figure 5.35, rate and position (unity) feedback are included with a forward path gain K. The root locus is in Figure 5.36(a). Here $H_c(s) = As + 1$, so if $A = 0.25$, a zero is placed at -4. To obtain the root locus, we must have $KG(s)H(s)$, which is

$$KG(s)H(s) = \frac{K(0.5)(s^2 + 10)(s + 4)}{(s^2 + 4)(s^2 + 16)}$$

Here $n = 4$, the number of open-loop poles, and $m = 3$, the number of open-loop zeros. Thus $n - m$ is one. There will be closed-loop zeros at $\pm j3.16$. Assume that the closed-loop poles near $\pm j3.16$ are nearly canceled by the closed-loop zeros there, then the dominant behavior belongs to the remaining two closed-loop poles. It is possible to maximize the relative stability of those two closed-loop poles by adjusting K until there are double real roots at -9.02. With $K = 36.24$ the closed-loop poles are at -9.02, -9.02, and $-0.04 \pm j3.11$.

Design results.

The step response of the closed-loop system is shown in Figure 5.36(b). The oscillations have been almost completely damped out. If K is increased or if additional filtering and compensation is added, the oscillation can be removed further. The relatively simple PD compensation has provided a significant improvement in stability, as we can see by comparing Figure 5.36(b) with Figure 5.34.

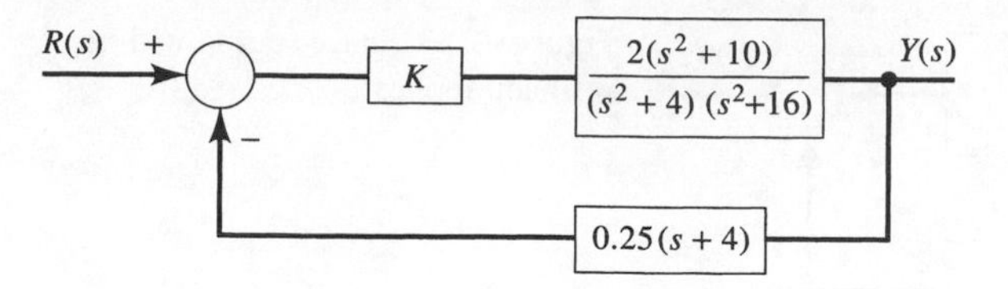

Figure 5.35 Space station with PD compensation.

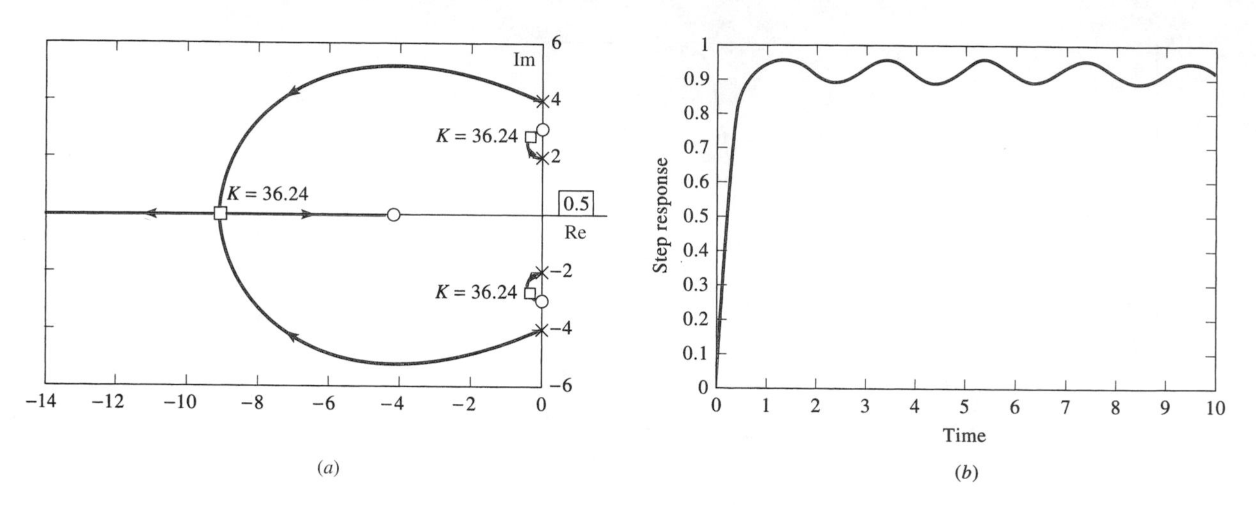

Figure 5.36 Improving stability of the space station. (a) Root locus with PD compensation. (b) Step response with $K = 36.24$.

The oscillation is nearly gone.

In general, if there are p resonant modes as in Figure 5.32, the plant will have $n = 2p$ open-loop poles and $2p - 2$ open-loop zeros. If we add one open-loop zero due to the combination of position and rate feedback, then $m = 2p - 1$ and we always have $n - m = 1$. Thus, PD compensation can always stabilize a flexible plant of this type.

5.15 Control of a Solar Furnace

Solar energy provides a readily available source of heat, but controlling that heat is a formidable task. Figure 5.37 shows an application of solar energy to heating some sample material for research purposes. The temperature of the material must be closely controlled. Too much heat can melt the sample, while insufficient heat would compromise the test being conducted.

In Figure 5.37, a mirror (heliostat) is positioned to receive maximum solar light intensity. That positioning is easily carried out by using the amount of light received and the known astronomical solar position. Our problem is to control the temperature of the sample by opening and closing a parallel set of shutters. Light progresses through the shutters, reflecting from a fixed parabolic mirror onto the sample. The design goal is to control temperature with a settling time of less than the 1–3 min. achievable by a skilled operator. There should be minimal overshoot to avoid melting

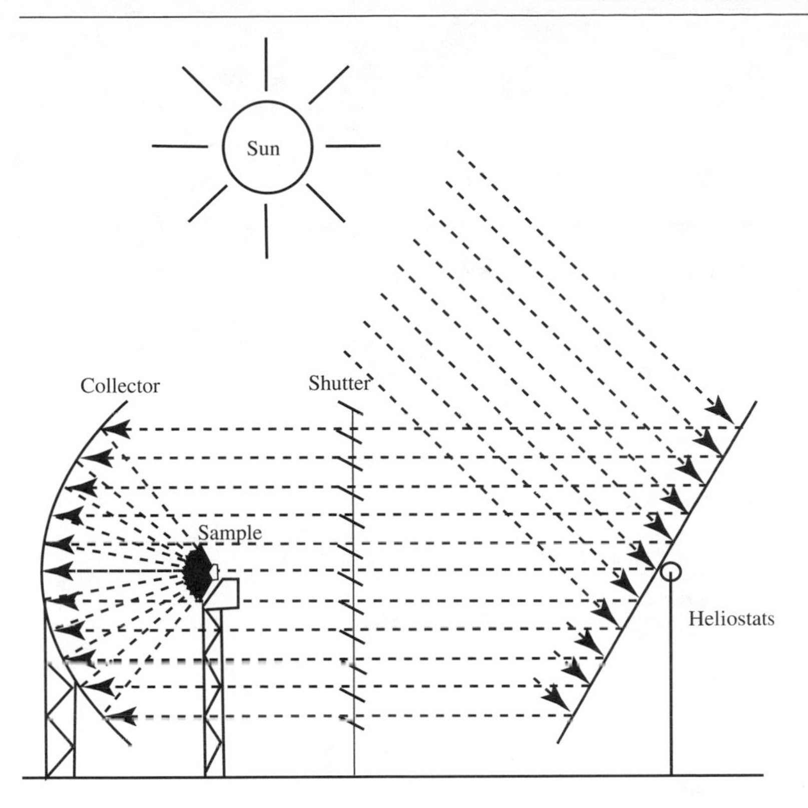

Figure 5.37 Solar furnace.

the sample. There should also be zero steady state step error (to obtain the desired temperature) and a small ramp error (to track changing light conditions throughout the day).

One such control system is shown in Figure 5.38, a unity feedback system. In the forward path are a compensator $G_C(s)$, an adaptive device $G_a(s)$, and the plant $G_p(s)$, consisting of the solar furnace. Light intensity, I acts as a disturbance affecting plant efficiency. That same light intensity is measured by the adaptive device, intended to render control system operation independent of I.

The plant may be modified by

$$G_p(s) = \frac{K'Ib}{s+b}$$

The adaptive device removes influence of solar intensity I by defining

$$G_a(s) = \frac{I_{\text{ref}}}{I}$$

The effect of I_{ref} may be removed by defining

$$K' = \frac{K_G}{I_{\text{ref}}}$$

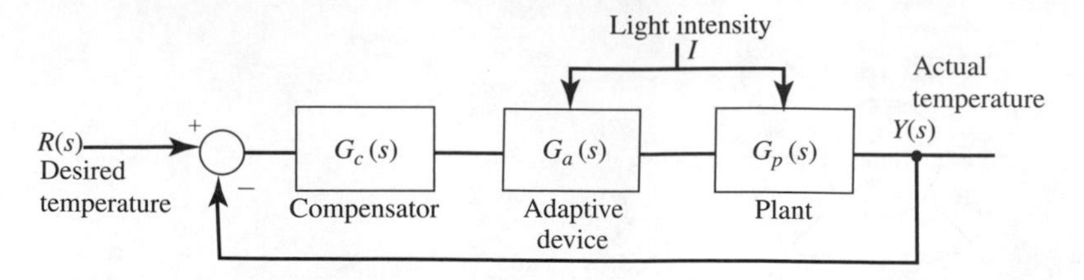

Figure 5.38 Solar furnace shutter control system.

If $G_c(s)$ is chosen to be a PI compensator, the system becomes type 1 and, as required, there is zero steady state error for a step input. Then

$$G_c(s) = \frac{K(s+a)}{s}$$

In the forward path we have

$$\begin{aligned} G_e(s) &= G_c(s)G_a(s)G_p(s) \\ &= \frac{K(s+a)}{s}\,\frac{I_{\text{ref}}}{I}\,\frac{(K_G/I_{\text{ref}})}{s+b}Ib \\ &= \frac{KK_G b(s+a)}{s(s+b)} \end{aligned}$$

It can be assumed that K_G and b are fixed by the operation of the solar furnace. We need to select K and a. The steady state ramp error follows from the ramp error coefficient

$$K_1 = \lim_{s\to 0} sG_E(s) = KK_G a$$

$$e_{\text{ramp}}(\infty) = \frac{1}{K_1}$$

Let's start by finding a range of acceptable values for a. If we choose $a < b$ as in Figure 5.39(a), one closed-loop pole must be to the right of $-b$, causing the closed-loop system to be slower than the open-loop system. That would not be desirable. If we choose $a > b$ as in Figure 5.39(b), the closed-loop system can be faster than the open-loop system.

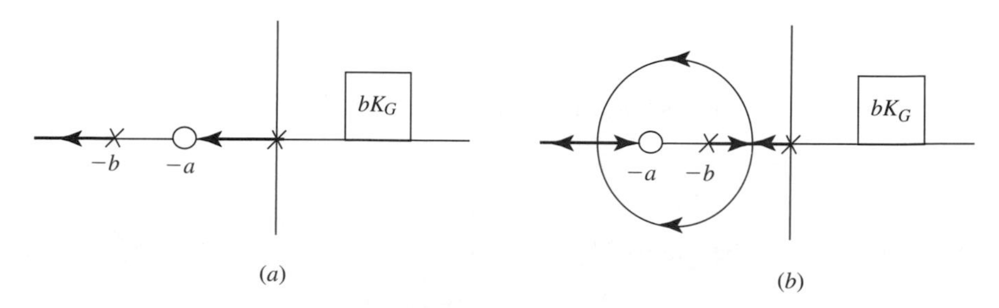

Figure 5.39 Root locus as K varies. (a) $a < b$. (b) $a > b$.

If we want to minimize overshoot, the closed-loop poles should be real-valued. There may be some overshoot due to the closed-loop zero at $-a$ but that overshoot would be less (for real closed-loop poles) than when the closed-loop poles are complex conjugates. To ensure real-valued closed-loop poles as a function of d, lets consider K to be fixed and get a root locus as a varies. To isolate a as the variable, we write

$$T(s) = \frac{G_E(s)}{1 + G_E(s)}$$

$$= \frac{KK_G b(s+a)}{s^2 + s[b(1+KK_G)] + KK_G ba}$$

$$T(s) = \frac{\dfrac{KK_G b(s+a)}{s[s+b(1+KK_G)]}}{1 + a\dfrac{KK_G b}{s[s+b(1+KK_G)]}}$$

Thus, the root locus for variable a follows from

$$aG(s)H(s) = a\frac{N(s)}{D(s)} = a\frac{KK_G b}{s[s+b(1+KK_G)]}$$

See Figure 5.40. To ensure real closed-loop poles, a must be no greater than the value at breakaway. From our earlier discussion the breaking point s_b is the value of s solving

$$N\frac{dD}{ds} + d\frac{dN}{ds} = 0$$

Since dN/ds is zero, we just require

$$\frac{dD}{ds} = 0$$

$$s_b = \frac{-b(1+KK_G)}{2}$$

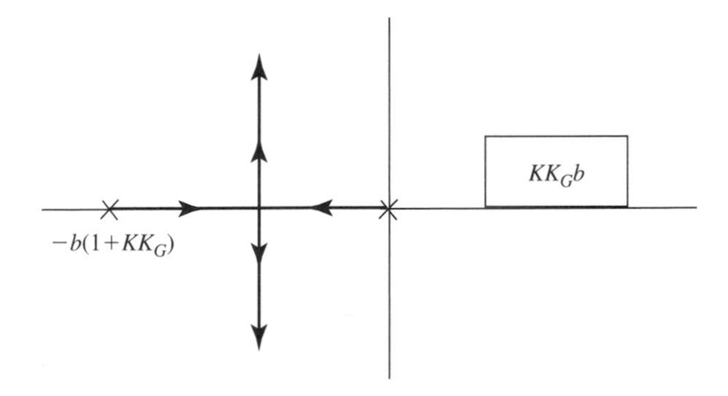

Figure 5.40 Root locus as a varies.

Then we require

$$a \leqslant -\frac{1}{G(s_b)H(s_b)} = -\frac{D(s_b)}{N(s_b)} = \frac{b(1+KK_G)^2}{4KK_G}$$

In summary, we require

$$b \leqslant a \leqslant \frac{b(1+KK_G)^2}{4KK_G}$$

Following Berenguel et al. (1999), we put

$$K_G = 50$$
$$b = 0.01$$
$$K = 0.5$$
$$0.01 \leqslant a \leqslant 0.068$$

If we choose $a = 0.02$ (and the values of K, K_G and b just listed) we get the step response of Figure 5.41. There is only 2% overshoot, the settling time is 10 s, and there is zero steady state step error, thus meeting the goals mentioned earlier. Also

$$e_{\text{ramp}}(\infty) = \frac{1}{(0.02)(0.5)(50)} = 2$$

The system will run for several hours. After one hour the ramp error is only $100 \times 2/3600$ or 0.06%, so that error is acceptably small, and the design goals are met.

The solar furnace control system has some interesting nonlinear properties that could render operation less than adequate. To protect the plant from overactivity, certain limiters are included. When an output is frozen at some value because of a limiter, the input to that device may build up to an unacceptably high level. That process is called windup. The paper by Berengmel et al. contains countermeasures to overcome windup so that the response (including nonlinear limiters) is close to

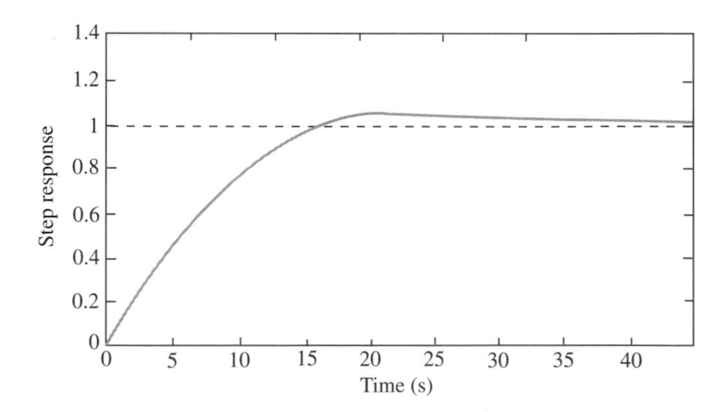

Figure 5.41 Step response of the solar furnace.

that shown in Figure 5.41 (assuming linear behavior). Nonlinear control methods are beyond the scope of this text, but the interested reader should explore the reference mentioned.

5.16 Summary

The simplest way to stabilize a plant is to place a gain amplifier K in the forward path and then to add unity feedback around the amplifier and the plant. That type of configuration is referred to as *uncompensated* in this text, although some engineers might also call this configuration *proportional* or *P-compensated*. Additional elements (compensators) may be added to the forward path and/or to the feedback path to improve steady state accuracy, stability, or both.

A five-step procedure is suggested for compensator design. First, the uncompensated system gain is selected to provide a trade-off between steady state accuracy and stability. Second, if that performance is not satisfactory, an appropriate compensator is selected to improve some undesirable trait of the compensated system. Third, compensator parameters are chosen to trade off steady state performance and stability. A root locus is useful for changing values of one parameter when all other parameters are fixed. Fourth, given the design from step 3, gain K is adjusted to improve the time response. Fifth, steps 3 and 4 are repeated as needed to improve performance.

The cascade proportional integral (PI) compensator adds a new pole to the forward path, which improves steady state accuracy by increasing the system type number. A new zero is also added so that the difference between the number of open-loop poles and open-loop zeros remains as with the uncompensated case. The gain K may be the same as for the uncompensated case, while the zero is adjusted for maximum relative stability. A cascade lag compensator adds a new pole and zero to the system, but the pole is not at the origin. The error coefficient may be improved by a desired multiplying factor. The gain K and the new zero may then be selected as for the PI compensator.

A cascade lead compensator also adds a new pole and zero to the system, but the new zero is closer to the origin than the new pole, opposite to their relationship in a cascade lag compensator. The primary purpose of cascade lead compensator is to improve stability. The gain K is usually higher than for the uncompensated case, while the zero is adjusted for maximum relative stability. A cascade lag–lead compensator improves steady state accuracy and stability, so the design rules for lag and lead compensation are combined.

A feedback rate (derivative) compensator can be added to the feedback path while a gain K is added to the forward path. The combination provides PD compensation, which primarily improves stability. The gain K is usually higher than for the uncompensated case, while one coefficient is adjusted for maximum relative stability. A PID compensator provides proportional, integral, and derivative operation so that the design combines PI design with PD design to improve steady state accuracy and stability. Chapter 5 uses one example system to compare and contrast the design rules.

Poles may also be placed at exact locations rather than in some region of the complex plane as with root locus design. Poles may be placed to shape steady state and transient performances. Two types of compensator may be used: fixed-structure compensators (like a PID compensator) of fixed form, or algebraic compensator, which may be quite complex (chosen to force all poles to be at predetermined locations).

The chapter concludes with three design examples: a high-performance aircraft that is open-loop unstable, control of a flexible space station, and control of a solar furnace. It is strongly suggested that a computer-aided approach be taken to fully explore the design options. The rltool option of MATLAB is a good example of one such aid.

REFERENCES

Compensator Design

Byrne, Raymond H., et al., "Experimental Results" in Robust Lateral Control of Highway Vehicles," *IEEE Control Syst. Mag.*, vol. 18, no. 2, April 1998, pp. 70–76.

Herget, D., "A Closed Loop PID Control Program for Process Control." *Int. J. Appl. Eng. Educ.* vol. 5, no. 1, (1989), pp. 83–88.

Ho, Wang Khuen, et al., "Self-Tuning PID Control of a Plant with Under-Damped Response with Specifications on Gain and Phase Margins." *IEEE Trans. Control Syst. Technol.*, vol. 5, no. 4, July 1997, pp. 446–452.

Mosch, H.E., and Steiner, H., "Robust PID Control for an Industrial Distillation Column." *IEEE Control Syst. Mag.*, vol. 15, no. 4, August 1995, pp. 46–55.

Ohta, T. et al., "A New Optimization Method of PID Control Parameters for Automatic Tuning by Process Computer," *Computer Aided Design of Control Systems, Proceedings of the IFAC Symposium*, Zurich, 1979, pp. 133–138.

Pole Placement

Chen, C.T. and Seo, B., "Applications of the Linear Algebraic Method for Control System Design." *IEEE Control Syst. Mag.* (January 1990): 43–47.

Chen, C.T., *Control System Design: Conventional, Algebraic and Optimal Methods.* New York: Holt, Rinehart & Winston, 1984.

Maffezoni, C. et al., "Robust Design of Cascade control." *IEEE Control Syst. Mag.* (January 1990): 21–25.

Aircraft Design

Chant, C., *Aviation: An Illustrated History.* Secaucus, NJ: Chartwell Books, 1978.

Culic, F.E.C., "The Origins of the First Powered. Man-Carrying Airplane." *Sci. Am.* (July 1979): 86–100.

Encyclopedia of Transportation. Chicago: Rand McNally, 1976.

Markowski, M.A. "Ultralight Airplanes." *Sci. Am.* (July 1982): 62–68.

Reed, F., "The Electric Jet." *Air Space* (December 1986/January 1987): 42–48.

Flexible Space Structures

Bar-Kana, I., Fischl, R., and Kalata, P., "Direct Position Plus Velocity Feedback Control of Large Flexible Space Structures." *IEEE Trans. Autom. Control* (October 1991): 1186–1188.

Solar Furnace

Berenguel, M. et al., "Temperature Control of a Solar Furnace." *IEEE Control Syst. Mag.*, vol. 19, no. 1, February 1999, pp. 8–24.

PROBLEMS

1. Each of the following systems consists of the plant driven an amplifier with gain K, as in Figure 5.1. For each of the following plant transmittances and amplifier gains, find the closed-loop system poles. Then find the step error coefficient K_0.

   (a) $G_p(s) = \dfrac{s+4}{s+10}$
   $K = 10$

   (b) $G_p(s) = \dfrac{s+10}{s+4}$
   $K = 10$

   Ans. -9.45; 25

   (c) $G_p(s) = \dfrac{s+10}{(s+2)^2}$
   $K = 10$

   (d) $G_p(s) = \dfrac{1}{(s+10)(s+4)}$
   $K = 58$

   Ans. $-7 \pm j7$; 1.45

2. Design cascade PI compensators for unity feedback systems with the following plant transmittances. For each design, find the steady state error coefficients and the dominant roots, if a is adjusted for maximum relative stability with $K = K_u$. Then adjust K to improve settling time, if possible.

   (a) $G_p(s) = \dfrac{6}{(s+4)^2}$
   $K_u = 2.67$

   (b) $G_p(s) = \dfrac{1}{(s+4)^3}$
   $K_u = 40$

   (c) $G_p(s) = \dfrac{s+9}{(s^2+4s+40)}$
   $K_u = 15$

$$\text{(d)} \quad G_p(s) = \frac{2s^2 + 10s + 13}{(s+4)(s^2 + 3s + 10)}$$
$$K_u = 1$$

Ans. (a) $a = 2.96,\ K = 2.67;\ K_0 = \infty,\ K_1 = 2.96;\ -2.67 \pm j3.27,\ -2.67$

3. Design cascade lag compensators for unity feedback systems with the following plant transmittances. For each design find the steady state error coefficients and the dominant roots, if a is adjusted for maximum relative stability with $K = K_u$. Then adjust K to improve settling time if possible.

$$\text{(a)} \quad G_p(s) = \frac{10}{s^2 + 4s + 10}$$
$$K_u = 10,\ d = 10$$

$$\text{(b)} \quad G_p(s) = \frac{s+4}{s^2 + 10}$$
$$K_u = 10,\ d = 10$$

$$\text{(c)} \quad G_p(s) = \frac{2s + 20}{s^2 + 9}$$
$$K_u = 10,\ d = 10$$

$$\text{(d)} \quad G_p(s) = \frac{6}{(s+5)(s^2 + 4s + 10)}$$
$$K_u = 10,\ d = 10$$

4. A plant to be controlled has transmittance

$$G_p(s) = \frac{s+8}{s(s^2 + 4s + 40)}$$
$$K_u = 20$$

Use the root locus methods to select a rate feedback compensator. Adjust K before selecting A so K_1 (comp) is at least as large as K_1 (uncomp).

5. Use root locus methods to select a PID compensator for

$$G_p(s) = \frac{1}{(s+4)(s^2 + 10s + 50)}$$
$$K_u = 600$$

6. A fine positioning system for an elevator when it is in the vicinity of the correct floor is modeled in Figure P5.6. Use root locus methods to locate the overall system poles as a function of $K_1 > 0$, when $K_2 = 0$. Then, with a specific choice of K_1, use root locus methods to determine the overall system poles as a function of $K_2 > 0$. Using several iterations of choosing K_1, then choosing K_2, find values of K_1 and K_2, if possible, for which the system has a pair of complex poles with damping ratio approximately 0.7.

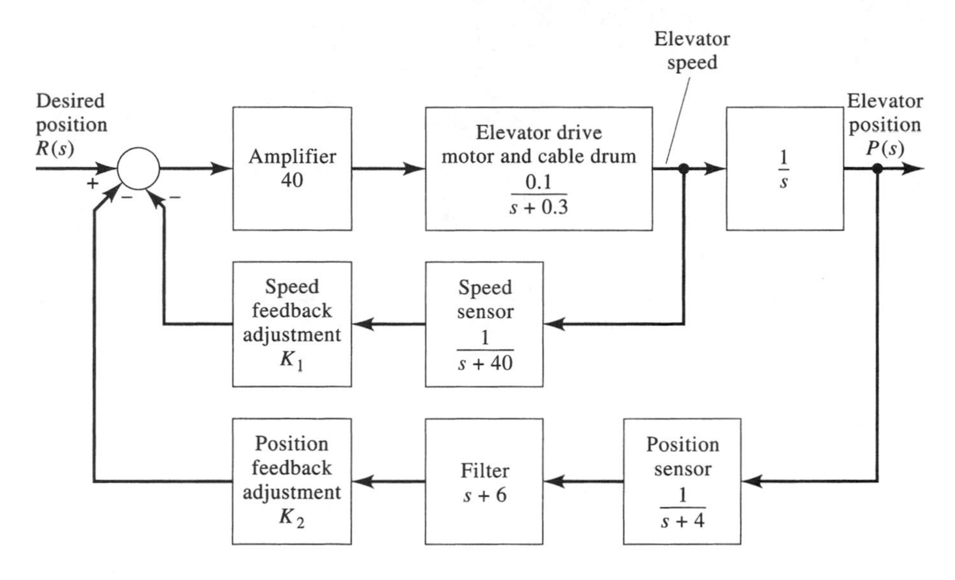

Figure P5.6

7. A linearized inventory control system model is given in Figure P5.7. Based upon the difference between desired and actual inventory levels, management sets the production quota, which in turn determines the production rate. Use root locus methods to determine the production time constant

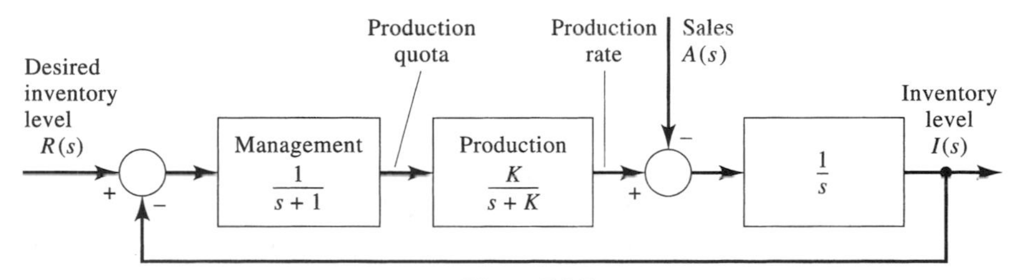

Figure P5.7

$$\tau = \frac{1}{K}$$

in the range

$$1 < \tau < 10$$

which gives greatest relative stability of the system.

Ans. $K = 1$

8. The pitch control system for high-altitude aircraft is modeled in Figure P5.8. Find the value of the adjustable constant $K > 0$ for which the potentially complex pole pair of the closed-loop system has critical damping.

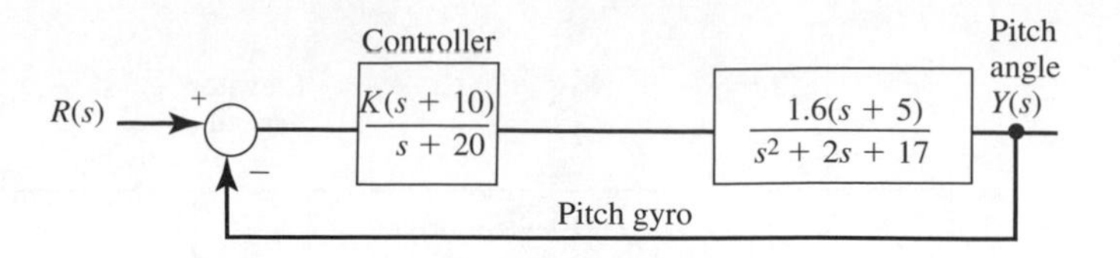

Figure P5.8

Ans. $K = 70.5$

9. Design K_u for plants with the following transmittances. Select a damping ratio of 0.707 for the dominant roots, if possible.

(a) $G_p(s) = \dfrac{3s}{s^2 + 4}$

(b) $G_p(s) = \dfrac{5}{s^2 + 4s + 5}$

(c) $G_p(s) = \dfrac{1}{s^3 + 6s^2 + 9s}$

(d) $G_p(s) = \dfrac{s + 10}{s(s^2 + 6s + 13)}$

Ans. (a) 0.943 (b) 0.6

10. It is desired that, for a plant with transmittance

$$G_p(s) = \frac{s + 3}{(s + 2)(s + 2 + 0.5j)(s + 2 - 0.5j)}$$

the steady state error to a step input be no more than 0.5 and the damping ratio of the dominants roots should be no less than 0.7. Find a range of values for $G_c(s) = K$ that achieves both goals.

11. The control of a medical patient's heart rate by an implanted pacemaker with a heart rate sensor is modeled in Figure P5.11. Use a root locus diagram to design the pacemaker gain K for response with critical damping. In terms of K, find the steady state error to an input

$$r(t) = (80 + 0.2t)u(t)$$

Ans. $25;\ 2/K$

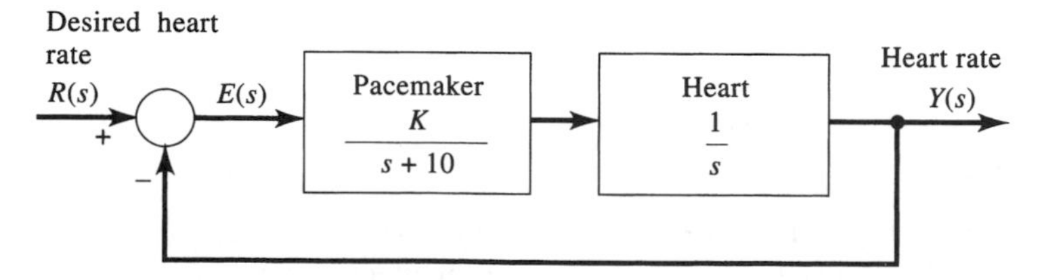

Figure P5.11

12. Design K_u for unity feedback systems with the following plant transmittances. Select K_u so that the dominant roots are 0.5 damped.

(a) $G_p(s) = \dfrac{9}{(s+3)^2}$

(b) $G_p(s) = \dfrac{s+20}{(s+3)^3}$

(c) $G_p(s) = \dfrac{s+5}{(s^2+4s+40)}$

13. Design cascade PI compensators for unity feedback systems with the following plant transmittances. For each design, find steady state errors to unit step and ramp inputs, the amount of relative stability, and the damping ratios of any complex closed-loop poles. Choose a for maximum relative stability. Leave $K = K_u$.

(a) $G_p(s) = \dfrac{10}{s^2+7s+10}$
$K_u = 1$

(b) $G_p(s) = \dfrac{s+2}{(s^2+4s+3)s}$
$K_u = 2$

(c) $G_p(s) = \dfrac{s+4}{s^2+5s+6}$
$K_u = 2$

(d) $G_p(s) = \dfrac{6s+30}{s^2+4s+13}$
$K_u = 1$

14. Design cascade lag compensators for unity feedback systems with the following plant transmittances. Find the steady state errors to unit step and ramp inputs, the amount of relative stability, and the damping ratios of any complex closed-loop poles. Choose a for maximum relative stability and leave $K = K_u$.

(a) $G_p(s) = \dfrac{3s+30}{s(s^2+4s+13)}$
$K_u = 4,\ d = 10$

(b) $G_p(s) = \dfrac{3}{s^2+10s}$
$K_u = 20,\ d = 10$

15. Design cascade lead compensators for unity feedback systems with the plant transmittances of Problem 14. Find the steady state errors to unit step and ramp inputs, the amount of relative stability, and the damping ratios of any complex closed-loop poles. Let $K = K_u d$; then choose a for maximum relative stability.

16. Design cascade lag–lead compensators for unity feedback systems with the plant transmittances of Problem 14. Find the steady state errors to step and ramp inputs, the amount of relative stability, and the damping ratios of any complex closed-loop poles.

17. Design feedback rate compensators for systems with the following plant transmittances. Use root locus plots to help select the compensator parameters. Let $K = 2K_u$; then select A for maximum relative stability. Find the steady state errors to step and ramp inputs, the amount of relative stability, and the damping ratios of any complex closed-loop poles.

(a) $$G_p(s) = \frac{1}{s(s+10)^2}$$ $$K_u = 250$$

(b) $$G_p(s) = \frac{10}{s^2 + 7s + 10}$$ $$K_u = 1$$

18. Design cascade PID compensators for unity feedback systems with the plant transmittances of Problem 17. Select a for the PI compensator to achieve maximum relative stability when $K = K_u$. For the final PID design let $K = 2K_u$. Find the steady state errors to step and ramp inputs, the amount of relative stability, and the damping ratios of any complex closed-loop poles.

19. A plant to be controlled has transmittance

$$G_p(s) = \frac{5}{s(s^2 + 4s + 5)}$$

Use root locus methods to design tracking systems of each of the following types:

(a) Uncompensated, using $G_c(s) = K = K_u$, so the damping ratio of the dominant roots is 0.707.

(b) Cascade lag compensated, $d = 10$, choosing a for maximum relative stability. Let $K = K_u$.

(c) Rate feedback. Let $K = 2K_u$. Choose A for maximum relative stability.

(d) Cascade lead compensation with $d = 10$. Let $K = K_u d$ and choose a for maximum relative stability.

For each design, find the steady state error to unit step, ramp, and parabolic inputs, the amount of relative stability, and the damping ratios of any complex closed-loop poles.

20. For a plant with transmittance

$$G_p(s) = \frac{1}{(s+1)(s^2 + 4s + 5)}$$

it is desired that the steady state error to a step input be no more than 5%, and with this requirement that the relative stability be maximum (or nearly so). If possible, design a cascade lag compensator with $d = 10$.

21. An electrohydraulic positioning mechanism is shown schematically in Figure P5.21(a). The block diagram of a linearized model of this system is shown in Figure P5.21(b). Use root locus methods to choose the gain K for a damping ratio of 0.7 for the dominant roots. For this value of K, what is the steady state error to a unit ramp input.

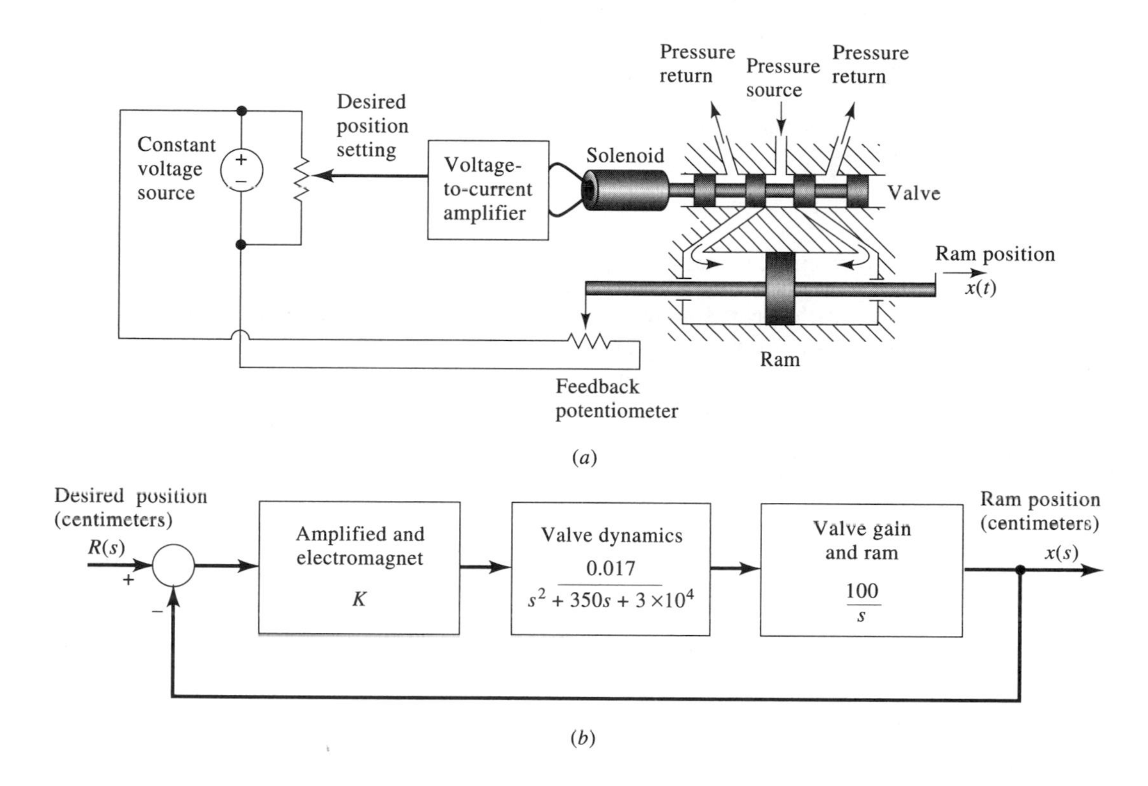

Figure P5.21

22. Without a feedback control system, most helicopters are unstable. For helicopter dynamics having unstable transmittance,

$$G_p(s) = \frac{10(s + 0.05)}{(s + 0.5)[(s - 0.2)^2 + (0.4)^2]}$$

$$K_u = 1$$

relating pitch angle to blade control (Figure P5.22), design a feedback compensator

$$H_c(s) = 1 + AS$$

Let $K = 2K_u$ and select A for maximum relative stability.

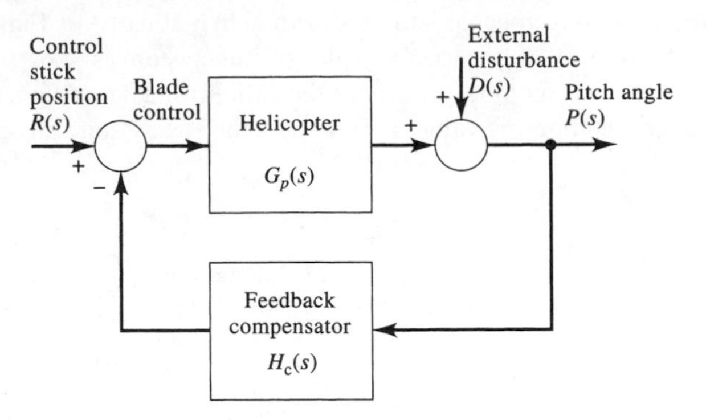

Figure P5.22

23. Suppose we want to place closed-loop poles exactly.

 (a) Select a second-order transfer function with a damping ratio of 0.7 and an undamped natural frequency of 4 rad/s. There should be zero steady state error for a step input.

 Ans. $T(s) = 16/(s^2 + 5.6s + 16)$

 (b) Suppose the plant in Figure 5.18 is

$$G_p(s) = \frac{1}{(s+2)^2}$$

 Select an algebraic cascade compensator as in Figure 5.18 to create the transfer function of (a).

 (c) Sketch the root locus of your design, where $K = 1$ should generate the desired closed-loop poles.

24. For the plant and transfer function of Problem 23, Select a PID compensator

$$G_c(s) = \frac{K_d s^2 + K_p s + K_i}{s}$$

 so that the third pole is at -20.

25. As in Problem 23, suppose we want to place closed-loop poles exactly.

 (a) Select a third-order transfer function with two critically damped closed-loop poles at $s = -5$ and one other closed-loop pole at $s = -50$. There should be zero steady state error for a step input.

$$\textbf{Ans.}\quad T(s) = \frac{1250}{(s+5)^2(s+50)}$$

 (b) Suppose the plant in Figure 5.18 is

$$G_p(s) = \frac{1}{(s+1)(s^2+4)}$$

 Select an algebraic cascade compensator as in Figure 5.18 to create the transfer function of (a).

(c) Sketch the root locus of your design, where $K = 1$ should generate the desired closed-loop poles.

26. If the plant can change (or if the model for the plant may be incorrect), it is desirable to design a robust compensator to reduce the impact of using the wrong $G_p(s)$.

(a) Suppose the plant of Problem 25(b) is actually

$$G'_p(s) = \frac{1}{(s+2)(s^2+4)}$$

but the compensator of Problem 25(b) is used. Find $T'(s)$. Do the desired closed-loop poles change?

(b) Refer to the design methodology of Figure 5.21. For the plant of Problem 25(b), assume the part that will not change is

$$G_{p2}(s) = \frac{1}{s^2+4}$$

while the part that might change is

$$G_{p1}(s) - \frac{1}{s+1}$$

Further, for the $T(s)$ of Problem 25(a), assume we do not want the closed-loop poles of

$$T_2(s) = \frac{25}{(s+5)^2}$$

to change but we will allow the closed-loop pole of

$$T_1(s) = \frac{50}{s+50}$$

to change if $G_{p1}(s)$ changes. Thus

$$G_p(s) = G_{p1}(s) \quad G_{p2}(s)$$
$$T(s) = T_1(s)T_2(s)$$

Find the robust compensator of Figure 5.21; that is, calculate

$$G_{c1}(s), G_{c2}(s), H_1(s)$$

(c) Suppose $G'_{p1}(s)$ changes to

$$G'_{p1}(s) = \frac{1}{s+2}$$

Find $T'_1(s)$, the changed version of $T_1(s)$.

27. Figure P5.27 shows a plant for a space station having two resonant modes. A rate feedback compensator has been chosen to place an open-loop zero at -10. Select K so that the system is critically damped. Find the step response for your design.

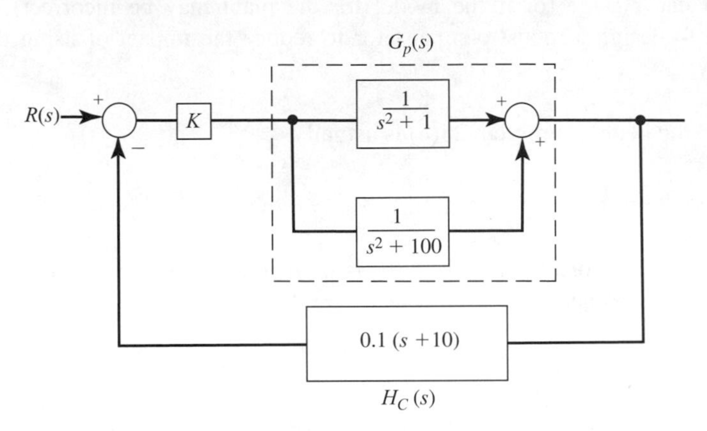

Figure P5.27

CHAPTER

6

Frequency Response Analysis

6.1 Preview

Whenever a radio or TV is tuned a particular station, some circuit's performance varies so as to tune in the desired signal and reject undesired signals. It is logical that a key to designing an effective communication system is to understand the behavior of that system as a function of the incoming frequencies.

Research by Nyquist in the 1930s led to a book by Bode in the 1940s in which the same frequency response methods used to understand electrical circuits were used to understand the stability of control systems. Once a system's performance is understood, remedies can be created to reduce unwanted effects.

The technology of the 1930s was based on analysis and design using slide rules and simple (by current standards) mechanical computational aids. Frequency response computations are fairly easy to perform, so these methods were applied with vigor. Further, the ease with which the frequency response can be measured experimentally was another reason for the acceptance of this method. While the frequency response is usually found for linear systems, the method can be applied experimentally to nonlinear systems, as we will see later in this chapter.

Three viewpoints are generally taken when designing and analyzing control system behavior. First, behavior as a function of s is evaluated using Routh–Hurwitz methods (dating to the 1890s when steam engines used mechanical governors) and root locus methods (dating to Walter Evans in the late 1940s). Second, time domain behavior can be determined by inverse Laplace-transforming or by applying techniques to be developed in Chapters 7 and 8. Third, behavior as a function of frequency requires use of $s = j\omega$; this approach is the topic of this chapter.

The three approaches are not mutually exclusive. In Chapter 5 it was pointed out that dominant pole locations in the s plane imply time domain response in terms of undamped natural frequency and damping ratio. Later in this chapter we will show that frequency domain figures of merit are also related to dominant pole locations (and therefore to the time domain).

Root locus, time domain, and frequency domain methods provide a comprehensive viewpoint of the strengths and weaknesses of a system. All three methods should be applied to fully understand (and improve) system performance.

As was mentioned in the preface and in the preview of Chapter 4, a number of computational packages are available. However, we again stress the importance of the user backing up and double-checking computer results with manual sketches and numerical values. In many cases, very accurate frequency response data and sketches can be arrived at without use of a PC or other large-scale computational device.

6.2 Frequency Response

6.2.1 Forced Sinusoidal Response

In the development to follow, a general symbol for a transmittance, $F(s)$ will be used. In later sections of this chapter, $F(s)$ is taken to represent the transmittance of a system component or the loop transmittance $G(s)H(s)$ of a feedback system, depending upon the application at hand.

The response of a system with transmittance $F(s)$ to a sinusoidal input signal,

$$r(t) = B\cos(\omega t + \beta)$$

generally consists of both forced and natural components. The forced part of the response is also sinusoidal, with the same frequency ω as the input. Generally, the amplitude and the phase angle of the forced sinusoidal response are different from those of the input and they depend upon the input frequency:

$$y_{\text{forced}}(t) = C\cos(\omega t + \gamma) = A(\omega)B\cos[\omega t + \beta + \Phi(\omega)]$$

The magnitude and phase of the frequency response is defined.

The magnitude of the transmittance when evaluated at $s = j\omega$ is the ratio of output amplitude to input amplitude:

$$A(\omega) = |F(s = j\omega)| = \frac{C}{B}$$

The angle of the transmittance, for $s = j\omega$, is the difference in phase angles between output and input:

$$\Phi(\omega) = \angle F(s = j\omega) = \gamma - \beta$$

As a numerical example, consider the transmittance

$$F(s) = \frac{6}{s + 4}$$

and the input signal

$$r(t) = 3\cos(7t + 20^\circ)$$

as in Figure 6.1. For $s = j7$, the transmittance is

$$F(s = j7) = \frac{6}{j7 + 4} = 0.74e^{-j60^\circ}$$

The forced sinusoidal output has amplitude and phase angle

The forced response follows easily from $A(\omega)$ and $\phi(\omega)$.

$$C = (0.74)(3) = 2.22 \quad \gamma = 20^\circ + (-60^\circ) = -40^\circ$$

giving

$$y_{\text{forced}}(t) = 2.22 \cos(7t - 40^\circ)$$

This important property only holds for linear time-invariant systems. In general, if a system is nonlinear or time varying, the response to sinusoids will not necessarily be sinusoids. For example, consider the following nonlinear differential equation (Van der Pol equation)

$$\ddot{y} = -y + \dot{y}(1 - y^2) + u$$

Because the derivative term is multiplied by $(1-y^2)$ the equation is both nonlinear and time-varying. The response of this system to two sinusoidal inputs of two frequencies are shown in Figure 6.1(b–c).

$$u = \sin(0.5t) \qquad \text{and} \qquad u = \sin(t)$$

Note that the output is not a sinusoid and, in addition, the form of the output changes when the input frequency is changed. In linear systems, only the amplitude and phase may change as a result of a change in frequency. For this reason, frequency response is an important tool for linear systems and cannot directly be used in nonlinear systems.

In general, the values of $A(\omega)$ and $\Phi(\omega)$ vary considerably as frequency varies. Figure 6.2 shows a hypothetical example in which the amplitude of $y(t)$ varies for different input frequencies because of the frequency characteristics of $F(s = j\omega)$. For a fixed input amplitude, we refer to the situation of the amplitude of the output being larger in some frequency range (as in Figure 6.2) as *band-pass behavior*. When low-frequency signals have larger amplitudes than high-frequency signals, the behavior is termed *low-pass*. Finally, when the situation is reversed so that the high-frequency signals have a larger amplitude, we have *high-pass behavior*. Communications engineers use frequency response methods so select circuits that shape the spectrum of signals for various purposes. Although our interest is primarily in the stability of control systems using frequency response methods, it is important to be able to measure the frequency response of system components accurately, sometimes by applying sinusoidal signals in much the same way as a communications engineer would test a transmitting device.

Low-pass, high-pass, and band-pass behaviors are defined.

6.2.2 Frequency Response Measurement

When a transmittance $F(s)$ is driven by a sinusoidal input signal, the ratio of the amplitude of the forced output to the amplitude of the input is

$$A(\omega) = \frac{\text{amplitude of sinusoidal output}}{\text{amplitude of sinusoidal input}} = |F(s = j\omega)| \qquad \textbf{[6.1]}$$

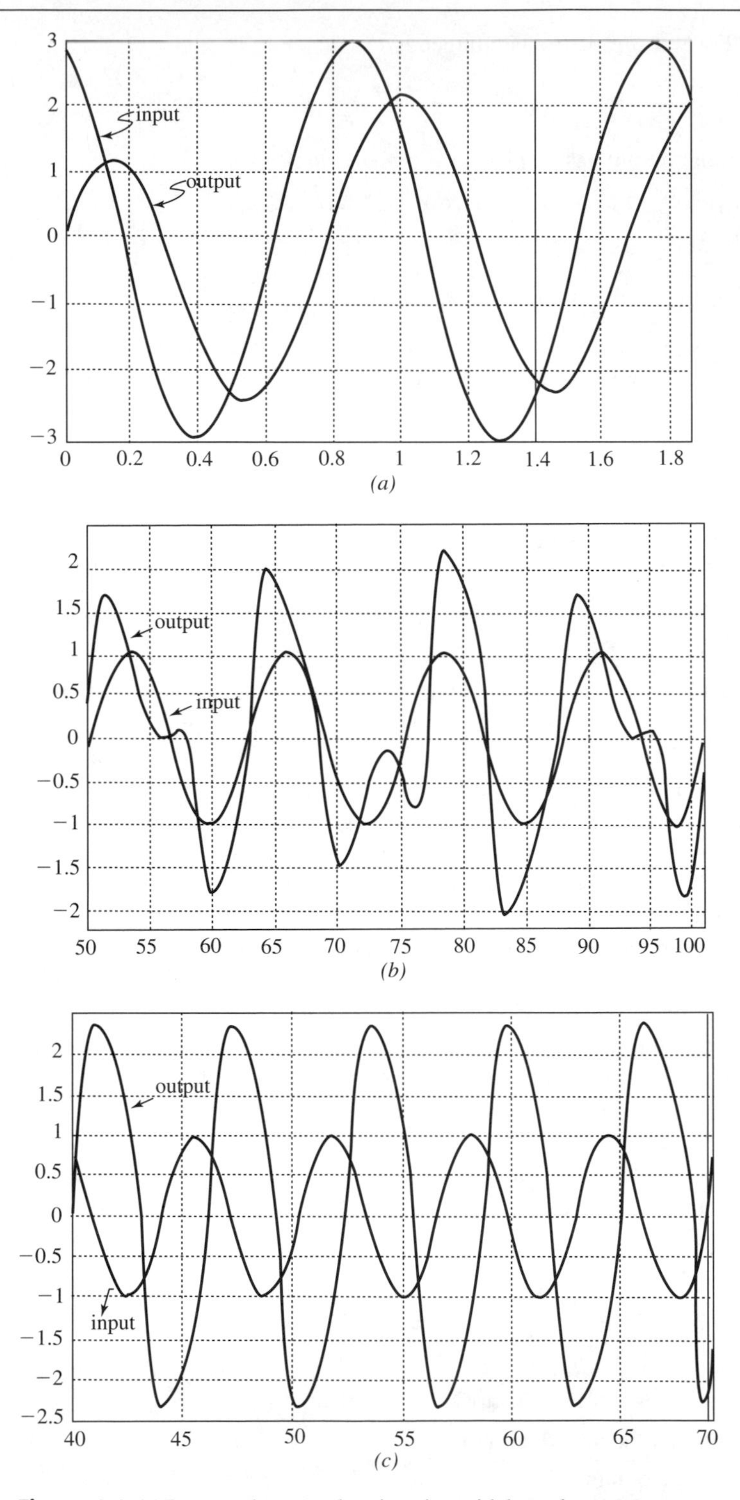

Figure 6.1 (a) Input and output showing sinusoidal steady state response. (b) Response of Van der Pol equation to input $u(t) = \sin(0.5t)$. (c) Response of Van der Pol equation to input $u(t) = \sin(t)$.

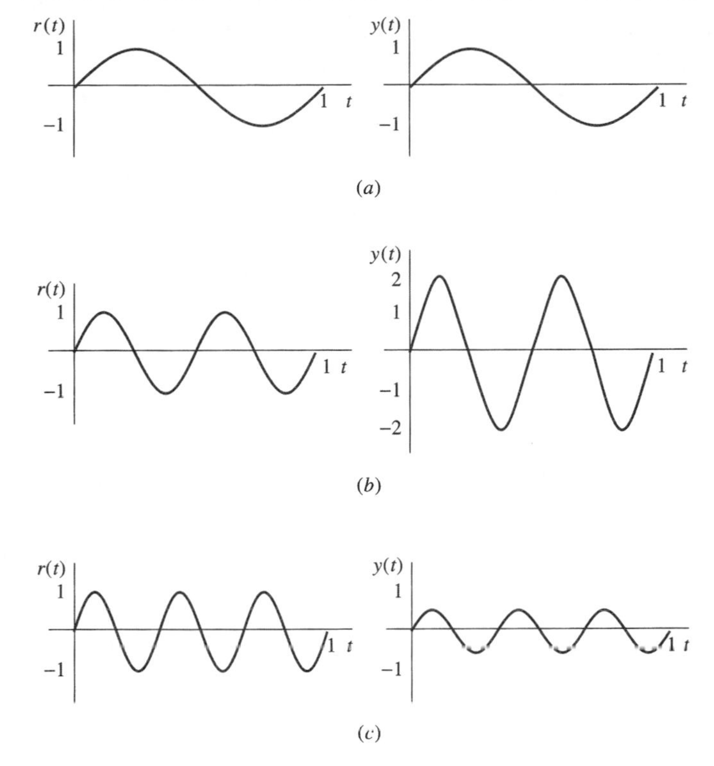

Figure 6.2 Forced response. (a) $\omega = 2\pi$, $A(2\pi) = 1$, $\Phi(2\pi) = 0°$. (b) $\omega = 4\pi$, $A(4\pi) = 2$, $\Phi(4\pi) = 0°$, (c) $\omega = 6\pi$, $A(6\pi) = 0.5$, $\Phi(6\pi) = 0°$.

which is a function of the radian frequency of the sinusoid, ω. The difference in the phase angles of the output and the input is

$$\begin{aligned}\Phi(\omega) &= \text{phase angle of sinusoidal output} - \text{ phase angle of sinusoidal input} \\ &= \angle F(s = j\omega) \end{aligned} \quad [6.2]$$

To measure the frequency response of a stable component or system at some frequency, apply a sinusoidal input of that frequency. Choose any convenient amplitude that is not so large as to overload the system, yet not so small that the system signals are masked by noise. Wait until the system natural behavior, arising from the connection of the input, dies out. Measure the amplitudes of the input and output sinusoids and form the ratio of Equation (6.1). The phase shift is the difference in phase between the forced-output sinusoid and the sinusoidal input signal. If the two signals can be plotted side by side, the phase shift will be found by comparing the plots. Stroboscopic light techniques are especially useful for comparing the phases of mechanical elements in rapid motion. If the system signals are electrical or can be monitored by electrical transducers, the elliptical pattern formed by plotting the input versus the output signal on an oscilloscope or instruments known as *phase meters* or *vector voltmeters* can be used to display the amount of phase shift.

With a number of such measurements of amplitude ratio and phase shift at various frequencies, curves for $A(\omega)$ and $\Phi(\omega)$ can be sketched.

6.2.3 Response at Low and High Frequencies

There are some considerations that make the determination of amplitude and phase shift curves from a limited number of measurements more accurate. If the transmittance is a rational function

$$F(s) = \frac{b_m s^m + b_{m-1}s^{m-1} + \cdots + b_1 s + b_0}{s^n + a_{n-1}s^{n-1} + \cdots + a_1 s + a_0}$$

then

$$F(j\omega) = \frac{b_m (j\omega)^m + b_{m-1}(j\omega)^{m-1} + \cdots + b_1 (j\omega) + b_0}{(j\omega)^n + a_{n-1}(j\omega)^{n-1} + \cdots + a_1 (j\omega) + a_0}$$

Approximating high frequency behavior.

For large values of ω, all but the highest powers of ω in the numerator and denominator can be ignored:

$$F(j\omega) \cong \frac{b_m (j\omega)^m}{(j\omega)^n} = b_m j^{m-n} \omega^{m-n}$$

The amplitude curve approaches a power of ω, and the phase curve approaches a multiple of 90°.

Approximating low frequency behavior.

For small ω, all but the lowest powers of ω in the transmittance numerator and denominator are negligible. If a_0 and b_0 are nonzero,

$$F(j\omega) \cong \frac{b_1 (j\omega) + b_0}{a_1 (j\omega) + a_0} \cong \frac{b_0}{a_0}$$

In general, at low frequencies, a rational transmittance's amplitude curve approaches a power of ω or a constant and the phase curve approaches a multiple of 90°, 0°, or 180°.

For example, consider

$$F(s) = \frac{6s^3 + 2s^2 + 3s}{s^5 + 4s^4 + 2s^3 + s^2 + s + 10}$$

At high frequencies,

$$F(j\omega) = \frac{6(j\omega)^3}{(j\omega)^5} = -6\omega^{-2}$$

$$A(\omega) = |F(j\omega)| \cong \frac{6}{\omega^2} = 6w^{-2} \quad \Phi(\omega) = \angle F(j\omega) \cong 180°$$

At low frequencies,

$$F(j\omega) \cong \frac{3(j\omega)}{10}$$

$$A(\omega) = |F(j\omega)| \cong \tfrac{3}{10}\omega \quad \Phi(\omega) = \angle F(j\omega) \cong 90°$$

When the experimenter is certain that measurements are being made in the small-ω or large-ω region, a few measurements will suffice for the entire region. It is most useful to make measurements at more closely spaced frequencies where the largest changes in amplitude or phase occur.

❑ DRILL PROBLEMS

D6.1 Find the forced sinusoidal response of each of the following transmittances to the indicated input signals:

(a)

$$F(s) = \frac{s}{s+3}$$

$$r(t) = 7\cos(3t - 40°)$$

Ans. $(7/\sqrt{2})\cos(3t + 5°)$

(b)

$$F(s) = \frac{4}{s+2}$$

$$r(t) = 6\cos(5t + 30°)$$

Ans. $4.46\cos(5t - 38°)$

(c)

$$F(s) = \frac{10}{s^2 + 3s + 10}$$

$$r(t) = 8\cos(2t + 70°)$$

Ans. $(40/3\sqrt{2})\cos(2t + 25°)$

D6.2 For transmittances with the following amplitude ratios and phase shift functions, find the forced sinusoidal response to the given input signal:

(a)

$$A(\omega) = \frac{1}{\sqrt{\omega^2 + 100}}$$

$$\Phi(\omega) = \tan^{-1}\frac{\omega}{10}$$

$$r(t) = 7\cos(6t + 80°)$$

Ans. $0.6\cos(6t + 111°)$

(b)

$$A(\omega) = 4/(\sqrt{\omega^2 + 4})$$

$$\Phi(\omega) = 90° - \tan^{-1}(\omega/2)$$

$$r(t) = 3\cos 2t$$

Ans. $(3\sqrt{2})\cos(2t + 45°)$

(c)

$$A(\omega) = 4$$

$$\Phi(\omega) = -3\omega(\text{rad})$$

$$r(t) = 10\cos(5t - 30°)$$

Ans. $40\cos(5t - 170°)$

6.2.4 Graphical Frequency Response Methods

$A(\omega)$ and $\phi(\omega)$ can be obtained graphically.

Given the pole–zero plot for a transmittance, its evaluation for various values of $s = j\omega$ can be done graphically by drawing sets of directed line segments to points of evaluation on the imaginary axis. With a little practice, the amplitude and phase shift curves can be roughly sketched from just a few evaluations.

For example, the transmittance with the pole–zero plot given in Figure 6.3(a) is

$$F(s) = \frac{s+3}{(s+1+j3)(s+1-j3)}$$

where the directed line segments are defined by vectors υ_1, υ_2, and υ_3. For s restricted to the imaginary axis

$$F(s = j\omega) = \frac{|\upsilon_1|\,(\angle\upsilon_1 - \angle\upsilon_2 - \angle\upsilon_3)}{|\upsilon_2|\,|\upsilon_3|}$$

This expression can be calculated mathematically, beginning with the general equation

$$F(j\omega) = \frac{j\omega+3}{(j\omega+1+j3)(j\omega+1-j3)} = \frac{j\omega+3}{[1+j(\omega+3)][1+j(\omega-3)]}$$

The magnitude and phase are

$$A(\omega) = \frac{\sqrt{\omega^2+9}}{\sqrt{1+(\omega+3)^2}\sqrt{1+(\omega-3)^2}}$$

$$\Phi(\omega) = \tan^{-1}(\omega/3) - \tan^{-1}(\omega+3) - \tan^{-1}(\omega-3)$$

For example, if $\omega = 3$, then

$$A(\omega) = \frac{\sqrt{18}}{\sqrt{37}\sqrt{1}} = 0.70$$

$$\Phi(\omega) = \tan^{-1}(1) - \tan^{-1}(6) - \tan^{-1}(0) = 45° - 80.5° - 0° = -35.5°$$

A number of magnitude and phase values can be computed and plotted as in Figure 6.3(b). That figure can be interpreted as the result of changes in the three vector magnitudes and phase angles in Figure 6.3(a) as $s = j\omega$ moves up the imaginary axis. As we shall soon see, it is also possible to plot magnitude and phase versus the logarithm of frequency, rather then versus linearly plotted frequency.

Visualizing the frequency response from the poles and zeros.

The amplitude curve for a transmittance can be easily visualized from the pole–zero plot by imagining a very flexible rubber sheet suspended over the complex plane. The sheet is poked up by thin rods at each pole location and tacked down to the plane at each zero location. The height of the rubber sheet will represent $|F(s)|$ for each value of s. The frequency response amplitude is a cross section of the sheet displacement along the imaginary axis, as indicated in Figure 6.4.

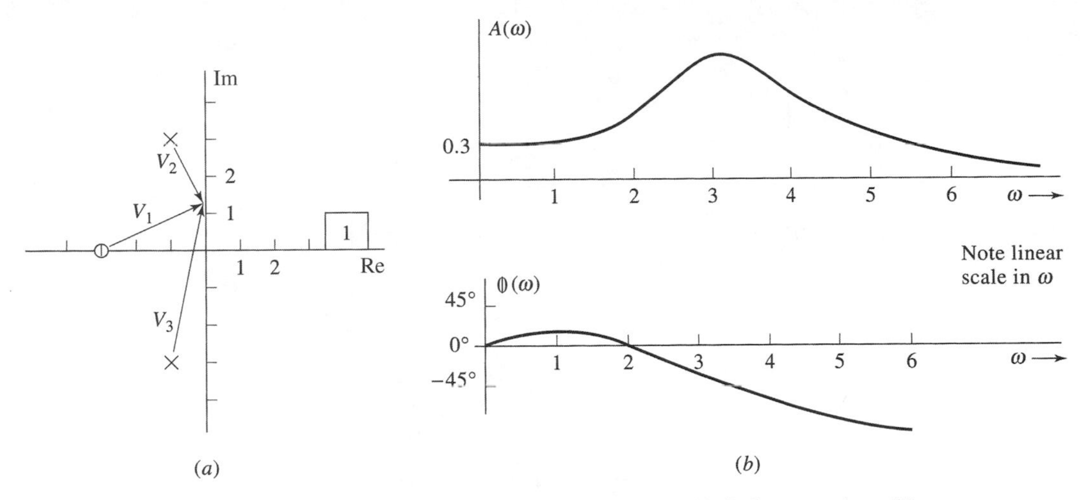

Figure 6.3 Sketching frequency response from a pole–zero plot. (a) Pole–zero plot. (b) Approximate frequency response.

To use the rubber sheet analogy effectively, the height of the sheet at large distances from the origin, that is, as $|s| \rightarrow \infty$, must be determined. If

$$F(s) = \frac{b_m s^m + b_{m-1} s^{m-1} + \cdots + b_1 s + b_0}{a_n s^n + a_{n-1} s^{n-1} + \cdots + a_1 s + a_0}$$

$$\lim_{|s| \rightarrow \infty} |F(s)| = \left| \frac{b_m}{a_n} \right| \frac{|s|^m}{|s|^n}$$

If the number of poles of $F(s)$ is greater than its number of zeros,

$$\lim_{|s| \rightarrow \infty} |F(s)| = 0$$

and the rubber sheet is visualized as being tacked down to the complex plane at large distances from the origin, as in Figure 6.5(a).

If $F(s)$ has an equal number of poles and zeros,

$$\lim_{|s| \rightarrow \infty} |F(s)| = \left| \frac{b_m}{a_n} \right|$$

and the rubber sheet approaches a fixed height above the complex plane, as in Figure 6.5(b). If there are more zeros than poles in $F(s)$,

$$\lim_{|s| \rightarrow \infty} |F(s)| = \infty$$

This case is unusual in practical applications because it means that the amplitude ratio of the transmittance increases with frequency without bound.

Imaginary axis zeros and poles give zero or infinite amplitude for the values of ω where they occur. The phase shift is discontinuous at these values of ω, but limiting values of the phase shift are found by considering imaginary axis points slightly below and slightly above the zero or pole. Complex conjugate poles on the imaginary axis mean that the corresponding system natural behavior is sinusoidal, neither decaying

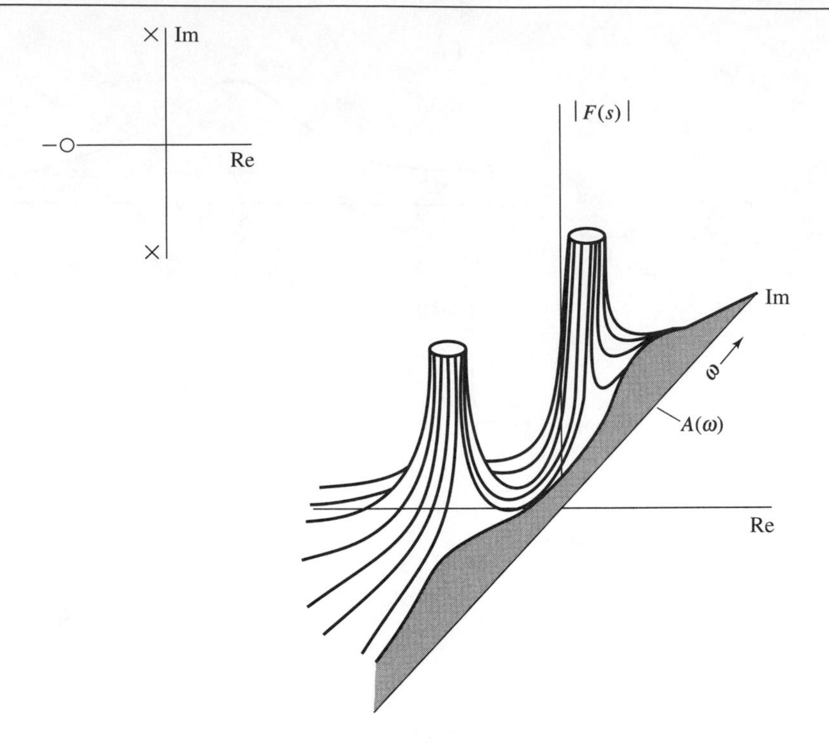

Figure 6.4 Frequency response amplitude as the height of a rubber sheet.

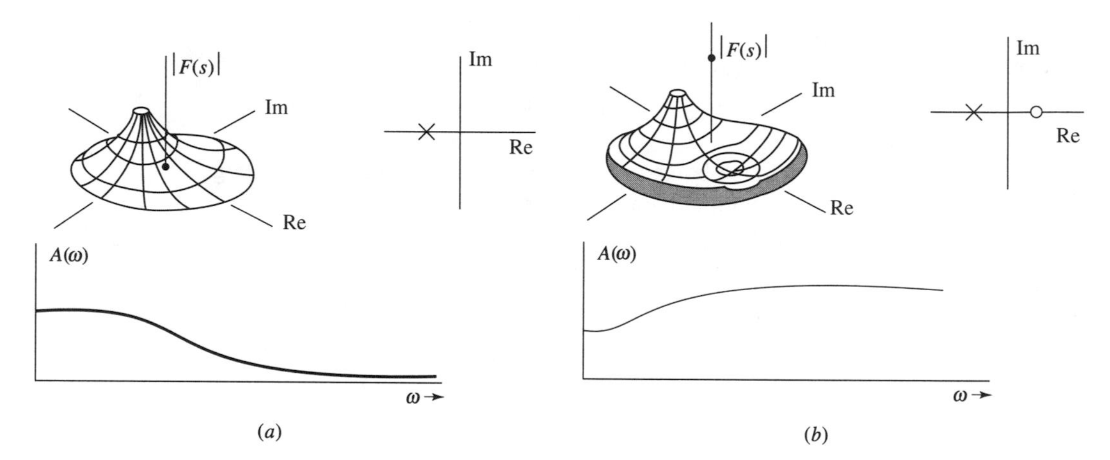

Figure 6.5 Using the rubber sheet analogy to visualize frequency response amplitude ratios. (a) More poles than zeros. (b) Equal number of poles and zeros.

nor expanding with time. If such a system is *driven* with a sinusoidal signal of the same frequency as this natural behavior, the output becomes larger and larger. Of course, it cannot really become infinite in a practical system, since nonlinearities will eventually be reached that limit the output. Possibly, the system will be destroyed or it will not be possible to continue to supply the ever-increasing energy needed to maintain the input signal. A linear, time-invariant model does not take these practical constraints into account.

Complex conjugate poles in the LHP near the imaginary axis represent slowly decaying oscillatory natural behavior. Driving such a system with a sinusoidal signal of the same or nearly the same frequency as the natural behavior oscillations results in a relatively large, but not infinite, forced output. That is, there is typically a corresponding peak in the system's frequency response, termed a *resonanse peak.*

❑ DRILL PROBLEM

D6.3 Sketch approximate frequency response curves (both amplitude and phase shift) for functions with the pole–zero plots shown in Figure D6.3.

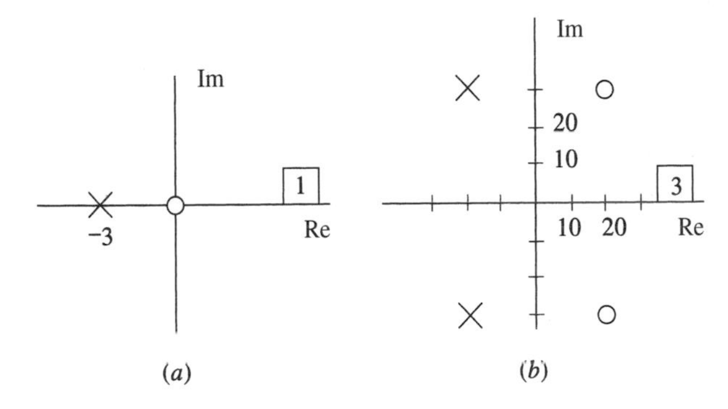

Figure D6.3

Ans.

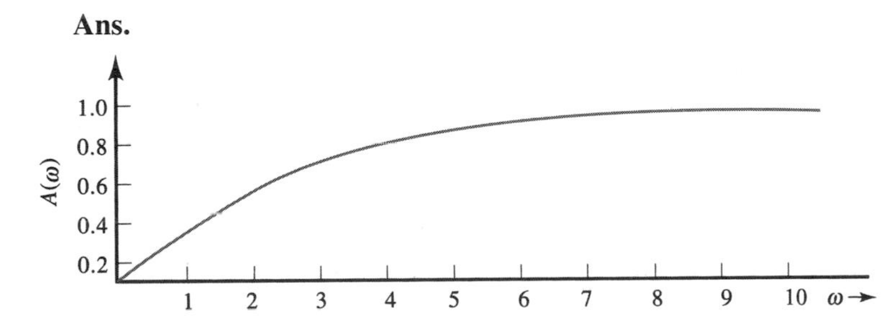

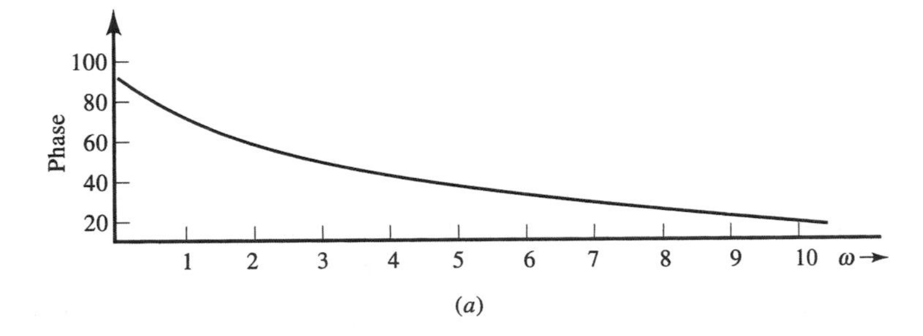

Figure DA6.3

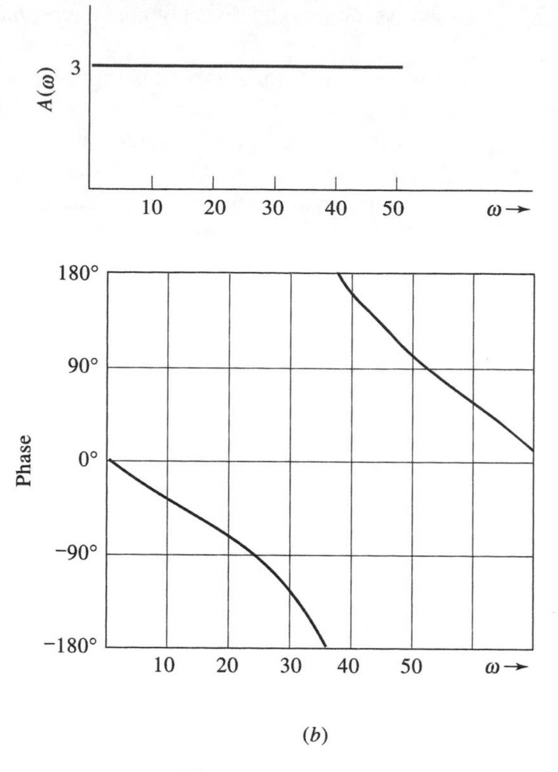

(*b*)

Figure DA6.3

Frequency response data are complex numbers, and as we vary the input frequency over a range, we obtain a vector of frequency response data (a complex vector). This complex vector can be manipulated and displayed in a variety of ways, as explained briefly in the following list and expanded upon in detail elsewhere in the chapter.

1. **Linear frequency response plots**. Two plots: one of magnitude of $G(j\omega)$ and another of the phase angle of $G(j\omega)$. The horizontal axis in both plots is frequency ω with linear scaling. This is shown in Figure 6.6.
2. **Linear frequency response plots versus log(ω)**. Two plots, similar to the preceding case, but the horizontal axes are logarithmically scaled (also known as *semilog* plots). This allows a larger frequency scale to be used. Notice how most of the change in the magnitude plot occurs between 0.1 to 10 rad/s, and this is amplified in log scaling.
3. **Bode plots**. Two plots: the magnitude M is in decibels [i.e., 20 log (M)]; phase is in degrees. The horizontal axes are logarithmic. Bode plots are most frequently used in control systems analysis and design.
4. **Polar (Nyquist) plot**. This is plot of the imaginary versus the real part of $G(j\omega)$.
5. **Nichols plot**. A Nichols plot is the plot of the magnitude (in dB) versus the phase in degrees.

To demonstrate these plots let us consider the transfer function

$$G(s) = \frac{500}{(s+1)(s+5)(s+10)}$$

Its frequency response is obtained by letting $s = j\omega$ to get $G(j\omega)$. We will evaluate the frequency response over the frequency range $\omega = 0.1$ to 100. The following data shows the vectors of frequency ω, frequency response $G(j\omega)$, its magnitude and angle, magnitude in decibels, and real and imaginary parts of $G(j\omega)$.

ω	$G(j\omega)$	$\lvert Gj\omega\rvert$	$\angle G(j\omega)$	$\lvert G(j\omega\rvert_{dB}$	Re $\{G(j\omega)\}$	Im $\{G(j\omega)\}$
1.0000e-001	9.8644e+000-1.2863e+000i	9.9479e+000	-7.4293e+000	1.9955e+001	9.8644e+000	-1.2863e+000
1.0000e-001	9.8644e+000-1.2863e+000i	9.9479e+000	-7.4293e+000	1.9955e+001	9.8644e+000	-1.2863e+000
1.2690e-001	9.7829e+000-1.6218e+000i	9.9165e+000	-9.4128e+000	1.9927e+001	9.7829e+000	-1.6218e+000
1.6103e-001	9.6539e+000-2.0370e+000i	9.8664e+000	-1.1915e+001	1.9883e+001	9.6539e+000	-2.0370e+000
2.0434e-001	9.4512e+000-2.5430e+000i	9.7873e+000	-1.5059e+001	1.9813e+001	9.4512e+000	-2.5430e+000
2.5929e-001	9.137e+000-3.14446e+000i	9.6636e+000	-1.8990e+001	1.9703e+001	9.1377e+000	-3.1446e+000
3.2903e-001	8.6636e+000-3.8324e+000i	9.4734e+000	-2.3863e+001	1.9530e+001	8.6636e+000	-3.8324e+000
4.1753e-001	7.9708e+000-4.5698e+000i	9.1879e+000	-2.9826e+001	1.9264e+001	7.9708e+000	-4.5698e+000
5.2983e-001	7.0081e+000-5.2806e+000i	8.7748e+000	-3.6998e+001	1.8865e+001	7.0081e+000	-5.2800e+000
6.7234e-001	5.7600e+000-5.8449e+000i	8.2062e+000	-4.5419e+001	1.8283e+001	5.7600e+0000	-5.8449e+000
8.5317e-001	4.2826e+000-6.1229e+000i	7.4720e+000	-5.5030e+001	1.7469e+001	4.2826e+0000	-6.1229e+000
1.0826e+000	2.7164e+000-6.0074e+000i	6.5930e+000	-6.5669e+001	1.6382e+001	2.7164e+0000	-6.0074e+000
1.3738e+000	1.2517e+000-5.4808e+000i	5.6219e+000	-7.7135e+001	1.4998e+001	1.2517e+000	-5.4808e+000
1.7433e+000	5.8822e-002-4.6281e+000i	4.6285e+000	-8.9272e+001	1.3309e+001	5.8822e-002	-4.6281e+000
2.2122e+000	-7.6570e-001-3.5973e+000i	3.6779e+000	-1.0202e+002	1.1312e+001	-7.6570e-001	-3.5973e+000
2.8072e+000	-1.2077e+000-2.5452e+000i	2.8172e+000	-1.1539e+002	8.9962e+000	-1.2077e+000	-2.5452e+000
3.5622e+000	-1.3160e+000-1.6025e+000i	2.0736e+000	-1.2939e+002	6.3344e+000	-1.3160e+0000	-1.6025e+000
4.5204e+000	-1.1807e+000-8.5888e-001i	1.4600e+000	-1.4397e+002	3.2872e+000	-1.1807e+000	-8.5888e-001
5.7362e+000	-9.1308e-001-3.5283e-001i	9.7888e+001	-1.5887e+002	-1.8544e-001	-9.1308e-001	-3.5283e-001
7.2790e+000	-6.1933e-001-6.7909e-002i	6.2304e-001	-1.7374e+002	-4.1096e+000	-6.1933e-0001	-6.7909e-002
9.2367e+000	-3.7262e-001+5.3173e-002i	3.7640e-001	-1.8812e+002	-8.4871e+000	-3.7262e-001	5.3173e-002
1.1721e+001	-2.0135e-001+7.9524e-002i	2.1649e-001	-2.0155e+022	-1.3291e+001	-2.0135e-001	7.9524e-002
1.4874e+001	-9.9271e-002+6.6101e-002i	1.1926e-001	-2.1366e+002	-1.8470e+001	-9.9271e-002	6.6101e-002
1.8874e+001	-4.5466e-002+4.4234e-002i	6.3434e-002	-2.2421e+002	-2.3954e+001	-4.5466e-002	4.4234e-002
2.3950e+001	-1.9697e-002+2.6287e-002i	3.2847e-002	-2.3315e+002	-2.9670e+001	-1.9697e-002	2.6287e-002
3.0392e+001	-8.2011e-003+1.4531e-002i	1.6685e-002	-2.4056e+002	-3.5553e+001	-8.2011e-003	1.4531e-002
3.8566e+001	-3.3232e-003+7.6763e-003i	8.3648e-003	-2.4659e+002	-4.1551e+001	-3.3232e-003	7.6763e-003
4.8939e+001	-1.3227e-003+3.9409e-003i	4.1570e-003	-2.5145e+002	-4.7624e+001	-1.3227e-003	3.9409e-003
6.2102e+001	-5.2034e-004+1.9872e-003i	2.0542e-003	-2.5533e+002	-5.3747e+001	-5.2034e-004	1.9872e-003
7.8805e+001	-2.0319e-004+9.9082e-004i	1.0114e-003	-2.5841e+002	-5.9901e+001	-2.0319e-004	9.9082e-004
1.0000e+002	-7.8978e-005+4.9056e-004i	4.9687e-004	-2.6085e+002	-6.6075e+001	-7.8978e-005	4.9056e-004

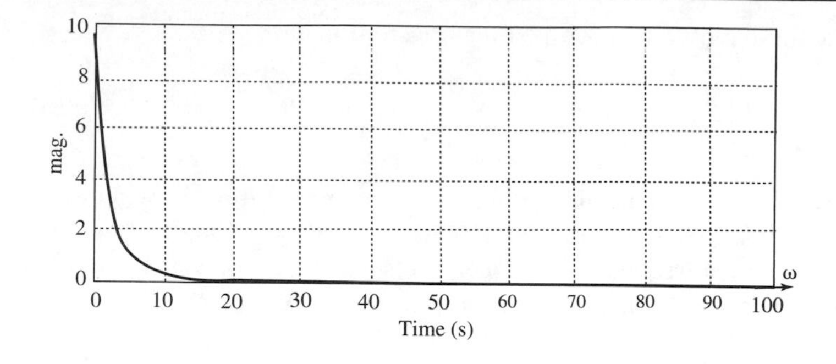

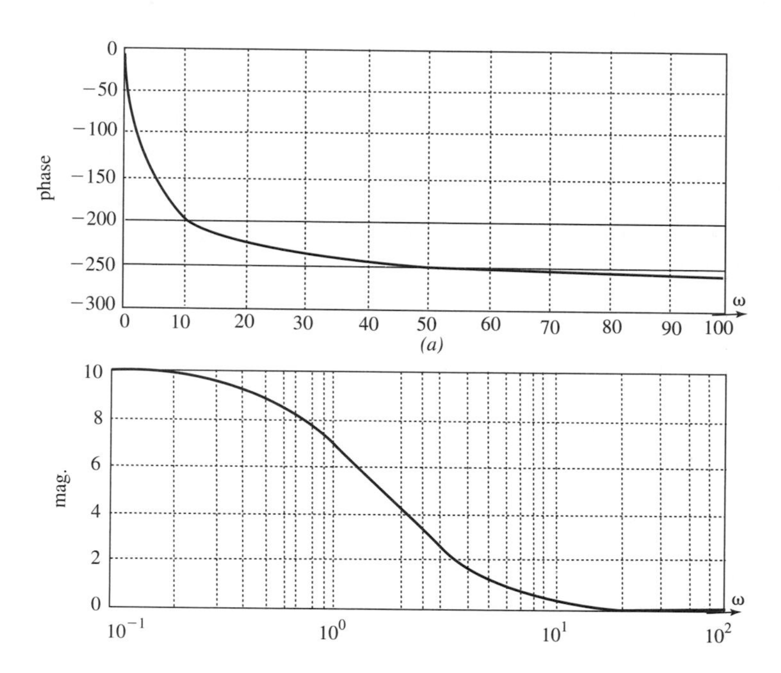

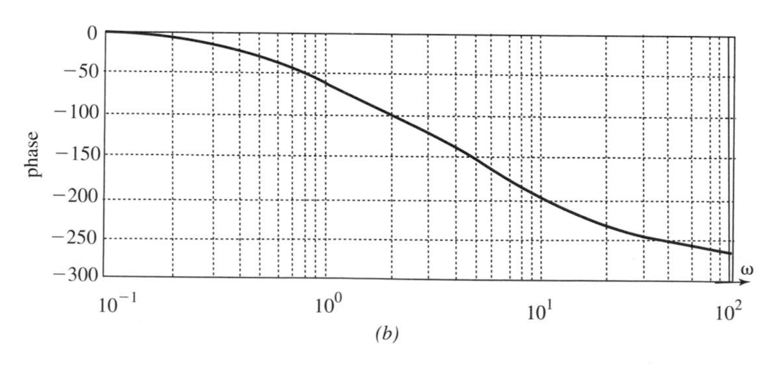

Figure 6.6 (a) Linear frequency axes, linear magnitude, and phase. (b) Logarithmic frequency axes, linear magnitude, and phase plots.

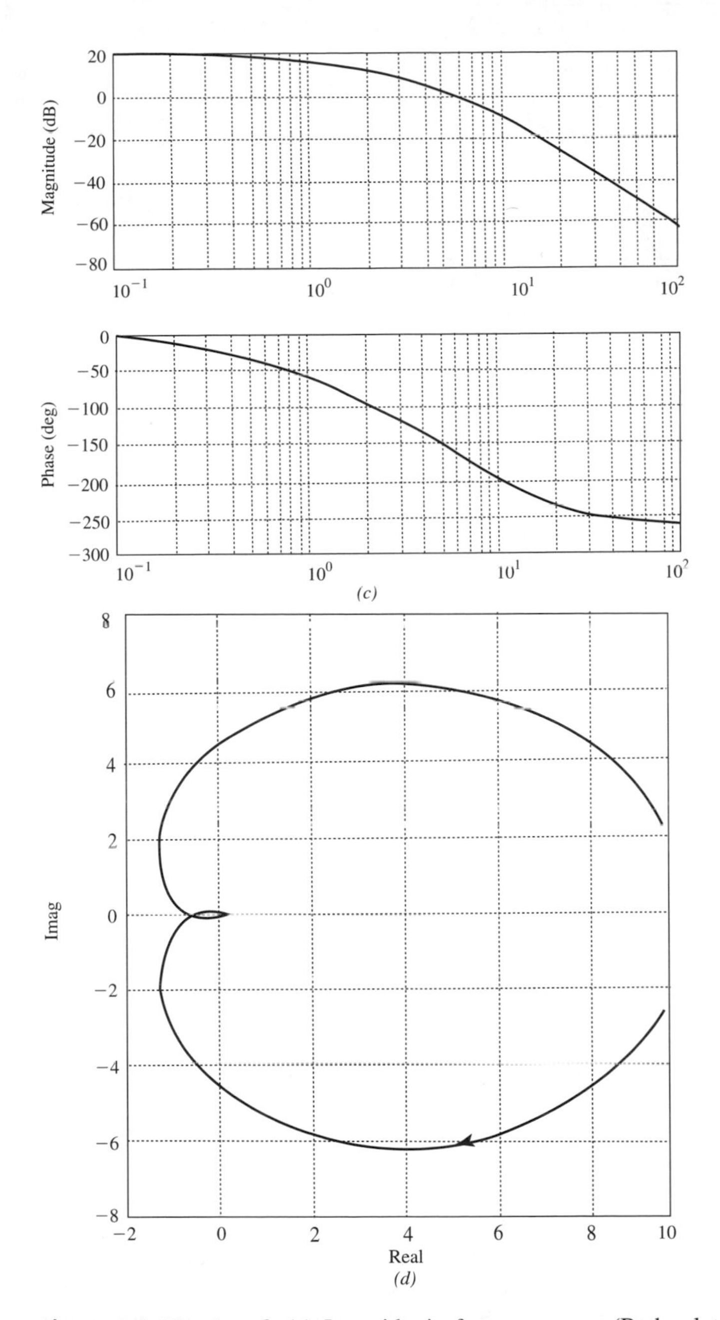

Figure 6.6 (*Continued*) (c) Logarithmic frequency axes (Bode plots). (d) Nyquist or polar plot: imaginary versus real part (arrow shows direction of increasing positive frequency).

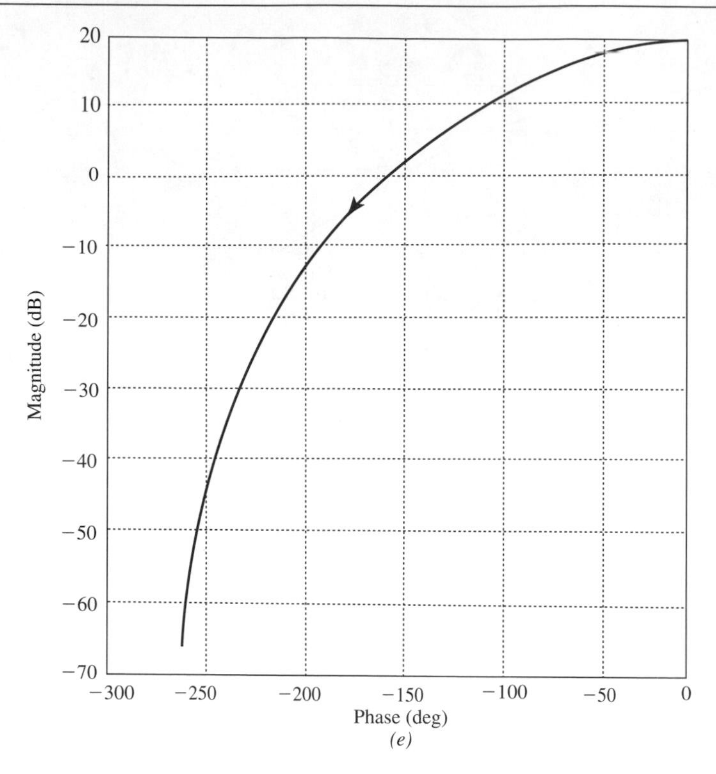

Figure 6.6 (*Continued*) (e) Nichols plot; arrow shows direction of increasing frequency.

The various plots referred are displayed in Figure 6.6. The sections, that follow give plotting techniques for Bode and Nyquist plots. But it is important to keep in mind that all the plots display the *same* information, (i.e., they are plots of complex numbers versus frequency); they are just various ways at looking the same data, and some plots are more convenient for some applications.

6.3 Bode Plots

6.3.1 Amplitude Plots in Decibels

The magnitude of the product of complex numbers is the product of the individual magnitudes, and the angle of a product is the sum of the individual angles. A transmittance that is the product of several simple terms thus has an amplitude curve that is the product of the amplitude curves for the individual terms. The overall phase shift curve is the sum of the individual phase shift curves.

If, instead of dealing with the amplitude curves directly, their logarithms are used, multiplication of individual amplitude curves is, in terms of logarithms, addition. Commonly, *decibels* (dB) are used for the description of frequency response amplitude ratios:

$$\text{dB} = 20\log_{10} A(\omega) = 20\log_{10}|F(s = j\omega)|$$

If

$$F(s) = F_1(s)F_2(s)F_3(s)\ldots$$

then

$$F(\text{in dB}) = F_1(\text{in dB}) + F_2(\text{in dB}) + F_3(\text{in dB}) + \cdots$$

There is little reason for this choice of common logarithm units; the choice dates from Alexander Graham Bell's study of the response of the human ear. In electromagnetics and some other applications, the natural logarithm is used, and the definition does not appear to be so contrived. The units then are *nepers*.

For rational functions $F(s)$, it is only necessary to be able to plot amplitude and phase shift for the following types of terms:

Types of function.

Constants
Poles and zeros at the origin of the complex plane
Real axis poles and zeros
Complex conjugate pairs of poles and zeros

A rational function can be factored into terms of these types and the individual dB and phase shift curves plotted. The complete dB curve is then the sum of the component dB curves, and the complete phase shift curve is the sum of the individual phase shift curves.

It is usually most convenient to plot dB and phase shift curves with a logarithmic scale for ω, in which event they are termed *Bode plots*, after Hendrik W. Bode. Bode plots display information much more clearly than the corresponding linear plots. Magnitude and frequency both vary over many powers of 10 so that most information would be compressed near the origin by a linear plot, while the dB scale expands magnitude in a Bode plot and the logarithmic frequency scale expands frequency. Since the vertical axes of Bode plots are linear in decibels and linear in phase, the Bode plots are termed *semilog* plots. For example, Table 6.1 shows Bode plots for constant and power-of-s transmittances.

The advantage of Bode plots over linear plots is a matter of clarity.

Commonly used dB values are 40 dB ($A = 100$), 20 dB ($A = 10$), 6 dB ($A \cong 2$), 3 dB ($A \cong \sqrt{2}$), 0 dB ($A = 1$), −20 dB ($A = 0.1$), and −40 dB ($A = 0.01$). The student should attempt to gain a perception of dB and phase trends to be able to catch obvious programming errors. A partnership must exist between the computer user and the computer. The user must be able to double-check computer output by (what used to be called) back-of-the-envelope analysis.

A positive constant k has constant amplitude in dB

$$20\log_{10} k$$

and angle 0°. A negative constant, $-k$, has dB amplitude

$$20\log_{10}|k|$$

Table 6.1 *Logarithmic Frequency Response (Bode) Plots for Some Basic Terms*

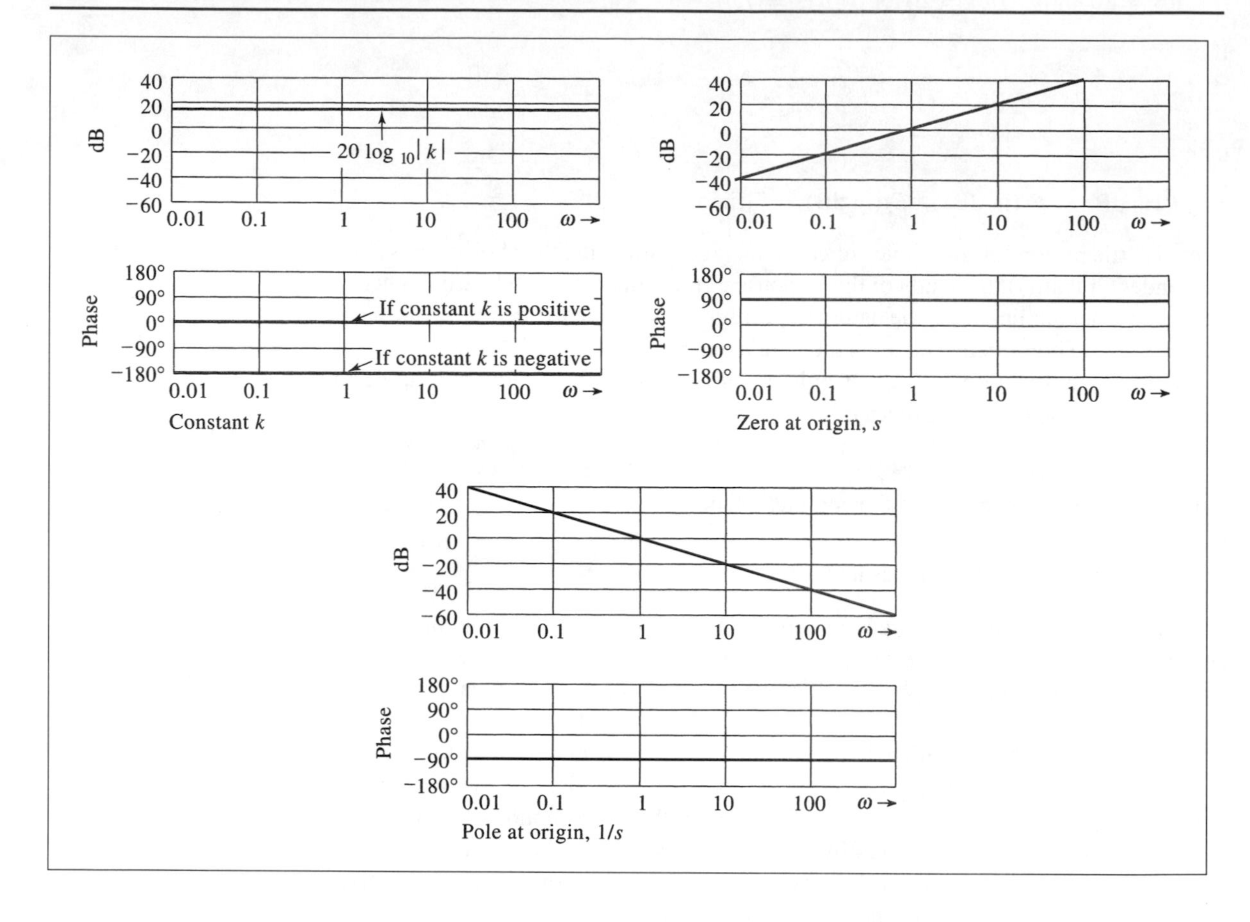

and angle 180°. For example.

$$F = 100$$

has constant amplitude ratio and constant phase shift

$$\text{dB} = 20\log_{10}(100) = 40 \text{ dB} \qquad \Phi = 0^\circ$$

as plotted in Figure 6.7(a). The transmittance

$$F = -\tfrac{1}{10}$$

has the curves of Figure 6.7(b).

For the transmittance

$$F(s) = s$$

the curves are

$$A(\omega) = |F(s = j\omega)| = \omega \qquad \text{dB} = 20\log_{10} A(\omega) = 20\log_{10}\omega$$

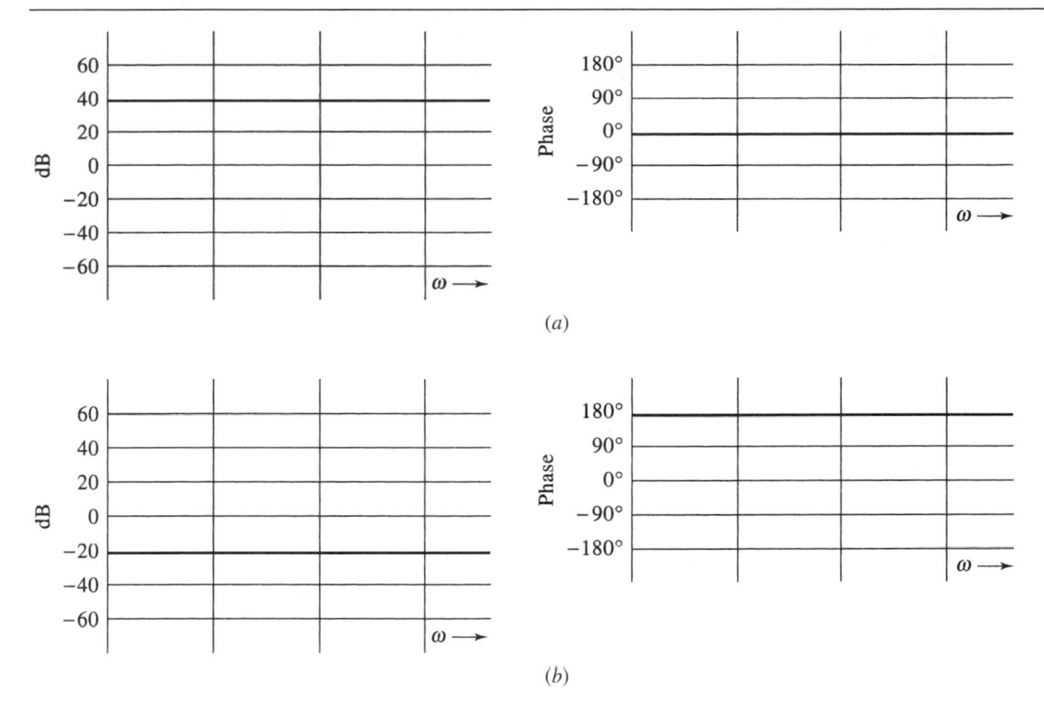

Figure 6.7 Frequency response curves for two constant transmittances. (a) $F = 100$. (b) $F = -1/10$.

and

$$\Phi(\omega) = \angle F(s = j\omega) = 90^\circ$$

Since dB is plotted versus the logarithm of ω, the Bode plot is a straight line with slope 20 dB per decade of ω, passing through 0 dB at $\omega = 1$ as shown in Table 6.1. The *dec* in decade comes from the Latin word meaning "ten." A decade is an increase in frequency by a factor of 10, so slope is usually measured in decibels per decade. Another slope measures is decibels per octave. The *oct* in octave comes from the Latin word meaning "eight." The use of octave is somewhat misleading, because eight notes in the western musical scale refers to doubling in frequency; hence, an octave is an increase in frequency by a factor of 2. A slope of 20 dB per decade implies that the amplitude doubles when the frequency doubles, making the slope 6 dB per octave. In the rest of this chapter, slope is measured in dB per decade, which we believe to be a more understandable slope measure than dB per octave.

Decade is defined so the slope is dB/decade.

Octave is defined so the slope is dB/octave.

The curves for

$$F(s) = s^2$$

shown in Figure 6.8(a) are just the sums of two $F(s) = s$ curves, that is, double the plots for s. In fact, the nth power of any transmittance has curves that are n times the

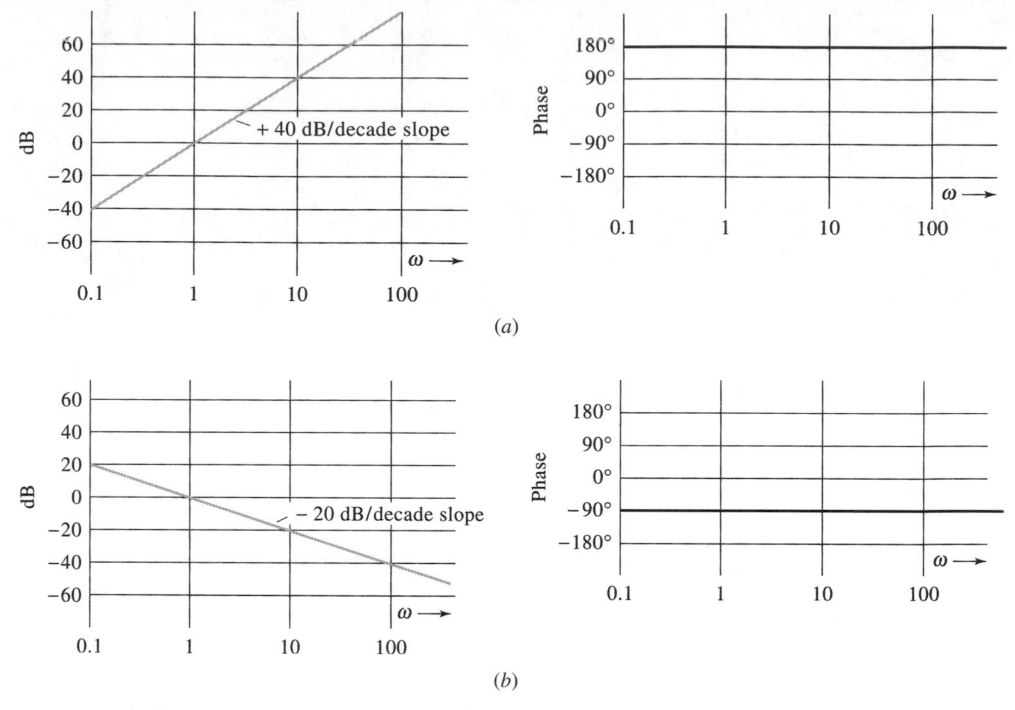

Figure 6.8 Bode plots for transmittances having poles and zeros at the origin of the complex plane. (a) Frequency response plots for $F(s) = s^2$. (b) Frequency response plots for $F(s) = 1/s$.

plots for the original transmittance. The Bode plots for

$$F(s) = \frac{1}{s} = s^{-1}$$

[Figure 6.8(b)] are the negatives of the dB and phase angle plots for $F(s) = s$.

6.3.2 Real Axis Roots

Consider a transmittance that has a left half-plane (LHP) zero, of the form

$$F(s) = \frac{s + a}{a} = 1 + \frac{s}{a}$$

The frequency response is plotted for LHP zero.

where a is a positive number. Instead of considering a single zero term, $s + a$, terms are put in this form because the relations are simplest when

$$F(s = j\omega) = 1 + j\frac{\omega}{a}$$

For this term,

$$|F(s = j\omega)| = \sqrt{1 + \left(\frac{\omega}{a}\right)^2} \qquad \angle F(s = j\omega) = \tan^{-1}\left(\frac{\omega}{a}\right)$$

and

$$\text{dB} = 20 \log_{10} |F(s = j\omega)| = 10 \log_{10} \left[1 + \left(\frac{\omega}{a} \right)^2 \right]$$

For $\omega \ll a$,

$$F(s = j\omega) \cong 1 \qquad \text{dB} \cong 10 \log_{10} 1 = 0 \qquad \angle F(s = j\omega) \cong \tan^{-1} 0 = 0^\circ$$

For $\omega \gg a$,

$$F(s = j\omega) \cong j\frac{\omega}{a} \qquad \text{dB} \cong 20 \log_{10} \frac{\omega}{a} = 20 \log_{10} \omega - 20 \log_{10} a$$

$$\angle F(s = j\omega) \cong 90^\circ$$

The actual dB and phase curves appear in Figure 6.9. The low-frequency and high-frequency approximations can be represented by straight lines called *asymptotes*. For the dB curve, the straight-line approximation at high frequency with a slope of 20 dB/decade can be intersected with the flat low-frequency approximation at a frequency called the *corner* or *break* frequency. At that frequency the dB curve changes slope by +20 dB/decade, as is noted below the frequencies in Figure 6.9. At the break frequency, the difference between the true curve and the approximate curve is the amplitude of the true curve, since the approximate curve is 0 dB. That error is

Slope changes are used in plotting magnitude and phase asymptotes.

$$|F(ja)| = 20 \log_{10} \sqrt{2} = 3.01 \text{ dB}$$

It is common practice in control system applications to approximate the phase by a three-segment curve that changes slope by +45°/decade at one-tenth of the break frequency and by −45°/decade at 10 times the break frequency. The approximate phase curve has a maximum error of less than 6°, as noted in Figure 6.9. The slope changes are also noted.

In the approximation, the angle is 0° up to one-tenth of the break frequency, rises 45° per decade through 45° at the break frequency, and continues at 45° per decade slope up to 10 times the break frequency. Beyond 10 times the break frequency, the approximate angle is 90°. At the break frequency a, the actual and approximate curves are equal, since

$$\angle F(ja) = \tan^{-1} \left(\frac{1}{1} \right) = 45^\circ$$

As a numerical example, the transmittance

$$F(s) = \frac{s + 10}{10}$$

has the specific Bode curves of Figure 6.10. Approximately, the dB curve is along 0 dB up to the break frequency, $\omega = 10$, then the dB rises 20 dB per decade of ω. The phase shift curve is approximately 0° up to one-tenth of the break frequency, and 90° beyond 10 times the break frequency. In between, it is approximately a straight line with slope 45° per decade through 45° at $\omega = 10$.

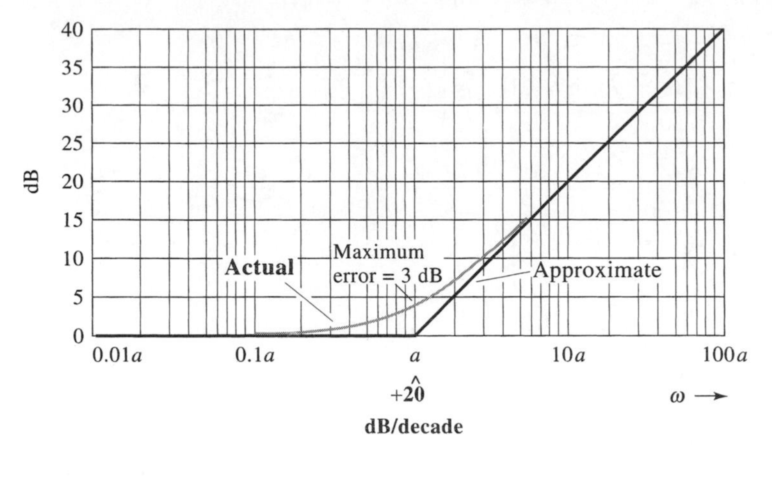

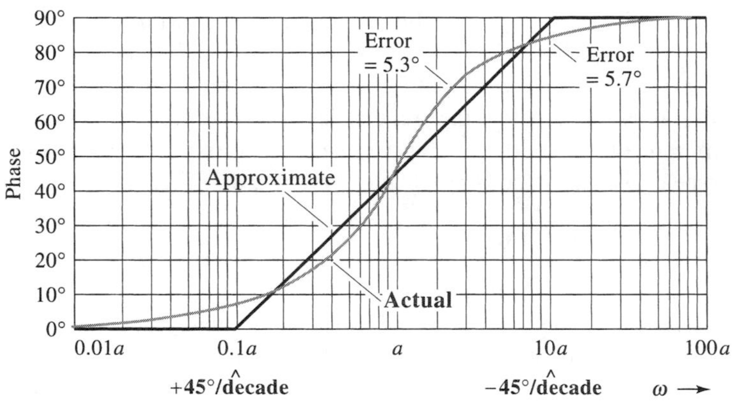

Figure 6.9 Bode plot for $F(s) = (s + a)/a$ showing actual and approximate curves with slope changes for dB and phase.

Bode plots for real axis LHP poles and for RHP (right half-plane) zeros and poles are given in Table 6.2. The Bode frequency response plots for a real axis LHP pole term.

$$F(s) = \frac{a}{s + a} = \frac{1}{1 + (s/a)}$$

The dB and phase for real LHP poles are the negatives of those for real LHP zeros.

have dB and phase shift curves that are negatives of the curves for a corresponding LHP zero. The dB curve is

$$\begin{aligned} \text{dB} &= 20 \log_{10} |F(s = j\omega)| = 20 \log_{10} \left| \frac{1}{1 + j(\omega/a)} \right| \\ &= -20 \log_{10} \left| 1 + j\left(\frac{\omega}{a}\right) \right| \end{aligned}$$

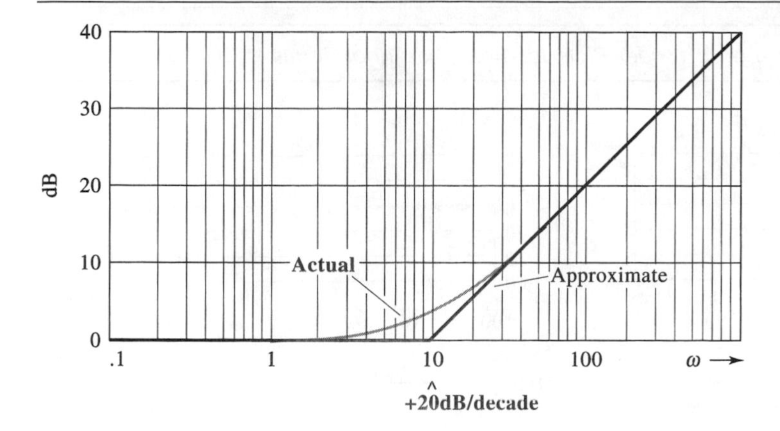

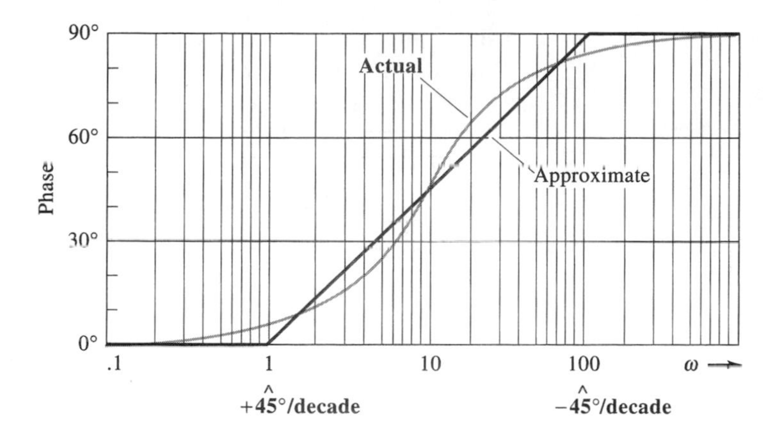

Figure 6.10 Bode plots for the transmittance $(s + 10)/10$.

which is the negative of that for a zero. The phase shift is, similarly,

$$\angle F(s = j\omega) = \Bigg/\!\underline{\frac{1}{1 + j(\omega/a)}} = -\Bigg/\!\underline{\left[1 + j\left(\frac{\omega}{a}\right)\right]}$$

RHP zeros and poles differ from their LHP counterparts by the algebraic sign of the phase shift; amplitude ratios are the same as for LHP roots of the same type.

For example, the transmittance

$$F(s) = \frac{-10}{s - 10}$$

with a RHP pole term has a dB curve that is the negative of that for the zero term $(s+10)/10$, a phase shift that is the same as that for the zero, as shown in Figure 6.11. Although this transmittance, having a RHP pole, is unstable, it could be part of an overall system of interest that is stable.

For RHP roots, change the sign on s. Magnitudes are the same but phase changes sign.

Table 6.2 *Logarithmic Frequency Response (Bode) Plots for Real Axis Root Terms*

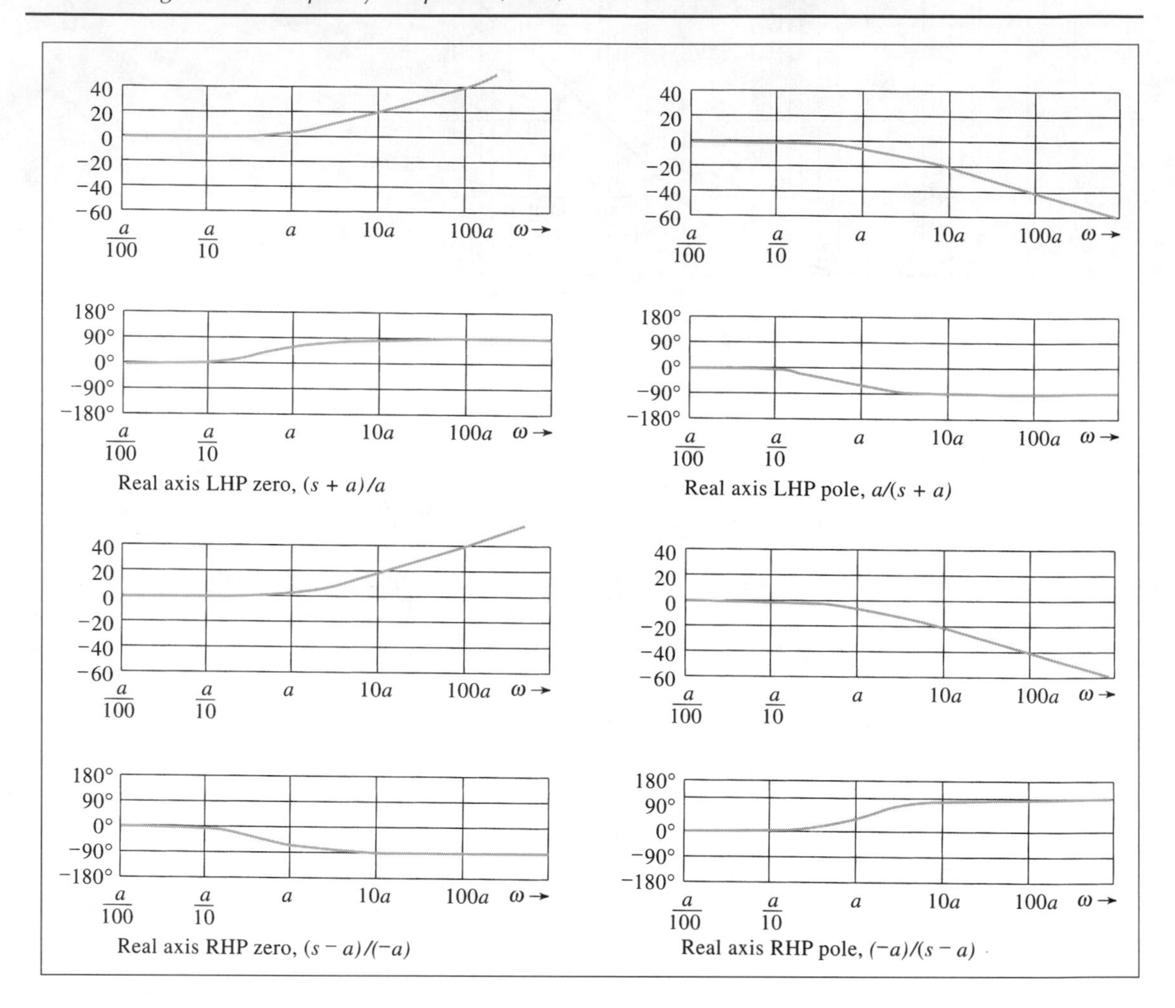

Real axis LHP zero, $(s + a)/a$

Real axis LHP pole, $a/(s + a)$

Real axis RHP zero, $(s - a)/(-a)$

Real axis RHP pole, $(-a)/(s - a)$

6.3.3 *Products of Transmittance Terms*

Consider plotting Bode frequency response curves for

$$F(s) = \frac{50(s + 2)}{s(s + 10)}$$

First, $F(s)$ is decomposed into factors for which the frequency response is known and can easily be sketched:

$$F(s) = 10\left(\frac{1}{s}\right)\left(\frac{s + 2}{2}\right)\left(\frac{10}{s + 10}\right)$$

There are two break frequencies: 2 rad/s from the real zero and 10 rad/s from the real pole. It is good practice to obtain a Bode plot in which the minimum frequency is

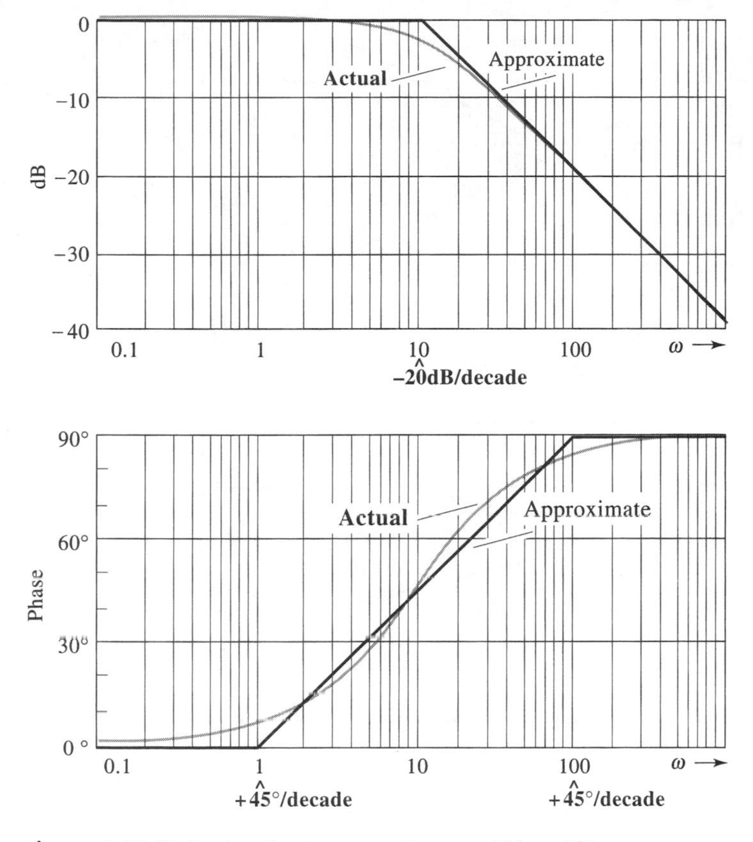

Figure 6.11 Bode plots for the transmittance $-10(s-10)$.

no higher than one-tenth of the lowest break frequency and the maximum frequency is no lower than 10 times the highest break frequency. That range will include the frequency region where the response changes. Here we should plot the response from 0.2 to 100 rad/s. Using powers of 10, we should therefore include the range from 0.1 to 100 rad/s.

When several factors are multiplied together, add the dB and phase curves. Slope changes facilitate the addition of curves.

In Figure 6.12(a), the straight-line approximations of each of the four components are shown along with the slope changes. To obtain an approximation for $F(s)$, an engineer can either add the components or just use slope change information. With practice, a notation of slope change is probably sufficient to obtain the composite plot. For the dB plot, the composite slope is -20 dB/decade at frequencies below 2 rad/s due to adding the flat slope of 10 to the -20 dB/decade of $1/s$. At 2 rad/s the composite slope changes to 0 dB/decade. At 10 rad/s, the composite slope changes to -20 dB/decade. The composite dB curve appears in Figure 6.12(b). It is necessary to calculate the dB value from 2 rad/s to 10 rad/s, where the curve does not change, and it may be useful to approximate the dB value at other frequencies. Notice that using the straight-line approximation results in a simple first-order equation at all frequencies, regardless of the order of the system. Of course, the result is only an approximation; but engineers should be able to quickly approximate values.

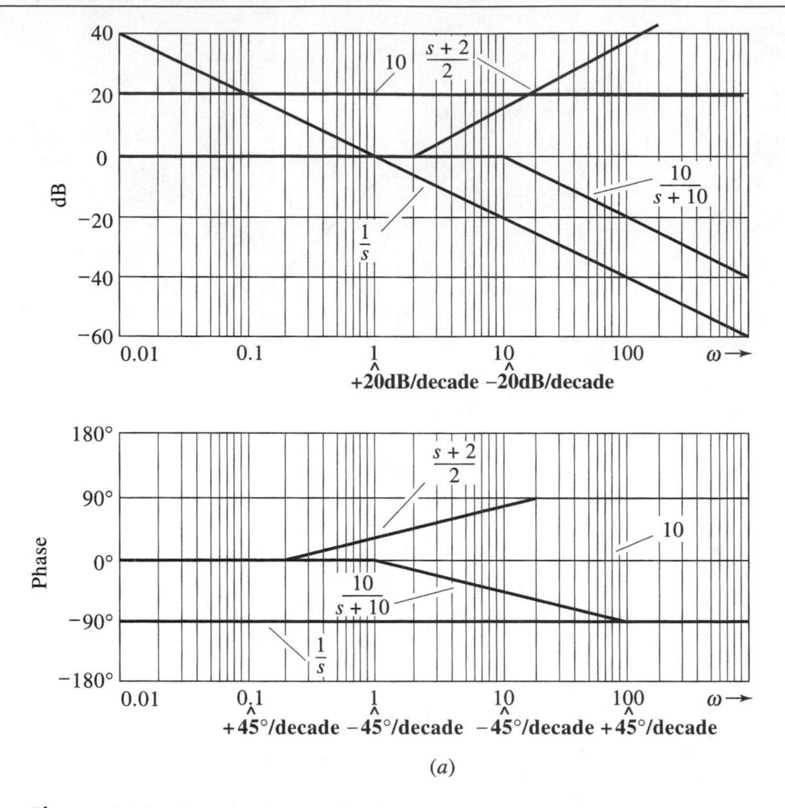

Figure 6.12 Example of Bode plot construction. (a) Bode plots of individual terms.

We can use standard straight-line equations, taking care that the horizontal axis is logarithmic. A straight-line equation can be written with y as the vertical axis variable and x as the horizontal axis variable and starting the straight-line equation at the point x_0, y_0.

$$y = y_0 + m(x - x_0)$$

For a dB plot, the slope is in units of dB per decade. For a phase plot, the slope is in units of degrees per decade. The distance in the horizontal (frequency) direction must be measured in decades. That can be accomplished (where $x = \omega$) by

$$\text{decades of } (\omega - \omega_0) = \log_{10} \frac{\omega}{\omega_0}$$

When $\omega = \omega_0$, there is no decade of change. When $\omega = 10\omega_0$, the equation yields 1 decade.

Straight-line equations may be derived for each linear segment.

Suppose a straight-line equation is needed for the Bode magnitude plot of Figure 6.12 for frequencies between 1 and 2 rad/s. At 1 rad/s, the amplitude is 20 dB and the slope is −20 dB/decade.

$$y_0 = 20 \text{ dB}$$

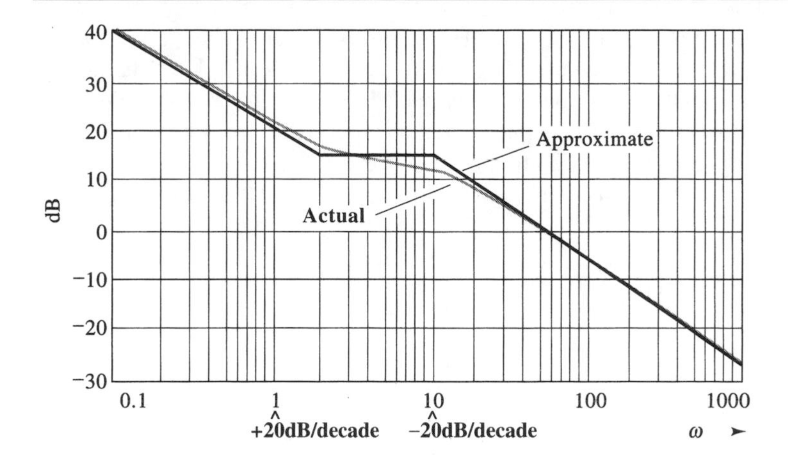

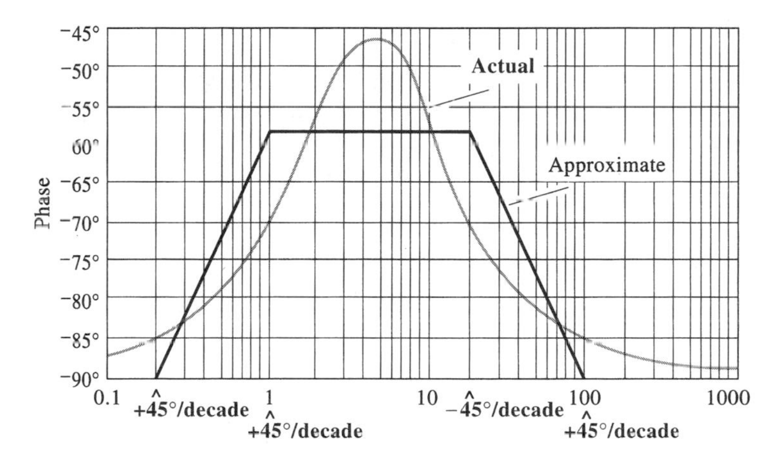

Figure 6.12 (*Continued*) Example of Bode plot construction. (b) Complete plots, consisting of sums of terms, showing slope changes.

$$\omega_0 = 1$$

$$m = -20 \text{ dB/decade}$$

$$y = 20 - 20\log_{10}\left(\frac{\omega}{1}\right)$$

When $\omega = 2$, then

$$y = 20 - 20\log_{10} 2 = 14 \text{ dB}$$

For frequencies above 10 rad/s, the approximate curve is

$$y = 14 - 20\log_{10}\left(\frac{\omega}{10}\right)$$

The composite phase curve can be approximated by using the phase changes from Figure 6.12(a). The approximate composite curve is flat at −90° until 0.2 rad/s, when the slope changes to +45°/decade. The curve is flat until the slope changes by −45°/decade at 1 rad/s, becoming 0°/decade. At 20 rad/s the slope changes to −45°/decade, and finally the slope changes back to 0°/decade at 100 rad/s. The approximate value when the curve is flat as well as other values may be found using the straight-line approach we applied to the dB curve.

For the phase curve where frequencies lie between 0.2 and 1 rad/s, the approximate phase is

$$y = -90^\circ + 45^\circ \log_{10}\left(\frac{\omega}{0.2}\right)$$

At 1 rad/s we have

$$y = -90^\circ + 45^\circ \log_{10}\left(\frac{1}{0.2}\right) = -58.5^\circ$$

For frequencies between 20 and 100 rad/s we have

$$y = -58.5^\circ - 45^\circ \log_{10}\left(\frac{\omega}{20}\right)$$

Straight-line equations can be used to approximate the frequency of 0 dB.

We can use the approximate curve to approximate the frequency at which zero dB occurs, and then we can compare that result with the true value found using the actual $F(s)$.

The frequency at which zero dB occurs can be approximated by

$$0 = 14 - 20 \log_{10} \frac{\omega}{10}$$

$$\log_{10}\left(\frac{\omega}{10}\right) = \left(\frac{14}{20}\right) = 0.7$$

$$\omega = 10 \times 10^{0.7} = 50 \text{ rad/s}$$

To find the frequency for 0 dB by trial and error from the expression

$$F(j\omega) = \frac{50(j\omega + 2)}{j\omega(j\omega + 10)}$$

$$A(\omega) = \frac{50\sqrt{(\omega^2 + 4)}}{\omega\sqrt{(\omega^2 + 100)}}$$

we can use a number of trial values of frequency near 50 rad/s. For example,

$$A(50) = \frac{50\sqrt{2504}}{50\sqrt{2600}} = 0.9814$$

$$A(49) = 1.0006$$

$$A(48) = 1.0207$$

The true value of frequency for 0 dB is therefore 49 rad/s rather than the approximate value of 50 rad/s. The approximate and true values are close together in this example because the frequency for 0 dB occurs fairly far from a break frequency, and thus the approximate and true magnitude curves are close together.

6.3.4 Complex Roots

Bode frequency response curves for complex conjugate pole or zero terms are plotted by hand, if desired, from straight-line asymptotes similar to simple real axis poles and zeros. Factor a set of complex conjugate poles or zeros to the form

$$F_1(s) = \frac{\omega_n^2}{s^2 + 2\zeta\omega_n s + \omega_n^2}$$

for a set of complex conjugate poles, or

$$F_2(s) = \frac{s^2 + 2\zeta\omega_n s + \omega_n^2}{\omega_n^2}$$

for a set of complex conjugate zeros.

First, we consider a complex conjugate set of poles $F_1(s)$. For small values of ω compared to ω_n,

$$F_1(s = j\omega) \cong 1$$

For large values of ω compared to ω_n,

$$F_1(s = j\omega) \cong -\frac{\omega_n^2}{\omega^2}$$

which is a dB curve with slope -40 dB/decade and a phase shift of 180°. At $\omega = \omega_n$,

$$F_1(j\omega_n) = \frac{1}{j2\zeta}$$

for which the dB curve has value

The magnitude when $\omega = \omega_n$ depends on ζ.

$$\text{dB} = 20\log_{10}\frac{1}{2\zeta}$$

and the phase shift is $-90°$.

Plots of amplitude and phase shift for various damping ratios for complex conjugate pole terms are given in Figure 6.13. For $\zeta \geq 1$, the poles are real, not complex, and so can be handled by the real axis pole methods discussed earlier. Figure 6.13(c) shows the true dB and phase values (for various actual damping ratios) minus the approximate curve (the asymptotic curve plotted with a damping ratio of 1). With a bit of practice, a designer can sketch the asymptotic curve for a damping ratio of 1 and then add the correctional curve [from Figure 6.13(c)] for the actual damping ratio. Curves of reasonable accuracy can be derived in a fairly short time.

Straight-line asymptotes are drawn for $\zeta = 1$. Corrective curves may be added to approximate the true curve.

For damping ratios between 0.5 and 1, the true dB curve lies generally below the asymptotic curve; so the correctional dB curve is generally negative. For damping ratios between 0 and 0.5, the true curve lies generally above the correctional curve, so the correctional dB values are generally positive. The true phase curve forms an S-shaped curve with the asymptotic curve; therefore, the correctional phase curves are of both signs, with magnitudes depending on the actual damping ratio.

As an example, consider the transmittance

$$F(s) = \frac{1000}{s^2 + 2s + 100} = (10)\left(\frac{100}{s^2 + 2s + 100}\right)$$

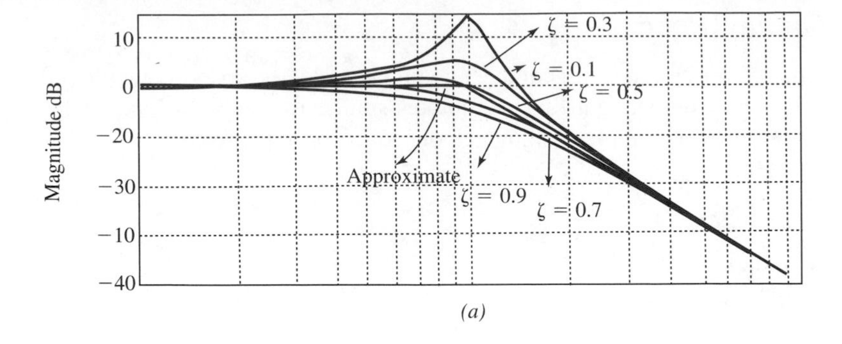

(a)

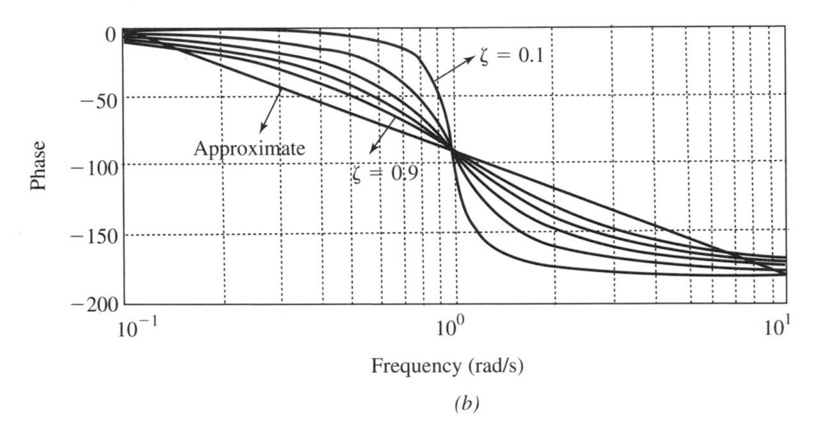

(b)

Figure 6.13 Bode plots for second-order terms. (a) Magnitude. (b) Phase.

For the complex conjugate pole term

$$\omega_n = \sqrt{100} = 10$$

$$2\zeta\omega_n = 20\zeta = 2 \qquad \zeta = \tfrac{1}{10}$$

First the straight-line approximations shown in Figure 6.14 are drawn. The conjugate pair of roots are first approximated as if they were critically damped, of the form

$$\frac{100}{s^2 + 20s + 100} = \left(\frac{10}{s + 10}\right)^2$$

The dB approximation is 0 dB (+20 dB for the constant factor of 10) to the break frequency of $\omega = 10$, then -40 dB/decade slope thereafter. The angle approximation is $0°$ to one-tenth of the break frequency, then $-90°$/decade slope, passing through $-90°$ at the break frequency and continuing until 10 times the break frequency.

The corrections of Figure 6.13, for $\zeta = \frac{1}{10}$, are then applied to the approximate curves to give the final result in Figure 6.14. Transferring several points from each curve and then drawing a smooth curve through the result will usually suffice. Since the

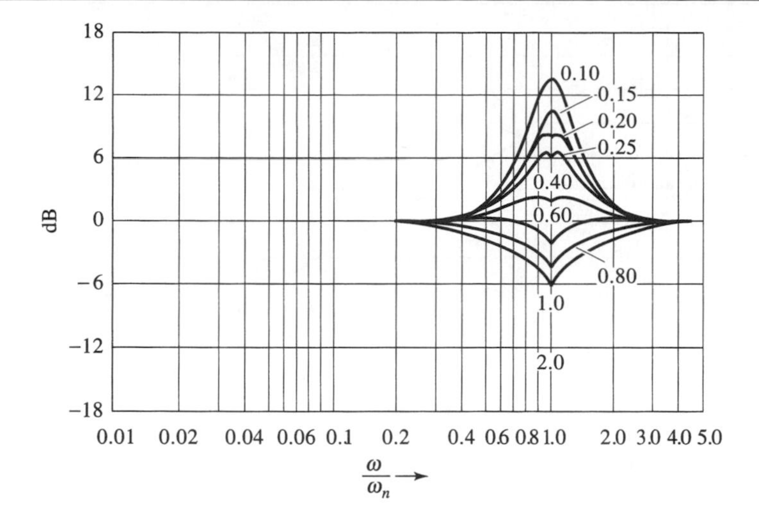

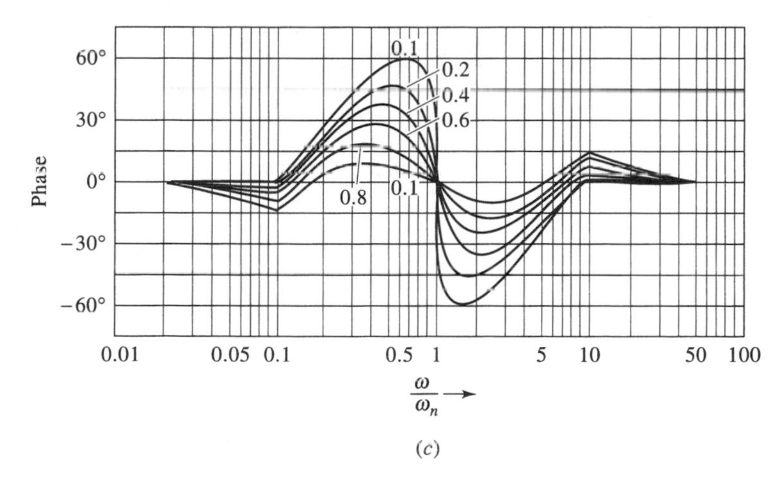

(c)

Figure 6.13 (*Continued*) Bode plots for second-order terms. (c) Corectional curves (true—asymptotic).

error of the straight-line approximation is sizable, it is important that the corrections be made.

A potential problem exists when the phase of an underdamped transmittance such as the one in Figure 6.14 is calculated using an inverse tangent function. To be specific, $F(s)$ can be written two different ways

$$F_1(s) = \frac{1000}{s^2 + 2s + 100}$$

$$F_2(s) = \frac{1000}{(s + 1 + j9.95)(s + 1 - j9.95)}$$

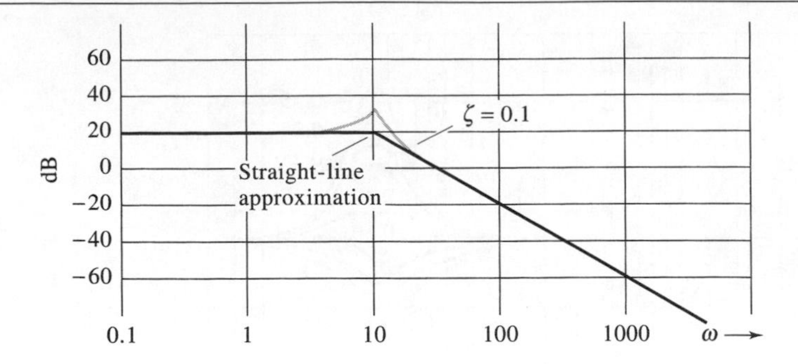

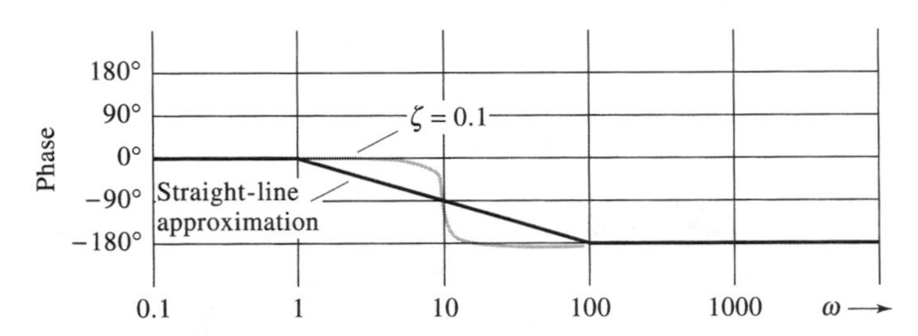

Figure 6.14 Frequency response plots for $F(s) = 1000/(s^2+2s+100)$.

If s is replaced with $j\omega$,

$$F_1(j\omega) = \frac{1000}{100 - \omega^2 + 2j\omega}$$

$$F_2(j\omega) = \frac{1000}{[1 + j(\omega + 9.95)]\,[1 + j(\omega - 9.95)]}$$

It is easier to calculate the magnitude from $F_1(j\omega)$

Use this form to calculate magnitude.

$$A_1(\omega) = \frac{1000}{\sqrt{(100 - \omega^2)^2 + 4\omega^2}}$$

because using $F_2(j\omega)$ would require more operations

$$A_2(\omega) = \frac{1000}{\sqrt{1 + (\omega + 9.95)^2}\sqrt{1 + (\omega - 9.95)^2}}$$

but both $A_1(\omega)$ and $A_2(\omega)$ would result in the correct magnitude.

Suppose the phase is calculated from $F_1(j\omega)$.

$$\Phi_1(\omega) = -\tan^{-1}\left[2\omega/(100 - \omega^2)\right]$$

Figure 6.14 shows that the phase angle is between 0° and –180°. Notice that if we use $\Phi_1(\omega)$, when the frequency exceeds 10 rad/s, the inverse tangent function operates on a positive numerator and a negative denominator, which should imply an angle between 90°and 180° and therefore an angle between −90° and −180° after multiplying by the minus sign. Unfortunately, when the frequency exceeds 10 rad/s, some computers that are not using a rectangular-to-polar transformation would calculate an angle between 0° and −90° for the inverse tangent function and a result between 0° and 90° after multiplying by the minus sign. Of course, when the frequency exceeds 10 rad/s, we could use

Use this form to calculate phase.

$$\Phi_1(\omega) = -180^\circ + \tan^{-1}[2\omega/(\omega^2 - 100)]$$

but it is less trouble to use $F_2(j\omega)$, so that

$$\Phi_2(\omega) = -\tan^{-1}(\omega + 9.95) - \tan^{-1}(\omega - 9.95)$$

which will result in an angle between 0° and −180° for any frequency. In summary, $A_1(\omega)$ is the easiest expression to use when calculating the magnitude, while $\Phi_2(\omega)$ provides the easiest computation of phase when dealing with an underdamped transmittance.

For example, when the frequency is 5 rad/s, neither procedure requires any special consideration:

$$\begin{aligned}
A_1(5) &= \frac{1000}{\sqrt{5725}} = 13.22 \\
A_2(5) &= \frac{1000}{\sqrt{224.5}\sqrt{25.5}} = 13.22 \\
\Phi_1(5) &= -\tan^{-1}(10/75) = -7.6^\circ \\
\Phi_2(5) &= -\tan^{-1}(14.95) - \tan^{-1}(-4.95) = -86.2^\circ + 78.6^\circ \\
&= -7.6^\circ
\end{aligned}$$

However, when the frequency is 12 rad /s, $\Phi_1(12)$ requires special handling, while $\Phi_2(12)$ does not.

$$\begin{aligned}
A_1(12) &= \frac{1000}{\sqrt{2512}} = 13.22 \\
A_2(12) &= \frac{1000}{\sqrt{482.8}\sqrt{5.2}} = 13.22 \\
\Phi_1(12) &= -\tan^{-1}(24/-44) \\
&= -180^\circ + \tan^{-1}(24/44) = -151.4^\circ \\
\Phi_2(12) &= -\tan^{-1}(21.95) - \tan^{-1}(2.05) = -87.4^\circ - 64.0^\circ \\
&= -151.4^\circ
\end{aligned}$$

Table 6.3 shows Bode plots for complex conjugate LHP and RHP pole and zero pairs. Conjugate LHP zeros of the form

LHP poles, LHP zeros, and RHP roots are compared as to their Bode Plots.

$$F(s) = \frac{s^2 + 2\zeta\omega_n s + \omega_n^2}{\omega_n^2}$$

have Bode plots that are the negatives of the curves for the corresponding conjugate poles. Conjugate RHP poles and zeros,

$$F(s) = \frac{\omega_n^2}{s^2 - 2\zeta\omega_n s + \omega_n^2}$$

Table 6.3 *Bode Plots for Complex Conjugate Pole and Zero Pairs*

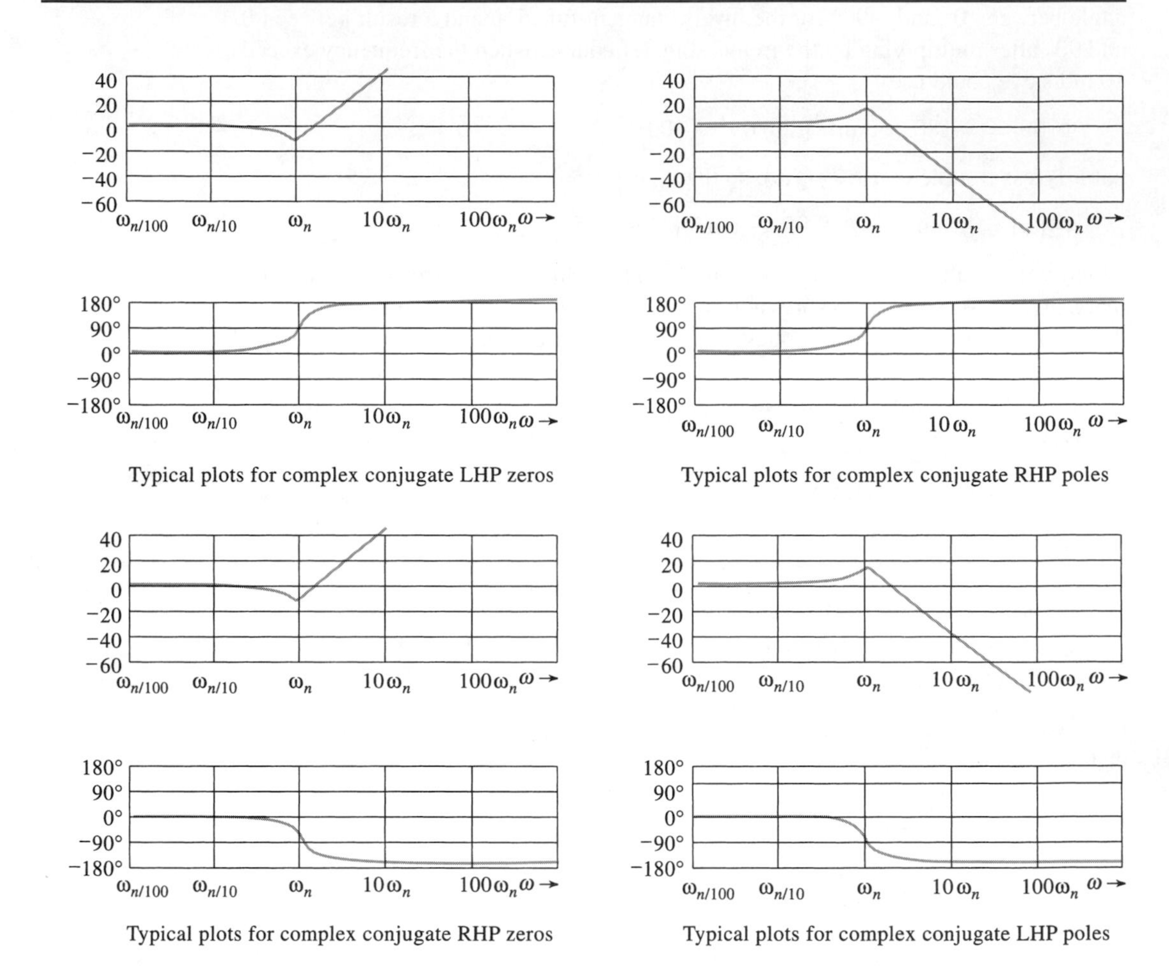

Typical plots for complex conjugate LHP zeros

Typical plots for complex conjugate RHP poles

Typical plots for complex conjugate RHP zeros

Typical plots for complex conjugate LHP poles

and

$$F(s) = \frac{s^2 - 2\zeta\omega_n s + \omega_n^2}{\omega_n^2}$$

differ from their LHP counterparts in algebraic sign of the phase shift; the dB amplitude ratios are the same as for LHP roots of the same type.

As another example, consider the transmittance

$$F(s) = \frac{s^2 + s + 8}{s^2}$$

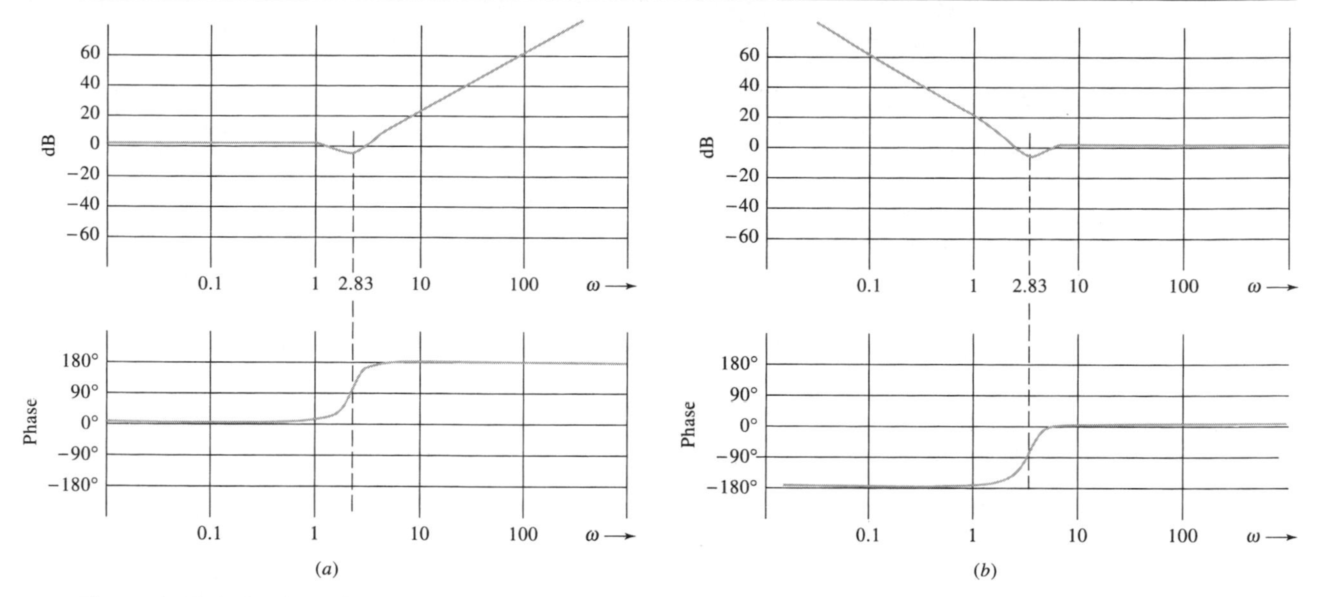

Figure 6.15 Bode plots of a transmittance involving complex roots. (a) Approximate frequency response curves for $F(s) = (s^2 + s + 8)/8$. (b) Approximate frequency response curves for $F(s) = (s^2 + s + 8)/s^2$.

Decomposing, we have

$$F(s) = (8)\left(\frac{s^2 + s + 8}{8}\right)\left(\frac{1}{s}\right)^2$$

The set of complex conjugate zeros has curves that are the negatives of the given curves for a pole, with

$$\omega_n = \sqrt{8} = 2\sqrt{2} = 2.83$$
$$2\zeta\omega_n = 5.66\zeta = 1 \qquad \zeta = 0.177$$

from which the curves of Figure 6.15 are sketched.

❑ DRILL PROBLEM

D6.4 Sketch frequency response curves (both amplitude in dB and phase shift) for the following transmittances:

(a)

$$F(s) = \frac{1}{s + 1000}$$

(b)

$$F(s) = \frac{1}{(s + 10)^3}$$

(c)

$$F(s) = \frac{s - 10}{s + 10}$$

Ans.

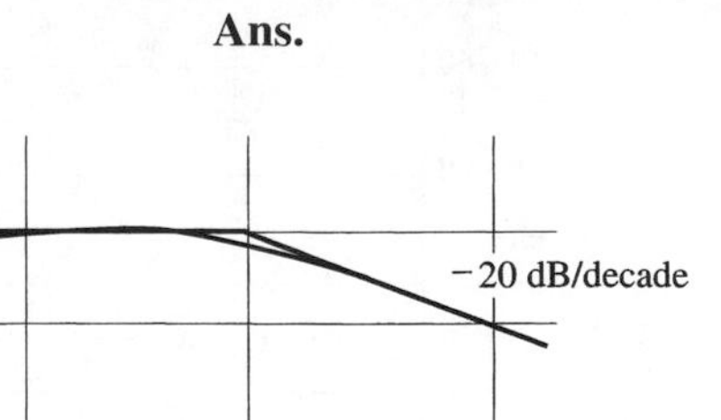

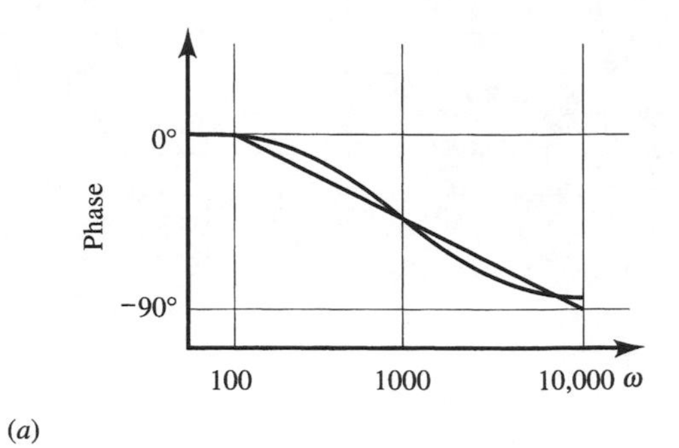

(*a*)

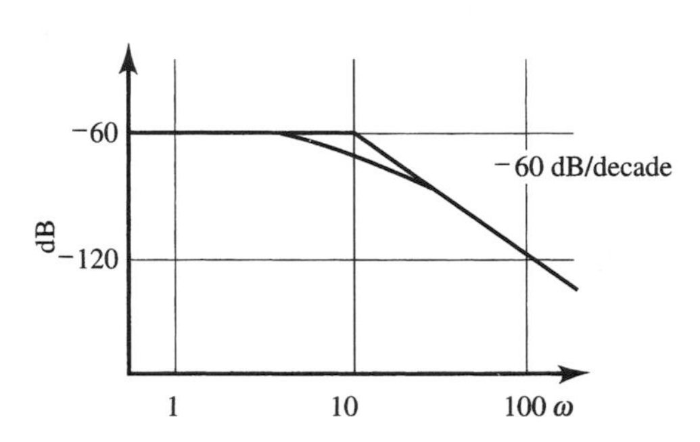

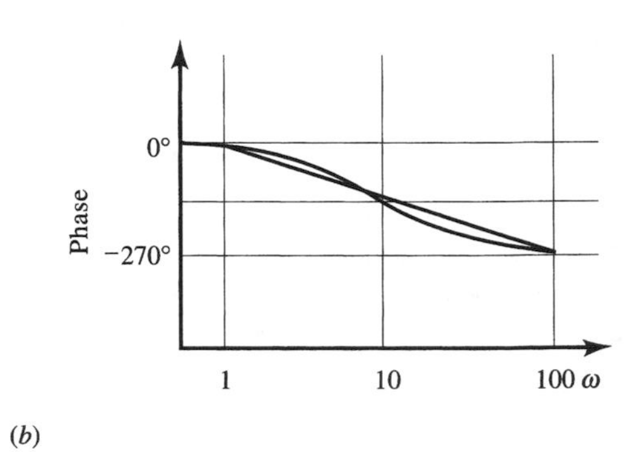

(*b*)

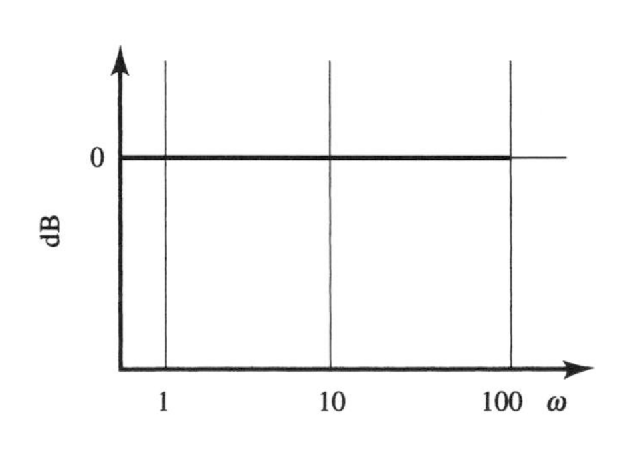

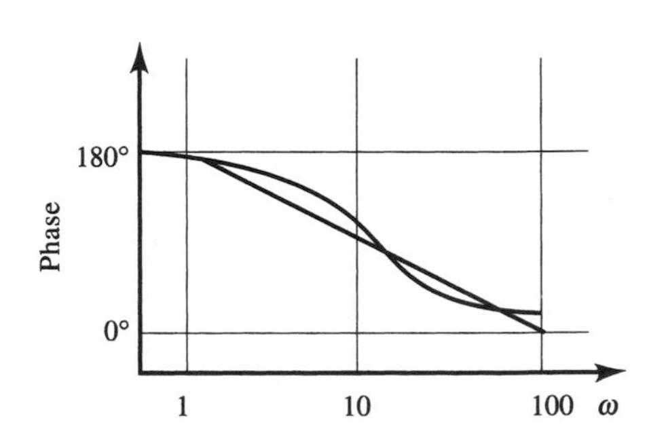

(*c*)

Figure D6.4

(d)

$$F(s) = \frac{10}{s^2 + s + 4}$$

(e)

$$F(s) = \frac{s^2 - 4s + 30}{(s + 10)^2}$$

Ans.

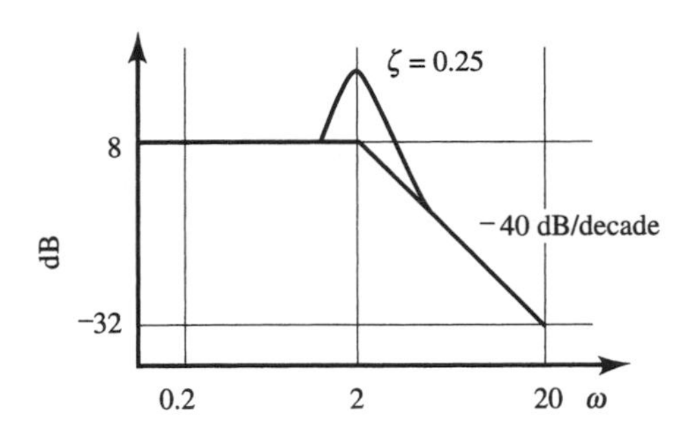

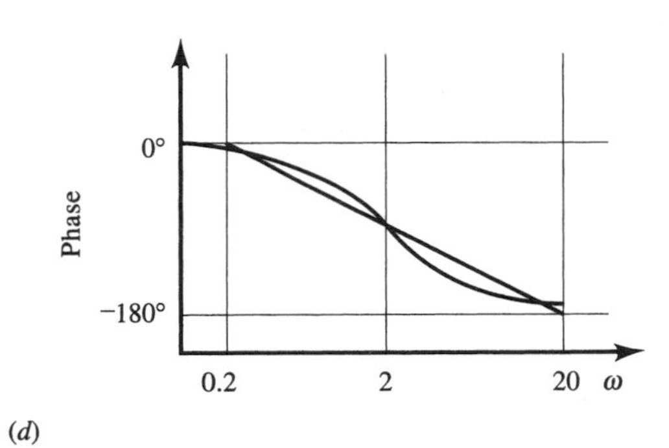

(*d*)

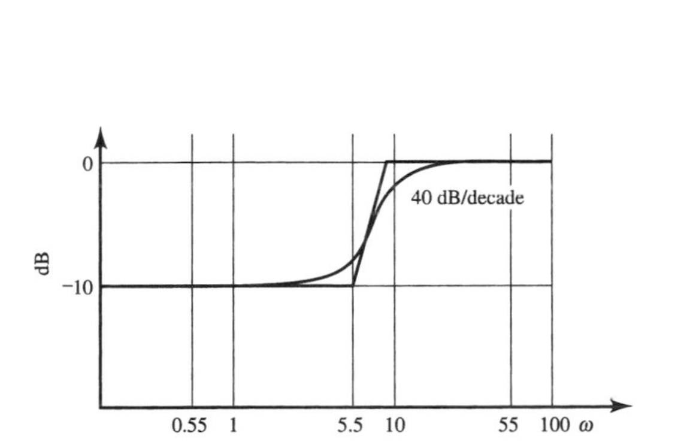

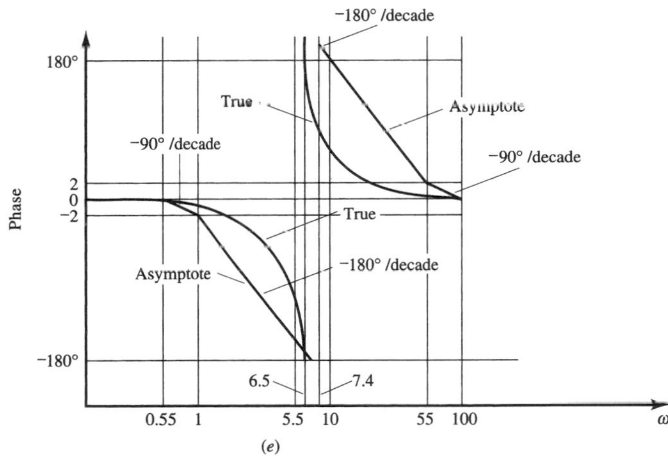

(*e*)

Figure D6.4

(f)

$$F(s) = \frac{s^2 + 2s + 100}{s^2 + 10s + 100}$$

(g)

$$F(s) = \frac{s}{s^2 + 20s + 100}$$

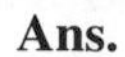

Ans.

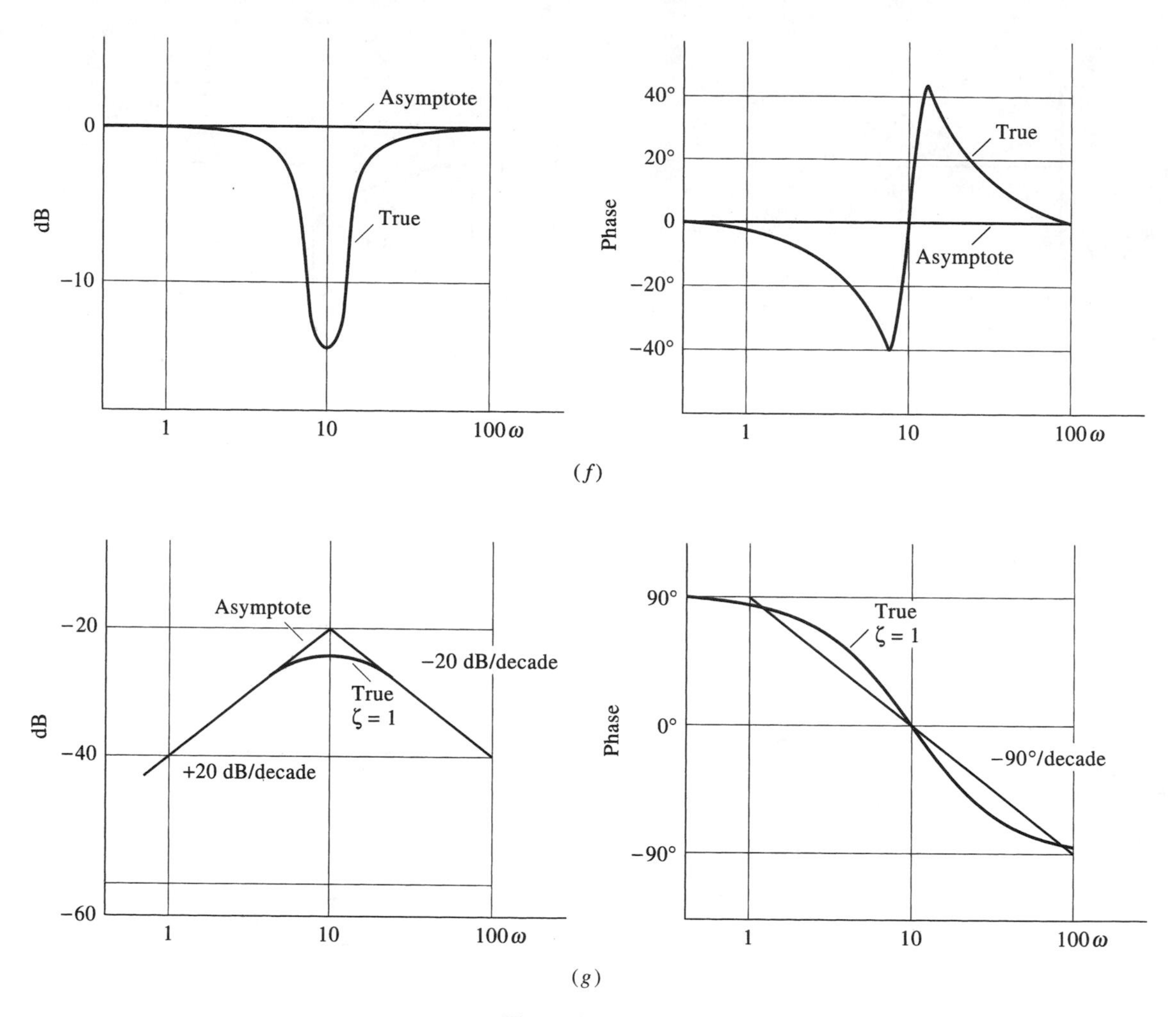

Figure D6.4

❑ Computer-Aided Learning

This section shows how MATLAB can be used for frequency response analysis. The main frequency response commands are as follows:

```
bode, nyquist, nichols, margin
```

We must first define the system transfer function as an object using the "tf" command. For instance, consider the system

$$G(s) = \frac{10}{s + 10}$$

```
g=tf(10,[1,10])
```

```
Transfer function:
  10
------
s + 10
```

To obtain a quick sketch of the Bode plot we use the "bode" command

```
bode(g)
```

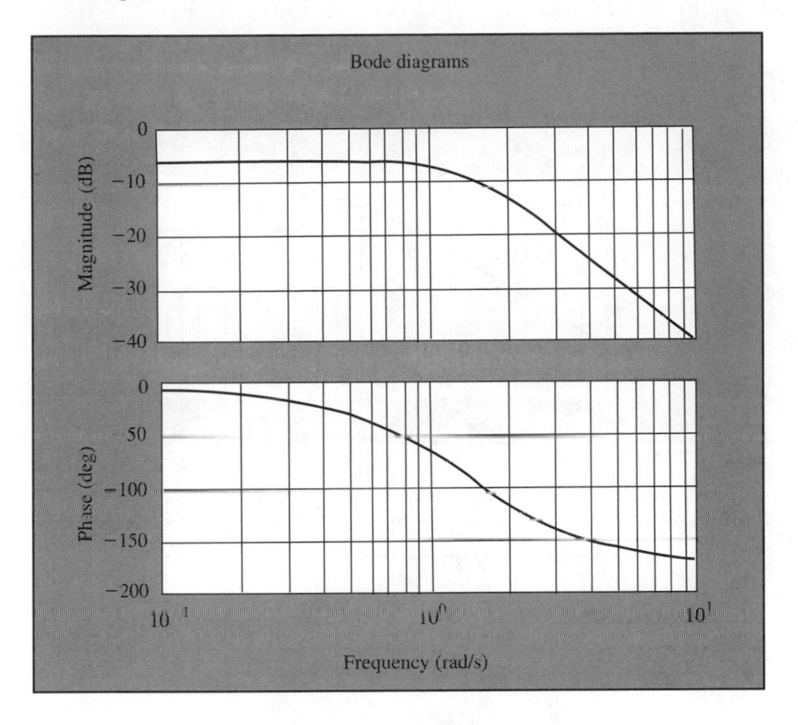

Figure C6.1

The quick form of the "bode" command goes automatically into the graphic mode. All plot options including the frequency axis are automatic. We also do not have the raw magnitude and phase data. Other forms of the bode command are:

```
bode(g,{wmin, wmax})
```

which draws the Bode plot for frequencies between wmin and wmax in radians per second, and

```
bode(g,w)
```

uses the user-supplied vector w of frequencies, in radians per second, at which the Bode response is to be evaluated. We can use the "logspace" to generate logarithmically spaced frequency vectors.

```
bode (g1,g2,g3, ... ,w)
```

plots the Bode response of several systems g1, g2, etc. on a single plot. The frequency vector w is optional. We can also specify a color, line style, and marker for each system, as in

```
bode(g1,'opt',g2,'opt',g3,'opt')
```

where 'opt' stands for plot options (type "help plot" to find out about plotting options). For instance, to plot $s/(s+1)$ in red with double-dashed lines and $1/(s+1)$ in magenta using the marker **x** we enter

```
bode(tf([1 0],[1 1]),'r-',tf([1],[1 1]),'mx')
```

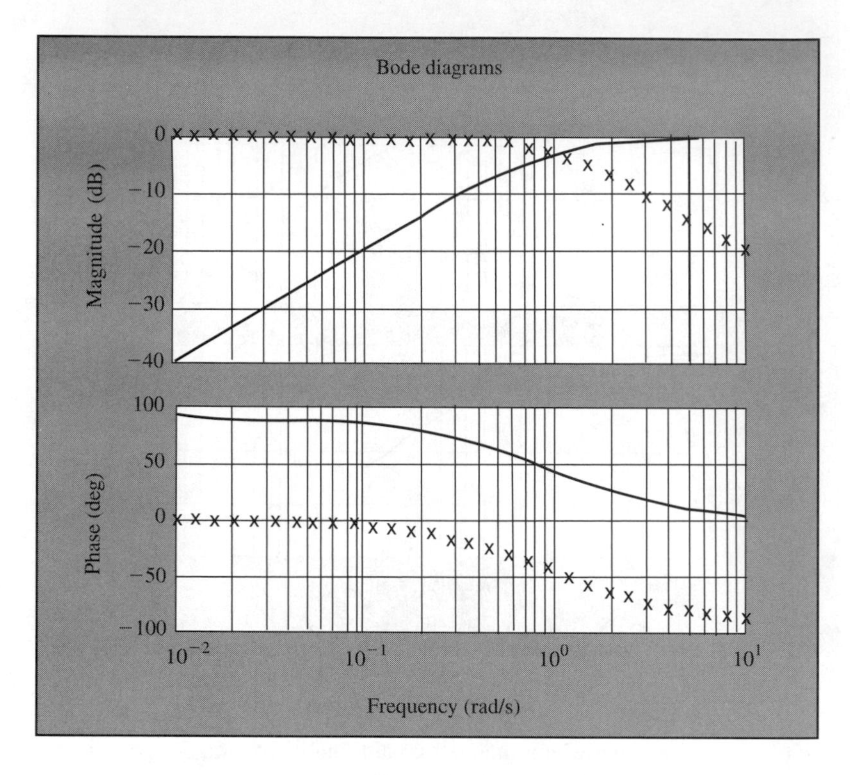

Figure C6.2

when invoked with left-hand arguments, such as the following,

```
[Mag,Phase,w]=bode(g,...)
```

the magnitude and phase are returned as three-dimensional arrays. We can then convert them to column vectors and plot them directly. The "subplot(m,n,p)" command splits the page in m rows and n columns and puts the current graph in the cell p. The "semilogx" command is used to create semilog plots with a logarithmic x-axis.

```
g=tf(1,[1,2,2]);
w=logspace(-1,2,100); % creates a vector of size 100
from 10^-1 to 10^2
[magn, phase] = bode (g,w);
mag=magn (:,:)'; ph=phase (:,:)'; % extract mag and
```

```
phase as column vectors
subplot(2,1,1), semilogx (w,20*log10(mag)), grid
subplot(2,1,2), semilogx (w,ph), grid
```

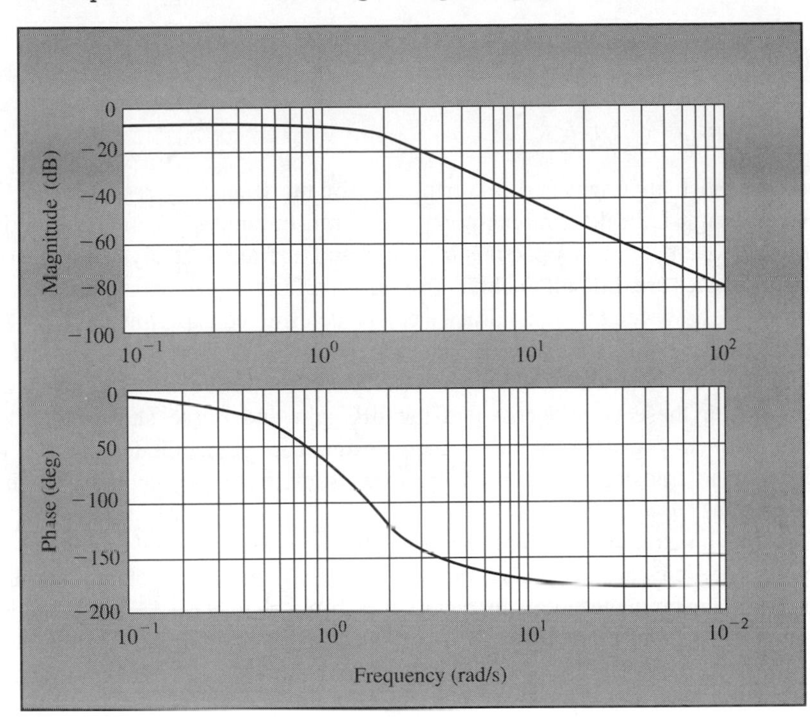

Figure C6.3

We can use this form to compute sinusoidal steady state response. For example, to compute the steady state response $y(t)$ of $G(s) = 1/(s+1)$ to $x(t) = \cos(5t)$, we enter

```
g=tf(1,[1 5]); [mag,ang]=bode(g, 50); mag,
ang=ang*pi/180
mag=
     0.0199
ang=
    1.4711
```

Therefore $y(t) = 0.0199\ \cos(5t - 1.47)$.

C6.1 Verify the results of Drill Problem 6.1 using the "bode" command.

Ans. (a) `g=tf([1 0],[1 3]);[m,p]=bode(g,3); m=7*m;`
`p=p-40`

We will use the "fprintf" command to display the answer.

```
fprintf('y(t)=%fcos(3t+%f)\n',m,p)
```

C6.2 Repeat Drill Problem 6.4 using the "bode" command.

Ans. (e) `f=tf([1-4 30],[1 2*10 10^2]); bode(f)`

6.4 Using Experimental Data

6.4.1 Finding Models

One of the most important and powerful uses of the frequency response method of system design is in determining component transmittances. For many practical system components, such as pneumatic values and airframes, analytic expressions for transmittances are difficult to obtain from theory. If a frequency response test can be performed, however, the transmittance can be determined experimentally.

Pneumatic valves, for example, do not readily lend themselves to analytic determination of their transfer functions. Very often, however, a frequency test can be run on these components and the dB gain and phase shift versus logarithmic frequency can be plotted. Then design can be facilitated by the use of approximate transfer functions determined from the experimentally obtained frequency plots.

Consider an example transmittance for which the experimental amplitude and phase characteristics are as tabulated in Table 6.4. It is desired to obtain a transfer function that approximates these characteristics. If the characteristics are plotted, as in Figure 6.16, a series of straight-line asymptotes can be fitted to these data for both amplitude and phase. By use of the slopes and corresponding break frequencies, a

Table 6.4 *Experimental Frequency Response Data*

f	ω	Gain (dB)	Phase Shift (deg)
60	377	−7.75	−155
50	314	−4.3	−150
40	251	−0.2	−145
35	219	0.75	−140
25	157	5.16	−135
20	126	7.97	−120
16	100	10.5	−110
10	63	15.0	−100
7	44	16.9	−85
2.5	16	20.4	−45
1.3	8	21.6	−30
0.22	1.38	24.0	−5
0.16	1.0	24.1	0

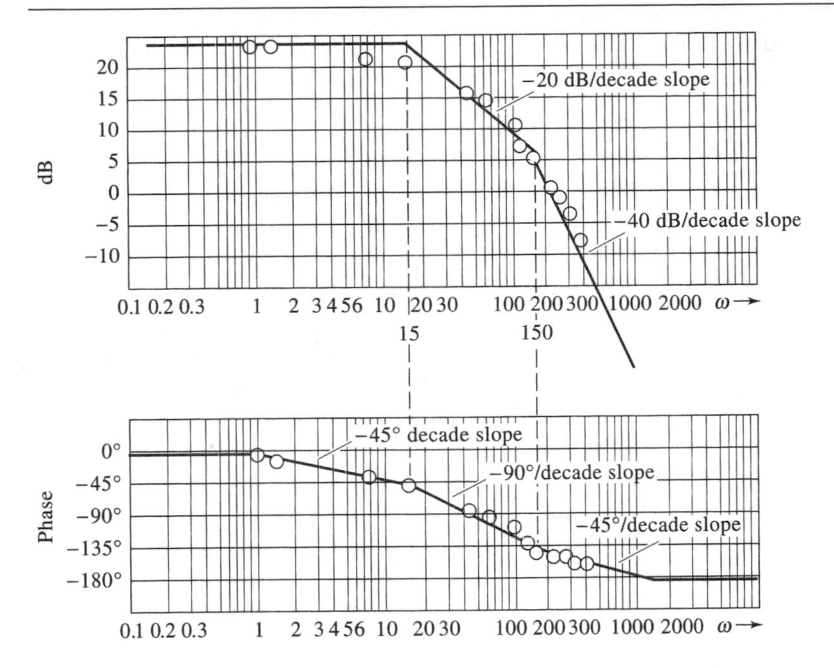

Figure 6.16 Incorporating experimental data for frequency response analysis.

transfer function is obtained. For the example given, an approximate transmittance is

$$F(s) = 16\left(\frac{15}{s+15}\right)\left(\frac{150}{s+150}\right) = \frac{16.0}{(0.07s+1)(0.007s+1)}$$

Often the phase versus logarithmic frequency, as calculated from the approximate transfer function, will not completely agree with the corresponding experimental curve. The problem of obtaining the best match for both amplitude and phase curves is simplified if linear asymptotes are used for both amplitude and phase curves.

Use of this approximated transfer function, in conjunction with the remaining analytically obtainable transfer functions, permits the engineer to analyze the system.

6.4.2 Irrational Transmittances

Another advantage of frequency response methods is that it is not necessary to restrict the type of transfer function to rational polynomials. Frequency response methods can be brought to bear on such irrational transfer functions as

$$F(s) = \sqrt{s}$$

and

$$F(s) = \cos s$$

An irrational transmittance of considerable practical importance is of the form

$$F(s) = e^{-\tau s}$$

where τ is a positive constant. This transmittance represents the time delay of the incoming signal by τ seconds. In the language of the Laplace transformation,

$$\mathcal{L}^{-1}\left[Y(s)e^{-\tau s}\right] = y(t-\tau)$$

One type of time delay scheme involves recording the input signal on magnetic tape. A time delay is obtained as the tape moves from the record to the playback head. This arrangement is often used on broadcast interview programs to allow censorship of the material before it is aired. Other simple time delay systems are transmission lines, digital shift registers, conveyor belts, and audio reverberation generators.

The transmittance for a system that is delayed only in time emerges with no change in amplitude versus frequency, but it does undergo a change in phase. The higher the frequency, the greater the phase shift for the same time delay For

$$F(s) = e^{-s\tau} \qquad F(j\omega) = e^{-j\omega\tau}$$

A delay term has a magnitude of 1 and a decreasing phase at all frequencies.

and

$$A(\omega) = |F(j\omega)| \qquad \Phi(\omega) = \angle F(j\omega) = -\omega\tau \text{ rad}$$

The frequency response of the time delay system

$$F(s) = e^{-(1/2)s}$$

is plotted on a linear scale of frequency in Figure 6.17. On a logarithmic frequency scale, the phase shift curve is more and more compressed for larger values of ω.

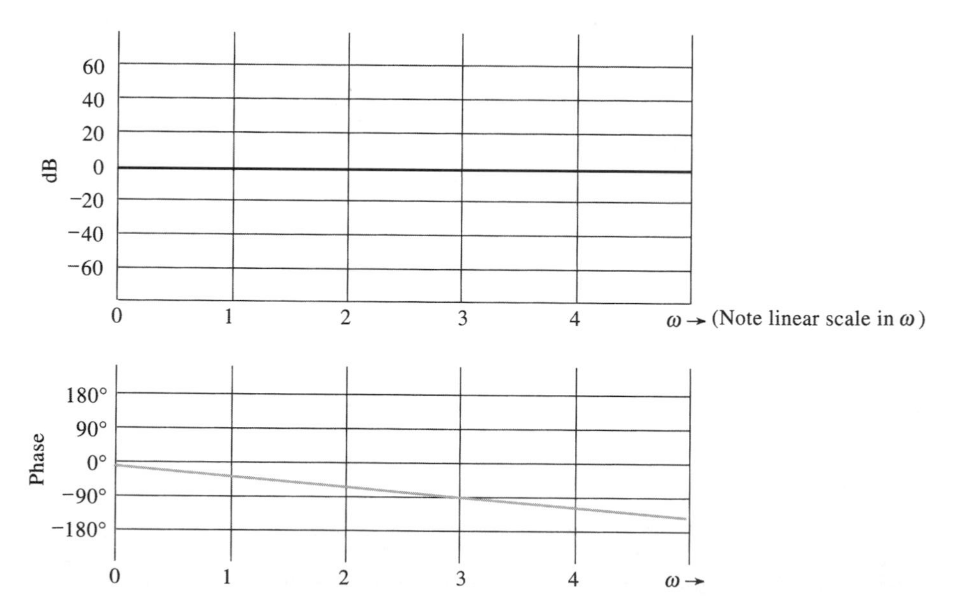

Figure 6.17 Frequency response plots for $F(s) = e^{-(1/2)s}$.

❑ DRILL PROBLEMS

D6.5 Determine an approximate transmittance from the following experimental frequency response data:

ω	dB	Phase
0.1	−20	0
0.5	−21	0
1.0	−21	−9°
2.0	−22	−54°
3.0	−24	−90°
5.0	−28	−135°
10.0	−40	−170°
30.0	−60	−178°
100.0	−84	−180°

Ans. $F(s) = \dfrac{0.1}{(s/3+1)^2}$

D6.6 Find and sketch the frequency response (both amplitude and phase shift) for the following irrational functions:

(a)

$F(s) = 6e^{-0.2s}$

Ans. $6,\ -0.2\omega$

(b)

$F(s) = \dfrac{e^{-4s}}{s}$

Ans. $1/\omega,\ -4\omega - \pi/2$

(c)

$F(s) = \sqrt{s}$

Ans. $\sqrt{\omega}$, 45° or 225°

6.5 Nyquist Methods

Nyquist's work in the 1930s led to a procedure that can determine whether a system is stable by using the frequency response of the open-loop transmittance $G(s)H(s)$. Since the frequency response could be calculated fairly easily (given the slide rule-based technology of that era), Nyquist used calculus to prove that the number of right half-plane poles could be found for an electrical network (radio and TV were under active development then) or for a control system (for which the method is now primarily used).

The origin of Nyquist's work is discussed.

We will discuss the Nyquist procedure in two parts. First we will show how to create a Nyquist (polar) plot, and then we will evaluate system stability by interpreting the Nyquist plot.

6.5.1 Generating the Nyquist (polar) Plot

Figure 6.18 contains a typical control system with gain K, plant $G(s)$, and feedback element $H(s)$. The closed-loop transfer function is

$$T(s) = \frac{KG(s)}{1 + KG(s)H(s)} \qquad [6.3]$$

The Nyquist plot is defined.

The Nyquist plot assumes $K = 1$.

The RHP boundary is influenced by open-loop IA roots.

The frequency response of a transmittance has important implications in a communication system, but our primary reason for examining the frequency response of a transmittance is to evaluate system stability for a system of the form of Equation (6.3). A Nyquist plot connects values of the open-loop transmittance $G(s)H(s)$ for those s that lie on the boundary of the right half-plane (RHP). The procedure results in counting the number of RHP closed-loop poles much as a fisherman might cast out a net and examine its contents for edible fish. It is likewise important to look at the resulting plot to determine whether any RHP closed-loop poles are caught as the RHP boundary is traversed. At first, it is assumed that $K = 1$. Later, other values of K are chosen to force the closed-loop system into marginal stability.

The RHP boundary is traversed in a clockwise direction, so the RHP is always on the right. Since the purpose of the Nyquist plot is to count closed-loop poles lying to the right of the boundary, the RHP boundary must not include imaginary axis (IA) poles of $G(s)H(s)$, otherwise there might be confusion about counting the IA poles as closed-loop RHP poles. The boundary of Figure 6.19(a) is used when $G(s)H(s)$ has no IA poles. That boundary consists of the imaginary axis and a semicircle of infinite radius that covers the open RHP.

If $G(s)H(s)$ contains an open-loop pole at the origin, the boundary of Figure 6.19(b) is used so that a semicircle of very small radius bypasses the pole of $G(s)H(s)$ at $s = 0$. Similarly, other IA poles of $G(s)H(s)$ are bypassed by small semicircles as in Figure 6.19(c). Four Nyquist plot examples are now considered in detail.

Suppose the open-loop transmittance is

$$G_1(s)H_1(s) = \frac{6}{(s+1)(s+2)}$$

Since there are no open-loop IA poles, the boundary of Figure 6.19(a) is used as in Figure 6.20(a). For, convenience, the pole-zero plot of $G_1(s)H_1(s)$ is included

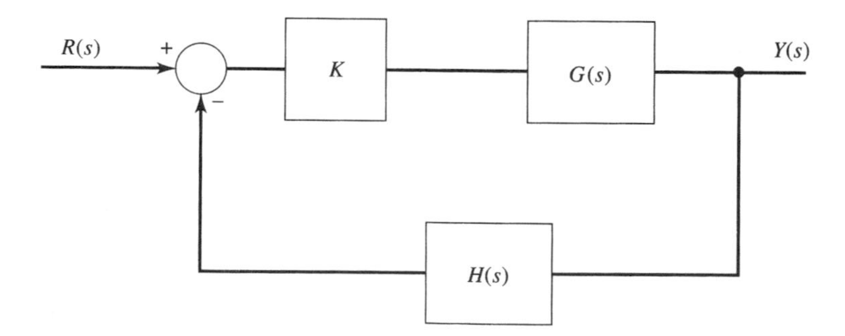

Figure 6.18 Closed-loop system.

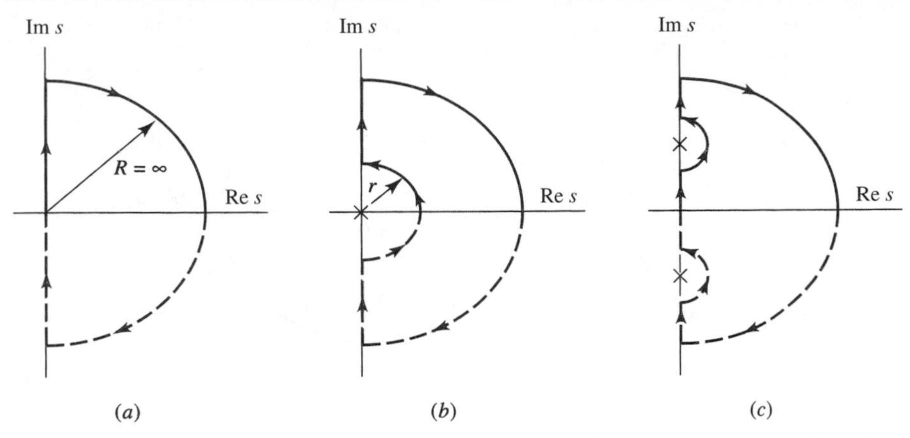

Figure 6.19 Right half-plane boundary. (a) $G(s)H(s)$ has no imaginary axis poles. (b) $G(s)H(s)$ has a pole at the origin. (c) $G(s)H(s)$ has imaginary axis poles.

with the RHP boundary. Point a represents $s = 0$, so we must map $s = 0$ using $G_1(s)H_1(s)$.

First, map near $s = 0$.

$$G_1(0)H_1(0) = 6/2 = 3$$

Next the upper IA (marked I on the boundary) must be mapped (when points are calculated and then plotted for some function, that process is called *mapping*). For the upper IA, $s = j\omega$ for $0 \leqslant \omega < \infty$. These are exactly the values of s used in a Bode plot. Here

Second, map I.

$$G_1(j\omega)H_1(j\omega) = \frac{6}{(j\omega + 1)(j\omega + 2)}$$

so that

$$A_1(\omega) = \frac{6}{\sqrt{(\omega^2 + 1)}\sqrt{(\omega^2 + 4)}}$$

$$\Phi_1(\omega) = -\tan^{-1}(\omega) - \tan^{-1}(\omega/2)$$

The phase $\Phi_1(\omega)$ varies from 0° to − 180° as frequency varies from 0 to infinity. Table 6.5 contains a number of points spanning the entire frequency range.

These points are plotted in Figure 6.20(b). The points c, d, and e on the boundary occur for values of s with infinite magnitude, so $G_1(s)H_1(s)$ is zero at each value.

Third, map infinite s.

Table 6.5 *Evaluation of $G_1(s)H_1(s)$ for s on the IA*

ω	0	.5	1	1.5	2	10	x
$A_1(\omega)$	3	2.6	1.89	1.33	0.95	0.06	0
$\Phi_1(\omega)$	0°	−41°	−72°	−93°	−108°	−163°	−180°

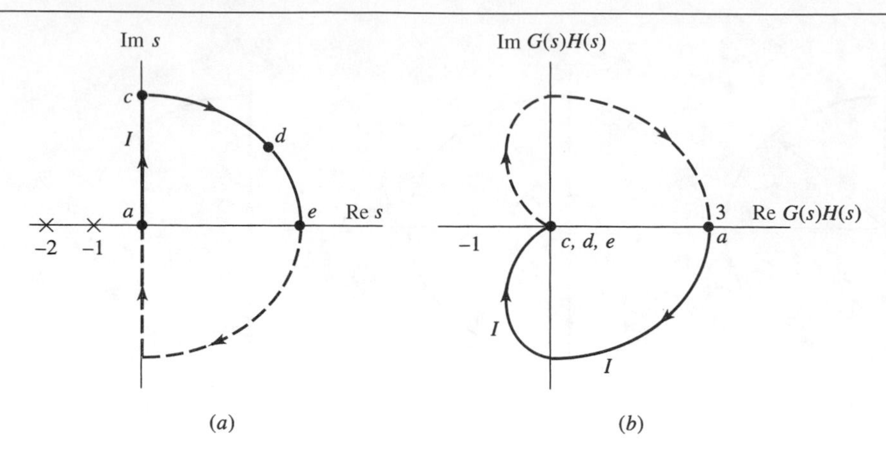

Figure 6.20 Nyquist plot for $G(s)H(s) = 6/(s+1)(s+2)$. (a) RHP boundary. (b) Nyquist plot.

The mapped values c, d, and e lie at the origin of the Nyquist plot. It is common practice to draw arrows to connect the points mapped in the same order as the boundary is traversed so that the mapped location resulting in a, I, c, d, and e is oriented in that order.

Finally, map the mirror image of the Nyquist plot for the upper RHP boundary.

It is not necessary to calculate the mapping of the lower half boundary Figure 6.20(a). Since the lower half (dashed-line) boundary is the mirror image of the upper half boundary reflected about the real (horizontal) axis, it can be shown that the Nyquist plot for the mapping of the lower half boundary is just the mirror image of the Nyquist plot mapping of the upper half boundary, where the mapping is reflected about the real (horizontal) axis and shown by a dashed line in Figure 6.20(b). The complete Nyquist plot is therefore a closed curve. Stability implications will be considered shortly. Three more examples are considered first.

As a second example, consider

$$G_2(s)H_2(s) = \frac{6}{s(s+2)}$$

Because of the open-loop pole at $s = 0$, the boundary of Figure 6.19(b) is used as in Figure 6.21(a). The mapping of the upper half boundary is done first. To create the Nyquist plot for the part of the boundary between a and b in Figure 6.21(a), s is considered to be $r\angle\theta$ where r is a very small number and the angle θ varies $0° \leqslant \theta \leqslant 90°$. For these values of s we can approximate $G_2(s)H_2(s)$ by

First map s between a and b.

$$G_2(s)H_2(s) \approx \frac{6}{(r\angle\theta)(2)} = \frac{3}{r\angle\theta} = R_1\angle -\theta$$

where R_1 is a very large number and $-\theta$ follows $0° \geqslant -\theta \geqslant -90°$. Thus the Nyquist plot for the mapping of $[a, b]$ becomes a semicircle of large radius R_1 moving through angles from 0° to − 90° as in Figure 6.21(b).

Second, map I. Third, map infinite s.

Next, we must map the upper IA boundary, denoted by I in Figure 6.21(a). Bode plot data could be used, or, a representative set of points as in Table 6.6 can be

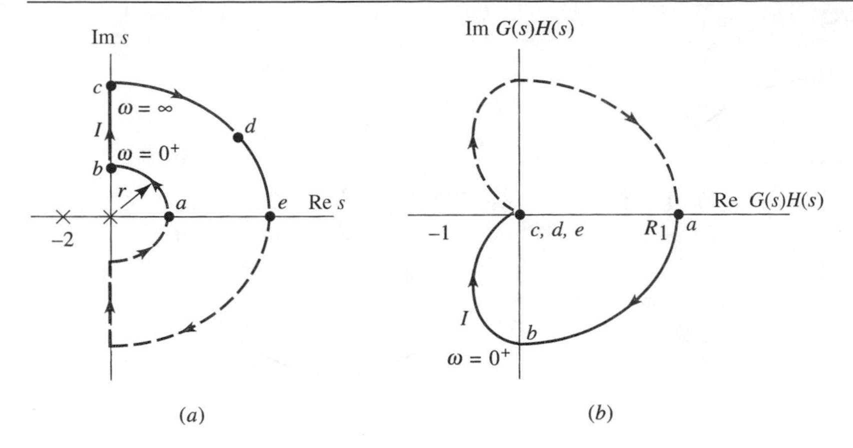

Figure 6.21 Nyquist plot for $G(s)H(s) = 6/s(s+2)$. (a) RHP boundary. (b) Nyquist plot. (Not to scale.)

Table 6.6 *Evaluation of* $G_2(s)H_2(s)$ *for* s *on the* IA

ω	0	.5	1	1.5	2	10	∞
$A_2(\omega)$	∞	5.82	2.68	1.60	1.06	0.06	0
$\Phi_2(\omega)$	$-90°$	$-104°$	$-117°$	$-127°$	$-135°$	$-169°$	$-180°$

calculated from

$$G_2(j\omega)H_2(j\omega) = \frac{6}{j\omega(j\omega + 2)}$$

so that

$$A_2(\omega) = \frac{6}{\omega\sqrt{(\omega^2 + 4)}}$$

$$\Phi_2(\omega) = -90° - \tan^{-1}(\omega/2)$$

Here the phase angle $\Phi_2(\omega)$ varies from $-90°$ to $-180°$. The frequency value at the point b is denoted by a small positive frequency 0^+ in both the boundary and the mapped Nyquist plot. The actual $\omega = 0$ frequency would map to infinity, which could not be plotted. Even so, Figure 6.21(b) cannot be drawn exactly to scale, so the general shape of the Nyquist is shown instead.

The dashed bottom-half boundary of Figure 6.21(a) is mapped by taking the mirror image of the Nyquist plot for the upper-half boundary and then reflecting about the horizontal axis in Figure 6.21(b).

Finally, take the mirror image.

The third example contains two open-loop poles at the origin.

$$G_3(s)H_3(s) = \frac{6}{s^2(s + 2)}$$

Because of these two open-loop poles at $s = 0$, the boundary of Figure 6.19(b) is used as in Figure 6.22(a). As before, the mapping of the upper half boundary is

First, map s between a and b.

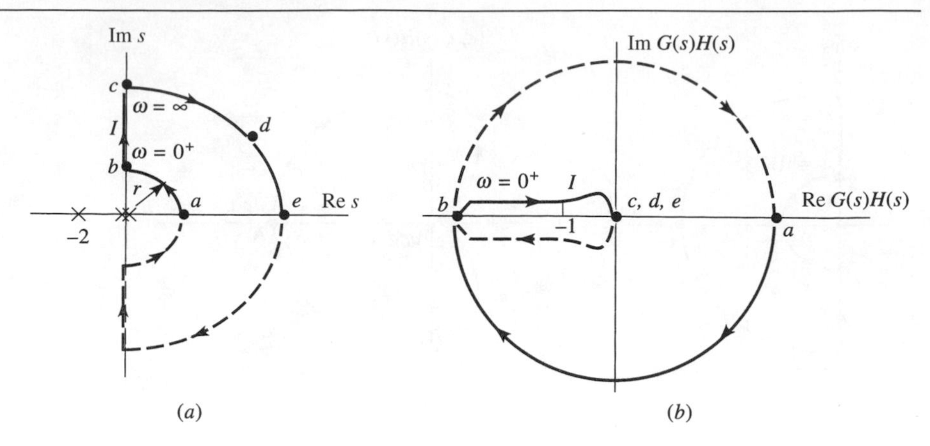

Figure 6.22 Nyquist plot for $G(s)H(s) = 6/s^2(s+2)$ (not to scale). (a) RHP boundary. (b) Nyquist plot.

done first. To create the Nyquist plot for the part of the boundary between a and b in Figure 6.22(a), s is considered to be $r\angle\theta$, where r is a very small number and the angle θ varies $0° \leqslant \theta \leqslant 90°$. For these values of s we can approximate $G_3(s)H_3(s)$ by

$$G_3(s)H_3(s) \approx \frac{6}{(r\angle\theta)^2(2)} = \frac{3}{r^2\angle 2\theta} = R_1\angle -2\theta$$

where R_1 is a very large number and -2θ follows $0° \geqslant -2\theta \geqslant -180°$. Thus the Nyquist plot for the mapping of $[a, b]$ becomes a semicircle of infinitely large radius R_1 moving through angles from 0° to − 180° as in Figure 6.22(b).

Second, map I. Third, map infinite s.

Next, we must map the upper IA boundary, denoted by I in Figure 6.22(a). Bode plot data can be used, or a representative set of points as in Table 6.7 can be calculated from

$$G_3(j\omega)H_3(j\omega) = \frac{6}{(j\omega)^2(j\omega+2)}$$

So that

$$A_3(\omega) = \frac{6}{\omega^2\sqrt{(\omega^2+4)}}$$

$$\Phi_3(\omega) = -180° - \tan^{-1}(\omega/2)$$

Here the phase angle $\Phi_3(\omega)$ varies from −180° to −270°.

Table 6.7 *Evaluation of $G_3(s)H_3(s)$ for s on the IA*

ω	0	.5	1	1.5	2	10	∞
$A_3(\omega)$	∞	11.64	2.68	1.07	0.53	0.006	0
$\Phi_3(\omega)$	−180°	−194°	−207°	−217°	−225°	−259°	−270°

The bottom-half dashed boundary of Figure 6.22(a) is mapped by taking the mirror image of the Nyquist plot for the upper-half boundary and then reflecting about the horizontal axis in Figure 6.22(b).

Finally, take the mirror image.

The fourth example has three open-loop poles, one of which is at the origin.

$$G_4(s)H_4(s) = \frac{4}{s(s+1)(s+2)}$$

Because of the open-loop pole at $s = 0$, the boundary of Figure 6.19(b) is used as in Figure 6.23(a). To create the Nyquist plot for the part of the boundary between a and b in Figure 6.23(a), s is considered to be $r\angle\theta$ where r is a very small number and the angle θ varies $0^\circ \leqslant \theta \leqslant 90^\circ$. For these values of s we can approximate $G_4(s)H_4(s)$ by

First, map s between a and b.

$$G_4(s)H_4(s) \approx \frac{4}{(r\angle\theta)(2)} = \frac{2}{r\angle\theta} = R_1\angle-\theta$$

Where R_1 is a very large number and $-\theta$ follows $0^\circ \geqslant -\theta \geqslant -90^\circ$. Thus the Nyquist plot for the mapping of $[a, b]$ becomes a semicircle of infinitely large radius R_1 moving through angles from 0° to -90° as in Figure 6.23(b).

Next, we must map the upper IA boundary denoted by I in Figure 6.23(a). Bode plot data can be used, or a representative set of points as in Table 6.8 can be calculated from

Second map I. Third, map infinite s.

$$G_4(j\omega)H_4(j\omega) = \frac{4}{j\omega(j\omega+1)(j\omega+2)}$$

so that

$$A_4(\omega) = \frac{4}{\omega\sqrt{(\omega^2+1)}\sqrt{(\omega^2+4)}}$$

$$\Phi_4(\omega) = -90^\circ - \tan^{-1}(\omega) - \tan^{-1}(\omega/2)$$

Here the phase angle $\Phi_4(\omega)$ varies from -90° to -270°.

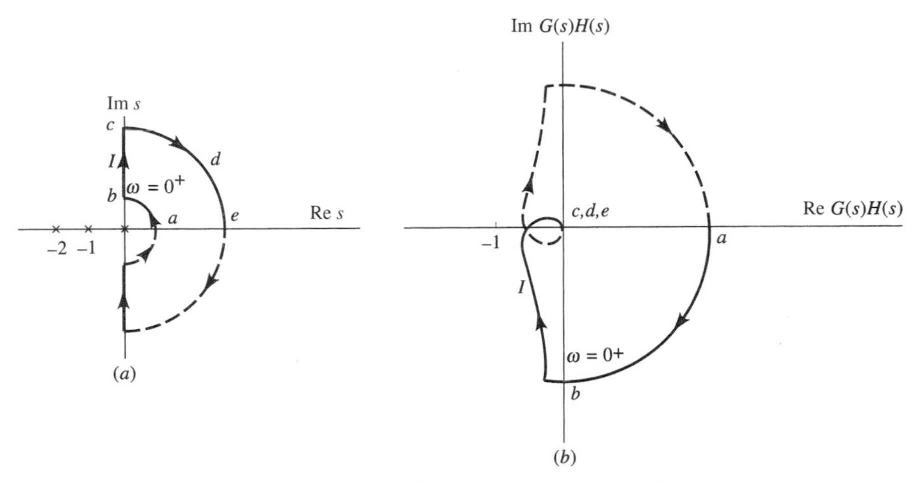

Figure 6.23 Nyquist plot $G(s)H(s) = 4/s(s+1)(s+2)$. (a) RHP boundary. (b) Nyquist plot. (Not to scale.)

Table 6.8 *Evaluation of $G_4(s)H_4(s)$ for s on the IA*

ω	0	.5	1	1.414	1.5	2	10	∞
$A_4(\omega)$	∞	3.47	1.26	0.67	0.59	0.31	0.004	0
$\Phi_4(\omega)$	$-90°$	$-131°$	$-162°$	$-180°$	$-183°$	$-198°$	$-253°$	$-270°$

Finally, take the mirror image.

The dashed bottom-half boundary of Figure 6.23(a) is mapped by taking the mirror image of the Nyquist plot for the upper-half boundary and then reflecting about the horizontal axis as in Figure 6.23(b).

The purpose in generating these plots is to evaluate the stability of each system. Stability is covered in the next section.

6.5.2 Interpreting the Nyquist Plot

CW encirclements are defined and calcualted.

Nyquist showed that a relationship exists between the way the Nyquist plot of the open-loop transmittance $G(s)H(s)$ encircles the -1 point and the number of RHP poles of the closed-loop system. An important value is the number of clockwise (CW) encirclements of the -1 point. In Figure 6.24 that value can be established by looking at a segment of the Nyquist plot. A vector can be drawn outward from the -1 point. If the vector is corssed by the Nyquist plot twice in a CW direction and once in a counterclockwise (CCW) direction, the result is interpreted as one CW encirclement. CCW encirclements are considered to be negative.

Nyquist proved that the number of RHP poles of the closed-loop transfer function $T(s)$, which is Equation (6.3) with $K = 1$,

$$T(s) = \frac{G(s)}{1 + G(s)H(s)} \tag{6.3$'$}$$

The Nyquist plot is interpreted.

is given by

$$\begin{pmatrix}\text{Number of RHP}\\ \text{poles of } T(s)\end{pmatrix} = \begin{pmatrix}\text{number of CW}\\ \text{encirclements of}\\ \text{the } -1 \text{ point on}\\ \text{the } GH \text{ plane}\end{pmatrix} + \begin{pmatrix}\text{number of RHP}\\ \text{poles of } G(s)H(s)\end{pmatrix} \tag{6.4}$$

Let us apply Nyquist's test of Equation (6.4) to the examples in Figures 6.20 to 6.23. The $G(s)H(s)$ of Figure 6.20(b) has no open-loop RHP pole. If Nyquist's test

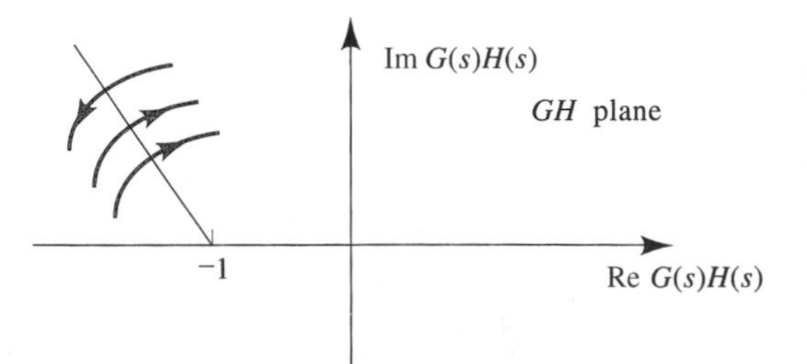

Figure 6.24 Segment of a Nyquist plot used to interpret the number of CW encirclements of -1. (Here there is one CW encirclement).

is applied to the Nyquist plot of Figure 6.20(b), a vector drawn in the second quadrant either is not crossed at all or is crossed once in each direction. The conclusion is that there are no CW encirclements, so

$$\text{Number of RHP poles of } T(s) = 0 + 0 = 0$$

System 1 is stable for $K = 1$.

and the system is stable. The conclusion is the same if Nyquist's test of Equation (6.4) is applied to the system of Figure 6.21(b). That system also has no open-loop RHP poles, and there is no CW encirclement of -1.

System 2 is stable for $K = 1$.

The system of Figure 6.22(b) has no open-loop RHP pole. However, the two open-loop poles at $s = 0$ cause the large radius R_1 to sweep through a clockwise-moving arc from 0° to 180° due to the boundary from $[a, b]$ and to sweep through an additional CW 180° when the lower boundary is mapped. The result from the complete Nyquist plot is that there are two CW encirclements of -1, as is demonstrated by drawing a vector from -1 into the second quadrant. The conclusion is that

System 3 is unstable for $K = 1$.

$$\text{Number of RHP poles of } T(s) = 2 + 0 = 2$$

and the closed-loop system is unstable, with two RHP poles.

The system of Figure 6.23(b) exhibits no CW encirclements of -1 and since there are no open-loop RHP poles,

System 4 is stable for $K = 1$.

$$\text{Number of RHP poles of } T(s) = 0 + 0 = 0$$

and the closed-loop system is stable.

Suppose that K is not necessarily one, so that $T(s)$ is given by Equation (6.3) instead of Equation (6.3′). To determine stability, Nyquist plots would be required for $KG(s)H(s)$. It is easy to visualize the case for K not equal to 1 since the plots of Figures 6.20(b) to 6.23(b) would expand for K greater than 1 and contract for K less than 1. The number of CW encirclement of -1 in Figures 6.20(b) and 6.21(b) would remain at zero and so no positive gain K could alter stability. Each system is stable for all positive K, which can easily be verified by plotting the root locus for each case.

System 1 and 2 are stable for all $K > 1$.

Conversely, the two CW encirclements in Figure 6.22(b) would remain for any K because the radius R_1 is always quite large. It follows that the system of Figure 6.22(b) is unstable for all positive K, which can also be demonstrated from the root locus.

System 3 is unstable for all $K > 1$.

If K is sufficiently large, the Nyquist plot of Figure 6.23(b) would expand and have two CW encirclements of -1. The system would become unstable, with two closed-loop RHP poles. The conclusion regarding the system of Figure 6.23(b) is that the system is stable (no CW encirclements of -1) for small K and unstable, with two CW encirclements of -1, for large K. The system of Figure 6.23(b) is the only one of the four examples whose stability is alterable by varying K, assuming that K takes on only positive values.

System 4 can be stable, marginally stable, or unstable as K increases.

The analysis indicates that CW encirclements are always to be avoided, since each additional encirclement indicates additional closed-loop RHP poles. Conversely, CCW encirclements are beneficial to stability. For example, a system with

This example has an open-loop RHP pole.

$$G(s)H(s) = \frac{2(s+3)}{(s+2)^2(s-1)} \qquad K = 1$$

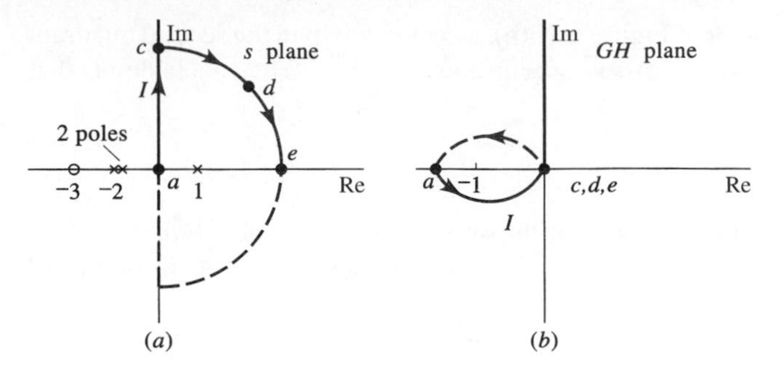

Figure 6.25 Applying the Nyquist criterion.

has the Nyquist plot given in Figure 6.25. $G(s)H(s)$ has one RHP pole, but the Nyquist plot circles the -1 point once in a CCW sense, so

$$(\text{Number of RHP poles of } T(s)) = (-1) + (1) = 0$$

and the overall system is stable.

Notice that if K is reduced sufficiently below 1, the Nyquist plot of Figure 6.25(b) shrinks, thus eliminating the CCW encirclement, in which case the open-loop RHP pole causes

$$\text{Number of RHP poles of } T(s) = 0 + 1 = 1$$

The system is unstable for small K and stable for large K.

and the closed-loop system is unstable, with one RHP pole. It is concluded that the system of Figure 6.25(b) is stable for large positive K and unstable for small positive K. A control system analyst will always obtain the same conclusion, whether using a root locus plot or a complete Nyquist plot.

❑ DRILL PROBLEMS

D6.7 Sketch Nyquist plots for feedback systems with the following loop transmittances, then use the plots to determine whether or not each system is stable:

(a)

$$G(s)H(s) = \frac{s}{s+4}$$

Ans. stable

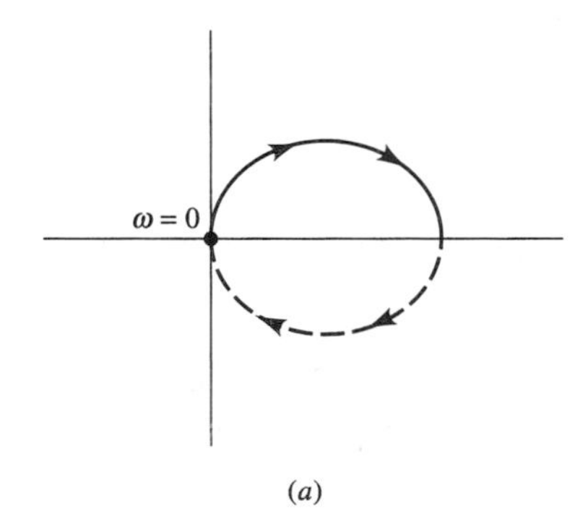

Figure D6.7

(b)

$$G(s)H(s) = \frac{10}{(s+2)(s+6)}$$

Ans. stable

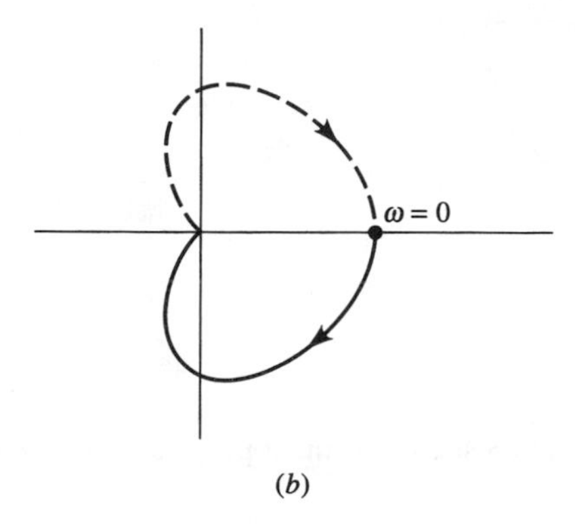

(b)

Figure D6.7

(c)

$$G(s)H(s) = \frac{s^2}{s^2 + 2s + 10}$$

Ans. stable

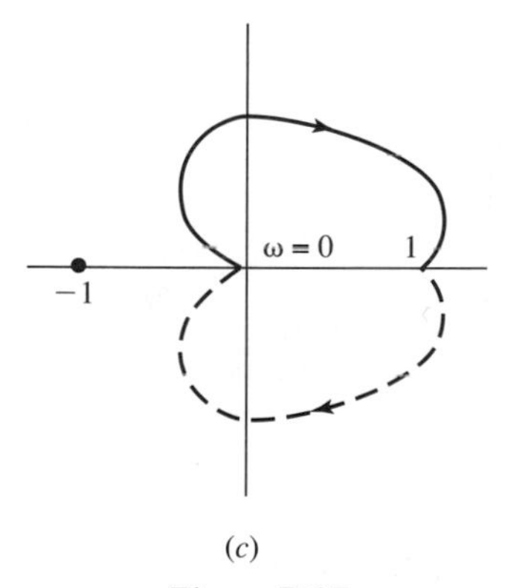

(c)

Figure D6.7

(d)

$$G(s)H(s) = \frac{2}{s^2(s+3)}$$

Ans. unstable

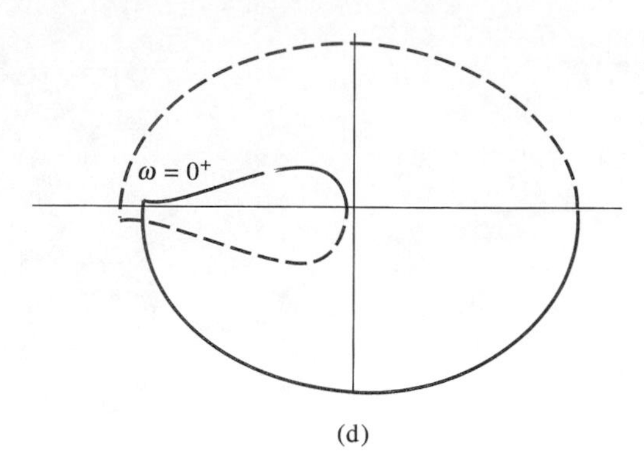

(d)

Figure D6.7

D6.8 For the feedback systems in Figure D6.8 sketch Nyquist plots and use them to determine whether the system is stable.

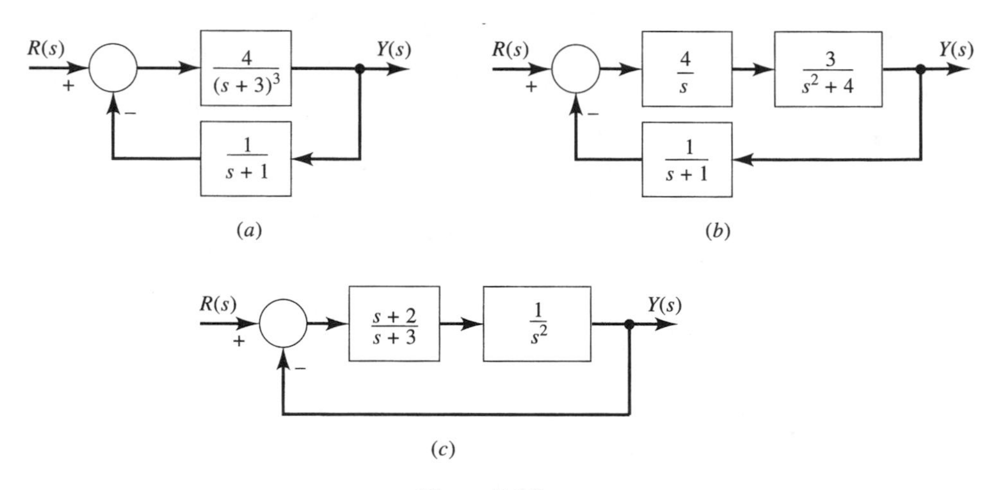

Figure D6.8

Ans. (a) stable; (b) unstable; (c) stable

The stabilities of the systems of Table 6.9 are evaluated.

Table 6.9 shows a number of complete Nyquist plots for systems with rational transmittances $G(s)H(s)$. None of these systems have RHP open-loop poles, so the number of CW encirclements of the -1 point equals the number of closed-loop RHP poles. In some cases, selecting K to be other than one in Figure 6.18 alters stability, while in other cases stability is the same for all positive K values. For example, the systems of a, b, c, and d are stable for any positive K. However, the systems of e and f are stable for small K, marginally stable for some intermediate K, and unstable for large K, in which case there are two closed-loop RHP poles.

The system of g is stable for all positive K. The system of h is unstable for all positive K, and there is one closed-loop RHP pole. The system of i is unstable for all positive K, and there are two closed-loop RHP poles. The system of j is stable for all positive K.

The main function of the complete Nyquist plot is to determine whether there are encirclements of the -1 point. Once it has been determined how the -1 point is (or is not) encircled, it is only necessary to consider enough of the Nyquist plot to identify encirclements. Generally, we only need to examine the part of the Nyquist plot resulting from mapping the upper imaginary axis, that is, mapping $G(s)H(s)$ for $s = j\omega,\ 0 \leqslant \omega < \infty$. Bode noticed that fact, and, as we shall see, stability can be determined from the Bode plot for all systems with no open-loop RHP poles or zeros.

For example, many practical systems may contain a delay due to the inclusion of a computing device or due to delays in sending information to the plant. Since a T-second delay may be modeled by its Laplace transform e^{-sT}, a partial Nyquist plot may be used to evaluate the stability of a plant $1/s$ containing a delay. That part of the Nyquist plot that determines the presence of an encirclement of -1 is examined.

The stability of a system with delay is evaluated.

Table 6.9 *A Collection of Nyquist Plots*

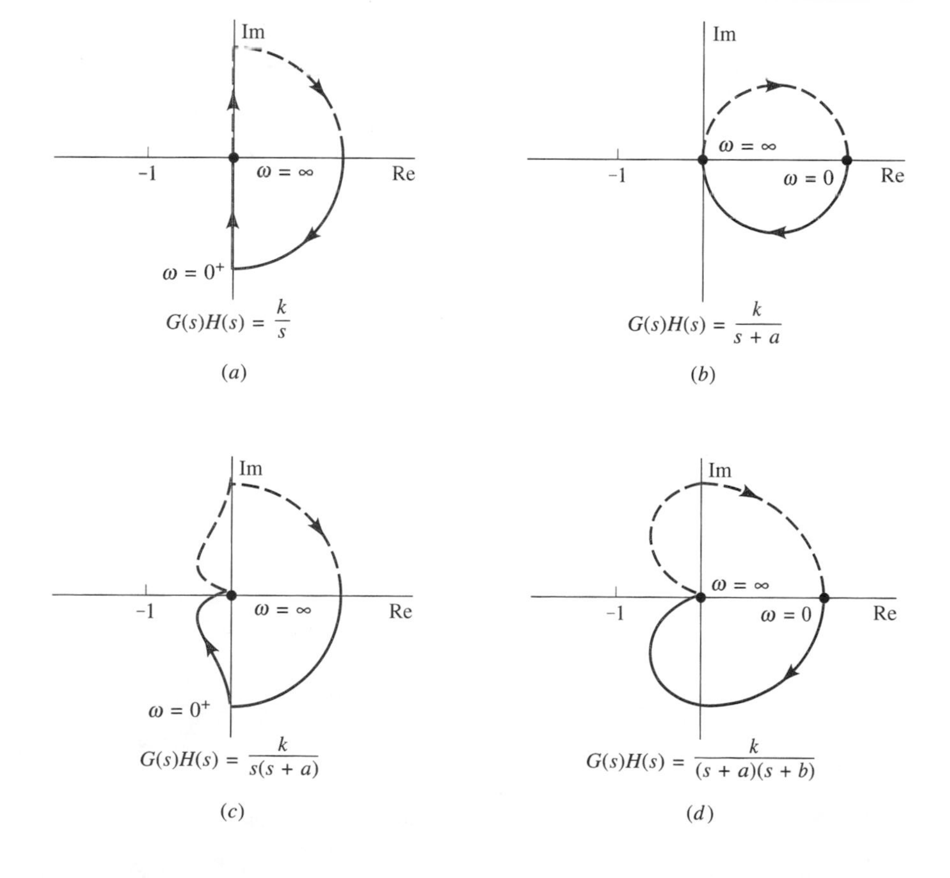

Table 6.9 *A Collection of Nyquist Plots (Continued)*

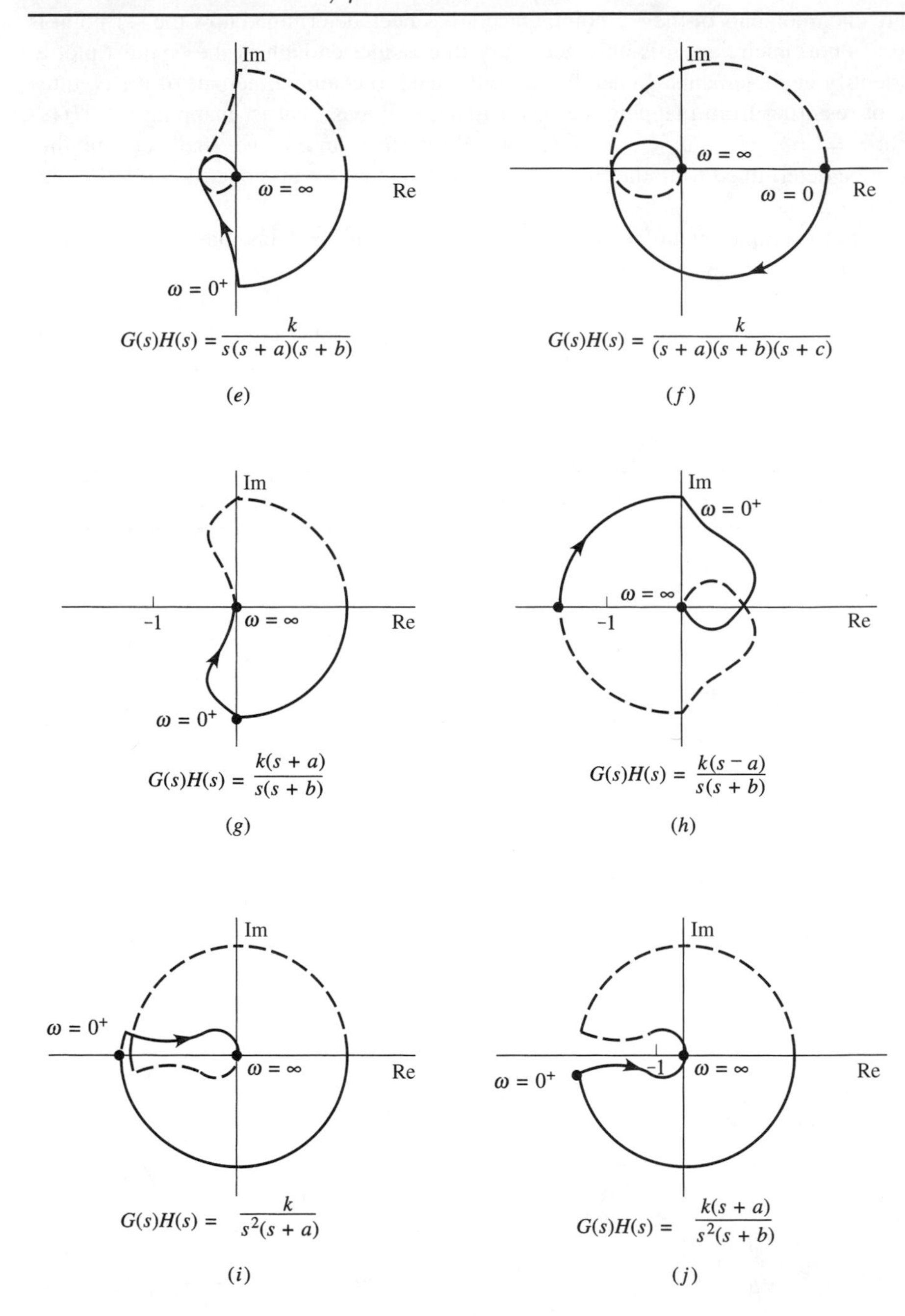

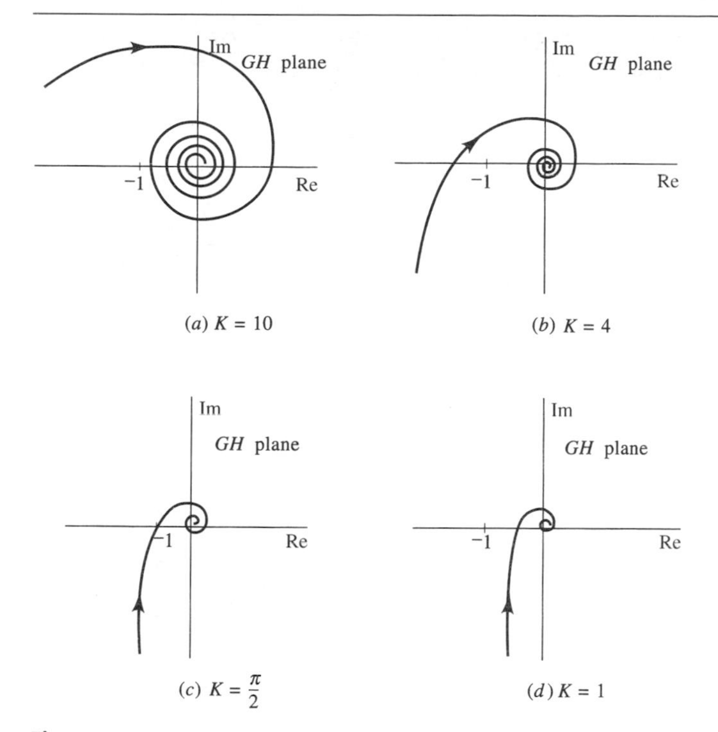

Figure 6.26 Nyquist plots for systems involving a time delay. For readability only the positive frequency part of the plot is drawn. (a) $K = 10$. (b) $K = 4$. (c) $K = \pi/2$. (d) $K = 1$.

With $T = 1$, Figure 6.26 shows Nyquist plots of

$$G(s)H(s) = \frac{Ke^{-s}}{s}$$

for various values of the positive constant K. For sufficiently large K, this system's Nyquist plot circles the -1 point of the GH plane several times, as in Figure 6.26(a), indicating the presence of several RHP poles in the overall system transfer function $T(s)$. For a smaller value of K, the Nyquist plot is given in Figure 6.26(b), with a single CW encirclement of the -1 point, indicating one RHP pole in $T(s)$. For $K = \pi/2$, Figure 6.26(c), the Nyquist curve passes through the -1 point and $T(s)$ has imaginary axis poles. For smaller K, the Nyquist plot is as in Figure 6.26(d) and $T(s)$ is stable.

The gain and phase margins of simple feedback systems can easily be determined from their Nyquist plots.

❑ Computer-Aided Learning

The Nyquist plot is obtained similarly using the “nyquist” command.

```
g=tf(10,[1,10]); nyquist(g), grid
```

With left hand arguments, we get

```
[re,im]=nyquist(g,w), re=re(:,:)': im=im(:,:)';
```

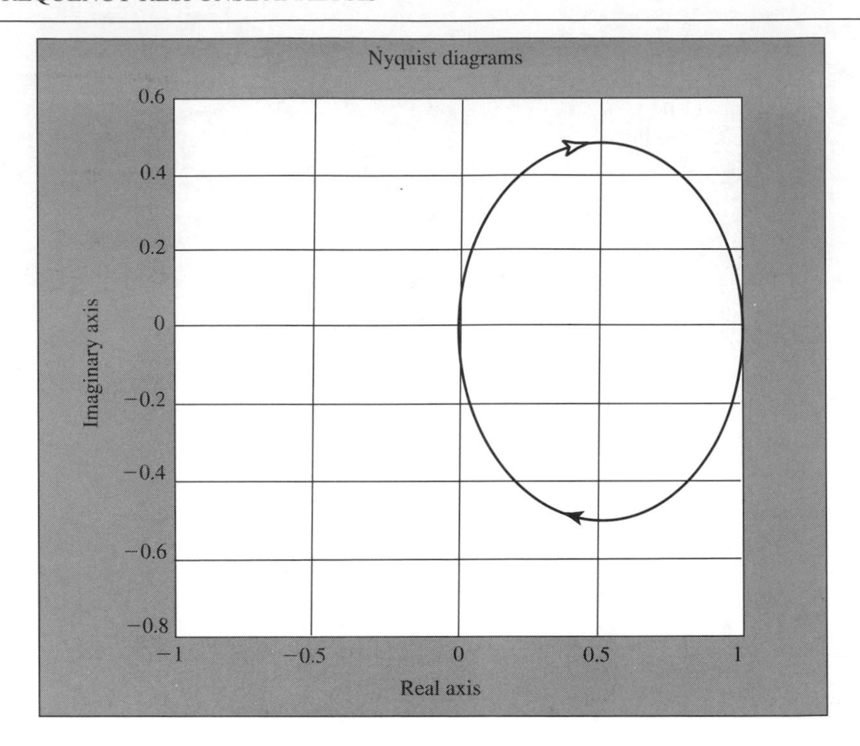

Figure C6.4

The Nyquist plot is the plot of the imaginary versus the real part of the frequency response.

C6.3 Redo Drill problem 6.7 using MATLAB, Note that the Nyquist plots in the text are drawn by artists for demonstration purposes and are not necessarily to scale. Because of scaling, the actual plots in MATLAB may appear very different from the ones in the text. By selecting the range of frequencies properly, you can get "better looking" Nyquist plots by trial and error. This is particularly so in systems with imaginary poles or poles at the origin (because the magnitude approaches ∞).

Ans. (c) `g=tf([1 0 0],[1 2 10]); nyquist(g)`

C6.4 Use MATLAB to redo Drill Problem 6.8.

6.6 Gain Margin

We now know that a system with no open-loop RHP poles should have no CW encirclements of the -1 point for stability and that a system with open-loop RHP poles should have as many CCW encirclements as there are open-loop RHP poles. Also, some systems with open-loop poles at the origin require CCW encirclements of the -1 point to cancel CW encirclements caused by the mapping of $[a, b]$ in Figure 6.19(b). In general, CW encirclements of -1 are always bad, while CCW encirclements are always good. It is often possible to determine whether the -1 point is encircled by looking at only that part of the Nyquist plot (or Bode plot) that identifies the presence of an encirclement. Generally, that means examining the mapping for $s = j\omega$, where $0 \leqslant \omega < \infty$, especially when there are no open-loop poles or zeros of $G(s)H(s)$ in the RHP.

A transfer function is called *minimum phase* when all the poles and zeros are LHP and *non-minimum-phase* when there are RHP poles or zeros. This means that stability is relatively easy to determine when $G(s)H(s)$ is minimum phase, but special care must be taken for the non-minimum-phase case.

Minimum and nonminimum phases are defined.

Recall that the Nyquist plot establishes the stability of a system of the form of Figure 6.18 with $K = 1$ by examining $G(s)H(s)$. One measure of stability arises from use of the *phase crossover frequency*, denoted ω_{PC}, and defined as the frequency at which the phase of $G(s)H(s)$ is $-180°$, that is,

Phase crossover frequency causes $\phi(\omega_{PC})$ to be $-180°$.

$$\angle G(j\omega_{PC})H(j\omega_{PC}) = -180° = \Phi(\omega_{PC})$$

The magnitude of $G(s)H(s)$ at the phase crossover frequency is denoted $A(\omega_{PC})$. The gain margin, *GM*, is defined as follows:

$$GM = \frac{1}{A(\omega_{PC})}$$

Suppose that, the gain K in Figure 6.18 is not selected to be one, rather K is selected to be

The gain margin is the value of K for marginal stability.

$$K = GM$$

then

$$\begin{aligned} KG(j\omega_{PC})H(j\omega_{PC}) &= KA(\omega_{PC})\angle - 180° \\ &= 1\angle - 180° \end{aligned}$$

and the system becomes marginally stable.

For example, let us reconsider the system of Figure 6.23 with frequency response summarized in Table 6.8. Figure 6.27(a) shows the root locus for variable K; Figure 6.27(b, c) contains the Nyquist plot of $G(s)H(s)$, and Figure 6.27 (d, e) shows the Bode plot of $G(s)H(s)$. The gain margin *GM* is the K for marginal stability in the root locus of Figure 6.27(a). $A(\omega_{PC})$ is the distance from the origin to where the Nyquist crosses the negative real axis in Figure 6.27(b), which occurs for $A(\omega_{PC}) = 0.67$. The gain margin is thus measured at $\omega_{PC} = 1.414$ rad/s.

Gain margin may be interpreted on a root locus. Nyquist plot, or Bode plot.

$$GM = 1/0.67 = 1.5$$

$$\text{dB}(GM) = -\text{dB}[A(\omega_{PC})] = 20\log_{10}(1.5) = 3.5 \text{ dB}$$

The *GM* in decibels is the distance on the Bode magnitude plot from the amplitude at the phase crossover frequency up to the 0 dB point.

For stability, every minimum-phase system should have no encirclements of -1 on the Nyquist plot. Thus, $A(\omega_{PC})$ should be a number less than 1 and the gain margin should exceed 1 (i.e., *GM* in dB should be positive). In general, for a minimum-phase $G(s)H(s)$—that is, a system with LHP open-loop poles and zeros—the system is stable if the dB(*GM*) is positive, marginally stable if dB(*GM*) is zero and unstable if dB(*GM*) is negative.

Stability is easy to find for a minimum-phase system.

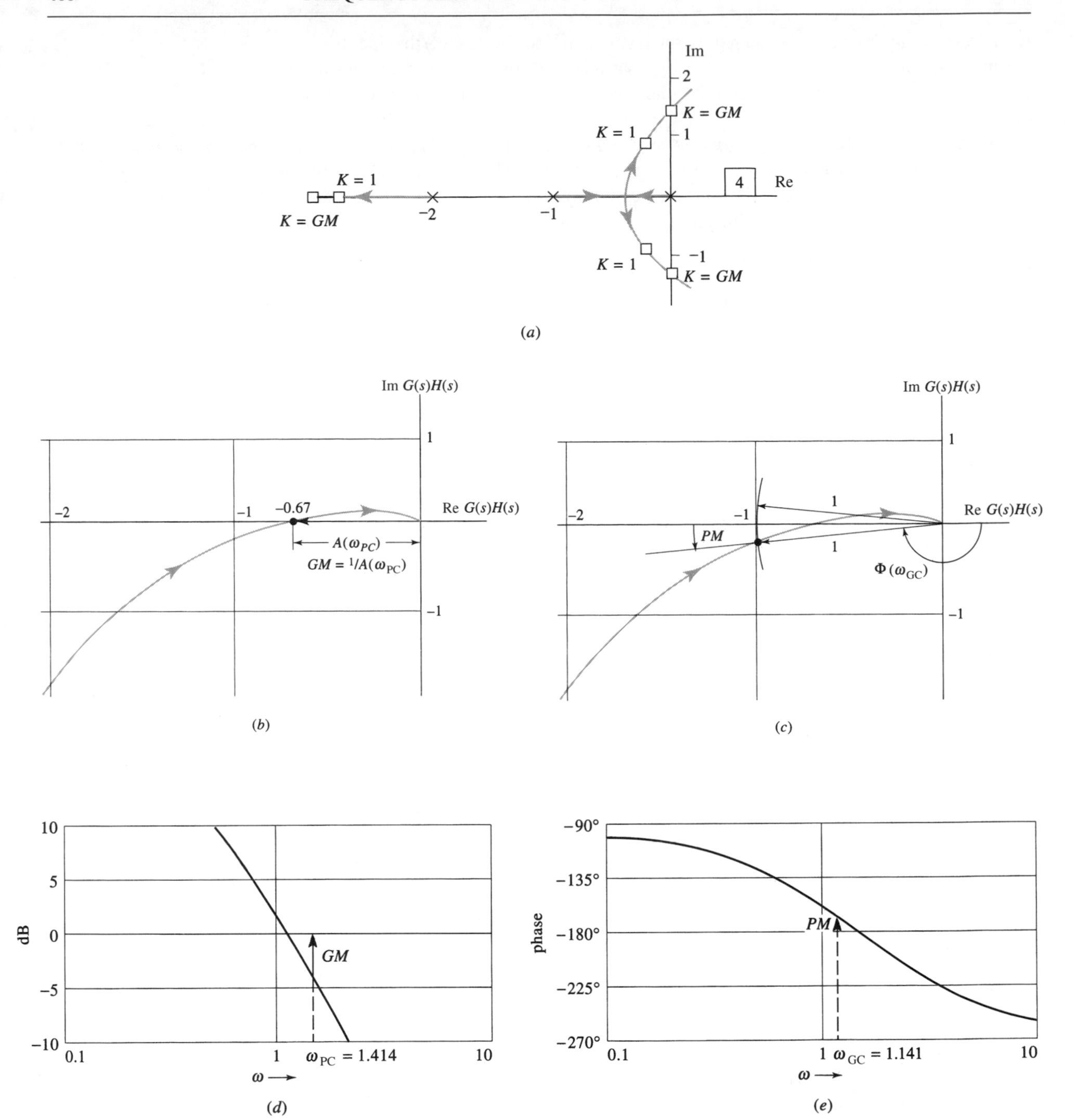

Figure 6.27 Stability of the system $G(s)H(s) = 4/s(s+1)(s+2)$. (a) Root locus for variable K. (b) Nyquist plot of $G(s)H(s)$ showing gain margin. (c) Nyquist plot of $G(s)H(s)$ showing phase margin. (d) Bode magnitude plot of $G(s)H(s)$. (e) Bode phase plot of $G(s)H(s)$.

When a system is stable for all positive K, as in Table 6.9(a–d), the phase crossover frequency is generally infinite, $A(\omega_{PC})$ is zero, and *GM* is infinite. Conversely, when a system is unstable for all positive K, as in Table 6.9(h, i), the phase crossover frequency is generally at 0 rad/s, $A(\omega_{PC})$ is infinite, and the *GM* is zero.

When $K = GM$, the system with open-loop transmittance $G(s)H(s)$ becomes marginally stable. That fact can be interpreted in two ways. First, the designer can purposely vary K to effect some level of stability. Second, the actual system open-loop transmittance is not actually $G(s)H(s)$, but when the uncertainty in $G(s)H(s)$ affects only the magnitude, the larger the gain margin is the greater the allowable margin of error in knowing $G(s)H(s)$ is before the system moves to marginal stability.

Notice that the system of Figure 6.25 has an open-loop RHP pole. The system is stable and the phase crossover frequency is at zero. Also $G(0)H(0)$ is -4, so $A(\omega_{PC})$ is 4 and $GM = 0.25$. In this case the system is stable with a dB(*GM*) that is negative. Care must be taken in interpreting stability based on dB(*GM*) for non-minimum-phase systems. There it is safest to examine the complete Nyquist or the root locus rather than to rely only on the sign of dB(*GM*).

Care must be taken for non-minimum-phase systems.

For example, the frequency response plots for

$$G(s)H(s) = \frac{100}{(s+1)^3}$$

the system of Figure 6.28(a), are given in Figure 6.28(b). To determine system stability, the dB value of *GM* can be used since the system has no RHP poles or zeros. The phase crossover frequency is easy to determine from

$$\Phi(\omega_{PC}) = -3\tan^{-1}(\omega_{PC}) = -180^\circ$$

$$\omega_{PC} = \tan(60^\circ) = 1.73\text{ rad/s}$$

Therefore

$$A(\omega_{PC}) = \frac{100}{\sqrt{(1.73^2+1)^3}} = 12.53$$

$$GM = 1/12.53 = 0.08$$

$$\text{dB}(GM) = -22$$

The system has a *GM* with negative dB; hence, the system of Figure 6.28 is unstable. A K of 0.08 in Figure 6.18 is required to move the system from instability to marginal stability.

When the transmittance *GH* is more complicated, the frequency for a phase of -180° can be found by trial and error. Consider the feedback system with irrational transmittance

$$G(s)H(s) = \frac{100e^{-0.07s}}{s+\frac{1}{10}}$$

The frequency response for *GH* is given in Figure 6.29. It is concluded that the overall system is unstable because the amplitude is larger than 0 dB when the phase is 180° and because the open-loop transmittance has only LHP poles.

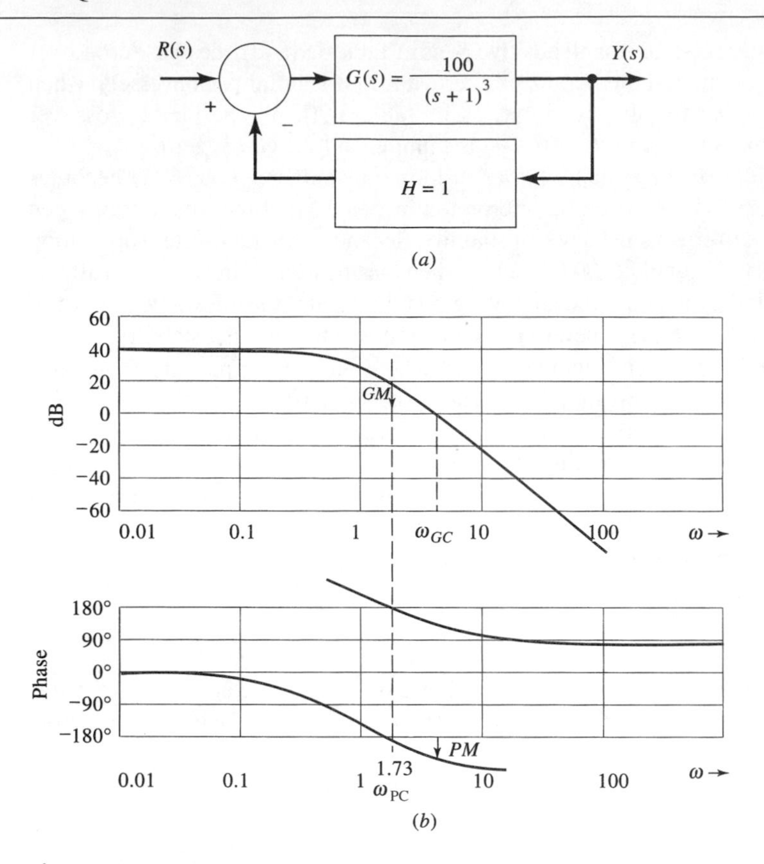

Figure 6.28 Feedback system example.

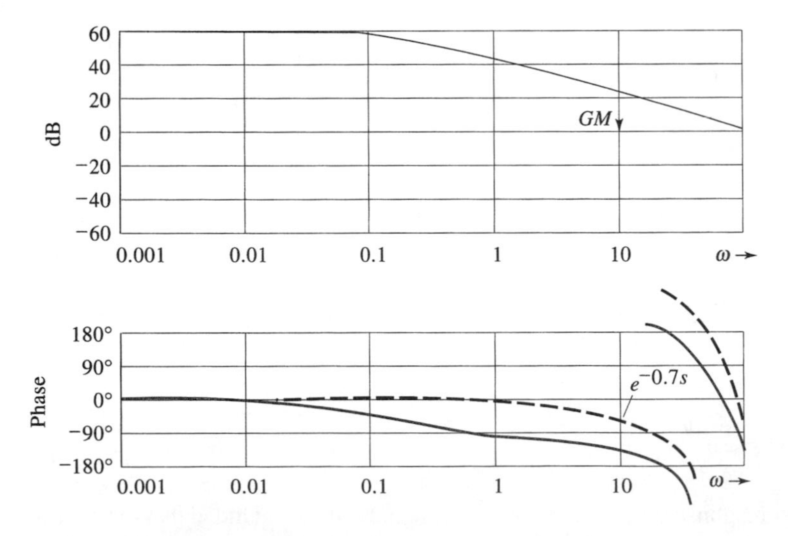

Figure 6.29 An irrational loop transmittance.

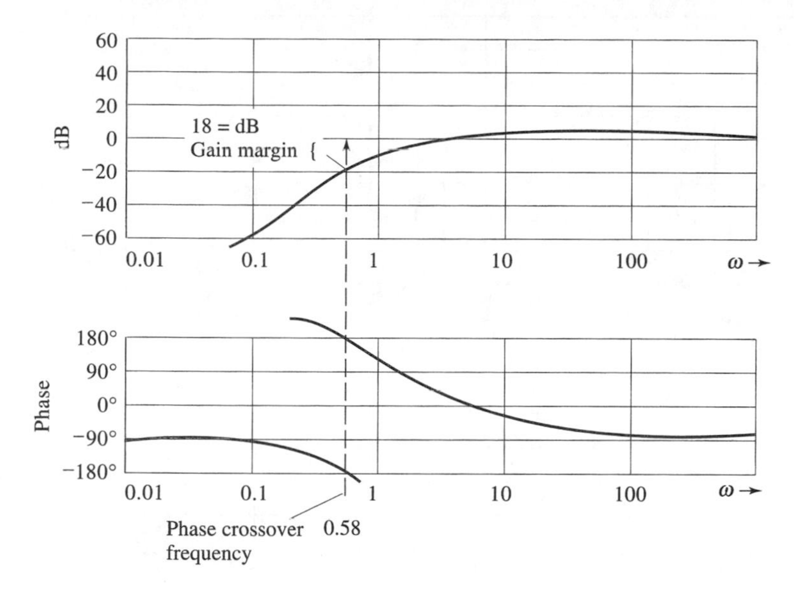

Figure 6.30 Gain margin of a simple feedback system. Frequency response plotted for $G(s)H(s)$.

For example, the frequency response for

$$G(s)H(s) = \frac{s^3}{(s+1)^3}$$

is given in Figure 6.30. At the phase crossover frequency the amplitude curve is at −18 dB, giving an 18 dB gain margin. The loop gain could be increased by 18 dB before instability would result.

For the system with

$$G(s)H(s) = \frac{1000s^3}{(s+1)^3(s+10)^4}$$

there are two phase crossover frequencies, as shown in Figure 6.31. The gain margin is the smaller of the two candidates. If there is no phase crossover frequency, the gain margin can be said to be infinite.

6.7 Phase Margin

In the section on gain margin, we learned that if K is chosen to be a positive real number equal to the gain margin, the closed-loop system becomes marginally stable. In that case, only the magnitude of $KG(s)H(s)$ is changed compared to that of $G(s)H(s)$, while the phase is not changed. Suppose instead that the gain K has unit magnitude and a nonzero phase angle, so that only the phase of $KG(s)H(s)$ is changed compared to that of $G(s)H(s)$.

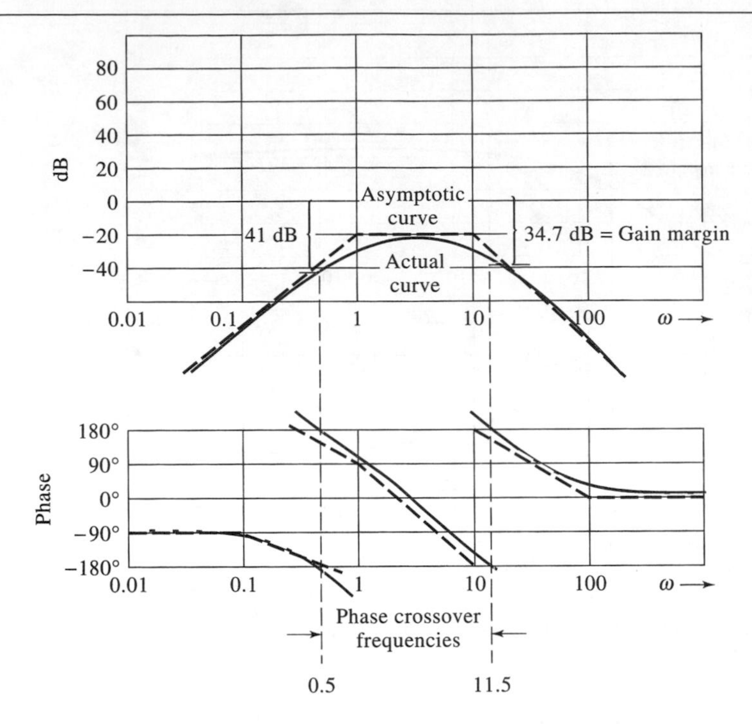

Figure 6.31 Gain margins when there is more than one phase crossover frequency.

Gain crossover frequency causes $A(\omega_{GC}) = 1$.

It is useful to define the *gain crossover frequency* ω_{GC} as the frequency at which the magnitude of $G(s)H(s)$ is one (0 dB). Thus

$$A(\omega_{GC}) = 1$$

The phase of $G(s)H(s)$ at the gain crossover frequency is denoted by $\Phi(\omega_{GC})$. The *phase margin* (*PM*) is defined by

$$PM = 180^\circ + \Phi(\omega_{GC})$$

A phase shift of $-PM$ causes marginal stability.

Suppose the gain K in Figure 6.18 is selected to be

$$K = 1\angle - PM$$

then at the gain crossover frequency

$$\begin{aligned} KG(j\omega_{GC})H(j\omega_{GC}) &= [1\angle - PM][1\angle\Phi(\omega_{GC})] \\ &= 1\angle[-PM + \Phi(\omega_{GC})] \\ &= 1\angle[-180^\circ - \Phi(\omega_{GC}) + \Phi(\omega_{GC})] \\ &= 1\angle - 180^\circ \end{aligned}$$

and the system is marginally stable.

For example, let us reconsider the system of Figure 6.23 with frequency response summarized in Table 6.8. Figure 6.27 (b, c) contains the Nyquist plot of $G(s)H(s)$ and Figure 6.27(d, e) shows the Bode plot of $G(s)H(s)$.

From Table 6.8 it is apparent that the gain crossover frequency is between 1 and 1.414 rad/s. By trial and error, the gain crossover frequency is found at 1.141 rad/s, so that

$$A_4(\omega_{GC}) = 1$$

$$\Phi_4(\omega_{GC}) = \Phi_4(1.141) = -168.6^\circ$$

Therefore

$$PM = 180^\circ - 168.6^\circ = 11.4^\circ$$

The phase margin *PM* is the angle in the Nyquist plot of Figure 6.27(c) drawn from the negative real axis to the point at which the Nyquist plot penetrates a circle of unit radius (called the *unit circle*). On the Bode phase plot of Figure 6.27(e), the *PM* is the distance from -180° up to the phase at the gain crossover frequency. The phase margin is therefore the negative of the phase shift through which the Nyquist plot can be rotated, and similarly the Bode plot can be shifted, so that the closed-loop system becomes marginally stable.

Phase margin may be interpreted on a Nyquist plot or Bode plot.

Every minimum phase system should have no encirclements of -1 on the Nyquist for stability. As in Figure 6.27(c), this implies that there should be a positive amount of *PM*. In general, for a minimum phase $G(s)H(s)$, that is, a system with LHP open-loop poles and zeros, the system is stable if the *PM* is positive, marginally stable if *PM* is zero and unstable if *PM* is negative.

Stability is easy to determine for a minimum-phase system.

To properly calculate *PM*, it is generally best to define $\Phi(\omega_{GC})$ as follows:

$$-270^\circ \leqslant \Phi(\omega_{GC}) \leqslant 90^\circ$$

For example, a third-quadrant $\Phi(\omega_{GC})$ would be written as -160°, so *PM* would be $+20^\circ$ (stable minimum-phase system), and a second-quadrant angle would be written as -200° with a *PM* of -20° (unstable minimum-phase system).

Notice that the minimum-phase systems of Table 6.9 that are stable—(a), (b), (c), (d), (g), and (j)—have positive values of *PM*; minimum-phase system (i), which is unstable, has a negative *PM*, and the remaining minimum phase systems, (e) and (f), can have either positive, zero, or negative *PM* values, so each can be stable, marginally stable, or unstable. The complete Nyquist plot for non-minimum-phase systems such as in Figure 6.25 and in Table 6.9(h) must be examined rather than relying on the sign of *PM* or dB(*GM*) to determine stability.

As noted earlier, when $K = 1\angle -PM$, the system with open-loop transmittance $G(s)H(s)$ becomes marginally stable. That fact can be interpreted in two ways. First, the designer can purposely vary K to effect some level of stability. Second, the actual system open-loop transmittance is not actually $G(s)H(s)$, but when the uncertainty in $G(s)H(s)$ affects only the phase, the larger the phase margin is, the greater the allowable margin of error is in knowing $G(s)H(s)$ before the system moves to marginal stability. In most systems, there is uncertainty in both the magnitude and phase of $G(s)H(s)$ so that substantial gain and phase margins are required to assure the designer that imprecise knowledge of $G(s)H(s)$ will not necessarily cause instability.

If there is more than one gain crossover frequency, there is more than a single phase margin. For

$$G(s)H(s) = \frac{10s}{(s+1)^2}$$

the two phase margins are shown in Figure 6.32. A decrease of 101° or an increase of $360° - 258° = 102°$ in the loop transmittance phase shift results in instability.

The design of a closed-loop system can be accomplished by using experimental data directly, without approximating the transmittances of complicated components from their experimental data. The experimental data are plotted and the frequency responses of analytically known components are added directly to the experimental data. The resulting combined loop transmittance is used in the conventional fashion to determine stability and gain and phase margins.

As an example of this powerful technique, consider the yaw control system for a ground-effect vehicle, modeled in the block diagram of Figure 6.33(a). Experimental data for the vehicle are used in part to construct a frequency response plot for the loop gain $G_1(s)G_2(s)$. The experimental data are not approximated with asymptotes or an equation here. Instead, the experimental data for $G_2(s)$ are plotted and the analytical response for $G_1(s)$ is added in Figure 6.33(b). The system of Figure 6.33 has significant gain and phase margins; hence, the system is stable.

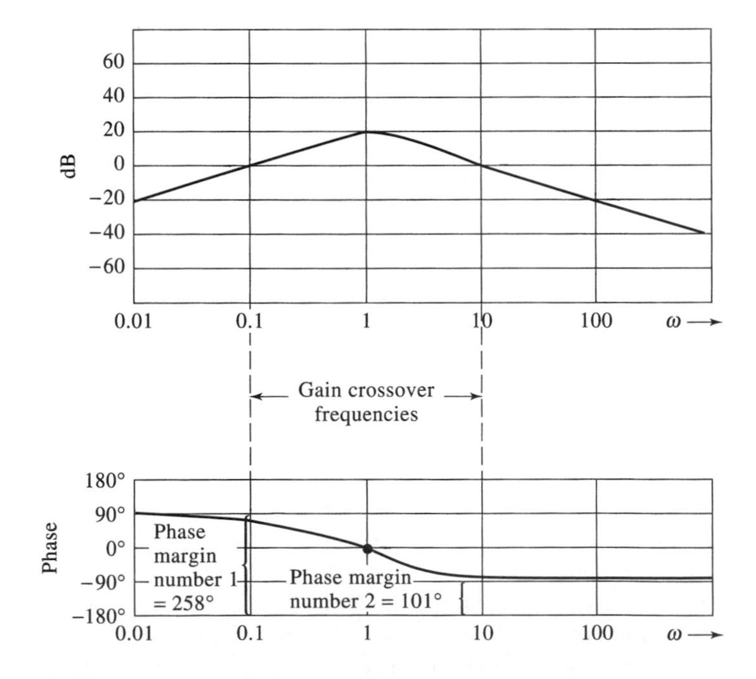

Figure 6.32 Phase margins when there are two gain crossover frequencies.

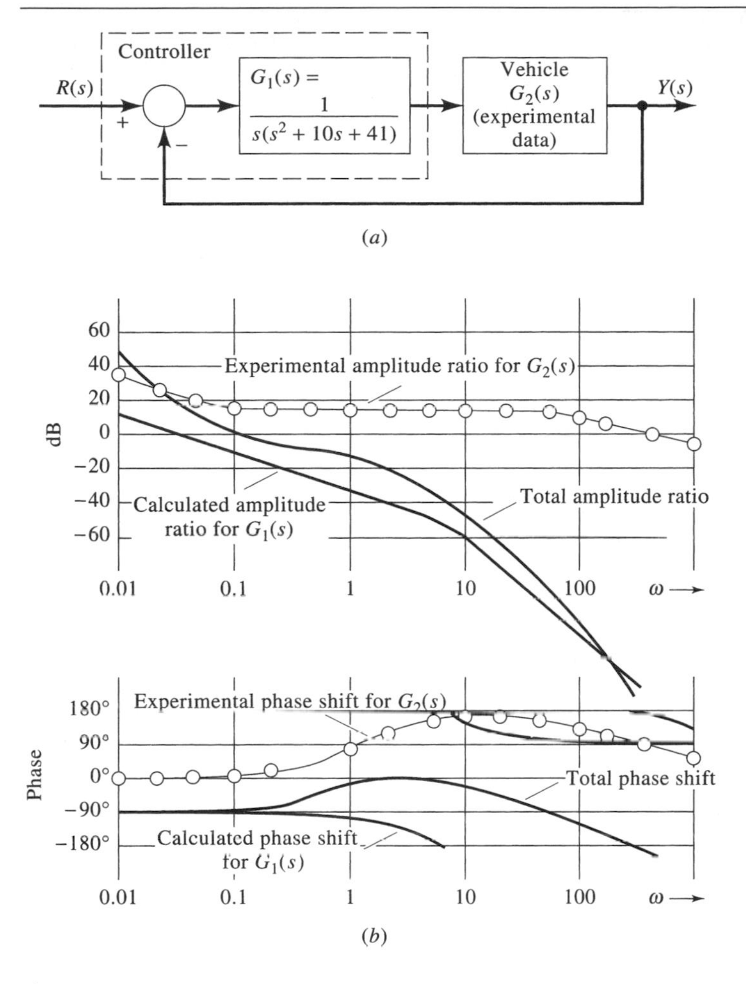

Figure 6.33 Incorporating experimental data into frequency response plots. (a) Model of yaw system. (b) Frequency response for the transmittance $G_1(s)G_2(s)$.

Before going on to more topics and applications, we will gather together the definitions from this section.

Phase crossover frequency (rad/s) = frequency at which the phase of GH is $-180°$.
Gain margin (dB) = −(dB of GH measured at the phase crossover frequency).
Gain crossover frequency (rad/s) = frequency at which the magnitude of GH is 0 dB.
Phase margin (degrees) = 180° + phase of GH measured at the gain crossover frequency (count first-quadrant angles as positive and other quadrant angles as negative).
A minimum-phase system has all the open-loop poles and zeros of $G(s)H(s)$ in the left half-plane. For stability both the *PM* and dB(*GM*) are positive. For marginal stability both the *PM* and dB(*GM*) are zero. For instability both the *PM* and dB(*GM*) are negative.

A non-minimum-phase system has open-loop poles and/or zeros of $G(s)H(s)$ in the right half-plane. Frequency response stability should be determined by examining the entire Nyquist plot.

❑ DRILL PROBLEMS

D6.9 Find gain margins and phase margins (if they exist) for feedback systems with the following loop transmittances:

(a)

$$G(s)H(s) = \frac{2000}{(s+2)(s+7)(s+16)}$$

Ans. 5.4 dB, 20°

(b)

$$G(s)H(s) = \frac{20}{s(s^2+7s+140)}$$

Ans. 33.8 dB, 89°

(c)

$$G(s)H(s) = \frac{-s}{(s+100)^3}$$

Note: the negative algebraic sign.

Ans. 88.5 dB, infinite phase margin

(d)

$$G(s)H(s) = \frac{e^{-0.1s}}{s}$$

Ans. 23.9 dB, 84°

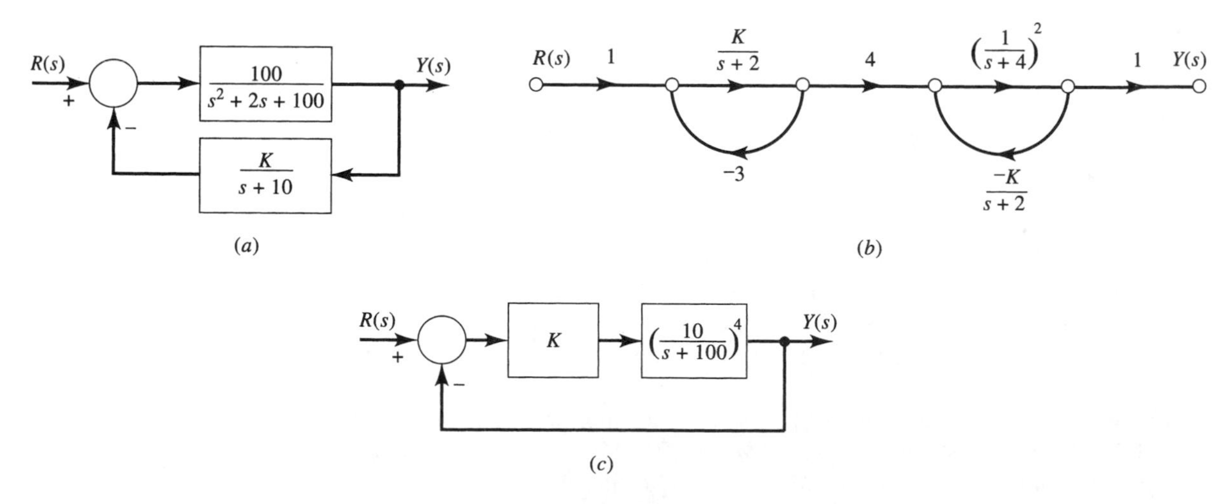

Figure D6.10

D6.10 Use frequency response methods to determine the range of the positive constant K for which the systems in Figure D6.10 are stable.

Ans. (a) $K < 4.4$; (b) $k < 288$; (c) $K < 40{,}000$

6.8 Relations Between Closed-Loop and Open-Loop Frequency Response

We have seen how Bode and Nyquist plots can be used to determine the stability of systems. In particular, we have used the open-loop transfer function GH in determining stability. Since one can also obtain the Bode plot of the closed- loop transfer function $T(s)$, we would like to see the relation between the open-loop and closed-loop frequency response plots. Consider the unity feedback system with closed-loop transfer function $T(s)$:

$$T(s) = \frac{G}{1+G}$$

$$T(j\omega) = T(s)|_{s=j\omega} = \frac{G(j\omega)}{1+G(j\omega)}$$

Because $G(j\omega)$ is a complex number, we can write it in rectangular form as follows:

$$G(j\omega) = X(j\omega) + jY(j\omega)$$

$$T(j\omega) = \frac{X+jY}{(X+1)+jY} \Rightarrow \quad |T|^2 = \frac{X^2+Y^2}{(X+1)^2+Y^2} \triangleq M^2$$

$$\Rightarrow (X+1)^2M^2 + Y^2M^2 = X^2+Y^2$$

$$X^2M^2 + M^2 + 2XM^2 + Y^2M^2 = X^2+Y^2$$

$$X^2(-M^2+1) + Y^2(-M^2+1) - 2M^2X = +M^2$$

$$X^2+Y^2 - 2\frac{M^2}{1-M^2}X = \frac{M^2}{1-M^2}$$

$$\text{add} \left(\frac{M^2}{1-M^2}\right)^2 \Rightarrow$$

$$X^2+Y^2-\frac{2M^2}{1-M^2}X + \left(\frac{M^2}{1-M}\right)^2 = \frac{M^2}{1-M^2} + \left(\frac{M^2}{1-M^2}\right)^2$$

$$\Rightarrow \left(X - \frac{M^2}{1-M^2}\right)^2 + Y^2 = \frac{M^2-M^4+M^4}{(1-M^2)^2} = \frac{M^2}{(1-M^2)^2}$$

$$\Rightarrow \boxed{\left(X - \frac{M^2}{1-M^2}\right)^2 + Y^2 = \left(\frac{M}{1-M^2}\right)^2}$$

This is the equation of a circle in the X-Y plane with its center at $X = M^2/(1-M^2)$ and $Y = 0$ and radius $M/(1-M^2)$, where M is the magnitude of the closed-loop transfer function $T(s)$.

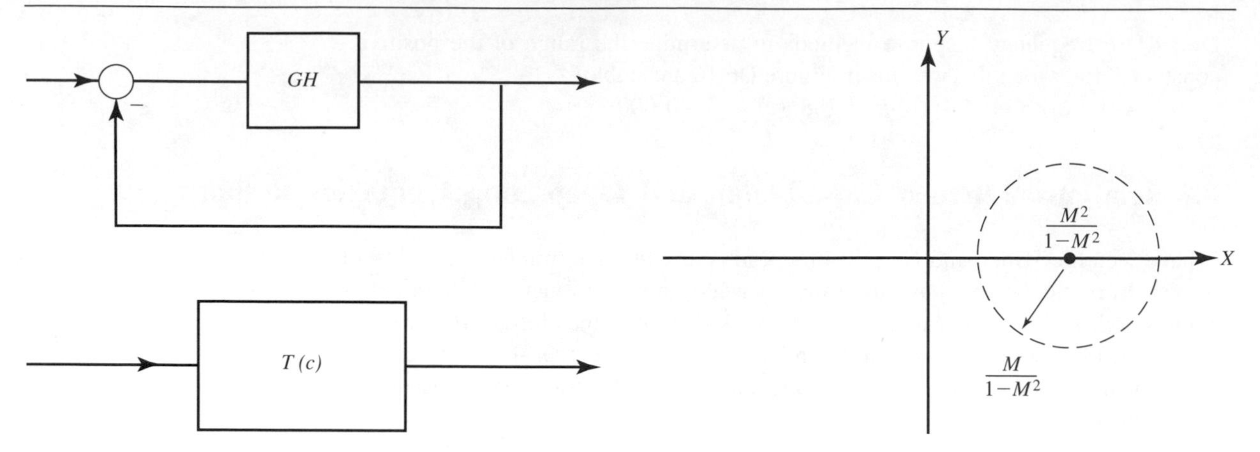

Figure 6.34 Open-loop and closed-loop system.

Figure 6.35 An M circle.

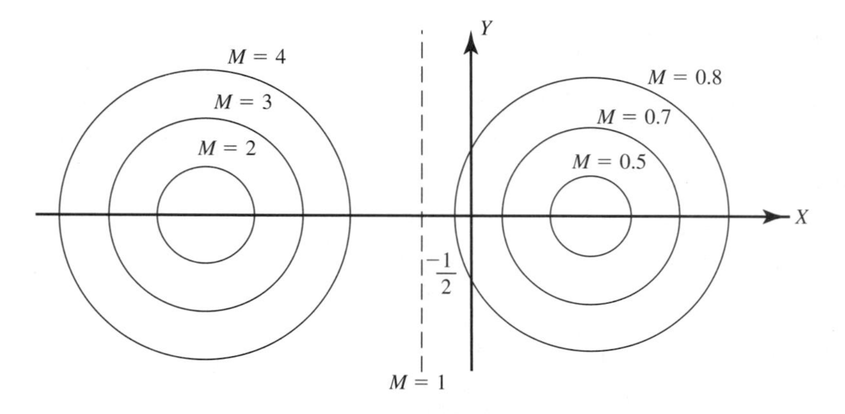

Figure 6.36 Family of M circles.

Over a frequency range, as $G(j\omega)$ changes, so will $T(j\omega)$ and M, and this will result in a family of circles of different radii and centers. These circles are called constant M circles. Note that as M tends to zero, the circles get smaller, with the center moving toward the origin. In the limit, we will have a point at the origin. Also, as $M \to 1$, $X \to -\frac{1}{2}$. Also, for $0 < M < 1$, $M^2/(1 - M^2) > 0$ and $X > 0$, and the circles are centered to the left of the $X = -\frac{1}{2}$ line.

We can use M circles (as shown in Figure 6.37) by superimposing the polar plot of $G(j\omega)$ on an M-circle template, and the M-circle gives the closed-loop magnitude. We can recreate the closed-loop magnitude plot by finding intersections at various frequencies. In fact, before computer-aided tools were available, this is exactly how engineers obtained the closed-loop magnitude plot. The alternative approach would have been to obtain the closed-loop transfer function, factor its numerator and denominator (difficult without computers), and draw Bode plots. Additionally, if the gain changed this whole process would have to be repeated, a very cumbersome and inefficient undertaking. It would be easy to use M circles to redraw the polar plot of $KG(j\omega)$

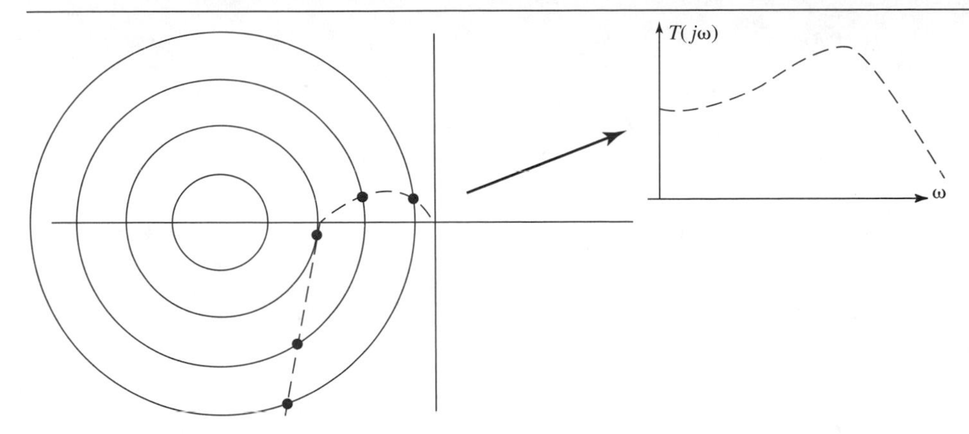

Figure 6.37 Using *GH* and *M* circles to generate closed-loop frequency response magnitude.

and read the new intersections. Now that we have the closed-loop magnitude plot, let us see how we can use it to obtain information about the system.

There are two important pieces of information obtained from the closed-loop magnitude plot: bandwidth and resonance peak.

Bandwidth (Figure 6.38) is the range of frequencies over which the closed-loop magnitude of the response to a unit amplitude input exceeds $\frac{1}{\sqrt{2}}$ (or $\sqrt{2}/2 = 0.707 = -3$ dB).
If $|T(j\omega)| = 1 = 0$ dB at low frequencies (dc), then ω_b corresponds to the −3 dB point. To find bandwidth, we set $|T(j\omega)|$ equal to $1/\sqrt{2}$ and solve for ω_b.

$$T(s) = \left.\frac{\omega_n^2}{s^2 + 2\zeta\omega_n s + \omega_n^2}\right|_{s=j\omega} = \frac{\omega_n^2}{-\omega^2 + j2\zeta\omega_n\omega + \omega_n^2}$$

$$|T(j\omega)| = \frac{\omega_n^2}{\sqrt{(\omega_n^2 - \omega)^2 + 4\zeta^2\omega_n^2\omega^2}} = \frac{1}{\sqrt{2}} \Rightarrow |T(j\omega)|^2 = \frac{1}{2}$$

solving for ω_b, we get

$$\omega_b = \omega_n\sqrt{1 - 2\zeta^2 + \sqrt{(2 - 4\zeta^2 + 4\zeta^4)}}$$

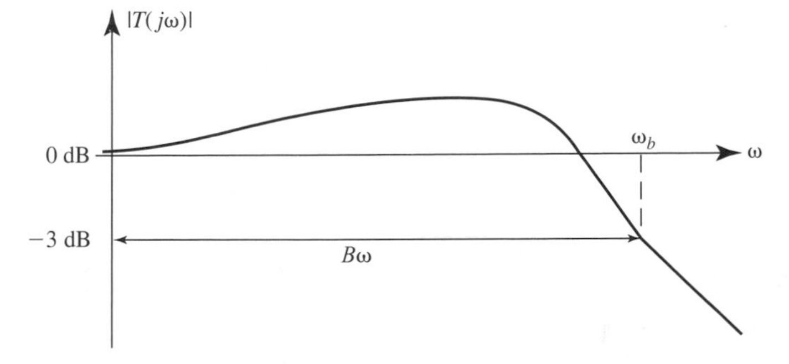

Figure 6.38 closed-loop magnitude response showing bandwidth.

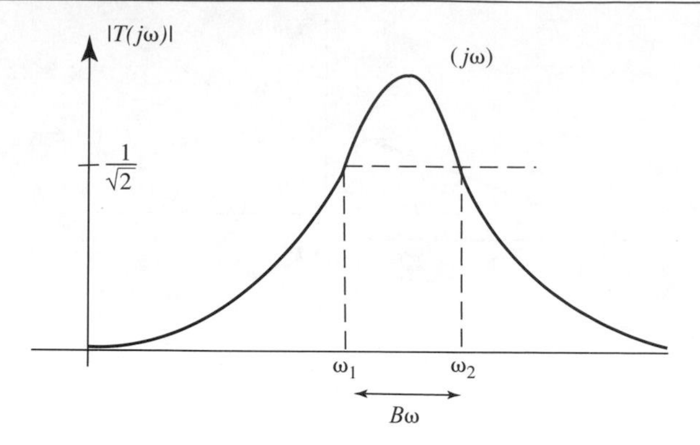

Figure 6.39 Band-pass filter.

The closed-loop magnitude of control systems typically has a low-pass filter characteristic, and bandwidth is defined as $B\omega = \omega_b - 0 = \omega_b$.

This is in contrast to communications filters that may be band-pass (see, e.g., Figure 6.39) and $B\omega = \omega_2 - \omega_1$.

Another feature of $|T(j\omega)|$ is its peak. This can be found by setting the derivative of $|T(j\omega)|$ equal to zero and solving for the frequency ω_r and evaluating $|T(j\omega)|$ at ω_r. The peak occurs at the resonant frequency ω_r (Figure 6.40) given by

$$\omega_r = \omega_n\sqrt{1 - 2\zeta^2}$$

and the resonance peak is

$$M_r = \frac{1}{2\zeta\sqrt{1 - \zeta^2}}$$

Note that we get a resonance peak only when $\zeta < 1/\sqrt{2}$; otherwise there is no peaking, and $|T(j\omega)|$ rolls off monotonically with frequency, as shown in Figure 6.41. Note that M_r is a function of ζ and hence similar to overshoot (which is also directly related to ζ). M_r is a measure of relative stability. Infact, the peak in the step response, M_p, and the peak in the magnitude Bode plot are directly related. Large overshoots (small

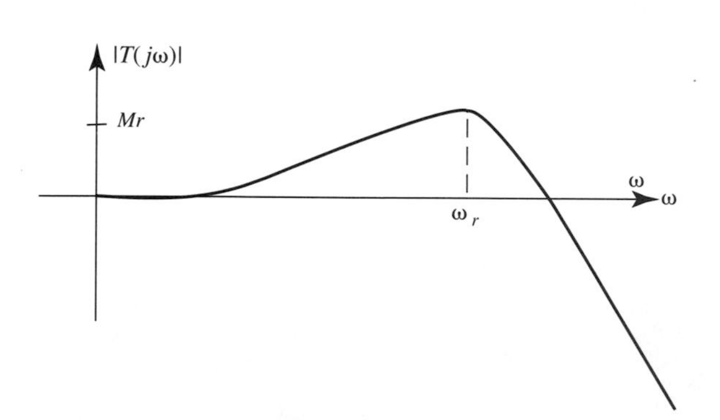

Figure 6.40 Resonant frequency.

Figure 6.41 Magnitude response when $\zeta > 1/\sqrt{2}$.

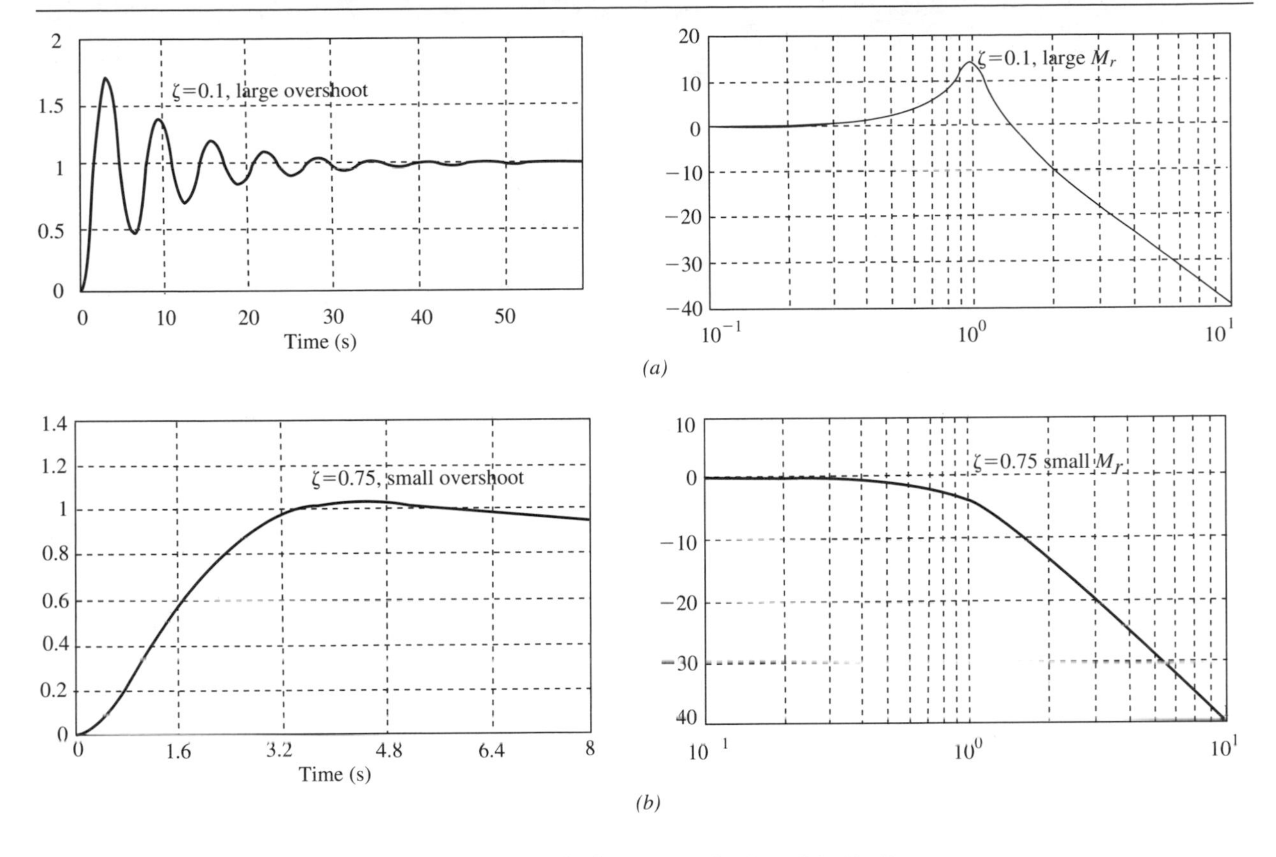

Figure 6.42 (a) Step response and closed-loop magnitude response for $\zeta = 0.1$. (b) Same plots as (a) but for $\zeta = 0.75$.

ζ) result in large M_r (see Figure 6.42). Recent results show that M_r is a very reliable measure of relative stability, also known as robustness.

The discussion about the magnitude of the closed-loop transfer function can be extended to its phase.

Let

$$\Phi = \angle T(j\omega) = \angle \frac{X + jY}{1 + (X + jY)}$$

where $G(j\omega) = X + jY$.

Then

$$\Phi = \tan^{-1} Y/X - \tan^{-1} Y/(1 + X)$$

If we let $N = \tan\Phi$ and use $\tan(a - b) = (\tan a - \tan b)/(1 + \tan a \cdot \tan b)$, We get

$$N = \tan\Phi = \frac{\dfrac{Y}{X} - \dfrac{Y}{1+X}}{1 + \dfrac{Y}{X}\left(\dfrac{Y}{1+X}\right)} = \frac{Y}{X^2 + X + Y^2}$$

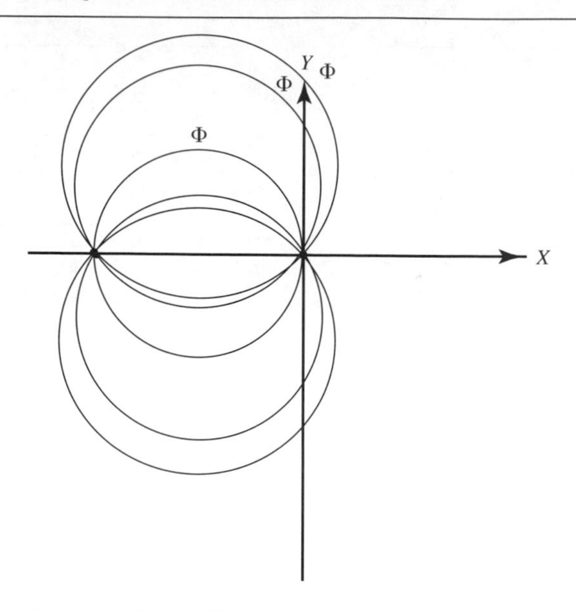

Figure 6.43 Families of N circles.

or

$$X^2 + X + Y^2 - \frac{1}{N}Y = 0$$

Adding $1/(2N)^2 + \frac{1}{4}$ to both sides yields

$$\boxed{\left(X + \frac{1}{2}\right)^2 + \left(Y - \frac{1}{2N}\right)^2 = \frac{1}{4} + \left(\frac{1}{2N}\right)^2}$$

This is an equation of a circle centered at $X = -1/2, Y = 1/2N$ with radius $r = \sqrt{1/4 + (1/2N)^2}$. For given values of Φ, we get what are known as families of N circles (Figure 6.43). Because the points $X = Y = 0$ and $X = -1,\ Y = 0$ satisfy the equation, all circles go through these points.

M and N circles are usually used in what is known as a Nichols chart. The plot of magnitude in decibel versus phase in degrees is called a Nichols plot (also called log-magnitude, phase diagram). A Nichols chart superimposes the M and N circles in a Nichols plot. Most computer programs for control have a built-in Nichols chart template. (In MATLAB, use the command ngrid after obtaining a Nichols plot by using the Nichols command. A modern and powerful technique for design of control systems called QFT (Quantitative Feedback Technique) is directly based on the Nichols chart. A MATLAB-generated Nichols chart is shown in Figure 6.44. Note that logarithmic scaling has distorted the circles.

6.9 Frequency Response of a Flexible Spacecraft

Chapter 4 contains a discussion of controlling a flexible spacecraft. Very large satellites extend solar panels to generate and store electrical energy. As a result, the satellite is

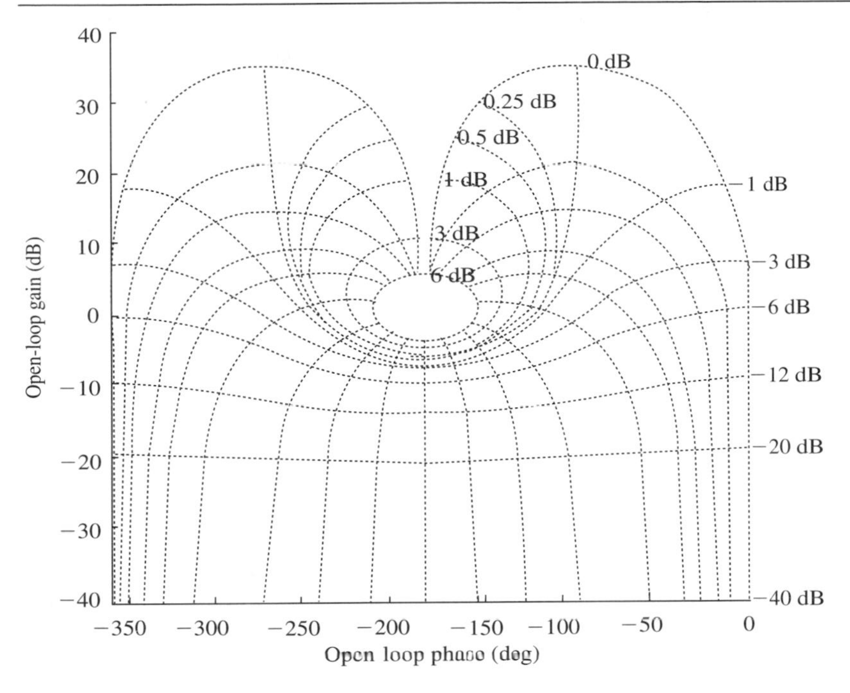

Figure 6.44 MATLAB generated Nichols chart obtained by using ngrid.

very flexible. When a thruster fires to cause the satellite to change position, the entire vehicle moves (rigid motion) about some center of gravity but individual parts of the vehicle bend and oscillate about local null points (flexible motion). Most aircraft and satellites have both rigid and flexible behaviors.

Figure 6.45 shows a hypothetical spacecraft where rigid behavior is included and flexible behavior is ignored. Figure 6.46 shows the spacecraft with flexible behavior included. The root locus for these systems concluded (in Chapter 4) that the flexible behavior has a strong destabilizing influence, which must be included. Otherwise, the root locus would appear overly optimistic as a source of stability analysis. Frequency domain analysis is equally revealing.

From Figure 6.45 (rigid behavior) a Bode plot must be constructed for

$$\frac{2K(s+4)}{s^2}$$

Suppose K is chosen to force the gain crossover frequency ω_{GC} to be 4. Then

$$K = \frac{\omega_{GC}^2}{2\sqrt{16+\omega_{GC}^2}} = \frac{16}{2\sqrt{32}} = \sqrt{2} = 1.414$$

The phase margin is

$$PM = 180^\circ + \tan^{-1}(\omega_{GC}/4) - 180^\circ = 45^\circ$$

The Bode plot appears in Figure 6.47.

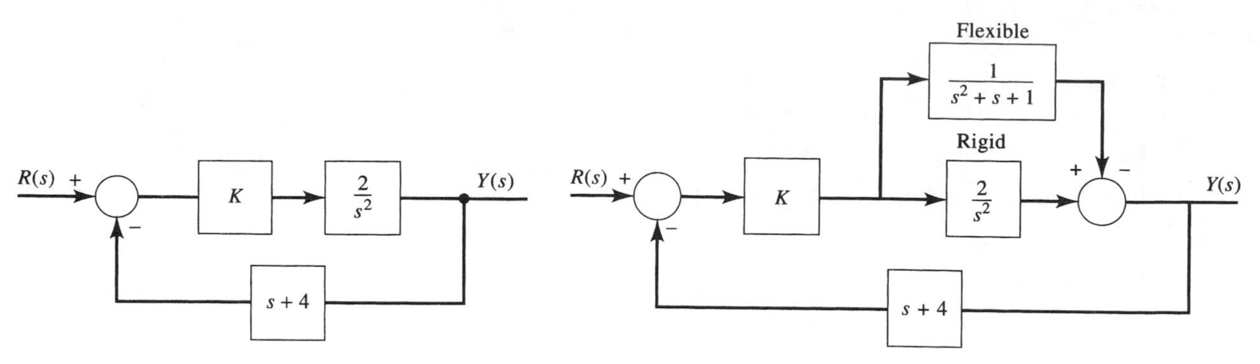

Figure 6.45 Rigid behavior.

Figure 6.46 Flexible and rigid behavior.

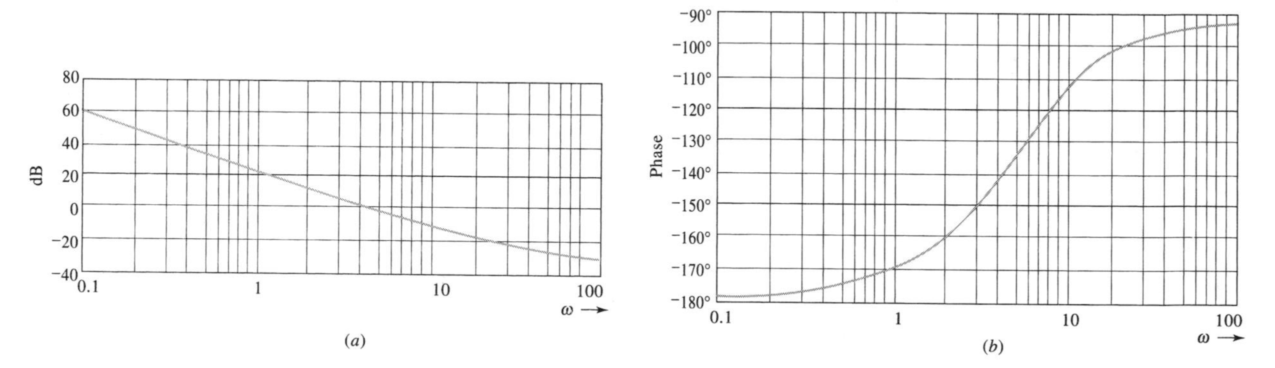

Figure 6.47 Bode plots for the system in Figure 6.34. (a) Amplitude versus frequency. (b) Phase versus frequency.

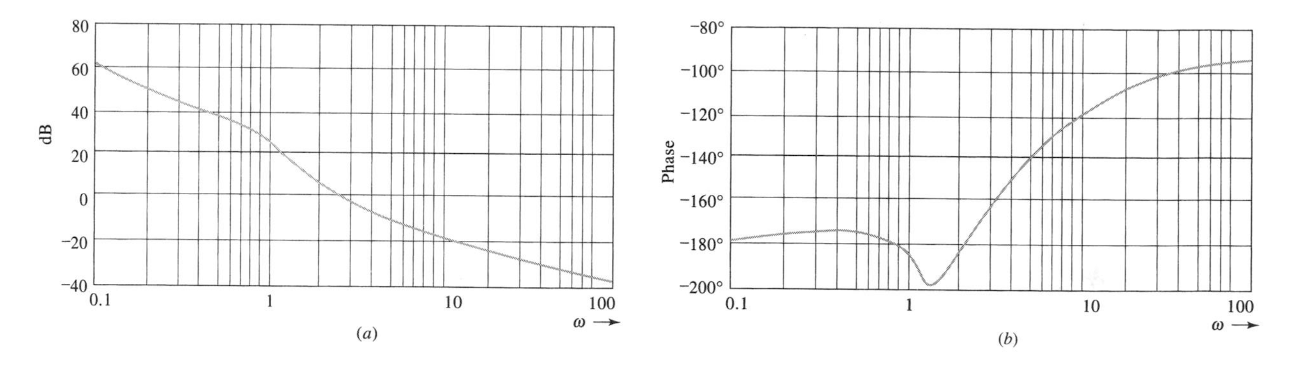

Figure 6.48 Bode plots for the system in Figure 6.46. (a) Amplitude versus frequency. (b) Phase versus frequency.

From Fig 6.46 (including flexible behavior) a Bode plot must be constructed for

$$\frac{1.414(s^2+2s+2)(s+4)}{s^2(s^2+s+1)}$$

The Bode plot appears in Figure 6.48. The gain crossover frequency is 2.75 (not 4 as would result from ignoring flexible behavior). The negative phase created by flexible behavior lowers the actual phase margin to 12.5°. The same conclusion results as before: Flexible behavior must be included to obtain a true picture of stability.

❑ Computer-Aided Learning

The Nichols plot, obtained using the nichols command, has similar syntax to 'bode' and 'nyquist'.

```
g=tf(10,[1 10]); nichols(g), grid
```

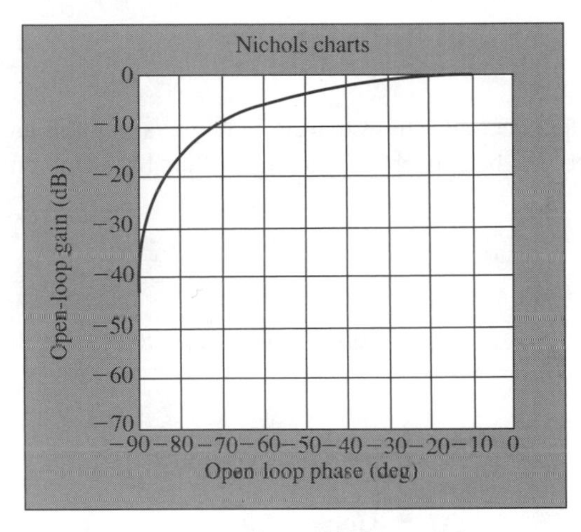

Figure C6.5

Moreover, we can add the M circles and N circles using the command

```
ngrid.
```

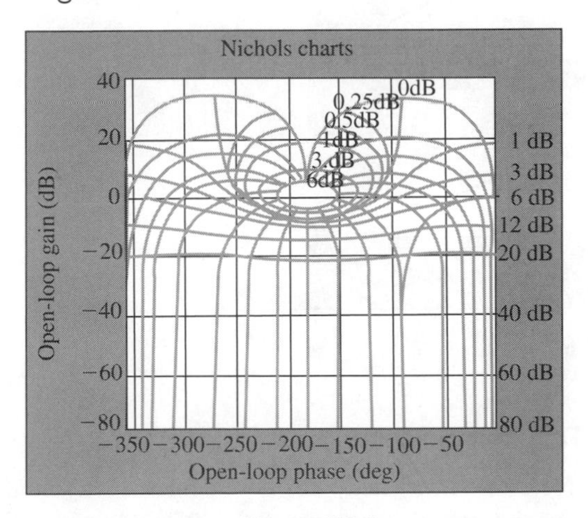

Figure C6.6

To obtain relative stability measures such as gain and phase margins, we can use the "margin" command.

```
g=zpk([], [0, -1, -2], 1)
zero/pole/gain:
      1
-----------
s(s + 1)(s + 2)
[Gm,Pm,Wcg,Wcp]=margin(g)
Gm=
    6.0000
Pm=
    53.4108
Wcg=
     1.4142
Wcp=
     0.4457
```

The frequencies are in radians per second, phase margin in degrees; but the gain margin is not in decibels and needs to be converted by using the "log10" command.

```
Gm=20*log10(Gm)
Gm=
     15.5630
```

Without left-hand arguments, the Bode plot is drawn with the margins indicated on the plot.

```
margin(g)
```

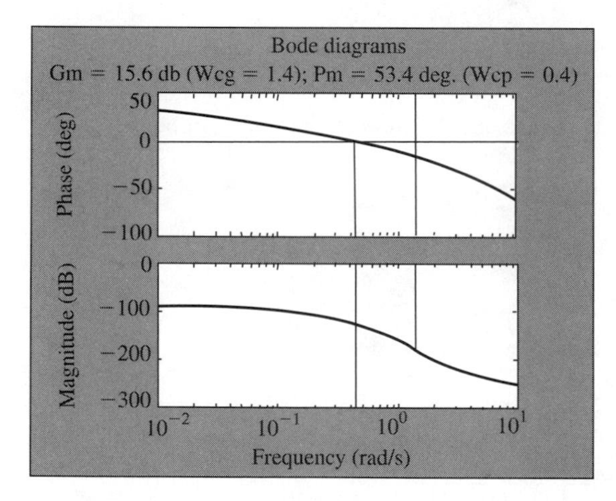

Figure C6.7

C6.5 Use Nichols plots to do Drill Problem 6.9(a–c).

C6.6 Do Drill Problem 6.9d. Note that time delay has no change in magnitude and only adds negative phase. Use the "bode" command with left-hand arguments and subtract from the phase to compute the stability margins. Use the frequency range $\omega = 0.1–16$ for accuracy.

Ans.
```
g=tf(1,[1 0]); w=logspace(-1,1.2,100)';
[m,p]=bode(g,w);
m=m(:,:)'; mdb=20*log10(m); p=p(:,:)';
p=p-.1*w*180/pi;
% subtract -0.1*w*180/π for time delay
subplot(211), semilogx(w,mdb); grid, subplot(212),
semilogx(w,p), grid
[gm,pm]=margin(m,p,w);
% use m for margin not mdb (should not be in dB)
gm=
  15.7050
pm=
  84.2704
gmdb=20*log10(gm)
gmdb=
  23.9207
```

Note that the phase margin is different from the book answer (84). If you look at the raw data, you will see that 84 is the correct answer.

```
[w mdb p]
ans=
 (partial list data)
 0.1000   20.0000    -90.5730
 0.9501   0.4444     -95.4438
 1.0000   0.0000     -95.7296   ← phase margin
 1.0525   0.4444     -96.0304
 15.0584  -23.5556   -176.2781
 15.8489  -24.0000   -180.8077  ← gain margin
```

6.10 SUMMARY

When a transmittance $F(s)$ is driven with a sinusoidal signal, its forced response is also sinusoidal and of the same frequency. The ratio of sinusoidal output amplitude to input amplitude is the amplitude ratio

$$A(\omega) = |F(s = j\omega)|$$

as a function of radian frequency ω. The phase shift between input and output is

$$\Phi(\omega) = \angle F(s = j\omega)$$

The frequency response amplitude curve may be visualized as the height along the imaginary axis of a rubber sheet laid over the complex plane, pushed up by poles and tacked down by zeros.

Bode plots are frequency response curves in a format that is especially convenient for rational transmittances. The amplitude ratio is plotted in decibels.

$$\text{dB} = 20 \log_{10} A(\omega)$$

and both amplitude ratio and phase shift are plotted on a logarithmic frequency scale. The frequency response contributions of real axis pole and zero terms are approximated well with straight-line segments. While the frequency response for complex conjugate pairs of roots may be constructed from standard, normalized curves.

The following process can be used to plot dB and phase shift curves for complicated transmittances.

1. Decompose the transmittance into simple factors.
2. Plot the dB and phase shift curves for each factor and/ or note slope changes.
3. Add the individual dB curves and/or slope changes to obtain the overall dB curve.
4. Add the individual phase shift curves and/or slope changes to obtain the overall phase shift. Multiples of 360° may be added to or subtracted from the phase shift to keep that curve within a convenient range of angle.

Frequency response methods apply also to systems with irrational transmittances. One such system, of considerable practical importance, is the time delay

$$F(s) = e^{-s\tau}$$

where τ, the delay time, is a constant. For $F(s)$, the frequency response amplitude ratio is a uniform 0 dB. The phase shift is proportional to frequency:

$$\Phi(\omega) = -\tau\omega$$

A major advantage of frequency response methods, in addition to their applicability to systems with irrational transmittances, is that experimentally derived data can be easily incorporated. Approximate transmittances may be determined from experimental data, or the data may be used directly in analysis and design.

A Nyquist plot consists of a curve on the complex plane representing the frequency response of the loop transmittance $G(s)H(s)$ of a simple feedback system.

A Nyquist plot may be mapped in four parts. First, map values of s near the origin in the s plane. A semicircle around the origin is mapped when one or more poles are located there. Second, the magnitudes and phases are calculated and mapped for a representative set of positive radian frequencies. Third, behavior at infinity is usually mapped to origin on the $G(s)H(s)$ plane. Finally, the mirror image of the mapping on the $G(s)H(s)$ plane completes the mapping of the lower right-half s plane. It is difficult to map the entire Nyquist plot to scale. It is best to examine the entire plot not to scale and then to create a scaled plot near the -1 point in the $G(s)H(s)$ plane, since that region influences stability. Four complete examples of Nyquist plots are presented.

A minimum-phase system has all the open-loop poles and zeros of $G(s)H(s)$ in the left half-plane. A non-minimum-phase system has open-loop poles and/or zeros of $G(s)H(s)$ in the right half-plane. A stable minimum-phase system must have no clockwise encirclements of the -1 point on the Nyquist plot. It is possible to determine the stability of a minimum-phase system by examining a portion of the Nyquist plot or a portion of the Bode plot to determine whether CW encirclements of -1 exist. To meet that goal, gain and phase margins are useful. Gain margin is defined as the algebraic inverse of the amplitude of $G(s)H(s)$ measured at the phase crossover frequency—that is, the frequency at which the phase of $G(s)H(s)$ is $-180°$. If a gain K is included in the system and if K equals the gain margin, the system becomes marginally stable. A stable minimum-phase system has a positive gain margin measured in dB. Phase margin is defined as 180° plus the phase of $G(s)H(s)$ measured at the gain crossover frequency—that is, the frequency at which the magnitude of $G(s)H(s)$ equals one. If a gain K is included in the system, and if K has unit magnitude at an angle equal to the negative of the phase margin, the system becomes marginally stable. A stable minimum-phase system has a positive phase margin. The stability of non-minimum-phase systems should generally be established by examining the entire Nyquist plot.

Stability for a minimum-phase system can be determined directly from that portion of the Bode plot that permits evaluation of phase margin. For example, the gain crossover frequency for the system of Figure 6.28 can be found from

$$A(\omega_{GC}) = \frac{100}{\sqrt{\left(\omega_{GC}{}^2 + 1\right)^3}} = 1$$

Therefore ω_{GC} is 4.53 rad/s. From

$$\Phi(\omega_{GC}) = -3\tan^{-1}(4.53) = -232.7°$$

is follows that

$$PM = 180° - 232.7° = -52.7°$$

so the system is unstable owing to negative phase margin.

For the system of Figure 6.30, 0 dB occurs at a gain crossover frequency of infinity. Since the phase is 0° at infinite frequency, PM is $180° + 0° = 180°$ and we conclude that the system is stable. When no gain crossover frequency exists, as in the system of Figure 6.31, the phase margin is defined as infinite.

If there is more than one gain crossover frequency there is more than one phase margin. The smallest candidate phase margin should be chosen. For

$$G(s)H(s) = \frac{10s}{(s+1)^2}$$

the two phase margins are shown in Figure 6.32. The smaller phase margin of 101° is chosen. Since both phase margins are positive and since the system is minimum-phase, the system is stable with $K = 1$.

The relation between open- and closed-loop frequency response is discussed, as are such closed-loop measures as bandwidth and resonant peak. The Nichols chart, which shows constant M and N circles is also discussed.

Frequency response methods are applied to a flexible spacecraft. The chapter ends with examples of using MATLAB to obtain frequency response results.

REFERENCES

Frequency Response Methods

Bode, H. W., *Network Analysis and Feedback Amplifier Design*. Princeton, NJ: Van Nostrand, 1945.

Nyquist, H., "Regeneration Theory." *Bell Syst. Tech. J.* (January 1932): 126–147.

PROBLEMS

1. For the following transmittances, find the forced sinusoidal response to the indicted input signals:

(a) $F(s) = \dfrac{1}{s+8}$
$r(t) = 5\cos 5t$

(b) $F(s) = \dfrac{s^2}{s+4}$
$r(t) = 100\ \cos(4t + 40°)$

(c) $F(s) = \dfrac{10}{s^2 + 2s + 10}$
$r(t) = 4\cos(5t - 70°)$

(d) $F(s) = \dfrac{s}{(s+4)(s+8)}$
$r(t) = 10\cos(4t + 120°)$

(e) $F(s) = \dfrac{10}{(s+6)^2}$
$r(t) = 5\sin 6t$

2. Sketch approximate frequency response curves (both amplitude ratio and phase shift) for functions with the pole–zero plots of Figure P6.2.

3. Draw Bode plots (both dB and phase shift) for the following transmittances:

(a) $F(s) = \dfrac{100}{s^2}$

(b) $F(s) = \dfrac{s}{s+10}$

(c) $F(s) = \dfrac{10(s+6)}{s+4}$

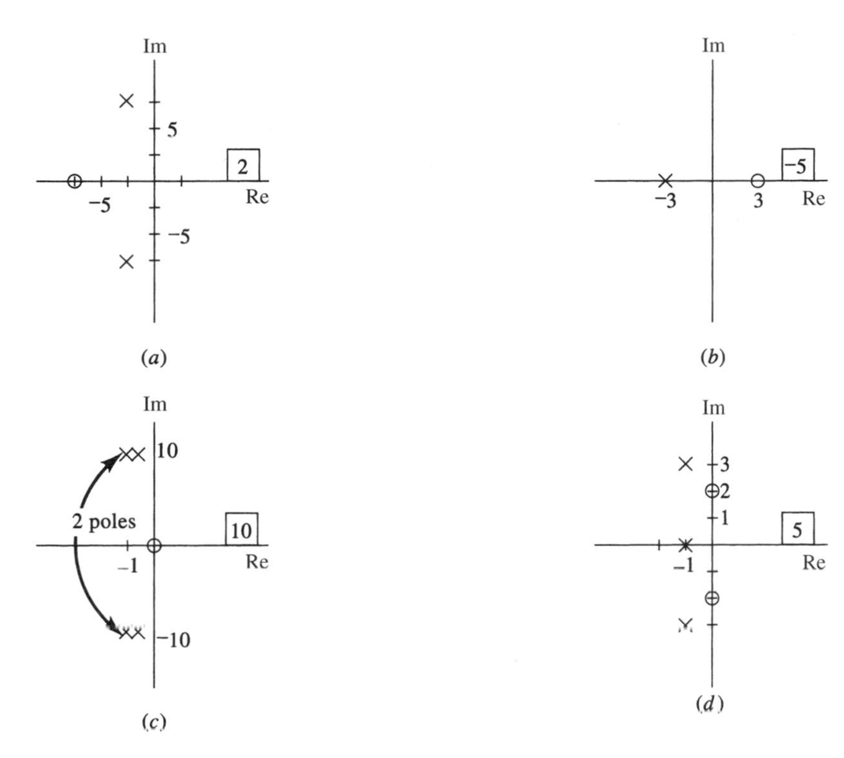

Figure P6.2

Ans.

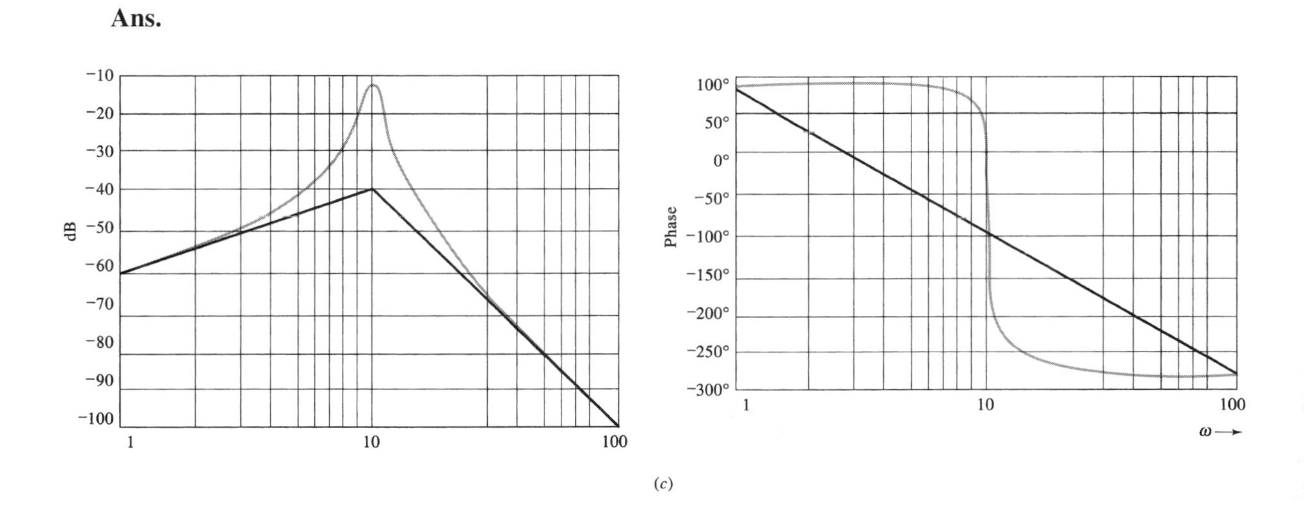

Figure P6.2

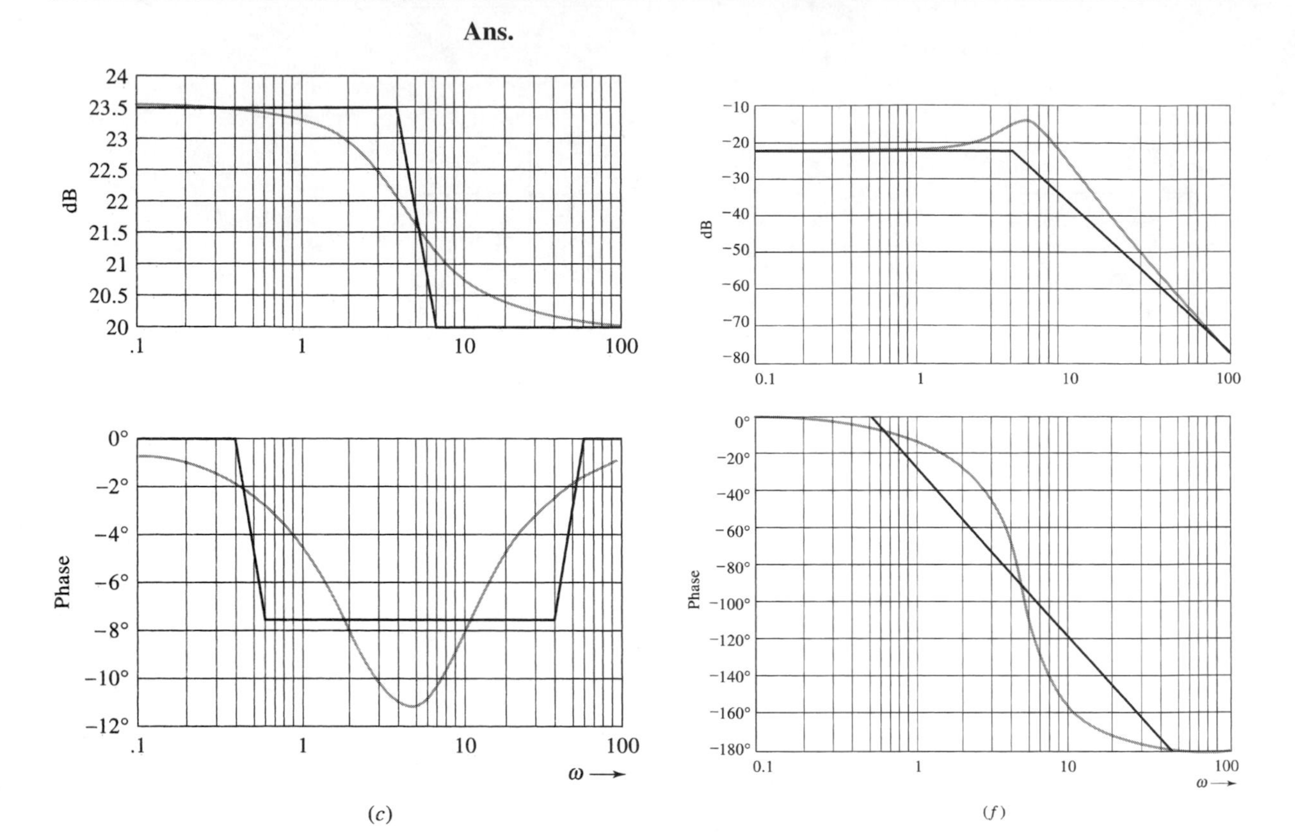

Figure P6.3

(d) $F(s) = \dfrac{-40}{s(s+40)}$

(e) $F(s) = \dfrac{s+10}{s(s+100)}$

(f) $F(s) = \dfrac{1}{s^2+2s+20}$

(g) $F(s) = \dfrac{1}{3s(s^2+s+10)}$

(h) $F(s) = \dfrac{s^2-s+4}{s^2+s+4}$

Ans.

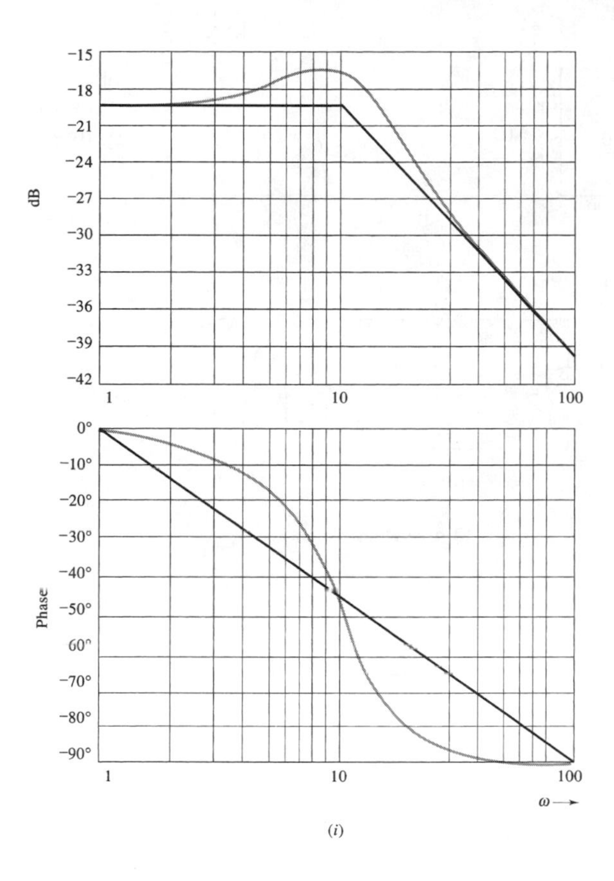

Figure P6.3

(i) $F(s) = \dfrac{s+10}{s^2+10s+100}$

(j) $F(s) = \dfrac{4s-1}{(s^2+s+4)^2}$

4. Draw Bode plots for the transmittances

$$F(s) = \left.\frac{E_0(s)}{E_i(s)}\right|_{\text{zero initial conditions}}$$

of the electrical networks of Figure P6.4.

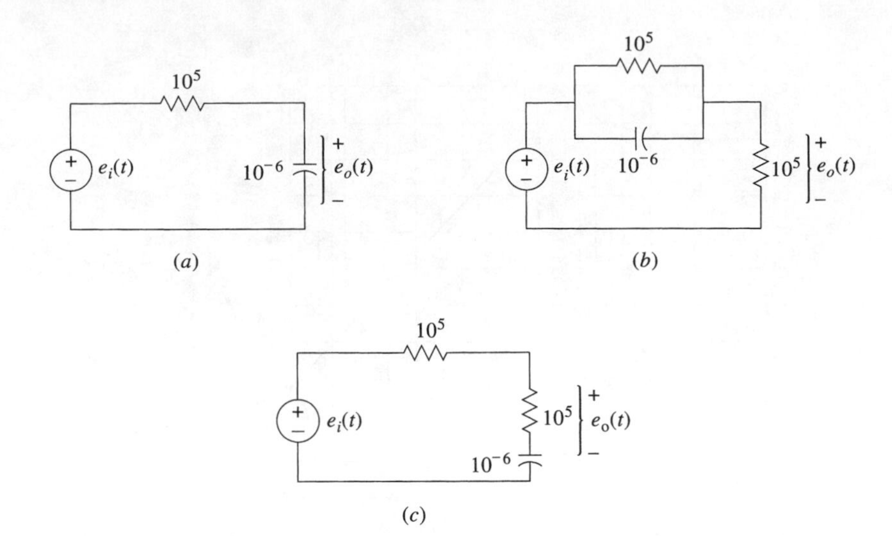

Figure P6.4

Ans.

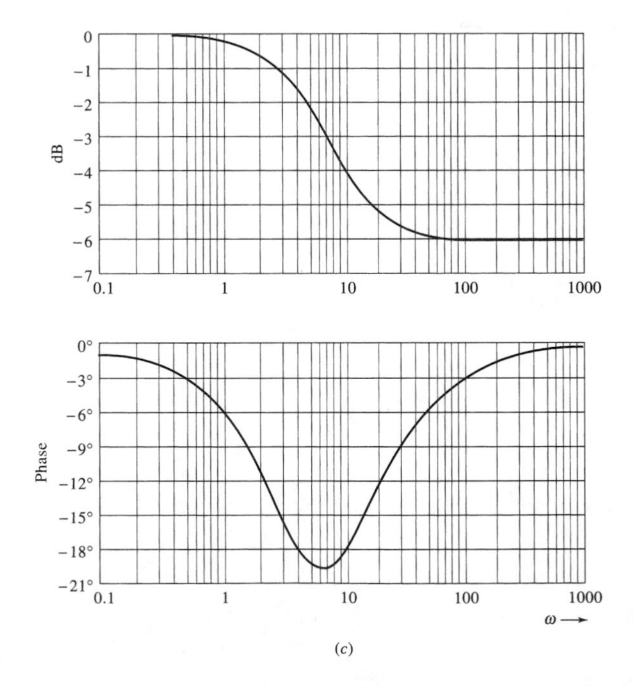

(c)

5. Draw Bode plots for the indicated transmittances

$$T(s) = \left.\frac{X(s)}{F(s)}\right|_{\text{zero initial conditions}}$$

for each of the translational mechanical networks of Figure P6.5.

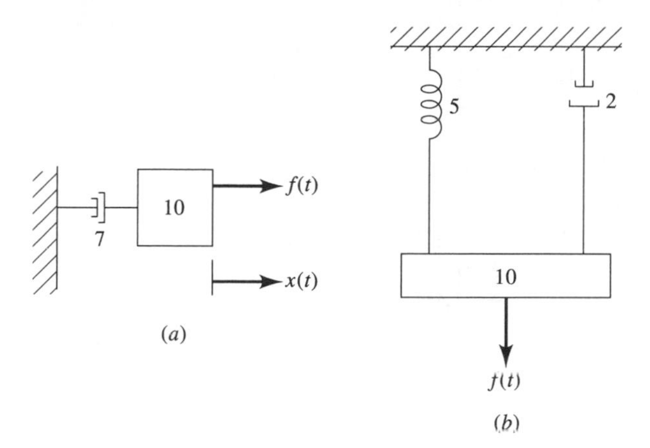

Figure P6.5

6. Draw Bode plots for the indicated transmittances

$$F(s) = \left.\frac{\Theta(s)}{\tau(s)}\right|_{\text{zero initial conditions}}$$

of each of the rotational mechanical networks of Figure P6.6.

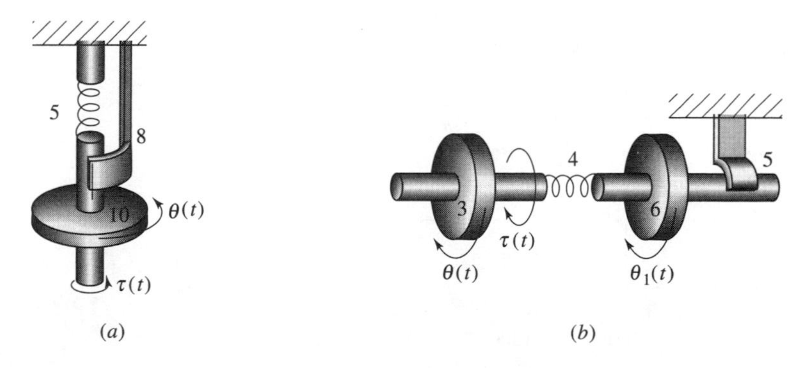

Figure P6.6

Ans.

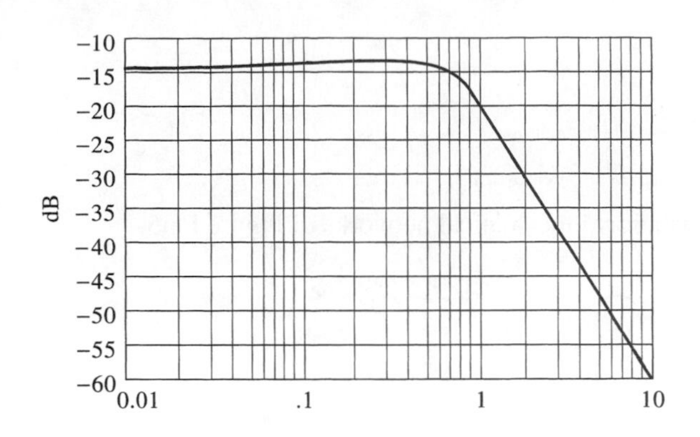

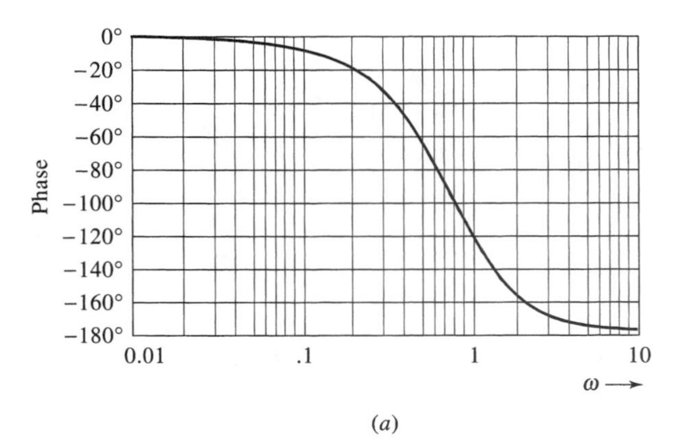

(*a*)

7. Find a transmittance that has the approximate dB curve as shown in Figure P6.7. A *minimum phase* transmittance has all poles and zeros in the left half of the complex plane. Find a minimum phase solution.

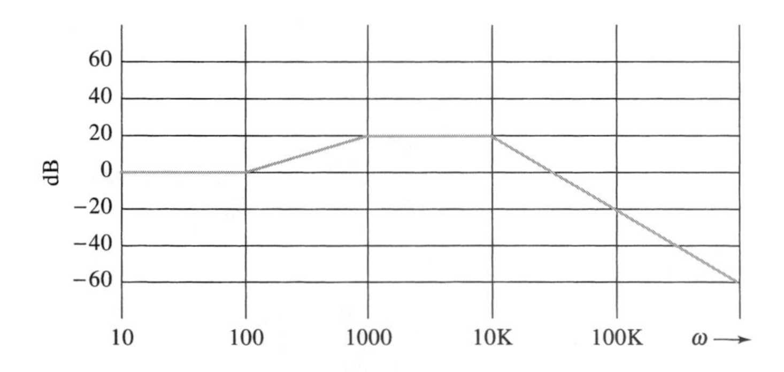

Figure P6.7

8. Find a stable transmittance that has the approximate phase shift curve of Figure P6.8.

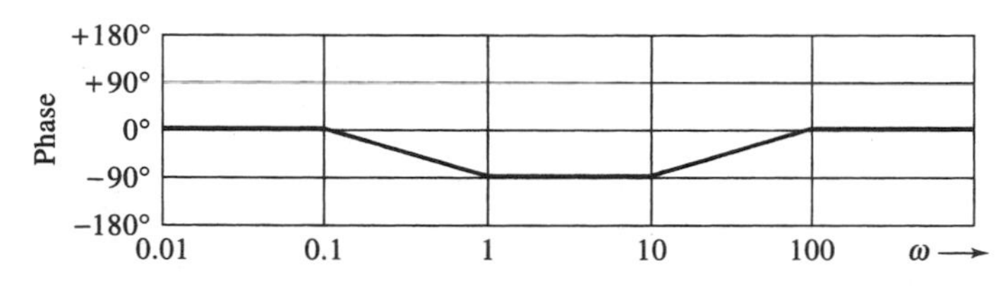

Figure P6.8

9. From the experimental frequency response data for two transmittances $G(s)$ given, find an approximate analytical expression for each of these $G(s)$ functions.

(a)

Radian Frequency ω	$G(s = j\omega)$
0.1	$0 - j20$
0.2	$-1.3 - j10$
0.4	$-0.8 - j5$
1.0	$-0.6 - j1.7$
2.0	$-0.4 - j0.6$
4.0	$-0.2 - j0.2$
8.0	$-0.05 - j0.008$
16.0	$-0.0022 + j0.0056$

(b)

Frequency f (Hz)	Amplitude Ratio $\lvert G\rvert$(dB)	Phase Shift $\angle G$(deg)
120	−7.8	−165
100	−4.3	−160
80	−0.2	−160
70	0.75	−150
50	5.2	−140
40	8.0	−130
32	10.5	−125
20	15.0	−90
14	16.9	−55
5	20.4	−40
2.5	22.0	−30
0.5	24.0	−10
0	24.0	0

Ans. (b) $G(s) \approx 158{,}000/(s + 100)^2$

10. Find gain margins and phase margins (if they exist) for feedback systems with the following loop transmittances:

(a) $$G(s)H(s) = \frac{100}{(s+7)^3}$$

(b) $$G(s)H(s) = \frac{100}{s(s+10)^2}$$

(c) $$G(s)H(s) = \frac{1}{(s+4)^3}$$

(d) $$G(s)H(s) = \frac{\sqrt{10}\,10^5}{(s+10)(s+100)^2}$$

(e) $$G(s)H(s) = \frac{1}{s(s+10)^2}$$

Ans. (a) gain margin $=$ 30 dB; (b) gain margin $= 25dB$, Phase margin $= 80°$

11. For the system of Figure P6.11, use frequency response methods to determine values of $K > 0$, if any, that result in marginal stability of the overall system.

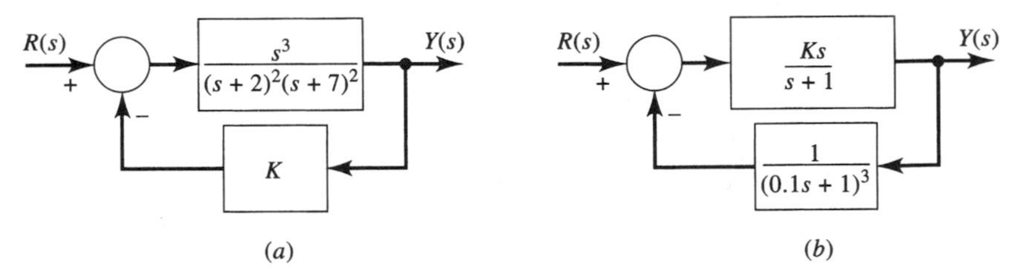

Figure P6.11

12. Use frequency response methods to find the ranges of the positive constant K (if any) for which each of the systems of Figure P6.12 is stable.

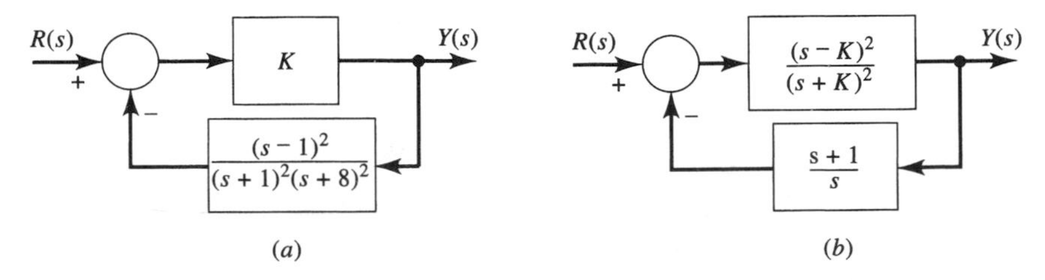

Figure P6.12

13. Construct Bode plots for feedback systems with the following loop transmittances, and then use the plots to determine whether each system is stable.

(a) $$G(s)H(s) = \frac{s+3}{(s+2)(s+7)}$$

(b) $$G(s)H(s) = \frac{10}{s(s+3)(s+10)}$$

(c) $$G(s)H(s) = \frac{6}{(s^2+1)(s+10)}$$

(d) $$G(s)H(s) = \frac{s^2+5}{s^2(s^2+4s+8)}$$

14. Repeat Problem 13 by sketching Nyquist plots for each of the GH functions. Determine whether each system is stable using the Nyquist plots.

Ans.

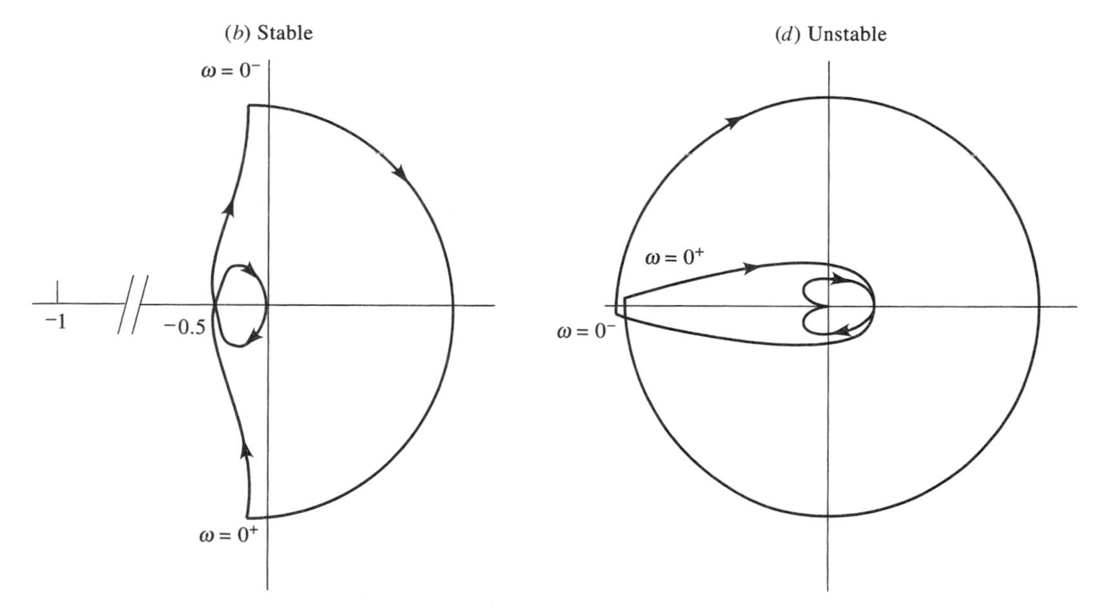

Figure P6.14

15. For the feedback systems of Figure P6.15, sketch Nyquist plots and use them to determine whether each system is stable.

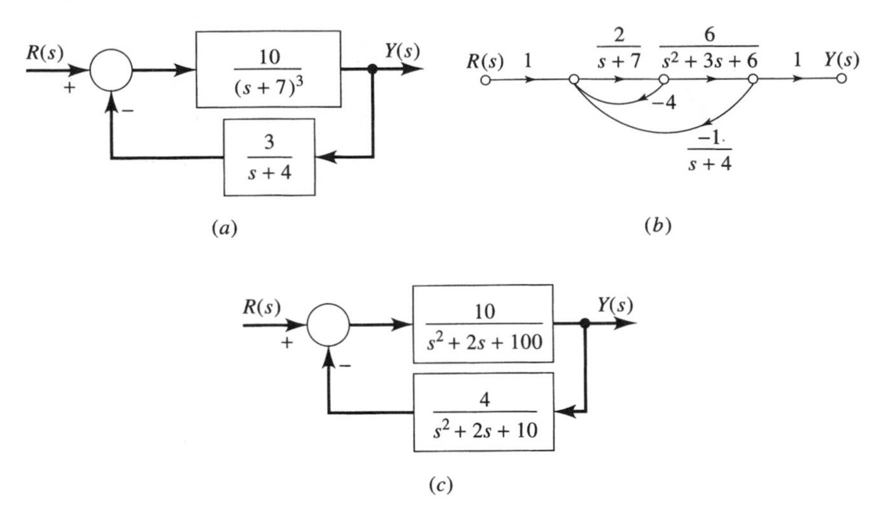

Figure P6.15

16. Use frequency response methods to determine the range of positive values of the constant a for which the system of Figure P6.16 is stable.

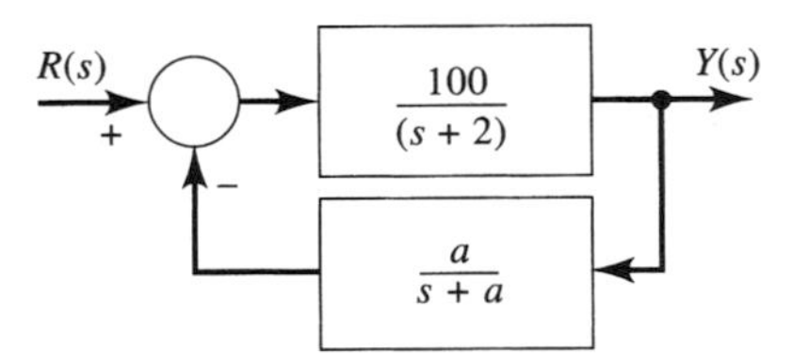

Figure P6.16

17. Figure P6.17 shows a *Nichols chart*, which shows the graphical conversion of the frequency response of the open-loop transmittance $G(s = j\omega)$ of a unity feedback system to its closed-loop frequency response,

$$T(s = j\omega) = \frac{G(s = j\omega)}{1 + G(s = j\omega)}$$

This form of the chart deals with the amplitude ratio rather than dB. Write and test a digital computer or pocket calculator program to perform this computation. Given the dB and phase of G, your program should produce the dB and phase of T.

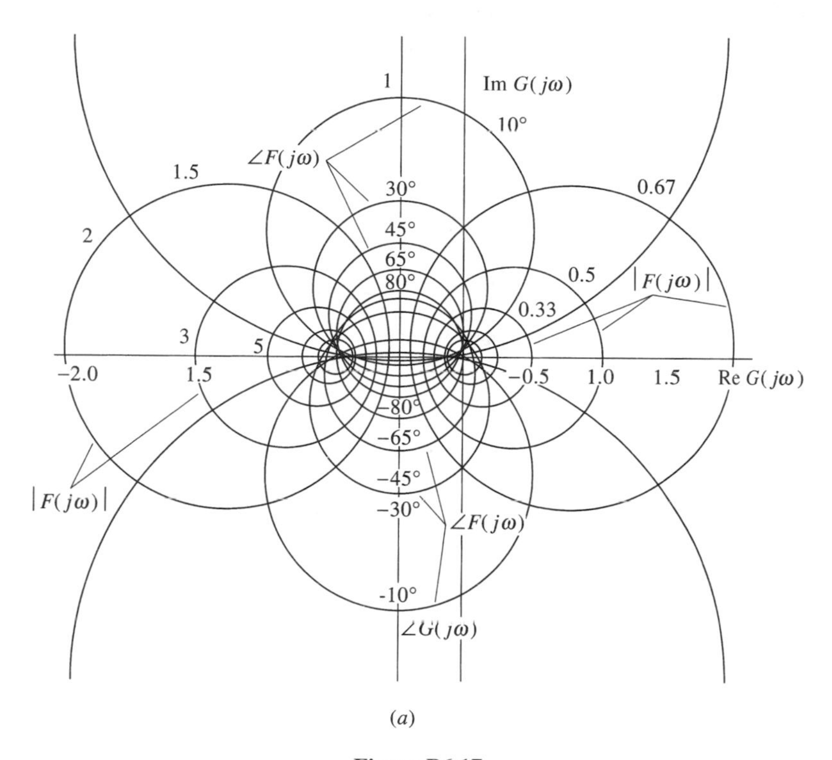

(*a*)

Figure P6.17

18. Carefully describe a method for measuring the frequency response of a control system component with an *unstable* transmittance by connecting it as part of a stable feedback system.

19. For the system of Figure P6.19, how large can the constant K be for each positive integer value of n if the overall system is to be stable?

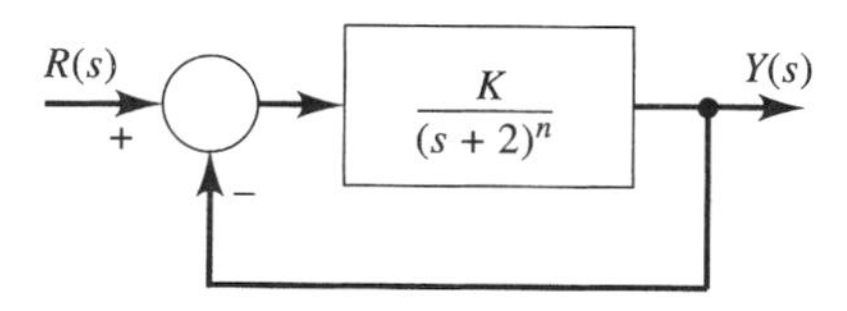

Figure P6.19

Ans. $0 < K < \dfrac{2^n}{\cos^n(180°/n)}$

20. An aircraft heading control system is diagrammed in Figure P6.20. Use frequency response methods to determine a suitable value of "pilot gain" K.

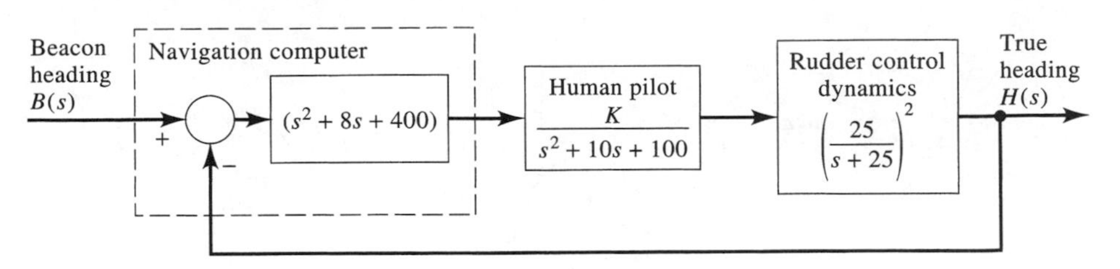

Figure P6.20

CHAPTER 7

Frequency Response Design

7.1 Preview

In Chapter 5 we focused on the selection of compensators to improve system performance as viewed from the perspective of the root locus. There are two main areas to be improved: steady state accuracy and stability. We used error coefficients to measure steady state accuracy, and we used the relative stability of the two dominant root locations to measure stability.

In Chapter 6 we learned that phase margin and gain crossover frequency are two important measures of stability from the frequency response viewpoint.

Chapter 7 begins by showing that phase margin is related to the damping ratio of the dominant roots. Also, the gain crossover frequency is related to the undamped natural frequency of the dominant roots. It follows that the time response, root locus, and frequency response are related even though the three are calculated by very different procedures.

We will discuss the design of six kinds of compensator, where each is selected because of the effect on steady state accuracy as measured using the error coefficient and the effect on stability as measured by gain crossover frequency and phase margin. The compensators are cascade PI, cascade lag, cascade lead, cascade lag–lead, rate feedback, and PID.

The chapter concludes with a design example.

7.2 Relation Between Root Locus, Time Domain, and Frequency Domain

A control system is usually designed using either the root locus (s domain), the time response (time domain), or the Bode/Nyquist plot (frequency domain). Although the three approaches portray three different views of a control system defined on three different domains, relationships exist among these methods.

Dominant roots are defined.

Figure 7.1 shows a second-order control system. A higher-order system will be far more complex. However, if two dominant roots exist that are much closer to the vertical axis than any other, Figure 7.1 approximately describes that higher-order system also.

The closed-loop transfer function for Figure 7.1 has the standard form including the damping ratio ζ and the undamped natural frequency ω_n.

$$T(s) = \frac{Y(s)}{R(s)} = \frac{\omega_n^2}{s^2 + 2\zeta\omega_n s + \omega_n^2}$$

If the damping ratio is between zero and 1, complex conjugate poles of $T(s)$ result.

$$s^2 + 2\zeta\omega_n s + \omega_n^2 = (s + \sigma + j\omega)(s + \sigma - j\omega)$$
$$= s^2 + 2\sigma s + \sigma^2 + \omega^2$$

Figure 7.2 shows the complex conjugate closed-loop poles for $T(s)$, denoted by a box. From the figure and the characteristic polynomial, the damping ratio and the angle are related.

$$\zeta = \frac{\sigma}{\omega_n} = \cos\phi$$

Notice that ω_n is the length of the vector from the origin to the complex conjugate poles. Figure 7.3 shows loci of constant ζ with increasing ω_n (straight lines projecting outward) and loci of constant ω_n with increasing ζ (semicircular arcs).

Effects upon rise time and settling time are discussed.

Figures 7.4 and 7.5 were discussed in Section 5.4, where they appeared as Figure 5.6. It was pointed out that when damping ratio is constant, both rise time and settling time decrease as undamped natural frequency increases. For fixed undamped natural frequency, it was stated that increased damping ratio causes rise time to increase. Settling time is a minimum for a damping ratio of 0.7.

The gain crossover frequency and the phase margin are calculated for a second-order type 1 model.

These trends may be related to the frequency domain. Figure 7.1 may be redrawn as in Figure 7.6 where

$$G(j\omega) = \frac{\omega_n^2}{j\omega(j\omega + 2\zeta\omega_n)}$$

The gain crossover frequency requires

$$|G(j\omega_\phi)| = 1$$

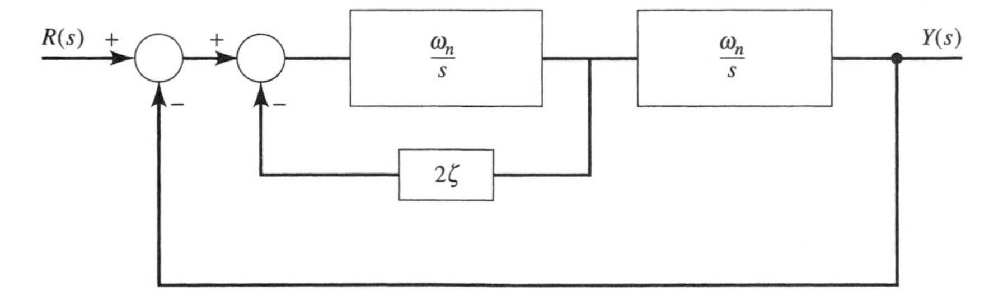

Figure 7.1 Second-order system.

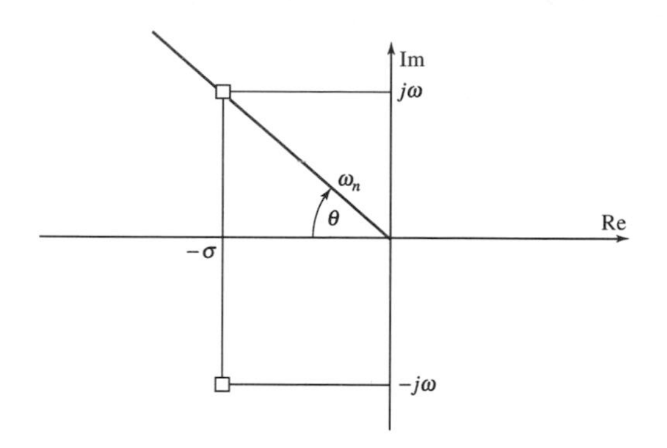

Figure 7.2 Complex conjugate poles.

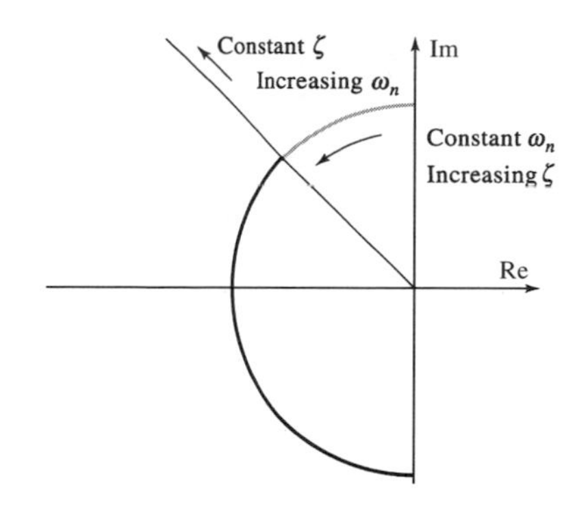

Figure 7.3 Changes in damping ratio and undamped natural frequency.

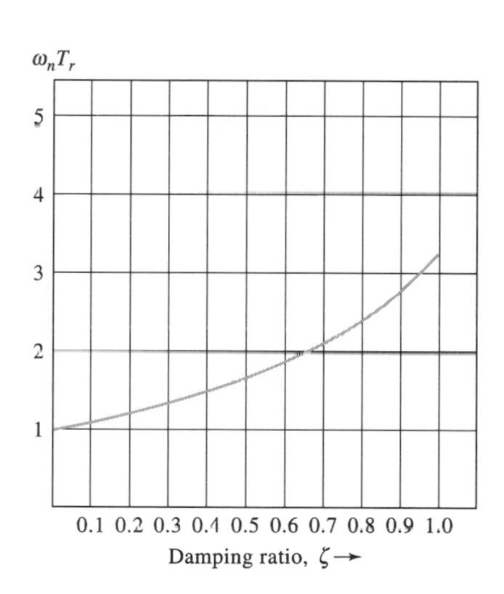

Figure 7.4 Rise time versus damping ratio.

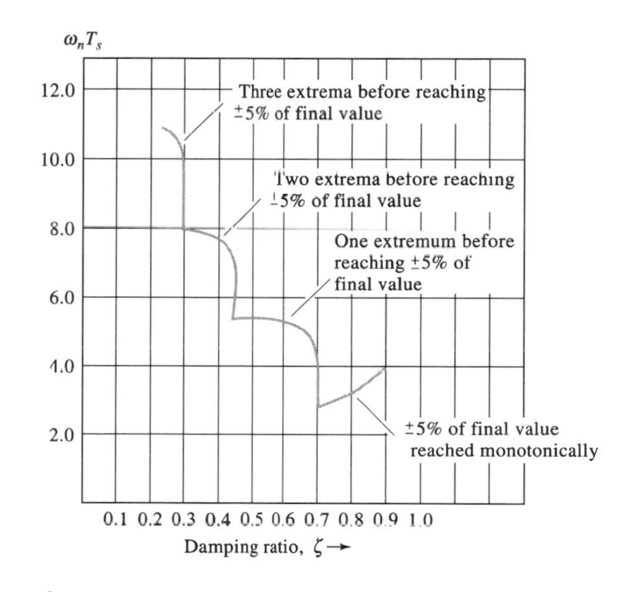

Figure 7.5 Settling time versus damping ratio.

The phase margin is then

$$PM = 180° + \text{phase of } G(j\omega_\phi)$$

From the magnitude condition, we write

$$\omega_n^2 = \omega_\phi\sqrt{\omega_\phi^2 + (2\zeta\omega_n)^2}$$

$$\omega_\phi^4 + 4\zeta^2\omega_n^2\omega_\phi^2 - \omega_n^4 = 0$$

The roots of the last expression follow by applying the quadratic formula in terms of ω_ϕ^2:

$$\omega_\phi^2 = \omega_n^2(-2\zeta^2 \pm \sqrt{4\zeta^4 + 1})$$

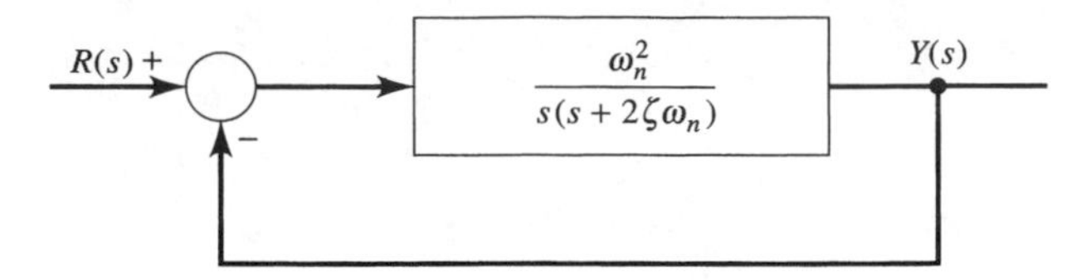

Figure 7.6 Second-order system redrawn.

For ω_ϕ to be real-valued, the positive root must be used so that

$$\omega_\phi = \omega_n k$$

$$k = \sqrt{\sqrt{4\zeta^4 + 1} - 2\zeta^2}$$

Figure 7.7 plots k versus ζ. For most values of ζ, k is near unity and thus the gain crossover frequency and the undamped natural frequency are closely related.

The phase margin becomes

$$PM = 180^\circ - 90^\circ - \tan^{-1}\left(\frac{k}{2\zeta}\right)$$

$$= 90^\circ - \tan^{-1}\left(\frac{k}{2\zeta}\right)$$

$$= \tan^{-1}\left(\frac{2\zeta}{k}\right)$$

Phase margin is about 100 times the damping ratio.

The last result is due to a trigonometric identity. Figure 7.8 shows phase margin in degrees versus ζ. Phase margin is directly proportional to damping ratio. For damping ratios less than 0.5, the phase margin (in degrees) is about 100 times the damping ratio. The slope of phase margin versus damping ratio diminishes as damping ratio varies from 0.5 to 1.

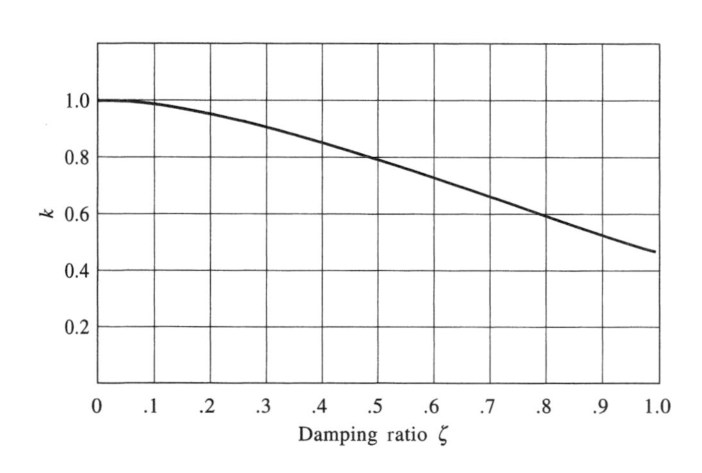

Figure 7.7 Plot of k versus damping ratio.

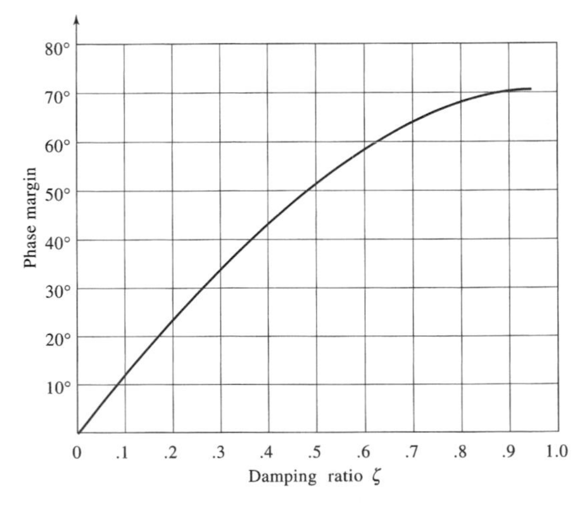

Figure 7.8 Phase margin versus damping ratio.

To summarize, for most values of damping ratio the gain crossover frequency and the undamped natural frequency are nearly equal. The phase margin is directly (and almost linearly) related to the damping ratio. Therefore, when a frequency domain design holds phase margin constant while increasing the gain crossover frequency, the resulting rise time and settling time would diminish in the time domain and the root locus would move outward along a line of constant damping ratio as in Figure 7.3.

Effect on the time domain

Good control design involves examination of all three evaluative tools: root locus, time domain, and frequency domain. The preceding discussion demonstrates (approximately) how frequency domain design influences the other two portraits.

❑ DRILL PROBLEM

D7.1 For each of the following unity feedback systems, a forward path transmittance is given. Compute the gain crossover frequency and the phase margin. Use Figures 7.7 and 7.8 to approximate the damping ratio and the undamped natural frequency for the dominant closed-loop roots. Compare with the exact values.

(a)

$$G(s) = \frac{9}{s(s+3)}$$

Ans. 2.36 rad/s, 51.8°; 0.5, 3 rad/s; 0.5, 3 rad/s

(b)

$$G(s) = \frac{100}{s(s+3)(s+9)}$$

Ans. 2.66 rad/s, 31.9°; 0.28, 2.89 rad/s; 0.27, 3.11 rad/s

(c)

$$G(s) = \frac{30(s+20)}{s(s+3)(s+9)}$$

Ans. 7.15 rad/s, 4°; 0.04, 7.2 rad/s; 0.03, 7.20 rad/s

7.3 Compensation Using Bode Plots

The components of Figure 7.9 appeared in Chapter 5 (Figures 5.6 and 5.7); in connection with our consideration of the influence of compensators on the root locus. The same compensators may be analyzed by using the Bode plot concepts of this chapter. A designer may decide to adjust a gain K to provide an acceptable error coefficient, gain crossover frequency, phase margin, phase crossover frequency, and gain margin. The resulting control system of Figure 7.9(a) is called uncompensated in that only a gain K and a unity feedback structure are used.

If the uncompensated system cannot be designed to meet specifications, a cascade compensator [Figure 7.9(b)] or a feedback compensator [Figure 7.9(c)] may be selected. Table 7.1 contains a representative sample of the types of compensators often employed. The same compensators were considered in Chapter 5 as each affected root locus behavior. Now interest is focused on Bode plot behavior. Of course, the influence

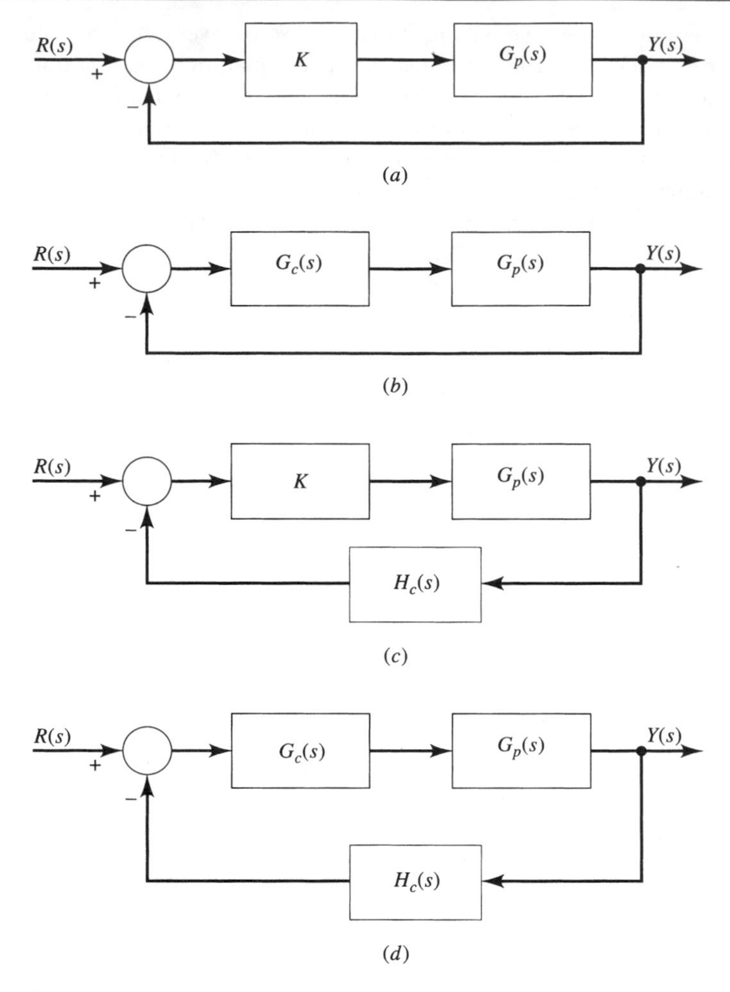

Figure 7.9 Compensator configurations. (a) Uncompensated unity feedback system. (b) Cascade-compensated system. (c) Feedback-compensated system. (d) Feedback- and cascade-compensated system.

on steady state accuracy should always be considered (see Chapter 5 for the steady state effects of these compensators).

The PI and cascade lag compensators create negative phase.

A cascade PI compensator or a cascade lag compensator provides a negative phase angle where s is replaced by $j\omega$. Since the phase angle of the compensated system becomes more negative, phase margin tends to be reduced for fixed gain crossover frequency or gain crossover frequency tends to be reduced for fixed phase margin. These compensators are primarily intended for improvement of steady state accuracy.

Cascade lead compensator.

The cascade lead compensator provides a positive phase angle. Therefore phase margin may be increased for fixed gain crossover frequency or gain crossover frequency may be increased for fixed phase margin.

The cascade lag–lead compensator combines the attributes of the lag and lead compensators so that improvements in steady state accuracy and in stability (through the gain crossover frequency and the phase margin) are possible.

Cascade lag–lead compensator.

The rate feedback compensator provides a positive phase angle. The same benefits derived from lead compensation result. Parameter selection follows by recognizing that the rate feedback compensator $H_c(s)$ represents the combination of two loops, one containing K and one containing both K and A.

Rate Feedback compensator.

The PID compensator combines the steady state improvement using a cascade PI compensator with the stability improvement of the rate feedback compensator. The design procedure thus combines the design procedure of a PI compensator with that of a rate feedback compensator. The parameters are K, a, and A.

PID Compensator.

7.4 Uncompensated System

The effects of the compensators of Table 7.1 will be evaluated using the same uncompensated system as in Chapter 5. See Table 7.2. That is

Uncompensated system example.

$$G_p(s) = \frac{1}{s(s^2 + 4s + 5)}$$

With K equal to 3, the ramp error coefficient is 0.6. Figure 7.10 shows the Bode plot for the uncompensated system. The gain crossover frequency is at 0.576 rad/s. The phase margin is 63.7°. Compensation may be applied to improve steady state accuracy by increasing the error coefficient and/or to improve stability by operating on the gain crossover frequency and the phase margin.

Table 7.1 *Common Types of Compensator*

Compensator	Transmittance	Typical Effect on Steady State Errors	Typical Effect on Phase Margin and Gain Crossover Frequency
Cascade PI	$G_c(s) = \frac{K(s+a)}{s}$	Greatly improved	Reduced
Cascade lag	$G_c(s) = \frac{K(s+a)}{s+b}$ $b < a$	Improved	Reduced
Cascade lead	$G_c(s) = \frac{K(s+a)}{s+b}$ $a < b$	Somewhat improved or somewhat worse	Increased
Cascade lag–lead	$G_c(s) = K\left(\frac{s+a}{s+b}\right)_{\text{lag}} \times \left(\frac{s+a}{s+b}\right)_{\text{lead}}$	Improved	Increased
Rate feedback (PD)	$H_c(s) = 1 + As$	Somewhat improved or somewhat worse	Increased
PID	$G_c(s) = \frac{K(s+a)}{s} + KAs$	Greatly improved	increased

Table 7.2 *Summary of Designs*

Compensator	Transmittance	Ramp Error Coefficient	Gain Crossover Frequency	Phase Margin
Uncompensated	$K = 3$	0.6	0.58	63.7°
Cascade PI	$G_c(s) = \dfrac{3(s+0.0576)}{s}$	∞	0.58	57.9°
Cascade lag	$G_c(s) = \dfrac{3(s+0.0576)}{s+0.00576}$	6	0.58	58.5°
Cascade lead	$G_c(s) = \dfrac{30(s+0.576)}{s+5.76}$	0.6	1.08	92.9°
Cascade lag–lead	$G_c(s) = \dfrac{30(s+0.0576)}{s+0.00576} \times \dfrac{s+0.576}{s+5.76}$	6	1.08	90.1°
Rate feedback (PD)	$H_c(s) = \frac{1}{2}(s+2)$ $K = 10$	1.0	1.81	55.5°
PID	$G_c(s) = \dfrac{5(s+1.94)(s+0.059)}{s}$	∞	1.79	54.9°

Throughout the design process it is expected that the engineer will make effective use of computer programs. It is important, however, that designers be able to sketch Bode plots and verify design values so as to validate computer results, and they will not become totally dependent on a computer.

❑ DRILL PROBLEM

D7.2 Each of the following uncompensated systems is of the form of Figure 7.9. Use computer programs to find the gain crossover frequency, phase margin, and appropriate error coefficient—that is, the error coefficient K_i for a type i plant.

(a)

$$G_p(s) = \frac{1}{s(s+2)} \qquad K = K_u = 2$$

Ans. $0.91,\ 65.5°,\ K_1 = 1$

(b)

$$G_p(s) = \frac{1}{s(s+2)(s+4)} \qquad K = K_u = 5.2$$

Ans. $0.61,\ 64.2°,\ K_1 = 0.65$

(c)

$$G_p(s) = \frac{s+8}{(s+1)(s+2)(s+6)} \qquad K = K_u = 1.64$$

Ans. $0.39,\ 147.1°,\ K_0 = 1.09$

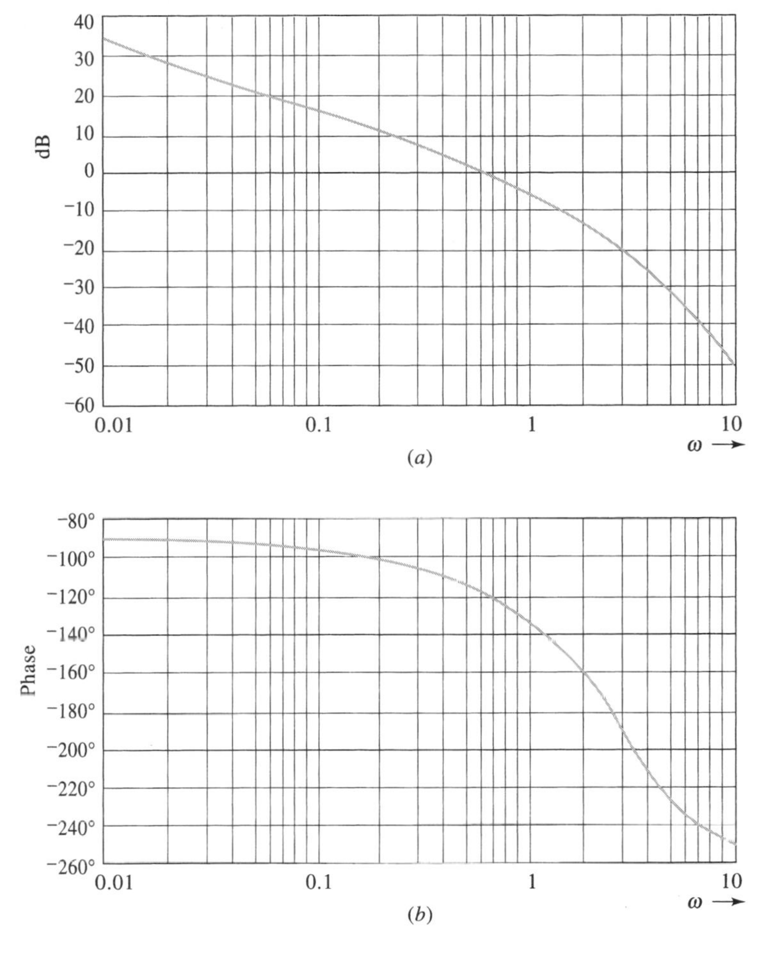

Figure 7.10 Uncompensated system Bode plots. (a) Amplitude versus frequency. (b) Phase versus frequency.

7.5 Cascade Proportional Plus Integral (PI) and Cascade Lag Compensations

Definition of the PI compensator

The Bode plot for the cascade proportional plus integral (PI) compensator follows by replacing s with $j\omega$:

$$G_c(j\omega) = \frac{K(1 + j\omega/a)}{j\omega/a}$$

Definition of the cascade compensator.

Similarly, the cascade lag compensator is as follows, where $b = a/d$ for $d > 1$

$$G_c(j\omega) = \frac{Kd(1 + j\omega/a)}{1 + j\omega d/a}$$

As d rises toward infinity, the two compensators become equal (although the RC elements of the lag compensator would be very difficult to accurately realize).

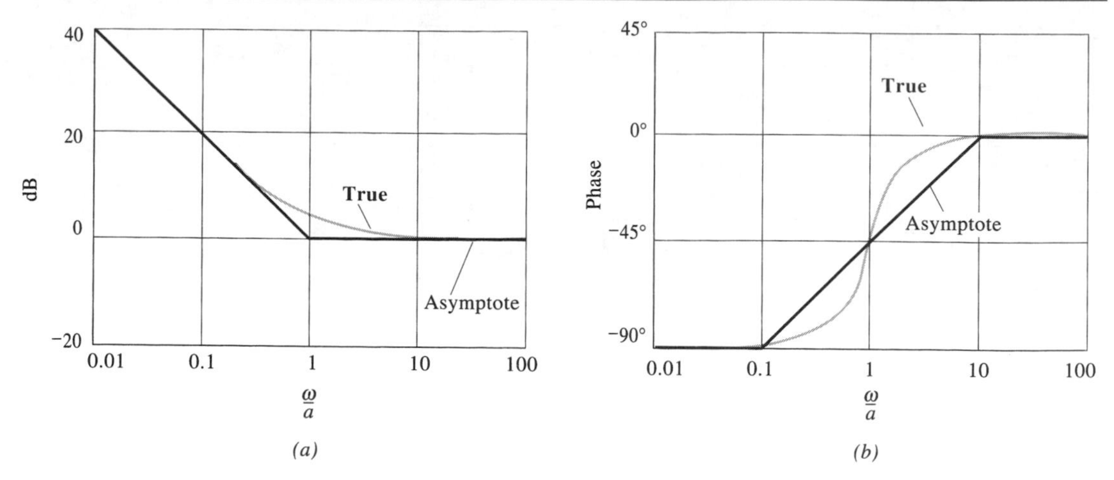

Figure 7.11 Proportional plus integral compensator Bode plots of $G'_c(j\omega) = (1 + j\omega/a)/j\omega/a$. (a) Amplitude versus frequency. (b) Phase versus frequency.

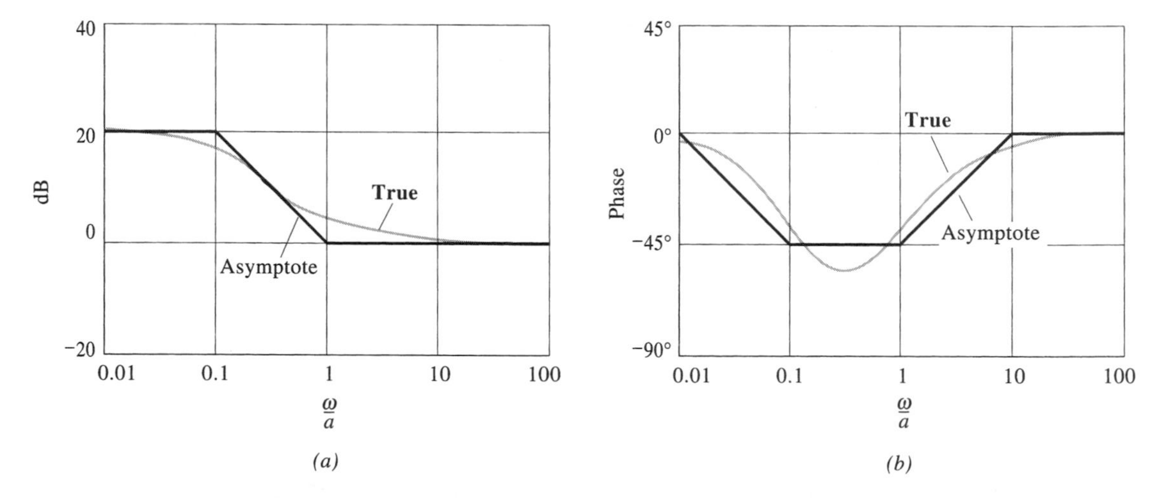

Figure 7.12 Lag compensator Bode plots ($d = 10$) of $G'_c(j\omega) = [10(1 + j\omega/a)]/(1 + j10\omega/a)$. (a) Amplitude versus frequency. (b) Phase versus frequency.

Figures 7.11 and 7.12 contain the Bode plots for these two compensators. For convenience, the frequency axis is normalized by using ω/a. Also, K is set equal to 1. It would be a simple matter to include other values of K because all dB values would be increased by the dB values of K.

Phase effects.

Each compensator provides negative phase angles. In fact, the term "lag" compensator results because negative phase angles provide a steady state sinusoidal output that "lags" behind a sinusoidal input of the same frequency. In this chapter, the gain crossover frequency is denoted by ω_ϕ.

Because each compensator creates a negative phase angle, the phase margin would be decreased for the combination of compensator and plant if K is adjusted so the gain crossover frequency remains unchanged. When the compensator is added, we put

$$\begin{aligned}\text{Phase of } G_c(j\omega_\phi)G_p(j\omega_\phi) &= \text{phase of } G_c(j\omega_\phi) + \text{phase of } G_p(j\omega_\phi)\\ \text{Phase margin (compensated)} &= 180^\circ + \text{phase of } G_p(j\omega_\phi) + \text{phase of } G_c(j\omega_\phi)\\ &= \text{phase margin (uncompensated)}\\ &\quad + \text{phase of } G_c(j\omega_\phi)\end{aligned}$$

If the phase margin must be maintained, then ω_ϕ must be reduced from the uncompensated value, so the phase angle of the plant is increased by the same amount as the negative phase angle of the compensator. In either case, the stability of the system is worse. However, these compensators are primarily intended to improve steady state accuracy (see Chapter 5).

Stability is worse.

One way to reduce the destabilizing influence that the compensators have upon the closed-loop system is to select

$$a = \frac{\omega_\phi(\text{uncompensated})}{10}$$

The PI compensator then has a phase angle at ω_ϕ of

Design concepts.

$$\tan^{-1} 10 - 90^\circ = -5.7^\circ$$

If $d = 10$, the lag compensator has a phase angle at ω_ϕ of

$$\tan^{-1} 10 - \tan^{-1} 100 = -5.1^\circ$$

If the gain crossover frequency remains unchanged, phase margin is reduced by less than 6°.

For the example system, we write

$$a = \frac{0.576}{10} = 0.0576$$

The resulting integral plus proportional compensator is

$$G_c(s) = \frac{K(s + 0.0576)}{s}$$

A value for K must somehow be chosen even though the Bode plot cannot be plotted without knowing K. One way out of that problem is to factor out K from the compensator

Factoring out K.

$$G_c(s) = KG'_c(s)$$

The compensated system

$$G_c(s)G_p(s) = KG'_c(s)G_p(s)$$

can be analyzed from the Bode plot of

$$G'_c(s)G_p(s)$$

Since K is usually positive, the phase with or without K is usually the same. If the designer selects a new gain crossover frequency ($\omega_{\phi 1}$) to create some phase margin, then K must be

$$K = \frac{1}{\left|G'_c(j\omega_{\phi 1})G_p(j\omega_{\phi 1})\right|}$$

If the designer selects K for some other reason (perhaps to create an error coefficient), then the new gain crossover frequency becomes such that

$$\left|G'_c(j\omega_{\phi 1})G_p(j\omega_{\phi 1})\right| = \frac{1}{K}$$

Design examples.

Figure 7.13 shows the PI compensated system Bode plot with K factored out. If K is chosen to remain at the uncompensated value of 3, the gain crossover frequency (where the magnitude is $\frac{1}{3}$ or -9.5 dB) becomes 0.58 rad/s with a phase margin of 57.9°. The phase margin is 6° less than for the uncompensated system, as in Table 7.2.

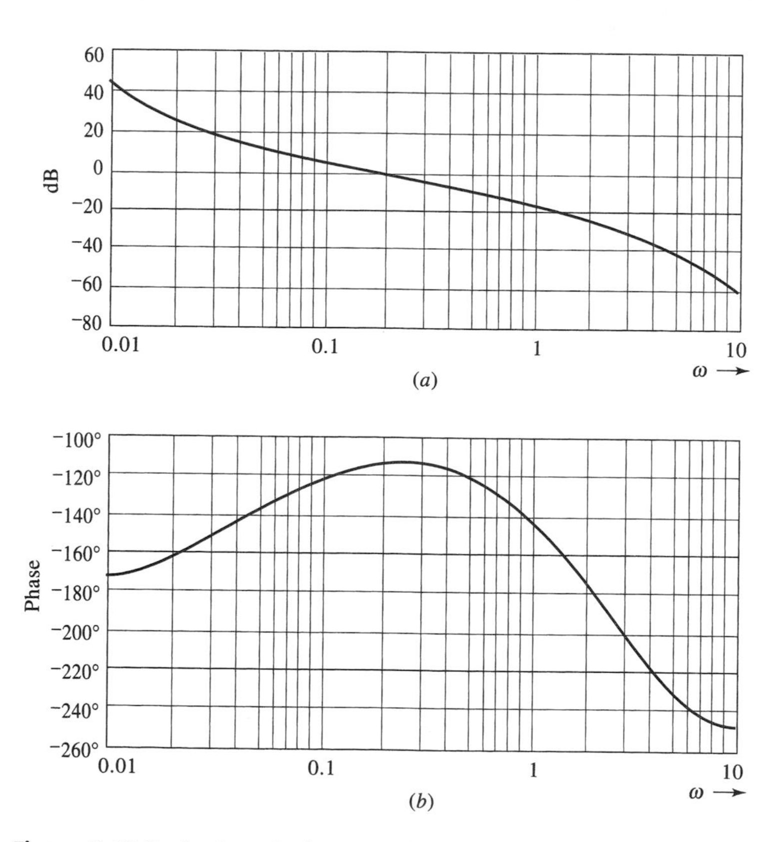

Figure 7.13 Bode plots of $G'_c(j\omega)G_p(j\omega)$ for PI compensation. (a) Amplitude versus frequency. (b) Phase versus frequency.

For the lag compensator, suppose d is chosen to be 10 (perhaps the error coefficient is to be increased by a factor of 10). Then we have

$$G_c(s) = \frac{K(s + 0.0576)}{s + 0.00576}$$

Figure 7.14 shows the Bode plot of $G_c'(s)G_p(s)$. If K is chosen to be 3, the ramp error coefficient becomes 6. The gain crossover frequency (where the magnitude is $\frac{1}{3}$ or -9.5 dB) is 0.58 rad/s with a phase margin of 58.5°. The phase margin is 5° less than for the uncompensated system, as in Table 7.2.

To summarize, the steps in the design procedure for cascade PI compensation are as follows.

1. Obtain the Bode plot of $KG_p(s)$ for the uncompensated system where K takes on a value K_u. Find the gain crossover frequency ω_ϕ, the phase margin, and the error coefficient.
2. Let $a = 0.1\,\omega_\phi$.
3. Let $K = K_u$.

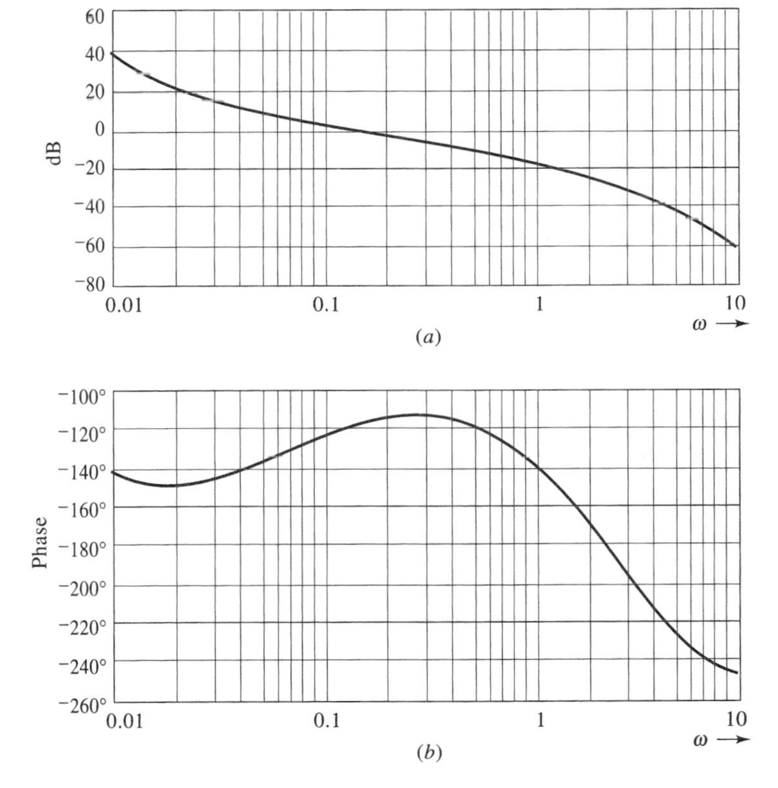

Figure 7.14 Bode plots of $G_c'(j\omega)G_p(j\omega)$ for lag compensation. (a) Amplitude versus frequency. (b) Phase versus frequency.

4. Obtain the Bode plot of $G_c'(j\omega)G_p(j\omega)$ where $G_c'(s)$ is the compensator with K factored out. The new gain crossover frequency $\omega_{\phi 1}$ is the frequency where $\left|G_c'(j\omega_{\phi 1})G_p(j\omega_{\phi 1})\right| = 1/K$. Find the phase margin.
5. Vary K and a as required to meet design requirements.

This procedure results in an infinite error coefficient K_i for a plant with type number i, while the phase margin is about 6° less than for the uncompensated case.

The steps in the design procedure for cascade lag compensation are as follows.

1. Obtain the Bode plot of $KG_p(s)$ for the uncompensated system where K takes on a value K_u. Find the gain crossover frequency ω_ϕ, the phase margin, and the error coefficient.
2. Let $d =$ the factor by which the error coefficient is to be increased.
3. Let $a = 0.1\,\omega_\phi$ and $b = a/d$.
4. Let $K = K_u$.
5. Obtain the Bode plot of $G_c'(j\omega)G_p(j\omega)$ where $G_c'(s)$ is the compensator with K factored out. The new gain crossover frequency $\omega_{\phi 1}$ is the frequency where $\left|G_c'(j\omega_{\phi 1})G_p(j\omega_{\phi 1})\right| = 1/K$. Find the phase margin.
6. Vary K and a as required to meet design requirements.

This procedure results in an error coefficient K_i for a plant with type number i that is d times that of the uncompensated system, while the phase margin is about 5° less than for the uncompensated case.

❑ DRILL PROBLEMS

D7.3 For each of the plants from Drill Problem D7.2 carry out design steps 1–4 for a cascade PI compensator. For step 4 show the value of a, K, the gain crossover frequency, phase margin, and appropriate error coefficient.

Ans. (a) $0.091, 2, 0.91, 59.8°, K_1 = \infty$; (b) $0.061, 5.2, 0.61, 58.4°, K_1 = \infty$; (c) $0.039, 1.64, 0.39, 142.0°, K_0 = \infty$

D7.4 For each of the plants from Drill Problem D7.2 carry out design steps 1–5 for a cascade lag compensator with $d = 10$. For step 5 show the value of a, b, K, the gain crossover frequency, phase margin, and appropriate error coefficient.

Ans. (a) $0.091, 0.0091, 2, 0.91, 60.4°, K_1 = 10$; (b) $0.061, 0.0061, 5.2, 0.61, 59.0°, K_1 = 6.5$; (c) $0.039, 0.0039, 1.64, 0.39, 142.6°, K_0 = 10.9$

7.6 Cascade Lead Compensation

Definition of the cascade lead compensators.

A cascade lead compensator has the following form if d is defined to be b/a and $d > 1$:

$$G_c(j\omega) = \frac{K}{d}\frac{(1 + j\omega/a)}{(1 + j\omega/(ad))}$$

Figure 7.15 shows the Bode plot of a lead compensator where K equals 1 and where the frequency axis is normalized to ω/a.

Phase effects.

The phase angle is always positive, which is why this compensator is called a "lead" compensator. A sinusoidal steady state output would "lead" a sinusoidal input of the same frequency. Since the phase angle is always positive, the phase margin is increased in the compensated case if the gain crossover frequency is unchanged. That fact follows from the same analysis as when lag compensators were discussed. If the same phase margin is needed, then the gain crossover frequency would increase.

Design concepts.

A designer must select values for K, a, and d. The system of Figure 7.16 is useful for providing some general clues as to Bode design procedure in the same way that the system of Figure 7.16 provided guidance for root locus design. Suppose a damping ratio of 0.7 occurs for the uncompensated system dominant roots. Figure 7.7 suggests (for the uncompensated case) that

$$\frac{\omega_\phi}{\omega_n} = 0.65$$

The ramp error coefficient with the lead compensator included remains the same if K equals d. The Bode plot of Figure 7.17 show $G_c(s)G_p(s)$ where $a/\omega_n = 0.63$, $K = d - 10$, and the frequency axis is normalized by using ω/ω_n. The compensated

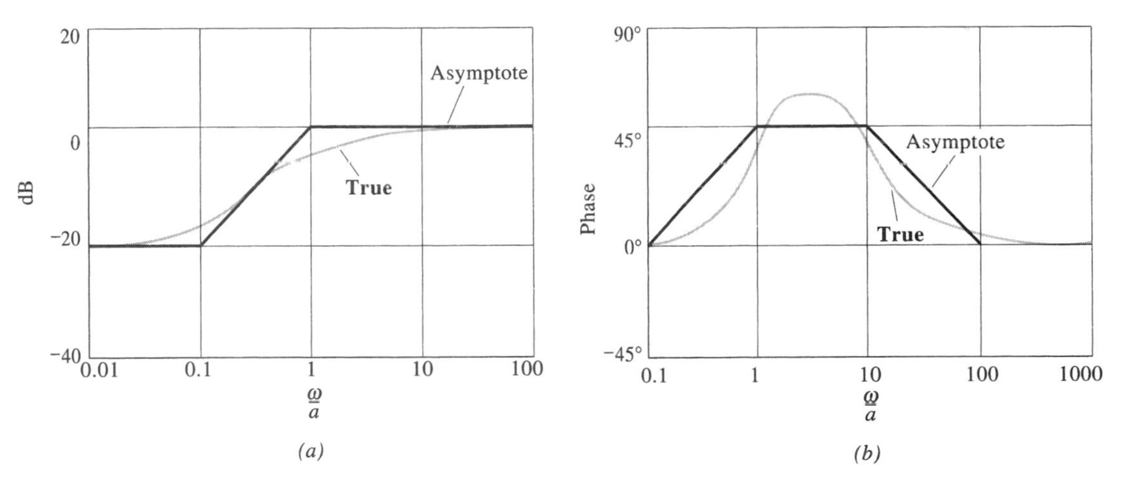

Figure 7.15 Lead compensator Bode plots ($d = 10$) of $G_c'(j\omega) = [0.1(1 + j\omega/a)]/(1 + j0.1\omega/a)$. (a) Amplitude versus frequency. (b) Phase versus frequency.

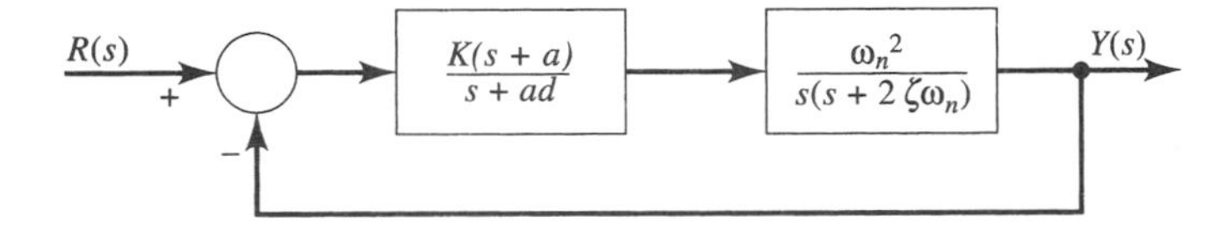

Figure 7.16 Lead compensation of the uncompensated dominant roots.

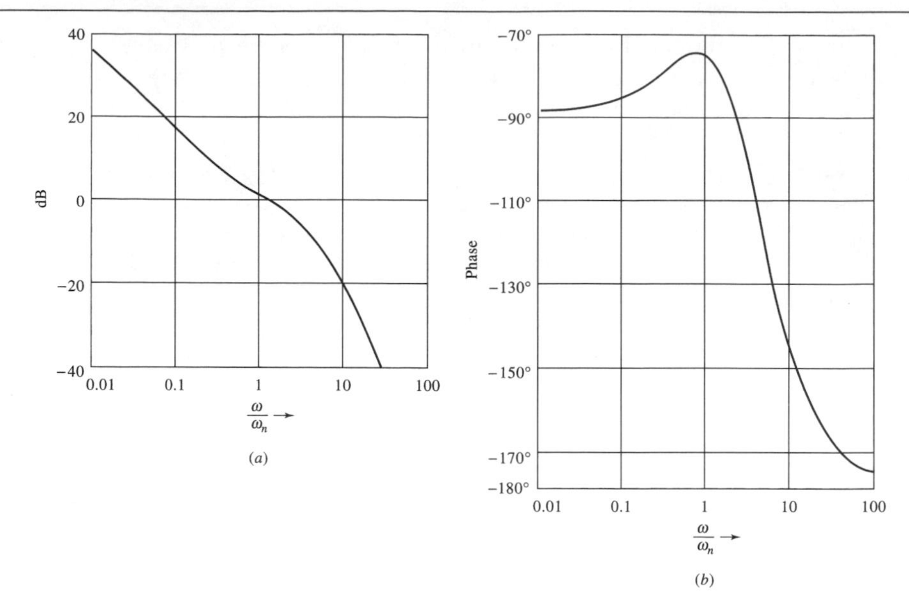

Figure 7.17 Bode plots for the system in Figure 7.16 where the damping ratio $= 0.7$; $a/\omega_n = 0.63$; $K = d = 10$. (a) Amplitude versus frequency. (b) Phase versus frequency.

normalized gain crossover frequency and phase margin are

$$\frac{\omega_{\phi 1}}{\omega_n} = 1.12$$

$$\text{Phase margin} = 101.9^\circ$$

If another value of a is chosen, a lower phase margin results. By trial and error, the best value of a (largest phase margin) may be chosen for various damping ratios. The results are in Table 7.3. In general the best choice of a is approximately

$$a \cong \omega_\phi(\text{uncompensated})$$

The steps in the design procedure for cascade lead compensation are as follows:

1. Obtain the Bode plot of $KG_p(s)$ for the uncompensated system where K takes on a value K_u. Find the gain crossover frequency ω_ϕ, the phase margin, and the error coefficient.
2. Select d. A large value may create problems in realizing an RC circuit with widely separated poles and zeros, so the range 5–20 is normal.
3. Let $a = \omega_\phi$ and $b = ad$.
4. Select $K = K_u \cdot d$. This maintains steady state accuracy compared to the uncompensated case.
5. Obtain the Bode plot of $G_c'(j\omega)G_p(j\omega)$, where $G_c'(s)$ is the compensator with K factored out. The new gain crossover frequency $\omega_{\phi 1}$ is the frequency where $\left|G_c'(j\omega_{\phi 1})G_p(j\omega_{\phi 1})\right| = 1/K$. Find the phase margin.
6. Vary K and a as required to meet design requirements.

Table 7.3 *Lead Compensation (best phase margin)*

Damping	Uncompensated System	Compensated System			a
Ratio	ω_ϕ/ω_n	a/ω_n	ω_ϕ/ω_n	Phase Margin	ω_ϕ(uncomp)
0.3	0.91	0.65	1.52	75.3°	0.71
0.4	0.85	0.67	1.41	82.3°	0.78
0.5	0.79	0.67	1.31	89.2°	0.85
0.6	0.72	0.66	1.20	95.8°	0.92
0.7	0.65	0.63	1.12	101.9°	0.97

If we apply the design procedure to the example system, the result is

1. $\omega_\phi = 0.576$ rad/s, $PM = 63.7°$, and $K_1 = 0.6$.
2. $d = 10$
3. $a = 0.576$, $b = 5.76$.
4. $K = 3 \cdot 10 = 30$

For step 5, the Bode plots for $G'_c(s)G_p(s)$ are shown in Figure 7.18. The gain crossover frequency (when the magnitude equals $1/30 = -29.5$ dB) is 1.08 rad/s (an improvement by a factor of 2 over the uncompensated case) with a phase margin of 92.9° (an improvement of 29.2°) as in Table 7.2.

Design Example.

If the design is not acceptable, we could increase K, which would make $1/K$ have a lower dB value, thus raising the gain crossover frequency and reducing the phase margin (since the phase is increasingly more negative at higher frequencies). However, for the foregoing design, stability has been substantially improved (more phase margin and higher gain crossover frequency), so the design goals are met and the process is terminated with step 5. The compensator is

$$G_c(s) = \frac{30(s + 0.576)}{s + 5.76}$$

❑ DRILL PROBLEM

D7.5 For each of the plants from Drill Problem D7.2 carry out design steps 1–5 for a cascade lead compensator with $d = 10$. For step 5 show the value of *a, b, K,* the gain crossover frequency, phase margin, and appropriate error coefficient.

Ans. (a) 0.91, 9.1, 20, 1.54, 102.3°, $K_1 = 1$; (b) 0.61, 6.1, 52, 1.13, 96.0°, $K_1 = .65$; (c) 0.39, 3.9, 16.4, 3.33, 84.0°, $K_0 = 1.09$

7.7 Cascade Lag–Lead Compensation

A cascade lag–lead compensator is given by

Definition of the cascade lag–lead compensator.

$$G_c(s) = K\left(\frac{s+a}{s+b}\right)_{\text{lag}}\left(\frac{s+a}{s+b}\right)_{\text{lead}}$$

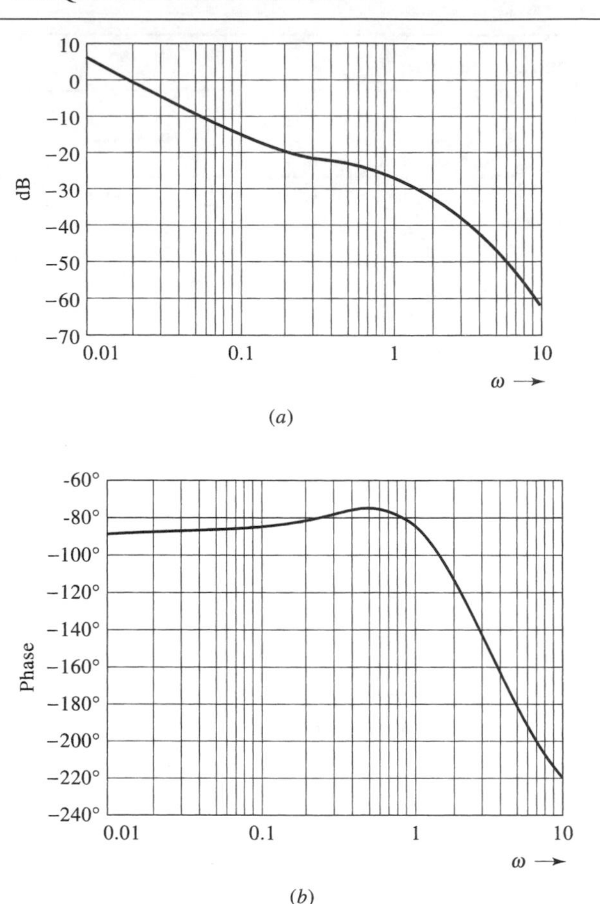

Figure 7.18 Bode plots of $G'_c(j\omega)G_p(j\omega)$ for lead compensation. (a) Amplitude versus frequency. (b) Phase versus frequency.

To simplify the design we can let $b_{\text{lag}} = a_{\text{lag}}/d$ and $b_{\text{lead}} = a_{\text{lead}}d$ for the same $d > 1$. The lag–lead compensator becomes

$$G_c(s) = K\left(\frac{s+a}{s+(a/d)}\right)_{\text{lag}}\left(\frac{s+a}{s+ad}\right)_{\text{lead}}$$

Design concepts.

The lag part of the compensator should influence the low-frequency response, to increase steady state accuracy, while the lead part should influence the higher-frequency response around the gain crossover frequency of the uncompensated system, to increase the phase margin. We can combine the lag design and lead design procedures in designing the lag–lead compensator to combine the best attributes of each. For the lag part of the compensator:

1. Obtain the Bode plot of $KG_p(s)$ for the uncompensated system in which K takes on a value K_u. Find the gain crossover frequency ω_ϕ, the phase margin and the error coefficient.

2. Let d = the factor by which the error coefficient is to be increased.
3. Let $a_{\text{lag}} = 0.1\,\omega_\phi,\ b_{\text{lag}} = a_{\text{lag}}/d$.

For the lead part of the compensator

4. Let $a_{\text{lead}} = \omega_\phi,\ b_{\text{lead}} = a_{\text{lead}}d$.
5. Select $K = K_u \cdot d$. This maintains steady state accuracy compared to the lag case.

For the lag–lead compensator

6. Obtain the Bode plot of $G'_c(j\omega)G_p(j\omega)$ where $G'_c(s)$ is the compensator with K factored out. The new gain crossover frequency $\omega_{\phi 1}$ is the frequency where $\left|G'_c(j\omega_{\phi 1})G_p(j\omega_{\phi 1})\right| = 1/K$. Find the phase margin.
7. Vary K, a_{lag}, and a_{lead} as required to meet design requirements.

The design procedure should cause the system to have somewhat less phase margin than for the lead compensated system with the same error coefficient as for the lag compensated system. Compared to the uncompensated system, the phase margin and gain crossover frequency should each be increased and the error coefficient is increased by a factor of d.

If we apply the design procedure to the example system,

Design example.

1. $\omega_\phi = 0.576$ rad/s, $PM = 63.7°$, and $K_1 = 0.6$
2. $d = 10$
3. $a_{\text{lag}} = 0.0576,\ b_{\text{lag}} = 0.00576$
4. $a_{\text{lead}} = 0.576,\ b_{\text{lead}} = 5.76$
5. $K = 3 \times 10 = 30$

For step 6, the Bode plot for $G'_c(j\omega)G_p(j\omega)$ where

$$G_c(s) = \frac{K(s + 0.0576)(s + 0.576)}{(s + 0.00576)(s + 5.76)}$$

is shown in Figure 7.19. The new gain crossover frequency $\omega_{\phi 1}$ is the frequency where $\left|G'_C(j\omega_{\phi 1})G_p(j\omega_{\phi 1})\right| = 1/30 = -29.5$ dB. Then $\omega_{\phi 1} = 1.08$ rad/s and the phase margin is $90.1°$, $26.4°$ better than for the uncompensated system and only $2.8°$ worse than for the lead-compensated system. The gain crossover frequency is twice that of the uncompensated system, while the error coefficient is 10 times higher. Thus both steady state accuracy and stability have been improved in comparison to the uncompensated case so that design can terminate with step 6. See Table 7.2. If we need to make further adjustments, then if K is increased above 30, the error coefficient increases and the gain crossover frequency increases, but phase margin becomes smaller.

❑ DRILL PROBLEM

D7.6 For each of the plants from Drill Problem D7.2 carry out design steps 1–6 for a cascade lag–lead compensator with $d = 10$. For step 6 show the value of $a_{\text{lag}}, b_{\text{lag}}, a_{\text{lead}}, b_{\text{lead}}, K$, the gain crossover frequency, phase margin, and appropriate error coefficient.

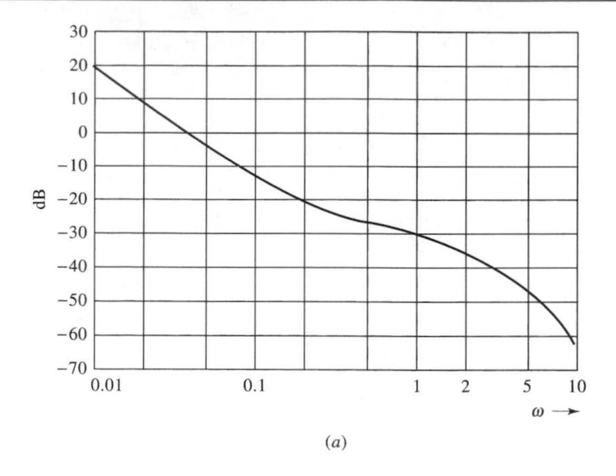

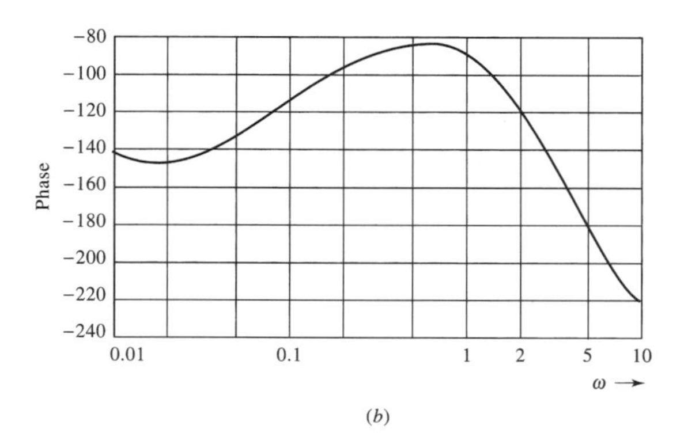

Figure 7.19 Bode plots of $G'_c(j\omega)G_p(j\omega)$ for lag–lead compensation. (a) Amplitude versus frequency. (b) Phase versus frequency.

Ans. (a) $0.091, 0.0091, 0.91, 9.1, 20, 1.54, 99.2°, \ K_1 = 10$;

(b) $0.061, 0.0061, 0.61, 6.1, 52, 1.13, 93.2°, \ K_1 = 6.5$;

(c) $0.039, 0.0039, 0.39, 3.9, 16.4, 3.33, 83.4°, \ K_0 = 10.9$

7.8 Rate Feedback Compensation

Definition of the rate feedback compensator.

The typical rate feedback system in Fig. 7.20 contains two feedback signals. The outer loop contains unity feedback while the inner loop provides rate of change of the output. Two parameters must be chosen: K and A.

One procedure with Bode methods is to stabilize the inner loop and then the entire system. The inner loop contains $KAsG_p(s)$, and the entire system then contains

$$KA\left(s + \frac{1}{A}\right)G_p(s)$$

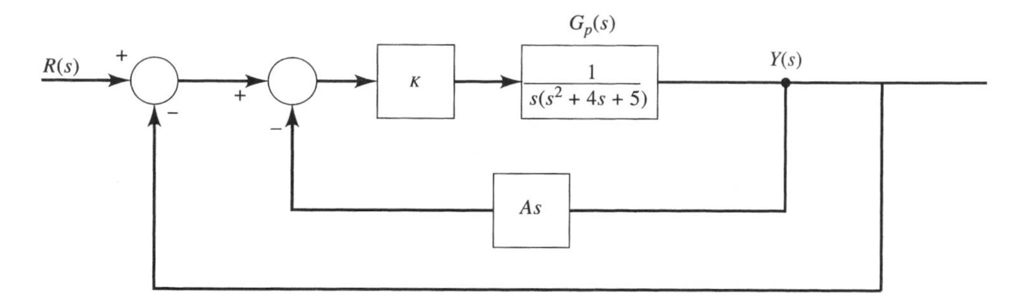

Figure 7.20 Rate feedback compensated system.

For the example system, analysis and then design of the inner loop requires a close look at

Design using the inner loop, then the outer loop.

$$\frac{KA}{s^2 + 4s + 5}$$

where one factor of s cancels. Figure 7.21 shows the Bode plot for the inner loop. Stability is ensured if

$$\frac{KA}{5} = 1$$

$$KA = 5$$

because the magnitude can never exceed unity. For this inner loop, there is no gain crossover frequency. The magnitude when the phase is $-180°$ is negative infinity so the gain margin is infinite.

For the entire system, the designer must evaluate

$$\frac{KA(s + 1/A)}{s(s^2 + 4s + 5)}$$

where KA equals 5. Once $1/A$ has been selected, the gain crossover frequency and the phase margin become fixed. The ramp error coefficient follows from

$$G_E(s) = \frac{KG_p(s)}{1 + AsKG_p(s)}$$

$$K_1(\text{comp}) = \lim_{s \to 0} sG_E(s) = \frac{K/5}{1 + AK/5}$$

Table 7.4 contains the gain crossover frequency, phase margin, and ramp error coefficient for several values of $1/A$.

As $1/A$ increases, the open-loop zero due to rate feedback moves to the left (higher break frequency). The ramp error coefficient and the gain crossover frequency increase but the phase margin decreases. If $1/A$ is 2, then K is 10, the gain crossover frequency is 1.81, the ramp error coefficient is 1, and the phase margin is 55.5°. In comparison to the uncompensated case, the gain crossover frequency is tripled. The ramp error coefficient is increased by more than 50%. The phase margin is reduced by 8°, as in Table 7.2. That design would appear to provide a good balance for the trade-offs

With $KA = 5$, the best trade-off from Table 7.4 is $K = 10.1$, $A = 2$.

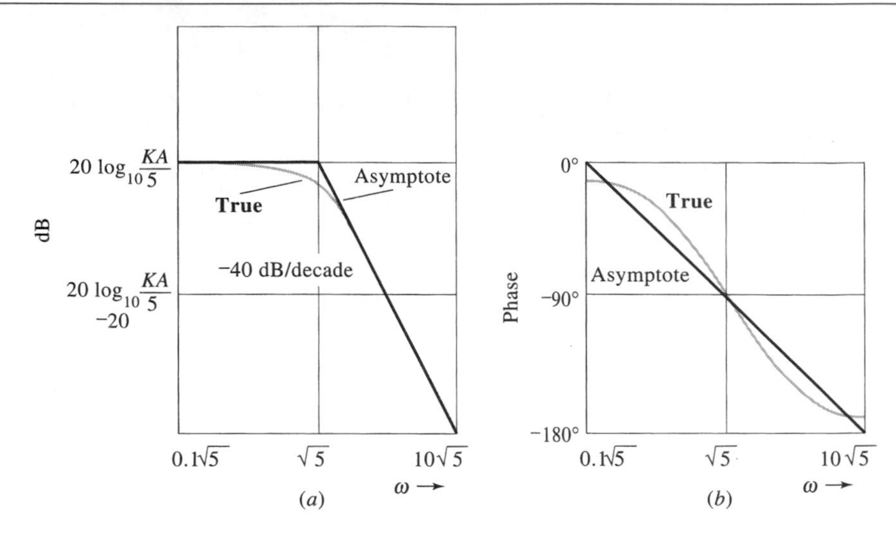

Figure 7.21 Bode plots for the inner loop of the system in Figure 7.20, with $KA/(s^2 + 4s + 5)$. (a) Amplitude versus frequency. (b) Phase versus frequency.

Table 7.4 *Rate Feedback Design*

1/A	K	Gain Crossover Frequency	Phase Margin	Ramp Error Coefficient
1	5	1.34	84.2°	0.5
2	10	1.81	55.5°	1.0
3	15	2.15	38.2°	1.5
4	20	2.41	26.2°	2.0
5	25	2.64	17.3°	2.5
6	30	2.83	10.4°	3.0

involved. Other considerations could favor other values of $1/A$, or the value of KA might have to be adjusted.

The design procedure for rate feedback compensation is to

1. Obtain the Bode plot of $KG_p(s)$ for the uncompensated system in which K takes on a value K_u. Find the gain crossover frequency ω_ϕ, the phase margin, and the error coefficient.
2. Obtain the Bode plot of $KAsG_p(s)$, or just $sG_p(s)$, and select the KA product to provide an acceptable gain crossover frequency and phase margin.
3. Obtain the Bode plot of $KA(s + 1/A)G_p(s)$ for various values of K and $A = KA/K$. Select appropriate K and A for error coefficient, gain crossover frequency, and phase margin.
4. Vary K and A as required to meet design requirements.

❑ **DRILL PROBLEM**

D7.7 For each of the plants from Drill Problem D7.2, carry out design steps 1–3 for a rate feedback compensator. For step 2, select KA as the largest value so that the peak Bode magnitude of $KAG_p(s)$ is less than 1. For step 3, select K so that the error coefficient with the rate feedback compensator is the same as the uncompensated value from Drill Problem D7.2. Show the value of A, K, the gain crossover frequency, phase margin and appropriate error coefficient.

Ans. (a) 0.500, 4.0, 2, 90°, $K_1 = 1$; (b) 0.769, 10.4, 1.49, 81.8°, $K_1 = 0.65$; (c) 0.612, 1.64, 0.46, 156.8°, $K_0 = 1.09$

7.9 Proportional-Integral-Derivative Compensation

A cascade proportional-integral-derivative (PID) compensator has a transmittance of the form

Definition of the PID compensator.

$$G_c(s) = A_p + A_i/s + A_d s$$

The A_p term provides proportional (P) action, where the compensator output is proportional to the compensator input. In a similar manner, the A_i/s provides integrator (I) action and the $A_d s$ term provides derivative (D) action; hence, the compensator is referred to as PID.

Instead of designing values for the three coefficients A_p, A_i, and A_d, it is convenient to rewrite the compensator by combining the PI parts so that

Rewriting as PI plus PD.

$$G_c(s) = \frac{K(s+a)}{s} + KAs \qquad [7.1]$$

The variables are K, a, and A. Now Equation (7.1) shows clearly that the compensator combines PI action (where $A = 0$) with PD action (where $a = 0$). Here the derivative action is in the forward path, while with a rate feedback compensator, the derivative action is in the feedback path. These two PD compensators have the same influence on the Bode plot. Thus, the PID compensator is seen to combine the steady state accuracy improvement of the PI compensator (with some loss of stability) with the improvement in stability of a PD compensator. We can therefore combine the design procedures of the PI compensator with those of the PD compensator in much the same way as we designed the cascade lag–lead compensator by combining the lag design, which improves steady state accuracy, with the lead design, which improves stability.

The compensator has a from the PI compensator and K and A from the PD compensator. The values can then be modified as required as a last design step. The design procedures are as follows.
For the PI compensator with $A = 0$:

1. Obtain the Bode plot of $KG_p(s)$ for the uncompensated system in which K takes on a value K_u. Find the gain crossover frequency ω_ϕ, the phase margin, and the error coefficient.
2. Let $a = 0.1\omega_\phi$.

For the PD compensator with $a = 0$:

3. Obtain the Bode plot of $KAsG_p(s)$, or just $sG_p(s)$, and select the *KA* product to provide an acceptable gain crossover frequency and phase margin.
4. Obtain the Bode plot of $KA(s + 1/A)G_p(s)$ for various values of K and $A = KA/K$. Select appropriate K and A for error coefficient, gain crossover frequency, and phase margin.

For the PID compensator

5. Select K and A from the PD compensator and select a from the PI compensator. Obtain the Bode plot of $G_c'(j\omega)G_p(j\omega)$ where $G_c'(s)$ is the compensator with K factored out. The new gain crossover frequency $\omega_{\phi 1}$ is the frequency where $\left|G_c'(j\omega_{\phi 1})G_p(j\omega_{\phi 1})\right| = 1/K$. Find the phase margin.
6. Vary K, a, and A as required to meet design requirements.

The preceding designs for the example system can now be used to obtain a PID compensator. At the conclusion of each step we have

PID example.

1. $\omega_\phi = 0.576$ rad/s, $PM = 63.7°$, and $K_1 = 0.6$.
2. $a = 0.0576$
3. $KA = 5$
4. $A = 0.5,\ \ K = 10$.

For step 5 of the design process,

$$G_c(s) = \frac{KA[s^2 + (1/A)s + a/A]}{s} = \frac{K(0.5)[s^2 + 2s + 0.115]}{s} = \frac{K(0.5)(s + 1.94)(s + 0.059)}{s}$$

Figure 7.22 shows the Bode plots of $G_c'(s)G_p(s)$ that result from using the PID compensator. The new gain crossover frequency $\omega_{\phi 1}$ is the frequency where $\left|G_c'(j\omega_{\phi 1})G_p(j\omega_{\phi 1})\right| = 1/10 = -20$ dB. Then $\omega_{\phi 1} = 1.79$ rad/s and the phase margin is 54.9°. See Table 7.2 for a comparison of all compensators. For the PID compensator, the gain crossover frequency and the phase margin are each nearly the same as for rate feedback (PD), while the ramp error coefficient is now infinite. In comparison to the uncompensated case, the gain crossover frequency is tripled with a modest loss of phase margin and an infinite ramp error coefficient. Since both stability and steady state accuracy are improved over the uncompensated case, the design goals are met and design terminates with step 5.

❑ DRILL PROBLEM

D7.8 For each of the plants from Design Problem D7.2 carry out design steps 1–5 for a PID compensator. For step 5 show the value of a, A, K, the gain crossover frequency, phase margin, and appropriate error coefficient.

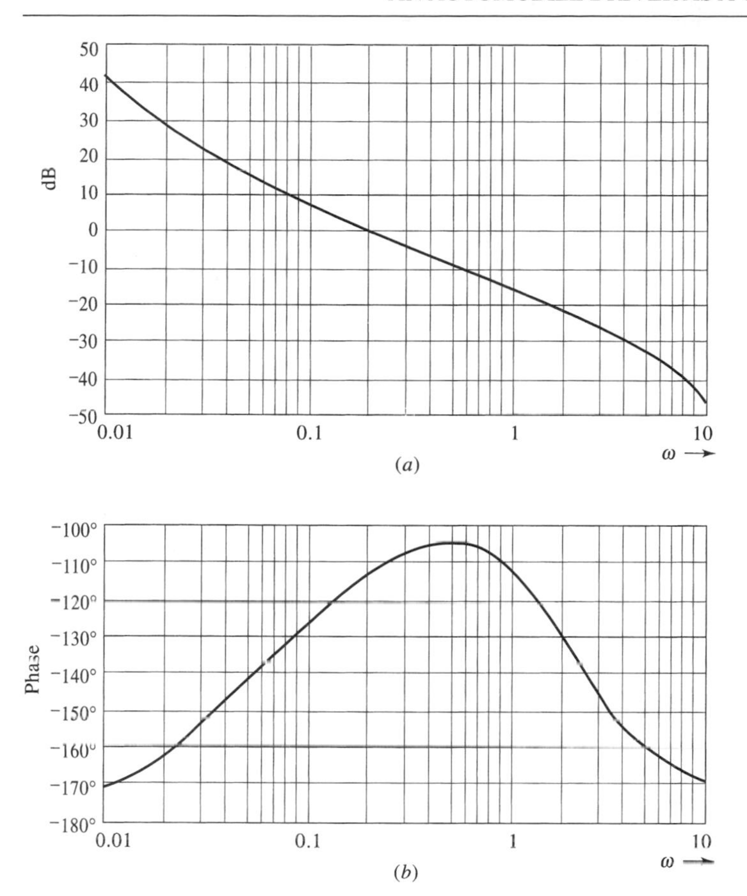

Figure 7.22 Bode plots of $G_c'(j\omega)G_p(j\omega)$ for PID compensation. (a) Amplitude versus frequency. (b) Phase versus frequency.

Ans. (a) 0.091, 0.5, 4.0, 2.0, 88.6°, $K_1 = \infty$; (b) 0.061, 0.769, 10.4, 1.45, 81.1°, $K_1 = \infty$; (c) 0.039, 0.612, 1.64, 0.41, 154.1°, $K_0 = \infty$

7.10 An Automobile Driver as a Compensator

In this chapter we have studied frequency response methods that can be used to select a compensator that is intended to improve the steady state and transient performances of a plant. The compensator is often produced by using active electronic components, such as operational amplifiers, and passive circuits, such as those using resistors and capacitors.

It is interesting that a human being acts as a compensator when performing everyday tasks. A person tracking a moving object such as a football, baseball, or tennis ball so as to catch it or hit it reacts in much the same way as the compensators we have studied. For example, Figure 7.23 can be used to understand how an automobile driver, given by the compensator transmittance $G_c(s)$, compensates for the dynamics of the automobile, given by the plant transmittance $G_p(s)$.

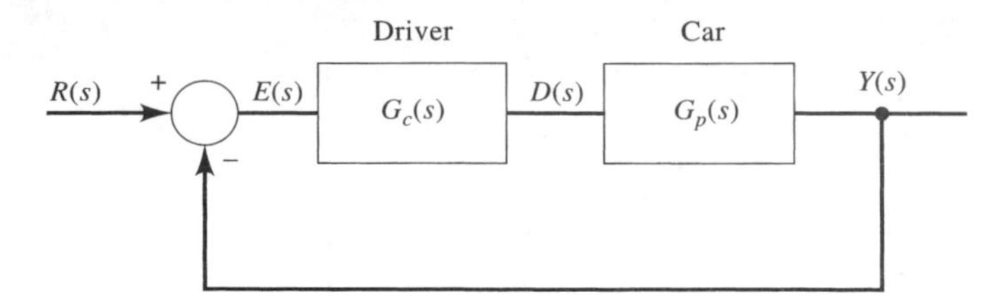

Figure 7.23 Cascade-compensated system.

Suppose the driver is steering an automobile from one lane to another, attempting to stay in the center of the new lane. In that case, $R(s)$ would be a step, the change in position by one lane. A skilled driver (the compensator) should be able to move to the new lane with zero steady state error, regardless of what car (plant) is being driven. The plant dynamics can be evaluated with the input $D(s)$ being the angular rotational deflection of the steering wheel and the output $Y(s)$ being the lateral right-to-left position of the auto with respect to some reference such as the edge of the road. Research into the handling characteristics of a full-sized car traveling at 30 mi/h (see the reference at the end of this chapter) has led to the following transmittance

Definition of plant input and output.

Model for the automobile.

$$G_p(s) = \frac{7.7(s^2 + 4.8s + 36)}{s^2(s + 4.44)(s + 5)} = \frac{Y(s)}{D(s)}$$

Figure 7.24 shows the Bode plots for the plant. The gain crossover frequency is 2.82 rad/s and the phase margin is $-36.1°$. The plant is minimum phase (LHP poles and zeros), and it is therefore unstable due to the negative phase margin. Because of this instability, if the driver simply rotates the wheel until the center of the lane is crossed and then steers in the other direction (simple unity feedback), the auto will weave back and forth across the center of the lane with increasingly larger movements. Obviously, a skilled driver does not weave from one side of the road to the other, so a skilled driver must employ much more than simple positioning unity feedback control.

The uncompensated system is unstable.

Figure 7.25 shows a model for the driver of an automobile where the driver is considered to act as a compensator in the sense of Figure 7.23. The driver will typically act to maintain the gain crossover frequency of the plant. In Figure 7.25, $G_{cPD}(s)$ is a PD compensation effect that the driver uses to improve the stability of the driven system. $G_{cd}(s)$ represents a Padé approximation for a pure delay of 0.15 s. This Padé approximation was discussed in Section 4.6. The remainder of the model contains a neuromuscular block $G_{cnm}(s)$ and compensation elements $G_{c1}(s)$ and $G_{c2}(s)$. Most of the blocks are relatively consistent for a variety of applications, while the parameters K, A, and b are tuned by the driver to fit the handling characteristics of the auto. The transmittance of the compensator is thus

The driver as a compensator.

$$G_c(s) = \frac{G_{cPD}(s)G_{cd}(s)G_{cnm}(s)}{1 + G_{cnm}(s)G_{c1}(s)[1 + G_{c2}(s)]}$$

$$= \frac{K(As + 1)(1 - 0.15s/2)100(s + 0.4)(s + b)}{1 + 0.15s/2}$$

$$[(s^2 + 14s + 100)(s + 0.4)(s + b) + 100s(s + 10 + b)]$$

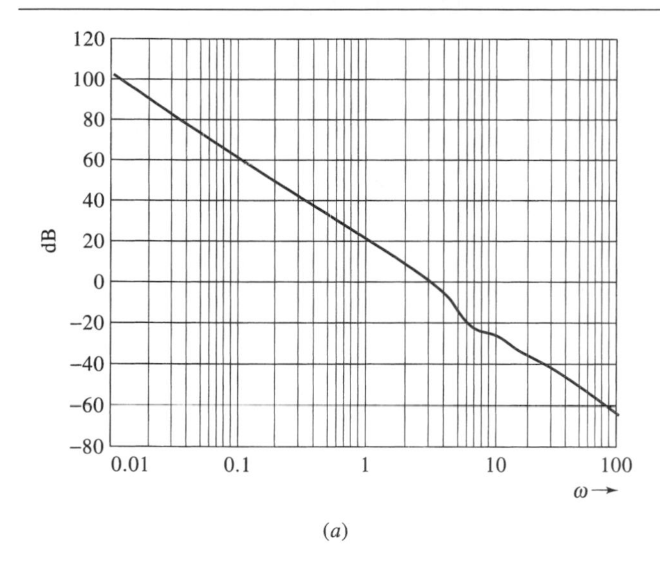

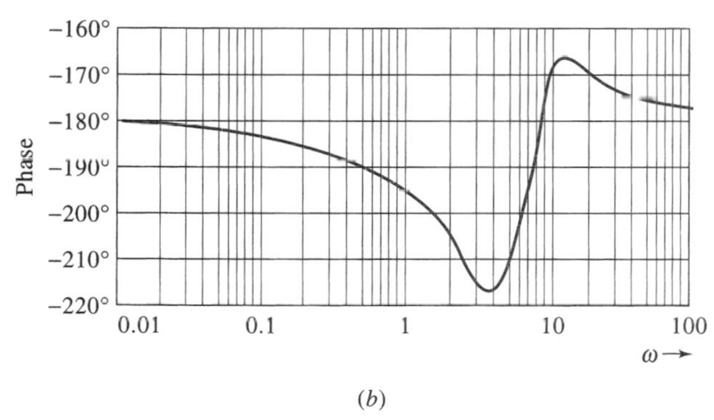

Figure 7.24 Bode plots of $G_p(s)$. (a) Magnitude. (b) Phase.

If the driver operates upon the plant just described, the compensation element $G_{cPD}(s)$ is selected to provide a significantly large positive phase angle at the gain crossover frequency of the plant. The driver's performance is equivalent to selecting

$$A = \frac{10}{\omega_{GC}} = \frac{10}{2.82} = 3.55$$

At the gain crossover frequency, the phase of $G_{cPD}(j\omega_{GC})$ is thus $\tan^{-1}(10)$, which is nearly 90°. The value of b may be selected to cancel a pole of the plant. For example, the driver may operate the auto so that, in effect, $b = 4.44$. Finally, K is adjusted to maintain the same gain crossover frequency for $G_c(s)G_p(s)$ as for $G_p(s)$. If $K = 0.36$, the Bode plot for $G_c(s)$ is shown in Figure 7.26 with a net phase of 63.7° at a frequency of 2.82 rad/s. The Bode plot for $G_c(s)G_p(s)$ in Figure 7.27 has a gain crossover frequency of 2.82 rad/s and a phase margin of 27.6° The system is now stable. At the gain crossover frequency, the Bode plot of $G_c(s)G_p(s)$ is

The driver-compensated system is stable.

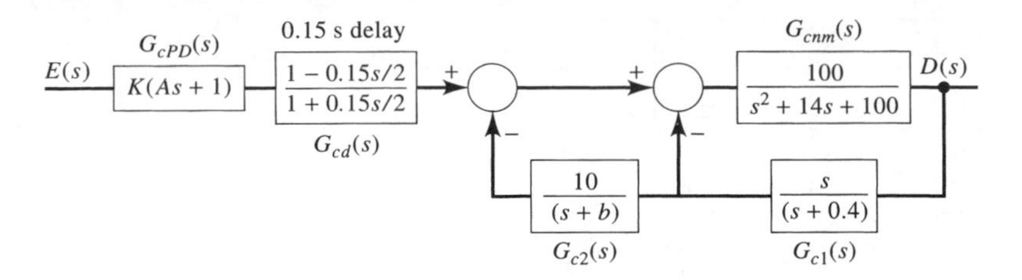

Figure 7.25 Model of an automobile driver as a compensator.

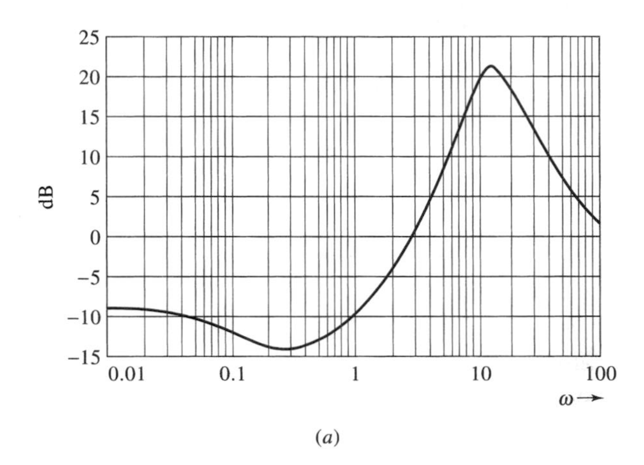

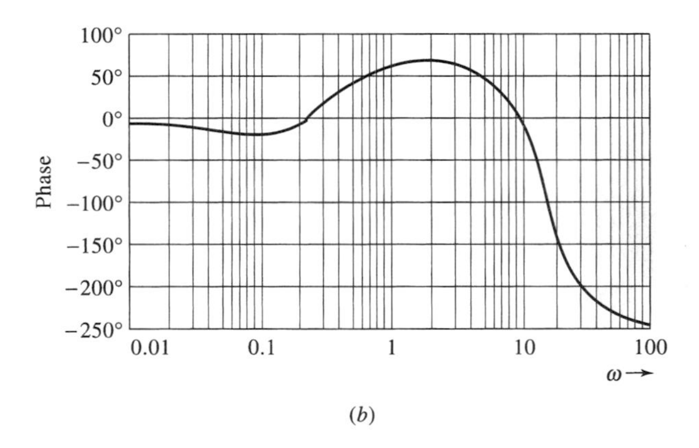

Figure 7.26 Bode plots of $G_c(s)$. (a) Magnitude. (b) Phase.

roughly equivalent to ω_{GC}/s, which many researchers believe is a good model for a skillfully driven automobile.

Without thinking in those terms, the driver of an automobile is actually selecting the parameters of a compensator to generate an appropriate phase margin at the gain crossover frequency of the automobile.

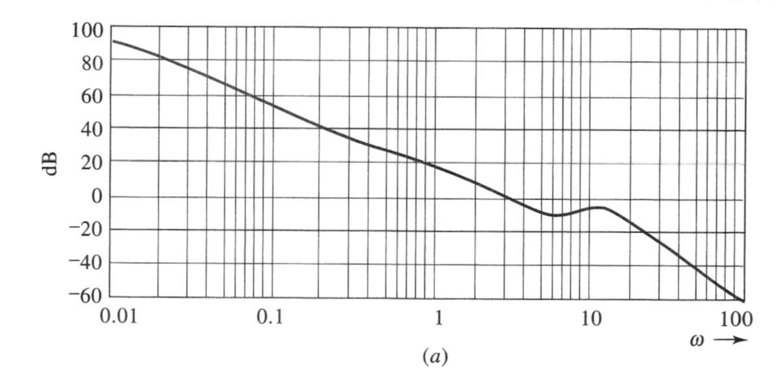

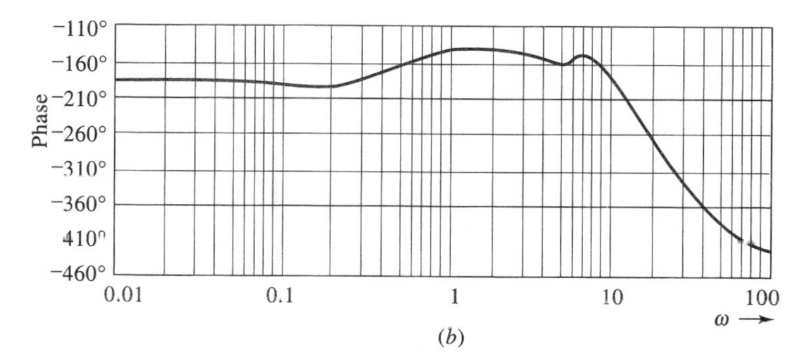

Figure 7.27 Bode plots of $G_c(s)G_p(s)$. (a) Magnitude. (b) Phase.

7.11 SUMMARY

The relationship between frequency domain figures of merit and time domain figures of merit is demonstrated. Gain crossover frequency can be related to the undamped natural frequency, and phase margin can be related to the damping ratio. Since the closed-loop roots of the characteristic polynomial can be used to predict the undamped natural frequency and the damping ratio, relationships exist between the frequency response, the time domain, and the root locus.

A designer may vary the gain of the uncompensated unity feedback system to provide an acceptable error coefficient, which provides reasonable steady state accuracy, and an acceptable gain crossover frequency and phase margin, which are measures of stability. If that uncompensated system cannot meet specifications, compensators may be added to the forward and/or feedback paths.

Two compensators affect low-frequency behavior (which, in turn, can improve steady state accuracy): the cascade PI and cascade lag compensators. Each provides a negative phase angle, which tends to reduce stability. Good design dictates reducing the gain crossover frequency and phase margin as little as possible.

Two compensators affect the frequency response near the gain crossover frequency while having little influence on steady state accuracy: the cascade lead and rate feedback compensators. Here good design is aimed at improving the gain crossover frequency and the phase margin.

Two compensators can improve steady state accuracy as well as the gain crossover frequency and the phase margin: the cascade lag–lead and the PID compensators. The cascade lag–lead compensator is designed by combining procedures for the cascade lag and cascade lead compensators. The PID compensator is designed by combining procedures for the PD (as with feedback rate) and the PI compensators.

REFERENCES

Frequency Response Methods

Bode, H. W., *Network Analysis and Feedback Amplifier Design.* Princeton, NJ: Van Nostrand, 1945.

Nyquist, H., "Regeneration Theory." *Bell Syst. Tech J.* (January 1932): 126–147.

Model of Driver Steering

Hess, R. A., "A Control Theoretic Model of Driver Steering Behavior." *IEEE Control Syst. Mag.* (August 1990): 3–8.

PROBLEMS

1. For each of the following uncompensated systems, find, where applicable, the gain crossover frequency, phase margin, phase crossover frequency, and gain margin.

(a) $$G_p(s) = \frac{s+4}{s+10}$$

$$K = K_u = 10$$

(b) $$G_p(s) = \frac{s+10}{s+4}$$

$$K = K_u = 10$$

(c) $$G_p(s) = \frac{s+10}{(s+2)^2}$$

$$K = K_u = 10$$

$$G_p(s) = \frac{1}{(s+10)(s+4)}$$

$$K = K_u = 58$$

2. Find cascade PI compensators for each of the following uncompensated systems. For each design, show the compensator, the steady state error coefficients, the gain crossover frequency, and the phase margin.

(a) $G_p(s) = \dfrac{6}{(s+4)^2}$

$K_u = 3$

(b) $G_p(s) = \dfrac{1}{(s+4)^3}$

$K_u = 70$

(c) $G_p(s) = \dfrac{s+9}{(s^2+4s+40)}$

$K_u = 15$

(d) $G_p(s) = \dfrac{2s^2+10s+13}{(s+4)(s^2+3s+10)}$

$K_u = 4$

3. Find cascade lag compensators for each of the following uncompensated systems. For each design, show the compensator, the steady state error coefficients, the gain crossover frequency, and the phase margin.

(a) $G_p(s) = \dfrac{10}{s^2+4s+10}$

$K_u = 10, \; d = 10$

(b) $G_p(s) = \dfrac{s+4}{s^2+10}$

$K_u = 10, \; d = 10$

(c) $G_p(s) = \dfrac{2s+20}{s^2+9}$

$K_u = 10, \; d = 10$

(d) $G_p(s) = \dfrac{6}{(s+5)(s^2+4s+10)}$

$K_u = 10, \; d = 10$

4. A plant to be controlled has transmittance

$$G_p(s) = \frac{s+8}{s(s^2+4s+40)}$$

$$K_u = 20$$

Use Bode design methods to select a rate feedback compensator.

5. Use Bode design methods to select a PID compensator for

$$G_p(s) = \frac{1}{(s+4)(s^2+10s+50)}$$

$$K_u = 600$$

6. Design K_u for plants with the following transmittances so the phase margin is 70°.

(a) $$G_p(s) = \frac{3s}{s^2+4}$$

(b) $$G_p(s) = \frac{5}{s^2+4s+5}$$

(c) $$G_p(s) = \frac{1}{s^3+6s^2+9s}$$

(d) $$G_p(s) = \frac{s+10}{s(s^2+6s+13)}$$

7. Design K_u so the phase margin is 60°.

$$G_p(s) = \frac{s+4}{(s+1)(s+2+j)(s+2-j)}$$

8. Design K_u for unity feedback systems with the following plant transmittances. Select K_u so $PM = 50°$.

(a) $$G_p(s) = \frac{9}{(s+3)^2}$$

(b) $$G_p(s) = \frac{s+20}{(s+3)^3}$$

(c) $$G_p(s) = \frac{s+5}{s^2+4s+40}$$

9. Find cascade PI compensators for each of the following uncompensated systems. For each design, show the compensator, the steady state error coefficients, the gain crossover frequency, and the phase margin.

(a) $$G_p(s) = \frac{10}{s^2+7s+10}$$

$$K_u = 2$$

(b) $$G_p(s) = \frac{s+2}{(s^2+4s+3)s}$$

$$K_u = 2$$

(c) $$G_p(s) = \frac{s+4}{s^2+5s+6}$$

$$K_u = 2$$

(d) $$G_p(s) = \frac{6s+30}{s^2+4s+13}$$

$$K_u = 1$$

10. Find cascade lag compensators for each of the following uncompensated systems. For each design, show the compensator, the steady state error coefficients, the gain crossover frequency, and the phase margin.

    (a) $G_p(s) = \dfrac{3s + 30}{s(s^2 + 4s + 13)}$

    $K_u = 4,\ d = 10$

    (b) $G_p(s) = \dfrac{3}{s^2 + 10s}$

    $K_u = 20,\ d = 10$

11. Find cascade lead compensators for each of the uncompensated systems of Problem 10. For each design, show the compensator, the steady state error coefficients, the gain crossover frequency, and the phase margin.

12. Find cascade lag–lead compensators for each of the uncompensated systems of Problem 10. For each design, show the compensator, the steady state error coefficients, the gain crossover frequency, and the phase margin.

13. Find rate feedback compensators for each of the following uncompensated systems. For each design, show the compensator, the steady state error coefficients, the gain crossover frequency, and the phase margin.

    (a) $G_p(s) = \dfrac{1}{s(s + 10)^2}$

    $K_u = 250$

    (b) $G_p(s) = \dfrac{10}{s^2 + 7s + 10}$

    $K_u = 1$

14. Find cascade PID compensators for each of the uncompensated systems of Problem 13. For each design, show the compensator, the steady state error coefficients, the gain crossover frequency, and the phase margin.

15. A plant to be controlled has transmittance

    $$G_p(s) = \frac{5}{s(s^2 + 4s + 40)}$$

    Use Bode design methods to design each of the following

    (a) K_u, so $PM = 70°$.

    (b) Cascade lag compensation, $d = 10$, using (a).

    (c) Rate feedback using (a).

    (d) Cascade lead compensation with $d = 10$, using (a).

16. For the electrohydraulic positioning mechanism of Figure P5.21(b) for Problem 5.21, with $K = 3.2 \times 10^6$, select a rate of feedback compensator. Show the compensator, the steady state error coefficients, the gain crossover frequency, and the phase margin.

17. For the flexible spacecraft of Figure 6.45 with $K = 1.414$, select a cascade lead compensator with $d = 10$. Show the compensator, the steady state error coefficients, the gain crossover frequency, and the phase margin.
18. For the unstable helicopter of Problem 5.22, select A so the system has a phase margin of $+30°$.

CHAPTER

8

State Space Analysis

8.1 Preview

The development of control system analysis and design can be divided into three eras. In the first era, we have classical control theory, which deals with techniques developed before 1950. Classical control embodies such methods as root locus, Bode, Nyquist, and Routh–Hurwitz. These methods have in common the use of transfer functions in the complex frequency (s) domain, emphasis on the use of graphical techniques, the use of feedback, and the use of simplifying assumptions to approximate the time response. Since computers were not available at that time, a great deal of emphasis was placed on developing methods that were amenable to manual computation and graphics. A major limitation of classical control methods was the use of single-input, single-output (SISO) methods. Multivariable (i.e., multiple-input, multiple-output, or MIMO) systems were analyzed and designed one loop at a time. Also, the use of transfer functions and the frequency domain limited one to linear time-invariant systems.

In the second era, we have modern control (which is not so modern any longer), which refers to state-space-based methods developed in the late 1950s and early 1960s. In modern control, system models are directly written in the time domain. Analysis and design are also done in the time domain. It should be noted that before Laplace transforms and transfer functions became popular in the 1920s, engineers were studying systems in the time domain. Therefore, the resurgence of time domain analysis was not unusual, but it was triggered by the development of computers and advances in numerical analysis. Because computers were available, it was no longer necessary to develop analysis and design methods that were strictly manual. An engineer could use computers to numerically solve or simulate large systems that were nonlinear and time-varying. State space methods removed the previously mentioned limitations of classical control. The period of the 1960s was the heyday of modern control.

That period did not last very long, however. For one thing, classical control was already well entrenched, tested, and established. Modern control methods initially enjoyed a great deal of success in academic circles, but they did not perform very well in real applications. Modern control provided a lot of insight into system structure and properties, but it masked other important feedback properties that could be studied and manipulated using classical control. For example, a basic problem in control theory is to design control systems that will work properly when the plant model is uncertain. This issue is tackled in classical control using gain and phase margins; most modern control design methods, however, inherently require a precise model of the plant. During the third era of the 1970s and 1980s, a body of methods finally emerged that tried to provide answers to the plant uncertainty problem. These techniques, commonly known as robust control, are a combination of modern state-space and classical frequency domain techniques.

For a thorough understanding of these new methods, we need to have a basic knowledge of state space analysis. Other advanced techniques in control, such as optimal and adaptive control, are also formulated in state space. Therefore, this chapter presents a brief introduction to state space. Since most of the mathematics associated with the modern approach relies heavily on matrix algebra, Appendix A provides a brief review of matrix algebra. To appreciate the material that follows, matrix algebra must be well understood.

8.2 State Space Representation

Up to this point in this textbook, all control systems have been represented by using transfer functions as functions of the complex frequency variable s. That approach is often called *classical* compared with the "modern" approach in which time domain (differential) equations are used. A fundamental apparatus needed to describe a control system in the time domain is the use of state variables. In general, a system that can be described by an n-order linear differential equation can be defined by creating n state variables. For example, a system whose transfer function has a second-order denominator would require two state variables, because if

$$\frac{Y(s)}{U(s)} = \frac{3(s+1)}{s^2 + 2s + 4}$$

that system can be described by

$$\frac{d^2y}{dt^2} + 2\frac{dy}{dt} + 4y = 3\frac{du}{dt} + 3u$$

which is a second-order linear differential equation in y. By obtaining the transfer function, the system order is determined from the denominator. That order identifies the number of state variables that are needed. The problem, then, is to determine those state variables so the n choices are independent of each other.

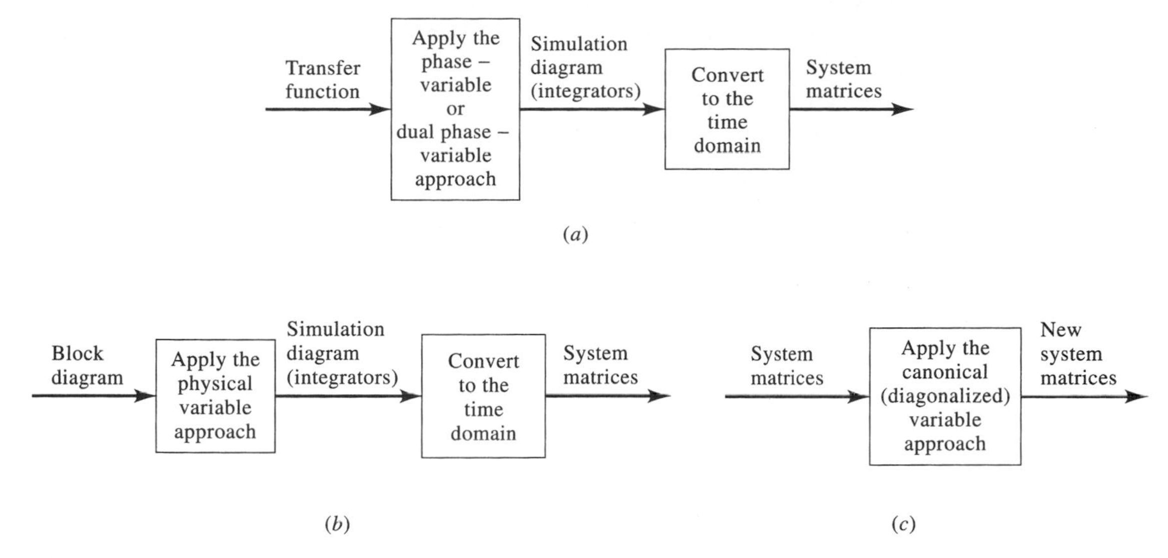

Figure 8.1 Creating a state-variable representation. (a) Phase and dual phase variables. (b) Physical variables. (c) Canonical (diagonalized) variables.

Figure 8.1 shows the three common methods for selecting state variables. If the system transfer function (but not the actual structure of the system) is available, then we see from Figure 8.1(a) that phase or dual phase variables can be used to create a linkage, called a *simulation diagram*, between the classical and the modern. Each integrator output (as we will soon see) is defined as a state variable. If the actual system structure is known in block diagram form (e.g., when a closed-loop system includes compensators), the resulting simulation diagram will follow the system structure, providing integrator outputs that have physical significance. These physical variables might include voltage, current, velocity, and position [e.g., Figure 8.1(b)].

In either case mentioned, the ultimate result is a set of matrices. Finally, if the matrices are available [Figure 8.1(c)], certain operations can be performed on them to create new system matrices of a particularly simple form (providing what are called a *set of canonical state variables*).

Physical variables are most closely related to real-world recognizable quantities, while canonical variables are least related to real-world quantities and are the most theoretical. The phase and dual-phase variables lie between the extremes of practical and theoretical.

8.2.1 Phase-Variable Form

An important problem in control system design is the synthesis of specific transfer functions through the interconnection of simple components, as is needed for many of the controllers (or compensators) of the preceding chapters. Synthesis is important also in the simulation of systems, where system behavior is predicted from a model governed by equivalent equations. Above all, the viewpoint of synthesis leads to fundamental techniques for system description, analysis, and design. These methods are systematic, compact, and suitable for computer analysis. They are also extendable to nonlinear and time-varying systems.

A basic component for synthesis is the integrator, a block or branch having transmittance $1/s$. A block diagram or signal flow graph composed only of constant transmittances and integrators is termed a *simulation diagram*. The order of such a system is simply the number of integrators present. Signal flow graphs are especially convenient for representing simulation diagrams because in many cases, system transfer functions are evident by inspection, making use of Mason's gain rule.

A transfer function that is the ratio of two polynomials in s is termed rational. If the numerator degree is less than or equal to the denominator degree, the transfer function is said to be *proper*. Any proper rational transfer function may be realized with a simulation diagram—that is, using only integration, multiplication by a constant, and summation operations. One very useful realization known as the *phase-variable* form, is described. The development, which is in terms of a specific numerical example for clarity, is applicable to any proper rational transfer function.

For the transfer function

Writing transfer function in Mason's form.

$$T(s) = \frac{-5s^2 + 4s - 12}{s^3 + 6s^2 + s + 3} = \frac{-5/s + 4/s^2 - 12/s^3}{1 + 6/s + 1/s^2 + 3/s^3} = \frac{P_1 + P_2 + P_3}{1 - L_1 - L_2 - L_3}$$

dividing the numerator and denominator by the highest power-of-s term in the denominator places a 1 in the denominator and results in other numerator and denominator terms that are inverse powers of s, representing multiple integrations. In this form the transfer function may be interpreted as a Mason's gain rule expression. The numerator terms

$$\frac{-5}{s} + \frac{4}{s^2} + \frac{-12}{s^3}$$

are each taken to be paths through integrators, and the paths are intermingled as in Figure 8.2(a) to require a minimum number of integrators—in this case, three. The denominator terms

$$\frac{6}{s} + \frac{1}{s^2} + \frac{3}{s^3}$$

are taken to be the negative of the loop gains. By placing each of these loops through the node to which $U(s)$ couples, all loops touch one another, so no product of loop gain terms is involved. All the loops touch each of the paths, so each path cofactor is unity.

In Figure 8.2(b), each integrator output signal has been labeled. These signals are termed the *state variables* of the system. This realization of the example transfer function is then described by the following Laplace-transformed equations:

$$X_1(s) = \frac{1}{s} X_2(s)$$

$$X_2(s) = \frac{1}{s} X_3(s)$$

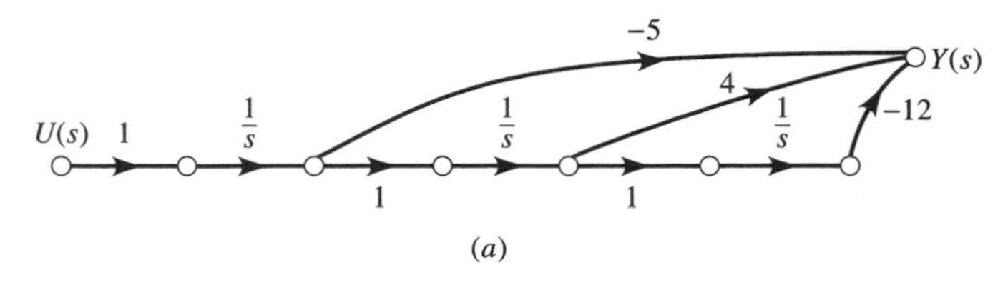

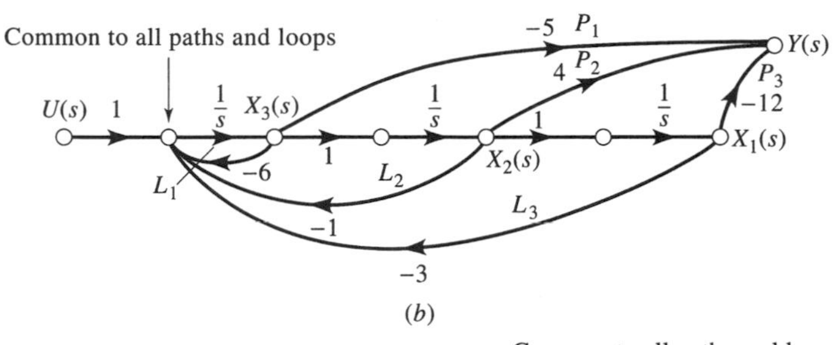

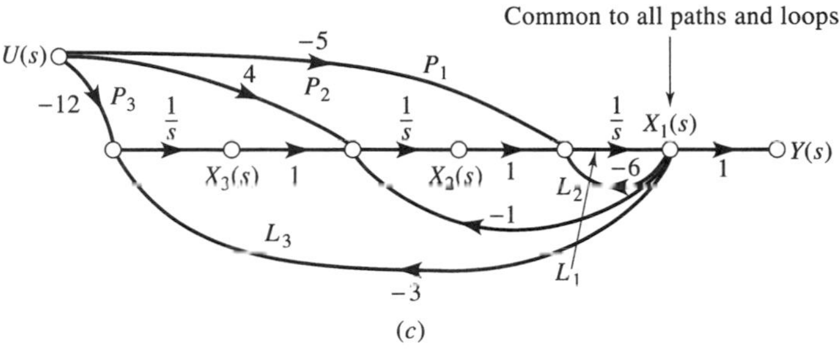

Figure 8.2 Phase-variable realization of a single-input, single-output system. (a) Paths in the simulation diagram. (b) Complete simulation diagram. (c) Realizing a transfer function in the dual phase-variable form.

$$X_3(s) = \frac{1}{s}[-3X_1(s) - X_2(s) - 6X_3(s) + U(s)]$$
$$Y(s) = -12X_1(s) + 4X_2(s) - 5X_3(s)$$

or

$$sX_1(s) = X_2(s)$$
$$sX_2(s) = X_3(s)$$
$$sX_3(s) = -3X_1(s) - X_2(s) - 6X_3(s) + U(s)$$
$$Y(s) = -12X_1 + 4X_2(s) - 5X_3(s)$$

As functions of time, the signals satisfy

$$\frac{dx_1}{dt} = x_2(t)$$
$$\frac{dx_2}{dt} = x_3(t)$$

$$\frac{dx_3}{dt} = -\,3x_1(t) - x_2(t) - 6x_3(t) + u(t)$$

$$y(t) = -\,12x_1(t) + 4x_2(t) - 5x_3(t)$$

which is a set of coupled first-order differential equations.

Denoting time derivatives by $\dot{x}$, and defining the vectors x and $\dot{x}$ by

$$x = \begin{bmatrix} x_1 \\ x_2 \\ x_3 \end{bmatrix} \qquad \dot{x} = \begin{bmatrix} \dot{x}_1 \\ \dot{x}_2 \\ \dot{x}_3 \end{bmatrix}$$

allows us to rewrite the foregoing equations in matrix–vector format as follows:

State equations in vector-matrix form.

$$\begin{bmatrix} \dot{x}_1 \\ \dot{x}_2 \\ \dot{x}_3 \end{bmatrix} = \begin{bmatrix} 0 & 1 & 0 \\ 0 & 0 & 1 \\ -3 & -1 & -6 \end{bmatrix} \begin{bmatrix} x_1 \\ x_2 \\ x_3 \end{bmatrix} + \begin{bmatrix} 0 \\ 0 \\ 1 \end{bmatrix} u$$

$$y = [-12 \quad 4 \quad -5] \begin{bmatrix} x_1 \\ x_2 \\ x_3 \end{bmatrix} + 0 \cdot u$$

or more compactly as follows:

$$\dot{x} = Ax + Bu$$

$$y = Cx + Du$$

where the set of four matrices $\{A,\ B,\ C,\ D\}$ are called a *quadruple*. In general, any proper transfer function can be converted to the general form just shown. In fact, any linear differential equation (possibly with variable coefficients) can be converted to this form (in this case, the coefficient matrices may be functions of time rather than constants).

8.2.2 Dual Phase-Variable Form

Another especially convenient way to realize a transfer function with integrators is to arrange the signal flow graph so that all the paths and all the loops touch an output node. For the previous transfer function

$$\begin{aligned} T(s) &= \frac{-5s^2 + 4s - 12}{s^3 + 6s^2 + s + 3} \\ &= \frac{-5/s + 4/s^2 - 12/s^3}{1 + 6/s + 1/s^2 + 3/s^3} \\ &= \frac{P_1 + P_2 + P_3}{1 - L_1 - L_2 - L_3} \end{aligned}$$

for example, the diagram of Figure 8.2(c) shows this *dual phase-variable* arrangement. The output signal is derived from a single node, while the input signal is coupled to each integrator.

The Laplace transform relations describing this system are, in terms of the indicated state variables,

$$\begin{aligned} sX_1(s) &= -6X_1(s) + X_2(s) - 5U(s) \\ sX_2(s) &= -X_1(s) + X_3(s) + 4U(s) \\ sX_3(s) &= -3X_1 - 12U(s) \\ Y(s) &= X_1(s) \end{aligned}$$

or in the time domain

$$\dot{x} = \begin{bmatrix} -6 & 1 & 0 \\ -1 & 0 & 1 \\ -3 & 0 & 0 \end{bmatrix} x + \begin{bmatrix} -5 \\ 4 \\ -12 \end{bmatrix} u$$

$$y = [1 \quad 0 \quad 0]x + 0 \cdot u$$

Phase-variable and dual phase-variable forms are called *canonical forms*. Our constructions always lead to a special form for the $\{A, \; B, \; C, \; D\}$ quadruplets. For example, in the phase-variable form, assuming that the state variables are defined from right to left, we can observe the following patterns. B is a column vector of zeros except the last element which is 1; C is a row vector that contains the coefficients of the transfer function numerator in ascending powers of s. The D term is a scalar, and is always 0 if the transfer function is *strictly proper* (i.e., the degree of the numerator is strictly less than the degree of the denominator). The A matrix has a special form, which can be partitioned as follows:

$$A = \left[\begin{array}{c:cc} 0 & 1 & 0 \\ 0 & 0 & 1 \\ \hdashline -3 & 1 & 6 \end{array}\right]$$

Note that the last-row elements are the negative of the coefficients of the transfer function denominator (in ascending powers of s, the highest-degree term is always assumed to have a coefficient of 1). The first column is all zeros (except the last element). The remaining submatrix is an identity matrix.

If you look closely at the matrices in the dual phase-variable form, you will see almost the same pattern. If fact, if you make the following substitutions in the preceding two paragraphs, you will get the dual phase-variable form matrices: replace B and C, row with column, first with last, ascending with descending. A substitution that allows us to go from one form to another is called a *dual*. In fact, you may recall such dualities from basic circuit theory. This explains the name, *dual phase-variable form*.

Because of the above-mentioned patterns, we can obtain these forms directly from the transfer functions. If a system transfer function is given by

$$G(s) = \frac{b_1 s^{n-1} + \cdots + b_{n-1}s + b_n}{s^n + a_1 s^{n-1} + \cdots + a_n}$$

The phase-variable form matrices are given by

Phase-variable form.

$$A = \begin{bmatrix} 0 & 1 & 0 & \cdot & \cdots & 0 \\ 0 & 0 & 1 & 0 & \cdots & 0 \\ & & & & & \cdot \\ \vdots & \vdots & & \ddots & & 0 \\ & & & & & 1 \\ -a_n & -a_{n-1} & \cdot & \cdots & & -a_1 \end{bmatrix} \qquad B = \begin{bmatrix} 0 \\ \cdot \\ \cdot \\ 0 \\ 1 \end{bmatrix}$$

$$C = [b_n \; b_{n-1} \cdots b_1] \qquad D = [0]$$

The dual phase-variable form matrices are given by

Dual phase-variable form.

$$A = \begin{bmatrix} -a_1 & 1 & 0 & \cdots & 0 \\ -a_2 & 0 & 1 & 0 \cdots & 0 \\ & & 0 & & \cdot \\ \vdots & \vdots & \vdots & \ddots & 0 \\ & & & & 1 \\ -a_n & 0 & 0 & \cdots & 0 \end{bmatrix} \qquad B = \begin{bmatrix} b_1 \\ b_2 \\ \cdot \\ \cdot \\ b_n \end{bmatrix}$$

$$C = [1 \quad 0 \quad \cdots \quad 0] \quad D = [0]$$

Matrices that have the special structure observed in the A matrices are called *companion* matrices in matrix algebra. An important property of companion matrices is that their characteristic equation can be obtained by inspection. In particular, the characteristic equation for the above A matrices is as follows:

$$\text{Characteristic equation} = s^n + a_1 s^{n-1} + \cdots + a_n$$

8.2.3 Multiple Inputs and Outputs

Additional system outputs may be easily derived from the phase-variable arrangement. For example, the single-input, two-output system of Figure 8.3(a) has the following transfer functions:

$$T_{11}(s) = \left.\frac{Y_1(s)}{U(s)}\right|_{\substack{\text{initial}\\ \text{conditions} = 0}} = \frac{-5/s + 4/s^2 + (-12/s^3)}{1 + 6/s + 1/s^2 + 3/s^3}$$

$$= \frac{-5s^2 + 4s - 12}{s^3 + 6s^2 + s + 3}$$

$$T_{21}(s) = \left.\frac{Y_2(s)}{U(s)}\right|_{\substack{\text{initial}\\ \text{conditions}=0}} = \frac{3/s + 1/s^2 + (-6/s^3)}{1 + 6/s + 1/s^2 + 3/s^3}$$

$$= \frac{3s^2 + s - 6}{s^3 + 6s^2 + s + 3}$$

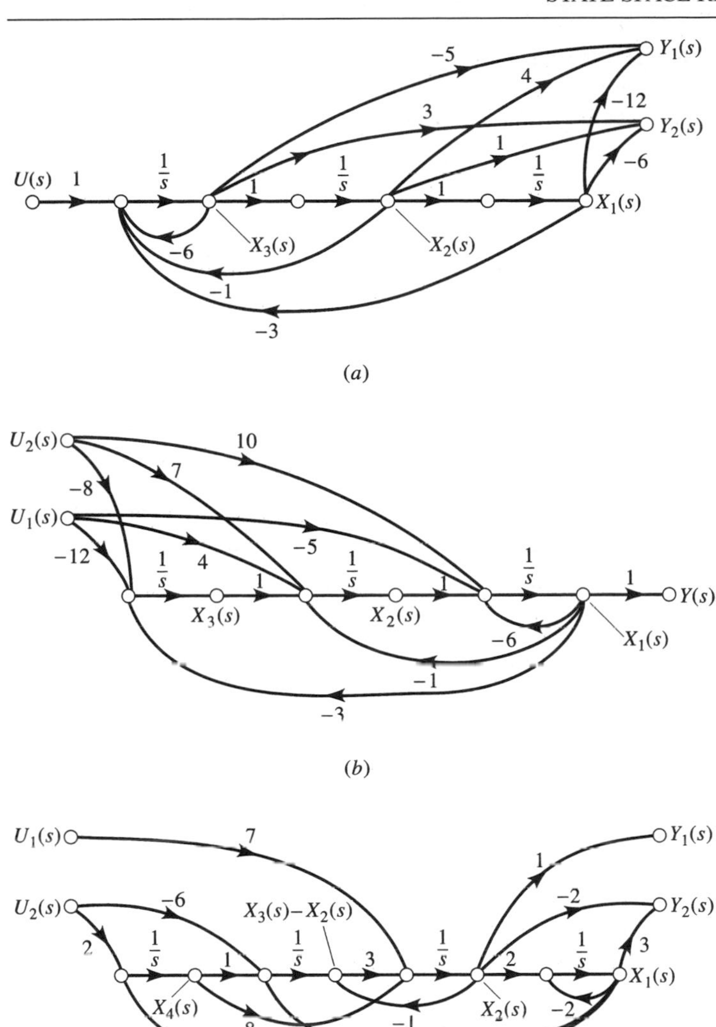

Figure 8.3 Achieving multiple outputs and multiple inputs. (a) Multiple outputs from the phase-variable arrangement. (b) Multiple inputs with the dual phase-variable arrangement. (c) System with both multiple inputs and multiple outputs.

The system transfer function can be written as a vector

$$T(s) = \begin{bmatrix} T_{11}(s) \\ T_{21}(s) \end{bmatrix} = \frac{1}{s^3 + 6s^2 + s + 3} \begin{bmatrix} -5s^2 + 4s - 12 \\ 3s^2 + s - 6 \end{bmatrix}$$

The state space realization for the system in phase-variable form can be obtained from the signal flow graph. Note that both transfer functions share the same denominator and input, hence A and B are the same as the single-input case. The realization is given by

A two-output system.

$$\dot{x} = \begin{bmatrix} 0 & 1 & 0 \\ 0 & 0 & 1 \\ -3 & -1 & -6 \end{bmatrix} x + \begin{bmatrix} 0 \\ 0 \\ 1 \end{bmatrix} u$$

$$y = \begin{bmatrix} -12 & 4 & -5 \\ -6 & 1 & 3 \end{bmatrix} x + \begin{bmatrix} 0 \\ 0 \end{bmatrix} u$$

Because there are two outputs, y is a two-dimensional vector; hence C and D must have two rows to make the dimensions match. Also note that if the transfer function is available, the phase-variable realization could have been written by inspection.

Additional inputs are easily added to the dual phase-variable arrangement. The example two-input, single-output system of Figure 8.3(b) has the following transfer functions:

$$T_{11}(s) = \left.\frac{Y(s)}{U_1(s)}\right|_{\substack{\text{initial} \\ \text{conditions} \\ \text{and } R_2=0}} = \frac{-5/s + 4/s^2 - 12/s^3}{1 + 6/s + 1/s^2 + 3/s^3}$$

$$= \frac{-5s^2 + 4s - 12}{s^3 + 6s^2 + s + 3}$$

$$T_{12}(s) = \left.\frac{Y(s)}{U_2(s)}\right|_{\substack{\text{initial} \\ \text{conditions} \\ \text{and } R_1=0}} = \frac{10/s + 7/s^2 - 8/s^3}{1 + 6/s + 1/s^2 + 3/s^3}$$

$$= \frac{10s^2 + 7s - 8}{s^3 + 6s^2 + s + 3}$$

You can use the same kind of reasoning as in the preceding case to verify that the dual phase-variable realization for the two-input case is given by

A two-input system.

$$\dot{x} = \begin{bmatrix} -6 & 1 & 0 \\ -1 & 0 & 1 \\ -3 & 0 & 0 \end{bmatrix} x + \begin{bmatrix} -5 & 10 \\ 4 & 7 \\ -12 & -8 \end{bmatrix} u$$

$$y = [1 \quad 0 \quad 0]x + [0 \quad 0]u$$

The system described by the simulation diagram of Figure 8.3(c) is in neither phase-variable nor dual phase-variable form. Its two inputs and two outputs are governed by the following Laplace-transformed equations:

$$sX_1(s) = -2X_1(s) + 2X_2(s)$$

$$sX_2(s) = -3X_2(s) + 3X_3(s) + 8X_4(s) + 7U_1(s)$$

$$sX_3(s) = -X_1(s) + X_4(s) - 6U_2(s)$$

$$sX_4(s) = -\tfrac{1}{3}X_1(s) + 2U_2(s)$$

$$Y_1(s) = X_2(s)$$

$$Y_2(s) = 3X_1(s) - 2X_2(s)$$

We can write the state space realization directly from these equations:

$$\dot{x} = \begin{bmatrix} -2 & 0 & 0 & 0 \\ 0 & -3 & 3 & 8 \\ -1 & 0 & 0 & 1 \\ -\frac{1}{3} & 0 & 0 & 0 \end{bmatrix} x + \begin{bmatrix} 0 & 0 \\ 7 & 0 \\ 0 & -6 \\ 0 & 2 \end{bmatrix} u$$

A two-input two-output system.

$$y = \begin{bmatrix} 0 & 1 & 0 & 0 \\ 3 & -2 & 0 & 0 \end{bmatrix} x + \begin{bmatrix} 0 & 0 \\ 0 & 0 \end{bmatrix} u$$

Observe that matrices A, B, C, and D are not in any particular form.

❑ DRILL PROBLEM

D8.1 Draw simulation diagrams in either the phase-variable or the dual phase-variable form for systems with the following transfer functions:

(a) $$T(s) = \frac{-4s + 3}{s^2 + 6s + 2}$$

(b) $$T(s) = \frac{-s^2 + 5s + 9}{3s^3 + 2s^2 + 4s + 1}$$

(c) $$T_{11}(s) = \frac{0.4s^2 + 1.4s + 0.8}{s^3 + 0.3s^2 + 1.7s + 0.2}$$

$$T_{12}(s) = \frac{-0.5s^2 + 0.7s - 1.9}{s^3 + 0.3s^2 + 1.7s + 0.2}$$

(d) $$T_{11}(s) = \frac{4s^2 - 1}{s^3 + 6s^2 + 2s + 5}$$

$$T_{21}(s) = \frac{3s + 6}{s^3 + 6s^2 + 2s + 5}$$

Phase-variable form is particularly convenient for the synthesis of single- and multiple-output systems, while in dual phase-variable form, single- and multiple-input systems are easily arranged. There are a whole spectrum of other ways of connecting integrators to achieve systems with desired transfer functions, including systems with both multiple inputs and multiple outputs. Moreover, the representation of systems in terms of integrators is useful not only for transfer function synthesis, but for the description of systems of all kinds, particularly those that are very complicated, for which a standard, compact notation is especially helpful.

A general state variable description of an nth-order system involves n integrators, the outputs of which are the state variables. The input of each of the integrators are driven with a linear combination of the state signals and the inputs:

$$
\begin{aligned}
sX_1(s) &= a_{11}X_1(s) + a_{12}X_2(s) + \cdots + a_{1n}X_n(s) + b_{11}u_1(s) + \cdots + b_{1i}u_i(s) \\
sX_2(s) &= a_{21}X_1(s) + a_{22}X_2(s) + \cdots + a_{2n}X_n(s) + b_{21}u_1(s) + \cdots + b_{2i}u_i(s) \\
&\vdots \\
sX_n(s) &= a_{n1}X_1(s) + a_{n2}X_2(s) + \cdots + a_{nn}X_n(s) + b_{n1}u_1(s) + \cdots + b_{ni}u_i(s)
\end{aligned}
\tag{8.1}
$$

In the time domain, these are a set of n first-order differential equations in the n state variables and the inputs:

$$
\begin{aligned}
\frac{dx_1}{dt} &= a_{11}x_1 + a_{12}x_2 + \cdots + a_{1n}x_n + b_{11}u_1 + \cdots + b_{1i}u_i \\
\frac{dx_2}{dt} &= a_{21}x_1 + a_{22}x_2 + \cdots + a_{2n}x_n + b_{21}u_1 + \cdots + b_{2i}u_i \\
&\vdots \\
\frac{dx_n}{dt} &= a_{n1}x_1 + a_{n2}x_2 + \cdots + a_{nn}x_n + b_{n1}u_1 + \cdots + b_{ni}u_i
\end{aligned}
$$

These state equations are compactly written in matrix notation as follows:

$$
\frac{d}{dt}\begin{bmatrix} x_1 \\ x_2 \\ \vdots \\ x_n \end{bmatrix} = \begin{bmatrix} \dot{x}_1 \\ \dot{x}_2 \\ \vdots \\ \dot{x}_n \end{bmatrix} = \begin{bmatrix} a_{11} & a_{12} & \cdots & a_{1n} \\ a_{21} & a_{22} & \cdots & a_{2n} \\ \vdots & & & \\ a_{n1} & a_{n2} & \cdots & a_{nn} \end{bmatrix} \begin{bmatrix} x_1 \\ x_2 \\ \vdots \\ x_n \end{bmatrix} + \begin{bmatrix} b_{11} & \cdots & b_{1i} \\ b_{21} & \cdots & b_{2i} \\ \vdots & & \\ b_{n1} & \cdots & b_{ni} \end{bmatrix} \begin{bmatrix} u_1 \\ u_2 \\ \vdots \\ u_i \end{bmatrix}
$$

or

$$
\frac{dx}{dt} = \dot{x} = Ax + Bu
$$

The column matrix of state variables

$$
x = \begin{bmatrix} x_1 \\ x_2 \\ \vdots \\ x_n \end{bmatrix}
$$

is called the *state vector.* The inputs are arranged to form the *input vector,*

$$u = \begin{bmatrix} u_1 \\ \vdots \\ u_i \end{bmatrix}$$

The system outputs are similarly arranged in an *output vector,*

$$y = \begin{bmatrix} y_1 \\ \vdots \\ y_m \end{bmatrix}$$

related linearly to the state variables through the output equations:

$$\begin{cases} y_1 = c_{11}x_1 + c_{12}x_2 + \cdots + c_{1n}x_n \\ \vdots \\ y_m = c_{m1}x_1 + c_{m2}x_2 + \cdots + c_{mn}x_n \end{cases}$$

or

$$\begin{bmatrix} y_1 \\ \vdots \\ y_m \end{bmatrix} = \begin{bmatrix} c_{11} & c_{12} & \cdots & c_{1n} \\ \vdots & & & \\ c_{m1} & c_{m2} & \cdots & c_{mn} \end{bmatrix} \begin{bmatrix} x_1 \\ \vdots \\ x_n \end{bmatrix}$$

or

$$y = Cx$$

The state equations describe how the system state vector evolves in time. One may imagine the tip of the vector tracing a curve, the *state trajectory*, in an n-dimensional space. The output equations describe how the output signals are related to the state.

For systems described by linear constant-coefficient integrodifferential equations, the state-variable arrangement is simply a standard form for the equations describing a system. Instead of dealing with a mixed collection of simultaneous system equations—some of first order, some of second order, some involving running integrals, and so on—additional manipulation of the original equations is done to place them in the standard form. The advantages of a standard form are that systematic methods may be easily brought to bear upon very involved problems and that a degree of unification results.

8.2.4 Physical State Variables

State space equations are sometimes written directly from first principles. For example, consider a standard series *RLC* circuit driven by a voltage source. From Kirchhoff's voltage law (KVL), we have

$$V_s = Ri(t) + L\frac{di}{dt} + \frac{1}{C}\int i(\tau)d\tau$$

There are many ways to convert this equation to state space form. It is possible, however, to obtain the state equations directly if we know how to choose the states. Voltage across capacitors and current through inductors are frequently chosen as (physical) state variables in circuit analysis. We therefore let

$$V_C = x_1 = \frac{1}{C}\int i(\tau)d\tau \qquad \text{and} \qquad I_L = x_2$$

Because the elements are in series, the loop current is equal to the inductor current. Therefore, from the foregoing definitions we have

$$\dot{x}_1 = \frac{1}{C}x_2$$

and using KVL, we also have

$$V_s = Rx_2 + L\dot{x}_2 + x_1$$

Rearranging, we get

$$\dot{x}_2 = -\frac{1}{L}x_1 - \frac{R}{L}x_2 + \frac{V_s}{L}$$

The output equation depends on what we desire to control, or what we can measure, or both. In fact, in the most general setting, the controlled variables and the measured variables might be different. Suppose we can measure only the loop current (measured output, y), but we want to control the capacitor voltage (controlled output, z); the complete equations in vector–matrix form are (u stands for V_s)

$$\dot{x} = \begin{bmatrix} 0 & \frac{1}{C} \\ \frac{-1}{L} & \frac{-R}{L} \end{bmatrix} x + \begin{bmatrix} 0 \\ \frac{1}{L} \end{bmatrix} u$$

$$y = [0 \quad 1]x$$

$$z = [1 \quad 0]x$$

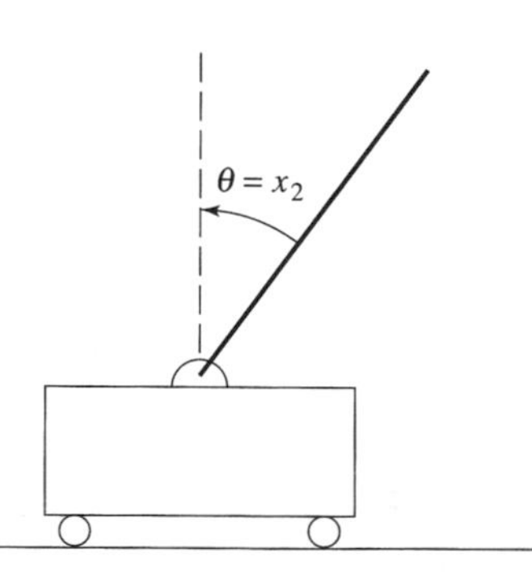

As another example, consider the well-known problem of balancing an inverted pendulum on a moving cart shown in the figure. The linearized equations of motion for this mechanical system are given by

$$(J + ml^2)\ddot{\theta} + ml\ddot{x} - mgl\theta = 0$$

$$(M + m)\ddot{x} + ml\ddot{\theta} = u$$

where $\theta =$ pendulum angle and $x =$ cart position. The system parameters are as follows: cart mass $M = 26/3$, pendulum mass $m = 4/3$, $l = 3/4 =$ (pendulum length)/2, gravity constant $g = 9.8$ m/s, and J, the pendulum moment of inertia about its center of gravity (assumed to be in the middle of the pendulum$= ml^2/3$. From these hypothetical values, we get

$$\ddot{\theta} + \ddot{x} - g\theta = 0 \quad \textbf{[8.2]}$$

$$10\ddot{x} + \ddot{\theta} = u \quad \textbf{[8.3]}$$

These equations are a set of two coupled second-order differential equations. How we define the state variables for this system depends on the control objectives. That is, the choice of the state variables and the control task are interrelated. We consider the following three situations.

Case I: The objective is to balance the pendulum (i.e., we do not care about the cart position or velocity). In this case, the pendulum angle and its angular velocity are chosen as the state variables. Solving for $\ddot{x}$ from Equation (8.2) and substituting in Equation (8.3), we get

$$\ddot{\theta} = \frac{10}{9}g\theta - \frac{1}{9}u$$

Letting $x_1 = \theta$ and $x_2 = \dot{\theta}$, and assuming we can measure the pendulum angle, we get

$$A = \begin{bmatrix} 0 & 1 \\ \frac{10g}{9} & 0 \end{bmatrix}, \quad B = \begin{bmatrix} 0 \\ \frac{-1}{9} \end{bmatrix} \quad \text{and} \quad C = [1 \quad 0]$$

Case II: The objective is to balance the pendulum and stop the cart. The position of the cart is not important here as long as its velocity is brought down to zero. In this case, we add another state variable for the cart velocity. Writing and solving Equation (8.3) in terms of cart velocity v, we get

$$\dot{v} = -\frac{g}{9}\theta + \frac{1}{9}u$$

Using v as the third state variable, we get the following third-order system equation

$$A = \begin{bmatrix} 0 & 1 & 0 \\ \frac{10g}{9} & 0 & 0 \\ \frac{-g}{9} & 0 & 0 \end{bmatrix}, \quad B = \begin{bmatrix} 0 \\ \frac{-1}{9} \\ \frac{1}{9} \end{bmatrix} \quad \text{and} \quad C = [1 \quad 0 \quad 0]$$

Case III: The objective is to balance the pendulum and position the cart (e.g., return it to its original position). In this case, we add the cart position as the fourth state variable.

$$A = \begin{bmatrix} 0 & 1 & 0 & 0 \\ \dfrac{10g}{9} & 0 & 0 & 0 \\ \dfrac{-g}{9} & 0 & 0 & 0 \\ 0 & 0 & 1 & 0 \end{bmatrix}, \qquad B = \begin{bmatrix} 0 \\ \dfrac{-1}{9} \\ \dfrac{1}{9} \\ 0 \end{bmatrix} \qquad \text{and} \qquad C = [1 \quad 0 \quad 0 \quad 0]$$

Note that as the task complexity increases, so will the order of the model used. Whether it is possible to accomplish the preceding tasks by measuring only the pendulum angle is another question. This is related to the notions of controllability and observability discussed later in the chapter.

❑ DRILL PROBLEMS

D8.2 Draw simulation diagrams for the given state space equations.

(a)
$$\begin{bmatrix} \dot{x} \\ \dot{x}_2 \end{bmatrix} = \begin{bmatrix} -2 & 5 \\ -4 & -3 \end{bmatrix} \begin{bmatrix} x_1 \\ x_2 \end{bmatrix} + \begin{bmatrix} -7 \\ 6 \end{bmatrix} u$$

$$y = [1 \quad 0] \begin{bmatrix} x_1 \\ x_2 \end{bmatrix}$$

(b)
$$\begin{bmatrix} \dot{x}_1 \\ \dot{x}_2 \end{bmatrix} = \begin{bmatrix} -2 & 1 \\ -3 & -1 \end{bmatrix} \begin{bmatrix} x_1 \\ x_2 \end{bmatrix} + \begin{bmatrix} 0 \\ 1 \end{bmatrix} u$$

$$y = [1 \quad 1] \begin{bmatrix} x_1 \\ x_2 \end{bmatrix}$$

(c)
$$\begin{bmatrix} \dot{x}_1 \\ \dot{x}_2 \\ \dot{x}_3 \end{bmatrix} = \begin{bmatrix} -2 & 0 & 0 \\ 0 & -3 & 0 \\ 0 & 0 & -4 \end{bmatrix} \begin{bmatrix} x_1 \\ x_2 \\ x_2 \end{bmatrix} + \begin{bmatrix} 1 \\ -6 \\ 10 \end{bmatrix} u$$

$$\begin{bmatrix} y_1 \\ y_2 \end{bmatrix} = \begin{bmatrix} 1 & 1 & 0 \\ 0 & 1 & 1 \end{bmatrix} \begin{bmatrix} x_1 \\ x_2 \\ x_3 \end{bmatrix}$$

D8.3 Draw simulation diagrams to represent the following systems:

(a)
$$\begin{bmatrix} \dot{x}_1 \\ \dot{x}_2 \end{bmatrix} = \begin{bmatrix} -2 & -3 \\ 4 & 1 \end{bmatrix} \begin{bmatrix} x_1 \\ x_2 \end{bmatrix} + \begin{bmatrix} 5 \\ -6 \end{bmatrix} u$$

$$y = [7 \quad -8] \begin{bmatrix} x_1 \\ x_2 \end{bmatrix}$$

(b)
$$\begin{bmatrix} \dot{x}_1 \\ \dot{x}_2 \\ \dot{x}_3 \end{bmatrix} = \begin{bmatrix} 0 & 10 & 3 \\ 5 & 8 & 0 \\ -2 & -7 & -3 \end{bmatrix} \begin{bmatrix} x_1 \\ x_2 \\ x_3 \end{bmatrix} + \begin{bmatrix} 0 & 1 \\ 2 & 0 \\ -1 & 3 \end{bmatrix} \begin{bmatrix} u_1 \\ u_2 \end{bmatrix}$$

$$y = [0 \quad 4 \quad -3] \begin{bmatrix} x_1 \\ x_2 \\ x_3 \end{bmatrix}$$

(c)
$$\begin{bmatrix} \dot{x}_1 \\ \dot{x}_2 \\ \dot{x}_3 \end{bmatrix} = \begin{bmatrix} -2 & 0 & -6 \\ 3 & 5 & 0 \\ -4 & 0 & 7 \end{bmatrix} \begin{bmatrix} x_1 \\ x_2 \\ x_3 \end{bmatrix} + \begin{bmatrix} 8 \\ -2 \\ 0 \end{bmatrix} u$$

$$\begin{bmatrix} y_1 \\ y_2 \end{bmatrix} = \begin{bmatrix} 0 & -1 & 1 \\ -1 & 1 & 0 \end{bmatrix} \begin{bmatrix} x_1 \\ x_2 \\ x_3 \end{bmatrix}$$

8.2.5 Transfer Functions

The transfer functions of a system represented in state variable form may be found by Laplace-transforming the state equations with zero initial conditions. In general, these are Equations (8.1). Collecting the terms involving $X(s)$, there results

$$\begin{bmatrix} (s-a_{11}) & -a_{12} & \cdots & -a_{1n} \\ -a_{21} & (s-a_{22}) & \cdots & -a_{2n} \\ \vdots & & & \\ -a_{n1} & -a_{n2} & \cdots & (s-a_{nn}) \end{bmatrix} \begin{bmatrix} X_1(s) \\ X_2(s) \\ \vdots \\ X_n(s) \end{bmatrix}$$

$$= \begin{bmatrix} b_{11} & b_{12} & \cdots & b_{1i} \\ b_{21} & b_{22} & \cdots & b_{2i} \\ \vdots & & & \\ b_{n1} & b_{n2} & \cdots & b_{ni} \end{bmatrix} \begin{bmatrix} U_1(s) \\ U_2(s) \\ \vdots \\ U_i(s) \end{bmatrix}$$

or

$$[sI \quad -A]X(s) = BU(s)$$

where I is the $n \times n$ identity matrix

$$I = \begin{bmatrix} 1 & 0 & \cdots & 0 & 0 \\ 0 & 1 & \cdots & 0 & 0 \\ \vdots & & & & \\ 0 & 0 & \cdots & 0 & 1 \end{bmatrix}$$

Solving for the Laplace transform of the state vector, we have

$$X(s) = [sI - A]^{-1} BU(s)$$

The output and state vectors are related by

$$\begin{bmatrix} Y_1(s) \\ Y_2(s) \\ \vdots \\ Y_m(s) \end{bmatrix} = \begin{bmatrix} c_{11} & c_{12} & \cdots & c_{1n} \\ c_{21} & c_{22} & \cdots & c_{2n} \\ \vdots & & & \\ c_{m1} & c_{m2} & \cdots & c_{mn} \end{bmatrix} \begin{bmatrix} X_1(s) \\ X_2(s) \\ \vdots \\ X_n(s) \end{bmatrix}$$

or

$$Y(s) = CX(s) = \left\{C[sI - A]^{-1}B\right\} U(s)$$

The $m \times i$ matrix in braces { } consists of the input–output transfer functions of the system, arranged as a matrix:

$$C[sI - A]^{-1}B = \begin{bmatrix} T_{11}(s) & T_{12}(s) & \cdots & T_{1i}(s) \\ T_{21}(s) & T_{22}(s) & \cdots & T_{2i}(s) \\ \vdots & & & \\ T_{m1}(s) & T_{m2}(s) & \cdots & T_{mi}(s) \end{bmatrix}$$

For example, a single-input, single-output system with state equations

$$\begin{bmatrix} \dot{x}_1 \\ \dot{x}_2 \end{bmatrix} = \begin{bmatrix} -3 & 1 \\ -2 & 0 \end{bmatrix} \begin{bmatrix} x_1 \\ x_2 \end{bmatrix} + \begin{bmatrix} 4 \\ -5 \end{bmatrix} u$$

$$y = [1 \quad -1] \begin{bmatrix} x_1 \\ x_2 \end{bmatrix}$$

has transfer function given by

$$T(s) = [1 \quad -1] \begin{bmatrix} s+3 & -1 \\ 2 & s \end{bmatrix}^{-1} \begin{bmatrix} 4 \\ -5 \end{bmatrix}$$

$$= [1 \quad -1] \frac{\begin{bmatrix} s & 1 \\ -2 & s+3 \end{bmatrix} \begin{bmatrix} 4 \\ -5 \end{bmatrix}}{s^2 + 3s + 2}$$

$$= \frac{[1 \quad -1] \begin{bmatrix} (4s - 5) \\ (-5s - 23) \end{bmatrix}}{s^2 + 3s + 2}$$

$$= \frac{9s + 18}{s^2 + 3s + 2}$$

The two-input, two-output system

$$\begin{bmatrix} \dot{x}_1 \\ \dot{x}_2 \end{bmatrix} = \begin{bmatrix} -3 & 1 \\ -2 & 0 \end{bmatrix} \begin{bmatrix} x_1 \\ x_2 \end{bmatrix} + \begin{bmatrix} 4 & 6 \\ -5 & 0 \end{bmatrix} \begin{bmatrix} u_1 \\ u_2 \end{bmatrix}$$

$$\begin{bmatrix} y_1 \\ y_2 \end{bmatrix} = \begin{bmatrix} 1 & -1 \\ 8 & 1 \end{bmatrix} \begin{bmatrix} x_1 \\ x_2 \end{bmatrix}$$

is described by the transfer function matrix given by

$$T(s) = \begin{bmatrix} 1 & -1 \\ 8 & 1 \end{bmatrix} \begin{bmatrix} s+3 & -1 \\ 2 & s \end{bmatrix}^{-1} \begin{bmatrix} 4 & 6 \\ -5 & 0 \end{bmatrix}$$

$$= \frac{\begin{bmatrix} 1 & -1 \\ 8 & 1 \end{bmatrix} \begin{bmatrix} s & 1 \\ -2 & s+3 \end{bmatrix} \begin{bmatrix} 4 & 6 \\ -5 & 0 \end{bmatrix}}{s^2+3s+2}$$

$$= \frac{\begin{bmatrix} 1 & -1 \\ 8 & 1 \end{bmatrix} \begin{bmatrix} (4s-5) & 6s \\ (-5s-23) & -12 \end{bmatrix}}{s^2+3s+2}$$

$$= \begin{bmatrix} \dfrac{9s+18}{s^2+3s+2} & \dfrac{6s+12}{s^2+3s+2} \\ \dfrac{27s-63}{s^2+3s+2} & \dfrac{48s-12}{s^2+3s+2} \end{bmatrix} = \begin{bmatrix} T_{11}(s) & T_{12}(s) \\ T_{21}(s) & T_{22}(s) \end{bmatrix}$$

Transfer function matrix.

where

$$T_{11}(s) = \frac{9s+18}{s^2+3s+2} = \left.\frac{Y_1(s)}{U_1(s)}\right|_{\substack{\text{initial} \\ \text{conditions} \\ \text{and } U_2=0}}$$

$$T_{12}(s) = \frac{6s+12}{s^2+3s+2} = \left.\frac{Y_1(s)}{U_2(s)}\right|_{\substack{\text{initial} \\ \text{conditions} \\ \text{and } U_1=0}}$$

$$T_{21}(s) = \frac{27s-63}{s^2+3s+2} = \left.\frac{Y_2(s)}{U_1(s)}\right|_{\substack{\text{initial} \\ \text{conditions} \\ \text{and } U_2=0}}$$

$$T_{22}(s) = \frac{48s-12}{s^2+3s+2} = \left.\frac{Y_2(s)}{U_2(s)}\right|_{\substack{\text{initial} \\ \text{conditions} \\ \text{and } U_1=0}}$$

All the transfer functions of a system share the denominator polynomial

$$|sI - A|$$

where A is the state coupling matrix for the system, since

$$[sI - A]^{-1} = \frac{\text{adjoint}[sI - A]}{|sI - A|}$$

The nth-degree polynomial

$$|sI - A| = 0$$

is termed the characteristic polynomial of an $n \times n$ matrix A, and the n roots of that polynomial are the eigenvalues, or characteristic roots, of the matrix. A system is stable if and only if the eigenvalues of the state coupling matrix are all in the left half of the complex plane.

❑ DRILL PROBLEM

D8.4 Find the transfer function matrices of the following systems:

(a)

$$\begin{bmatrix} \dot{x}_1 \\ \dot{x}_2 \end{bmatrix} = \begin{bmatrix} -2 & 3 \\ -1 & -1 \end{bmatrix} \begin{bmatrix} x_1 \\ x_2 \end{bmatrix} + \begin{bmatrix} 4 & 0 \\ -5 & 6 \end{bmatrix} \begin{bmatrix} u_1 \\ u_2 \end{bmatrix}$$

$$y = [7 \quad 8] \begin{bmatrix} x_1 \\ x_2 \end{bmatrix}$$

Ans. $\begin{bmatrix} \dfrac{-12s - 189}{s^2 + 3s + 5} & \dfrac{48s + 222}{s^2 + 3s + 5} \end{bmatrix}$

(b)

$$\begin{bmatrix} \dot{x}_1 \\ \dot{x}_2 \end{bmatrix} = \begin{bmatrix} -3 & 4 \\ -2 & 0 \end{bmatrix} \begin{bmatrix} x_1 \\ x_2 \end{bmatrix} + \begin{bmatrix} 2 \\ 1 \end{bmatrix} u$$

$$\begin{bmatrix} y_1 \\ y_2 \end{bmatrix} = \begin{bmatrix} -4 & 6 \\ 5 & -1 \end{bmatrix} \begin{bmatrix} x_1 \\ x_2 \end{bmatrix}$$

Ans. $\begin{bmatrix} \dfrac{-2s - 22}{s^2 + 3s + 8} \\ \dfrac{9s + 21}{s^2 + 3s + 8} \end{bmatrix}$

(c)

$$\begin{bmatrix} \dot{x}_1 \\ \dot{x}_2 \\ \dot{x}_2 \end{bmatrix} = \begin{bmatrix} -4 & 0 & 2 \\ -1 & -1 & 0 \\ 3 & 0 & -3 \end{bmatrix} \begin{bmatrix} x_1 \\ x_2 \\ x_3 \end{bmatrix} + \begin{bmatrix} -3 & -2 \\ 4 & 1 \\ 0 & 0 \end{bmatrix} \begin{bmatrix} u_1 \\ u_2 \end{bmatrix}$$

$$\begin{bmatrix} y_1 \\ y_2 \end{bmatrix} = \begin{bmatrix} 1 & 1 & 1 \\ -1 & 0 & 1 \end{bmatrix} \begin{bmatrix} x_1 \\ x_2 \\ x_3 \end{bmatrix}$$

Ans. $\begin{bmatrix} \dfrac{s^2 + 10s + 15}{(s+1)(s^2 + 7s + 6)} & \dfrac{-s(s+5)}{(s+1)(s^2 + 7s + 6)} \\ \dfrac{3s}{s^2 + 7s + 6} & \dfrac{2s}{s^2 + 7s + 6} \end{bmatrix}$

8.3 State Transformations and Diagonalization

You have already observed that for a given system there is more than one state space representation. We have introduced two canonical forms, namely, the phase-variable and dual phase-variable forms. Hence, state space representation is not unique. In fact, if you change the way you label the states in a simulation diagram (number the states from left to right instead), you will obtain other forms. In general, there are infinitely many representations. These are generally called (state space) *realizations* of a system. Because these realizations correspond to the same transfer function, we want to determine how we can go from one state space realization to another.

The answer is that all state space realizations of the same system are related to each other via a linear transformation.

$$x(t) = Pz(t) \quad P = \text{nonsingular matrix}$$

State transformation.

where $x(t)$ represents the old state, and $z(t)$, the new state vector. How do we obtain the matrices corresponding to the new realization? We do this by differentiating both sides of the equation

$$\dot{x} = P\dot{z} = Ax + Bu = APz + Bu$$

Multiplying the equation on the left by the inverse of P, we get

$$\dot{z} = P^{-1}APz + P^{-1}Bu$$

The transformed equation.

The output equation becomes

$$y = Cx + Du = CPz + Du$$

Summarizing, we have that if the original realization is $\{A, B, C, D\}$, and the new realization is $\{\bar{A}, \bar{B}, \bar{C}, \bar{D}\}$, then the relation between the realizations is given by

$$\bar{A} = P^{-1}AP \qquad \bar{B} = P^{-1}B \qquad \bar{C} = CP \qquad \bar{D} = D$$

Let us demonstrate the procedure by an example. Consider the system shown in Figure 8.4:

$$\dot{x}_1 = x_1 + x_2$$
$$\dot{x}_2 = x_1 + x_2$$
$$y = x_1 + x_2$$

Define new state variables as

$$z_1 = \frac{x_1 - x_2}{2} \quad \text{and} \quad z_2 = \frac{x_1 + x_2}{2}$$

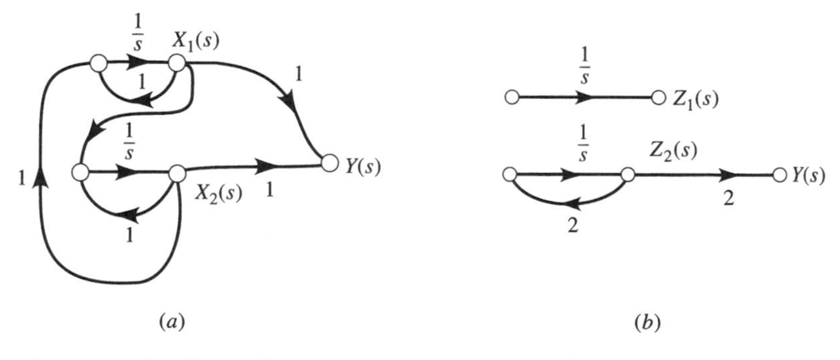

Figure 8.4 (a) Signal flow graph of a system. (b) Signal flow graph of the transformed system.

This can be written in matrix form as

$$z = \frac{1}{2}\begin{bmatrix} 1 & -1 \\ 1 & 1 \end{bmatrix} x \quad \text{or} \quad x = Pz = \begin{bmatrix} 1 & 1 \\ -1 & 1 \end{bmatrix} z$$

Therefore the state space matrices will be transformed accordingly

$$\begin{aligned} A &= \begin{bmatrix} 1 & 1 \\ 1 & 1 \end{bmatrix} \rightarrow \bar{A} = P^{-1}AP \\ &= \frac{1}{2}\begin{bmatrix} 1 & -1 \\ 1 & 1 \end{bmatrix}\begin{bmatrix} 1 & 1 \\ 1 & 1 \end{bmatrix}\begin{bmatrix} 1 & 1 \\ -1 & 1 \end{bmatrix} = \begin{bmatrix} 0 & 0 \\ 0 & 2 \end{bmatrix} \\ C &= [1 \quad 1] \rightarrow \bar{C} = CP \\ &= [1 \quad 1]\begin{bmatrix} 1 & 1 \\ -1 & 1 \end{bmatrix} = [0 \quad 2] \end{aligned}$$

B and D are zero since the system has no inputs. Writing the new state equations, we get

$$\begin{aligned} \dot{z}_1 &= 0 \\ \dot{z}_2 &= 2z_2 \\ y &= 2z_2 \end{aligned}$$

Observe that the system in this form is much easier to work with. In the original realization, the second-order system can be viewed as two coupled first-order systems. In the new realization, the same system appears as two uncoupled first-order systems. Every state space realization represents the same system, but each one allows us to look at the system from a different perspective. In modern control theory, state space transformations are used quite frequently for numerical purposes. Some realizations have superior numerical properties to others.

In the case of our present example, we note that even though we have not yet discussed how to solve state space equations, the solution in the new decoupled realization is almost trivial. The solution is

$$\begin{aligned} z_1(t) &= z_1(0) \\ z_2(t) &= e^{2t} z_2(0) \end{aligned}$$

To obtain the solution to our original equation, we note that

$$x(t) = Pz(t) = \begin{bmatrix} 1 & 1 \\ -1 & 1 \end{bmatrix}\begin{bmatrix} z_1(0) \\ e^{2t} z_2(0) \end{bmatrix}$$

Also note that the initial conditions have to be transformed as

$$z(0) = P^{-1}x(0)$$

Suppose the initial conditions are given by $x(0) = \begin{bmatrix} 1 \\ 1 \end{bmatrix}$; then $z(0) = \begin{bmatrix} 0 \\ 1 \end{bmatrix}$ and the solution is

$$x(t) = \begin{bmatrix} e^{2t} \\ e^{2t} \end{bmatrix}$$

The input–output relations of a system are unchanged by a nonsingular change of state variables; it is only the internal description, in terms of its state, that is changed. This can be proved by the following line of argument

$$\bar{T}(s) = \bar{C}(sI - \bar{A})^{-1}\bar{B}$$

Substituting for $\bar{A}$, $\bar{B}$, and $\bar{C}$, we get

$$\bar{T}(s) = CP(sI - P^{-1}AP)^{-1}P^{-1}B$$

Let $I = P^{-1}P$, then

$$\bar{T}(s) = CP(sP^{-1}P - P^{-1}AP)^{-1}P^{-1}B = CP[P^{-1}(sI - A)P]^{-1}P^{-1}B$$

Recall from matrix algebra the identity: $(XYZ)^{-1} = Z^{-1}Y^{-1}X^{-1}$

$$\bar{T}(s) = CPP^{-1}(sI - A)^{-1}PP^{-1}B = C(sI - A)^{-1}B = T(s)$$

State transformations leave the transfer function unchanged.

❑ DRILL PROBLEM

D8.5 Make the indicated change of state variables, finding the new set of state and output equations in terms of z.

(a)

$$\begin{bmatrix} \dot{x}_1 \\ \dot{x}_2 \end{bmatrix} = \begin{bmatrix} -2 & 1 \\ -3 & 0 \end{bmatrix} \begin{bmatrix} x_1 \\ x_2 \end{bmatrix} + \begin{bmatrix} 4 \\ 5 \end{bmatrix} u$$

$$y = [1 \quad 0] \begin{bmatrix} x_1 \\ x_2 \end{bmatrix}$$

$$\begin{bmatrix} x_1 \\ x_2 \end{bmatrix} = \begin{bmatrix} 2 & 1 \\ 4 & 3 \end{bmatrix} \begin{bmatrix} z_1 \\ z_2 \end{bmatrix}$$

Ans.

$$\begin{bmatrix} \dot{z}_1 \\ \dot{z}_2 \end{bmatrix} = \begin{bmatrix} 3 & 3 \\ -6 & -5 \end{bmatrix} \begin{bmatrix} z_1 \\ z_2 \end{bmatrix} + \begin{bmatrix} 3.5 \\ -3 \end{bmatrix} u$$

$$y = [2 \quad 1] \begin{bmatrix} z_1 \\ z_2 \end{bmatrix}$$

(b)

$$\begin{bmatrix} \dot{x}_1 \\ \dot{x}_2 \end{bmatrix} = \begin{bmatrix} 2 & -6 \\ 12 & 16 \end{bmatrix} \begin{bmatrix} x_1 \\ x_2 \end{bmatrix} + \begin{bmatrix} 0 & -1 \\ 2 & 1 \end{bmatrix} \begin{bmatrix} u_1 \\ u_2 \end{bmatrix}$$

$$y = \begin{bmatrix} 1 & 1 \end{bmatrix} \begin{bmatrix} x_1 \\ x_2 \end{bmatrix}$$

$$\begin{bmatrix} x_1 \\ x_2 \end{bmatrix} = \begin{bmatrix} 1 & 2 \\ 1 & 4 \end{bmatrix} \begin{bmatrix} z_1 \\ z_2 \end{bmatrix}$$

Ans.

$$\begin{bmatrix} \dot{z}_1 \\ \dot{z}_2 \end{bmatrix} = \begin{bmatrix} -74 & -54 \\ 128 & 92 \end{bmatrix} \begin{bmatrix} z_1 \\ z_2 \end{bmatrix} + \begin{bmatrix} -2 & -3 \\ 1 & 1 \end{bmatrix} \begin{bmatrix} u_1 \\ u_2 \end{bmatrix}$$

$$y = \begin{bmatrix} 2 & 6 \end{bmatrix} \begin{bmatrix} z_1 \\ z_2 \end{bmatrix}$$

(c)

$$\begin{bmatrix} \dot{x}_1 \\ \dot{x}_2 \\ \dot{x}_3 \end{bmatrix} = \begin{bmatrix} -2 & 1 & 1 \\ -3 & 0 & 0 \\ 0 & 0 & 0 \end{bmatrix} \begin{bmatrix} x_1 \\ x_2 \\ x_3 \end{bmatrix} + \begin{bmatrix} 1 \\ 1 \\ 1 \end{bmatrix} u$$

$$\begin{bmatrix} y_1 \\ y_2 \end{bmatrix} = \begin{bmatrix} 2 & -2 & 1 \\ 0 & -1 & 1 \end{bmatrix} \begin{bmatrix} x_1 \\ x_2 \\ x_3 \end{bmatrix}$$

$$\begin{bmatrix} x_1 \\ x_2 \\ x_3 \end{bmatrix} = \begin{bmatrix} 1 & 0 & 1 \\ 0 & 1 & 3 \\ 0 & 0 & 4 \end{bmatrix} \begin{bmatrix} z_1 \\ z_2 \\ z_3 \end{bmatrix}$$

Ans.

$$\begin{bmatrix} \dot{z}_1 \\ \dot{z}_2 \\ \dot{z}_3 \end{bmatrix} = \begin{bmatrix} -2 & 1 & 5 \\ -3 & 0 & -3 \\ 0 & 0 & 0 \end{bmatrix} \begin{bmatrix} z_1 \\ z_2 \\ z_3 \end{bmatrix} + \begin{bmatrix} 3/4 \\ 1/4 \\ 1/4 \end{bmatrix} u$$

$$y = \begin{bmatrix} 2 & -2 & 0 \\ 0 & -1 & 1 \end{bmatrix} \begin{bmatrix} z_1 \\ z_2 \\ z_3 \end{bmatrix}$$

8.3.1 Diagonal Forms

As shown in Figure 8.1(a), a simulation diagram results in a set of system matrices when a transfer function is properly decomposed. Phase variables or dual phase variables result in a recognizable form for the A matrix. In this section, a properly chosen change of variables can create a set of A, B, and C matrices in which the A

matrix has a very simple (diagonal) form. The state variables [Figure 8.1(c)] are often called *canonical* when the A matrix becomes diagonal.

When a nonsingular change of state variables in a system representation is made,

$$z = P^{-1}x, \qquad x = Pz$$

the new state coupling matrix $\bar{A}$ is related to the original one A by

$$\bar{A} = P^{-1}AP$$

Such an operation on a matrix is termed a *similarity transformation*. One of the most important results of matrix algebra is that, provided a square matrix A has no repeated eigenvalues, a similarity transformation P may be found for which

$$\bar{A} = P^{-1}AP$$

is diagonal, with the eigenvalues as the diagonal elements.

A similarity transformation that diagonalizes A can be created using a set of eigenvectors, one for each eigenvalue. The German word "eigen" means "characteristic." These eigenvectors are not unique. Each eigenvector can be multiplied by a constant that works just as well. The following procedure can be used to get P.

1. Find the eigenvalues s_i where

$$|sI - A| = 0$$

2. Find an eigenvector x_i for each s_i

$$[s_i I - A]x_i = 0$$

3. Let P be a matrix consisting of the eigenvectors

$$P = [x_1 : x_2 : \cdots : x_n]$$

$$P^{-1}AP = \begin{bmatrix} s_1 & 0 & 0 & . \\ 0 & s_2 & 0 & . \\ \text{etc.} & & & \end{bmatrix}$$

In the example just analyzed, the P matrix was actually computed by means of the same procedure. You can verify that the eigenvalues of the system are 0 and 2, and the eigenvectors are the columns of P.

As another example, consider

$$\begin{bmatrix} \dot{x}_1 \\ \dot{x}_2 \\ \dot{x}_3 \end{bmatrix} = \begin{bmatrix} -1 & -2 & 0 \\ 1 & 2 & 0 \\ -2 & -1 & -3 \end{bmatrix} \begin{bmatrix} x_1 \\ x_2 \\ x_3 \end{bmatrix} + \begin{bmatrix} 1 \\ 0 \\ 0 \end{bmatrix} u = Ax + bu$$

$$\begin{bmatrix} y_1 \\ y_2 \end{bmatrix} = \begin{bmatrix} 1 & 0 & 1 \\ 1 & -1 & 0 \end{bmatrix} \begin{bmatrix} x_1 \\ x_2 \\ x_3 \end{bmatrix} = Cx$$

The characteristic polynomial is

$$\begin{vmatrix} s+1 & 2 & 0 \\ -1 & s-2 & 0 \\ 2 & 1 & s+3 \end{vmatrix} = s(s-1)(s+3)$$

The eigenvalues can be selected in any order. If

$$s_1 = 0 \quad s_2 = 1 \quad s_3 = -3$$

Then the first eigenvector is found by solving

$$\begin{bmatrix} s_1+1 & 2 & 0 \\ -1 & s_1-2 & 0 \\ 2 & 1 & s_1+3 \end{bmatrix} \begin{bmatrix} x_{11} \\ x_{21} \\ x_{31} \end{bmatrix} = \begin{bmatrix} 0 \\ 0 \\ 0 \end{bmatrix}$$

$$x_1 = \begin{bmatrix} x_{11} \\ x_{21} \\ x_{31} \end{bmatrix}$$

With the first eigenvalue being zero

$$\begin{aligned} x_{11} + 2x_{21} \quad\quad &= 0 \\ -x_{11} - 2x_{21} \quad\quad &= 0 \\ 2x_{11} + x_{21} + 3x_{31} &= 0 \end{aligned}$$

The first two equations are equivalent, so an infinite number of solutions exist. It is possible to select x_{21} arbitrarily (say -1). Then x_{11} is 2 and x_{31} is -1. By proceeding in a similar way, three eigenvectors result:

$$x_1 = \begin{bmatrix} 2 \\ -1 \\ -1 \end{bmatrix} \quad x_2 = \begin{bmatrix} 4 \\ -4 \\ -1 \end{bmatrix} \quad x_3 = \begin{bmatrix} 0 \\ 0 \\ 1 \end{bmatrix}$$

Any nonzero multiple of any eigenvector also works. Collecting these eigenvectors into P, the transformation

$$P = \begin{bmatrix} 2 & 4 & 0 \\ -1 & -4 & 0 \\ -1 & -1 & 1 \end{bmatrix} \quad P^{-1} = \begin{bmatrix} 1 & 1 & 0 \\ -\frac{1}{4} & -\frac{1}{2} & 0 \\ \frac{3}{4} & \frac{1}{2} & 1 \end{bmatrix}$$

gives a state-variable representation for which the state coupling matrix is diagonal:

$$\bar{A} = P^{-1}AP = \begin{bmatrix} 1 & 1 & 0 \\ -\frac{1}{4} & -\frac{1}{2} & 0 \\ \frac{3}{4} & \frac{1}{2} & 1 \end{bmatrix} \begin{bmatrix} -1 & -2 & 0 \\ 1 & 2 & 0 \\ -2 & -1 & -3 \end{bmatrix} \begin{bmatrix} 2 & 4 & 0 \\ -1 & -4 & 0 \\ -1 & -1 & 1 \end{bmatrix}$$

$$= \begin{bmatrix} 1 & 1 & 0 \\ -\frac{1}{4} & -\frac{1}{2} & 0 \\ \frac{3}{4} & \frac{1}{2} & 1 \end{bmatrix} \begin{bmatrix} 0 & 4 & 0 \\ 0 & -4 & 0 \\ 0 & -1 & -3 \end{bmatrix} = \begin{bmatrix} 0 & 0 & 0 \\ 0 & 1 & 0 \\ 0 & 0 & -3 \end{bmatrix}$$

$$\bar{B} = P^{-1}B = \begin{bmatrix} 1 & 1 & 0 \\ -\frac{1}{4} & -\frac{1}{2} & 0 \\ \frac{3}{4} & \frac{1}{2} & 1 \end{bmatrix} \begin{bmatrix} 1 \\ 0 \\ 0 \end{bmatrix} = \begin{bmatrix} 1 \\ -\frac{1}{4} \\ \frac{3}{4} \end{bmatrix}$$

$$\bar{C} = CP = \begin{bmatrix} 1 & 0 & 1 \\ 1 & -1 & 0 \end{bmatrix} \begin{bmatrix} 2 & 4 & 0 \\ -1 & -4 & 0 \\ -1 & -1 & 1 \end{bmatrix} = \begin{bmatrix} 1 & 3 & 1 \\ 3 & 8 & 0 \end{bmatrix}$$

The system described by the new state variables,

$$\begin{bmatrix} \dot{z}_1 \\ \dot{z}_2 \\ \dot{z}_3 \end{bmatrix} = \begin{bmatrix} 0 & 0 & 0 \\ 0 & 1 & 0 \\ 0 & 0 & -3 \end{bmatrix} \begin{bmatrix} z_1 \\ z_2 \\ z_3 \end{bmatrix} + \begin{bmatrix} 1 \\ -\frac{1}{4} \\ \frac{3}{4} \end{bmatrix} u$$

$$\begin{bmatrix} y_1 \\ y_2 \end{bmatrix} = \begin{bmatrix} 1 & 3 & 1 \\ 3 & 8 & 0 \end{bmatrix} \begin{bmatrix} z_1 \\ z_2 \\ z_3 \end{bmatrix}$$

has the same relation between u and y. Because the state coupling matrix is diagonal, however, the state equations are decoupled from one another. The system is represented in the form of three separate first-order systems, as in the simulation diagram of Figure 8.5.

Finding a transformation (eigenvalues) matrix that diagonalizes a square matrix A with distinct characteristic roots is a fundamental technique of linear algebra. It is termed the *characteristic value problem* and is discussed in detail in most texts on linear algebra, including those cited in the references at the end of this chapter.

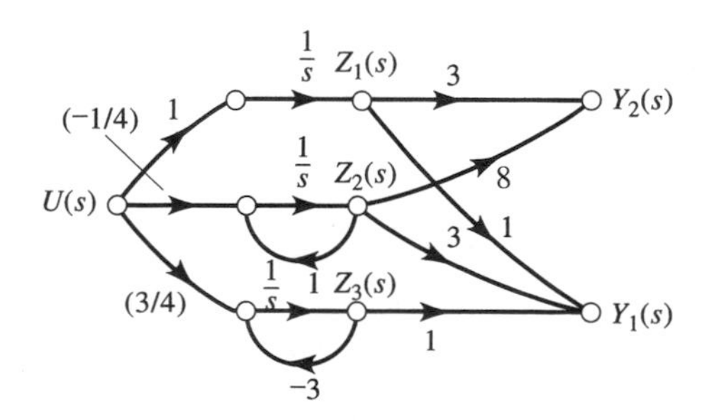

Figure 8.5 Simulation diagram for the diagonalized example system.

8.3.2 Diagonalization Using Partial Fraction Expansion

Another method of determining a diagonal form for a system involves partial fraction expansion. For a single-input, single-output system such as

$$\begin{bmatrix} \dot{x}_1 \\ \dot{x}_2 \\ \dot{x}_3 \end{bmatrix} = \begin{bmatrix} -1 & -2 & 0 \\ 1 & 2 & 0 \\ -2 & -1 & -3 \end{bmatrix} \begin{bmatrix} x_1 \\ x_2 \\ x_3 \end{bmatrix} + \begin{bmatrix} 1 \\ 0 \\ 0 \end{bmatrix} u$$

$$y = [1 \quad 0 \quad 1] \begin{bmatrix} x_1 \\ x_2 \\ x_3 \end{bmatrix}$$

the transfer function is

$$\begin{aligned} T(s) &= C(sI - A)^{-1}B \\ &= [1 \quad 0 \quad 1] \begin{bmatrix} s+1 & 2 & 0 \\ -1 & s-2 & 0 \\ 2 & 1 & s+3 \end{bmatrix}^{-1} \begin{bmatrix} 1 \\ 0 \\ 0 \end{bmatrix} \\ &= [1 \quad 0 \quad 1] \frac{\begin{bmatrix} s^2+s-6 & -2s-6 & 0 \\ s+3 & s^2+4s+3 & 0 \\ -2s+3 & -s+3 & s^2-s \end{bmatrix}}{s^3+2s^2-3s} \begin{bmatrix} 1 \\ 0 \\ 0 \end{bmatrix} \\ &= \frac{[1 \quad 0 \quad 1]}{s^3+2s^2-3s} \begin{bmatrix} (s^2+s-6) \\ (s+3) \\ (-2s+3) \end{bmatrix} = \frac{s^2-s-3}{s^3+2s^2-3s} \end{aligned}$$

Upon expanding this transfer function in partial fractions, there results

$$T(s) = \frac{s^2-s-3}{s(s-1)(s+3)} = \frac{1}{s} + \frac{-\frac{3}{4}}{s-1} + \frac{\frac{3}{4}}{s+3}$$

which may be considered as the tandem (or parallel) connection of first-order systems shown in Figure 8.6(a). Each of these first-order subsystems is drawn in state-variable form in Figure 8.6(b), where the three integrator output signals are labeled as state-variables. The state-variable equations for this alternate system representation, which has the same transfer function as the original system, are

$$\dot{z}_1 = u$$

$$\dot{z}_2 = z_2 + u$$

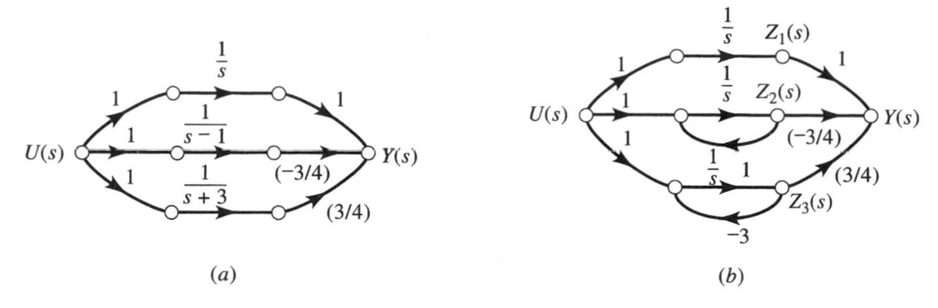

Figure 8.6 Diagonalizing a single-input, single-output system. (a) Tandem first-order sub systems from the partial fraction expansion of the transfer function. (b) Subsystems in simulation diagram form.

$$\dot{z}_3 = -3z_3 + u$$

$$y = z_1 - \tfrac{3}{4}z_2 + \tfrac{3}{4}z_3$$

or

$$\begin{bmatrix} \dot{z}_1 \\ \dot{z}_2 \\ \dot{z}_3 \end{bmatrix} = \begin{bmatrix} 0 & 0 & 0 \\ 0 & 1 & 0 \\ 0 & 0 & -3 \end{bmatrix} \begin{bmatrix} z_1 \\ z_2 \\ z_3 \end{bmatrix} + \begin{bmatrix} 1 \\ 1 \\ 1 \end{bmatrix} u$$

$$y = \begin{bmatrix} 1 & -\tfrac{3}{4} & \tfrac{3}{4} \end{bmatrix} \begin{bmatrix} z_1 \\ z_2 \\ z_3 \end{bmatrix}$$

which is diagonal.

❑ DRILL PROBLEMS

D8.6 Use the partial fraction method to find diagonal state equations for single–input, single–output systems with the following transfer functions:

(a)

$$T(s) = \frac{-5s + 7}{s^2 + 7s + 12}$$

Ans.

$$\begin{bmatrix} \dot{x}_1 \\ \dot{x}_2 \end{bmatrix} = \begin{bmatrix} -3 & 0 \\ 0 & -4 \end{bmatrix} \begin{bmatrix} x_1 \\ x_2 \end{bmatrix} + \begin{bmatrix} 22 \\ -27 \end{bmatrix} u$$

$$y = \begin{bmatrix} 1 & 1 \end{bmatrix} \begin{bmatrix} x_1 \\ x_2 \end{bmatrix}$$

(b)

$$T(s) = \frac{3s^2 - 2}{(s + 1)(s + 4)(s + 10)}$$

Ans.

$$\begin{bmatrix} \dot{x}_1 \\ \dot{x}_2 \\ \dot{x}_3 \end{bmatrix} = \begin{bmatrix} -1 & 0 & 0 \\ 0 & -4 & 0 \\ 0 & 0 & -10 \end{bmatrix} \begin{bmatrix} x_1 \\ x_2 \\ x_3 \end{bmatrix} + \begin{bmatrix} \frac{1}{27} \\ -\frac{23}{9} \\ \frac{149}{27} \end{bmatrix} u$$

$$y = [1 \quad 1 \quad 1] \begin{bmatrix} x_1 \\ x_2 \\ x_3 \end{bmatrix}$$

(c)

$$T(s) = \frac{4}{s^3 + 3s^2 + 2s}$$

Ans.

$$\begin{bmatrix} \dot{x}_1 \\ \dot{x}_2 \\ \dot{x}_3 \end{bmatrix} = \begin{bmatrix} 0 & 0 & 0 \\ 0 & -1 & 0 \\ 0 & 0 & -2 \end{bmatrix} \begin{bmatrix} x_1 \\ x_2 \\ x_3 \end{bmatrix} + \begin{bmatrix} 1 \\ 1 \\ 1 \end{bmatrix} u$$

$$y = [2 \quad -4 \quad 2] \begin{bmatrix} x_1 \\ x_2 \\ x_3 \end{bmatrix}$$

D8.7 Draw a simulation diagram for each of the following state equations.

(a)

$$\begin{bmatrix} \dot{z}_1 \\ \dot{z}_2 \end{bmatrix} = \begin{bmatrix} -3 & 0 \\ 0 & -4 \end{bmatrix} \begin{bmatrix} z_1 \\ z_2 \end{bmatrix} + \begin{bmatrix} 8 \\ 4 \end{bmatrix} u$$

$$y = [1 \quad 1] \begin{bmatrix} z_1 \\ z_2 \end{bmatrix}$$

(b)

$$\begin{bmatrix} \dot{z}_1 \\ \dot{z}_2 \end{bmatrix} = \begin{bmatrix} -2 & 0 \\ 0 & -4 \end{bmatrix} \begin{bmatrix} z_1 \\ z_2 \end{bmatrix} + \begin{bmatrix} 1 \\ 1 \end{bmatrix} u$$

$$y = [6 \quad -9] \begin{bmatrix} z_1 \\ z_2 \end{bmatrix}$$

8.3.3 Complex Conjugate Characteristic Roots

In general, diagonalized state equations for systems with complex characteristic roots involve state equations with complex coefficients. For example, the single-input, single-output system with transfer function

$$T(s) = \frac{6s^2 + 26s + 8}{(s+2)(s^2+2s+10)} = \frac{-2}{s+2} + \frac{4+j}{s+1+j3} + \frac{4-j}{s+1-j3}$$

may be represented in terms of state variables as in the simulation diagram of Figure 8.7(a). The gains associated with the complex characteristic roots are generally complex numbers. The state equations, in terms of the indicated state variables, are given by

$$sX_1(s) = -2X_1(s) + U(s)$$
$$sX_2(s) = (-1 - j3)X_2(s) + U(s)$$
$$sX_3(s) = (-1 + j3)X_3(s) + U(s)$$
$$Y(s) = -2X_1(s) + (4 + j)X_2(s) + (4 - j)X_3(s)$$

or

$$\begin{bmatrix} \dot{x}_1 \\ \dot{x}_2 \\ \dot{x}_3 \end{bmatrix} = \begin{bmatrix} -2 & 0 & 0 \\ 0 & -1 - j3 & 0 \\ 0 & 0 & -1 + j3 \end{bmatrix} \begin{bmatrix} x_1 \\ x_2 \\ x_3 \end{bmatrix} + \begin{bmatrix} 1 \\ 1 \\ 1 \end{bmatrix} u$$

$$y = [-2 \quad 4 + j \quad 4 - j] \begin{bmatrix} x_1 \\ x_2 \\ x_3 \end{bmatrix}$$

Although the individual physical components of this representation involve complex numbers, hence cannot be assembled, the mathematical relationships are valid. To build such a system, or to represent it in a convenient form that does not involve complex numbers, the two complex conjugate component parts may be combined just as one commonly combines the corresponding conjugate partial fraction terms:

$$\frac{4 + j}{s + 1 + j3} + \frac{4 - j}{s + 1 - j3} = \frac{8s + 14}{s^2 + 2s + 10}$$

This portion of the system may be represented in phase-variable form, giving the real-number simulation diagram of Figure 8.7(b). The state equations for this alternative arrangement are given by

$$sZ_1(s) = -2Z_1(s) + U(s)$$
$$sZ_2(s) = Z_3(s)$$
$$sZ_3(s) = -10Z_2(s) - 2Z_3(s) + U(s)$$
$$Y(s) = -2Z_1(s) + 14Z_2(s) + 8Z_3(s)$$

or

$$\begin{bmatrix} \dot{z}_1 \\ \dot{z}_2 \\ \dot{z}_3 \end{bmatrix} = \begin{bmatrix} -2 & 0 & 0 \\ 0 & 0 & 1 \\ 0 & -10 & -2 \end{bmatrix} \begin{bmatrix} z_1 \\ z_2 \\ z_3 \end{bmatrix} + \begin{bmatrix} 1 \\ 0 \\ 1 \end{bmatrix} u$$

$$y = [-2 \quad 14 \quad 8] \begin{bmatrix} z_1 \\ z_2 \\ z_3 \end{bmatrix}$$

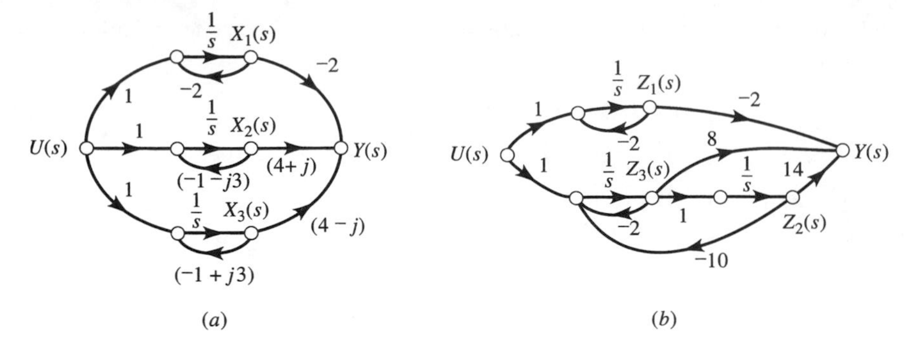

Figure 8.7 A system with complex characteristic roots. (a) Diagonalized system. (b) Alternative form for the diagonalized system, where the complex conjugate root terms have been combined and placed in phase-variable form.

It is thus possible to represent systems with one or more pairs of complex conjugate characteristic roots with diagonalized state equations involving complex numbers or in block diagonal form involving real numbers. For example, the following state equations

$$\begin{bmatrix} \dot{x}_1 \\ \dot{x}_2 \\ \dot{x}_3 \\ \dot{x}_4 \\ \dot{x}_5 \\ \dot{x}_6 \end{bmatrix} = \begin{bmatrix} 3 & 0 & 0 & 0 & 0 & 0 \\ 0 & -4 & 0 & 0 & 0 & 0 \\ 0 & 0 & 0 & 1 & 0 & 0 \\ 0 & 0 & -17 & -2 & 0 & 0 \\ 0 & 0 & 0 & 0 & 0 & 1 \\ 0 & 0 & 0 & 0 & -10 & 3 \end{bmatrix} \begin{bmatrix} x_1 \\ x_2 \\ x_3 \\ x_4 \\ x_5 \\ x_6 \end{bmatrix} + \begin{bmatrix} 1 \\ 1 \\ 0 \\ 1 \\ 0 \\ 1 \end{bmatrix} u$$

$$y = \begin{bmatrix} 6 & -8 & 1 & -5 & 0 & 7 \end{bmatrix} \begin{bmatrix} x_1 \\ x_2 \\ x_3 \\ x_4 \\ x_5 \\ x_6 \end{bmatrix}$$

are in block diagonal form and represent a system with transfer function

$$T(s) = \frac{6}{s-3} + \frac{-8}{s+4} + \frac{-5s+1}{s^2+2s+17} + \frac{7s}{s^2-3s+10}$$

❑ DRILL PROBLEMS

D8.8 The following transfer functions for single-input, single-output systems involve complex characteristic roots. Find diagonal state equations for these systems. Then find an alternative block diagonal representation that does not involve complex numbers.

(a)

$$T(s) = \frac{10}{s^3 + 2s^2 + 5s}$$

Ans.

$$\begin{bmatrix} \dot{x}_1 \\ \dot{x}_2 \\ \dot{x}_3 \end{bmatrix} = \begin{bmatrix} 0 & 0 & 0 \\ 0 & 0 & 1 \\ 0 & -5 & -2 \end{bmatrix} \begin{bmatrix} x_1 \\ x_2 \\ x_3 \end{bmatrix} + \begin{bmatrix} 2 \\ 0 \\ 1 \end{bmatrix} u$$

$$y = [1 \quad -4 \quad -2] \begin{bmatrix} x_1 \\ x_2 \\ x_3 \end{bmatrix}$$

(b)

$$T(s) = \frac{3s^2 - 1}{(s^2 + 4)(s^2 + 4s + 5)}$$

Ans.

$$\begin{bmatrix} \dot{x}_1 \\ \dot{x}_2 \\ \dot{x}_3 \\ \dot{x}_4 \end{bmatrix} = \begin{bmatrix} 0 & 1 & 0 & 0 \\ -4 & 0 & 0 & 0 \\ 0 & 0 & 0 & 1 \\ 0 & 0 & -5 & -4 \end{bmatrix} \begin{bmatrix} x_1 \\ x_2 \\ x_3 \\ x_4 \end{bmatrix} + \begin{bmatrix} 0 \\ 1 \\ 0 \\ 1 \end{bmatrix} u$$

$$y = \begin{bmatrix} -\frac{1}{5} & \frac{4}{5} & 0 & -\frac{4}{5} \end{bmatrix} \begin{bmatrix} x_1 \\ x_2 \\ x_3 \\ x_4 \end{bmatrix}$$

(c)

$$T(s) - \frac{s^2 - 4s + 10}{(s + 2)(s^2 + 6s + 13)}$$

Ans.

$$\begin{bmatrix} \dot{x}_1 \\ \dot{x}_2 \\ \dot{x}_3 \end{bmatrix} = \begin{bmatrix} -2 & 0 & 0 \\ 0 & 0 & 1 \\ 0 & -13 & -6 \end{bmatrix} \begin{bmatrix} x_1 \\ x_2 \\ x_3 \end{bmatrix} + \begin{bmatrix} 1 \\ 0 \\ 1 \end{bmatrix} u$$

$$y = \begin{bmatrix} \frac{22}{5} & \frac{118}{5} & -\frac{17}{5} \end{bmatrix} \begin{bmatrix} x_1 \\ x_2 \\ x_3 \end{bmatrix}$$

8.3.4 Repeated Characteristic Roots

The state equations for a system with repeated characteristic roots may not necessarily be diagonalized. A block diagonal form, termed a *Jordan canonical form*, is commonly used when a simple representation is desired. For example, the single-input, single-output system with transfer function

$$T(s) = \frac{10s^2 + 51s + 56}{(s+4)(s+2)^2} = \frac{3}{s+4} + \frac{-6}{s+2} + \frac{7}{(s+2)^2}$$

may be represented as in Figure 8.8(a). A simplification results when the $1/(s+2)$ transmittance is used in common by two paths, as shown in Figure 8.8(b). A corresponding state-variable representation is given in Figure 8.8(c).

The state equations are given by

$$\begin{aligned} sX_1(s) &= -4X_1(s) + U(s) \\ sX_2(s) &= -2X_2(s) + X_3(s) \\ sX_3(s) &= -2X_3(s) + U(s) \\ Y(s) &= 3X_1(s) + 7X_2(s) - 6X_3(s) \end{aligned}$$

or

Jordan form.

$$\begin{bmatrix} \dot{x}_1 \\ \dot{x}_2 \\ \dot{x}_3 \end{bmatrix} = \begin{bmatrix} -4 & 0 & 0 \\ 0 & -2 & 1 \\ 0 & 0 & -2 \end{bmatrix} \begin{bmatrix} x_1 \\ x_2 \\ x_3 \end{bmatrix} + \begin{bmatrix} 1 \\ 0 \\ 1 \end{bmatrix} u$$

$$y = [3 \quad 7 \quad -6] \begin{bmatrix} x_1 \\ x_2 \\ x_3 \end{bmatrix}$$

For three repetitions of a characteristic root, the corresponding transfer function partial fraction terms are

$$\frac{k_1}{s+a} + \frac{k_2}{(s+a)^2} + \frac{k_3}{(s+a)^3}$$

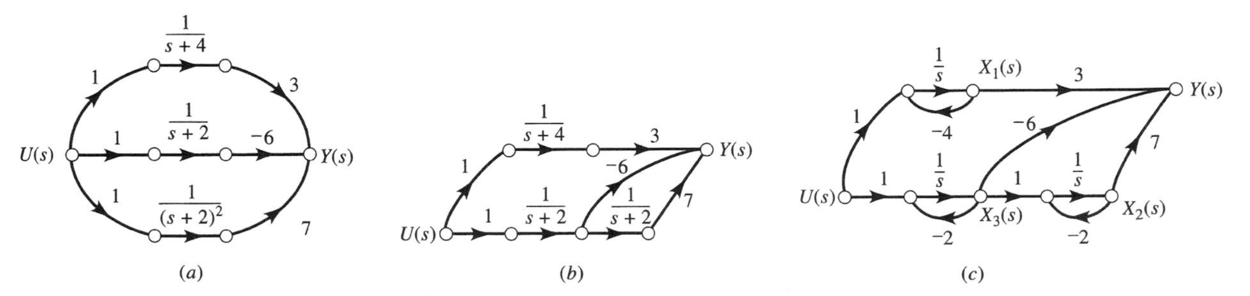

Figure 8.8 State equations for a system with repeated characteristic roots. (a) Diagram showing each partial fraction term. (b) Diagram with common signal path through a repeated transmittance. (c) Diagram showing state variables.

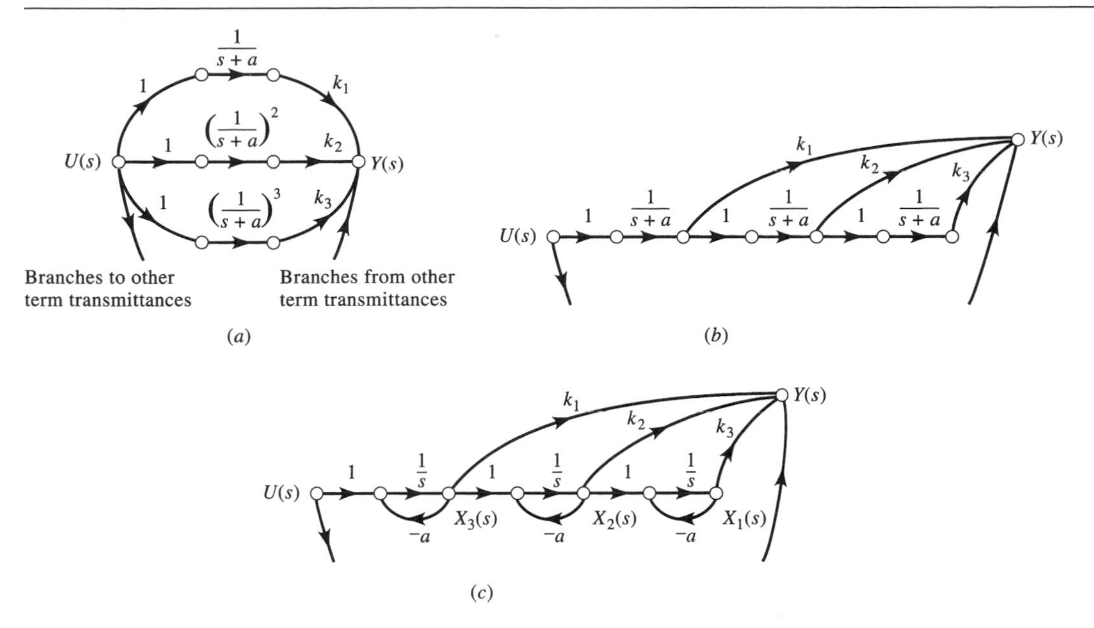

Figure 8.9 State variables for repeated roots. (a) Diagram showing each partial fraction term. (b) Diagram using common signal paths. (c) Diagram showing state variables.

and the state variables may be defined as in Figure 8.9. The resulting Jordan block has the following structure:

$$\begin{bmatrix} \dot{x}_1 \\ \dot{x}_2 \\ \dot{x}_3 \\ \vdots \end{bmatrix} = \begin{bmatrix} -a & 1 & 0 & 0 & 0 & \cdots \\ 0 & -a & 1 & 0 & 0 & \cdots \\ 0 & 0 & -a & 0 & 0 & \cdots \\ 0 & 0 & 0 & & & \\ \vdots & \vdots & \vdots & \vdots & \vdots & \end{bmatrix} \begin{bmatrix} x_1 \\ x_2 \\ x_3 \\ \vdots \end{bmatrix} + \begin{bmatrix} 0 \\ 0 \\ 1 \\ \vdots \end{bmatrix} u,$$

$$y = [k_3 \quad k_2 \quad k_1 \quad \cdots] \begin{bmatrix} x_1 \\ x_2 \\ x_3 \\ \vdots \end{bmatrix}$$

The state-variable equations

$$\begin{bmatrix} \dot{x}_1 \\ \dot{x}_2 \\ \dot{x}_3 \\ \dot{x}_4 \\ \dot{x}_5 \\ \dot{x}_6 \end{bmatrix} = \begin{bmatrix} -2 & 1 & 0 & 0 & 0 & 0 \\ 0 & -2 & 0 & 0 & 0 & 0 \\ 0 & 0 & -3 & 0 & 0 & 0 \\ 0 & 0 & 0 & 4 & 1 & 0 \\ 0 & 0 & 0 & 0 & 4 & 1 \\ 0 & 0 & 0 & 0 & 0 & 4 \end{bmatrix} \begin{bmatrix} x_1 \\ x_2 \\ x_3 \\ x_4 \\ x_5 \\ x_6 \end{bmatrix} + \begin{bmatrix} 0 \\ 1 \\ 1 \\ 0 \\ 0 \\ 1 \end{bmatrix} u$$

$$y = [4 \quad -5 \quad 6 \quad 7 \quad -8 \quad 9] \begin{bmatrix} x_1 \\ x_2 \\ x_3 \\ x_4 \\ x_5 \\ x_6 \end{bmatrix}$$

for example, are in Jordan canonical form. They represent a system with transfer function

$$T(s) = \frac{-5}{s+2} + \frac{4}{(s+2)^2} + \frac{6}{s+3} + \frac{9}{s-4} + \frac{-8}{(s-4)^2} + \frac{7}{(s-4)^3}$$

❑ DRILL PROBLEM

D8.9 The following systems have repeated characteristic roots. Find an alternate set of state equations in Jordan canonical form.

(a)

$$\begin{bmatrix} \dot{x}_1 \\ \dot{x}_2 \end{bmatrix} = \begin{bmatrix} 2 & 9 \\ -1 & -4 \end{bmatrix} \begin{bmatrix} x_1 \\ x_2 \end{bmatrix} + \begin{bmatrix} 4 \\ -3 \end{bmatrix} u$$

$$y = [2 \quad -6] \begin{bmatrix} x_1 \\ x_2 \end{bmatrix}$$

Ans.

$$\begin{bmatrix} \dot{z}_1 \\ \dot{z}_2 \end{bmatrix} = \begin{bmatrix} -1 & 1 \\ 0 & -1 \end{bmatrix} \begin{bmatrix} z_1 \\ z_2 \end{bmatrix} + \begin{bmatrix} 0 \\ 1 \end{bmatrix} u$$

$$y = [-60 \quad 26] \begin{bmatrix} z_1 \\ z_2 \end{bmatrix}$$

(b)

$$\begin{bmatrix} \dot{x}_1 \\ \dot{x}_2 \\ \dot{x}_3 \end{bmatrix} = \begin{bmatrix} -6 & 1 & 0 \\ -9 & 0 & 1 \\ 0 & 0 & 0 \end{bmatrix} \begin{bmatrix} x_1 \\ x_2 \\ x_3 \end{bmatrix} + \begin{bmatrix} 1 \\ 2 \\ -1 \end{bmatrix} u$$

$$y = [3 \quad -2 \quad 1] \begin{bmatrix} x_1 \\ x_2 \\ x_3 \end{bmatrix}$$

Ans.
$$\begin{bmatrix} \dot{z}_1 \\ \dot{z}_2 \\ \dot{z}_3 \end{bmatrix} = \begin{bmatrix} -3 & 1 & 0 \\ 0 & -3 & 1 \\ 0 & 0 & -3 \end{bmatrix} \begin{bmatrix} z_1 \\ z_2 \\ z_3 \end{bmatrix} + \begin{bmatrix} 0 \\ 0 \\ 1 \end{bmatrix} u$$

$$y = \begin{bmatrix} 2 & -2 & 0 \end{bmatrix} \begin{bmatrix} z_1 \\ z_2 \\ z_3 \end{bmatrix}$$

❑ Computer-Aided Learning

To create a state space model, we use the "ss" command with the following syntax:

```
SYS=SS(A,B,C,D)
```

For example, to define the following system:

$$\dot{x} = \begin{pmatrix} 1 & 2 \\ 3 & 4 \end{pmatrix} x + \begin{pmatrix} -1 \\ 6 \end{pmatrix} u$$

$$y = (0 \quad -14)x + 8u$$

we enter the following commands

```
>> a=[1, 2; 3, 4]; b=[-1; 6]; c=[0, -14]; d=8;
>> g=ss(a,b,c,d)
```

and MATLAB responds

```
a=
             x1          x2
     x1   1.00000     2.00000
     x2   3.00000     4.00000
b=
             u1
     x1  -1.00000
     x2   6.00000
c=
             x1          x2
     y1      0       -14.00000
d=
             u1
     y1   8.00000
```

Continuous-Time System

We can convert from state space to transfer function form using the "tf" command.

```
>> g_s=tf(g)
```

Transfer function:

```
  8 s^2-124 s+110
-----------------
     s^2-5 s-2
```

We can then transfer back to state space form using the "ss" command

```
g_ss=ss(g_s)
a=
              x1          x2
      x1    5.00000    0.50000
      x2    4.00000          0
b=
              u1
      x1    8.00000
      x2          0
c=
              x1          x2
      y1  -10.50000    3.93750
d=
              u1
      y1    8.00000
```

Continuous-Time System

Note that the A, B, C, D matrices that MATLAB returns are different from the ones we used to define the original system we called g. The reason as explained in the text is that the state space representation is not unique.

Moreover the "tf" and "ss" commands create system objects that MATLAB commands can interpret. If we want direct access to the object features such as numerator, denominator, and the $\{A, B, C, D\}$ matrices, we need to extract them. To extract the $\{A, B, C, D\}$ matrices, we use the "ssdata" command.

```
>>[a,b,c,d]=ssdata(g_ss)
   a=
     5.0000     0.5000
     4.0000          0
   b=
     8
     0
   c=
     -10.5000   3.9375
   d=  8
```

We can now use these matrices to determine stability, controllability and other system properties.

The system can be transformed to other forms using the "canon" and "ss2ss" commands. The "canon" command has the following syntax:

```
sc=canon(sys,'type')
```

where type is either 'companion' or 'modal.' The latter diagonalizes the system (also known as the *modal realization*), and the former converts it to companion form (a variation of the phase or dual-phase variable forms)

```
>>sd=canon(g_ss,'modal')
 a=
              x1           x2
      x1   5.37228          0
      x2      0        -0.37228
 b=
              ul
      x1   9.32759
      x2  -5.59456
 c=
              x1           x2
      yl  -6.07043     4.89359
 d=
              ul
      yl   8.00000
>>[sd]=canon(g_ss,'com')
 a=
              x1           x2
      x1      0         2.00000
      x2   1.00000      5.00000
 b=
              ul
      x1   1.00000
      x2      0
 c=
              x1           x2
      yl  -84.00000   -294.00000
 d=
              ul
      yl   8.00000
```

The "ss2ss" command has the syntax

`sys2=ss2ss(sysl,P)`   or   `[a2,b2,c2,d2]=ss2ss(a,b,c,d,P)`

where P is the nonsingular state transformation matrix. Note that our definition of P is different from that of MATLAB, which defines P by

`z(t)=P x(t)`   or   `new state=P.old state`

We, on the other hand, use

`x(t)=P z(t)`   or   `old state=P.new state`

Hence our P is the inverse of what MATLAB uses. To get the answers in the book using the "ss2ss" command use inv(p) instead of p.

Here is an example:

$$x = \begin{pmatrix} -2 & 1 & 1 \\ -3 & 0 & 0 \\ 0 & 0 & 0 \end{pmatrix} x + \begin{pmatrix} 1 \\ 1 \\ 1 \end{pmatrix} u$$

$$y = \begin{pmatrix} 2 & -2 & 1 \\ 0 & -1 & 1 \end{pmatrix} x$$

Use the transformation matrix $P = \begin{pmatrix} 1 & 0 & 1 \\ 0 & 1 & 3 \\ 0 & 0 & 4 \end{pmatrix}$

```
>>p=[1 0 1;0 1 3;0 0 4];
>>a=[-2 1 1;-3 0 0;0 0 0];
>>b=[1;1;1];
>>c=[2 -2 1;0 -1 1]; d=[0; 0];
>>g=ss(a,b,c,d);
>>tf(g) % the transfer function of the 2-output system
```

Transfer function from input to output

```
       s^2+8s+9
#1:  ----------
      s^3+2s^2+3s
         3s+6
#2:  ----------
       s^3+2s^2+3s
```

To transform using the given P matrix we use "inv (p)" to get the matrix P^{-1}. We then extract the new $\{A, B, C, D\}$ matrices to verify our answers:

```
[a2,b2,c2,d2]=ss2ss(a,b,c,d,inv(p))

a2=
    -2    1    5
    -3    0   -3
     0    0    0
b2=
     0.7500
     0.2500
     0.2500
c2=
    2    -2    0
    0    -1    1
d2=
    0
    0
```

Compare the foregoing answers with the ones from the formulas: $\bar{A} = P^{-1}AP$, $\bar{B} = P^{-1}B$, $\bar{C} = CP$

```
>>p\a*p
ans=
   -2    1    5
   -3    0   -3
    0    0    0
>>p\b
ans=
         0.7500
         0.2500
         0.2500
>>c*p
ans=
    2   -2    0
    0   -1    1
```

C8.1

(a) Find the transfer function of the systems defined in Drill Problem D8.4.

(b) Redo Drill Problem D8.5.

(c) Convert the systems defined in Drill Problem D8.6 to state space form, and then diagonalize using the "canon" command to verify the answers.

8.4 Time Response from State Equations

8.4.1 Laplace Transform Solution

One method of calculating the state of a system as a function of time is to Laplace-transform the equations, solve for the transform of the signals of interest, then invert the transforms. The system outputs, being linear combinations of the state signals, are easily found from the state.

For example, consider the system

$$\begin{bmatrix} \dot{x}_1 \\ \dot{x}_2 \end{bmatrix} = \begin{bmatrix} -6 & 1 \\ -5 & 0 \end{bmatrix} \begin{bmatrix} x_1 \\ x_2 \end{bmatrix} + \begin{bmatrix} 0 \\ 1 \end{bmatrix} u$$

$$y = [1 \quad -2] \begin{bmatrix} x_1 \\ x_2 \end{bmatrix}$$

with initial state

$$\begin{bmatrix} x_1(0^-) \\ x_2(0^-) \end{bmatrix} = \begin{bmatrix} -3 \\ 1 \end{bmatrix}$$

and input

$$u(t) = 7$$

where 7 represents a step function starting at $t = 0$.

The Laplace-transformed state equations are as follows:

$$\begin{cases} sX_1(s) + 3 = -6X_1(s) + X_2(s) \\ sX_2(s) - 1 = -5X_1(s) + \dfrac{7}{s} \end{cases}$$

$$\begin{cases} (s+6)X_1(s) - X_2(s) = -3 \\ 5X_1(s) + sX_2(s) = 1 + \dfrac{7}{s} = \dfrac{s+7}{s} \end{cases}$$

$$X_1(s) = \frac{\begin{vmatrix} -3 & -1 \\ (s+7)/s & s \end{vmatrix}}{\begin{vmatrix} s+6 & -1 \\ 5 & s \end{vmatrix}} = \frac{-3s + (s+7)/s}{s^2 + 6s + 5}$$

$$= \frac{-3s^2 + s + 7}{s(s+1)(s+5)} = \frac{\frac{7}{5}}{s} + \frac{-\frac{3}{4}}{s+1} + \frac{-\frac{73}{20}}{s+5}$$

$$x_1(t) = \tfrac{7}{5} - \tfrac{3}{4}e^{-t} - \tfrac{73}{20}e^{-5t} \qquad t \geqslant 0$$

8.4.2 Time Domain Response of First-Order Systems

In many situations, it is advantageous to have an expression for the solution of a set of state equations as functions of time rather than in terms of Laplace transforms. For a first-order state-variable system, we write

$$\frac{dx}{dt} = ax + bu$$

$$sX(s) - x(0^-) = aX(s) + bu(s)$$

$$X(s) = \frac{x(0^-)}{s-a} + bu(s)\frac{1}{s-a}$$

$$x(t) = \mathcal{L}^{-1}\left\{\frac{x(0^-)}{s-a} + bu(s)\frac{1}{s-a}\right\}$$

$$= e^{at}x(0^-) + \text{convolution } [bu(t), e^{at}]$$

$$= e^{at}x(0^-) + \int_{0^-}^{t} e^{a(t-\tau)}bu(\tau)d\tau$$

the inverse transform of a product of Laplace transforms being the convolution of the corresponding time functions.

As a numerical example, consider the first-order system

$$\dot{x} = -2x + 3$$

$$y = 4x$$

The general solution for $x(t)$ is

$$x(t) = e^{-2t}x(0^-) + \int_{0^-}^{t} 3e^{-2(t-\tau)}u(\tau)d\tau$$

If

$$x(0^-) = 10 \quad and\ u = 5$$

then

$$\begin{aligned} x(t) &= 10e^{-2t} + \int_{0^-}^{t} 15e^{-2(t-\tau)}d\tau \\ &= 10e^{-2t} + 15e^{-2t}\left.\frac{e^{2\tau}}{2}\right|_{0^-}^{t} \\ &= 10e^{-2t} + 15e^{-2t}\frac{e^{-2t}-1}{2} \\ &= \tfrac{5}{2}e^{-2t} + \tfrac{15}{2} \quad t \geqslant 0 \end{aligned}$$

and the system output is

$$y(t) = 4x(t) = 10e^{-2t} + 30 \quad t \geqslant 0$$

8.4.3 Time Domain Response of Higher-Order Systems

In general, a state-variable system

$$\dot{x} = Ax + Bu$$

has state response given by

$$\begin{aligned} &sX(s) - x(0^-) = AX(s) + BU(s) \\ &[sI - A]X(s) = x(0^-) + BU(s) \\ &X(s) = [sI - A]^{-1}x(0^-) + [sI = -A]^{-1}BU(s) \end{aligned}$$

Denoting the *resolvent matrix* by $\Phi(s) = (sI - A)^{-1}$ and its inverse Laplace transform, the *state transition matrix*, by

$$\Phi(t) = \mathcal{L}^{-1}\{[sI - A]^{-1}\}$$

then

Solution of state equations.

$$\begin{aligned} x(t) &= \Phi(t)x(0^-) + \text{convolution } [Bu(t), \Phi(t)] \\ &\boxed{= \Phi(t)x(0^-) + \int_0^t \Phi(t-\tau)Bu(\tau)d\tau} \end{aligned}$$

For example, for the system

$$\begin{bmatrix} \dot{x}_1 \\ \dot{x}_2 \end{bmatrix} = \begin{bmatrix} -3 & 1 \\ -2 & 0 \end{bmatrix}\begin{bmatrix} x_1 \\ x_2 \end{bmatrix} + \begin{bmatrix} 2 \\ -1 \end{bmatrix}u$$

the state transition matrix is given by

$$\begin{aligned} \Phi(t) &= \mathcal{L}^{-1}\left\{[sI - A]^{-1}\right\} \\ &= \mathcal{L}^{-1}\left\{\begin{bmatrix} s+3 & -1 \\ 2 & s \end{bmatrix}^{-1}\right\} \end{aligned}$$

$$= \mathcal{L}^{-1}\begin{bmatrix} \frac{s}{s^2+3s+2} & \frac{1}{s^2+3s+2} \\ \frac{-2}{s^2+3s+2} & \frac{s+3}{s^2+3s+2} \end{bmatrix}$$

$$= \mathcal{L}^{-1}\begin{bmatrix} \frac{-1}{s+1}+\frac{2}{s+2} & \frac{1}{s+1}+\frac{-1}{s+2} \\ \frac{-2}{s+1}+\frac{2}{s+2} & \frac{2}{s+1}+\frac{-1}{s+2} \end{bmatrix}$$

$$= \begin{bmatrix} -e^{-t}+2e^{-2t} & e^{-t}-e^{-2t} \\ -2e^{-t}+2e^{-2t} & 2e^{-t}-e^{-2t} \end{bmatrix}$$

The system state is, in terms of initial conditions and the inputs,

$$\begin{bmatrix} x_1(t) \\ x_2(t) \end{bmatrix} = \begin{bmatrix} -e^{-t}+2e^{-2t} & e^{-t}-e^{-2t} \\ -2e^{-t}+2e^{-2t} & 2e^{-t}-e^{-2t} \end{bmatrix}\begin{bmatrix} x_1(0^-) \\ x_2(0^-) \end{bmatrix}$$

$$+ \int_{0^-}^{t} \begin{bmatrix} -e^{-(t-\tau)}+2e^{-2(t-\tau)} & e^{-(t-\tau)}-e^{-2(t-\tau)} \\ -2e^{-(t-\tau)}+2e^{-2(t-\tau)} & 2e^{-(t-\tau)}-e^{-2(t-\tau)} \end{bmatrix}$$

$$\times \begin{bmatrix} 2 \\ -1 \end{bmatrix} u(\tau)d\tau$$

❑ DRILL PROBLEMS

D8.10 Use Laplace transform methods to find the outputs of the following systems for $t \geqslant 0$ with the given inputs and initial conditions:

(a)

$$\dot{x} = -2x + u(t)$$

$$y = 10x$$

$$x(0^-) = 3$$

$$u(t) = 4e^{5t}$$

Ans. $\frac{170}{7}e^{-2t} + \frac{40}{7}e^{5t}$

(b)

$$\begin{bmatrix} \dot{x}_1 \\ \dot{x}_2 \end{bmatrix} = \begin{bmatrix} 0 & 1 \\ -12 & -7 \end{bmatrix}\begin{bmatrix} x_1 \\ x_2 \end{bmatrix} + \begin{bmatrix} 1 \\ 1 \end{bmatrix} u$$

$$y = [1 \quad -1]\begin{bmatrix} x_1 \\ x_2 \end{bmatrix}$$

$$\begin{bmatrix} x_1(0^-) \\ x_2(0^-) \end{bmatrix} = \begin{bmatrix} 10 \\ 0 \end{bmatrix}$$

where the input $u(t)$ is the unit step function.

Ans. $Y(s) = (10s^2 + 190s + 20)/s(s+3)(s+4)$

$y(t) = \frac{5}{3} + \frac{460}{3}e^{-3t} - 145e^{-4t} \qquad t \geqslant 0$

(c)

$$\begin{bmatrix} \dot{x}_1 \\ \dot{x}_2 \\ \dot{x}_3 \end{bmatrix} = \begin{bmatrix} -5 & 1 & 0 \\ -6 & 0 & 1 \\ 0 & 0 & 0 \end{bmatrix} \begin{bmatrix} x_1 \\ x_2 \\ x_3 \end{bmatrix} + \begin{bmatrix} 0 \\ 0 \\ 1 \end{bmatrix} u$$

$$y = [1 \quad 0 \quad 0] \begin{bmatrix} x_1 \\ x_2 \\ x_3 \end{bmatrix}$$

$x(0) = 0$

$u(t) = \delta(t)$, where $\delta(t)$ is the unit impulse.

Ans. $\frac{1}{6} + \frac{1}{3}e^{-3t} - \frac{1}{2}e^{-2t} \quad t \geqslant 0$

D8.11 Calculate state transition matrices for system with the following state coupling matrices A, using

$$\Phi(t) = \mathcal{L}^{-1}\left\{[sI - A]^{-1}\right\}:$$

(a)

$$\begin{bmatrix} -9 & 1 \\ -14 & 0 \end{bmatrix}$$

Ans.
$$\begin{bmatrix} -\frac{2}{5}e^{-2t} + \frac{7}{5}e^{-7t} & \frac{1}{5}e^{-2t} - \frac{1}{5}e^{-7t} \\ -\frac{14}{5}e^{-2t} + \frac{14}{5}e^{-7t} & \frac{7}{5}e^{-2t} - \frac{2}{5}e^{-7t} \end{bmatrix}$$

(b)

$$\begin{bmatrix} 1 & -1 \\ 2 & -4 \end{bmatrix}$$

Ans.
$$\begin{bmatrix} 1.11e^{0.56t} - 0.11e^{-3.56t} & -0.24e^{0.56t} + 0.24e^{-3.56t} \\ 0.48e^{0.56t} - 0.48e^{-3.56t} & -0.11e^{0.56t} + 1.11e^{-3.56t} \end{bmatrix}$$

8.4.4 System Response Computation

One advantage of placing system equations in a state-variable form is that it is well suited to digital computer calculations. Computers are not particularly efficient at equation manipulation, Laplace transformation, and the like, but they excel at such repetitive tasks as matrix addition and multiplication. The capability of simulating a system, that is, investigating and testing its performance by modeling, is important to the designer, particularly for the common situation in which the plant is expensive and the design must be correct when it is first installed.

The state transition matrix can be approximated by an $(m+1)$-term Taylor series

$$\Phi(\Delta t) = I + A(\Delta t) + \frac{1}{2}A^2(\Delta t)^2 + \cdots + \left(\frac{1}{m!}\right) A^m (\Delta t)^m$$

The convolution integral depends on the state transition matrix and on the input, both of which can be functions of time. However, if Δt is a very short time, then $u(t)$ can be removed from the integral so that

$$\text{Convolution integral} = D(\Delta t)Bu(t)$$

$$D(\Delta t) = I\Delta t + \frac{1}{2}A(\Delta t)^2 + \left(\frac{1}{3!}\right)A^2(\Delta t)^3 + \cdots + \left(\frac{1}{(m+1)!}\right)A^m(\Delta t)^{m+1}$$

For sufficiently small time increments Δt, one can start with the initial state $x(0)$ and calculate $x(\Delta t)$ as follows:

$$x(\Delta t) \cong (I + A\Delta t)\,x(0) + (B\Delta t)\,u(0)$$

then $x(2\Delta t)$ may be calculated from $x(\Delta t)$,

$$x(2\Delta t) \cong (I + A\Delta t)\,x(\Delta t) + (B\Delta t)\,u(\Delta t)$$

and so on, obtaining approximate solutions for the state,

$$x\{(k+1)\,\Delta t\} \cong (I + A\Delta t)\,x\,(k\Delta t) + (B\Delta t)\,u\,(k\Delta t)$$

For example, the response of the first-order system

$$\dot{x} = -2x + u$$

$$y = x$$

with

$$x(0^-) = 10$$

$$u(t) = 3\sin t$$

is approximated by

$$x\{(k+1)\,\Delta t\} \cong (1 - 2\Delta t)\,x\,(k\Delta t) + 3\Delta t\ \sin(k\Delta t)$$

with

$$x(0\cdot\Delta t) = 10$$

Representative computer-generated plots of $x(t)$ are given in Figure 8.10 for various choices of Δt. For a sufficiently small time increment Δt, the approximate response is very nearly the actual system response.

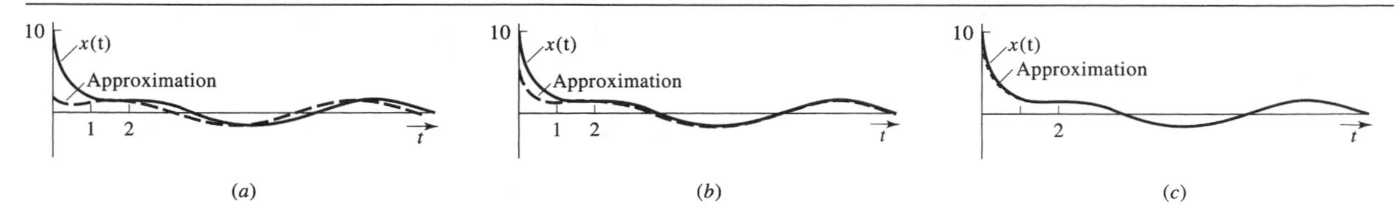

Figure 8.10 Computer-generated response plots for a first-order system. (a) Step size $\Delta t = 0.4$. (b) Step size $\Delta t = 0.2$. (c) Step size $\Delta t = 0.05$.

Another example system is the following:

$$\begin{bmatrix} \dot{x}_1 \\ \dot{x}_2 \end{bmatrix} = \begin{bmatrix} -2 & 1 \\ -3 & 0 \end{bmatrix} \begin{bmatrix} x_1 \\ x_2 \end{bmatrix} + \begin{bmatrix} 2 \\ -1 \end{bmatrix} u$$

$$y = \begin{bmatrix} 1 & -\frac{1}{2} \end{bmatrix} \begin{bmatrix} x_1 \\ x_2 \end{bmatrix}$$

with

$$\begin{bmatrix} x_1(0^-) \\ x_2(0^-) \end{bmatrix} = \begin{bmatrix} -4 \\ 5 \end{bmatrix}$$

$$u(t) = \cos 0.25t$$

It is approximated by

$$\begin{bmatrix} x_1\{(k+1)\,\Delta t\} \\ x_2\{(k+1)\,\Delta t\} \end{bmatrix} = \begin{bmatrix} 1-2\Delta t & \Delta t \\ -3\Delta t & 1 \end{bmatrix} \begin{bmatrix} x_1(k\,\Delta t) \\ x_2(k\,\Delta t) \end{bmatrix} + \begin{bmatrix} 2 \\ -1 \end{bmatrix} \Delta t \cos(0.25k\,\Delta t)$$

$$y\{(k+1)\,\Delta t\} = \begin{bmatrix} 1 & -\frac{1}{2} \end{bmatrix} \begin{bmatrix} x_1\{(k+1)\,\Delta t\} \\ x_2\{(k+1)\,\Delta t\} \end{bmatrix}$$

or

$$\begin{cases} x_1\{(k+1)\,\Delta t\} = (1-2\Delta t)\,x_1(k\,\Delta t) + \Delta t x_2(k\,\Delta t) + 2\Delta t \cos(0.25k\,\Delta t) \\ x_2\{(k+1)\,\Delta t\} = -3\Delta t x_1(k\,\Delta t) + x_2(k\,\Delta t) - \Delta t \cos(0.25k\,\Delta t)) \\ y\{(k+1)\,\Delta t\} = x_1\{(k+1)\,\Delta t\} - \frac{1}{2}x_2\{(k+1)\,\Delta t\} \end{cases}$$

with

$$\begin{cases} x_1(0 \cdot \Delta t) = -4 \\ x_2(0 \cdot \Delta t) = 5 \end{cases}$$

Computer-generated response plots for this system are given in Figure 8.11, where $\Delta t = 0.05$.

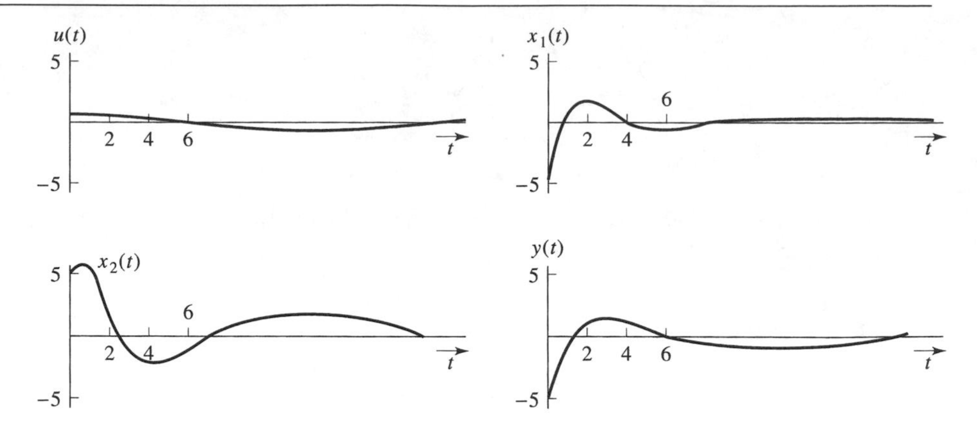

Figure 8.11 Computer-generated response plots for a second-order system.

Improved accuracy and reduced computation time may result from using more involved approximations—for example, matrix power series, predictor correctors, or Runge–Kutta methods.

❑ DRILL PROBLEMS

D8.12 For the following systems, develop discrete-time approximation equations using the indicated time steps Δt.

(a)

$$\begin{bmatrix} \dot{x}_1 \\ \dot{x}_2 \end{bmatrix} = \begin{bmatrix} 3 & 1 \\ 2 & -1 \end{bmatrix} \begin{bmatrix} x_1 \\ x_2 \end{bmatrix} + \begin{bmatrix} 1 \\ 4 \end{bmatrix} u$$

$$y = [-3 \quad -1] \begin{bmatrix} x_1 \\ x_2 \end{bmatrix}$$

$$\Delta t = 0.2$$

Ans.

$$x\,[(k+1)\ \Delta t] = \begin{bmatrix} 1.6 & 0.2 \\ 0.4 & 0.8 \end{bmatrix} x(k\ \Delta t) + \begin{bmatrix} 0.2 \\ 0.8 \end{bmatrix} u(k\ \Delta t)$$

$$y = [-3 \quad -1]\, x(k\ \Delta t)$$

(b)

$$\begin{bmatrix} \dot{x}_1 \\ \dot{x}_2 \\ \dot{x}_3 \end{bmatrix} = \begin{bmatrix} 1 & 2 & 3 \\ 7 & -2 & -3 \\ 6 & 0 & 4 \end{bmatrix} \begin{bmatrix} x_1 \\ x_2 \\ x_3 \end{bmatrix} + \begin{bmatrix} 1 & -2 \\ -1 & 3 \\ 0 & 4 \end{bmatrix} \begin{bmatrix} u_1 \\ u_2 \end{bmatrix}$$

$$y = [5 \quad -2 \quad 1] \begin{bmatrix} x_1 \\ x_2 \\ x_3 \end{bmatrix}$$

$$\Delta t = 0.01$$

Ans.

$$x\,[(k+1)\,\Delta t] = \begin{bmatrix} 1.01 & 0.02 & 0.03 \\ 0.07 & 0.98 & -0.03 \\ 0.06 & 0 & 1.04 \end{bmatrix} x(k\ \Delta t) + \begin{bmatrix} 0.01 & -0.02 \\ -0.01 & 0.03 \\ 0 & 0.04 \end{bmatrix} u(k\ \Delta t)$$

$$y = [\,5 \quad -2 \quad 1\,]\,x(k\ \Delta t)$$

D8.13 For the set of state equations

$$\begin{bmatrix} \dot{x}_1 \\ \dot{x}_2 \end{bmatrix} = \begin{bmatrix} -2 & 1 \\ -3 & 0 \end{bmatrix} \begin{bmatrix} x_1 \\ x_2 \end{bmatrix} + \begin{bmatrix} 1 \\ 4 \end{bmatrix} u$$

$$y = [\,1 \quad -1\,] \begin{bmatrix} x_1 \\ x_2 \end{bmatrix}$$

a discrete-time approximation is

$$\begin{bmatrix} x_1\,\{(k+1)\,\Delta t\} \\ x_2\,\{(k+1)\,\Delta t\} \end{bmatrix} = \begin{bmatrix} (1-2\Delta t) & \Delta t \\ -3\Delta t & 1 \end{bmatrix} \begin{bmatrix} x_1(k\ \Lambda t) \\ x_2(k\ \Delta t) \end{bmatrix} + \begin{bmatrix} \Lambda t \\ 4\Delta t \end{bmatrix} u(k\ \Delta t)$$

$$y(k\ \Delta t) = [\,1 \quad -1\,] \begin{bmatrix} x_1(k\ \Delta t) \\ x_2(k\ \Delta t) \end{bmatrix}$$

If

$$\begin{bmatrix} x_1(0^-) \\ x_2(0^-) \end{bmatrix} = \begin{bmatrix} 10 \\ 0 \end{bmatrix} \quad \text{and} \quad u(t) = 2 \quad t \geqslant 0$$

calculate approximate values for $x(\Delta t)$, $x(2\Delta t)$, and $x(3\Delta t)$ for the following:

(a)

$\Delta t = 0.2$

Ans. $$\begin{bmatrix} 6.4 \\ -4.4 \end{bmatrix}, \begin{bmatrix} 3.36 \\ -6.64 \end{bmatrix}, \begin{bmatrix} 1.09 \\ -7.06 \end{bmatrix}$$

(b)

$\Delta t = 0.1$

Ans. $$\begin{bmatrix} 8.2 \\ -2.2 \end{bmatrix}, \begin{bmatrix} 6.54 \\ -3.86 \end{bmatrix}, \begin{bmatrix} 6.37 \\ -5.02 \end{bmatrix}$$

(c)

$$\Delta t = 0.02$$

Ans. $\begin{bmatrix} 9.84 \\ -0.44 \end{bmatrix}, \begin{bmatrix} 9.67 \\ -0.87 \end{bmatrix}, \begin{bmatrix} 9.503 \\ -1.29 \end{bmatrix}$

8.5 Stability

A system is stable if its eigenvalues (or characteristic values) are in the left half-plane (LHP). We want to expand upon this issue and formally define various notions of stability.

8.5.1 Asymptotic Stability

Consider a system represented in state space:

$$\dot{x} = Ax \qquad x(0) = x_0$$

The system is said to be *asymptotically stable* if all the states approach zero with time—that is,

$$x(t) \to 0 \quad \text{as} \quad t \to \infty$$

Now, let us diagonalize the system. This simplifies the task as it will allow us to look at the components of the system one at a time.

$$x(t) = Pz(t) \to \dot{z}(t) = \Lambda z(t)$$

where Λ is the diagonal matrix of eigenvalues of A—that is,

$$\Lambda = \begin{bmatrix} \lambda_1 & 0 & \cdot & 0 \\ 0 & \lambda_2 & \cdot & \cdot \\ \cdot & 0 & \cdot & \cdot \\ \cdot & \cdot & \cdot & 0 \\ 0 & 0 & \cdot & \lambda_n \end{bmatrix}$$

The individual subsystems are

$$\dot{z}_i(t) = \lambda_i z_i(t)$$

For asymptotic stability eigenvalues must be in the LHP.

The solution to this first-order system is

$$z_i(t) = e^{\lambda_i t} z_i(0)$$

Clearly, if the eigenvalues of the system have negative real parts (i.e., they are in the LHP), the individual states go to zero asymptotically with time: $z_i(t) \to 0$ implies that $z(t) \to 0$. Because $x(t) = Pz(t)$, we conclude that $x(t) \to 0$. Hence, it has been proved that the system is asymptotically stable if its eigenvalues are in the LHP.

8.5.2 BIBO Stability

Another notion of stability that has been used throughout the book, starting from Chapter 2, is input–output stability. This concept is also referred to as *bounded-input, bounded-output* (or *BIBO*) stability. BIBO stability means that the system output is bounded for all bounded inputs. This is,

$$|u(t)| \leqslant N < \infty \rightarrow |y(t)| \leqslant M < \infty \quad \text{for all bounded inputs}$$

where M and N are some finite bounds for u and y. Examples of bounded functions are negative exponentials and sinusoids. For example,

$$\left|e^{-at}\right| \leqslant 1 \quad \text{or} \quad |\sin(t)| \leqslant 1 \rightarrow M = 1 \quad \text{or} \quad N = 1$$

The condition that guarantees BIBO stability is the familiar condition that all the transfer function poles be in the LHP.

What is the difference between the two definitions? Does one imply the other? The answer is obtained from looking at the expression for the transfer function in the terms of state space matrices:

$$T(s) = \frac{N(s)}{D(s)} = C(sI - A)^{-1}B = \frac{C \,\text{adjoint}\,[sI - A]\, B}{|sI - A|}$$

Recall that eigenvalues of the system are found from the equation

$$|sI - A| = 0$$

Because this is also the polynomial that appears in the denominator of the transfer function, we can draw the following the conclusion:

In the absence of pole–zero cancellations, transfer function poles are identical to the system eigenvalues, hence BIBO stability and asymptotic stability are equivalent.

The pole zero cancellation condition is important for the equivalence. This is because, poles are formally to be identified from the simplified transfer function (i.e., after all the numerator and denominator common terms have been canceled out). For instance, consider the following situation,

$$T(s) = \frac{s - 1}{(s - 1)(s + 2)}$$

in which $T(s)$ has one pole at $s = -2$. The term $(s - 1)$ must be canceled out before the poles and zeros are identified. To prove that the transfer function does not have a pole at 1, take the limit of $T(s)$ as s approaches 1.

$$\lim_{s\to 1} T(s) = \lim_{s\to 1} \frac{s - 1}{(s - 1)(s + 2)} = \lim_{s\to 1} \frac{1}{2s + 1} = \frac{1}{3} \neq \infty$$

This system does not have a pole at $s = 1$.

where we have used L'Hôpital's rule. It is concluded that the system is BIBO stable because it has no poles in the RHP or on the imaginary axis. To determine asymptotic stability, we need to obtain a state space realization. One realization is given by

$$\dot{x} = \begin{bmatrix} -1 & 2 \\ 1 & 0 \end{bmatrix} x + \begin{bmatrix} 1 \\ 0 \end{bmatrix} u$$

$$y = [\,1 \quad -1\,]x$$

The system eigenvalues are at 1 and −2. Because of the eigenvalue of 1, we conclude that the system is unstable in the asymptotic sense. Such contradictory answers to system stability occur only when the system transfer function has pole–zero cancellations in the RHP or on the imaginary axis. What do we make of the system stability in such cases?

We can make some progress toward the answer by looking at the diagonalized system given by

$$\dot{x} = \begin{bmatrix} 1 & 0 \\ 0 & -2 \end{bmatrix} x + \frac{1}{3}\begin{bmatrix} 1 \\ -1 \end{bmatrix} u$$

$$y = [\,0 \quad -3\,]x$$

Observing the signal flow graph of the system shown in Figure 8.12, we note that the unstable mode at 1 is not connected to the output. Since the transfer function describes the input–output properties of the system, it is not surprising that it does not detect the offending mode. Imagine an experiment where you apply an input to the system, and connect your measurement instrument to the output. Even though the first state is practically "blowing up," you will not be aware of it. This is what is happening here. Fortunately, such pathological cases do not happen very often in practice. After all, because of modeling uncertainties and component tolerances, it is practically impossible for a system to have its poles and zeros exactly at the same place. We will see in the next section that this issue is related to system controllability and observability properties, and that it can be detected and avoided.

State space description of systems are called *internal representation* because they allow us to observe the internal structure of systems, whereas transfer function description is an external representation. In the case of our example, the state space representation allowed a more accurate description of the system.

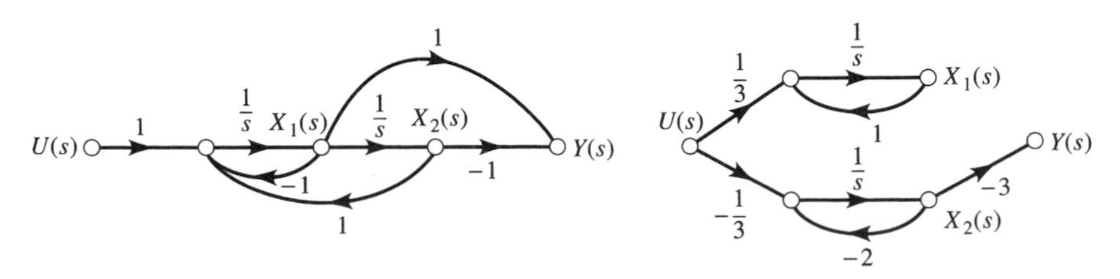

Figure 8.12 (a) Signal flow graph of a system with pole–zero cancellation. (b) Diagonal representation of the system.

Going back to the question of stability, we conclude that the preceding system is practically unstable, because any nonzero initial conditions on the first state will grow indefinitely.

8.5.3 Internal Stability

Asymptotic stability is a notion that is based on the state space description of systems, whereas BIBO stability is based on transfer function description. There is another notion of stability, called *internal stability*, that is based on transfer function description and is stronger than BIBO stability. To define it, we need to consider the general block diagram in Figure 8.13. This figure is a more realistic diagram of feedback control systems. The blocks represent the plant, the controller, and the sensor dynamics. Besides the usual reference (or command) input, the ever-present noise and disturbance inputs are explicitly included.

Internal stability requires that all signals within the feedback system remain bounded for all bounded inputs. This is equivalent to the requirement that all possible transfer functions between all inputs and outputs be stable. It can be shown that only nine transfer functions between the three inputs (R, D, N) and the three outputs taken at the output of the summing junctions (U, V, W) are sufficient. In practice, determining the nine transfer functions and checking them for stability is still a major task. The following result, a necessary and sufficient condition for internal stability, will be used as a test for internal stability. The feedback system is internally stable if and only if the transfer function $1 + KGH$ has no zeros in the RHP (including the imaginary axis), and the product of KGH has no pole–zero cancellations in the RHP (including the imaginary axis).

As an example, consider the case of

$$G(s) = \frac{1}{s-1} \qquad H(s) = \frac{s-1}{s+1} \qquad \text{and} \qquad K = 1$$

Because there is an RHP pole–zero cancellation in KGH, the system is not internally stable. To show that at least one signal will blow up, note the following transfer functions:

$$\frac{Y(s)}{R(s)} = \frac{1}{s+2} \quad \frac{Y(s)}{D(s)} = \frac{s+1}{(s-1)(s+2)}$$

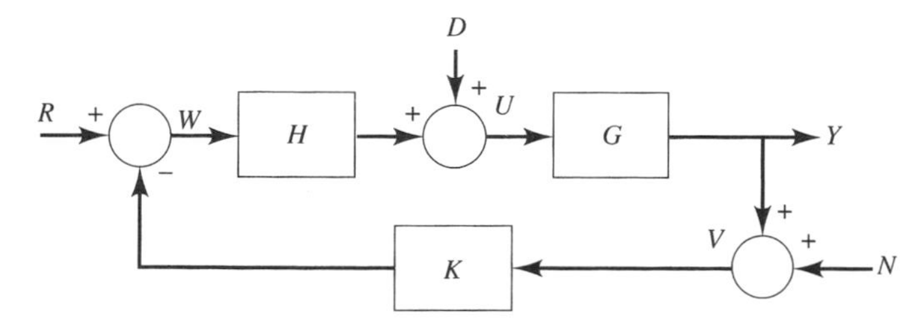

Figure 8.13 Block diagram of a feedback system showing disturbance and noise inputs.

Note that in traditional BIBO stability, it is the transfer function between R and Y that is examined. This transfer function is clearly stable. However, the transfer function between the disturbance D and output Y is unstable. Hence, the slightest disturbances in the system will grow unbounded. For all practical purposes, the system is unstable.

The summary of our discussion of stability is that internal stability is the true stability requirement that must be imposed on feedback control systems. A design lesson that can be drawn from our discussion is the following. We must never cancel the unstable (RHP) plant poles by unstable (RHP) compensator zeros, for this will render the closed-loop system internally unstable. This caution is warranted because canceling poles in undesirable locations by compensator zeros and replacing them by poles in more desirable locations is commonly used by control system designers. This is an acceptable and effective technique, but only for poles that are in the LHP.

❑ DRILL PROBLEMS

D8.14 Determine stability of the following systems. Check for BIBO, asymptotic, and internal stability.

(a)
$$\dot{x} = \begin{bmatrix} 1 & 0 \\ 1 & -1 \end{bmatrix} x + \begin{bmatrix} 0 \\ 1 \end{bmatrix} u$$
$$y = [1 \quad 1]x$$

(b)
$$\dot{x} = \begin{bmatrix} -1 & 1 & 0 \\ 0 & -1 & 0 \\ 0 & 0 & 0 \end{bmatrix} x + \begin{bmatrix} 1 \\ 1 \\ 1 \end{bmatrix} u$$
$$y = [1 \quad 1 \quad 1]x$$

(c)
$$\dot{x} = \begin{bmatrix} 1 & 0 \\ 0 & 1 \end{bmatrix} x + \begin{bmatrix} -1 \\ 1 \end{bmatrix} u$$
$$y = [1 \quad 1]x + u$$

Ans. (a) $T(s) = 1/(s+1)$, eigenvalues $= 0, -1$; BIBO stable but not asymptotically stable; (b) $T(s) = (3s^2 + 5s + 1)/[s(s+1)]^2$; eigenvalues $= 0, -1, -1$; neither BIBO nor asymptotically stable; (c) $T(s) = 1$; eigenvalues $= 1, 1$; BIBO stable but not asymptotically stable

D8.15 Consider the feedback control system of Figure 8.15.

Let $G(s) = 1/(s-1)$. Determine if the system is internally stable in each case.

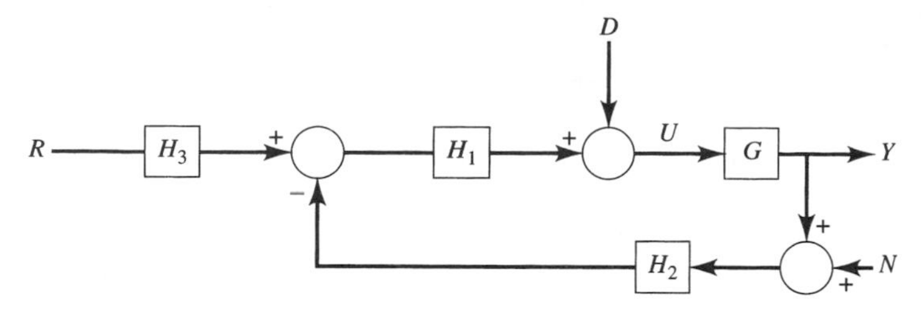

Figure D8.15

(a) $H_1 = \dfrac{s-1}{s+1},\ H_3 = 1,\ H_2 = 1$

(b) $H_1 = 1,\ H_2 = \dfrac{s-1}{s+1},\ H_3 = 1$

(c) $H_1 = 1,\ H_2 = 1,\ H_3 = \dfrac{s-1}{s+1}$

Ans. (a) no; (b) no; (c) no

8.6 Controllability and Observability

In the preceding chapters, a control system was operated upon to provide acceptable phase margin, rise time, or other figures of merit. Perhaps the system is constructed in a way that conspires to thwart efforts at improving performance. By using state-variable methods it is possible to answer fundamental questions about the ability of the control-system designer to effect meaningful improvement in performance and to generate needed sensor measurements. The terms *controllability* and *observability,* respectively, address those needs.

A system is completely controllable if the system state $x(t_f)$ at time t_f can be forced to take on any desired value by applying a control input $u(t)$ over a period of time from t_0 until t_f.

The definition does not restrict the choice of $u(t)$. The idea is that it is possible to move the system state to any desired destination. Perhaps the system is (or is not) constructed in a way that allows control to take place. A test for controllability can easily be constructed.

A system is completely observable if any initial state vector $x(t_0)$ can be reconstructed by examining the system output $y(t)$ over some period of time from t_0 until t_f.

There are no restrictions placed on the output. The definition indicates that any earlier value of the state vector is determinable by watching the output $y(t)$. An automobile would be considered completely observable if, by monitoring speedometer (for speed), odometer (for distance), and steering wheel position (for turning), it is possible to determine where the car was parked before being driven.

For systems of certain kinds (with diagonal A matrices), the tests for controllability and observability are easy to apply. For a nondiagonal system, a test can also be constructed. For a system that is completely controllable, methods will be developed by which an appropriate control can be derived. Similarly, for a system that is completely observable, an observer will be designed to carry out that task of state reconstruction.

Figure 8.14 shows that controllability is tested assuming a zero-state response and that observability is tested assuming a zero-input response. The tests provide a worst-case scenario, where the initial condition does not necessarily aid in control and an input does not necessarily aid in reconstruction of an earlier state. A system that passes the controllability test is usually applied in an environment that has a nonzero initial condition. Similarly, a system that passes the observability test (observers will be considered shortly) is usually applied in an environment that includes an input and control.

Consider the following system:

$$\dot{x}_1 = x_1$$

$$\dot{x}_2 = 2x_2 + u$$

The objective is to force the system states to go to zero. This is another way of stating that we want to make the system asymptotically stable, a common objective. According to definition of controllability, this is an achievable objective if the system is controllable. The solution of the system is

$$x_1(t) = e^t x_1(0)$$

$$x_2(t) = e^{2t} x_2(0) + \int_0^t e^{2(t-\tau)} u(\tau) d\tau$$

Observe that by appropriate choice of the control signal u, the second state can be driven to zero. The first state, however, is uncontrollable. It will always blow up, unless its initial condition is zero. Upon examining the system signal flow graph, shown in Figure 8.15, it is clear that the control signal is not even connected to the first state, so it cannot affect it in any way. Because the system is in decoupled form, its eigenvalues can be obtained by inspection; they are 1 and 2. These are also called system *modes*. When the system is in decoupled form, we can be more specific: The first mode is uncontrollable, whereas the second mode is controllable. Because the objective was to drive all states to zero, the system is declared to be uncontrollable.

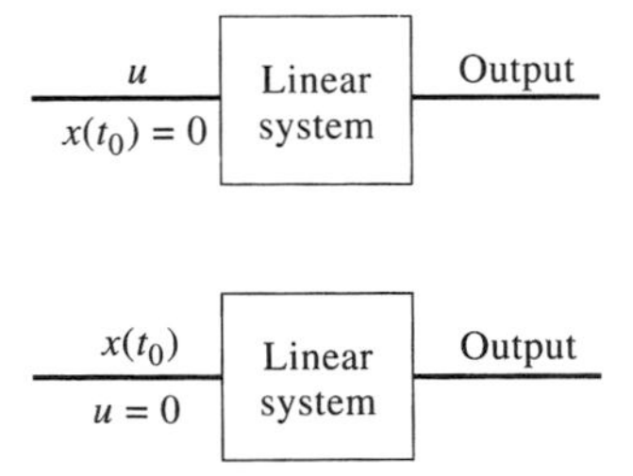

Figure 8.14 Significance of controllability and observability tests. (a) Test for controllability. (b) Test for observability.

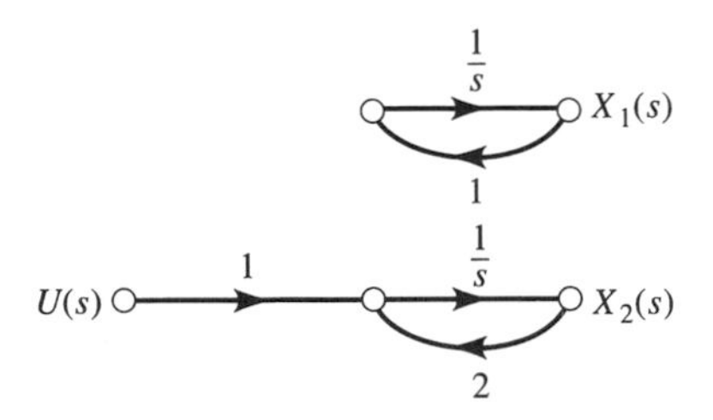

Figure 8.15 Signal flow graph of an uncontrollable system.

How to determine the control signal u to drive the states to arbitrary values. will be discussed later. Driving the states to zero (i.e., stabilization), however, is easy. For instance, in the preceding example, if we choose u as

$$u = -kx_2$$

we get

$$\dot{x}_2 = 2x_2 - kx_2 = (2 - k)x_2$$

Clearly if the gain k is greater than 2, this state will indeed go to zero. The idea behind this choice is to feed the state back to the input using an appropriate gain. This is called *state feedback*, and more will be said about it in the next chapter. That this scheme works is not surprising—after all, feedback has been used for stabilization throughout the book.

Determining controllability and observability is easy when the system is in diagonalized form. To see this, consider the following general example for SISO systems.

$$\begin{bmatrix} \dot{x}_1 \\ \dot{x}_2 \\ \dot{x}_3 \\ \dot{x}_4 \end{bmatrix} = \begin{bmatrix} \lambda_1 & 0 & 0 & 0 \\ 0 & \lambda_2 & 0 & 0 \\ 0 & 0 & \lambda_3 & 0 \\ 0 & 0 & 0 & \lambda_4 \end{bmatrix} \begin{bmatrix} x_1 \\ x_2 \\ x_3 \\ x_4 \end{bmatrix} + \begin{bmatrix} 1 \\ 1 \\ 0 \\ 0 \end{bmatrix} u$$

$$y = [1 \quad 0 \quad 1 \quad 0]x$$

Observing the input connections in Figure 8.16, we conclude that modes 3 and 4 are uncontrollable because they are not connected to the control input. Also, modes 2 and 4 are unobservable because they are not connected to the output. In general we can always categorize the system modes into four categories: controllable and observable

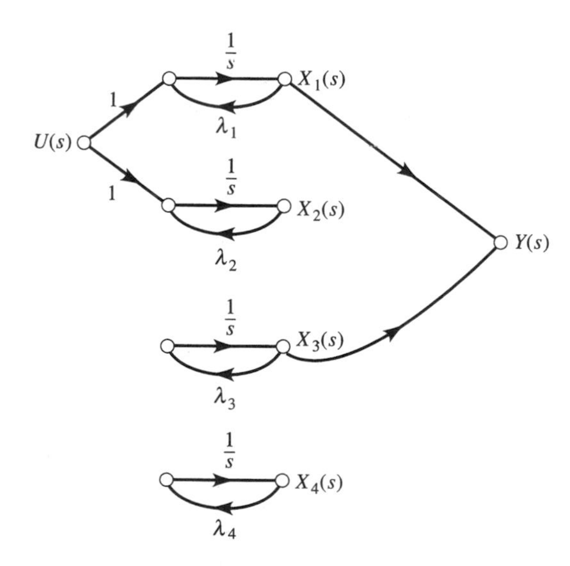

Figure 8.16 Signal flow graph of a system showing controllable, observable, uncontrollable, and unobservable modes.

(as in mode 1), controllable but unobservable (as in mode 2), uncontrollable but observable (as in mode 3), and uncontrollable and unobservable (as in mode 4).

This information is also available from examining the rows and columns of B and C matrices. Uncontrollable modes correspond to zero rows of B: unobservable modes correspond to zero columns of C. The latter applies to multiple-input, multiple-output systems with distinct (i.e., nonrepeated) modes in diagonalized form.

For systems in general (nondiagonalized) form, these properties cannot be determined from the signal flow graph (or block diagram). Similarly, zero rows in B or zero columns in C do not imply anything in general.

8.6.1 The Controllability Matrix

Fortunately, there is a much simpler method of determining system controllability than diagonalization. It can be shown that an nth-order system, with or without repeated modes (eigenvalues),

$$\dot{x} = Ax + Bu$$

is completely controllable if and only if its controllability matrix

$$M_c = \left[B \;\vdots\; AB \;\vdots\; \cdots \;\vdots\; A^{n-1} \quad B \right]$$

is of full rank. The controllability matrix consists of the columns of B followed by the columns of AB, and so on.

For the system

$$\begin{bmatrix} \dot{x}_1 \\ \dot{x}_2 \\ \dot{x}_3 \end{bmatrix} = \begin{bmatrix} -2 & 1 & 2 \\ 4 & 0 & 3 \\ 1 & -1 & 0 \end{bmatrix} \begin{bmatrix} x_1 \\ x_2 \\ x_3 \end{bmatrix} + \begin{bmatrix} 0 & 4 \\ -5 & 0 \\ 0 & 0 \end{bmatrix} \begin{bmatrix} u_1 \\ u_2 \end{bmatrix}$$

A is 3×3, so

$$M_c = \left[B \;\vdots\; AB \;\vdots\; A^2 B \right]$$

Upon using

$$AB = \begin{bmatrix} -2 & 1 & 2 \\ 4 & 0 & 3 \\ 1 & -1 & 0 \end{bmatrix} \begin{bmatrix} 0 & 4 \\ -5 & 0 \\ 5 & 0 \end{bmatrix} = \begin{bmatrix} -5 & -8 \\ 0 & 16 \\ 5 & 4 \end{bmatrix}$$

$$A^2 B = A(AB) = \begin{bmatrix} -2 & 1 & 2 \\ 4 & 0 & 3 \\ 1 & -1 & 0 \end{bmatrix} \begin{bmatrix} -5 & -8 \\ 0 & 16 \\ 5 & 4 \end{bmatrix} = \begin{bmatrix} 20 & 40 \\ -5 & -20 \\ -5 & -24 \end{bmatrix}$$

we can write

$$M_c = \begin{bmatrix} 0 & 4 & -5 & -8 & 20 & 40 \\ -5 & 0 & 0 & 16 & -5 & -20 \\ 0 & 0 & 5 & 4 & -5 & -24 \end{bmatrix}$$

To be of full rank, the controllability matrix must have three linearly independent columns, which it does, since

$$\begin{vmatrix} 0 & 4 & -5 \\ -5 & 0 & 0 \\ 0 & 0 & 5 \end{vmatrix} \neq 0$$

The system

$$\begin{bmatrix} \dot{x}_1 \\ \dot{x}_2 \end{bmatrix} = \begin{bmatrix} 2 & 3 \\ 6 & -1 \end{bmatrix} \begin{bmatrix} x_1 \\ x_2 \end{bmatrix} + \begin{bmatrix} 1 \\ -2 \end{bmatrix} u$$

has controllability matrix

$$M_c = [\, B \;\vdots\; AB \,]$$

where

$$AB = \begin{bmatrix} 2 & 3 \\ 6 & -1 \end{bmatrix} \begin{bmatrix} 1 \\ -2 \end{bmatrix} = \begin{bmatrix} -4 \\ 8 \end{bmatrix}$$

Thus we have

$$M_c = \begin{bmatrix} 1 & -4 \\ -2 & 8 \end{bmatrix}$$

which is not of rank 2, since

$$\begin{vmatrix} 1 & -4 \\ -2 & 8 \end{vmatrix} = 0$$

This system is not completely controllable.

Note that for SISO systems the controllability matrix is square. From matrix algebra, we recall that a square matrix has full rank if and only if its determinant is not zero. Therefore, in the SISO case, the system is controllable if and only if the determinant of M_c is nonzero. The controllability matrix, although easy to apply, provides only a "yes or no" answer on system controllability. To get specific information on individual modes, you have to diagonalize the system.

8.6.2 The Observability Matrix

To determine whether a nondiagonalized nth-order system is completely observable, its observability matrix

$$M_o = \begin{bmatrix} C \\ \text{-----} \\ CA \\ \text{-----} \\ \vdots \\ \text{-----} \\ CA^{n-1} \end{bmatrix}$$

may be formed. The system is completely observable if and only if the observability matrix is of full rank, that is, if M_o has n linearly independent rows.

For example, the system

$$\begin{bmatrix} \dot{x}_1 \\ \dot{x}_2 \\ \dot{x}_3 \end{bmatrix} = \begin{bmatrix} 2 & 1 & 0 \\ -3 & 0 & 1 \\ 4 & 0 & 0 \end{bmatrix} \begin{bmatrix} x_1 \\ x_2 \\ x_3 \end{bmatrix} + \begin{bmatrix} 1 \\ 1 \\ 1 \end{bmatrix} u$$

$$y = [0 \quad 0 \quad 1] \begin{bmatrix} x_1 \\ x_2 \\ x_3 \end{bmatrix}$$

is completely observable:

$$CA = [0 \quad 0 \quad 1] \begin{bmatrix} 2 & 1 & 0 \\ -3 & 0 & 1 \\ 4 & 0 & 0 \end{bmatrix} = [4 \quad 0 \quad 0]$$

$$CA^2 = (CA)\,A = [4 \quad 0 \quad 0] \begin{bmatrix} 2 & 1 & 0 \\ -3 & 0 & 1 \\ 4 & 0 & 0 \end{bmatrix}$$

$$= [8 \quad 4 \quad 0]$$

$$M_o = \begin{bmatrix} 0 & 0 & 1 \\ 4 & 0 & 0 \\ 8 & 4 & 0 \end{bmatrix}$$

As another example, consider the system

$$\begin{bmatrix} \dot{x}_1 \\ \dot{x}_2 \end{bmatrix} = \begin{bmatrix} 1 & 0 \\ 1 & 1 \end{bmatrix} \begin{bmatrix} x_1 \\ x_2 \end{bmatrix} + \begin{bmatrix} 1 \\ 1 \end{bmatrix} u$$

$$\begin{bmatrix} y_1 \\ y_2 \end{bmatrix} = \begin{bmatrix} 1 & -1 \\ -2 & 2 \end{bmatrix} \begin{bmatrix} x_1 \\ x_2 \end{bmatrix}$$

$$M_o = \begin{bmatrix} 1 & -1 \\ -2 & 2 \\ 0 & -1 \\ 0 & 2 \end{bmatrix}$$

The observability matrix has two linearly independent rows (1 and 3).

8.6.3 Controllability, Observability, and Pole–Zero Cancellation

It can be shown, in general, that uncontrollable or unobservable (SISO) systems will have pole–zero cancellations in their transfer functions. We will not prove this fact, but it will be investigated by an example. Consider the second-order system

$$\dot{x} = \begin{bmatrix} 1 & 0 \\ 0 & 2 \end{bmatrix} x + \begin{bmatrix} b_1 \\ b_2 \end{bmatrix} u$$

$$y = \begin{bmatrix} c_1 & c_2 \end{bmatrix} x$$

The transfer function of the system is given by

Lack of controllability or observability leads to pole zero cancellation in the transfer function.

$$T(s) = \frac{(b_1c_1 + b_2c_2)s - (2b_1c_1 + b_2c_2)}{(s-1)(s-2)}$$

Now, if $b_1 = 0$ (or $c_1 = 0$), the mode at 1 becomes uncontrollable (or unobservable) and the pole term $(s-1)$ gets canceled in the transfer function. Similarly, when $b_2 = 0$ (or $c_2 = 0$), the mode at 2 becomes uncontrollable (or unobservable) and the pole term $(s-2)$ will get canceled out.

The example demonstrates that lack of either controllability or observability will lead to pole–zero cancellation in the transfer function. Conversely, pole–zero cancellation in a transfer function implies either uncontrollability or unobservability. As another example, consider the system

$$\dot{x} = \begin{bmatrix} 1 & 1 \\ 0 & 2 \end{bmatrix} x + \begin{bmatrix} 1 \\ b_2 \end{bmatrix} u$$

$$y = \begin{bmatrix} c_1 & 1 \end{bmatrix} x$$

$$T(s) = \frac{(b_2 + c_1)s + (-2c_1 - b_2 + b_2c_1)}{(s-1)(s-2)}$$

When $b_2 = 0$, the term $(s-2)$ is canceled. Hence, the corresponding mode is either uncontrollable or unobservable. Let us see if the system is observable.

$$M_o = \begin{bmatrix} c_1 & 1 \\ c_1 & 2 + c_1 \end{bmatrix}$$

Because the determinant of the observability matrix is nonzero, the system is observable. Consequently, the term$(s-2)$ corresponds to an uncontrollable mode. Also, observe that the transfer function becomes 0 when $b_2 = c_1 = 0$. This strange case occurs because there is no path from the input to the output when both parameters are zero.

Never cancel RHP poles or zeros.

Our earlier caution against unstable pole–zero cancellation is worth repeating here. When canceled by a zero, an unstable pole does not really disappear, it simply becomes either uncontrollable or unobservable. In the first case, you will observe the state blowing up, but you cannot do anything about it. In the second case, you will not even be aware that something is wrong because the unstable state does not appear at the output. In either case, the results are disastrous.

8.6.4 Causes of Uncontrollability

What are some of the causes of uncontrollability or unobservability? One cause, as has been indicated, is pole–zero cancellation. Another source of problem is symmetry in the system. For instance, consider the second-order system

$$\dot{x} = \begin{bmatrix} 1 & 1 \\ 1 & 1 \end{bmatrix} x + \begin{bmatrix} 1 \\ 1 \end{bmatrix} u$$

$$y = [\,1 \quad 1\,]x$$

The system is neither controllable, nor observable. In fact, the diagonal realization and transfer function of the system indicate that the pole at the origin is canceled out.

$$\dot{x}_1 = 0$$

$$\dot{x}_2 = 2x_2 + u \qquad T(s) = \frac{2s}{s(s-2)} = \frac{2}{s-2}$$

$$y = 2x_2$$

In physical systems, such symmetry is rare. The preceding two cases either can be avoided or are unlikely to occur in practice. Another common cause is redundant state variables. During the process of modeling complex systems, one may introduce unnecessary or redundant state variables. In this case, lack of controllability/observability indicates modeling errors and can be corrected by proper system modeling. The following example demonstrates the case of modeling error.

Consider the simple RL circuit shown in Figure 8.17. The circuit input is a voltage source, and the output is the current flowing through the inductor. From basic circuit theory, we know that the correct equation is given by

$$u(t) = i(t) + \frac{di(t)}{dt}$$

If we let $x = i(t)$, then we have

$$\dot{x} = -x + u$$

If the output is the current, we have

$$y = x = i$$

so

$$T(s) = \frac{Y(s)}{U(s)} = \frac{1}{s+1}$$

Now, it is also known that current is the rate of flow of charge. Suppose electrical charge is selected as a state variable—that is,

An incorrect model can result in an uncontrollable/unobservable realization.

$$x_1 = q$$
$$x_2 = \dot{q} = i$$
$$\dot{x}_1 = x_2$$
$$\dot{x}_2 = -x_2 + u$$
$$y = x_2$$

so that

$$A = \begin{bmatrix} 0 & 1 \\ 0 & -1 \end{bmatrix} \quad B = \begin{bmatrix} 0 \\ 1 \end{bmatrix} \quad C = [0 \quad 1]$$

The transfer function between the current and the input is given by

$$\frac{Y(s)}{U(s)} = C(sI - A)^{-1}B = \frac{s}{s(s+1)} = \frac{1}{s+1}$$

The observability matrix of this second-order model is

$$M_o = \begin{bmatrix} 0 & 1 \\ 0 & -1 \end{bmatrix} \rightarrow |M_o| = 0$$

which is singular (i.e., the model is not observable). This does not mean that the circuit is unobservable. It simply indicates that the model is not good. In fact, the extra state variable defined for charge is redundant for our purposes.

Another common cause is inappropriate or insufficient control actuators or sensors. The latter cause is an important system design issue. For a given control objective, we need an appropriate model and a sufficient number of control actuators and sensors that are appropriately positioned. To illustrate this issue, consider the classic problem of stabilizing an inverted pendulum on a moving cart shown in Figure 8.18.

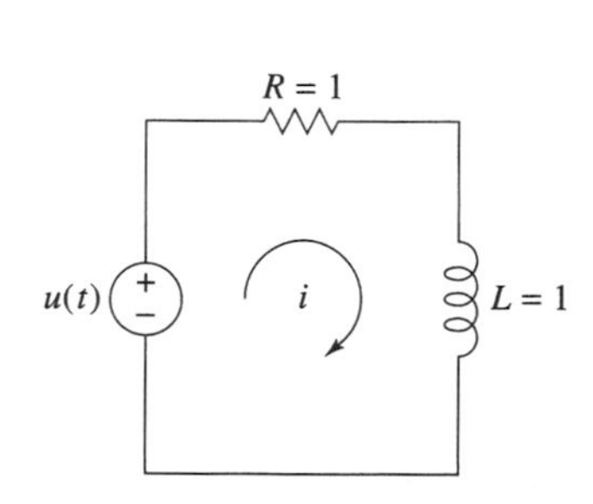

Figure 8.17 *RL* circuit: $R = 1, L = 1$.

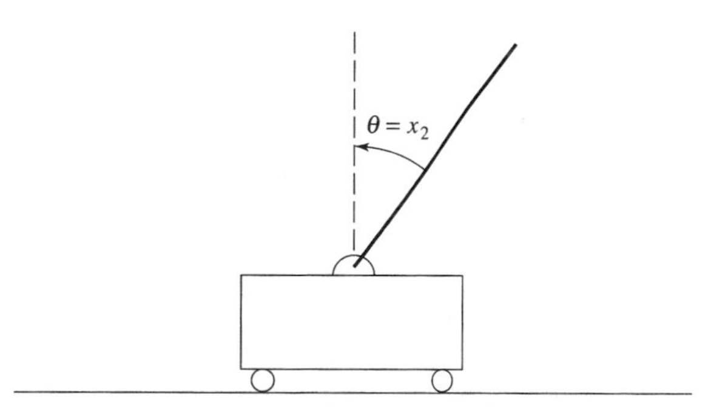

Figure 8.18 Inverted pendulum on a moving cart.

Inverted pendulum on a cart example.

Suppose the objective is to balance the pendulum and stop the cart. A linearized model is given by

$$\dot{x} = \begin{bmatrix} 0 & -a & 0 \\ 0 & 0 & 1 \\ 0 & a & 0 \end{bmatrix} x + \begin{bmatrix} b_1 \\ 0 \\ -b_2 \end{bmatrix} u$$

The state vector components correspond to cart velocity, pendulum angle, and pendulum angular velocity, respectively. To meet the objective, one of the state variables is measured and fed back. That signal is used to drive a motor that will move the cart to stabilize the system. Note that asymptotic stability implies that all states will approach zero, which means that the pendulum will be balanced and the cart will stop moving.

We can show that this system is controllable. Now the question is where to place the sensor—that is, which state should be measured. If we measure the pendulum angle and use that as the feedback signal, we get

$$y = \begin{bmatrix} 0 & 1 & 0 \end{bmatrix} x$$

The system is not observable in this case because

$$|M_n| = \begin{vmatrix} 0 & 1 & 0 \\ 0 & 0 & 1 \\ 0 & a & 0 \end{vmatrix} = 0$$

Intuitively, we can imagine that it is possible to balance the pendulum while the cart is still moving. Hence, it is obvious that using the pendulum angle as the feedback signal would not be a good choice. Similar arguments can be made against measuring the pendulum angular velocity. Finally, by using the cart velocity as the measured signal—that is,

$$y = \begin{bmatrix} 1 & 0 & 0 \end{bmatrix} x$$

we can verify that the system is observable because

$$|M_o| = \begin{vmatrix} 1 & 0 & 0 \\ 0 & -a & 0 \\ 0 & 0 & -a \end{vmatrix} = a^2 \neq 0$$

Therefore, it is theoretically possible to meet the design objectives by measuring the cart velocity.

The preceding example demonstrated one common cause of uncontrollability/unobservability. As system complexity increases, we may no longer have the benefit of our intuition, and we must resort to system concepts.

In general, if system design issues are well thought out, and adequate models are obtained, we need not worry about controllability/observability issues in practice. These issues will appear frequently as theoretical conditions for state space and optimal design, however.

❑ Computer-Aided Learning

System stability can be obtained by looking at the poles or eigenvalues of systems. MATLAB has two commands to find poles and zeros. They are "pole" and "tzero." The "pole" command returns the poles and "tzero" returns the so-called transmission zeros of systems. If the A matrix is available, we can determine asymptotic stability by finding the eigenvalues of A using the "eig" command. For the system defined by

$$\dot{x} = \begin{pmatrix} -2 & 1 & 1 \\ -3 & 0 & 0 \\ 0 & 0 & 0 \end{pmatrix} x + \begin{pmatrix} 1 \\ 1 \\ 1 \end{pmatrix} u$$

$$y = \begin{pmatrix} 2 & -2 & 1 \end{pmatrix} x$$

we get

```
>>pole(g)
ans=
-1.0000+1.4142i
-1.0000-1.4112i
 0
>>cig(a)   % same as the poles
ans=
-1.0000+1.4142i
-1.0000-1.4142i
 0
>>tzero(g)
ans=
-6.6458
-1.3542
```

The system controllability and observability can be determined by examining the corresponding matrices. MATLAB has ctrb and obsv to obtain them. For the above-defined system, we get

```
>>CO=ctrb(a,b)
CO=
  1   0  -3
  1  -3   0
  1   0   0
>>OB=obsv(a,c)
OB=
  2  -2  1
  2   2  2
-10   2  2
```

We can then use det and rank to find the determinant and rank of these matrices. Sometimes systems have pole–zero cancellations (owing lack of controllability and/or observability). To obtain the so-called minimal realization of the system,

MATLAB has the "mineral" command. As an example, consider the third-order system where all matrices are all Ts (similar to the example in the book).

```
>>a=ones(3,3);b=ones(3,1);c=ones(1,3);d=0;
>>g=ss(a,b,c,d);
>>rank(ctrb(a,b))  % system not controllable
ans=
  1
>>rank(obsv(a,c))  % system not observable
ans=
  1
>>tf(g)  % transfer function has 2 poles and zeros at the origin

Transfer function:
           3s^2
-------------------------
s^3-3s^2-3.077e-015s+1.972e-031

>>eig(a)  % system not asymptotically stable
                (double eigenvalues at the origin)
ans=
  0.0000
  0.0000
  3.0000
>>gmin=mineral(g);  % after pole–zero cancellation
2 state(s) removed
>>tf(gmin)

Transfer function:
3
------
s – 3
```

C8.2

(a) Determine the stability of the systems defined in Drill Problem 8.14.

(b) Determine the stability, controllability, and observability of

```
a=[-1-2 0;1 2 0;-2-1-3]; b=[1;0;0]; c=[1 0 1]; d=0;
```

(c) Use MATLAB to do drill Problem 8.17.

Solution of state equation can be obtained using the "lsim" and "initial" commands.

```
LSIM(SYS, U, T, X0)
```

plots the response of a state space system with input $U(T)$ and initial state X_0. The time sequence T must be defined as a vector, and the input must be of the same size as T. Here is an example:

```
g=tf(1, [1 1.4 1])
```

Transfer function:

```
            1
   ---------------
   s^2 + 1.4 s + 1
   gs=ss(g);
   t=linspace(0,pi,100); u=sin(2*t); x0=[-2 3];
   lsim(gs,u,t,x0)
```

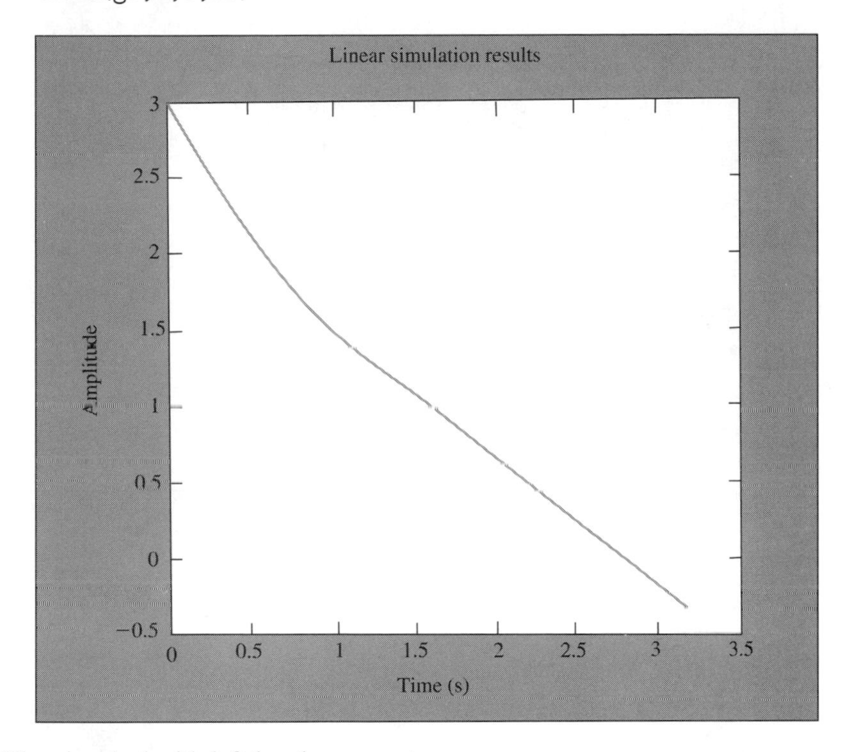

When invoked with left-hand arguments,

```
[Y,T,X]=LSIM(SYS,U,T,X0)
```

returns the output Y, the state vector X, and time vector T used for simulation, and no plot is drawn on the screen.

```
   [y,t,x]=lsim(gs,u,t,x0);
   subplot(211), plot(t,x(:,1)), subplot(212), plot(t,x(:,2))
```

The "initial" command is used to solve for the zero input response (ZIR) of state space systems: $\dot{x} = Ax$ subject to initial conditions:

INITIAL(SYS,XO,TF) simulates and plots the time response from $t = 0$ to the final time $t = TF$

INITIAL(SYS,X0,T) specifies a time vector T to be used for simulation

[Y,T,X]=INITIAL(SYS,X0,$\cdots$) returns the left hand side vectors but does not plot.

```
initial(gs, [-1 2])
```

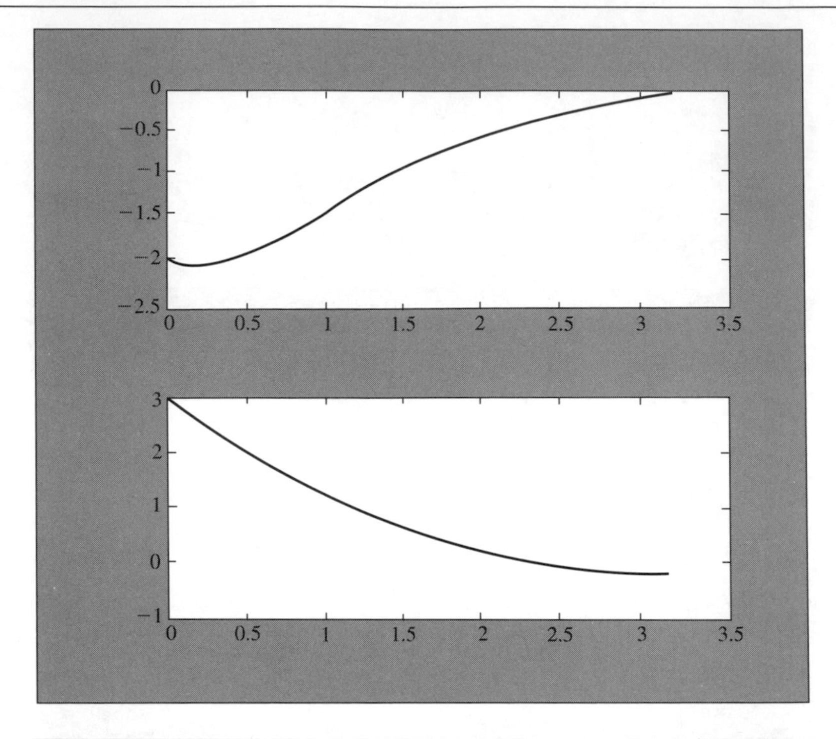

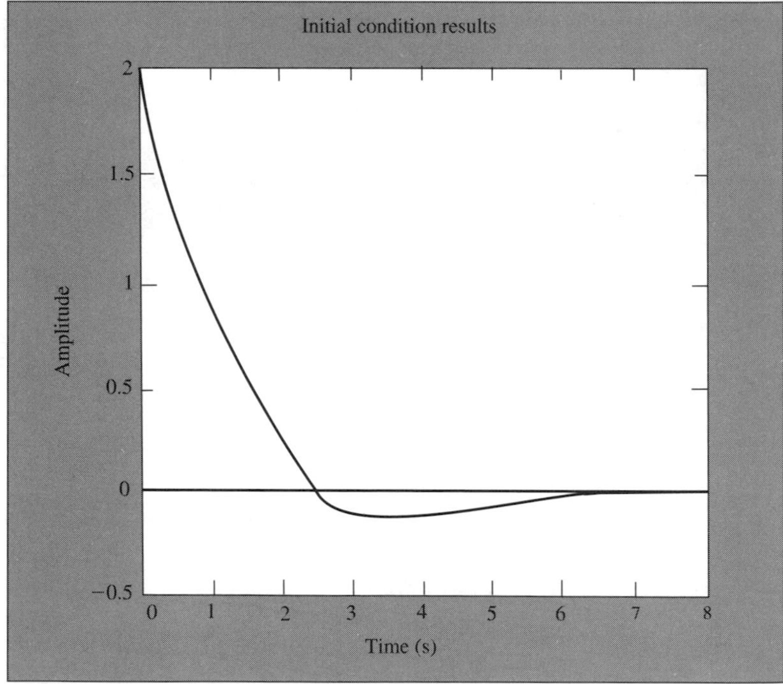

C8.3 Repeat Drill Problem D8.10 using the "lsim" command.

8.7 Inverted Pendulum Problems

One of the most celebrated and well-publicized problems in control is the inverted pendulum (broom balancer) problem. This is an unstable system that may model a rocket before launch. Almost all known and novel control techniques have been tested on the inverted pendulum (IP) problem. In this section we discuss models of a variety of IP-type problems. The IP problem is also highly nonlinear, but it can easily be controlled by using linear controllers in an almost vertical position.

Some of the varieties of IP are single pendulum, single-rotary pendulum, double side-by-side pendulum, double-pendulum, double-rotary, and triple-pendulum problems.

We start with a derivation for a single IP problem (see Figure 8.19).

The horizontal and vertical coordinates of center of gravity of the mass are given by

$$y_1 = x + l \sin \theta$$

$$y_2 = l \cos \theta$$

Newton's law in the horizontal direction gives us

$$u = M\ddot{x} + m\ddot{y}_1 = M\ddot{x} + m\ddot{x} + ml(-\sin\theta \cdot \dot{\theta}^2 + \cos\theta \cdot \ddot{\theta})$$

or

$$\boxed{(M + m)\ddot{x} - ml \sin\theta \cdot (\dot{\theta}^2) + \cos\theta \cdot \ddot{\theta} = u}$$

Newton's law for the rotational motion about the pivot gives

$$m\ddot{y}_1 l \cos\theta - m\ddot{y}_2 l \sin\theta = mgl \sin\theta$$

or

$$(\cos\theta) \cdot ml\ddot{x} + ml^2(-\sin\theta \cdot \dot{\theta}^2 + \cos\theta \cdot \ddot{\theta}) \cos\theta$$
$$+ ml^2(\ddot{\theta} \sin\theta + \dot{\theta}^2 \cos\theta) \sin\theta = mgl \sin\theta$$

which simplifies to

$$\boxed{m\ddot{x} \cdot \cos\theta + ml\ddot{\theta} = mg \sin\theta}$$

These are the nonlinear IP models. If we assume that θ is small (i.e., we want to control the IP near its vertical equilibrium position), we can linearize these equations. Recall that for small θ, $\sin\theta \approx \theta$ and $\cos\theta \approx 1$, we get

$$\ddot{x} + l\ddot{\theta} = g\theta$$

$$(M + m)\ddot{x} - ml\dot{\theta}^2 + \ddot{\theta} = u$$

Assuming that for small θ, $\dot{\theta}^2$ is negligible, we get our final IP equation.

Linear IP equations.

$$\boxed{\begin{aligned} &(M + m)\ddot{x} + \ddot{\theta} = u \\ &\ddot{x} + l\ddot{\theta} = g\theta \end{aligned}}$$

Obtaining transfer functions, we get

$$\begin{bmatrix} (M+m)\,s^2 & s^2 \\ s^2 & ls^2 - g \end{bmatrix} \begin{bmatrix} X(s) \\ \theta(s) \end{bmatrix} = \begin{bmatrix} u(s) \\ 0 \end{bmatrix}$$

$$\begin{bmatrix} X(s) \\ \theta(s) \end{bmatrix} = \begin{bmatrix} ls^2 - g & -s^2 \\ -s^2 & (M+m)s^2 \end{bmatrix} \begin{bmatrix} u(s) \\ 0 \end{bmatrix} \frac{1}{\Delta(s)}$$

$$\Delta(s) = (ls^2 - g)(M+m)s^2 - s^4$$

$$= s^2[(M+m)(\ ls^2 - g) - 1]$$

$$\frac{X(s)}{U(s)} = \frac{ls^2 - g}{\Delta(s)}, \qquad \frac{\theta(s)}{U(s)} = \frac{-s^2}{\Delta(s)}$$

The poles are

$$[(M+m)\quad g+1] = (M+m)\quad ls^2 \Rightarrow \text{s} = \pm\sqrt{\frac{(M+m)\,g+1}{l(M+m)}}, 0, 0$$

If $M \gg m$, the poles are at $s = \pm\sqrt{g/l}, 0, 0$ (Figure 8.20). Also if l is small (short pendulum), it is more unstable (RHP pole farther into the plane), Simple experiment confirms that longer pendulums are easier to control, too.

Also notice that the transfer function $\theta(s)/U(s)$ has an unstable pole–zero cancellation (double pole at the origin, s^2 term, cancels out). This indicates that the system cannot be controlled by measuring θ alone. The transfer function $X(s)/U(s)$ does not have this problem, but it has RHP zeros (this makes it more difficult to control the system). Hence, stabilization is possible by measuring cart position.

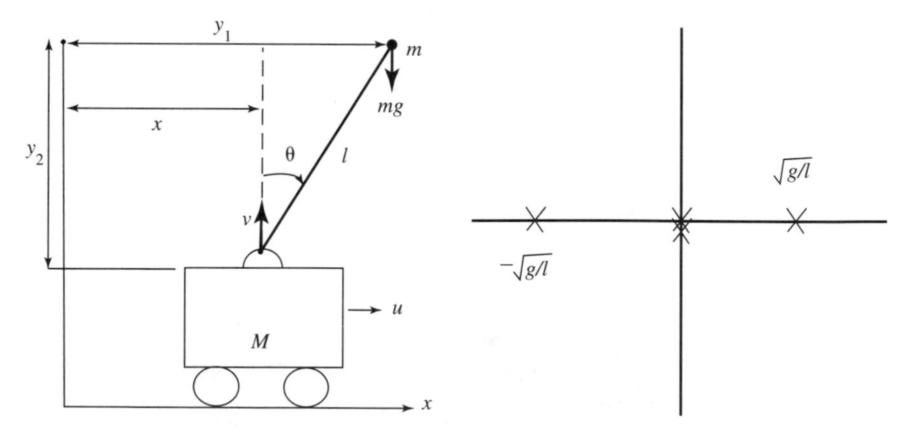

Figure 8.19 The inverted pendulum.

Figure 8.20 Poles of the inverted pendulums

Let us now obtain the state space representation of the IP, rewriting the differential equations:

$$(M+m)\ddot{x} + \left(\frac{g\theta - \ddot{x}}{l}\right) = u \Rightarrow l(M+m)\ddot{x} + g\theta - \ddot{x} = lu$$

$$\left(\frac{u - \ddot{\theta}}{M+m}\right) + l\ddot{\theta} = g\theta \Rightarrow u - \ddot{\theta} + l(M+m)\ddot{\theta} = g(M+m)\theta$$

Simplifying, we have

$$\ddot{x}\,[lc(M+m) - 1] + g\theta = lu$$

$$\ddot{\theta}\,[l(M+m) - 1] - g(M+m)\theta = -u$$

and

$$\boxed{\begin{aligned} \ddot{x} + \alpha\theta &= \beta l u \\ \ddot{\theta} - \gamma\theta &= -\beta u \end{aligned}}$$

where

$$\alpha = \frac{g}{l(M+m) - 1}, \qquad \beta = \frac{1}{l(M+m) - 1}, \qquad \gamma = (M+m)\alpha$$

Defining $x_1 = x, \quad x_2 = \alpha, \quad x_3 - \dot{x}, \quad x_4 - \dot{\alpha}$, we write

$$\dot{x}_1 = x_3$$

$$\dot{x}_2 = x_4$$

$$\dot{x}_3 = -\alpha x_2 + \beta l u \Rightarrow \begin{bmatrix} \dot{x}_1 \\ \dot{x}_2 \\ \dot{x}_3 \\ \dot{x}_4 \end{bmatrix} = \begin{bmatrix} 0 & 0 & 1 & 0 \\ 0 & 0 & 0 & 1 \\ 0 & -\alpha & 0 & 0 \\ 0 & \gamma & 0 & 0 \end{bmatrix} \begin{bmatrix} x_1 \\ x_2 \\ x_3 \\ x_4 \end{bmatrix} + \begin{bmatrix} 0 \\ 0 \\ \beta l \\ -\beta \end{bmatrix} u$$

$$\dot{x}_4 = \gamma x_2 - \beta u$$

The output equation depends on what we measure. If we measure the cart position and the pendulum angle (single input, multi output) problem, we get

$$\begin{bmatrix} y_1 \\ y_2 \end{bmatrix} = \begin{bmatrix} 1 & 0 & 0 & 0 \\ 0 & 1 & 0 & 0 \end{bmatrix} \begin{bmatrix} x_1 \\ x_2 \\ x_3 \\ x_4 \end{bmatrix}$$

Measuring only the pendulum angle (single input, single output), we have

$$y = [\,0 \quad 1 \quad 0 \quad 0] \begin{bmatrix} x_1 \\ x_2 \\ x_3 \\ x_4 \end{bmatrix}$$

and sensing only the cart position gives

$$y = [\ 1 \quad 0 \quad 0 \quad 0] \begin{bmatrix} x_1 \\ x_2 \\ x_3 \\ x_4 \end{bmatrix}$$

We can verify system controllability by noting that $\det \mathcal{C} \neq 0$

$$\mathcal{C} = \begin{bmatrix} b & A \cdot b & A^2 \cdot b & A^3 \cdot b \end{bmatrix} = \begin{bmatrix} 0 & \beta l & 0 & -\alpha\beta l \\ 0 & -\beta & 0 & \gamma\beta l \\ \beta l & 0 & -\alpha\beta l & 0 \\ -\beta & 0 & \gamma\beta l & 0 \end{bmatrix}, \det \mathcal{C} \neq 0$$

Checking observability: is a two-step procedure.

1. Sensing pendulum angle: $C = [0 \quad 1 \quad 0 \quad 0]$

$$\mathcal{O} = \begin{bmatrix} 0 & 1 & 0 & 0 \\ 0 & 0 & 0 & 1 \\ 0 & -\alpha & 0 & 0 \\ 0 & 0 & 0 & -\alpha \end{bmatrix}, \det \mathcal{O} = 0$$

   This is reflected in pole–zero cancellation in the $\theta(s)/u(s)$ transfer function. We cannot achieve stabilization (all states going to zero asymptotically) by measuring only the pendulum angle.

2. Sensing cart position: $C = [1 \quad 0 \quad 0 \quad 0]$

$$\mathcal{O} = \begin{bmatrix} 1 & 0 & 0 & 0 \\ 0 & 0 & 1 & 0 \\ 0 & -\alpha & 0 & 0 \\ 0 & 0 & 0 & -\alpha \end{bmatrix}, \det \mathcal{O} \neq 0$$

In this case, the system is both controllable and observable [no pole–zero cancellation in the $X(s)/U(s)$ transfer function], and we can stabilize the system.

Note that it makes a big difference which state variable we measure. This is called the "sensor location" problem. In practice, sensitive potentiometers are used to measure both θ and x.

We now consider a slight variation of the IP known as the rotary inverted pendulum (RIP) problem (Figure 8.21). The pendulum, standing on a short arm, can rotate in one plane about a hinge, with a possible potentiometer to measure its angle ϕ. The arm itself can also rotate through an angle of θ.

This version of IP is easier to build very compactly. Regular IP is usually built on a moving car platform or on a track. The track version is large and bulky because the track must be long enough to allow the pendulum to move a distance sufficient to stabilize. The RIP replaces the linear track with a rotating arm, and the arm can rotate

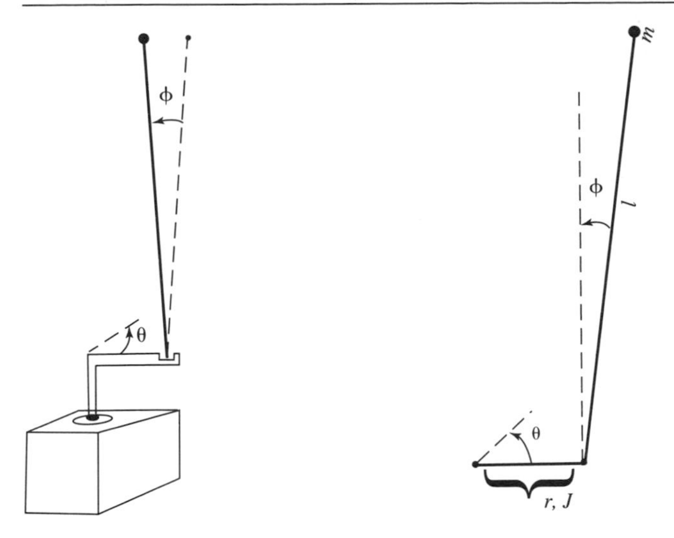

Figure 8.21 The rotary inverted pendulum. **Figure 8.22** Parameter definition for RIP.

as many degrees as desired to stabilize the system. In state space form the linearized equations for the RIP (see Figure 8.22) are given by

$$\begin{bmatrix} \dot{x}_1 \\ \dot{x}_2 \\ \dot{x}_3 \\ \dot{x}_4 \end{bmatrix} = \begin{bmatrix} 0 & 0 & 1 & 0 \\ 0 & 0 & 0 & 1 \\ 0 & -\alpha & 0 & 0 \\ 0 & \gamma & 0 & 0 \end{bmatrix} \begin{bmatrix} x_1 \\ x_2 \\ x_3 \\ x_4 \end{bmatrix} + \begin{bmatrix} 0 \\ 0 \\ (r/l)\beta \\ -\beta \end{bmatrix} u$$

where the states are $x_1 = \theta$, $x_2 = \phi$, $x_3 = \dot{\theta}$, $x_4 = \dot{\phi}$, and $\beta = 1/J$, $\alpha = mrg/J$, and $\gamma = \left(J + mr^2/Jl\right) g$, in which

J = moment of inertia of the arm
r = length of the arm
m = pendulum mass
l = length of pendulum
g = gravity constant

Note that the equations are dynamically similar to the rectilinear IP system, and similar system properties are expected. It is possible to control the system theoretically by using measurements of θ (i.e., x_1). In practice, it is relatively easy to control the system by measuring both θ and ϕ.

In the double side-by-side inverted pendulum problem (Figure 8.23), we have two pendulums of equal mass m with lengths l_1 and l_2 mounted on a moving cart. The hinge lines of the pendulums are parallel.
The equations of motion are given by

$$\begin{cases} M\ddot{x} = -mg\theta_1 - mg\theta_2 + u \\ m\ddot{x} = mg\theta_i - ml_i\ddot{\theta}_i \quad i = 1, 2 \end{cases}$$

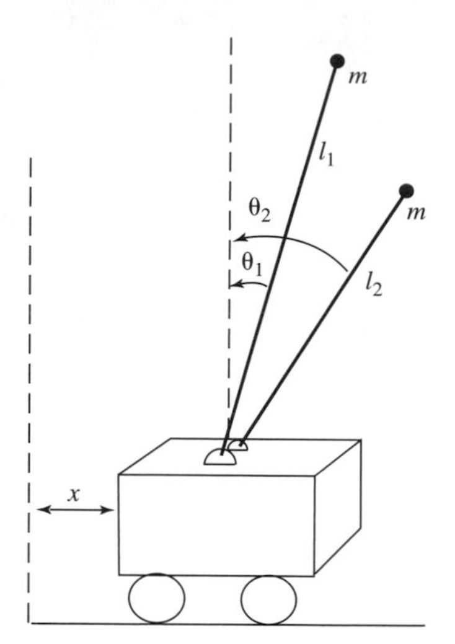

Figure 8.23 Double side-by-side inverted pendulum.

Substituting for $\ddot{x}$ from the first equation into the second set of equations (to eliminate $\ddot{x}$ in the second set) gives

$$\begin{cases} \ddot{\theta}_1 = \dfrac{\alpha}{l_1}\theta_2 + \dfrac{\gamma}{l_1}\theta_1 - \dfrac{1}{l_1}\beta u & \alpha = \dfrac{mg}{M}, \quad \gamma = \alpha + g \\ \ddot{\theta}_2 = \dfrac{\alpha}{l_2}\theta_1 + \dfrac{\gamma}{l_2}\theta_2 - \dfrac{1}{l_2}\beta u & \beta = \dfrac{1}{M} \\ \ddot{x} = -\alpha\theta_1 - \alpha\theta_2 + \beta u \end{cases}$$

Upon defining state variables as $x_1 = x,\ x_2 = \theta_1,\ x_3 = \theta_2,\ x_4 = \dot{x},\ x_5 = \dot{\theta}_1$, and $x_6 = \dot{\theta}_2$, we get the following state equations:

$$\begin{bmatrix} \dot{x}_1 \\ \dot{x}_2 \\ \dot{x}_3 \\ \dot{x}_4 \\ \dot{x}_5 \\ \dot{x}_6 \end{bmatrix} = \begin{bmatrix} 0 & 0 & 0 & 1 & 0 & 0 \\ 0 & 0 & 0 & 0 & 1 & 0 \\ 0 & 0 & 0 & 0 & 0 & 1 \\ 0 & -\alpha & -\alpha & 0 & 0 & 0 \\ 0 & \dfrac{\gamma}{l_1} & \dfrac{\alpha}{l_1} & 0 & 0 & 0 \\ 0 & \dfrac{\alpha}{l_2} & \dfrac{\gamma}{l_2} & 0 & 0 & 0 \end{bmatrix} \begin{bmatrix} x_1 \\ x_2 \\ x_3 \\ x_4 \\ x_5 \\ x_6 \end{bmatrix} + \begin{bmatrix} 0 \\ 0 \\ 0 \\ \beta \\ -\dfrac{\beta}{l_1} \\ -\dfrac{\beta}{l_2} \end{bmatrix} u$$

Note that the A, B matrices can be partitioned as follows:

$$A = \left[\begin{array}{c|c} 0 & I \\ \hline \bar{A} & 0 \end{array}\right], \quad B = \begin{bmatrix} 0 \\ \hline \bar{B} \end{bmatrix}$$

where I and 0 stand for identity and zero matrices of appropriate size. This makes it easier to compute the controllability matrix

$$\mathfrak{C} = \begin{bmatrix} 0 & \bar{B} & 0 & \bar{A}\bar{B} & 0 & \bar{A}^2\bar{B} \\ \bar{B} & \bar{0} & \bar{A}\bar{B} & 0 & \bar{A}^2\bar{B} & 0 \end{bmatrix}$$

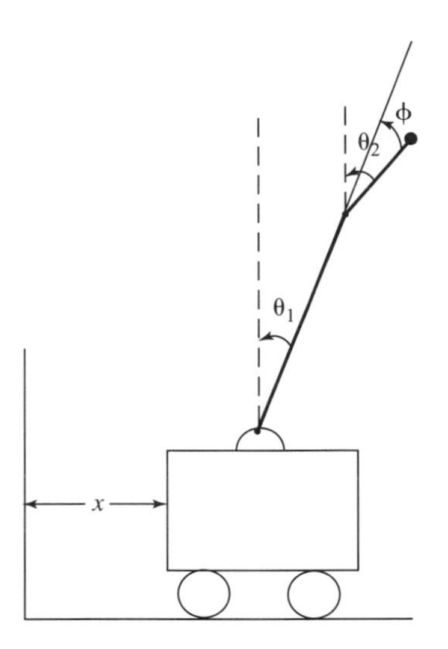

Figure 8.24 The double inverted pendulum.

It is left as an exercise to show that the system is controllable only when $l_1 \neq l_2$. The reader can also verify conditions for observability under various sensor assumptions (sensing x, x and θ_1; θ_1, θ_1 and θ_2; x and θ_1 and θ_2).

We now consider two pendulums, one on top of the other, the double inverted pendulum problem (DIP) (Figure 8.24). Possible sensor variables are the cart position, the angles of the pendulums with respect to the vertical plane (θ_1, θ_2), and also measuring the angle difference between the pendulums (angle ϕ). Researchers have indicated successful control measuring x, θ_1, and ϕ, and even measuring only x and θ_1 (which means stabilizing both pendulums without measuring the apparently crucial angle θ_2 or ϕ). The linearized equations of motion are given by:

$$\begin{cases} r_1\ddot{x} + r_2\ddot{\theta}_1 + r_3\ddot{\theta}_2 = u \\ r_4\ddot{\theta}_1 + r_5\ddot{\theta}_2 + r_2\ddot{x} = r_7\theta_1 \\ r_5\ddot{\theta} + r_6\ddot{\theta}_2 + r_3\ddot{x} = r_8\theta_2 \end{cases}$$

where
$$\begin{aligned} r_1 &= M + m_1 + m_2 \qquad (l_i = \text{distance to center of mass for pendulum } i) \\ r_2 &= m_1 l_1 + m_2 l \qquad l = \text{length of pendulum 1} \\ r_3 &= m_2 l_2 \\ r_4 &= J_1 + m_1 l_1^2 + m_2 l^2 \\ r_5 &= m_2 l_2 l \\ r_6 &= J_2 + m_2 l_2^2 \\ r_7 &= m_1 l_1 g + m_2 l g, \\ r_8 &= m_2 l_2 g \end{aligned}$$

These implicit differential equations can be solved explicitly for θ_1, θ_2, and u:

$$\begin{bmatrix} r_1 & r_2 & r_3 \\ r_2 & r_4 & r_5 \\ r_3 & r_5 & r_6 \end{bmatrix} \begin{bmatrix} \ddot{x} \\ \ddot{\theta}_1 \\ \ddot{\theta}_2 \end{bmatrix} = \begin{bmatrix} u \\ r_7\theta_1 \\ r_8\theta_2 \end{bmatrix}$$

Solving for $\ddot{x}, \ddot{\theta}_1, \ddot{\theta}_2$, we get

$$\ddot{x} = \alpha_{11}\theta_1 + \alpha_{12}\theta_2 + \beta_1 u$$

$$\ddot{\theta}_1 = \alpha_{21}\theta_1 + \alpha_{22}\theta_2 + \beta_2 u$$

$$\ddot{\theta}_2 = \alpha_{31}\theta_1 + \alpha_{32}\theta_2 + \beta_3 u$$

Letting the state variables be

$$x_1 = x, \quad x_2 = \theta_1, \quad x_3 = \theta_2$$

$$x_4 = \dot{x}, \quad x_5 = \dot{\theta}_1, \quad x_6 = \dot{\theta}_2$$

we write

$$\begin{bmatrix} \dot{x}_1 \\ \dot{x}_2 \\ \dot{x}_3 \\ \dot{x}_4 \\ \dot{x}_5 \\ \dot{x}_6 \end{bmatrix} = \begin{bmatrix} 0 & 0 & 0 & 1 & 0 & 0 \\ 0 & 0 & 0 & 0 & 1 & 0 \\ 0 & 0 & 0 & 0 & 0 & 1 \\ 0 & \alpha_{11} & \alpha_{12} & 0 & 0 & 0 \\ 0 & \alpha_{21} & \alpha_{22} & 0 & 0 & 0 \\ 0 & \alpha_{31} & \alpha_{32} & 0 & 0 & 0 \end{bmatrix} \begin{bmatrix} x_1 \\ x_2 \\ x_3 \\ x_4 \\ x_5 \\ x_6 \end{bmatrix} + \begin{bmatrix} 0 \\ 0 \\ 0 \\ \beta_1 \\ \beta_2 \\ \beta_3 \end{bmatrix} u$$

The A, B matrices can be partitioned as follows.

$$A = \left[\begin{array}{c|c} 0 & I \\ \hline \bar{A} & 0 \end{array} \right], \quad B = \begin{bmatrix} 0 \\ - \\ \bar{B} \end{bmatrix}$$

which is similar in structure to the double side-by-side problem.

The controllability of the system is left as an exercise. The reader can also determine observability under a variety of sensor decisions (measuring x, θ_1, θ_2, x and θ_1, x and θ_2, θ_1 and θ_2, x and θ_1 and θ_2). It is important to appreciate the use of symbolic math programs (such as Symbolic Math Toolbox of MATLAB, Mathematica, Maple, and Macsyma, to name a few) in the various IP problems because they create rather large symbolic matrices that soon test the patience of humans.

8.8 SUMMARY

At least three distinct procedures exist for converting a system describable by an nth-order linear differential equation into a system having n state variables: phase/dual phase variables, canonical (diagonalized) variables, and physical (block diagram) variables. The first of these translates a system transfer function into system matrices, the second provides diagonalized system matrices, while the third preserves actual system quantities (e.g., velocity, current, temperature).

The phase-variable form was shown to be especially convenient for single-and multiple-output system transfer function synthesis, and the dual phase-variable form is convenient for multiple-input system transfer functions. Simulation diagrams are not only useful in transfer function synthesis, they also give a standard, systematic and compact description of a system.

The relationships between signals in a simulation diagram were shown to be a set of coupled first-order differential state equations and linear algebraic output equations relating the system outputs to the state variables. These state-variable equations are compactly expressed using matrix notation:

$$\dot{x} = Ax + Bu$$

$$y = Cx$$

System transfer functions were calculated systematically from the state-variable equations by using matrix algebra:

$$T(s) = C[sI - A]^{-1} B$$

It was seen that all transfer functions of a system share a common characteristic polynomial,

$$|sI - A|$$

A nonsingular change of state variables gives a new representation for a system but leaves the system's input–output relations, its transfer functions, unchanged. Hence, a system characterized by a set of transfer functions may be represented in countless different ways, each differing in the choice of state variables.

A very special set of state variables for a system are those for which the state equations, each of first order, are decoupled from one another. A system so represented is said to bc in *normal or diagonal* form:

$$A = \begin{bmatrix} s_1 & 0 & 0 & \cdots & 0 \\ 0 & s_2 & 0 & \cdots & 0 \\ \vdots & & & & \\ 0 & 0 & 0 & & s_n \end{bmatrix}$$

Determining the change of variables that places a system in diagonal form is the *characteristic value problem* of matrix algebra. An alternative transformation method involves expansion of a system transfer function into partial fractions. Systems with repeated characteristic roots cannot be diagonalized: however, they may be placed in a related *Jordan* form, where the repeated root terms involve a distinctive nonzero "block" along the diagonal of the state coupling matrix.

A fundamental system property is stability. It can be studied from transfer function or state space points of view. A system with a given transfer function is BIBO stable if for all bounded inputs, the outputs are bounded. The condition for BIBO stability is that the poles be in the LHP. A system with a given state space realization is asymptotically stable if the states approach 0 as time approaches infinity. The condition for asymptotic stability is that the eigenvalues of the A matrix be in the LHP. These two notions of stability are equivalent when the transfer function of the system contains no common poles and zeros (i.e., there are no pole–zero cancellations). Since most control systems are of the feedback type and are always subject to external disturbances and noise, a more appropriate condition is that all transfer functions between all inputs and outputs be stable. This is called *internal stability*, and it ensures that all signals within the system remain bounded.

Two fundamental properties of a system are controllability and observability. A system is completely controllable if the system state $x(t_f)$ at time t_f can be forced to take on any desired value by applying a control input $u(t)$ over a period of time from t_0 until t_f. A system is completely observable if any initial state vector $x(t_0)$ can be reconstructed by examining the system output $y(t)$ over some period of time from t_0 until t_f.

For a system represented in diagonal form, controllability and observability are apparent from inspection of the input and output coupling matrices. For systems in other than diagonal form, simple rank tests of the controllability and observability matrices,

$$M_c = \left[B \;\vdots\; AB \;\vdots\; \cdots \;\vdots\; A^{n-1}B \right]$$

$$M_o = \begin{bmatrix} C \\ \hline CA \\ \hline \vdots \\ \hline CA^{n-1} \end{bmatrix}$$

may be used.

Whenever the system's transfer function has pole–zero cancellations, its state space realizations will be either uncontrollable or unobservable or both. The uncontrollable or unobservable modes correspond to the canceled poles. If a canceled pole is in the RHP, contradictory answers between BIBO and asymptotic stability result. Most physical systems that are properly modeled satisfy controllability and observability conditions, hence checking the poles for stability is sufficient in most cases.

For an nth-order system

$$x(t) = \Phi(t)\,x\left(0^-\right) + \int_{0-}^{t} \Phi(t-\tau)\,Br(\tau)d\tau$$

where $\Phi(t)$ is the $n \times n$ state transition matrix, we have

$$\Phi(t) = \mathcal{L}^{-1}\left\{(sI - A)^{-1}\right\}$$

The time response can be approximated by using a Taylor series to compute the state transition matrix and the convolution integral. The classic inverted pendulum problem and its variations are discussed at the end of this chapter to demonstrate state space models and system concepts such as stability, controllability, and observability.

REFERENCES

Simulation Diagrams

Jackson, A. S., *Analog computation*. New York: McGraw-Hill, 1960.

Korn, G. A., and Korn, T. M., *Electronic Analog Computers*. New York: McGraw-Hill, 1952.

State Variables

Brockett, R. W., "Poles, Zeros and Feedback: State Space Interpretation." *IEEE Trans. Autom. Control* (April 1965).

Brogan, W. L., *Modern Control Theory*, 3rd ed. Englewood Cliffs, NJ: Prentice-Hall, 1991.

DeCarlo, R. A., *Linear Systems: A State Variable Approach with Numerical Implementation*. Englewood Cliffs, NJ: Prentice-Hall, 1989.

De Russo, P. M., Roy, R. J., and Close, C. M., *State Variables for Engineers*. New York: Wiley, 1965.

Friedland, B., *Control Systems Design: An Introduction to State-Space Methods*. New York: McGraw-Hill, 1986.

Gupta, S. C., *Transform and State Variable Methods in Linear Systems*. New York: Wiley, 1966.

Horowitz, I. C., and Shaked, U., "Superiority of Transfer Function Over State Variable Methods in Linear Time-Invariant Feedback System Design." *IEEE Trans. Autom. Control* (February1975).

Kailath, T., *Linear Systems*. Englewood Cliffs, NJ: Prentice-Hall, 1980.

Kalman, R. E., "Mathematical Description of Linear Dynamical Systems." *SIAM J. Control ser.* A, (1): (1963).

Luenberger, D. G., *Introduction to Dynamic Systems: Theory, Models, & Applications*. New York: Wiley, 1979.

Ogata, K., *State Space Analysis of Control Systems*. Englewood Cliffs, NJ: Prentice-Hall, 1967.

Timothy, L. K., and Bona, B. E., *State Space Analysis: An Introduction*. New York: McGraw-Hill, 1968.

Zadeh, L. A., and Desoer, C. A., *Linear System Theory: The State Space Approach*. New York: McGraw-Hill, 1963.

Matrix Algebra and the Characteristic Value Problem

Bellman, R., *Introduction to Matrix Analysis*. New York: McGraw-Hill, 1960.

Noble, B., *Applied Linear Algebra*. Englewood Cliffs, NJ: Prentice-Hall, 1969.

Strang, G., *Linear Algebra and Its Applications*, 2nd ed. New York: Academic Press, 1980.

Strang, G., *Introduction to Applied Mathematics*.: Wellesley-Cambridge Press, 1986.

Controllability and Observability

Gillbert, E. G., "Controllability and Observability in Multivariable Control Systems." *J. SIAM*, ser. A (1963): 128–151.

Kalman, R. E., "Canonical Structure of Linear Dynamical Systems." *Proc. Natl. Acad. Sci. USA* (April 1962): 596–600.

Stubberud, A. R., "A Controllability Criterion for a Class of Linear Systems." *IEEE Trans. Appl. Ind.* 68 (1964): 411–413.

Computational Methods

Faddeeva, D. K., and Faddeeva, V. N., *Computational Methods of Linear Algebra.* San Fransisco: Freeman, 1963.

Shahian, B., and Hassul, M., *Control System Design Using MATRIX$_X$*. Englewood Cliffs, NJ: Prentice-Hall, 1992.

Shahian, B., and Hassul, M., *Control System Design Using MATLAB.* Englewood Cliffs, NJ: Prentice-Hall, 1993.

PROBLEMS

1. Draw phase-variable form simulation diagrams for systems with the following transfer functions. Then write the state-variable equations in matrix form.

(a) $T(s) = \dfrac{-2s+8}{s^2+8}$

(b) $T(s) = \dfrac{10s}{s^3+12s^2+7s+2}$

Ans.

$$\begin{bmatrix} \dot{x}_1 \\ \dot{x}_2 \\ \dot{x}_2 \end{bmatrix} = \begin{bmatrix} 0 & 1 & 0 \\ 0 & 0 & 1 \\ -2 & -7 & -12 \end{bmatrix} \begin{bmatrix} x_1 \\ x_2 \\ x_3 \end{bmatrix} + \begin{bmatrix} 0 \\ 0 \\ 1 \end{bmatrix} u$$

$$y = \begin{bmatrix} 0 & 10 & 0 \end{bmatrix} \begin{bmatrix} x_1 \\ x_2 \\ x_3 \end{bmatrix}$$

(c) $T(s) = \dfrac{7s^3-2s^2+s}{s^4+3s^3+9s^2+s+1}$

(d) Two outputs:

$$T_{11}(s) = \frac{-s^2+9}{s^3+3s^2+s+4}$$

$$T_{21}(s) = \frac{s^2+s+10}{s^3+3s^2+s+4}$$

Ans.

$$\begin{bmatrix} \dot{x}_1 \\ \dot{x}_2 \\ \dot{x}_3 \end{bmatrix} = \begin{bmatrix} 0 & 1 & 0 \\ 0 & 0 & 1 \\ -4 & -1 & -3 \end{bmatrix} \begin{bmatrix} x_1 \\ x_2 \\ x_3 \end{bmatrix} + \begin{bmatrix} 0 \\ 0 \\ 1 \end{bmatrix} u$$

$$\begin{bmatrix} y_1 \\ y_2 \end{bmatrix} = \begin{bmatrix} 9 & 0 & -1 \\ 10 & 1 & 1 \end{bmatrix} \begin{bmatrix} x_1 \\ x_2 \\ x_3 \end{bmatrix}$$

2. Draw dual phase-variable form simulation diagrams for systems with the following transfer functions. Then write the state-variable equations in matrix form.

(a) $T(s) = \dfrac{-2s + 8}{s^2 + 8}$

(b) $T(s) = \dfrac{2s + 8}{3s^3 + 7s^2 + 8s + 2}$

Ans.

$$\begin{bmatrix} \dot{x}_1 \\ \dot{x}_2 \\ \dot{x}_3 \end{bmatrix} = \begin{bmatrix} -\frac{7}{3} & 1 & 0 \\ -\frac{8}{3} & 0 & 1 \\ -\frac{2}{3} & 0 & 0 \end{bmatrix} \begin{bmatrix} x_1 \\ x_2 \\ x_3 \end{bmatrix} + \begin{bmatrix} 0 \\ \frac{2}{3} \\ \frac{8}{3} \end{bmatrix} u$$

$$y = [1 \quad 0 \quad 0] \begin{bmatrix} x_1 \\ x_2 \\ x_3 \end{bmatrix}$$

(c) $T(s) = \dfrac{-s^3 + 4s^2 - 9s + 4}{s^4 + 8s^3 + 2s^2 + s + 9}$

(d) Two inputs:

$$T_{11}(s) = \frac{3s^2 + 9}{s^3 + 3s^2 + s + 9}$$

$$T_{12}(s) = \frac{s - 4}{s^3 + 3s^2 + s + 9}$$

Ans.

$$\begin{bmatrix} \dot{x}_1 \\ \dot{x}_2 \\ \dot{x}_3 \end{bmatrix} = \begin{bmatrix} -3 & 1 & 0 \\ -1 & 0 & 1 \\ -9 & 0 & 0 \end{bmatrix} \begin{bmatrix} x_1 \\ x_2 \\ x_3 \end{bmatrix} + \begin{bmatrix} 3 & 0 \\ 0 & 1 \\ 9 & -4 \end{bmatrix} \begin{bmatrix} u_1 \\ u_2 \end{bmatrix}$$

$$y = [1 \quad 0 \quad 0] \begin{bmatrix} x_1 \\ x_2 \\ x_3 \end{bmatrix}$$

3. Draw simulation diagrams to represent the following systems:

(a)

$$\begin{bmatrix} \dot{x}_1 \\ \dot{x}_2 \\ \dot{x}_3 \end{bmatrix} = \begin{bmatrix} 1 & 8 & -1 \\ 2 & 0 & 4 \\ -2 & 1 & 8 \end{bmatrix} \begin{bmatrix} x_1 \\ x_2 \\ x_3 \end{bmatrix} + \begin{bmatrix} 1 \\ 0 \\ 0 \end{bmatrix} u$$

$$y = [1 \quad -4 \quad 0] \begin{bmatrix} x_1 \\ x_2 \\ x_3 \end{bmatrix}$$

(b)
$$\begin{bmatrix} \dot{x}_1 \\ \dot{x}_2 \\ \dot{x}_3 \end{bmatrix} = \begin{bmatrix} 0 & 4 & 0 \\ -1 & -1 & 4 \\ 8 & 0 & 3 \end{bmatrix} \begin{bmatrix} x_1 \\ x_2 \\ x_3 \end{bmatrix} + \begin{bmatrix} 1 \\ 8 \\ -5 \end{bmatrix} u$$
$$y = [0 \quad -3 \quad 6] \begin{bmatrix} x_1 \\ x_2 \\ x_3 \end{bmatrix}$$

(c)
$$\begin{bmatrix} \dot{x}_1 \\ \dot{x}_2 \end{bmatrix} = \begin{bmatrix} 2 & 1 \\ -8 & 0 \end{bmatrix} \begin{bmatrix} x_1 \\ x_2 \end{bmatrix} + \begin{bmatrix} 1 & 0 & 8 \\ 2 & 4 & 1 \end{bmatrix} \begin{bmatrix} u_1 \\ u_2 \\ u_3 \end{bmatrix}$$
$$y = [1 \quad 0] \begin{bmatrix} x_1 \\ x_2 \end{bmatrix}$$

(d)
$$\begin{bmatrix} \dot{x}_1 \\ \dot{x}_2 \\ \dot{x}_3 \end{bmatrix} = \begin{bmatrix} 0 & 0 & 1 \\ -1 & 2 & 4 \\ -8 & 1 & 0 \end{bmatrix} \begin{bmatrix} x_1 \\ x_2 \\ x_3 \end{bmatrix} + \begin{bmatrix} 4 & 0 \\ 0 & 0 \\ -1 & 7 \end{bmatrix} \begin{bmatrix} u_1 \\ u_2 \end{bmatrix}$$
$$\begin{bmatrix} y_1 \\ y_2 \\ y_3 \end{bmatrix} = \begin{bmatrix} 4 & 0 & 0 \\ 0 & 1 & 0 \\ 1 & 4 & 3 \end{bmatrix} \begin{bmatrix} x_1 \\ x_2 \\ x_3 \end{bmatrix}$$

4. Draw a simulation diagram to represent the following state equation.

$$\begin{bmatrix} \dot{x}_1 \\ \dot{x}_2 \\ \dot{x}_3 \end{bmatrix} = \begin{bmatrix} -1 & 3 & 0 \\ 0 & -4 & 8 \\ 0 & 0 & -3 \end{bmatrix} \begin{bmatrix} x_1 \\ x_2 \\ x_3 \end{bmatrix} + \begin{bmatrix} 0 & 8 \\ -3 & 0 \\ 0 & 1 \end{bmatrix} \begin{bmatrix} u_1 \\ u_2 \end{bmatrix}$$
$$y = \begin{bmatrix} 9 & 2 & 0 \end{bmatrix} \begin{bmatrix} x_1 \\ x_2 \\ x_3 \end{bmatrix}$$

5. For the following systems, find the transfer function matrices:

(a)
$$\begin{bmatrix} \dot{x}_1 \\ \dot{x}_2 \end{bmatrix} = \begin{bmatrix} -2 & 1 \\ -8 & 0 \end{bmatrix} \begin{bmatrix} x_1 \\ x_2 \end{bmatrix} + \begin{bmatrix} 2 \\ -4 \end{bmatrix} u$$
$$\begin{bmatrix} y_1 \\ y_2 \end{bmatrix} = \begin{bmatrix} 1 & -1 \\ 0 & 4 \end{bmatrix} \begin{bmatrix} x_1 \\ x_2 \end{bmatrix}$$

(b) $$\begin{bmatrix} \dot{x}_1 \\ \dot{x}_2 \end{bmatrix} = \begin{bmatrix} 0 & 1 \\ -3 & 8 \end{bmatrix} \begin{bmatrix} x_1 \\ x_2 \end{bmatrix} + \begin{bmatrix} 0 & 1 \\ 4 & -7 \end{bmatrix} \begin{bmatrix} u_1 \\ u_2 \end{bmatrix}$$

$$y = [-1 \quad 4] \begin{bmatrix} x_1 \\ x_2 \end{bmatrix}$$

Ans. $\begin{bmatrix} \dfrac{16s-4}{s^2-8s+3} & \dfrac{-29s+3}{s^2-8s+3} \end{bmatrix}$

(c) $$\begin{bmatrix} \dot{x}_1 \\ \dot{x}_2 \\ \dot{x}_3 \end{bmatrix} = \begin{bmatrix} 0 & 4 & 1 \\ 0 & 0 & 4 \\ -2 & 8 & -4 \end{bmatrix} \begin{bmatrix} x_1 \\ x_2 \\ x_3 \end{bmatrix} + \begin{bmatrix} 4 & 2 \\ 0 & 1 \\ 3 & -8 \end{bmatrix} \begin{bmatrix} u_1 \\ u_2 \end{bmatrix}$$

$$\begin{bmatrix} y_1 \\ y_2 \end{bmatrix} = \begin{bmatrix} 1 & 0 & 0 \\ 0 & 4 & -8 \end{bmatrix} \begin{bmatrix} x_1 \\ x_2 \\ x_3 \end{bmatrix}$$

6. Find the characteristic equations of the following systems. Then determine whether each is stable.

(a) $$\begin{bmatrix} \dot{x}_1 \\ \dot{x}_2 \end{bmatrix} = \begin{bmatrix} -1 & 3 \\ -3 & 8 \end{bmatrix} \begin{bmatrix} x_1 \\ x_2 \end{bmatrix} + \begin{bmatrix} 4 & 1 \\ -3 & 3 \end{bmatrix} \begin{bmatrix} u_1 \\ u_2 \end{bmatrix}$$

$$\begin{bmatrix} y_1 \\ y_2 \end{bmatrix} = \begin{bmatrix} 1 & 2 \\ 0 & 7 \end{bmatrix} \begin{bmatrix} x_1 \\ x_2 \end{bmatrix}$$

(b) $$\begin{bmatrix} \dot{x}_1 \\ \dot{x}_2 \\ \dot{x}_3 \end{bmatrix} = \begin{bmatrix} 0 & -2 & 3 \\ 0 & -3 & -1 \\ 0 & 1 & -8 \end{bmatrix} \begin{bmatrix} x_1 \\ x_2 \\ x_3 \end{bmatrix} + \begin{bmatrix} 2 \\ -8 \\ 4 \end{bmatrix} u$$

$$y = [4 \quad 1 \quad 6] \begin{bmatrix} x_1 \\ x_2 \\ x_3 \end{bmatrix}$$

Ans. $s^3 + 11s^2 + 25s$, marginally stable

(c) $$\begin{bmatrix} \dot{x}_1 \\ \dot{x}_2 \\ \dot{x}_3 \end{bmatrix} = \begin{bmatrix} -1 & -2 & 2 \\ 2 & 0 & 6 \\ -1 & 2 & -4 \end{bmatrix} \begin{bmatrix} x_1 \\ x_2 \\ x_3 \end{bmatrix} + \begin{bmatrix} 2 & -1 \\ 0 & 0 \\ 2 & 8 \end{bmatrix} \begin{bmatrix} u_1 \\ u_2 \end{bmatrix}$$

$$y = [4 \quad 0 \quad -1] \begin{bmatrix} x_1 \\ x_2 \\ x_3 \end{bmatrix}$$

7. Although the following transfer functions do not share a common denominator polynomial, they may be made to have a common denominator by multiplying their numerators and denominators by appropriate factors. Find a simulation

diagram and a matrix state-variable diagram and a matrix state-variable representation for a single-input, two-output system with the following two transfer functions:

$$T_{11}(s) = \frac{4s+1}{(s+2)(s+4)}$$

$$T_{21}(s) = \frac{10s}{(s+1)(s+4)}$$

The best solutions will involve only three integrators.

8. Use the partial fraction method to find diagonal state equations for single-input, single-output systems with the following transfer functions:

(a) $T(s) = \dfrac{-7s+4}{s^2+8s+12}$

(b) $T(s) = \dfrac{2s^2+3s-7}{(s+2)(s+8)(s+5)}$

Ans.
$$\begin{bmatrix} \dot{z}_1 \\ \dot{z}_2 \\ \dot{z}_3 \end{bmatrix} = \begin{bmatrix} -2 & 0 & 0 \\ 0 & -8 & 0 \\ 0 & 0 & -5 \end{bmatrix} \begin{bmatrix} z_1 \\ z_2 \\ z_3 \end{bmatrix} + \begin{bmatrix} 1 \\ 1 \\ 1 \end{bmatrix} u$$

$$y = \begin{bmatrix} -\frac{5}{18} & \frac{97}{18} & -\frac{28}{9} \end{bmatrix} \begin{bmatrix} z_1 \\ z_2 \\ z_3 \end{bmatrix}$$

(c) $T(s) = \dfrac{10}{s^3+8s^2+15s}$

9. The following transfer functions for single-input, single-output systems involve complex characteristic roots. Find diagonal state equations for these systems. Then find an alternative block-diagonal representation that does not involve complex numbers.

(a) $T(s) = \dfrac{4s}{s^2+2s+7}$

(b) $T(s) = \dfrac{s^2+3s-8}{(s+8)(s+3+j)(s+3-j)}$

(c) $T(s) = \dfrac{4}{(s+2)\left(s^2+2s+17\right)}$

10. The following transfer functions for single-input, single-output systems involve repeated characteristic roots. Find block diagonal Jordan canonical form state equations for these systems.

(a) $T(s) = \dfrac{3s-1}{s^2+4s+4}$

(b) $T(s) = \dfrac{s^3-4s^2+s-2}{(s+2)(s+3)^3}$

Ans.

$$\begin{bmatrix} \dot{z}_1 \\ \dot{z}_2 \\ \dot{z}_3 \\ \dot{z}_4 \end{bmatrix} = \begin{bmatrix} -2 & 0 & 0 & 0 \\ 0 & -3 & 1 & 0 \\ 0 & 0 & -3 & 1 \\ 0 & 0 & 0 & -3 \end{bmatrix} \begin{bmatrix} z_1 \\ z_2 \\ z_3 \\ z_4 \end{bmatrix} + \begin{bmatrix} 1 \\ 0 \\ 0 \\ 1 \end{bmatrix} u$$

$$y = [-28 \quad 68 \quad 16 \quad 29] \begin{bmatrix} z_1 \\ z_2 \\ z_3 \\ z_4 \end{bmatrix}$$

(c) $T(s) = \dfrac{7s^3}{(s+2)^2 (s+6)^2}$

11. The following systems have real characteristic roots. Find alternative diagonal state equations.

(a)
$$\begin{bmatrix} \dot{x}_1 \\ \dot{x}_2 \end{bmatrix} = \begin{bmatrix} -9 & 1 \\ -20 & 0 \end{bmatrix} \begin{bmatrix} x_1 \\ x_2 \end{bmatrix} + \begin{bmatrix} 1 \\ 4 \end{bmatrix} u$$

$$y = [2 \quad -3] \begin{bmatrix} x_1 \\ x_2 \end{bmatrix}$$

(b)
$$\begin{bmatrix} \dot{x}_1 \\ \dot{x}_2 \end{bmatrix} = \begin{bmatrix} 0 & 1 \\ -6 & -5 \end{bmatrix} \begin{bmatrix} x_1 \\ x_2 \end{bmatrix} + \begin{bmatrix} 1 & 0 \\ 0 & 1 \end{bmatrix} \begin{bmatrix} u_1 \\ u_2 \end{bmatrix}$$

$$y = [0 \quad 1] \begin{bmatrix} x_1 \\ x_2 \end{bmatrix}$$

Ans.

$$\begin{bmatrix} \dot{z}_1 \\ \dot{z}_2 \end{bmatrix} = \begin{bmatrix} -2 & 0 \\ 0 & -3 \end{bmatrix} \begin{bmatrix} z_1 \\ z_2 \end{bmatrix} + \begin{bmatrix} 3 & 1 \\ -2 & -1 \end{bmatrix} u$$

$$y = [-2 \quad -3] \begin{bmatrix} z_1 \\ z_2 \end{bmatrix}$$

(c)
$$\begin{bmatrix} \dot{x}_1 \\ \dot{x}_2 \end{bmatrix} = \begin{bmatrix} -5 & 1 \\ -4 & 0 \end{bmatrix} \begin{bmatrix} x_1 \\ x_2 \end{bmatrix} + \begin{bmatrix} 1 & 0 \\ 1 & -1 \end{bmatrix} \begin{bmatrix} u_1 \\ u_2 \end{bmatrix}$$

$$\begin{bmatrix} y_1 \\ y_2 \end{bmatrix} = \begin{bmatrix} 1 & 1 \\ -1 & 0 \end{bmatrix} \begin{bmatrix} x_1 \\ x_2 \end{bmatrix}$$

(d) $$\begin{bmatrix} \dot{x}_1 \\ \dot{x}_2 \\ \dot{x}_3 \end{bmatrix} = \begin{bmatrix} -3 & 1 & 0 \\ -2 & 0 & 1 \\ 0 & 0 & 0 \end{bmatrix} \begin{bmatrix} x_1 \\ x_2 \\ x_3 \end{bmatrix} + \begin{bmatrix} 1 \\ 1 \\ 1 \end{bmatrix} u$$

$$\begin{bmatrix} y_1 \\ y_2 \end{bmatrix} = \begin{bmatrix} 1 & 1 & 0 \\ 0 & 1 & 1 \end{bmatrix} \begin{bmatrix} x_1 \\ x_2 \\ x_3 \end{bmatrix}$$

12. The following system has a set of complex conjugate characteristic roots. Find an alternative diagonal set of state equations. Then find another alternative set of state equations where the complex root terms are placed in real number block diagonal form.

$$\begin{bmatrix} \dot{x}_1 \\ \dot{x}_2 \\ \dot{x}_3 \end{bmatrix} = \begin{bmatrix} 0 & 1 & 0 \\ 0 & 0 & 1 \\ 0 & -17 & -2 \end{bmatrix} \begin{bmatrix} x_1 \\ x_2 \\ x_3 \end{bmatrix} + \begin{bmatrix} 2 \\ -1 \\ 0 \end{bmatrix} u$$

$$y = \begin{bmatrix} 1 & 1 & 1 \end{bmatrix} \begin{bmatrix} x_1 \\ x_2 \\ x_3 \end{bmatrix}$$

13. The following system has a repeated characteristics root. Find an alternative set of state equations in Jordan form:

$$\begin{bmatrix} \dot{x}_1 \\ \dot{x}_2 \\ \dot{x}_3 \end{bmatrix} = \begin{bmatrix} 0 & 1 & 0 \\ 0 & 0 & 1 \\ -9 & -15 & -7 \end{bmatrix} \begin{bmatrix} x_1 \\ x_2 \\ x_3 \end{bmatrix} + \begin{bmatrix} 0 \\ 2 \\ 3 \end{bmatrix} u$$

$$y = \begin{bmatrix} 1 & 1 & 0 \\ 0 & 1 & 1 \end{bmatrix} \begin{bmatrix} x_1 \\ x_2 \\ x_3 \end{bmatrix}$$

14. Find diagonal state equations for systems with the following transfer function matrices:

(a) $$T(s) = \begin{bmatrix} \dfrac{-6s}{s^2 + 4s + 3} & \dfrac{4}{s^2 + 4s + 3} \end{bmatrix}$$

(b) $$T(s) = \begin{bmatrix} \dfrac{s^2 - 4}{s^3 + 3s^2 + 2s} & \dfrac{4s - 8}{s^3 + 3s^2 + 2s} & \dfrac{s^2 + 3s - 4}{s^3 + 3s^2 + 2s} \end{bmatrix}$$

Ans. One possibility is the following:

$$\begin{bmatrix} \dot{x}_1 \\ \dot{x}_2 \\ \dot{x}_3 \end{bmatrix} = \begin{bmatrix} 0 & 0 & 0 \\ 0 & -1 & 0 \\ 0 & 0 & -2 \end{bmatrix} \begin{bmatrix} x_1 \\ x_2 \\ x_3 \end{bmatrix} + \begin{bmatrix} -2 & -4 & -2 \\ 3 & 12 & 6 \\ 0 & -8 & -3 \end{bmatrix} \begin{bmatrix} u_1 \\ u_2 \\ u_3 \end{bmatrix}$$

$$y = \begin{bmatrix} 1 & 1 & 1 \end{bmatrix} \begin{bmatrix} x_1 \\ x_2 \\ x_3 \end{bmatrix}$$

(c)

$$T(s) = \begin{bmatrix} \dfrac{3s-1}{s^2+4} \\ \dfrac{-s+8}{s^2+4} \end{bmatrix}$$

(d)

$$T(s) = \frac{\begin{bmatrix} s \\ -3s^2-4 \\ 8 \end{bmatrix}}{s^3+3s^2+2s}$$

Ans. One possibility is the following:

$$\begin{bmatrix} \dot{x}_1 \\ \dot{x}_2 \\ \dot{x}_3 \end{bmatrix} = \begin{bmatrix} 0 & 0 & 0 \\ 0 & -1 & 0 \\ 0 & 0 & -2 \end{bmatrix} \begin{bmatrix} x_1 \\ x_2 \\ x_3 \end{bmatrix} + \begin{bmatrix} 1 \\ 1 \\ 1 \end{bmatrix} u$$

$$\begin{bmatrix} y_1 \\ y_2 \\ y_3 \end{bmatrix} = \begin{bmatrix} 0 & 1 & -1 \\ -2 & 7 & -8 \\ 4 & -8 & 4 \end{bmatrix} \begin{bmatrix} x_1 \\ x_2 \\ x_3 \end{bmatrix}$$

15. Find a simulation diagram and a matrix state-variable representation for a two-input, two-output system with the following transfer function matrix:

$$T(s) = \begin{bmatrix} \dfrac{4s}{s^2+3s+2} & \dfrac{s-3}{s^2+3s+2} \\ \dfrac{-6}{s^2+3s+2} & \dfrac{s+4}{s^2+3s+2} \end{bmatrix}$$

16. A transfer function with equal numerator and denominator polynomial degrees may be expanded as a constant plus a proper remainder, as in the following example:

$$T(s) = \frac{3s^2+2s-4}{s^2+3s+2} = 3 + \frac{-7s-10}{s^2+3s+2}$$

It may be realized by adding to the system output a term that is proportional to the system input. The resulting state-variable equations have the form

$$\dot{x} = Ax + Bu$$

$$y = Cx + Du$$

Find matrices A, B, C, and D for a system with the transfer function above. For such a second-order single-input, single-output system, the state-variable equations will be of the form

$$\begin{bmatrix} \dot{x}_1 \\ \dot{x}_2 \end{bmatrix} = \begin{bmatrix} a_{11} & a_{12} \\ a_{21} & a_{22} \end{bmatrix} \begin{bmatrix} x_1 \\ x_2 \end{bmatrix} + \begin{bmatrix} b_1 \\ b_2 \end{bmatrix} u$$

$$y = [c_1 \quad c_2] \begin{bmatrix} x_1 \\ x_2 \end{bmatrix} + du$$

17. Use controllability and observability matrices to determine whether the following systems are completely controllable and whether these systems are completely observable. In addition, determine BIBO and asymptotic stability in each case.

(a)
$$\begin{bmatrix} \dot{x}_1 \\ \dot{x}_2 \end{bmatrix} = \begin{bmatrix} 2 & -4 \\ 0 & 1 \end{bmatrix} \begin{bmatrix} x_1 \\ x_2 \end{bmatrix} + \begin{bmatrix} 1 \\ 0 \end{bmatrix} u$$

$$y = [1 \quad 1] \begin{bmatrix} x_1 \\ x_2 \end{bmatrix}$$

(b)
$$\begin{bmatrix} \dot{x}_1 \\ \dot{x}_2 \\ \dot{x}_3 \end{bmatrix} = \begin{bmatrix} 3 & 0 & -5 \\ -2 & 1 & 5 \\ 0 & 0 & -2 \end{bmatrix} \begin{bmatrix} x_1 \\ x_2 \\ x_3 \end{bmatrix} + \begin{bmatrix} 1 & 0 \\ 2 & 0 \\ 0 & -1 \end{bmatrix} \begin{bmatrix} u_1 \\ u_2 \end{bmatrix}$$

$$\begin{bmatrix} y_1 \\ y_2 \end{bmatrix} = \begin{bmatrix} 4 & 1 & -3 \\ 3 & 2 & -1 \end{bmatrix} \begin{bmatrix} x_1 \\ x_2 \\ x_3 \end{bmatrix}$$

Ans. completely controllable but not completely observable

(c)
$$\begin{bmatrix} \dot{x}_1 \\ \dot{x}_2 \\ \dot{x}_3 \end{bmatrix} = \begin{bmatrix} 1 & 0 & -2 \\ 3 & -3 & 0 \\ 0 & 0 & 1 \end{bmatrix} \begin{bmatrix} x_1 \\ x_2 \\ x_3 \end{bmatrix} + \begin{bmatrix} 1 & -1 \\ 2 & 0 \\ 0 & 0 \end{bmatrix} \begin{bmatrix} u_1 \\ u_2 \end{bmatrix}$$

$$\begin{bmatrix} y_1 \\ y_2 \end{bmatrix} = \begin{bmatrix} 0 & 4 & 1 \\ 0 & -2 & 3 \end{bmatrix} \begin{bmatrix} x_1 \\ x_2 \\ x_3 \end{bmatrix}$$

18. Write state equations for systems—each with modes $e^{2t}, e^{-3t}, e^{(-4+j)t}$, and $e^{(-4-j)t}$—that have the following properties:

(a) The mode e^{2t} is uncontrollable.

(b) The mode e^{-3t} is unobservable.

(c) The mode e^{2t} is both uncontrollable and unobservable.

(d) The modes $e^{(-4+j)t}$ and $e^{(-4-j)t}$ are uncontrollable.

(e) The mode e^{-3t} is uncontrollable and the mode e^{2t} is unobservable.

19. The system

$$\begin{bmatrix} \dot{x}_1 \\ \dot{x}_2 \end{bmatrix} = \begin{bmatrix} -1 & 1 \\ 2 & 0 \end{bmatrix} \begin{bmatrix} x_1 \\ x_2 \end{bmatrix} + \begin{bmatrix} 0 \\ 1 \end{bmatrix} u$$

$$y = [\,1 \quad 1\,] \begin{bmatrix} x_1 \\ x_2 \end{bmatrix}$$

is unstable. Can the instability be detected from input–output measurements? Determine whether the system is completely observable. Then calculate the system transfer function. A common factor in the numerator and the denominator should cancel.

Repeat if instead the output equation is

$$y = [\,-2 \quad 1\,] \begin{bmatrix} x_1 \\ x_2 \end{bmatrix}$$

20. Find a third-order system, if possible, in phase-variable form, that is not completely controllable.

21. Show that an nth-order system with n outputs is completely observable if its $(n \times n)$-output coupling matrix is nonsingular.

22. Use the time domain method involving convolution to solve

$$\dot{x} = -2x + u\,(t)$$

with

$$x\left(0^-\right) = 7$$
$$u\,(t) = 3e^{3t}$$

23. Use Laplace transform methods to find the state response of the following systems for $t \geqslant 0$ with the given inputs and initial conditions. Also find the system output.

(a)
$$\dot{x} = -3x + 2r\,(t)$$
$$y = 4x$$
$$x\left(0^-\right) = 7$$
$$r\,(t) = 5u(t), \text{ where } u\,(t) \text{ is the unit step function}$$

(b)

$$\begin{bmatrix} \dot{x}_1 \\ \dot{x}_2 \end{bmatrix} = \begin{bmatrix} -3 & 1 \\ 7 & 0 \end{bmatrix} \begin{bmatrix} x_1 \\ x_2 \end{bmatrix} + \begin{bmatrix} 1 \\ 0 \end{bmatrix} u$$

$$y = [1 \quad 0] \begin{bmatrix} x_1 \\ x_2 \end{bmatrix}$$

$$\begin{bmatrix} x_1(0^-) \\ x_2(0^-) \end{bmatrix} = \begin{bmatrix} 3 \\ 0 \end{bmatrix}$$

$u(t) = \delta(t)$, the unit impulse function

Ans. $\begin{bmatrix} 1.01e^{1.54t} & 2.99e^{-4.54t} \\ 4.61e^{1.54t} & -4.61e^{-4.54t} \end{bmatrix}$; $y(t) = 1.01e^{1.54t} + 2.99e^{-4.54t}$

(c)

$$\begin{bmatrix} \dot{x}_1 \\ \dot{x}_2 \end{bmatrix} = \begin{bmatrix} 0 & 1 \\ -3 & -8 \end{bmatrix} \begin{bmatrix} x_1 \\ x_2 \end{bmatrix} + \begin{bmatrix} 0 & 4 \\ 1 & -1 \end{bmatrix} \begin{bmatrix} u_1 \\ u_2 \end{bmatrix}$$

$$y = [1 \quad 1] \begin{bmatrix} x_1 \\ x_2 \end{bmatrix}$$

$$\begin{bmatrix} x_1(0^-) \\ x_2(0^-) \end{bmatrix} = \begin{bmatrix} 7 \\ -3 \end{bmatrix}$$

$$\begin{bmatrix} u_1(t) \\ u_2(t) \end{bmatrix} = \begin{bmatrix} e^t \\ 5 \end{bmatrix}$$

(d)

$$\begin{bmatrix} \dot{x}_1 \\ \dot{x}_2 \\ \dot{x}_3 \end{bmatrix} = \begin{bmatrix} -3 & 1 & 0 \\ -2 & 0 & 1 \\ 0 & 0 & 0 \end{bmatrix} \begin{bmatrix} x_1 \\ x_2 \\ x_3 \end{bmatrix} + \begin{bmatrix} 0 \\ 1 \\ 0 \end{bmatrix} u$$

$$\begin{bmatrix} y_1 \\ y_2 \end{bmatrix} = \begin{bmatrix} 1 & 0 & 1 \\ 0 & 0 & 1 \end{bmatrix} \begin{bmatrix} x_1 \\ x_2 \\ x_3 \end{bmatrix}$$

$$\begin{bmatrix} x_1(0^-) \\ x_2(0^-) \\ x_3(0^-) \end{bmatrix} = \begin{bmatrix} 1 \\ 0 \\ 0 \end{bmatrix}$$

$u(t) = 2$

24. The state transition matrix for a certain system is

$$\Phi(t) = \begin{bmatrix} \frac{1}{2}e^{-t} + \frac{1}{2}^{-2t} & 2(e^{-t} - e^{-2t}) \\ 3e^{-t} - 3e^{-2t} & -e^{-t} + 2e^{-2t} \end{bmatrix}$$

Find the state $x(t)$ for $t \geqslant 0$ if all system inputs are zero and

$$x\left(0^-\right) = \begin{bmatrix} x_1\left(0^-\right) \\ x_2\left(0^-\right) \end{bmatrix} = \begin{bmatrix} -7 \\ 2 \end{bmatrix}$$

25. Calculate state transition matrices for systems with the following state coupling matrices A, using

$$\Phi(t) = \mathcal{L}^{-1}\left\{[sI - A]^{-1}\right\}$$

(a) $\begin{bmatrix} 0 & 1 \\ -6 & -4 \end{bmatrix}$

(b) $\begin{bmatrix} -1 & 4 \\ 8 & -2 \end{bmatrix}$

Ans. $\begin{bmatrix} 0.4e^{-7.1t} + 0.6e^{4.1t} & 0.3e^{-7.1t} - 0.3e^{4.1t} \\ 0.6e^{-7.1t} - 0.6e^{4.1t} & 0.54e^{-7.1t} + 0.46e^{4.1t} \end{bmatrix}$

(c) $\begin{bmatrix} -5 & 1 & 0 \\ -4 & 0 & 5 \\ 0 & 0 & 0 \end{bmatrix}$

26. Show that the state transition matrix for a diagonalized system is diagonal, with the system modes along the diagonal.

27. Find state-variable equations for each of the systems of Fig. P8.27. Then find the transfer function(s) from the original drawing and compare with the transfer function(s) of the state-variable model.

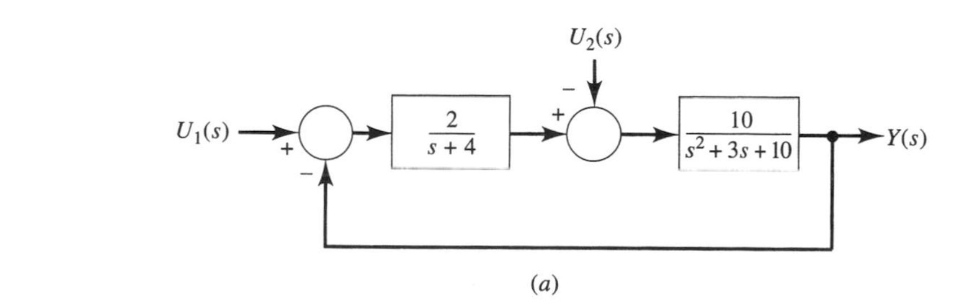

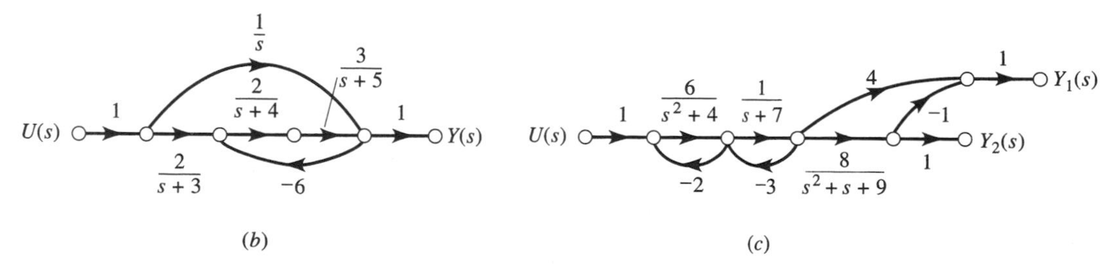

Figure P8.27

CHAPTER 9

State Space Design

9.1 Preview

Feedback along with many of its properties has been the underlying theme in control engineering. In classical design, the plant output is fed back and processed by standard compensators (lead–lag, PID) to modify system dynamics with a view to satisfying stability and performance requirements. Typically, a design engineer tries to reshape the system root locus, or Bode plot, to meet the requirements. These methods allow limited control of the closed loop poles.

In state space design, the basic idea of feedback is maintained. Rather than feeding back one or two outputs, we feedback the complete system state vector to modify system dynamics. We will see that feeding back the complete state vector gives the designer total control over the closed-loop poles. Such complete control is possible because, ideally, the system's state contains all the available information about the system. In a sense, we are using *full information* about the system to control its behavior.

9.2 State Feedback and Pole Placement

Consider the simple first-order system described by

$$\dot{x} = x + u \qquad x(0) = x_0$$

$$y = x$$

where u is the input to the plant. The pole of this open loop system is 1, indicating that the open-loop system is unstable. To stabilize the system, we may feed the state back using some gain k where

$$u = -kx$$

Therefore, the compensated system becomes

$$\dot{x} = x - kx = (1 - k)x$$

The closed-loop pole is $1 - k$, which results in an asymptotically stable system for $k > 1$. In fact, by a suitable choice of k, we can place the closed-loop pole anywhere on the real axis.

As another example of placing the system poles at desired locations with state feedback, consider the following single-input, single-output system, which is described in phase-variable form:

$$\begin{bmatrix} \dot{x}_1 \\ \dot{x}_2 \\ \dot{x}_3 \end{bmatrix} = \begin{bmatrix} 0 & 1 & 0 \\ 0 & 0 & 1 \\ -5 & -7 & -3 \end{bmatrix} \begin{bmatrix} x_1 \\ x_2 \\ x_3 \end{bmatrix} + \begin{bmatrix} 0 \\ 0 \\ 1 \end{bmatrix} u$$

$$y = [\,-2 \quad 4 \quad 3\,] \begin{bmatrix} x_1 \\ x_2 \\ x_3 \end{bmatrix}$$

This system is shown in the simulation diagram of Figure 9.1(a) and, from Mason's gain rule, we find that it has transfer function

$$\begin{aligned} T(s) &= \frac{3/s + 4/s^2 + -2/s^3}{1 + 3/s + 7/s^2 + 5/s^3} \\ &= \frac{3s^2 + 4s - 2}{s^3 + 3s^2 + 7s + 5} \end{aligned}$$

Since the characteristic equation factors as follows:

$$s^3 + 3s^2 + 7s + 5 = (s + 1 + j2)(s + 1 - j2)(s + 1)$$

its poles are at $s = -1 - j2, -1 + j2$, and -1.

With state feedback,

$$u(t) = -k_1 x_1 - k_2 x_2 - k_3 x_3 + r(t)$$

the state equations are of the form

$$\begin{bmatrix} \dot{x}_1 \\ \dot{x}_2 \\ \dot{x}_3 \end{bmatrix} = \begin{bmatrix} 0 & 1 & 0 \\ 0 & 0 & 1 \\ -k_1 - 5 & -k_2 - 7 & -k_3 - 3 \end{bmatrix} \begin{bmatrix} x_1 \\ x_2 \\ x_3 \end{bmatrix} + \begin{bmatrix} 0 \\ 0 \\ 1 \end{bmatrix} r(t)$$

$$y = [\,-2 \quad 4 \quad 3\,] \begin{bmatrix} x_1 \\ x_2 \\ x_3 \end{bmatrix}$$

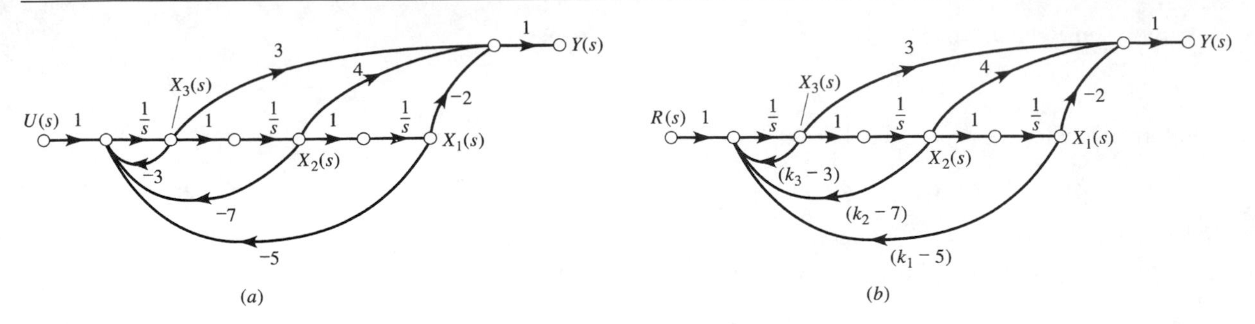

Figure 9.1 State feedback example. (a) Open-loop system. (b) System with state feedback.

as diagrammed in Figure 9.1(b). The feedback system has transfer function, in terms of the feedback gain constants k,

$$T(s) = \frac{3/s + 4/s^2 - 2/s^3}{1 + (3 + k_3)/s + (7 + k_2)/s^2 + (5 + k_1)/s^3}$$
$$= \frac{3s^2 + 4s - 2}{s^3 + (3 + k_3)s^2 + (7 + k_2)s + (5 + k_1)}$$

The coefficients of the characteristic equation may be chosen at will, by appropriately selecting k_1, k_2, and k_3. If, for instance, it is desired that the system poles be located at $s = -4, -4$, and -5, the characteristic polynomial should be

$$(s + 4)(s + 4)(s + 5) = s^3 + 13s^2 + 56s + 80$$
$$= s^3 + (3 + k_3)s^2 + (7 + k_2)s + (5 + k_1)$$

which will be the case for

$$k_1 = +75$$
$$k_2 = +49$$
$$k_3 = +10$$

In general, given a system in state space form

$$\dot{x} = Ax + Bu$$

Using state feedback, we get

$$u = -Kx \rightarrow \dot{x} = Ax - BKx = (A - BK)x$$

The closed-loop system matrix has been modified from A to $A - BK$. The closed-loop characteristic polynomial is given by

$$\Delta_c(s) = |sI - (A - BK)|$$

The closed-loop poles (or eigenvalues) are the roots of the foregoing polynomial. If the original system is represented in phase-variable form, the A matrix will be in companion form. An important property of companion form matrices is that their characteristic polynomial can be written by inspection. In fact the coefficients of the characteristic polynomial can be read off the last row of the A matrix. For instance, in the preceding example, the last row of the $A - BK$ matrix is

$$\text{Last row} = [-k_1 - 5 \quad -k_2 - 7 \quad -k_3 - 3]$$

and the characteristic polynomial is

$$\text{Characteristic polynomial} = s^3 + (3 + k_3)s^2 + (7 + k_2)s + (5 + k_1)$$

It is clear from this equation that, in this case, any desired polynomial can be achieved by selecting the feedback gains.

Can all systems be stabilized by using state feedback? To answer this question, consider the following example.

$$\dot{x} = \begin{bmatrix} 1 & 0 \\ 0 & 2 \end{bmatrix} x + \begin{bmatrix} b_1 \\ b_2 \end{bmatrix} u$$

Using state feedback, the closed-loop characteristic equation becomes

$$\Delta_c(s) = \left| \begin{bmatrix} s & 0 \\ 0 & s \end{bmatrix} - \begin{bmatrix} 1 & 0 \\ 0 & 2 \end{bmatrix} + \begin{bmatrix} b_1 \\ b_2 \end{bmatrix} [k_1 \quad k_2] \right|$$

$$= s^2 + (-3 + b_2k_2 + b_1k_1)s + (2 - b_2k_2 - 2b_1k_1)$$

To see whether we can place the closed-loop poles anywhere in the complex plane, consider an arbitrary second-order polynomial

$$\text{Desired characteristic polynomial} = \Delta_d(s) = s^2 + \alpha s + \beta$$

To find the state feedback gains, these polynomials must be identical. Thus

$$b_1k_1 + b_2k_2 - 3 = \alpha$$

$$-b_2k_2 - 2b_1k_1 + 2 = \beta$$

Writing the equations in matrix form, we get

$$\begin{bmatrix} b_1 & b_2 \\ -2b_1 & -b_2 \end{bmatrix} \begin{bmatrix} k_1 \\ k_2 \end{bmatrix} = \begin{bmatrix} 3 + \alpha \\ -2 + \beta \end{bmatrix}$$

We can solve for the unknown gains for any right-hand side (i.e., for any desired polynomial) if and only if the determinant of the coefficient matrix is nonzero.

$$\begin{vmatrix} b_1 & b_2 \\ -2b_1 & -b_2 \end{vmatrix} = -b_1b_2 + 2b_1b_2 = b_1b_2$$

Hence, both b_1 and b_2 must be nonzero. But note that this is the same condition as complete controllability of the system (refer to Section 8.6). We conclude that to place the closed-loop poles of this system arbitrarily, the system must be controllable. Let us see what happens if this condition is violated. Suppose b_1 is zero. In this case, the equations will be consistent when

$$3 + \alpha = 2 - \beta \qquad \text{or} \qquad \beta = -1 - \alpha$$

Imposing this condition on the desired polynomial

$$\text{Achievable desired polynomial} = s^2 + \alpha s - 1 - \alpha$$

But this polynomial is unstable. Hence, we can never stabilize this system if it is uncontrollable.

The results of the previous example can be generalized as follows:

The closed-loop poles of a system can be arbitrarily placed anywhere in the complex plane if and only if the system is completely controllable.

Controllability was defined earlier as the ability to move the system states from any initial state to any final state. We observe here that this is equivalent to the ability of placing (or shifting) the system poles anywhere in the complex plane. The foregoing result is commonly called the *pole placement* (or *pole-shifting*) theorem. If a system is not controllable, we may still be able to move some of the poles but not all of them. In general, controllable modes can be shifted, whereas uncontrollable modes are fixed. For instance, in the preceding example where b_1 was 0, the mode at $s = 1$ was uncontrollable (could not be moved), but the mode at $s = 2$ could be placed anywhere (we can show that to move the mode from 2 to -2, $k_2 = 4/b_2$).

9.2.1 Stabilizability

Defining stabilizabilty.

Careful study of the pole placement theorem reveals that controllability is too strong a condition. It allows arbitrary pole placement, even in the RHP. But we normally are not interested in placing system poles in the RHP. It turns out in practice that a weaker notion than controllability is sufficient for most purposes. This notion is called *stabilizability.* It refers to the ability to move only the unstable modes of the system. Therefore, we say a system is *stabilizable* if the unstable modes are controllable or, equivalently, if the uncontrollable modes are stable. The easiest way to check this is to convert the system to modal form, then check each mode and the corresponding row in the B matrix. The next example illustrates the notion.

$$\dot{x} = \begin{bmatrix} 2 & 0 \\ 0 & -1 \end{bmatrix} x + \begin{bmatrix} 1 \\ 0 \end{bmatrix} u, \quad \dot{z} = \begin{bmatrix} 2 & 0 \\ 0 & 1 \end{bmatrix} z + \begin{bmatrix} 1 \\ 0 \end{bmatrix} u$$

Neither of these systems is controllable. In the first system, the stable mode at -1 is not controllable, whereas the unstable mode at 2 is controllable. Hence, the system is stabilizable. For instance, by using state feedback control—$u = -kx_1$, with $k > 2$—the system is stabilized. In the second system, observe that the unstable mode at 1 is not controllable; therefore, the system is not stabilizable. Note that either stability or controllability implies that the system is stabilizable. For control system design, stabilizability is the minimum condition the system must satisfy for any problem. A system model that does not satisfy this condition is a poor model. Either the system must be remodeled or its structure must be modified to render it stabilizable.

Several formulas exist for computation of the state-feedback gain. *Ackermann's formula* is an example of one (for SISO controllable systems). Given the desired characteristic equation

$$\Delta_d(s) = s^n + \alpha_1 s^{n-1} + \cdots + \alpha_n$$

The state feedback gain vector is given by

Ackermann's formula.

$$k = [0, 0, \ldots, 1]M_c^{-1}\Delta_d(A)$$

where $\Delta_d(A) = A^n + \alpha_1 A^{n-1} + \cdots + \alpha_{n-1}A + \alpha_n I$ and M_c is the controllability matrix.

As an application of this formula, consider the following double-integrator plant.

$$G(s) = \frac{1}{s^2}$$

or in state space form

State feedback control of the double-integrator system.

$$\dot{x} = \begin{bmatrix} 0 & 1 \\ 0 & 0 \end{bmatrix} x + \begin{bmatrix} 0 \\ 1 \end{bmatrix} u$$

$$y = \begin{bmatrix} 1 & 0 \end{bmatrix} x$$

Let us place the system poles at $-1 \pm j$. That is, $\Delta_d(s) = s^2 + 2s + 2$. Then, k is given by

$$k = [0 \quad 1]\begin{bmatrix} 0 & 1 \\ 1 & 0 \end{bmatrix}^{-1}\left\{\begin{bmatrix} 0 & 1 \\ 0 & 0 \end{bmatrix}^2 + 2\begin{bmatrix} 0 & 1 \\ 0 & 0 \end{bmatrix} + 2\begin{bmatrix} 1 & 0 \\ 0 & 1 \end{bmatrix}\right\} = [2 \quad 2]$$

Let the plant input be

$$u = -kx + r$$

Then the compensated system block diagram will be as shown in Figure 9.2(a).

Rearranging the block diagram, as in Figure 9.2(b), shows that the state feedback controller is equivalent to a feedback PD compensator of the form

$$H(s) = 2(1 + s)$$

The open-loop transfer function is given by

$$G(s)H(s) = \frac{2(s+1)}{s^2}$$

This can be used to perform classical root locus and Bode analysis on the system. We can also obtain stability margins for the system. The open-loop and closed-loop transfer functions can also be obtained directly in terms of the state space matrices as shown next.

The plant is represented by

$$\dot{x} = Ax + Bu$$

Laplace-transforming, we have

$$sX(s) = AX(s) + BU(s) \qquad X(s) = (sI - A)^{-1}BU(s)$$

Let $\Phi(s) = (sI - A)^{-1}$; then $X(s) = \Phi(s)BU(s)$, and

$$U(s) = -KX(s) = -K\Phi(s)BU(s)$$

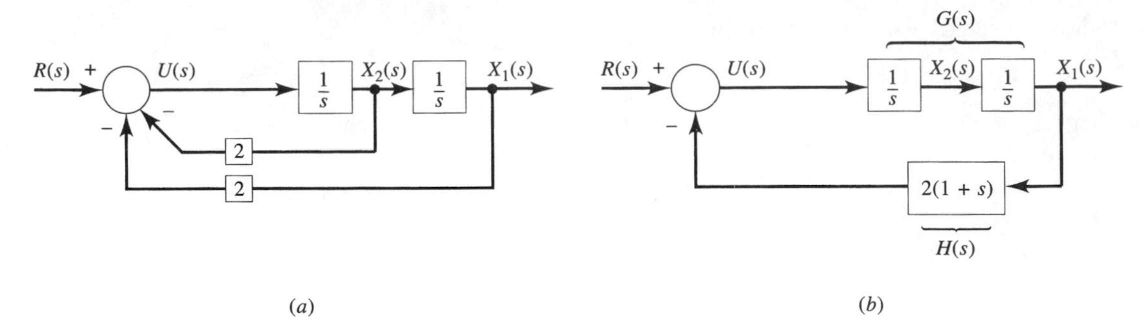

Figure 9.2 (a) Compensated system block diagram. (b) Rearranged block diagram.

State feedback open-loop transfer function.

Figure 9.3(c) shows that if the plant input is u, the signal fed back to the summing junction is $-K\Phi B$. Hence, the open-loop transfer function is $K\Phi B$. If this is compared to a classical feedback configuration, shown in Figure 9.3(d), we get

$$G(s)H(s) = K\Phi(s)B$$

The closed-loop transfer function of the state feedback system, shown in Figure 9.4, is given by

$$u = -Kx + r$$

Applying the input to the system (i.e., closing the loop), Laplace-transforming, and solving for the state vector, we get

$$\dot{x} = (A - BK)x + Br \rightarrow X(s) = (sI - A + BK)^{-1}BR(s)$$

Closed-loop transfer function under state feedback.

Substituting the state into the output equation gives the closed-loop transfer function

$$Y(s) = CX(s) = C(sI - A + BK)^{-1}BR(s) \rightarrow$$

Applying these to the present example, we get

$$k\Phi(s)B = [2 \quad 2]\begin{bmatrix} s & -1 \\ 0 & s \end{bmatrix}^{-1}\begin{bmatrix} 0 \\ 1 \end{bmatrix} = 2\frac{s+1}{s^2}$$

$$T(s) = [1 \quad 0]\left\{s\begin{bmatrix} 1 & 0 \\ 0 & 1 \end{bmatrix} - \begin{bmatrix} 0 & 1 \\ -2 & -2 \end{bmatrix}\right\}^{-1}\begin{bmatrix} 0 \\ 1 \end{bmatrix} = \frac{1}{s^2 + 2s + 2}$$

9.2.2 Choosing Pole Locations

State feedback gives the designer the option of relocating all system closed-loop poles. This is in contrast with classical design, where the designer can only hope to achieve a pair of complex conjugate poles that are dominant. Because all other poles and zeros may fall anywhere, meeting the design specifications becomes a matter of trial and error. With the freedom of choice rendered by state feedback comes the responsibility of selecting these poles judiciously.

Although there is no magic choice, there are some guidelines we can follow. Moving the poles around is costly. Suppose a first-order system has a pole at -1.

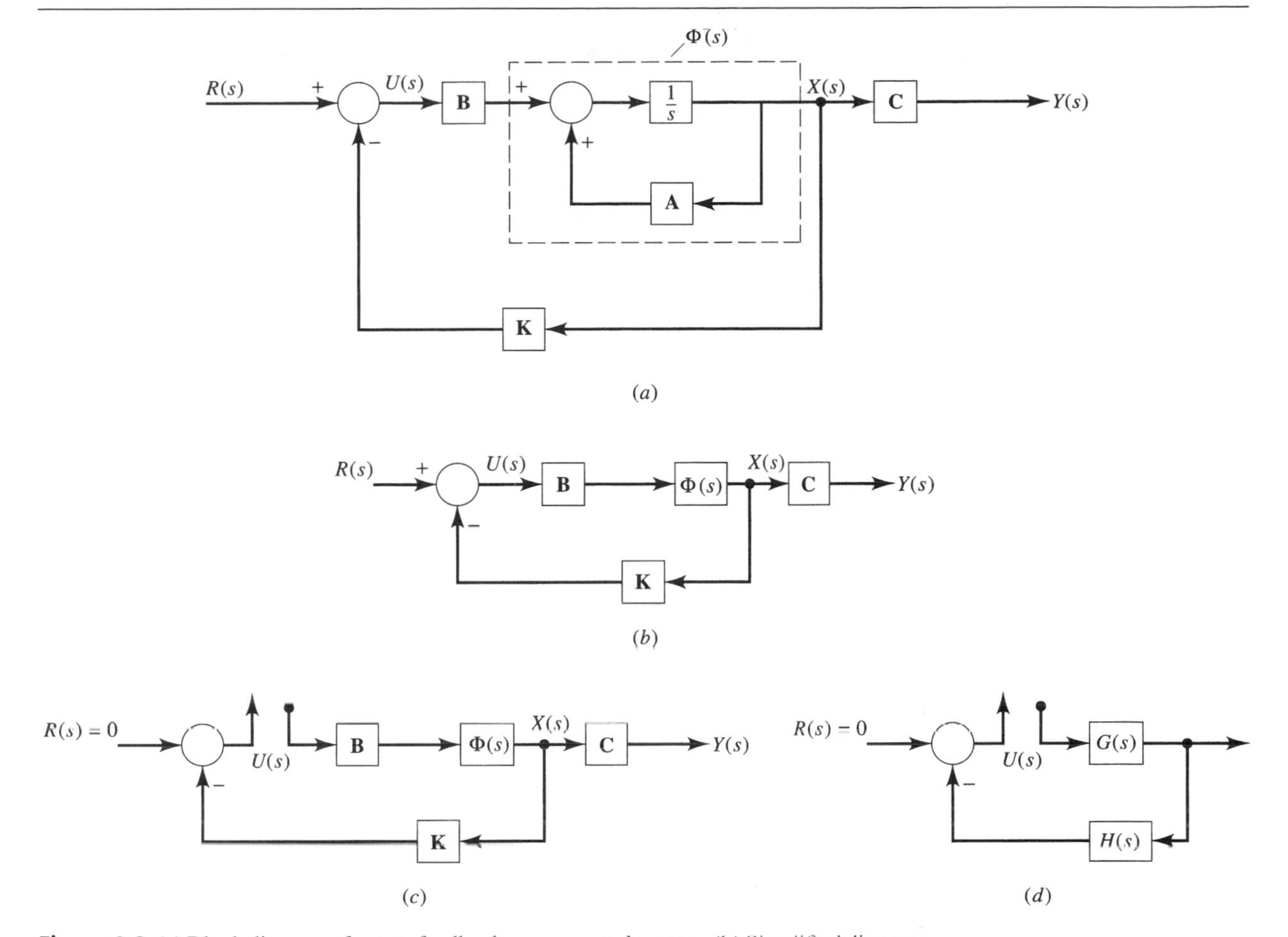

Figure 9.3 (a) Block diagram of a state feedback compensated system. (b) Simplified diagram (c) Diagram for computing the open-loop transfer function with the loop broken at the input. (d) Classical feedback configuration.

If this pole is moved to -10, the time constant is reduced, resulting in a faster system. The system output may be voltage, pressure, position, velocity, temperature, and so on. Sensors and an actuator are needed to do the job. A faster system may require a more accurate sensor and a larger or stronger actuator (such as a motor) to perform the task. These are some of the obvious costs that may be associated with pole shifting. Therefore, one guideline is that if an LHP pole has an acceptable location, leave it alone. Poles in the RHP, or poles on or close to the imaginary axis must be moved. A rule, suggested by optimal control, is that RHP poles must be reflected about the imaginary axis to minimize control energy (i.e., a pole at 2 must be shifted to -2). A pair of complex conjugate poles can be placed to meet transient response requirements. One should also be cautious about the temptation to push the poles too far into the LHP. The consequence of this is that the system bandwidth increases, and the system becomes sensitive to noise.

Guidelines from optimal control.

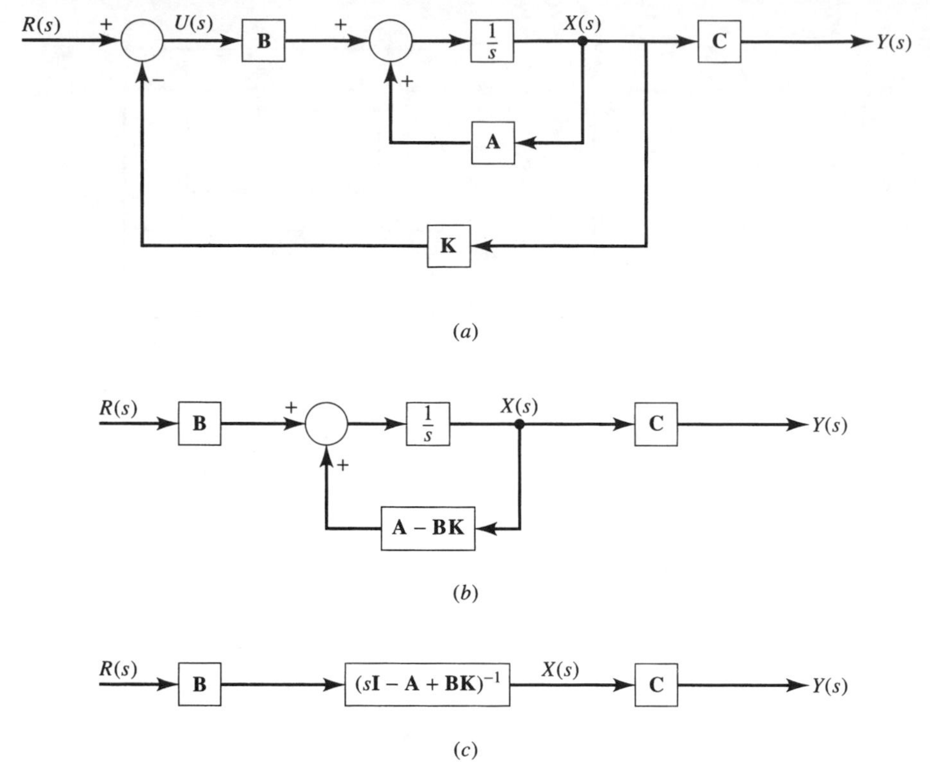

Figure 9.4 (a) State feedback compensated system. (b) The system after closing the feedback loop. (c) Diagram for the closed-loop transfer function.

Suppose an unstable system has poles at $\{2, -10, -0.1 - j10, -0.1 + j10\}$. It is specified that overshoot be less than 10%, and the settling time be less than 5. The pole at 2 can be shifted to -2. The pole at -10 can remain. The complex poles have a very small damping ratio. They could be moved to $-1 - j$ and $-1 + j$ to meet the transient response specifications. Hence, the desired pole locations are $\{-2, -10, -1 - j, -1 + j\}$. Figure 9.5 shows the resulting step response; clearly, the specifications are satisfied.

State feedback does not affect plant zeros.

Note that the closed-loop zero locations were not specified. The reason is that state feedback does not affect the system zeros. Therefore, if they are at undesirable locations, nothing can be done about them by using state feedback. Because steady state tracking properties depend on poles and zeros, this means that tracking properties cannot be helped by state feedback alone. For instance, the plant in the preceding example was a double integrator, hence it is a type 2 system with zero steady state error to unit step and ramp inputs. After state feedback, it became a type 0 system. For example, we put

$$T_E(s) = 1 - T(s) = \frac{s^2 + 2s + 1}{s^2 + 2s + 2}$$

and for a unit step input, $T_E(0) = \frac{1}{2} = 0.5$; hence, the steady state error is 0.5.

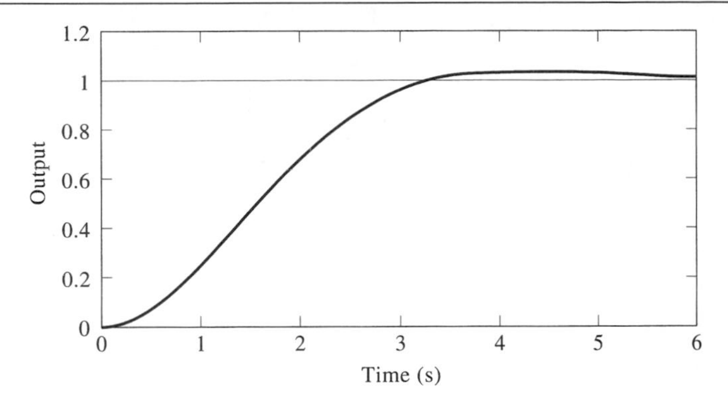

Figure 9.5 Step response of a fourth-order system with specified poles.

9.2.3 Limitations of State Feedback

State feedback is not usually practical.

The major limitation of state feedback is that it is not usually practical. It is not practical for two reasons. One is that state feedback leads to PD-type compensators, which have infinite bandwidth, whereas real components and compensators always have finite bandwidth. Another reason is that it is simply not possible or practical to sense all the states and feed them back. In reality, only certain states or combinations of them are measurable as outputs. Consequently, any practical compensator must rely on system outputs and inputs only for compensation; that is called *output feedback* and is discussed in the following sections.

❑ DRILL PROBLEMS

D9.1 For the state-feedback systems described by the following equations, choose the feedback gain constants k_i to place the closed-loop system poles at the indicated locations:

(a)

$$\begin{bmatrix} \dot{x}_1 \\ \dot{x}_2 \\ \dot{x}_3 \end{bmatrix} = \begin{bmatrix} 0 & 1 & 0 \\ 0 & 0 & 1 \\ -3 & -6 & -7 \end{bmatrix} \begin{bmatrix} x_1 \\ x_2 \\ x_3 \end{bmatrix} + \begin{bmatrix} 0 \\ 0 \\ 1 \end{bmatrix} u$$

$$u = \begin{bmatrix} -k_1 & -k_2 & -k_3 \end{bmatrix} \begin{bmatrix} x_1 \\ x_2 \\ x_3 \end{bmatrix} + r$$

$$y = \begin{bmatrix} 2 & 0 & -1 \end{bmatrix} \begin{bmatrix} x_1 \\ x_2 \\ x_3 \end{bmatrix}$$

Closed-loop poles at $s = -3, -4$, and -5

Ans. $k_1 = +57, k_2 = +41, k_3 = +5$

(b)

$$\begin{bmatrix} \dot{x}_1 \\ \dot{x}_2 \\ \dot{x}_3 \end{bmatrix} = \begin{bmatrix} -2 & 1 & 0 \\ 4 & 0 & 1 \\ 0 & 0 & 0 \end{bmatrix} \begin{bmatrix} x_1 \\ x_2 \\ x_3 \end{bmatrix} + \begin{bmatrix} 0 \\ 0 \\ 1 \end{bmatrix} u$$

$$u = \begin{bmatrix} -k_1 & -k_2 & -k_3 \end{bmatrix} \begin{bmatrix} x_1 \\ x_2 \\ x_3 \end{bmatrix} + r$$

$$y = \begin{bmatrix} 2 & 0 & -1 \\ 1 & 1 & 0 \end{bmatrix} \begin{bmatrix} x_1 \\ x_2 \\ x_3 \end{bmatrix}$$

Closed-loop poles at $s = -3 \pm j3, -3$

Ans. $k_1 = +30, k_2 = +26, k_3 = +7$

D9.2 Design a state-feedback controller for the following systems. Determine the controller gains, open-loop transfer functions, and closed-loop transfer functions. Use the open-loop transfer functions to obtain root locus, Bode plots, and gain and phase margins.

(a)

$$\dot{x} = \begin{bmatrix} 2 & 0 \\ 1 & 0 \end{bmatrix} x + \begin{bmatrix} 1 \\ 0 \end{bmatrix} u$$

$$y = \begin{bmatrix} 1 & -1 \end{bmatrix} x$$

$$u = -kx + r$$

Closed-loop poles at $s = -1 \pm j$

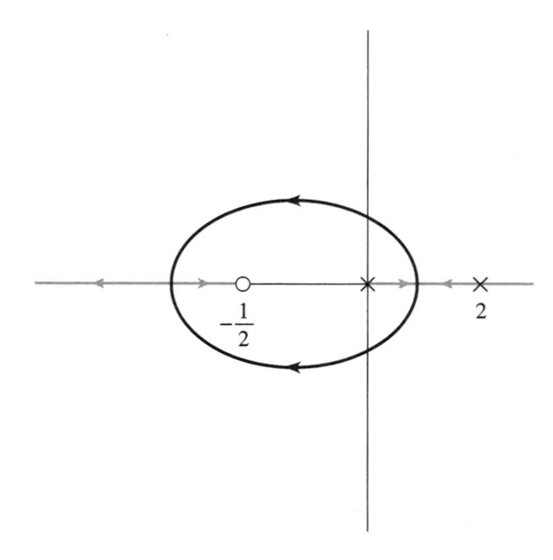

Figure D9.2a

Ans. $k = [4 \quad 2]$, $G(s)H(s) = (4s+2)/s(s-2)$, gain margin $= -6$ dB, phase margin $= 52°$, $T(s) = (s-1)/(s^2+2s+2)$

(b)

$$\dot{x} = \begin{bmatrix} 0 & 1 \\ 1 & 0 \end{bmatrix} x + \begin{bmatrix} 1 \\ 0 \end{bmatrix} u$$

$$y = [0 \quad 1]x$$

$$u = -kx + r$$

Closed-loop poles at $s = -1 \pm j$

Ans. $k = [2 \quad 3]$, $G(s)H(s) = (2s+3)/(s^2-1)$, gain margin $= -9.5$ dB, phase margin $= 53°$, $T(s) = 1/(s^2+2s+2)$

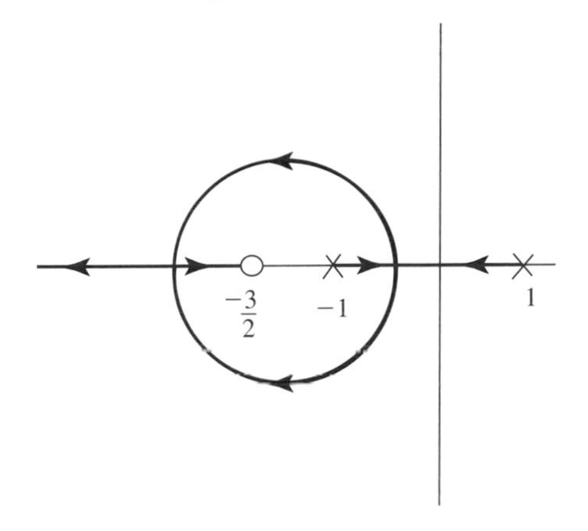

Figure D9.2b

9.3 Tracking Problems

The state feedback compensator design discussed so far has been a regulator problem. The command input has been ignored or set to zero. The objective has been to design a system that is stable and rejects disturbances. The issue of steady state error was not dealt with. In fact, if you check the step response of the double-integrator example, you will discover some steady state error. This is true even though the original plant is a type 2 system. The compensator has changed the system type. How do we incorporate the command input into the system, and design for a given steady state accuracy?

For a step input, this is easy. We simply add a summing junction with a gain, such as

$$u = -Kx + \bar{N}r$$

where r is the reference (or command) input. The constant, $\bar{N}$, can be computed to ensure zero steady state error to step inputs. Let us derive this for the case of state feedback.

$$\dot{x} = Ax + Bu = Ax + B(-Kx + \bar{N}r) = (A - BK)x + B\bar{N}r$$

By definition of steady state, the states and output must reach a constant value—that is,

$$\dot{x}_{ss} = 0 \qquad \text{which implies that} \qquad 0 = (A - BK)x_{ss} + B\bar{N}r$$

Solving for the steady state output

$$x_{ss} = -(A - BK)^{-1}B\bar{N}r \rightarrow y_{ss} = -C(A - BK)^{-1}B\bar{N}r$$

This inverse exists because $(A - BK)$ is a stable matrix. The steady state error to a constant input is the difference between the input and the output. It is therefore given by

$$e_{ss} = r - y_{ss} = r + C(A - BK)^{-1}B\bar{N}r = [1 + C(A - BK)^{-1}B\bar{N}]r$$

For zero step tracking error, the steady state output must be equal to the command input, therefore

$$\bar{N} = \frac{-1}{C(A - BK)^{-1}B}$$

9.3.1 Integral Control

The preceding technique places a gain outside of the feedback loop. As you know, when elements of a control system are not within a feedback loop, the overall system will be quite sensitive to elements outside the loop. An alternate method that allows us to achieve zero steady state error to step inputs is integral control. The idea is a classical one; we place an integrator in the forward path in series with the system, thereby increasing its system type. The block diagram of a state feedback controller using integral control is shown in Figure 9.6

Because the integral term increases the order of the system by 1, we need to augment the plant model with an added state variable to account for this. Define the new state variable as

$$x_i = \int e\,dt = \int (r - y)dt = \int (r - Cx)dt$$

Therefore, $\dot{x}_i = r - Cx$, and the augmented plant equation becomes

$$\begin{bmatrix} \dot{x} \\ \dot{x}_i \end{bmatrix} = \begin{bmatrix} A & 0 \\ -C & 0 \end{bmatrix} \begin{bmatrix} x \\ x_i \end{bmatrix} + \begin{bmatrix} B \\ 0 \end{bmatrix} u + \begin{bmatrix} 0 \\ 1 \end{bmatrix} r$$

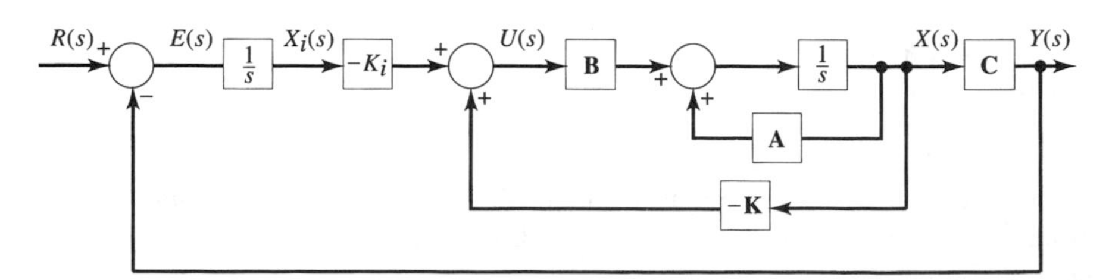

Figure 9.6 Block diagram of a state feedback compensated system with integral control.

where the zero matrices have compatible dimensions. The controller is modified to

$$u = -Kx - K_i x_i = -[\,K \quad K_i\,]\begin{bmatrix} x \\ x_i \end{bmatrix}$$

Using this controller, the compensated system becomes

$$\begin{bmatrix} \dot{x} \\ \dot{x}_i \end{bmatrix} = \begin{bmatrix} A - BK & -BK_i \\ -C & 0 \end{bmatrix}\begin{bmatrix} x \\ x_i \end{bmatrix} + \begin{bmatrix} 0 \\ 1 \end{bmatrix} r$$

The characteristic polynomial of the compensated system is then equated with the desired one to solve for the controller gains.

Double-integrator example with integral control.

Let us design a state feedback controller for the double-integrator system incorporating integral control action. First we need to augment the plant

$$\begin{bmatrix} \dot{x} \\ \dot{x}_i \end{bmatrix} = \left[\begin{array}{cc|c} 0 & 1 & 0 \\ 0 & 0 & 0 \\ \hline -1 & 0 & 0 \end{array}\right]\begin{bmatrix} x \\ \hline x_i \end{bmatrix} + \left[\begin{array}{c} 0 \\ 1 \\ \hline 0 \end{array}\right] u$$

$$y = [1 \quad 0 \,|\, 0]\begin{bmatrix} x \\ \hline x_i \end{bmatrix}$$

The poles of the system will be shifted to $\{-1 \pm j, -5\}$. Note that an extra pole needs to be selected because of the extra state. Solving for K, we get

$$K = [\,12 \quad 7 \quad -10\,]$$

The steady state output due to unit step input is

$$y_{ss} = -C(A - BK)^{-1}B = -\left[1 \quad 0 \,|\, 0\right]\left[\begin{array}{cc|c} 0 & 1 & 0 \\ -12 & -7 & 10 \\ \hline -1 & 0 & 0 \end{array}\right]^{-1}\left[\begin{array}{c} 0 \\ 0 \\ \hline 1 \end{array}\right] = 1$$

❑ DRILL PROBLEMS

D9.3 For the plants described in Drill Problem D9.2, use state feedback to design an integral controller. The desired pole locations are as indicated.

(a) Closed-loop poles at $s = -1 \pm j, -5$

Ans. $k = [\,9 \quad 22\,]$, $K_i = 10$

(b) Closed-loop poles at $s = -1 \pm j, -5$

Ans. $k = [\,7 \quad 13\,]$, $K_i = -10$

D9.4 For the plant described by $G(s) = 1/(s - 1)$, use state feedback to design an integral controller that will place closed-loop poles at $s = -1 \pm j$. Also, draw a block diagram or signal flow graph of the system.

Ans. $k = 3$, $K_i = 2$

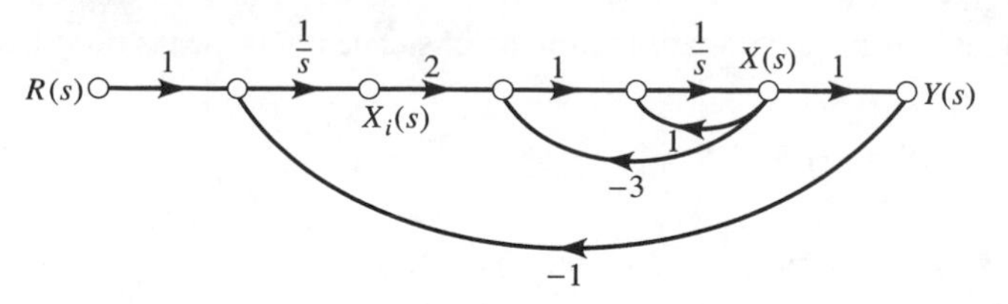

Figure D9.4

9.4 Observer Design

An observer estimates the states.

To fully implement the advantages of state feedback, all states should be fed back. Typically, some of the states are measured by sensors and the rest must be estimated by another device. An observer, shown in Figure 9.7, is a device that uses the inputs and outputs of a system to produce estimates of its states. Observers either are built using electronic components (hardware) or are equivalent computer (or microprocessor) programs (software) in digital control implementations. The word "device" implies either implementation. The idea of an observer is that if we have all system parameters, we can always simulate the model on an analog or digital computer. Even though we do not have access to system states, we have full access to the states of our simulation. Therefore, an observer is a device which simulates the original system. Letting $\hat{x}$ denote the state estimates in Figure 9.8, we have

$$\dot{x} = Ax + Bu \qquad x(0) = x_0$$
$$y = Cx$$

The observer dynamics are

$$\dot{\hat{x}} = A\hat{x} + Bu + L(y - C\hat{x}) \quad \hat{x}(0) = \hat{x}_0$$

The observer design proceeds by defining the error between the states and their estimates. Let

$$\tilde{x} = x - \hat{x}$$

To see how the error evolves with time, a differential equation for the error must be obtained.

$$\dot{\tilde{x}} = \dot{x} - \dot{\hat{x}} = Ax + Bu - A\hat{x} - Bu - L(y - C\hat{x}) = (A - LC)\tilde{x}$$
$$\tilde{x}(0) = \tilde{x}_0$$

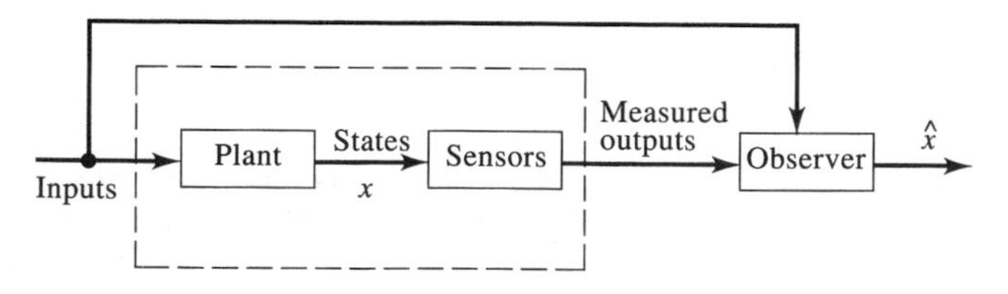

Figure 9.7 Block diagram of system and its observer.

The observer error will go to zero asymptotically if and only if the matrix $(A - LC)$ is a stable matrix (i.e., its eigenvalues are in the LHP) . But this matrix contains the matrix L, which is to be determined. It turns out that if the system is completely observable, L can be chosen such that the eigenvalues of $(A - LC)$ are arbitrary. The L matrix, which is a column vector for single-output systems, is called the *observer gain*. Also note that the observer equation can be written as

The observer works if the system is observable.

$$\dot{\hat{x}} = (A - LC)\hat{x} + Bu + Ly \quad \hat{x}(0) = \hat{x}_0$$

It can be seen that the eigenvalues of $(A - LC)$ are the observer poles. Hence, controllability allows arbitrary plant pole placement, whereas observability allows arbitrary observer pole placement.

Recall from Chapter 8 that there was a duality between phase-variable and dual phase-variable forms, in that one form could be converted into the other form via an algorithm. There is a similar duality between the notions of controllability and observability. For instance, if we transpose the controllability matrix M_c (rank remains the same), and replace A' with A and B' with C, we get M_0. Because stabilizability is a weaker version of controllability, we can also define its dual. The dual notion of *stabilizability* is called *detectability* . We say a system is *detectable* if the unstable modes are observable, or equivalently, the unobservable modes are stable.

Detectability.

Consider the following example.

$$\dot{x} = \begin{bmatrix} 2 & 0 \\ 0 & -1 \end{bmatrix} x + \begin{bmatrix} 1 \\ 0 \end{bmatrix} u \qquad \dot{z} = \begin{bmatrix} 2 & 0 \\ 0 & 1 \end{bmatrix} z + \begin{bmatrix} 1 \\ 0 \end{bmatrix} u$$

$$y = [1 \quad 0]x \qquad\qquad w = [1 \quad 0]z$$

In the first system, the mode at -1 is unobservable but detectable. In the second system, the mode at 1 is unobservable and undetectable. Let us design an observer for the double-integrator plant $G(s) = 1/s^2$.

An observer for the double-integrator system.

$$\dot{x} = \begin{bmatrix} 0 & 1 \\ 0 & 0 \end{bmatrix} x + \begin{bmatrix} 0 \\ 1 \end{bmatrix} u$$

$$y = [1 \quad 0]x$$

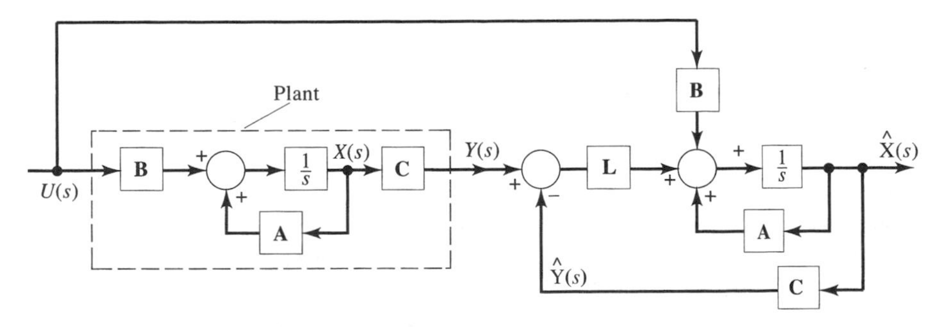

Figure 9.8 Block diagram of system with observer.

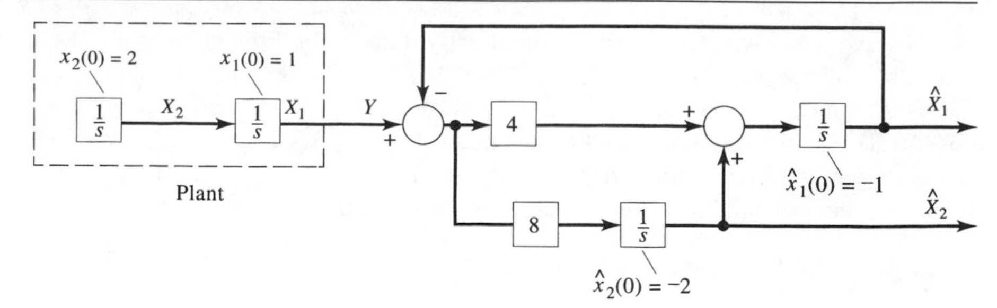

Figure 9.9 Structure of the observer for the double-integrator plant. Plant initial conditions are $(1, 2)$, observer initial conditions are $(-1, -2)$.

You can verify that the system is observable. We will design an observer with poles at $\{-2 \pm j2\}$. This choice is arbitrary, and more will be said about it later. The design starts with setting the observer characteristic polynomial equal to our desired polynomial, and solving for L.

$$|sI - (A - LC)| = \left| s\begin{bmatrix} 1 & 0 \\ 0 & 1 \end{bmatrix} - \begin{bmatrix} 0 & 1 \\ 0 & 0 \end{bmatrix} + \begin{bmatrix} l_1 \\ l_2 \end{bmatrix} [1 \quad 0] \right|$$
$$= s^2 + l_1 s + l_2$$

$$s^2 + l_1 s + l_2 = s^2 + 4s + 8 \rightarrow l_1 = 4, l_2 = 8, \text{ i.e., } L = \begin{bmatrix} 4 \\ 8 \end{bmatrix}$$

The observer equations are (with the input u set to zero).

$$\dot{\hat{x}}_1 = \hat{x}_2 + 4(y - \hat{x}_1)$$
$$\dot{\hat{x}}_2 = 8(y - \hat{x}_1)$$

The structure of this observer is shown in Figure 9.9.

This observer was simulated to verify its convergence. The zero-input response plots are shown in Figure 9.10. The plant initial conditions are $(1, 2)$ and the observer initial conditions are $(-1, -2)$. Because there are no inputs, the second state stays at a constant value of 2 and the first state is a ramp starting at 1. The observer estimates have an initial error, but they converge to true state values in about 2 s. The convergence time is the settling time of the observer, which is controlled by the real part of its poles. Because the real parts of observer poles are -2, the observer is expected to converge in about 4 time constants (i.e., 2 s).

❑ DRILL PROBLEM

D9.5 For each of the following systems, x_1 is measured while x_2 must be estimated by an observer. Select the observer gain L so both eigenvalues are as required.

Write the observer equations and create a block diagram or signal flow graph for the interconnected system and observer. Also use computer software to simulate the system and the observer, verifying the convergence of the observer.

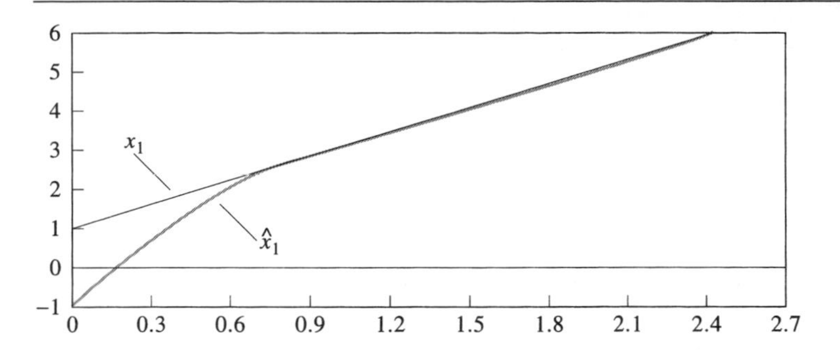

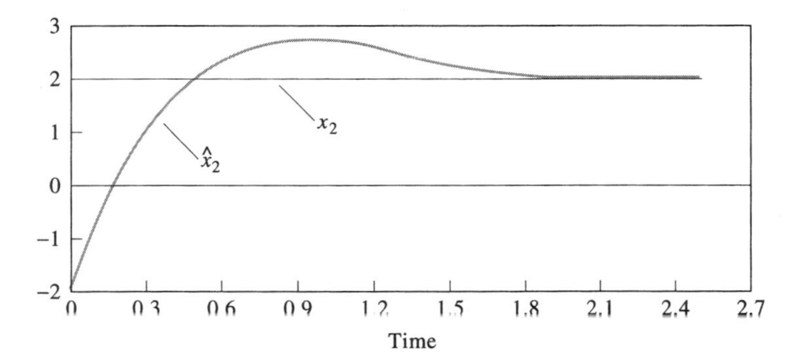

Figure 9.10 Simulation of the double-integrator plant and its observer. (a) Plot of the first state and its estimate. (b) Plot of the second state and its estimate.

(a)

$$A = \begin{bmatrix} -2 & -4 \\ 1 & -4 \end{bmatrix} \quad B = \begin{bmatrix} 2 \\ 0 \end{bmatrix} \quad C = [1 \quad 0]$$

Observer eigenvalues should be $\{-50, -50\}$

Ans.

$$L = \begin{bmatrix} 94 \\ -528 \end{bmatrix}$$

$$\dot{\hat{x}}_1 = -2\hat{x}_1 - 4\hat{x}_2 + 2u + 94(x_1 - \hat{x}_1)$$

$$\dot{\hat{x}}_2 = \hat{x}_1 - 4x_2 - 528(x_1 - \hat{x}_1)$$

(b)

$$A = \begin{bmatrix} -4 & -4 \\ 1 & -2 \end{bmatrix} \quad B = \begin{bmatrix} 0 \\ 2 \end{bmatrix} \quad C = [1 \quad 0]$$

Observer eigenvalues should be $\{-10, -10\}$.

Ans.

$$L = \begin{bmatrix} 14 \\ -15 \end{bmatrix}$$

$$\dot{\hat{x}}_1 = -4\hat{x}_1 - 4\hat{x}_2 + 14(x_1 - \hat{x}_1)$$

$$\dot{\hat{x}}_2 = \hat{x}_1 - 2x_2 + 2u - 15(x_1 - \hat{x}_1)$$

(c)

$$A = \begin{bmatrix} 0 & 1 \\ -10 & -3 \end{bmatrix} \quad B = \begin{bmatrix} 0 \\ 10 \end{bmatrix} \quad C = \begin{bmatrix} 1 & 0 \end{bmatrix}$$

Observer eigenvalues should be $\{-20, -20\}$.

Ans.

$$L = \begin{bmatrix} 37 \\ 279 \end{bmatrix}$$

$$\dot{\hat{x}} = \hat{x}_2 + 37(x_1 - \hat{x}_1)$$

$$\dot{\hat{x}}_2 = -10\hat{x}_1 - 3\hat{x}_2 + 10u + 279(x_1 - \hat{x}_1)$$

9.4.1 Control Using Observers

It was pointed out earlier that state feedback requires that all states be available for feedback. What happens when state estimates obtained using observers replace the actual states? Will we still be able to stabilize the system, or even to place its poles arbitrarily? The answer is yes. To see this, we need to obtain the equations of the closed-loop system, and examine its eigenvalues.

If the system is of order n, the observer will also be of the same order. When the observer estimates are fed back into the system, the order of the closed-loop system will be $2n$. The states of the composite system are the original plant states, x, and their estimates, $\hat{x}$. The composite state equations are obtained as follows.

$$\dot{x} = Ax + Bu \quad x(0) = x_0$$

$$y = Cx$$

Estimated state feedback:

$$u = -K\hat{x}$$

Observer:

$$\dot{\hat{x}} = A\hat{x} + Bu + L(y - C\hat{x}), \quad \hat{x}(0) = \hat{x}_0$$

Closing the loop (i.e., feeding the control back into the system and the observer), we get

$$\dot{x} = Ax - BK\hat{x}$$

$$\dot{\hat{x}} = (A - LC)\hat{x} - BK\hat{x} + Ly = (A - LC - BK)\hat{x} + LCx$$

Combining these two equations gives the closed-loop composite system

Closed-loop system.

$$\begin{bmatrix} \dot{x} \\ \dot{\hat{x}} \end{bmatrix} = \begin{bmatrix} A & -BK \\ LC & A - BK - LC \end{bmatrix} \begin{bmatrix} x \\ \hat{x} \end{bmatrix}$$

The system will be closed-loop stable if the eigenvalues of this block matrix are in the LHP. Although, this is not obvious from the matrix, we will soon show that it is indeed the case. For now, let us return to our example and see whether the double-integrator system is stabilized when the observer-based control is used. We recall that the control and observer gains were

$$k = [2 \quad 2] \qquad L = \begin{bmatrix} 4 \\ 8 \end{bmatrix}$$

The closed-loop system becomes

$$\begin{bmatrix} \dot{x} \\ \dot{\hat{x}} \end{bmatrix} = \begin{bmatrix} 0 & 1 & 0 & 0 \\ 0 & 0 & -2 & -2 \\ 4 & 0 & -4 & 1 \\ 8 & 0 & -10 & -2 \end{bmatrix} \begin{bmatrix} x \\ \hat{x} \end{bmatrix} \qquad \begin{bmatrix} x(0) \\ \hat{x}(0) \end{bmatrix} = \begin{bmatrix} 1 \\ 1.5 \\ -1 \\ -2 \end{bmatrix}$$

The eigenvalues of this matrix are $\{-1 \pm j, -2 \pm j2\}$. These are the plant and observer pole locations selected earlier. The zero-input response of the system due to the specified initial conditions is shown in Figure 9.11.

The system is clearly asymptotically stable. Therefore, the system has been stabilized by using state estimates instead of actual states. Figure 9.12, compares the

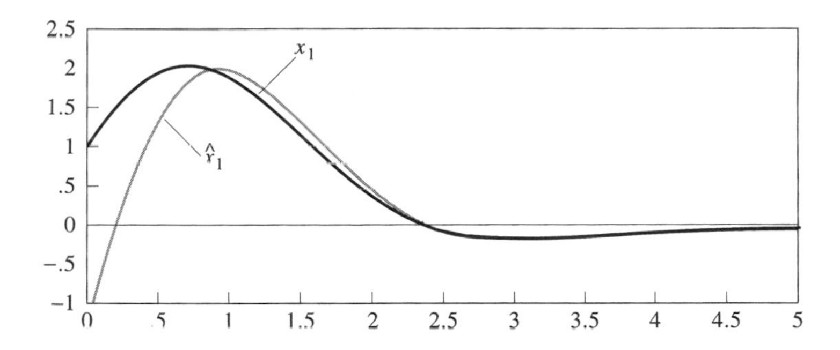

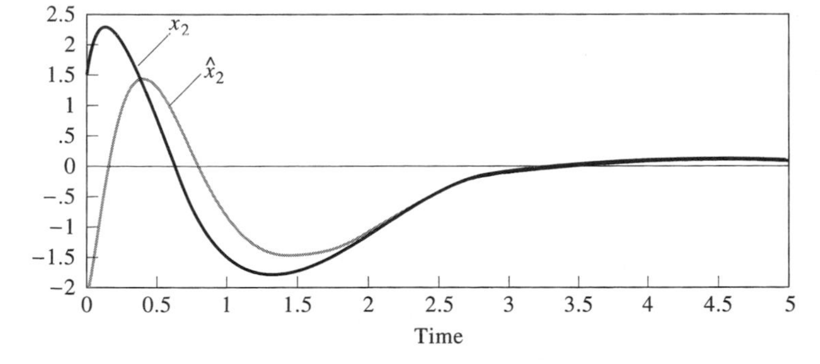

Figure 9.11 States and their estimates of the observer-based compensated double-integrator system. (a) First state and its estimate. (b) Second state and its estimate.

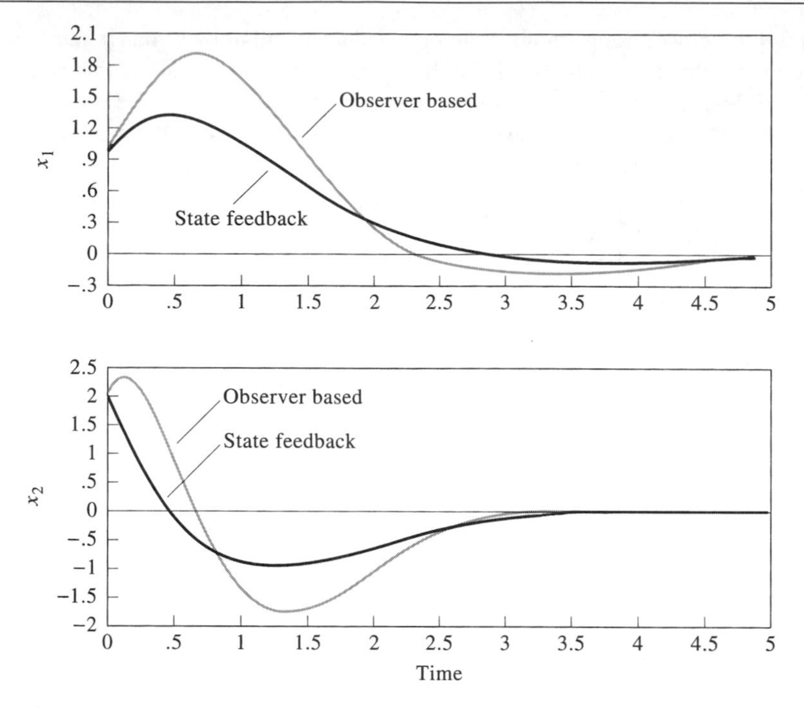

Figure 9.12 Comparison of state feedback and observer-based compensated systems. (a) First state. (b) Second state.

two designs: the zero-input responses of the system using full state feedback and using observer-based control. The figure indicates some performance degradation due to estimated states. The observer-based response has a longer settling time and a larger overshoot.

9.4.2 Separation Property

We want to show that the observer-based control results in a stable system. Recall from Section 8.3 that state space transformations allow us to look at the system in a different, and possibly more informative, way. Let us introduce the following state transformation.

$$\begin{bmatrix} x \\ \hat{x} \end{bmatrix} = P \begin{bmatrix} z \\ w \end{bmatrix} \qquad \text{where } P = \begin{bmatrix} I_n & 0_n \\ I_n & -I_n \end{bmatrix}$$

It turns out that the P matrix is its own inverse; hence the new states are

$$\begin{bmatrix} z \\ w \end{bmatrix} = \begin{bmatrix} I_n & 0_n \\ I_n & -I_n \end{bmatrix} \begin{bmatrix} x \\ \hat{x} \end{bmatrix} = \begin{bmatrix} x \\ x - \hat{x} \end{bmatrix} = \begin{bmatrix} x \\ \tilde{x} \end{bmatrix}$$

This transformation allows us to look at the states and observer errors. Recall that under state transformation, the "A" matrix becomes "$P^{-1}AP$". Therefore the transformed system becomes

$$\begin{bmatrix} \dot{x} \\ \dot{\tilde{x}} \end{bmatrix} = \begin{bmatrix} A - BK & BK \\ 0 & A - LC \end{bmatrix} \begin{bmatrix} x \\ \tilde{x} \end{bmatrix}$$

Note that, in this new realization, the matrix is block triangular (a block matrix is a matrix whose elements are matrices, as in our case). Now, we use a fact from matrix algebra.

FACT: Eigenvalues of a block triangular matrix are equal to the eigenvalues of the matrices along the diagonal blocks.

Using this fact, and the knowledge that system eigenvalues remain invariant under state transformations, we conclude that the closed-loop poles (or eigenvalues) of an observer-based control system are the union of the observer poles, and the poles of the system selected under state feedback (also known as *controller poles*). Because controllability allows us to place the eigenvalues of $A - BK$ arbitrarily, and observability does the same for the eigenvalues of $A - LC$, we see that under these two conditions we have complete freedom in controller and observer pole selections. Moreover, these selections can be made independently of each other because of the block triangular structure of the closed-loop system matrix. This property is known as the *separation property*.

Closed-loop poles = controller poles plus observer poles.

The separation between the control and the observer problem implies that we can find the controller gain, assuming the states are available, design an observer to estimate the states, and then use the estimates in place of the actual states.

9.4.3 Observer Transfer Function

It is instructive to obtain the transfer function of the observer-based compensator and compare it with classical designs. For instance, we note that state feedback resulted in a feedback PD-type compensator. The compensator output is the plant input u, and the compensator input is the plant output y. The observer also has a feedback from u, but this can be eliminated by substituting u into the observer equation. The derivation follows.

$$u = -K\hat{x}$$

$$\dot{\hat{x}} = A\hat{x} + Bu + L(y - C\hat{x}) = A\hat{x} - BK\hat{x} + Ly - LC\hat{x}$$

$$= (A - BK - LC)\hat{x} + Ly$$

Taking the Laplace transform of this equation and solving it for $\hat{X}(s)$, we get

$$\hat{X}(s) = (sI - A + BK + LC)^{-1}LY(s)$$

Substituting this into the control equation gives the transfer function

$$U(s) = -K(sI - A + BK + LC)^{-1}LY(s)$$

$$U(s) = -H(s)Y(s) \rightarrow H(s) = K(sI - A + BK + LC)^{-1}L$$

Transfer function for observer-based compensator.

The observer-based compensator for the double-integrator system is

$$H(s) = \frac{24s + 16}{s^2 + 6s + 18}$$

The open-loop transfer function is given by

$$G(s)H(s) = \frac{24s + 16}{s^2(s^2 + 6s + 18)}$$

The root locus and Bode plots are shown in Figures 9.13 and 9.14.

Stable but no guaranteed stability margins.

From the Bode plots, we read the gain and phase margins. They are approximately 10 dB and 36°, respectively. Observe that the gain margin is fairly small (i.e., raising the gain by a factor of about 3.16 would be destabilizing). One limitation of observer-based design is that we have no direct control over the stability margins. Thus designs that are perfect on paper might not work in a real situations. This is because we usually have imperfect models of our systems, and stability margins provide some protection against model uncertainties. System designed with low margins are inherently sensitive to model errors and may become unstable in actual operation.

No control over compensation poles and zeros.

Observing the resulting compensator, we note that it has a pair of complex conjugate poles and one zero, so it has no classical counterpart. In fact, the compensator poles and zeros could end up anywhere in the complex plane, including the RHP, and we have no control over this issue.

In the preceding example, the observer initial conditions were chosen rather arbitrarily. How do we choose them in general? The best choice is to use the plant initial conditions. This we do not have, however. If we did, given the plant model, we could have numerically solved for the states. From the convergence point of

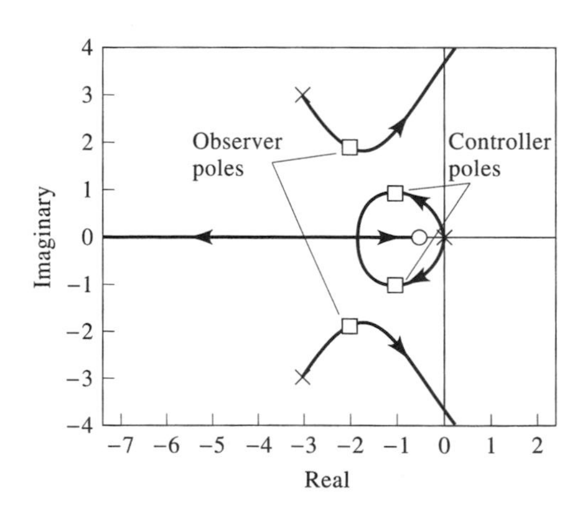

Figure 9.13 Root locus of the observer-based compensated double-integrator system.

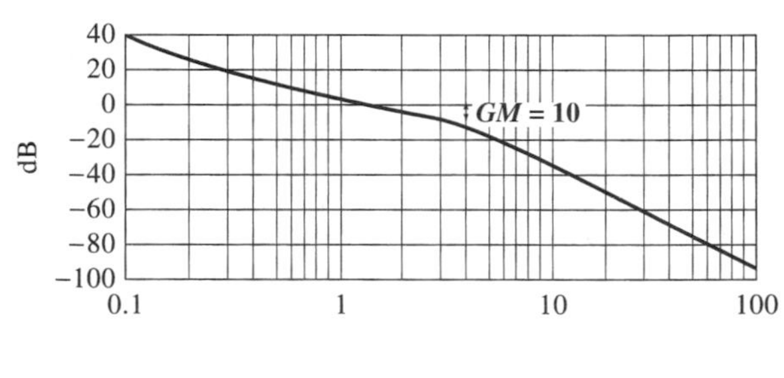

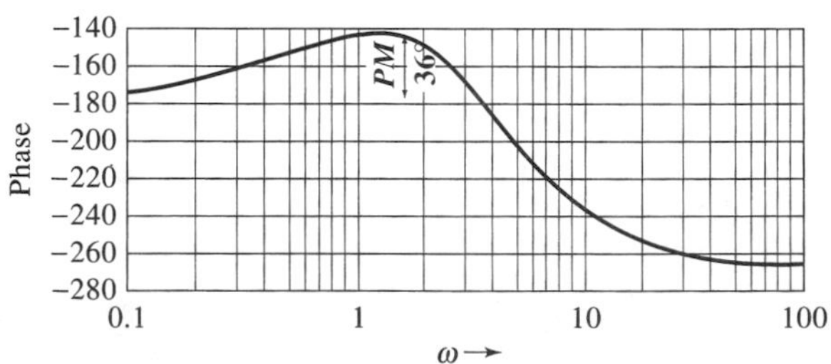

Figure 9.14 Bode plots of an observer-based compensated double-integrator system.

view, this is not an issue because the matrix $(A - LC)$ is a stable matrix, and the observer errors will go to zero independently of the initial conditions. It can be shown that because the initial output is known, an optimal choice (*in a mathematical sense, not necessarily optimal in a physical sense*) for observer initial conditions is

Choosing observer initial conditions.

$$\hat{x}(0) = C'(CC')^{-1}y(0)$$

What about the location of observer poles? There are several guidelines available. One guideline is to choose the observer poles that are faster than the controller poles. This choice ensures that the observer converges faster than the system, so the controller will be using more accurate estimates, thereby reducing the degradation caused by the observer. Again, one must caution against pushing the observer poles too far into the LHP, because this increases the system bandwidth and makes it more susceptible to noise. An alternate guidelines suggested by results from robust control is to choose the observer poles at the plant zeros (if the system has RHP zeros, use their LHP images).

Choosing observer poles.

❑ DRILL PROBLEMS

D9.6 For the system of Drill Problem D9.1, choose control gain k to place the closed-loop system poles at the indicated locations. In addition, design observers with the indicated poles. Use MATLAB to verify that the closed-loop system is stable using estimated states.

(a) Controller poles at $\{-6, -1 \pm j\}$

observer poles at $\left\{-5 \pm j5, -\sqrt{2}\right\}$.

Ans. $k = [9 \quad 8 \quad 1] \quad L = \begin{bmatrix} 2.336 \\ 2.3037 \\ 0.2579 \end{bmatrix}$

(b) Controller poles at $\{-3, -1 \pm j\}$

observer poles at $\{-3, -5 \pm j5\}$

Ans. $k = [6 \quad 6 \quad 3], \quad L = \begin{bmatrix} -2.7619 & 5.1429 \\ 1.1905 & 1.7143 \\ -9.6667 & 14.0000 \end{bmatrix}$

D9.7 For the systems of Drill Problem D9.2, design observers with indicated poles. Using the same control gains simulate the closed-loop systems to verify stability. Obtain compensator transfer functions, and root locus and Bode plots. Use these to determine stability margins.

(a) observer poles $= \{-1, -2\}$

Ans. $L = \begin{bmatrix} 12 \\ 7 \end{bmatrix}, \; H(s) = \dfrac{62s - 4}{s^2 + 7s - 38}$

(b) observer poles $= \{-1, -2\}$

$$\textbf{Ans.}\quad L = \begin{bmatrix} 3 \\ 3 \end{bmatrix},\ H(s) = \frac{15(s+1)}{s^2+5s+11}$$

9.5 Reduced-Order Observer Design

There is no need to estimate states that are directly measured.

The observer introduced in Section 9.4 has the same order as the system and is referred to as the *identity* or *full-order observer*. If the system has n states and m measurements are available, it is possible to build an observer that estimates the states that are not measured, hence reducing the order of the observer. It seems reasonable that an observer of order $(n-m)$ should be sufficient. This was first introduced by D. Luenberger and is referred to as *reduced-order* (or *Luenberger*) *observer.* The reduction in order leads to simpler and more economical compensators. If the number of measurements m is large, the benefits could be substantial.

Before we drive this observer, we make an assumption on the structure of the measurement matrix C. We will assume that C has the form

$$C = [\,I \quad 0\,]$$

where $I = m \times m$ and $0 = m \times (n-m)$

The consequence of this is that it allows us to divide the states into two categories: measured, unmeasured. That is,

$$y = [\,I \quad 0\,]\begin{bmatrix} x_m \\ x_u \end{bmatrix} = x_m$$

We can then partition the system accordingly as

$$\begin{bmatrix} \dot{x}_m \\ \dot{x}_u \end{bmatrix} = \begin{bmatrix} A_{11} & A_{12} \\ A_{21} & A_{22} \end{bmatrix}\begin{bmatrix} x_m \\ x_u \end{bmatrix} + \begin{bmatrix} B_1 \\ B_2 \end{bmatrix} u$$

$$y = [\,I \quad 0\,]\begin{bmatrix} x_m \\ x_u \end{bmatrix}$$

The unmeasured portion of the system is

$$\dot{x}_u = A_{22}x_u + (A_{21}x_m + B_2 u)$$

The terms within the parentheses are known quantities, so they are collected together. Because there are m measured states, the number of unmeasured states is $n-m$, so we will build an observer of order $n-m$ to estimate these states. The observer structure is given by the following procedure (this is the same procedure used for the full-order observer): copy the system equation, replace unknown quantities by their estimates, and add a correction term multiplied by the observer gain. The correction term is the difference between the plant output and the observer output.

$$\dot{\hat{x}}_u = A_{22}\hat{x}_u + (A_{21}x_m + B_2u) + L(\text{correction term})$$

The correction term in the full-order observer case was $(y - C\hat{x})$. In the present case, it is

$$y - [I \quad 0]\begin{bmatrix} x_m \\ \hat{x}_u \end{bmatrix} = y - x_m = 0$$

Therefore, using the output will not provide any correction. However, we note that if the outputs are available, we can assume that their derivatives are also available. Now, observe that the derivative of the plant output is equal to the measured portion of the system, that is,

$$\dot{y} = \dot{x}_m = A_{12}x_u + (A_{11}x_m + B_1u) \rightarrow \dot{y} - A_{11}x_m - B_1u = A_{12}x_u$$

where we have collected the known, or measured, quantities on the left-hand side. We can use the known quantities on the left as a substitute for plant output (we are basically using all the information available from the system), and the right-hand side as the observer output. Substituting this in the observer equation, we get

$$\dot{\hat{x}}_u = A_{22}\hat{x}_u + (A_{21}x_m + B_2u) + L(\dot{y} - A_{11}x_m - B_1u - A_{12}\hat{x}_u)$$

To verify that this scheme works, we need to show that the error dies out. Define the error as

$$\tilde{x}_u = x_u - \hat{x}_u$$

and derive a differential equation for the error.

Reduced-order observer error.

$$\dot{\tilde{x}}_u = (A_{22} - LA_{12})\tilde{x}_u$$

This is similar to the full-order observer error equation

Full-order observer error.

$$\dot{\tilde{z}} = (A - LC)\tilde{z}$$

Recall that if the pair (C, A) is observable (i.e., the system is observable), L can be chosen to place observer poles anywhere in the complex plane. Comparing the two error equations, by analogy we conclude that the same would be true for the reduced-order case under the condition that the pair of matrices (A_{12}, A_{22}) are observable. Luenberger showed that this condition is equivalent to the observability of the original system—that is, (C, A).

We are almost done. The last step is to eliminate the output derivative in the observer. Differentiation is to be avoided in system design because it is a noise-enhancing operation. Looking at the observer equation suggests that a simple change of variable will eliminate the derivative term. This change of variable is given by

$$z = \hat{x}_u - Ly \qquad \text{or} \qquad \hat{x}_u = z + Ly$$

A bit of algebra results in the final form of the reduced-order observer

Final form of the reduced-order observer.

$$\boxed{\begin{aligned} \dot{z} &= Dz + Fy + Gu \\ \hat{x}_u &= z - Ly \end{aligned}}$$

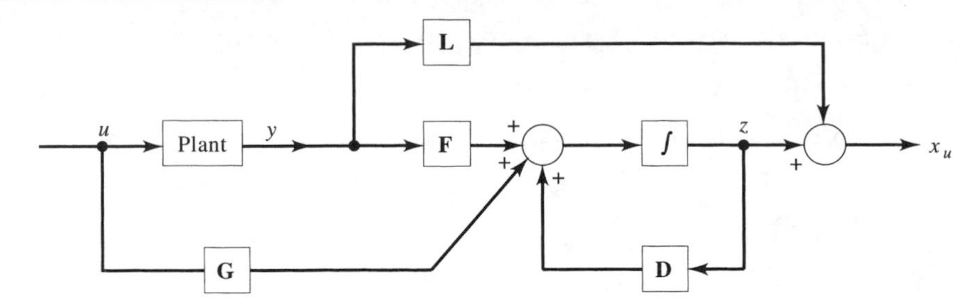

Figure 9.15 Block diagram of the reduced-order observer.

where

$$\boxed{\begin{aligned} D &= A_{22} - LA_{12} \\ F &= DL + A_{21} - LA_{11} \\ G &= B_2 - LB_1 \end{aligned}}$$

The block diagram of the observer is shown in Figure 9.15.

Designing a reduced-order observer for the double-integrator system.

Let us design a reduced-order observer for the double-integrator system. Because there are two states and one measurement, we require a first-order observer. The plant equations are repeated:

$$\dot{x} = \begin{bmatrix} 0 & 1 \\ 0 & 0 \end{bmatrix} x + \begin{bmatrix} 0 \\ 1 \end{bmatrix} u$$

$$y = [1 \quad 0]x$$

Although, we can simply plug into the formulas, we will repeat the observer derivation for this simple example from scratch. The unmeasured portion of the system is given by

$$\dot{x}_2 = u \rightarrow A_{21} = A_{22} = 0, \ B_2 = 1$$

The measured portion of the system, to be used for the correction term, is given by

$$\dot{x}_1 = \dot{y} = x_2 \rightarrow (\dot{y} - \hat{x}_2) \quad \text{(correction term)}$$

Now, we copy the unmeasured portion, and add the correction term to get the observer.

$$\dot{\hat{x}}_2 = u + L(\dot{y} - \hat{x}_2)$$

The observer error equation becomes

$$\dot{\tilde{x}}_2 = \dot{x}_2 - \dot{\hat{x}}_2 = u - u - L(\dot{y} - \hat{x}_2) = -L(x_2 - \hat{x}_2) = -L\tilde{x}_2$$

Choosing the observer pole at -2 yields $L = 2$. To eliminate the derivative term, let

$$z = \hat{x}_2 - Ly = \hat{x}_2 - 2y$$

First-order observer to estimate x_2.

The final form of the observer becomes

$$\begin{aligned} \dot{z} &= -2z - 4y + u \\ \hat{x}_2 &= z + 2y \end{aligned}$$

Figure 9.16 is a diagram of this reduced-order observer.

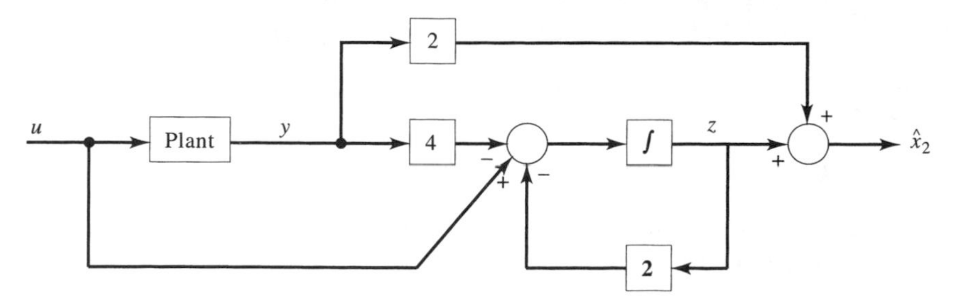

Figure 9.16 Block diagram of the reduced-order observer for the double-integrator example.

Note that in this example, the system was second-order, and we had one measurement. Solving for L involved solving a first-order algebraic equation, so we obtained a unique solution. In general, when there are n states and m measurements, there are $(m - 1)$ degrees of freedom in solving for L, and the solution is not unique.

❑ DRILL PROBLEM

D9.8 For the systems of Drill Problem D9.5, design a reduced-order, first-order observer to estimate only x_2. Select the observer eigenvalue as indicated. Write the observer equation and use computer software to verify its operation.

(a) The observer eigenvalue should be -50.

Ans. $D = -50,\ F = 553,\ G = 23,\ L = -11.5$

$$\begin{cases} \dot{z} = -50z + 23u + 553y \\ \hat{x}_2 = z - 11.5y \end{cases}$$

(b) The observer eigenvalue should be -10.

Ans. $D = -10,\ F = 13,\ G = 2,\ L = -2$

$$\begin{cases} \dot{z} = -10z + 2u + 13y \\ \hat{x}_2 = z - 2y \end{cases}$$

9.5.1 Separation Property

The separation property also holds in the reduced-order case. It can be derived by combining the closed-loop system and error system equations

$$\begin{bmatrix} \dot{x} \\ \dot{\tilde{x}}_u \end{bmatrix} = \begin{bmatrix} A - BK & BK_2 \\ 0 & A_{22} - LA_{12} \end{bmatrix} \begin{bmatrix} x \\ \tilde{x}_u \end{bmatrix}$$

Therefore, the closed-loop eigenvalues are the union of the controller and observer eigenvalues. In the above, the control gain vector has been partitioned as

$$u = -[K_1 \quad K_2] \begin{bmatrix} x_m \\ \hat{x}_u \end{bmatrix} = -K_1 x_m - K_2 \hat{x}_u$$

9.5.2 Reduced-Order Observer Transfer Function

The compensator transfer function is derived by substituting for u in the observer equation

$$H(s) = K_2\,[sI - (D - GK_2)]^{-1}\,(F - GK_1 - GK_2L) + (K_1 + K_2L)$$

The same techniques used for the full-order case can be used to handle tracking problems and integral control.

A first-order compensator will be obtained for the double-integrator problem. The control gain, observer gain, and observer parameters, obtained earlier, are as follows:

$$k = [k_1 \quad k_2] = [2 \quad 2]\,, \; L = 2, \; D = -2, \; F = -4, \; G = 1$$

First-order compensator for the double-integrator system.

By using the preceding parameters, we get the compensator transfer function

$$H(s) = \frac{6s + 4}{s + 4}$$

The compensator is recognized as a classical lead compensator. Root locus and Bode plots of the compensated system are shown in Figures 9.17 and 9.18, respectively.

The compensated system has 45° phase margin and infinite gain margin. Comparing this with the full-order observer-based compensator, we note that the reduced-order case has resulted in a simpler (first-order) compensator with better stability margins. The zero-input response of the compensated system is shown in Figure 9.19. The zero-input responses of all three designs (i.e., state feedback, observer-based, and reduced-order observer-based compensators) are shown in Figure 9.20 for comparison.

At the beginning of Section 9.5 we assumed that the C matrix is of the form

$$C = [I \quad 0]$$

The case when C is not of the form $[I \quad 0]$

This is not a restrictive assumption, because through a linear transformation, we can always convert C to this form. The transformation is given as follows. Choose any

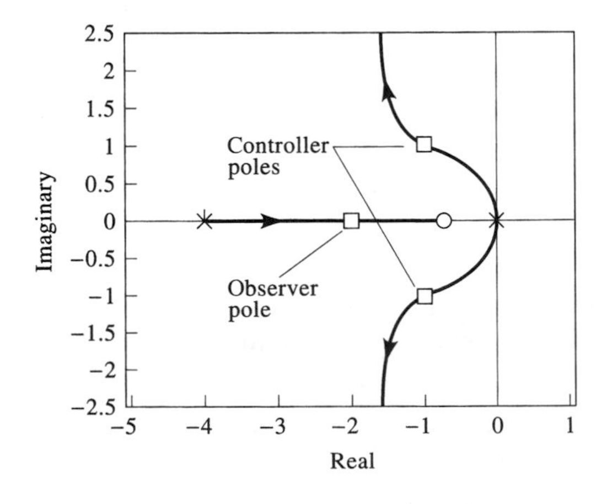

Figure 9.17 Root locus of the reduced-order, observer-based, compensated double-integrator system.

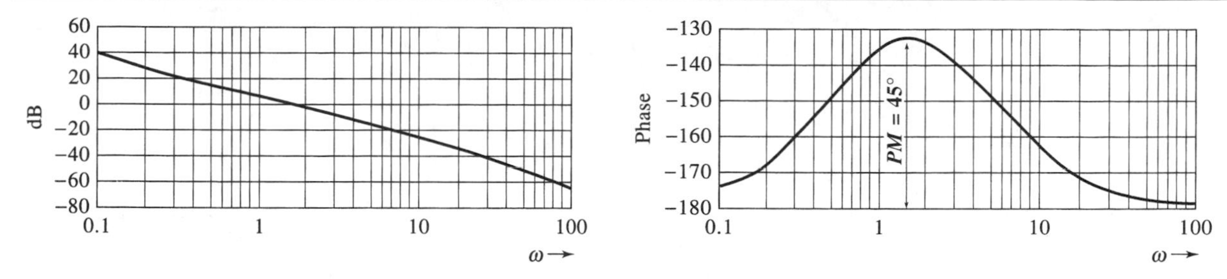

Figure 9.18 Bode plots of the reduced-order, observer-based, compensated double-integrator system.

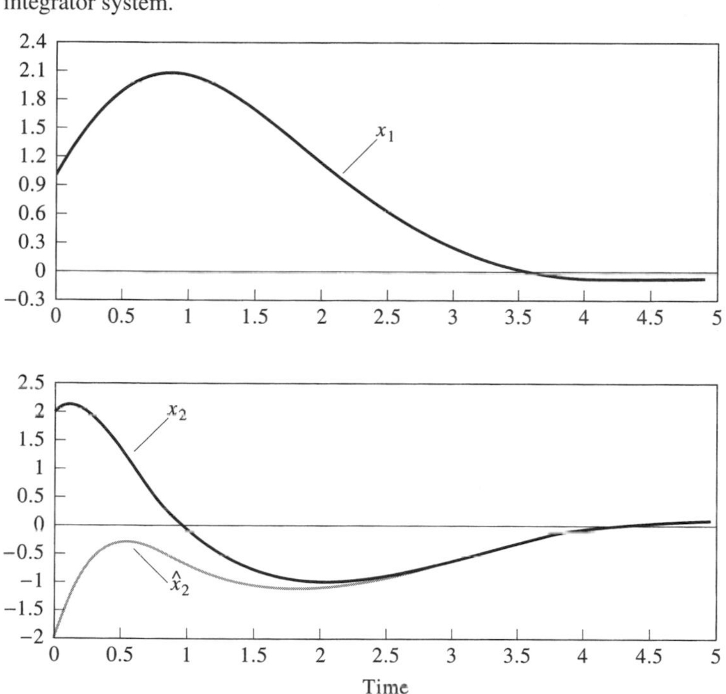

Figure 9.19 Using a reduced-order observer to obtain zero-input response of the double-integrator system.

arbitrary matrix, T, such that when it is stacked on top of C, the result is a nonsingular matrix—that is,

$$P = \begin{bmatrix} C \\ T \end{bmatrix} \text{ is nonsingular}$$

The inverse of this matrix, called Q, is our sought-after transformation. This is because P and Q are inverses of each other, and therefore satisfy

$$PQ = I \rightarrow \begin{bmatrix} C \\ T \end{bmatrix} [Q_1 \quad Q_2] = \begin{bmatrix} CQ_1 & CQ_2 \\ TQ_1 & TQ_2 \end{bmatrix} = \begin{bmatrix} I & 0 \\ 0 & I \end{bmatrix}$$

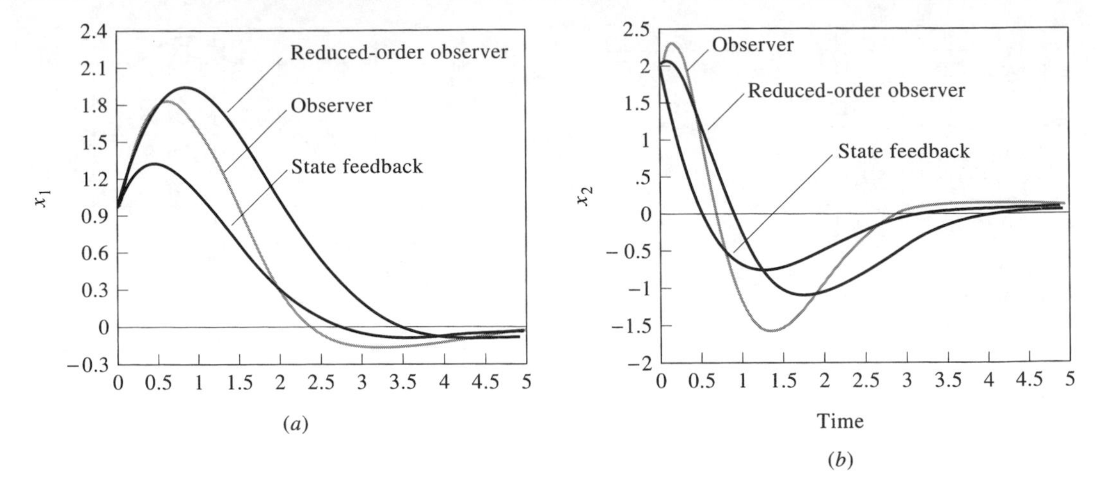

Figure 9.20 Comparison of the zero-input responses of the double-integrator system under state feedback, full-order observer, and reduced-order observer.

where the identity and zero matrices have compatible dimensions. Recall from Chapter 8 that under a linear transformation, the new C matrix becomes

$$\bar{C} = CQ = C\,[Q_1 \quad Q_2] = [CQ_1 \quad CQ_2] = [I \quad 0]$$

As an example, consider the following equivalent model for the double-integrator system (obtained by labeling the states from left to right):

$$\dot{x} = \begin{bmatrix} 1 & 0 \\ 0 & 0 \end{bmatrix} x + \begin{bmatrix} 0 \\ 1 \end{bmatrix} u$$

$$y = [0 \quad 1]\,x$$

Choosing T as shown below results in

$$T = [1 \quad 0] \rightarrow P = \begin{bmatrix} 0 & 1 \\ 1 & 0 \end{bmatrix} \rightarrow Q = P^{-1} = \begin{bmatrix} 0 & 1 \\ 1 & 0 \end{bmatrix} \rightarrow \bar{C} = CQ = [1 \quad 0]$$

$\bar{A}$ and $\bar{B}$ will turn out to be the same matrices as in the original example (this is just a coincidence in this simple example, usually the new matrices will be quite different, but the C matrix will have the required form).

❑ DRILL PROBLEM

D9.9 Consider the third-order system given by

$$G(s) = \frac{3}{s(s^2 + 4s + 5)}$$

or, in state-space form,

$$\dot{x} = \begin{bmatrix} 0 & 1 & 0 \\ 0 & 0 & 1 \\ 0 & -5 & -4 \end{bmatrix} x + \begin{bmatrix} 0 \\ 0 \\ 3 \end{bmatrix} u$$

Suppose the states x_1 and x_2 are measured but the state x_3 is to be reconstructed using a first-order observer.

(a) Find the observer gain and parameters. The observer pole is to be at -20. Also write down the observer equation.

(b) Find the controller gain to place poles at -10 and $-2 \pm j2$.

(c) Find the transfer function of the compensator (note that the compensator is a two-input, one-output system).

Ans. (a) $D = -20,\ G = 3,\ L = [0 \quad 16],\ F = [0 \quad -325]$

$$\dot{z} = -20z + [0 \quad -325]\begin{bmatrix} x_1 \\ x_2 \end{bmatrix} + 3u = -20z - 325x_2 + 3u$$

$$\hat{x}_3 = z + [0 \quad 16]\begin{bmatrix} x_1 \\ x_2 \end{bmatrix} = z + 16x_2$$

(b) $k = [80 \quad 43 \quad 10]\frac{1}{3}$

(c) $H(s) = \dfrac{1}{s+30}[-26.66s - 533.33,\ 30s + 2660]$

9.6 A Magnetic Levitation System

Beginning in 1969, West Germany sought to develop a high-speed electric train system to span central Europe. State space analysis and aircraft technology were used to design, build, and test such a train for operation at speeds as high as 400 km/h (248 mi/h).The train is suspended in midair by magnetic fields. This type of suspension is called magnetic levitation or MAGLEV.

Figure 9.21, shows the cross section of a MAGLEV vehicle. The track is a T-shaped concrete guideway. Once under way, the train does not touch the guideway, resulting in greatly reduced friction and reduced guideway construction costs. Electromagnets are distributed along the guideway and along the length of the train in matched pairs. The magnetic attraction of the vertically paired magnets balances the force of gravity and levitates the vehicle above the guideway. The horizontally paired magnets stabilize the vehicle against sideways forces. Forward propulsion is produced by linear induction motor action between train and guideway. Only the vertical motion and control of the suspended vehicle will be considered here.

The equations characterizing the train's vertical motion are now developed. It is desired to control the gap distance d within a close tolerance in normal operation of the train. The gap distance d between the track and the train magnets is

$$d = z - h$$

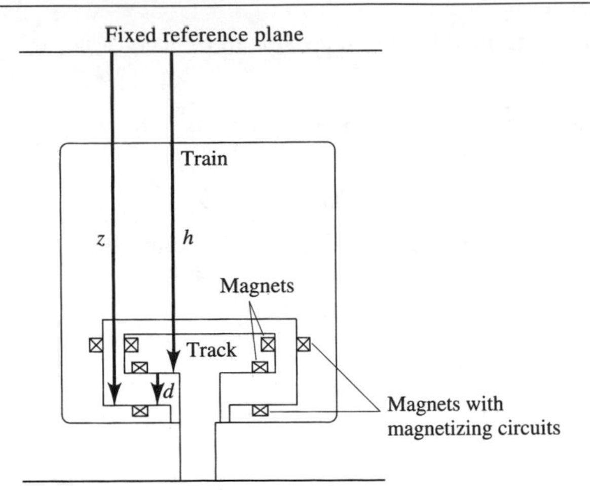

Figure 9.21 Cross section of a MAGLEV train.

Then

$$\dot{d} = \dot{z} - \dot{h}$$
$$\ddot{d} = \ddot{z} - \ddot{h}$$

where the dots denote time derivatives. The magnet produces a force that is dependent upon residual magnetism and upon the current passing through the magnetizing circuit. For small changes in the magnetizing current i and the gap distance d, that force is approximately

$$f_1 = -Gi + Hd$$

where G and H are positive constants. That force acts to accelerate the mass M of the train in a vertical direction, so

$$f_1 = M\ddot{z} = -Gi + Hd$$

For increased current, the distance z diminishes, reducing d as the vehicle is attracted to the guideway.

A network model for the magnetizing circuit is given in Figure 9.22. This circuit represents a generator driving a coil wrapped around the magnet on the vehicle. The voltage induced in the coil by the vehicle motion is represented by the term $(LH/G)\dot{d}$, for which it is assumed that the magnetic flux loss is negligible. For that circuit

$$Ri + L\dot{i} - \frac{LH}{G}\dot{d} = v$$

The three state variables

$$x_1 = d$$
$$x_2 = \dot{d}$$
$$x_3 = i$$

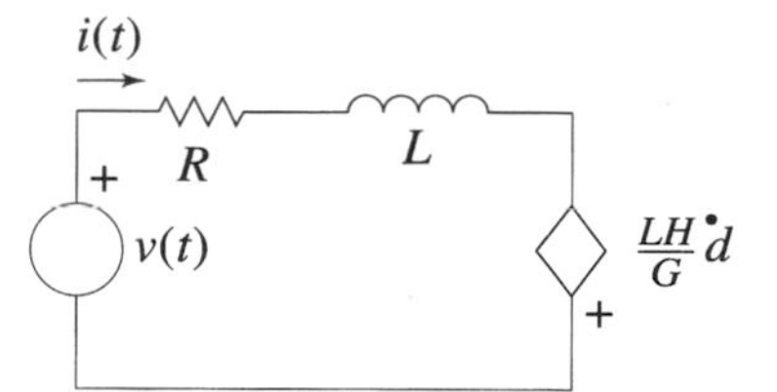

Figure 9.22 Magnetizing circuit model.

are convenient, and in terms of them the vertical motion state equations are

$$\begin{bmatrix} \dot{x}_1 \\ \dot{x}_2 \\ \dot{x}_3 \end{bmatrix} = \begin{bmatrix} 0 & 1 & 0 \\ \frac{H}{M} & 0 & -\frac{G}{M} \\ 0 & \frac{H}{G} & -\frac{R}{L} \end{bmatrix} \begin{bmatrix} x_1 \\ x_2 \\ x_3 \end{bmatrix} + \begin{bmatrix} 0 & 0 \\ 0 & -1 \\ \frac{1}{L} & 0 \end{bmatrix} \begin{bmatrix} v \\ f \end{bmatrix}$$

where

$$f = \ddot{h}$$

If the gap distance d is considered to be the system output, then the state variable output equation is

$$d = x_1$$

The voltage v is considered to be a control input, while guideway irregularities $f = \ddot{h}$ constitute a disturbance. Figure 9.23 shows block diagram and signal flow graph representations of the state equations.

The characteristic equation for the system, the roots of which are the transfer function poles, is given by

$$|sI - A| = \begin{vmatrix} s & -1 & 0 \\ -\frac{H}{M} & s & \frac{G}{M} \\ 0 & -\frac{H}{G} & s + \frac{R}{L} \end{vmatrix} = 0$$

$$= s \begin{vmatrix} s & \frac{G}{M} \\ -\frac{H}{G} & s + \frac{R}{L} \end{vmatrix} + \begin{vmatrix} -\frac{H}{M} & \frac{G}{M} \\ 0 & s + \frac{R}{L} \end{vmatrix}$$

$$= s\left(s^2 + \frac{R}{L}s + \frac{H}{M}\right) - \frac{H}{M}\left(s + \frac{R}{L}\right)$$

$$= s^3 + \frac{R}{L}s^2 - \frac{HR}{ML} = 0$$

The system is thus unstable, since its characteristic polynomial has coefficients with differing algebraic signs. Also, the coefficient of s in the characteristic equation is zero. The system instability is quite understandable when one considers the action

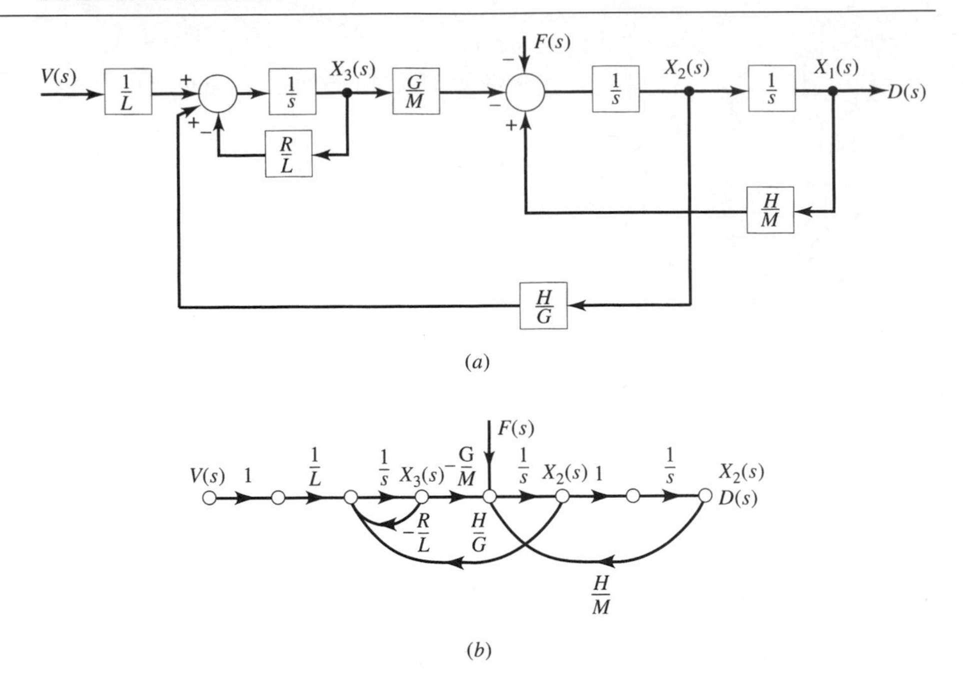

Figure 9.23 Diagrams of the state equations. (a) Block diagram. (b) Signal flow graph.

of the magnets. If the gap distance d should increase slightly, the magnetic attraction will decrease, tending to further increase the gap, and so on.

To control the system, the magnetizing circuit voltage is chosen to be a linear combination of the state signals plus a tracking input $u_1(t)$:

$$v = k_1x_1 + k_2x_2 + k_3x_3 + u_1(t)$$

The feedback signals are produced from sensors that monitor the state variables, namely, gap distance d, gap velocity $\dot{d}$, and magnetizing current i. The resulting feedback system is described by

$$\begin{bmatrix} \dot{x}_1 \\ \dot{x}_2 \\ \dot{x}_3 \end{bmatrix} = \begin{bmatrix} 0 & 1 & 0 \\ \dfrac{H}{M} & 0 & -\dfrac{G}{M} \\ \dfrac{k_1}{L} & \dfrac{H}{G} + \dfrac{k_2}{L} & -\dfrac{R}{L} + \dfrac{k_3}{L} \end{bmatrix} \begin{bmatrix} x_1 \\ x_2 \\ x_3 \end{bmatrix} + \begin{bmatrix} 0 & 0 \\ 0 & -1 \\ \dfrac{1}{L} & 0 \end{bmatrix} \begin{bmatrix} u_1 \\ f \end{bmatrix}$$

$$d = x_1$$

Appropriate choice of the feedback gain constants k_1, k_2, and k_3, that is, the feedback gain matrix,

$$K = [\,k_1 \quad k_2 \quad k_3\,]$$

will place the system poles at any desired locations.

To proceed with state-variable design methods, the parameters M, G, L, and R must be estimated. The following values do not necessarily represent those of any specific existing system, but the methods and values are representative of the design process in general.

Suppose an engineer finds that each train car weighs 8000 kg. Each car is supported by four magnets, each of which must therefore support 2000 kg. Each subsystem can be analyzed using $M = 2000$ kg.

A static test is performed without control. The air gap is clamped shut, causing d to be zero. A -120 V source is applied to the magnetizing circuit. With a time constant of $\frac{1}{30}$ s, -8 A eventually flows at steady state. A resultant force of 4000 N is measured (in addition to that of gravity).

The static test is concluded and the voltage is carefully varied until, at equilibrium, the car levitates with $d = 10$ mm under the influence of 8 A of current.

If the magnetizing circuit is at steady state, the static test can be used to get R and L, since

$$R = \frac{v}{i} = \frac{-120}{-8} = 15\ \Omega$$

and, from the time constant during the static test

$$T = \frac{L}{R} \qquad \text{so} \qquad L = RT = \frac{15}{30} = 0.5H$$

The data from when the air gap was clamped shut ($d = 0$) permit G to be computed.

$$f = -Gi + H \times 0$$

$$G = \frac{-f}{i} = \frac{-4000}{-8} = 500\ \text{N/A}$$

The data from when the car was levitated to equilibrium provides H:

$$0 = -500 \times 8 + H \times 10$$

$$H = 400\ \text{N/mm}$$

The parameter values are, therefore,

$$M = 2000 \qquad H = 400$$

$$G = 500 \qquad L = 0.5$$

$$R = 15$$

For these, the feedback system equations are

$$\begin{bmatrix} \dot{x}_1 \\ \dot{x}_2 \\ \dot{x}_3 \end{bmatrix} = \begin{bmatrix} 0 & 1 & 0 \\ 0.2 & 0 & -0.25 \\ 2k_1 & 0.8 + 2k_2 & -30 + 2k_3 \end{bmatrix} \begin{bmatrix} x_1 \\ x_2 \\ x_3 \end{bmatrix} + \begin{bmatrix} 0 & 0 \\ 0 & -1 \\ 2 & 0 \end{bmatrix} \begin{bmatrix} u_1 \\ f \end{bmatrix}$$

$$d = x_1$$

The characteristic equation for the feedback system is given by

$$\begin{vmatrix} s & -1 & 0 \\ -0.2 & s & 0.25 \\ -2k_1 & -0.8 - 2k_2 & s + 30 - 2k_3 \end{vmatrix}$$

$$= s \begin{vmatrix} s & 0.25 \\ -0.8 - 2k_2 & s + 30 - 2k_3 \end{vmatrix} + \begin{vmatrix} -0.2 & 0.25 \\ -2k_1 & s + 30 - 2k_3 \end{vmatrix}$$

$$= s^3 + (30 - 2k_3)s^2 + (0.5k_2)s + 0.4k_3 + 0.5k_1 - 6$$

The feedback gains k_1, k_2, and k_3 may be chosen to give any desired coefficients of the characteristic equation of the feedback system. For example, if it is desired to have the system poles at $s = -1 + j2, \ -1 - j2,$ and -3, the characteristic polynomial would be

$$(s + 1 - j2)(s + 1 + j2)(s + 3) = s^3 + 5s^2 + 11s + 15$$
$$= s^3 + c_2 s^2 + c_1 s + c_0$$

which is achieved with

$$k_3 = 0.5(30 - c_2) = 12.5$$
$$k_2 = 2c_1 = 22$$
$$k_1 = 2(c_0 + 0.2c_2) = 32$$

For this choice of feedback gains, the feedback system model is

$$\begin{bmatrix} \dot{x}_1 \\ \dot{x}_2 \\ \dot{x}_3 \end{bmatrix} = \begin{bmatrix} 0 & 1 & 0 \\ 0.2 & 0 & -0.25 \\ 64 & 44.8 & -5 \end{bmatrix} \begin{bmatrix} x_1 \\ x_2 \\ x_3 \end{bmatrix} + \begin{bmatrix} 0 & 0 \\ 0 & -1 \\ 2 & 0 \end{bmatrix} \begin{bmatrix} u_1 \\ f \end{bmatrix}$$

$$d = [1 \quad 0 \quad 0] \begin{bmatrix} x_1 \\ x_2 \\ x_3 \end{bmatrix}$$

The steady state output d due to a unit step disturbance input f is given by

$$0 = [A + BK]x + Bu$$
$$d = x_1 \qquad u_1 = 0 \qquad f = 1$$

So that

$$[A + BK]x = -Bu = - \begin{bmatrix} 0 \\ -1 \\ 0 \end{bmatrix} = \begin{bmatrix} 0 \\ 1 \\ 0 \end{bmatrix}$$

Cramer's rule can be used to obtain $d = x_1$. It is instructive to write the gain values in terms of the desired characteristic polynomial coefficients.

$$d = x_1 = \frac{\begin{vmatrix} 0 & 1 & 0 \\ 1 & 0 & -0.25 \\ 0 & 4c_1 + 0.8 & -c_2 \end{vmatrix}}{\begin{vmatrix} 0 & 1 & 0 \\ 0.2 & 0 & -0.25 \\ 4(c_0 + 0.2c_2) & 4c_1 + 0.8 & -c_2 \end{vmatrix}} = \frac{c_2}{-c_0} = -\frac{1}{3}$$

which depends only on the desired performance. Further study would be needed to determine whether this amount of disturbance rejection from track irregularities is sufficient. The negative algebraic sign simply means that a positive step in $f = \ddot{h}$ results in a steady state decrease in the gap distance. Types of disturbance other than constant ones should also be considered in the design.

The reference input u_1 would normally be a constant that sets the nominal gap distance. The steady state gap distance d due to a constant reference input u_1 where f is zero (level track) is given by

$$0 = [A + BK]x + Bu \qquad u = \begin{bmatrix} u_1 \\ 0 \end{bmatrix}$$

$$[A + BK]x = -Bu = \begin{bmatrix} 0 \\ 0 \\ -2u_1 \end{bmatrix}$$

$$d = x_1$$

Again, Cramer's rule can be used to obtain $d = x_1$ where it is instructive to write the gain values in terms of the desired characteristic polynomial coefficients

$$d = x_1 = \frac{\begin{vmatrix} 0 & 1 & 0 \\ 0 & 0 & -0.25 \\ -2u_1 & 4c_1 + 0.8 & -c_2 \end{vmatrix}}{-c_0}$$

$$= -\frac{0.5u_1}{c_0}$$

For a nominal gap of 10 mm, the reference input should be

$$u_1 = -(20)c_0 = -300$$

which depends only on the nominal gap and on a coefficient of the desired characteristic polynomial.

Figure 9.24 shows calculated system response where the train accelerates from a standstill and traverses an irregular guideway with a downgrade followed by an upgrade. In the German system, the nominal air gap distance is 14 mm (about $\frac{1}{2}$ in.). Improvement in disturbance rejection is obtained by modeling the track irregularities by differential equations that are included as part of an observer. Three levels of complexity are used depending on whether the track is level (actually somewhat curved between towers), following a hill, or following a curve.

Space does not permit a complete discussion of the system; however, one feature is of interest. The rate of change of the air gap is also estimated by using an observer. The state vector may be reordered as follows:

$$x = \begin{bmatrix} d \\ i \\ \cdots \\ \dot{d} \end{bmatrix} = \begin{bmatrix} y \\ x_u \end{bmatrix}$$

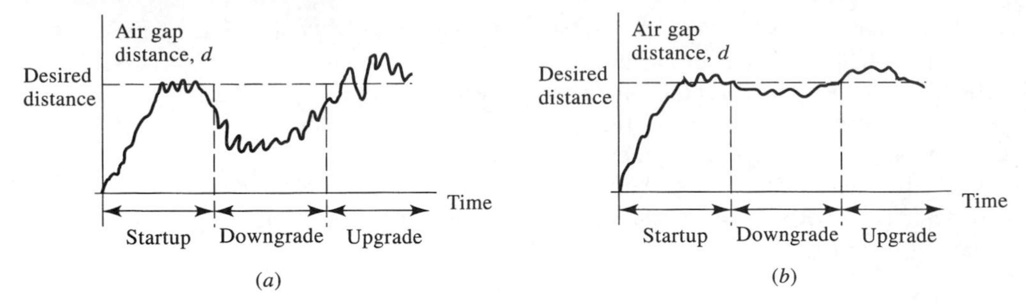

Figure 9.24 MAGLEV system response. (a) Response of the system with state feedback. (b) Improved response with disturbance modeling and feedback.

because d and i can be measured and the rate of change of d must be estimated by a reduced-order, first-order observer. The concepts of control and estimation can be separated (as mentioned earlier in this chapter); therefore, the open-loop system matrix is used to compute the observer dynamics. As a result of reordering, we have

$$A = \left[\begin{array}{cc|c} 0 & 0 & 1 \\ 0 & -30 & 0.8 \\ \hline 0.2 & -0.25 & 0 \end{array}\right] \qquad B = \left[\begin{array}{c} 0 \\ 2 \\ \hline 0 \end{array}\right]$$

The reordering is the result of augmenting the C matrix by T and transforming the state equations.

$$C = \begin{bmatrix} 1 & 0 & 0 \\ 0 & 0 & 1 \end{bmatrix} \qquad \text{choose } T \text{ as } T = [0 \quad 1 \quad 0] \text{ then}$$

$$P = \begin{bmatrix} C \\ T \end{bmatrix} = \begin{bmatrix} 1 & 0 & 0 \\ 0 & 0 & 1 \\ 0 & 1 & 0 \end{bmatrix} \quad \text{and} \quad Q = P^{-1} \quad \text{and} \quad \bar{C} = CQ = \begin{bmatrix} 1 & 0 & 0 \\ 0 & 1 & 0 \end{bmatrix}$$

Suppose the observer pole is selected at -10. Following the given design procedures, we get

$$A_{22} - LA_{12} = 0 - \begin{bmatrix} l_1 & l_2 \end{bmatrix} \begin{bmatrix} 1 \\ 0.8 \end{bmatrix} = -l_1 - 0.8 l_2 = -10$$

This results in one equation and two unknowns. We can set l_1 to zero so that $l_2 = 12.5$. Computing the remaining observer parameters, we get

$$F = \begin{bmatrix} 0.2 & -249.25 \end{bmatrix} \qquad G = -25 \qquad D = -10$$

Therefore, the observer equation is given by

$$\dot{z} = -10z + 0.2x_1 + 249.75x_2 - 25u$$

$$\hat{x}_3 = z + 12.5x_2$$

Figure 9.24(b) shows the improved performance when an observer estimates track motion. The closed-loop system also uses the observer estimate of d for feedback control.

❑ Computer-Aided Learning

To find the state feedback gain we use the "place" command with the following syntax:

```
K=place(A,B,DP)
```

where DP stands for the vector of desired pole locations. The only restriction is that desired poles must be distinct (not repeated). The "place" command also works in the multiinput case.

For example, let us place the poles of the following system in $-1 \pm j$.

$$\dot{x} = \begin{pmatrix} 1 & 2 \\ 3 & 4 \end{pmatrix} x + \begin{pmatrix} -1 \\ 6 \end{pmatrix} u$$

$$y = (0 \quad -14)x + 8u$$

We enter the following commands:

```
a=[1,2;3,4];b=[ 1;6];c=[0, 14];d=8;
g=ss(a,b,c,d);
dp=[-1+j, -1-j];
k=place(a,b,dp)
place: ndigits=15
k=
  1.0     1.3333
```

We can verify our work by finding the closed-loop poles by

```
eig(a-b*k)
ans=
  -1.0000+1.0000i
  -1.0000-1.0000i
```

MATLAB also has an implementation of the Ackerman formula under the "acker" command. This command only works for single-input systems but does not require that the poles be distinct. Here is an example:

```
k=acker(a,b,dp)
k=
  1.0000  1.3333
```

We can place both poles at $-1, -1$:

```
dp2=[-1,-1];
k2=acker(a,b,dp2)
k2=
  0.9310  1.3218
```

C9.1 Redo Drill Problems D9.1–9.3 using the "place" and "acker" commands.

Observer Design

Because of the duality that exists between control and observer problems, the same commands can be used for observer design with modified inputs as shown next.

For full-order observer, use

```
L=place(A',C',ODP)'
```

For reduced-order observers, use

```
Lr=place(A22',A12',ODP)'
```

Where ODP are the observer desired poles. Note that in the single-output cases we can use the "acker" command if the observer poles are repeated.

We will design a full-order observer with poles at $-5 \pm j5$ for the system just described.

```
odp=5*[-1+j, -1-j]; l=place(a',c',odp)'
Place: ndigits=15
l=
  -1.5952
  -1.0714
```

C9.2 (a) Use MATLAB to solve Drill Problem 9.5.

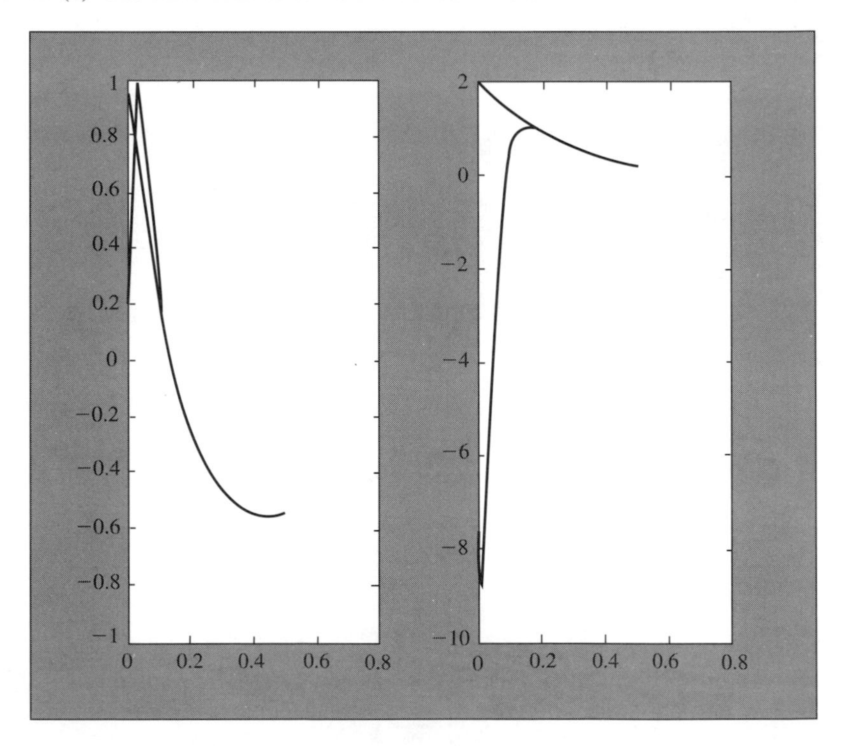

Figure C9.2

Ans. (a)
```
a=[-2,-4;1-4]; b=[2;0]; c=[1 0];
odp=[-50 -50]; l=acker(a',c',odp)'
l=
    94
  -528
ac=[a zeros (2, 2); l*c a-l*c] ;
bc=zeros (4,1); cc=eye (4); d=zeros (4,1);
[y,x,t]=initial (ac, bc, cc, zeros (4,1), [1 2 -1 -2], .5);
subplot (1, 2, 1), plot (t, y(: , 1), t, y (:,3)),
subplot (1, 2, 2), plot (t, y(: , 2), t, y(: , 4))
```

(b) Use MATLAB to solve Drill Problem 9.6, verify closed-loop stability.

(c) Use MATLAB to solve Drill Problem D9.7.

(d) Use MATLAB to solve Drill Problem D9.9.

9.7 SUMMARY

The techniques discussed in this chapter are very powerful and have expanded the range of problems that can be solved. If all the states are available, the advantages of state feedback become apparent. All closed-loop poles can be placed at desired locations in the complex plane as long as the system is controllable. If the system is not controllable, we can still stabilize the system as long as the system is stabilizable (i.e., the unstable modes are controllable). If some of the states are not available for measurement, for technological or economic reasons, observers can be implemented to estimate the states. Full-order or identity observer has a simple structure; its structure is the copy of the system plus a correction term, multiplied by observer gains. The observer gains can be computed to place the observer poles anywhere in the complex plane as long as the system is observable. Reduced-order observers reconstruct only the states that are not measured. For linear systems, there exists a seperation between the control and the estimation problems. This means that the poles of the closed-loop system (the interconnection of the controlled plant and the observer) are the union of the poles of the controlled plant and the observer.

Integral control, which allows us to eliminate steady state errors to constant inputs can be designed by using state space methods. This is done by augmenting the plant model by an integrator. Although not discussed here, it is also possible to track general command inputs using the methods discussed in this chapter. A major limitation of observer-based controllers is the lack of guaranteed stability margins. The methods rely heavily on the plant models. Because we rarely have accurate models of our plants, adequate stability margins are required to protect against these model uncertainties. In some situations we may design a controller that works perfectly under computer simulations but turns out to be unstable in practice. Therefore, any design must be tested thoroughly to prevent disastrous results.

The design of a magnetically levitated train exemplifies state space representation and controller and observer design.

REFERENCES

State Feedback

Anderson, B. D. O., and Moore, J. B., *Optimal Control*. Englewood Cliffs, NJ: Prentice-Hall, 1990.

Davison, E. J., "On Pole Assignment in Multivariable Linear Systems." *IEEE Trans. Autom. Control* AC-13 (December 1968): 747–748.

Wonham, W. M., "On Pole Assignment in Multi-Input Controllable Linear Systems." *IEEE Trans. Autom. Control* AC-12 (December 1967): 660–665.

Observers

Doyle. J. C., and Stein, G., "Robustness with Observers." *IEEE Trans. Autom. Control* (August 1979): 607–611.

Krogh, B., and Cruz. J. B., "Design of Sensitivity-Reducing Compensators Using Observers." *IEEE Trans. Autom. Control* (December 1978): 1058–1062.

Luenberger, D. G., "Observers for Multivariable Systems." *IEEE Trans. Autom. Control* AC-11 (April 1966): 190–197.

———, "An Introduction to Observers." *IEEE Trans. Autom. Control* AC-16 (December 1971): 596–602.

Nuyan, S., and Carroll, R. L., "Minimum Order Arbitrarily Fast Adaptive Observers and Identifiers." *IEEE Trans. Autom. Control* (April 1979): 289–297.

Sage, A. P., and White, C. C.,*Optimum Systems Control*. Englewood Cliffs. NJ: Prentice-Hall, 1977.

Stefani, R. T., "Reducing the Sensitivity to Parameter Variations of a Minimum-Order Reduced-Order Observer" *Int. J. Control* (1982): 983–995.

Stefani, R. T., "Observer Steady State Errors Induced by Errors in Realization." *IEEE Trans. Autom. Control* (April 1976): 280–282.

Magnetic Levitation of Trains

Brock, K. H., Gottzein, E., Pfefferl, J., and Schneider, E., "Control Aspects of a Tracked Magnetic Levitation High Speed Test Vehicle." *Automatica*. vol. 13. no. 3, 1977, pp. 205–233.

Glatzel, K., Khurdok, G., and Rogg, D., "The Development of the Magnetically Suspended Transportation System in the Federal Republic of Germany." *IEEE Trans. Vehic. Technol.* (February 1980): 3–17.

Glatzel, K., and Schulz, H., "Transportation: The Promise of MAGLEV." *IEEE Spectrum* (March 1980): 63–66.

Gottzein, E., Meisinger, R., and Miller, L., " The Magnetic Wheel in the Suspension of High Speed Ground Transportation Vehicles." *IEEE Trans. Vehic. Technol.* (February 1980): 17–22.

Kaplan, G., "Rail Transportation." *IEEE Spectrum* (January 1984): 82–85.

———. " Transportation." *IEEE Spectrum* (January 1985): 81–84.

PROBLEMS

1. For the state feedback systems described by the following equations, choose the feedback gain constants k_i to place the closed-loop system poles at the indicated locations. Then, for the feedback system, find the steady state outputs due to a unit step input.

(a)
$$\begin{bmatrix} \dot{x}_1 \\ \dot{x}_2 \\ \dot{x}_3 \\ \dot{x}_4 \end{bmatrix} = \begin{bmatrix} 0 & 1 & 0 & 0 \\ 0 & 0 & 1 & 0 \\ 0 & 0 & 0 & 1 \\ -8 & -3 & -7 & -5 \end{bmatrix} \begin{bmatrix} x_1 \\ x_2 \\ x_3 \\ x_4 \end{bmatrix} + \begin{bmatrix} 1 \\ 0 \\ 0 \\ 1 \end{bmatrix} u$$

$$u = -\begin{bmatrix} k_1 & k_2 & k_3 & k_4 \end{bmatrix} \begin{bmatrix} x_1 \\ x_2 \\ x_3 \\ x_4 \end{bmatrix} + r$$

$$y = \begin{bmatrix} 2 & -1 & 0 & 3 \\ 1 & 0 & 1 & -2 \end{bmatrix} \begin{bmatrix} x_1 \\ x_2 \\ x_3 \\ x_4 \end{bmatrix}$$

Closed-loop poles at $s = -5 \pm j3, \ -4 \pm j4$

(b)
$$\begin{bmatrix} \dot{x}_1 \\ \dot{x}_2 \\ \dot{x}_3 \end{bmatrix} = \begin{bmatrix} 0 & 1 & 0 \\ 0 & 0 & 1 \\ -10 & -5 & -2 \end{bmatrix} \begin{bmatrix} x_1 \\ x_2 \\ x_3 \end{bmatrix} + \begin{bmatrix} 0 & -1 \\ 0 & 0 \\ 1 & 7 \end{bmatrix} \begin{bmatrix} u \\ r \end{bmatrix}$$

$$u = -\begin{bmatrix} k_1 & k_2 & k_3 \end{bmatrix} \begin{bmatrix} x_1 \\ x_2 \\ x_3 \end{bmatrix}$$

$$y = [1 \quad 0 \quad 0] \begin{bmatrix} x_1 \\ x_2 \\ x_3 \end{bmatrix}$$

Closed-loop poles at $s = -4$ and $-4 \pm j2$

Ans. $k_1 = +70, \ k_2 = +47, \ k_3 = +10, \ y(\infty) = -0.5625$

2. Design first-order observers of the following plants. Choose the observer eigenvalues to be at $s = -30$.

(a)
$$\begin{bmatrix} \dot{x}_1 \\ \dot{x}_2 \\ \dot{x}_3 \end{bmatrix} = \begin{bmatrix} -2 & 1 & 0 \\ -4 & 0 & 1 \\ -2 & 0 & 0 \end{bmatrix} \begin{bmatrix} x_1 \\ x_2 \\ x_3 \end{bmatrix} + \begin{bmatrix} 0 \\ 1 \\ -1 \end{bmatrix} u$$
$$y = \begin{bmatrix} 1 & 0 & 0 \\ 0 & 1 & 0 \end{bmatrix} \begin{bmatrix} x_1 \\ x_2 \\ x_3 \end{bmatrix}$$

(b)
$$\begin{bmatrix} \dot{x}_1 \\ \dot{x}_2 \end{bmatrix} = \begin{bmatrix} 0 & 1 \\ -4 & -2 \end{bmatrix} \begin{bmatrix} x_1 \\ x_2 \end{bmatrix} + \begin{bmatrix} 0 \\ 3 \end{bmatrix} u$$
$$y = x_1$$

3. Design identity observers for the following plants. Choose the observer eigenvalues to be at -20.

(a)
$$A = \begin{bmatrix} 0 & 1 \\ -4 & -4 \end{bmatrix} \qquad B = \begin{bmatrix} 0 \\ 3 \end{bmatrix}$$
$$y = x_1$$

(b)
$$A = \begin{bmatrix} 0 & 1 \\ -3 & -8 \end{bmatrix} \qquad B = \begin{bmatrix} 0 \\ 2 \end{bmatrix}$$
$$y = x_1$$

4. For the systems of Problem 3, design control gains to place the desired closed-loop poles at -5 and -8, assuming the measurements are available. Next, close the loop by using a reduced-order observer to furnish an estimate of x_2. Show that the characteristic polynomial of the closed-loop system including the observer contains the desired closed-loop roots and the observer root.

5. Consider the system given by

$$\dot{x} = \begin{bmatrix} 2 & 0 \\ -1 & 1 \end{bmatrix} x + \begin{bmatrix} 1 \\ -1 \end{bmatrix} u$$
$$y = \begin{bmatrix} c_1 & c_2 \end{bmatrix} x$$

(a) Determine whether the system is controllable.

(b) Determine whether the system is stabilizable.

(c) Find the transfer function, $T(s)$, if $c_1 = c_2 = 1$.

(d) Repeat (c) for $c_1 = 1,\ c_2 = -1$.

(e) Can state feedback be used to stabilize the system?

Ans. (a) uncontrollable; (b) not stabilizable; (c) $T(s) = 0$; (d) $T(s) = 2(s-1)/[(s-1)(s-2)] = 2/(s-2)$ (e) no.

6. Consider the system given by

$$\dot{x} = \begin{bmatrix} 0 & 1 & -1 \\ -2 & -3 & 0 \\ p & 1 & 1 \end{bmatrix} x + \begin{bmatrix} 1 \\ 0 \\ 0 \end{bmatrix} u$$

$$y = [0 \quad 0 \quad 1]x$$

(a) Determine the values of the parameter p for which the system is controllable (observable).

(b) Find the transfer function of the system.

(c) Determine for what values of the parameter p the system is stabilizable (detectable).

7. Consider the system

$$\dot{x}_2 = \begin{bmatrix} 2 & 1 & 0 \\ 1 & 0 & 1 \\ -2 & 0 & 0 \end{bmatrix} x + \begin{bmatrix} 1 \\ 3 \\ 2 \end{bmatrix} u$$

$$y = [1 \quad 0 \quad 0]x$$

(a) Determine its controllability and observability.

(b) Diagonalize the system; that is, find $\{\bar{A},\ \bar{B},\ \bar{C}\}$.

(c) Identify the modes that are either uncontrollable or unobservable.

(d) Find the system transfer function.

(e) Determine the stabilizability and detectability of the system.

8. Consider the linear system

$$\dot{x} = \begin{bmatrix} 1 & 2 & -1 \\ 0 & 1 & 0 \\ 1 & -4 & 3 \end{bmatrix} x + \begin{bmatrix} 0 \\ 0 \\ 1 \end{bmatrix} u$$

$$y = [1 \quad -1 \quad 1]x$$

Determine the controllability, observability, stabilizability, and detectability of each mode. Also find the system transfer function and note any pole–zero cancellations.

9. Consider the following plant and answer the following questions.

$$\dot{x} = \begin{bmatrix} -1 & 0 \\ 0 & 2 \end{bmatrix} x + \begin{bmatrix} 0 \\ 1 \end{bmatrix} u$$

$$y = \begin{bmatrix} c_1 & c_2 \end{bmatrix} x$$

   (a) Is the system controllable?
   (b) Is the system observable?
   (c) Find a control gain vector, k, to place plant poles at $\{-1, \quad -2\}$, if possible.
   (d) Repeat (c) for poles at $\{-2, \quad -2\}$.
   (e) Explain any discrepancy in answers to (c) and (d).
   (f) Find the transfer function when $c_1 = 1, \ c_2 = 0$.
   (g) Repeat (f) for $c_1 = 0, \ c_2 = 1$.
   (h) Given the answers in (f) and (g), which state variable should be measured to stabilize the system using observers?

10. Consider the plant, $G(s)$.

$$G(s) = \frac{1}{s(s-2)}$$

   Use state feedback to move the closed-loop poles to $s = -1, \ -1$.

   (a) Find control gain, k.
   (b) Design a reduced-order observer with pole at $s = -2$.
   (c) Find the open-loop transfer function of the compensated system and use it to plot the root locus.

11. Design a first-order observer for the following system. Place the observer pole at -1. Also design a controller to place system poles at $-1 \pm j$. Obtain the compensator transfer function and draw the root locus of the open-loop transfer function.

   Use computer software to simulate the system. For simulation purposes, the plant initial conditions are 1 and 2. Set the initial condition for $\hat{x}_2$ at -2.

$$\dot{x} = \begin{bmatrix} 1 & 1 \\ 1 & 1 \end{bmatrix} x + \begin{bmatrix} 1 \\ 0 \end{bmatrix} u, \qquad x(0) = \begin{bmatrix} 1 \\ 2 \end{bmatrix}$$

$$y = [1 \quad 0]x$$

12. It was mentioned in this chapter that an optimal choice for observer initial conditions is given by $\hat{x}(0) = C'(CC')^{-1}y(0)$. Repeat Problem 11 by designing a full-order observer with poles at $-2 \pm 2j$. Simulate the closed-loop system using observer initial conditions of $\hat{x}(0) = \begin{bmatrix} -1 \\ -2 \end{bmatrix}$. Then, to repeat the simulation, use optimal initial conditions [note that $y(0) = 1$]. Compare the zero-input responses in both cases.

13. Consider the following plant: $\dot{x} = Ax + Bu$, where the state space matrices are given by

$$A = \begin{bmatrix} 0 & 1 & 0 \\ 0 & 0 & 1 \\ 2 & 0 & -1 \end{bmatrix} \quad B = \begin{bmatrix} 1 \\ 2 \\ 0 \end{bmatrix} \quad C = [1 \quad 0 \quad 0] \quad D = 0$$

(a) We want to place the closed-loop poles at $\{-10, \ -1 + j, \ -1 - j\}$. Find the state feedback gain vector.

(b) Obtain the equivalent transfer function of the compensator, root locus, Bode plots, and closed-loop step response. Tabulate the step response and frequency response features such as percent overshoot, rise time, settling time, and phase and gain margins.

(c) Design a full-order observer. Choose observer poles at $\{-40, -4 + j4, -4 - j4\}$. Repeat (b).

(d) Design a reduced-order observer. Choose observer poles at $\{-4 + j4, -4 - j4\}$. Repeat (b).

(e) Repeat (c) with the observer poles at $\{-40, \ -1, \ -2\}$.

(f) Repeat (d) with the observer poles at $\{-1, \ -2\}$

Note: In (e) and (f), the observer poles are chosen at the plant zeros. It is known that such a choice increases the robustness of the system. Because phase and gain margins are classical measures of robustness (protection against uncertainty), compare the margins in all cases. Does the choice of observer poles in (e) and (f) really improve the margins?

(g) To verify robustness, let us assume that the A matrix of the true plant model is

$$A_{\text{true}} = \begin{bmatrix} 0 & 1 & 0 \\ 0 & 0 & 1 \\ -2 & 0 & -1 \end{bmatrix}$$

Use the control gain vector of (a) to obtain the step responses and frequency responses for the true system, using the observers designed in (c)–(f). Compare the responses, and determine which observer design is more robust to parameter uncertainty and variation.

14. For the MAGLEV system choose instead values of the feedback gain constants k_1, k_2, and k_3 to place all three of the overall system poles at $s = -5$. For this system, find the steady state response d to a unit step disturbance f and the value of constant reference input u_1 to give a nominal gap distance $d = 15$.

15. For the open-loop MAGLEV system, suppose the vertical track elevation varies sinusoidally with time as the train is in motion, according to

$$\dot{h}(t) = 0.2 \sin \frac{\pi t}{10}$$

Find the second-order differential equation satisfied by $\dot{h}(t)$, then augment the original state equations with two more equations and two more state variables

$$x_4 = \dot{h}(t)$$

$$x_5 = \ddot{h}(t)$$

in place of the disturbance input $f = \ddot{h}$. With additional sensors for the signals x_4 and x_5 and feedback of the form

$$u = +k_1x_1 + k_2x_2 + k_3x_3 + k_4x_4 + k_5x_5 + u_1$$

find the state equations for the feedback system in terms of the k constants.

CHAPTER 10

Advanced State Space Methods

10.1 Preview

In the preceding chapters, compensators were designed to satisfy specified requirements for steady state error, transient response, stability margins, or closed-loop pole locations. Meeting all objectives is usually difficult because of various trade-offs that must be made and because of the limitations of the design techniques. For example, classical Bode design allows us to satisfy phase margin and steady state error requirements, but the step response characteristics may not be desirable. State space observer-based techniques allow arbitrary pole placement, but the stability margins cannot be controlled directly. Also, none of the techniques discussed so far address practical issues such as plant model uncertainty or actuator signal limits. In addition, none of the techniques result in the best possible performance. This chapter addresses some of these issues. In particular, we present optimization-based techniques that result in an optimal solution.

Optimization refers to the science of maximizing or minimizing objectives. Optimization requires a measure of performance. When mathematically formulated, this measure of performance is called the *objective* (or *cost*) *function*. Optimization of control systems is called *optimal control*. Optimization problems are either constrained or unconstrained. For example, finding the minimum of a parabola is an unconstrained optimization problem. Finding the minimum of a parabola in a given interval of its domain is a constrained optimization problem. You have seen examples of these types in calculus. Calculus-type problems are usually static problems because the constraints are algebraic equations. Optimal control problems are usually constrained dynamic optimization problems because the constraints are the system equations, which are differential equations (i.e., they are dynamic). Simple examples of optimal control appear in Section 3.5, where a gain or damping ratio is found to minimize the tracking error in a control system. In this chapter a more general and systematic treatment of optimal control is presented.

A typical optimal control problem formulation is the following:

$$\min_u J = \int_0^T L(x, u, t)dt \qquad \text{subject to} \qquad \dot{x} = f(x, u, t)$$

Here the plant is a nonlinear system, which is the constraint, and the cost function is the integral of some nonlinear function of the state x, control u, and time. The objective is to find a control function u that will minimize the cost function. Examples of optimal control problems are minimum time, minimum fuel, and minimum energy (more examples appear in Section 3.5). This formulation, in general, leads to controllers that are time-varying and nonlinear. Analog implementation of these nonlinear controllers is not usually practical or worthwhile.

10.2 The Linear Quadratic Regulator Problem

We will now restrict our attention to linear systems (or linearized versions of nonlinear systems) and choose a cost function that is a quadratic function of states and controls. In this case, the solution is a linear controller that is easily implemented. Hence, we will consider a special optimal control problem, called the *linear quadratic regulator* (*LQR*) problem. The formulation of the problem follows. Given the linear system

$$\boxed{\begin{aligned} \dot{x} &= Ax + Bu \\ y &= Cx \end{aligned}}$$

find a control function *u(t)* that will minimize the cost function J given by

$$\boxed{J = \frac{1}{2}\int_o^\infty \left(x'Qx + u'Ru\right) dt}$$

The function inside the integral is a quadratic form and the matrices Q and R are usually symmetric (see Appendix A for a brief review of quadratic forms). It is assumed that R is positive definite (i.e., it is symmetric and has positive eigenvalues) and Q is positive semi definite (i.e., it is symmetric and its eigenvalues are nonnegative). These assumptions imply that the cost is nonnegative, so its minimum value is zero. For the cost function to achieve its minimum value, both x and u must go to zero. This type of control problem is called a *regular problem*. When the state vector is to track nonzero values, J can be redefined to create an optimal *servomechanism* (tracking) problem. Many of the control systems considered in earlier chapters were servomechanisms. Regulator behavior is important for control systems of many types (e.g., attitude control of satellites or spacecraft, where a zero reference should be maintained in spite of disturbances).

A simple interpretation of the cost function is as follows. If the system is scalar (i.e., a first-order system), the cost function becomes

Scalar LQR problem.

$$J = \frac{1}{2}\int_0^\infty \left(qx^2 + ru^2\right) dt$$

Now we see that J represents the weighted sum of energy of the state and control. Small q and r are used, respectively, when x and u are scalars. If r is very large relative to q, which implies that the control energy is penalized heavily, the control effort will diminish at the expense of larger values for the state. This physically translates into smaller motors, actuators, and amplifier gains needed to implement the control law. Likewise if q is much larger than r, which means that the state is penalized heavily, the control effort rises to reduce the state, resulting in a damped system. In the general case, Q and R represent respective weights on different states and control channels. For example, if

$$Q = \begin{bmatrix} 10 & 0 \\ 1 & 0 \end{bmatrix} \qquad \text{and} \qquad R = r \text{ (a scalar)}$$

Then

$$x'Qx + u'Ru - 10x_1^2 + x_2^2 + ru^2$$

By putting a larger weight on the first state, we are putting more emphasis on controlling this state and restricting its fluctuations.

Several procedures are available to solve the LQR problem. Since optimization can easily become the subject for several textbooks, we will present only the main results. The work of mathematicians and engineers such as Hamilton, Euler, Lagrange, Jacobi, Pontryagin, Kalman, and Bellman have resulted in a rather complete understanding of the optimal control problem. Actual implemented control systems that have been designed by these methods were few in number as of the early 1990s, but they are now more popular.

One approach to finding a controller that minimizes the LQR cost function is based on finding the positive-definite solution of the following *algebraic Riccati equation* (ARE).

LQR Solution.

$$A'P - PA + Q - PBR^{-1}B'P = 0$$
$$u = -Kx \qquad K = R^{-1}B'P$$

It turns out that under the conditions stated shortly, the positive-definite solution of the ARE results in an asymptotically stable closed-loop system. The conditions are the following. The system is controllable, R is positive definite (this ensures that its inverse exists), and Q can be factored as $Q = C_q'C_q$, where C_q is any matrix such that $(C_{q,} A)$ is observable. These conditions are necessary and sufficient for the existence and uniqueness of the optimal controller that will asymptotically stabilize the system (these assumptions can be relaxed to stabilizability and detectability). Note that the assumption on Q allows us to define another output vector z as

$$z = C_qx$$

therefore

$$x'Qx = x'C_q'C_qx = z'z$$

the vector z is called the *controlled* or *regulated output*, and may differ from the *measured output* y.

Manually solving the Riccati equation is tedious and almost impossible for third- or higher-order systems; the second-order example, however, can be solved. Let us consider the double-integrator example considered in Chapter 9. The system matrices are

Solving the double integrator problem by LQR.

$$A = \begin{bmatrix} 0 & 1 \\ 0 & 0 \end{bmatrix} \quad B = \begin{bmatrix} 0 \\ 1 \end{bmatrix} \quad C_q = [1 \quad 0]$$

Let us assume Q and R are given by $Q = \begin{bmatrix} 1 & 0 \\ 0 & 0 \end{bmatrix}$ and $R = 1$.

First, we will check to see if the conditions are satisfied. The system is controllable because the matrix $[B \quad AB]$ has rank 2. Q can be factored as

$$Q = \begin{bmatrix} 1 & 0 \\ 0 & 0 \end{bmatrix} = \begin{bmatrix} 1 \\ 0 \end{bmatrix} [1 \quad 0] = C_q'C_q$$

The observability condition is also satisfied because the matrix $\begin{bmatrix} C_q \\ C_qA \end{bmatrix}$ has rank 2. Therefore, the ARE will have a stabilizing solution. Now, the ARE becomes

$$\begin{bmatrix} 0 & 0 \\ 1 & 0 \end{bmatrix} \begin{bmatrix} p_1 & p_2 \\ p_2 & p_3 \end{bmatrix} + \begin{bmatrix} p_1 & p_2 \\ p_2 & p_3 \end{bmatrix} \begin{bmatrix} 0 & 1 \\ 0 & 0 \end{bmatrix} + \begin{bmatrix} 1 & 0 \\ 0 & 0 \end{bmatrix}$$
$$- \begin{bmatrix} p_1 & p_2 \\ p_2 & p_3 \end{bmatrix} \begin{bmatrix} 0 & 0 \\ 0 & 1 \end{bmatrix} \begin{bmatrix} p_1 & p_2 \\ p_2 & p_3 \end{bmatrix} = \begin{bmatrix} 0 & 0 \\ 0 & 0 \end{bmatrix}$$

Multiplying and adding the matrices, and setting the sides of the equation equal to each other, element by element, we get three coupled algebraic-quadratic equations. In this case, the equations are very simple (usually, they are quite horrendous) because of the number of zero elements in the matrices. The equations are

$$p_2^2 = 1$$
$$p_1 = p_2 p_3$$
$$2p_2 - p_3^2 = 0$$

Therefore,

$$P = \begin{bmatrix} \sqrt{2} & 1 \\ 1 & \sqrt{2} \end{bmatrix} \quad \text{and} \quad K = R^{-1}B'P = [1 \quad \sqrt{2}]$$

The closed-loop system matrix becomes

$$A - BK = \begin{bmatrix} 0 & 1 \\ -1 & \sqrt{2} \end{bmatrix}$$

The closed-loop characteristic equation and its roots (closed-loop eigen values) are

$$\lambda^2 + \sqrt{2}\lambda + 1 = 0$$

$$\lambda = \frac{\sqrt{2}}{2}(-1 \pm j)$$

Therefore, the system has been stabilized and it has a damping ratio of 0.707. Observe that the optimal controller is of the state feedback form (i.e., we are assuming all states are available for feedback). We discuss an observer design in the next section. Let us obtain the open-loop transfer function and use it to obtain Bode plots and determine the stability margins. The open-loop transfer function L(s) is given by (refer to Section 9.2 and Figure 9.3)

$$L(s) = K\Phi(s)B = K(sI - A)^{-1}B = \frac{\sqrt{2}[s + (\sqrt{2}/2)]}{s^2}$$

The Bode plots in Figure 10.1 show that the system has 65° of phase margin and infinite gain margin.

❑ DRILL PROBLEMS

D10.1 For each of the following systems described by A and B matrices and LQR performance criteria measured by Q and R, solve the associated algebraic Riccati equation and find the optimal control gains.

(a) $A = -2,\ B = 4,\ Q = 4,\ R = 1$

(b) $A = 2,\ B = 4,\ Q = 4,\ R = 1$

(c) $A = \begin{bmatrix} 0 & 1 \\ -10 & -2 \end{bmatrix},\ B = \begin{bmatrix} 0 \\ 2 \end{bmatrix},\ Q = \begin{bmatrix} 1 & 0 \\ 0 & 0 \end{bmatrix},\ R = 1$

(d) $A = \begin{bmatrix} 0 & 1 \\ -10 & -2 \end{bmatrix},\ B = \begin{bmatrix} 0 \\ 2 \end{bmatrix},\ Q = \begin{bmatrix} 0 & 0 \\ 0 & 1 \end{bmatrix},\ R = 1$

Ans. (a) $P = 0.3904,\ k = +1.462$;

(b) $P = 0.6404,\ k = +2.562$;

(c) $P = \begin{bmatrix} 0.3455 & 0.0495 \\ 0.0495 & 0.0242 \end{bmatrix},\ k = [0.1 \quad 0.048]$;

(d) $p = \begin{bmatrix} 2.07 & 0 \\ 0 & 2.07 \end{bmatrix},\ k = [0 \quad -0.414]$

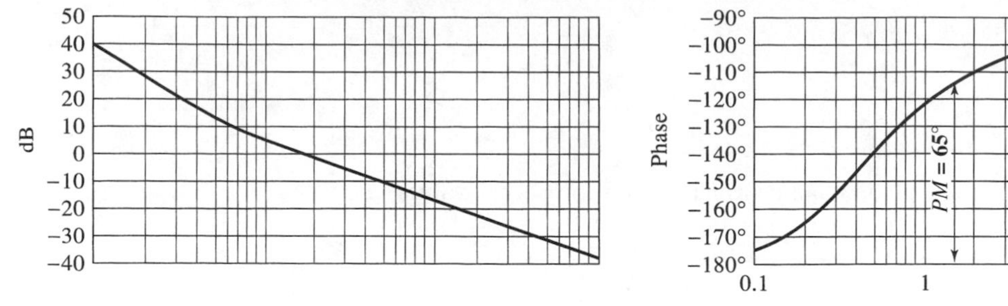

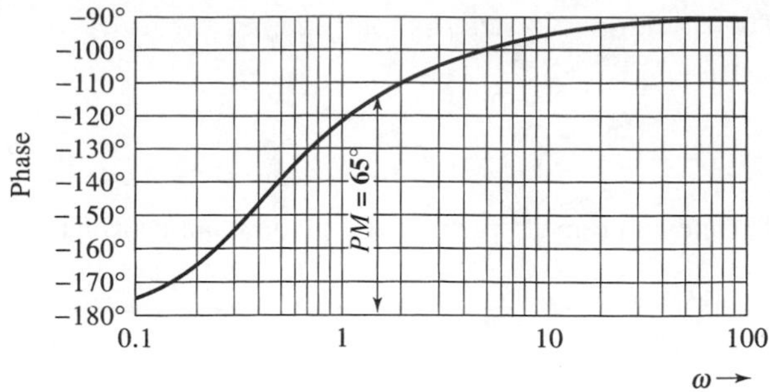

(a)

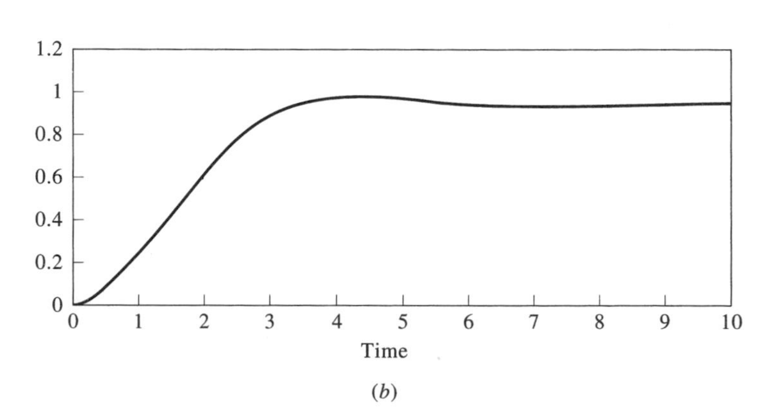

(b)

Figure 10.1 (a) Bode plots for the LQR design of the double-integrator plant. (b) The step response for LQR design.

10.2.1 Properties of the LQR Design

LQR has many desirable properties. Among them are good stability margins and sensitivity properties. We will also discuss the effects of weights in the LQR setting. Most of these properties can be derived using the *return-difference inequality* first derived by Kalman.

10.2.2 Return Difference Inequality

The algebraic Riccati equation can be manipulated to arrive at the following relation:

$$|1 + L\,(j\omega)|^2 = 1 + \frac{1}{\rho}\left|G_q(j\omega)\right|^2$$

Where $L(s)$ is the loop gain (open-loop transfer function) given by

$$L\,(s) = K\Phi\,(s)\,B$$

where $\Phi\,(s) = (sI - A)^{-1}$, relation assumes that $Q = C_q' C_q$ and $G_q\,(s) = C_q\Phi\,(s)\,B$.

Because the right-hand side of the return-difference equality (RDE) is always greater than 1, the following inequality holds:

$$|1 + L(j\omega)| \geqslant 1$$

The preceding return-difference inequality (RDI) implies that for all frequencies, the Nyquist plot of the open-loop transfer function of an LQR-based design always stays outside a unit circle centered at $(-1, 0)$. A typical Nyquist plot is shown in Figure 10.2. The term *return difference*, introduced by Bode, means the following. Suppose a feedback loop is broken at a given point; inject a 1-volt signal at the entrance of the point and measure the signal returned at the exit of that point; the difference between the injected signal and the returned signal is called the *return difference*. If the gain around the loop is $-GH$, the return difference is $1 - (-GH)$ or $1 + GH$. The return difference is a measure of the amount of feedback in a feedback loop. It is an important quantity and appears in many expressions, such as the denominator of the closed-loop transfer function and the sensitivity function defined in earlier chapters.

The return-difference inequality, along with simple geometric arguments, can be used to show that the LQR solution, in the SISO case, has at least 60° phase margin, infinite gain margin, and a gain reduction tolerance of −6 dB. The latter means that the gain can be reduced by a factor of $\frac{1}{2}$ before instability occurs. Therefore, an LQR design behaves quite well from a classical control point of view. It not only always results in an asymptotically stable system but also provides guaranteed stability margins. This is to be compared with the state feedback pole placement technique discussed in Chapter 9, where stability margins are not known or guaranteed ahead of time. Finally, observe that the LQR margins are a bit excessive in that lower gain and phase margins are generally acceptable in most designs.

Stability margins of LQR.

Another frequency domain property of the LQR solution is its high-frequency roll-off rate. Recall that the closed-loop transfer function of state feedback design is

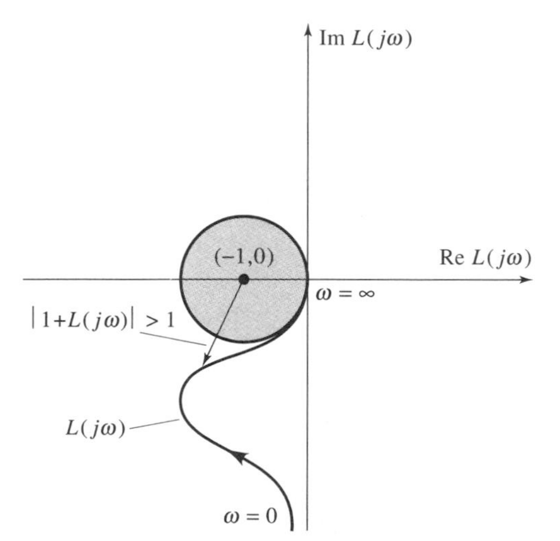

Figure 10.2 Nyquist plot of an LQR-based design. The plot always stays outside the unit circle centered at $(-1, 0)$.

given by

$$T\,(j\omega) = -K\,(j\omega I - A + BK)^{-1}\,B$$

it can be shown that

$$\lim_{\omega\to\infty} T\,(j\omega) = \frac{-1}{j\omega} KB = \frac{-1}{j\omega} R^{\;1} B'PB < 0$$

LQR Bode plot rolls off at −20 dB at high frequencies.

The preceding implies that $|T\,(j\omega)|$ drop as $1/(j\omega)$ in the SISO case, indicating a roll-off rate (i.e., a slope) of −20 dB per decade at high frequencies [see Figure 10.1(a)]. This, of course, affects the noise suppression properties of the optimal system and as such is not very good. It can be argued that this defect is the result of the excessive stability margins of the LQR solution.

10.2.3 Optimal Root Locus

We will see that a special choice of Q and R allows us to investigate the effects of weights on the location of closed-loop poles. Let us assume that Q and R are given by

$$Q = C_q' C_q \qquad \text{and} \qquad R = \rho I$$

where ρ is a positive scalar. Then

$$x'Qx = z'z$$

where $z = C_q x$
and the cost function becomes

$$J = \frac{1}{2}\int_0^{\alpha} \left(z'z + \rho u'u\right) dt$$

This means that we are minimizing the system output and control energy, and by increasing ρ, we can put more emphasis on minimizing control energy. Definite the following matrix, called the *Hamiltonian* matrix

The Hamiltonian matrix for LQR.

$$\mathcal{H} = \begin{bmatrix} A & -\frac{1}{\rho}BB' \\ -C_q'C_q & -A' \end{bmatrix}$$

Eigenvalues of $\mathcal{H}$ are symmetric with respect to the imaginary axis.

Because of the special structure of the Hamiltonian matrix, its characteristic polynomial is an even polynomial (i.e., if s is a root, so is $-s$). Therefore, it can be factored as a polynomial with only LHP roots and a polynomial with only RHP roots (the Hamiltonian has no roots on the imaginary axis). The Hamiltonian matrix is used in formal proofs of the LQR problem, and the eigenvalues and eigenvectors of the Hamiltonian matrix are used to solve the ARE (Potter's method).

The optimal closed-loop poles will be the stable (i.e., LHP) eigenvalues of the Hamiltonian matrix. If we denote the characteristics polynomial of the Hamiltonian matrix by $\Delta_c = |sI - \mathcal{H}|$, after a series of matrix manipulations, we arrive at the following equation (n = number of poles, m = number of zeros, with $m < n$, $r = n - m$)

$$\Delta_c\,(s) = (-1)^n\,\Delta\,(s)\,\Delta\,(-s)\left[1 + \frac{1}{\rho} G_q\,(s)\,G_q\,(-s)\right]$$

where $\Delta(s) = |sI - A|$ and preceding $G_q(s) = n_q(s)/d(s)$ is the transfer function from u to z (the regulated output). The preceding equation simplifies to

$$(-1)^n \Delta_c(s) = \Delta(s)\Delta(-s)\left[1 + \frac{1}{\rho}G_q(s)G_q(-s)\right]$$
$$= d(s)d(-s) + \frac{1}{\rho}n_q(s)n_q(-s)$$

This has the standard root locus form. It implies that the optimal closed-loop poles can be obtained from the root locus of $G_q(s)G_q(-s)$. Such root loci are generally called a *symmetric root locus* or *root-square locus.* We will discuss the effects of limiting values of ρ.

Root locus of $G_q(s)G_q(-s)$ gives the optimal pole locations.

Minimum Energy Control (or Expensive Control) Case

$$As \quad \rho \to \infty, \quad (-1)^n \Delta_c(s) \to d(s)d(-s)$$

Because the optimal closed-loop poles are always in the LHP, we conclude that as the control weighting is increased, the stable open-loop poles will remain where they are and the unstable ones will be reflected about the imaginary axis. This property can be used as a guideline for pole placement.

Cheap Control Case

$$\text{As } \rho \to 0, \quad (-1)^n \Delta_c(s) \to n_q(s)n_q(-s) \quad \text{for finite } s$$

Hence, the closed-loop poles approach the plant finite zeros or their stable images. For values of s approaching , the closed-loop poles will approach zeros at infinity in the so-called *Butterworth pattern.*

$$\text{for } |s| \to \infty \quad s = \left(\frac{\alpha_m^2}{\rho}\right)^{1/2r} \exp\left[j\frac{\pi k(r+1)}{2r}\right] \qquad k = \text{odd integer}$$

where α_m is the coefficient of the highest-order term in $n_q(s)$.

As an example of root-square locus (RSL) consider the following system

RSL example.

$$G_q(s) = \frac{5}{s^2 + s + 5}$$

Then the optimal characteristic equation is given by

$$1 + \frac{1}{\rho}\frac{25}{(s^2 + s + 5)(s^2 - s + 5)} = 0$$

The optimal closed-loop poles are along the LHP branches of the RSL shown in Figure 10.3. Note that the RSL is symmetric with respect to both the imaginary and the real axes. The RSL shows what happens to the poles as the control cost weight (ρ) increases from 0 to infinity (note that the actual root locus gain $1/\rho$). When ρ approaches infinity (i.e., the root locus gain goes to 0), the closed-loop poles approach the plant open-loop poles; therefore, when control is expensive and the

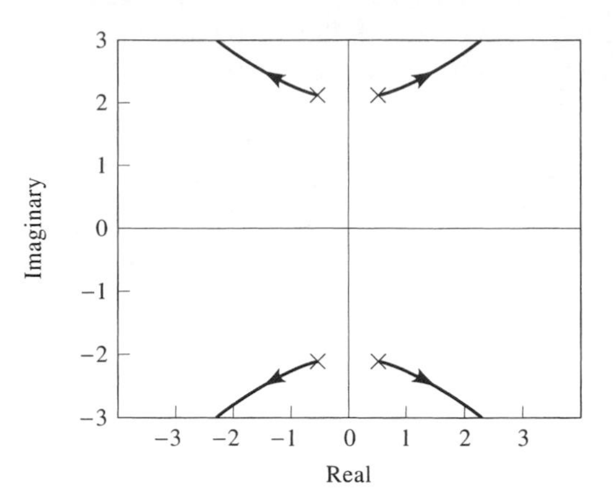

Figure 10.3 Root-Square locus for $G_q(s)$.

plant is stable, the best choice is to do nothing and leave the poles where they are. When ρ approaches 0 (i.e., the root locus gain goes to infinity), the closed-loop poles approach the plant open-loop zeros at infinity.

Now suppose that the plant is unstable and is given by

Another RSL example.

$$G_q(s) = \frac{5}{s^2 - s + 5}$$

We obtain the same RSL as shown in Figure 10.3. The interpretation is slightly different here. In the minimum-energy control case (expensive control), the best choice is to reflect the unstable poles about the imaginary axis. In either of these examples, the optimal characteristic equation for the minimum-energy control case is

$$s^2 + s + 5$$

To find the optimal control gain K using RSL, we first determine the optimal pole location from the RSL, form the characteristic polynomial, and set this equal to the characteristic polynomial of $A - BK$ and solve for K. For example, in the preceding case, we get

$$\begin{aligned} |sI - (A - BK)| &= s^2 + (-1 + 5k_2)\,s + 5 + 5k_1 \\ &= s^2 + s + 5 \end{aligned}$$

Therefore,

$$k_1 = 0 \quad \text{and} \quad k_2 = 0.4$$

Let us consider another example.

RSL example.

$$G_q(s) = \frac{s}{s^2 + 2s + 10} \quad \text{and} \quad 1 + \frac{1}{\rho}\frac{-s^2}{\left(s^2 + 2s + 10\right)\left(s^2 - 2s + 10\right)} = 0$$

The RSL for positive ρ is shown in Figure 10.4(a). Note that for positive ρ the locus has imaginary poles, and this cannot correspond to an optimal system. Therefore, we have to use negative-gain sketching rules for root locus. The optimal locus is shown in Figure 10.4(b). In this example, this was clear because of the negative sign in the numerator. In general, if r (the number of poles minus the number of zeros in the plant

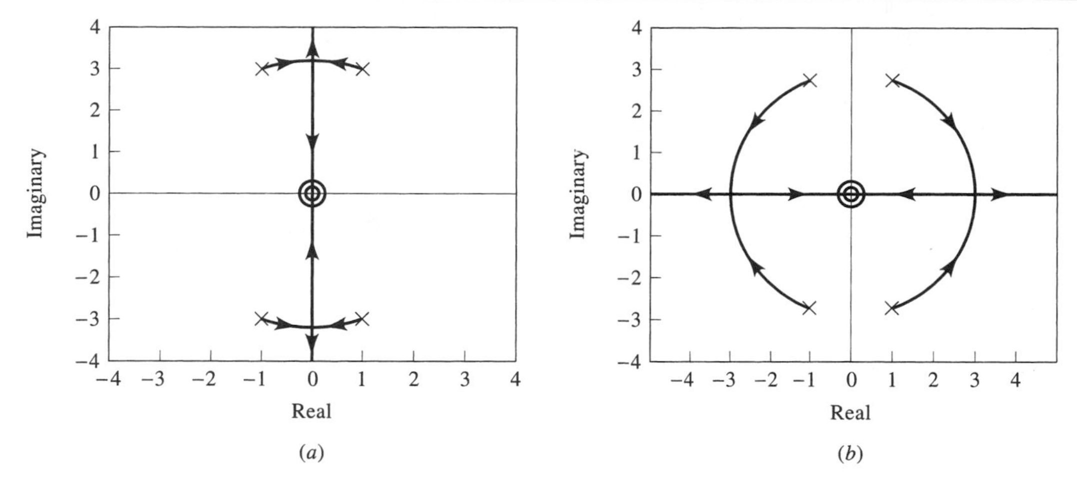

Figure 10.4 (a) Root-square locus using positive gain. (b) Root-square locus using negative gain.

transfer function) is odd, we have to use negative-gain sketching rules; otherwise we use positive-gain sketching rules for plotting the root locus.

❑ DRILL PROBLEMS

D10.2 Consider the systems presented in Drill Problem 10.1. In each case, factor $Q = C_q' C_q$ and let $R = \rho$. Then find $G_q(s)\,G_q(-s)$ and plot the appropriate root-square locus.

Ans. (a) $Q = 4 = 2\,(2)\,,\, C_q = 2,\, G_q(s)\,G_q(-s) =$ $[8/(s+2)]\,[8/(-s+2)]$

(b) $Q = (2)(2),\, C_q = 2,\, G_q(s) = [8/(s-2)]\,[8/(-s-2)]$

(c) $Q = \begin{bmatrix} 1 \\ 0 \end{bmatrix} [1 \quad 0],\, C_q = [1 \quad 0],\, G_q(s) =$ $\left[2/(s^2+2s+10)\right]\left[2/(s^2-2s+10)\right]$

(d) $Q = \begin{bmatrix} 0 \\ 1 \end{bmatrix} [0 \quad 1],\, C_q = [0 \quad 1],$ $G_q(s) = \left[2s/(s^2+2s+10)\right]\left[-2s/(s^2-2s+10)\right]$

See Figure 10.4(b) for the RSL

10.3 Optimal Observers—the Kalman Filter

The LQR solution is basically a state-feedback type of controller—i.e., it requires that all states be available for feedback. It was argued in the previous chapter that this is usually an unreasonable assumption and some form of state estimations necessary.

In addition, the concept of the observer was introduced and it was shown that the combination of the state feedback and observer will always result in a stable closed-loop system. In chapter 9, however, the designer was left with the responsibility of choosing the controller and observer poles. We have seen in the previous section that the controller performance can be optimized according to some quadratic cost function, resulting in optimal controller pole locations. The next obvious question is whether the observer design can also be done in an optimal manner. The answer is affirmative, provided the problem is formulated in a probabilistic (or stochastic) sense. The formulation of the state estimation problem is as follows

$$\dot{x} = Ax + Bu + \omega$$

$$y = Cx + v$$

where ω represents random noise disturbance input (process noise) and ν represents random measurement (sensor) noise; we also have to assume some statistical knowledge of the noise processes. For instance, in the case of an aircraft, the plant is subject to random wind disturbances (or process noise), and the measurement instrumentations (sensors) are not always accurate and may include random errors (sensor noise). Ships and other marine vessels are subjected to random wave motions (which may also have strong periodic components), and in general, most systems are subject to both kinds of random inputs.

The state-space solution to this problem was first provided by R.E. Kalman and R.S. Bucy. The optimal observer (commonly known as the *Kalman filter*) is given by

$$\dot{\hat{x}} = A\hat{x} + Bu + L(y - C\hat{x})$$

where $\hat{x}$ is the estimate of x. The observer gain is computed from

$$L = \Sigma C' R^{-1}$$

and Σ is found as the positive semi-definite solution of

$$A\Sigma + \Sigma A' + Q_o - \Sigma C' R_o^{-1} C \Sigma = 0$$

Note that the equation for the filter gain and Σ are very similar to the equations for the LQR solution. In particular, the equation for Σ is an algebraic Riccati equation. There are two matrices that appear in the filter equation that require some explanation. They are Q_o and R_o. These matrices represent the intensity of the process and sensor noise inputs and are the only parameters that are to be provided by the user. In the mathematical subject of random processes, these matrices are known as *co-variance matrices*. Their size (usually measured by their trace, the sum of the diagonal elements) is a measure of how strong the noise is—the larger the size, the more random or intense the noise—hence we refer to it as *noise intensity*.The Kalman filter attempts to minimize the size of the estimation error intensity (the intensity of the estimation error is given by Σ). Finally, we note that the mathematical conditions that are needed for the solution of the Kalman filter problem to exist are the following: $Q_o$and R_o must be positive semidefinite and positive definite respectively, and the system must be observable.

Estimation theory and, in particular, Kalman filter theory are vast and important areas that are common to control and communications. There are many reported successful applications of Kalman filters in a wide range of areas (many more than LQR implementations). Because our interest lies in control system design rather than pure state estimation, we will return to the control problem without pursuing this subject any further. Hence, we view the Q_o and R_o matrices as design parameters, not necessarily related to physical noise processes.

10.4 The Linear Quadratic Gaussian (LQG) Problem

Given an optimal filter to estimate the states, the next question is whether the closed-loop system remains stable and optimal when we combine the LQR controller of Section 10.2 and the Kalman filter of Section 10.3. This problem is known as the *linear quadratic Gaussian* (or *LQG*) problem. The term *Gaussian* refers to the statistical distribution of the noise processes. The plant equations and the problem solution are repeated:

$$\dot{x} = Ax + Bu + W$$
$$y = Cx + v$$

The controller portion is given by

$$u = -K\hat{x}(t)$$
$$K = R^{-1}B'P$$
$$A'P + PA - PBR^{-1}B'P + Q = 0$$

The observer (or filter) portion is given by

$$\dot{\hat{x}} = A\hat{x} + Bu + L\left(y - C\hat{x}\right)$$
$$L = \Sigma C'R^{-1}$$
$$A\Sigma + \Sigma A' + Q_o - \Sigma C'R_o^{-1}C\Sigma = 0$$

It can be shown that the LQG solution results in an asymptotically stable closed-loop system. In addition, the controller minimizes the average of the LQR cost function (i.e., the weighted variance of the state and input), resulting in an optimal solution. Because the structure of the controller and the Kalman filter are similar to the observer-based compensator discussed in Chapter 9 (the major difference is how the control and filter gains are computed), the LQG compensator will also exhibit the separation property (the mathematical proof of this fact is actually quite involved). Hence, the closed-loop poles will be the union of the controller poles and the filter poles, and the controller and the filter can be designed independently of each other (this means that the filter equation do not contain K or P, and the control equations do not depend on L or Σ).

The transfer function of the LQG compensator is similar to the observer-based compensator, and is given by

LQG compensator.

$$H(s) = K\left(sI - A + BK + LC\right)^{-1}L$$

Let us obtain an LQG compensator for the double-integrator plant. The controller portion has already been found, so we will design the Kalman filter. We will assume the noise intensities are

$$Q_o = \begin{bmatrix} 1 & 0 \\ 0 & 1 \end{bmatrix} \quad \text{and} \quad R_o = 1$$

The filter Riccati equation results in three coupled algebraic nonlinear equations:

$$a^2 = 2b + 1$$

$$ab = c$$

$$b^2 = 1$$

where $\Sigma = \begin{bmatrix} \text{a} & \text{b} \\ \text{b} & \text{c} \end{bmatrix}$

Therefore,

$$\Sigma = \begin{bmatrix} \sqrt{3} & 1 \\ 1 & \sqrt{3} \end{bmatrix} \quad \text{and} \quad L = \Sigma C' R_o^{-1} = \begin{bmatrix} \sqrt{3} \\ 1 \end{bmatrix}$$

LQG design for the double-integrator system.

The transfer function of the compensator is given by

$$H(s) = \frac{3.14\,(s + 0.31)}{(s + 1.57 + j1.4)\,(s + 1.57 - j1.4)}$$

Further computation shows that the closed-loop poles are at the locations of the controller and filter eigenvalues, respectively:

$$\text{Closed-loop poles} = \frac{\sqrt{2}}{2}(-1 \pm j)\,,\ \frac{-\sqrt{3} \pm j}{2}$$

The root locus, open-loop magnitude and phase plots, closed-loop magnitude plot, and closed-loop step response of the system are shown in Figure 10.5. The Bode plots indicate a gain margin of 10.7 dB and phase margin of 34.5°. Let us now compare the LQR and LQG designs.

LQR and LQG comparison for the double-integrator.

1. LQR has much higher stability margins.
2. The low-frequency gain in LQR is 40 dB and in LQG is 27 dB. Hence, LQR will have better steady state tracking properties (recall that error coefficients are obtained by letting s approach 0, so as low-frequency gain in the open-loop magnitude plot determines steady state error properties).
3. The gain-crossover frequency is higher in LQR. This means that LQR has a higher bandwidth, so it passes more noise into the system. Also, since gain crossover frequency is inversely related to the speed of response, this indicates a faster response in LQR, as can be seen in the step response.
4. The high-frequency roll-off rates, approximated by the slope of Bode magnitude plot in Figure 10.5(b), are -60 dB and -20 dB in LQG and LQR, respectively. This means that LQG has better noise suppression properties.

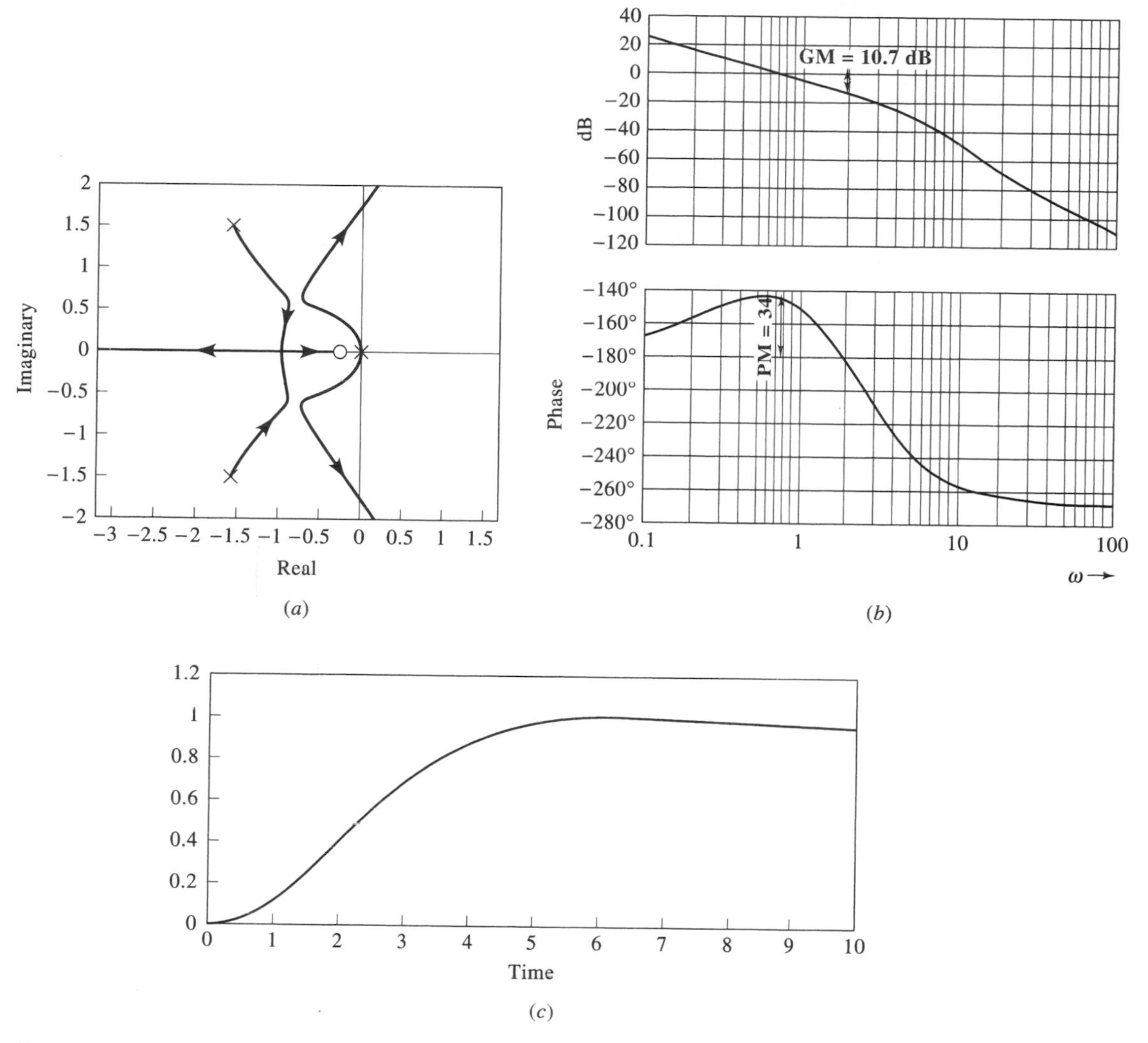

Figure 10.5 Classical root locus for LQG design. (b) Open-loop Bodc plots for LQG design. (c) Closed-loop step response for LQG design.

In this example, we see the trade-offs involved in control system design. Stability margin is traded off with high-frequency roll-off rate. Gain crossover frequency (or bandwidth) is traded off with speed of response.

We will now examine the Nyquist plots for both cases shown in Figure 10.6. As predicted by the return-difference inequality, the LQR plot avoids the unit circle centered at $(-1, 0)$, whereas the LQG plot enters it. This shows that the LQG open-loop transfer function does not satisfy the return-difference inequality. This has very important implications because the RDI is the basis for the guaranteed stability margins of LQR. In fact, it has been shown by counterexamples that LQG has no guaranteed stability margins and its margins can be dangerously low.

LQG has no guaranteed stability margins.

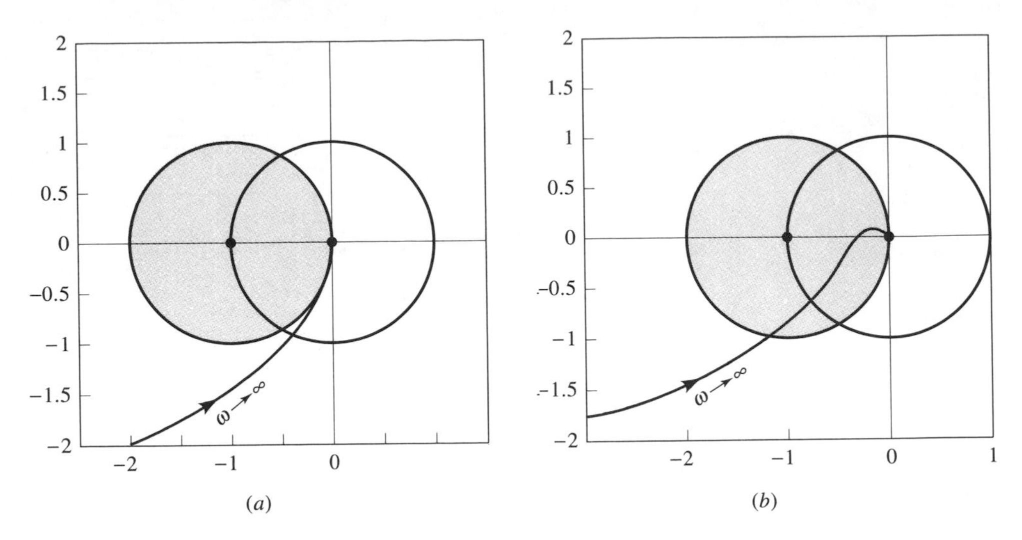

Figure 10.6 (a) Nyquist plot for LQR design. (b) Nyquist plot for LQG design. The unit circles centered at (0, 0) and (-1, 0) are shown.

You can experiment by changing the design parameters Q and R and the noise intensities, and you will observe that some parameters can have drastic effects on the system properties. But how does one choose these parameters? We will try to investigate this question in the subsequent sections.

10.4.1 Critique of LQG

Early pioneers of control, particularly H. W. Bode and I. M. Horowitz, studied and delineated most of the properties of feedback. In the early 1960s, with the birth of modern control, optimality and the design of optimal control systems became the dominant concern. The solution of the LQG problem was probably the highlight of this era. However, the LQG paradigm failed to meet the main objectives of control system designers. That is LQG control failed to work in real environments. The major problem with the LQG solution was lack of robustness. In a series of papers, researchers showed that LQG-based designs can become unstable in practice as more realism it added to the plant model. The same kinds of failure were also observed in industrial experiments with LQG. It became apparent that the main culprit was too much emphasis on optimality and not enough attention to the model uncertainty issue. During the 1980s, much of the attention was shifted back to feedback properties and frequency domain techniques (which were the main features of classical control), and their generalization to multivariable systems.

Section 10.6 discusses the *loop transfer recovery* (LQG/LTR) technique. This method maintains the LQG machinery but modifies the design procedure to address some of the shortcomings of the original LQG approach.

❑ DRILL PROBLEM

D10.3 Consider the following system (Doyle, J. and Stein, G., *IEET Trans. (Auto— Control,* August 1979).

$$\dot{x} = \begin{bmatrix} 1 & 1 \\ 0 & 1 \end{bmatrix} x + \begin{bmatrix} 0 \\ 1 \end{bmatrix} u + \begin{bmatrix} 1 \\ 1 \end{bmatrix} w$$

$$y = [1 \quad 0]x + v$$

Let the LQR parameters be $Q = qC'C$, $R = 1$, and let the filter parameters be $Q_o = q_o \begin{bmatrix} 1 \\ 1 \end{bmatrix} [1 \quad 1]$, $R_o = 1$. For each case given, compute gain and phase margins and draw the Nyquist plot overlaid on the unit circle centered at the origin.

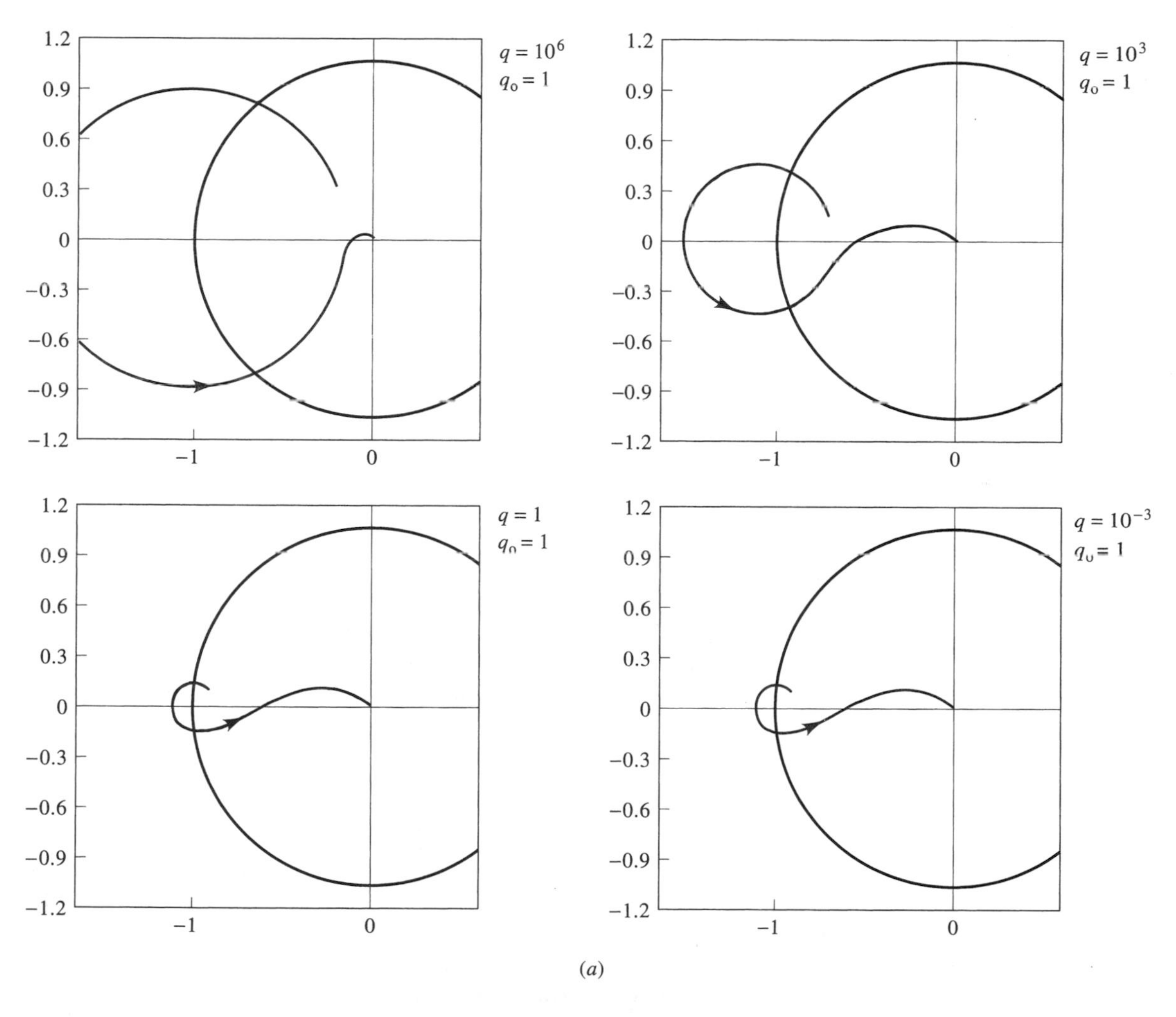

(*a*)

Figure D10.3

(a) $q = 10^6, 10^3, 1, 10^{-3}$ and $q_o = 1$

(b) $q_o = 10^6, 10^3, 1, 10^{-3}$ and $q = 1$

Ans. (a) $GM = \begin{bmatrix} -5.7 & -3.4 & -0.8 & -0.7 \\ 21.3 & 9.8 & 5.7 & 5.6 \end{bmatrix}$

$PM = \begin{bmatrix} -51 & -25.6 & -4.5 & -3.5 \\ 51 & 26.5 & 9.2 & 8.5 \end{bmatrix}$

(b) $GM = \begin{bmatrix} -1.2 & -1.2 & -0.8 & -0.8 \\ 49.7 & 20.7 & 5.7 & 5.3 \end{bmatrix}$

$PM = \begin{bmatrix} 19 & 17.7 & -4.5 & -4.6 \\ - & - & 9.2 & 8.6 \end{bmatrix}$

10.5 Robustness

The ultimate goal of a control system designer is to build a system that will work in real environment. Since the real environment may change with time (as components age or their parameters vary with temperature or other environmental conditions) or operating conditions may vary (load changes, disturbances), the control system must be able to withstand these variations. Assuming that the environment does not change, the second fact of life is the issue of model uncertainty. A mathematical representation of a system often involves simplifying and sometimes wishful assumptions. Nonlinearities are either unknown, hence unmodeled, or modeled and later ignored to simplify analysis. Different components of systems (actuators, sensors, amplifiers, motors, gears, belts, etc.) are sometimes modeled by constant gains, even though they may have dynamics or nonlinearities. Dynamic structures (e.g., aircrafts, satellites, missiles) have complicated dynamics in high frequencies, and these may initially be ignored. Since control systems are typically designed using much simplified models of systems, they may not work on the real plant in real environments.

The particular property a control system must possess to operate properly in realistic situations is called *robustness.* Mathematically this means that the controller must perform satisfactorily not just for one plant, but for a family (or set) of plants. Let us be more specific. Suppose the following plant is to be stabilized.

$$G(s) = \frac{1}{s-a}$$

It is suspected that the value of the parameter a is equal to 1, but this value could be off by 50%. If we design a controller that will stabilize the system for all values of $0.5 \leqslant a \leqslant 1.5$,we say the system has *robust stability.* If in addition the system is to satisfy performance specifications such as steady state tracking, disturbance rejection, and speed of response requirements, and the controller satisfies all requirements for all values of a in the specified range, we say the system possesses *robust performance.* The problem of designing controllers that satisfy robust stability and performance requirements is called *robust control.* This problem was investigated intensely during the 1980s and is still under investigation by many researchers following a variety of approaches. We will present a brief introduction to the robust control problem in the ensuring sections.

The underlying concept within control theory that has made it into a field of science is feedback. The study of feedback and its properties is responsible for the rapid growth of this field. What are these properties and why do we use feedback? The answer is that feedback has many properties that are discussed, either implicitly or explicitly, in this book. But there are two properties that a feedback system possesses that an open-loop system cannot have: sensitivity and disturbance rejection. By *sensitivity* it is meant that feedback reduces the sensitivity of the closed-loop system with respect to uncertainties or variations in elements located in the forward path of the system. *Disturbance rejection* refers to the fact that feedback can eliminate or reduce the effects of unwanted disturbances occuring within the feedback loop. It is mainly for these reasons that feedback is used. An open-loop system (i.e., a system with no feedback) does not have these properties. Of course, an open-loop system can also eliminate certain disturbances, but it requires full knowledge of the disturbance, which is not always available. Feedback is also used to stabilize unstable systems, but feedback itself is frequently the cause of instability. The stabilizing effects of feedback are emphasized so much in most texts that its other important properties are forgotten by beginning (or even experienced) students of control.

Two very important properties of feedback.

10.5.1 Feedback Properties

A feedback control system must satisfy certain performance specifications, and it must tolerate model uncertainties. We will study these issues from a frequency domain perspective. Consider the feedback system in Figure 10.7. The system has the following inputs:

$R(s)$ = command (or reference) input. This is the input that the system must be able to follow or track.

$D(s)$ = disturbance input. Disturbances are known or unknown inputs that the system must be able to reject. Disturbances may represent actual physical disturbances acting on the system such as wind gusts disturbing aircraft, disturbances owing to actuators such as motors, or uncertainties resulting from model errors in plant or actuator. Model uncertainties include neglected nonlinearities in plant or actuator and neglected or unknown modes in the system.

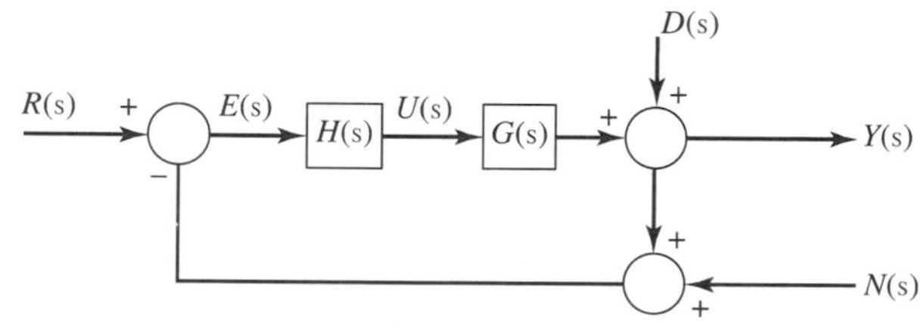

Figure 10.7 Block diagram of a feedback control system including disturbance and measured noise inputs.

$N(s)$ = sensor or measurement noise. This is introduced into the system via sensors, which are usually random high-frequency signals.

A properly designed control system must track reference inputs with small error and reject disturbance and noise inputs. The contribution of general disturbances to the output must be small. The total output of the closed-loop system is

$$Y(s) = \frac{G(s)H(s)}{1+G(s)H(s)}R(s) + \frac{1}{1+G(s)H(s)}D(s) - \frac{G(s)H(s)}{1+G(s)H(s)}N(s)$$

If we define the tracking error as $e = r - y$, we get

$$E(s) = \frac{1}{1+G(s)H(s)}R(s) - \frac{1}{1+G(s)H(s)}D(s) + \frac{G(s)H(s)}{1+G(s)H(s)}N(s)$$

Finally, the actuator output (i.e., the plant input) is given by

$$U(s) = \frac{H(s)}{1+G(s)H(s)}[R(s) - D(s) - N(s)]$$

Several quantities appear frequently in these relationships, they are

$$J(s) = 1 + G(s)H(s) \qquad \textit{return difference}$$

$$S(s) = \frac{1}{1+G(s)H(s)} = \frac{1}{J(s)} \qquad \text{sensitivity}$$

$$T(s) = \frac{G(s)H(s)}{1+G(s)H(s)} \qquad \textit{complementary sensitivity}$$

It can be seen that, for all frequencies, the following equality holds:

$$S(s) + T(s) = 1$$

Using the earlier definitions, we can write

$$Y(s) = S(s)D(s) + T(s)[R(s) - N(s)]$$
$$E(s) = S(s)[R(s) - D(s)] + T(s)N(s)$$
$$U(s) = H(s)S(s)[R(s) - D(s) - N(s)]$$

We are now ready to draw the following conclusions from these relations:

1. **Disturbance rejection:** $S(s)$ must be kept small to minimize the effects of disturbances. From the definition of S, this can be met if the loop gain (i.e., GH) is large.
2. **Tracking:** $S(s)$ must be kept small to keep tracking errors small.
3. **Noise suppression:** $T(s)$ must be kept small to reduce the effects of measurement noise on the output and errors. From the definition of T, this is met if the loop gain is small.
4. **Actuator limits:** $H(s)S(s)$ must be bounded to ensure that the actuating signal driving the plant does not exceed plant tolerances. Another reason for taking this relation into consideration is to reduce the control energy so that we can use smaller actuators (such as motors).

Tracking and disturbance rejection require small sensitivity. Noise suppression requires small complementary sensitivity. Because these two transfer functions add up to unity, we cannot reduce both transfer functions to zero simultaneously. We can, however, avoid this conflict by noticing that, in practice, command inputs and disturbances are low-frequency signals (i.e., they vary slowly with time), whereas measurement noise is a high-frequency signal. Therefore, we can meet both objectives by keeping S small in the low-frequency range and T small in high frequencies. The control energy constraint requires keeping HS small. Note that

$$H(s)S(s) = \frac{H(s)}{1 + G(s)H(s)} = \frac{T(s)}{G(s)}$$

Hence, by keeping T small we can reduce control energy. Putting together these effects, we arrive at a general desired shape for the open-loop transfer function (or loop gain) of a properly designed feedback system. This is shown in Figure 10.8. The general feature of this loop gain is that it has high gain at low frequencies (for good tracking and disturbance rejection) and low gain at high frequencies (for noise suppression). The intermediate frequencies typically control the gain and phase margins. Bode has shown that for a stable system, the slope of the magnitude plot should not exceed −40 dB/decade; that is, the transition from low- to high-frequency range must be smooth (e.g., −20 dB/decade). Desirable shapes for sensitivity and complementary sensitivity transfer functions are shown in Figure 10.9. Note that S must be small at low frequencies and roll off to 1 (0 dB) at high frequencies, whereas T must be at 1 (0 dB) at low frequencies and diminish at high frequencies. These properties are summarized in Table 10.1.

10.5.2 Uncertainty Modeling

The preceding performance specifications apply to a stable feedback system. As we discussed earlier, a stable system is not our final objective; rather, the stability must

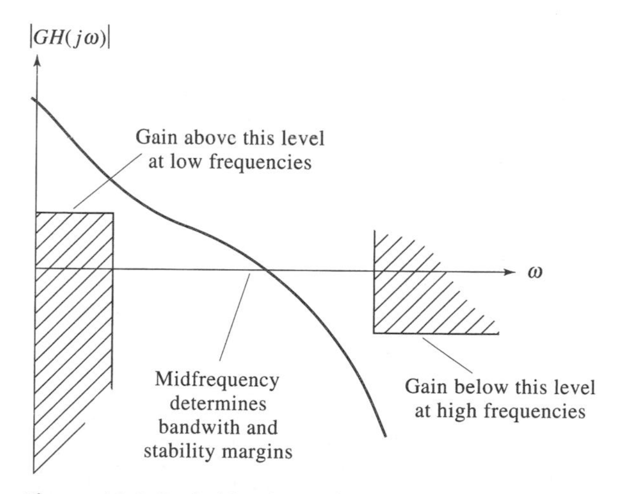

Figure 10.8 Desirable shape for the open-loop transfer function of a feedback system.

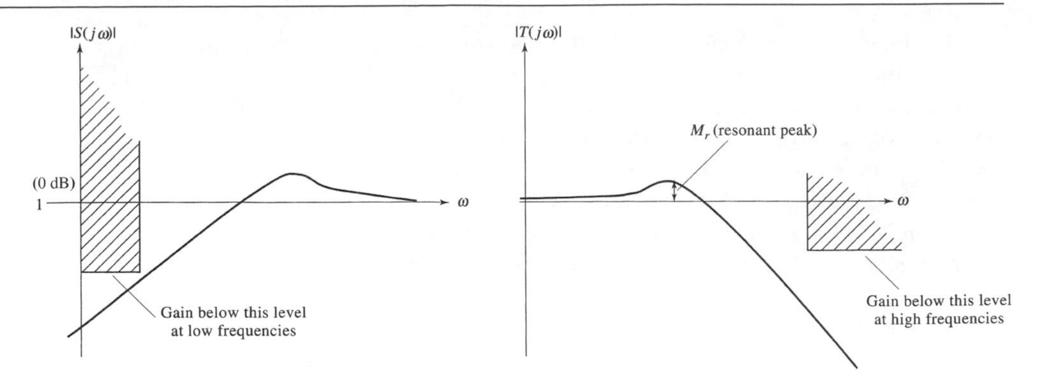

Figure 10.9 Desirable shape for the sensitivity and complementary sensitivity of a feedback system.

Table 10.1 *Loop Transfer Function Properties*

	Low Frequency	Mid Frequency	High Frequency
Performance (R)	High gain	Smooth transition	
Disturbance rejection (D)	High gain		
Noise suppression (N)			Low gain

be maintained despite model uncertainties. Model uncertainty is generally divided into two categories: *structured uncertainty* and *unstructured uncertainty*. Structured uncertainty assumes that the uncertainty is modeled and we have ranges and bounds for uncertain parameters in the system. For example, we may have a valid transfer function model of a system but have some uncertainty about the exact location of the poles, zeros, or gain of the system. In the case of an *RLC* circuit we know that it can be adequately modeled by a second-order transfer function (in a given frequency range), but the components may have up to 20–30% tolerance. These kinds of uncertainties are structured. Unstructured uncertainties assume less knowledge of the system. We only assume that the frequency response of the system lies between two bounds. Both kinds of uncertainties are usually present in most applications. We will discuss only unstructured uncertainty. (Dealing with structured uncertainty is still under investigation; owing to the complexity of the problem and space limitations, we will not discuss this case.)

Unstructured uncertainty can be modeled in different ways. We will discuss *additive* and *multiplicative* uncertainty. Suppose we model a system by $G(s)$, where the actual system is given by $\tilde{G}(s)$—That is,

$$\tilde{G}(s) = G(s) + \Delta_a(s)$$

Therefore, the model error, or the additive uncertainty, is given by

$$\Delta_a(s) = \tilde{G}(s) - G(s)$$

In the multiplicative uncertainty case, we assume the true model, $\tilde{G}(s)$ is given by

$$\tilde{G}(s) = [1 + \Delta_m(s)]G(s)$$

The uncertainty, or the model error, is given by

$$\Delta_m(s) = \frac{\tilde{G}(s) - G(s)}{G(s)}$$

Block diagram representations of these uncertainty models are shown in Figure 10.10. Because multiplicative uncertainty represents the relative error in the model, whereas the additive model represents absolute error, the multiplicative model is used more often.

As an example, consider the flexible spacecraft example in Section 4.10. The nominal plant model consists of the rigid mode, and is given by $G(s)$. The true plant model, $\tilde{G}(s)$, must also include the flexible mode.

$$G(s) = \frac{2}{s^2} \quad \tilde{G}(s) = \frac{s^2 + 2s + 2}{s^2(s^2 + s + 1)}$$

Modeling the flexible mode as additive uncertainty, we get

$$\Delta_u(s) - \tilde{G}(s) - G(s) - \frac{-1}{s^2 + s + 1}$$

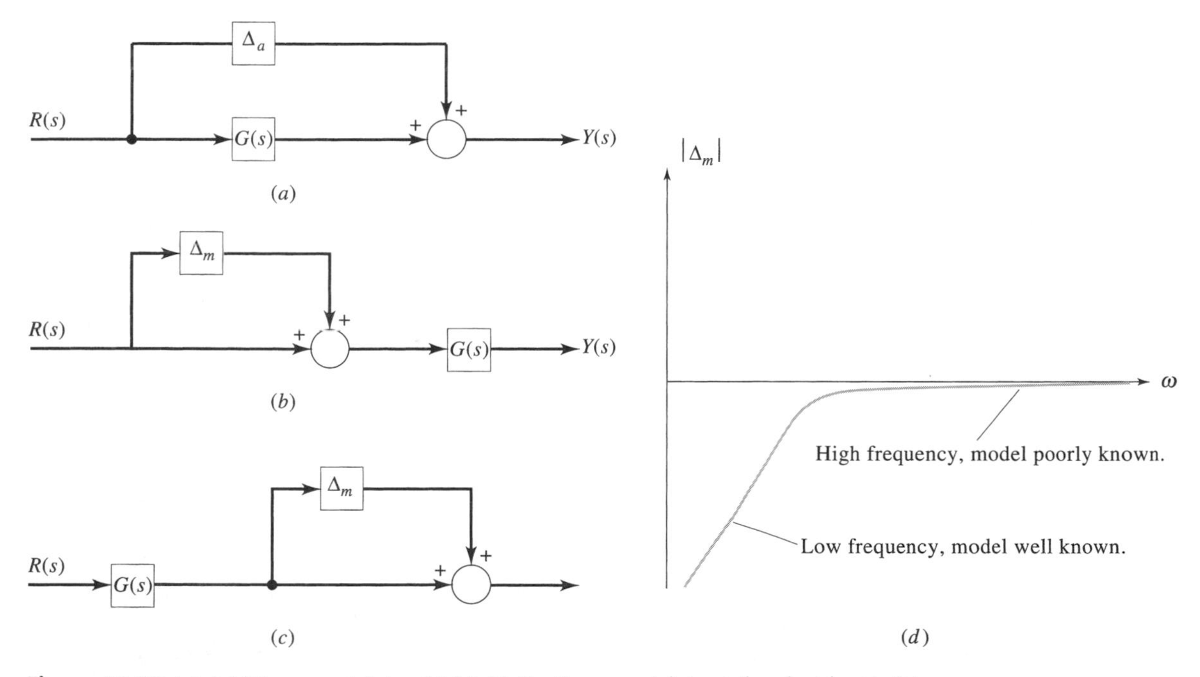

Figure 10.10 (a) Additive uncertainty. (b) Multiplicative uncertainty at the plant input. (c) Multiplicative uncertainty at the plant output. (d) Typical shape for multiplicative uncertainty.

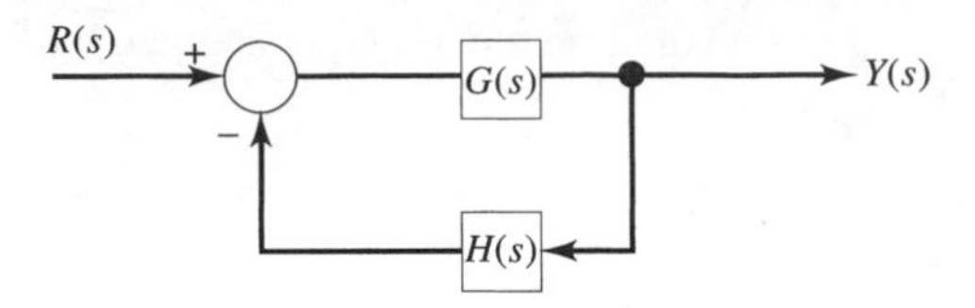

Figure 10.11 Block diagram of a feedback control system.

Using the multiplicative model, we have

$$\Delta_m(s) = \frac{\tilde{G}(s) - G(s)}{G(s)} = \frac{-s^2}{2(s^2 + s + 1)}$$

10.5.3 Robust Stability

Consider a feedback system containing a plant and a compensator. Suppose the compensator stabilizes the nominal plant model $G(s)$. We say that the compensator *robustly stabilizes* the system if the closed-loop system remains stable for the true plant $\tilde{G}(s)$. Most of the results and conditions for robust stability can be derived from variations of the Nyquist stability criterion or the following very powerful result, called the *small-gain theorem.*

Small-Gain Theorem

Consider the feedback system in Figure 10.11. Assume the plant and the compensator are stable. Then the closed-loop system will remain stable if

$$|G(s)H(s)| < 1$$

Also, because of the following inequality

$$|G(s)H(s)| \leqslant |G(s)|\,|H(s)|$$

We can also guarantee closed-loop stability if

$$|G(s)|\,|H(s)| < 1$$

In essence, the small-gain theorem states that for closed-loop stability, the loop gain must be small. The Nyquist stability criterion can be used to justify the validity of this theorem. Because we are requiring the open-loop transfer function to be inside the unit circle, there can be no encirclements of the $(-1,\ 0)$ point. In addition, we are assuming that the system is open-loop stable; it follows from the Nyquist stability criterion that the system has no closed-loop RHP poles and is therefore closed-loop stable. We should also add that the small-gain theorem guarantees internal stability, so all possible closed-loop transfer functions are stable and all internal signals will remain bounded for bounded inputs.

We can use the small-gain theorem to answer two kinds of question about robust stability. First, given that the uncertainty is stable and bounded, will the closed-loop system be stable for the given uncertainty? Second, for a given system, what is the smallest uncertainty that will destabilize the system? To use the small-gain theorem, it is helpful to convert our system block diagram to a two-block structure, shown in

Figure 10.11. Let us now derive the condition for robust stability under multiplicative uncertainty. Consider the feedback system shown in Figure 10.12(a). To obtain the two-block structure in Figure 10.11, we need to find the transfer function seen by the uncertainty block. The input and output of this transfer function are shown at the indicated points in Figure 10.12(b). It is given by [see Figure 10.12(c)]

$$M(s) = \frac{-G(s)H(s)}{1 + G(s)H(s)}$$

By the small-gain theorem, if the transfer function and the uncertainty transfer function are stable, the closed-loop system will be robustly stable if

$$|\Delta_m| < \frac{1}{\left|GH(1 + GH)^{-1}\right|}$$

Condition for robust stability.

Observe that the denominator of the right-hand side of the inequality is the complementary sensitivity, T, so the robust stability condition becomes

$$|\Delta_m| < \frac{1}{|T|}$$

We can use this result to answer the two questions posed earlier. Suppose the stable uncertainty is bounded by

$$|\Delta_m| < \gamma$$

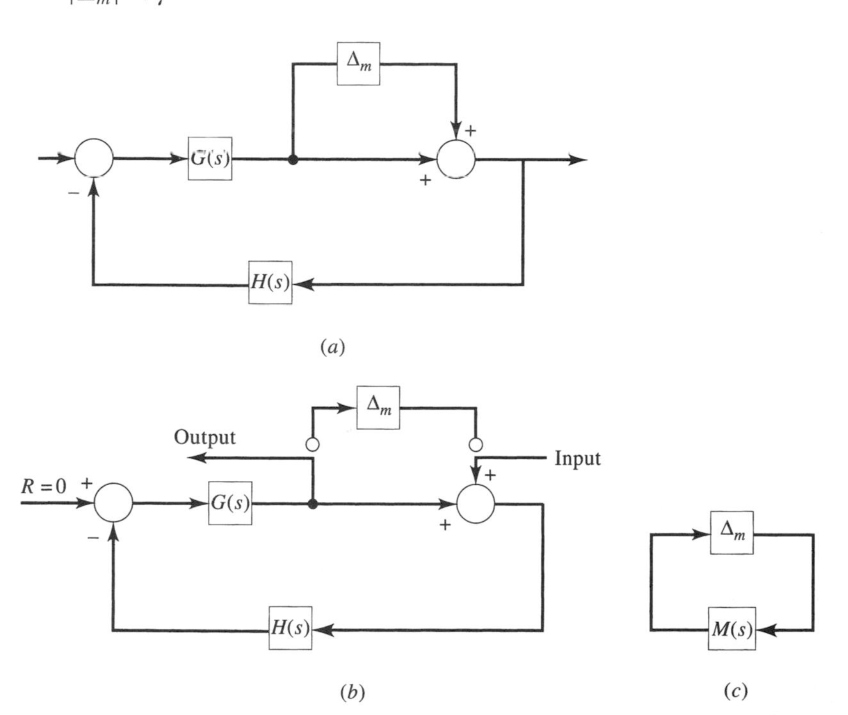

Figure 10.12 (a) Feedback system with multiplicative uncertainty. (b) Obtaining the transfer function seen by the uncertainty. (c) The system as seen by the uncertainty.

Then the closed-loop system will be stable if

$$|T| < \frac{1}{\gamma} \quad \text{or} \quad |\gamma T| < 1$$

To answer the second question, we are interested in finding the size of the smallest stable uncertainty that will destabilize the system. Because the uncertainty must be smaller than $1/T$, it must be smaller than the minimum of $1/T$. To minimize the right-hand side of the inequality, we must maximize T. The maximum of T over all frequencies is its peak value (also called the *resonant peak* in second-order systems: see Figure 10.9). Hence, the smallest destabilizing uncertainty (we call this the *multiplicative stability margin* or MSM) is given by

$$\text{MSM} = \frac{1}{M_r}$$

where

$$M_r = \sup_{\omega} |T(j\omega)|$$

and the symbol "sup" stands for the *supremum* of the function. The supremum (or least upper bound) of a function is its maximum value, even if it is not attained. This is needed for mathematical reasons. We frequently encounter transfer functions that have no maximum. For instance, the following transfer function (a lead network)

G(s) has no maximum, but it has a supremum.

$$G(s) = \frac{s+1}{s+5}$$

has no maximum (if you take the derivative of its magnitude and set it equal to zero, you will get the minimum value of 0.2). However, a glance at its frequency response shows that it approaches the value of 1 as the frequency approaches infinity. But because we never reach the infinite frequency, we never reach the maximum value (although we get very close to it). That is why it does not have a maximum. In these situations, we use the notion of the supremum (or sup for short). We have

$$\sup_{\omega} \frac{|j\omega + 1|}{|j\omega + 5|} = 1$$

The condition for robust stability under additive uncertainty modeling can be derived using the same approach. The transfer function seen by the uncertainty, in this case, is given by (see Figure 10.13)

$$M(s) = \frac{-H(s)}{1 + G(s)H(s)}$$

Hence, the closed-loop system will be robustly stable if

Condition for robust stability (additive case).

$$|\Delta_a| < \frac{1}{\left|H(1 - GH)^{-1}\right|} \quad \text{or} \quad |\Delta_a| < \frac{1}{|HS|}$$

If the uncertainty is stable and bounded by

$$|\Delta_a| < \gamma$$

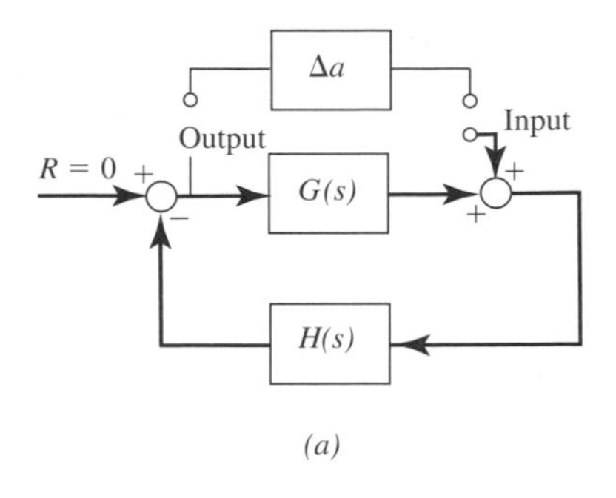

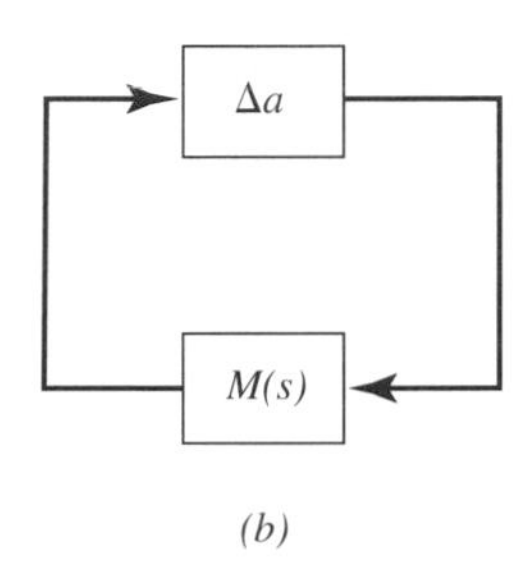

Figure 10.13 (a) Obtaining transfer function seen by the additive uncertainty. (b) The system as seen by the uncertainty.

then we can guarantee closed-loop stability if

$$|HS| < \frac{1}{\gamma} \qquad \text{or} \qquad |\gamma HS| < 1$$

We can also define the *additive stability margin* (ASM) by

$$\text{ASM} = \frac{1}{\sup_{\omega} |H(j\omega)S(j\omega)|}$$

Note that for increased protection against destabilizing multiplicative uncertainties, MSM must be large, implying that the complementary sensitivity must be small. This is compatible with good noise suppression but conflicts with tracking and disturbance rejection. Therefore, small loop gain at high frequencies will protect against multiplicative uncertainties in the high-frequency range. Similarly, observe that the appropriate transfer function for ASM is the same transfer function that determines control energy (actuator limits). Therefore, these requirements are compatible.

Let us apply these results to an example. Consider the following plant and compensator (the compensator is in the feedback path).

$$G(s) = \frac{(5 - s)}{(s + 5)(s^2 + 0.2s + 1)} \qquad \text{and} \qquad H(s) = \frac{5(s + 0.1)}{s}\frac{s + 0.2}{s + 5}$$

The open-loop Bode plot of the system is shown in Figure 10.14. The system has a phase margin of 38° and a gain margin of 9 dB. This means that a phase lag of 38° or a gain increase factor of 2.8 (9 dB) will destabilize the system. Let us compute the ASM and MSM for the system. For the MSM, we need to obtain the complementary sensitivity and find its peak value. The plot is shown in Figure 10.15. The peak value is 1.52, resulting in an MSM of 0.65. The interpretation of MSM is the following: the

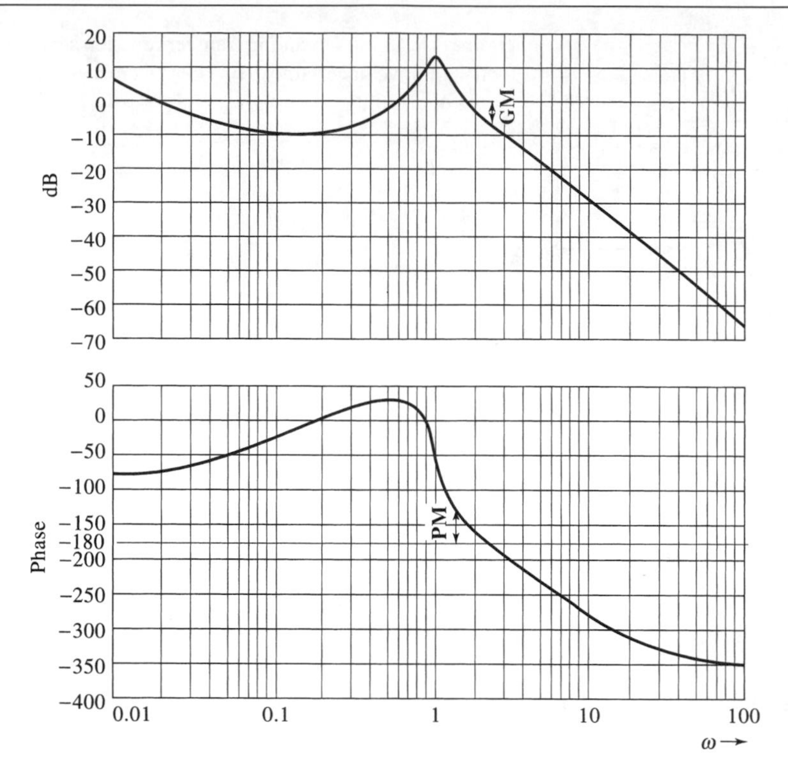

Figure 10.14 Bode plots of $G(s)H(s)$.

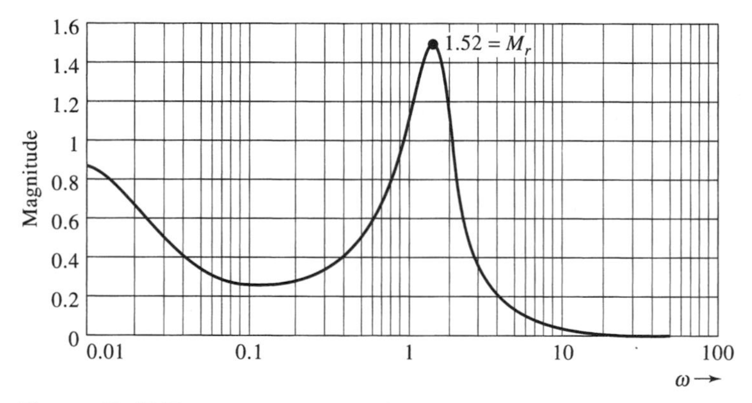

Figure 10.15 Frequency response of the complementary sensitivity for determining the MSM.

system will be robustly stable against unmodeled multiplicative uncertainties with transfer function magnitudes below 0.65. Two points need to be emphasized:

1. The uncertainty can be *any* stable transfer function, provided its magnitude is below our bound.

2. The small-gain theorem is only a sufficient condition (i.e., even if it is violated, the system can still be stable). These stability margins (ASM, MSM) are sometimes very conservative. Hence, the system may be able to tolerate uncertainties that violate the bounds.

To check the conservativeness of the MSM, we modeled the uncertainty by a first-order transfer function and varied its gains. With

$$\Delta_m(s) = \text{MSM}\frac{1 + 0.2k}{s + 1}$$

for $k = 0$, the gain of the uncertainty is the upper limit guaranteed by our theory. We varied k from 0 to 10 in steps of 2, and it was discovered that the system becomes unstable for $k = 8$ or

$$\Delta_m(s) = \frac{1.7}{s + 1} \quad \text{and} \quad 1 + \Delta_m(s) = \frac{s + 2.7}{s + 1}$$

The step responses for these values of k (along with the nominal system (i.e., with no uncertainty) are shown in Figure 10.16. The figure shows that the uncertainty causes oscillations that will eventually lead to instability. Note that the transfer function that is actually in series with the plant is $[1 + \Delta_m(s)]$, which has a maximum destabilizing gain of 2.7. Now, you may wonder why the system is unstable for a gain of 2.7 even though it has gain margin of 2.8. To answer this question, we must be specific about the meaning of *gain*, *phase*, and *multiplicative stability margins*. The gain margin is the factor by which the gain can be increased before instability occurs. This assumes no phase change, which implies that the gain margin is a measure of tolerance of pure gain uncertainty. Likewise, the definition of phase margin assumes that the gain is fixed, so phase margin is a measure of tolerance of pure phase uncertainty. MSM, however, allows simultaneous gain *and* phase changes. For example, the gain and phase of $[1 + \Delta_m(s)]$ near the gain phase crossover frequencies are:

At the gain crossover frequency: $\omega = 1.57$, gain $= 1.67$, phase $= -27$
At the phase crossover frequency: $\omega = 2.8$, gain $= 1.30$, phase $= -24$

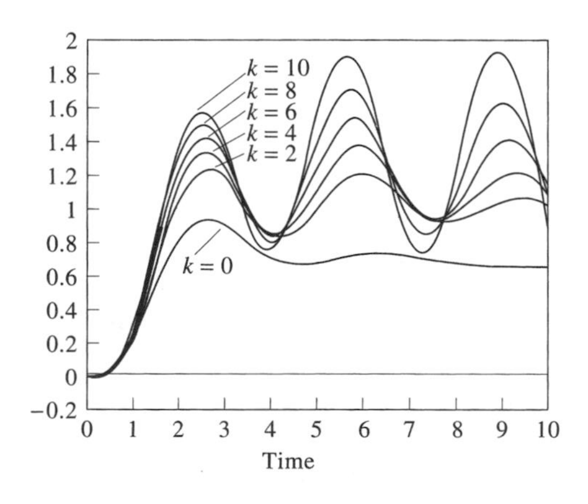

Figure 10.16 Step responses for the uncertain (or perturbed) system.

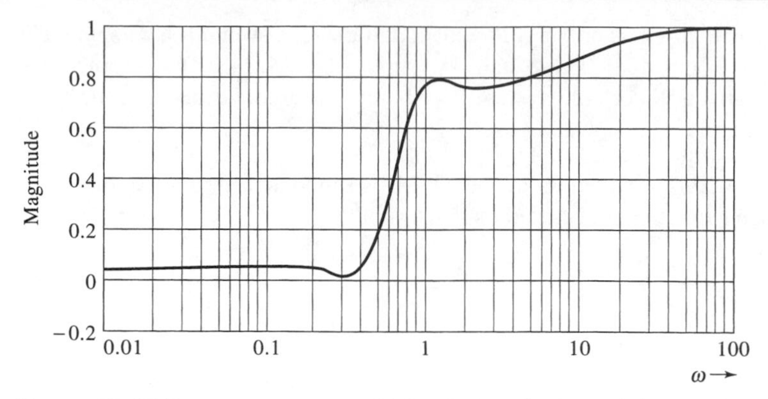

Figure 10.17 Frequency response of $0.2\,H(s)S(s)$ for determining additive robust stability.

Therefore, at the gain crossover frequency, the uncertainty introduces a phase lag of 27° (the phase margin is 38) in addition to a gain increase of 1.67. Also, at the phase crossover frequency, it multiplies the gain by 1.3, and adds 24° of phase lag. That is why it is destabilizing: it is adding both gain and phase lag near the critical frequencies of the system. In a sense, MSM is more general than gain and phase margins. For this reason, MSM is sometimes called *gain-phase margin*.

We can also study the system tolerance to additive uncertainty. We ask whether the system can withstand additive uncertainty transfer functions with magnitude less then $0.2(\gamma = 0.2)$. To answer this question, we obtain the frequency response of HS, and robust stability is guaranteed if

$$|0.2H(j\omega)S(j\omega)| < 1 \qquad \text{for} \qquad |\Delta_a(s)| < 0.2$$

Figure 10.17 shows the response. Because, its peak is less than 1, we conclude that the system is stable in the robust sense, and ASM = 1.

❑ DRILL PROBLEM

D10.4 Consider the double-integrator plant compensated by a feedback lead compensator.

$$G(s) = \frac{1}{s^2} \qquad H(s) = \frac{20(s+1)}{s+10}$$

Suppose the actual plant model contains an additive uncertainty given by

$$\Delta_a(s) = \frac{-1}{s^2 + s + 1}$$

(a) Determine if the compensator $H(s)$ stabilizes the plant $G(s)$.

(b) Determine if the compensator $H(s)$ stabilizes the actual plant given by $G_a(s) = G(s) + \Delta_a(s)$.

(c) Find $M(s)$, the transfer function "seen" by the uncertainty.

(d) Draw Bode plots of $\Delta_a(s)$ and $M(s)$ to determine the robust stability of the system.

Ans. (a) characteristic equation $= s^3 + 10s^2 + 20s + 20$, stable;
(b) characteristic equation $= s^5 + 11s^4 + 11s^3 + 50s^2 + 80s + 40$, unstable;
(c) $M(s) = [20s^2(1 + s)]/(s^3 + 10s^2 + 20s + 20)$;
(d) system not robustly stable

10.6 Loop Transfer Recovery (LTR)

It was discussed earlier that the LQR solution has excellent stability margins (infinite gain margin and 60° phase margin); we know that LQR is usually, but not always, considered impractical because it requires that all states be available for feedback. Doyle and Stein showed that under certain conditions, the LQG can asymptotically recover the LQR properties. One of the problems with LQG is that it requires statistical information of the noise processes. In most cases, however, this information is not available or is impractical to obtain. Mathematical arguments and simulations had shown that the LQG design parameters (Q, R, Q_o, and R_o) have a strong influence on the performance of the system. It was suggested that because Q_o, and R_o are not usually available, they should be used as tuning parameters to improve system performance.

Consider the block diagrams in Figure 10.18. With the loop broken at the indicated point, the (open) loop transfer function of the LQR is given by

LQR.

$$L(s) = K_{\Phi(s)}B$$

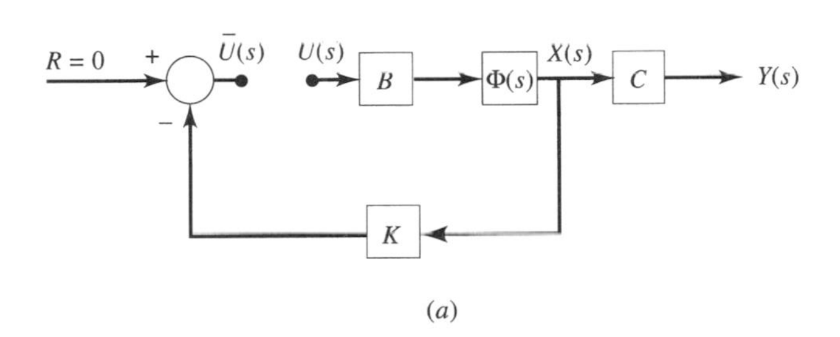

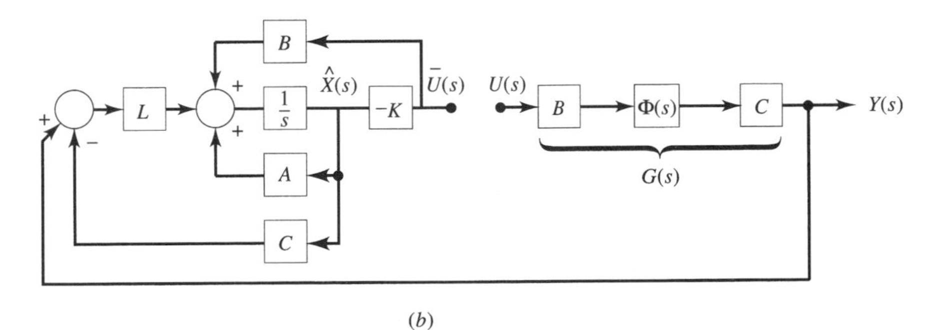

Figure 10.18 (a) Block diagram of an LQR controller. (b) Block diagram of an LQG controller.

where

$$\Phi(s) = (sI - A)^{-1}$$

The loop transfer function for LQG is likewise given by

LQG.

$$\underset{LQG}{L(s)} = K(sI - A + BK + LC)^{-1} LC\Phi(s)B$$

Under the following two conditions

LTR assumptions.

$G(s)$ is minimum-phase (i.e., it has no zeros in the RHP)
$R_o = 1$ and $Q_o = q^2 BB'$

it can be shown that

$$\lim_{q\to\infty} \underset{LQG}{L(s)} = L(s)$$

The following procedure for design is suggested by the foregoing conditions. Choose the LQR parameters such that the LQR loop transfer function (also called the *target feedback loop* or *TFL*) has desirable time and/or frequency domain properties. Design an observer with parameters specified in condition 2. Increase the tuning parameter q until the resulting loop transfer function is as close as possible to the TFL.

In many situations, the variable that is measured is different from the variable we want to control. For example, we may desire to control thrust in a jet engine, but we can sense only temperature and turbine speed. Let y denote the measured states, and z denote the controlled states, then

$$y = Cx \quad \text{and} \quad z = C_q x$$

Because the loop transfer function of LQG approaches that of LQR, it will asymptotically recover its properties. A more detailed procedure follows.

Loop Shaping Step

1. Determine the controlled variable and set

$$Q = C'C \quad \text{or} \quad Q = C_q' C_q$$

2. Convert the design specifications into a desired TFL. At this stage, if the system is type 0 and we want a type 1 system, we can add an integrator to the system.
3. Vary the parameter R until the resulting loop transfer function is similar to the TFL. One may use the RSL approach here. Also, check the sensitivity and complementary sensitivity transfer functions (S and T), to make sure they have desirable shapes.

Recovery Step

4. Select a scalar, q, and solve the filter Riccati equation

$$A\Sigma + \Sigma A' + q^2 BB' - \Sigma C' C \Sigma = 0 \quad \text{and set } L = \Sigma C'$$

Increase q until the resulting loop transfer function is close to the TFL.

The higher the value of q, the closer the LQG system comes to LQR performance. It should be noted that the value of q should not be increased indefinitely, because this may lead to unreasonably large values for the filter gain L. Also, because LQR has -20 dB slope at high frequencies, large values for q will also recover this slow roll-off rate. Smaller values for q will tend to trade off lower stability margins with higher roll-off rates at high frequencies.

LTR solution of the double integrator system.

We will now use LTR on the double-integrator system to recover the LQR properties. Because LTR requires solving the Riccati equation a number of times, the problem must be solved on the computer. First, we chose the LQR loop transfer function as the TFL. Therefore, our objective is to recover the Bode plots shown in Figure 10.1. We next let the parameter q vary over the range (1, 10, 100, 1,000). The plots for the closed-loop step response and open-loop Bode plots for the LQR case and LTR, for the specified values of q, are shown in Figure 10.19.

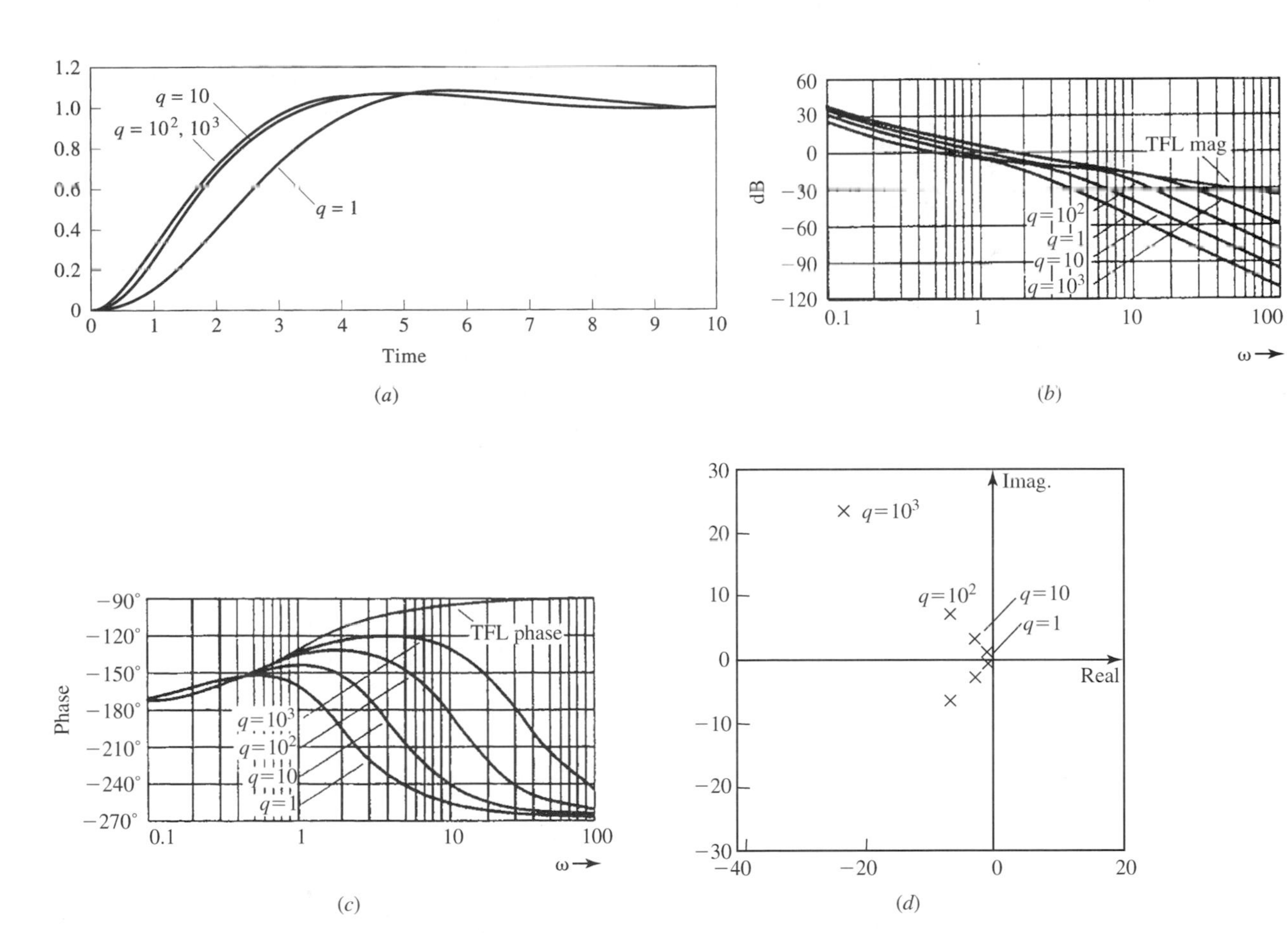

Figure 10.19 Step response, Bode plots, and filter oples for LTR using $q = (1, 10, 100, 1000)$. (a) Closed-loop step response. (b) Open-loop magnitude Bode plot.(c) Open-loop phase Bode plot. (d) Filter poles.

Table 10.2 *Results of LTR Design*

q	1	10	100	1000
PM	32.6	41.9	55.0	61.7
GM	9.5	13.0	21.1	30.4
L	1.4	4.5	14.1	44.7
	1.0	10.0	100.0	1000.0
Filter	$-0.7+0.7j$	$-2.2+2.2j$	$-7.0+7.0j$	$-22.3+22.3j$
poles	$-0.7-0.7j$	$-2.2-2.2j$	$-7.0-7.0j$	$-22.3-22.3j$

Note how the step response approaches the LQR case for increasing values of q. Also, as q increases, the low frequency gain of the system goes from 28 dB to 40 dB while the high frequency gain goes from -110 dB to -40. The values for the filter gain L, its eigenvalues, and the stability margins (*GM* and *PM*) are given in Table 10.2. The data show that the LQR phase margin is recovered. The gain margin increases from 10 dB to 30 dB. This can clearly be increased by increasing q. But note that increasing the margins will cost us in terms of higher values for the filter gain L, higher gain crossover frequency, and lower high-frequency gain. This will make the system more sensitive to noise and uncertainties at high frequencies. It appears that a value of q between 100 to 1000 is a reasonable compromise.

Note that the procedure uses the machinery of LQG (i.e., two Riccati equations) and its guaranteed stability. However, it allows us to work strictly with Bode plots of various transfer functions and to satisfy frequency domain measures (similar to classical control). Therefore, it can be considered a frequency domain design procedure that uses state-space equations for computation. This is the common feature of control system design after LQR/LQG, sometimes called *postmodern control* (i.e., frequency domain techniques that use state space machinery for computation).

❑ DRILL PROBLEM

D10.5 Consider the following system:

$$\dot{x} = Ax + Bu + \omega$$

$$y = cx + v$$

$$A = \begin{bmatrix} 0 & 1 \\ -3 & -4 \end{bmatrix} \quad B = \begin{bmatrix} 0 \\ 1 \end{bmatrix} \quad c = [2 \quad 1]$$

Let $Q = E'E$ where $E = 4\sqrt{5}[\sqrt{35} \quad 1]$

$$R = 1$$

$$Q_o = \begin{bmatrix} 35 \\ -61 \end{bmatrix} [35 \quad -61]$$

$$R_o = 1$$

(a) Find an LQR compensator $H(s)$. Compute K, the closed-loop poles, and the gain and phase margins.

(b) Design an LQG compensator. Compute the gain and phase margins in this case. Compare with (a).

(c) Design an LTR compensator by solving the following Riccati equation:

$$A\Sigma + \Sigma A' + (Q_o + q^2 BB') - \Sigma C' C \Sigma = 0$$

Increase the value of q to 10^6 and observe the effects. Compute the gain and phase margins for $q = 100$.

Ans. (a) $K = [50 \quad 10], -7 \pm j2, \; GM = \infty, \; PM = 85°$;

(b) $PM < 15°, \; GM = 6.7$ dB;

(c) $PM = 74°, \; GM = 37$ dB

10.7 H_∞ Control

10.7.1 A Brief History

One of the major challenges in control has been the analysis and design of multivariable (multiple-input, multiple-output or MIMO) control systems. This is a difficult problem, because the transfer function of a MIMO system is a matrix. Even very basic concepts such as system order, poles, and zeros run into difficulty in this case. For instance, there are at least five to ten different definitions of zeros of a multivariable system! Successful concepts and tools of classical control such as root locus, Bode plots, Nyquist stability criterion, and gain and phase margins ran into difficulty. State space techniques, based in the time domain, avoided the complexities of transfer function matrices and provided tools for analysis and design of MIMO systems. Within the state space framework, the only difference between a SISO system and a MIMO system is the number of columns of the B matrix (number of inputs) and the number of rows in the C matrix (number of outputs). Note that in all the techniques we have discussed, these dimensions play no part. In fact, the most important feature of LQR/LQG is that they are systematic methods for designing MIMO systems.

At the about the same time that most researchers were developing, extending, and refining time domain optimal control methods. Other researchers, mostly in Britain, (A. G. J. MacFarlane and H. H. Rosenbrock), were busy extending classical control tools to the multivariable case. They were largely successful in these endeavors. Classical tools such as root locus (renamed *characteristic locus*), Nyquist techniques (renamed *Nyquist arrays*), and Bode plots (renamed *singular value plots*) were extended to the multivariable case. As the shortcomings of LQG methods became more apparent in the 1970s, more attention was paid to classical control concepts and concerns.

During the 1980s a new paradigm emerged, H_∞ control. This control problem was first formulated by G. Zames. It was essentially a frequency domain optimization method for designing robust control systems. Robustness became the main concern in the control community, and other techniques for designing robust multivariable control systems soon followed. They are H_∞ control, μ synthesis (by J. Doyle, also called k_m synthesis by M. Safonov), QFT (quantitative feedback theory) by I. Horowitz, and methods based on Kharitonov's theorem for structured uncertainty. All these techniques are still being developed and refined today.

Our purpose in this section is to present a brief introduction to H_∞ control. Although this is a powerful technique for the MIMO case, our presentation is limited to the SISO case. The transition to the MIMO case is straightforward in theory but not necessarily in practice.

10.7.2 Some Preliminaries

H_∞ control has developed its own terminology, notation, and paradigm. For example, the classical block diagram has been modified to handle problems of more general types. Also, because the design equations are very lengthy, some shorthand notation is introduced to simplify the presentation. Because these notations have become standard in the literature, and because they could be confusing to the novice, we will introduce and use them in this discussion to ease the transition to more advanced books and the literature for the readers.

We first discuss the name. H_∞ refers to the space of stable and proper transfer functions. We generally desire that the closed-loop transfer functions be proper (i.e., the degree of the denominator $\geqslant$ the degree of the numerator) and stable (poles strictly in the LHP). Instead of repeating these requirements, we say $G(s)$ is in H_∞. The basic object of interest in H_∞ control is a transfer function. In fact, we will be optimizing over the space of transfer functions. Optimization presupposes a cost (or objective) function, because we want to compare different transfer functions and choose the best one in the space. In H_∞ control, we compare transfer functions according to their H_∞ norm (a mathematical term for the concept of size). The H_∞ norm of a transfer function is defined by

Measuring size of transfer functions.

$$\|G\|_\infty = \sup_\omega |G(j\omega)|$$

This is easy to compute graphically: it is simply the peak in the Bode magnitude plot of the transfer function (it is finite when the transfer function is proper with no imaginary poles). We have already seen this quantity before in the section on robust stability. For instance, the multiplicative stability margin (MSM) can be written

$$\text{MSM} = \frac{1}{\|T\|_\infty}$$

As an example

$$\left\|\frac{1}{s+1}\right\|_\infty = 1$$

In H_∞ control, the objective is to minimize the H_∞ norm of some transfer function, so we will try to minimize the peak in the Bode magnitude plots.

A notation that is rapidly becoming popular is the *packed-matrix* notation for representing transfer functions in state space. Recall that the transfer function of a system with state-space matrices $\{A, B, C, D\}$ is given by

$$G(s) = C(sI - A)^{-1}B + D$$

This transfer function in packed-matrix notation is written

$$G(s) = \left[\begin{array}{c|c} A & B \\ \hline C & D \end{array}\right]$$

New notation: packed-matrix.

We emphasize that this is not a matrix in the ordinary sense; it is just a shorthand notation for the foregoing expression for $G(s)$. For example, the transfer function of an LQG compensator, given in Section 10.4, can be expressed as follows:

$$H(s) = \left[\begin{array}{c|c} A - BK - LC & L \\ \hline K & 0 \end{array}\right]$$

The solution to the H_∞ control problem contains very messy Riccati equations, so the following notation is introduced to simplify solution representation. Consider the following Riccati equation:

$$A'X + XA - XRX + Q = 0$$

The stabilizing solution of this equation will be denoted by $X = \text{Ric}\ (H)$, where H is the following *Hamiltonian* matrix:

The Ric notation.

$$H = \begin{bmatrix} A & R \\ -Q & -A' \end{bmatrix} \quad \text{and} \quad (A - RX) \text{ is stable}$$

Instead of writing the Riccati equation, we will specify its associated Hamiltonian matrix and the reader can create the appropriate Riccati equation.

Finally, we introduce a more general block diagram representation of control systems shown in Figure 10.20. This new diagram is able to represent a variety of problems of interest. The diagram contains two main blocks, the plant and the controller. The plant section has two inputs and two outputs. The plant inputs are classified as control inputs and exogenous inputs. The control input u is the output of the controller, which becomes the input to the actuators driving the plant. The exogenous input w is actually a collection of inputs (a vector). The main distinction between w and u is that the controller cannot manipulate these exogenous inputs. Typical inputs that are lumped into w are external disturbances, noise from the sensors, and tracking or command signals.

The plant outputs are also categorized in two groups. The first group, y, are signals that are measured and fed back. These become the inputs to the controller. The second group, z, are the regulated outputs. These are all the signals we are interested to control or regulate. They could be states, error signals, and control signals. Even if the original

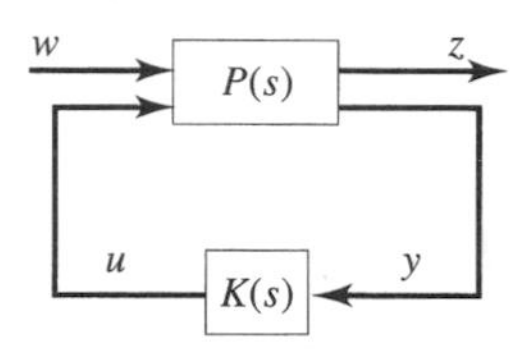

Figure 10.20 Generic block diagram for H_∞ control.

system is SISO (i.e., u and$y$ are scalars), the new formulation is essentially MIMO. Most realistic control system problem formulations are of the MIMO kind.

A transfer function representation of the system is given by

$$z = P_{zw}w + P_{zu}u$$
$$y = P_{yw}w + P_{yu}u$$
$$u = Ky$$

(Note the change in notation: although the inputs and outputs are s-domain quantities, we write them in lowercase letters and omit the dependency on s; as a rule we will use lowercase italic letters for scalars and vectors, and capital letters for matrices in our presentation.)

The closed-loop transfer function between the regulated outputs and the exogenous inputs is obtained as follows. First, we substitute for u in the equation for y.

$$y = P_{yw}w + P_{yu}Ky$$

then, we solve for y (note that all capital letters are matrices, so we have to use matrix inversion and watch for the order of multiplication)

$$(I - P_{yu}K)y = P_{yw}w \rightarrow y = (I - P_{yu}K)^{-1}P_{yw}w$$

Therefore, u becomes

$$u = Ky = K(I - P_{yu}K)^{-1}P_{yw}w$$

Substituting this into the equation for z, we get

$$z = P_{zw}w + P_{zu}K(I - P_{yu}K)^{-1}P_{yw}w = [P_{zw} + P_{zu}K(I - P_{yu}K)^{-1}P_{yw}]w$$

Finally

The linear fractional transformation.

$$z = T_{zw} \qquad \text{where } \mathrm{T}_{zw} = P_{zw} + P_{zu}K(I - P_{yu}L)^{-1}P_{yw}$$

This expression for the closed-loop transfer function of P and K is called the *linear fractional transformation.*

The plant can also be represented in state-space form as follows:

$$\dot{x} = Ax + B_1w + B_2u$$
$$z = C_1x + D_{11}w + D_{12}u$$
$$y = C_2x + D_{21}w + D_{22}u$$

Using the packed-matrix notation, we get

New plant representation.

$$P(s) = \left[\begin{array}{c|cc} A & B_1 & B_2 \\ \hline C_1 & D_{11} & D_{12} \\ C_2 & D_{21} & D_{22} \end{array}\right]$$

10.7.3 H_∞ Control: Solution

The H_∞ control problem is formulated as follows. Consider the generic block diagram in Figure 10.20 and find an internally stabilizing controller, $K(s)$, for the plant $P(s)$, such that the infinity norm of the closed-loop transfer function, T_{zw}, is below a given level γ (a positive scalar). This problem is called the *standard H_α control problem.* The *optimal H_∞ control problem* is

$$\begin{array}{lll} \text{Optimal problem} & \underset{K(s)\text{stabilizing}}{\text{Min}} & \|T_{zw}\|_\infty \\ \text{Standard problem} & \underset{K(s)\text{stabilizing}}{\text{Find}} & \|T_{zw}\|_\infty \leqslant \gamma \end{array}$$

The standard problem is more practical. In practice, control system design is more like a balancing act and trade-offs, and a mathematically optimal solution may not be so desirable after all the other real-life constraints have been taken into account. To solve the optimal problem, we start with a value for γ and reduce it until the problem fails to have a solution. As a starting value for γ, we can solve an LQG problem; find the peak in the resulting closed-loop transfer function and use this value. To lower γ, we can use a search algorithm (such as a binary search) to reach the optimal value. This procedure is called *γ-iteration.*

For the problem to have a solution, certain assumptions must be satisfied. They are listed below. The dimensions of various variables are listed first.

Dimensions: $\dim x = n$, $\dim w = m_1$, $\dim u - m_2$, $\dim z = p_1$, $\dim y = p_2$

H_∞ problem assumptions.

1. The pair (A, B_2) is stabilizable and (C_2, A) is detectable. Recall from Chapter 8 that these are weaker versions of controllability and obsevability conditions. This assumption is necessary for a stabilizing controller to exist. It simply guarantees that the controller can reach all unstable states and these states show up on the measurements.
2. Rank $D_{12} = m_2$, rank$D_{21} = p_2$. These conditions are needed to ensure that the controllers are proper. They also imply that the transfer function from w to y is nonzero at high frequencies. Unlike the first assumption, which is usually satisfied, this assumption is frequently violated (e.g., if the original plant is strictly proper: i.e., if it has more poles than zeros, this condition will be violated) unless the problem is formulated in a way that ensures its satisfaction.
3. Rank$\begin{bmatrix} A - j\omega I & B_2 \\ C_1 & D_{12} \end{bmatrix} = n + m_2$ for all frequencies.
4. Rank$\begin{bmatrix} A - j\omega I & B_1 \\ C_2 & D_{21} \end{bmatrix} = n + p_2$ for all frequencies.
5. $D_{11} = 0$ and $D_{22} = 0$. This assumption is not needed, but it will simplify the equations for the solution. It also implies that the transfer functions from w to z and u to y roll off at high frequencies, respectively.

Before we present the solution, it should be pointed out that the solutions of the H_∞ and LQG problems are very similar. Both use a state estimator and feed back the estimated states. The controller and estimator gains are also computed from two

Riccati equations. The differences are in the coefficients of the Ricatti equations, and the H_∞ state estimator contains an extra term. The compensator equations follow.

The controller is given by, K_c corresponds to K, the controller gain, in the LQG case

$$u = -K_c \hat{x}$$

and the state estimator is given by

$$\dot{\hat{x}} = A\hat{x} + B_2 u + B_1 \hat{w} + Z_\infty K_e (y - \hat{y})$$

where

$$\hat{w} = \gamma^{-2} B_1' X_\infty \hat{x}$$

$$\hat{y} = C_2 \hat{x} + \gamma^{-2} D_{21} B_1' X_\infty \hat{x}$$

We can also write this in packed-matrix notation as follows:

And finally the solution.

$$K(s) = \left[\begin{array}{c|c} A - B_2 K_c - Z_\infty K_e C_2 + \gamma^{-2}(B_1 B_1' - Z_\infty K_e D_{21} B_1') X_\infty & Z_\infty K_e \\ \hline -K_c & 0 \end{array}\right]$$

The extra term, $\hat{w}$, is an estimate of the worst-case input disturbance to the system, and $\hat{y}$ is the output of the estimator. The estimator gain is $Z_\infty K_e (K_e$ corresponds to L in the LQG case). The controller gain K_c and estimator gain K_e are given by

$$K_c = \tilde{D}_{12}(B_2' X_\infty + D_{12}' C_1) \quad \text{where } \tilde{D}_{12} = (D_{12}' D_{12})^{-1}$$

$$K_e = (Y_\infty C_2' + B_1 D_{21}') \tilde{D}_{21} \quad \text{where } \tilde{D}_{21} = (D_{21} D_{21}')^{-1}$$

The term Z_∞ is given by

$$Z_\infty = (I - \gamma^{-2} Y_\infty X_\infty)^{-1}$$

The terms X_∞ and Y_∞ are solutions to the controller and estimator Riccati equations—that is,

The Riccati equations.

$$X_\infty = \text{Ric}\begin{bmatrix} A - B_2 \tilde{D}_{12} D_{12}' C_1 & \gamma^{-2} B_1 B_1' - B_2 \tilde{D}_{12} B_2' \\ -\tilde{C}_1' \tilde{C}_1 & -(A - B_2 \tilde{D}_{12} D_{12}' C_1') \end{bmatrix}$$

$$Y_\infty = \text{Ric}\begin{bmatrix} (A - B_1 D_{21}' \tilde{D}_{21} C_2)' & \gamma^{-2} C_1' C_1 - C_2' \tilde{D}_{21} C_2 \\ -\tilde{B}_1 \tilde{B}_1' & -(A - B_1 D_{21}' \tilde{D}_{21} C_2) \end{bmatrix}$$

where $\tilde{C}_1 = (I - D_{12} \tilde{D}_{12} D_{12}') C_1$ and $\tilde{B}_1 = B_1 (I - D_{21}' \tilde{D}_{21} D_{21})$.

The closed-loop system becomes

$$\begin{bmatrix} \dot{x} \\ \dot{\hat{x}} \end{bmatrix} = \begin{bmatrix} A & -B_2 K_c \\ Z_\infty K_e C_2 & \begin{array}{c} A - B_2 K_c + \gamma^{-2} B_1 B_1' X_\infty \\ -Z_\infty K_e (C_2 + \gamma^{-2} D_{21} B_1' X_\infty) \end{array} \end{bmatrix} \begin{bmatrix} x \\ \hat{x} \end{bmatrix} + \begin{bmatrix} B_1 \\ Z_\infty K_e D_{21} \end{bmatrix} w$$

$$\begin{bmatrix} z \\ y \end{bmatrix} = \begin{bmatrix} C_1 & -D_{12}K_c \\ C_2 & 0 \end{bmatrix} \begin{bmatrix} x \\ \hat{x} \end{bmatrix} + \begin{bmatrix} 0 \\ D_{21} \end{bmatrix} w$$

Closed-loop system.

As we had promised, the equations are quite complicated and messy!

Finally, it can be proved that there exists a stabilizing compensator if and only if there exist positive semidefinite solutions to the two Riccati equations and the following condition:

$$\rho(X_\infty Y_\infty) < \gamma^2$$

where $\rho(A)$ = spectral radius of A = largest eigenvalue of $A = \lambda_{\max}(A)$.

The block diagrams of the LQG and H_∞ control systems are shown in Figure 10.21. Compare these diagrams to see the similarities and differences between them.

It should be fairly obvious that H_∞ problems cannot be solved manually. Computer programs such as Program CC, MATRIX$_X$, and MATLAB have special functions and utilities for solving these problems. For every value of γ, two Riccati equations must be solved. In addition, even if the plant is first-order, we still may need to add weights to the system either to satisfy design requirements or to satisfy the necessary assumptions for a feasible solution. This increases the order of the equations and makes manual solution almost impossible. The steps can be summarized as follows.

Solution procedure.

1. Set up the problem to obtain the state space representation for $P(s)$.
2. Check to see whether the assumptions (the rank conditions) are satisfied. If they are not, reformulate the problem by adding weights or adding (fictitious) inputs or outputs.
3. Select a large positive value of γ.
4. Solve the two Riccati equations. Determine if the solutions are positive semidefinite; also, verify that the spectral radius condition is met.
5. If all the conditions given are satisfied, lower the value of γ. Otherwise, increase it. Repeat steps 4 and 5 until either an optimal or satisfactory solution is obtained.

10.7.4 Weights in H_∞ Control Problems

Practical control problems require weighting the inputs and outputs. There are a few reasons for using weights. Constant weights are used for scaling inputs and outputs, they are also used for unit conversions. Transfer function weights are used to shape the various measures of performance in the frequency domain. In H_∞ control problems, weights are also used to satisfy the rank conditions. These assumptions are frequently violated unless appropriate weights are selected. In fact, the weights are the only parameters that the designer must specify. Proper selection of these weights depends a great deal on the experience of the user and on his or her understanding of the physics of the problem and other practical engineering constraints.

Tracking and disturbance rejection require that the sensitivity transfer function be small in the low-frequency range. This can be formulated as specifying that the

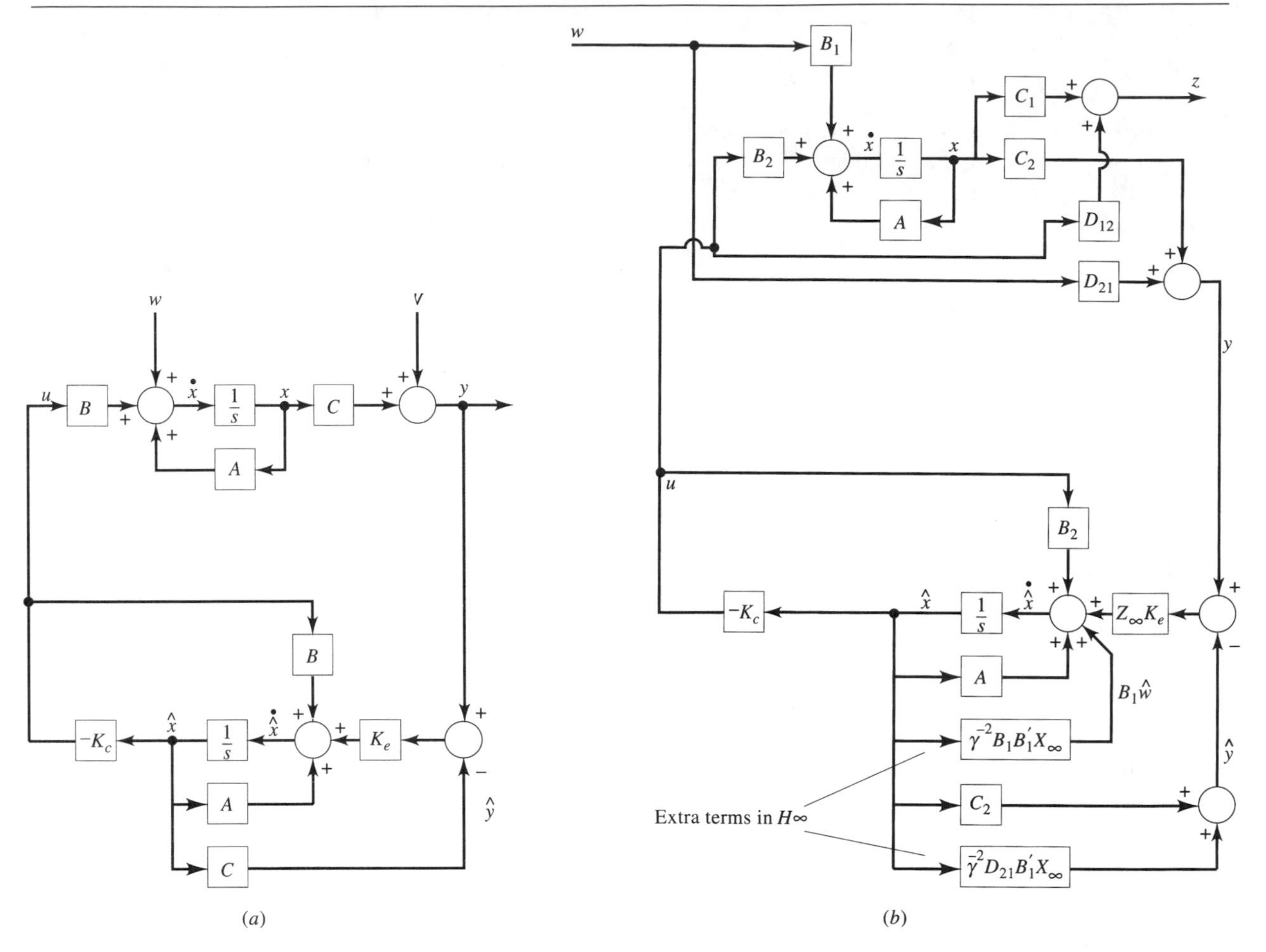

Figure 10.21 (a) The LQG controller block diagram. (b) Block diagram showing the structure of the H_∞ control system.

Weights give us frequency domain control over system behavior.

sensitivity remain below a given frequency-dependent weight—that is,

$$|S| \leqslant W_s^{-1} \qquad \text{or} \qquad |W_s S| \leqslant 1$$

Similarly we can specify that the complementary sensitivity be kept below a given weight in the high-frequency range—that is,

$$|T| \leqslant W_t^{-1} \qquad \text{or} \qquad |W_t T| \leqslant 1$$

Finally, both requirements can be satisfied by solving what is called the *mixed-sensitivity problem*. Typical plots for both cases including weights are shown in Figure 10.22.

As an example, we will use the H_∞ approach to design a controller for the double-integrator system. The first step is setting up the problem appropriately. The

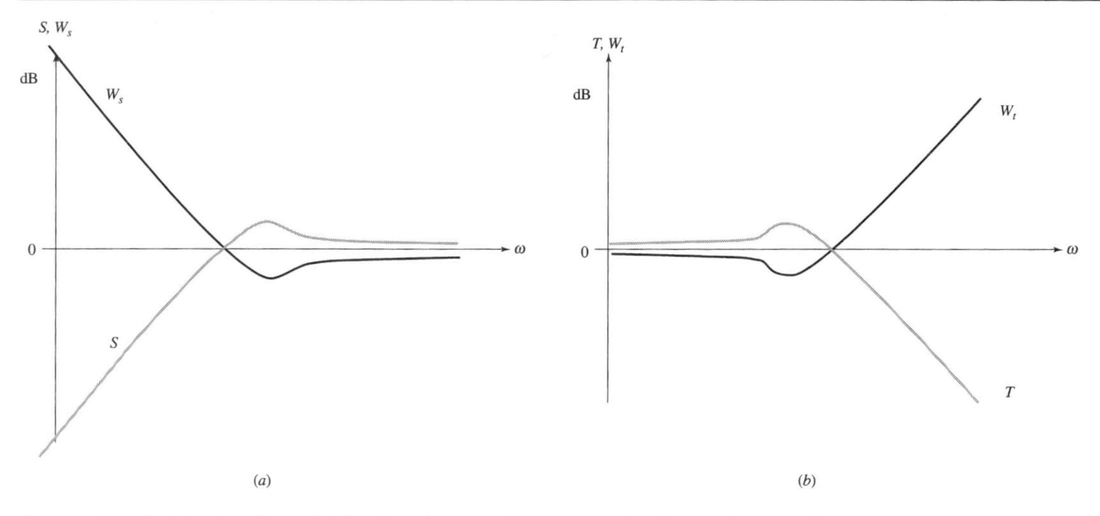

Figure 10.22 (a) Plot of sensitivity function and its weight. (b) Plot of the complementary sensitivity and its corresponding weight.

plant equations are given by

$$\dot{x}_1 = d + u$$
$$\dot{x}_2 = x_1$$

The new term that we have added is the disturbance term, d; this term corresponds either to an actual disturbance or to unmodeled dynamics in the system. The regulated outputs are given by

H_∞ solution of the double-integrator.

$$z = \begin{bmatrix} x_2 \\ u \end{bmatrix}$$

It is important that the control signal be included in the regulated outputs so that we can bound its magnitude. This is also needed to ensure that the rank condition on D_{12} is satisfied. The measurement equation is given by

$$y = x_2 + n$$

The noise term, n, is either actual sensor noise or, perhaps, it represents high-frequency unmodeled dynamics. It is also needed to ensure that the rank condition on D_{21} is met. Collecting these equations, we obtain the system equations in packed-matrix notation as follows:

$$P(s) = \left[\begin{array}{c|cc} A & B_1 & B_2 \\ \hline C_1 & D_{11} & D_{12} \\ C_2 & D_{21} & D_{22} \end{array}\right] = \left[\begin{array}{cc|cc|c} 0 & 0 & 1 & 0 & 1 \\ 1 & 0 & 0 & 0 & 0 \\ \hline 0 & 1 & 0 & 0 & 0 \\ 0 & 0 & 0 & 0 & 1 \\ \hline 0 & 1 & 0 & 1 & 0 \end{array}\right]$$

The block diagram of the system is shown in Figure 10.23 in the usual form and in the generic H_∞ form.

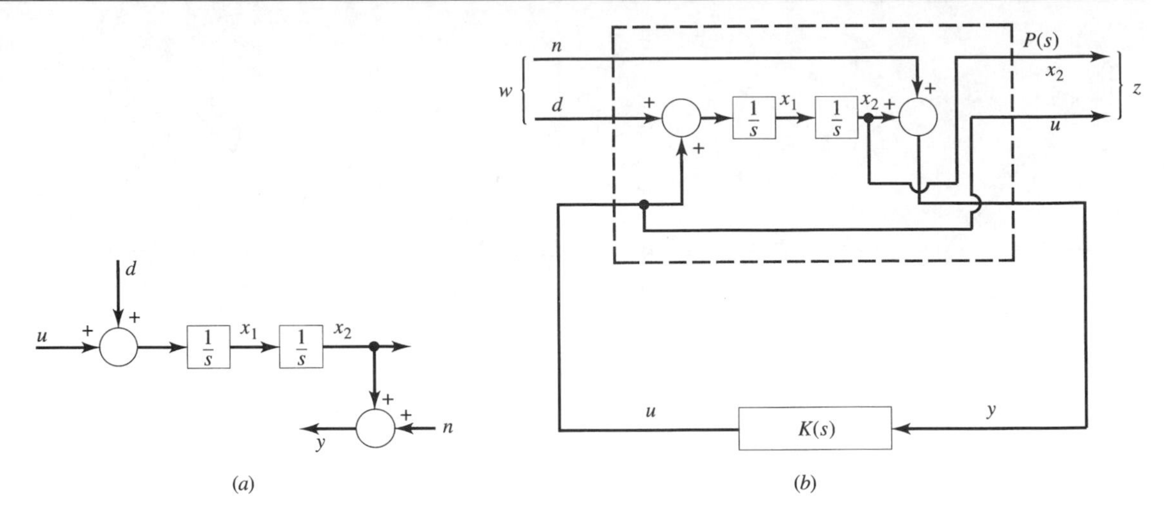

Figure 10.23 (a) The block diagram for the double-integrator system. (b) The generic H_∞ block system.

This problem was solved on the computer, and after several trials we found that the value of γ could not be reduced below 2.62. Hence, we conclude that 2.62 is the optimal value (note that the solution of the optimal H_∞ control problem involves a search over γ and we can get as close to it as possible but not achieve it). The following are the relevant data obtained:

$$\gamma = 2.62 \qquad X_\infty = \begin{bmatrix} 1.59 & 1.08 \\ 1.08 & 1.47 \end{bmatrix} \qquad Y_\infty = \begin{bmatrix} 1.47 & 1.08 \\ 1.08 & 1.59 \end{bmatrix}$$

$$K_c = [\, 1.59 \quad 1.08 \,] \qquad K_e = \begin{bmatrix} 1.08 \\ 1.59 \end{bmatrix}$$

The transfer function of the compensator and the resulting closed-loop poles are given by

The H_∞ compensator. Looks like a lead plus an extra pole.

$$K(s) = \frac{-578.3(s + 0.39)}{(s + 2.33)(s + 220.72)}$$

Closed-loop poles $= \{-0.71, -0.81 \pm 0.91j, -200.7\}$

The Bode and Nyquist plots of the system are shown in Figure 10.24(a, b). We have obtained a gain margin of 44 dB and phase margin of 45°. The responses of the system to a unit-pulse disturbance and random sensor noise also is shown in Figure 10.24 (c). As expected, both responses approach zero asymptotically.

We will end our brief introduction to H_∞ control at this point. We point out that this subject is still very novel and is rapidly progressing. We have also limited our discussion to the treatment of unstructured uncertainty and have presented only one of the approaches to robust control. The different approaches, however, have one feature in common: They are all frequency domain, computer-assisted tools for

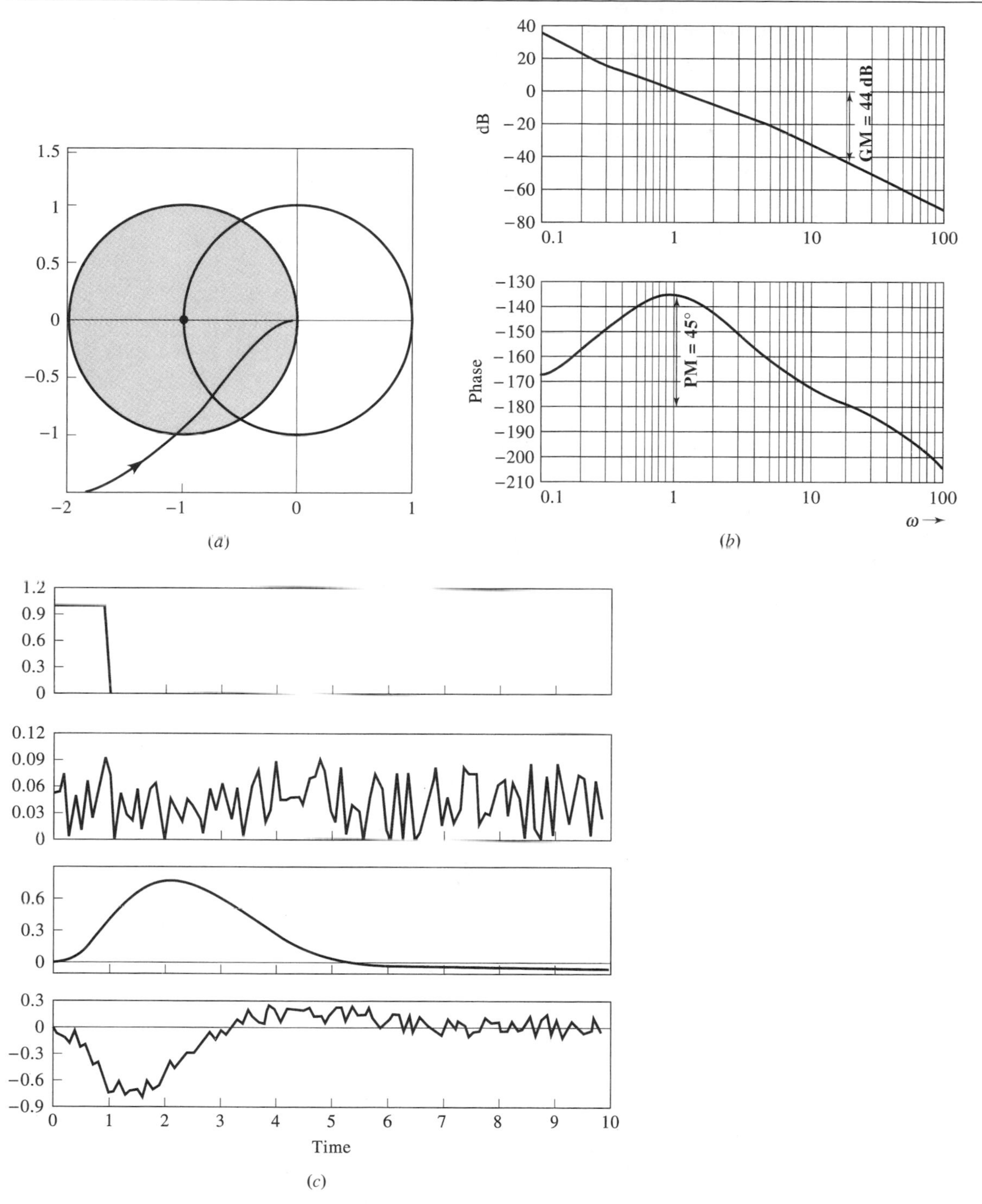

Figure 10.24 (a) The Nyquist plot of the H_∞ compensated system. (b) The Bode plots of the system. (c) Unit pulse, random noise, and plant response to these inputs.

design of MIMO systems satisfying practical constraints. For this reason, H_∞ control is expected to find a permanent place in the control engineer's toolbox.

❑ DRILL PROBLEMS

D10.6 (*Note:* As much as possible, try to solve this problem manually.) Consider the first-order system given by

$$\begin{aligned} \dot{x} &= x + u + d \\ y &= x + n \end{aligned}$$

Our objective is to regulate the state and control signals (x, u) in presence of disturbance and noise inputs (d, n).

Let $w = \begin{bmatrix} d \\ n \end{bmatrix}$ and $z = \begin{bmatrix} x \\ u \end{bmatrix}$

(a) Obtain the plant equation, $P(s)$, in packed-matrix notation.

(b) Draw the generic H_∞ block diagram of the system.

(c) Verify that all the rank conditions are met.

(d) Compute $\tilde{D}_{12}, \tilde{D}_{21}, \tilde{C}_1, \tilde{B}_1$.

(e) Find the controller Hamiltonian matrix and solve for X_∞ in terms of γ.

(f) Repeat (e) for the estimator, and solve for Y_∞.

(g) Compute $(X_\infty Y_\infty)$ in terms of γ. Find out if the spectral radius condition is met for $\gamma = 2$. Repeat for $\gamma = 3$.

For the rest of the drill, let $\gamma = 3$.

(h) Compute $X_\infty, Y_\infty, Z_\infty, K_c, K_e$.

(i) Find the compensator transfer function and the closed-loop poles.

(j) Draw the Nyquist plot and obtain gain and phase margins.

Ans.

(a) $\dot{x} = x + [1 \quad 0]w + u$

$$z = \begin{bmatrix} 1 \\ 0 \end{bmatrix} x + \begin{bmatrix} 0 \\ 1 \end{bmatrix} u \qquad P(s) = \left[\begin{array}{c|cc:c} 1 & 1 & 0 & 1 \\ \hline 1 & 0 & 0 & 0 \\ 0 & 0 & 0 & 1 \\ \hdashline 1 & 0 & 1 & 0 \end{array}\right]$$

$y = x + [0 \quad 1]w$

(b)

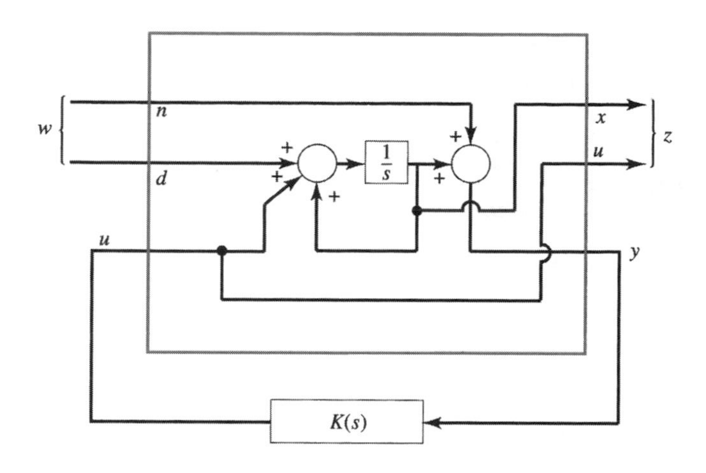

(c) Rank D_{12} =rank $\begin{bmatrix} 0 \\ 1 \end{bmatrix} = 1$, rank D_{21} = rank $[0 \quad 1] = 1$,

$D_{11} = 0, D_{22} = 0$ rank $\begin{bmatrix} A - j\omega & B_2 \\ C_1 & D_{12} \end{bmatrix}$

$= \text{rank} \begin{bmatrix} j\omega - 1 & 1 \\ 1 & 0 \\ 0 & 1 \end{bmatrix} = 2$ for all ω rank $\begin{bmatrix} A - j\omega & B_1 \\ C_2 & D_{21} \end{bmatrix}$

$= \text{rank} \begin{bmatrix} j\omega - 1 & 1 & 0 \\ 1 & 0 & 1 \end{bmatrix} = 2$ for all ω

(d) $\tilde{D}_{12} = \tilde{D}_{21} = 1,\ \tilde{C}_1 = \begin{bmatrix} 1 \\ 0 \end{bmatrix},\ \tilde{B}_1 - [1 \quad 0]$

(e) $X_\infty = \text{Ric} \begin{bmatrix} 1 & \gamma^{-2} - 1 \\ -1 & -1 \end{bmatrix} = \dfrac{-\gamma^2 - \gamma\sqrt{2\gamma^2 - 1}}{1 - \gamma^2}$

(f) $Y_\infty = X_\infty$ for this problem

(g) $X_\infty Y_\infty = 1/(1 - \gamma^2)^2 \left[\gamma^4 + \gamma^2(2\gamma^2 - 1) + 2\gamma^3\sqrt{2\gamma^2 - 1}\right]$
for $\gamma = 2$, $X_\infty Y_\infty > 4$, violated
for $\gamma = 3$, $X_\infty Y_\infty < 9$, satisfied

(h) $X_\infty = Y_\infty = 2.67$, $Z_\infty = 4.82$, $K_c = K_e = 2.67$

(i) $K(s) = -34.43 \quad 1/(s + 14.26)$, poles at $s = -1.75, -11.51$

(j) phase margin= 56.5°, infinite gain margin

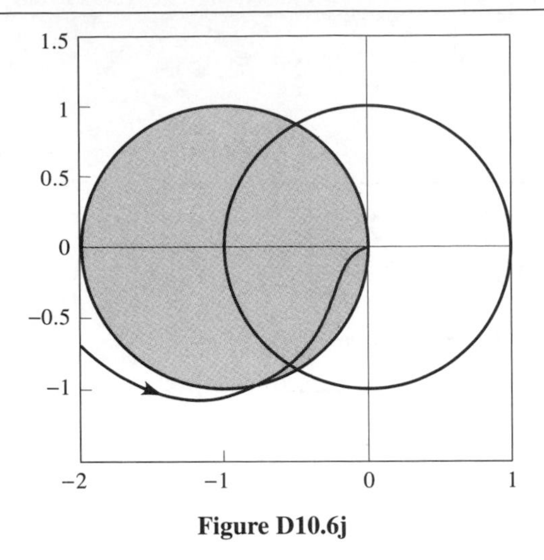

Figure D10.6j

10.8 SUMMARY

Linear quadratic methods for control system design have been discussed. These techniques lead to linear controllers that are easy to implement. Another important feature is guaranteed closed-loop stability. The LQR technique requires that all states be available for measurement. If the system is controllable (or at least stabilizable), this method gives excellent stability margins. The guaranteed margins are 60° phase margin, infinite gain margin, and−6 dB gain reduction margin. The design involves selection of the state and control weights, Q and R matrices, and solution of the Riccati equation. This can be also accomplished by using the root-square locus approach.

If all the states are not available for feedback, a Kalman filter (observer or estimator) can be designed. The combination of the filter and the LQ controller is called the LQG compensator. The design starts with selection of the state control weights Q and R, selection (or determination) of the process and measurement noise intensities, and solution of two Riccati equations. The compensator structure is of the observer-based type seen in Chapter 9, and the solution satisfies the separation principle. Although the closed-loop system is guaranteed to be stable, it will have no guaranteed stability margins. The design requires perfect knowledge of the system model and consequently is not robust.

Feedback properties of systems can be adequately characterized by the sensitivity or complementary sensitivity transfer functions. Performance specifications impose bounds of the open-loop transfer function of the system in various frequency ranges. Stability or performance of a system is robust if it is maintained in spite of model uncertainty. Model uncertainty can be modeled as either structured or unstructured uncertainty. Unstructured uncertainty can be modeled as either additive or multiplicative uncertainty. The small-gain theorem can be used to determine if the system is robustly stable under model uncertainties.

Loop transfer recovery (LTR) is a modification of the LQG technique; it allows recovery of the LQR stability margins. One begins with selection the LQR parameters until a desired open-loop transfer function (the target feedback loop) is obtained. The recovery step involves iterating on the filter design parameters until the desired loop transfer function shape has been obtained. The LTR methodology converts the LQG technique from a rigid time domain method to a flexible frequency domain design technique. The computations are still based on state space techniques, but they can remain hidden from the user.

H_∞ control is the newest tool for control-system design. It is a computer-aided frequency domain method for design of multivariable systems. The exogenous inputs (disturbances, command inputs, sensor noise) are collected into one vector; the regulated outputs (control signals, errors) are collected into another vector. This will result in a dual-input, dual-output block diagram, called the *generic* (or *synthesis*) *block diagram.* The objective is to maintain the peak in the closed-loop frequency response of the system below a specified level γ. The optimal problem can be solved by iteration on γ. The solution involves selecting weights (possibly frequency-dependent weights) and solving two Riccati equations. The compensator structure is similar to LQG with some added terms. It can be shown that if the value of γ is allowed to go to infinity, the solution approaches the LQG solution.

REFERENCES

LQR/LQG/LTR

Anderson, B. D. O., and Moore, J. B., *Linear Optimal Control.* Englewood Cliffs, NJ: Prentice-Hall, 1990.

Bryson, A. E, Jr., and Ho, Y. C., *Applied Optimal Control: Optimization, Estimation and Control.* Washington, DC: Hemisphere Publishing Corporation, 1975.

Chen, C. T., *Control System Design: Conventional, Algebraic and Optimal Methods.* Pond Woods Press, 1987.

Doyle, J. C., and G. Stein. Robustness with Observers. *IEEE Trans. Autom. Control,* vol. 24, 1979, pp. 607–611.

B. Friedland. *Control Systems Design: An Introduction to State Space Methods.* New York: McGraw-Hill, 1986.

IEEE Trans. Autom. Control. Special issue on linear multivariable control systems, (February 1981).

R. E. Kalman. When Is a Linear Control System Optimal? *J. Basic Eng. Trans. ASME D*, 86 (1964): 51–60.

Kirk, D. E., *Optimal Control Theory: An Introduction.* Englewood Cliffs, NJ: Prentice-Hall, 1970.

Kwakernaak, H., and Sivan, R., *Linear Optimal Control Systems.* New York: Wiley-Interscience, 1972.

Lewis, F. L., *Optimal Control.* New York: Wiley, 1986.

H_∞ Control

Boyd, S, P., and Barratt, C. H., *Linear Controller Design.* Englewood Cliffs, NJ: Prentice-Hall, 1991.

Doyle, J. C., Francis, B. A., and Tannenbaum, A. R., *Feedback Control Theory.* New York: Macmillan, 1992.

Hauser, Frank, Lecture Notes for Modern Control Theory for the Practitioner, UCLA Extension course, Los Angeles, 1992.

Levine, W.S., and Reichert, R. T., "An Introduction to H_∞ Control System Design." *Proceedings of the 29th Conference on Decision and Control,* Honolulu, December 1990.

Maciejowski, J. M., *Multivariable Feedback Design.* Reading, MA: Addison-Wesley, 1989.

Safanov, M. G., Lecture Notes for a Course in H_∞ Control offered at the University of Southern California, Los Angeles 1992.

Computational Techniques

Balas, G. J., Doyle, J. C., Glover, K., Packard, A., and Smith, R., *User's Guide to the μ-Analysis and Synthesis Toolbox for MATLAB.* MUSYN Inc. and Math Works, Inc., S. Natick, MA, 1991.

Balas, G. J., Packard, A., Doyle, J. C., Glover, K., and Smith, R., "Development of Advanced Control Design Software for Researchers and Engineers. *Proceedings of the American Control Conference*, Boston, June 26–28, 1991.

Chiang, R. Y., and Safonov, M. G., "A Hierarchical Data Structure and New Capabilities of the Robust-Control Toolbox." *Proceedings of the American Control Conference*, Boston, June 26–28, 1991.

Chiang, R. Y., and Safonov, M. G., *User's Guide to the Robust Control Toolbox for MATLAB*. South Natick, MA: MathWorks, Inc., 1988.

Integrated Systems Inc., *User's Guide to the Robust Control Module for $MATRIX_X$*. Santa Clara, CA:, Integrated Systems Inc., 1990.

Shahian, B., and Hassul, M., *Control System Design Using MATLAB*. Englewood Cliffs, NJ: Prentice-Hall, 1993.

Shahian, B., and Hassul, M., *Control System Design Using $MATRIX_X$*. Englewood Cliffs, N J: Prentice-Hall, 1992.

Thompson, Peter, *Tutorial and User's Guide for program CC, Version 4.* Systems Technology, Hawthorn, CA: 1988.

PROBLEMS

1. For each of the following systems, obtain the optimal control gain using LQR by solving the Riccati equation.

(a) $A = \begin{bmatrix} 0 & 1 \\ -4 & -4 \end{bmatrix}$, $B = \begin{bmatrix} 0 \\ 3 \end{bmatrix}$, $Q = \begin{bmatrix} 1 & 0 \\ 0 & 0 \end{bmatrix}$, $R = 1$

(b) $A = \begin{bmatrix} 0 & 1 \\ -4 & -4 \end{bmatrix}$, $B = \begin{bmatrix} 0 \\ 3 \end{bmatrix}$, $Q = \begin{bmatrix} 0 & 0 \\ 0 & 1 \end{bmatrix}$, $R = 1$

(c) $A = \begin{bmatrix} 0 & 1 \\ -4 & -4 \end{bmatrix}$, $B = \begin{bmatrix} 0 \\ 3 \end{bmatrix}$, $Q = \begin{bmatrix} 1 & 0 \\ 0 & 1 \end{bmatrix}$, $R, = 1$

Ans. (a) $K = [0.33 \quad 0.08]$

2. Obtain the root-square locus in each case for Problem 1. Let $R = \rho$

3. Obtain the root-square locus for each of the following systems.

(a) $A = \begin{bmatrix} 0 & 1 \\ -3 & -8 \end{bmatrix}$, $B = \begin{bmatrix} 0 \\ 4 \end{bmatrix}$, $Q = \begin{bmatrix} 1 & 0 \\ 0 & 0 \end{bmatrix}$, $R = \rho$

(b) same system as (a) but let $Q = \begin{bmatrix} 1 & 0 \\ 0 & 1 \end{bmatrix}$, $R = \rho$

(c) $A = \begin{bmatrix} 0 & 1 \\ -4 & 0 \end{bmatrix}$, $B = \begin{bmatrix} 1 \\ 1 \end{bmatrix}$, $Q = \begin{bmatrix} 1 & 0 \\ 0 & 0 \end{bmatrix}$, $R = \rho$

4. Use the Riccati equation to solve Problem 3. Find the gain for $\rho = 0.1, \ 1, \ 10$ in each case.

5. Obtain the compensator transfer function, draw Bode plots, and compute gain and phase margins for each system in Problem 1.

Let $C = [1 \quad 0]$ in each case.

6. Consider the system given by

$$A = \begin{bmatrix} 0 & 1 \\ 0 & 0 \end{bmatrix}, B = \begin{bmatrix} 0 \\ 1 \end{bmatrix} \quad Q = \begin{bmatrix} 1 & 0 \\ 0 & q \end{bmatrix} \quad R = 1$$

Find the optimal control gain and the closed-loop poles as a function of q. Discuss what happens to these quantities as q increases ($q > 0$).

7. Consider the following system:

$$A = \begin{bmatrix} 0 & 1 & 0 & 0 \\ 0 & -1 & 0 & 0 \\ 0 & 0 & 0 & 1 \\ -12 & 0 & 12 & 0 \end{bmatrix} \quad B = \begin{bmatrix} 0 \\ 1 \\ 0 \\ 0 \end{bmatrix} \quad C = \begin{bmatrix} 1 & 0 & 0 & 0 \\ -1 & 0 & 1 & 0 \end{bmatrix}$$

$$\text{Let } Q = \begin{bmatrix} 0 & 0 & 0 & 0 \\ 0 & 0 & 0 & 0 \\ 0 & 0 & 1 & 0 \\ 0 & 0 & 0 & 0 \end{bmatrix} \quad R = 10^{-6}$$

(a) Find the control gain vector K and the optimal closed-loop poles

(b) Let $Q_o = \begin{bmatrix} 0 & 0 & 0 & 0 \\ 0 & 0.5 & 0 & 0 \\ 0 & 0 & 0 & 0 \\ 0 & 0 & 0 & 0 \end{bmatrix}$ and $R_o = \begin{bmatrix} 2 \times 10^{-6} & 0 \\ 0 & 10^{-6} \end{bmatrix}$

Design a Kalman filter. Find the filter gain L and filter eigenvalues.

(c) Find the equation for the LQG compensator of (a) and (b).

(d) Find the eigenvalues of the closed-loop system.

(e) Plot the impulse and step response of the system.

8. Consider the system given by

$$A = \begin{bmatrix} -54 & 2 & 10 \\ 2 \times 10^{-4} & -10^{-3} & -5 \times 10^{-3} \\ -10^{-3} & -24 \times 10^{-3} & -0.14 \end{bmatrix}$$

$$B = \begin{bmatrix} -10^4 \\ 0.25 \\ -2 \end{bmatrix} \qquad C = [0 \quad 3 \quad 0.05]$$

The open-loop transfer function must meet the following specification:

(i) $|GH| > 20$ dB for $\omega \leqslant 0.1$ rad/s

for good disturbance rejection and command tracking.

(ii) $|1 + GH| \geqslant 25$ dB for $\omega \leqslant 0.05$ rad/s

for insensitivity to parameter variations.

(iii) $|GH| \leqslant -20$ dB for $\omega \geqslant 5$ rads

for good immunity to noise.

(a) Design an LQG compensator and record its performance with respect to the foregoing specifications.

(b) Use LTR to meet the specification as closely as possible.

9. Consider the problem in Drill Problem D10.4.

$$G(s) = \frac{1}{s^2} \qquad H(s) = \frac{20(s+1)}{(s+10)}$$

Suppose the actual plant is given by

$$\tilde{G}(s) = \frac{2(s+1)}{s^2(s^2+s+1)}$$

(a) Find a multiplicative uncertainty model for the system.

(b) Find $M(s)$, the transfer function as seen by the multiplicative uncertainty.

(c) Determine if the closed-loop system is robustly stable under the multiplicative uncertainty computed in (a).

Ans. (a) $\Delta_m(s) = (-s^2 + s + 1)/(s^2 + s + 1)$;

(b) $M(s) = -20(s + 1)/(s^3 + 10s^2 + 20s + 20)$

10. This problem is adapted from a paper by W. S. Levine and R. T. Reichert (*Proceedings of Conference on Decision and Control*, December, 1990).

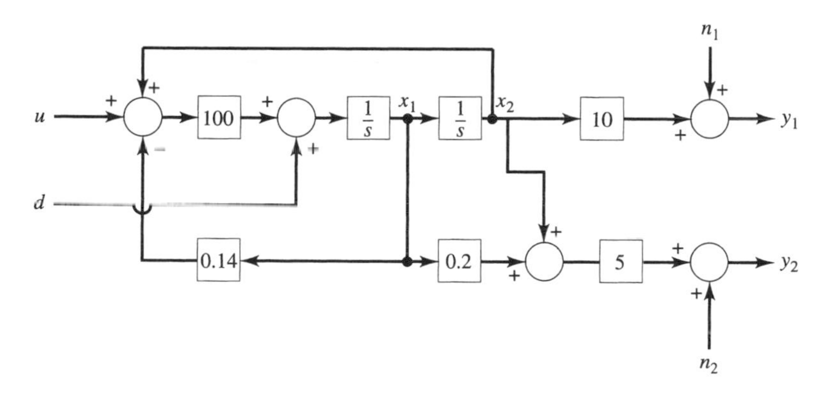

Figure P10.10

Consider the single-input, dual output plant, in Figure P10.10(a). where u = control, d = disturbance, n_1 and n_2 = sensor noise sources. The specifications are to have integral tracking performance for output y_1 with a time constant of 0.6 s. The second output y_2, is also available for feedback.

It has been suggested that for integral tracking, one should regulate the integral of the tracking error. Hence, we will introduce a command input y_c, and an integrator at $y_1 - y_c$(before the entry of the noise source n_1).

In addition, all regulated outputs and disturbance and noise inputs will be weighted. The new block diagram is shown in Figure P10.10 (b).

The following weights are suggested: $W_d = 0.1$, $W_{n1} = W_{n2} = 1$, $W_e = 0.5$, $W_u = 1$.

(By varying these weights, one can manipulate the performance of the system).

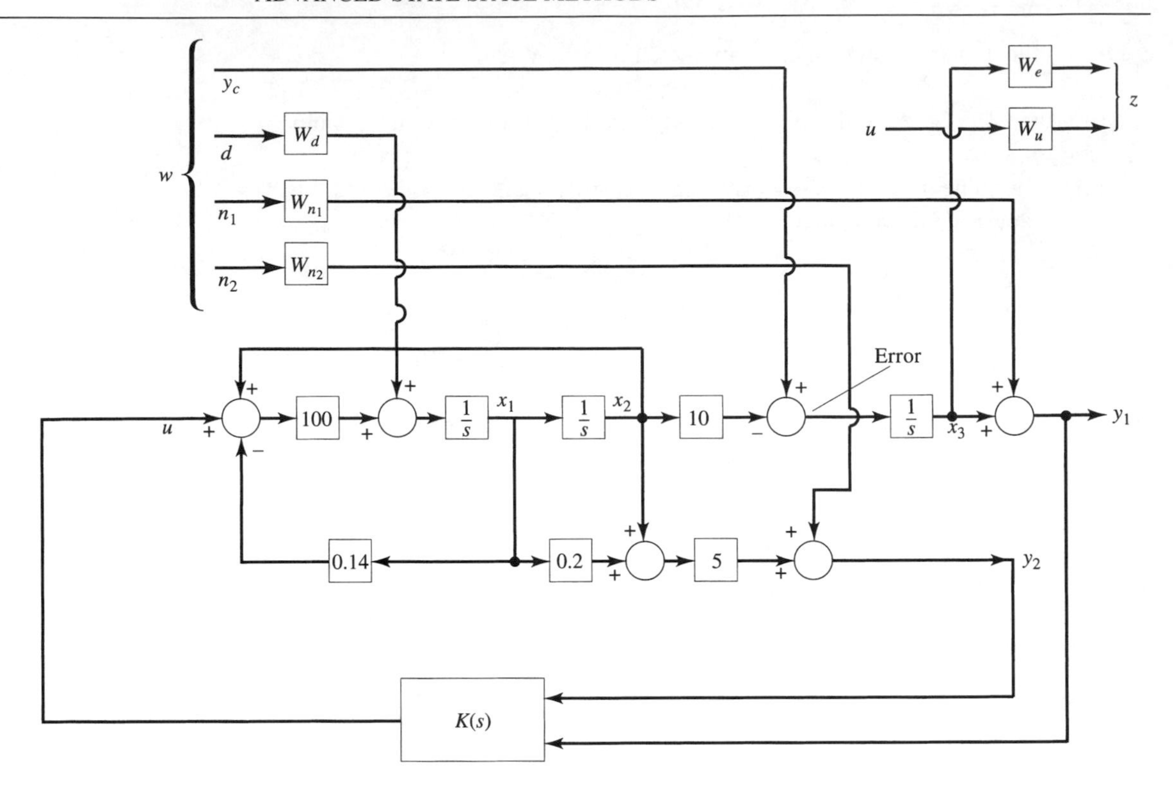

Figure P10.10

(a) Verify that the plant equations are given by

$$\dot{x}_1 = -14x_1 + 100x_2 + 0.1d + 100u$$

$$\dot{x}_2 = x_1$$

$$\dot{x}_3 = -10_{x2} + y_c$$

$$z = \begin{bmatrix} 0.5x_3 \\ u \end{bmatrix}$$

$$y = \begin{bmatrix} x_3 + n_1 \\ x_1 + 5x_2 + n_2 \end{bmatrix}$$

(b) Obtain the plant matrix $P(s)$.

(c) For a value of $\gamma = 0.5922$ (optimal value obtained by computer) find K_c, K_e, and the compensator transfer function.

(d) Draw the Nyquist plot, and compute the stability margins.

(e) Plot the unit step response of the system (set all inputs to zero and let $y_c = 1$, then plot y_1).

(f) Use γ-iteration to verify that the $\gamma = 0.5922$ is optimal.

Ans. (b) $P(s) = \left[\begin{array}{ccc|cccc:c} -14 & 100 & 0 & 0 & 0.1 & 0 & 0 & 100 \\ 1 & 0 & 0 & 0 & 0 & 0 & 0 & 0 \\ 0 & -10 & 0 & 1 & 0 & 0 & 0 & 0 \\ \hline 0 & 0 & 0.5 & 0 & 0 & 0 & 0 & 0 \\ 0 & 0 & 0 & 0 & 0 & 0 & 0 & 1 \\ \hdashline 0 & 0 & 1 & 0 & 0 & 1 & 0 & 0 \\ 1 & 5 & 0 & 0 & 0 & 0 & 1 & 0 \end{array}\right]$

Note: The (1, 3) element of C_2 is incorrect in the original source.

(c) $K_c = [\ 0.14 \quad 3 \quad -0.52]$

$$K_e = \begin{bmatrix} -0.89 & 5.26 \\ -0.17 & 1.01 \\ 2.16 & -1.76 \end{bmatrix}$$

$$K(s) = \frac{1}{(s+4.9)(s-16.6)(s+318.2)}[(s+3.6)(s+19.2)$$

$$(s+0.5)(s+19.2)]$$

(d) $PM = 51, GM = -5.5.$

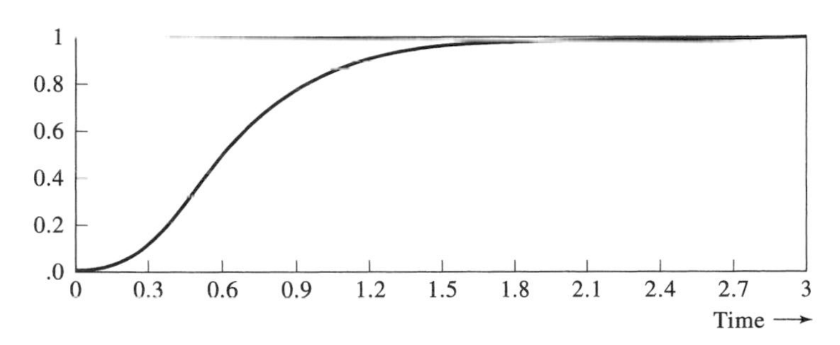

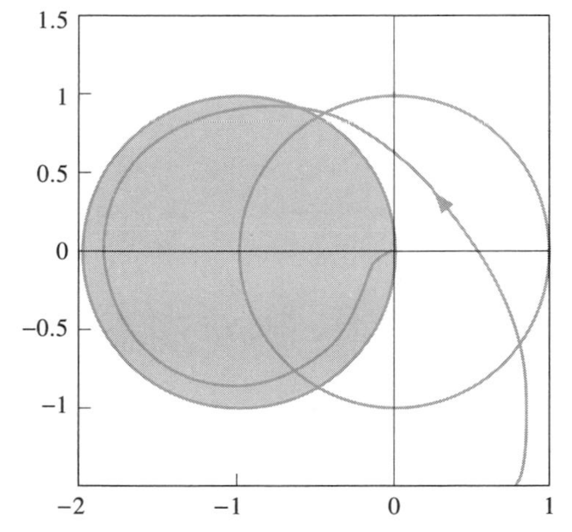

Figure P10.10

11. Consider the plant given by $G(s) = (s-1)/[(s+1)(s^2+s+1)]$. The specification is $|S(j\omega)| \leqslant 0.1$ for $\omega \leqslant 0.01$ rad/s. This is a sensitivity minimization problem. The requirement means that we want to reject low-frequency disturbances (or equivalently, to reduce the system sensitivity to parameters variations or model uncertainties). To set up the problem for H_∞, consider Figure P10.11(a).

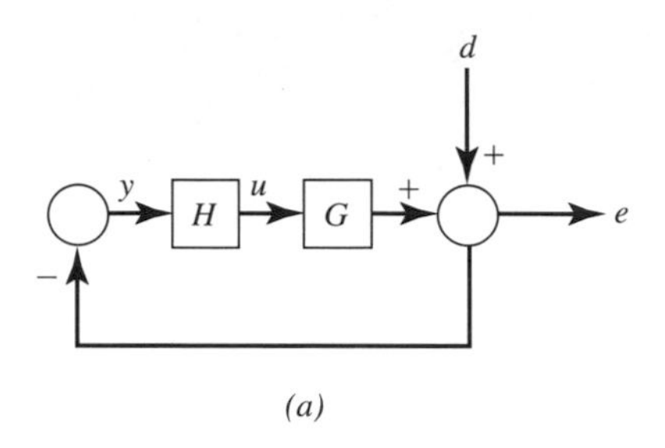

Figure P10.11

The transfer function from disturbance d to error e is given by

$$\frac{E(s)}{D(s)} = \frac{1}{1+GH} = S$$

Now, we redraw the diagram as in Figure P 10.11(b). (We have changed the notation to correspond to Section 10.7.)

Obtain the system matrix, $P(s)$, and use γ-iteration to solve the problem. Plot the sensitivity, $S(j\omega)$, and vary γ until the specification is met. For the final design, display $|S(j\omega)|$, the compensator transfer function $K(s)$, and the step response of the system.

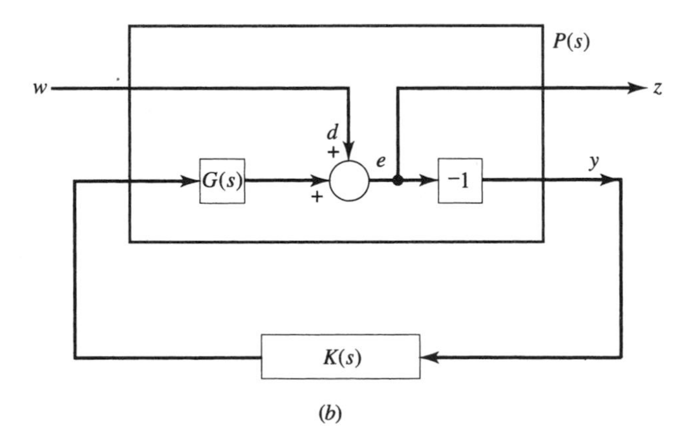

Figure P10.11

12. Consider the H_∞ set up in Figure P10.12.

$$G(s) = \frac{1}{(s + 10^{-3})^2}$$

$$G(s) = \frac{1}{(s + 10^{-3})^2}$$

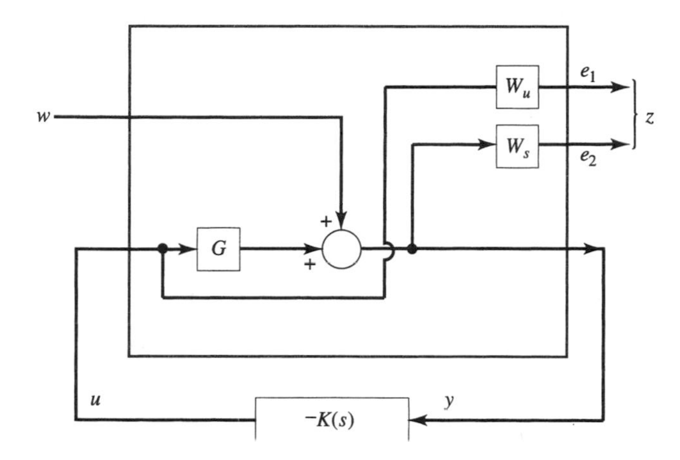

Figure P10.12

The objective is to minimize $\|T_{zw}\|_\infty$. This corresponds to minimizing the effects of the disturbance w. Note that the transfer function from w to e_2 is the sensitivity weighted by W_s, and the transfer function from w to e_1 is related to the ASM (Section 10.5).

$$\frac{e_2}{w} = \frac{1}{1 + KG} W_s \qquad \frac{e_1}{w} = \frac{-K}{1 + KG} W_u$$

Hence by reducing the $\|e_1/w\|_\infty$, we are also increasing the additive stability margin.

Let

$$W_s = \frac{1}{(s + 10^{-3})^2}$$

and

$$W_u = \begin{cases} 1 & \text{Case I} \\ 10^{-4} & \text{Case II} \end{cases}$$

Find the H_∞ optimal compensator in each case. In each case display the Bode magnitude plot for GK, S, and T. Also give the optimal value of γ. Discuss the effects of the weight W_u and compare the two cases.

13. Repeat Problem 12 for $G(s) = (s-1)/[(s+1)(s+10^{-3})^2]$.

$$W_u = 10^{-4} \qquad W_s = \begin{cases} \dfrac{1}{(s+10^{-3})^2} & \text{Case I} \\ \dfrac{s^2+1.4s+1}{(s+10^{-3})^2(1+10^{-3}s)} & \text{Case I} \\ 1 & \text{Case III} \end{cases}$$

14. It is possible to use the Nyquist plot and some geometry to obtain formulas for gain and phase margins in terms of return difference and sensitivity.

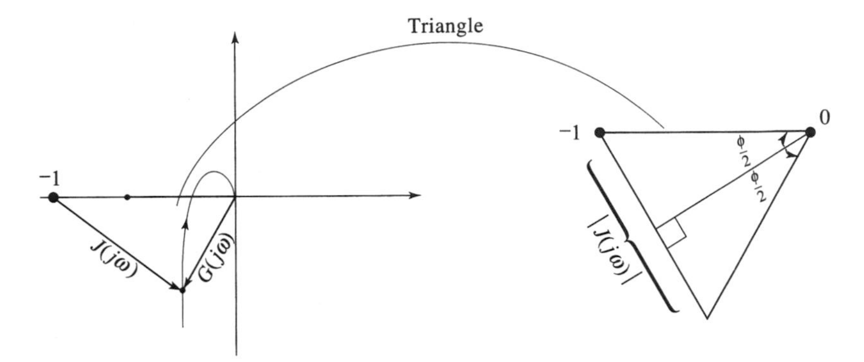

Figure P10.14

In Figure P10.14, if

$$J(\omega) = 1 + G(j\omega) = \text{return difference}$$

and

$$S(j\omega) = \frac{1}{1+G(j\omega)} = \text{sensitivity}$$

use the figure as a hint to show that

$$PM = 2\sin^{-1}\left|\frac{J(j\omega_{gc})}{2}\right| \qquad \omega_{gc} = \text{gain crossover frequency}$$

$$GM = -20\ \log\left(1 - \frac{1}{|S(j\omega)|}\right)$$

$$PM = 2\sin^{-1}\ \frac{1}{2\,|S(j\omega)|}$$

Tabulate $|S(j\omega)|$, GM, PM for $|S(j\omega)|$ from 1 to 3 in 0.2 increments. [Note: for $|S| = 1$, we get LQR stability margins.]

CHAPTER 11

Digital Control

11.1 Preview

The terms *continuous-time* and *analog* are identical in meaning when applied to signals and systems. Analog signals are functions of a continuous-time variable, and analog systems are systems described in terms of such signals. Similarly, *discrete-time* and *digital* have the same meaning; they refer to signals that are defined only for specified instants of time. The use of the words *analog* and *digital* in this sense dates from an era when large-scale analog computers were common.

Past years have seen an exponential growth in the capability and application of digital computers, and there is every indication that a high rate of growth will continue far into the future. In the field of control systems, digital computers were first applied in military and space applications, where the high costs were justified by new capabilities. As computer costs dropped, their use for control began in large-scale industry such as chemical processing, heavy manufacturing, and telecommunications plants, which could afford the large investment. Later, minicomputers, faster, more powerful, and much less expensive than their predecessors, began to revolutionize industry everywhere. No longer was a huge central computing installation necessary; general- and special-purpose computers could be economically distributed and tailored to specific tasks. Now, with the widespread availability and low cost of microcomputers, every process is a candidate for digital control, and sophisticated control systems that were impractically expensive only a few years ago are feasible.

It is no longer far-fetched to imagine a digital-computer-based control system that monitors plant behavior to determine a mathematical model of the plant, then through repeated simulation or other calculation determines an optimum control strategy and proceeds to effect control according to programmed objectives.

One advantage of digital systems is that the time response can be computed by applying simple long division to a special transform, the z transform. Conversely, when continuous-time systems are analyzed, more difficult methods are needed, such as the inverse Laplace transform or the state transition matrix approach. Stability, control, observability, and controllability can all be defined and applied in the context of discrete-time control.

This chapter introduces the main concepts of discrete-time control. This material may be used as a brief terminating study or as an introduction to a more in-depth discussion of digital control systems in another course to follow.

11.2 Computer Processing

11.2.1 Computer History and Trends

A brief digital computer chronology is given Table 11.1. Although their basic concepts are credited largely to Charles Babbage (c. 1830), today's computers became practical only after the invention and development of the transistor. Rapid technological advances in solid state physics have been the primary driving force behind the rapid evolution of digital computers since 1960. The emphasis here is upon stored-program general-purpose mini- and microcomputers rather than special-purpose digital logic (which will accomplish the same purpose) because it appears that the vast majority of future control applications will be with low-cost, mass-produced hardware.

The charts given in Figure 11.1 show three dramatic trends in computer evolution. Their speed of operation has increased by about an order of magnitude every 5 years since 1955. The density of their electronic circuits has increased by roughly the same factor of 10 every 5 years since the early use of solid state components in 1960. With this compactness has come a large decrease in power requirements. The total costs of digital computation have similarly plummeted.

Whenever a digital computer becomes part of an otherwise analog system, signal conversion takes place. Each analog signal that is to be operated upon by a computer must be converted from analog form to digital form by a A/D converter. Similarly, each digital value that it to influence the analog system must be converted to analog form by a D/A converter. Since the computer output does not change until the next set of calculations and D/A conversions are completed, the analog resultant of some D/A process may be held constant during each cycle by a sample-and-hold (S/H) device. See Figure 11.2.

Analog signals are input, analog signals are output, and in between may be placed powerful computational ability. Operations such as square-rooting, correlation, function generation, and spectral analysis, which are a nightmare in analog hardware, are simply and routinely done digitally. Furthermore, if a general-purpose programmable digital computer is used, changes in objectives and improvements in design may require only changes in the stored program (the software), not changes to the equipment itself (the hardware).

Table 11.1 *Digital Computer Chronology*

Date	Development
B.C.	Abacus is in use. It becomes widespread in Europe and Asia.
c.1650	Pascal builds a mechanical desk calculator for addition.
c.1670	Leibniz builds a calculator also capable of subtraction, multiplication, division, and root extraction.
c.1800	Jacquard perfects automatic looms, which weave designs programmed by punched cards.
c.1830	Babbage develops modern computer principles, including memory, program control, and branching capabilities.
1890	Hollerith uses a punched card system for the U.S. census. Hollerith's company later becomes IBM.
1940	Aiken builds an electromechanical programmed computer. Mark I, used for ballistics calculations by the U.S. Army and capable of several additions per second. Similar work is done at Bell Telephone Laboratories by Stibitz, who notices similarity between telephone switching and computation.
1946	Eckert and Mauchly build the first vacuum tube digital computer. ENIAC, at the University of Pennsylvania. It is capable of several thousand additions per second.
1948	Von Neumann directs construction of the IAS stored-program computer at Princeton. Memory involves charge storage on cathode-ray-tube targets and a rotating magnetic drum. Addition is performed in Approximately 65 μs EDSAC at Cambridge University is completed first, becoming the first stored-program computer. IBM also builds a stored-program computer, the SSEC.
1950	Sperry Rand Corporation builds the first commercial data processing computer, using semiconductor diodes and vacuum tubes, the UNIVAC I.
1952	IBM begins marketing the digital computer model 701 commercially.
1960	The "second generation" of computers is introduced to the market. These machines use solid state components in place of vacuum tubes.
1964	The "third generation" of computers begins. Integrated circuit hardware predominates as in the IBM System 360.
1965	Digital Equipment Corporation markets the PDP-8 at about $50,000. The minicomputer industry begins.
1973	Intel markets the first microcomputer system using the 8080 microprocessor chip.
1975	Ten different microprocessors are on the market, including the Fairchild F8, the Intel 8080A, the Motorola 6800, and the Signetics 2650. Others soon enter the field.
1977	Motorola, Texas Instruments, Intel, and Zilog each begin to market competing 16-bit microprocessors; 32-bit microprocessors are announced.
1978	Rockwell International markets the AIM-65 microprocessor system, priced at about $500. System includes alphanumeric display, small printer, cassette tape interface, and input–output ports. ROM-based software includes system monitor, assembler, and BASIC compiler.

Table 11.1 *Digital Computer Chronology (Continued)*

Date	Development
1982	Very-large-scale integration (VLSI) technology is capable of integrating a million devices on a tiny semiconductor chip, approximately the number of devices in a very large computer processing unit and roughly the complexity level of primitive organisms.
1983	Personal desktop computers are comparable in storage and capability to the mainframes of the 1960s.
1984	Handheld devices are programmable in BASIC.
1986	Intense competition occurs among duplicates (clones) of IBM PCs (personal computers).
1988	The distinction between mainframes and minicomputers and between minicomputers and microcomputers blurs as capacity and architectural improvements continue.
1992	32-bit microprocessor-based PCs with large memory at affordable prices reach workstation capability.
2001	Powerful PCs become pervasive, intelligent and Internet-centric devices (Palm PCs, wireless devices, networked appliances) threaten the market of PCs.

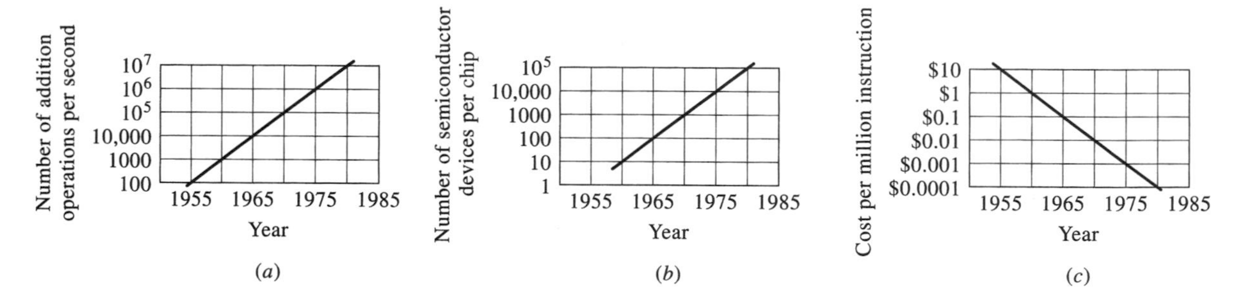

Figure 11.1 Digital computer trends. (a) Approximate speed increase of high-speed computers. (b) Approximate semiconductor device density increase. (c) Approximate decrease in computation cost.

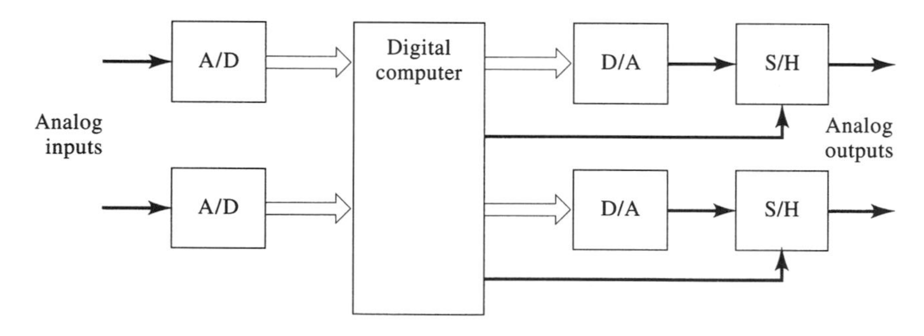

Figure 11.2 A simple digital controller.

Digital computation is subject to numerical *roundoff* or *truncation errors* because numbers are represented by a finite number of bits. For example, the product of two 16-bit numbers involves potentially 32 bits. To represent such a product as another 16-bit number, the least significant 16 bits of the product must be eliminated. Ordinarily, finite arithmetic precision is not of great concern, since the number of bits used to represent a number far exceeds the required numerical precision. However, the power of a computer to perform huge numbers of calculations in a short time means that it is possible for tiny errors to quickly accumulate to large proportions. One must also be aware that certain kinds of calculations, such as forming differences between large but nearly identical numbers, are especially susceptible to error.

11.3 A/D and D/A Conversion

11.3.1 Analog-to-Digital Conversion

An analog signal such as a voltage can be expressed as a binary number, suitable for computer processing, by assigning weights to each bit position. Table 11.2 gives a 4-bit coding of an analog signal that may range between 0 to 10 V. Each binary increment represents $2^{-4} = \frac{1}{16}$, the maximum representable voltage of 10 V. The information in the table is shown in graphical form in Figure 11.3.

Each binary number represents a *range* of analog voltage; hence there is a *quantization error* associated with the conversion. For a 4-bit conversion, the maximum

Table 11.2 *Representing a Nonnegative Analog Voltage with a Binary Number*

Analog Voltage	Binary Representation
0–0.625	0000
0.625–1.25	0001
1.25–1.875	0010
1.875–2.5	0011
2.5–3.125	0100
3.125–3.75	0101
3.75–4.375	0110
4.375–5.0	0111
5.0–5.625	1000
5.625–6.25	1001
6.25–6.875	1010
6.875–7.5	1011
7.5–8.125	1100
8.125–8.75	1101
8.75–9.375	1110
9.375–10.0	1111

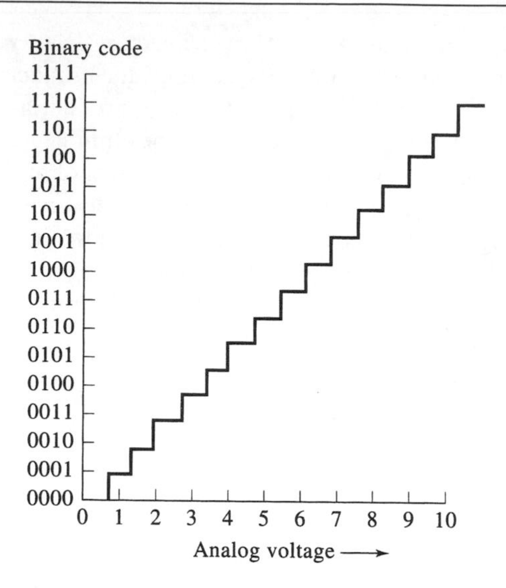

Figure 11.3 Binary coding of an analog voltage.

quantization error is $2^{-4} = 6.25$ %. Table 11.3 shows quantization error percentages for various numbers of bits in the digital representation. The quantization error in 16-bit conversion, for example, corresponds to a signal-to-noise ratio (SNR) of

$$\text{SNR (in dB)} = 20 \log_{10} 2^{16} = 96.3 \text{ dB}$$

By comparison, typical signal-to-noise ratios in quality audio recording and reproduction are 60 to 70 dB, which may be accurately portrayed by only 12-bit coding.

For bipolar signals, the three types of binary code shown in Table 11.4 are the most commonly used. In the sign and magnitude arrangement, the most significant bit of the binary code represents the algebraic sign of the signal, with a zero meaning a positive number. The remaining bits are the binary representation

Table 11.3 *Quantization Error For Analog Digital Conversion*

Number of Bits	Maximum Percentage Error
1	50
2	25
4	6.25
6	1.56
8	0.391
10	0.0977
12	0.0244
14	0.0061
16	0.0015

Table 11.4 *Bipolar Analog Voltage Representations*

Analog Voltage	Sign and Magnitude	Offset Binary	Two's Complement
−5.0 to −4.375	1111	0000	1001
−4.375 to −3.75	1110	0001	1010
−3.75 to −3.125	1101	0010	1011
−3.125 to −2.5	1100	0011	1100
−2.5 to −1.875	1011	0100	1101
−1.875 to −1.25	1010	0101	1110
−1.25 to −0.625	1001	0110	1111
0.625–0	1000	0111	1000
0–0.625	0000	1000	0000
0.625–1.25	0001	1001	0001
1.25–1.875	0010	1010	0010
1.875–2.5	0011	1011	0011
2.5–3.125	0100	1100	0100
3.125–3.75	0101	1101	0101
3.75–4.375	0110	1110	0110
4.375–5.0	0111	1111	0111

of the signal's magnitude. The offset binary code is equivalent to adding a fixed constant (or bias) to the signal to be converted so that the sum is always nonnegative. In two's complement coding, negative signals are represented as the two's complement of their magnitude, in the same manner as negative numbers are commonly manipulated in digital computers. In applications involving digital displays, binary-coded-decimal (BCD) coding may be used, where the signal is representated as *decimal* digits and then each decimal digit is individually converted to a 4-bit binary equivalent.

❑ DRILL PROBLEM

D11.1 What is the maximum percentage error if a binary number is truncated to 10 bits? What if the number is rounded?

Ans. 0.098 %; 0.049 %

11.3.2 Sample and Hold

A/D and D/A converters are generally used to repetitively perform conversions. For analog-to-digital conversion it is often desirable to "freeze" the anlog signal while the conversion is taking place. A sample-and-hold device, with symbol given in Figure 11.4(a), may be used to hold an analog signal steady while conversion proceeds, as in Figure 11.4(b).

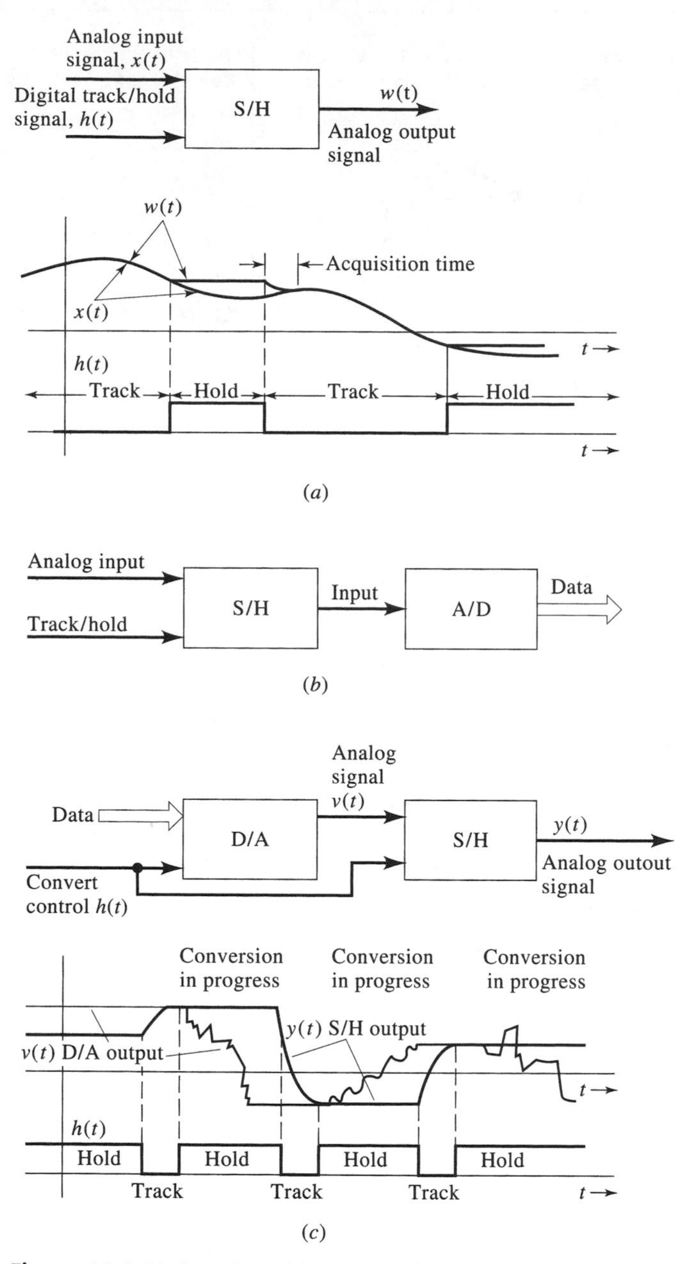

Figure 11.4 (a) Sample-and-hold used in analog/digital conversion. (b) Sample-and-hold symbol and representative symbols. (c) S/H used to "freeze" an analog signal while D/A conversion takes place.

11.3.3 Digital-to-Analog Conversion

In converting from digital to analog, a D/A converter output may fluctuate wildly while conversion is taking place. A sample-and-hold device is conveniently used to hold the previously converted signal while a new conversion takes place, as in Figure 11.4(c). The result is an output signal that changes in nearly a stepwise fashion each time a conversion occurs.

❑ **DRILL PROBLEMS**

D11.2 A 12-bit D/A converter has minimum output voltage -10 and maximum output voltage $+10$. After the binary code 010110101001 is applied, what is the output voltage if the converter is of each of the following types?

(a) Sign and magnitude

(b) Offset binary

(c) Two's complement

Ans. (a) 7.08; (b) -2.92; (c) 7.08

D11.3 The sinusoidal signal

$$f(t) = 10 \sin t$$

is tracked for $t < 1$, held for $1 \leqslant t < 2$, tracked for $2 \leqslant t < 5$, then held thereafter to form the signal $g(t)$. Sketch both $f(t)$ and $g(t)$.

11.4 Discrete-Time Signals

11.4.1 Representing Sequences

Periodic samples of a continuous-time signal, as are generated by an A/D converter, form a sequence of numbers termed a *discrete-time signal*. Figure 11.5 shows a continuous-time signal $f(t)$ and the corresponding sequence of samples

$$f(t = 0),\ f(t = T),\ f(t = 2T),\ f(t = 3T), \ldots$$

Although it results in an ambiguity that is resolved only by context, it is common practice to denote the sequence by $f(k)$, where k is the sample number.

Some important sequences are shown in Figures 11.6 to 11.10. All sequences consist of zero samples prior to $k = 0$. The unit pulse sequence, δk in Figure 11.6, has unit sample value for $k = 0$ and all other values zero. The unit step sequence $u(k)$, Figure 11.7, has samples that are all unity for $k = 0$ and thereafter. The ramp, Figure 11.8, consists of samples of the continuous-time unit ramp function.

The sampled exponential, Figure 11.9, has samples that are progressive powers of the number

$$c = e^{-aT}$$

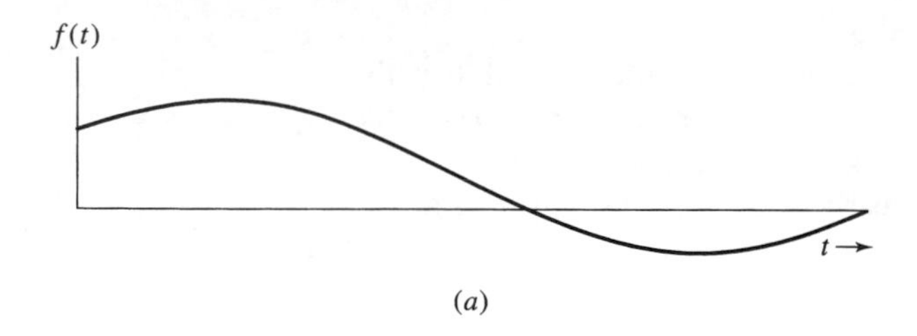

(a)

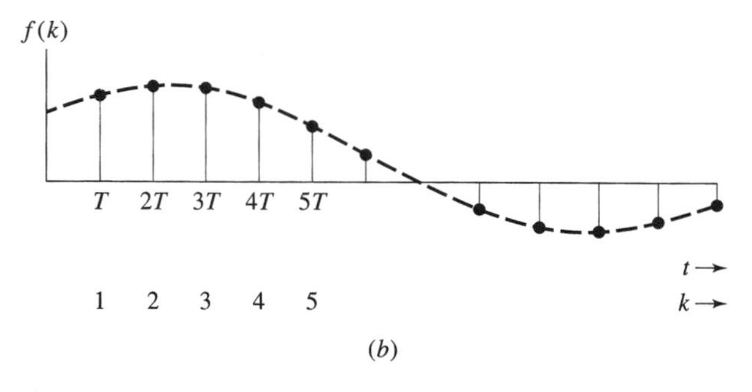

(b)

Figure 11.5 Sampling a continuous-time signal. (a) A continuous-time signal. (b) Samples of the continuous-time signal.

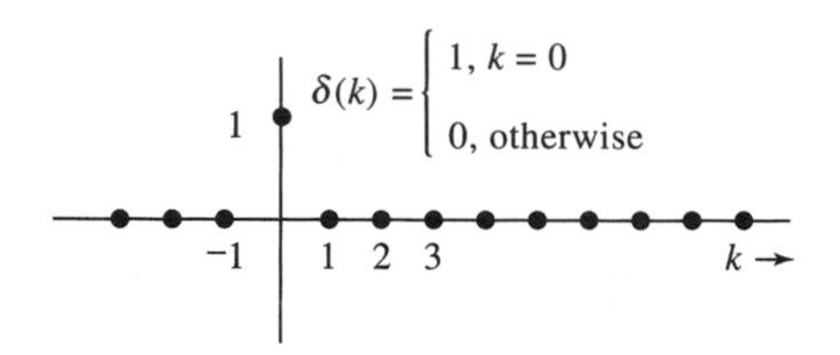

Figure 11.6 Unit impulse (or pulse) sequence.

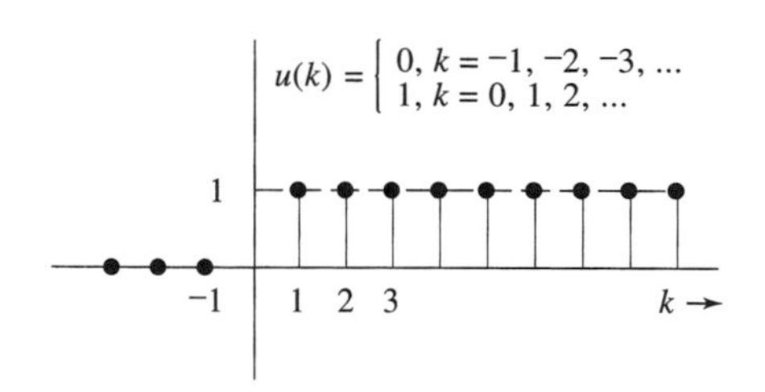

Figure 11.7 Unit step.

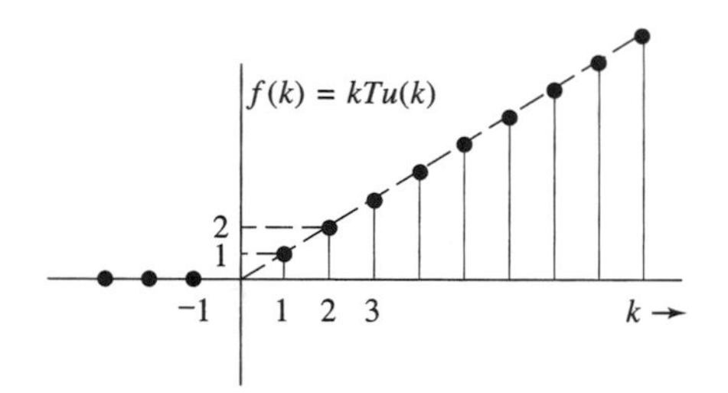

Figure 11.8 Ramp.

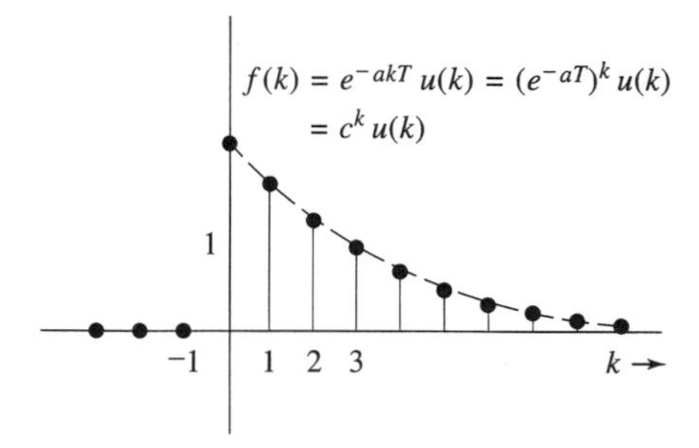

Figure 11.9 Exponential (or geometric).

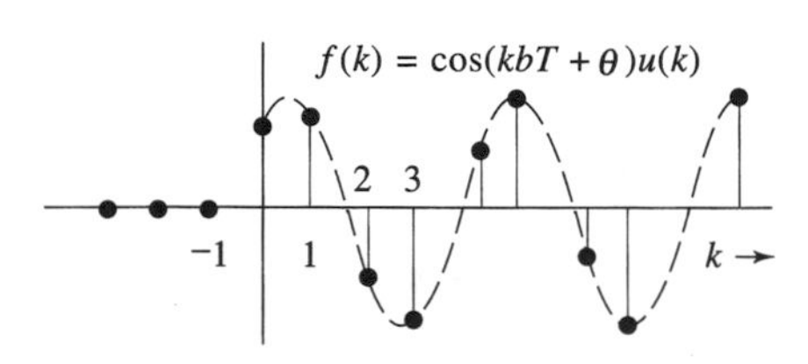

Figure 11.10 Sinusoidal.

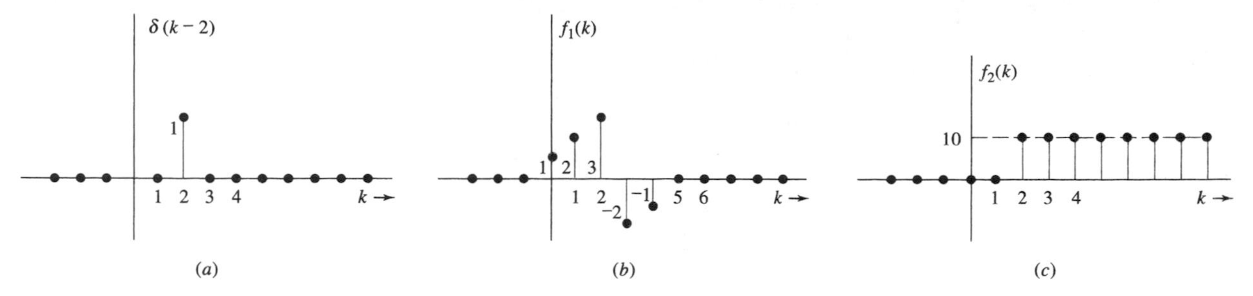

Figure 11.11 Shifting and summing sequences. (a) A shifted unit pulse sequence. (b) A finite sum of shifted pulses. (c) Shifted step sequence or a step sequence with samples at $k = 0$ and $k = 1$ canceled by pulses.

and so forms a geometric sequence:

$$f(0) = c^0 = 1$$

$$f(1) = c$$

$$f(2) = c^2$$

$$f(n) = c^n$$

Geometric sequences (or sampled exponential functions) have fundamental importance to discrete time systems in the same way that exponential functions are basic to continuous-time system. A sampled sinusoidal function as in Figure 11.10 is termed a *sinusoid sequence*.

More complicated sequences may often be represented as shifts and sums of the basic sequences. For example, $\delta(k-2)$ is the unit pulse sequence shifted two samples to the right, as in Figure 11.11 (a). The sequence of Figure 11.11(b) is thus

$$f_1(k) = \delta(k) + 2\delta(k-1) + 3\delta(k-2) - 2\delta(k-3) - \delta(k-4)$$

The sequence

$$f_2(k) = \begin{cases} 10 & k = 2,\ 3,\ 4, \ldots \\ 0 & \text{otherwise} \end{cases}$$

drawn in Figure 11.11(c), is

$$f_2(k) = 10u(k) - 10\delta(k) - 10\delta(k-1)$$

Alternatively, we write

$$f_2(k) = 10u(k-2)$$

❑ DRILL PROBLEMS

D11.4 Sketch the function $f(t)$ and the samples $f(k)$ for a sampling period $T = 0.5$ s.

(a)

$$f(t) = e^{-0.5t}u(t)$$

Ans. $f(k) = e^{-0.25k}u(k)$

(b)

$$f(t) = (\sin \pi t)u(t)$$

Ans. $f(k) = (\sin k\pi/2)u(k)$

(c)

$$f(t) = (\cos 2\pi t)u(t)$$

Ans. $f(k) = (-1)^k u(k)$

(d)

$$f(t) = (\sin 2\pi t)u(t)$$

Ans. $f(k) = 0$

11.4.2 z-Transformation and Properties

The z transform of a sequence $f(k)$ is defined as the infinite series

$$Z[f(k)] = F(z) = \sum_{k=0}^{\infty} f(k)z^{-k}$$

It plays much the same role in the description of discrete-time signals as the Laplace transform does with continuous-time signals.

Table 11.5 lists basic z-transform pairs together with the corresponding continuous-time function that gives the sequence when sampled with period T. When we use the z-transform definition, the transform of the unit pulse is

$$Z[\delta(k)] = \sum_{k=0}^{\infty} \delta(k)z^{-k} = z^{-0} = 1$$

It should be noted that the unit pulse sequence $\delta(k)$, while analogous to the unit impulse $\delta(t)$,does not consist of samples of $\delta(t)$, which is infinite for $k = t = 0$.

The z transform of the unit step sequence is as follows:

$$Z[u(k)] = \sum_{k=0}^{\infty} 1 \cdot z^{-k}$$

using the fact that for a geometric series,

The geometric series formula is very useful.

$$\sum_{k=0}^{\infty} x^k = \frac{1}{1-x} \qquad |x| < 1$$

then

$$Z[u(k)] = \sum_{k=0}^{\infty} z^{-k} = \sum_{k=0}^{\infty} \left(\frac{1}{z}\right)^k = \frac{1}{1-1/z} = \frac{z}{z-1} \qquad \left|\frac{1}{z}\right| < 1$$

Table 11.5 *Some Laplace and z-Transform Pairs*

$f(t)$	$F(s)$	$f(k)$	$F(z)$
		$\delta(k)$, unit pulse	1
$u(t)$, unit step	$\frac{1}{s}$	$u(k)$, unit step	$\frac{z}{z-1}$
$tu(t)$	$\frac{1}{s^2}$	$kTu(k)$	$\frac{Tz}{(z-1)^2}$
$e^{-at}u(t)$	$\frac{1}{s+a}$	$(e^{-aT})^k u(k) = c^k u(k)$ where $c = e^{-aT}$	$\frac{z}{z-e^{-aT}} = \frac{z}{z-c}$
$te^{-at}u(t)$	$\frac{1}{(s+a)^2}$	$KT(e^{-at})^k u(k) = kTc^k u(k)$	$\frac{Tze^{-aT}}{(z-e^{-aT})^2} = \frac{Tcz}{(z-c)^2}$
$(\sin bt)u(t)$	$\frac{b}{s^2+b^2}$	$(\sin kbT)u(k)$	$\frac{z \sin bT}{z^2 - 2z \cos bT + 1}$
$(\cos bt)u(t)$	$\frac{s}{s^2+b^2}$	$(\cos kbT)u(k)$	$\frac{z(z-\cos bT)}{z^2 - 2z \cos bT + 1}$
$e^{-at}(\sin bt)u(t)$	$\frac{b}{(s+a)^2+b^2}$	$(e^{-aT})^k(\sin kbT)u(k)$ $= c^k(\sin kbT)u(k)$	$\frac{z(e^{-aT} \sin bT)}{(z-e^{(-a+jb)T})(z-e^{(-a-jb)T})}$ $= \frac{zc \sin bT}{z^2 - (2c \cos bT)z + c^2}$
$e^{-at}(\cos bt)u(t)$	$\frac{s+a}{(s+a)^2+b^2}$	$(e^{-aT})^k(\cos kbT)u(k)$ $= c^k(\cos kbT)u(k)$	$\frac{z(z-e^{-aT} \cos bT)}{(z-e^{(-a+jb)T})(z-e^{(-a-jb)T})}$ $= \frac{z(z-c \cos bT)}{z^2 - (2c \cos bT)z + c^2}$

Conditions for z-transform convergence are satisfied by all but the most pathological sequences and so are not emphasized here.

A sampled exponential function, whether decaying or expanding, is of the form

$$f(k) = e^{-kaT} = (e^{-aT})^k$$

Its z transform is given by

$$Z\left[e^{-kaT}\right] = \sum_{k=0}^{\infty} e^{-kaT} z^{-k} = \sum_{k=0}^{\infty} \left(\frac{1}{ze^{aT}}\right)^k = \frac{1}{1 - 1/ze^{aT}} = \frac{z}{z - e^{-aT}}$$

This z transform is worth memorizing.

Samples of an exponential function form a geometric series, since by definition

$$c = e^{-aT}$$
$$f(k) = e^{-kaT} = (e^{-aT})^k = c^k$$

In terms of c, the z transform is

$$Z\left[c^k\right] = \frac{z}{z - c}$$

The transforms of sampled sinusoids given in Table 11.5 follow easily from expanding the sinusoidal function into Euler components and applying the result for sampled exponentials. For the sampled sine, we have

$$\begin{aligned}
Z[\sin kbT] &= \sum_{k=0}^{\infty} \left(\frac{e^{jkbT} - e^{-jkbT}}{2j}\right) z^{-k} \\
&= \frac{1}{2j} \sum_{k=0}^{\infty} e^{jkbT} z^{-k} - \frac{1}{2j} \sum_{k=0}^{\infty} e^{-jkbT} z^{-k} \\
&= \frac{z/2j}{z - e^{jbT}} - \frac{z/2j}{z - e^{-jbT}} \\
&= \frac{(z/2j)(z - e^{-jbT} - z + e^{jbT})}{z^2 - (e^{jbT} + e^{-jbT})z + 1} \\
&= \frac{z(e^{jbT} + e^{-jbT})/2j}{z^2 - 2\left[(e^{jbT} + e^{-jbT}/2)\right] z + 1} \\
&= \frac{z \sin bT}{z^2 - 2z \cos bT + 1}
\end{aligned}$$

Basic z-transform properties are listed in Table 11.6. The transform of a sequence scaled by a multiplicative constant is that constant times the original z transform. The z transform of a sample-by-sample sum of sequences is the sum of their individual z transforms. A sequence weighted by the step number k has z transform

$$\begin{aligned}
Z[kf(k)] &= \sum_{k=0}^{\infty} kf(k) z^{-k} = \sum_{k=0}^{\infty} f(k)(kz^{-k}) \\
&= \sum_{k=0}^{\infty} f(k) \frac{zd}{dz}(-z^{-k}) = -\frac{zd}{dz} \sum_{k=0}^{\infty} f(k) z^{-k} \\
&= -\frac{zd}{dz} F(z)
\end{aligned}$$

A sequence weighted by successive powers of a constant c has z transform as follows:

$$\begin{aligned}
Z\left[c^k f(k)\right] &= \sum_{k=0}^{\infty} c^k f(k) z^{-k} = \sum_{k=0}^{\infty} f(k) \left(\frac{z}{c}\right)^{-k} \\
&= F\left(\frac{z}{c}\right)
\end{aligned}$$

Table 11.6 *Some z-Transform Properties*

$Z[cf(k)] = cF(z)$, c a constant

$Z[f(k) + g(k)] = F(z) + G(z)$

$Z[kf(k)] = -\frac{zdF(z)}{dz}$

$Z[c^k f(k)] = F\left(\frac{z}{c}\right)$, c a constant

$Z[f(k-1)] = f(-1) + z^{-1}F(z)$

$Z[f(k-2)] = f(-2) + z^{-1}f(-1) + z^{-2}F(z)$

$Z[f(k-n)] = f(-n) + z^{-1}f(1-n) + z^{-2}f(2-n) + \cdots + z^{-n+1}f(-1) + z^{-n}F(z)$

$Z[f(k+1)] = zF(z) - zf(0)$

$Z[f(k+2)] = z^2F(z) - z^2f(0) - zf(1)$

$Z[f(k+n)] = z^nF(z) - z^nf(0) - z^{n-1}f(1) - \cdots - z^2f(n-2) - zf(n-1)$

$f(0) = \lim_{z\to\infty} F(z)$

If $\lim_{k\to\infty} f(k)$ exists and is finite,

$\lim_{k\to\infty} f(k) - \lim_{z\to 1}\left[\frac{z-1}{z}F(z)\right]$

Figure 11.12(a) shows an example of a sequence that is shifted one step to the right, its z transform is given by

$$\begin{aligned} Z[f(k-1)] &= \sum_{k=0}^{\infty} f(k-1)z^{-k} = \sum_{k=-1}^{\infty} f(k)z^{-(k+1)} \\ &= f(-1) + z^{-1}\sum_{k=0}^{\infty} f(k)z^{-k} \\ &= f(-1) + z^{-1}F(z) \end{aligned}$$

Then

$$\begin{aligned} Z[f(k-2)] &= f(-2) + z^{-1}Z[f(k-1)] \\ &= f(-2) + z^{-1}\left[f(-1) + z^{-1}F(z)\right] \\ &= f(-2) + z^{-1}f(-1) + z^{-2}F(z) \end{aligned}$$

and, similarly,

$$Z[f(k-n)] = f(-n) + z^{-1}f(1-n) + \cdots + z^{-n+1}f(-1) + z^{-n}F(z)$$

For a left shift of the sequence, Figure 11.12(b), we write

$$Z[f(k+1)] = \sum_{k=0}^{\infty} f(k+1)z^{-k} = \sum_{k=1}^{\infty} f(k)z^{-(k-1)}$$

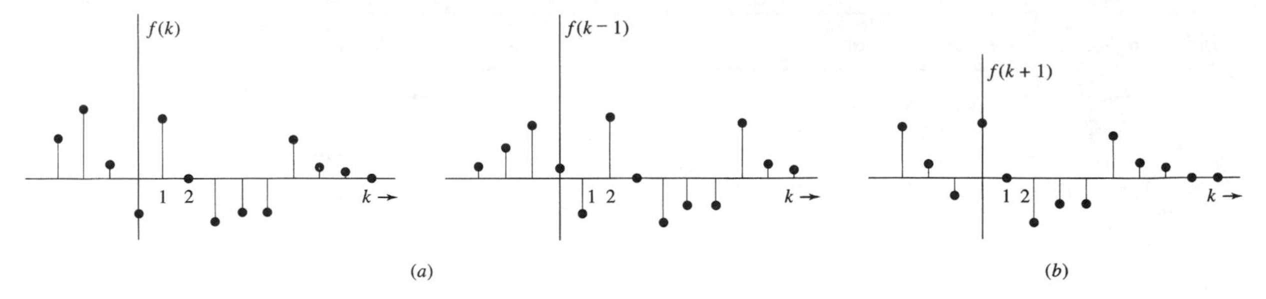

Figure 11.12 Right- and left-shifted sequences. (a) A sequence and the same sequence shifted right one step. (b) The sequence shifted left one step.

$$= z\sum_{k=1}^{\infty} f(k)z^{-k} = z\sum_{k=0}^{\infty} f(k)z^{-k} - zf(0)$$

$$= zF(z) - zf(0)$$

These properties are useful for solving difference equations.

Similarly,

$$Z[f(k+2)] = z[zF(z) - zf(0)] - zf(1)$$
$$= z^2F(z) - z^2f(0) - zf(1)$$

and

$$Z[f(k+n)] = z^nF(z) - z^nf(0) - z^{n-1}f(1) - \cdots - z^2f(n-2) - zf(n-1)$$

❑ DRILL PROBLEM

D11.5 Find the z transforms of the following sequences:

(a) $f(k) = \left[(-0.5)^k - 4(0.2)^k\right]u(k)$

Ans. $\dfrac{z}{z+0.5} - \dfrac{4z}{z-0.2}$

(b) $f(k) = \begin{cases} (-1)^k & k = 3, 4, 5, \ldots \\ 0 & \text{otherwise} \end{cases}$

Ans. $-z^{-2}/(1+z)$

(c) $f(1) = 2,\ f(4) = -3,\ f(7) = 8$, and all other samples are zero

Ans. $2z^{-1} - 3z^{-4} + 8z^{-7}$

(d) $f(k) = u(k)\sin 3k - 2u(k-4)\sin 3(k-4)$

Ans. $\left[(1 - 2z^{-4})z\sin 3\right]/(z^2 - 2z\cos 3 + 1)$

11.4.3 Inverse z Transform

The sequence of samples represented by a rational z transform may be obtained, if desired, by long division. Consider the z transform

$$F(z) = \frac{4z}{z^2 - z + 0.5}$$

For example:

$$\begin{array}{r|l} & 4z^{-1} + 4z^{-2} + 2z^{-3} + \cdots \\ \hline z^2 - z + 0.5 & 4z \\ & \underline{4z - 4 + 2z^{-1}} \\ & \quad 4 - 2z^{-1} \\ & \quad \underline{4 - 4z^{-1} + 2z^{-2}} \\ & \qquad 2z^{-1} - 2z^{-2} \end{array}$$

Since

$$F(z) = \frac{4z}{z^2 - z + 0.5} = 0z^0 + 4z^{-1} + 4z^{-2} + 2z^{-3} + \cdots$$

then we have

$$f(k) = 4\delta(k-1) + 4\delta(k-2) + 2\delta(k-3) + \cdots$$

and

$$f(0) = 0$$
$$f(1) = 4$$
$$f(2) = 4$$
$$f(3) = 2$$
$$\vdots$$

Repeated steps of long division do not give a closed-form expression for the sequence represented by a z transform, although in principle as many terms in the sequence as desired may be found in this manner. To find a formula for the sequence of samples, partial fraction expansion may be used. Rather than expanding a z transform $F(z)$ directly in partial fractions, $F(z)/z$ is expanded so that terms of the form

$$\frac{z}{z - e^{-aT}}$$

result. For example, for the z transform

To use partial fractions, divide F(z) by z first.

$$F(z) = \frac{-2z^2 + 2z}{z^2 + 4z + 3}$$

$$\frac{F(z)}{z} = \frac{-2z + 2}{(z+1)(z+3)} = \frac{2}{z+1} + \frac{-4}{z+3}$$

gives

$$F(z) = \frac{2z}{z+1} + \frac{-4z}{z+3}$$

$$f(k) = 2(-1)^k - 4(-3)^k \qquad k = 0,\ 1,\ 2,\ 3, \cdots$$

Another example is the following:

$$F(z) = \frac{z^3 - 3}{z(z-0.25)(z-0.5)}$$

$$\frac{F(z)}{z} = \frac{z^3 - 3}{z^2(z-0.25)(z-0.5)}$$

$$= \frac{-144}{z} + \frac{-24}{z^2} + \frac{191}{z-0.25} + \frac{-46}{z-0.5}$$

Therefore,

$$F(z) = -144z^0 - 24z^{-1} + \frac{-191z}{z-0.25} + \frac{-46z}{z-0.5}$$

and then

$$f(k) = -144\delta(k) - 24\delta(k-1) - 191(0.25)^k - 46(0.5)^k$$

A set of complex conjugate root terms should be manipulated into the form of the last two entries of Table 11.5:

$$F(z) = \frac{4z^2 - 3z}{z^2 + 2z + 2}$$

$$= \frac{K_1 zc \ \sin\ bT}{z^2 - 2c\ \cos\ bT + c^2} + \frac{K_2 z(z - c\ \cos\ bT)}{z^2 - 2c\ \cos\ bT + c^2}$$

Equating

$$c^2 = 2 \qquad c = \sqrt{2}$$

$$2c \cos bT = 2\sqrt{2} \cos bT = -2 \qquad bT = \frac{3\pi}{4}$$

$$\sin bT = \sin \frac{3\pi}{4} = \frac{1}{\sqrt{2}}$$

we write

$$F(z) = \frac{4z^2 - 3z}{z^2 + 2z + 2} = \frac{K_1 z + K_2 z(z+1)}{z^2 + 2z + 2}$$

giving

$$K_2 = 4$$

$$K_1 + K_2 = -3 \qquad K_1 = -7$$

$$f(k) = 4(\sqrt{2})^k\ \cos\ \frac{3\pi k}{4} - 7(\sqrt{2})^k \sin\ \frac{3\pi k}{4} \qquad k = 0,\ 1,\ 2, \ldots$$

❑ DRILL PROBLEMS

D11.6 Use long division to show that the inverse z transform of

$$F(z) = \frac{10z}{(z-1)^2}$$

is $f(k) = 10k$.

D11.7 Find the inverse z transforms for $k \geqslant 0$:

(a)

$$F(z) = \frac{1}{z+0.3}$$

Ans. $\frac{10}{3}\delta(k) - \frac{10}{3}(-0.3)^k$

(b)

$$F(z) = \frac{-6z^2 + z}{z^2 + 5z + 6}$$

Ans. $-19(-3)^k + 13(-2)^k$

(c)

$$F(z) = \frac{4z^2 - 3z + 2}{z^2 + 4z + 4}$$

Ans. $\frac{1}{2}\delta(k) + \frac{7}{2}(-2)^k + 6k(-2)^k$

(d)

$$F(z) = \frac{3z^2 - z}{z^2 + 2z + 10}$$

Ans. $3(\sqrt{10})^k \cos 1.89k - 1.33(\sqrt{10})^k \sin 1.89k$

11.5 Sampling

When an analog signal $f(t)$ is sampled to form the sequence $f(k)$, there is a direct relationship between the Laplace transform $F(s)$ of the analog signal and the z transform $F(z)$ of the sequence. If a rational Laplace transform is expanded into a sum of terms of the type given in Table 11.5, the z transform of the sample sequence is obtained by simply summing the corresponding z-transform terms from the table.

For example, for a continuous-time signal with Laplace transform

$$F(s) = \frac{4s^2 + 13s + 18}{s^3 + 5s^2 + 6s}$$

$$= \frac{3}{s} + \frac{-4}{s+2} + \frac{5}{s+3}$$

from which we get

$$f(t) = (3 - 4e^{-2t} + 5e^{-3t})u(t)$$

For a sampling interval $T = 0.2$, (we let $t = 0.2k$) to get

$$f(k) = \left(3 - 4e^{-0.4k} + 5e^{-0.6k}\right) u(k)$$
$$= \left[3(1)^k - 4(e^{-0.4})^k + 5(e^{-0.6})^k\right] u(k)$$
$$F(z) = 3\left(\frac{z}{z-1}\right) - 4\left(\frac{z}{z - e^{-0.4}}\right) + 5\left(\frac{z}{z - e^{-0.6}}\right)$$

When the Laplace transform involves delay operations in multiples of the sampling interval T, the remainder of the transform is expanded into partial fraction terms, as in the following example, for which $T = 0.1$:

$$F(s) = \frac{e^{-0.1s} + 2}{s(s+3)} = (e^{-0.1s} + 2)\left(\frac{1/3}{s} + \frac{-1/3}{s+3}\right)$$

Upon denoting

$$G(s) = \frac{1/3}{s} + \frac{-1/3}{s+3}$$
$$F(s) = (e^{-0.1s} + 2)G(s)$$
$$f(t) = g(t - 0.1)u(t - 0.1) + 2g(t)u(t)$$

we have

$$f(k) = g(k-1)u(k-1) + 2g(k)u(k)$$
$$F(z) = (z^{-1} + 2)G(z)$$
$$= (z^{-1} + 2)\left(\frac{1}{3}\frac{z}{z-1} - \frac{1}{3}\frac{z}{z - e^{-0.3}}\right)$$
$$= \frac{0.173z^2 + 0.086z}{z(z-1)(z-0.74)}$$

Finding the z transform of the corresponding sequence thus involves separating the time-delay operations from the rational part of the Laplace transform, then substituting z^{-1} for each unit of time delay in the delay operation and substituting from the entries of Table 11.5 for each term in the partial fraction expansion of the remainder. Another example is the following, for which the sampling period is $T = 0.05$:

$$F(s) = \frac{10}{s^2 + 4} + \frac{6e^{-0.2s}}{s^2 + 3s}$$
$$= \frac{10}{4}\left(\frac{4}{s^2 + 4}\right) + e^{-0.2s}\left(\frac{2}{s} + \frac{-2}{s+3}\right)$$
$$F(z) = \frac{10}{4}\left(\frac{z \sin 0.1}{z^2 - 2z\cos 0.1 + 1}\right) + z^{-4}\left(\frac{2z}{z-1} - \frac{2z}{z - e^{-0.15}}\right)$$
$$= \frac{10}{4}\left(\frac{0.0998z}{z^2 - 1.99z + 1}\right) + \frac{1}{z^4}\left(\frac{2z}{z-1} - \frac{2z}{z - 0.86}\right)$$

❑ **DRILL PROBLEM**

D11.8 For each of the analog signals with the given Laplace transform $F(s)$, find the z transform $F(z)$ of the corresponding sample sequence with the given sampling interval T:

(a)

$$F(s) = \frac{-4s+1}{s^2+7s+12} \qquad T = 0.2$$

Ans. $(-4z^2+3.48)/(z-0.5488)(z-0.45)$

(b)

$$F(s) = \frac{100}{s^2+2s+10} \qquad T = 0.5$$

Ans. $20.2z/(z^2-0.086z+0.368)$

(c)

$$F(s) = \frac{e^{-0.1s}-3e^{-0.3s}+2}{s^2}\left(\frac{10}{s+1}\right) \qquad T = 0.1$$

Ans. $(2z^3+z^2-3)(0.05z+0.0045)/z^2(z-1)^2(z-0.905)$

(d)

$$F(s) = \frac{se^{-s}+4}{s^2+9} \qquad T = 0.2$$

Ans. $(z-0.825)/z^4(z^2-1.65z+1)+0.753z/(z^2-1.65z+1)$

11.6 Reconstruction of Signals from Samples

11.6.1 Representing Sampled Signals with Impulses

Description of A/D conversion involves discrete-time representation of continuous-time signals, the process termed *sampling*. Conversely, describing D/A conversion requires continuous-time representation of discrete-time signals. This process is termed *reconstruction.*

Although very often the analog signal reconstructed from digital samples is a sampled-and-held waveform, that is not always the case. A more fundamental continuous-time signal related to a sequence of samples is a train of impulses, timed periodically at intervals T, with strengths equal to the corresponding samples. Figure 11.13 shows an original analog signal $f(t)$, the corresponding sample sequence $f(k)$, and the impulse train $f^*(t)$, which is a useful representation of the samples in the analog domain. The objective of reconstruction is to recover from the samples $f(k)$, the analog signal $f(t)$ or a sufficiently close approximation to it. Of course, the signal $f(t)$ may not actually exist anywhere in the system. Typically the samples $f(k)$ are computed from combinations of samples of other signals, present and delayed. From the samples $f(k)$, it is desired to generate an analog signal that *could have been* sampled to obtain $f(k)$.

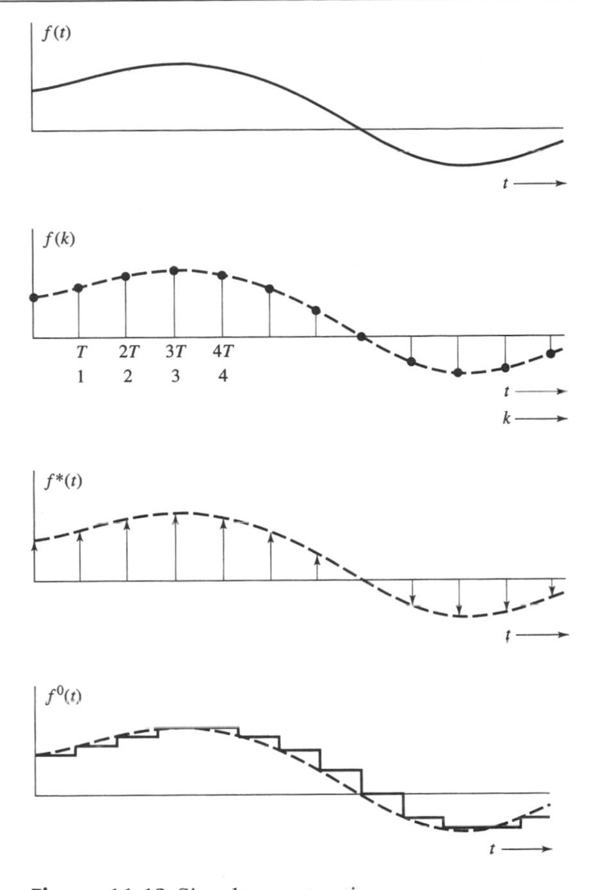

Figure 11.13 Signal reconstruction.

The sampled-and-held waveform $f^\circ(t)$ in Figure 11.14 is a reconstructed signal with samples $f(k)$. It may be derived from the impulse train by passing the impulses through an appropriate transmittance, termed a *zero-order hold*. The impulse train is related to the samples by

An impulse train.

$$f^*(t) = f(0)\delta(t) + f(1)\delta(t-T) + f(2)\delta(t-2T) + \cdots$$
$$= \sum_{k=0}^{\infty} f(k)\delta(t-kT)$$

To obtain the sample-and-hold waveform from $f^*(t)$ requires a linear, time-invariant analog system with the impulse response given in Figure 11.14(a). A unit impulse input to this transmittance causes a unit rectangular pulse output of duration T as shown. A delayed impulse with different amplitude produces a delayed pulse with that amplitude, as in Figure 11.14(b), and an impulse train, Figure 11.14(c), results in the desired sampled-and-held reconstruction.

The required impulse response of the zero-order hold is

$$y_{\text{impulse}}(t) = u(t) - u(t-T)$$

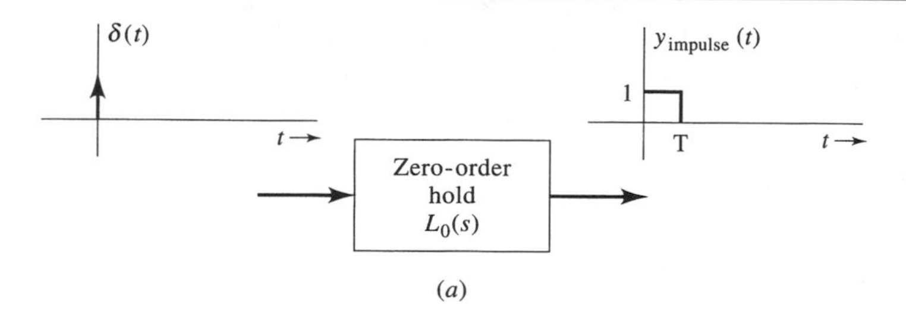

(a)

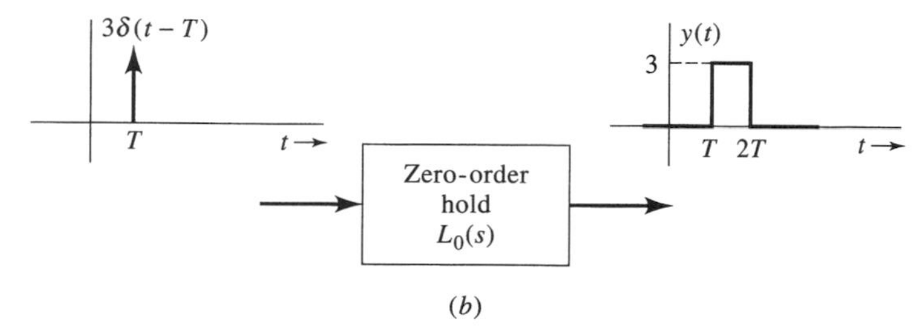

(b)

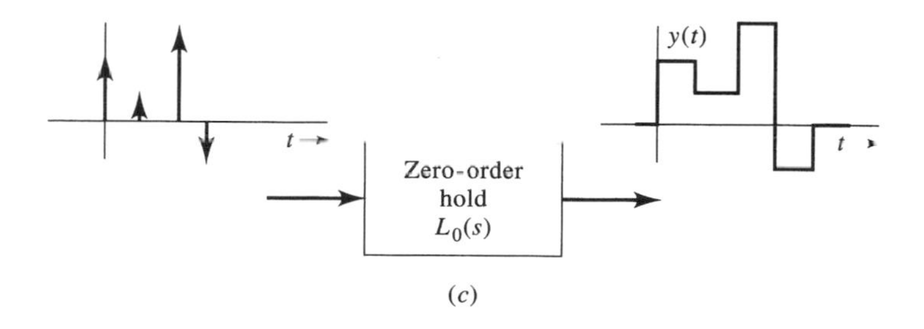

(c)

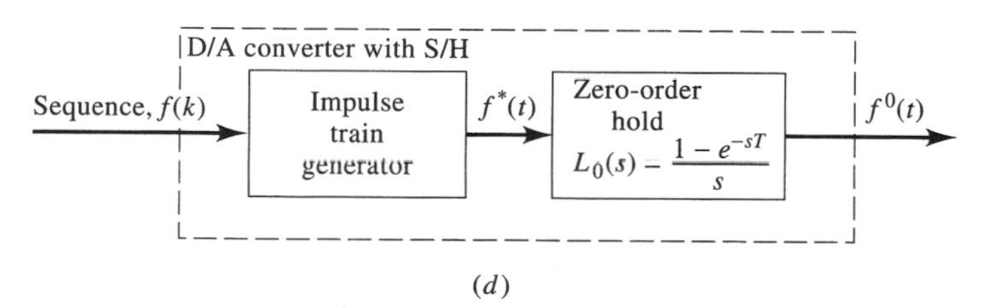

(d)

Figure 11.14 Response of the zero-order hold transmittance. (a) Impulse response. (b) esponse to a scaled and delayed impulse. (c) Response to an impulse train. (d) Model of a D/A converter with S/H output.

and its Laplace transform is

$$Y_{\text{impulse}}(s) = \frac{1 - e^{-sT}}{s}$$

Since the Laplace transform of the unit impulse is unity, the zero-order hold transmittance, which is the ratio of the transforms, is

$$L_0(s) = Y_{\text{impulse}}(s) = \frac{1 - e^{-sT}}{s}$$

Transfer function of ZOH.

A D/A converter with sample-and-hold output may thus be modeled by the idealized impulse train generator followed by $L_0(s)$, as in Figure 11.14(d).

11.6.2 Relation Between the z Transform and the Laplace Transform

The impulse train associated with a sequence $f(k)$ is

$$f^*(t) = f(0)\delta(t) + f(1)\delta(t-T) + f(2)\delta(t-T) + f(3)\delta(t-3T) + \cdots$$
$$= \sum_{k=0}^{\infty} f(k)\delta(t-kT)$$

This looks like the z transform.

The Laplace transform of the impulse train is

$$F^*(s) = \mathcal{L}[f^*(t)]$$
$$= f(0) + f(1)e^{-sT} + f(2)e^{-2sT} + f(3)e^{-3sT} + \cdots$$
$$= \sum_{k=0}^{\infty} f(k)(e^{sT})^{-k}$$

Letting

$$z = e^{sT}$$

there results

$$F^*(s)\Big|_{e^{sT}=z} = \sum_{k=0}^{\infty} f(k)z^{-k} = Z[f(k)] = F(z)$$

One interpretation of the z transformation is that it is the Laplace transform of the impulse train with e^{sT} replaced by z.

A sequence $f(k)$ with a rational z transform $F(z)$ has a corresponding impulse train $f^*(t)$ with Laplace transform that may be obtained simply by substitution:

$$F^*(s) = F(z)\Big|_{z=e^{sT}}$$

For example, the sequence with z transform

$$F(z) = \frac{-4z^3 + 5z^2 - 6z}{z^3 + 2z^2 - z + 3}$$

has the following related impulse train transform when the sampling period $T = 0.1$:

$$F^*(s) = \frac{-4e^{0.3s} + 5e^{0.2s} - 6e^{0.1s}}{e^{0.3s} + 2e^{0.2s} - e^{0.1s} + 3}$$

11.6.3 The Sampling Theorem

In applications such as communications, it is especially important to establish conditions for which a signal $g(t)$ is completely specified by (and thus recoverable from) its samples. Communication signals are typically band-limited, or nearly so, meaning that they contain no frequencies higher than a certain band-limit frequency f_B. The frequency content of a signal $g(t)$ is given by its Fourier transform

$$G(\omega) = \int_{-\infty}^{\infty} g(t)e^{-j\omega t}\,dt$$

a calculation similar to the Laplace transformation with $s = j\omega$ but extending over all time, not just from $t = 0$ and thereafter. A signal band-limited beyond frequency f_B is one for which

$$G(\omega) = 0 \quad |\omega| > 2\pi f_B$$

A statement of the sampling theorem is as follows:

Sampling theorem.

> *A signal $g(t)$ that is band-limited above (hertz) frequency f_B can be recovered from an infinite sequence of its periodic samples $g(k)$ if and only if the sampling interval T is less than $\frac{1}{2}f_B$.*

That is, a band limited signal must be sampled at a rate over twice that of its highest component frequency in order for the samples to be unique. The rate $2f_B$, where f_B is the highest frequency in a band-limited signal, is termed the *Nyquist rate* for that signal.

For a single sinusoidal signal of radian frequency b,

$$g(t) = A\cos(bt + \theta)$$

the sample sequence is, in terms of T,

$$g(k) = A\cos(kbT + \theta)$$

If the sampling interval is less than $\frac{1}{2}f_B$

$$T < \frac{1}{2f_B} = \frac{2\pi}{2b}$$
$$bT < \pi$$

then the samples are unique, there being at least two per cycle of $g(t)$. If this condition is not met, then any higher-frequency sinusoid

$$h(t) = A\cos\,(b't + \theta)$$

for which

$$b'T = bT + n2\pi \quad n = 1, 2, 3, \ldots$$

could be present, as it produces precisely the same sample sequence:

$$\begin{aligned} h(k) &= A\cos[k(bT + n2\pi) + \theta] \\ &= A\cos(kbT + kn2\pi + \theta) \\ &= A\cos(kbT + \theta) = g(k) \end{aligned}$$

The effects of any of these higher frequencies, being indistinguishable from those below the presumed band limit, are termed *aliasing distortion.*

A constructive statement of how to recover a band-limited signal from its samples is as follows:

Signal reconstruction.

> *To recover a suitably band-limited signal $g(t)$ from its samples $g(k)$, form the impulse train $g^*(t)$ and pass it through a low-pass filter that passes, unchanged, all frequencies in $g(t)$ below its band-limit frequency f_B and removes all frequencies above $\frac{1}{2}T$.*

This arrangement is shown in Figure 11.15(a). The required low-pass filter has the frequency response shown in Figure 11.15(b) if the analog signal is to be reconstructed without delay. In practice, a phase shift proportional to frequency, representing a time delay in the reconstruction, is approximated. The frequency response of the zero-order hold is given in Figure 11.15(c) and is seen to approximate a low-pass characteristic with time delay.

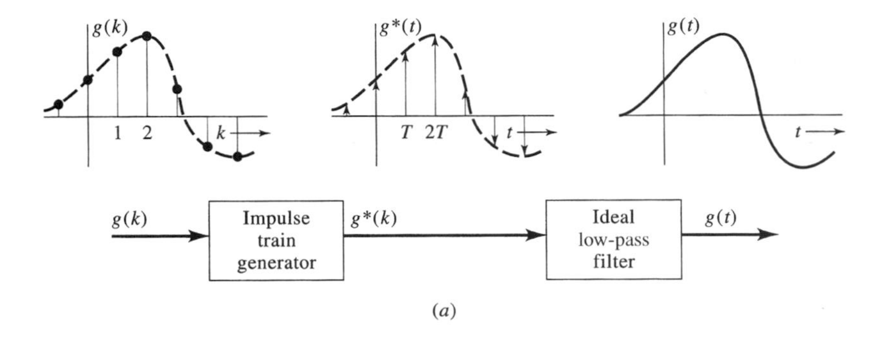

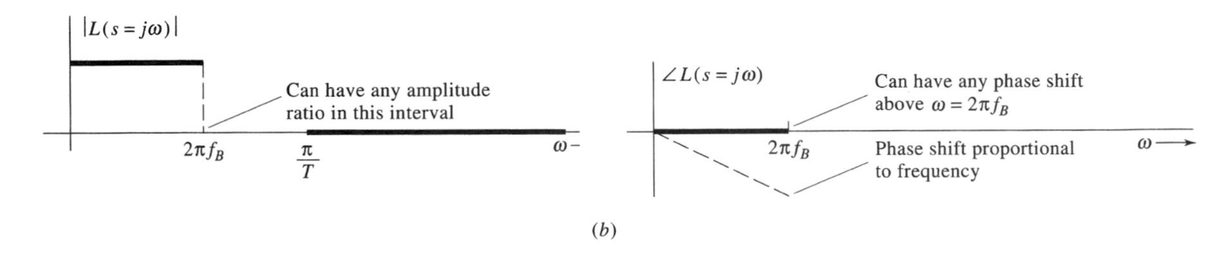

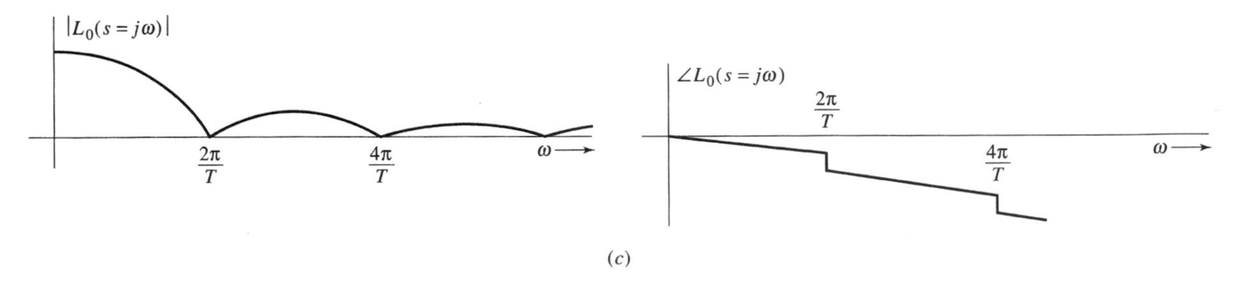

Figure 11.15 Reconstruction from impulses. (a) Reconstruction of a band-limited signal. (b) Ideal low-pass filter frequency response. (c) Frequency response of the zero-order hold.

The sampling theorem does not apply directly to most control system design problems for the following reasons:

1. Many of the signals used in control system analysis, such as those involving step changes in amplitude and slope, are not band-limited.
2. For a band-limited signal, perfect reconstruction requires an infinite number of samples. Another way of stating this fact is to note that the low-pass filter required for reconstruction is a physical impossibility. It can be approximated only if a delay is introduced into the signal processing. Better approximation requires longer delays.
3. In control, good reconstruction of signals from their samples is only occasionally of primary interest in comparison to such concerns as stability, relative stability, and steady state errors.

Nevertheless, the sampling theorem is useful to digital control because it shows important properties of the analog-digital interface such as the following:

1. Samples of a signal uniquely determine that signal only under special circumstances, the sampling theorem stating one such situation. In particular, large transients and high-frequency oscillations in an analog signal such as a system output may not be evident from relatively widely spaced samples of that signal.
2. When A/D conversion is done on a signal, say from a sensor, containing significant frequency components above half the sampling rate, the high frequencies produce errors equivalent to the presence of lower-frequency sinusoidal components. It is thus possible for a poorly designed digital feedback system to attempt to correct presumed low-frequency errors when in fact high-frequency sensor noise is the culprit. For this reason, low-pass filters, termed *antialiasing filters*, are commonly placed before the A/D converters to greatly reduce high-frequency sensor noise in many applications.
3. To improve the smoothing of reconstructed signals, the equivalent impulse-train-to-output transmittance should have a frequency response that better approximates an appropriate low-pass filter with time delay. In practice, this is achieved with analog low-pass filters and/or higher-order hold circuits. The higher-order holds produce outputs based upon more than the single sample used by the zero-order hold.

In practice, control system designers use the following rule of thumb for sample rate selection:
Sampling frequency $\approx 10 \times BW$ where BW is the desired closed-loop bandwidth.

❑ DRILL PROBLEMS

D11.9 For the following sequences $f(k)$, find the corresponding impulse-train Laplace transforms $F^*(s)$ for a sampling interval T:

(a)

$$f(k) = [1 - 3(-1)^k + 4\left(\tfrac{1}{2}\right)^k]u(k) \qquad T = 1$$

Ans. $(2e^{3s} + 5e^{2s} - 3e^s)/(e^{2s} - 1)\left(e^s - \frac{1}{2}\right)$

(b)

$$f(k) = \left[(-1)^k - \sin\frac{\pi k}{2}\right]u(k) \qquad T = 0.2$$

Ans. $(e^{0.6s} - e^{0.4s})/(e^{0.2s} + 1)(e^{0.4s} + 1)$

(c)

$$f(k) = \left(\tfrac{1}{2}\right)^k (10\sin 4k - 8\cos 4k)u(k) \qquad T = 0.1$$

Ans. $(-6.38e^{0.1s} - 8e^{0.2s})/(e^{0.2s} + 0.65e^{0.1s} + 0.25)$

D11.10 For the continuous-time function

$$f(t) = 10 + 3\cos\pi t - 7\sin 6t$$

determine which of the following functions have the same sample sequence as $f(k)$ for a sampling interval of $T = 0.2$:

(a) $g_1(t) = 10\cos 10\pi t + 3\cos 11\pi t - 7\sin 6t$

(b) $g_2(t) = 10 + \sin 5\pi t + 3\cos\pi t - 7\sin[(6 + 10\pi)t]$

(c) $g_3(t) = 10\cos 20\pi t - 3\cos 6\pi t + 7\sin[(6 + 5\pi)t]$

(d) $g_4(t) = 5 + 6\cos 10\pi t - 2\cos 20\pi t + \cos 30\pi t + 6\cos 11\pi t - 3\cos\pi t - 7\sin 6t$

(e) $g_5(t) = 10\sqrt{2}\sin(170\pi t/8) + 3\cos 51\pi t - 8\sin 40\pi t - 7\sin[(6 + 30\pi)t]$

Then find five other functions that, when sampled at this rate, have the same sample sequence.

11.7 Discrete-Time Systems

Computer processing of input signal samples to produce output signal samples may be described by difference equations, analogous to the differential equations that characterize continuous-time systems. In this introductory treatment, only linear, step-invariant (or constant-coefficient) difference equations are considered.

11.7.1 Difference Equations and Response

Discrete-time systems are described by difference equations, of the form

$$\begin{aligned} y(k+n) &+ a_{n-1}y(k+n-1) + a_{n-2}y(k+n-2) \\ &+ \cdots + a_1 y(k+1) + a_0 y(k) \\ &= b_m r(k+m) + b_{m-1}r(k+m-1) + \cdots \\ &+ b_1 r(k+1) + b_0 r(k) \end{aligned}$$

Where $y(k)$ is the output sequence and $r(k)$ is the input sequence. Solving this nth-order difference equation for $y(k+n)$ gives

$$y(k+n) = -a_{n-1}y(k+n-1) - \cdots - a_1 y(k+1) - a_0 y(k)$$
$$+ b_m r(k+m) + \cdots + b_0 r(k)$$

In other words, the $(k+n)$th sample of the output is a linear combination of the previous output samples through $y(k)$ and of the input samples from step $(k+m)$ through step k.

For example,

$$y(k+2) = 3y(k+1) - 2y(k) + 2r(k+1) - r(k)$$

is a discrete-time system described by a second-order difference equation. Given the input sequence $r(k)$ and two initial values of the sequence $y(k)$, the entire output sequence can be calculated recursively. If $r(k) = u(k)$, the unit step sequence, and

$$y(0) = 1$$
$$y(1) = 4$$

then

$$y(2) = 3y(1) - 2y(0) + 2u(1) - u(0)$$
$$= 12 - 2 + 2 - 1 = 11$$
$$y(3) = 3y(2) - 2y(1) + 2u(2) - u(1)$$
$$= 33 - 8 + 2 - 1 = 26$$
$$y(4) = 3y(3) - 2y(2) + 2u(3) - u(2)$$
$$= 78 - 22 + 2 - 1 = 57$$
$$\vdots$$

and so forth.

A closed-form expression for the response of a discrete-time system may be obtained by z-transform methods. For the discrete-time system

$$y(k+1) = -0.5y(k) + 3r(k)$$

for example, suppose that

$$y(0) = 4$$
$$r(k) = u(k)$$

is the unit step sequence. Then the sequence shift relation of Table 11.6, upon z transformation, gives

$$zY(z) - zy(0) = -0.5Y(z) + 3R(z)$$

$$(z+0.5)Y(z) = 4z + \frac{3z}{z-1}$$

$$Y(z) = \frac{4z^2 - z}{(z + 0.5)(z - 1)}$$

$$\frac{Y(z)}{z} = \frac{4z - 1}{(z + 0.5)(z - 1)} = \frac{2}{z + 0.5} + \frac{2}{z - 1}$$

$$Y(z) = \frac{2z}{z + 0.5} + \frac{2z}{z - 1}$$

$$y(k) = 2(-0.5)^k + 2u(k) \quad k = 0, 1, 2, \ldots$$

11.7.2 z-Transfer Functions

The z-transfer function of a discrete-time system is the ratio of the z transform of the output to the z transform of the input when all initial conditions are zero:

$$D(z) = \left.\frac{Y(z)}{R(z)}\right|_{\substack{\text{initial} \\ \text{conditions}=0}}$$

For the discrete-time system

$$y(k + 3) = -0.3y(k + 2) + y(k + 1) - 0.5y(k) + 4r(k + 3) - r(k + 1) - 0.6r(k)$$

z-transforming with zero initial conditions gives

$$(z^3 + 0.3z^2 - z + 0.5)Y(z) = (4z^3 - z - 0.6)R(z)$$

$$D(z) = \left.\frac{Y(z)}{R(z)}\right|_{\substack{\text{initial} \\ \text{conditions}=0}} = \frac{4z^3 - z - 0.6}{z^3 + 0.3z^2 - z + 0.5}$$

A discrete-time system with z-transfer function

$$D(z) = \frac{3z + 1}{z^2 - z + 2}$$

is described by the difference equation

$$y(k + 2) - y(k + 1) + 2y(k) = 3r(k + 1) + r(k)$$

If the initial conditions are zero and

$$R(z) = \frac{5}{z + 4}$$

The z transform of the output is

$$Y(z) = D(z)R(z) = \left(\frac{3z + 1}{z^2 - z + 2}\right)\left(\frac{5}{z + 4}\right)$$

Nonzero initial conditions and multiple inputs and outputs may be accommodated in a manner analogous to the transfer function development for continuous-time systems given in Chapter 1.

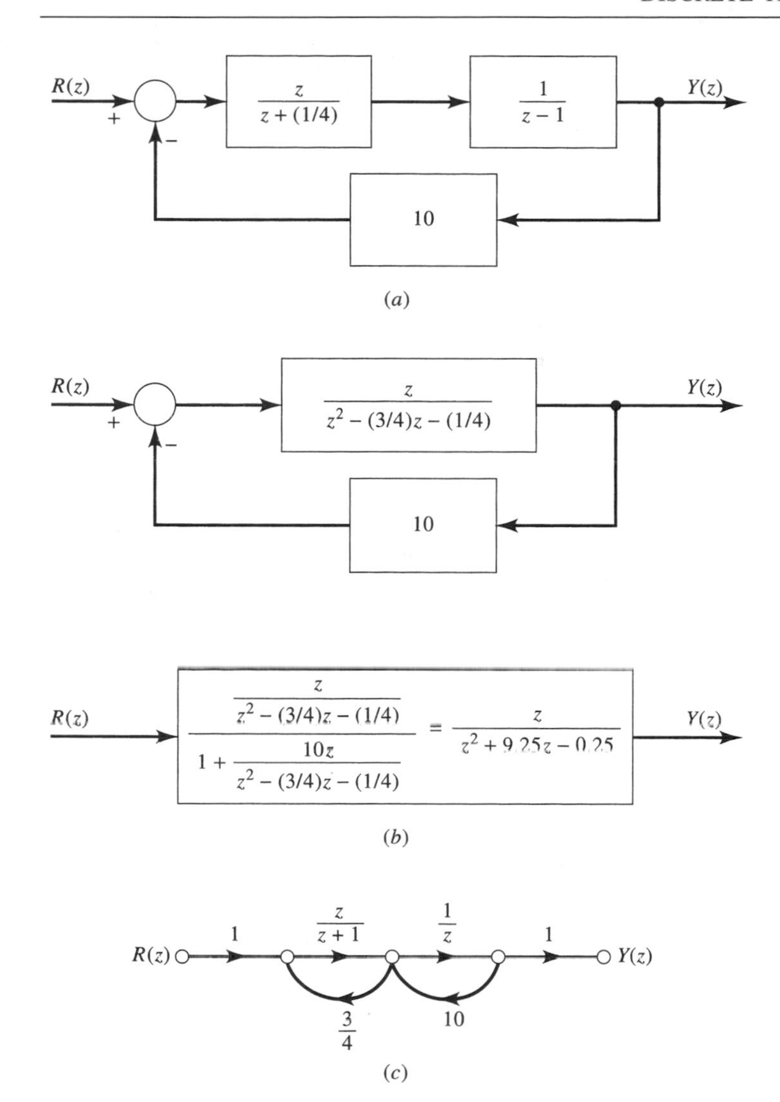

Figure 11.16 Discrete-time systems represented by block diagrams and signal flow graphs. (a) Block diagram of discrete-time system. (b) Reduction of the block diagram. (c) Signal flow graph of a discrete-time system.

11.7.3 Block Diagrams and Signal Flow Graphs

The block diagram and signal flow graph manipulations used for continuous-time system components also apply to discrete-time systems. For example, Figure 11.16(a) shows the description of a discrete-time system by a block diagram. Reduction of the block diagram to determine the overall system z-transfer function is shown in Figure 11.16(b). A discrete-time system signal flow graph is given in Figure 11.16(c).

By using Mason's gain rule, we find that the system z-transfer function is

$$D(z) = \frac{1/(z+1)}{1 - \frac{3}{4}z/(z+1) + 10/z} = \frac{4z}{z^2 + 44z + 40}$$

11.7.4 Stability and the Bilinear Transformation

Table 11.7 shows the characters of sequences corresponding to various complex plane locations of the denominator roots of a z-transformed sequence $F(z)$. Denominator roots within the unit circle on the complex plane give rise to sequences that decay with k, while roots outside the unit circle represent response terms that grow in magnitude with k.

A discrete-time system is said to be stable if and only if its unit pulse response decays with k. If a system with z-transfer function $D(z)$ has a unit pulse input

$$r(k) = \delta(k) \quad R(z) = 1$$

the system output has z-transform equal to the z-transfer function

$$Y(z) = D(z) \cdot 1 = D(z)$$

Hence the stability of a discrete-time system hinges upon whether all the poles of its z-transfer function are within the unit circle on the complex plane.

Stability testing for a discrete-time system involves determining whether all the poles of the system's z-transfer function are within the unit circle on the complex plane. One stability-testing method that avoids factoring the denominator polynomial of $D(z)$ involves a change of variables from z to W, for which the region within the unit circle on the complex plane is mapped to the left half of the complex plane. Then, Routh–Hurwitz testing may be applied to determine stability. The change of variable involved is known as the *bilinear transformation*:

Bilinear transformation allows use of continuous techniques to discrete systems.

$$z = \frac{1+W}{1-W}$$

For example, when the bilinear change of variables is made on the z-transfer function

$$D(z) = \frac{8z^3 - 3z^2 + z}{z^3 + 0.4z^2 - 0.25z - 0.1}$$

there results

$$\begin{aligned}
D(W) &= \frac{8\left(\dfrac{1+W}{1-W}\right)^3 - 3\left(\dfrac{1+W}{1-W}\right)^2 + \left(\dfrac{1+W}{1-W}\right)}{\left(\dfrac{1+W}{1-W}\right)^3 + 0.4\left(\dfrac{1+W}{1-W}\right)^2 - 0.25\left(\dfrac{1+W}{1-W}\right) - 0.1} \\
&= \frac{(12W^3 + 26W^2 + 20W + 6)/(1-W)^3}{(0.45W^3 + 2.55W^2 + 3.95W + 1.05)/(1-W)^3} \\
&= \frac{12W^3 + 28W^2 + 22W + 6}{0.45W^3 + 2.55W^2 + 3.95W + 1.05}
\end{aligned}$$

Table 11.7 *Sequences Corresponding to Various z-Transform Denominator Polynomial Root Locations*

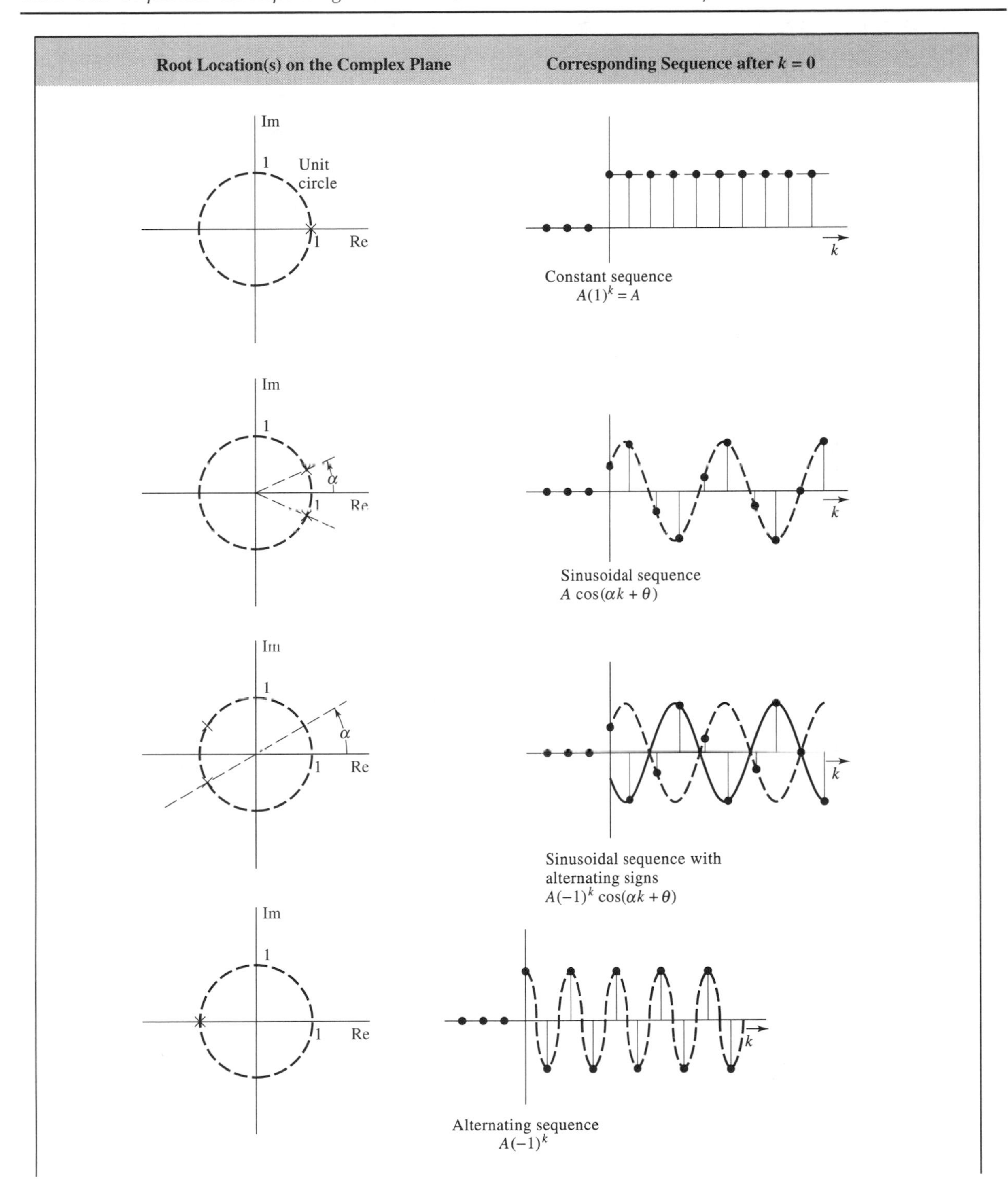

Table 11.7 *(continued)*

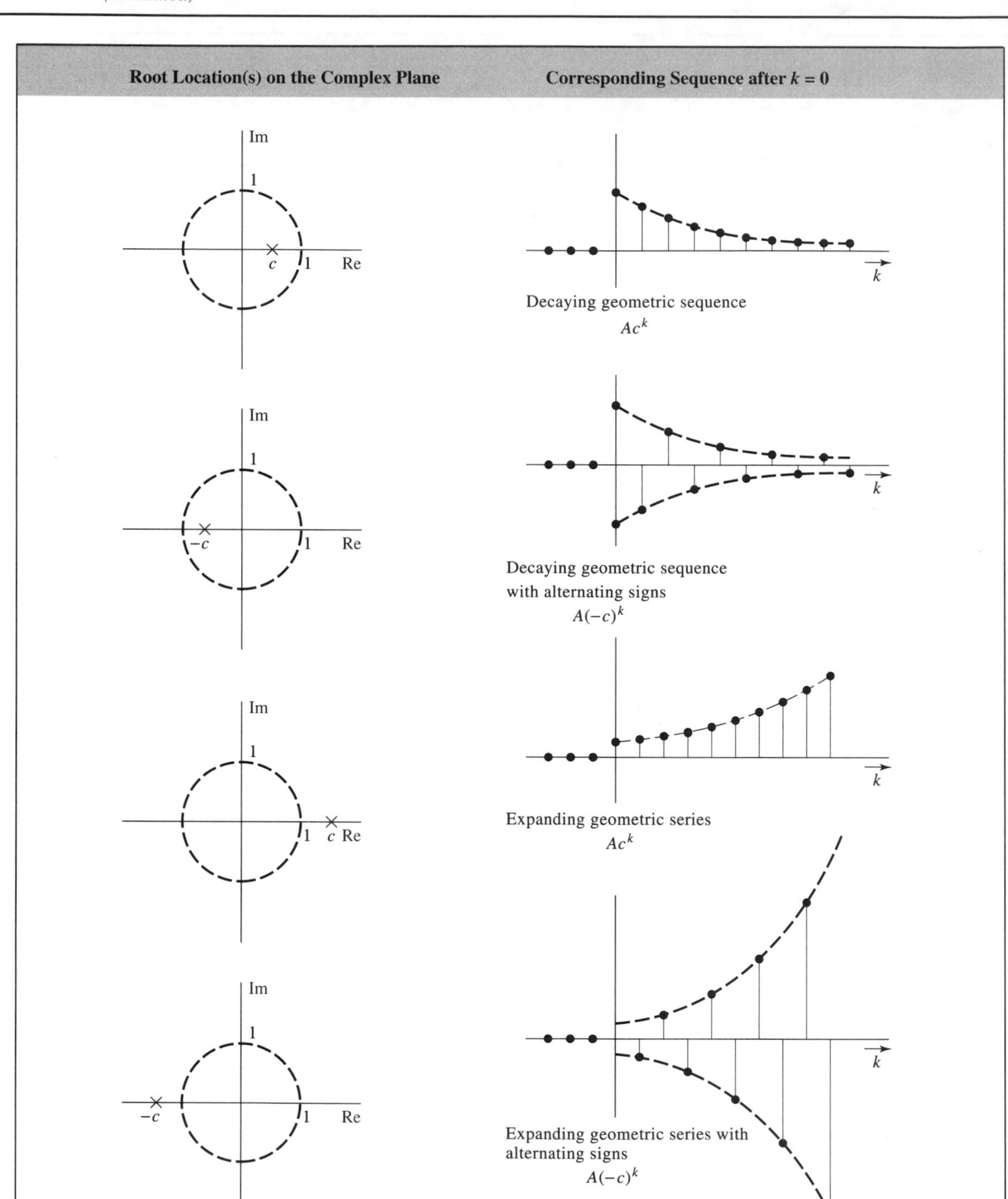

Table 11.7 *(continued)*

Root Location(s) on the Complex Plane	Corresponding Sequence after $k = 0$
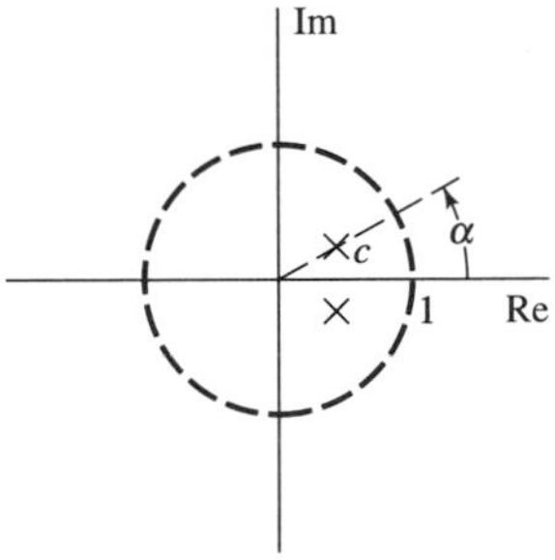	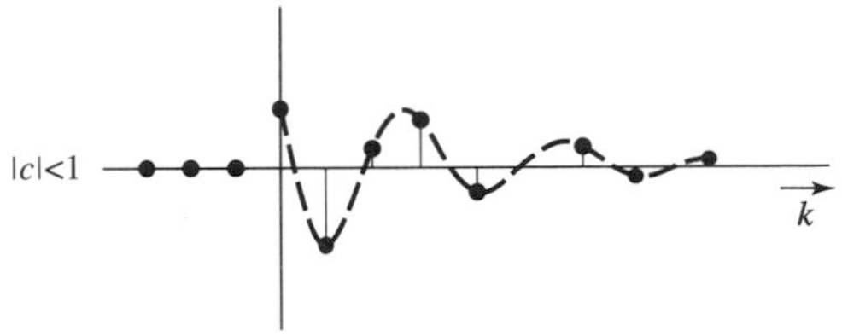 $\lvert c\rvert<1$ Damped sinusoidal sequence $Ac^k \cos(\alpha k + \theta)$
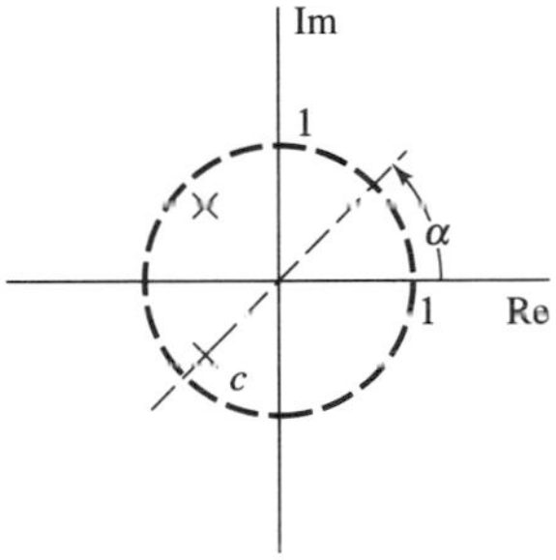	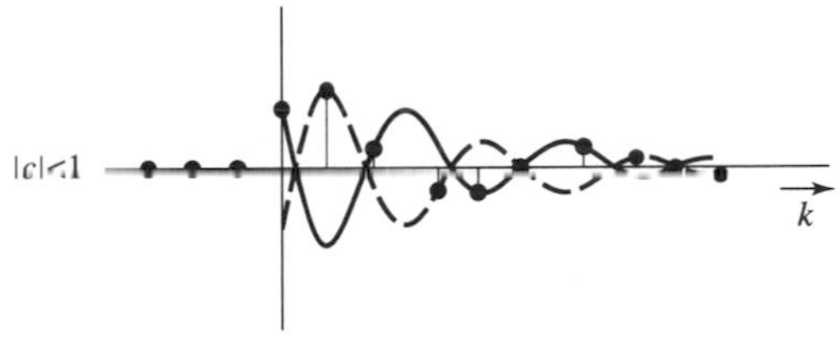 $\lvert c\rvert<1$ Damped sinusoidal sequence with alternating signs $A(-c)^k \cos(\alpha k + \theta)$
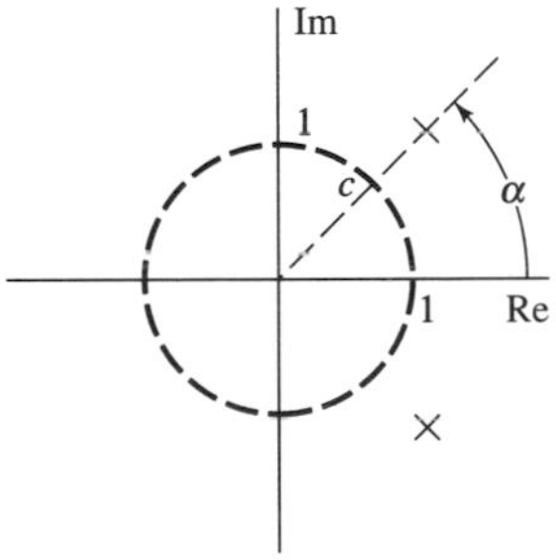	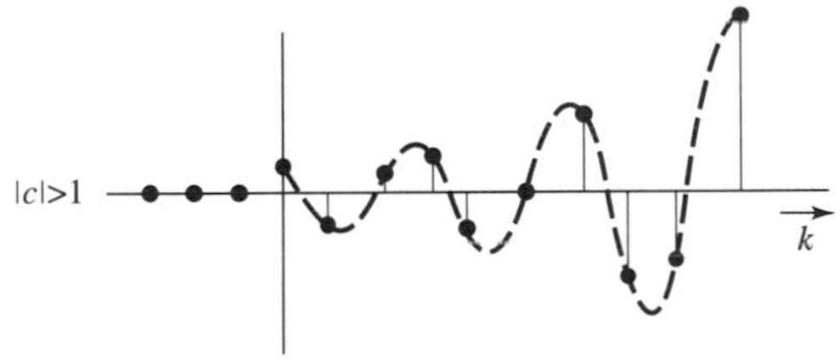 $\lvert c\rvert>1$ Exponentially expanding sinusoidal sequence $Ac^k \cos(\alpha k + \theta)$
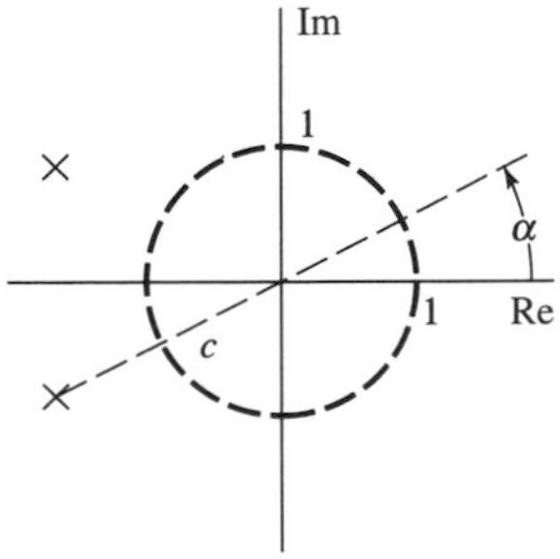	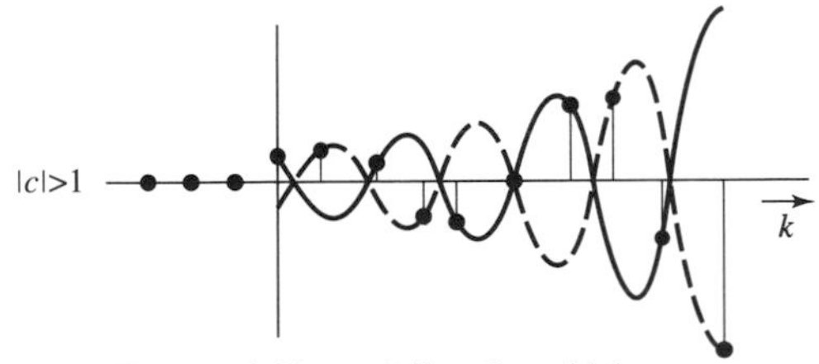 $\lvert c\rvert>1$ Exponentially expanding sinusoidal sequence with alternating signs $A(-c)^k \cos(\alpha k + \theta)$

Poles and zeros of $D(z)$ within the unit circle are mapped to the LHP in $D(W)$, roots of $D(z)$ outside the unit circle are mapped to the RHP in terms of $D(W)$, and roots of $D(z)$ located precisely on the unit circle are mapped to the imaginary axis in $D(W)$.

A Routh–Hurwitz test of the poles of $D(W)$ is as follows:

W^3	0.45	3.95
W^2	2.55	1.05
W^1	3.76	
W^0	1.05	

There are no left-column sign changes in the array, so all poles of $D(W)$ are in the LHP. All poles of $D(z)$ are then within the unit circle on the complex plane. The system represented by $D(z)$ is thus stable.

Because it converts a digital problem to a related analog one, the bilinear transformation is also very useful for applying root locus and frequency response methods to digital systems.

11.7.5 Computer Software

Programming a digital computer with A/D and D/A capability as discrete-time system is straightforward. For example, a system with z-transfer function

$$D(z) = \frac{2z^2 + 5}{z^2 + 3z + 2}$$

is described by the difference equation

$$y(k + 2) = -3y(k + 1) - 2y(k) + 2r(k + 2) + 5r(k)$$

A Fortran program for this system is outlined in the flow diagram of Figure 11.17 and listed in Table 11.8. The variables $Y2$, $Y1$, and $Y0$ are used for $y(k + 2)$, $y(k + 1)$, and $y(k)$, respectively, while $R2$, $R1$, and $R0$ represent $r(k+2)$, $r(k+1)$, and $r(k)$. The initial conditions $Y0$, $Y1$, $R0$, and $R1$ are first set to zero. Then an input value $R2$ is read from the A/D converter. $Y2$ is calculated and the values of $Y0$, $Y1$, $R0$, and $R1$ are updated for the next calculation cycle. $Y2$ is then output to the D/A converter. Assuming sufficient time between samples to perform the calculations, the program waits for a new input sample, computes the next $Y2$ sample, and so forth.

There are, of course, many other functions the computer could perform. It could limit the output signal, check that the input signal samples are "reasonable" (i.e., within some predetermined bounds), and trigger alarms in the event that a malfunction is detected.

❑ DRILL PROBLEMS

D11.11 Find the z-functions of the following discrete-time systems:

(a)

$$y(k + 3) + 3y(k + 2) - 2y(k + 1) + y(k) = r(k + 2) + r(k + 1) - 4r(k)$$

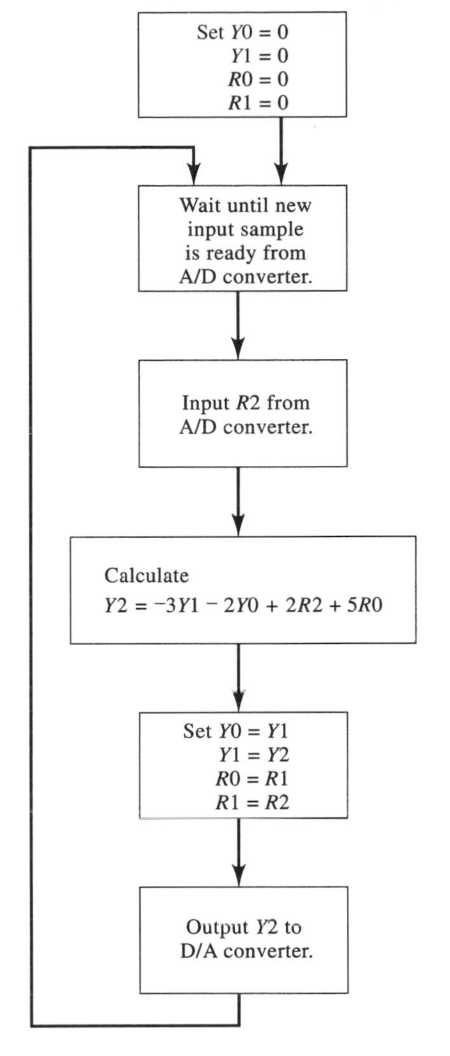

Figure 11.17 Flow diagram for a program to realize a discrete-time system.

Ans. $D(z) = (z^2 + z - 4)/(z^3 + 3z^2 - 2z + 1)$

(b)

$$y(k+4) = 0.2r(k+4) - 0.3r(k+3) + 0.1r(k+2) + 0.7r(k+1) - 0.5r(k)$$

Ans. $D(z) = (0.2z^4 - 0.3z^3 + 0.1z^2 + 0.7z - 0.5)/z^4$

(c)

$$y(k+3) = 0.5y(k+2) - y(k+1) + 0.125y(k) + 10r(k+3)$$

Ans. $D(z) = 10z^3/(z^3 - 0.5z^2 + z - 0.125)$

Table 11.8 *Fortran Program to Realize a Discrete-Time System*

```
100 FORMAT (F16.8)
110 Y0 = 0.
120 Y1 = 0.
130 R1 = 0.
140 R0 = 0.
a150 READ (1, 100)R2
160 Y2 = 3.*Y1-2.*Y0+2.*R2+5.*R0
170 Y0 = Y1
180 Y1 =Y2
190 R0 = R1
200 R1 = R2
b210 WRITE (2,100)Y2
220 GO TO 150
230 STOP
240 END
```

[a]The A/D converter is taken to be device number 1 with F16.8 format. It is here assumed that the processor waits at step 150 at each looping until a new sample is ready, just as it would wait for a character to be input from a keyboard device.

[b]The D/A converter is taken to be device number 2 with F16.8 format. It is here assumed that the D/A device contains a buffer that stores each new output sample for conversion at the sample time.

D11.12 For

$$y(k+2) - 5y(k+1) - 6y(k) = 2r(k)$$

recursively find $y(2)$, $y(3)$, $y(4)$, and $y(5)$ if

$$r(k) = (-1)^k$$

$$y(0) = -4$$

$$y(1) = 7$$

Ans. 13, 105, 605, 3653

D11.13 For discrete-time systems with the following z-transfer functions and input sequences, find the output sequences for $k = 0$ and thereafter if the initial conditions are zero:

(a)

$$D(z) = \frac{4}{z+3}$$

$$r(k) = 5\delta(k)$$

Ans. $\frac{20}{3}\delta(k) - \frac{20}{3}(-3)^k$

(b)

$$D(z) = \frac{z}{z - \frac{1}{10}}$$

$$r(k) = u(k)$$

Ans. $-\frac{1}{9}\left(\frac{1}{10}\right)^k + \frac{10}{9}$

(c)

$$D(z) = \frac{-8}{z + \frac{1}{3}}$$

$$r(k) = \left(\frac{1}{4}\right)^k$$

Ans. $-\frac{96}{7}\left(\frac{1}{4}\right)^k + \frac{96}{7}\left(-\frac{1}{3}\right)^k$

(d)

$$D(z) = \frac{z}{2z - 1}$$

$$r(k) = \delta(k - 3)$$

Ans. $\frac{1}{2}\left(\frac{1}{2}\right)^{k-3} u(k - 3)$

D11.14 Determine whether each of the following discrete-time systems is stable:

(a)

$$D(z) = \frac{-3z^2 + 1}{4z^2 + 2z - 1}$$

Ans. denominator roots 0.31 and -0.81; stable

(b)

$$D(z) = \frac{z^3}{z^3 + 0.3z^2 - 0.25z - 0.075}$$

Ans. $D(W) = \dfrac{W^3 + 3W^2 + 3W + 1}{0.525W^3 + 2.725W^2 + 3.775W + 0.975}$; stable

(c)

$$D(z) = \frac{5(z - 0.2)}{z^3 - 2.8z^2 + 1.75z - 0.3}$$

Ans. $D(W) = \dfrac{1.026(W + 0.667)(1 - W)^2}{W^3 + 0.538W^2 - 0.111W - 0.060}$; unstable

11.8 State-Variable Descriptions of Discrete-Time Systems

11.8.1 Simulation Diagrams and Equations

Simulation diagrams for discrete-time systems involve as a basic element blocks or branches having transmittance $1/z$. A simulation diagram, in phase-variable canonical form, for a system with z-transfer function

$$D(z) = \frac{3z^2 - 2z + 8}{z^3 + 0.5z^2 - 0.25z + 0.75}$$
$$= \frac{3/z - 2/z^2 + 8/z^3}{1 + 0.5/z - 0.25/z^2 + 0.75/z^3}$$

is given in Figure 11.18(a). In terms of the indicated state variables, the z-transformed equations describing the system are as follows:

$$zX_1(z) = X_2(z)$$
$$zX_2(z) = X_3(z)$$
$$zX_3(z) = -0.75X_1(z) + 0.25X_2(z) - 0.5X_3(z) + U(z)$$
$$Y(z) = 8X_1(z) - 2X_2(z) + 3X_3(z)$$

In terms of the step, k, these equations are

$$x_1(k+1) = x_2(k)$$
$$x_2(k+1) = x_3(k)$$
$$x_3(k+1) = -0.75x_1(k) + 0.25x_2(k) - 0.5x_3(k) + u(k)$$
$$y(k) = 8x_1(k) - 2x_2(k) + 3x_3(k)$$

or

$$\begin{bmatrix} x_1(k+1) \\ x_2(k+1) \\ x_3(k+1) \end{bmatrix} = \begin{bmatrix} 0 & 1 & 0 \\ 0 & 0 & 1 \\ -0.72 & 0.25 & -0.5 \end{bmatrix} \begin{bmatrix} x_1(k) \\ x_2(k) \\ x_3(k) \end{bmatrix} + \begin{bmatrix} 0 \\ 0 \\ 1 \end{bmatrix} u(k)$$

$$y(k) = [8 \quad -2 \quad 3] \begin{bmatrix} x_1(k) \\ x_2(k) \\ x_3(k) \end{bmatrix}$$

A simulation diagram for a system with the same transfer function but using the dual phase-variable form is given in Figure 11.18(b). For this system,

$$X_1(z) = \frac{1}{z}[-0.5X_1(z) + X_2(z) + 3U(z)]$$
$$X_2(z) = \frac{1}{z}[0.25X_1(z) + X_3(z) + 2U(z)]$$
$$X_3(z) = \frac{1}{z}[-0.75X_1(z) + 8U(z)]$$
$$Y(z) = X_1(z)$$

or

$$\begin{bmatrix} x_1(k+1) \\ x_2(k+1) \\ x_3(k+1) \end{bmatrix} = \begin{bmatrix} -0.5 & 1 & 0 \\ 0.25 & 0 & 1 \\ -0.75 & 0 & 0 \end{bmatrix} \begin{bmatrix} x_1(k) \\ x_2(k) \\ x_3(k) \end{bmatrix} + \begin{bmatrix} 3 \\ -2 \\ 8 \end{bmatrix} u(k)$$

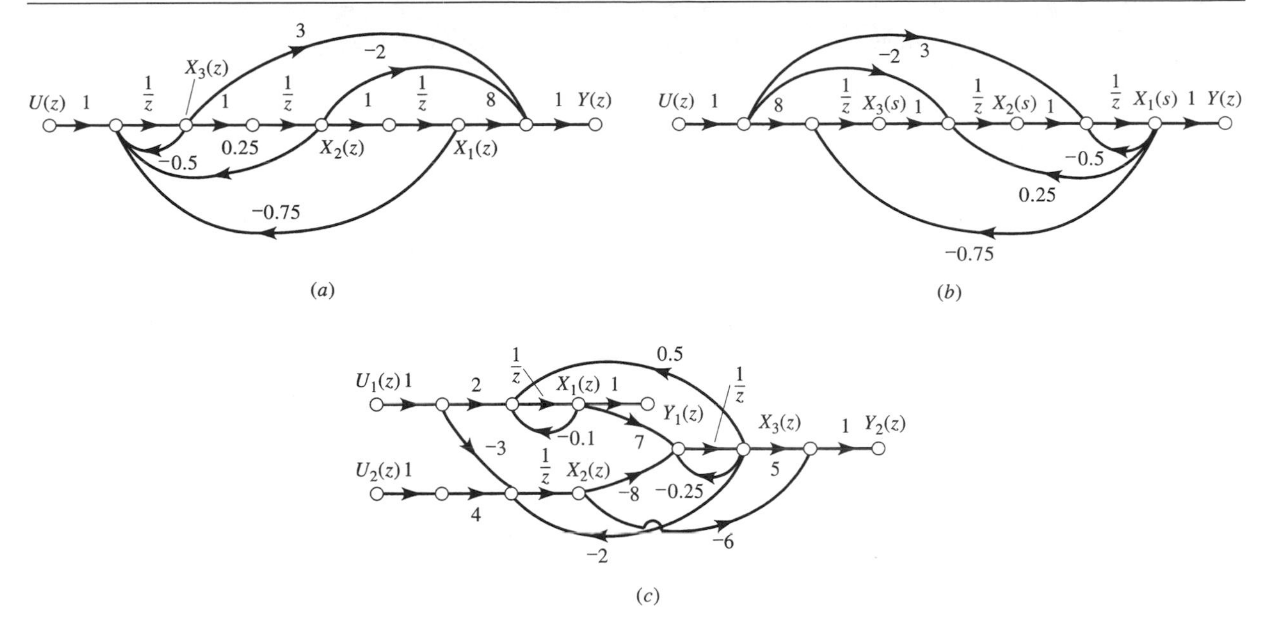

Figure 11.18 Simulation diagrams for discrete-time systems. (a) A system in phase-variable form. (b) A system in dual phase-variable form. (c) A multiple-input, multiple-output system.

$$y(k) = [1 \quad 0 \quad 0] \begin{bmatrix} x_1(k) \\ x_2(k) \\ x_3(k) \end{bmatrix}$$

A simulation diagram for a multiple-input, multiple-output system is given in Figure 11.18(c). It consists of constant transmittances and unit delay transmittances $1/z$. The output of each delay is a state variable, and each delay input consists of a linear combination of the inputs and the state variables. For the example system,

$$\begin{cases} zX_1(z) &= -0.1X_1(z) + 0.5X_3(z) + 2U_1(z) \\ zX_2(z) &= -2X_3(z) - 3U_1(z) + 4U_2(z) \\ zX_3(z) &= 7X_1(z) - 8X_2(z) - 0.25X_3(z) \\ Y_1(z) &= X_1(z) \\ Y_2(z) &= -6X_2(z) + 5X_3(z) \end{cases}$$

$$\begin{bmatrix} x_1(k+1) \\ x_2(k+1) \\ x_3(k+1) \end{bmatrix} = \begin{bmatrix} -0.1 & 0 & 0.5 \\ 0 & 0 & -2 \\ 7 & -8 & -0.25 \end{bmatrix} \begin{bmatrix} x_1(k) \\ x_2(k) \\ x_3(k) \end{bmatrix} + \begin{bmatrix} 2 & 0 \\ -3 & 4 \\ 0 & 0 \end{bmatrix} \begin{bmatrix} u_1(k) \\ u_2(k) \end{bmatrix}$$

$$\begin{bmatrix} y_1(k) \\ y_2(k) \end{bmatrix} = \begin{bmatrix} 1 & 0 & 0 \\ 0 & -6 & 5 \end{bmatrix} \begin{bmatrix} x_1(k) \\ x_2(k) \\ x_3(k) \end{bmatrix}$$

In general, the state equations for a discrete-time system are of the form

$$\begin{aligned} x(k+1) &= Fx(k) + Gu(k) \\ y(k) &= Hx(k) \end{aligned} \qquad [11.1]$$

When written out, these are

$$\begin{bmatrix} x_1(k+1) \\ x_2(k+1) \\ x_3(k+1) \\ \vdots \\ x_n(k+1) \end{bmatrix} = \begin{bmatrix} f_{11} & f_{12} & f_{13} & \cdots & f_{1n} \\ f_{21} & f_{22} & f_{23} & & f_{2n} \\ f_{31} & f_{32} & f_{33} & & f_{3n} \\ \vdots & & & & \\ f_{n1} & f_{n2} & f_{n3} & \cdots & f_{nn} \end{bmatrix} \begin{bmatrix} x_1(k) \\ x_2(k) \\ x_3(k) \\ \vdots \\ x_n(k) \end{bmatrix} + \begin{bmatrix} g_{11} & g_{12} & \cdots & g_{1i} \\ g_{21} & g_{22} & \cdots & g_{2i} \\ g_{31} & g_{32} & \cdots & g_{3i} \\ \vdots & & & \\ g_{n1} & g_{n2} & \cdots & g_{ni} \end{bmatrix} \begin{bmatrix} u_1(k) \\ u_2(k) \\ \vdots \\ u_i(k) \end{bmatrix}$$

$$\begin{bmatrix} y_1(k) \\ y_2(k) \\ \vdots \\ y_m(k) \end{bmatrix} = \begin{bmatrix} h_{11} & h_{12} & \cdots & h_{1n} \\ h_{21} & h_{22} & \cdots & h_{2n} \\ \vdots & & \vdots & \\ h_{m1} & h_{m2} & \cdots & h_{mn} \end{bmatrix} \begin{bmatrix} x_1(k) \\ x_2(k) \\ \vdots \\ x_n(k) \end{bmatrix}$$

11.8.2 Response and Stability

The response of a discrete-time system may be calculated recursively, starting with an initial state and repeatedly using the state equations (11.1). From $x(0)$ and $u(0)$, $x(1)$ may be calculated:

$$x(1) = Fx(0) + Gu(0)$$

Then, using $x(1)$ and $u(1)$, $x(2)$ is calculated:

$$\begin{aligned} x(2) &= Fx(1) + Gu(1) \\ &= F^2x(0) + FGu(0) + Gu(1) \end{aligned}$$

Continuing, we write

$$\begin{aligned} x(3) &= Fx(2) + Gu(2) \\ &= F^3x(0) + F^2Gu(0) + FGu(1) + Gu(2) \\ &\vdots \\ x(k) &= F^kx(0) + F^{k-1}Gu(0) + F^{k-2}Gu(1) \end{aligned}$$

$$+\cdots + FGu(k-2) + Gu(k-1)$$

$$= F^k x(0) + \sum_{n=1}^{k} F^{k-n} Gu(n-1)$$

As a numerical example, the system

$$\begin{bmatrix} x_1(k+1) \\ x_2(k+2) \end{bmatrix} = \begin{bmatrix} 0 & -2 \\ -1 & 3 \end{bmatrix} \begin{bmatrix} x_1(k) \\ x_2(k) \end{bmatrix} + \begin{bmatrix} 2 \\ 1 \end{bmatrix} u(k)$$

$$y(k) = [\,3 \quad -2\,] \begin{bmatrix} x_1(k) \\ x_2(k) \end{bmatrix} \qquad [\mathbf{11.2}]$$

with

$$\begin{bmatrix} x_1(0) \\ x_2(0) \end{bmatrix} = \begin{bmatrix} 5 \\ -7 \end{bmatrix}$$

and

$$u(k) = \delta(k)$$

has response as follows:

$$y(0) = [\,3 \quad -2\,] \begin{bmatrix} 5 \\ -7 \end{bmatrix} = 29$$

$$\begin{bmatrix} x_1(1) \\ x_2(1) \end{bmatrix} = \begin{bmatrix} 0 & -2 \\ -1 & 3 \end{bmatrix} \begin{bmatrix} 5 \\ -7 \end{bmatrix} + \begin{bmatrix} 2 \\ 1 \end{bmatrix} = \begin{bmatrix} 16 \\ -25 \end{bmatrix}$$

$$y(1) = [\,3 \quad -2\,] \begin{bmatrix} 16 \\ -25 \end{bmatrix} = 98$$

$$\begin{bmatrix} x_1(2) \\ x_2(2) \end{bmatrix} = \begin{bmatrix} 0 & -2 \\ -1 & 3 \end{bmatrix} \begin{bmatrix} 16 \\ -25 \end{bmatrix} + \begin{bmatrix} 0 \\ 0 \end{bmatrix} = \begin{bmatrix} 50 \\ -91 \end{bmatrix}$$

$$y(2) = [\,3 \quad -2\,] \begin{bmatrix} 50 \\ -91 \end{bmatrix} = 332$$

$$\begin{bmatrix} x_1(3) \\ x_2(3) \end{bmatrix} = \begin{bmatrix} 0 & -2 \\ -1 & 3 \end{bmatrix} \begin{bmatrix} 50 \\ -91 \end{bmatrix} + \begin{bmatrix} 0 \\ 0 \end{bmatrix} = \begin{bmatrix} 182 \\ -323 \end{bmatrix}$$

$$y(3) = [\,3 \quad -2\,] \begin{bmatrix} 182 \\ -323 \end{bmatrix} = 1192$$

$$\vdots$$

A discrete-time system's z-transfer function is found by z-transforming the state equations (11.1) with zero initial conditions and solving for the ratio of output to input transforms:

$$\begin{cases} zX(z) = FX(z) + GU(z) \\ Y(z) = HX(z) \end{cases}$$

$$(zI - F)X(z) = GU(z)$$

$$X(z) = (zI - F)^{-1}GU(z)$$

$$Y(z) = HX(z) = H(zI - F)^{-1}GU(z)$$

The z-transfer function matrix of the system is thus

$$D(z) = H(zI - F)^{-1}G = \frac{H \operatorname{adj}(zI - F)G}{|zI - F|}$$

and the system is stable if and only if all the roots of the characteristic polynomial

$$|zI - F| = 0$$

are within the unit circle on the complex plane.

For the single-input, single-output example system of Equation (11.2), the transfer function is

$$\begin{aligned} D(z) &= \begin{bmatrix} 3 & -2 \end{bmatrix} \begin{bmatrix} z & 2 \\ 1 & z-3 \end{bmatrix}^{-1} \begin{bmatrix} 2 \\ 1 \end{bmatrix} \\ &= \frac{\begin{bmatrix} 3 & -2 \end{bmatrix} \begin{bmatrix} z-3 & -2 \\ -1 & z \end{bmatrix} \begin{bmatrix} 2 \\ 1 \end{bmatrix}}{z^2 - 3z - 2} \\ &= \frac{\begin{bmatrix} 3 & -2 \end{bmatrix} \begin{bmatrix} (2z-8) \\ (z-2) \end{bmatrix}}{z^2 - 3z - 2} \\ &= \frac{4z - 20}{z^2 - 3z - 2} \end{aligned}$$

This system is unstable, since the roots of the characteristic equation are

$$\begin{aligned} z^2 - 3z - 2 = (z-1)(z-2) &= 0 \\ z_1,\ z_2 &= 1,\ 2 \end{aligned}$$

which do not lie within the unit circle on the complex plane.

11.8.3 Controllability and Observability

For discrete-time systems with nonrepeated characteristic roots, an appropriate change of state variables

$$x = P\bar{x} \qquad \bar{x} = P^{-1}x$$

determined as for continuous-time systems with the methods of Section 8.3, will decouple the state equations:

$$\bar{x}(k+1) = P^{-1}FP\bar{x}(k) + P^{-1}Gu(k) = \bar{F}x(k) + \bar{G}u(k)$$
$$y(k) = HP\bar{x}(k) = \bar{H}\bar{x}(k)$$

where

$$\bar{F} = \begin{bmatrix} z_1 & 0 & 0 & \cdots & 0 \\ 0 & z_2 & 0 & \cdots & 0 \\ \vdots & & & & \\ 0 & 0 & 0 & \cdots & z_n \end{bmatrix}$$

In terms of the new state variables $\bar{x}$, the state coupling matrix $\bar{F}$ is diagonal, with the system's characteristic roots along the diagonal.

If, when the state equations for a discrete-time system are diagonalized, any row of the new input coupling matrix

$$\bar{G} = P^{-1}G$$

is zero, the corresponding discrete-time system mode is uncontrollable. If any column of

$$\bar{H} = HP$$

is zero, the corresponding system mode is unobservable. The same rank tests for complete controllability and complete observability that apply to continuous-time systems apply here because controllability and observability are algebraic properties of the system matrices.

An nth-order discrete-time system of Equations (11.1) is completely controllable if and only if its controllability matrix

$$M_c = [G \,|\, FG \,|\, F^2G \,|\cdots|\, F^{n-1}G]$$

is of full rank. The system is completely observable if and only if the observability matrix

$$M_o = \begin{bmatrix} H \\ \hdashline HF \\ \hdashline HF^2 \\ \hdashline \vdots \\ \hdashline HF^{n-1} \end{bmatrix}$$

is of full rank.

❑ DRILL PROBLEMS

D11.15 Draw discrete-time simulation diagrams in phase-variable canonical form for systems with the following z-transfer functions:

(a)

$$D(z) = \frac{4z}{z^2 + z + 0.5}$$

(b)

$$D(z) = \frac{10z^3 - 4z^2 + 5z}{z^3 + 0.5z^2 - 0.2z + 0.3}$$

D11.16 Find state equations in matrix form for the systems with the discrete-time simulation diagram of Figure D11.16.

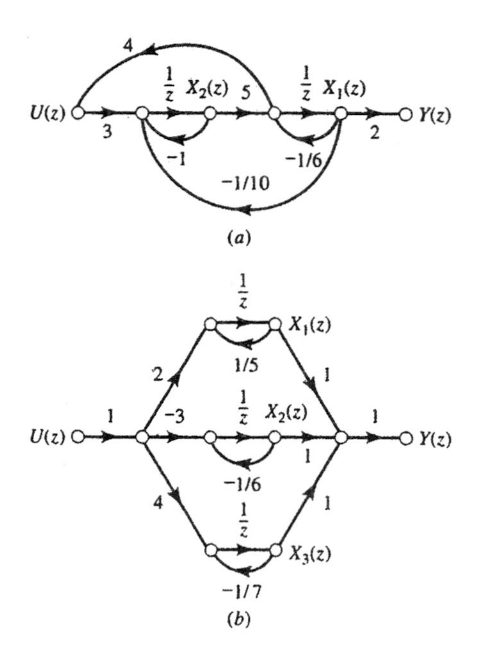

Figure D11.16

Ans.

(a)

$$\begin{bmatrix} x_1(k+1) \\ x_2(k+1) \end{bmatrix} = \begin{bmatrix} \frac{1}{6} & 5 \\ -\frac{1}{10} & -1 \end{bmatrix} \begin{bmatrix} x_1(k) \\ x_2(k) \end{bmatrix} + \begin{bmatrix} 4 \\ 3 \end{bmatrix} u(k)$$

$$y(k) = [2 \quad 0] \begin{bmatrix} x_1(k) \\ x_2(k) \end{bmatrix}$$

(b)

$$\begin{bmatrix} x_1(k+1) \\ x_2(k+1) \\ x_3(k+1) \end{bmatrix} = \begin{bmatrix} \frac{1}{5} & 0 & 0 \\ 0 & -\frac{1}{6} & 0 \\ 0 & 0 & -\frac{1}{7} \end{bmatrix} \begin{bmatrix} x_1(k) \\ x_2(k) \\ x_3(k) \end{bmatrix} + \begin{bmatrix} 2 \\ -3 \\ 4 \end{bmatrix} u(k)$$

$$y(k) = [1 \quad 1 \quad 1] \begin{bmatrix} x_1(k) \\ x_2(k) \\ x_3(k) \end{bmatrix}$$

D11.17 For the system

$$\begin{bmatrix} x_1(k+1) \\ x_2(k+1) \end{bmatrix} = \begin{bmatrix} -2 & 1 \\ 1 & -3 \end{bmatrix} \begin{bmatrix} x_1(k) \\ x_2(k) \end{bmatrix} + \begin{bmatrix} 1 \\ 2 \end{bmatrix} u(k)$$

$$y(k) = [-1 \quad 1] \begin{bmatrix} x_1(k) \\ x_2(k) \end{bmatrix}$$

with

$$\begin{bmatrix} x_1(0) \\ x_2(0) \end{bmatrix} = \begin{bmatrix} 2 \\ 0 \end{bmatrix}$$

and $r(k) = 1$, find $y(0)$, $y(1)$, $y(2)$, and $y(3)$.

Ans. -2, 7, -24, 86

11.9 Digitizing Control Systems

11.9.1 Step-Invariant Approximation

An important technique in the design of digital control systems is to require that the unit step sequence response of a digital transmittance be samples of the continuous-time unit step response of a model analog transmittance. In practice, the analog transmittance is often a working component of the system and it is desired to replace the analog component with a digital one that performs similarly. In Figure 11.19(a), an analog transmittance $G(s)$ and its unit step response $f_{\text{step}}(t)$ are indicated. The step-invariant digital approximation to $G(s)$, Figure 11.19(b), has a unit step response sequence $f_{\text{step}}(k)$ that consists of samples of the analog step response $f_{\text{step}}(t)$.

The conversion from $F_{\text{step}}(s)$ to $F_{\text{step}}(z)$, where the sequence consists of samples of the time function, is the sampling conversion of Section 11.5. For example, the analog transmittance

$$G(s) = \frac{-s+2}{s^2+3s+2}$$

has unit step response given by

$$F_{\text{step}}(s) = \frac{1}{s}G(s) = \frac{-s+2}{s(s^2+3s+2)}$$
$$= \frac{1}{s} - 3\left(\frac{1}{s+1}\right) + 2\left(\frac{1}{s+2}\right)$$

Samples of this step response at a sampling interval $T = 0.1$ are given by

$$F_{\text{step}}(z) = \frac{z}{z-1} - 3\left(\frac{z}{z-e^{-0.1}}\right) + 2\left(\frac{z}{z-e^{-0.2}}\right)$$
$$= \frac{z}{z-1} + \frac{-3z}{z-0.905} + \frac{2z}{z-0.82}$$

so the step-invariant digital system with $T = 0.1$ is to have unit step sequence response

$$F_{\text{step}}(z) = U(z)D(z) = \frac{z}{z-1}D(z)$$

The step-invariant z-transmittance for $T = 0.1$ is thus

$$D(z) = 1 + \frac{-3(z-1)}{z-0.905} + \frac{2(z-1)}{z-0.82}$$
$$= \frac{-0.07z + 0.092}{(z-0.905)(z-0.82)}$$

Step-invariant discretization method is designed to preserve the quality of the step response.

The digital system with analog input and analog output in Figure 11.19(c) then approximates the performance of $G(s)$. Generally, the smaller the sampling interval, the better the approximation. Further improvement in the approximation may be obtained by further smoothing of the output waveform rather than simply holding it between samples. Step-invariance design has the properties that the digital transmittance is stable if the analog model is stable and that the resulting z-transmittance is of the same order as the original continuous-time transmittance.

Step invariance is but one of several useful approximations of an analog transmittance by a digital one. Other commonly used approximations include impulse invariance, ramp invariance, matched z-transformation, and bilinear transformation.

❑ DRILL PROBLEM

D11.18 Find the step-invariant approximations, for the given sampling interval T, to the following continuous-time transmittances:

(a)

$$G(s) = \frac{1}{s} \qquad T = 0.5$$

Ans. $0.5/(z-1)$

(b)

$$G(s) = \frac{2}{s+4} \qquad T = 0.1$$

Ans. $0.165/(z-0.67)$

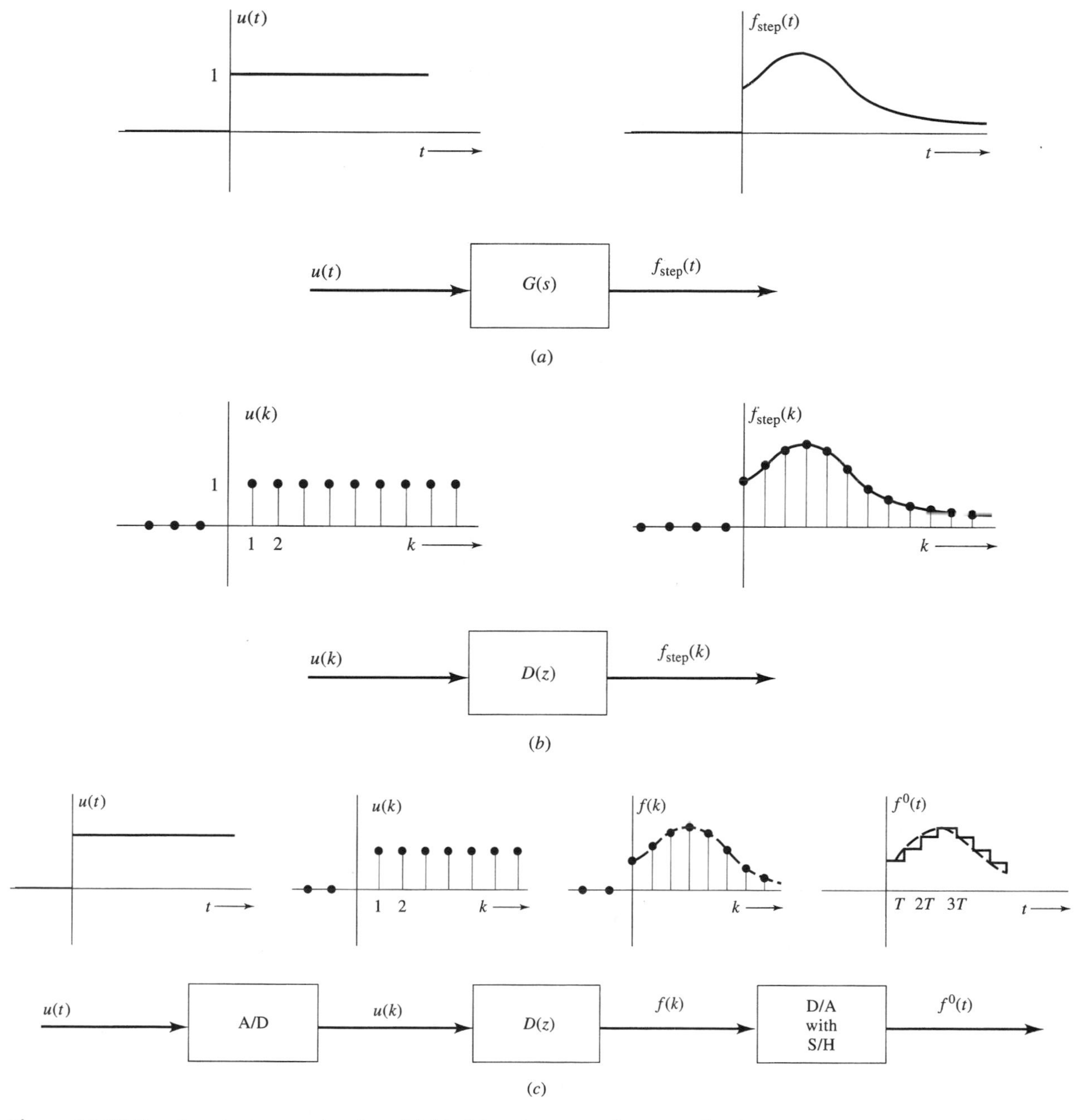

Figure 11.19 Step-invariant approximation. (a) Model analog transmittance and unit step response. (b) Digital system derived from $G(s)$ by using the step-invariant approximation.

(c)

$$G(s) = \frac{2}{s+4} \qquad T = 0.03$$

Ans. $0.057/(z - 0.887)$

(d)

$$G(s) = \frac{e^{-s}}{s+3} \qquad T = 0.1$$

Ans. $0.086/z(z - 0.74)$

11.9.2 *z*-Transfer Functions of Systems with Analog Measurements

Often in digital control system analysis and design the situation of Figure 11.20(a) occurs, where it is desired to find the z-transfer function of a system or subsystem with discrete-time input and output but intervening analog components. For this basic situation, the Laplace transform of the impulse train is

$$F^*(s) = F(z)|_{z=e^{sT}}$$

and the analog output $y(t)$ is given by

$$Y(s) = F^*(s)D(s) = [F(z)|_{z=e^{sT}}]D(s)$$

To obtain the z-transform $Y(z)$ of the output sequence, the time-shift terms are separated from the rational part of the transform and the usual z-transform term substitutions made. The e^{sT} terms involved in $F^*(s)$ are part of the time-shift portion of $Y(s)$, so in forming $Y(z)$, the e^{sT} terms in $F^*(s)$ are simply converted back to $e^{sT} = z$, giving

$$Y(z) = F(s)D(z)$$

The z-transmittance of the basic subsystem is thus

$$\frac{Y(z)}{F(z)} = D(z)$$

That is, the analog transmittance $D(s)$ is converted to the corresponding z-transmittance just as in the sampling process where a signal given by $D(s)$ is sampled, yielding a z-transform $D(z)$.

The example digital-input, digital-output subsystem of Figure 11.20(b) involves sample-and-hold and so has the intervening analog transmittance

$$D(s) = (1 - e^{-sT})\left[\frac{12}{s(s+3)}\right] = (1 - e^{-sT})\left(\frac{4}{s} - \frac{4}{s+3}\right)$$

for which

$$D(z) = (1 - z^{-1})\left[4\left(\frac{z}{z-1}\right) - 4\left(\frac{z}{z - e^{-3T}}\right)\right]$$

where T is the sampling interval.

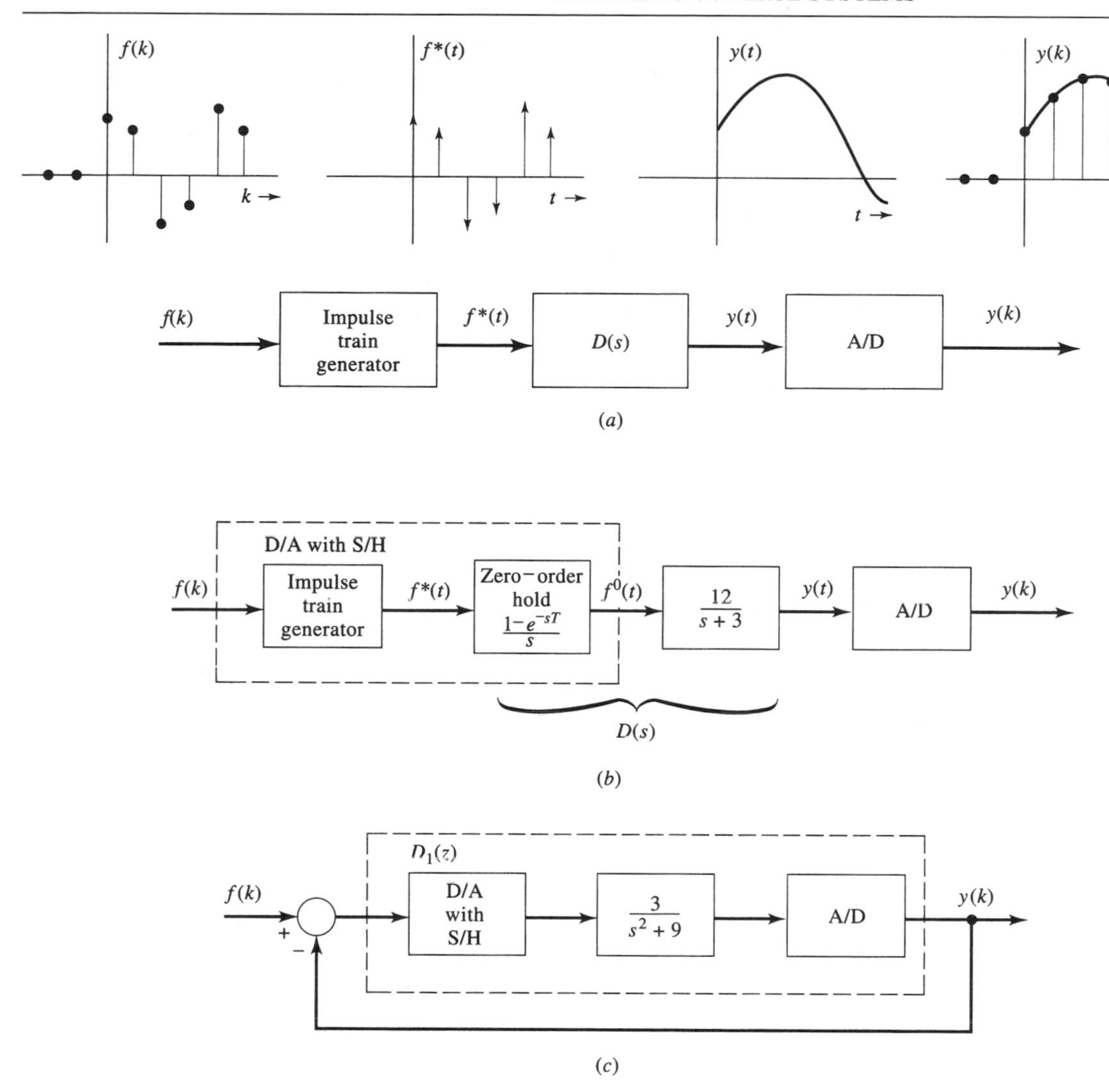

Figure 11.20 Digital subsystems with analog components. (a) Basic arrangement. (b) An example subsysem. (c) An example system with digital feedback.

For the system of Figure 11.20(c), the z-transmittance

$$D_1(z) = \frac{Y(z)}{E(z)}$$

is given by

$$D_1(s) = \frac{1 - e^{-sT}}{s}\left(\frac{3}{s^2+9}\right)$$

$$= (1 - e^{-sT})\left[\frac{\frac{1}{3}}{s} + \frac{-\frac{1}{3}s}{s^2+9}\right]$$

$$D_1(z) = (1 - z^{-1})\left\{\frac{1}{3}\left(\frac{z}{z-1}\right) - \frac{1}{3}\left[\frac{z(z - \cos 3T)}{z^2 - 2z\cos 3T + 1}\right]\right\}$$

$$= \frac{1}{3}\left(\frac{1-z}{z}\right)\frac{(1 - \cos 3T)(z^2 + z)}{(z-1)(z^2 - 2z\cos 3T + 1)}$$

$$= \frac{\frac{1}{3}(1 - \cos 3T)(z+1)}{z^2 - 2z\cos 3T + 1}$$

The overall feedback system z-transfer function is

$$D(z) = \frac{Y(z)}{F(z)} = \frac{D_1(z)}{1 + D_1(z)}$$

$$= \frac{\frac{1}{3}(1 - \cos 3T)(z+1)}{z^2 + [\frac{1}{3} - \frac{7}{3}\cos 3T]z + [\frac{4}{3} - \frac{1}{3}\cos 3T]}$$

❑ DRILL PROBLEM

D11.19 Find the z-transfer functions of the systems shown in Figure D11.19 in terms of the sampling interval T.

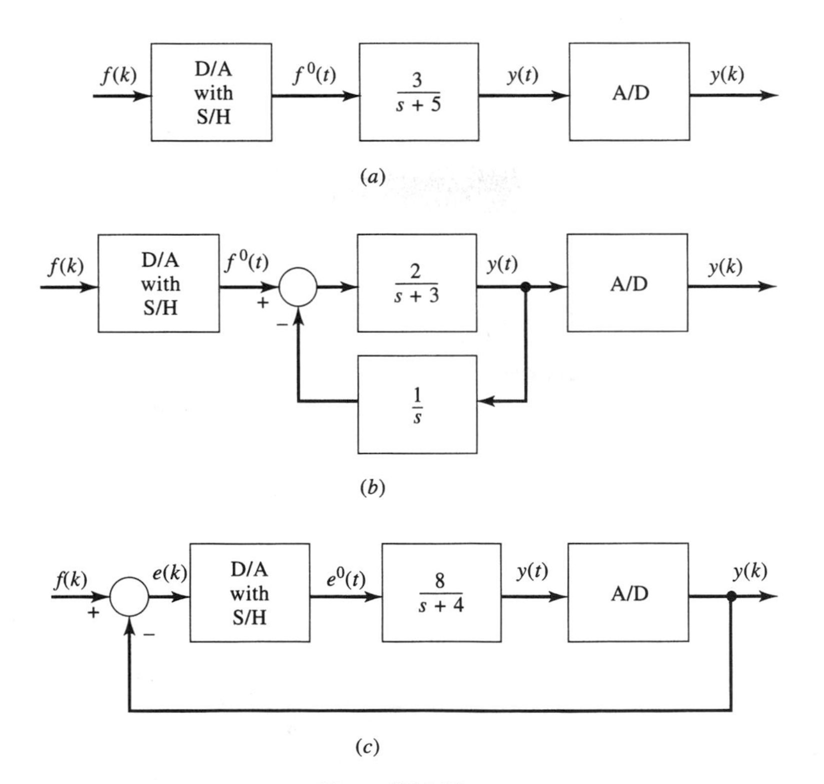

Figure D11.19

Ans. (a) $0.6(e^{-5T} - 1)/(z - e^{5T})$

(b) $\dfrac{2(z-1)(e^{-T} - e^{-2T})}{(z - e^{-T})(z - e^{-2T})}$

(c) $(2e^{-4T} - 2)/(-z + 3e^{-4T} - 2)$

11.9.3 A Design Example

Figure 11.21(a) shows a unity feedback continuous-time system consisting of a controller with compensator $G_1(s)$ and the plant $G_2(s)$. The system transfer function is

$$T(s) = \frac{G_1(s)G_2(s)}{1 + G_1(s)G_2(s)} = \frac{1}{s^2 + s + 1}$$

Its unit step response is given by

$$Y_{\text{step}}(s) = \frac{1}{s}T(s) = \frac{1}{s(s^2 + s + 1)}$$

$$= \frac{1}{s} + \frac{0.58e^{-j150^\circ}}{s + 0.5 + j0.866} + \frac{0.58e^{j150^\circ}}{s + 0.5 - j0.866}$$

$$Y_{\text{step}}(t) = [1 + 1.16e^{-0.5t}\cos(0.866t + 150^\circ)]u(t)$$

To convert the analog controller to a digital one, analog-to-digital converters are used to sample the input and the output signals, a digital-to-analog converter drives the plant, and the compensator transmittance is replaced by a z-transmittance as in Figure 11.21(b). The unit step response of the original analog is given by

$$F_{\text{step}}(s) = \frac{1}{s}\left(\frac{1}{s+1}\right) = \frac{1}{s} - \frac{1}{s+1}$$

$$f_{\text{step}}(t) = (1 - e^{-t})u(t)$$

Requiring that the unit step response of the discrete-time compensator consist of samples of the analog unit step response gives, in terms of the sampling interval T,

$$f_{\text{step}}(k) = (1 - e^{-kT})u(k) = u(k) - (e^{-T})^k u(k)$$

$$F_{\text{step}}(z) = \frac{z}{z-1} - \frac{z}{z - e^{-T}}$$

$$= \frac{z}{z-1}\left(\frac{1 - e^{-T}}{z - e^{-T}}\right) = U(z)D_1(z)$$

Hence the step-invariant design for $D_1(z)$ is

$$D_1(z) = 1 - \frac{z-1}{z - e^{-T}} = \frac{1 - e^{-T}}{z - e^{-T}}$$

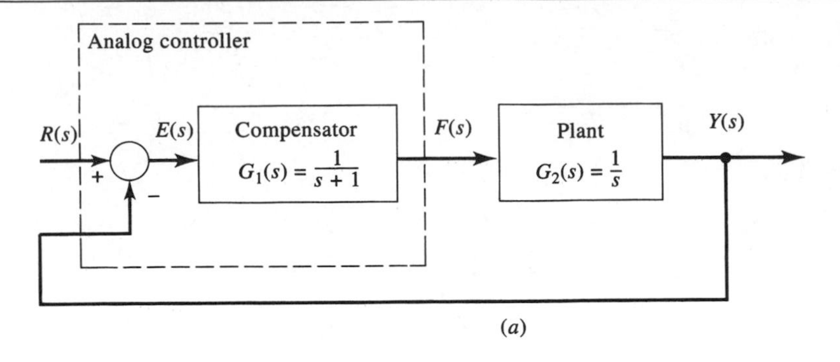

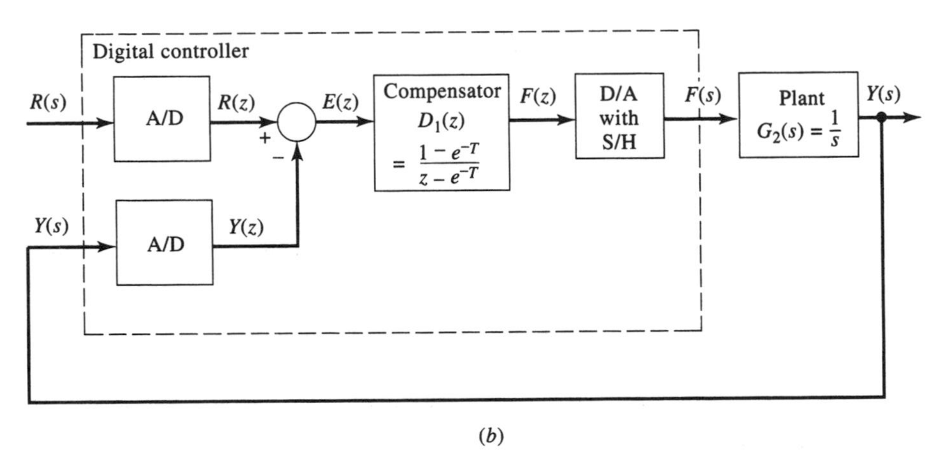

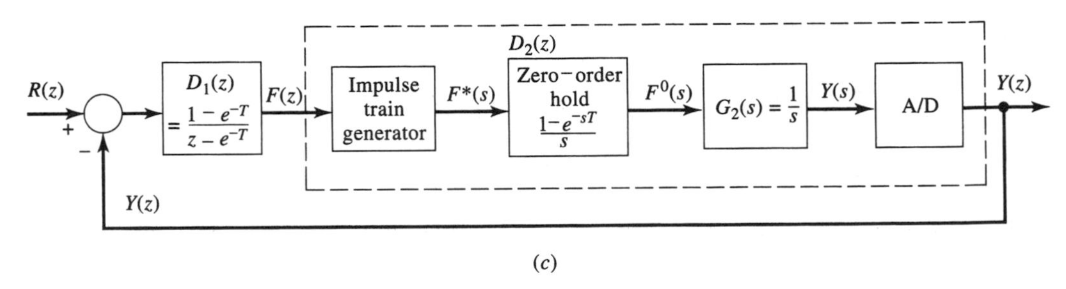

Figure 11.21 Converting analog-to-digital control. (a) A continuous-time feedback system. (b) Conversion to digital control. (c) Relations between the digital signals.

The relationships between the digital signals in the system with digital control are shown in Figure 11.21(c). The z-transmittance relating the sequence $y(k)$ to $f(k)$ is given by

$$D_2(s) = \frac{1 - e^{-sT}}{s}\left(\frac{1}{s}\right) = (1 - e^{-sT})\frac{1}{s^2}$$

$$D_2(z) = (1 - z^{-1})\frac{T_z}{(z-1)^2} = \frac{T}{z-1}$$

where T is the sampling interval. The feedback system thus has overall z-transfer function

$$D(z) = \frac{Y(z)}{R(z)} = \frac{D_1(z)D_2(z)}{1 + D_1(z)D_2(z)}$$

$$= \frac{\dfrac{1-e^{-T}}{z-e^{-T}}\left(\dfrac{T}{z-1}\right)}{1 + \dfrac{1-e^{-T}}{z-e^{-T}}\left(\dfrac{T}{z-1}\right)}$$

$$= \frac{T(1-e^{-T})}{z^2 - (1+e^{-T})z + T(1-e^{-T}) + e^{-T}}$$

Response samples of the step response of the digitized system for various sampling rates are plotted in Figure 11.22, where they are compared with the continuous-time step response of the original analog system. For a half-second sampling interval, $T = 0.5$,

$$Y(z) = U(z)D(z) = \frac{0.197z}{z^3 - 2.607z + 2.41z - 0.804}$$
$$= 0.197z^{-2} + 0.51z^{-3} + 0.86z^{-4} + \cdots$$

For $T = 0.2$,

$$Y(z) = U(z)D(Z) = \frac{0.036z}{z^3 - 2.819z^2 + 2.674z - 1.655}$$
$$= 0.036z^{-2} + 0.102z^{-3} + 0.191z^{-4} + \cdots$$

For $T - 0.05$,

$$Y(z) = U(z)D(z) = \frac{0.0024z}{z^3 - 2.95z^2 + 1.95z + 0.954}$$
$$= 0.0024z^{-2} + 0.00708z^{-3} + 0.014z^{-4} + \cdots$$

The higher the sampling rate, the more closely the digital system approximates the behavior of the original analog system.

Since $G_1(s)$ is an integrator driven by the piecewise constant sample-and-hold (S/H) waveform $f^0(t)$ and since the continuous-time output $y(t)$ passes through the sample points $y(k)$, $y(t)$ is a piecewise linear signal through the sample points.

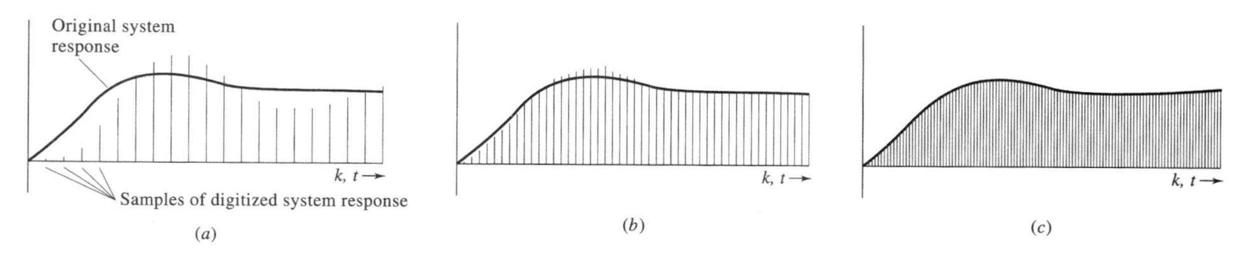

Figure 11.22 Response of the digitized system. (a) $T = 0.5$. (b) $T = 0.2$. (c) $T = 0.05$.

The final value of the unit step response is, from the formula in Table 11.6,

$$\begin{aligned}\lim_{k\to\infty} y(k) &= \lim_{z\to 1}\left[\frac{z-1}{z}Y(z)\right]\\ &= \lim_{z\to 1}\left[\frac{z-1}{z}\frac{z}{z-1}D(z)\right]\\ &= \lim_{z\to 1}\left[\frac{T(1-e^{-T})}{z^2-(1+e^{-T})z+T(1-e^{-T})+e^{-T}}\right]\\ &= \frac{T(1-e^{-T})}{T(1-e^{-T})} = 1\end{aligned}$$

11.10 Direct Digital Design

11.10.1 Steady State Response

If a discrete-time signal $f(k)$ reaches a finite, constant steady state value, it is given by

$$\lim_{k\to\infty} f(k) = \lim_{z\to 1}\frac{z-1}{z}F(z)$$

which is the discrete-time version of the final value theorem. Let us expand $F(z)/z$ into partial fractions:

$$\frac{F(z)}{z} = \frac{K_1}{z-1} + \cdots$$

Now the residue of the $(z-1)$ pole represents the constant term in the sequence $f(k)$, which may be found by multiplying $F(z)/z$ by $(z-1)$ and evaluating at $z=1$.

For example, the sequence with z-transform

$$\begin{aligned}F(z) &= \frac{3z^2+4}{z^2-\frac{1}{2}z-\frac{1}{2}}\\ &= \frac{3z^2+4}{(z-1)(z+\frac{1}{2})} = \frac{K_1}{z-1}+\frac{K_2}{z+\frac{1}{2}}\end{aligned}$$

has steady state value given by

$$\begin{aligned}\lim_{k\to\infty} f(k) &= \lim_{z\to 1}\frac{z-1}{z}F(z) = \lim_{z\to 1}\frac{3z^2+4}{z^2+\frac{1}{2}z}\\ &= \tfrac{14}{3}\end{aligned}$$

The sequence with z transform

$$F(z) = \frac{z^3-4z+1}{z^3+z^2-2z} = \frac{K_1}{z-1}+\frac{K_2}{z}+\frac{K_3}{z+2}$$

does not reach a finite steady state because the $K_3/(z+2)$ term in its partial fraction expansion represents an expanding term in the sequence $f(k)$. The limit exists and is finite, however:

$$\lim_{z\to 1} \frac{z-1}{z} F(z) = \lim_{z\to 1} \frac{z^3 - 4z + 1}{z^3 + 2z^2} = -\frac{2}{3}$$

For a system with stable z-transfer function

$$D(z) = \frac{4z^2 + 8z - 1}{(3z+2)(4z-1)}$$

the error between input and output, when the input is a unit step sequence, is

$$E(z) = R(z) - Y(z) = R(z)[1 - D(z)]$$
$$= \frac{z}{z-1}\left(1 - \frac{4z^2 + 8z - 1}{12z^2 + 5z - 2}\right)$$

which has a final value

$$\lim_{k\to\infty} e(k) = \lim_{z\to 1} \frac{z-1}{z} E(z)$$
$$= \lim_{z\to 1} \frac{8z^2 - 3z - 1}{12z^2 + 5z - 2}$$
$$= \frac{4}{15}$$

❑ DRILL PROBLEM

D11.20 Each of the systems in Figure D11.20 is stable. Find the steady state error between input and output for each when the input is a unit step sequence.

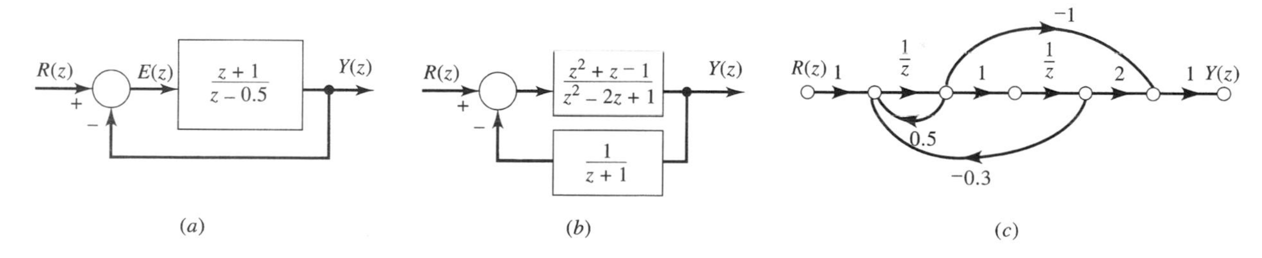

Figure D11.20

Ans. (a) 0.2; (b) 0; (c) −0.25

11.10.2 Deadbeat Systems

A digital system with all of its poles at $z = 0$ is termed *deadbeat*. Such systems have a remarkable property: their pulse response is zero after n steps, where n is the order of the system. For example, the system with z-transfer function

$$D(z) = \frac{z^3 + 3z^2 - 2z + 4}{z^3}$$

Deadbeat means all poles at the origin.

is deadbeat. Its unit pulse response is

$$Y_{\text{pulse}}(z) = 1 \cdot D(z)$$
$$= 1 + 3z^{-1} - 2z^{-2} + 4z^{-3}$$
$$y_{\text{pulse}}(k) = \delta(k) + 3\delta(k-1) - 2\delta(k-2) + 4\delta(k-3)$$

Deadbeat systems are also commonly termed *finite-duration impulse response* (FIR) systems.

As pulse response is representative of the natural component of a system's response in general, deadbeat systems have a natural response that goes to zero after n steps. There is no counterpart in analog systems; analog natural response may only decay asymptotically to zero. In many practical situations, digital systems are designed to be deadbeat if possible.

❑ DRILL PROBLEM

D11.21 Find the response $y(k)$ of each of the systems of Figure D11.21 if the input is a unit step sequence and the initial conditions are zero.

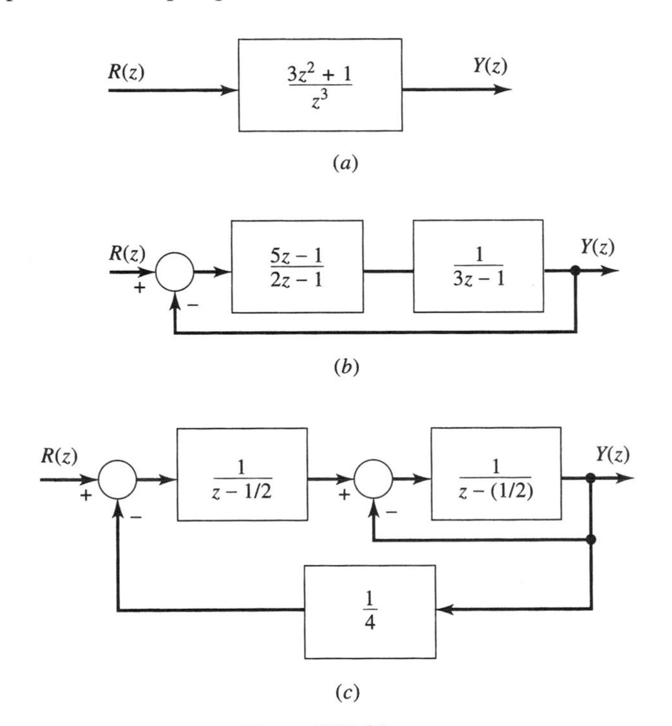

Figure D11.21

Ans. (a) $3u(k-1) + u(k-3)$; (b) $\frac{5}{6}u(k-1) - \frac{1}{6}u(k-2)$; (c) $u(k-2)$

11.10.3 A Design Example

Figure 11.23(a) shows a simplified model of a satellite tracking control system. Computer-generated digital commands, based on orbital calculations, are applied to the system, which consists of a digital controller driving the analog positioning subsystem. The diagram is rearranged in Figure 11.23(b) to show the relations between digital signals. The sampling interval is $T = 0.1$. It is desired to design the controller transmittance $D_1(z)$ so that, if possible,

1. The overall digital system $D(z) = Y(z)/R(z)$ is deadbeat.
2. The error between output and input, $E(z)$, to a step input $R(z)$ is zero in steady state.

In addition, it is important that the analog position output $y(t)$ be well behaved between samples, particularly as a constant steady state is approached.

The analog transmittance $D_2(s)$ in Figure 11.23(b) is

$$D_2(s) = \frac{10(1 - e^{-0.1s})}{s^2(s+10)}$$

$$= (1 - e^{-0.1s})\left(\frac{-\frac{1}{10}}{s} + \frac{1}{s^2} + \frac{\frac{1}{10}}{s+10}\right)$$

hence the corresponding z-transmittance is

$$D_2(z) = (1 - z^{-1})\left[\frac{-\frac{1}{10}z}{z-1} + \frac{0.1z}{(z-1)^2} + \frac{\frac{1}{10}z}{z-0.368}\right]$$

$$= \frac{(z-1)(0.0368z^2 + 0.0264z)}{z(z-1)^2(z-0.368)}$$

$$= \frac{0.0368z + 0.0264}{(z-1)(z-0.368)}$$

In terms of the controller transmittance $D_1(z)$, the overall system transfer function is

$$D(z) = \frac{D_1(z)\,D_2(z)}{1 + D_1(z)\,D_2(z)}$$

$$= \frac{D_1(z)(0.0368z + 0.0264)/[(z-1)(z-0.368)]}{1 + D_1(z)(0.0368z + 0.0264)/[(z-1)(z-0.368)]}$$

Let $D(z)$ have n poles, all of which are at $z = 0$. Then

$$1 + D_1(z)\frac{0.0368z + 0.0264}{(z-1)(z-0.368)} = \frac{z^n}{\text{polynomial}}$$

where "polynomial" denotes an as-yet-unspecified polynomial in z. Solving for $D_1(z)$,

$$D_1(z) = \frac{(z-1)(z-0.368)}{0.0368z + 0.0264}\left(\frac{z^n}{\text{polynomial}} - 1\right)$$

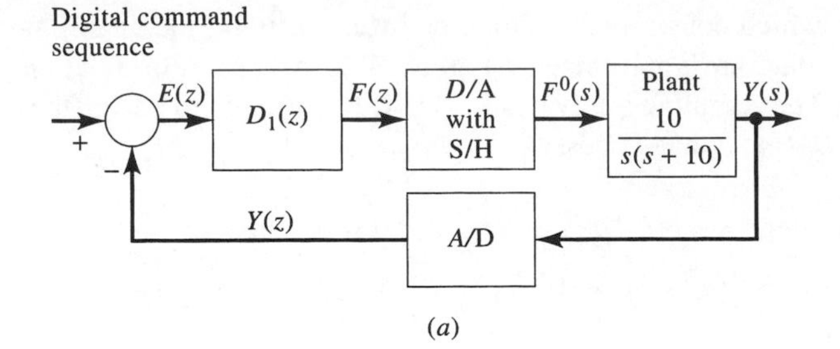

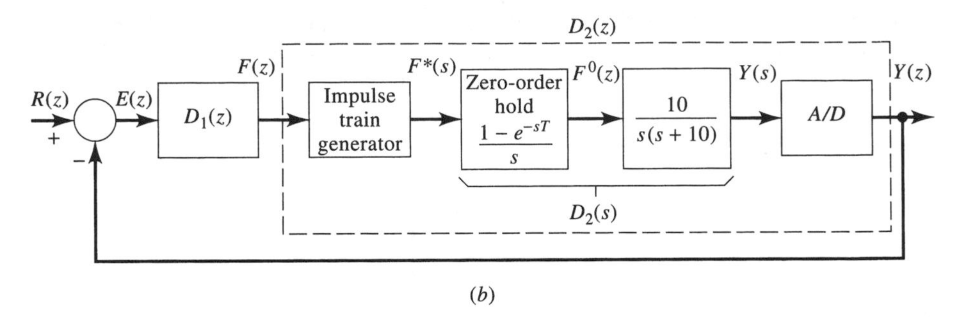

Figure 11.23 Satellite tracking system. (a) System block diagram. (b) Block diagram rearranged to show digital relations.

Any such z-transmittance $D_1(z)$ will result in an overall system that is deadbeat:

$$D(z) = \frac{z^n/\text{polynomial} - 1}{1 + (z^n/\text{polynomial} - 1)}$$

$$= \frac{z^n - \text{polynomial}}{z^n}$$

The error between input and output is

$$E(z) = R(z) - Y(z) = R(z)\,[1 - D(z)]$$

$$= R(z)\left(1 - \frac{z^n - \text{polynomial}}{z^n}\right)$$

$$= R(z)\left(\frac{\text{polynomial}}{z^n}\right)$$

For a step input

$$R(z) = \frac{z}{z - 1}$$

the error signal has z-transform

$$E(z) = \frac{1}{z - 1}\left(\frac{\text{polynomial}}{z^n}\right)$$

and steady state value given by

$$\lim_{k\to\infty} e(k) = \lim_{z\to 1}\left(\frac{\text{polynomial}}{z^n}\right)$$
$$= \lim_{z\to 1}(\text{polynomial})$$

To summarize:

1. For the overall system to be deadbeat, $D_1(z)$ must be of the form

$$D_1(z) = \frac{(z-1)(z-0.368)}{0.0368z+0.0264}\left(\frac{z^n}{\text{polynomial}} - 1\right)$$

2. For the system's steady state step error to be zero, the unspecified polynomial must have the property

$$\lim_{z\to 1}(\text{polynomial}) = 0$$

The choice of $n = 1$ and polynomial $= z - 1$ is simple, has property 2, and results in a simplification of $D_1(z)$:

$$D_1(z) = \frac{z - 0.0368}{0.0368z + 0.0264}$$

The overall transfer function for this choice is

$$D(z) = \frac{y(z)}{R(z)} = \frac{1}{z}$$

The digital step response of this system is given by

$$Y(z) = R(z)D(z) = \frac{z}{z-1}\left(\frac{1}{z}\right)$$
$$\frac{y(z)}{z} = \frac{1}{z(z-1)} = \frac{-1}{z} + \frac{1}{z-1}$$
$$y(z) = -1 + \frac{1}{z-1}$$
$$y(k) = -\delta(k) + u(k)$$

For a step command input, the output reaches its final value in one step, as shown in Figure 11.24(a).
For the step input, the controller's digital signal $F(z)$ is given by

$$F(z) = [R(z) - Y(z)]\,D_1(z)$$
$$= \left(\frac{z}{z-1} - \frac{1}{z-1}\right)\frac{z-0.368}{0.0368z+0.0264}$$
$$= \frac{z-0.368}{0.0368z+0.0264}$$

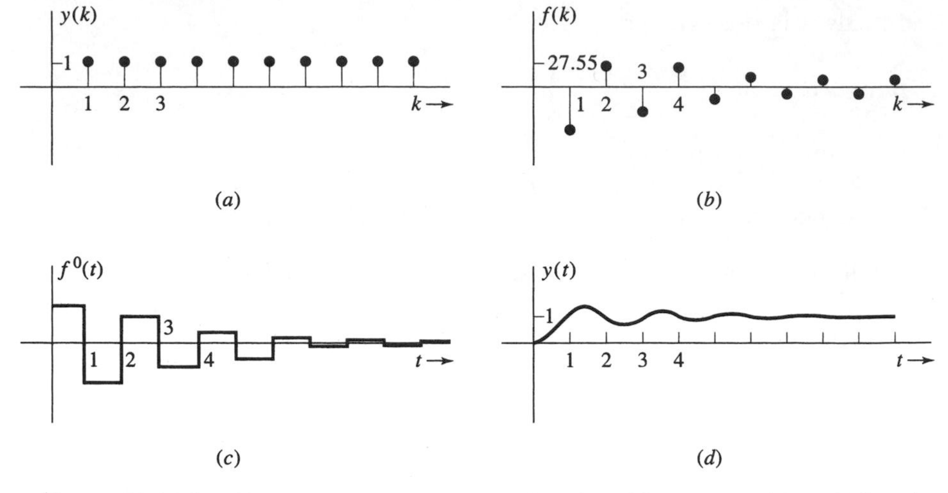

Figure 11.24 Tracking system step response. (a) Samples of the output position. (b) Samples of the controller output. (c) Sample-and-hold controller output. (d) Output position as a function of time.

$$\frac{F(z)}{z} = \frac{27.17z - 10}{z\,(z + 0.717)} = \frac{-13.95}{z} + \frac{+41.1}{z + 0.717}$$

$$F(z) = -13.95 + 41.1\frac{z}{z + 0.717}$$

❑ Computer-Aided Learning

MATLAB handles digital systems by including a sample period parameter in the system definitions. For example, to enter the discrete transfer function

$$G(z) = \frac{z - 1}{z^2 - 3z + 4} \quad \text{sample period} \qquad T = 20$$

we can use the "tf" command with a third parameter for the sample period.

```
g=tf([1 -1], [1 -3 4], 20)
```

Transfer function:

```
z-1
-------
z^2-3z+4
Sampling time: 20
```

or we can convert this system to state space using the "ss" command. Notice that since $G(z)$ is defined as a discrete transfer function object, MATLAB converts it to a discrete state space equation:

```
gs=ss(g)
a=
                 x1          x2
        x1   3.00000   -2.00000
        x2   2.00000          0
```

```
b=
              u1
    x1   1.00000
    x2         0

c=
              x1          x2
    y1   1.00000   - 0.50000

d=
              u1
    y1         0

Sampling time: 20
Discrete-time system.
```

We can also enter the system directly in state space using the matrices and an optional sample period.

```
   a=[1 2; 3 4]; b=[1; 2]; c=[1 2];d=0; g2=ss (a,b,c,d,20)
a=
    x1         x2
    x1    1.00000    2.00000
    x2    3.00000    4.00000

b=
    u1
    x1    1.00000
    x2    2.00000

c=
               x1         x2
    y1    1.00000    2.00000

d=
               u1
    y1          0

Sampling time: 20
Discrete-time system.
```

When we transform this to transfer function form we get a discrete transfer function.

```
    g2z=tf(g2)
    Transfer function:
    5z+2
    --------
    z^2-5z-2

    Sampling time: 20
```

We can convert continuous systems to discrete systems (discretization) using the "c2d" command. MATLAB uses the zero-order hold method for approximation.

For example, consider the continuous plant $G(s)$:

$$G(s) = \frac{10}{s^2 + 2s + 10}$$

Let us discretize this plant. To select the sample period, we obtain the Bode plot to find the bandwidth (3 dB below the low-frequency magnitude):

```
g=tf(10,[1 2 10])
Transfer function:
     10
---------
s^2+2s+10
bode(g)
```

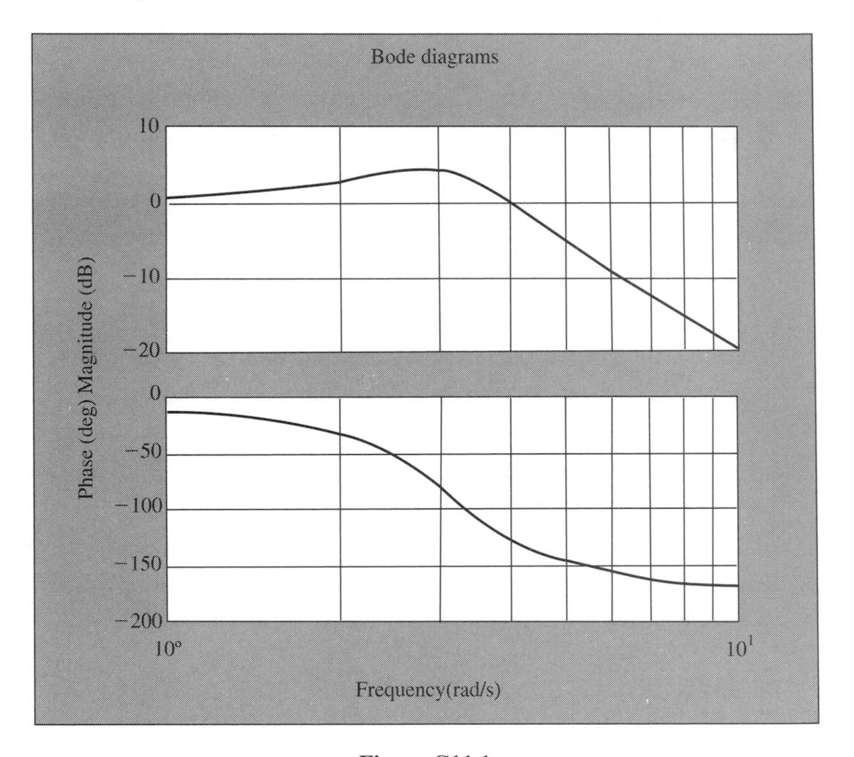

Figure C11.1

The system has a bandwidth of approximately 4.5. A rule of thumb for sample rate selection for control systems is to use 10 times the closed-loop bandwidth. So we use $\omega_s = 45$, and $T = 2\pi/45 = 0.14$.

```
T=2*pi/45; gz=c2d(g,T)
Transfer function:
0.08759 z+0.07976
---------
z^2-1.589 z+0.7563
```

```
Sampling time: 0.13963
step(g, 'k'), grid, hold, step(gz, 'r')
```

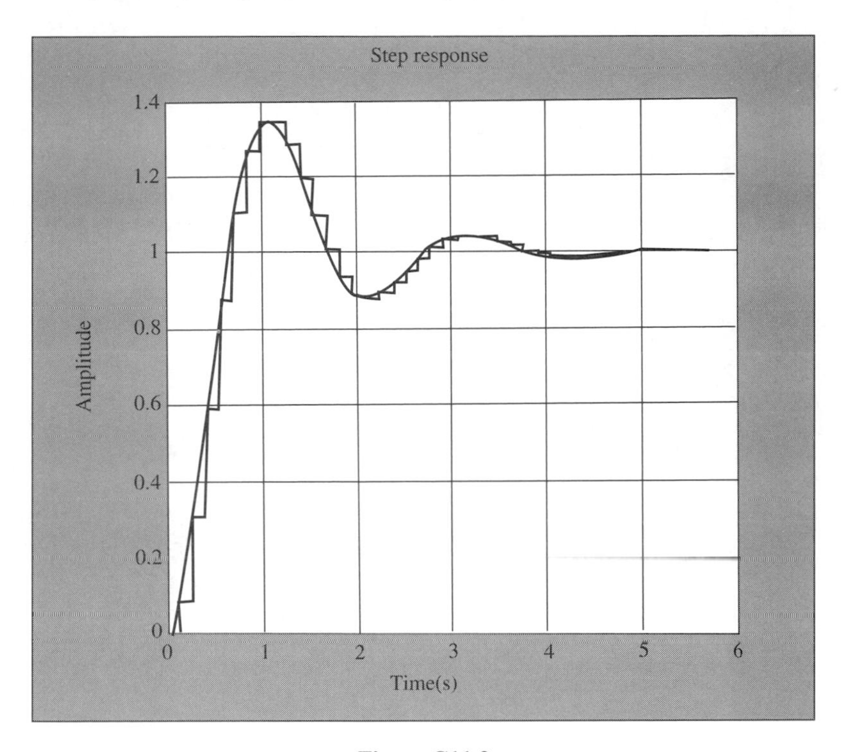

Figure C11.2

```
T=2*pi/10;  gz=c2d(g,T);
step(g, 'k'), grid, hold, step (gz, 'r'),
Current plot held
```

Notice how a small sample period affects the quality of the discrete step response. The full syntax for the "c2d" command is

```
SYSD=C2D(SYSC,TS,METHOD)
Where METHOD is a string that allows you to choose among a
number of discretization methods such as:
  'zoh'  Zero-order hold on the inputs.
  'foh' Linear interpolation of inputs.(triangle appx.)
  'tustin'  Bilinear (Tustin) approximation.
  'prewarp' Tustin approximation with frequency prewarping.
  'matched' Matched pole-zero method (for SISO systems
only).
The default is 'zoh' when METHOD is omitted.
```

To determine stability once the closed-loop transfer function has been given, we can use the "pole" and "abs" commands to find the magnitude of the poles.

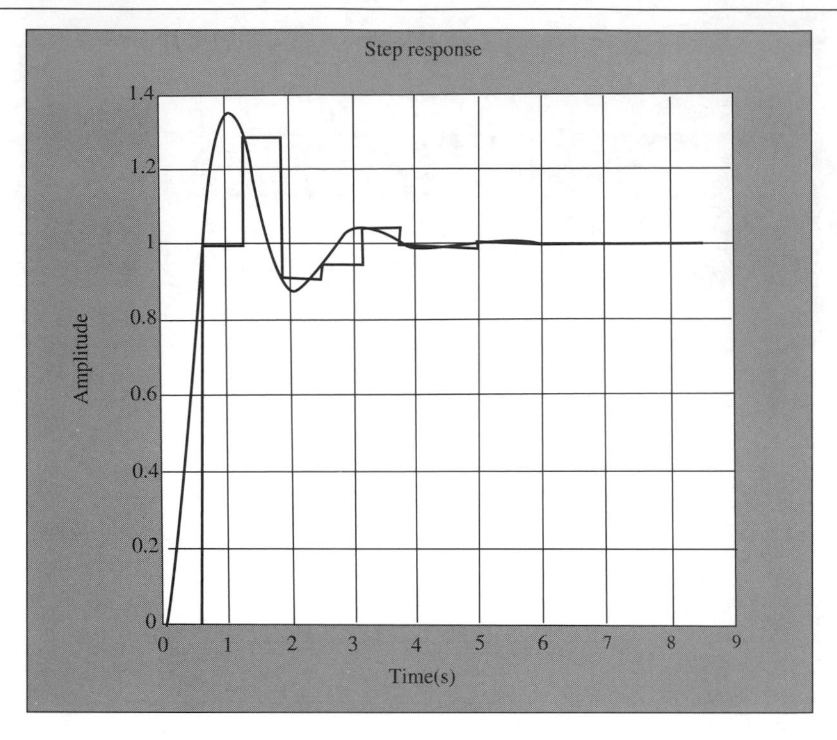

Figure C11.3

```
abs (pole (gz))
ans=
  5.3349e-001
  5.3349e-001
```

All other Control Systems commands, such as step, impulse, rlocus, bode, and nyquist work identically for discrete systems with no change.

11.11 SUMMARY

The relation between an analog signal *f(t)* and its sample sequence

$$f(k) = F(t = kT)$$

is found by relating the z-transform $F(z)$ of the sequence to the Laplace transform $F(s)$ of the analog signal. Each term in the partial fraction expansion of $F(s)$ is replaced by the corresponding z-transform expansion term from Table 11.5 to form $F(z)$.

Approximate reconstruction of an analog signal $f(t)$ from its samples is commonly performed by a D/A converter with sample-and-hold. The S/H waveform $f^0(t)$ is conveniently represented by the conversion of digital samples to an analog impulse train, followed by a zero-order hold transmittance. More accurate reconstruction may be done with higher-order holds, involving more than a single sample, and low-pass filtering. The sampling theorem states conditions for which a band-limited signal is uniquely represented by its samples.

Discrete-time systems are described by difference equations,

$$\begin{aligned} y(k+n) &+ a_{n-1}y(k+n-1) + a_{n-2}y(k+n-2) \\ &+ \cdots + a_1 y(k+1) + a_0 y(k) \\ &= b_m r(k+m) + b_{m-1}r(k+m-1) + \ldots + b_1 r(k+1) + b_0 r(k) \end{aligned}$$

where $r(k)$ is the input sequence and $y(k)$ is the output sequence. The z-transfer function of such a system is

$$\begin{aligned} D(z) &= \left.\frac{Y(z)}{R(z)}\right|_{\text{initial conditions}=0} \\ &= \frac{b_m z^m + b_{m-1}z^{m-1} + \ldots + b_1 z + b_0}{z^n + a_{n-1}z^{n-1} + a_{n-2}z^{n-2} + \ldots + a_1 z + a_0} \end{aligned}$$

Block diagrams and signal flow graphs for discrete-time systems, described in terms of z-transforms, are manipulated just like continuous-time system descriptions in terms of Laplace transforms.

A discrete-time system is termed *stable* if and only if all poles of its z-transfer function are within the unit circle on the complex plane. The bilinear transformation

$$z = \frac{1+W}{1-W}$$

maps the unit circle to the left half-plane on the complex plane and so is very useful in relating discrete-time situations to equivalent continuous-time ones. With the above substitution for z in a rational z-transfer function. Routh–Hurwitz methods may be applied to determine stability.

State equations offer a systematic and powerful method of system representation. In terms of these,

$$\begin{aligned} x(k+1) &= Fx(k) + Gu(k) \\ y(k) &= Hx(k) \end{aligned}$$

The z-transfer function matrix of a system is

$$D(z) = H(zI - F)^{-1}G$$

and the system is stable if and only if all roots of the characteristic equation

$$|zI - F| = 0$$

are within the unit circle on the complex plane. An nth-order system is completely controllable if and only if

$$M_c = \left[G \mid FG \mid F^2G \mid \cdots \mid F^{n-1}G\right]$$

is of full rank and is completely observable if and only if

$$M_o = \begin{bmatrix} H \\ \hdashline HF \\ \hdashline HF^2 \\ \hdashline \vdots \\ \hdashline HF^{n-1} \end{bmatrix}$$

is of full rank.

Change of stable variables

$$x(k) = P\bar{x}(k)$$

gives alternative state-variable representations of a system and shows available design freedom. If the system characteristic roots are not repeated, the methods of Section 8.3 may be used to find a set of state variables for which the state equations are decoupled from one another.

A common analysis problem involves a system containing analog components but with discrete-time input and output. To find the z-transmittance, the system is converted to the form where an impulse train generator drives a continuous-time transmittance $D(s)$, followed by a sampler. The transmittance $D(s)$ is separated into delay terms and rational terms, with the rational terms expanded into partial fractions. Substituting

$$e^{sT} \to z$$

into the delay terms and

$$\frac{K}{s+a} \to \frac{Kz}{z - e^{aT}}$$

for the partial fractions yields the z-transmittance $D(z)$.

One design method for digital control involves replacing a model analog subsystem with a digital subsystem consisting of A/D converter, z-transmittance, and D/A converter. For the step-invariant approximation, the z-transmittance is required to have a step response consisting of samples of the step response of the analog subsystem.

Digital control system design may also be done directly, in terms of overall system performance requirements and objectives.

REFERENCES

Computer Processing

Gothmann, W. H., *Digital Electronics, an Introduction to Theory and Practice.* Englewood Cliffs. NJ: Prentice-Hall, 1977.

Mano, M. M., *Digital Logic and Computer Design.* Englewood Cliffs, NJ: Prentice-Hall, 1979.

Osborne. A., *An Introduction to Microcomputers*, 2nd ed. New York: McGraw-Hill, 1980.

Peatman, J. B., *Microcomputer-Based Design.* New York: McGraw-Hill, 1977.

Sampling and Reconstruction

Cadzow, J. A, *Discrete-Time Systems, an Introduction with Interdesciplinary Applications.* Englewood Cliffs, NJ: Prentice-Hall, 1973.

Hamming, R. W., *Digital Filters.* Englewood Cliffs, NJ: Prentice-Hall, 1977.

Jury, E. I., *Theory and Application of the z-Transform Method.* New York: Wiley, 1964.

Oliver, B. M., Pierce, J. R., and Shannon, C. E., "The Philosophy of PCM." *Proc. IRE* 36 (November 1948): 1324–1331.

Stearns, S. D., *Digital Signal Analysis.* Rochelle Park, NJ: Hayden, 1975.

Tretter, S. A., *Introduction to Discrete-Time Signal Processing.* New York: Wiley, 1976.

Discrete-Time Systems and Filtering

Chen, C.-T., *One-Dimensional Digital Signal Processing.* New York: Marcel Dekker, 1979.

Oppenheim, A. V., and Schafer, R. W., *Digital Signal Processing.* Englewood Cliffs, NJ: Prentice-Hall, 1975.

Peled, A., and Liu, B., *Digital Signal Processing: Theory, Design and Implementation.* New York: Wiley, 1976.

Rabiner, L. R., and Gold, B., *Theory and Application of Digital Signal Processing.* Englewood Cliffs, NJ: Prentice-Hall, 1975.

Schwartz, M., and Shaw, L., *Signal Processing.* New York: McGraw-Hill, 1975.

Stanley, W. D., *Digital Signal Processing.* Reston, VA: Reston, 1975.

Digital Control

Cadzow, J. A., and Martens, H. R., *Discrete-Time and Computer Control Systems.* Englewood Cliffs, NJ: Prentice-Hall, 1970.

Franklin, G. F., and Powell, J. D., *Digital Control of Dynamic Systems.* Reading, MA: Addison-Wesley, 1980.

Freeman, H., *Discrete-Time Systems.* New York: Wiley, 1965.

Kuo, B. C., *Digital Control Systems.* New York: Holt, Rinehart & Winston, 1980.

Monroe, A. J., *Digital processes for Sampled Data systems.* New York: Wiley, 1962.

Ragazzini, J. R., and Franklin, G. F., *Sampled Data Control Systems.* New York: McGraw-Hill, 1958.

Schwartz, R. J., and Friedland, B., *Linear systems.* New York: Wiley, 1965.

Tou, J. T., *Digital and Sampled-Data Control Systems.* New York: McGraw-Hill, 1959.

Van Landingham, H. F., *Introduction to Digital Control Systems.* New York: Macmillan, 1985.

Digital Processing and Control Applications

Allan. R., "Busy Robots Spur Productivity." *IEEE Spectrum* (September 1979): 31–36.

———, "The Microcomputer Invades the Production Line." *IEEE Spectrum* (January 1979): 53–57.

Andrews, H. C., and Hunt, B. R., *Digital Image Restoration*. Englewood Cliffs, NJ: Prentice-Hall, 1977.

Kahne, s., "Automatic Control by Distributed Intelligence." *Sci. Amr.* (June 1979): 78–109.

Oppenheim, A. V., ed., *Applications of Digital Signal Processing.* Englewood Cliffs, NJ: Prentice-Hall, 1978.

Rabiner, L. R., and Schafer, R. W., *Digital Processing of Speech Signals.* Englewood Cliffs, NJ: Prentice-Hall, 1978.

PROBLEMS

1. List five home appliances for which microprocessor control is today useful and economically feasible. For each, describe the functions performed by the digital system.
2. Carefully describe functions that a microprocessor-based control system might perform in each of the following. Specify the signals to be sensed and the quantities to be controlled.
   (a) An automobile
   (b) A hotel
   (c) Aboard a ship
   (d) An electric power generating plant
   (e) A hospital operating room.
3. The analog signal

$$f(t) = 3 + 4\cos 50t$$

is sampled at 0.01 s intervals by an A/D converter preceded by a sample and-hold (S/H) that freezes the sample while conversion takes place. Then the samples are reconverted to an analog signal with S/H. Sketch your visualization of signal $f(t)$, the first S/H waveform, the A/D samples, and the output wave-form.

4. A bank account pays (9%) interest per year, compounded monthly. Initially a deposit of $1000 is made, and thereafter $65 per month is deposited into the account each month. Describe the monthly bank balance as a function of the month k after the initial deposit.

5. Sketch the functions $f(t)$ and samples $f(k)$ with the given sampling interval T:

   (a) $f(t) = 3e^{-10t}u(t) \qquad T = 0.1$

   (b) $f(t) = \left(3e^{-2t}\cos\dfrac{5\pi t}{4}\right)u(t) \qquad T = 0.2$

   (c) $f(t) = \left(2\sin\dfrac{\pi t}{8}\right)u(t) \qquad T = 1$

   (d) $f(t) = \left(4 + 3e^{0.1t}\right)u(t) \qquad T = 0.5$

6. Find the z transforms of the following sequences:

   (a) $f(k) = u(k) - u(k-4)$

   (b) $f(k) = 0.1ku(k) - 0.1ku\,(k-4)$

   Ans. $0.1(1 - z^{-4})z/(z-1)^2$

   (c) $f(k) = \left[\cos\left(0.1k + \dfrac{\pi}{4}\right)\right]u(k)$

   (d) $f(k) = \left[(-1)^k \sin 0.5k\right]u(k)$

   Ans. $-0.48z/\left(z^2 + 1.76z + 1\right)$

7. Find the z transforms of the sequences consisting of samples of the following functions $f(t)$ with the given sampling interval T:

   (a) $f(t) = 5tu(t) \qquad T = 0.5$

   (b) $f(t) = 3u(t) - 4u(t)\sin 3t \qquad T = 0.2$

   Ans. $3z/(z-1) - 2.26z/(z^2 - 1.65z + 1)$

   (c) $15t^2e^{-10t}u(t) \qquad T = 4$

   (d) $5u(t)e^{-3t}\cos(\pi t + 45°) \qquad T = 0.1$

   Ans. $3.54\left(z^2 - 0.93z\right)/(z^2 - 1.4z + 0.55)$

8. Use long division to find $f(0)$, $f(1)$, $f(2)$ and $f(3)$ for each of the discrete-time signals with the following z-transforms:

   (a) $F(z) = \dfrac{4z^2 - 3z + 6}{2z^2 + z - 1}$

   (b) $F(z) = \dfrac{2z - 3}{z^2 - 0.5z + 1}$

   Ans. $0, 2, -2, -3$

   (c) $F(z) = \dfrac{z}{0.3z^3 - 0.1z^2 + 0.2z + 1}$

9. Find the inverse z-transforms:

   (a) $F(z) = \dfrac{4}{z + \frac{1}{3}}$

(b) $F(z) = \dfrac{3z^2 - 2z}{4z^2 + 5z + 1}$

Ans. $-\left(\frac{11}{3}\right)\left(-\frac{1}{4}\right)^k u(k) + \frac{5}{3}(-1)^k u(k)$

(c) $F(z) = \dfrac{z^2}{(2z + 1)^2}$

(d) $F(z) = \dfrac{3z - 2}{z^2 + 1.5z + 0.5}$

Ans. $-4\delta(k) - 10(-1)^k u(k) + 14\left(-\frac{1}{2}\right)^k u(k)$

(e) $F(z) = \dfrac{1}{z^3 - 0.25z}$

10. For the following functions $f(t)$ and sampling periods T, find $f(k)$:

(a) $f(t) = 3e^{-10t}u(t) \qquad T = 0.1$

(b) $f(t) = 3e^{-2t}u(t)\cos\dfrac{5\pi t}{4} \qquad T = 0.2$

Ans. $3(0.67)^k u(k)\cos(0.785k)$

(c) $f(t) = 2u(t)\sin\dfrac{\pi t}{8} \qquad T = 1$

(d) $f(t) = \left(t^2 e^{-7t} + 5e^{-6t}\sin 10t\right)u(t) \qquad T = 0.01$

Ans. $\left[10^{-4}k^2(0.93)^k + 5(0.94)^k \sin 0.1k\right]u(k)$

11. For each of the following analog signals with given Laplace transform $F(s)$, find the z transform $F(z)$ of the corresponding sample sequence with the given sampling interval T:

(a) $F(s) = \dfrac{16}{s^2 + 4} \qquad T = 0.1$

(b) $F(s) = \dfrac{-2s + 20}{s^3 + 9s^2 + 20s} \qquad T = 0.05$

Ans. $z/(z - 1) - 7z/(z - 0.819) + 6z/(z - 0.789)$.

(c) $F(s) = \dfrac{4e^{-0.5s}}{s^2 + 2s + 1} \qquad T = 0.25$

(d) $F(s) = \dfrac{3e^{-s}}{s} - \dfrac{4}{s + 2} + \dfrac{se^{-2s}}{s^2 + 9} \qquad T = 0.2$

Ans. $3/(z^5 - z^4) - 4z/(z - 0.67) + (z - 0.825)/\left(z^{11} - 1.65z^{10} + z^9\right)$

(e) $F(s) = \dfrac{1 - e^{-2s} - e^{-3s}}{s^2 + 2s + 10} \qquad T = 0.5$

12. For the following sequence $f(k)$, find the corresponding impulse train Laplace transform $F^*(s)$ for the given sampling interval T:

(a) $f(k) = (3\cos 2k - 4\sin 2k)u(k) \qquad T = 0.5$

(b) $f(k) = \left[(-1)^k - 1\right]u(k) \qquad T = 0.1$

Ans. $-2e^{-0.1s}/\left(e^{-0.2s} - 1\right)$

(c) $f(k) = 6(-0.5)^k u(k)\cos 3k \qquad T = 1$

(d) $f(k) = 10ke^{-0.1k}u(k-3) \qquad T = 0.5$

Ans. $\left(22.22e^{-0.5s} - 13.4e^{-s}\right) \Big/ \left(e^{0.5s} - 0.905\right)^2$

(e) $f(k) = ku(k) - ku(k-5) \qquad T = 0.3$

13. The continuous-time function

$$f(t) = 20 - 15\cos 1000t$$

is sampled at the rate of 100 samples/s. It is then reconstructed from the impulse train $f^*(t)$, according to the sampling theorem, as if it were a signal band-limited at 50 Hz. Find the signal that results.

14. Find the z-transfer functions of the following discrete-time systems:

(a) $y(k+2) + 0.2y(k+1) - 0.5y(k) = r(k)$

(b) $y(k+3) - y(k) = 4r(k+2) - 3r(k)$

Ans. $\left(4z^2 - 3\right) / \left(z^3 - 1\right)$

(c) $y(k+3) = 0.75r(k+3) + 0.25r(k+2) - 0.25r(k+1) - 0.75r(k)$

15. For

$$y(k+2) - 0.5y(k) = r(k+1) - 2r(k)$$

recursively find $y(2)$, $y(3)$, $y(4)$, and $y(5)$ if

$$r(k) = \left(\tfrac{1}{2}\right)^k$$

$$y(0) = 0$$

$$y(1) = 3$$

16. For discrete-time systems with the following z-transfer functions and input sequences, find the output sequences for $k = 0$ and thereafter if the initial conditions are zero:

(a) $D(z) = \dfrac{z}{z + \frac{1}{2}}$

$r(k) = 2\delta(k)$

(b) $D(z) = \dfrac{1}{z - 2}$

$r(k) = 3u(k)$

Ans. $-3u(k) + 3(2)^k u(k)$

(c) $D(z) = \dfrac{z}{z^2 + 1}$

$r(k) = (-1)^k$

(d) $D(z) = \dfrac{1}{z^2 - \frac{1}{2}z}$

$r(k) = \left(\tfrac{1}{2}\right)^k$

Ans. $4\delta(k) - 4\left(\frac{1}{2}\right)^k u(k) + 4k\left(\frac{1}{2}\right)^k u(k)$

(e) $D(z) = \dfrac{z^2}{8z^2 + 6z + 1}$

$r(k) = 3\left(-\frac{1}{2}\right)^k$

17. Reduce the block diagram of Figure P11.17, finding the system z-transfer functions. Figure P11.17.

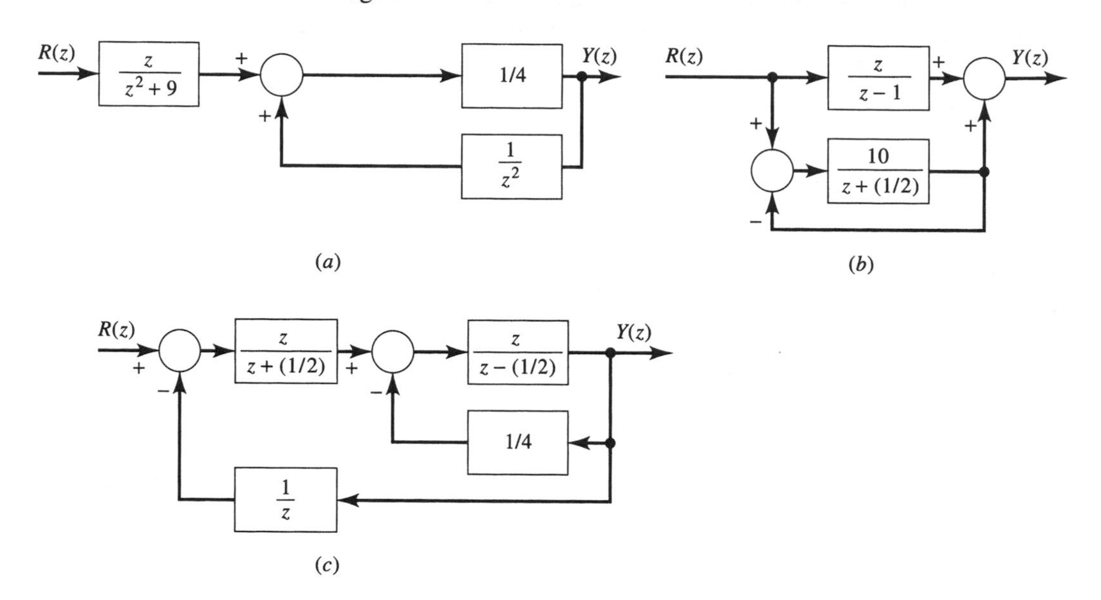

Figure P11.17

Ans. (b) $(2z^2 + 41z - 20)/(2z^2 + 19z - 21)$

18. Use Mason's gain rule to find the z-transfer functions of the systems of Figure P11.18.

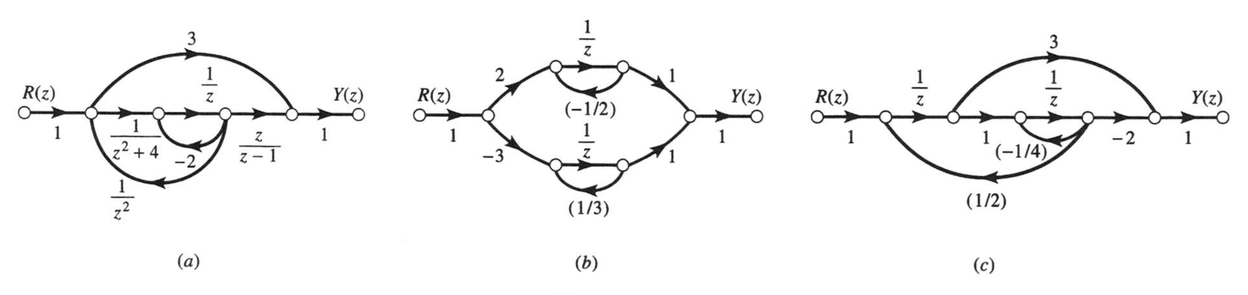

Figure P11.18

Ans. (b) $(-6z - 13)/\left(6z^2 + z - 1\right)$

19. Use the bilinear transformation and Routh–Hurwitz tests to determine whether each of the discrete-time systems with the given z-transfer functions is stable:

(a) $D(z) = \dfrac{4z^3 - 3z}{z^3 - 0.8z^2 + 0.17z - 0.01}$

(b) $D(z) = \dfrac{0.63z^2 - 8.73z + 1}{z^3 - 2.3z^2 + 0.62z - 0.04}$

Ans. unstable

(c) $D(z) = \dfrac{0.02}{z^4 - 2.8z^3 + 1.77z^2 - 0.35z + 0.02}$

(d) $D(z) = \dfrac{z^4 + 0.005}{z^4 - 1.3z^3 + 0.57z^2 - 0.095z + 0.005}$

Ans. stable

20. Draw discrete-time simulation diagrams in phase-variable canonical form for systems with the following z-transfer functions:

(a) $D(z) = \dfrac{z^2 - z + 2}{z^3 - 0.5z^2 + z}$

(b) $D(z) = \dfrac{10z}{z^4 + 0.1z^3 - 0.2z^2 + 0.3z + 0.4}$

(c) $D(z) = \dfrac{4z^2 - 6z + 8}{5z^4 + 3z^3 - 2z^2 + z + 1}$

21. Find state equations in phase-variable matrix form for the discrete-time systems with the following z-transfer functions:

(a) $D(z) = \dfrac{3z + 2}{z^2 + z + 4}$

(b) $D(z) = \dfrac{8}{6z^3 + z^2 - z}$

Ans.
$$\begin{bmatrix} x_1(k+1) \\ x_2(k+1) \\ x_3(k+1) \end{bmatrix} = \begin{bmatrix} 0 & 1 & 0 \\ 0 & 0 & 1 \\ 0 & \frac{1}{6} & -\frac{1}{6} \end{bmatrix} \begin{bmatrix} x_1(k) \\ x_2(k) \\ x_3(k) \end{bmatrix} + \begin{bmatrix} 0 \\ 0 \\ 1 \end{bmatrix} u(k)$$

$$y(k) = \begin{bmatrix} \frac{4}{3} & 0 & 0 \end{bmatrix} \begin{bmatrix} x_1(k) \\ x_2(k) \\ x_3(k) \end{bmatrix}$$

(c) $D(z) = \dfrac{4z^2 - 3}{z^3 - 0.5z^2 + 1}$

22. For each of the following discrete-time systems, with the indicated input and initial conditions, recursively find the state vectors $x(1)$, $x(2)$, and $x(3)$ and the

outputs $y(0)$, $y(1)$, $y(2)$, $y(3)$:

(a) $$\begin{bmatrix} x_1(k+1) \\ x_2(k+1) \end{bmatrix} = \begin{bmatrix} 1 & -2 \\ -1 & 1 \end{bmatrix} \begin{bmatrix} x_1(k) \\ x_2(k) \end{bmatrix}$$

$$y(k) = [3 \quad 0] \begin{bmatrix} x_1(k) \\ x_2(k) \end{bmatrix}$$

$$\begin{bmatrix} x_1(0) \\ x_2(0) \end{bmatrix} = \begin{bmatrix} -4 \\ 0 \end{bmatrix}$$

(b) $$\begin{bmatrix} x_1(k+1) \\ x_2(k+1) \\ x_3(k+1) \end{bmatrix} = \begin{bmatrix} 0 & 1 & 0 \\ -1 & 0 & 1 \\ 2 & 1 & -2 \end{bmatrix} \begin{bmatrix} x_1(k) \\ x_2(k) \\ x_3(k) \end{bmatrix} + \begin{bmatrix} 0 \\ 1 \\ -1 \end{bmatrix} (-1)^k$$

$$y(k) = [3 \quad 0 \quad 4] \begin{bmatrix} x_1(k) \\ x_2(k) \\ x_3(k) \end{bmatrix}$$

$$\begin{bmatrix} x_1(0) \\ x_2(0) \\ x_3(0) \end{bmatrix} = \begin{bmatrix} 0 \\ 0 \\ 0 \end{bmatrix}$$

Ans. $\begin{bmatrix} 0 \\ 1 \\ -1 \end{bmatrix}, \begin{bmatrix} 1 \\ -2 \\ 4 \end{bmatrix} \begin{bmatrix} -2 \\ 4 \\ -9 \end{bmatrix}$, 0, −4, 19, −42

23. Find step-invariant discrete-time system approximations to the following analog transmittances, using the indicated sampling intervals T:

(a) $G(s) = \dfrac{1}{s} \qquad T = 0.01$

(b) $G(s) = \dfrac{1}{s+1} \qquad T = 0.5$

Ans. $0.393/(z - 0.607)$

(c) $G(s) = \dfrac{s}{s^2+1} \qquad T = 0.02$

(d) $G(s) = \dfrac{1 - e^{-0.5s}}{s+4} \qquad T = 0.1$

Ans. $(0.0825)\left(z^5 - 1\right) / \left(z^6 - 0.67z^5\right)$

(e) $G(s) = \dfrac{10\left(1 - e^{-0.1s}\right)}{s^2 + 4s} \qquad T = 0.01$

24. An *impulse-invariant* approximation of an analog transmittance by a digital system is the digital system with unit pulse response, that consists of samples of the unit impulse response of the analog transmittance. Find the z-transfer function of the impulse-invariant approximation to

$$G(s) = \frac{10}{s + 4}$$

with sampling interval $T = 0.2$.

25. Find the z-transfer functions of the systems of Figure p11.25. For each system, the sampling interval is $T = 0.3$.

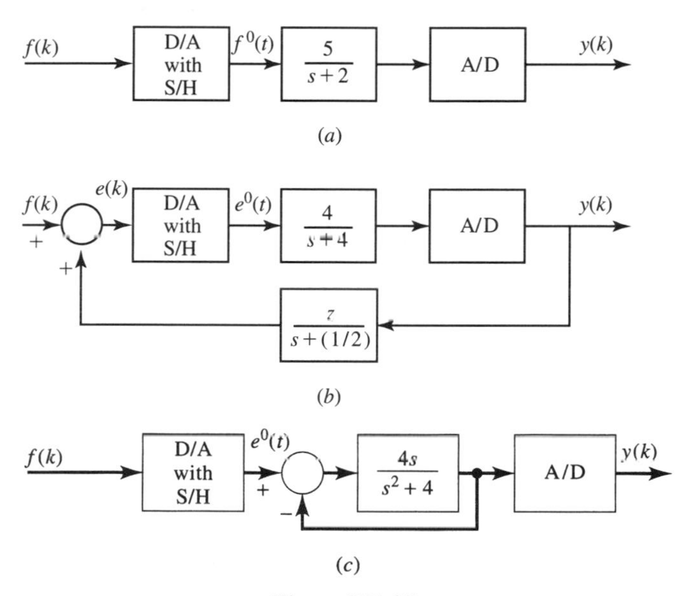

Figure P11.25

Ans. (b) $(0.7z + 0.35) / \left(z^2 - 0.5z - 0.15\right)$

26. For the system of Figure P11.26, with sampling interval $T = 0.2$, it is desired that the z-transfer function be

$$D(z) = \frac{Y(z)}{F(z)} = \frac{z}{z - 1}$$

Find, if possible the necessary analog transmittance $G(s)$.

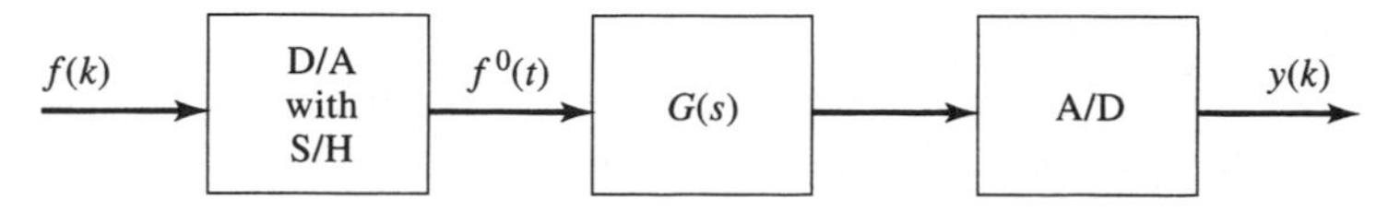

Figure P11.26

27. For each of the systems of Figure P11.27, find the z-transfer function relating $y(k)$ to $r(k)$. The sampling interval is $T = 0.2$. In figure P11.27 (b), (c), note that an equivalent system is one in which $Y(s)$ and $R(s)$ are first A/D-converted, then summed, as in Figure P11.27(a).

Ans. (b) $\dfrac{1.69\,(z-1)^2}{z\left(z^2-1.65z+1\right)+1.69\,(z-1)^2}$

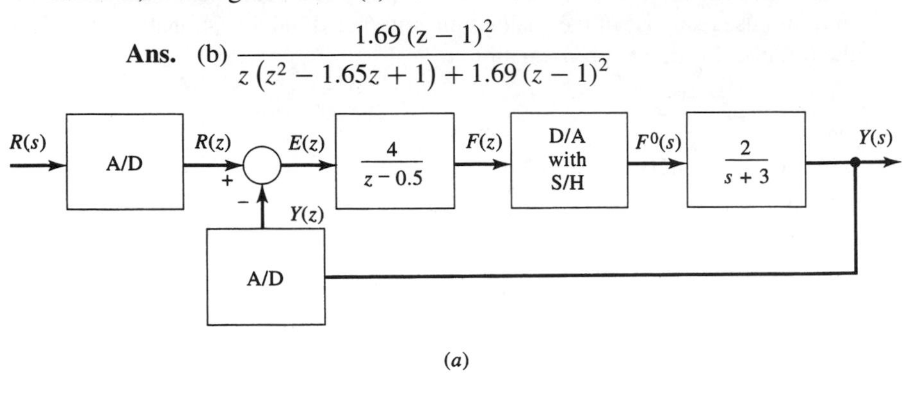

(a)

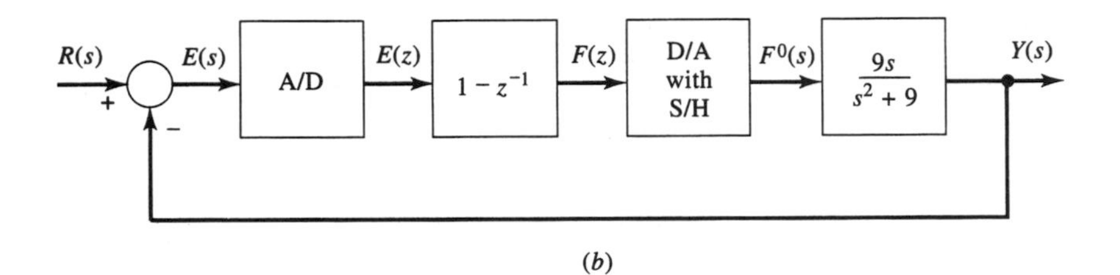

(b)

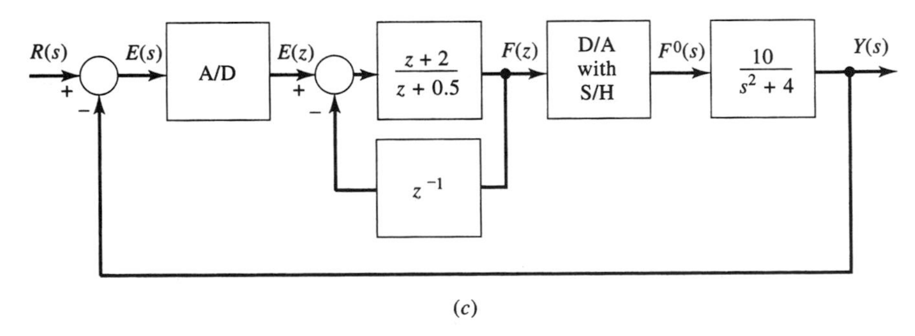

(c)

Figure P11.27

28. Determine whether the systems of Figure P11.28 are stable. For each system, the sampling interval is $T = 0.5$.

Ans. (b) stable

29. For systems with the following z-transfer functions, find the normalized steady state error, between output and input, for a step input:

(a) $D(z) = \dfrac{20z^2}{12z^2+7z+1}$

(b) $D(z) = \dfrac{0.25z^2 - 3z}{z^2 + 1.6z + 0.48}$

Ans. infinite error; system unstable

(c) $D(z) = \dfrac{2}{2z^3 - z^2 + z}$

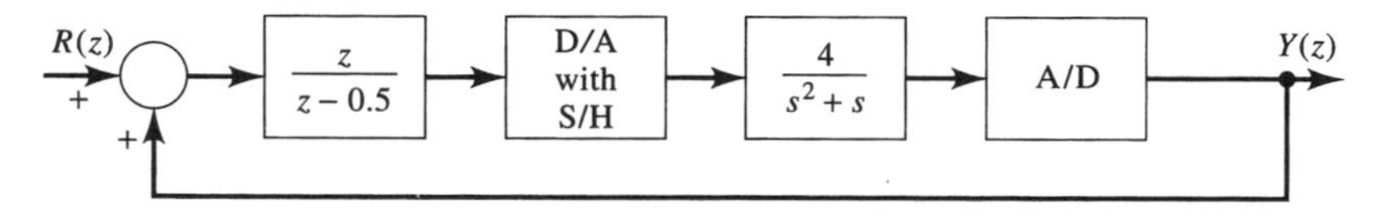

(a)

(b)

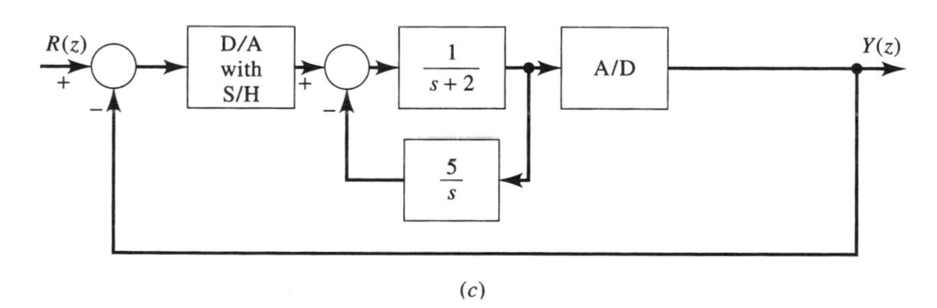

(c)

Figure P11.28

30. For the system of Figure P11.30, find a transmittance $H(z)$ so that the overall transfer function $D(z) = Y(z)/R(z)$ is deadbeat.

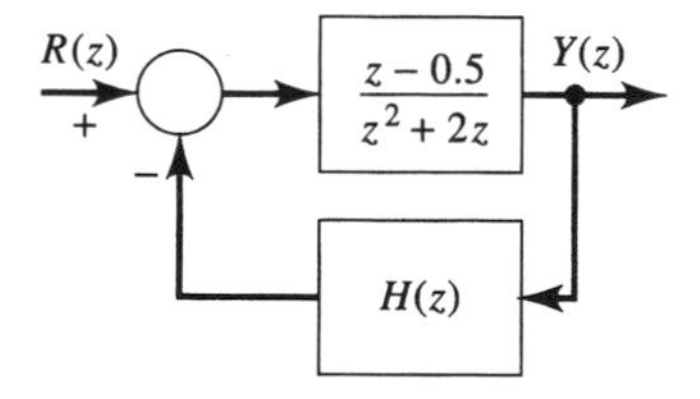

Figure P11.30

APPENDIX

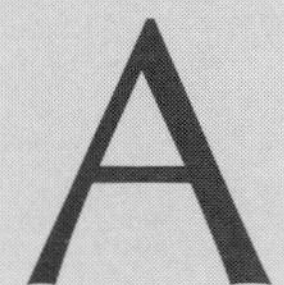

Matrix Algebra

A.1 Preview

The modern approach to control system analysis depends heavily on the manipulation of matrices and functions of matrices. This appendix reviews matrix addition, subtraction, multiplication, transposition, inversion, and differentiation. This appendix is not intended to be a primary source of knowledge regarding matrix algebra. Instead, this review provides a handy reference to operations required in several chapters of the text.

A.2 Nomenclature

Uppercase letters (A, B, C) denote matrices of dimension $n \times m$ (n rows and m columns) where both n and m exceed 1. Lowercase letters (x, y) denote column vectors of dimension $n \times 1$. A row vector would be denoted by x'. Where n is 1, the vector becomes a scalar. The symbol ($'$) denotes transpose, while -1 signifies inversion. The general element of matrix A is denoted by a_{ij} (the element in row i and column j).

A.3 Addition and Subtraction

Two matrices or two column vectors may be added if and only if both components are of the exact same dimension (the number of rows and columns are equal). If two matrices A and B are to be added to yield C, then

$$C = A + B$$

$$c_{ij} = a_{ij} + b_{ij}$$

For example, let

$$A = \begin{bmatrix} 1 & 2 \\ 3 & 4 \end{bmatrix}$$

$$B = \begin{bmatrix} 2 & 1 \\ 2 & 3 \end{bmatrix}$$

Then

$$C = \begin{bmatrix} 3 & 3 \\ 5 & 7 \end{bmatrix}$$

Subtraction requires

$$C = A - B$$

$$c_{ij} = a_{ij} - b_{ij}$$

For the foregoing A and B matrices

$$C = \begin{bmatrix} -1 & 1 \\ 1 & 1 \end{bmatrix}$$

A.4 Transposition

The transpose of a matrix follows by interchanging rows and columns. The general element for the transpose of A follows from

$$a'_{ij} = a_{ji}$$

For A and B of Section A.3, we write

$$A' = \begin{bmatrix} 1 & 3 \\ 2 & 4 \end{bmatrix}$$

$$B' = \begin{bmatrix} 2 & 2 \\ 1 & 3 \end{bmatrix}$$

A.5 Multiplication

The product C of two matrices A and B has a general element in row i and column j of C that results from multiplying each element of row i in A by each element in row j of B and summing the products. If one wants to compute

$$C = AB$$

then the number of columns of A must equal the number of rows of B (so that an equal number of elements may be multiplied together). In general,

$$c_{ij} = \sum_{k=1}^{m} a_{ik} b_{kj}$$

If A is $n \times m$ then B must be $m \times$ (say p) so that C is $n \times p$. Thus for A (2×2) and B (2×2), $C = AB$ becomes 2×2.

$$C = \begin{bmatrix} 1 \times 2 + 2 \times 2 & 1 \times 1 + 2 \times 3 \\ 3 \times 2 + 4 \times 2 & 3 \times 1 + 4 \times 3 \end{bmatrix} = \begin{bmatrix} 6 & 7 \\ 14 & 15 \end{bmatrix}$$

❑ DRILL PROBLEM

DA.1 Use A, B, and C to compute the indicated matrix functions.

$$A = \begin{bmatrix} 1 & 2 & 3 \\ 0 & 1 & 2 \end{bmatrix} \quad B = \begin{bmatrix} 4 & 6 & 5 \\ 1 & 0 & 3 \end{bmatrix} \quad C = \begin{bmatrix} 2 & 1 \\ 0 & 3 \end{bmatrix}$$

(a) $A + B$

(b) $B - A$

(c) A'

(d) $A'B$

(e) $B'C$

(f) $\frac{1}{2}(C + C')$

Ans.

$$\text{(a)} \begin{bmatrix} 5 & 8 & 8 \\ 1 & 1 & 5 \end{bmatrix}; \quad \text{(b)} \begin{bmatrix} 3 & 4 & 2 \\ 1 & -1 & 1 \end{bmatrix}; \quad \text{(c)} \begin{bmatrix} 1 & 0 \\ 2 & 1 \\ 3 & 2 \end{bmatrix};$$

$$\text{(d)} \begin{bmatrix} 4 & 6 & 5 \\ 9 & 12 & 13 \\ 14 & 18 & 21 \end{bmatrix}; \quad \text{(e)} \begin{bmatrix} 8 & 7 \\ 12 & 6 \\ 10 & 14 \end{bmatrix}; \quad \text{(f)} \begin{bmatrix} 2 & \frac{1}{2} \\ \frac{1}{2} & 3 \end{bmatrix}$$

A.6 Determinants and Cofactors

The determinant of a matrix A (denoted by $|A|$) is computed by adding all possible products of combinations of the matrix elements where each combination can contain only one element from each row and one element from each column. The determinant of a 2×2 matrix follows from the pattern

$$\begin{bmatrix} a_{11} & a_{12} \\ a_{21} & a_{12} \end{bmatrix} \quad \begin{matrix} \nearrow x(-1) \\ \searrow x(+1) \end{matrix}$$

$$= a_{11}a_{22} - a_{21}a_{12}$$

For A and B as in Section A.3, we have

$$|A| = 1 \times 4 - 3 \times 2 = 4 - 6 = -2$$

$$|B| = 2 \times 3 - 2 \times 1 = 6 - 2 = 4$$

Where multiplication is valid,

$$|C| = |AB| = |A|\,|B|$$

Thus the determinant of a product (of matrices) is the product of the determinant. From section A.5, we write

$$|C| = 6 \times 15 - 14 \times 7 = 90 - 98 = -8$$

$$= -2 \times 4 = |A|\,|B|$$

In the determinant of a 3 × 3 matrix that follows, we use a pattern of multiplications similar to the 2 × 2 pattern where the first two columns of the 3 × 3 matrix are repeated. If one needs

$$|E|$$

$$E = \begin{bmatrix} 1 & 3 & 2 \\ 0 & 1 & 2 \\ 1 & 0 & 4 \end{bmatrix}$$

then rewrite E and follow the indicated pattern.

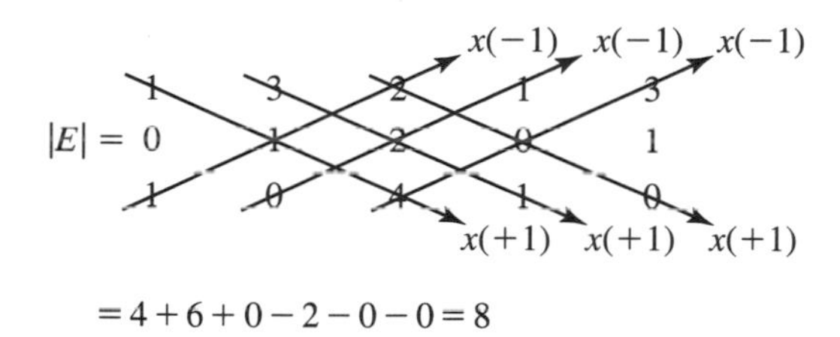

$$= 4 + 6 + 0 - 2 - 0 - 0 = 8$$

The determinant of a higher-order matrix (and, as we shall see, the inverse of a matrix) requires the use of cofactors and minors. Let $De(i, j, A)$ denote a matrix resulting from A where row i and column j are deleted. The determinant of each De is a minor of A. Each such minor must be multiplied by $(-1)^{i+j}$ to yield a cofactor of A (a signed minor). Denote each cofactor by $\mathrm{Co}(i, j, A)$. Thus for E we have

$$De(1, 1, E) = \begin{bmatrix} 1 & 2 \\ 0 & 4 \end{bmatrix}$$

$$\mathrm{Co}(1, 1, E) = (-1)^{1+1}\,|De(1, 1, E)| = 1 \times 4 = 4$$

Similarly,

$$De(1, 2, E) = \begin{bmatrix} 0 & 2 \\ 1 & 4 \end{bmatrix}$$

$$\mathrm{Co}(1, 2, E) = (-1)^{1+2}\,|De(1, 2, E)| = -(-2) = 2$$

The remaining cofactors are

$$\begin{aligned}
&\text{Co}(1, 3, E) = -1 \\
&\text{Co}(2, 1, E) = -12 \\
&\text{Co}(2, 2, E) = +2 \\
&\text{Co}(2, 3, E) = +3 \\
&\text{Co}(3, 1, E) = +4 \\
&\text{Co}(3, 2, E) = -2 \\
&\text{Co}(3, 3, E) = +1
\end{aligned}$$

The determinant of E may be found by expanding E along any row or column. By expanding along row i is meant the computation of

$$|E| = \sum_{k=1}^{m} e_{ik}\text{Co}(i, k, E)$$

For row 2, we have

$$\begin{aligned}
|E| &= e_{21}\text{Co}(2, 1, E) + e_{22}\text{Co}(2, 2, E) + e_{23}\text{Co}(2, 3, E) \\
&= 0 \times (-12) + 1 \times 2 + 2 \times 3 = 2 + 6 = 8
\end{aligned}$$

This agrees with the determinant computed earlier. Upon expanding along the rightmost column, we find

$$|E| = 2 \times (-1) + 2 \times 3 + 4 \times 1 = 8$$

A.7 Inverse

The inverse of a matrix A (if the inverse exists) must satisfy

$$AA^{-1} = A^{-1}A = I$$

where I is an identity matrix. A must be square. The determinant of A must not be zero; hence A must be nonsingular (a singular matrix is one with a zero-valued determinant). An identity matrix has ones along the main diagonal and zeros elsewhere. For example, a 3×3 identity matrix would be

$$I = \begin{bmatrix} 1 & 0 & 0 \\ 0 & 1 & 0 \\ 0 & 0 & 1 \end{bmatrix}$$

Denote the adjoint of a matrix A by $\text{Adj}(A)$, the general element of which is the cofactor $\text{Co}(j, i, A)$. The inverse of A is given by

$$A^{-1} = \frac{\text{Adj}(A)}{|A|}$$

For the matrix E,

$$\text{Adj}(E) = \begin{bmatrix} 4 & -12 & 4 \\ 2 & 2 & -2 \\ -1 & 3 & 1 \end{bmatrix}$$

$$E^{-1} = \frac{\text{Adj}(E)}{|E|} = \begin{bmatrix} \frac{1}{2} & -\frac{3}{2} & \frac{1}{2} \\ \frac{1}{4} & \frac{1}{4} & -\frac{1}{4} \\ -\frac{1}{8} & \frac{3}{8} & \frac{1}{8} \end{bmatrix}$$

Note that $E^{-1}E$ and EE^{-1} both equal I. This formula is not very efficient computationally. Better methods such as conversion to reduced row echelon form via row operations are to be preferred. (See textbooks on matrices or linear algebra.)

A.8 Simultaneous Equations

Suppose a set of simultaneous equations is given by

$$Ax = y$$

where x is an unknown vector, y is a known vector, and A is a matrix of coefficients (also known). These exists a unique solution where A is $n \times n$ with n independent equations (the determinant is not zero). If both sides of the matrix equation are multiplied byA^{-1},

$$x = A^{-1}y$$

Suppose

$$A = \begin{bmatrix} 1 & 2 \\ 3 & 4 \end{bmatrix} \quad y = \begin{bmatrix} 2 \\ 1 \end{bmatrix}$$

Then

$$\text{Adj}(A) = \begin{bmatrix} 4 & -2 \\ -3 & 1 \end{bmatrix}$$

$$|A| = -2$$

$$A^{-1} = \begin{bmatrix} -2 & 1 \\ \frac{3}{2} & -\frac{1}{2} \end{bmatrix}$$

$$x = A^{-1}y = \begin{bmatrix} -3 \\ \frac{5}{2} \end{bmatrix}$$

Cramer's rule provides an alternate way to solve the equation for x.

$$x_1 = \frac{\begin{vmatrix} 2 & 2 \\ 1 & 4 \end{vmatrix}}{\begin{vmatrix} 1 & 2 \\ 3 & 4 \end{vmatrix}} = \frac{8-2}{-2} = -3$$

$$x_2 = \frac{\begin{vmatrix} 1 & 2 \\ 3 & 1 \end{vmatrix}}{\begin{vmatrix} 1 & 2 \\ 3 & 4 \end{vmatrix}} = \frac{-5}{-2} = \frac{5}{2}$$

In general x_i is computed using a determinant ratio. The numerator matrix follows by replacing column i of A by y. The denominator is always the determinant of A. The x vector becomes

$$x = \begin{bmatrix} x_1 \\ x_2 \end{bmatrix}$$

❑ DRILL PROBLEMS

DA.2 Perform the indicated operations on

$$A = \begin{bmatrix} 1 & 2 & 1 \\ 0 & 2 & 3 \\ 1 & 0 & 4 \end{bmatrix}$$

(a) Find Co(1, 1, A).

(b) Find Co(3, 1, A).

(c) Find the determinant of A by expanding down the first column.

(d) Find the determinant of A using the pattern for a 3×3 matrix in Section A.6.

(f) Find adj(A).

(g) Find A^{-1} using the adjoint.

Ans. (a) 8; (b) 4; (c) $1 \times 8 + 1 \times 4 = 12$

(d) 12; (e) $\begin{bmatrix} 8 & -8 & 4 \\ 3 & 3 & -3 \\ -2 & 2 & 2 \end{bmatrix}$; (f) $\begin{bmatrix} \frac{2}{3} & -\frac{2}{3} & \frac{1}{3} \\ \frac{1}{4} & \frac{1}{4} & -\frac{1}{4} \\ -\frac{1}{6} & \frac{1}{6} & \frac{1}{6} \end{bmatrix}$

DA.3 Perform the indicated operations on

$$A = \begin{bmatrix} 3 & 0 & 1 \\ 2 & 0 & 3 \\ 0 & 1 & 1 \end{bmatrix}$$

(a) Find the determinant of A by expanding down the second column.

(b) Find Co(2, 3, A).

(c) Find A^{-1} using the adjoint.

Ans. (a) $1 \times (-7) = -7$; (b) -3; (c) $\begin{bmatrix} \frac{3}{7} & -\frac{1}{7} & 0 \\ -\frac{2}{7} & -\frac{3}{7} & 1 \\ -\frac{2}{7} & \frac{3}{7} & 0 \end{bmatrix}$

DA.4 For the A matrix of Drill Problem DA.2, solve

$$A \begin{bmatrix} x_1 \\ x_2 \\ x_3 \end{bmatrix} = \begin{bmatrix} 0 \\ 2 \\ 3 \end{bmatrix}$$

(a) by using the inverse computed earlier.

(b) by using Cramer's rule.

Ans. $\begin{bmatrix} -\frac{1}{3} \\ -\frac{1}{4} \\ \frac{5}{6} \end{bmatrix}$; $x_2 = \dfrac{\begin{vmatrix} 1 & 0 & 1 \\ 0 & 2 & 3 \\ 1 & 3 & 4 \end{vmatrix}}{\begin{vmatrix} 1 & 2 & 1 \\ 0 & 2 & 3 \\ 1 & 0 & 4 \end{vmatrix}}$

DA.5 Repeat Drill Problem DA.4, using A from Drill Problem DA.3.

Ans. $\begin{bmatrix} -\frac{2}{7} \\ 2\frac{1}{7} \\ \frac{6}{7} \end{bmatrix}$; $x_1 = \dfrac{\begin{vmatrix} 0 & 0 & 1 \\ 2 & 0 & 3 \\ 3 & 1 & 1 \end{vmatrix}}{\begin{vmatrix} 3 & 0 & 1 \\ 2 & 0 & 3 \\ 0 & 1 & 1 \end{vmatrix}}$

A.9 Eigenvalues and Eigenvectors

It may be of interest to find a scalar λ and a nonzero vector x associated with a matrix A so that

$$\lambda x = Ax$$

The values of λ satisfying that equation are those λ (eigenvalues) for which

$$|\lambda I - A| = 0$$

Associated with each eigenvalue is a vector x (an eigenvector). If A is an $n \times n$ matrix, there are n eigenvalues. Associated with eigenvalue λ_i is an eigenvector x_i where

$$[\lambda_i I - A]x_i = 0$$

No eigenvector is unique. Any eigenvector can be multiplied by a scalar and the result also satisfies the eigenvector equation. The fact that the x_i are not unique follows from the fact that $|\lambda_i I - A| = 0$.

Suppose

$$A = \begin{bmatrix} -1 & 0 \\ 1 & -2 \end{bmatrix}$$

Then

$$|\lambda I - A| = \begin{vmatrix} \lambda + 1 & 0 \\ -1 & \lambda + 2 \end{vmatrix} = (\lambda + 1)(\lambda + 2)$$

The two eigenvalues are $\lambda_1 = -1$ and $\lambda_2 = -2$. (One could also write $\lambda_1 = -2$ and $\lambda_2 = -1$.) For the former choice eigenvectors are sought. For $\lambda_1 = -1$,

$$\begin{bmatrix} 0 & 0 \\ -1 & 1 \end{bmatrix} x_1 = \begin{bmatrix} 0 & 0 \\ -1 & 1 \end{bmatrix} \begin{bmatrix} x_{11} \\ x_{21} \end{bmatrix} = \begin{bmatrix} 0 \\ 0 \end{bmatrix}$$

As long as the elements of x_1 are such that as $x_{11} = x_{12,}$ any such x_1 vector is an eigenvector for λ_1. For example,

$$x_1 = \begin{bmatrix} 1 \\ 1 \end{bmatrix}$$

is an eigenvector of λ_1 (or any multiple of x_1). For $\lambda_2 = -2$,

$$\begin{bmatrix} -1 & 0 \\ -1 & 0 \end{bmatrix} x_2 = \begin{bmatrix} -1 & 0 \\ -1 & 0 \end{bmatrix} \begin{bmatrix} x_{12} \\ x_{22} \end{bmatrix} = \begin{bmatrix} 0 \\ 0 \end{bmatrix}$$

As long as x_{12} is 0, an x_2 vector with any x_{22} will be an eigenvector for λ_2. For example,

$$x_2 = \begin{bmatrix} 0 \\ 1 \end{bmatrix}$$

(or any multiple thereof) is acceptable.

If the eigenvalues are distinct, a matrix can be formed by writing the eigenvectors side by side. If one defines

$$P = [\, x_1 \mid x_2 \,]$$

then (for distinct values of λ_1 and λ_2) one has

$$P^{-1}AP = \begin{bmatrix} \lambda_1 & 0 \\ 0 & \lambda_2 \end{bmatrix}$$

In this example

$$P = \begin{bmatrix} 1 & 0 \\ 1 & 1 \end{bmatrix}$$

Therefore,

$$P^{-1} = \begin{bmatrix} 1 & 0 \\ -1 & 1 \end{bmatrix}$$

$$P^{-1}AP = \begin{bmatrix} 1 & 0 \\ -1 & 1 \end{bmatrix}\begin{bmatrix} -1 & 0 \\ 1 & -2 \end{bmatrix}\begin{bmatrix} 1 & 0 \\ 1 & 1 \end{bmatrix} = \begin{bmatrix} -1 & 0 \\ 0 & -2 \end{bmatrix}$$

❑ DRILL PROBLEMS

DA.6 Perform the indicated operations on

$$A = \begin{bmatrix} 1 & 2 \\ 0 & 3 \end{bmatrix}$$

(a) Obtain the characteristic polynomial $|\lambda I - A| = 0$.

(b) Substitute A for λ in the characteristic polynomial and show that A satisfies its own characteristic polynomial. This is called the Cayley-Hamilton Theorem. Zero here implies a 2×2 matrix of zeros. Multiply the λ^0 coefficient by I.

(c) Solve the characteristic polynomial for λ_1 and λ_2.

(d) Obtain any eigenvector x_1 for λ_1 and x_2 for λ_2.

Ans.

(a) $\lambda^2 - 4\lambda + 3$; (b) $A^2 - 4A + 3I = \begin{bmatrix} 0 & 0 \\ 0 & 0 \end{bmatrix}$;

(c) $\lambda_1 = +1, \lambda_2 = +3$ (or $\lambda_1 = +3$ and $\lambda_2 = +1$);

(d) $x_1 =$ any multiple of $\begin{bmatrix} 1 \\ 0 \end{bmatrix}$; $x_2 =$ any multiple of $\begin{bmatrix} 1 \\ 1 \end{bmatrix}$

DA.7 Repeat Drill Problem DA.6 with

$$A = \begin{bmatrix} 3 & 0 \\ 1 & -2 \end{bmatrix}$$

(a) $\lambda^2 - \lambda - 6 = 0$

(b) $A^2 - A - 6I = \begin{bmatrix} 0 & 0 \\ 0 & 0 \end{bmatrix}$

(c) $\lambda_1 = 3, \lambda_2 = -2$

(d) $x_1 =$ any multiple of $\begin{bmatrix} 5 \\ 1 \end{bmatrix}$

$x_2 =$ any multiple of $\begin{bmatrix} 0 \\ 1 \end{bmatrix}$

A.10 Derivative of a Scalar with Respect to a Vector

The gradient vector is defined as the derivative of a scalar function of a vector taken with respect to that vector. Suppose the scalar is called $f(x)$, where x is an $n \times 1$

vector. Then

$$\frac{df(x)}{dx} = \begin{bmatrix} \frac{df(x)}{dx_1} \\ \frac{df(x)}{dx_2} \\ \vdots \\ \frac{df(x)}{dx_n} \end{bmatrix}$$

For example, if

$$f(x) = x_1^2 + 2x_1x_2 + 2x_2^3$$

then the gradient vector is

$$\frac{df(x)}{dx} = \begin{bmatrix} 2x_1 + 2x_2 \\ 2x_1 + 6x_2^2 \end{bmatrix}$$

A special case of the gradient pertains to

$$f(x) = [y_1 \quad y_2] \begin{bmatrix} x_1 \\ x_2 \end{bmatrix} = y'x = y_1x_1 + y_2x_2$$

where y is a vector also. From the definition of the gradient vector,

$$\frac{df(x)}{dx} = \begin{bmatrix} y_1 \\ y_2 \end{bmatrix} = y$$

Notice also that

$$\frac{df(y)}{dy} = \begin{bmatrix} x_1 \\ x_2 \end{bmatrix} = x$$

In general then, for

$$f(x) = y'x$$

$$\frac{df(x)}{dx} = y$$

$$\frac{df(y)}{dy} = x$$

This result may be applied to quadratic forms.

❑ DRILL PROBLEMS

DA.8 What is the gradient vector for $f(x) = 2x_1x_2 + e^{3x_2}$ computed for $x_1 = 1$, $x_2 = -1$?

Ans. $$\begin{bmatrix} 2x_2 \\ 2x_1 + 3e^{3x_2} \end{bmatrix}_{\begin{bmatrix} 1 \\ -1 \end{bmatrix}} = \begin{bmatrix} -2 \\ 2 + 3e^{-3} \end{bmatrix}$$

DA.9 What is $df(x)/dx$ for $f(x) = y'x$, where

$$y = \begin{bmatrix} 1 \\ 3 \\ 4 \end{bmatrix} \quad x = \begin{bmatrix} x_1 \\ x_2 \\ x_3 \end{bmatrix}$$

Ans. $\begin{bmatrix} 1 \\ 3 \\ 4 \end{bmatrix}$

A.11 Quadratic Forms and Symmetry

A quadratic form is a special scalar function of a vector x where

$$f(x) = x'Qx$$

To be able to compute $f(x)$ where x is an $n \times 1$ vector, the matrix Q must be $n \times n$ (i.e., square). It can be shown that no loss of generality ensues by requiring Q to be symmetrical.

A symmetrical matrix is equal to its own transpose, that is Q, is symmetrical if and only if

$$Q' = Q$$

The general element requires

$$q'_{ij} = q_{ji} = q_{ij}$$

For example,

$$Q = \begin{bmatrix} 1 & 2 \\ 2 & 3 \end{bmatrix}$$

is symmetrical because $q_{12} = q_{21}$. However,

$$Q = \begin{bmatrix} 1 & -2 \\ 4 & 3 \end{bmatrix}$$

is not symmetrical because $q_{12} \neq q_{21}$. A matrix

$$Q_s = \tfrac{1}{2}(Q + Q')$$

is always symmetrical because

$$Q'_s = Q_s$$

Notice that any Q can be written as

$$\begin{aligned} Q &= \tfrac{1}{2}(Q + Q') + \tfrac{1}{2}(Q - Q') \\ &= Q_s + Q_{\text{sk}} \end{aligned}$$

Q_s is symmetrical. Q_{sk} is called a skew symmetric matrix because

$$q_{\text{skij}} = \tfrac{1}{2}(q_{ij} - q_{ji}) = -\tfrac{1}{2}(q_{ji} - q_{ij})$$
$$= -q_{\text{skji}}$$
$$q_{skii} = 0$$

That is, the diagonal elements of Q_{sk} are 0, while the off-diagonal elements are equal and opposite. For example,

$$\begin{matrix} Q = & Q_s & + & Q_{\text{sk}} \end{matrix}$$
$$\begin{bmatrix} 1 & -2 \\ 4 & -3 \end{bmatrix} = \begin{bmatrix} 1 & 1 \\ 1 & 3 \end{bmatrix} + \begin{bmatrix} 0 & -3 \\ 3 & 0 \end{bmatrix}$$

It can be shown for any Q that

$$x'Qx = x'\,[Q_s + Q_{\text{sk}}]\,x = x'Q_s x$$

since the skew symmetric part of Q does not contribute to $f(x)$. For example,

$$\begin{bmatrix} x_1 & x_2 \end{bmatrix} \begin{bmatrix} 0 & -3 \\ 3 & 0 \end{bmatrix} \begin{bmatrix} x_1 \\ x_2 \end{bmatrix} = 0$$

For this reason, Q is always a symmetric matrix when $f(x)$ is a quadratic form.

To differentiate $f(x)$ with respect to x, it is useful to form the derivative with respect to each occurrence of x while holding the other part constant. We write

$$\frac{df(x)}{dx} = \frac{d(y'x)}{dx} + \frac{d(x'y_1)}{dx}$$

where

$$y = Q'x$$
$$y_1 = Qx$$

Using the results from the end of Section A.10, we then have

$$\frac{df(x)}{dx} = y + y_1 = (Q' + Q)x$$

Assuming that Q is symmetric,

$$\frac{df(x)}{dx} = 2Qx$$

A.12 Definiteness

The definiteness of a scalar quadratic form $f(x)$ depends on potential values of $f(x)$ for all possible nonzero values of the vector x. A nonzero value of x implies that at least one element of x is nonzero.

$f(x)$	Type of Definiteness
> 0	Positive definite
$\geqslant 0$	Positive semidefinite
< 0	Negative definite
$\leqslant 0$	Negative semidefinite
None of above	Indefinite

The definiteness of a symmetric matrix Q (associated with a quadratic form $x'Qx$) depends on the real part σ_i of eigenvalues $\lambda_i = \sigma_i + j\omega_i$.

σ_i	Type of Definiteness
> 0	Positive definite
$\geqslant 0$	Positive semidefinite
< 0	Negative definite
$\leqslant 0$	Negative semidefinite
None of above	Indefinite

The definiteness of an $n \times n$ symmetric matrix Q may be tested by examining the value of a sequence of n determinants, $\det(i, Q)$, each defined as the determinant of the upper left-hand $i \times i$ submatrix of Q.

The index i varies from 1 to n.

Test	Type of Definiteness
$\det(i, Q) > 0$	Positive definite
$\det(i, Q) \geqslant 0$	Positive semidefinite
$\det(i, -Q) > 0$	Negative definite
$\det(i, -Q) \geqslant 0$	Negative semidefinite
None of above	Indefinite

Suppose

$$Q = \begin{bmatrix} -1 & 0 & 0 \\ 0 & -\frac{5}{2} & \frac{1}{2} \\ 0 & \frac{1}{2} & -\frac{5}{2} \end{bmatrix}$$

$$|\lambda I - Q| = (\lambda + 1)(\lambda + 2)(\lambda + 3)$$

The test of the real parts of the λ_i indicates the Q is negative definite, since the real parts are all negative. Using the determinant test, one first examines

$$\det(1, Q) = |-1| = -1$$

Since this determinant is negative, the matrix Q cannot be positive definite or positive semidefinite. The determinant test now examines $-Q$.

$$\det(1, -Q) = |1| = 1$$

$$\det(2, -Q) = \begin{vmatrix} -1 & 0 \\ 0 & -\frac{5}{2} \end{vmatrix} = \frac{5}{2}$$

$$\det(3, -Q) = |-Q| = 6$$

Since all three determinants are positive, the matrix is identified as negative definite.

❑ DRILL PROBLEMS

DA.10 Use the determinant test to establish the definiteness of

(a)

$$\begin{bmatrix} 6 & 3 & 0 \\ 3 & 2 & 0 \\ 0 & 0 & 4 \end{bmatrix}$$

(b)

$$\begin{bmatrix} -2 & 0 & -2 \\ 0 & -3 & 0 \\ 2 & 0 & -4 \end{bmatrix}$$

Ans. (a) positive definite; (b) negative definite

DA.11 What is $df(x)/dx$, where

$$f(x) = \begin{bmatrix} x_1 \\ x_2 \end{bmatrix}^T \begin{bmatrix} 6 & 2 \\ 2 & 1 \end{bmatrix} \begin{bmatrix} x_1 \\ x_2 \end{bmatrix}$$

Ans. $\begin{bmatrix} 12 & 4 \\ 4 & 2 \end{bmatrix}$

A.13 Rank

The rank of a matrix A indicates the number of linearly independent pieces of information contained in A. If A is a square $n \times n$ matrix, the maximum possible rank is n. If A is not square with dimension $n \times m$ where n and m are not equal, the maximum possible rank is the smaller of n or m denoted by min (n, m).

To formulate a test for rank, it is useful to define $\text{Pe}(i, A, j)$ to be permutation j, where an $i \times i$ submatrix of A is generated. A submatrix occurs when the rows and columns of A are (or are not) exchanged and a contiguous $i \times i$ matrix is selected. For a 3×3 matrix, several 2×2 submatrices can be generated but only one 3×3 would be used. The rank of A is the largest value of i for which the determinant of at least one Pe is nonzero. The test begins by testing for the largest possible rank [n or $\min(n, m)$].

Should all Pe yield a zero-valued determinant for that i, i is decremented (reduced by 1) and the test continues. Suppose

$$A = \begin{bmatrix} 1 & 2 & 1 \\ 0 & 1 & 3 \\ 0 & 2 & 6 \end{bmatrix}$$

Since A is square and 3×3, the test first examines the determinant of A since, for all j $|\text{Pe}(3, A, j)|$ will be the same. However, $|A|$ is zero, so the rank is not 3. For the first permutation of a 2×2 submatrix let

$$\text{Pe}(2, A, 1) = \begin{bmatrix} 1 & 2 \\ 0 & 1 \end{bmatrix}$$

Obviously that determinant is 1; hence that rank is 2, since at least on 2×2 submatrix has a nonzero determinant.

❑ DRILL PROBLEMS

DA.12 What is the rank of

$$A = \begin{bmatrix} 1 & 2 & 1 \\ 1 & 2 & 3 \\ 1 & 2 & 2 \end{bmatrix}$$

Ans. 2

DA.13 What is the rank of

$$A = \begin{bmatrix} 1 & 0 & 0 \\ 0 & 2 & 0 \\ 0 & 0 & 3 \end{bmatrix}$$

Ans. 3

DA.14 What is the rank of

$$A = \begin{bmatrix} 1 & 0 & 2 \\ 0 & 0 & 0 \\ 2 & 0 & 4 \end{bmatrix}$$

Ans. 1

A.14 Partitioned Matrices

When a matrix is partitioned into component parts, the normal rules of addition and multiplication may be applied to the components (including compatibility of dimensions). If

$$A = \begin{bmatrix} A_{11} & A_{12} \\ A_{21} & A_{22} \end{bmatrix} \quad B = \begin{bmatrix} B_{11} \\ B_{21} \end{bmatrix}$$

$$C = \begin{bmatrix} C_{11} & C_{12} \\ C_{21} & C_{22} \end{bmatrix}$$

then

$$A + C = \begin{bmatrix} A_{11} + C_{11} & A_{12} + C_{12} \\ A_{21} + C_{21} & A_{22} + C_{22} \end{bmatrix}$$

$$AB = \begin{bmatrix} A_{11}B_{11} + A_{12}B_{21} \\ A_{21}B_{11} + A_{22}B_{21} \end{bmatrix}$$

For example, if

$$A = \left[\begin{array}{c|cc} 1 & 2 & 0 \\ \hline 4 & 5 & 6 \\ 7 & 8 & 9 \end{array}\right] \qquad B = \left[\begin{array}{c} 1 \\ \hline 2 \\ 3 \end{array}\right]$$

$$C = \left[\begin{array}{c|cc} 9 & 8 & 7 \\ \hline 6 & 5 & 4 \\ 3 & 2 & 1 \end{array}\right]$$

then the foregoing rules result in the following values for $A + C$ and AB:

$$A + C = \left[\begin{array}{c|cc} 10 & 10 & 7 \\ \hline 10 & 10 & 10 \\ 10 & 10 & 10 \end{array}\right]$$

$$AB = \left[\begin{array}{c} 1 + 4 \\ \hline \begin{bmatrix} 4 \\ 7 \end{bmatrix} + \begin{bmatrix} 28 \\ 43 \end{bmatrix} \end{array}\right] = \left[\begin{array}{c} 5 \\ \hline 32 \\ 50 \end{array}\right]$$

The determinant of a A can be found using the components.

$$|A| = |A_{11}| \left| A_{22} - A_{21}A_{11}^{-1}A_{12} \right|$$

The inverse of A is given by

$$D = \begin{bmatrix} D_{11} & D_{12} \\ D_{21} & D_{22} \end{bmatrix} = A^{-1}$$

$$D_{22} = \left[A_{22} - A_{21}A_{11}^{-1}A_{12} \right]^{-1}$$

$$D_{12} = -A_{11}^{-1}A_{12}D_{22}$$

$$D_{21} = -D_{22}A_{21}A_{11}^{-1}$$

$$D_{11} = A_{11}^{-1}\left[I - A_{12}D_{21} \right]$$

For example, the determinant and inverse of a 3×3 matrix can be found by partitioning

$$A = \left[\begin{array}{c|cc} 1 & 2 & 0 \\ \hline 4 & 5 & 6 \\ 7 & 8 & 9 \end{array}\right]$$

then

$$|A| = |1| \left| -\begin{bmatrix} 4 \\ 7 \end{bmatrix} (1)[2 \quad 0] + \begin{bmatrix} 5 & 6 \\ 8 & 9 \end{bmatrix} \right|$$

$$= \left| -\begin{bmatrix} 8 & 0 \\ 14 & 0 \end{bmatrix} + \begin{bmatrix} 5 & 6 \\ 8 & 9 \end{bmatrix} \right| = \begin{vmatrix} -3 & 6 \\ -6 & 9 \end{vmatrix} = -27 + 36 = +9$$

The inverse follows because

$$D_{22} = \left[\begin{bmatrix} 5 & 6 \\ 8 & 9 \end{bmatrix} - \begin{bmatrix} 4 \\ 7 \end{bmatrix} (1)[2 \quad 0] \right]^{-1}$$

$$= \begin{bmatrix} -3 & 6 \\ -6 & 9 \end{bmatrix}^{-1} = \begin{bmatrix} 1 & -\frac{2}{3} \\ \frac{2}{3} & -\frac{1}{3} \end{bmatrix}$$

$$D_{12} = -\,(1)[2 \quad 0] \begin{bmatrix} 1 & -\frac{2}{3} \\ \frac{2}{3} & -\frac{1}{3} \end{bmatrix} = \begin{bmatrix} -2 & \frac{4}{3} \end{bmatrix}$$

$$D_{21} = -\begin{bmatrix} 1 & -\frac{2}{3} \\ \frac{2}{3} & -\frac{1}{3} \end{bmatrix} \begin{bmatrix} 4 \\ 7 \end{bmatrix} (1) = \begin{bmatrix} \frac{2}{3} \\ -\frac{1}{3} \end{bmatrix}$$

$$D_{11} = (1) \left[1 - [2 \quad 0] \begin{bmatrix} \frac{2}{3} \\ -\frac{1}{3} \end{bmatrix} \right] = -\frac{1}{3}$$

$$A^{-1} = \left[\begin{array}{c|cc} -\frac{1}{3} & -2 & \frac{4}{3} \\ \hline \frac{2}{3} & 1 & -\frac{2}{3} \\ -\frac{1}{3} & \frac{2}{3} & -\frac{1}{3} \end{array}\right]$$

As a check, notice that

$$AA^{-1} = I$$

The partitioned approach reduces the dimension of the inversion process. For example, the 3×3 inversion example was reduced to a 2×2 inversion. A 4×4 inversion can also be reduced to a 2×2 inversion.

❑ DRILL PROBLEMS

DA.15 $A = \left[\begin{array}{c|cc} 2 & 1 & 3 \\ \hline 0 & 1 & 0 \\ 1 & 0 & -2 \end{array}\right] \quad B = \left[\begin{array}{c} 2 \\ \hline 1 \\ 6 \end{array}\right]$

Answer the following where A and B are partitioned as just shown.

(a) What is AB?

(b) What is $|A|$?

(c) What is A^{-1}?

Ans. (a) $\left[\begin{array}{c} 23 \\ \hline 1 \\ -10 \end{array}\right]$; (b) $2\begin{vmatrix} 1 & 0 \\ -\frac{1}{2} & -\frac{7}{2} \end{vmatrix} = -7$; (c) $\left[\begin{array}{c|cc} \frac{2}{7} & -\frac{2}{7} & \frac{3}{7} \\ \hline 0 & 1 & 0 \\ \frac{1}{7} & -\frac{1}{7} & -\frac{2}{7} \end{array}\right]$

DA.16 Repeat Drill Problem DA.15 (c), where

$$A = \left[\begin{array}{cc|c} 2 & 1 & 3 \\ 0 & 1 & 0 \\ \hline 1 & 0 & -2 \end{array}\right]$$

Ans. $A^{-1} = \left[\begin{array}{cc|c} \frac{2}{7} & -\frac{2}{7} & \frac{3}{7} \\ 0 & 1 & 0 \\ \hline \frac{1}{7} & -\frac{1}{7} & -\frac{2}{7} \end{array}\right]$

PROBLEMS

1. Use A and x to compute the indicated matrix functions.

$$A = \begin{bmatrix} 3 & 2 \\ 1 & 0 \end{bmatrix} \quad x = \begin{bmatrix} 1 \\ 2 \end{bmatrix}$$

(a) $A'A$

(b) Ax

(c) $x'A'Ax$

(d) $x'x$

Ans. (a) $\begin{bmatrix} 10 & 6 \\ 6 & 4 \end{bmatrix}$; (b) $\begin{bmatrix} 7 \\ 1 \end{bmatrix}$; (c) 50; (d) 5

2. Use A, B, x to compute the indicated matrix functions.

$$A = \begin{bmatrix} 4 & 3 \\ 6 & 2 \end{bmatrix} \quad B = \begin{bmatrix} 2 & 1 \\ 3 & 6 \\ 4 & 2 \end{bmatrix} \quad x = \begin{bmatrix} 1 \\ 3 \\ 6 \end{bmatrix}$$

(a) $B'B$

(b) BA

(c) $x'B$

(d) $x'BA$

3. Perform the indicated operations on

$$A = \begin{bmatrix} 2 & 1 & 6 \\ 3 & 1 & 2 \\ 4 & 2 & 12 \end{bmatrix}$$

(a) Co(3, 1, A)

(b) Co(2, 2, A)

(c) determinant of A

(d) adjoint of A

(e) Does A^{-1} exist? If so, find A^{-1}

Ans. (a) -4; (b) 0; (c) 0, (d) $\begin{bmatrix} 8 & 0 & -4 \\ -28 & 0 & 14 \\ 2 & 0 & -1 \end{bmatrix}$; (e) no

4. Repeat Problem 3, where

$$A = \begin{bmatrix} 1 & 0 & 2 \\ 1 & 1 & 0 \\ 0 & 2 & 3 \end{bmatrix}$$

5. Find A^{-1} for

$$A = \begin{bmatrix} 3 & 0 & 2 \\ 1 & 0 & 0 \\ 0 & 1 & 0 \end{bmatrix}$$

Ans. $\begin{bmatrix} 0 & 1 & 0 \\ 0 & 0 & 1 \\ \frac{1}{2} & -\frac{3}{2} & 0 \end{bmatrix}$

6. Use Cramer's rule to find x.

$$3x_1 + 2x_2 = 4$$

$$x_1 - 3x_2 = 2$$

7. Use the inverse from Problem 5 to find x.

$$A \begin{bmatrix} x_1 \\ x_2 \\ x_3 \end{bmatrix} = \begin{bmatrix} 2 \\ 1 \\ 2 \end{bmatrix}$$

8. Does the following have a unique solution?

$$x_1 + 3x_2 + x_3 = 7$$
$$x_1 - x_2 = 3$$
$$2x_1 - 2x_2 = 2$$

9. What is the characteristic polynomial for

$$A = \begin{bmatrix} 2 & 1 \\ 3 & 2 \end{bmatrix}$$

Ans. $\lambda^2 - 4\lambda + 1$

10. What are the characteristic polynomials for A, B, AB, where

$$A = \begin{bmatrix} 1 & 2 \\ 1 & 0 \end{bmatrix} \quad B = \begin{bmatrix} 2 & 0 \\ 0 & 2 \end{bmatrix}$$

11. Show that the matrix in problem 9 satisfies its own characteristic polynomial.

12. Give the eigenvalues of

$$A = \begin{bmatrix} 2 & 0 & 0 \\ 0 & 1 & 2 \\ 0 & 3 & 2 \end{bmatrix}$$

13. Give the eigenvectors of

$$A = \begin{bmatrix} 1 & 0 \\ 1 & 2 \end{bmatrix}$$

Ans. 1, 2; for $\lambda = 1$ any multiple of $\begin{bmatrix} 1 \\ -1 \end{bmatrix}$; for $\lambda = 2$ any multiple of $\begin{bmatrix} 0 \\ 1 \end{bmatrix}$

14. Find a matrix P such that $P^{-1}AP$ is diagonal using the matrix A from Problem 13.

15. What is the gradient vector of $f(x) = \sin 2x_1 + x_2 e^{2x_1}$?

Ans. $\begin{bmatrix} 2\cos 2x_1 + 2x_2 e^{2x_1} \\ e^{2x_1} \end{bmatrix}$

16. Show that

$$\frac{df(x)}{dx} = y$$
$$f(x) = y'x = y_1x_1 + y_2x_2 + y_3x_3$$

17. For the quadratic form

$$f(x) = 2x_1^2 + 3x_1x_2 + 4x_2^2$$

find the following.

(a) Q for $f(x) = x'Qx$

(b) $\dfrac{df(x)}{dx}$

(c) Definiteness of Q

Ans. (a) $Q = \begin{bmatrix} 2 & 1.5 \\ 1.5 & 4 \end{bmatrix}$; (b) $\begin{bmatrix} 4 & 3 \\ 3 & 8 \end{bmatrix}$; (c) positive-definite

18. What is the definiteness of

$$Q = \begin{bmatrix} 1 & 2 & 0 \\ 2 & 1 & 0 \\ 0 & 0 & 8 \end{bmatrix}$$

19. For the A matrix of Problem 5, show that

$$x'Ax = x'A_s x$$
$$A_s - \tfrac{1}{2}[A + A']$$

20. What is the rank of the A matrix in Problem 3?

21. What is the rank of the A matrix in Problem 5?
Ans. 3

22. Use the indicated partitioned matrices to answer the following questions.

$$A = \left[\begin{array}{c|cc} 1 & 0 & 1 \\ \hline 2 & 1 & 1 \\ 3 & 2 & 6 \end{array}\right] \qquad B = \left[\begin{array}{c|cc} 1 & 3 & 2 \end{array}\right]$$

(a) What is BA?

(b) What is $A'B'$?

(c) What is $|A|$?

(d) What is A^{-1}?

23. Obtain the inverse of the following 4×4 matrix by partitioning into 2×2 matrices

$$A = \begin{bmatrix} 0 & 1 & 2 & 3 \\ 2 & 1 & 0 & 0 \\ 1 & 0 & 2 & 1 \\ 0 & 1 & 1 & 2 \end{bmatrix}$$

Ans. $A^{-1} = \left[\begin{array}{cc|cc} 3 & 1 & -1 & -4 \\ -6 & -1 & 2 & 8 \\ \hline -4 & -1 & 2 & 5 \\ 5 & 1 & -2 & -6 \end{array}\right]$

APPENDIX

B

Laplace Transform

B.1 Preview

The analysis and design of linear control systems depends heavily upon Laplace transform methods. The Laplace transform variable s is found throughout the text in reference to mathematical models, block diagrams, steady state errors, root loci, and so on. Most students studying the subject of control system design have already had a course in Laplace transforms. However, because of the importance of this topic, we review and summarize the method here. This appendix, however, is not intended to be a primary source of knowledge on Laplace transforms. Instead, it should serve as a handy reference for several of the chapters in this text.

B.2 Definition and Properties

The expression of a system description in the language of mathematics is termed *modeling*. Modeling involves idealizations whereby the important aspects of a problem are isolated and the minor ones ignored, so that simplicity with sufficient accuracy is obtained. When the system models consist of linear, constant-coefficient integrodifferential equations, Laplace transform methods can be used to advantage. The Laplace transform of a function $f(t)$ is

$$\mathcal{L}[f(t)] = F(s) = \int_{0^-}^{\infty} f(t)e^{-st}\,dt$$

The inverse Laplace transform recovers the original function for $t \geqslant 0$ and gives zero for times prior to $t = 0$:

$$\mathcal{L}^{-1}[F(s)] = \frac{1}{2\pi j}\int_{\sigma-j\infty}^{\sigma+j\infty} F(s)e^{st}\,ds = \begin{Bmatrix} f(t) & t \geqslant 0 \\ 0 & t < 0 \end{Bmatrix} = f(t)u(t)$$

where $u(t)$ is the unit step function. When functions with discontinuities at $t = 0$ are involved, it is inconvenient to have $t = 0$ as a limit to the Laplace transform integral. The most useful and common definition begins the integration just before $t = 0$, at $t = 0^-$.

Table B.1 is a collection of common functions $f(t)$ and their Laplace transforms, $f(s)$. Since these Laplace transforms are unique, the table can also be used to find time functions from transforms. Table B.2 gives important Laplace transform properties.

B.3 Solving Differential Equations

By Laplace-tramsforming linear constant-coefficient integrodifferential equations, one obtains linear algebraic equations, which can then be solved for the transforms of the solutions; Initial conditions can be included when the equations are transformed. For example, transforming

$$\frac{d^2y}{dt^2} + 9\frac{dy}{dt} + 2y = 6e^{-4t}$$

with

$$y(0^-) = 2$$
$$y'(0^-) = -4$$

gives

$$s^2Y(s) - 2s + 4 + 9[sY(s) - 2] + 2Y(s) = \frac{6}{s+4}$$

$$Y(s) = \frac{6}{(s+4)(s^2+9s+2)} + \frac{2s+14}{s^2+9s+2}$$

The Laplace transform method is also easily applied to systems described by simultaneous integrodiffierential equations. Transforming the simultaneous equations

$$\frac{dy_1}{dt} + 2y_1 - \frac{3dy_2}{dt} = r(t)$$

$$y_1 + \frac{dy_2}{dt} + 4\int_0^t y_2 dt = 0$$

gives

$$sY_1(s) - y_1(0^-) + 2Y_1(s) - 3sY_2(s) + 3y_2(0^-) = R(s)$$

$$Y_1(s) + sY_2(s) - y_2(0^-) + \frac{4}{s}Y_2(s) = 0$$

Table B.1 *Selected Laplace Transforms*

$f(t)$	$F(s)$
$\delta(t)$ (unit impulse)	1
$u(t)$ (unit impulse)	$\frac{1}{s}$
$tu(t)$	$\frac{1}{s^2}$
$t^n u(t)$	$\frac{n!}{s^{n+1}}$
$e^{-at}u(t)$	$\frac{1}{s+a}$
$te^{-at}u(t)$	$\frac{1}{(s+a)^2}$
$t^n e^{-at}u(t)$	$\frac{n!}{(s+a)^{n+1}}$
$(\sin bt)\, u(t)$	$\frac{b}{s^2+b^2}$
$(\cos bt)u(t)$	$\frac{s}{s^2+b^2}$
$(t \sin bt)u(t)$	$\frac{2bs}{(s^2+b^2)^2}$
$(t \cos bt)u(t)$	$\frac{s^2-b^2}{(s^2+b^2)^2}$
$(e^{-at} \sin bt)u(t)$	$\frac{b}{(s+a)^2+b^2}$
$(e^{-at} \cos bt)u(t)$	$\frac{(s+a)}{(s+a)^2+b^2}$
$Ae^{-at} \cos(bt+\theta)u(t)$	$\frac{(A/2)e^{j\theta}}{s+a-jb} + \frac{(A/2)e^{-j\theta}}{s+a+jb} = \frac{\text{first-degree numerator}}{(s+a)^2+b^2}$
$Ae^{-at} \sin(bt+\theta)u(t)$	$\frac{s+a}{(s+a)^2+b^2}$ $\quad A = \frac{1}{b}\sqrt{(\alpha-a)^2+b^2}$ $\theta = \tan^{-1} \frac{b}{\alpha - a}$

or

$$
\begin{aligned}
(s+2)Y_1(s) - 3sY_2(s) &= R(s) + y_1(0^-) - 3y_2(0^-) \\
Y_1(s) + \left(s + \frac{4}{s}\right) Y_2(s) &= y_2(0^-)
\end{aligned}
$$

which can be solved for $Y_1(s)$ and $Y_2(s)$, given the input $R(s)$ and the initial conditions $y_1(0^-)$ and $y_2(0^-)$.

Table B.2 *Fundamental Laplace Transform Properties*

$\mathcal{L}[kf(t)] = kF(s)$, k a constant
$\mathcal{L}[f_1(t) + f_2(t] = F_1(s) + F_2(s)$
$\mathcal{L}[f_1(t)f_2(t)]$ does *not* equal $F_1(s)F_2(s)$
$\mathcal{L}[f(t-T)] = e^{-sT}F(s)$, T a constant, provided $f(t)$ and $f(t-T)$ are both zero prior to $t = 0$
$\mathcal{L}[f(at)] = \frac{1}{a}F\left(\frac{s}{a}\right)$, a is a positive constant
$\mathcal{L}[e^{-at}f(t)] = F(s+a)$
$\mathcal{L}\left[\frac{df}{dt}\right] = sF(s) - f(0^-)$
$\mathcal{L}\left[\frac{d^2f}{dt^2}\right] = s^2F(s) - sf(0^-) - f'(0^-)$
$\mathcal{L}\left[\frac{d^nf}{dt^2}\right] = s^nF(s) - s^{n-1}f(0^-) - s^{n-2}f'(0^-) - \cdots - sf^{[n-2]}(0^-) - f^{[n-1]}(0^-)$
$\mathcal{L}\left[\int_{0^-}^{t} f(t)dt\right] = \frac{F(s)}{s}$
$\mathcal{L}\left[\int_{\infty}^{t} f(t)dt\right] = \frac{F(s)}{s} + \frac{1}{s}\int_{-\infty}^{0^-} f(t)dt$
$\mathcal{L}[tf(t)] = -\frac{dF(s)}{ds}$
$\mathcal{L}[t^2f(t)] = \frac{d^2F(s)}{ds^2}$
$\mathcal{L}[t^nf(t)] = (-1)^n\frac{d^nF(s)}{ds^n}$

B.4 Partial Fraction Expansion

Determining inverse Laplace transforms of rational functions involves expansion into partial fractions. If the numerator polynomial is of lower order than the denominator polynomial and if the denominator polynomial has no repeated roots, then the residues $K_1,\ K_2,\ K_3, \ldots$ can be found such that

$$Y(s) = \frac{\text{numerator polynomial}}{(s+a)(s+b)(s+c)\cdots} = \frac{K_1}{s+a} + \frac{K_2}{s+b} + \frac{K_3}{s+c} + \cdots$$

The individual terms in the expansion represent exponential functions of time after $t = 0$:

$$y(t) = K_1e^{-at} + K_2e^{-bt} + K_3e^{-ct} + \cdots \qquad t \geqslant 0$$

A residue other than for a repeated root is found by multiplying each side of the expansion by the denominator term, then evaluating the result at the value of s

which makes that denominator term zero:

$$\frac{(s+a)\text{ numerator}}{(s+a)(s+b)(s+c)\cdots} = \frac{(s+a)K_1}{s+a} + \frac{(s+a)K_2}{s+b} + \frac{(s+a)K_3}{s+c} + \cdots$$

$$\left.\frac{\text{numerator}}{(s+b)(s+c)\cdots}\right|_{s=-a} = K_1$$

Let

$$Y(s) = \frac{-3s^2+4}{s(s+2)(s+3)}$$

Applying this method to $Y(s)$ gives the residues quite easily:

$$K_1 = \left.\frac{-3s^2+4}{(s+2)(s+3)}\right|_{s=0} = \frac{2}{3} \qquad K_2 = \left.\frac{-3s^2+4}{s(s+3)}\right|_{s=-2} = 4$$

$$K_3 = \left.\frac{-3s^2+4}{s(s+2)}\right|_{s=-3} = -\frac{23}{3}$$

Hence

$$Y(s) = \frac{2/3}{s} + \frac{4}{s+2} - \frac{23/3}{s+3} \quad \text{and} \quad y(t) = \frac{2}{3} + 4e^{-2t} - \frac{23}{3}e^{-3t}, \quad t \geqslant 0$$

If the numerator polynomial is not of lower order than the denominator polynomial, the denominator must be divided into the numerator until a remainder polynomial of lower order than the denominator is obtained, giving

$$\frac{\text{numerator polynomial}}{(s+a)(s+b)(s+c)\cdots} = \frac{\text{dividend}}{\text{polynomial}} + \frac{\text{remainder polynomial}}{(s+a)(s+b)(s+c)\cdots}$$

For example,

$$Y(s) = \frac{3s^2-4s+1}{s^2+5s+6} = 3 + \frac{-19s-17}{s^2+5s+6} = 3 + \frac{21}{s+2} + \frac{-40}{s+3}$$

$$y(t) = 3\delta(t) + 21e^{-2t} - 40e^{-3t} \qquad t \geqslant 0$$

A constant term in the Laplace transform corresponds to an impulsive time function.

If denominator roots are repeated, the corresponding terms in the partial function expansion are as follows:

$$\frac{\text{numerator}}{(s+a)^n D(s)} = \frac{K_1}{s+a} + \frac{K_2}{(s+a)^2} + \cdots + \frac{K_n}{(s+a)^n} + \text{ other}$$

The inverse Laplace transform of a repeated root is of the form

$$\mathcal{L}^{-1}\left[\frac{K_n}{(s+a)^n}\right] = \frac{K_n}{(n-1)!}t^{n-1}e^{-at}u(t)$$

For a repeated root, the residue evaluation method works only for the highest-order repeated term in the expansion. For example, with

$$Y(s) = \frac{4s^2-1}{(s+2)^3} = \frac{K_1}{s+2} + \frac{K_2}{(s+2)^2} + \frac{K_3}{(s+2)^3}$$

evaluation gives

$$\frac{(4s^2-1)(s+2)^3}{(s+2)^3} = K_1(s+2)^2 + K_2(s+2) + K_3$$

$$K_3 = (s+2)^3 Y(s)\big|_{s=-2} = (4s^2-1)\big|_{s=-2} = 15$$

Multiplying both sides of the expansion by $(s+2)^2$ or $(s+2)$ will leave $(s+2)$ denominator factors, however, so K_1 and K_2 cannot be determined in this manner.

K_1 and K_2 are found by multiplying both sides of the expansion by the repeated root term and differentiating with respect to s:

$$\frac{d}{ds}\left[\frac{(4s^2-1)(s+2)^3}{(s+2)^3}\right] = \frac{d}{ds}\left[K_1(s+2)^2 + K_2(s+2) + K_3\right]$$

$$8s = 2K_1(s+2) + K_2 \qquad \textbf{[B.1]}$$

Evaluating at $s=-2$ gives

$$8s\big|_{s=-2} = K_2 = -16$$

Differentiating Equation (B.1) a second time with respect to s and evaluating, we have

$$\frac{d}{ds}(8s)\bigg|_{s=-2} = 2K_1$$

$$K_1 = 4$$

In general, the coefficients of the repeated root terms of a partial fraction expansion, repeated p times,

$$Y(s) = \frac{K_1}{s+a} + \frac{K_2}{(s+a)^2} + \cdots + \frac{K_p}{(s+a)^p} + \cdots \text{ terms for other different roots}$$

are given by

$$K_i = \frac{1}{(p-i)!}\frac{d^{p-i}}{ds^{p-i}}\left\{(s+a)^p Y(s)\right\}\big|_{s=-a} \qquad i = 1,\ 2, \ldots,\ p$$

For complex root terms, the corresponding residues are complex numbers that are complex conjugates of one another. It is thus necessary to calculate only one of the residues for a set of conjugate terms. For the transform

$$Y(s) = \frac{-s+8}{s(s^2+2s+10)}$$

$$= \frac{-s+8}{s(s+1-j3)(s+1+j3)}$$

$$= \frac{K_1}{s} + \frac{K_2}{s+1-j3} + \frac{K_3 = K_2^*}{s+1+j3}$$

then

$$K_1 = \left.\frac{-s+8}{s^2+2s+10}\right|_{s=0} = 0.8$$

$$K_2 = \left.\frac{-s+8}{s(s+1+j3)}\right|_{s=-1+j3} = \frac{9-j3}{(-1+j3)(j6)} = 0.5e^{j143^\circ}$$

so that

$$Y(s) = \frac{0.8}{s} + \frac{0.5e^{j143^\circ}}{s+1-j3} + \frac{0.5e^{-j143^\circ}}{s+1+j3}$$

From Table B.1, the corresponding time function is

$$y(t) = 0.8 + e^{-t}\cos(3t + 143^\circ) \qquad t \geqslant 0$$

❑ DRILL PROBLEMS

DB.1 For the following Laplace-transformed signals, find $y(t)$ for $t \geqslant 0$:

(a)

$$Y(s) = \frac{s}{s+2}$$

Ans. $\delta(t) - 2e^{-2t}$

(b)

$$Y(s) = \frac{3s-5}{s^2+4s+2}$$

Ans. $5.39e^{-3.41t} - 2.39e^{-0.586t}$

(c)

$$Y(s) = \frac{3-6e^{-2s}}{(s+2)(s+3)}$$

Ans. $3e^{-2t} - 3e^{-3t} - \left[6e^{-2(t-2)} - 6e^{-3(t-2)}\right]u(t-2)$

(d)

$$Y(s) = \frac{10}{s^3+2s^2+5s}$$

Ans. $2 + e^{-t}(-2\cos 2t - \sin 2t) = 2 + 2.24e^{-t}\cos(2t + 153^\circ)$

(e)

$$Y(s) = \frac{4(s+1)}{(s+2)(s+3)^2}$$

Ans. $-4e^{-2t} + 4e^{-3t} + 8te^{-3t}$

DB.2 Use Laplace transform methods to solve the following differential equations for $t \geqslant 0$ with the indicated boundary conditions:

(a)

$$\frac{dy}{dt} + 4y = 6e^{2t}$$

$$y(0^-) = 3$$

Ans. $e^{2t} + 2e^{-4t}$

(b)

$$\frac{dy}{dt} + y = 3\cos 2t$$

$$y(0^-) = 0$$

Ans. $-\frac{3}{5}e^{-t} + \frac{3}{5}\cos 2t + \frac{6}{5}\sin 2t = -\frac{3}{5}e^{-t} + 1.34\cos(2t - 63.4°)$

(c)

$$\frac{d^2y}{dt^2} + +7\frac{dy}{dt} + 12y = 10$$

$$y(0^-) = 3$$

$$y'(0^-) = 0$$

Ans. $\frac{10}{12} - \frac{13}{2}e^{-4t} + \frac{26}{3}e^{-3t}$

(d)

$$\frac{d^2y}{dy^2} + 4\frac{dy}{dt} + 20y = 4$$

$$y(0^-) = -2$$

$$y'(0^-) = 0$$

Ans. $\frac{1}{5} + e^{-2t}\left(-\frac{11}{5}\cos 4t - \frac{11}{10}\sin 4t\right) = \frac{1}{5} + 2.46e^{-2t}\cos(4t + 153°)$

(f)

$$\frac{d^3y}{dt^3} + 5\frac{d^2y}{dt^2} + 6\frac{dy}{dt} = 0$$

$$y(0^-) = 3$$

$$y'(0^-) = -2$$

$$y''(0^-) = 7$$

Ans. $\frac{15}{6} + e^{-3t} - \frac{1}{2}e^{-2t}$

B.5 Additional Properties of the Laplace Transform

Included in this section are some input–output properties of the Laplace transform.

B.5.1 Real Translation

A translation in the time domain is shown for this function $f(t)$ in Figure B.1. The shifted function has the form $f(t-a)u(t-a)$ when $u(t)$ is the unit step function. With the Laplace transform, it can be shown that a time shift can be transferred into the s plane. The following transform pair

$$\mathcal{L}[f(t-a)u(t-a)] = e^{-as}F(s) \qquad \textbf{[B.2]}$$

results when $f(t) = F(s)$. This function is often termed a *transport lag* because the entire function of time is translated down the time axis. As an example, let us use this property to find the Laplace transform of one cycle of the sine wave, $A\sin\omega t$. This function can be formed by subtracting from $A\sin\omega t$, a sine wave that has been translated in time by $2\pi/\omega$. Hence one cycle is given by

$$f_1(t) = A\sin\omega t\; u(t) - A\sin\omega\left(t - \frac{2}{\pi}\omega\right) u\left(t - \frac{2}{\pi}\omega\right) \qquad \textbf{[B.3]}$$

The Laplace transform of $f_1(t)$ is found with the help of Equation (B.2).

$$\mathcal{L}f_1(t) = \frac{A\omega}{(s^2+\omega^2)} - \frac{A\omega}{(s^2+\omega^2)} \cdot e^{-2\pi s/\omega}$$

$$= \frac{A\omega}{(s^2+\omega^2)} \cdot (1 - e^{-2\pi s/\omega}) \qquad \textbf{[B.4]}$$

B.5.2 Second Independent Variable

A useful theorem results when a second independent variable exists in the function of time. Suppose that the function depends on the time t and another independent variable a. The function is expressed as $f(t, a)$, and the Laplace transform is found with respect to t.

$$\mathcal{L}f(t, a) = F(s, a) \qquad \textbf{[B.5]}$$

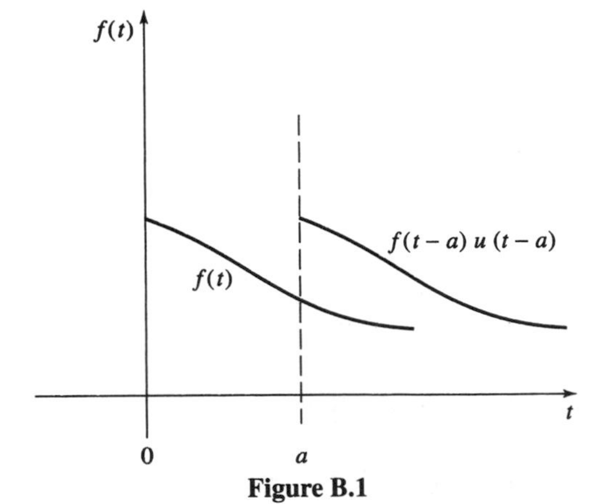

Figure B.1

Several useful expressions result:

1. Differentiation with respect to a,

$$\mathcal{L}[df(t, a)/da] = dF(s, a)/da \qquad \textbf{[B.6]}$$

2. Taking the limit, we write

$$\mathcal{L}\left[\lim_{a \to a_0} = f(t, a)\right] = \lim_{a \to a_0} F(s, a) \qquad \textbf{[B.7]}$$

As an example of Equation (B.6), the transform pair is

$$\mathcal{L}e^{-at} = \frac{1}{(s+a)}$$

and if we differentiated with respect to a

$$\frac{\mathcal{L}de^{-at}}{da} = \mathcal{L}_t(-te^{-at}) = \frac{d[1/(s+a)]}{da} = \frac{-1}{(s+a)^2}$$

which results in

$$\mathcal{L}te^{-at} = \frac{1}{(s+a)^2} \qquad \textbf{[B.8]}$$

Equation (B.7) is used to develop another transform pair from Equation (B.8):

$$\lim_{a \to 0}[\mathcal{L}te^{-at}] = \lim_{a \to 0} \frac{1}{(s+a)^2}$$

When a is set equal to zero, another transform results:

$$\mathcal{L}t = \frac{1}{s^2} \qquad \textbf{[B.9]}$$

B.5.3 Final-Value and Initial-Value Theorems

These two theorems are valuable in finding steady state errors. They are unique in that the value of the function of time at either $t \to 0$ $t \to \infty$ can be found from the function of s as $s \to \infty$ or $s \to 0$. These theorems are stated here, without proof.

Final-Value Theorem

$$\lim_{t \to \infty} f(t) = \lim_{s \to 0} sF(s) \qquad \textbf{[B.10]}$$

provided no poles of $sF(s)$ lie in the right half-plane.

Initial-Value Theorem

$$\lim_{t \to 0^-} f(t) = \lim_{s \to \infty} sF(s) \quad \textbf{[B.11]}$$

provided the limit exists.

As an example, Equation (B.8) is used to find the final value of the function:

$$f(t) = te^{-at}$$

Since the Laplace transform is $F(s) = 1/(s+a)^2$ and since $sF(s)$ is stable, then the final value of $f(t)$ as $t \to \infty$ is

$$\lim_{t \to \infty} te^{-at} = \lim_{s \to 0} s/(s+a)^2 = 0$$

B.5.4 Convolution Integral

This theorem provides a means to evaluate the product of two Laplace transformed functions. If $F_1(s) = \mathcal{L}f_1(t)$ and $F_2(s) = \mathcal{L}f_2(t)$, then

$$\mathcal{L}\left[\int_0^t f_1(\tau)f_2(t-\tau)d\tau\right] = F_1(s)F_2(s) \quad \textbf{[B.12]}$$

The integral on the left-hand side of Equation (B.12) is termed the *convolution integral.* Hence, Equation (B.12) states that the Laplace transform of the convolution of two time functions results in the product of the individual Laplace-transformed functions.

The theorem expressed in Equation (B.12) is the basis for the use of block diagrams in control system design. $G(s)$ is defined as the Laplace-transform of the response of the system to an input impulse, so $G(s)$ is $\mathcal{L}g(t)$. Now the output of the system, $y(t)$ for any input $f(t)$ is found from the convolution integral

$$y(t) = \int_0^t g(t-\tau)f(\tau)d\tau \quad \textbf{[B.13]}$$

When Equation (B.13) is Laplace-transformed, we have

$$Y(s) = G(s)F(s) \quad \textbf{[B.14]}$$

As a result, the Laplace transform of the oputput $Y(s)$ is the product of the Laplace transform of the input, $F(s)$, and the Laplace transform of the impulse response, $G(s)$. Block diagrams express the concept shown in Equation (B.14).